Microstrip Antennas

IEEE PRESS
445 Hoes Lane, P.O. Box 1331
Piscataway, NJ 08855-1331

Microstrip Antennas

The Analysis and Design of Microstrip Antennas and Arrays

Edited by
David M. Pozar
Daniel H. Schaubert
University of Massachusetts at Amherst

IEEE Antennas and Propagation Society, *Sponsor*

IEEE
The Institute of Electrical and Electronics Engineers, Inc., New York

WILEY-INTERSCIENCE
A JOHN WILEY & SONS, INC., PUBLICATION

Published by John Wiley & Sons, Inc., Hoboken, New Jersey.

For general information on our other products and services please contact our Customer Care Department within the U.S. at 877-762-2974, outside the U.S. at 317-572-3993 or fax 317-572-4002.

Wiley also publishes its books in a variety of electronic formats. Some content that appears in print, however, may not be available in electronic format.

ISBN 0-7803-1078-0

Library of Congress Cataloging-in-Publication Data

Microstrip antennas : the analysis and design of microstrip antennas and arrays / edited by David M. Pozar, Daniel H. Schaubert.

p. cm.

"A Selected reprint volume."

"IEEE Antennas and Propagation Society, sponsor."

Includes bibliographical references and index.

ISBN 0–7803–1078–0

1. Microstrip antennas. I. Pozar, David M. II. Schaubert, D.

TK7871.6.M512 1995

621.381′331—dc20 95–1229

CIP

Contents

INTRODUCTION ix

CHAPTER 1 REVIEW ARTICLES 1

Microstrip Antenna Technology 3
K. R. Carver and J. W. Mink (*IEEE Transactions on Antennas and Propagation*, Jan. 1981)
Research on Planar Antennas and Arrays: Structures Rayonnates 26
J. P. Daniel et al. (*IEEE Antennas and Propagation Magazine*, Feb. 1993)
A Review of CAD for Microstrip Antennas and Arrays 51
D. M. Pozar and J. R. James

CHAPTER 2 BASIC MICROSTRIP ANTENNA ELEMENTS AND FEEDING TECHNIQUES 57

A Review of Some Microstrip Antenna Characteristics 59
D. H. Schaubert
Conformal Microstrip Antennas and Microstrip Phased Arrays 68
R. E. Munson (*IEEE Transactions on Antennas and Propagation*, Jan. 1974)
An Experimental Investigation of Electrically Thick Rectangular Microstrip Antennas 73
E. Chang, S. A. Long, and W. F. Richards (*IEEE Transactions on Antennas and Propagation*, June 1986)
The Effect of Various Parameters of Circular Microstrip Antennas on Their Radiation Efficiency and the Mode Excitation 79
A. A. Kishk and L. Shafai (*IEEE Transactions on Antennas and Propagation*, Aug. 1986)
Crosspolarisation Characteristics of Circular Patch Antennas 87
K. F. Lee, K. M. Luk, and P. Y. Tam (*Electronics Letters*, March 1992)
Guidelines for Design of Electromagnetically Coupled Microstrip Patch Antennas on Two-Layer Substrates 90
G. Splitt and M. Davidovitz (*IEEE Transactions on Antennas and Propagation*, July 1990)
Design of Microstrip Antennas Covered with a Dielectric Layer 95
I. J. Bahl, P. Bhartia, and S. S. Stuchly (*IEEE Transactions on Antennas and Propagation*, March 1982)
The Finite Ground Plane Effect on the Microstrip Antenna Radiation Patterns 100
J. Huang (*IEEE Transactions on Antennas and Propagation*, July 1983)

CHAPTER 3 DUAL AND CIRCULARLY POLARIZED ELEMENTS 105

Review of Techniques for Dual and Circularly Polarised Microstrip Antennas 107
P. S. Hall
Analysis and Optimized Design of Single Feed Circularly Polarized Microstrip Antennas 117
P. C. Sharma and K. C. Gupta (*IEEE Transactions on Antennas and Propagation*, Nov. 1983)
A Circularly Polarized Microstrip Antenna Using Singly-Fed Proximity Coupled Feed 124
H. Iwasaki, H. Sawada, and K. Kawabata (*Proceedings of the 1992 International Symposium on Antennas and Propagation*, Sept. 1992)
Dual Aperture-Coupled Microstrip Antenna for Dual or Circular Polarization 128
A. Adrian and D. H. Schaubert (*Electronics Letters*, Nov. 1987)
Design of Wideband Circularly Polarized Aperture-Coupled Microstrip Antennas 130
S. D. Targonski and D. M. Pozar (*IEEE Transactions on Antennas and Propagation*, Feb. 1993)
Wideband Circularly Polarized Array with Sequential Rotations and Phase Shift of Elements 136
T. Teshirogi, M. Tanaka, and W. Chujo (*Proceedings of the 1985 International Symposium on Antennas and Propagation*, Aug. 1985)
Gain of Circularly Polarized Arrays Composed of Linearly Polarized Elements 140
P. S. Hall et al. (*Electronics Letters*, Jan. 1989)
Optimised Feeding of Dual Polarised Broadband Aperture-Coupled Printed Antenna 142
M. Edimo, A. Sharaiha, and C. Terret (*Electronics Letters*, Sept. 1992)

Feed Circuits of Double-Layered Self-Diplexing Antenna for Mobile Satellite Communications 145
M. Nakano et al. (*IEEE Transactions on Antennas and Propagation*, Oct. 1992)
Microstrip Antennas with Frequency Agility and Polarization Diversity 148
D. H. Schaubert et al. (*IEEE Transactions on Antennas and Propagation*, Jan. 1981)

CHAPTER 4 TECHNIQUES FOR IMPROVING ELEMENT BANDWIDTH 155

A Review of Bandwidth Enhancement Techniques for Microstrip Antennas 157
D. M. Pozar
An Impedance-Matching Technique for Increasing the Bandwidth of Microstrip Antennas 167
H. F. Pues and A. R. Van de Capelle (*IEEE Transactions on Antennas and Propagation*, Nov. 1989)
Probe Compensation in Thick Microstrip Patches 176
P. S. Hall (*Electronics Letters*, May 1987)
Increasing the Bandwidth of a Microstrip Antenna by Proximity Coupling 178
D. M. Pozar and B. Kaufman (*Electronics Letters*, April 1987)
Characteristics of a Two-Layer Electromagnetically Coupled Rectangular Patch Antenna 180
R. Q. Lee, K. F. Lee, and J. Bobinchak (*Electronics Letters*, Sept. 1987)
The SSFIP: A Global Concept for High Performance Broadband Planar Antennas 182
J.-F. Zurcher (*Electronics Letters*, Nov. 1988)
Millimeter-Wave Design of Wide-Band Aperture-Coupled Stacked Microstrip Antennas 185
F. Croq and D. M. Pozar (*IEEE Transactions on Antennas and Propagation*, Dec. 1991)
Multioctave Bandwidth Log-Periodic Microstrip Antenna Array 192
P. S. Hall (*IEE Proceedings*, April 1986)

CHAPTER 5 MODELING TECHNIQUES FOR MICROSTRIP ANTENNA ELEMENTS 203

Accurate Transmission-Line Model for the Rectangular Microstrip Antenna 205
H. Pues and A. Van de Capelle (*IEE Proceedings*, Dec. 1984)
CAD-Oriented Cavity Model for Rectangular Patches 212
D. Thouroude, M. Himdi, and J. P. Daniel (*Electronics Letters*, June 1990)
Analysis of Aperture-Coupled Microstrip Antenna Using Cavity Method 215
M. Himdi, J. P. Daniel, and C. Terret (*Electronics Letters*, March 1989)
Analysis of Arbitrarily Shaped Microstrip Patch Antennas Using Segmentation Technique and Cavity Model 217
V. Palanisamy and R. Garg (*IEEE Transactions on Antennas and Propagation*, Oct. 1986)
Fundamental Superstrate (Cover) Effects on Printed Circuit Antennas 223
N. G. Alexópoulos and D. R. Jackson (*IEEE Transactions on Antennas and Propagation*, Aug. 1984)
General Integral Equation Formulation for Microstrip Antennas and Scatterers 232
J. R. Mosig and F. E. Gardiol (*IEE Proceedings*, Dec. 1985)
A Reciprocity Method of Analysis for Printed Slot and Slot-Coupled Microstrip Antennas 241
D. M. Pozar (*IEEE Transactions on Antennas and Propagation*, Dec. 1986)
Multiport Scattering Analysis of General Multilayered Printed Antennas Fed by Multiple Feed Ports: Part II—Applications 249
N. K. Das and D. M. Pozar (*IEEE Transactions on Antennas and Propagation*, May 1992)
Accurate Characterization of Planar Printed Antennas Using Finite-Difference Time-Domain Method 259
C. Wu et al. (*IEEE Transactions on Antennas and Propagation*, May 1992)

CHAPTER 6 MICROSTRIP ANTENNA ARRAY DESIGN 267

Review of Microstrip Antenna Array Techniques 269
D. H. Schaubert
The Synthesis of Shaped Patterns with Series-Fed Microstrip Patch Arrays 274
B. B. Jones, F. Y. M. Chow, and A. W. Seeto (*IEEE Transactions on Antennas and Propagation*, Nov. 1982)
Coplanar Corporate Feed Effects in Microstrip Patch Array Design 280
P. S. Hall and C. M. Hall (*IEE Proceedings*, June 1988)

A Study of Microstrip Array Antennas with the Feed Network 287
E. Levine et al. (*IEEE Transactions on Antennas and Propagation*, April 1989)
Design Considerations for Low Sidelobe Microstrip Arrays 295
D. M. Pozar and B. Kaufman (*IEEE Transactions on Antennas and Propagation*, Aug. 1990)
A Parallel-Series-Fed Microstrip Array with High Efficiency and Low Cross-Polarization 305
J. Huang (*Microwave and Optical Technology Letters*, May 1992)

CHAPTER 7 ANALYSIS OF ARRAYS AND MUTUAL COUPLING 309

Input Impedance and Mutual Coupling of Rectangular Microstrip Antennas 311
D. M. Pozar (*IEEE Transactions on Antennas and Propagation*, Nov. 1982)
Phased Array Simulation Using Circular Patch Radiators 317
K. Solbach (*IEEE Transactions on Antennas and Propagation*, Aug. 1986)
Finite Phased Arrays of Rectangular Microstrip Patches 323
D. M. Pozar (*IEEE Transactions on Antennas and Propagation*, May 1986)
Analysis of a Series-Fed Aperture-Coupled Patch Array Antenna 331
C. Wu et al. (*Microwave and Optical Technology Letters*, Feb. 1991)
Performance of Probe-Fed Microstrip-Patch Element Phased Arrays 335
C. Liu, A. Hessel, and J. Shmoys (*IEEE Transactions on Antennas and Propagation*, Nov. 1988)
Analysis of Infinite Arrays of One- and Two-Probe-Fed Circular Patches 344
J. T. Aberle and D. M. Pozar (*IEEE Transactions on Antennas and Propagation*, April 1990)
Modelling of Wideband Proximity Coupled Microstrip Array Elements 356
J. S. Herd (*Electronics Letters*, Aug. 1990)
Scanning Characteristics of Infinite Arrays of Printed Antenna Subarrays 359
D. M. Pozar (*IEEE Transactions on Antennas and Propagation*, June 1992)

CHAPTER 8 OTHER TOPICS 369

Microstrip Antennas for Commercial Applications 371
J. Huang
Design of Low Cost Printed Antenna Arrays 380
J. P. Daniel et al. (*Proceedings of the 1985 International Symposium on Antennas and Propagation*, Aug. 1985)
Low-Cost Flat-Plate Array with Squinted Beam for DBS Reception 384
A. Henderson and J. R. James (*IEE Proceedings*, Dec. 1987)
Microstrip Yagi Antenna for Mobile Satellite Vehicle Application 390
J. Huang and A. C. Densmore (*IEEE Transactions on Antennas and Propagation*, July 1991)
Post Loaded Microstrip Antenna for Pocket Size Equipment at UHF 397
H. Kuboyama et al. (*Proceedings of the 1985 International Symposium on Antennas and Propagation*, Aug. 1985)
A Conformal Cylindrical Microstrip Array for Producing Omnidirectional Radiation Pattern 401
I. Jayakumar et al. (*IEEE Transactions on Antennas and Propagation*, Oct. 1986)
Radiation and Scattering from a Microstrip Patch on a Uniaxial Substrate 405
D. M. Pozar (*IEEE Transactions on Antennas and Propagation*, June 1987)
Analysis and Design of a Microstrip Reflectarray Using Patches of Variable Size 414
S. D. Targonski and D. M. Pozar (*IEEE Symposium on Antennas and Propagation Digest*, June 1994)
Active Subarray Module Development for Ka Band Satellite Communication Systems 417
S. Sanzgiri et al. (*IEEE Symposium on Antennas and Propagation Digest*, June 1994)
An MMIC Aperture-Coupled Microstrip Antenna in the 40GHz Band 421
H. Ohmine et al. (*Proceedings of the 1992 International Symposium on Antennas and Propagation*, Sept. 1992)

AUTHOR INDEX 425

SUBJECT INDEX 427

EDITORS' BIOGRAPHIES 431

Introduction

MICROSTRIP antenna technology has been the most rapidly developing topic in the antenna field in the last fifteen years, receiving the creative attentions of academic, industrial, and government engineers and researchers throughout the world. During this period there have been over 1500 published journal articles, five books, and innumerable symposia sessions and short courses devoted to the subject of microstrip antennas and arrays. As a result, microstrip antennas have quickly evolved from academic novelty to commercial reality, with applications in a wide variety of microwave systems. In fact, rapidly developing markets in personal communications systems (PCS), mobile satellite communications, direct broadcast television (DBS), wireless local area networks (WLANs), and intelligent vehicle highway systems (IVHS), suggest that the demand for microstrip antennas and arrays will increase even further.

The approaching maturity of microstrip antenna technology, coupled with the increasing demand and applications for such antennas, has led us to compile this reprint book on microstrip antennas. Since it is possible to include only a small fraction of the huge volume of work that has been published on the subject of microstrip antennas, we have had to be very selective in the choice of reprinted papers. We took as our guiding principle the goals of giving a thorough and up-to-date overview of the microstrip antenna art, and of selecting articles that would be most useful to the reader, with an emphasis on practical microstrip antenna designs, design data, and experimental prototypes. Therefore, we have not selected papers on the basis of historical precedence, preferring instead later papers if they were more complete or notable compared to the original coverage of a particular topic. Also, since our intended audience includes working antenna engineers and researchers as well as academic researchers and students, we have tried to avoid an overemphasis on theory and computer analysis.

Although microstrip antennas have proven to be a significant advance in the established field of antenna technology, it is interesting to note that it is usually their nonelectrical characteristics that make microstrip antennas preferred over other types of radiators. Microstrip antennas have a low profile and are light in weight, they can be made conformal, and they are well suited to integration with microwave integrated circuits (MICs). If the expense of materials and fabrication is not prohibitive, they can also be low in cost. When compared to traditional antenna elements such as reflectors, horns, slots, or wire antennas, however, the electrical performance of the basic microstrip antenna or array suffers from a number of serious drawbacks, including very narrow bandwidth, high feed network losses, poor cross polarization, and low power handling capacity. Many of the papers included in this volume address these issues directly, with the result that most of these drawbacks can be avoided, or at least alleviated to some extent, with innovative designs and configurations.

To ensure that we are presenting the most current state-of-the-art ideas in this still-evolving field, we are happy to be able to include six original review articles written solely for this book by experts in the field. These articles include overviews of CAD for microstrip antennas by Dave Pozar and Jim James, microstrip antenna characteristics by Dan Schaubert, dual and circularly polarized elements by Peter Hall, bandwidth enhancement techniques by Dave Pozar, microstrip array design by Dan Schaubert, and microstrip antennas for commercial applications by John Huang.

We have divided the book into eight chapters, beginning with three review articles on microstrip antennas in Chapter 1, and coverage of the basic element characteristics and feeding techniques in Chapter 2. The articles in these first two chapters should provide a good starting point for the reader new to the field. Methods for designing dual and circularly polarized microstrip elements are discussed in Chapter 3, with many practical results and design data provided. The important practical problem of increasing microstrip element bandwidth is the subject of Chapter 4; each paper in this chapter discusses an experimentally demonstrated technique. Chapter 5 presents a summary of the most popular modeling methods for microstrip antennas; this particular subject has received a great deal of attention, so we are only able to provide a small sampling of the variety of analysis methods that have been developed for the difficult problems encountered with microstrip antennas and arrays. One of the main advantages of microstrip antenna technology is the ease with which an array feed network can be fabricated in microstrip form, and some of the many popular variations of microstrip array design are presented in Chapter 6. Related to array design is the topic of mutual coupling, its calculation, and its effects in large arrays of microstrip elements; this topic is discussed in the papers included in Chapter 7. Finally, we close with Chapter 8, which includes several important topics that did not fit into the above chapters. These articles discuss the subjects of low-cost microstrip antenna fabrication, application to commercial communications systems, radar cross-section of microstrip antennas, design of microstrip reflectarrays, and integration of microstrip elements with active circuits.

Each chapter has been organized with the same format, beginning with an introduction to the subject and a discussion of the papers that make up that chapter. Each chapter also includes a list of additional references. We have tried to give some attention to the historical record of developments in the field in these introductions, as well as giving our view as to the relative merits and drawbacks of the methods and designs covered by the articles in that chapter. The reader may also note that we have included a substantial number of articles from the European and Japanese literature; this inclusion is a reflection of the fact that many developments in the field of microstrip antennas are being done outside the U.S.

We would like to extend our gratitude to Dr. Jim James, Dr. Peter Hall, and Dr. John Huang for contributing review papers written especially for this book. We hope that the inclusion of these articles, and the other original material in this book, will make it as current as possible. We would also like to thank our many colleagues who have contributed to the advancement of microstrip antenna design, analysis, and understanding. We are only sorry that we could not include more of their papers.

David M. Pozar
Daniel H. Schaubert

Amherst, Mass.

Microstrip Antennas

Chapter 1
Review Articles

WITH more than 1500 journal articles published on microstrip antennas, there has been ample opportunity and considerable demand for review articles that summarize work in this area. Hundreds of articles, both review and specific-topic papers, have been considered for this book, but only a small number could be selected for reprinting. Although many noteworthy review articles were considered, only two were selected for inclusion. Because the amount of published work in the area is so great, many review articles are either very long, or are focused on specific types of microstrip antennas or specific applications. We have selected the 1981 review paper by Carver and Mink as a useful article for this book. Although it is rather old, this article is still frequently referenced because it contains many fundamental design concepts that have spawned the various configurations in use today. This review paper was written after the 1979 Workshop on Printed Circuit Antenna Technology, at which designers and theoreticians gathered to exchange information about the fledgling but already well established field of microstrip antennas. It appeared in a special issue of the IEEE Transactions on Antennas and Propagation devoted to printed antennas, and was the earliest comprehensive survey of the field to appear in a major journal. It has stood the test of time as a useful reference for designers, and has thus been selected for this book. The second review article, by Daniel, Dubost, Terret, Citerne, and Drissi, appeared more recently and contains information about a variety of microstrip and related printed-circuit antennas. This article is also noteworthy because it gives a European perspective of developments in the field of microstrip antennas. A third review article, on CAD for microstrip antennas and arrays, was specially written for this book by D.M. Pozar and J.R. James. The authors present a current view of the topic and include unique commentaries, on the roles of CAD, and on their engineering experience in successfully designing microstrip antennas and arrays.

A sampling of other review articles and books is given in the reference list that follows. A cursory look at the index of most journals in the fields of antennas, microwaves, and electromagnetics will reveal other articles, and many conferences have included invited papers that provide snapshots of topics of importance at the time the article was prepared. In addition, the list below includes all of the books published to date on microstrip antennas.

Further Reading

Bahl, I. J., and Bhartia, P. *Microstrip Antennas.* Canton, Mass.: Artech House, 1980.

Bhartia, P., Rao, K., and Tomar, R. *Millimeter-Wave Microstrip and Printed Circuit Antennas,* Canton, Mass.: Artech House, 1991.

Gupta, K. C., and Benalla, A., eds. *Microstrip Antenna Design,* Canton, Mass.: Artech House, 1988.

Hall, P. S. "Review of practical issues in microstrip antenna design." *Dig. 1990 Journees Internationales de Nice sur les Antennes,* JINA '90, pp. 266–273, Nov. 1990.

James, J. R. "What's new in antennas." *IEEE Antennas and Prop.,* vol. 32, no. 1, pp. 6–18, February 1990.

James, J. R., and Hall, P. S., eds. *Handbook of Microstrip Antennas,* London: Peter Peregrinus (IEE), 1989.

James, J. R., Hall, P. S., and Wood, C. *Microstrip Antenna Theory and Design,* London: Peter Peregrinus (IEE), 1981.

Mailloux, R. J., McIlvenna, J. F., and Kernweis, N. P. "Microstrip array technology," *IEEE Trans. Antennas and Prop.,* vol. AP-29, pp. 25–37, Jan. 1981.

Pozar, D. M. "Microstrip antennas," *IEEE Proc.,* vol. 80, pp. 79–91, Jan. 1992.

Schaubert, D. H. "Microstrip antennas," *Electromagnetics,* vol. 12, pp. 381–401, 1992.

Microstrip Antenna Technology

KEITH R. CARVER, MEMBER, IEEE, AND JAMES W. MINK, MEMBER, IEEE

***Abstract*—A survey of microstrip antenna elements is presented, with emphasis on theoretical and practical design techniques. Available substrate materials are reviewed along with the relation between dielectric constant tolerance and resonant frequency of microstrip patches. Several theoretical analysis techniques are summarized, including transmission-line and modal-expansion (cavity) techniques as well as numerical methods such as the method of moments and finite-element techniques. Practical procedures are given for both standard rectangular and circular patches, as well as variations on those designs including circularly polarized microstrip patches. The quality, bandwidth, and efficiency factors of typical patch designs are discussed. Microstrip dipole and conformal antennas are summarized. Finally, critical needs for further research and development for this antenna are identified.**

INTRODUCTION

THE PURPOSES of this paper are to describe analytical and experimental design approaches for microstrip antenna elements, and to provide a comprehensive survey of the state of microstrip antenna element technology. A companion paper [1] discussed microstrip array design techniques. Taken together, these papers provide a reference for the current state of development of microstrip elements and arrays of elements at a time when advancements in this relatively new technology are being reported primarily in a wide variety of technical reports and private communications, and to a lesser extent in this TRANSACTIONS and other journals. This paper begins with a review of the state of printed circuit materials technology as it affects the design of microstrip antennas, and then describes several theoretical approaches to the analysis of rectangular and circular patches, as well as patches of other shapes and microstrip dipoles. Design curves are presented for both rectangular and circular patch shapes, and for linearly and circularly polarized elements. A discussion of the bandwidth and efficiency of the elements is presented with the patch size, shape, substrate thickness, and material properties as parameters. Several practical techniques are outlined for modifying the basic element for such special purpose applications as conformal arrays, feeds for dishes, dual-frequency communication systems, etc. The paper concludes with suggestions for future critical needs in the further development of the antenna.

The microstrip antenna concept dates back about 26 years to work in the U.S.A. by Deschamps [2] and in France by Gutton and Baissinot [3]. Shortly thereafter, Lewin [99] investigated radiation from stripline discontinuities. Additional studies were undertaken in the late 1960's by Kaloi, who studied basic rectangular and square configurations. However, other than the original Deschamps report, work was not reported in the literature until the early 1970's, when a conducting strip radiator separated from a ground plane by a dielectric substrate was described by Byron [4]. This half-wavelength wide and several-wavelength long strip was fed by coaxial connections at periodic intervals along both radiating edges, and was used as an array for Project Camel. Shortly thereafter, a microstrip element was patented by Munson [5] and data on basic rectangular and circular microstrip patches were published by Howell [6]. Weinschel [7] developed several microstrip geometries for use with cylindrical *S*-band arrays on rockets. Sanford [8] showed that the microstrip element could be used in conformal array designs for *L*-band communication from a KC-135 aircraft to the ATS-6 satellite. Additional work on basic microstrip patch elements was reported in 1975 by Garvin *et al.* [9], Howell [10], Weinschel [11], and Janes and Wilson [12]. The early work by Munson on the development of microstrip antennas for use as low-profile flush-mounted antennas on rockets and missiles showed that this was a practical concept for use in many antenna system problems, and thereby gave birth to a new antenna industry.

Mathematical modeling of the basic microstrip radiator was initially carried out by the application of transmission-line analogies to simple rectangular patches fed at the center of a radiating wall [13], [14]. The radiation pattern of a circular patch was analyzed and measurements reported by Carver [15]. The first mathematical analysis of a wide variety of microstrip patch shapes was published in 1977 by Lo *et al.* [16], who used the modal-expansion technique to analyze rectangular, circular, semicircular, and triangular patch shapes. Similar comprehensive reports on advanced analysis techniques were published by Derneryd [14], [17], Shen and Long [18], and Carver and Coffey [19]. By 1978 the microstrip patch antenna was becoming much more widely known and used in a variety of communication systems. This was accompanied by increased attention by the theoretical community to improved mathematical models which could be used for design. In October 1979, the first international meeting devoted to microstrip antenna materials, practical designs, array configurations, and theoretical models was held at New Mexico State University (NMSU), Las Cruces, under cosponsorship of the U.S. Army Research Office and NMSU's Physical Science Laboratory [20].

The terms stripline and microstrip are often encountered in the literature, in connection with both transmission lines and antennas. A *stripline* or *triplate* device is a sandwich of three parallel conducting layers separated by two thin dielectric substrates, the center conductor of which is analogous to the center conductor of a coaxial transmission line. If the center conductor couples to a resonant slot cut orthogonally in the upper conductor, the device is said to be a stripline radiator [21]. Although there are many variations on this printed-circuit stripline slot antenna, these are outside the scope of this paper and will not be considered further.

By contrast a *microstrip* device in its simplest form con-

Reprinted from *IEEE Trans. Antennas Propaga.*, vol. AP-29, no. 1, pp. 2–24, Jan. 1981.

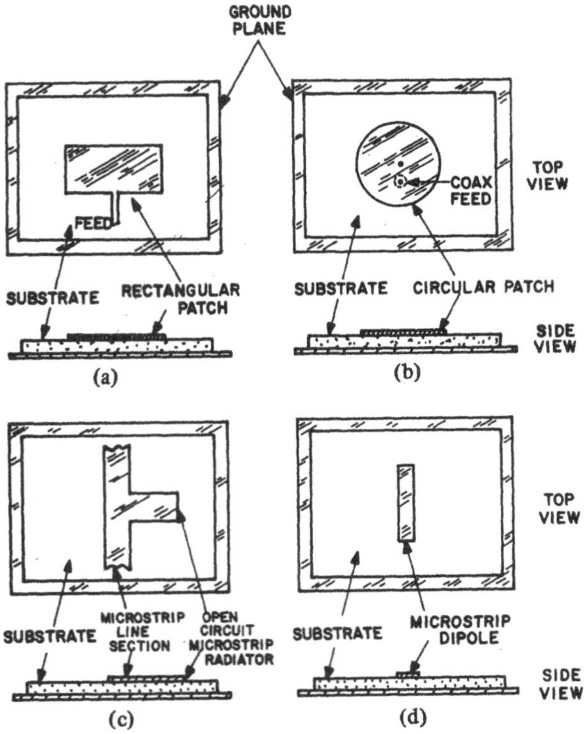

Fig. 1. (a) Rectangular microstrip patch antenna. (b) Circular microstrip patch antenna. (c) Open-circuit microstrip radiator. (d) Microstrip dipole antenna.

sists of a sandwich of two parallel conducting layers separated by a single thin dielectric substrate [22]. The lower conductor functions as a ground plane, and the upper conductor may be a simple resonant rectangular or circular patch, a resonant dipole, or a monolithically printed array of patches or dipoles and the associated feed network. Since arrays of microstrip patches and dipoles were considered in the companion article on microstrip arrays [1], this paper will concentrate on basic microstrip patches and dipoles. Fig. 1 shows a representative collection of microstrip patch and dipole shapes and their associated dielectric substrates and ground planes. Practical microstrip antennas have been developed for use from 400 MHz to 38 GHz, and it can be expected that the technology will soon extend to 60 GHz and beyond. Since mutual coupling between microstrip elements is considered elsewhere in [88], it will not be discussed in this paper.

II. MATERIALS FOR PRINTED CIRCUIT ANTENNAS

The propagation constant for a wave in the microstrip substrate must be accurately known in order to predict the resonant frequency, resonant resistance, and other antenna quantities. Antenna designers have found that the most sensitive parameter in microstrip antenna performance estimation is the dielectric constant of the substrate material, and that the manufacturer's tolerance on ϵ_r is sometimes inadequate.

The change in operating frequency of a thin substrate microstrip antenna due solely to a small tolerance-related change of the substrate dielectric constant may be expressed as

$$\frac{\delta f}{f_0} = -\frac{1}{2}\frac{\delta \epsilon_r}{\epsilon_r}, \tag{1}$$

where f_0 is the resonant frequency of a microstrip antenna assuming a magnetic wall boundary condition, ϵ_r is the relative dielectric constant, δf is the change in resonant frequency, and $\delta \epsilon_r$ is the change in relative dielectric constant. For example, if the operating frequency of the antenna is to be predicted to ±0.5 percent using $\epsilon_r = 2.55$, the required accuracy is $\delta \epsilon_r = 0.025$. However a typical quoted dielectric constant accuracy for materials of this type is $\delta \epsilon_r = \pm 0.04$. The relative frequency change for small dimensional changes may be expressed in terms of linear dimensions or in terms of temperature changes as follows:

$$\frac{\delta f}{f_0} = -\frac{\delta l}{l} = -\alpha_t \delta T, \tag{2}$$

where α_t is the thermal expansion coefficient, T is the temature in degrees Celsius, l is the frequency-determining length of the microstrip antenna. An uncertainty of less than 0.5 percent in the operating frequency with a temperature variation of 100°C would require the thermal expansion coefficient α_t to be less than $50 \times 10^{-6}/°C$. Commonly used materials are adequate in terms of thermal expansion. While thickness variation in the substrate material can have an effect upon the operating frequency, this factor is much less important than the dielectric constant tolerance. With this background one can determine the suitability of various dielectric materials for use in printed circuit antennas.

Available Microwave Substrates

There are many substrate materials on the market today with dielectric constants ranging from 1.17 to about 25 and loss tangents from 0.0001 to 0.004 [102]–[104]. Comparative data on most substrates ($2.1 < \epsilon_r < 25$) are given in Table I [23], [24]. Polytetrafluoroethylene (PTFE) substrates reinforced with either glass woven web or glass random fiber are very commonly used because of their desirable electrical and mechanical properties, and because of a wide range of available thicknesses and sheet sizes. For woven web materials, thicknesses range from 0.089 mm to 12.7 mm and sheet sizes up to 91.4 cm × 91.4 cm. Glass random fiber is available in thicknesses from 0.508 mm to 3.175 mm and in sheet sizes up to 40.64 cm × 101.6 cm. The discontinuous nature of the fiber and the relatively soft and deformable polymer matrix allow one to form this material on complex surfaces. Stress relief may be accelerated by heating the material. Also, this material is available in shapes other than sheets, such as rods or cylinders. For applications requiring high dielectric constants, alumina ceramic substrates ($9.7 < \epsilon_r < 10.3$) are frequently used. Typical commercially available substrates are K-6098 teflon/glass cloth ($\epsilon_r \cong 2.5$), RT/duroid-5880 PTFE ($\epsilon_r \cong 2.2$), and Epsilam-10 ceramic-filled teflon ($\epsilon_r \cong 10$).

Anisotropy

In order to obtain the necessary mechanical properties of PTFE, fill materials are introduced into the polymer matrix [23], [24]. This fill material is commonly glass fiber although it may also be a ceramic. In either case these filler materials take on preferred orientations during the manufacturing process. Composites containing fibrous reinforcement material oriented in the plane of the sheet will show a dependence of the dielectric constant on the electric field orientation with a higher value for electric fields in the plane of the sheet

TABLE I
AN OVERVIEW OF MAJOR MICROWAVE SUBSTRATES (AFTER [23])

Product	ϵ_r (X-Band)	tan δ (X-Band)	Dimensional Stability	Temperature Range in °C
PTFE unreinforced	2.10	0.0004	poor	−27 to +260
PTFE glass woven web	2.17	0.0009	excellent	−27 to +260
	2.33	0.0015		
	2.45	0.0018	very good	−27 to +260
	2.55	0.0022		
PTFE glass random fiber	2.17	0.0009	fair	−27 to +260
	2.35	0.0015		
PTFE quartz reinforced	2.47	0.0006	excellent	−27 to +260
Cross linked poly styrene/ woven quartz	2.65	0.0005	good	−27 to +260
Cross linked poly styrene/ ceramic powder-filled	3 to 15	from 0.00005 to 0.0015	fair	−27 to +110
Cross linked poly styrene/ glass reinforced	2.62	0.001	good	−27 to +110
Irradiated polyolefin	2.32	0.0005	poor	−27 to +100
Irradiated polyolefin/ glass reinforced	2.42	0.001	fair	−27 to +100
Polyphenylene oxide (PPO)	2.55	0.00016	poor	−27 to +193
Silicone resion ceramic powder-filled	3 to 25	from 0.0005	fair to medium	−27 to +268
Sapphire	9.0	0.0001	excellent	−24 to +371
Alumina ceramic	9.7 to 10.3	0.0004	excellent	to 1600 unclad
Glass bonded mica	7.5	0.0020	excellent	−27 to +593 unclad
Hexcell (laminate)	1.17 to 1.40 at 1.4 GHz	–	excellent	−27 to +260
Air with/rexolite standoffs				
Fused quartz	3.78	0.001	excellent	–

TABLE II
TYPICAL DIELECTRIC CONSTANT VERSUS MAJOR AXIS ORIENTATION OF THE ELECTRIC FIELD

Material \ ϵ_r	X Direction	Y Direction	Z Direction	Quoted Value	$\frac{\delta\epsilon_r}{\epsilon_r}$ (Percent)
Random fiber PTFE	2.454	2.432	2.347	2.35 ± 0.04	1.7
Ceramic PTFE	10.68	10.70	10.40	10.5 ± 0.25	2.4
Glass cloth PTFE	2.88	2.88	2.43	2.45 ± 0.04	1.6

than when the field is transverse to the sheet. The magnitude of this effect is a function of the difference in dielectric constants between the fiber orientation and the volume ratio of the fiber to polymer. Typical examples of this effect are shown in Table II.

As one can see from Table II, the value of the dielectric constant quoted by the manufacturer is essentially the value for the case where the electric field is perpendicular to the sheet. Usually this orientation of the electric field is the one needed for antenna engineers. However, the designer needs to be aware of this material property to insure the proper operation of the antenna system or for the proper interpretation of material measurements. In the microwave region, dielectric constant measurements are typically made using stripline resonator techniques. Because of fringing fields around the strip, there is an uncertainty associated with the measurements. The dielectric constant of PTFE-based substrate materials tends to decrease with increasing temperature as shown in Fig. 2. For this material the average change in dielectric constant over the temperature range −75°C to +100°C is about $\alpha_E = 96$ ppm/°C. An abrupt transition change of about $\delta\epsilon = 0.011$, which occurs at a temperature between zero and 20°C, is characteristic of PTFE-based materials. The exact temperature at which this change occurs is a function of the rate at which the temperature is changing. Over the temperature range of −75°C to 100°C the relative change in operating frequencies is about 0.8 percent due to the change of dielectric constant. It turns out that changes in linear dimensions due to thermal expansion tend to compensate the effect of a changing dielectric constant. Combining (1) and (2) one obtains

$$\frac{\delta f}{f_0} = (-\alpha_r + \tfrac{1}{2}\alpha_E)\delta T. \tag{3}$$

Over the temperature range from −75°C to 100° a typical net change of resonant frequency is 0.03 percent. Thus, with proper selection of materials, it is possible to almost eliminate temperature effects on the resonant frequency of a microstrip patch antenna.

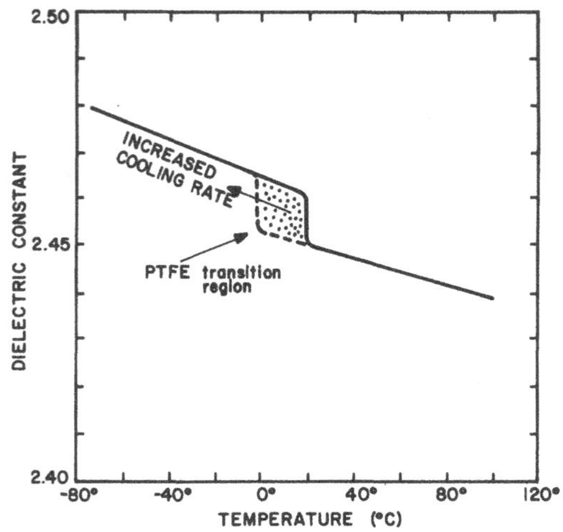

Fig. 2. Dependence of dielectric constant on temperature for polytetrafluoroethylene (PTFE) substrates. After Nowicki [23].

Fig. 3. Composite microstrip square patch using 0.0065-in PTFE substrate bonded on both sides of 0.25-in Hexcell honeycomb dielectric. Substrate is cut away to show both Hexcell and white adhesive on bottom PTFE layer.

Specialized Substrate Material

While the material most frequently used for printed antenna elements is PTFE, there are other materials used for specialized applications. Composite materials find applications where weight is important, such as for spacecraft antennas, or where large physical separation between the antenna element and the ground plane is required. One such substrate consists of two thin layers of PTFE bonded on each side of hexcell (honeycomb) material as shown in Fig. 3 [25], [26]. Depending upon the thickness of the dielectric layers, the dielectric constant ranges from 1.17 to about 1.40 for a composite substrate thickness of 0.25 in.

A second approach to achieve lightweight antenna structures is to support the radiating elements on dielectric spacers between the ground plane and the radiating element. If these spacers are placed at regions within the antenna where the electric field is small, the change in operating parameters from an air dielectric antenna will be small and can easily be computed using perturbation theory [27].

It is expected that PTFE will continue to be the dominant substrate material for printed circuit antennas. The dimensional stability, ease of processing, relatively low electrical losses, good copper adhesion, and availability of large sheets as well as preformed shapes make this class of materials very attractive. A primary limiting factor for this material is the relative uncertainty of the dielectric constant from batch to batch. As systems move to higher frequencies, other substrate materials with lower losses will need to be developed. One approach may be to employ syntactic foams with a combination of bubbles and PTFE.

III. ANALYSIS TECHNIQUES FOR MICROSTRIP ELEMENTS

Transmission-Line Models

The simplest analytical description of a rectangular microstrip patch utilizes transmission-line theory and models the patch as two parallel radiating slots [13] as shown in Fig. 4. Each radiating edge of length a is modeled as a narrow slot radiating into a half-space, with a slot admittance given by [27, p. 183]

$$G_1 + jB_1 \cong \frac{\pi a}{\lambda_0 z_0}[1 + j(1 - 0.636 \ln k_0 w)], \tag{4}$$

where λ_0 is the free-space wavelength, $z_0 = \sqrt{\mu_0/\epsilon_0}$, $k_0 = 2\pi/\lambda_0$, and w is the slot width, approximately equal to the substrate thickness t. Since the slots are identical (except for fringing effects associated with the feed point on edge 1), an identical expression holds for the admittance of slot 2. Assuming no field variation along the direction parallel to the radiating edge, the characteristic admittance is given by

$$Y_0 = \frac{a\sqrt{\epsilon_r}}{t z_0}, \tag{5}$$

where t is the substrate thickness and ϵ_r is the relative dielectric constant. Since it is desired to excite the slots 180° out of phase, the dimension b is set equal to slightly less than $\lambda_d/2$, where $\lambda_d = \lambda_0/\sqrt{\epsilon_r}$, i.e., $b = 0.48\lambda_d$ to $0.49\lambda_d$. This slight reduction in resonant length is necessary because of the fringing fields at the radiating edges. By properly choosing this length reduction factor q, the admittance of slot 2 after transformation becomes [90]

$$\tilde{G}_2 + j\tilde{B}_2 = G_1 - jB_1, \tag{6}$$

so that the total input admittance at resonance becomes

$$Y_{in} = (G_1 + jB_1) + (\tilde{G}_2 + j\tilde{B}_2) = 2G_1. \tag{7}$$

In a typical design, $a = \lambda_0/2$ so that $G_1 = 0.00417$ mhos, i.e.,

$$R_{in} = (1/2G_1) = 120 \quad \Omega. \tag{8}$$

The resonant frequency is found from

$$f_r = \frac{c}{\lambda_d\sqrt{\epsilon_r}} = q\,\frac{c}{2b\sqrt{\epsilon_r}}. \tag{9}$$

The advantage of this model lies in its simplicity, i.e., the resonant frequency and input resistance are given by the simple formulas (8) and (9). The fringe factor q determines the accuracy of the resonant frequency and in practice is

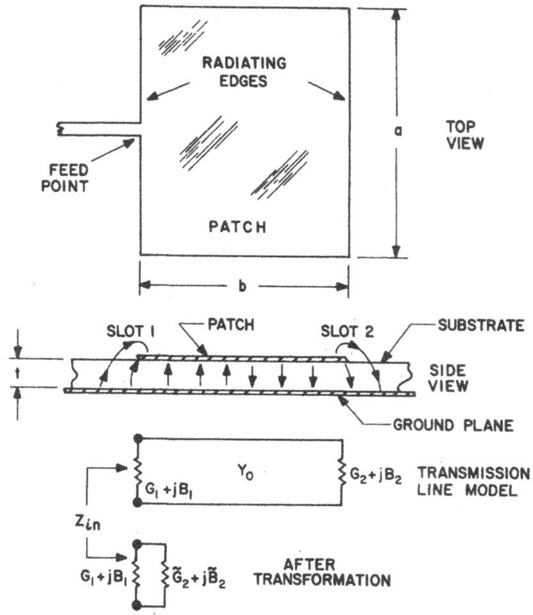

Fig. 4. Transmission-line model of rectangular microstrip antenna. After Munson [13].

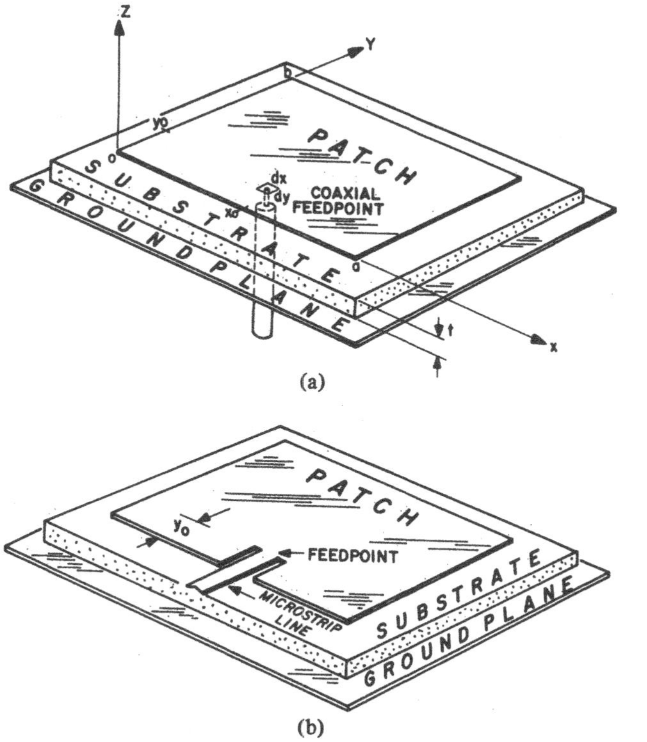

Fig. 5. (a) Rectangular microstrip patch with inset coaxial feedpoint. (b) Patch with inset microstrip transmission-line feed.

determined by measuring f_r for a rectangular patch on a given substrate. It is then assumed that the same q value holds for patches of other sizes on this same substrate and in the same general frequency range.

Modal-Expansion Cavity Models

Although the preceding transmission-line model is easy to use, it suffers from numerous disadvantages. It is only useful for patches of rectangular shape, the fringe factor q must be empirically determined, it ignores field variations along the radiating edge, it is not adaptable to inclusion of the feed, etc. These disadvantages are eliminated in the modal-expansion analysis technique whereby the patch is viewed as a thin TM_z-mode cavity with magnetic walls [16], [19], [28]-[34]. The field between the patch and the ground plane is expanded in terms of a series of cavity resonant modes or eigenfunctions along with its eigenvalues or resonant frequencies associated with each mode. The effect of radiation and other losses is represented in terms of either an artifically increased substrate loss tangent [16] or by the more elegant method of an impedance boundary condition at the walls [28], [29]. This results in a much more accurate formulation for the input impedance, resonant frequency, etc, for both rectangular and circular patches at only a modest increase in mathematical complexity.

Rectangular Patch

Consider a rectangular patch of width a and length b over a ground plane with a substrate of thickness t and a dielectric constant ϵ_r, as shown in Fig. 5. So long as the substrate is electrically thin, the electric field will be z-directed and the interior modes will be TM_{mn} to z so that

$$E_z(x, y) = \sum_m \sum_n A_{mn} e_{mn}(x, y), \tag{10}$$

where A_{mn} are the mode amplitude coefficients and e_{mn} are the z-directed orthonormalized electric field mode vectors. For the elementary case of a nonradiating cavity with perfect open-circuit walls,

$$e_{mn}(x, y) = \frac{\chi_{mn}}{\sqrt{\epsilon abt}} \cos k_n x \cos k_m y \tag{11}$$

with

$$\chi_{mn} = \begin{cases} 1, & m = 0 \text{ and } n = 0 \\ \sqrt{2}, & m = 0 \text{ or } n = 0 \\ 2, & m \neq 0 \text{ and } n \neq 0. \end{cases} \tag{12}$$

The mode vectors satisfy the homogeneous wave equation, and the eigenvalues satisfy the separation equation

$$k_{mn}^2 = \omega_{mn}^2 \mu\epsilon = k_n^2 + k_m^2. \tag{13}$$

For the nonradiating cavity, $k_n = (n\pi/a)$ and $k_m = (m\pi/b)$. The magnetic field orthonormalized mode vectors are found from Maxwell's equations as

$$\mathbf{h}_{mn} = \frac{1}{j\omega\mu} \frac{\chi_{mn}}{\sqrt{\epsilon abt}} \cdot \{\mathbf{x} k_m \cos k_n x \sin k_m y - \mathbf{y} k_n \sin k_n x \cos k_m y\}. \tag{14}$$

For this nonradiating case it is seen that the boundary condition $\mathbf{n} \times \mathbf{h}_{mn} = 0$ is satisfied on each perimeter wall.

As the cavity is now allowed to radiate, the eigenvalues become complex, corresponding to complex resonant frequencies, so that $|k_n|$ is slightly less than $n\pi/a$ and $|k_m|$ is slightly less than $m\pi/b$. The magnetic field mode vectors

$\mathbf{h}_{mn}$ no longer have a zero tangential component on the cavity sidewalls. However a perturbational solution shows that the electric field mode vectors are still very accurately given by (11).

Consider now the effect of a z-directed current probe I_0 of small rectangular cross section $(d_x d_y)$ at (x_0, y_0) as shown in Fig. 5(a). The coefficients of each electric mode vector are found from [27]:

$$A_{mn} = \frac{j\sqrt{\mu\epsilon}k}{k^2 - k_{mn}^2} \iiint \mathbf{J} \cdot \mathbf{e}_{mn}^* \, dv, \tag{15}$$

which then reduces to

$$A_{mn} = jI_0 \sqrt{\frac{\mu t}{ab}} \frac{k \chi_{mn}}{k^2 - k_{mn}^2} G_{mn} \cos k_m y_0 \cos k_n x_0 \tag{16}$$

where

$$G_{mn} = \frac{\sin (n\pi d_x/2a)}{n\pi d_x/2a} \cdot \frac{\sin (m\pi d_y/2b)}{m\pi d_y/2b} \tag{17}$$

and

$$k_{mn} = \tilde{\omega}_{mn}\sqrt{\mu\epsilon}. \tag{18}$$

In (18) $\tilde{\omega}_{mn}$ is the complex resonant frequency of the mnth mode as found from (13). The relation (15) for the coefficients is based on the orthogonality of the mode vectors; although the introduction of the radiation condition means that these mode vectors are no longer orthogonal in the strict sense, for electrically thin substrates the error due to this assumption is negligible. The factor G_{mn} accounts for the width of the feed; for coaxial feeds $d_x = d_y$ and the cross-section area $d_x d_y$ is set equal to the effective cross-section area of the probe. For patches fed by a microstrip transmission line at $y_0 = 0$, set $d_y = 0$ and use d_x as the feed line width as a zero-order approximation ignoring junction capacitance effects.

Substituting (16) into (10) we obtain

$$E_z(x, y) = jI_0 Z_0 k \sum_{m=0}^{\infty} \sum_{n=0}^{\infty} \frac{\psi_{mn}(x, y)\psi_{mn}(x_0, y_0)}{k^2 - k_{mn}^2} G_{mn}, \tag{19}$$

where $Z_0 = \sqrt{\mu/\epsilon}$, $k = \omega\sqrt{\mu\epsilon}$, $k_{mn}^2 = k_m^2 + k_n^2$, and

$$\psi_{mn} = \frac{\chi_{mn}}{\sqrt{ab}} \cos k_n x \cos k_m y \cong \frac{\chi_{mn}}{\sqrt{ab}} \cos \frac{n\pi x}{a} \cos \frac{m\pi y}{b}. \tag{20}$$

The voltage at the feed is now computed as

$$V_{\text{in}} = -tE_z(x_0, y_0) = -jI_0 Z_0 kt \sum_{m=0}^{\infty} \sum_{n=0}^{\infty} \frac{\psi_{mn}^2(x_0, y_0)}{k^2 - k_{mn}^2} G_{mn}. \tag{21}$$

Therefore the input impedance is

$$Z_{\text{in}} = \frac{V_{\text{in}}}{I_0} = -jZ_0 kt \sum_{m=0}^{\infty} \sum_{n=0}^{\infty} \frac{\psi_{mn}^2(x_0, y_0)}{k^2 - k_{mn}^2} G_{mn}. \tag{22}$$

The (0, 0) term with $k_{00} = 0$ is the static capacitance term with a shunt resistance to represent loss in the substrate. The (1, 0) term represents the dominant RF mode and is identical to the transmission-line mode discussed in the previous section; for this mode, (11) shows that there is no field variation in the x direction and a $\cos (\pi y/b)$ variation in the y direction. This mode is equivalent to a parallel R-L-C network where R represents radiation, substrate, and copper losses. If the patch is square or nearly so, the (0, 1) mode can also be excited as a degenerate mode. All the higher order modes have negligible losses and sum to form a net inductance L.

Fig. 6(a) shows a general network representation of the input impedance, and Fig. 6(b) shows a network model over a narrow band about an isolated TM_{10} mode, where the net series inductance is L_T. The feed probe diameter as expressed by the factor G_{mn} is the major factor in determining L_T, since it governs the convergence of the series. Equation (22) can be written as

$$Z_{\text{in}} = jX_L - \frac{j(\omega/C_{10})}{\omega^2 - (\omega_r + j\omega_i)^2}, \tag{23}$$

where

$$(\omega_r + j\omega_i)^2 = \omega_{10}^2(1 + j/Q) \tag{24}$$

$$C_{10} = \tfrac{1}{2} C_{\text{dc}} \cos^{-2} (\pi y_0/b) \tag{25}$$

with C_{dc} being the dc patch capacitance $(\epsilon ab/t)$, Q the quality factor for the TM_{10} mode, and ω_{10} the radian frequency at resonance. A simple means for determining both ω_{10} and Q will be given in a subsequent paragraph. The series inductive reactance is given by

$$X_L = -\frac{1}{\omega C_{\text{dc}}} + \frac{\omega}{C_{\text{dc}}} \sum_{\substack{mn \neq 10 \\ mn \neq 00}}^{M} \sum^{N} \frac{\chi_{mn}^2 \cos^2 \left(\frac{n\pi x_0}{a}\right) \cos^2 \left(\frac{m\pi y_0}{b}\right)}{\omega_{mn}^2 - \omega^2} \cdot G_{mn}, \tag{26}$$

which shows that the series reactance is proportional to the substrate thickness.

The next problem is to find the complex eigenvalues k_{mn}. Except near the TM_{10} mode resonant frequency (or also the TM_{01} resonant frequency for nearly square patches), $k_{mn}^2 \cong (m\pi/b)^2 + (n\pi/a)^2$. The complex eigenvalue k_{10} may be found by either lumping all the losses into an effective dielectric loss tangent [32], or by incorporating the losses into the conductance of the radiating walls and imposition of impedance-type boundary conditions [28], which leads to a complex transcendental eigenvalue equation [29] which

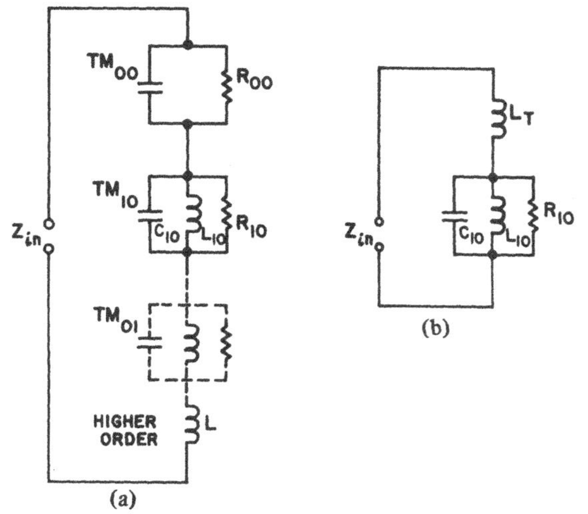

Fig. 6. (a) General network model representing microstrip antenna. (b) Network model over narrowband about isolated TM_{10} mode. After Richards *et al.* [32].

holds for thin substrates:

$$\tan k_{10}b = \frac{2k_{10}\alpha_{10}}{k_{10}^2 - \alpha_{10}^2}, \tag{27}$$

where

$$\alpha_{10} = j\frac{2\pi Z_0}{\lambda_0}\frac{t}{a} Y_w \tag{28}$$

with Y_w being the admittance of the radiating walls at $y = 0$ and $y = b$. A simple iterative algorithm has been developed [29] for finding the complex eigenvalue, i.e.,

$$k_{10} = (\pi/b) - (\Delta_4/b), \tag{29}$$

where

$$\Delta_{p+1} = \frac{2(\alpha_{10}b)(\pi - \Delta_p)}{(\alpha_{10}b)^2 + 2\Delta_p\pi - \Delta_p^2 - \pi^2} - \frac{\Delta_p^3}{3} \tag{30}$$

with $\Delta_0 = 0$ as a seed value. Equation (30) is derived from (27) with $\tan k_{10}b$ expanded in the first two terms of a Taylor series about π. By using (27), k_{10} is found as a complex pole whose real part is typically from 96 to 98 percent of (π/b), and whose imaginary part is positive and proportional to the power lost through radiation. This is equivalent to rigorously solving for the fringing factor q. The radiation quality factor is then found for thin substrates by [29]

$$Q_r = \frac{\text{Re}\,(k_{10})}{2\,\text{Im}\,(k_{10})}, \tag{31}$$

from which the radiation resistance at resonance (referred to the input) is found by

$$R_{\text{rad}} = \frac{Q_r}{\omega C_{10}}, \tag{32}$$

where

$$C_{10} = (1/2)C_{\text{dc}} \cos^{-2}(\pi y_0/b). \tag{33}$$

In addition to radiation losses, the cavity also sustains losses through the external surface wave (caused by the presence of the substrate) as well as heat losses associated with the copper (and adhesive film used to bond to the substrate) and the substrate itself; for thin substrates, these losses are small at resonance in comparison to the radiation loss. It may be shown that the loss conductance referred to the input voltage is given by

$$G_{\text{cu}} = R_s \frac{\pi^2 a}{2\omega^2\mu^2 bt} \quad (\mho), \tag{34}$$

where $R_s = \sqrt{\mu\omega/2\sigma}$ is the wave resistance of the conductor. The substrate loss conductance is given by

$$G_{\text{di}} = \omega C_{10} \tan\delta \quad (\mho) \tag{35}$$

where $\tan\delta$ is the substrate loss tangent (typically 0.001 or less). The total Q for thin substrates is therefore given by

$$Q = \frac{\omega W}{P_{\text{in}}} = \omega \frac{C_{10}}{G_{\text{in}}}, \tag{36}$$

where G_{in} is the input conductance given by

$$G_{\text{in}} = G_{\text{rad}} + G_{\text{cu}} + G_{\text{di}} = \frac{1}{R_{\text{rad}}} + G_{\text{cu}} + G_{\text{di}}. \tag{37}$$

In a practical design for an edge-fed patch, the input resistance ranges from 100–200 Ω; this value can be reduced by insetting the feed point for either coaxial inputs [19] or microstripline inputs [35] by noting through (32) and (33) that the radiation resistance varies as $\cos^2(\pi y_0/b)$. The antenna efficiency is given by

$$\eta = \frac{G_{\text{rad}}}{G_{\text{in}}} \tag{38}$$

and ranges typically from 95 to 99 percent, i.e., from 0.2 to 0.05 dB.

Wall Admittance of Rectangular Patch

Radiated and reactively stored power in the region exterior to the patch cavity is represented as the wall admittance Y_w, as used in (28). No rigorous solutions for the wall admittance of a microstrip patch as yet have been found, although several approximate solutions have been suggested, including the admittance of a slot in a ground plane [36], a parallel-plate TEM waveguide radiating into a half-space [19], the fringe admittance of a microstrip transmission line [37], [98], [99], and a Green's function for a long rectangular microstrip patch [38]. None of these analogous geometries is completely satisfactory, and a solution with full generality awaits current work in progress based on the Wiener–Hopf method [39], [40]. In the absence of a rigorous solution, a reasonable approach is to assume that the wall conductance is that of a wave normally incident on a parallel-plate TEM waveguide slot radiating into a half-space [27]; for electrically small slot widths, the conductance per unit length is given by $\pi/(376\lambda_0)$ mho/m. If it is further assumed that only the dominant TM_{10} mode is excited, then the wave is normally incident on the radiating edges with the field intensity being uniform across both of these edges. In this case the total wall conductance is given by

$$G_w = (\pi/376)(a/\lambda_0) \qquad (\mho). \tag{39}$$

The wall susceptance may be approximated from Hammerstad's formula for the capacitance of an open microstrip circuit [37] and assumes the form

$$B_w = 0.01668\,(\Delta l/t)(a/\lambda_0)\epsilon_e \qquad (\mho), \tag{40}$$

where

$$\Delta l/t = 0.412\left[\frac{\epsilon_e + 0.300}{\epsilon_e - 0.258}\right]\left[\frac{\dfrac{a}{t} + 0.262}{\dfrac{a}{t} + 0.813}\right] \tag{41}$$

and ϵ_e is an effective dielectric constant given by [41]

$$\epsilon_e = \frac{\epsilon_r + 1}{2} + \frac{\epsilon_r - 1}{2}\left[1 + \frac{10t}{a}\right]^{-1/2} \tag{42}$$

so that the TM_{10} lumped wall admittance of the radiating edges is

$$Y_w = G_w + jB_w. \tag{43}$$

It should be noted that the susceptance given by (40) is based on Hammerstad's nondispersive static capacitance relation and disagrees with the susceptance given in (4) which is based on a dynamic capacitance. Neither formula is rigorously correct for the microstrip antenna, and better relations await theoretical work in progress.

It will be shown in a subsequent section that (39) and (40) lead to a prediction of resonant input resistance and resonant frequency which is in good agreement with measured results for the aspect ratios $1 < a/b < 2$; for larger aspect ratios, the assumption of a uniform field and normal incidence on the radiating edges is no longer very good, so that (39) and (40) are insufficiently accurate. The advantage to this impedance boundary condition method of representing the exterior field through Y_w is that it explicitly provides (through the eigenvalue equation (27)) for improved solutions to the exterior problem, when these are published in future literature.

It should be mentioned that the mode vectors of (11) may be viewed as spatial harmonics resulting from the resonance of quasi-TEM plane waves launched from the feed which, by zig-zagging off the cavity parameter wall, travel a total distance and experience phase shifts at the walls so as to produce constructive interference. An analysis of this resonance condition as a function of the patch aspect ratio a/b has been provided by Chang and Kuester [42], who have shown that an optimum range for the aspect ratio exists in the sense of low-Q operation. The Wiener–Hopf technique was used to obtain the wall reflection coefficient (as a function of incidence angle, substrate thickness, and dielectric constant) which may in principle then be used to obtain the wall admittance. The reflection coefficient involves two infinite integrals, the evaluation of which reveals both LSE and LSM surface-wave modes which can propagate on the exterior grounded substrate. Importantly, this analysis shows that the wall admittance is a function of both frequency and angle of incidence, which then shows that Y_w cannot be rigorously represented by the approximate expressions (39) or (40) which assume normal incidence. We may therefore anticipate that Y_w will depend on both dimensions a and b. Carver [29], by near-field probing of the fields near the wall, has shown empirically that the wall admittance expressions (39), (40), and (43) may be modified by multiplying Y_w by an aspect ratio factor $F_y(a/b)$ given by

$$F_y(a/b) = 0.7747 + 0.5977\,(a/b - 1) - 0.1638\,(a/b - 1)^2, \tag{44}$$

which leads to better agreement of the predicted resonant resistance and resonant frequency versus aspect ratio with measured results at L-band and S-band than by assuming that $F_y = 1$. Nonetheless, (44) is empirical, and the upper frequency limit to its validity is unknown; clearly more work in the numerical evaluation of the Wiener–Hopf solution is needed, perhaps reducing this to curve-fit polynomials such as given in (44).

Radiation Pattern of Rectangular Patch

The far-field radiation pattern of a rectangular microstrip patch operating in the TM_{10} mode is broad in both the E and H planes. The pattern of a patch over a large ground plane may be calculated by modeling the radiator as either two parallel uniform magnetic line sources of length a, separated by distance b [96], or as two equivalent electric current sources as suggested in Fig. 7. The effect of the ground plane and substrate is handled by imaging the slot at an electrical distance kt. If the slot voltage across either radiating edge is taken as V_0, the calculated fields are

$$E_\theta = -\frac{jV_0k_0ae^{-jk_0r}}{\pi r}\,[\cos(kt\cos\theta)] \cdot \left[\frac{\sin\left[k_0\dfrac{a}{2}\sin\theta\sin\phi\right]}{k_0\dfrac{a}{2}\sin\theta\sin\phi}\right] \cdot \left[\cos\left(k_0\frac{b}{2}\sin\theta\cos\phi\right)\right]\cos\phi, \quad \left(0 \leqslant \theta \leqslant \frac{\pi}{2}\right) \tag{45}$$

$$E_\phi = \frac{jV_0k_0ae^{-jk_0r}}{\pi r}\,[\cos(kt\cos\theta)] \cdot \left[\frac{\sin\left[k_0\dfrac{a}{2}\sin\theta\sin\phi\right]}{k_0\dfrac{a}{2}\sin\theta\sin\phi}\right] \cdot \left[\cos\left(k_0\frac{b}{2}\sin\theta\cos\phi\right)\right]\cos\theta\sin\phi, \quad \left(0 \leqslant \theta \leqslant \frac{\pi}{2}\right), \tag{46}$$

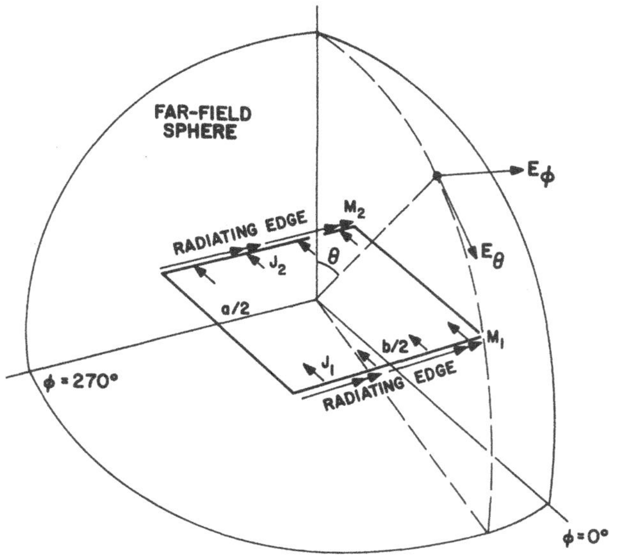

Fig. 7. Geometry for far-field pattern of rectangular microstrip patch.

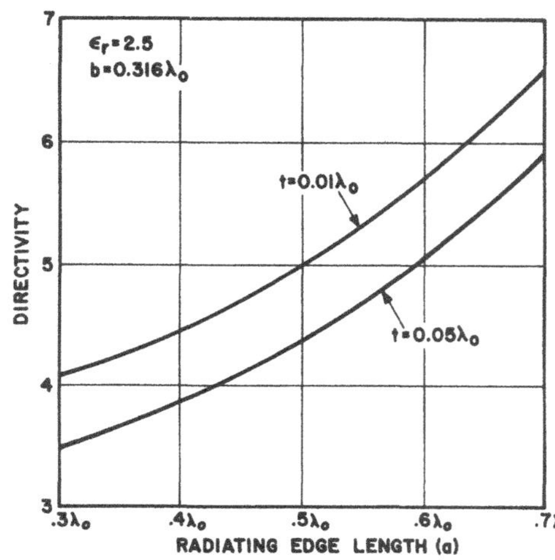

Fig. 8. Calculated directivity for a rectangular microstrip patch over a large ground plane.

where $k = k_0\sqrt{\epsilon_r}$. The image factor $\cos(kt \cos\theta)$ is obtained by assuming that the slot is imbedded in a half-space of dielectric constant ϵ_r. Although a more rigorous expression for the image factor is desirable, the image effect is small for thin PTFE substrates so that the image factor error in (45) and (46) is small for these cases. The directivity of a single element over an idealized infinite ground plane can be found by the numerical integration of the far-field power pattern as computed from the fields above. The computed directivity as a function of the radiating edge length with the substrate thickness as a parameter is shown in Fig. 8. As expected, an increase in the edge length causes an increase in directivity, so long as the TM_{10} mode alone is excited. Thicker substrates cause a decrease in directivity as a result of destructive interference between patch and image currents. A single patch mounted on a small ground plane will have less directivity than shown here, as a result of spillover into the region behind the ground plane.

Circular Patch

The modal-expansion technique may also be used for the analysis of a circular patch, along the same general lines as used for the rectangular patch [18], [29]. Trigonometric functions are replaced by Bessel functions, and a transcendental equation for the eigenvalues (complex resonant frequencies) analogous to (27) may be obtained. The solution for these eigenvalues is dependent on the expression used for the wall admittance Y_w, and approximate expressions for the admittance are available [29]. It has been shown by Mink [43] that there is an approximately linear relationship between percent error in the wall susceptance and percent error in the predicted resonant frequency. Typically, an eight-percent change in wall susceptance corresponds to a one-percent change in resonant frequency; this frequency f_r may be calculated from the complex eigenvalue k_{10}' by the equation $f_r = c\,\mathrm{Re}(k_{10}')/(2\pi a\sqrt{\epsilon_r})$, where $k_{10}' = 1.84118 - \Delta_p$ and where Δ_p is a complex correction to the zeroth-order eigenvalue 1.84118. As in the case of the rectangular patch, the Q may be calculated by (31) and the radiation resistance by (32); the patch capacitance may be calculated using an expression given in [29] or alternatively by the expression given by Shen, Long, Alldering, and Walton [44]. The appropriate diameter of the circular patch may be roughly estimated by using the above equation with k_{10}' reduced from 98 to 94 percent, depending on the substrate thickness. More accurate expressions are available in the literature, although none to date produce consistent agreement to within 1 MHz of the measured results for patches in the *L*-, *S*-, or *C*-band regions; 1-MHz agreement is often required in order to meet practical design requirements, and current theoretical work in progress may soon produce more accurate design formulas and graphs.

One example of a more rigorous approach to the circular microstrip patch has been provided by Butler [38], who has solved the canonical problem of a center-fed circular microstrip patch in the form of a radiating annular slot of inner radius a and outer radius b in the upper plate. In the limit, as b becomes large, this becomes a circular microstrip antenna with a null on the axis. Fig. 9 shows the variation of the radially directed slot electric field E_ρ^A as a function of the radial distance for an air dielectric, a substrate thickness of $0.1\ \lambda_0$, and an inner disk diameter of $1\ \lambda$. It is apparent from Fig. 9 that the rapid decay in the radial electric field in the slot is not appreciably affected by the slot width and that coupling to the radial waveguide beyond the slot is very small. Butler and Yung [45] have used a similar technique to that presented here for the analysis of a long rectangular microstrip radiator.

Numerical Analysis Techniques

The basic rectangular or circular microstrip patch has been modified for some applications to other shapes, including a five-sided patch producing circular polarization [11], a quarter-wave shorted patch [46], and a rectangular patch with clipped edges or diagonal center slots [47]. For these geometries, the modal-expansion technique is a more cumbersome analysis method than a direct numerical analysis, due to the difficulty in finding the appropriate orthogonal mode vectors. In recent years several numerical techniques applied to the microstrip antenna have been proposed, including the method of moments [48], [49], the unimoment–Monte Carlo method [50], [55], [56], the finite-elements technique [19], and the direct form of network analogs (DFNA) method [51]. Each of these techniques has certain advantages and disadvantages.

Method of Moments

In this technique advanced by Newman [49], the method of moments is used in connection with Richmond's reaction

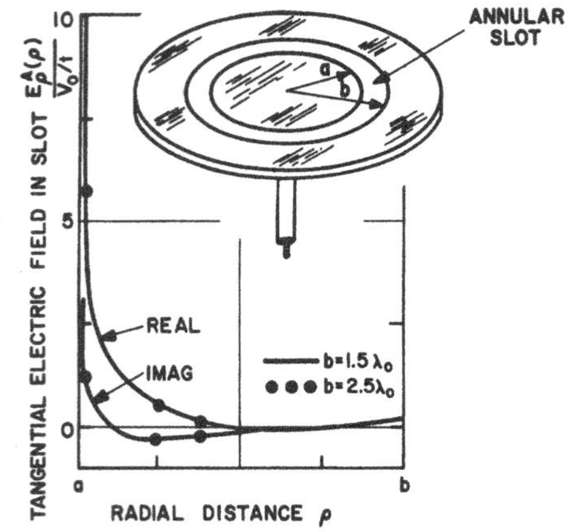

Fig. 9. Tangential electric field in annular slot versus radial distance for $a = 0.5\ \lambda_0$ and $t = 0.1\ \lambda_0$. After Butler [38].

method [52] to determine unknown surface currents ($\mathbf{J}_s$, $\mathbf{M}_s$) flowing on the walls forming the microstrip patch, ground plane, and magnetic walls. This begins with the reaction integral equation

$$\iint_s (\mathbf{J}_s \cdot \mathbf{E}_T - \mathbf{M}_s \cdot \mathbf{H}_T)\, ds + \iiint_v (\mathbf{J}_i \cdot \mathbf{E}_T - \mathbf{J}_i \cdot \mathbf{H}_T)\, dv = 0, \tag{47}$$

where ($\mathbf{E}_T$, $\mathbf{H}_T$) are the fields of an electric test source placed in the interior region, and the volume integral is over the source volume. For perfect conductors, $\mathbf{M}_s = 0$. The integral equation (47) is solved using the method of moments as a Galerkin method in which both expansion and testing functions are taken as a surface subpatch mode or as a wire attachment mode [48]. Thus the unknown current $\mathbf{J}_s$ is expanded in a set of N expansion functions $\mathbf{J}_n$, and (47) is enforced for N electric test sources (producing the fields $\mathbf{E}_m$, $\mathbf{H}_m$) placed inside the surface S bounding the microstrip antenna. This procedure reduces the reaction integral equation to a system of N simultaneous linear equations, with coefficients given by an impedance matrix Z_{mn}. The near fields of a suitable flat subpatch used as a testing function have been found [53], thus enabling the evaluation of the elements of the Z_{mn} matrix and the V_m source voltage vector. When a wire is attached to the surface of the microstrip patch, a special attachment mode consisting of a z-directed wire and a disk is introduced. The effect of the microstrip substrate is taken into account by using the volume equivalence theorem $\mathbf{J}_v = j\omega(\epsilon - \epsilon_0)\mathbf{E}$, where $\mathbf{E}$ is the electric field in the substrate. The Z_{mn} matrix is then modified by adding an incremental ΔZ_{mn} matrix, as described elsewhere [49]. Although the application of the method of moments to the microstrip antenna appears to be straightforward, there are several cautionary notes. First, the surface current $\mathbf{J}_s$ which is found is that on the interior side of both the patch and the associated ground plane; it is *not* the surface current on the exterior side of the patch and cannot be used directly to find the exterior field. Second, the method of moments applied to the reaction integral equation does not shed any new light on the mathematical connection between the interior and exterior fields, except insofar as the magnetic surface current $\mathbf{M}_s$ on the radiating perimeter walls is correctly formulated. Finally, the evaluation of the integrals for the impedance matrix and source voltage vector elements requires detailed attention to the geometry and polarization of any given microstrip patch and is not necessarily a trivial exercise. Due to the fact that the currents are inversely proportional to the difference between impedance elements, the method requires unusually precise computation of the impedance matrix [54]. The method of moments technique has been successfully used to find the input impedance of a quarter-wavelength shorted microstrip antenna and can be adapted to other microstrip antennas of nonstandard patch shape.

Finite-Element Technique

The numerical analysis of the fields interior to the microstrip antenna cavity can also be carried out using a finite-element approach [19]. This is a variational method in which a minimization process automatically seeks out the solution which is closest to the true analytical solution. The interior region of the microstrip antenna is mathematically decoupled from the exterior region through the use of an equivalent aperture admittance as the boundary condition, in an analogous fashion to that used by Carver [28] for the modal analysis of microstrip patches. The interior electric field E_z satisfies the inhomogeneous wave equation along with an impedance boundary condition on the perimeter walls. The variational formulation equivalent to solving these equations is to minimize a functional $I(v)$ [57, pp. 70–71] for all permissible functions $v(x, y)$. The particular function $v^*(x, y)$ which minimizes the functional is the "best" solution to the problem. This problem may be solved on a computer, via the eigenvalue problem

$$\mathbf{K}_1\mathbf{Q} = -k^2\mathbf{K}_2\mathbf{Q} \tag{48}$$

where $\mathbf{Q}$ is the column matrix of coefficients and $k = \omega\sqrt{\mu\epsilon}$.

The calculation of the $\mathbf{K}_1$ and $\mathbf{K}_2$ matrices for a general polygonal microstrip antenna has been implemented in a computer code MICRO, a listing of which is available in [58]. This technique, including the use of the code MICRO, has been successfully used to analyze the interior fields and polarization states of a pentagonal microstrip antenna developed by Weinschel [11] for which the classical technique of separation of variables cannot be used to find the mode vectors [19]. Since most of the entries in the $\mathbf{K}_1$ and $\mathbf{K}_2$ matrices are zero, a linked-list sparse matrix routine can be used to effect savings of up to 90 percent of the computer storage required to invert the K matrix. It should be pointed out that, by contrast, the method of moments generates full dense matrices so that sparse matrix techniques cannot be used. This is because the moment method is applied to the reaction integral equation, whereas the finite-element problem arises from the inhomogeneous wave equation.

IV. DESIGN PROCEDURES FOR MICROSTRIP ANTENNAS

This section presents design procedures for rectangular and circular microstrip patch antennas. For patches of simple rectangular or circular shape, the theoretical models presented earlier are used to generate design curves. In addition empirically derived procedures for modification of the basic patch shapes to yield enhanced or special performance characteristics are given. The material given here relates the antenna geometry (patch shape, size, substrate thickness, dielectric constant, and feed point type and location) to antenna performance (resonant frequency, resonant resistance, bandwidth, efficiency, polarization, and gain).

Rectangular Microstrip Antennas

The design of a rectangular microstrip antenna begins by recognizing that the desired TM_{10} mode is excited by making the patch dimension b slightly less than one-half wavelength in the substrate, $\lambda_d = \lambda_0/\sqrt{\epsilon_r}$, thus causing the two parallel radiating edges of length a to behave effectively as a two-element broadside array. The length a is chosen to be approximately $\lambda_0/2$ in a typical design. If there were no fringing, the resonant frequency would be given by $f_{ro} = c/(2b\sqrt{\epsilon_r})$. However, in practice, the fringing capacitance effect associated with the radiating edges causes the effective distance between the radiating edges to be slightly greater than b, so that the actual resonant frequency is slightly less than f_{ro} by a factor q as discussed earlier in (9). By using the modal-expansion analysis technique and solving the transcendental equation (27), the factor q may be found from the real part of the complex eigenvalue k_{10}. This is shown in Fig. 10 as a function of the electrical thickness of the substrate and for several values of the aspect ratio a/b. As the substrate becomes thicker, the fringing effect increases the effective distance between radiating edges, so that the resonant frequency decreases approximately linearly with increasing substrate thickness. According to (40) the radiating wall susceptance and thus the fringing capacitance is approximately proportional to the radiating edge length a. Thus for a given substrate thickness, an increase in length a will cause a decrease in resonant frequency as shown in Fig. 10.

It was pointed out in (8) that a simple transmission-line model analysis yields an input resonant resistance of approximately 120 Ω for a rectangular patch with a radiating edge length of $\lambda_0/2$; this assumed that the radiating edges were separated by one-half wavelength in the substrate. In the modal-expansion analysis, (32) was derived for the resonant resistance of the patch in terms of the radiation Q_r and the patch capacitance. This analysis reveals that the resonant resistance is also a function of the substrate thickness and the feed point location y_0/b. The calculated resonant resistance is shown in Fig. 11 for an edge-fed patch ($y_0 = 0$) with a substrate dielectric constant of 2.5, i.e., $b = 0.316\ \lambda_0$. The resonant resistance for an edge-fed patch varies typically between 100 and 200 Ω, depending on the aspect ratio a/b. It is not a strong function of substrate thickness except for very thin substrates where the radiation resistance for nearly square patches falls off rapidly with decreasing thickness. It is seen from (32) and (33) that a patch with an inset feed point has a resonant resistance given by

$$R_{\text{rad}} = R_{\text{rad}}{}^e \cos^2 (\pi y_0/b), \tag{49}$$

i.e., insetting the feed point causes a decrease in resistance. The use of (49) to determine the feed point location can be valuable in controlling the resonant resistance, particularly for square patches where the edge-fed resonant resistance is $R_{\text{rad}}{}^e \cong 260$ Ω. In this case a match to 50 Ω can be obtained by choosing $y_0 = 0.36\ b$. Tolerance here is important; for example, an error in feed point location of $0.01\ b$ ($y_0 = 0.37\ b$) would yield a resonant resistance of 41 Ω. The resonant resistance can be decreased by increasing the length a of the radiating edge, so long as the dimension b is held to one-half the substrate wavelength. However, ratios of a/b greater than about 2 are not advisable, since the aperture

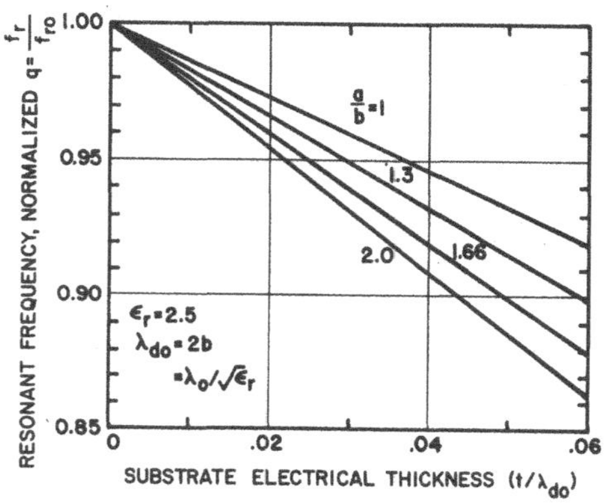

Fig. 10. Dependence of resonant frequency on substrate thickness and aspect ratio for TM_{10} mode edge-fed rectangular patch.

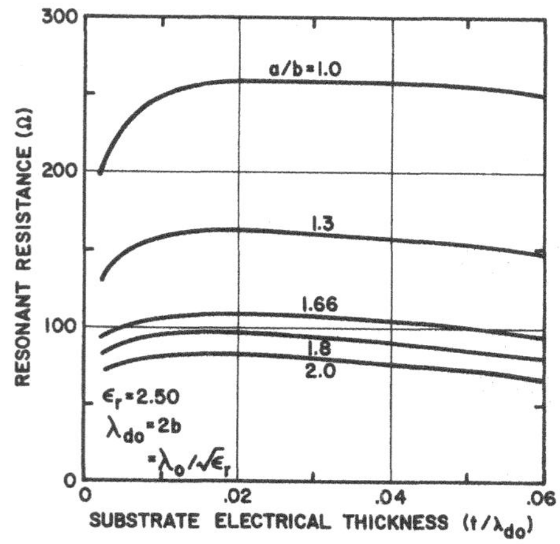

Fig. 11. Dependence of resonant resistance on substrate thickness and aspect ratio for TM_{10} mode edge-fed rectangular patch.

efficiency of a single patch begins to drop; this problem can be circumvented by the use of multiple feed points spaced one-half wavelength [4], [13].

The feed for a microstrip patch is usually a coaxial through-the-substrate connection or a microstrip transmission-line connection printed monolithically on the same board, as shown in Fig. 5(b). Fig. 11 and (49) can be used to determine the resonant resistance for either of these cases. Weinschel [35] has shown that (49) agrees fairly well with experimentally measured values of the resonant resistance as a function of inset distance for microstrip-fed microstrip patch antennas. However, due to junction capacitance effects associated with the inset notch, the resonant frequency may vary by about one percent from that associated with the edge-fed (no notch) case, depending on the inset distance and notch width [35].

The input impedance to the rectangular microstrip patch is given by (23). On the Smith chart the input impedance is approximately a circle whose center lies on the constant inductive reactance X_L line, as shown in Fig. 12. If X_L were zero, then the impedance circle would be symmetrically disposed about the zero-reactance horizontal line, and the resonant frequency would be determined as that frequency where the right-half portion of the circle crosses the zero-reactance line. However, since X_L is greater than zero, the resonant frequency must be determined as the point where the right-half portion of the impedance circle crosses the constant X_L line, as shown. The series inductive reactance X_L may be calculated rigorously from (26), although the series converges slowly. The series reactance depends on the substrate thickness, the probe diameter, and the probe inset distance

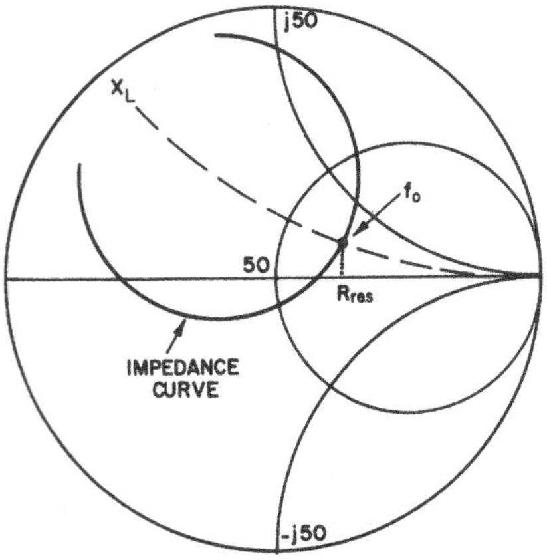

Fig. 12. Typical Smith chart display of microstrip antenna impedance circle, showing resonant frequency and resonant resistance locations.

[59]. The inductive reactance may be simply approximated by the formula

$$X_L \cong \sqrt{\mu/\epsilon}\ \tan\left(2\pi t/\lambda\right). \tag{50}$$

Fig. 13 is a Smith chart display of the input impedance to a typical *S*-band rectangular microstrip patch ($\epsilon_r = 2.5$) for both the edge-fed case ($y_0 = 0$) and an inset-fed case ($y_0 = 0.245\ b$). Both theoretical (solid line) and measured (dashed line) impedance curves are shown. As noted in (79), the effect of insetting the feed point is to lower the resonant resistance. In Fig. 13 it is seen that a good match to 50 Ω is obtained at 2200 MHz when the feed point is inset as shown. Because the field does not vary with location along the x axis (over the 2140–2300 MHz bandwidth), the feed point can be located at any point x_0 without changing the impedance curves.

Variations on the Rectangular Patch

In addition to the standard rectangular patch there are numerous variations on the design which have been used for special purposes. As an example, Fig. 14 shows a dual-frequency shepherd's crook feed developed by Kerr [60] for a 1.22-m dish. A linearly polarized *L*-band microstrip patch is mounted at the flange of an *X*-band waveguide which illuminates the dish through a hole cut in the center of the *L*-band patch. The inset measured antenna patterns for both 1250 and 9500 MHz are the secondary patterns for the 1.22-m dish and show good sidelobe control.

Another technique for designing a dual-frequency dish feed is to use an element which resonates at one frequency imbedded within another element which resonates at a lower frequency, as suggested by Kerr [60] and shown in Fig. 15. In this design an *X*-band horizontally polarized notched rectangular patch is etched within a rectangular hole in the center of an *L*-band vertically polarized patch etched on the same substrate. When this is used to feed a 1.22-m dish reflector, 1250- and 9500-MHz low-sidelobe patterns are obtained as shown in Fig. 15. A single rectangular patch with two feed points can be used as a two-port radiator with impedance loading on one port used to effect a measure of frequency control. Fig. 16 shows such a design described by Kerr [60] where a variable-length short circuit on port 2 can be adjusted to produce an input voltage standing-wave ratio (VSWR) of 1.5 or less at port 1 at frequencies from 1275 to 1500 MHz.

Dual-frequency operation can also be obtained by stacking

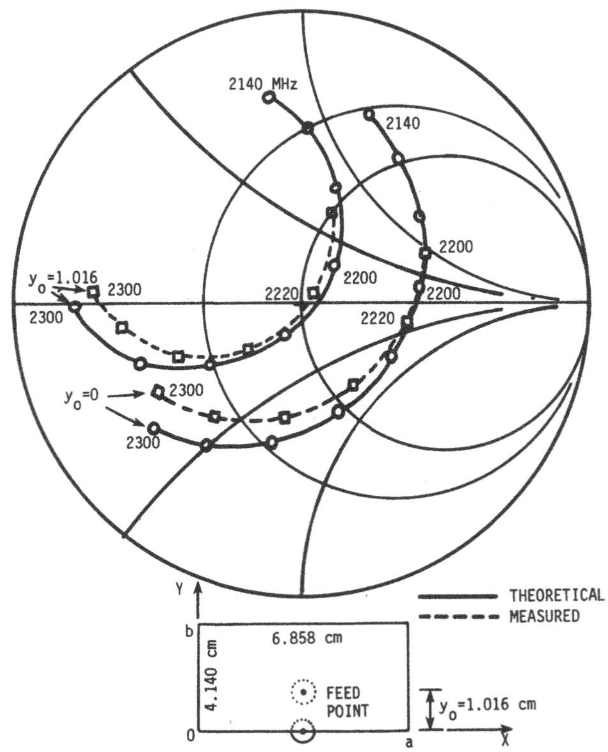

Fig. 13. Measured and calculated impedance curves for a rectangular microstrip antenna with both edge-feed and inset-feed locations. After Carver and Coffey [19].

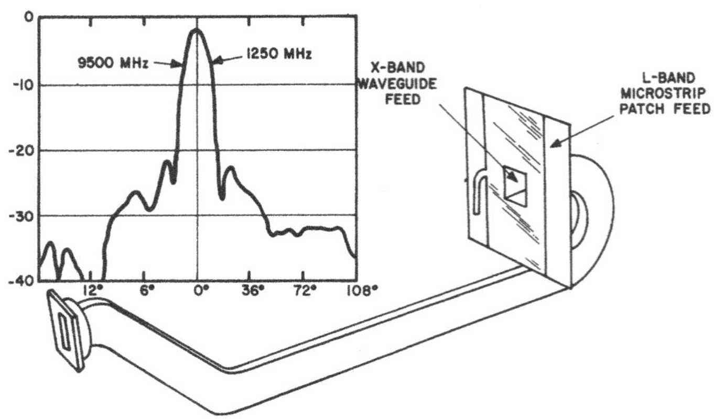

Fig. 14. Dual-frequency (*L*- and *X*-band) shepherd's crook feed for 1.22-m diameter dish. Inset: right-half pattern measured at 1250 MHz, left-half pattern measured at 9500 MHz, both *E*-plane patterns. After Kerr [60].

one element on another, for pentagon patches [61], circular patches [62], and trapezoidal patches [46]. Fig. 17 illustrates a piggyback antenna developed by Schaubert and Farrar [46] consisting of a λ/4 length shorted parallel-plate radiator resonant at 1140 MHz mounted over a 990-MHz λ/2 resonant microstrip patch. The microstrip patch acts as a ground plane for the λ/4 parallel-plate radiator. With a 1.6-mm substrate thickness, a 0.5-percent bandwidth (VSWR = 2) was obtained for the microstrip element, with an isolation between elements of 20 dB at 990 MHz and 37 dB at 1140 MHz. Parasitic strips placed parallel fo the nonradiating edges of a square patch may be used to improve the match to 50 Ω and to increase the bandwidth, as summarized in Fig. 18 from the work of Schaubert and Farrar [46] for a UHF microstrip antenna. A novel design proposed by Dubost [63] and illustrated in Fig. 19 has impedance bandwidths in excess of 20 percent at a VSWR =

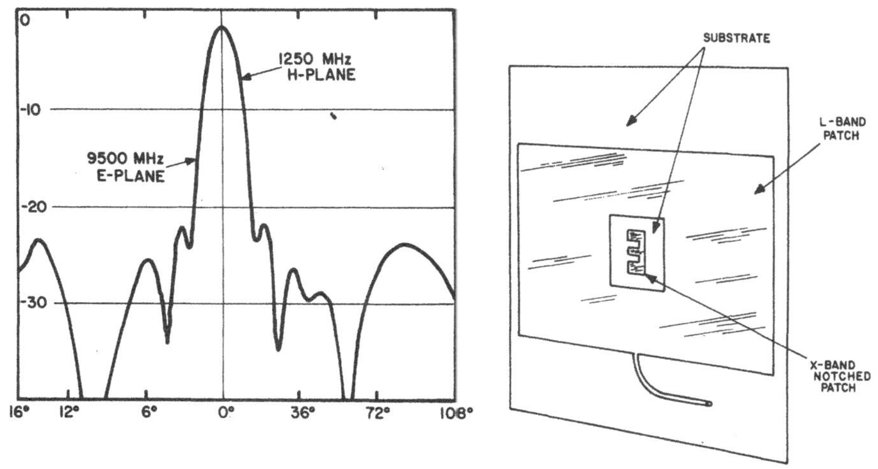

Fig. 15. Dual-frequency (*L*- and *X*-band) orthogonal and concentric microstrip patches with *L*-band patch vertically polarized and centered *X*-band notched patch horizontally polarized. After Kerr [60].

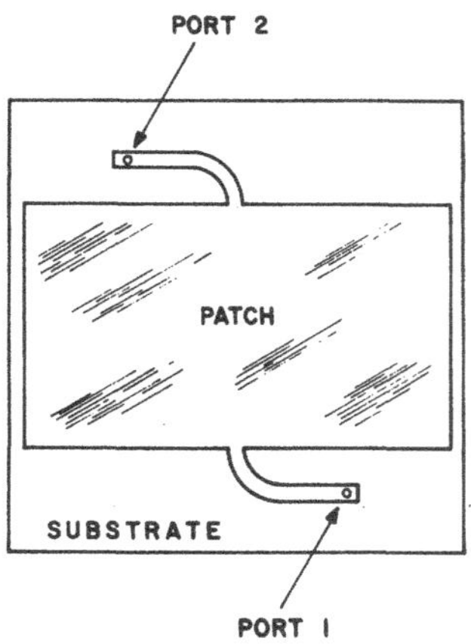

Fig. 16. Two-port rectangular microstrip patch. After Kerr [60].

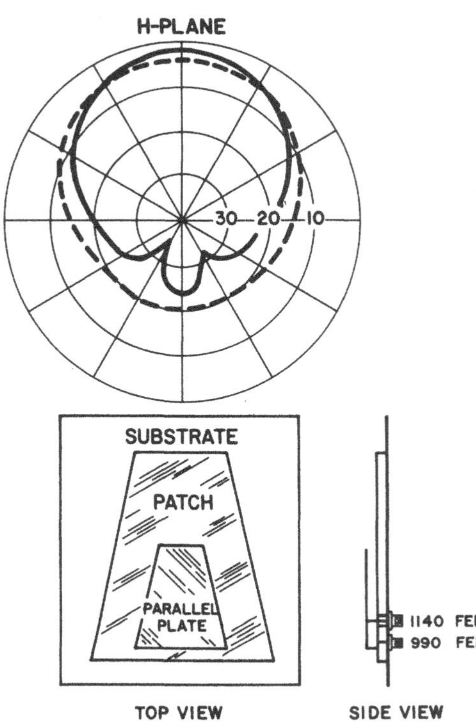

Fig. 17. Piggyback antenna consisting of parallel-plate radiator on top of microstrip patch. After Schaubert and Farrar [46]. —— Patch, 990 MHz. --- Parallel plate, 1140 MHz.

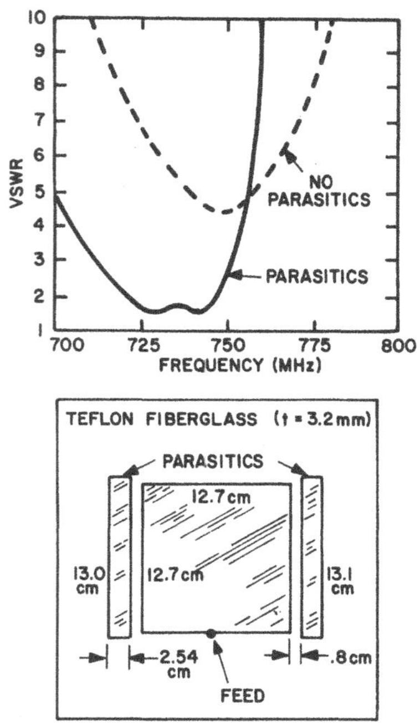

Fig. 18. Parasitic-tuned microstrip patch and VSWR characteristics. After Schaubert and Farrar [46].

3:1 level; to convert the VSWR = 3:1 bandwidth to a VSWR = 2:1 bandwidth, multiply by 0.612. This microstrip antenna is essentially a half-dipole [92] which radiates as an open circuit from a patch of width W and length h. The patch is printed on the underside of a thin substrate and is shorted at the feed end to the ground plane by a bar of height H. It is fed by a microstrip feed line printed on the upper side of the substrate, so that both the printed patch and its image are excited. An *X*-band microstrip patch on a 0.625-mm thick alumina ceramic substrate ($\epsilon_r = 9.8$) has a bandwidth (VSWR = 1.9) of 1.1 percent, whereas a polyguide substrate ($\epsilon_r = 2.32$) 1.59 mm thick produces a 6.6-percent bandwidth [64]. However the alumina ceramic substrate is often desirable in order to decrease the patch size. Hall, Wood, and Garrett [64] have shown that an *X*-band 3.9 mm × 3.9 mm patch on an alumina ceramic substrate gives a 13-percent bandwidth when the patch is covered by an 8 mm × 8 mm × 1.59 mm polyguide substrate, which then serves as a matching

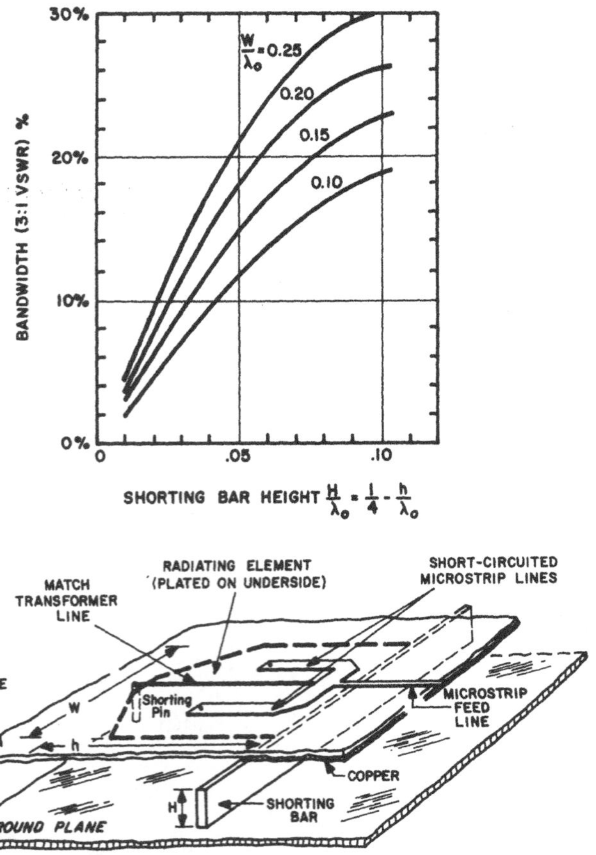

Fig. 19. Microstrip half-dipole radiator of width W and length h printed on underside of substrate and separated by H from ground plane. Top: 3:1 VSWR bandwidth versus height H for four radiator widths. After Dubost [63].

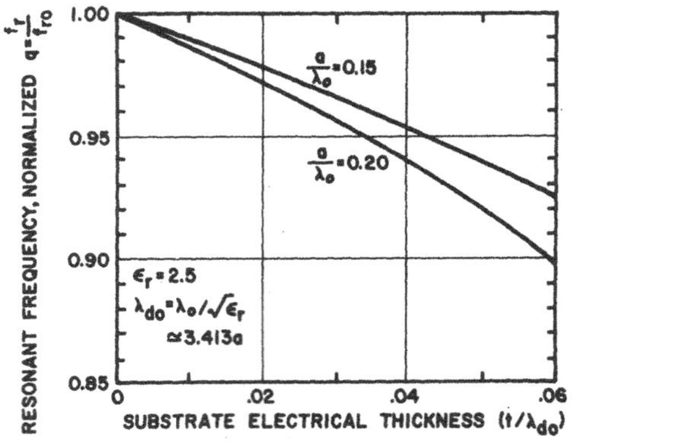

Fig. 20. Dependence of normalized resonant frequency on substrate thickness for a dominant-mode circular microstrip patch of radius a and $\epsilon_r = 2.5$.

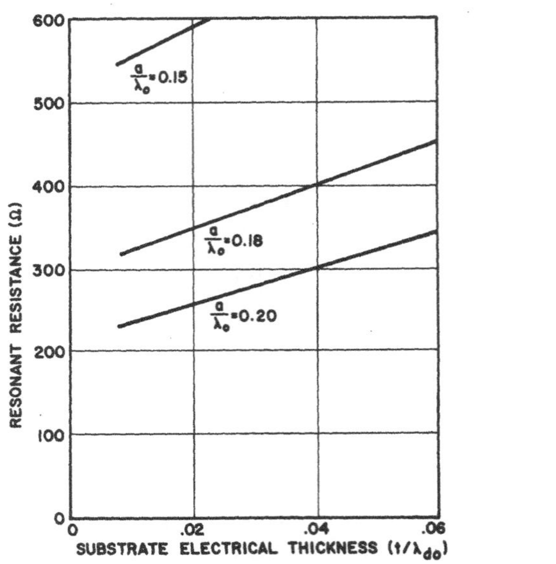

Fig. 21. Dependence of resonant radiation resistance on substrate thickness for circular microstrip patch.

transformer to free space. By using three layers, bandwidths of 18 percent were obtained with an element gain greater than 5.3 dB. In the preceding microstrip antenna configurations, the ground plane was much larger than the radiating element so that the pattern is roughly cardioid in shape with a peak broadside to the patch. Kaloi [65] has developed electrically small microstrip patches with small ground planes which produce nearly omnidirectional patterns and which have low resonant resistances.

Circular Microstrip Antennas

A circular microstrip patch of radius a and with a nonradiating zero-admittance wall has a dominant RF mode whose resonant frequency is given by $f_{ro} = ck_{10}'/(2\pi a\sqrt{\epsilon_r})$, where $k_{10}' = 1.84118$ [29]. In this case the resonant wavelength in the dielectric is therefore $\lambda_{do} = 3.413\ a$. For the case of a radiating circular patch, the wall admittance is complex so that the resonant frequency becomes complex, as discussed in [29]. The real resonant frequency f_r is therefore less than f_{ro}. Fig. 20 illustrates the dependence of f_r/f_{ro} on the substrate electrical thickness for an edge-fed patch with a substrate dielectric constant of 2.5. This calculated curve is based on the validity of the wall admittance given in [29], which is only approximate. As was previously mentioned, slight errors in the wall susceptance lead to errors in the computed resonant frequency. Thus the behavior illustrated in Fig. 20 should be taken as simply illustrating the parametric dependence of the resonant frequency on substrate thickness and patch radius. More accurate calculations can be made when more accurate values of the wall susceptance are made available. Fig. 21 is a graph of the resonant resistance versus substrate thickness for an edge-fed circular patch with the same dielectric constant as in the previous figure; and it shows that for the range of parameters listed, the resonant resistance increases with both patch radius and substrate thickness. The absolute accuracy of this curve is dependent on how accurately the wall conductance is given by the Re (Y_w) in [29]. In an analogous fashion to (49) for a rectangular patch, it may be shown that for an inset-fed circular patch the resonant resistance is related to $R_{rad}{}^e$ for an edge-fed patch by

$$R_{rad} = R_{rad}{}^e \frac{J_1{}^2\left(\mathrm{Re}\,(k_{10}')\dfrac{\rho_0}{a}\right)}{J_1{}^2\,(\mathrm{Re}\,(k_{10}'))}, \tag{51}$$

where ρ_0 is the radius of the feed point. This means that for TM_{10} mode operation, insetting the feed causes a decrease in resonant resistance.

The dominant mode electric field can be written as

$$E_{10} = E_0 J_1\left(k_{10}'\frac{\rho}{a}\right)\cos\phi, \tag{52}$$

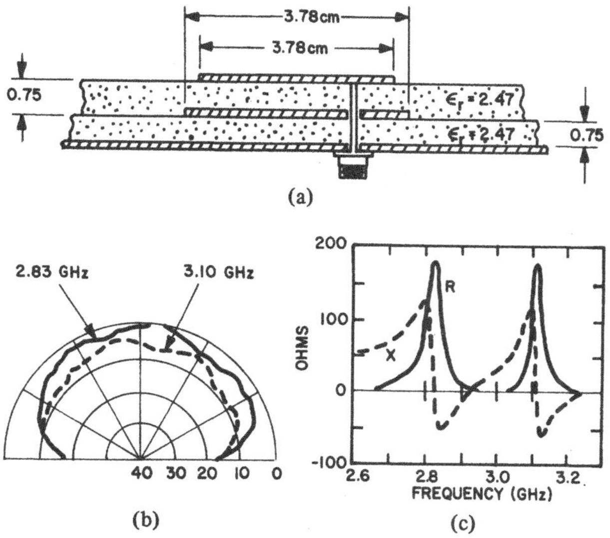

Fig. 22. (a) Cross section of typical stacked circular disk microstrip antenna. (b) Measured E_θ patterns at 2.83 and 3.10 GHz. (c) Measured input impedance showing resonance at 2.83 and 3.10 GHz. After Long and Walton [62].

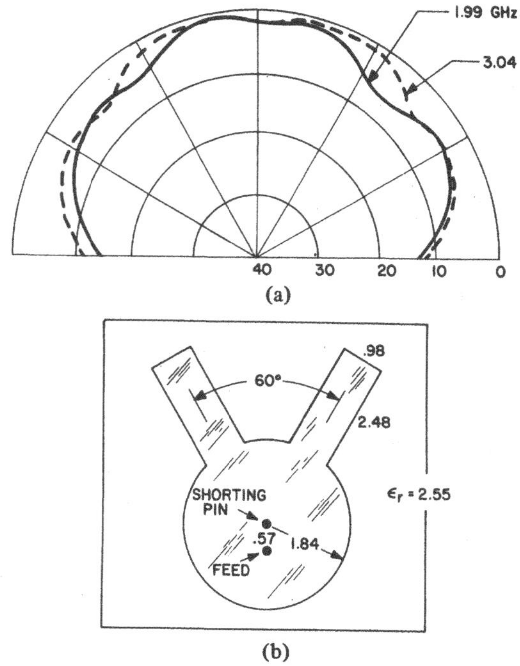

Fig. 23. (a) *S*-band patterns measured at two frequencies. (b) Circular disk microstrip with two ears (bunny antenna). All dimensions are in centimeters. After McIlvenna and Kernweis [69].

i.e., the field is zero at the center and is maximum at the edge; the fields are oppositely directed at the $\phi = 0$ and $\phi = 180^\circ$ locations on the edge, where $\phi = 0$ is identified with the feed point. By contrast, the static mode has a field which is uniform throughout the patch. In order to reject this mode and other higher order modes, and to retain the above TM_{10} mode, a shorting pin is usually placed at the center of the circular patch. If a match to 50 Ω is desired, (51) can be used in conjunction with Fig. 25 to calculate the location of the feed point.

Several theoretical analyses of the circular microstrip antennas are available in the literature, in addition to the one presented previously in this paper. Several authors [66], [74], [93], [95] have given formulas for the capacitance and resonant frequency of a circular microstrip disk. Shen, Long, Allerding, and Walton [44] have used these relations and other analytical results to obtain an improved formula for the capacitance of a radiating microstrip circular disk antenna. Derneryd [68] has calculated the radiation conductance, antenna efficiency, and quality factor associated with the circular disk antenna radiating in lower order TM modes. These papers are particularly helpful in the prediction of the performance of specific microstrip circular disk antenna designs.

Variations of the Circular Patch

The basic circular patch can be modified to reproduce resonance at more than one frequency close to the nominal resonance frequency. One technique is to stack one circular patch on top of another in a sandwich construction such as is shown in Fig. 22 [62]. For the geometry shown, resonant frequencies of 2.83 GHz and 3.10 GHz were obtained; a single patch of the same radius would produce a resonant frequency of 2.95 GHz, about halfway between the two resonances with the two patches. Another technique for achieving dual resonance is to use two ears at 60° angular separation, as developed by McIlvenna and Kernweis [69] and shown in Fig. 23. Good pattern and impedance characteristics were obtained at 1.99 GHz and 3.04 GHz with the bunny patch having the dimensions shown. A basic unadorned patch of the same radius and the same substrate thickness (0.159 cm) had a resonant frequency of 2.88 GHz [69]. A half-disk microstrip patch can also be excited at either a point along the disk bisector line or at a point along the circular arc [16]. As an example, a half-disk of radius 6.75 cm, substrate thickness 0.158 cm, dielectric constant 2.65, and feed point at 3.38 cm from the disk center and on the bisector line has a resonant resistance of approximately 90 Ω at 1323 MHz. Another circular microstrip antenna is a ring or annulus patch which, for a given maximum size, has a substantially lowered operating frequency [70].

V. CIRCULARLY POLARIZED MICROSTRIP ELEMENTS

Microstrip antennas may be designed for circular polarization by adjusting their physical dimensions so as to produce two degenerate orthogonal modes within the cavity region. This in turn results in the radiation of two orthogonally polarized waves near the broadside direction. Thus circularly polarized radiation is obtained when two orthogonal modes are excited with equal amplitude and in-phase quadrature. In this section we will discuss some of the techniques employed to achieve circularly polarized radiation from printed antenna elements.

The most direct approach to obtaining circular polarization is to use two separate and spatially orthogonal feeds excited with a relative phase shift of ±90°. This configuration then provides two orthogonal linearly polarized waves which are in time phase quadrature. The antenna can be excited from a single feed point by the use of a 90° hybrid or power splitter with unequal lengths of transmission line to obtain the necessary phase shift [71], [10], as shown in Fig. 24.

Several methods have been proposed to provide circular polarization without the complexities inherent in the dual-feed devices. One approach is to attach a single feed point at a location so as to excite two equal amplitude degenerate orthogonal modes and then to introduce some asymmetry in

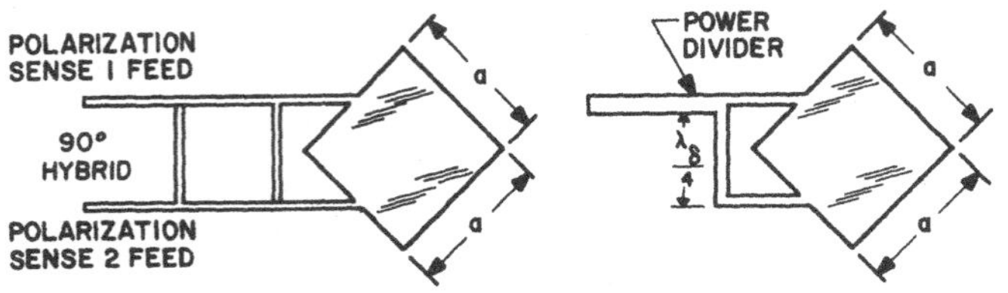

Fig. 24. Single-feed circularly polarized symmetrical microstrip antennas.

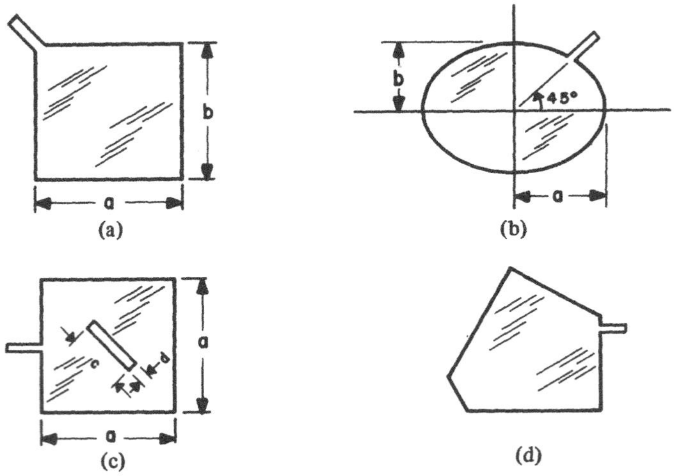

Fig. 25. Single-feed circularly polarized asymmetrical microstrip antennas. (a) Almost square. (b) Elliptical. (c) Square with 45° slot. $C = a/2.72, d = c/10$. (d) Pentagon.

the cavity so that the degeneracy of the modes is removed. If the asymmetry is properly applied, one mode will decrease in frequency by a specified amount while the orthogonal mode will increase by an equal amount. The equivalent circuit for such a configuration is two uncoupled parallel resonant circuits excited by a common current source, as suggested in Fig. 6. Since the equivalent circuits have different resonant frequencies, with proper design the phase of one mode voltage can lead the impressed current by 45°, with the other mode voltage lagging by 45°, thus producing circularly polarized radiation. Several geometrical arrangements have been devised to achieve this goal. Examples of the technique are the corner-fed rectangle [72], [32], [19], as shown in Fig. 25(a), the slightly elliptical patch [73], [74], [75], as shown in Fig. 25(b), the square patch with a 45° center slot [47], as shown in Fig. 25(c), and the pentagon-shaped patch [11], as shown in Fig. 25(d).

An illustration of the spatially orthogonal degenerate modes is found in a corner-fed rectangular patch where the dimensions a and b are approximately the same. For a nearly square patch with dimension a slightly less than dimension b, the TM_{10} and TM_{01} modes will have closely spaced resonant frequencies so that only these two modes need be considered.

It can be shown [19] that the magnetic surface current on the $y = 0$ wall of a rectangular patch (Fig. 5) is given by

$$\mathbf{M}_{sx} = \mathbf{x}\frac{j\omega I_{in}}{\epsilon ab} \cdot \left\{ \frac{1}{\omega^2 - \omega_{10}^2\left(1 + \dfrac{j}{Q_{10}}\right)} + \frac{\cos\dfrac{\pi x}{a}}{\omega^2 - \omega_{01}^2\left(1 + \dfrac{j}{Q_{01}}\right)} \right\} \tag{53}$$

and the magnetic surface current on the $x = 0$ wall is given by

$$\mathbf{M}_{sy} = \mathbf{y}\frac{j\omega I_{in}}{\epsilon ab} \cdot \left\{ \frac{\cos\dfrac{\pi y}{b}}{\omega^2 - \omega_{10}^2\left(1 + \dfrac{j}{Q_{10}}\right)} + \frac{1}{\omega^2 - \omega_{01}^2\left(1 + \dfrac{j}{Q_{01}}\right)} \right\}. \tag{54}$$

The second term in (53) and the first term in (54) have no net effect on the polarization in the broadside direction.

To achieve circular polarization, $|M_{sx}| = |M_{sy}|$ and the phase angles must differ by ±90°. It has been shown by Richards, Lo, Simon, and Harrison [32] that an optimum choice for a and b results when $b = a(1 + 1/Q)$. This relationship will separate the two resonant frequencies by f_0/Q and results in the largest band of frequencies over which good axial ratio is achieved.

A corner-fed nearly square patch was constructed by Carver [19] and used to verify this theory. The dimension of the patch was a = 4.14 cm, b = 4.26 cm, and t = 1.588 mm, with a dielectric constant of 2.62. The theoretical and measured axial ratios versus frequency are shown in Fig. 26, and the measured impedance diagram is shown in Fig. 27. The agreement between theoretical and experimental axial ratios is very good, and the excitation of two modes is in evidence from the cusp in the impedance diagram at 2200 MHz. It is also noted that the cusp occurs where the axial ratio is minimum.

Circular polarization may also be obtained from slightly elliptical (nearly circular) patches and has been investigated theoretically by Shen [74] who solved the Helmholtz equation for E_z in elliptical coordinates using Mathieu functions to express the modal spectrum. The results of his investigation are that circular polarization in the broadside direction is obtained when the eccentricity of the ellipse is 0.98, with the feed point on a line at 45° relative to the semimajor axis. This work has been experimentally verified by Long *et al.* [75]. Their results showed that the best circular polarization was obtained with an eccentricity of 0.976 and that an axial ratio of 6 dB or less was obtained over a bandwidth of about 1.5 percent. Other elliptical microstrip resonators have also been investigated [76].

The pentagon patch and patches with diagonal slots have been shown to produce circular polarization. However, modal-expansion techniques for the interior fields cannot be carried out by the classical separation of variable method since the boundary condition causes coupling between the modes. These structures may be analyzed using computer techniques such as the method of finite elements; however these techniques may be time consuming. As a result these structures have been studied primarily through experimental investigations. The result of one investigation by Kerr [47] is shown in Fig. 28. In this study the diagonal slot was employed to obtain cir-

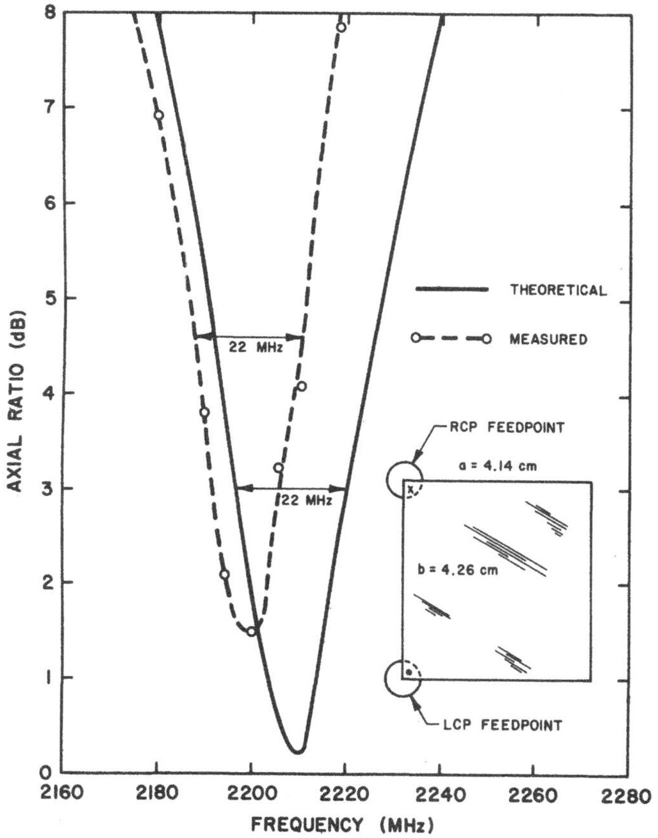

Fig. 26. On-axis axial ratio versus frequency for a corner-fed microstrip patch. After Carver and Coffey [19].

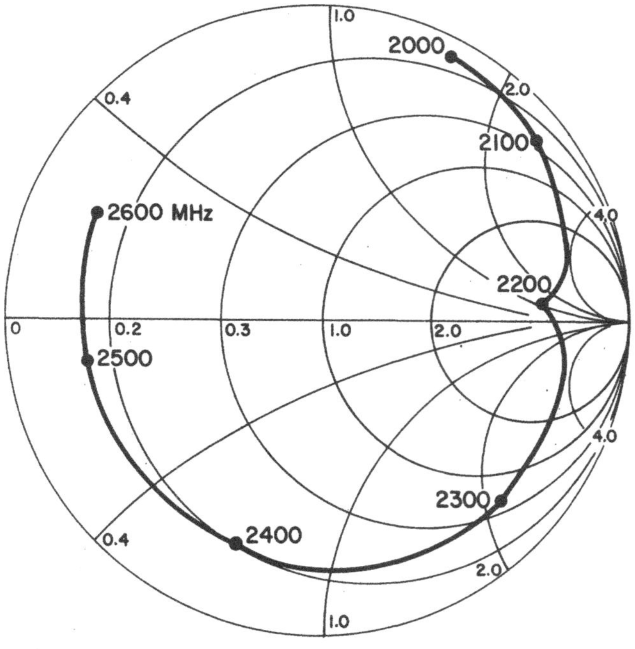

Fig. 27. Measured input impedance to corner-fed microstrip patch with dimensions given in Fig. 26.

cularly polarized radiation from both square and circular patches. Fig. 28 also shows the L-band radiation patterns obtained with a spinning dipole. Very good axial ratios were obtained over at least 120° segments of the radiation pattern. The bandwidth over which the axial ratio was not greater than 6 dB was about two percent.

VI. QUALITY FACTOR, BANDWIDTH, AND EFFICIENCY OF MICROSTRIP ELEMENTS

The usual goal of an antenna design is to produce an antenna system which has high efficiency and large bandwidth. However, these parameters are interrelated and one does not have complete freedom to independently set these parameters. The stored energy in the cavity region, including that energy stored in the fringing fields around the structure, may be calculated and then compared with the various losses to compute the Q factor associated with each. The steps required to perform these loss calculations were outlined previously. There are four loss mechanisms to be considered, namely, radiation, the loss associated with surface wave propagation on a dielectric coated conductor, the loss due to heating in the conducting elements and the ground plane, and the loss due to heating within the dielectric medium. The total Q of the antenna is given by

$$\frac{1}{Q} = \frac{1}{Q_{\mathrm{rad}}} + \frac{1}{Q_{\mathrm{sw}}} + \frac{1}{Q_{\mathrm{di}}} + \frac{1}{Q_{\mathrm{cu}}}. \tag{55}$$

The term involving Q_{sw} associated with the surface wave is negligible for thin substrates. For thicker substrates, techniques are available to estimate the surface wave contribution [77], [32]. The Q factors may then be calculated assuming that energy stored in the fringing fields is negligible, and the field distribution within the cavity region does not depend on thickness. Formulas for the Q factors due to conductor loss and dielectric loss for circular microstrip antennas have been obtained [18], [78], [97]. It can be shown that these relationships apply in general to thin microstrip antennas of arbitrary shape, i.e.,

$$Q_{\mathrm{di}} = \frac{1}{\tan \delta} \tag{56}$$

$$Q_{\mathrm{cu}} = \frac{t}{d_s}, \tag{57}$$

where $d_s = (\pi f \mu \sigma)^{-1/2}$ is the skin depth associated with the conductor. Providing that the field distribution along the radiating aperture or within the cavity region of the antenna does not change as the thickness is varied, it can be shown that the radiation quality factor Q_{rad} has the following form:

$$Q_{\mathrm{rad}} = \frac{2\omega\epsilon_r}{tG/l} K, \tag{58}$$

where G/l is the conductance per unit length of the radiating aperture and

$$K = \frac{\iint_{\mathrm{area}} |E|^2 \, dA}{\oint_{\mathrm{perimeter}} |E|^2 \, dl}. \tag{59}$$

For a rectangular patch operating in the TM_{10} mode, K becomes $b/4$ and G/l becomes G_{rad}/a. Equation (58) shows that for a microstrip antenna whose Q is dominated by the radiation term, the Q of the antenna is proportional to the reciprocal of the substrate thickness.

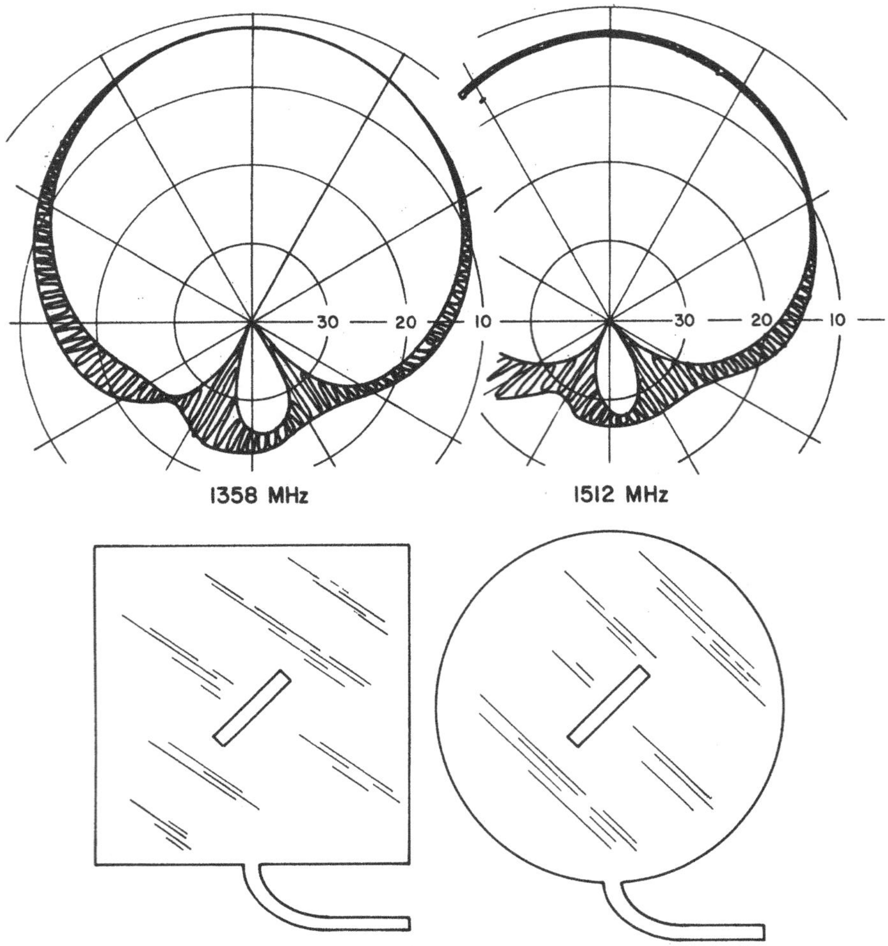

Fig. 28. Circularly polarized square and circular microstrip patches with thin diagonal center slot and associated measured patterns. After Kerr [47].

The various Q factors for typical square and rectangular microstrip antennas are shown in Fig. 29. As can be seen by these curves, the overall antenna Q will be dominated by the radiation Q_{rad} for substrate thicknesses greater than about $t/\lambda_{do} = 0.01$, and that the Q_{rad} of a square patch of area $0.0906\ \lambda_0{}^2$ is somewhat higher than for the rectangular patch of area $0.1504\ \lambda_0{}^2$. Fig. 30 shows the calculated total Q of a typical circular microstrip antenna of area $0.1057\ \lambda_0{}^2$. This curve takes into account all loss mechanisms present except those associated with the surface wave.

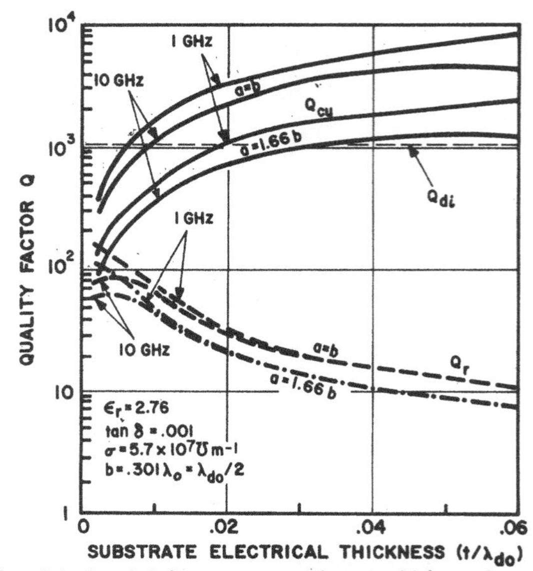

Fig. 29. Calculated Q factors versus substrate thickness for a rectangular microstrip patch antenna.

Bandwidth as referred to microstrip antennas may take one of several meanings. The usual definition of the bandwidth, $\Delta f = Q/f_0$, is not extremely useful by itself. There is usually an impedance matching network between the antenna radiating element and its input port which must be considered. A more meaningful measure of bandwidth is that band of frequencies where the input VSWR is less than a specified value, usually 2:1, assuming that a unity VSWR is obtained at the design frequency. The bandwidth may then be expressed in terms of Q and maximum allowable VSWR as follows [100]:

$$\mathrm{BW} = \frac{\mathrm{VSWR} - 1}{Q\sqrt{\mathrm{VSWR}}}. \tag{60}$$

Fig. 31 shows the VSWR = 2:1 bandwidth of various square and rectangular microstrip antennas. As shown by this curve, obtainable bandwidths range from about one percent for a substrate thickness of 0.01 t/λ_{do} to about six to ten percent for a substrate thickness of 0.06 t/λ_{do}, depending upon the aspect ratio of the antenna. One would expect the bandwidth of a circular patch to be slightly less because of its higher Q.

The bandwidth for circularly polarized microstrip patches usually must be defined in terms of the band of frequencies over which the axial ratio of the radiated energy is within

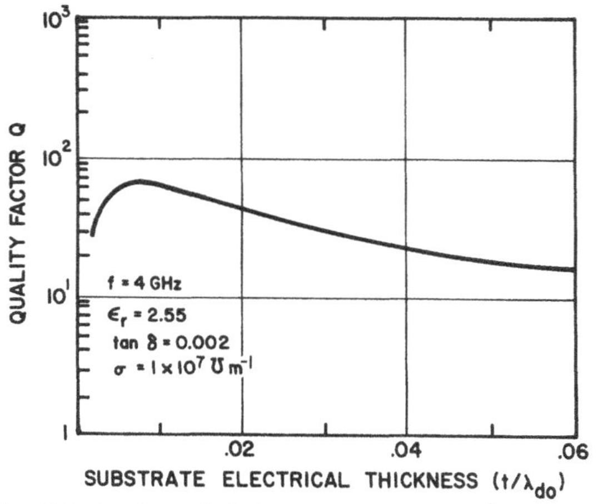

Fig. 30. Calculated total Q factor versus substrate thickness for a circular microstrip patch antenna. After Long, Shen, and Morel [101].

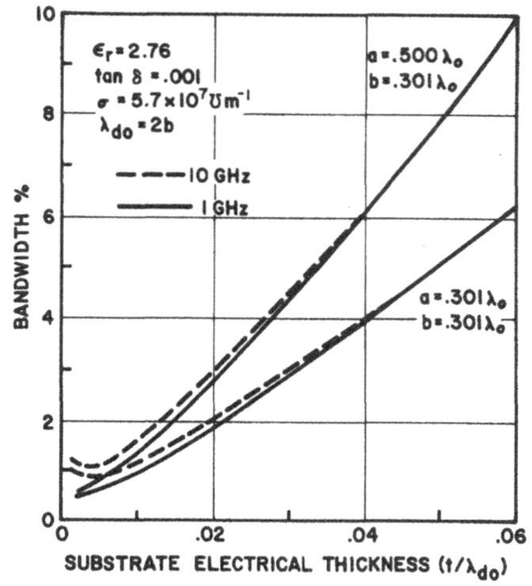

Fig. 31. Calculated VSWR = 2:1 bandwidth versus substrate thickness at 1 GHz and 10 GHz for both a square and a rectangular microstrip patch.

certain limits, typically 3 dB. This bandwidth is usually much less than the previously defined impedance bandwidth. For the case of a single-feed-point circularly polarized antenna, there is a relationship between the Q of the antenna and bandwidth over which good circular polarization (CP) results [32]. Richards *et al.* [32] have shown that for an axial ratio of 3 dB, one is limited to a bandwidth of about 35 percent of the frequency difference between the two resonant frequencies or about $35/Q$ percentage bandwidth. Thus the bandwidth of single-feed circularly polarized antennas is extremely limited.

The antenna efficiency (power radiated/power input) may be calculated from (38) with the help of (34), (35), and (32). However it may also be expressed in terms of the quality factors given by (55), (56), (57), and (58) as follows:

$$\eta = \frac{Q}{Q_{\text{rad}}}. \tag{61}$$

Antenna engineers usually express this as the antenna loss, i.e., $10 \log (1/\eta)$, in decibels. A typical graph of antenna loss versus substrate thickness is shown in Fig. 32 for a rectangular patch with $\epsilon_r = 2.5$ and with $a/b = 1.58$. Since the copper loss increases with increasing frequency, there is more loss for an X-band patch than for an L-band patch of the same electrical size. For both curves, the loss decreases with increasing substrate thickness.

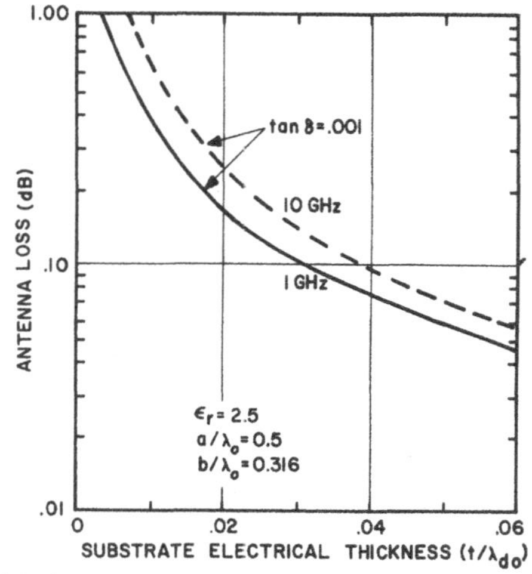

Fig. 32. Calculated antenna loss versus substrate electrical thickness for an edge-fed rectangular microstrip patch.

VII. MICROSTRIP DIPOLE ANTENNAS

In the preceding discussion of the rectangular microstrip patch antenna it was assumed that both patch side dimensions were appreciable fractions of a wavelength and that the patch was excited as a two-dimensional cavity resonator. By making one side length resonant and the other patch dimension very thin, a microstrip dipole is formed. By printing one arm of the dipole on one side of the substrate and the other arm on the opposite side, and spacing the substrate dipole one-quarter wavelength from a ground plane, either a bow-tie dipole [79] or a thin rectangular dipole printed circuit element [80] is formed which can then be used in an array configuration. However, because the spacing to the ground plane is a quarter-wavelength, these printed dipoles are not properly classified as microstrip printed dipoles which use electrically thin substrates. Oltman [81] has pointed out that thin resonant microstrip dipoles can be efficiently excited by electromagnetic proximity coupling to a microstrip transmission line imbedded in the substrate. Huebner [82] has successfully used this technique in the development of a 24-element X-band array of electromagnetically coupled microstrip dipoles. In addition linear resonant or traveling-wave arrays of printed monopoles or open-circuit microstrip radiators have been shown [83] to be efficient and relatively broadband antennas. One advantage of the printed dipole or open-circuit radiator is an inherent capability of larger bandwidth than is obtainable from a simple microstrip patch.

A rigorous analysis of thin-wire printed microstrip dipoles and coupled dipoles has been provided by Rana and Alexopoulos [84]. It has been found that the input impedance is not critically dependent on the gap length so long as that length is less than 0.1 λ_0. Fig. 33(a) shows the computed input impedance versus printed dipole length for a substrate with $\epsilon_r = 3.25$, $t = 0.1016\ \lambda_0$, and 0.0001 λ_0 wire diameter [84]. The first resonance is obtained for a dipole length of 0.317 λ_0, for which the input resistance is 34.5 Ω. At a length of 0.500 λ_0, the input resistance is $330 - j880$ Ω. The relatively gentle slope of the input reactance at the first resonance is evidence of the low Q behavior of the microstrip dipole for this length.

The same technique presented above can be easily extended to compute the mutual impedance between parallel, collinear, and echelon microstrip dipoles. Fig. 33(b) is a graph of the mutual impedance between two parallel broad-

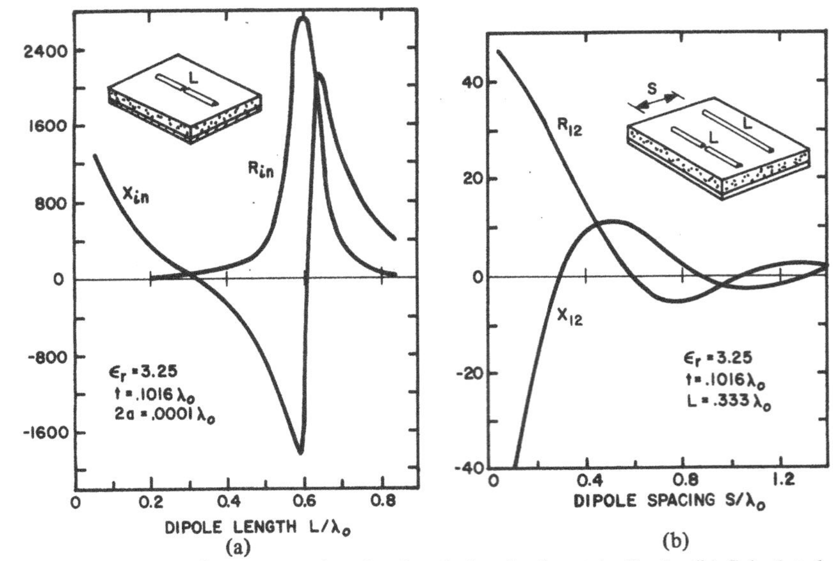

Fig. 33. (a) Calculated input impedance versus length of an isolated microstrip dipole. (b) Calculated mutual impedance between two broadside microstrip dipoles. After Rana and Alexopoulos [84].

side dipoles of 0.333 λ_0 length, and with the same substrate as used for Fig. 33(a). The graph is similar to that found for mutual coupling between two free-space broadside dipoles, except that when the dipoles are in free space, the mutual reactance for close spacing is inductive whereas it is capacitive when the dipoles are on a microstrip substrate.

The input impedance of a strip dipole of width w and length L on a microstrip substrate may also be calculated by variational techniques and the use of the appropriate Green's function [85]; the Green's function may be evaluated for far fields by using a stationary phase integration, and from this the far-field pattern may be computed. As an example a dipole of length 0.75 cm and width 0.05 cm on a substrate of 0.1 cm thickness and $\epsilon_r = 9.9$ has a broadside pattern at 10 GHz with a half-power beamwidth of 54° in the plane of the dipole axis and 90° in the orthogonal plane [85].

VIII. CONFORMAL PRINTED CIRCUIT ANTENNAS

There are numerous examples of conformal printed circuit antennas, although space limitations precluded a detailed discussion of these antennas. Conformal microstrip antennas can be mounted on a relatively small body such as a rocket so that the antenna elements in turn excite currents on the body so as to produce the desired radiation pattern. The pattern is then dependent upon the location of the antenna on the vehicle and the geometry of the vehicle. An example of such an antenna is the spiral-slot antenna [46] which was used to excite the dipole mode on a small missile with a bandwidth of about two percent at 238 MHz. A second example is a dielectric-filled edge-slot antenna [86]. This antenna consists of a dielectric disk with conductors on both surfaces which is coaxial with the conducting body of revolution so that its aperture coincides with the surface. The disk is excited by a coaxial feed, and the device is tuned to the proper operating frequency by means of a series of inductive posts across the dielectric disk. By proper location of these inductive posts and choice of the dielectric thickness, it is possible to tune the antenna over a six to one range. Instantaneous bandwidths of 12 percent have been observed at X band. Again, the radiation pattern is largely determined by the body upon which the antenna is mounted.

A spherical antenna was developed by DeSantis and Schwering [87] to produce greater than hemispheric coverage for satellite-to-aircraft reception of navigation signals. The antenna consisted of a conducting sphere which had an azimuthal slot cut along its 30° meridian. The slot was excited by a parallel-plate dielectric-filled resonator, which excited currents on the sphere surface. The radiation field was that of a low-order spherical harmonic, and the device produced adequate coverage when the antenna was isolated from the aircraft on a short mast.

Microstrip radiating elements have been employed to excite radiation from cones with very good results when the cone dimensions are of the order of the wavelength [88], [89], [91]. Again the radiation results from currents induced on the cone due to the microstrip element. It has been shown experimentally that good coverage in the forward direction may be obtained by properly exciting two elements mounted in the base of the cone. As the cone becomes longer in terms of wavelength, it becomes necessary to install elements near the apex of the cone to maintain good coverage in the forward direction.

IX. CONCLUSION

This paper has provided a comprehensive review of the state of microstrip antenna element technology as it exists in 1981. A wide variety of substrate materials suitable for element plating has been found to exist, with mechanical, thermal, and electrical properties which are attractive for use in both planar and conformal antenna configurations. However tolerance control of the dielectric constant remains a problem for accurate designs, particularly at higher microwave and millimeter frequencies. The mathematical analysis of the microstrip patch can be undertaken at several levels of sophistication, with the choice of the method dependent on the need for design accuracy as well as the shape of the patch. The simplest design technique for rectangular patches is based on a resonant half-wave transmission-line analogy and leads to very simple formulas for the resonant frequency and resonant resistance which are in approximate agreement with measured results. A more powerful, although slightly more complicated, approach is to use the modal-expansion technique in which the patch and ground plane are viewed as a resonant cavity with leaky magnetic walls. This is particularly well adapted to rectangular and circular patches and leads to design formulas for the resonant frequency, input impedance, bandwidth, efficiency, and directivity which are considerably more accurate than the simpler technique using transmission-line theory. Nonetheless, the modal-expansion technique is limited in its accuracy by the accuracy of the wall admittance, for which better formulas are needed. Patches of other shapes such as pentagonal or trapezoidal may be more conveniently analyzed by numerical techniques such as the method of moments, the finite-element approach, or by the unimoment–Monte Carlo method.

Most practical microstrip antenna designs use either the rectangular or circular patch, although other configurations such as the open-circuit microstrip radiator or the microstrip dipole are being used with increasing success. Design procedures and graphical presentations of typical microstrip patch performance data have been discussed, with emphasis on the rectangular and circular patches. In addition several practical modifications to these patches for special applications have been presented with typical performance data.

The microstrip antenna has typical bandwidths from one to six percent, although greater bandwidths may be achieved by using increased substrate thickness or larger patch sizes. A discussion of both the quality factor, the bandwidth, and the efficiency reveals that increased substrate thicknesses produce increased bandwidth and efficiency. The microstrip patch can be excited so as to produce either right-hand or left-hand circular polarization. Several practical methods for achieving circular polarization operation have been presented. In addition approaches to using the microstrip antenna on conformal bodies have been discussed.

Exclusive of the problems in microstrip arrays, there is a critical need for attention to the development of key improvements in the microstrip element itself. The first and most pressing of these is the need for better substrate dielectric constant tolerance control, as discussed in the first part of this paper. The second is the need for more detailed attention to rigorous solutions for the radiating wall admittance for various microstrip antenna geometries, including electrically thicker substrates, since this is crucial to improving design procedures and formulas. The third requirement is for the development of a larger class of layered microstrip element configurations which can be used for the design of multifrequency elements. Related to this is the need for more design approaches which can produce greater bandwidth. Finally, but certainly not the least of these, is the need for greatly expanded efforts in the development of monolithically integrated microstrip elements and associated active components.

X. ACKNOWLEDGMENT

The authors wish to gratefully acknowledge Prof. David Chang of the University of Colorado for his suggestion that this comprehensive review be undertaken, and for his encouragement of the project.

REFERENCES

[1] R. J. Mailloux, J. McIlvenna, and N. Kernweis, "Microstrip array technology," *IEEE Trans. Antennas Propagat.*, vol. AP-29, no. 1, pp. 25–38, Jan. 1981.

[2] G. A. Deschamps, "Microstrip microwave antennas," presented at the 3rd USAF Symp. on Antennas, 1953.

[3] H. Gutton and G. Baissinot, "Flat aerial for ultra high frequencies," French Patent No. 703113, 1955.

[4] E. V. Byron, "A new flush-mounted antenna element for phased array application," in *Proc. Phased-Array Antenna Symp.*, 1970, pp. 187–192.

[5] R. E. Munson, "Single slot cavity antennas assembly," U.S. Patent No. 3 713 162, Jan. 23, 1973.

[6] J. Q. Howell, "Microstrip antennas," in *Dig. Int. Symp. Antennas Propagat. Soc.*, Williamsburg, VA, Dec. 1972, pp. 177–180.

[7] H. D. Weinschel, "Progress report on development of microstrip cylindrical arrays for sounding rockets," Physic. and Sci. Lab., New Mexico State Univ., Las Cruces, 1973.

[8] G. G. Sanford, "Conformal microstrip phased array for aircraft tests with ATS-6," in *Proc. Nat. Electronics Conf.*, vol. 29, Oct. 1974, pp. 252–257.

[9] G. W. Garvin, R. E. Munson, L. T. Ostwald, and K. G. Schroeder, "Low profile electrically small missile base mounted microstrip antennas," in *Dig. Int. Symp. Antennas Propagat. Soc.*, Urbana, IL, June 1975, pp. 244–247.

[10] J. Q. Howell, "Microstrip antennas," *IEEE Trans. Antennas Propagat.*, vol. AP-23, no. 1, pp. 90–93, Jan. 1975.

[11] H. D. Weinschel, "A cylindrical array of circularly polarized microstrip antennas," in *Dig. Int. Symp. Antennas Propagat. Soc.*, Urbana, IL, June 1975, pp. 177–180.

[12] J. R. James and G. J. Wilson, "New design techniques for microstrip antenna arrays," in *Proc. 5th European Micro. Conf.*, Hamburg, Sept. 1975, pp. 102–106.

[13] R. E. Munson, "Conformal microstrip antennas and microstrip phased arrays," *IEEE Trans. Antennas Propagat.*, vol. AP-22, no. 1, pp. 74–77, Jan. 1974.

[14] A. G. Derneryd, "Linear microstrip array antennas," Chalmer Univ. Technol., Goteborge, Sweden, Tech. Rep. TR 7505, Oct. 1975.

[15] K. R. Carver, "The radiation pattern of a microstrip disc antenna," Physic. and Sci. Lab., New Mexico State Univ., Las Cruces, Tech. Memo., Nov. 29, 1976.

[16] Y. T. Lo, D. D. Harrison, D. Solomon, G. A. Deschamps, and F. R. Ore, "Study of microstrip antennas, microstrip phased arrays, and microstrip feed networks," Rome Air Development Center, Tech. Rep. TR-77-406, Oct. 21, 1977.

[17] A. G. Derneryd, "A theoretical investigation of the rectangular microstrip antenna element," Rome Air Development Center, Tech. Rep. TR-77-206, June, 1977.

[18] L. C. Shen and S. A. Long, "Low profile printed circuit antennas," Dept. Elec. Eng., Univ. Houston, Houston, TX, Contract DAAG-29-75-0187, Final Rep., Oct. 1977.

[19] K. R. Carver and E. L. Coffey, "Theoretical investigation of the microstrip antenna," Physic. and Sci. Lab., New Mexico State Univ., Las Cruces, Tech. Rep. PT-00929, Jan. 23, 1979.

[20] *Proc. Workshop on Printed Circuit Antenna Technology*, 31 papers, 480 pp., New Mexico State Univ., Las Cruces, Oct. 17–19, 1979.

[21] A. Waterman and D. G. Henry, "Stripline strap-on antenna array," in *Abstracts 21st USAF Antenna Symp.*, Allerton Park, IL, Oct. 12–14, 1971.

[22] *Reference Data for Radio Engineers*, 5th ed. Indianapolis, IN: Howard W. Sams, Oct. 1968, ch. 22, pp. 25–27.

[23] T. E. Nowicki, "Microwave substrates, present and future," in *Proc. Workshop Printed Circuit Antenna Tech.*, New Mexico State Univ., Las Cruces, Oct. 1979, pp. 26/1–22.

[24] G. R. Traut, "Clad laminates of PTFE composites for microwave antennas," *Proc. Workshop Printed Circuit Antenna Tech.*, New Mexico State University, Las Cruces, pp. 27/1-17, Oct., 1979.

[25] L. R. Murphy, "SEASAT and SIR-A microstrip antennas," in *Proc. Workshop Printed Circuit Antenna Tech.*, New Mexico State Univ., Las Cruces, Oct. 1979, pp. 18/1–20.

[26] K. R. Carver, "Description of a composite hexcell microstrip antenna," private communication to J. W. Mink, Dec. 1979.

[27] R. F. Harrington, *Time-Harmonic Electromagnetic Waves*. New York: McGraw-Hill, 1961.

[28] K. R. Carver, "A modal expansion theory for the microstrip antenna," in *Dig. Int. Symp. Antennas Propagat. Soc.*, Seattle, WA, June 1979, pp. 101–104.

[29] K. R. Carver, "Practical analytical techniques for the microstrip antenna," in *Proc. Workshop Printed Circuit Antenna Tech.*, New Mexico State Univ., Las Cruces, Oct. 1979, pp. 7/1–20.

[30] Y. T. Lo, D. Solomon, and W. F. Richards, "Theory and

experiment on microstrip antennas," *IEEE Trans. Antennas Propagat.*, vol. AP-27, no. 2, pp. 137–145, Mar. 1979.

[31] W. F. Richards and Y. T. Lo, "An improved theory for microstrip antennas and applications," in *Dig. Int. Symp. Antennas Propagat. Soc.*, Seattle, WA, June 1979, pp. 113–116.

[32] W. F. Richards, Y. T. Lo, P. Simon, and D. D. Harrison, "Theory and applications for microstrip antennas," in *Proc. Workshop Printed Circuit Antenna Tech.*, New Mexico State Univ., Las Cruces, Oct. 1979, pp. 8/1–23.

[33] A. R. Van de Capelle, "Theoretical investigations of microstrip antennas," in *Proc. Workshop Printed Circuit Antenna Tech.*, New Mexico State Univ., Las Cruces, Oct. 1979, pp. 11/1–8.

[34] E. Van Lil, R. Van Loock, and A. Van de Capelle, "Design models for rectangular microstrip resonator antennas," in *Dig. Int. Symp. Antennas Propagat. Soc.*, College Park, MD, May 1978, pp. 264–267.

[35] H. D. Weinschel, "Measurements of various microstrip parameters," in *Proc. Workshop Printed Circuit Antenna Tech.*, New Mexico State Univ., Las Cruces, Oct. 1979. pp. 2/1–15.

[36] A. G. Derneryd, "Linearly polarized microstrip antennas," *IEEE Trans. Antennas Propagat.*, vol. AP-24, no. 6, pp. 846–850, Nov. 1976.

[37] E. O. Hammerstad, "Equations for microstrip circuit design, in *Proc. 5th European Micro. Conf.*, Hamburg, Sept. 1975, pp. 268–272.

[38] C. M. Butler, "Analysis of a coax-fed circular microstrip antenna," in *Proc. Workshop Printed Circuit Antenna Tech.*, New Mexico State Univ., Las Cruces, Oct. 1979, pp. 13/1–17.

[39] N. G. Alexopoulos and I. E. Rana, "Mutual impedance computation between printed dipoles," *IEEE Trans. Antennas Propagat.*, vol. AP-29, no. 1, pp. 106–111, Jan. 1981.

[40] D. C. Chang and E. F. Kuester, "Total and partial reflection from the end of a parallel-plate waveguide with an extended dielectric loading," *Radio Sci*, to be published.

[41] M. V. Schneider, "Microstrip dispersion," *Proc. IEEE*, vol. 60, no. 1, pp. 144–146, Jan. 1972.

[42] D. C. Chang and E. F. Kuester, "Resonance characteristics of an rectangular microstrip antenna," in *Proc. Workshop Printed Circuit Antenna Tech.*, New Mexico State Univ., Las Cruces, Oct. 1979, pp. 28/1–18.

[43] J. W. Mink, "Sensitivity of microstrip antennas to admittance boundary variations," *IEEE Trans. Antennas Propagat.*, vol. AP-29, no. 1, pp. 143–145, Jan. 1981.

[44] L. C. Shen, S. A. Long, M. R. Allerding, and M. D. Walton, "Resonant frequency of a circular disc printed circuit antenna," *IEEE Trans. Antennas Propagat.*, vol. AP-25, no. 4, pp. 595–596, July 1977.

[45] C. M. Butler and E. K. Yung, "Analysis of a terminated parallel-plate waveguide with a slot in its upper plate," *Ann. Telecommun.*, tome 34, no. 9, 10, Sept.–Oct. 1979.

[46] D. H. Schaubert and F. G. Farrar, "Some conformal printed circuit antenna designs," in *Proc. Workshop Printed Circuit Antenna Tech.*, New Mexico State Univ., Las Cruces, Oct. 1979, pp. 5/1–21.

[47] J. L. Kerr, "Microstrip polarization techniques," in *Proc. 1978 Antenna Applications Symp.*, Allerton Park, IL, Sept. 1978.

[48] E. H. Newman and D. M. Pozar, "Electromagnetic modeling of composite wire and surface geometries," *IEEE Trans. Antennas Propagat.*, vol. AP-26, no. 6, pp. 784–789, Nov. 1978.

[49] E. L. Newman, "Strip antennas in a dielectric slab," *IEEE Trans. Antennas Propagat.*, vol. AP-26, no. 5, pp. 647–653, Sept. 1978.

[50] E. L. Coffey and K. R. Carver, "Towards the theory of microstrip antenna patterns," in *Proc. 1977 Antenna Application Symp.*, Allerton Park, IL, Apr. 1977.

[51] E. L. Coffey and T. H. Lehman, "A new analysis technique for calculating the self and mutual impedance of microstrip antennas," in *Proc. Workshop Printed Circuit Antenna Tech.*, New Mexico State Univ., Las Cruces, pp. 31/1–21.

[52] J. H. Richmond, "A wire-grid model for scattering by conducting bodies," *IEEE Trans. Antenna Propagat.*, vol. AP-14, pp. 782–786, Nov. 1966.

[53] J. H. Richmond, D. M. Pozar, and E. H. Newman, "Rigorous near-zone field expressions for rectangular sinusoidal monopole," *IEEE Trans. Antennas Propagat.*, vol. AP-26, p. 509–510, May 1978.

[54] E. H. Newman and P. Tulyathan, "Analysis of microstrip antennas using moment methods," *IEEE Trans. Antennas Propagat.*, vol. AP-29, no. 1, pp. 47–53, Jan. 1981.

[55] K. K. Mei, "Unimoment method of solving antenna and scattering problems," *IEEE Trans. Antennas Propagat.*, to be published.

[56] E. L. Coffey and W. L. Stutzman, "Hybrid computer solutions of electromagnetic scattering problems," in *Dig. Int. Symp. Antennas Propagat.*, Amherst, MA, Oct. 1976, pp. 82–85.

[57] G. Strang and G. J. Fix, *An Analysis of the Finite Element Method*. Englewood Cliffs, NJ: Prentice-Hall, 1973.

[58] R. W. Campbell, "Documentation of MICRO for microstrip antennas," Physic. and Sci. Lab., New Mexico State Univ., Las Cruces, Tech. Memo., Aug. 1979.

[59] K. R. Carver, "Input impedance to probe-fed microstrip antennas," in *Dig. Int. Symp. Antennas Propagat. Soc.*, Quebec City, Canada, June 1980.

[60] J. L. Kerr, "Microstrip antenna developments," in *Proc. Workshop Printed Circuit Antenna Tech.*, New Mexico State Univ., Las Cruces, Oct. 1979, pp. 3/1–20.

[61] H. D. Weinschel and K. R. Carver, "A medium-gain circularly polarized macrostrip UHF antenna for marine DCP communication to the GOES satellite system," in *Dig. Int. Symp. Antennas Propagat. Soc.*, Amherst, MA, Oct. 1976, pp. 391–394.

[62] S. A. Long and M. D. Walton, "A dual-frequency stacked circular disc antenna," in *Dig. Int. Symp. Antennas Propagat. Soc.*, College Park, MD, May 1978, pp. 260–263.

[63] G. Dubost, "Theory and experiments of a broadband short-circuited microstrip dipole at resonance," in *Proc. Workshop Printed Circuit Antenna Tech.*, New Mexico State Univ., Las Cruces, Oct. 1979, pp. 32/1–13.

[64] P. S. Hall, C. Wood and C. Garrett, "Wide bandwidth microstrip antennas for circuit integration," *Electron. Lett.*, vol. 15, no. 15, pp. 458–459, July 19, 1979.

[65] C. M. Kaloi, "Microstrip antennas, experimental results," in *Proc. Workshop Printed Circuit Antenna Tech.*, New Mexico State University, Las Cruces, Oct. 1979, p. 6/1.

[66] J. Watkins, "Circular resonant structures in microstrip," *Electron. Lett.*, vol. 5, pp. 524–525, Oct. 1974.

[67] I. Wolff and N. Knoppik, "Rectangular and circular microstrip disk capacitors and resonators," *IEEE Trans. Microwave Theory Technol.*, vol. MTT-22, no. 10, pp. 857–864, Oct. 1974.

[68] A. G. Derneryd, "Analysis of the microstrip disk antenna element," *IEEE Trans. Antennas Propagat.*, vol. AP-27, no. 5, pp. 660–664, Sept. 1979.

[69] J. McIlvenna and N. Kernweis, "Variations on the circular microstrip antenna element," in *Proc. Workshop Printed Circuit Antenna Tech.*, New Mexico State Univ., Las Cruces, Oct. 1979, pp. 25/1–15.

[70] J. W. Mink, "Circular ring microstrip antenna elements," in *Dig. Int. Symp. Antennas Propagat. Soc.*, Quebec City, Canada, June 1980.

[71] G. G. Sanford, "Conformal microstrip phased array for aircraft tests with ATS-6," *IEEE Trans. Antennas Propagat.*, vol. AP-26, no. 5, pp. 642–646, Sept. 1978.

[72] R. E. Munson, private communication, June 1975.

[73] L. C. Shen and S. A. Long, "Theoretical analysis of printed-circuit antennas," in *Proc. Workshop Printed Circuit Antenna Tech.*, New Mexico State Univ., Las Cruces, Oct. 1979, p. 14/1.

[74] L. C. Shen, "The elliptical microstrip antenna with circular polarization," *IEEE Trans. Antennas Propagat.*, vol. AP-29, no. 1, pp. 90–94, Jan. 1981.

[75] S. A. Long, L. C. Shen, D. H. Schaubert, and F. G. Farrar, "An experimental study of the circular-polarized, elliptical, printed-circuit antenna," *IEEE Trans. Antennas Propagat.*, vol. AP-29, no. 1, pp. 95–99, Jan. 1981.

[76] R. T. Irish, "Elliptic resonator and its use in microcircuit systems," *Electron. Lett.*, vol. 7, pp. 149–150, Apr. 1971.

[77] R. J. Collier and P. D. White, "Surface waves in microstrip circuits," in *Proc. 6th European Micro. Conf.*, 1976, pp. 632–636.

[78] A. G. Derneryd and A. G. Lind, "Extended analysis of rectangular microstrip resonator antennas," *IEEE Trans. Antennas Propagat.*, vol. AP-27, no. 6, pp. 846–849, Nov. 1979.

[79] W. C. Wilkinson, "A class of printed circuit antennas," in *Dig. Int. Symp. Antennas Propagat. Soc.*, Atlanta, GA, June 1974, pp. 270–273.

[80] R. W. Olthius, "Printed circuit antenna for radar altimeter," in

Dig. Int. Symp. Antennas Propagat. Soc., Amherst, MA, Oct. 1976, pp. 554–557.

[81] H. G. Oltman, Jr., U.S. Patent No. 4054874, Oct. 18, 1977.

[82] D. A. Huebner, "An electrically small microstrip dipole planar array," in *Proc. Workshop Printed Circuit Antenna Tech.*, New Mexico State Univ., Las Cruces, Oct. 1979, pp. 17/1–16.

[83] J. R. James, P. S. Hall, C. Wood, and A. Henderson, "Microstrip antenna research at the Royal Military College of Science," in *Proc. Workshop Printed Circuit Antenna Tech.*, New Mexi State Univ., Las Cruces, Oct. 1979, pp. 1/1–10.

[84] I. E. Rana and N. G. Alexopoulos, "Printed wire antenna," *Proc. Workshop Printed Circuit Antenna Tech.*, New Mexi State Univ., Las Cruces, Oct. 1979, pp. 30/1–38.

[85] N. K. Uzunoglu, N. G. Alexopoulos, and J. G. Fikior "Radiation properties of microstrip dipoles," *IEEE Trai Antennas Propagat.*, vol. AP-27, no. 6, pp. 853–858, Nov. 1979.

[86] H. S. Jones, Jr., D. H. Schaubert, and F. Reggia, "A conformal, dielectric-filled edge-slot antennas for bodies of revolution," in *Proc. 1977 Antenna Applications Symp.*, Allerton Park, IL, Apr. 1977.

[87] C. DeSantis and F. Schwering, "Circularly polarized spherical antenna," U.S. Patent No. 4185289, Jan. 22, 1980.

[88] R. P. Jedlicka, M. T. Poe, and K. R. Carver, "Measured mutual coupling between microstrip antennas," *IEEE Trans. Antennas Propagat.*, vol. AP-29, no. 1, pp. 148–150, Jan. 1981.

[89] A. R. Sindoris and F. G. Farrar, "A four-element conformal array antenna," in *Proc. 1977 Antenna Applications Symp.*, Allerton Park, IL, Apr. 1977.

[90] A. G. Derneryd, "A theoretical investigation of the rectangular microstrip antenna," *IEEE Trans. Antennas Propagat.*, vol. AP-26, no. 4, pp. 532–535, July 1978.

[91] H. S. Jones, "Integrated randome-antenna designs," *Microwaves*, pp. 34–39, Sept. 1967.

[92] G. Dubost, M. Nicolas, and H. Havot, "Theory and applications of broadband microstrip antennas," in *Proc. 6th European Micro. Conf.*, Sept. 1976, pp. 275–279.

[93] T. Itoh and R. Mittra, "A new method for calculating the capacitance of a circular disk for microwave integrated circuits," *IEEE Trans. Microwave Theory Technol.*, vol. MTT-21, pp. 431–432, June 1973.

[94] T. Itoh and W. Menzel, "A high frequency analysis method for open microstrip structures," in *Proc. Workshop Printed Circuit Antenna Tech.*, New Mexico State Univ., Las Cruces, Oct. 1979, pp. 10/1–16.

[95] S. Coen and G. M. L. Gladwell, "A Legendre approximation method for the circular microstrip disk problem," *IEEE Trans. Microwave Theory Technol.*, vol. MTT-25, no. 1, pp. 1–6, Jan. 1977.

[96] P. Hammer, D. Van Bouchaute, D. Verschraeven, and A. Van de Capelle, "A model for calculating the radiation field of microstrip antennas," *IEEE Trans. Antennas Propagat.*, vol. AP-27, no. 2, pp. 267–270, Mar. 1979.

[97] E. Belohoubek and E. Denlinger, "Loss considerations for microstrip resonators," *IEEE Trans. Microwave Theory Technol.*, vol. MTT-23, pp. 522–526, June 1975.

[98] H. Sobol, "Radiation conductance of open-circuit microstrip," *IEEE Trans. Microwave Theory Technol.*, vol. MTT-19, pp. 885–887, Nov. 1971.

[99] L. Lewin, "Radiation from discontinuities in stripline," in *Proc. Inst. Elec. Eng.*, vol. 107, pt. C, Feb. 1960, pp. 163–170.

[100] A. G. Derneryd, "The circular microstrip antenna element," in *Proc. Int'l. Conf. Antennas Propagation (IEE)*, Part 1, Nov. 1978, pp. 307–310.

[101] S. A. Long, L. C. Shen, and P. D. Morel, "A theory of the circular-disk printed-circuit antenna," *Proc. IEE*, vol. 125, no. 10, pp. 925–928, Oct. 1978.

[102] G. R. Traut, "Clad laminates of PTFE composites for microwave antennas," *Microwave J.*, vol. 23, no. 11, pp. 47–51, Nov. 1980.

[103] M. Olyphant, Jr. and T. E. Nowicki, "Microwave substrates support MIC technology," *Microwaves*, Part I, vol. 19, no. 12, pp. 74–80, Nov. 1980.

[104] ——, "Microwave substrates support MIC technology," *Microwaves* Part II, vol. 19, no. 13, pp. 47–52, Dec. 1980.

Research on planar antennas and arrays: "Structures Rayonnantes"

J. P. Daniel, G. Dubost, C. Terret
URA CNRS 834
Université de Rennes I - INSA de Rennes
Université 35042
Rennes, France

J. Citerne, M. Drissi
URA CNRS 834
Université de Rennes I - INSA de Rennes
INSA 35043
Rennes, France

1. Presentation

During the last ten years a strong evolution has appeared in the antenna area: both new kinds of radiating structures (flat-microstrip and wire antennas) and arrays have emerged, with different associated mathematical models analysis. Among the wide world of radiating elements, the laboratory "Structures Rayonnantes," of the University of Rennes I, France, has developed some original structures related to flat and wire antennas, while new applications are under development.

The initial laboratory, "Antennes et Rayonnement," was created by Professor Dubost, in 1965, at the University of Rennes. The activity increased progressively, and four teams joined the CNRS to yield the final laboratory, "Structures Rayonnantes" (URA 834). Three of the teams came from the University of Rennes I, and one came from the INSA (Rennes), all of them being located on the same campus.

Today, the basic research on radiowave systems deals with three main domains: Electromagnetism, electronics, and signal processing. The fusion among these different domains occurs for studies concerning systems where radiating structures, active devices and interfaces, and signal processing play a fundamental role, and can't be separated (e.g., in telecommunications and radar). The studies are divided into four parts:

1. Antennas and arrays: design of basic radiating elements, CAD of antennas and arrays, technology
2. Diffraction: frequency-selective surfaces, RCS
3. Electronics and antennas: active antennas, MMIC design
4. Propagation, communications systems, and signal processing.

The investigated radio frequencies spread from HF up to the millimeter band (1 MHz - 110 GHz), and researchers are engaged in both theoretical and experimental aspects. The laboratory owns the basic equipment for radio measurements (e.g., HP 8510 network analyzers, spectrum analyzers, four anechoic chambers). Radar measurements are carried out at the open experimental site (30 hectacres) located outside the campus.

Circuit computations are performed on various local workstations, using commercial circuit software such as HP *MDS* and *Touchstone*. In contrast, mathematical models of antennas and communication links are developed in the laboratory, and codes are implemented on PCs or workstations, or on larger computers, depending on the required memory size, speed, and available routines.

In most radio systems, the antenna appears to be a key component, even if it is strongly coupled to electronics and signal processing. Only the first topic related to antennas and arrays is presented here: more precisely, flat antennas (most of them being printed on copper-clad laminate), and arrays studied recently (during the last three years) are described, with their typical applications.

2. Basic planar radiating sources

Different geometries and methods have been developed for various types of patches, dipoles, and slot antennas, and for mixed structures (slots on a patch), as shown in Figure 1. A classification scheme, depending on the nature of the radiating sources and the

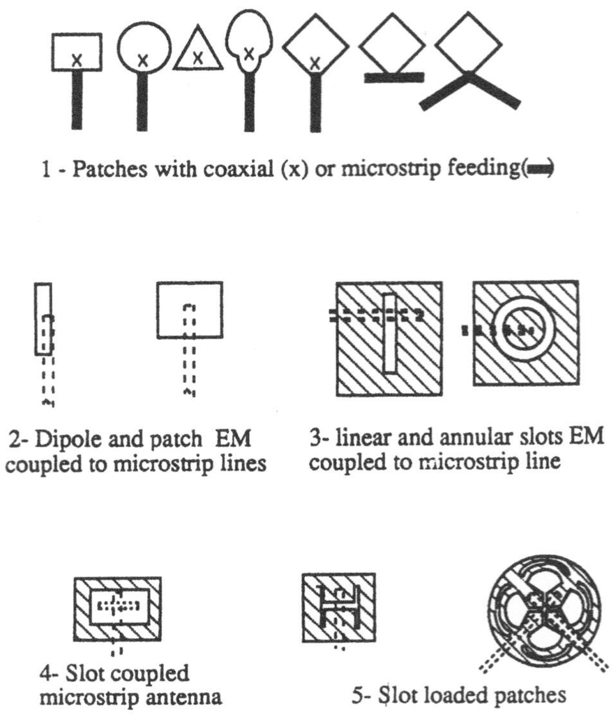

Figure 1. Five types of flat antennas.

Reprinted from *IEEE Antennas Propaga. Mag.* vol. 35, no. 1, pp. 14–38, Feb. 1993.

feeding technique, leads to the definition of five groups. An additional, parasitic element, located above the driven patch, leads to a second class of antennas, which are named "stacked microstrip patches." Performance factors, such as band width and gain, are improved. For each group, only the latest, interesting (or the most representative) results will be detailed here, while the previous, published methods will be mentioned, with references.

Two kinds of analysis have been considered:

–The first group starts from initial physical assumptions, which generally offers simple and analytical formulas, well suited for a physical understanding of phenomena and for future antenna CAD. These methods are known as transmission-line models (radiation losses are included in the attenuation coefficient of the propagation constant) and cavity models (radiation losses are included in the effective loss tangent of the dielectric).

–The second group one is based on an EM boundary problem, which leads to expression as an integral equation, using proper Green functions, either in the spectral domain (the SDA method), or directly in the space domain, using moment methods. Without any initial assumption, the choice of test functions and the path integration appear to be more critical during the final, numerical solution.

2.1 Patch with coaxial or microstrip feed

The most typical shapes of printed patches are the rectangle, the circle, the triangle, or the more general geometry, which preserves a main symmetry axis, passing through the feed point. The usual feed generally is a microstrip line, connected to the edge, or a coaxial line, suitably located inside the patch, in order to get a proper match to 50 ohms. On the other hand, a more original feed technique is also given in Figure 1, for the square patch (the parallel-patch antenna, or PPA, and the series-patch antenna, or SPA).

2.1.1 Transmission-line model ([1]...[8]). This transmission model differs from previous models and, particularly, from the work of Van De Capelle (c.f. Chapter 10 of ref. [17]), who considers radiation losses due to the radiation of slots, located around the patches. A TEM approximation, along the axis of each symmetrical patch, is assumed first. Next, equivalent ohmic and radiation loses are considered to be distributed along the whole transmission line. As far as a general, symmetrical shape (Figure 2) is concerned, the width, $a(s)$, the characteristic admittance, $G_c(s)$, and the transmis-

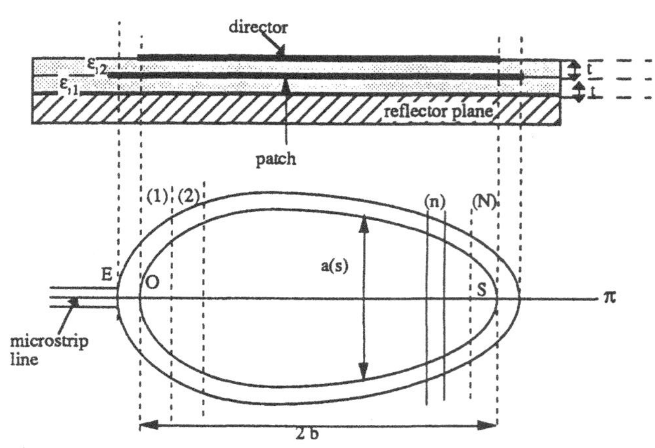

Figure 2a. Stacked microstrip antennas of arbitrary shape.

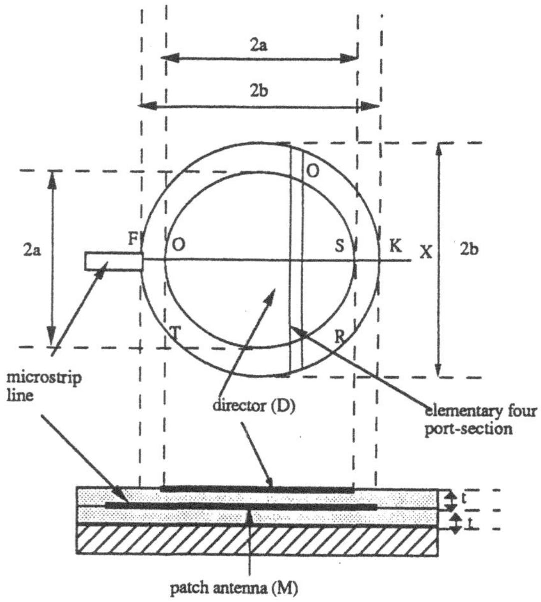

Figure 2b. Stacked microstrip antennas of circular shape.

sion coefficient, $\gamma(s)$, depend on the s variable along the axis. The normalized radiation admittance, Y_r, related to $G_0 = (\varepsilon_0/\mu_0)^{1/2}$, is given by the Riccati equation:

$$\frac{dY_r(y)}{dy} + \left[\frac{4\pi^3}{5}\frac{t_r^2}{\sqrt{\varepsilon_e}}g_c(y) + \pi\sqrt{\varepsilon_e}\left(\tan\delta + \frac{D_S}{t_r}\right) + j2\pi\sqrt{\varepsilon_e}\right] \times \left[g_c(y) - \frac{Y_r^2(y)}{g_c(y)}\right] = 0 \tag{1}$$

with $y = s/\lambda_0$, $t_r = t/\lambda_0$, $D_S = \sqrt{|(\varepsilon_0 f)/(\pi\sigma)|}$. $g_c(y) = G_c(y)/(\varepsilon_0/\mu_0)^{1/2}$ is the normalized characteristic admittance, and it depends on the $t/a(y)$ ratio and on ε_r. t_r is the normalized dielectric thickness, δ is the dielectric loss angle, D_S the normalized skin depth, σ is the (finite) conductivity, and $\varepsilon_e(y)$ is the effective relative permittivity. The method used to solve the Riccati equation (1) is justified by the following property: with any homographic application, a Riccati equation is converted into a Riccati equation. We cut the patch into N sections. A numerical solution is then obtained with the boundary condition $Y_r(2b/\lambda_0) = 0$. Finally, the radiation admittance, $Y_r(0)$, at the input, can be expressed as $Y_r(0) = (Y_r)_0$. The lowest resonance frequency is obtained where the real part of $Y_r(0)$ is a maximum for the first time.

For a circular patch with $t_r \ll 0.586/\sqrt{\varepsilon_r}$ [ε_r is the relative permittivity] [6, 7, 8, 9] and $A = 2b/\lambda_0$, we have

$$g_c(y) = 2\sqrt{\varepsilon_r}\sqrt{[y(A-y))]}/t_r.$$

Then, the right resonant diameter, which corresponds to

$$A_r = 0.586/\sqrt{\varepsilon_r},$$

can be compared with the lowest-order resonance, calculated with the transmission-line model (TLM), as is shown in Table 1.

Table 1. Theoretical resonant diameter (normalized to free-space wavelength) of circular patches, compared with transmission-line model results.

ε_r	1.00	2.32	2.47	3.80	7.00	9.80
A_r	0.586	0.385	0.373	0.300	0.221	0.187
A_r(TLM)	0.592	0.398	0.386	0.314	0.223	0.198

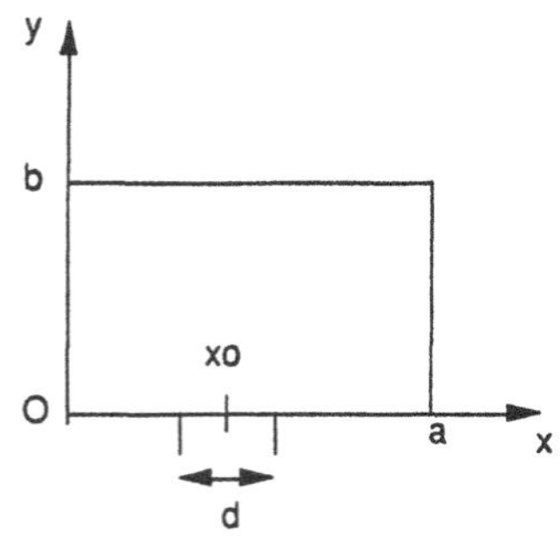

Figure 3. A rectangular patch fed with a microstrip line.

For a rectangular shape, of length b and width a, the Riccati equation (1) has an analytical solution,

$$Y_r = G_c \tanh[\gamma(b-s)],$$

and we obtained the radiation conductance when $a >> t$:

$$\frac{1}{R_r} = G_r = \sqrt{\frac{\varepsilon_0}{\mu_0}}\left[\frac{2\pi^3}{5}\left(\frac{b}{\lambda_0}\right)^2 + \frac{\pi}{2}\frac{b}{t}\sqrt{\varepsilon_e}\left(\tan\delta + \frac{D_S}{t_r}\right)\right] \tag{2}$$

Results and comparisons with others methods and experiments are given in Table 2.

2.1.2 Cavity model and effective loss tangent (15). This is a classical analysis for patch antennas of simple geometries, such as rectangular, circular, and triangular patches. For instance, the rectangular microstrip patch (Figure 3) is treated as a cavity bounded by four magnetic walls. The field is expressed as a series, using a mode-matching technique. All the loses (the dielectric loss, P_d, metallic loss, P_c, and radiated loss, P_r) in the antenna are represented by means of an effective loss tangent, δ_{eff}:

$$\delta_{eff} = \frac{P_r + P_d + P_c}{\omega(w_e + w_h)} = \frac{1}{Q} \tag{3}$$

where w_e and w_h are the electric and magnetic stored energy. The radiated power and δ_{eff} depend on the value of the electric field of the cavity, computation of which requires knowledge of δ_{eff}. The problem can be solved by an iterative procedure. The initial data are the dimensions, the frequency, and the electric properties of the dielectric substrate. Starting from $\delta_{eff} = \delta$, the internal and radiated fields are computed, leading to a new value of δ_{eff}, which is the starting point of a second iteration. The iterative sequence is stopped after the ith loop when

$$\frac{(\delta_{eff})_{i+1} - (\delta_{eff})_i}{(\delta_{eff})_i} < P\%,$$

where P is chosen arbitrarily (usually, less than 1 to 5). In fact, it has been shown [13] that δ_{eff} does not depend strongly on k (the wave number) or d. So, choosing an initial value of δ_{eff} very close to the real value is highly recommended, in order to reduce the computing time.

2.1.2a Effective loss tangent determination. According to (3), near the resonant frequency, the effective loss tangent may be written as

$$\delta_{eff} = \delta + \frac{\Delta}{t} + \frac{P_r}{2\omega_0 w_e} \tag{4}$$

where Δ is the skin depth. In order to find a proper initial value of δ_{eff}, δ_{eff0}, w_e and P_r have to be calculated for the dominant mode. The stored electric energy, at resonance, is

$$w_e = \frac{\varepsilon_0 \varepsilon_r ab V_0^2}{8t}$$, where V_0 is the input voltage.

The far fields may be calculated by modeling the radiator as four radiating slots. In order to carry out simple integrations, these fields are expressed in well-suited coordinates x',y',z', as explained in [15]. Then, using polynomial expressions for the electric-potential components $F_{y'}$ and $F_{z'}$, analytical integration for the radiated power, P_r, can be performed:

$$P_r = \frac{V_0^2}{R_0}\frac{A\pi^5}{192}\left[(1-B)\left(1 - \frac{A}{15} + \frac{A^2}{420}\right) + \frac{B^2}{5}\left(2 - \frac{A}{7} + \frac{A^2}{189}\right)\right] \tag{5}$$

with $A = [\pi(a + 2\Delta a)/\lambda_0]^2$ and $B = [\pi(b + 2\Delta b)/\lambda_0]^2$, where Δa and Δb are obtained using the Hammerstad formula, and R_0 is the intrinsic impedance of free space.

Equation (5) gives the radiated power with an accuracy better than 2.5% for $b/\lambda_0 = 0.3$ (with a typical limit of $a/\lambda_0 = 0.6$), and better than 4% for $b/\lambda_0 = 0.15$ (and a similar typical limit of $a/\lambda_0 = 0.3$).

2.1.2b Results and comparisons among models (radiating edge feeds). The iterative procedure is similar to the previous one, except for the initial value of the first step, where $\delta_{eff} = \delta_{eff0}$. Table 2 presents measured and computed values of the resonant frequency and resonant resistance, obtained using the cavity model (with δ_{eff0} and $\delta = 2.4 \times 10^{-3}$ for $\varepsilon_r = 10.2$, and $\delta = 10^{-3}$ for $\varepsilon_r = 2.2$), the moment method [16], and the previous transmission-line model. Measured values are those given by Schaubert et al. [16], obtained using a microstrip feed line connected to the middle of the radiating edge (Figure 3). The cavity- and transmission-line models give reasonable results for the resonant frequency and the resistance. The resonant frequencies are obtained with about 2% or less error for low permittivity. However, when ε_r is high or when the substrate is thick, there is some discrepancy between the measured and calculated resonant resistances. Three reasons can explain these differences:

- the surface-wave effect is assumed to be negligible

- the width of the feed line is considered to be small enough to keep identical radiating slot lengths both on the edge

Table 2. Comparison of resonant frequencies and input resistance (experimental and theoretical results) of printed rectangular patches.

					Measured (ref.16)		Cavity model (ref.15)				Moment method (ref.16)				Transmission line (ref.4)			
εr	t(mm)	b(mm)	a(mm)	d(mm)	fr	Rr	fr	%	Rr	%	fr	%	Rr	%	fr	%	Rr	%
10.2	1.27	20	30	1.19	2.26	335	2.314	+ 2.6	306	-8.86	2.25	- 0.4	350	+4.4	2.31	+ 2.4	306	-8.6
10.2	1.27	9.5	15	1.19	4.43	339	4.59	+ 3.6	358	+5.6	4.50	+ 1.6	350	+3.2	4.81	+ 8.6	398	+17.4
10.2	2.54	19	30	2.38	2.18	363	2.29	+ 5.0	364	+0.2	2.33	+ 6.9	420	+15.	2.35	+ 7.8	423	+16.5
2.22	0.79	25	40	2.42	3.92	136	3.95	+ 0.7	139	+2.0	3.92	0	130	+4.4	3.94	+ 0.5	102	- 25
2.22	0.79	12.5	20	2.42	7.56	152	7.64	+ 1	154	+1.3	7.60	+ 0.5	160	+5.3	7.70	+ 1.85	112	- 26
2.22	1.52	25	40	4.66	3.82	119	3.83	+ 0.2	153	28.	3.80	- 0.5	143	20.	3.86	+ 1	112	- 5.9
2.22	1.52	12	20	4.66	7.72	69	7.5	- 2.8	150	117.	7.75	0.38	145	110.	7.64	- 1	117	+ 69

connected to the microstrip feed line and on the opposite edge

– the value of the loss tangent of the dielectric is quite critical for the resistance value. For example, when δ changes from 10^{-3} to 2.4×10^{-3}, R_r changes from 354 Ω to 306 Ω (see the computed results of the cavity model and the dimensions in the first line of Table 2).

The best results are obtained with the moment method, at the expense of a large computing time. However, using the cavity model, the computation time is less than 20 seconds for a range of 60 frequencies, on a '286-type PC with a numeric coprocessor. Transmission-line and cavity models are well suited as initial CAD tools.

2.1.3 Stacked-microstrip antennas. Electromagnetically-coupled microstrip patches with stacked configurations have recently gone through a great deal of development, due to their performance features. Mainly, these include large bandwidth, higher gain, and/or dual-frequency operation. A quasi-TEM two-coupled-transmission-lines model is used to express the radiation admittance, the bandwidth, the current distributions, and the radiation patterns of an arbitrarily-shaped symmetrical-patch antenna, coupled with a director [5, 6]. The rectangular-, square-, or circular-shaped patches are particular cases.

We also studied such rectangular- and square-patch antennas, with different configurations of feeds, applying a Galerkin method in the spectral domain [18]. In order to reduce the computation time, we have used Legendre polynomials as basis functions. Their Fourier transform decays more rapidly, as they are combinations of spherical Bessel functions, rather than the cylindrical ones commonly reported in the literature. We perform the numerical integration in the complex plane, on an integration path far away from surface-wave poles. This makes the integrand more regular and smooth [19].

2.1.3.1 Transmission-line model analysis of arbitrarily-shaped symmetrical-patch antennas coupled with a director. The conduction and polarization currents are taken into account, to express the radiation admittance [5] and the radiation pattern [6] of the antenna (an arbitrarily-shaped symmetrical patch with a director).

Model definition: From the origin, O (Figure 2 a), and in the right direction, the radiating system of arbitrary shape, but symmetrical about the (π) plane, is equivalent to two coupled transmission lines between O and S, and acts according to a quasi-TEM mode.

Table 3. Resonant frequency and bandwidth of circular patch with a director (dimension per Figure 4b). Comparison between theory, transmission-line model, and experience (after [5]).

	f_M GHz	f_m GHz	B, % (VSWR<2)	f_o, GHz	$(VSWR)_{min}$
Theory (N = 500)	8.50	7.15	17.2	8.15	1.1
Experiments	8.38	7.08	16.8	8.10	1.3

The patch antenna is fed at (E) by a microstrip line. The equivalent four-port structure, of length $2b$, is divided into N elementary four-port sections, each of equal length. The nth elementary four-port section is equivalent to two short transmission lines, coupled with a capacitance. Moreover, the radiation and the ohmic losses are taken into account, as explained in [5] and [6].

In a particular case, related to a circular patch with a director (Figure 2b), a model has been tested with the following parameters: $2b = 13.7$ mm; $2a = 12.9$ mm; $\varepsilon_{r1} = \varepsilon_{r2} = 2.17$; $t_1 = t_2 = 1.6$ mm. Table 3 shows the bandwidth, B, related to the mean radiation resistance of 170 Ω, for a VSWR below two. We also indicate the maximum (f_M) and minimum (f_m) frequencies which limit the bandwidth. The $(VSWR)_{min}$ is obtained for the frequency f_o.

In Figure 4 (a and b) we present the current distributions calculated at f = 7.1 GHz. The modulus of the director conduction current is zero at $x_n = 0$ and $x_n = 2a$, and maximum at $x_n = a$, and its phase is constant. The modulus for the patch conduction current is equal to one ampere at $x_n = 0$, and zero at $x_n = 2a$. Its phase is not constant, and the asymmetry of the radiation pattern in the "E plane" is hence explained. The polarization currents for the director and the patch have their maximum modulus for $x_n/2a$ near to 0.2 and 0.8. In Figure 4c we present different patterns at 7.1 GHz, measured or calculated in the "E plane." Curve 2 is related to a circular reflector of diameter, $2d$, of 12 cm, with GTD corrections. In the "E plane," the measured cross-polarization pattern level is very low (curve 4).

2.1.3.2 Broad-band probe-fed EMCP antennas. In this configuration, the effect of the feeding probe results in an additional inductive component to the antenna input impedance. This probe inductance has been accounted for only through use of a simple formula [21, 18].

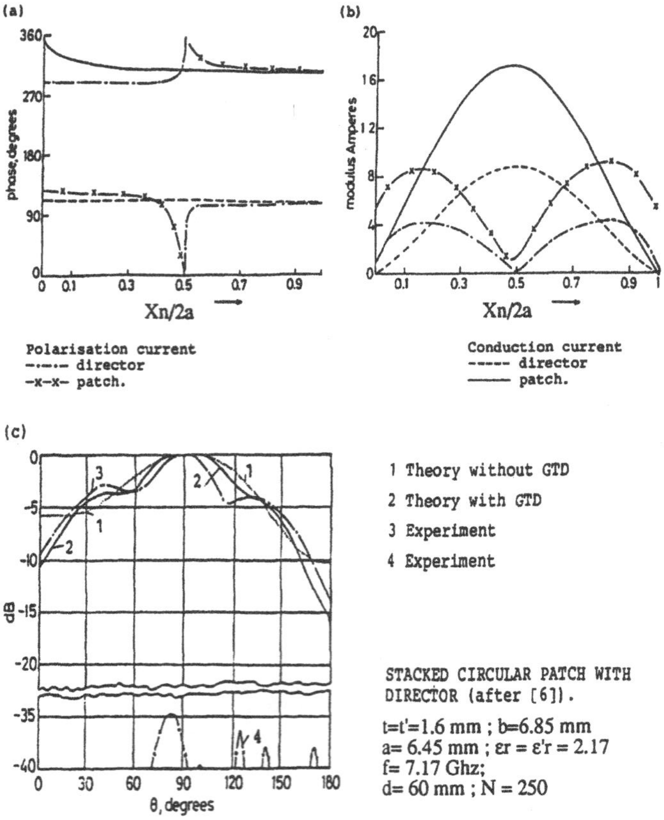

Figure 4. Stacked circular patches: The current distributions (conduction and polarization) on the director - - - and on the patch ——. (a) Phase (b) Modulus (c) Far-field pattern in E plane.

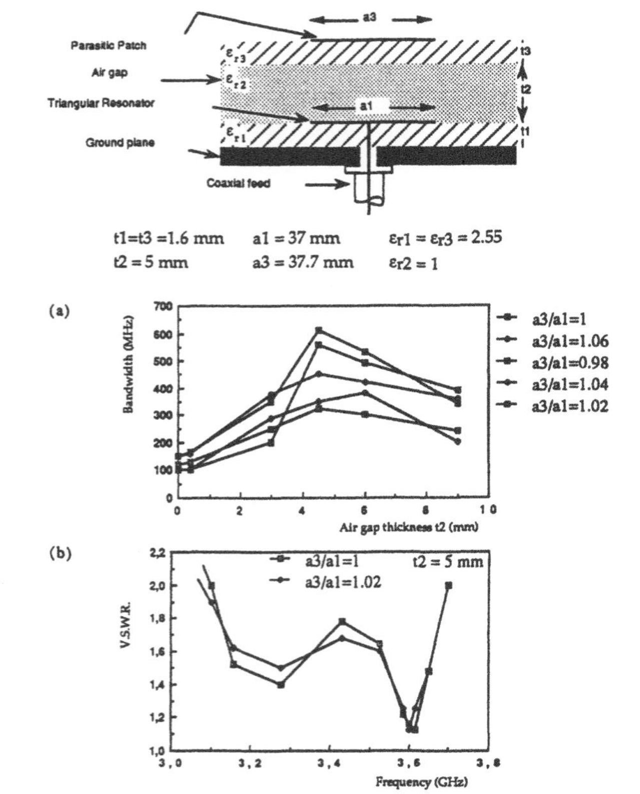

Figure 5. Probe-fed stacked triangular microstrip antenna: a) Bandwidth variation versus air-gap thickness; b) Frequency dependence of measured VSWR.

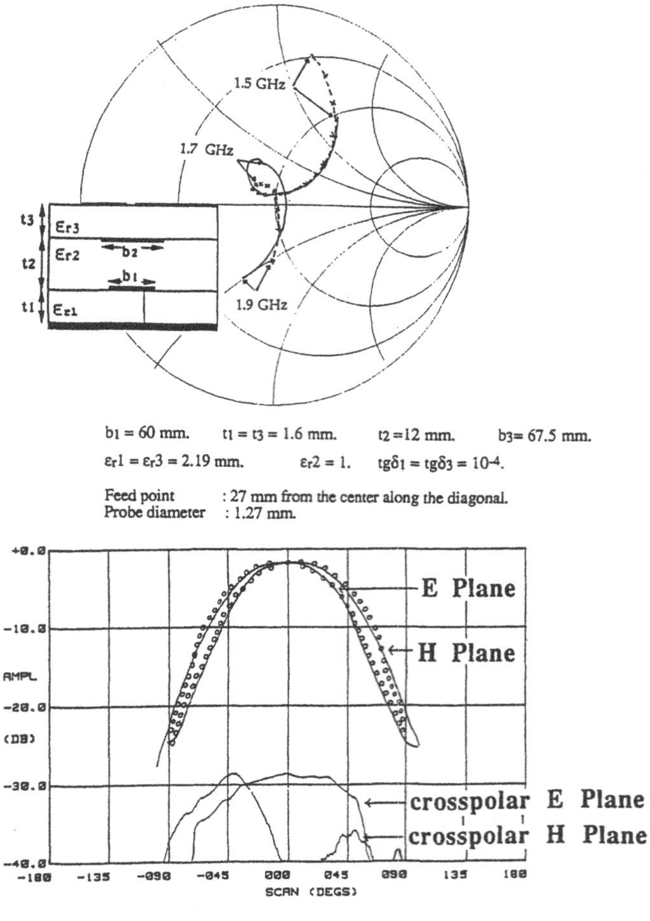

Figure 6. A stacked square-patch antenna. a) Input impedance versus frequency (increment 20 MHz):
—— Experiment; -x--x- Theory.
b) Radiation pattern at the center frequency f = 1.56 GHz of the stacked square-patch antenna:
—— Experiment; o o o o o Theory.

2.1.3.2.1 Triangular microstrip antenna. The stacked triangular microstrip antenna, presented in Figure 5, has been examined experimentally at frequencies in the S band [22, 23]. The behavior of the characteristic impedance, for varying air-gap thickness (1.5 mm to 9 mm), and varying sizes of the parasitic element (a_3/a_1 = 0.98 to 1.06), is presented in Figures 5a and 5b. An optimized bandwidth (about 19% for VSWR < 2) was obtained. The beamwidth measured over the full bandwidth varies from 55° to 65° in the E plane, and from 75° to 85° in the H plane. The cross-polarization level is less than -17 dB.

2.1.3.2.2 Square-patch antennas. This antenna is composed of two electromagnetically-coupled patches, deliberately fed on their diagonal by a coaxial probe. In this configuration, the upper dielectric slab additionally provides a protective layer, as shown in Figure 6. Cavity modes TM_{01} and TM_{10} are simultaneously excited [24], causing the following interesting property: the input impedance of the single patch is high (about 300 ohms), so it is easier to match this antenna when it becomes multi-layered. The polarization is linear and oriented along the diagonal. Cross polarization is mainly due to the TM_{11} cavity mode.

A broad-band microstrip antenna, with low cross polarization and high gain. This antenna was designed and experimented with in the L band. [25]. As shown in Figure 6a, we got a maximum bandwidth of 14.4%, for a VSWR < 1.7, around a central frequency of 1.67 GHz. The radiation patterns have been measured, and are compared to the theory in Figure 6b. The cross-polarization level is less than -27 dB over the whole frequency range. The beamwidth is narrower than that of the single patch, in both the E and H planes (67° to 80°). Thus, this element presents good efficiency, and its average gain is about 9 dB +/-1 dB over the frequency range from 1.56 to 1.80 GHz.

A broad-band dual fed circularly polarized antenna in S band. Another square-patch antenna has been designed for array applications in the S band. The influence of physical parameters has been theoretically investigated [18] (Figure 7). The air-layer thickness, t_2, is one of the most important parameters to adjust to increase the bandwidth. The chart plotted in Figure 7a shows the bandwidth characteristics (for a VSWR less than 1.7) of the double-patch antenna, as a function of the ratio b_2/b_1, and the distance, t_2/λ_0, between the upper and lower patches. We've used these charts to provide a stacked antenna CAD. We observe that the larger the required bandwidth, the shorter is the usable region over which the parameters can be varied. The maximum bandwidth (17% for a VSWR < 1.7) is obtained for an air-layer thickness of 8.8 mm, that is, 0.075 λ_0. The comparison with experiment is shown in Figure 7b. Only two basis functions were used on each patch, and this could explain the discrepancies between theory and experiment. All over the optimum bandwidth, the radiation patterns present quite a low cross polarization (less than -25 dB), and the gain is up to 8.5 dB [26].

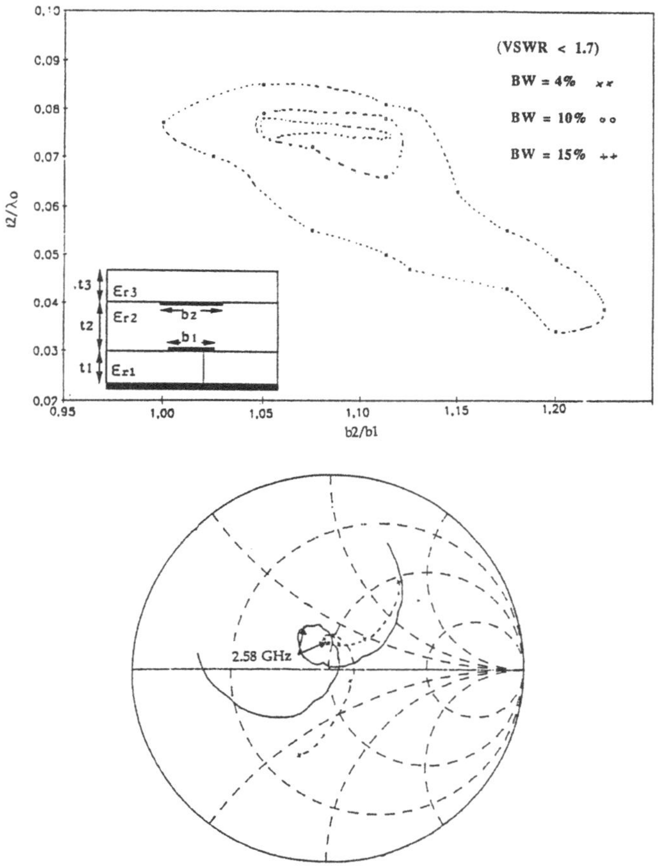

Figure 7. Broadband stacked square-patch antenna, 2.3 - 3.0 GHz.
a) Calculated bandwidths versus parameters
$t_1 = t_3 = 1.6$ mm $\quad \varepsilon_{r1} = \varepsilon_{r3} = 2.2$ mm
$\varepsilon_{r2} = 1$ $\quad \tan\delta_1 = \tan\delta_3 = 10^{-4}$
b) Input impedance versus frequency.
—— Experiment; -x--x- Theory (increment = 10 MHz).

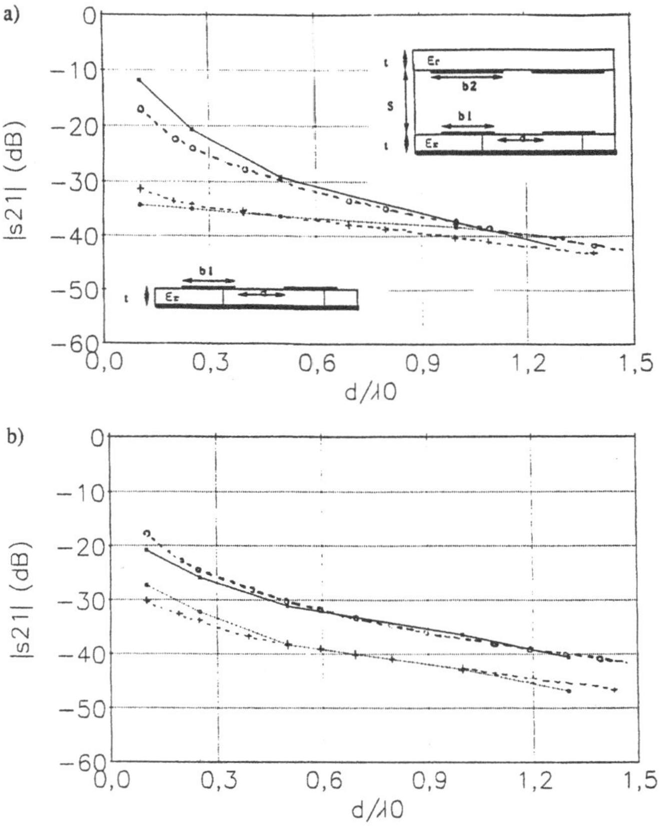

Figure 8. $|S_{12}|$ in dB plotted versus edge spacing for stacked and unstacked antennas: a) E-plane; b) H-plane.
Stacked antennas: *-*-*-* Experiment; θ-θ-θ-θ Theory.
Unstacked antennas: o o o o Experiment; + + + + Theory.

The previous stacked antenna is fed at two orthogonal points in phase quadrature. A good, broad-band, circular-polarized antenna can be achieved. The axial ratio is better than 1.2 dB over the whole bandwidth, for a range of thirty degrees around the broadside direction. The measured coupling between the two feeding points is less than -15 dB.

2.1.3.2.3 Mutual coupling between stacked antennas. We have investigated the mutual coupling between two stacked antennas, and compared stacked as well as unstacked configurations [81], [84]. The mutual coupling coefficient, S_{21}, between the antennas was calculated from the port impedance matrix, as usual [80]. The antennas studied are those presented previously in Section 1.3.2.2, and they operate in the S band [82]. In Figures 8a and 8b, we compare the measured coupling coefficient between stacked and unstacked antennas in the E and H planes, respectively. The solid curve corresponds to experimental results, and the dashed curve represents the calculated ones. Obviously, in the stacked case, the coupling should be stronger. But the measurement shows that it is much stronger than what we expected. This is probably due to the effects of the reinforced fringing fields between upper and lower patches. For instance, in the E plane, at $d = 0.3\ \lambda_0$, which corresponds to 0.78 λ_0 between patch centers, the difference is about 13 dB. However, the coupling between stacked antennas decreases very rapidly, and becomes comparable to that of the unstacked case, after a center spacing of about 1.5 λ_0. In the H plane, the coupling for the stacked structure is also higher. But the difference is only 7 dB, whatever the spacing may be.

2.1.4 Corner-fed square patch: influence of the feed-line geometry. When a square patch is excited at one corner, the

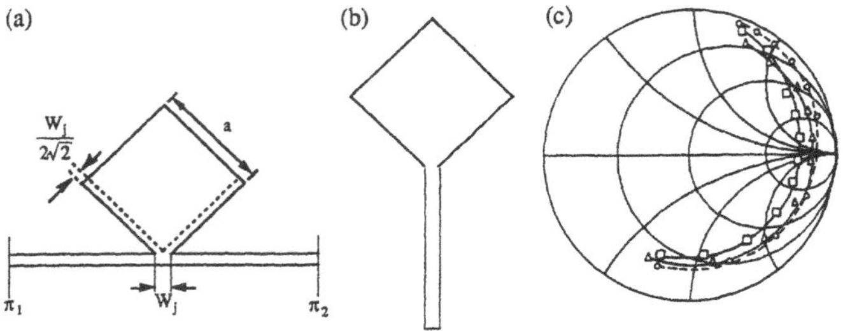

Figure 9. Corner-fed square-patch antennas: (a) Serial-patch antenna (SPA); (b) Parallel-patch antenna (PPA); (c) Input impedance locus: Frequency start = 4.5 GHz, Frequency stop = 5.5 GHz.
-□-□-□ SPA (measured) R_r = 360Ω, f_r = 5.20 GHz;
-Δ - Δ - Δ PPA (measured) R_r = 580Ω, f_r = 5.09 GHz;
-o-o-o Calculated (cavity method) R_r = 600Ω, f_r = 5.09 GHz.

cavity model shows that the internal field is the sum of two degenerate modes with equal amplitude, i.e., (0,1) and (1,0) [23]. However, the computation of the input impedance (using the cavity model) does not take into account the significant influence of the feed-line geometry. As a matter of fact, measurements show that the input resistance, R_r, at the resonance of a serial patch antenna (SPA), Figure 9a, is roughly one half that of a parallel patch antenna (PPA), Figure 9b, which result is close to the computed values of the cavity model, Fig 9c. The resonant frequency, f_r, measured from the SPA source, is also slightly higher than that of the PPA's. No theoretical analysis of the corner discontinuity is available today. An experimental analysis, up to K band, for the SPA, has been done, to obtain practical expressions for the resonant frequency and input impedance, taking into account the variation of the width of the patch-line junction and the proximity of the microstrip feed line [31].

2.1.4a SPA fed by a straight line. Measurements over 16 SPAs, fed by straight microstrip lines in three frequency bands, were carried out to test the influence of the feed junction. At first, the results show that the resonant frequency increases with widening of the feed junction. The presence of the line acts on the effective dimensions of the SPAs, reducing them by a term $w_j/2\sqrt{2}$ (Figure 9a), where w_j is the line patch junction width. Then, starting from the cavity model [14], the resonant frequency is given by

$$f_r = \frac{c}{2\sqrt{\varepsilon_{eff}}\left(a - \dfrac{w_j}{2\sqrt{2}} + 2\Delta_a\right)} \tag{6}$$

where c is the velocity of electromagnetic waves in free space, a is the patch's physical length, e_{eff} is the effective dielectric constant, and Δ_a is the edge extension. An error less than 1.8% in resonant frequency, up to K band, was observed.

The same measurements also show that the input impedances of the SPAs decrease with widening of the patch-line junction. At resonance, the equivalent circuit can be simply described by a resistance, R_r. Measurement results allow us to develop an exponential formula, where R_r is a function of f_r, w_j, and h, given by the following empirical expression:

$$R_r = 113\frac{\left(f_r[\text{GHz}]\right)^{0.354}}{\left(\dfrac{w_j}{h}\right)^{0.24}} \tag{7}$$

Figure 10 shows the measured and the calculated input resistances of 16 SPAs at resonance, as a function of w_j/h. w_j, and Values calculated from equation 7 show an accuracy better than 7 %, when compared to measurements.

2.1.4b SPA fed by a bent line. The electrical behavior of the SPA is also dependent on the geometry of the microstrip feed line. The electromagnetic coupling, between the radiating source and the line, leads to strong variations in the input impedance. Using the same experimental setup described above, measurements were made of four SPAs with different patch-line angles, ϕ_j. For this purpose, the line was bent at the feed junction. The results allow us to establish a modified empirical expression from equation 7, also a function of ϕ_j.

$$R_{r\phi} = \frac{86.5\left(f_r[\text{GHz}]\right)^{0.354}}{\left(\dfrac{w_j}{h}\right)^{0.24}\cos\left(\phi_j - 5^\circ\right)} \tag{8}$$

Figure 11 shows the input resistances calculated using equation (8), and compares them to the measured values for the SPAs. An error less than 7% is obtained between theoretical and experimental results.

2.2 Printed slots

Two shapes have been studied:

– the linear slot, which is very flexible in being fed by a microstrip line, suitably positioned to get the required impedance

– the annular slot, which is very attractive for obtaining circular-polarized waves, using two orthogonal feed lines.

Slot antennas are usually printed on a the top side of a double-sided board, while the microstrip feed lines are located below (Figure 12). Two analyses have been developed. The first one uses a simple model of a lossy transmission line [32, 33, 34]. The guided wavelength and attenuation coefficient are derived, and then the resonant frequency, input impedance, and radiation pattern are

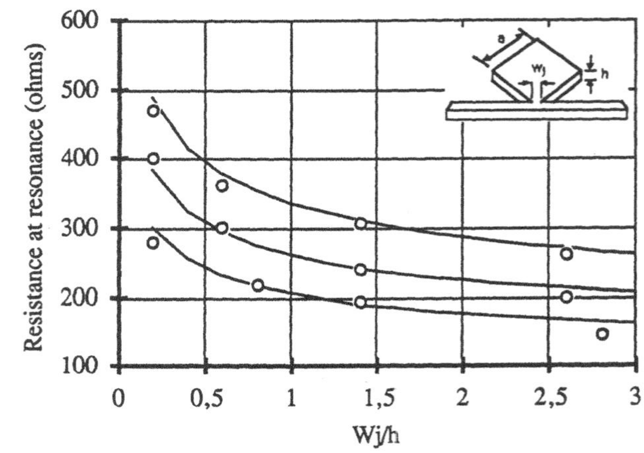

Figure 10. Resonant input resistance behavior of the SPA as a function of the ratio w_j/h: Comparison between R_r calculated by equation (3) and measured values. ε_r = 2.2.
K band, a = 4.5 mm, h = 0.38 mm;
X band, a = 9 mm, h = 0.79 mm;
C band, a = 18 mm, h = 1.57 mm;
—— Model (equation 3); o o o Measurements.

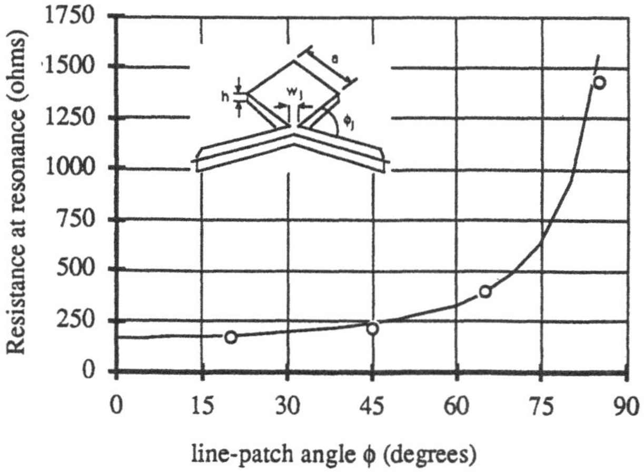

Figure 11. Resonant input resistance behavior of the SPA as a function of the angle ϕ_j: Comparison between $R_{r\phi}$ calculated by equation (4) and measured values. ε_r = 2.2, a = 9 mm, h = 0.79 mm, w_j = 1 mm.
—— Model (equation 4); o o o Measurements.

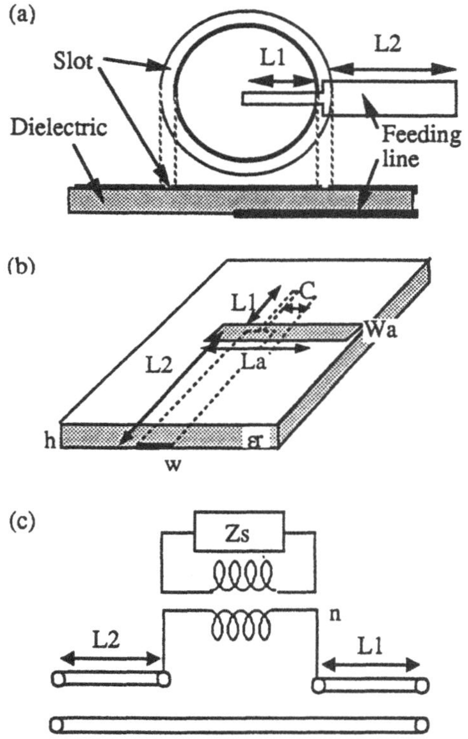

Figure 12. Feed and matching networks of radiating slots. (a) annular slot, (b) linear slot (with offset), (c) equivalent circuit.

computed. The second one considers the annular slot to be the complementary structure of an equivalent metallic loop (Babinet's principle).

2.2.1. Transmission line model of a radiating slot [32]: Model. The analysis of radiating slots has been developed using a lossy transmission-line model. It requires the computation of a propagation constant $\alpha + j\beta$, and a characteristic impedance, Z_c. β and Z_c are obtained using Cohn's method [35], and α is the solution of the numerical equation, $P_1(\alpha) = P_r(\alpha)$, where $P_1(\alpha)$ equals the power delivered to the lossy transmission line, and $P_r(\alpha)$ is the power radiated from the antenna. Finally, the impedance observed along the microstrip line (Figure 12) is calculated, using a transformation proposed by Knorr [36], and the reactance of the open microstrip line must be taken into account. The input impedance is

$$Z_{in} = n^2 Z_s - jZ_{cl} \cot(k^l L_s),$$

where Z_s is the aperture impedance, n is the transformation ratio, L_s is the length of an open-circuited stub, k^l is the wave number, and Z_{cl} is the characteristic impedance of the microstrip line.

Results. The antenna design is described in Figure 12, for linear and annular slots. The quarter-wave-length open stub, L_1, insures the coupling between the slot and the feed line. A second section of the line, L_2, is often used as a matching network. An offset position of the feed line is also possible with the linear slot. The slot exhibits a series impedance, $R + jX$, along the feed line. R and X are plotted in Figure 13b, for a centered-feed linear slot (and a 50Ω feed line). The input impedance of an annular slot, matched to 50Ω through a quarter-wave transformer, is shown on a Smith chart in Figure 13a. However, some problems occur when a ground plane is added to get a directional antenna. Further work is necessary to understand the parasitic effects of guided waves between the ground plane and the metallic plane of the antenna.

2.2.2 A slot as a loop-complementary structure. We deduced the radiation resistance of an isolated slot resonator from that of an equivalent circular loop of the same size, when Babinet's principle is applied. Then, the radiation resistance and the bandwidth of a stripline circular-slot resonator were deduced. Theoretical results, which are expressed by means of analytical forms, are in good agreement with experiment. The slot resonator, of radii r_1 and r_2, is made in the single metallic sheet of a printed circuit, with

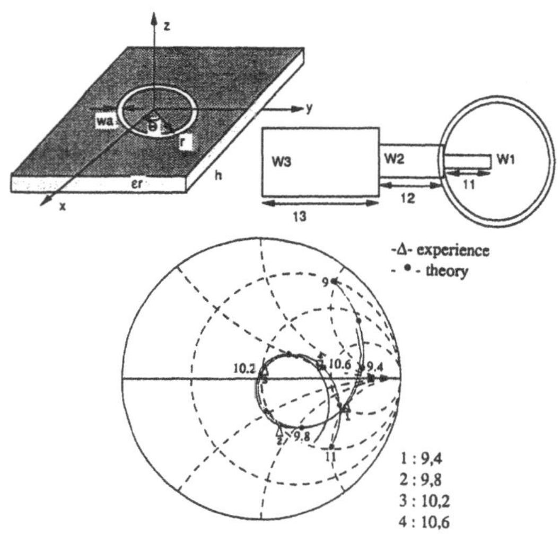

(a) annular slot : r = 4.09 mm, Wa = 0.154 mm, h = 0.78 mm, W1 = W2 = 0.373 mm, W3 = 2.31 mm l1 = 6.28 mm, l2 = 5 mm, l3 = 21.5 mm

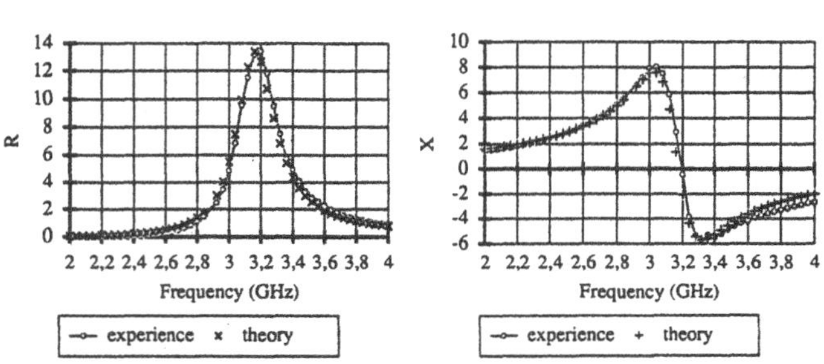

(b) linear slot : La= 40.2 mm, Wa = 0.7 mm, εr = 2.2, h= 1.587 mm

Figure 13. Impedance curves of annular slot (a), and linear slot (b).

Table 4. Comparison of experimental and theoretical input resistances for different printed slots.

r1,mm	77	77	4.00	4.02	3.51	0.45	0.325	18.45	30.48
r2,mm	82	79.5	4.14	4.17	5.21	0.70	0.375	18.75	33.02
t, mm	0	0	0.78	0.78	0.78	1.59	0.30	0.635	6.35
f_r, GHz			10	10	10	65.2	65.2	1.5	0.72
ε_r	1	1	2.17	2.17	2.17	2.23	9.6	10.2	12
ε_e	1	1	1.38	1.36	1.20	1.62	4.38	2.93	4.36
$(R_r)_S$,Ω Eqn. (9)	221	221	291	286	255	330	824	566	822
Theoretical Dusseux [32], [34]	270	271	330	330	300			606	860
Stephan et al [41]	240	244				390	413		590
$(R_r)_S$,Ω Dusseux [32], [34]			307	285	245			575	
Experimental Stephan et al [41]	235 +/−10	232 +/−10							518 +/−20

dielectric thickness t and relative permittivity ε_r. From the radiation resistance, $(R_r)_M$, of a metallic ring at the first resonance, we deduced the radiation resistance, $(R_r)_S$, of the complementary slot ring, applying Babinet's principle as in [39]. Finally, we obtained the following results (in SI units):

$$(R_r)_S = \frac{180\varepsilon_e}{1-\frac{1}{5\varepsilon_e}+\frac{1}{28\varepsilon_e^2}-\frac{1}{180\varepsilon_e^3}+\frac{1}{1320\varepsilon_e^4}-\cdots} \tag{9}$$

with

$$\sqrt{\varepsilon_e} = \frac{\lambda_0}{\pi(r_1+r_2)}.$$

In Table 4, we compare theoretical results obtained from equation (9), the spectral method of Stephan et al. [41], and the transmission-line model analysis of Dusseux et al. [32, 34].

The bandwidth is deduced from the radiation input admittance, with the expansion parameter [38, 40]

$$\Omega_S = 2\log_e\left(4\pi\frac{r_2+r_1}{r_2-r_1}\right), \tag{10}$$

and the antenna input radiation resistance is given by the expression

$$(R_r)_S^H = (R_r)_S\left[1+3\frac{\cos(2k_0t)}{(2k_0t)^2}-3\frac{\sin(2k_0t)}{(2k_0t)^3}\right] \tag{11}$$

Table 5 shows the theoretical bandwidth, B%, deduced by applying Babinet's principle to the radiation input admittance of a conducting circular loop [40]. A model with r_1 = 4.5 mm, r_2 = 5.5 mm, ε_r = 2.45, and t = 6.4 mm has been studied. It was fed by means of a symmetrical stripline. From equation (10) we deduced Ω_S = 9.7, then B = 19%. The measured bandwidth for a VSWR < 2 was 15.4%. The theoretical resonance frequency was 8.13 GHz, to be compared with the experimental value of 8.25 GHz.

2.3 Slot-fed patches

The structure Figure 14 has been previously described, and analyzed by the moment method, in references [37] and [44]. Simpler analyses, using a transmission-line or cavity model, have been developed in order to get an equivalent electric circuit well suited for CAD, and better understanding of the properties of this kind of antenna [46, 50]. The spectral-domain approach has also been considered [18].

Table 5. Theoretical bandwidth of annular slots for different values of the expansion parameter Ω.

Ω_S (10)	8.5	10	12	15	20
Bandwidth % (V.S.W.R. < 2)	26.1	18.2	12.6	8.6	5.9

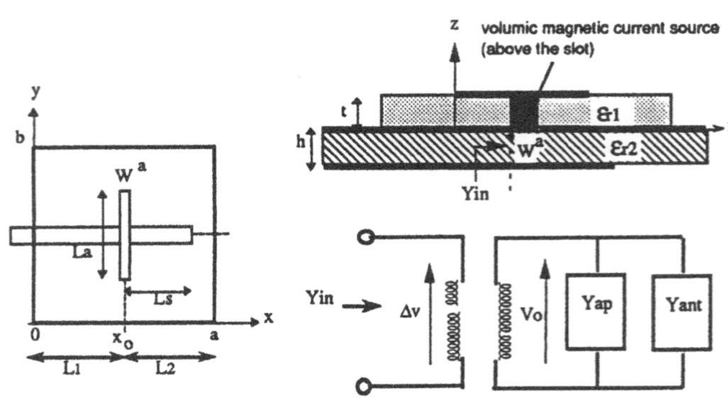

Figure 14. Aperture-coupled microstrip antenna and its equivalent circuit.

2.3.1 Cavity model [46, 51, 52]. The cavity model assumes a magnetic-current excitation. For thin substrates, the computations need only the dominant mode, TM_{10}, in a first approximation. More-general formulas have been obtained by taking into account the higher-order modes [51, 52].

2.3.1.1 Analysis. The analysis needs three steps to obtain the input impedance along the microstrip line:

a. The first step is to consider the microstrip antenna as a cavity, bounded by four perfect magnetic walls, and two electric walls, at $z = 0$ and t (Figure 14). To obtain the main, z component of the electric field in the over-all cavity volume, a magnetic-current source is assumed to be uniformly distributed in the volume area above the slot (Figure 14). The solution of the time-harmonic Maxwell's equations is easily obtained with a magnetic source which exhibits only a y component.

b. The second step deals with the radiation of the magnetic currents at the edges of the cavity. These currents are allowed to radiate into space, and the radiative losses, P_r, together with metallic losses, dielectric losses, and stored energy, can be computed in the usual manner. Then, Y_{ant} is easily deduced. The other electric-field components, near the slot, lead to slot-line reactive power. The susceptance, $Y_{ap} = jB_{ap}$ can be obtained simply from two short-circuited lines (with characteristic impedance and wave number determined by Cohn's method [35, 43]). Then, the total admittance at the aperture is

$$Y_{at} = Y_{ap} + Y_{ant} \tag{12}$$

Moreover, the analytical expression for Y_{ant} shows the frequency variation of an R-L-C series circuit for each mode [52].

c. The last step is to transform the impedance along the microstrip line (Figure 14). The discontinuity, ΔV, in modal voltage in the microstrip line, due to the slot cut on its ground plane, may be determined by [45]. Then, $Y_l = z_l^{-1} = Y_{at}\Delta V^{-2}$. Finally, the normalized input impedance is given by

$$Z_{in} = Z_l - jZ_c \cot\left(k^l L_s\right), \tag{13}$$

where L_s is the length of the open-circuited stub, and k^l is the wave number of the microstrip line.

2.3.1.2 Resonant behavior of slot-coupled microstrip antennas [48, 51]. Among all the parameters, the slot-length effects have been analyzed carefully. Similarly to the results in [44], Figure 15 shows that the resonant frequency decreases slightly with an increasing slot length, while the input resistance increases. One of the most interesting results concerns two opposite resonant behaviors of the slot-coupled patches, which can exhibit the impedance curves of a series R-L-C circuit [47], or the more usual parallel R-L-C properties [37, 44]. The proposed cavity model gives a simple explanation of these two aspects.

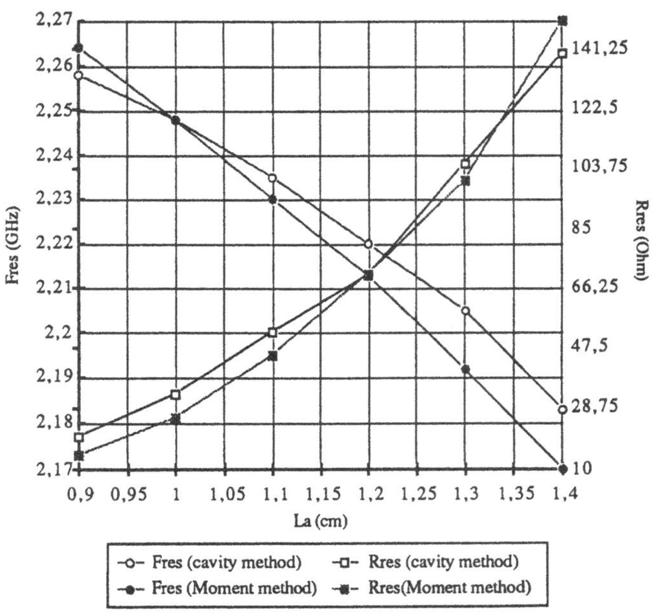

Figure 15. Resonant frequency and input resistance at resonance versus slot length: $a = 4.0$ cm, $b = 3.0$ cm, $\varepsilon_{r1} = 2.54$, $t = 0.16$ cm, $W = 0.495$ cm, $\varepsilon_r = 2.54$, $h = 0.16$ cm, $L_s = 2.0$ cm, $x_0 = a/2$, $y_0 = b/2$, $W_a = 0.11$ cm.

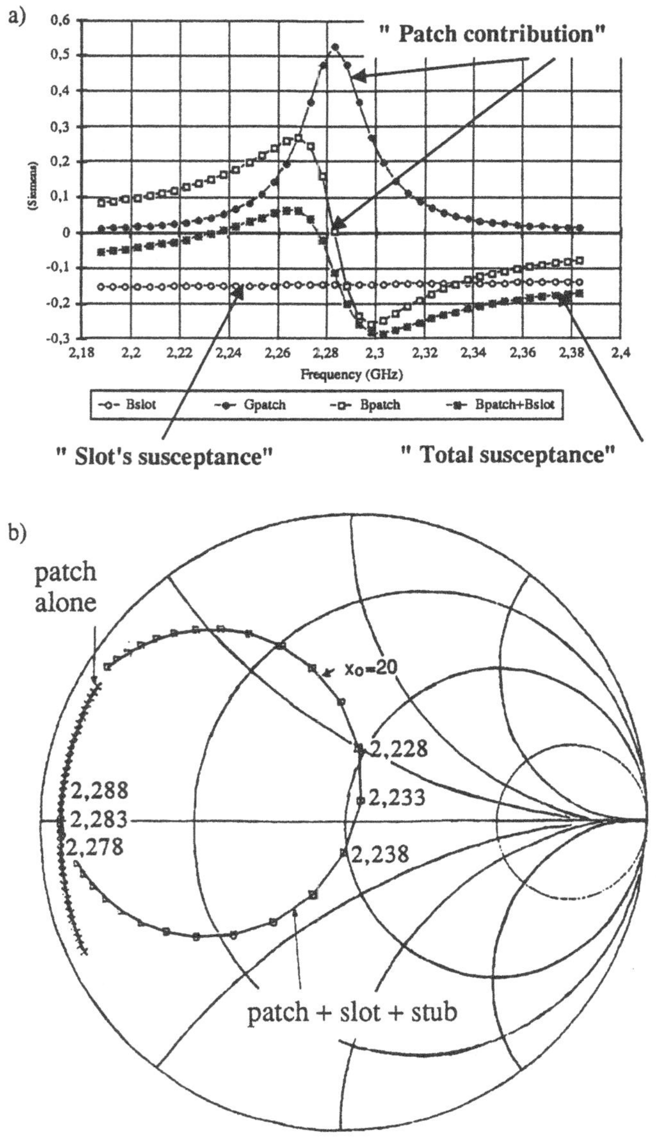

Figure 16. Behavior of antenna characteristics before and after addition of aperture susceptance. a) Values of conductance and susceptance of patch at the slot (with and without slot contribution). b) Values of the input impedance on microstrip line. c) The same values as in a) and b) on a Smith chart.

Parallel-type RLC. The antenna parameters are those given in [44]. The difference between the resonant frequency of the cavity, and the resonant frequency of the slot-fed patch, proves that the slot excitation has a fundamental role in both impedance values and resonant frequency. To give an idea of this effect, the admittances Y_{ant} (patch alone), Y_{ap}, and Y_{total} are plotted in Figure 16a. Transformer and stub contributions have been omitted. It is very clear from Figure 16b that the slot-coupled patch does not have the same resonant frequency as the patch alone (whose resistance at resonance is also very small).

Series-type RLC. Some previous papers [47] have described slot-coupled patches exhibiting typical series-RLC impedance. Theoretical and experimental results have been plotted in Figure 17. The curves are in good agreement, although a frequency shift remains. The cavity model uses the James [11] effective length, well suited for a low dielectric constant. It must first be noticed that the slot length is nearly equal to the width of the patch, and reaches a half-guided-slot wavelength in the frequency range 4.2 to 4.45 GHz. This means that the parallel reactive susceptance, Y_{ap}, equals zero, or remains very small. Then, the frequency variation of Z_{in} looks like the Z_{ant} of the cavity excited with a magnetic current. As explained in [49], this cavity will exhibit series-RLC impedance.

Conclusion. Despite its mathematical simplicity, the cavity model yields good results, which agree with previous theoretical and experimental data. Moreover, the model gives physical insight, and explains two opposite electrical properties (behavior like series and parallel circuits) of the same type of radiating element. Printed patches, on thin dielectric substrates, are well represented with only one fundamental mode. Therefore, a transmission-line analysis has also been developed, taking into account the different impedance transformations along each line section, or through the slot itself [50, 51].

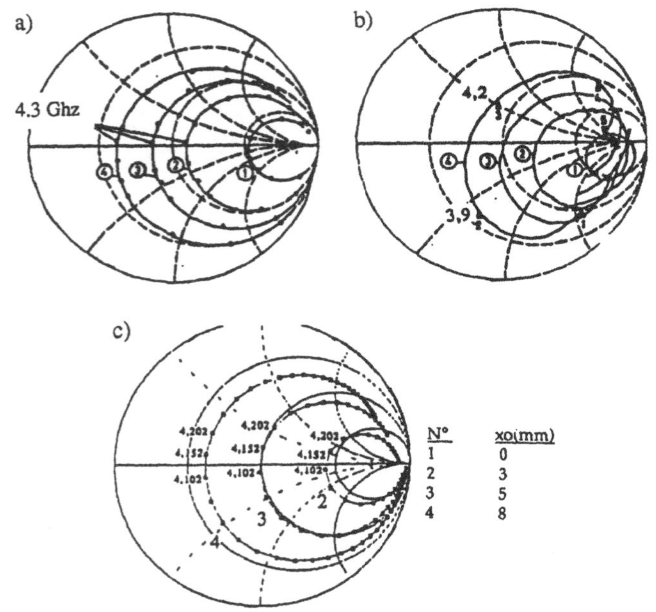

Figure 17. Input impedance variations versus frequency for different slot positions (series equivalent circuit). $a = 2.8$ cm, $b = 3.0$ cm, $\varepsilon_{r1} = 1$, $t = 3.15$ mm, $W = 0.92$ mm, $\varepsilon_r = 6$, $h = 0.635$ mm, $L_s = 8$ mm, $y_0 = b/2$, $L_a = 26.5$ mm, $W_a = 1$ mm. Curves: 1) $x_0 = 0$ mm, 2) $x_0 = 3$ mm, 3) $x_0 = 5$ mm, 4) $x_0 = 8$ mm.
a) Theory of [47]; b) Cavity method of [46]; c) Measurement of [47].

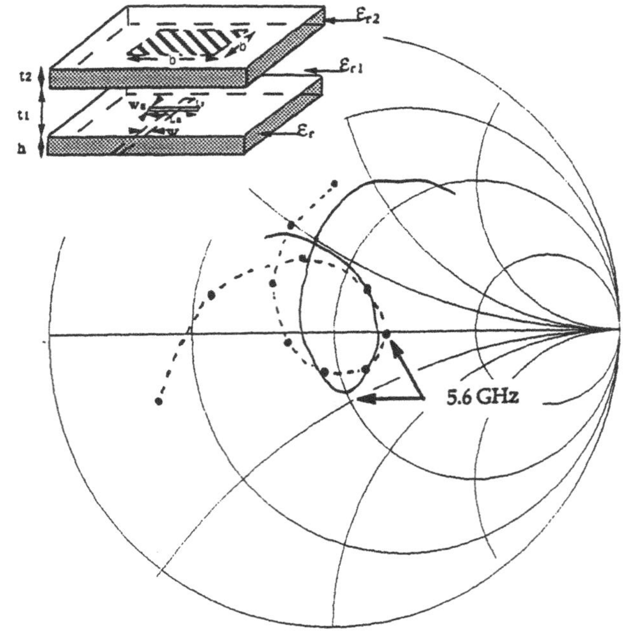

Figure 18. Input impedance versus frequency of aperture-coupled square-patch antenna, 4.8 - 6.8 GHz.
—— Experiment; o- - - -o Theory (increment 200 MHz).

$t_1 = 4.5$ mm	$\varepsilon_{r1} = 1$	$b = 17$ mm
$t_2 = 1.6$ mm	$\varepsilon_{r2} = 2.55$	$\tan\delta_2 = 10^{-3}$
$h = 0.8$ mm	$\varepsilon_r = 2.45$	$\tan\delta = 10^{-3}$
$W = 2.32$ mm	$L_s = 3$ mm	$W_a = 0.8$ mm $L_a = 15$ mm

2.3.2 Spectral-domain approach (optimization of bandwidth) [18]. Another useful alternative, to increase the bandwidth of a microstrip-patch antenna, is to use the mutual coupling between the radiating patch and the resonant feed slot, which transfers the power from the feed line to the radiator, through the ground plane [27]. The effect of the variation of parameters on the input impedance of such antennas has been investigated, by using the moment-method technique in the spectral domain, as proposed by Pozar [37]. The results for a rectangular patch can be found in [18].

2.3.2.1 Broad-band slot-fed rectangular-patch antenna. An aperture-coupled antenna has been optimized to get a bandwidth of 16% for a VSWR < 1.6 in the C band. Figure 18 presents a comparison between the theory and experiment of the antenna's input impedance. The measured radiation patterns show a cross-polarization level less than -25 dB, in both the E and H planes. The maximum cross polarization is found in the diagonal planes, as previously observed in [27]. The maximum back-to-front level is about -15 dB, over the full bandwidth.

2.3.2.2 Broad-band slot-fed triangular-patch antenna.

The structural configuration is presented in Figure 19. The variation of the input impedance versus the physical parameters has been studied experimentally. The stub, together with the aperture length or patch size, can be used to control the input impedance over a wide range of values. Around the central frequency of 6.5 GHz, a 24% impedance bandwidth (VSWR < 2) was obtained, as shown in Figure 19. In the E-plane, the 3 dB beamwidth varies from 75° to 85°, and the cross-polarization level is less than -22 dB over the full band of frequencies, covering the impedance bandwidth of 1500 MHz. In the H-plane, the 3 dB beamwidth varies from 60° to 80°, and the cross-polarization level is less than -18 dB.

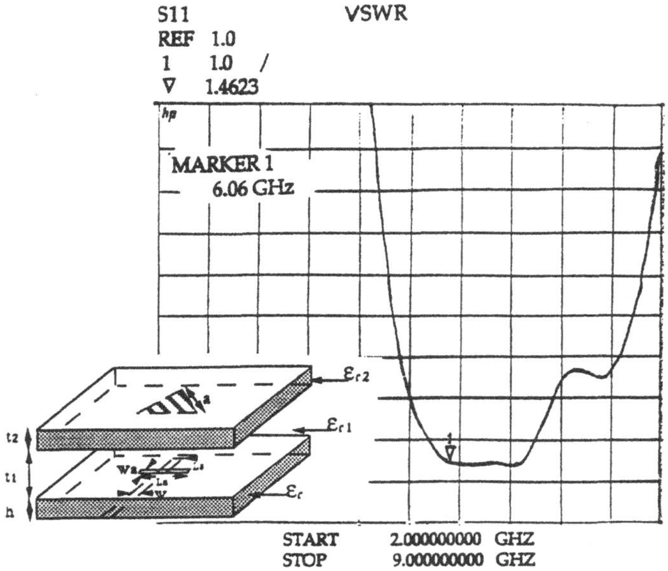

Figure 19. VSWR versus frequency of aperture-coupled triangular microstrip antenna.

$t_1 = 5$ mm	$\varepsilon_{r1} = 1$	
$t_2 = 0.8$ mm	$\varepsilon_{r2} = 2.2$	$a = 19.5$ mm
$h = 0.8$ mm	$\varepsilon_r = 2.2$	
$W = 2.32$ mm	$L_s = 3$ mm	$W_a = 0.8$ mm $L_a = 13.5$ mm

2.3.3 Dual-polarized slot-fed patches [53]. In Figure 20, we present a square metallic patch, fed by two orthogonal striplines, through two orthogonal slots. This antenna is advantageously realized by means of three stacked printed circuits (CI). Each stripline (P_1) simultaneously feeds the two slots (F_1) and (F_2), so that the wave radiated by the patch, P, is linearly polarized, following the electric moment, M_1. The opened end (C) of each stripline is located at a distance of a quarter-phase wavelength from the antenna center. The two points, A and B, are fed in opposite phase, so that the conducting currents induced on the patch can be designed as in Figure 20. Two models have been tested in the X and Ku bands [53]. The dimensions and electrical properties are shown in Tables 6 and 7, where the resonant frequency is f_0. The 3 dB half-beamwidth in the E and H planes, and the sidelobe level, are shown in the same tables. The isolation is measured between the two inputs.

We can note the small antenna size, due to the large length of the strength-lines of the conducting currents, and the large bandwidth, when compared with a traditional patch (respectively, 4.2% and 3.1%, for a VSWR < 1.5).

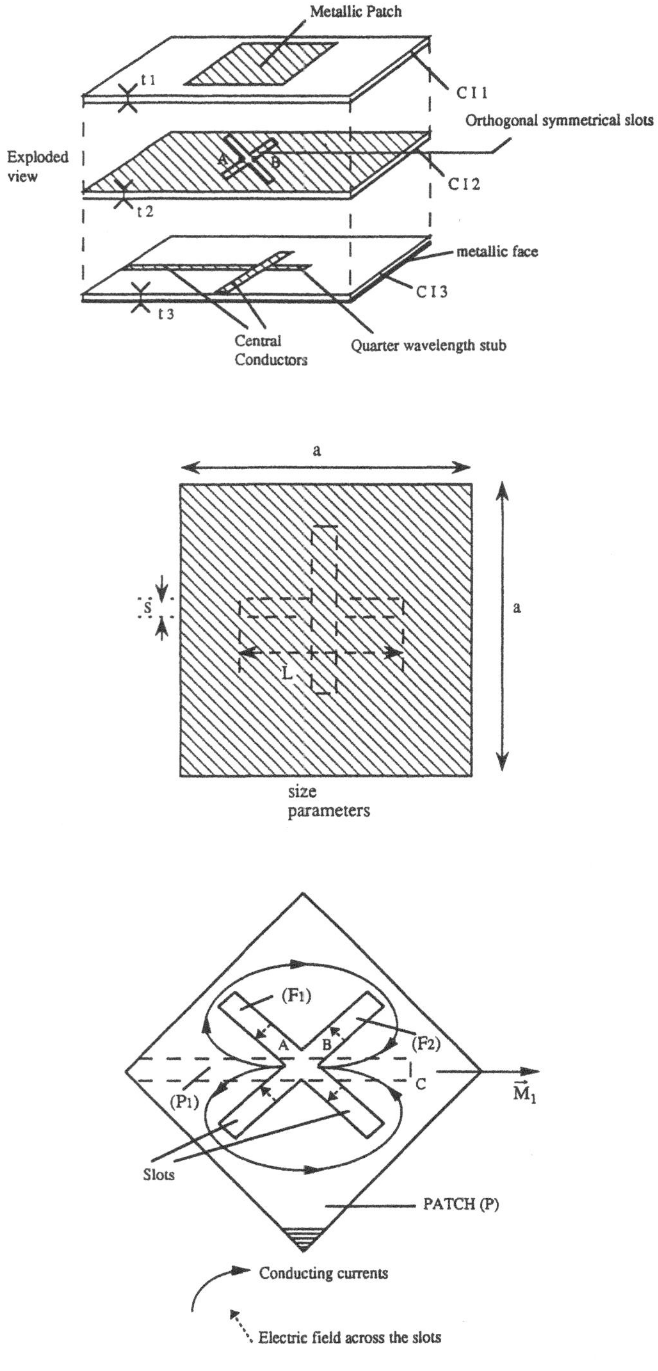

Figure 20. Dual-polarized slot-fed patches.

2.4 Slot-loaded patches

Two models have been studied, tested, and patented. One is linearly polarized [55, 1], and the other has a double, crossed polarization [54, 1, 4].

2.4.1 Mono-polarized flat-folded dipoles (see Figure 21a). The slot-microstrip antenna is used as a folded-slot dipole, symmetrically fed across a gap, g, by means of a stripline. A theoretical model, equivalent to several lossy transmission lines, has been described [55]. The coupling between the two equivalent, radiating lines, of every slot, is taken into account, and explains a fourth resonance near the third one. Theoretical results and experiments are in good agreement. In Figure 22, we present the theoretical radiation admittance related to points A and B. The antenna is

Table 6. Dimensions and electrical parameters of dual-polarized slot-fed patches.

	a(mm)	s(mm)	a/λ_o	L(mm)	L/λ	$t_1=t_2$ (mm) $= t_3$	$(\varepsilon_r)_1=(\varepsilon_r)_2 = (\varepsilon_r)_3$	t_1/λ_o
X band f_o=9.65 GHz	8	0.6	0.26	6.6	0.21	1.6	2.17	0.051
Ku Band f_o=14 GHz	5.2	0.3	0.23	4.5	0.21	0.8	2.17	0.037

Table 7. Electrical properties of two models of dual-polarized slot-fed patches (dimensions given in Table 6).

	Bandwidth R.O.S. < 1.5	Isolation (dB)	$(\theta_{3dB})_E$	$(\theta_{3dB})_H$	Sidelobe level (dB)
X Band $f_0 = 9.65$ GHz	5.2 %	< -24	110°	90°	-18
Ku Band $f_0 = 14$ GHz	6.1 %	< -16	70°	72°	-18

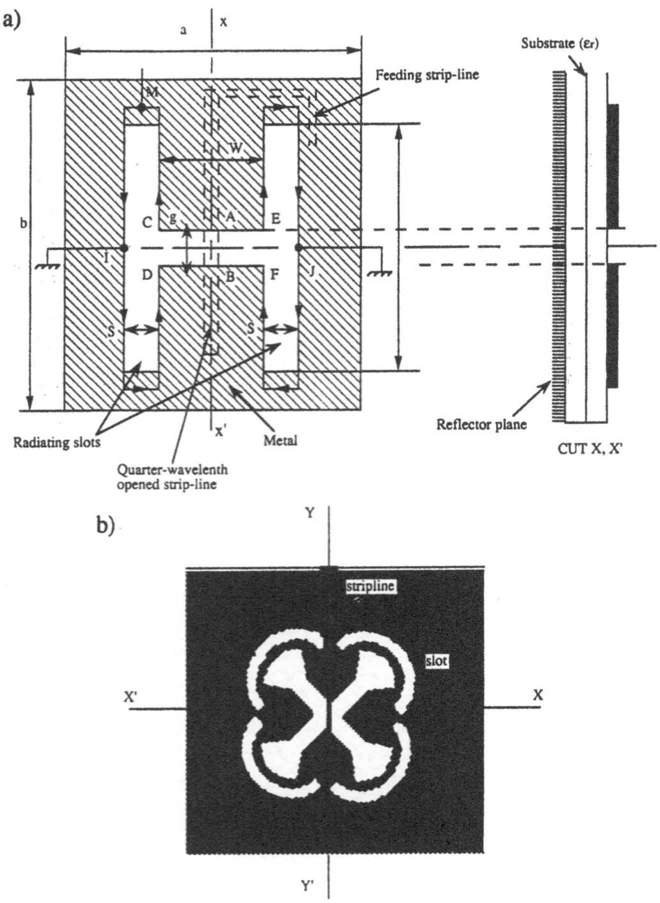

Figure 21. Slot-loaded patches: (a) flat-folded dipole, linearly polarized; (b) dual-polarized flat-folded dipoles.

operating near the third and fourth resonances, that is, between 10 and 11.5 GHz. It is equivalent to a four-port network, which is equivalent to *N* four-port identical sections, in a chain arrangement. Several antennas have been tested in the S, C, X, and Ku bands. Flat-folded dipoles are used in numerous arrays, as described later (see Figures 31 and 32).

A very-large-bandwidth special slot-loaded patch antenna has been studied and tested [2]. It operates in air. It is a microstrip antenna which is linearly polarized, with directional radiation. Moment and finite-difference methods have been applied to determine the radio-electric properties, using the electrical-image principle. The flat metallic radiating surface of the microstrip antenna is equivalent to a grate, which is composed of a set of cylindrical metallic cylinders, having the same length and orthogonally crossed. Thus, the antenna is equivalent to an array of conducting square meshes. Comparisons between theoretical and experimental properties have been done on two models, one operating between 8 GHz and 16 GHz, and the other, between 1.1 GHz and 2 GHz [2].

2.4.2 Dual-polarized flat-folded dipoles [see Figure 21b]. This model has been patented [54]. The radiating portion of the antenna is formed from two similar radiating folded dipoles, which are located in a single plane and are orthogonal, with the slots between driven elements of the folded dipoles crossing one another at the center (C) of the unit. The two folded dipoles are associated with the central conductors of striplines which are orthogonal, with their extensions crossing one another beneath the center (C) of the antenna, similarly to the first model of Figure 21a. Table 8 summarizes the experimental results obtained in the pass band of a single dipole, the other being matched on an adjusted load of 50 ohms. The 3 dB beamwidth and the crossed component ("c.c") are given. The isolation ("Dec") between the two inputs is better than -20 dB. The antenna operates over a relative large bandwidth. This model has been used in several arrays (see Figures 33, 34).

2.5 Electromagnetic coupled dipoles and patches

Electromagnetic coupling (EMC) was first introduced by Oltman, in 1981 [56], as an original means to feed microstrip antennas (Figure 23). It appeared to be able to provide essential matching facilities and reduced line radiation. The laboratory activities in EMC began in 1983, in connection with CNES (Space Studies National Center), which was interested in the development of a full-wave analysis tool.

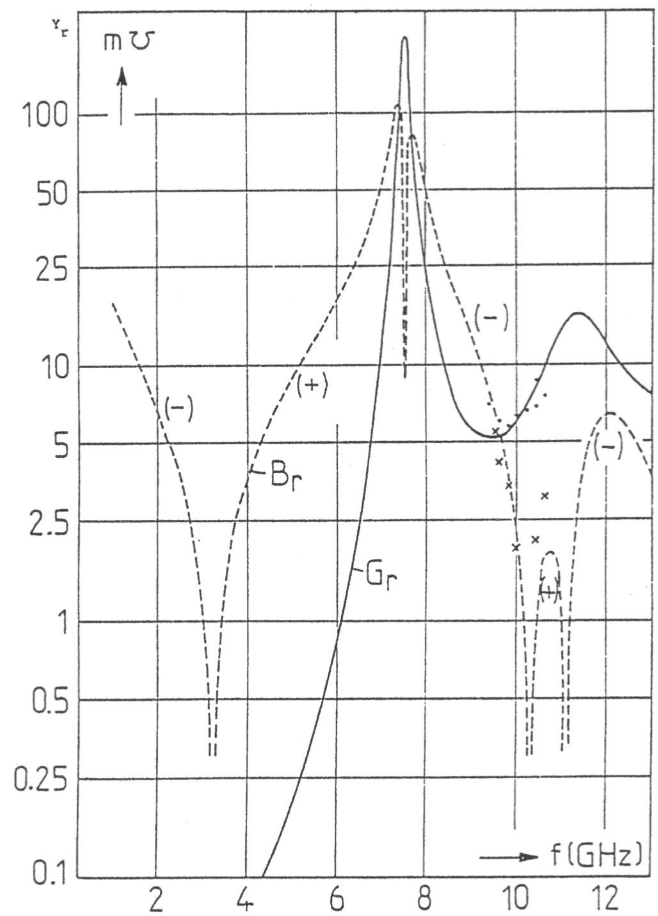

Figure 22. Theoretical and experimental radiation admittance of the flat-folded dipole. $Y_r = G_r + B_r$ and experimental points in a large frequency band (N = 250). G_e experimental points : . . . . ; B_e experimental points: x x x .

2.5.1. Theory. The integral-equation technique was chosen, due to its rigorous and general formulation. The theory is based on the solution of the electric-field integral equation (EFIE), using the mixed-potential formulation (scalar and vector). In this approach, the Green's functions for the multi-layer microstrip structure, which are defined in terms of Sommerfeld integrals, are first determined, using the boundary conditions. The integration is performed numerically for both a horizontal and a vertical Hertzian dipole [57, 58]. This integration requires special techniques, to insure convergence and to decrease the computing time. The surface-current densities are computed, using the moment method, which transforms the integral equation into a linear system of algebraic equations, solved by matrix inversion [59]. Once the currents on the antenna and its feed line are known, transmission-line theory is applied to determine the scattering parameters of the structure. This assumes a mono-modal propagation on the microstrip feed line. The radiated fields are derived from the obtained current densities, using an asymptotic evaluation of the Sommerfeld integrals. It is to be noted that the radiation from each conductor is calculated separately, which provides an efficient tool to evaluate the feed-line parasitic radiation.

Table 8. Electrical properties of a dual-polarized flat-folded dipole (described in Figure 21b).

f(GHz)	3.6	3.7	3.8	3.9	4	4.1
θ_E (degrees)	94	71	70	74	82	92
θ_H (degrees)	73	59	53	49	70	66
Max. gain (dB)	7.4	8.1	8.7	7.3	7.4	5.9
SWR/50 ohms	1.33	1.22	1.32	1.22	1.78	
c.c (dB)		-23	-18	-16	-16	-16
Dec (dB)	-20.0	-25.2	-22.0	-21.6	-22.5	-20

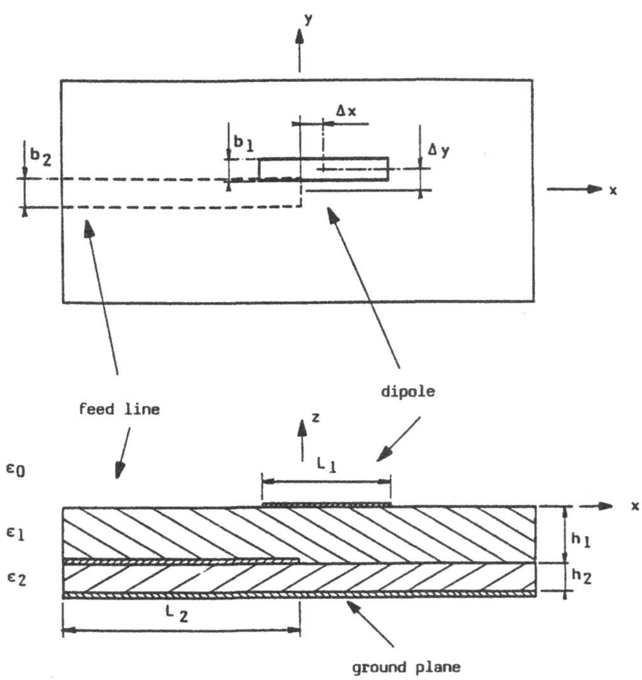

Figure 23. CEM dipole with longitudinal excitation.

The numerical computations are carried out on a HP 9000-835 computer. Particular effort was made to reduce both computing time and required storage space. Arbitrarily-shaped structures have been analyzed, such as longitudinal and transverse dipoles, patches and notched patches, crossed dipoles, etc. They are fabricated and tested to validate the theoretical model, and good agreement is observed.

2.5.2 Results. The structure studied (shown in Figure 23) has the following parameters. It has a 76 mm × 76 mm ground plane, and the dielectric is composed of two layers with $\varepsilon_r = 2.17$. The upper layer has a height of 1.6 mm, and the height of the lower one is 0.8 mm. The microstrip line is designed to have a 50Ω characteristic impedance. The model is applied to a longitudinal dipole, of 12.4 mm length and 1.0 mm width. The longitudinal and transverse positions of the dipole (relative to the end of the line) are chosen to obtain a good match [60]. Figure 24 presents the calculated and measured reflection coefficients on a Smith chart. From this figure, one can see the good agreement between theoretical and experimental results.

A rectangular patch of 11 mm length and 9 mm width is realized on the same structure, shown in Figure 23. The middle of the patch is located just above the end of the line, to provide a good match and to prevent the excitation of the transverse mode. Longitudinal and transverse currents on the patch, at the resonant frequency, are shown in Figure 25. As shown in Figure 26a, the resonant frequency is located at 8.03 GHz, with a 1.2% frequency shift from the measured value, while a 1 dB difference in the match level is observed. Again, the calculated results are in good agreement with the measured values. Figures 26b and 26c show the calculated and measured radiation patterns in the E and H planes. A non-negligible influence of parasitic-line radiation is observed in the E plane, which is only 13 dB less than the antenna radiation level. The difference, which appears at large angles, is due to diffraction from the finite-ground-plane edges.

Another interesting example, for antenna arrays, is the notched patch, which is designed in order to provide circular

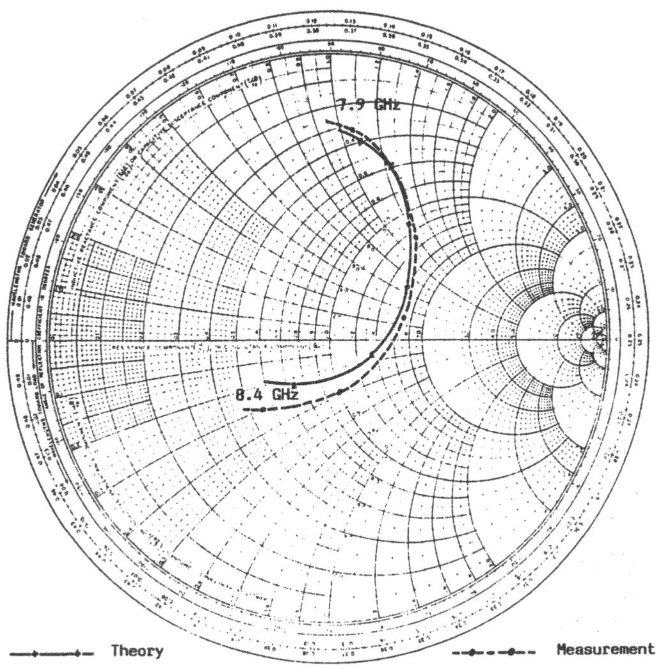

Figure 24. Input impedance comparison between theory and experiment at X-band.

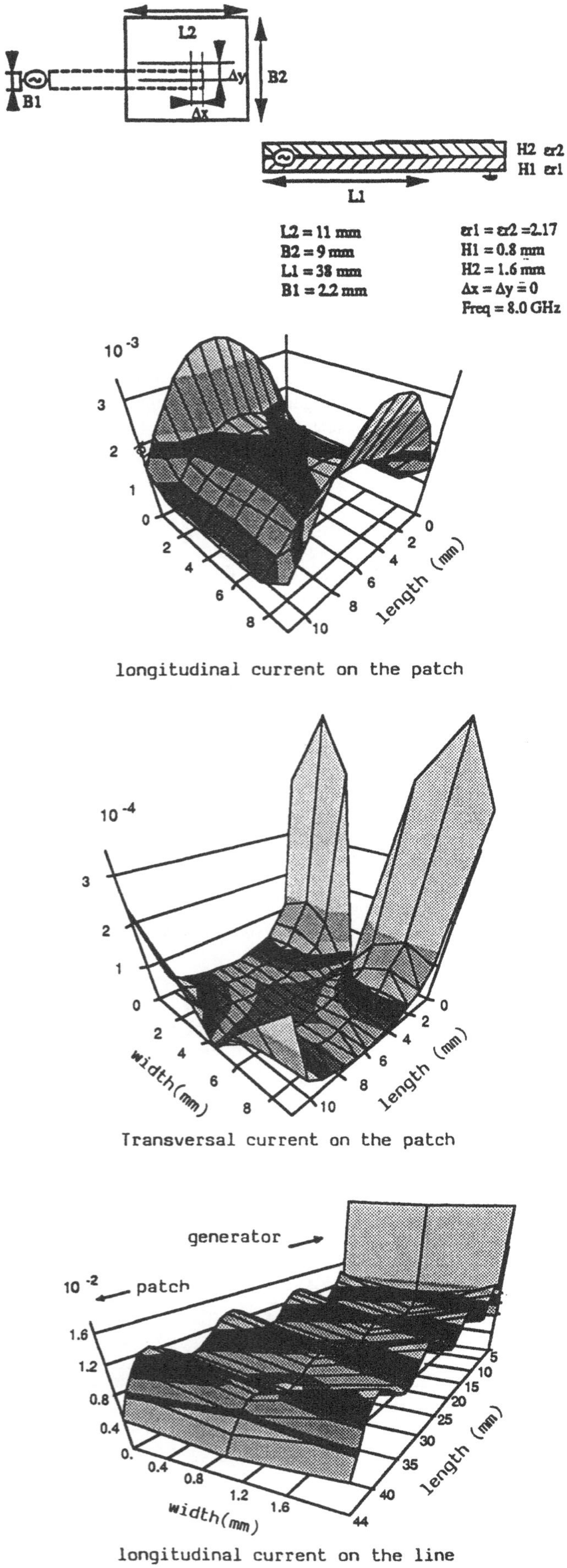

Figure 25. The current distribution on a patch electromagnetically coupled to a microstrip line.

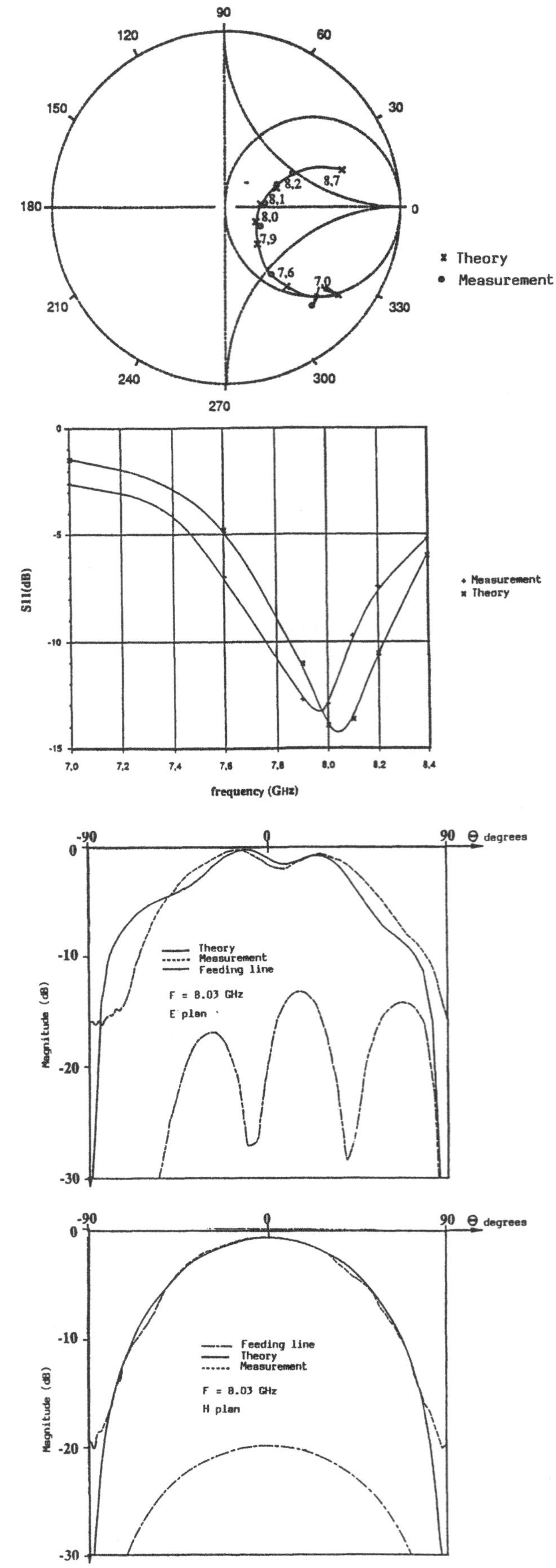

Figure 26. A rectangular patch electromagnetically coupled to a microstrip line. L_2 = 11 mm, B_2 = 9 mm, H_1 = 0.8 mm, H_2 = 1.6 mm, $\varepsilon_{r1} = \varepsilon_{r2}$ = 2.17.
a) input impedance b) E-plane pattern. c) H-plane pattern.

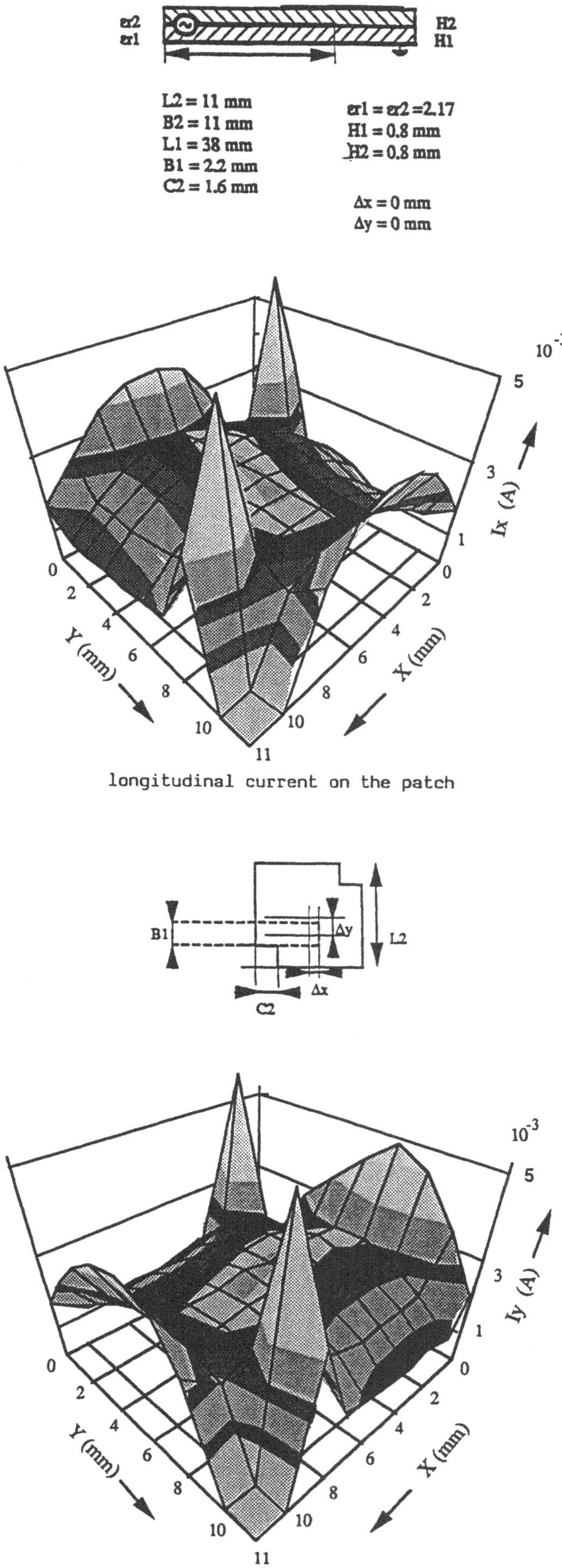

Figure 27. The current distribution of the notched patch.

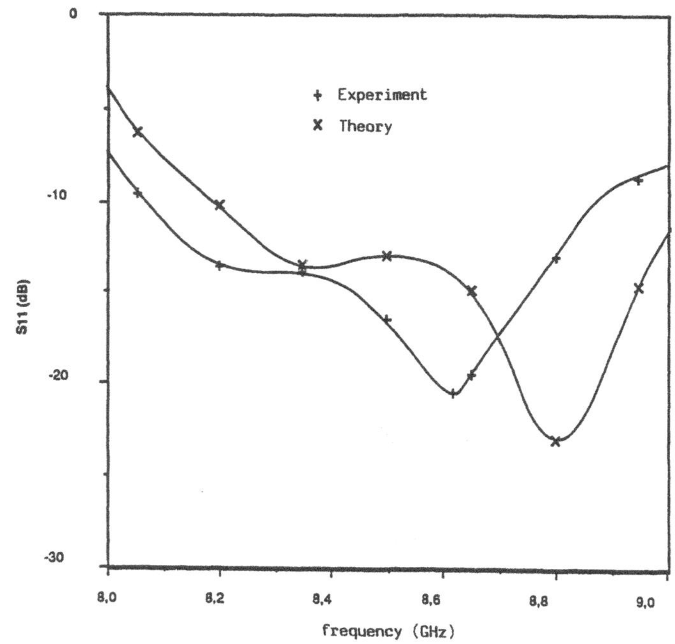

Figure 28. Return loss as a function of frequency for the notched patch.

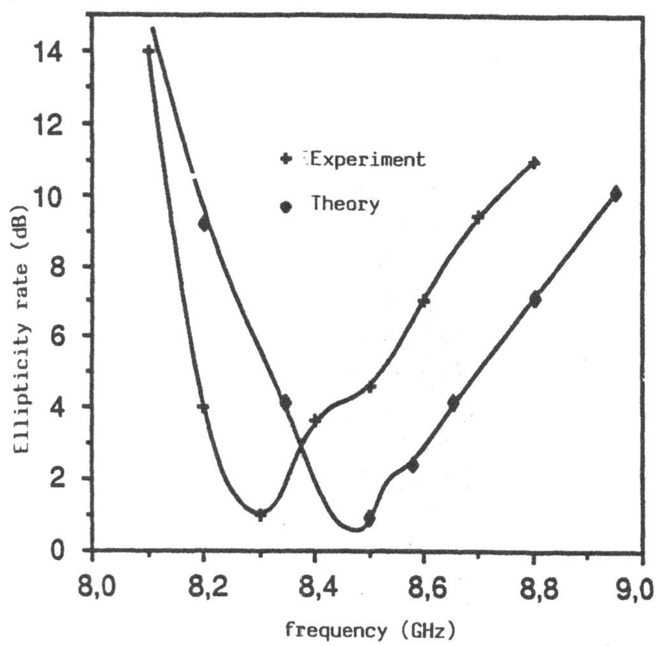

Figure 29. Polarization ellipticity rate as a function of frequency for the notched patch.

polarization using only one microstrip feed line [61]. The studied configuration is built, for a patch of 11 mm × 11 mm, with the same ground-plane size and line as those in Figure 22. The notched patch is obtained by removing two square notches, as shown in Figure 27. The dimensions of the notches are optimized, to permit the excitation of two orthogonal degenerate modes. The best elliptical ratio is obtained for notches whose lengths are about $\lambda_g/15$. Figure 28 shows the measured and the calculated return loss, as a function of frequency, wherein good agreement is obtained, within 2% frequency shift. It appears that two resonant frequencies exist, due to the presence of both longitudinal and transverse modes. Figure 29 presents the elliptical ratio as a function of frequency. The bandwidth, in which the elliptical ratio remains less than 3 dB, is

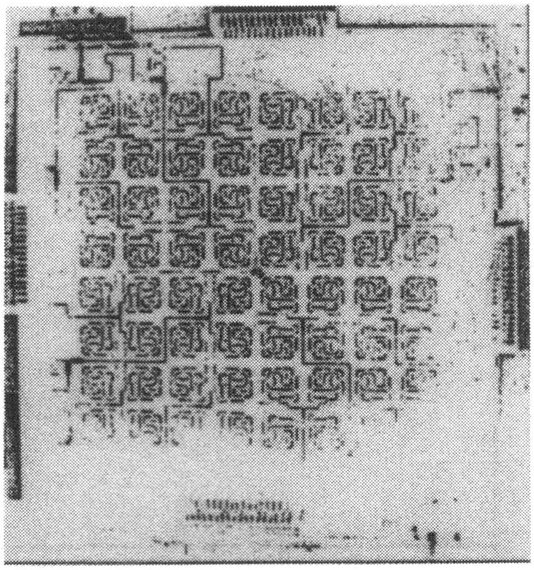

Sixty four 3 bit digital phase shifters

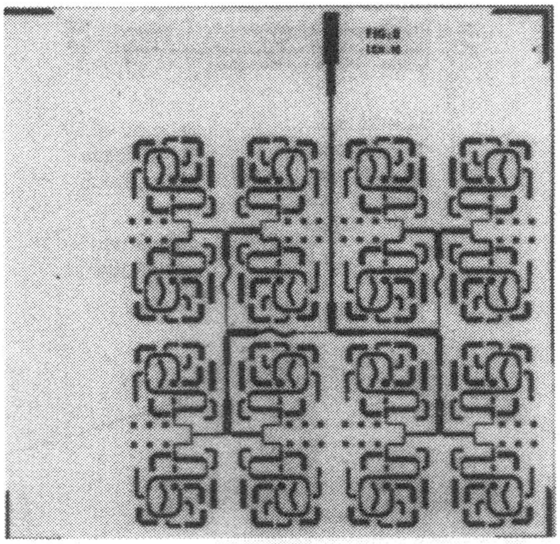

Sixteen 3 bit digital phase shifters of one quarter

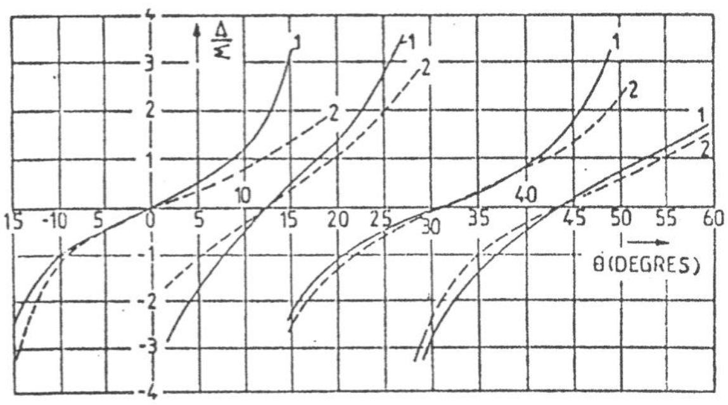

Interferometry curves for different deflections

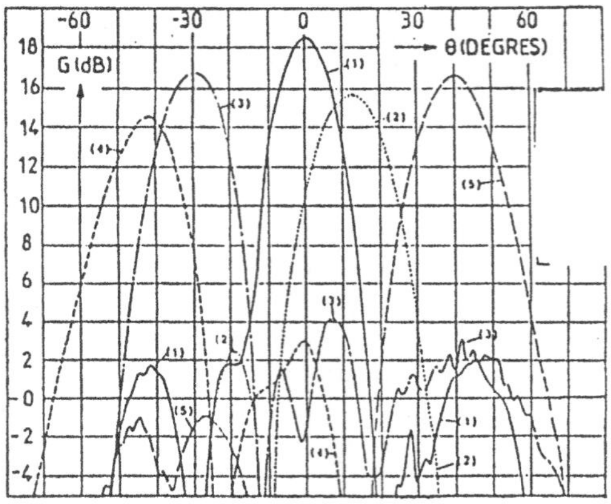

Measured gains for different deflections

Figure 30. Mono-polarized phased array with beam steering and variable directivity in the Ku band (digital phase shifter and feed arrangement).

about 3%, over a range of ±20°. The integral-equation technique is also used to develop charts for quick design of microstrip-patch antennas [62]. Moreover, this has recently been used to analyze similar structures with via hole [63] and active radiating elements [64].

3. Planar Arrays

Different kinds of arrays have been designed, with active or passive feed networks. The analysis includes mutual-coupling effects, especially when beam steering is considered with small element spacing. Most available synthesis methods (Fourier, Tchebyschev, etc.) are defined for arrays with uniform spacing. Moreover, they are not well suited when the pattern is specified. Consequently, different numerical methods have been developed:

–The relaxation method, which enables real excitation coefficients. As an example, a dual-beam pattern with low sidelobe level has been analyzed by this method.

–The synthesis of EM dipoles, based on previous works by R. S. Elliott and G. J. Stern.

3.1 Planar phased array

Figure 30 shows a mono-polarized phased array, with steerable beam and variable directivity in the Ku band. The radiating elements are 64 short-circuited quarter-wavelength microstrip patches, in an air medium. The distance between adjacent sources is equal to 0.5 wavelength. A first model was described in [4, 69]. A new, improved model, with variable directivity, is presented in the same figure. Feed through is used to excite each radiating element from the output of the associated phase shifter. A total of eight PIN diodes is required, for each three-bit digital phase shifter. So, for every deflected beam, we can obtain the sum and the difference patterns. The feed system, which is composed of splitters, branch lines, corporate feeds, and DC bias, is photo etched onto fused-quartz substrate plates, using a microstrip technique. The antenna efficiency is about 20% between 14.5 and 15.5 GHz, whatever the deflected beam angle, located in a solid angle of two steradians. The mean linear isotropic gain is 15 dB. A splitting system is used to change the antenna directivity.

RADIATING AREA

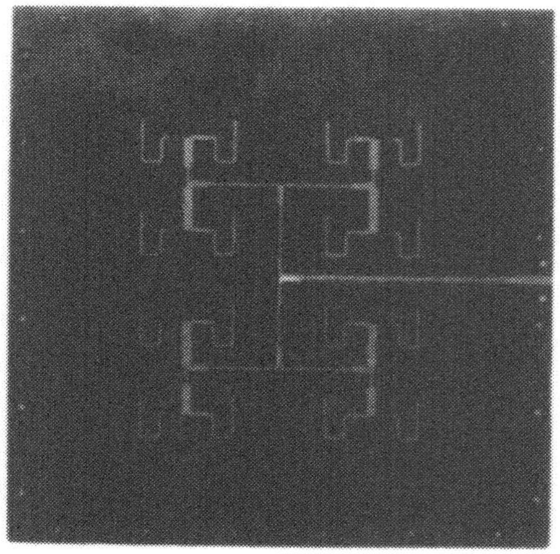

FEEDING NETWORK

Figure 31. Mono-polar array for a beacon at X band (9.1 to 10.1 GHz).

3.2 Planar passive array

3.2.1 Mono-polar array. Figure 31 shows a mono-polar array for a beacon in the X band (9.1 to 10.1 GHz). It is composed of 16 flat-folded dipoles, fed by a set of dividers using a strip-line technique [1, 4].

A flat-folded dipole radiating source [1, 2, 65] has been used in an array with 36 such elements (Figure 32). It is a large-bandwidth passive mono-polar array. The strip-line feed network is printed on a substrate of relative permittivity near to 1. While the measured, mean isotropic linear maximum gain is equal to 21 dB, the efficiency is equal to 60 percent. The array frequency bandwidth is 35%, between 1.3 and 1.85 GHz, when the flat-folded dipole bandwidth is equal to 37% for a VSWR lower than 2. The large bandwidth and good efficiency are obtained with a low relative permittivity, because the propagation of the dominant lowest-order TM_0 mode can be neglected [3, 67].

3.2.2 Dual-polarized flat array. Figure 33 shows an array at X band (9.5 to 9.8 GHz). These kinds of arrays are used, with two scatterometers, mounted on helicopters or aircraft. One array was operating at C band (5.25 to 5.45 GHz), and the other at X band (9.5 to 9.8 GHz) (Figure 33). Each array was composed of 96 dual-polarized multi-layer microstrip antennas. Two interlaced feed networks–stripline in concept–are used to excite the whole set of radiating sources equivalent to a dual-polarized array. The 3 dB beamwidths of the main lobe are equal to 3 and 30 degrees. The linear isotropic gain, measured at the mean frequency, is equal to 20 dB. The coupling between the two polarizations is lower than -14 dB [68].

Another recent dual-polarized flat array, operating at Ku band frequencies, is shown in Figure 34 [70, 71]. The array is composed of 52 double flat-folded dipoles, locating along several columns or subarrays. Each sub-array is equivalent to a stationary-wave and transverse-radiation plane array. For each polarization, the various transmission lines of the subarrays are fed, by means of a divider, from a central feed point. The whole array is manufactured with four stacked printed circuits. The array size is a circle, 7 wavelengths in diameter. The maximum linear gain is better than 21 dB, and the mean efficiency is 40%, in a bandwidth of 4 percent. The coupling between the two arrays inputs is less than -30 dB.

3.3 Dual-beam printed antennas

Dual-beam antennas are mainly of interest for the acquisition of information in navigation systems. Low-cost flat printed antenna arrays, associated with planar microwave components, can be designed, in order to get accurate speed measurements using the Doppler effect, for automotive applications (Figure 35). For such applications, pattern-synthesis methods are required to improve radiation diagrams, with small values of beamwidth and low side-lobe levels. The relaxation method is a well-suited synthesis technique, as it takes into account the envelope specification and the directivity pattern of the source. It directly enables the determination of the real excitation coefficients of the sources, with unequal spacing between them [76]. This method has already been used for arrays exhibiting directive or sectoral patterns [75].

3.3.1 The radiating element's pattern. The corner-fed square patch has been chosen as an elementary source, as it provides a high input impedance, well suited for series arrays (Figure 36a). The input impedance and the resonant frequency of the square patch are dependent on the sizes of the patch and the width of the feed-corner junction [31]. When the patch is excited at one corner (Figure 36b), the cavity model shows that the main part of the internal field is the sum of two degenerate modes of equal amplitudes, i.e., modes (1,0) and (0,1).

If the higher modes are neglected, the E_x and E_y fields, along the edges, exhibit the variations shown in Figure 36c. The far fields are linearly polarized, in either the E plane (y = 0°), or in the H plane (y = 90°). It is well known [77] that the E plane presents a weak-level spatial variation of the far field (around 6 dB). However, in the H plane, the same field decreases rapidly beyond an inclination angle greater than 60°. Therefore, it is necessary to take into account the directivity pattern of the source in the synthesis of the array pattern.

3.3.2 The relaxation method for dual-beam patterns. Let us consider a linear array of $2N$ equal-phase square-patch antennas, fed at their corners by a microstrip line. A coaxial probe is located in the middle of the microstrip line (Figure 37). Considering the patches located alternatively at each side of the feed line (and as a result, with a 180° phase shift), the physical distance, d_s, between such sources defines the angular positions of the two main lobes at $\pm\theta_0$:

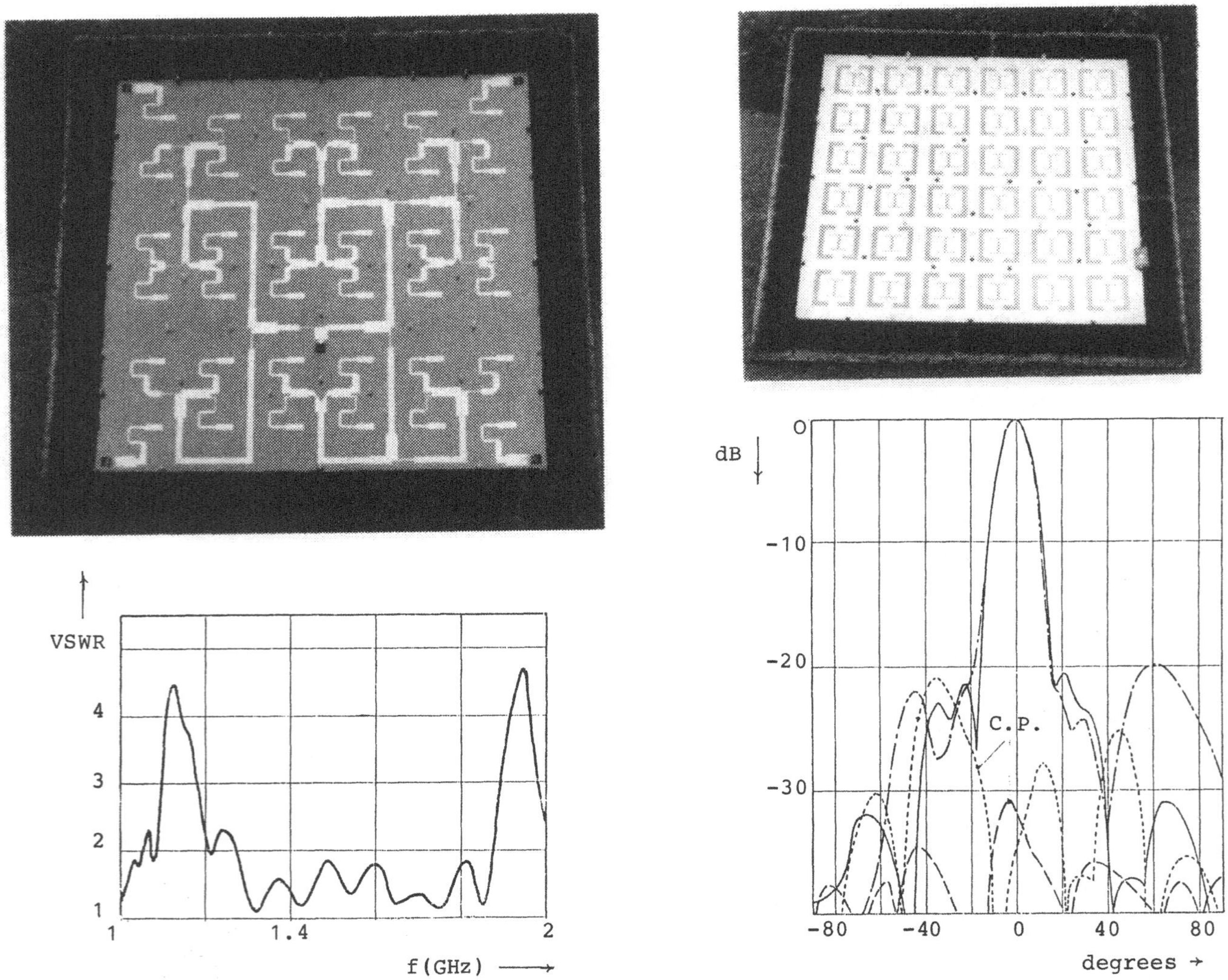

Figure 32. Large-bandwidth mono-polar array at L band (1.3 to 1.8 GHz).

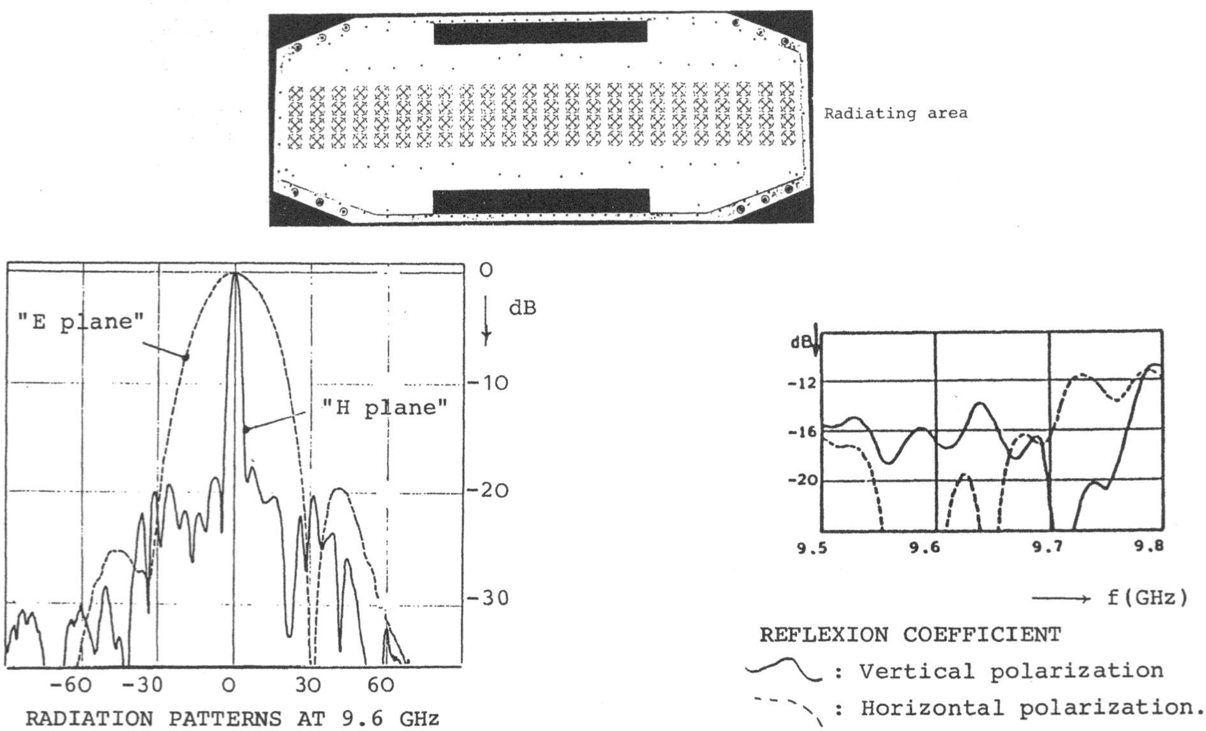

Figure 33. Dual-polarized flat array at X band (9.5 to 9.8 GHz).

Figure 34. Dual-polarized flat array at Ku band.

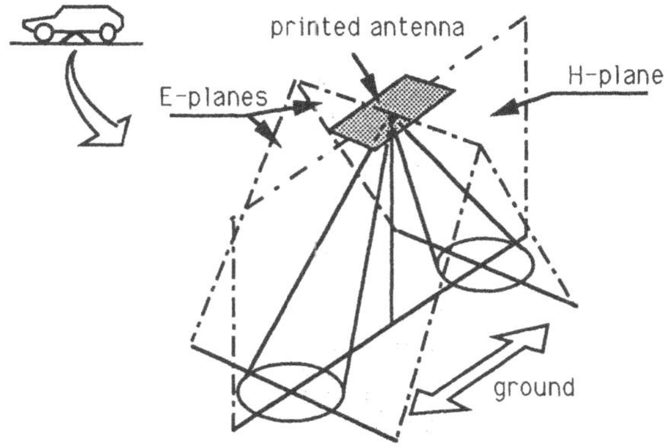

Figure 35. Dual-beam printed antenna for automotive Doppler radar sensor.

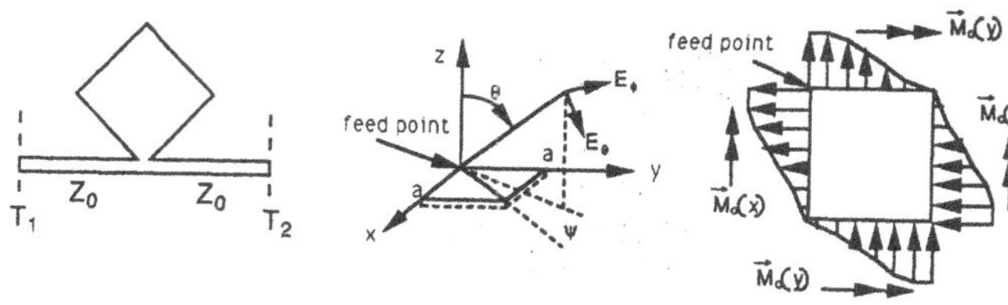

Figure 36. Geometry and magnetic currents of the corner-fed square patch.

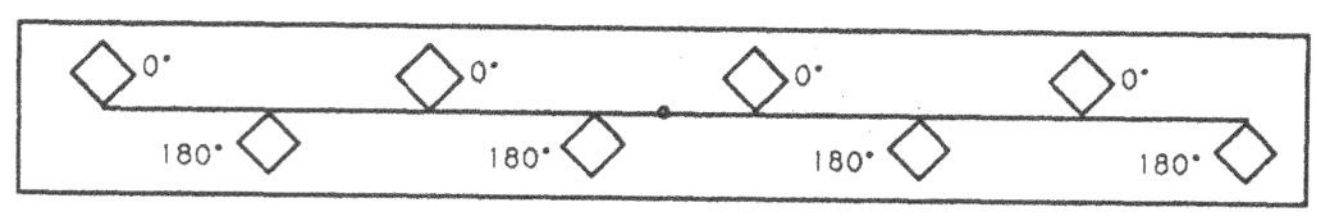

Figure 37. Linear series array.

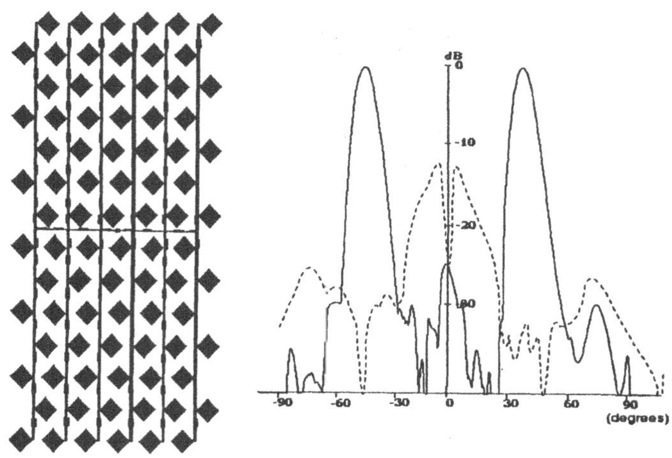

Figure 38. A 6 by 14 element dual-beam planar array: (a) schematic; (b) measured H-plane pattern (24 GHz).

$$\theta_0 = \arcsin\left(\frac{\lambda_0}{2d_s}\right), \tag{14}$$

where λ_0 is the free-space wavelength.

A non-uniform amplitude distribution along the patches is necessary, to obtain a desired beamwidth, and to achieve the prescribed sidelobe level. Quarter-wave transformers along the line joining the different patches allow tapered amplitudes.

Let us suppose that the $2N$ elements of the linear array are symmetrically fed, with an unknown excitation vector $[\mathbf{a}]$ given by

$$[\mathbf{a}] = [a_j]_{j=1}^{N} \qquad (15)$$

The dual-beam array pattern, $p(\theta)$ will be given by the following relation:

$$p(\theta) = \sum_{j=1}^{N} a_j \sin\left[\frac{\pi d_s}{\lambda_0}(1+2j)\sin(\theta_j)\right] g_\phi(\theta) \qquad (16)$$

where $g_\phi(\theta)$ is the directivity pattern of the isolated source. The physical distance, d_s, between the sources can be obtained as a function of the angular position of the two main lobes. The functional to be minimized is $J(\mathbf{a}) = p - p_d$ (where p_d is the desired pattern), in the sense of a given criterion. The search directions for the vector $\mathbf{a}$ are the co-ordinate axes, each of them being taken periodically.

Let us name BW_{3dB} the 3 dB main-lobe beamwidth, and SLL_{lim} the maximum sidelobe level outside the main-lobe pattern outline. The functional, $J(\mathbf{a})$ is replaced by two functionals, $BW(\mathbf{a})$ and $SLL(\mathbf{a})$, and minimizing $J(\mathbf{a})$ can be expressed in terms of the following improvement criteria, c1 and c2 (let $\mathbf{a}^+$ be the actual vector of the iteration):

c1: $[BW(\mathbf{a}^+) < BW(\mathbf{a}) \text{ and } SLL(\mathbf{a}^+) \leq SLL_{lim}]$

c2: $[BW(\mathbf{a}^+) \leq BW_{3dB} \text{ and } SLL(\mathbf{a}^+) < SLL(\mathbf{a})]$

The c1 criterion gives the best beamwidth for a given sidelobe level (SLL_{lim}), and c2 gives the best sidelobe level for a given beamwidth (BW_{3dB}). The c1 and c2 criteria can be applied successively, depending on the beamwidth and the sidelobe requirements.

3.3.3 Application. A 6 by 14-element dual-beam planar-antenna array (Figure 38a) was constructed on a polypropylene substrate [78, 79] ($\varepsilon_r = 2.2$, $\tan\delta = 10^{-3}$, thickness = 0.4 mm). According to equation (14),the H lane gives a directivity pattern with two main lobes at ±41.8° [77]. The synthesis method previously developed was used for the H lane. In this plane, the sidelobe level is lower than -30 dB, and a 3 dB beamwidth of 10° is obtained. The measured H plane pattern is plotted in Figure 38b.

Figure 39. A view of a 24 GHz radar sensor using the 6 by 14 element square-patch planar antenna array.

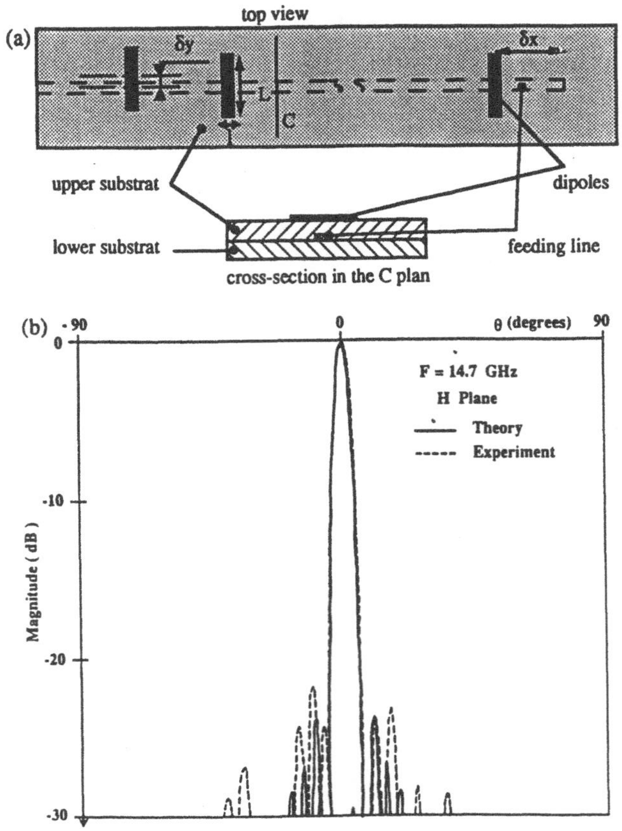

Figure 40. A linear series-fed array of microstrip dipoles. (a) Geometry of the array and feed network; (b) H-plane pattern.

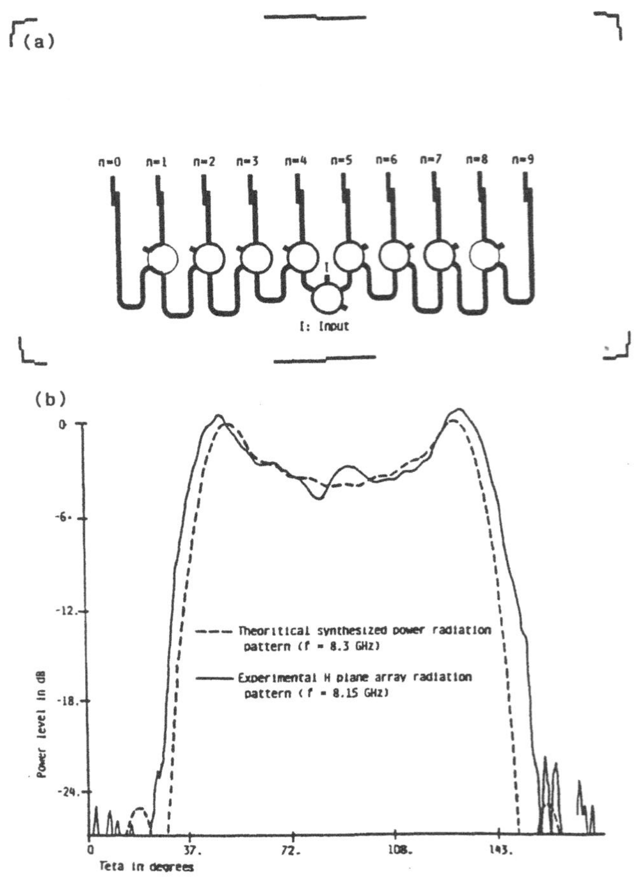

Figure 41. Linear parallel-fed array of microstrip dipoles. (a) Geometry of the array and feed network. (b) H-plane pattern.

Figure 39 presents a view of a 24 GHz radar sensor using this dual-beam planar printed antenna. The planar microwave electronics are placed on the backside of the ground plane of the antenna, while the signal processing is implemented on a second, parallel (usually epoxy) board.

4. Arrays of microstrip dipoles

The design procedure for antenna arrays fed by proximity is based on previous works by R. S. Elliott and G. J. Stern [85]. This method includes the mutual coupling between the array elements. The principle of operation of such arrays is controlled by both the electromagnetic energy transfer from the feed line to each array element, and by the participation of the mutual coupling, which represents the influence of other radiating elements. The array synthesis consists of two main steps [86]. First, self- and mutual-coupling impedances are obtained, using the integral-equation technique previously described. This takes into account the mutual coupling caused by both space and surface waves. A database is then constructed for different geometrical parameters, for elementary radiating sources. The radiation pattern is determined by making use of the array factor. By the application of the matching condition [87] on all active self and mutual impedances of the array, the configuration of the array can be calculated. The iterative re-matching process is then used to obtain the optimum array configuration.

This approach is applied to a microstrip dipole array, as shown in Figures 40 and 41. Figure 40 shows the computed and the measured radiating field patterns of a linear series-fed array of microstrip dipoles. Good agreement is observed between theoretical and experimental results. This type of array has the advantage of using only one microstrip line as the feed network.

The radiation pattern given in Figure 41 is obtained using a linear parallel-fed array of 10 elements, uniformly spaced by $\lambda_g/2$. The feed network for this array is realized with a compact power divider [88], composed of asymmetrical hybrid rings (Figure 41). The comparison with measurements given in the same figure shows the accuracy of the calculated radiation pattern of the array.

5. Conclusion

Availability of computers for antenna analysis has made possible precise knowledge of performance for simple geometries of microstrip structure, such as dipoles and rectangular patches (stacked or unstacked). However, innovative structures (slot-loaded patches, slot-fed patches, etc.) and their physical understanding often require an experimental start, and development of empirical models (transmission-line, cavity). These simple models also appear to be well suited for the analysis of microstrip antennas, with moderate accuracy and short CPU time on PC-level computers. Today, antenna designers need more-complete CAD software, including synthesis procedures for both antenna elements and arrays. Further research will be developed based on this idea.

Acknowledgments

Much of this paper originated through the collaborative efforts of our colleagues, the technical team, and many students, during their theses. The authors most gratefully acknowledge their help.

References

1. G. Dubost, "Méthode d'analyse et de synthèse de quelques microantennes à large bande en mode quasi transversal électromagnétique," *Annales des Télécom.*, **42**, 9-10, September-October 1987, pp. 588-605.

2. P. Poey, G. Dubost, M. Bahram, P.L. Guigue, "Analyse d'une microantenne à double fente à très large bande passante," *L'Onde Electrique n° 1*, **68**, pp. 60-66, January 1988.

3. G. Dubost, "Forme analytique du rendement d'une antenne plaque rectangulaire à la résonance demi-onde imputable à l'onde de surface du mode dominant TM_0". Annales des Télécom., juillet-août 1990, vol. 43, n° 7-8, pp. 429-436.

4. G. Dubost, "Wide band flat dipole and short-circuit microstrip patch antennas and arrays," Chapter 7, in J. R. James and P. S. Hall (eds.), *Handbook of Microstrip Antennas*, London, Peter Perigrinus Ltd., 1989, pp. 353-392.

5. G. Dubost, A. Zerguerras, "Transmission line model analysis of arbitrary shape symmetrical patch antenna coupled with a director," *Electronics Letters*, **26**, pp. 952-954, June 1990.

6. G. Dubost, S. Desclos, A. Zerguerras, "Radiation of arbitrary shape symmetrical patch antenna coupled with a director," *Electronics Letters*, **26**, 18, pp. 1539-1540, August 1990.

7. G. Dubost, S. Desclos, A. Zerguerras, "Current distributions and far field radiated by an arbitrary shape large bandwidth microstrip antenna," ANTEM'90 *Digest*, Winnipeg, Canada, 15-17 August, 1990, pp. 3-8.

8. G. Dubost, S. Desclos, A. Zerguerras, "Analyse d'antennes imprimées multicouches de forme quelconque à axe de symétrie en mode quasi TEM," *L'Onde Electrique*, **71**, 1, pp. 48-57, January-February 1991.

9. J. R. James, P.S. Hall, Wood C., *Microstrip Antenna Theory and Design*, London, Peter Perigrinus Ltd. (Electric Waves Series).

10. I. J. Bahll, and P. Bhartia, *Microstrip Antennas*, Dedham (MA), Artech house, 1980.

11. J. R. James, P. S. Hall, and C. Wood, *Microstrip Antenna Theory and Design*, London, Peter Peregrinus Ltd. (IEE Electromagnetic Waves Series 12), 1981.

12. Y. T. Lo, D. Solomon, and W. R. Richards, "Theory and experiment on microstrip antennas," *IEEE Trans. Ant. Prop.*, **AP-27**, pp.137-145, 1979.

13. K. R. Carver, and E. L. Coffey, "Theoretical investigation of the microstrip antenna," Technical report 00929, Physical Science Laboratory, New Mexico State University, Las Cruses (New Mexico), 1979.

14. E. Penard, "Etude d'antennes imprimées par la méthode de la cavité," Thesis, Rennes, 1982.

15. D. Thouroude, M. Himdi, J. P. Daniel, "CAD-oriented cavity model for rectangular patches," *Electronics Letters*, **26**, 13, pp. 842-844, June 21, 1990.

16. D. Schaubert, D. Pozar, A. Adrian, "Effect of microstrip antenna substrate thickness and permittivity: Comparison of theories with experiment," *IEEE Trans. Ant. Prop.*, **AP-37**, pp. 677-682, 1989.

17. Jr. James, P.S. Hall, *Handbook of Microstrip Antennas*, Longdon, Peter Peregrinus Ltd., 1989, Chapter 11.

18. S. Assailly, "Contribution à l'étude des antennes imprimées multicouches par une approche dans le domaine spectral. Application au couplage," DSc Thesis, University of Rennes, France, 1990.

19. S. Assailly, C. Terret, J. P. Daniel, G. Besnier, J. Mosig, B. Roudot, "Spectral domain approach applied to open resonators: application to microstrip antennas," *Electronics Letters*, **24**, n°2, pp. 105-106, January 22, 1988.

20. G. Dubost, G. Beauquet, J. Rocquencourt, G. Bonnet, "Patch antenna bandwidth increase by means of a director," *Electronics Letters*, **22**, 25, pp. 1345-1347, December 4, 1986.

21. M. C. Bailey, M. D. Deshpande, "Input impedance of microstrip antennas," *IEEE Trans. Ant. Prop.*, **AP-30**, 4, pp. 645-650, April, 1982.

22. P. S. Bhatnagar, C. Terret, J. P. Daniel, K. Mahdjoubi, "Experimental study on broadband triangular microstrip array antenna," in *Proceedings* of the International Symposium on Antennas and Propagation, 22-25 August 1989, Tokyo, Japan.

23. J. R James, P. S. Hall, *Handbook of microstrip antennas*, London, Peter Peregrinus Ltd., Chapter 3, pp. 311-351, 1989.

24. S. Assailly, C. Terret, J. P. Daniel, "Some results on broadband microstrip antenna with low crosspolar and high gain" *IEEE Trans. Ant. Prop.*, **AP-39**, 3, pp. 413-415, March 1991.

25. S. Assailly, C. Terret, J. P. Daniel, K. Mahdjoubi, "Low cost stacked circular polarized microstrip antenna," *Digest* of IEEE Antennas and Propagation Society International Symposium, vol 2, San Jose CA, June 1989, pp. 628-631.

26. C. Terret, S. Assailly, K. Mahdjoubi, A. Sharaiha, "Stacked microstrip antennas, advantages and drawbacks," presented at PIERS'91, Cambrige, Massachussets, July 1-5, 1991.

27. F. Croq, A. Papiernik, "Large bandwidth aperture coupled microstrip antenna, *Electronics Letters*, **26**, 16, pp. 1293-1294, August 2, 1990.

28. K. R. Carver, "Pratical Analytical Techniques for the Microstrip," *Proceedings* of Workshop on Printed Circuit Antenna Technology, Cruces, Mexico, pp. 7.1-7.20, October.

29. N. M. Martin, "Improved Cavity Model Parameters for Calculation Frequency of Rectangular Microstrip Antennas," *Electronic Letters*, **24**, 11, pp. 680-681, 1988.

30. E. Motta Cruz, J. P. Daniel, "Modèle d'antenne carrée imprimée alimentée en coin, application aux réseaux linéaires," JINA-90 Conference, November 13-15, 1990, Nice, France, pp. 297-300.

31. E. Motta Cruz, J. P. Daniel, "Experimental analysis of cornerfed patch antennas," *Electronic Letters*, **27**, 16, pp. 1410-1411, August 1, 1991.

32. T. Dusseux, "Etude d'antennes fentes annulaires imprimées, applications antennes melangeuses, réseaux," Thèse de Docteur Ingénieur, Université de Rennes I, May 4, 1987.

33. T. Dusseux, J. P. Daniel, Terret C., "Theoretical and experimental results ofguided wavelength of a slot on a low permittivity substrate," *Electronics Letters*, **22**, pp. 589-590, 1986.

34. T. Dusseux, J. P. Daniel, Terret C., "Analyse et réalisation d'antennes fentes imprimées annulaires," *Proceedings* JINA'86 Conference, November 1986, Nice, France, pp. 199-202.

35. S. B. Cohn, "Slot-line on a dielectric substrate," *IEEE Trans. Microwav. Theo. Techn.*, **MTT-17**, 10, pp.768-778, October 1969.

36. J. B. Knorr, "Slot-line transition," *IEEE Trans. Microwav. Theo. Techn.*, **MTT-22**, pp. 548-554, 1974.

37. D. M. Pozar, "A reciprocity method of analysis for printed slot and slot coupled microstrip antennas," *IEEE Trans. Ant. Prop.*, **AP-34**, pp. 1439-1446, December 1986.

38. G. Dubost, J. Grosset, "Large bandwidth dual polarized circular slot antenna. Application to an array in Ku band," *Proceedings* of International Symposium on Antennas and Propagation, August 22-25, 1989, Tokyo, Japan.

39. G. Dubost, "Theoretical Radiation Resistance of an isolated slot ring resonator," *Electronics Letters*, **23**, 18, pp. 928-930, August 1987.

40. G. Dubost, "Large bandwidth slot at resonance with directional radiation," *Electronics Letters*, **24**, 23, pp. 1449-1450, November 1988.

41. K. D. Stefan, N. Camillieri, T. Itoh, "A quasi-optical polarization duplexed balanced mixer for millimeter-wave applications," *IEEE Trans. Microwav. Theo. Techn.*, **MTT-31**, pp. 164, 1983.

42. D. M. Pozar, "Input impedance and mutual coupling of rectangular microstrip antennas," *IEEE Trans. Ant. Prop.*, **AP-30**, 6, pp. 1191-1196, November 1982.

43. M. Himdi, J. P. Daniel, "Characteristics of sandwich-slot lines in front of a parallel metallic strip," *Electronics Letters*, **27**, 5, pp. 455-456, February 28, 1991.

44. P. L. Sullivan, D. H. Schaubert, "Analysis of an aperture coupled micro-strip antenna," *IEEE Trans. Ant. Prop.*, **AP-34**, pp. 977-984, August 1986.

45. J. S. Rao, K. K. Joshi, B. N. Das, "Analysis of small aperture coupling between rectangular waveguide and microstripline," *IEEE Trans. Microwav. Theo. Techn.*, **MTT-29**, 2, February 1981.

46. M. Himdi, J. P. Daniel, C. Terret, "Analysis of aperture-coupled microstrip antenna using cavity method," *Electronics Letters*, **25**, 6, pp. 391-392, March 16, 1989.

47. G. Gronau, I. Wolff, "Aperture coupling of rectangular microstrip resonator," *Electronics Letters*, **22**, 10, pp. 554-556, 1986.

48. M. Himdi, J. P. Daniel, C. Terret, "Resonant behavior of slot coupled microstrip antennas," *Proceedings* International Symposium on Antennas and Propagation, Tokyo, Japan, August 22-25, 1989, pp. 217-220.

49. R. F. Harrington, *Time Harmonic Electromagnetic Fields*," New York, McGraw-Hill, 1968.

50. M. Himdi, J. P. Daniel, C. Terret, "Transmission line analysis of aperture coupled microstrip antenna". *Electronics Letters* Vol. 25, N°18, pp. 1229-1230, 31st August 1989.

51. M. Himdi, "Analyse et Synthèse d'antenne imprimée alimentée par fente. Application aux réseaux," Doctorat de l'Université de Rennes I, 1990.

52. M. Himdi, J. P. Daniel, D. Thouroude, "Antenne imprimée alimentée par fente (circuit équivalent)," *Proceedings* JINA'90 Conférences, November 13-15, 1990, Nice, France, pp. 293-296.

53. D. Beguin, G. Dubost, L. Dupont, "Antenne pastille carrée à double polarisation croisée alimentée par deux lignes triplaques à travers deux fentes orthogonales," Patent n° 9115515, December 13, 1991.

54. "Antenne plaque à double polarisations croisées," Titulaire Etat Français (C.N.E.T. et T.D.F.), Inventeurs, Dubost Gérard et Frin Roger, Date de publication de la demande, October 28, 1987, Numéro de publication du Brevet Européen (BE, DE, FR, GB, IT, NL, SE), June 19, 1991, n° 0243289B1, Numéro de publication aux US, Patent number, 4 922 263, May 1, 1990, "Plate antenna with double crossed polarizations".

55. G. Dubost, A. Rabbaa, "Analysis of a slot microstrip antenna," *IEEE Trans. Ant. Prop.*, **AP-34**, 2, pp. 155-163, February 1986.

56. Jr. Oltman, "Microstrip Dipole Antenna Elements and Arrays Thereof," United States Patent n° 703113.

57. Ph. Lepeltier, J. Citerne, and J. M. Floc'h, "Electromagnetically coupled microstrip dipole, Analysis by means of integral equations," *Digest* IEEE Antennas and Propagation Society International Symposium, Vancouver, Canada, June, 1985, pp. 777-780.

58. M. Drissi, V. Fouad Hanna and J. Citerne, "Analysis of coplanar waveguide radiating end effects using integral equation technique," *IEEE Trans. Microw. Theo. Techn.*, **39**, 1, pp. 112-116, January 1991.

59. R. F. Harrington, *Field Computations by Moment Methods*," New York, MacMillan, 1968.

60. Ph. Lepeltier, J. Citerne and J. M. Floc'h, "On the EMC dipole feedline parasistic radiation," *IEEE Trans. Ant. Prop.*, **38**, 6, pp. 878-883, June 1990.

61. H. Legay, J. M. Floc'h, J. Citerne and Ph. Lepeltier, "Theoritecal and experimental analysis of circularly polarized notched antenna fed electromgneticaly by one microstrip line," *Digest* IEEE Antennas and Propagation Society International Symposium, pp. 644-647, 1990.

62. H. Legay, R. Gillard, J. Citerne, G. Piton, "Charts for a quick design of microstrip line fed patch antennas," *Digest* IEEE Antennas and Propagation Society International Symposium, 1991.

63. H. Legay, R. Gillard, J. Citerne, G. Piton, "Via-hole effects in a patch microstrip antenna. Application to active antennas," *Proceedings* JINA'90, Nice, France, 1990, pp. 317-320.

64. R. Gillard, H. Legay, J.M. Floc'h, J. Citerne, "A rigorous analysis of an active receiving microstrip antenna," *Electronics Letters*, **27**, 25, pp. 2357-2359, 1991.

65. G. Dubost, S. Gueho, "Réseau d'antennes à large bande du type doublet replié plaque," *L'Onde Electrique*, **69**, 3, pp. 32-38, May-June 1989.

66. S. Gueho, "Contribution à l'étude des réseaux phasés. Antenne à balayage électronique de faisceaux à directivité variable dans la bande des 15 GHz," Doctorat de l'Université de Rennes I. Juillet, 1990.

67. G. Dubost, "Influence of surface wave upon efficiency and mutual coupling between rectangular microstrip antennas," *Digest* IEEE Antennas and Propagation Society International Symposium, Dallas, Texas, May 1990.

68. G. Dubost, R. Frin, S. Touimer, D. Beguin, L. Dupont, "Réseaux plats à double polarisations croisées dans la bande C et S," Colloque International sur le Radar, Paris April 24-28, 1989, Section B3, pp. 201-206.

69. G. Dubost, "Printed phased array with steering beams," *Proceedings* Military Microwaves'88, London , July 5-7, 1988, pp. 308-313.

70. G. Dubost, R. Frin, J. Grosset, "Dual polarized flat folded-dipoles. Application to an array in Ku band," Third A.P.M.C., Tokyo, 1990, pp. 79-82.

71. G. Dubost, R. Frin, "Dual polarized microstrip arrays in S, C, X and Ku bands," PIERS July 1-5, 1991, Cambridge, MA.

72. G. Dubost, J. Grosset, "Large bandwidth dual polarized circular slot antenna. Application to an array in Ku band," International Symposium on Antennas and Propagation, August 22-25, 1989, Tokyo, Japan.

73. G. Dubost, R. Frin, "Plate antenna with double crossed polarizations," US Patent number 4 922 263, May 1, 1990.

74. J. Grosset, "Etude d'un réseau de doublets repliés plaques à double polarisation croisée fonctionnant dans la bande Ku," Doctorat de l'Université de Rennes I, May, 1991.

75. M. Boguais, J. P. Daniel, C. Terret, "Antenna Pattern Synthesis using a Relaxation Method, Application to Printed Antennas," *Electronics Letters*, **22**, 7, pp. 375-376, March 27, 1986.

76. E. Motta Cruz, J. P. Daniel, "Pattern Synthesis of Dual-beam Printed Antennas," MIKON-91, Rydzyna, Pologne, May 1991, pp. 219-223.

77. E. Motta Cruz, J. P. Daniel, "Arrays of corner-fed square patch antennas, analysis and applications," SBMO, Rio de Janeiro, pp. 566-571, July 1991.

78. Ph. Dupuis, J. P. Daniel, J. L. Alanic, "Antenne plaque microonde notamment pour Radar Doppler," C.R.I.T.T./C.N.R.S. patent 87 12579.

79. CNET Patent "Support métallisé à base de polypropylène et procédé de fabrication de ce support," n° 84 402 7078.

80. D. M. Pozar, "Input impedance and mutual coupling of rectangular microstrip antennas," *IEEE Trans. Ant. Prop.*, **AP-30**, 6, pp. 1191-1196, November 1982.

81. C. Terret, S. Assailly, K. Mahdjoubi, "Mutual coupling between stacked microstrip antennas," ANTEM'90, Winnipeg, Canada, August 15-17, 1990

82. C. Terret, S. Assailly, K. Mahdjoubi, "Theoretical and experimental results on mutual coupling between stacked microstrip antennas," Asia Pacific Microwave Conference, Tokyo, Japan, August 1990.

83. R. Q. Lee, T. Talty, K. F. Lee, "Mutual coupling between antennas electromagnetically coupled rectangular patch antennas," *Electronics Letters*, **27**, 6, pp. 532-533, March 14, 1991.

84. C. Terret, S. Assailly, K. Mahdjoubi, M. Edimo, "Mutual coupling between stacked microstrip antennas," *IEEE Trans. Ant. Prop.*, **AP-39**, 7, July 1991.

85. G. J. Stern, R. S. Elliott, "The Design of Microstripdipole arrays including mutuel coupling," *IEEE Trans. Ant. Prop.*, **AP-29**, September 1991.

86. Yang, N. G. Alexopoulos, Ph. Lepeltier, P. M. Stern, "Design of transversly fed EMC microstrip dipole arrays including mutual coupling," *IEEE Trans. Ant. Prop.*, **AP-38**, pp. 145-151, February 1991.

87. J. M. Floch, J. Citerne, Ph. Lepeltier, "Standing Wave Array of EMC Tranverse Dipole Arrays," *Proceedings* of European Microwave Conference, pp. 1517-1522, September 1990.

88. D. Martin, J. M. Floch, J. Citerne, Ph. Lepeltier, "EMC dipole array synthesis including mutual coupling using the Elliott-Stern method," *Digest* IEEE Antennas Propagation Society International Symposium, pp. 952-955, 1988.

A Review of CAD for Microstrip Antennas and Arrays

D. M. POZAR
ECE DEPARTMENT
UNIVERSITY OF MASSACHUSETTS
AMHERST, MA 01003

J. R. JAMES
SCHOOL OF ELECTRICAL ENGINEERING AND SCIENCE
ROYAL MILITARY COLLEGE OF SCIENCE
CRANFIELD UNIVERSITY
SWINDON SN6 8LA ENGLAND

***Abstract*—While CAD software has reached fairly sophisticated levels in the areas of circuit analysis, solid-state device modeling, and microwave circuit analysis and optimization, microstrip antenna CAD software is substantially less advanced. This paper will discuss possible reasons for this present state of affairs, and present our view of the most desirable features of CAD software for microstrip antenna design and manufacturing. We will point out where microstrip antenna CAD is important and where it is not, and emphasize the fact that CAD software is not a panacea, or a substitute for experience and fundamental understanding of the technology and its physics.**

I. INTRODUCTION

Computer-aided design (CAD) software seems to be one of the most ubiquitous topics in the fields of microwave and antenna engineering today, perhaps because of the perception among engineers that such software will not only make their jobs easier but provide a tool to do work that would not otherwise be possible. In its ideal form antenna CAD software would combine a user-friendly interface with a computationally efficient set of accurate and versatile theoretical models [1]. Software with such features has reached a fairly high level of refinement in areas such as the analysis of low-frequency circuits (SPICE, etc.), and the analysis and optimization of passive and active microwave circuits (Touchstone, SuperCompact, etc.). With these software products, user confidence is high and prototype designs can be manufactured with an acceptable level of trial-and-error adjustment, if necessary. In contrast, microstrip antenna CAD software lags far behind, often committing the designer to costly experimental iterations, sometimes even for a single radiating element. For a large array of elements the cost of trial-and-error design soars with array size and complexity, and there is no guarantee that such a process is convergent!

For many years it has been stated that microstrip antenna technology is associated with manufacturing simplicity and low cost, but it is clear that for arrays at least, this will only be realized with drastically improved CAD software. Since the technical journals [2] have no shortage of papers describing a wide variety of numerical solutions to microstrip antenna problems, one may wonder why CAD embodying this work have not advanced further than they have at the present time.

The economic reality that the market for antenna software in general is relatively small perhaps explains why there is very little commercially available software for any type of antenna design. Another reason for the slow pace of microstrip CAD software development is the fact that such antennas are relatively new, receiving serious attention only during the last fifteen years. Furthermore, microstrip antenna geometries are relatively difficult to model because of the presence of dielectric inhomogeneities and a wide variety of feeding techniques and other geometrical features. This last consideration makes the development of a general-purpose microstrip antenna analysis package extremely difficult. Finally, there may be intrinsic sources of error within the mathematical formulations themselves, as well as the more obvious ones in the numerical algorithms, and the totality of these errors sets a ceiling to the achievable design accuracy.

Academic researchers have been prolific in generating analytical and numerical solutions for a wide variety of microstrip antennas and arrays, often with a high degree of originality as well as rigor. But this work is generally performed primarily for a graduate student thesis or publication, and the software is seldom written, validated, or documented for other users. Researchers in industry may be more pragmatic when developing comparable solutions for a specific antenna geometry, but such software is often considered proprietary. Antenna software that is available, such as the NEC code for wire antennas, the ESP code for modeling wires and plates, the TICRA software for reflector antenna design, and some others, constitute the few exceptions to this point.

This brief introduction has noted some of the many aspects of microstrip CAD software that we wish to bring into sharper focus in this paper. We begin with a brief qualitative overview of the various models that have been most successful for microstrip element and array design, including a list of some commercially available microstrip antenna CAD packages. Section III will present a fairly specific view of design methodologies for various types of microstrip antennas and arrays. Overall, we propose to demonstrate the following premises that CAD software is

- of less importance for many basic microstrip antenna and array designs,
- a more vital requirement for large arrays of microstrip antennas,
- not a substitute for experience and understanding of fundamentals, and

- still in need of extensive research and development to raise the operational accuracy and versatility to the level demanded by antenna manufacturers.

II. Discussion of CAD Models and Commercially Available Software

A good antenna model or theory can be used to calculate all necessary electrical parameters of the antenna under consideration, with enough accuracy for the intended purposes, in a computationally efficient and user-friendly manner. Further attractive features may include the versatility to treat variations in the basic antenna geometry, and to provide for the optimization of a particular performance variable against one or more design parameters. In general, a more sophisticated CAD model will be more accurate and versatile, but will involve higher product cost, and require more computer resources, as compared to CAD software based on a simpler model. As discussed above, the analytical treatment of microstrip antennas and arrays is complicated by several factors, and there have been many solutions and variations proposed for the treatment of many different structures, but most models can be divided into two groups: simplified (or reduced) analyses, and full-wave methods. Analytical models will be discussed within these two categories, drawn mostly from [2], but the reader is referred to the articles in Chapters 5 and 7 of this book for more technical detail. Figure 1 shows results for the input impedance of a probe-fed rectangular microstrip element computed from several different models and compared with experimental data.

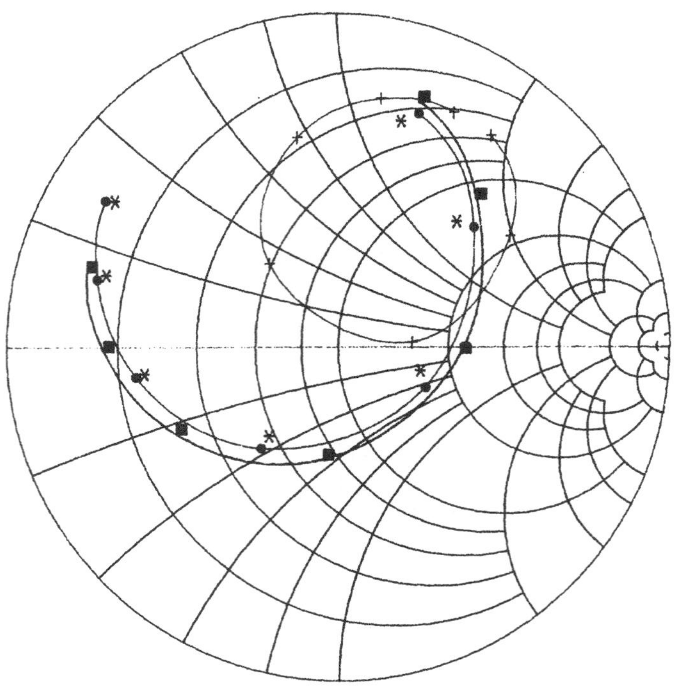

Fig. 1. Comparison of various models for the input impedance of a probe-fed microstrip antenna. *—measured data, ●—calculations using a full-wave moment method with a rigorous feed model, ■—calculations using a full-wave moment method with an idealized feed model, +—calculations using a cavity model. (Substrate thickness = 0.79mm; substrate dielectric constant = 2.20; patch length = 1.25 cm; patch width = 2.0 cm; feed probe is 0.4 cm from a radiating edge.) The frequency sweep starts at 7.3 GHz, in steps of 0.2 GHz.

II.1 Reduced Analyses

By **reduced** analyses we mean microstrip antenna models that introduce one or more significant approximations to simplify the problem. These include cavity models, which use a magnetic wall boundary condition approximation for the periphery of the patch; transmission line models, which model the element as a transmission line section with lumped loads at the radiating edges; and multiport network models, which can be viewed as a generalization of the cavity model. These models were the first to be developed for microstrip antennas, and have been useful for practical design as well as providing a good intuitive explanation of the operation of the microstrip antenna. Drawbacks of such models include limited accuracy for substrates that are not thin, and a limited capacity to handle related problems such as mutual coupling, feed network effects, surface wave effects, and multilayer substrate configurations.

II.2 Full-Wave Analyses

Microstrip antenna models that account for the dielectric substrate in a rigorous manner are referred to as **full-wave** solutions. Such models include moment method solutions that use the exact Green's function for the dielectric substrate(s), as well as solutions based on the finite difference time domain (FDTD) method, and the finite element (FE) method. Features of full-wave models include high accuracy and the ability to calculate all relevant electrical parameters for a wide range of antenna geometries, including multilayer configurations, arrays with feed networks, and various element-coupling configurations. The main disadvantages of full-wave solutions at the present time are their high computational cost, and a low level of user confidence when experimental or other independent validation is absent [1].

Of these, moment method solutions have received the most attention to date, with a large variety of solutions developed for specific antenna geometries. These solutions generally assume the substrate to be of infinite extent, and model the electric current on the patch elements and feed network in terms of subsectional basis functions. Such models are very time-consuming computationally, due to numerical integrations of Sommerfeld-type integrals, in either spectral domain or space domain form. Much of the current research in this area is devoted to improving computational performance.

FDTD and FE solutions take a more "brute force" approach by modeling the entire antenna, including dielectric and metal components, and some of the surrounding volume. This approach allows a very high degree of versatility for treating arbitrary geometries, including multilayer and inhomogeneous dielectrics, but the price is paid in terms of computer time, which is typically much longer than moment method solutions for a comparable geometry. For geometries for which they can be applied, moment method solutions usually have a computational advantage over FDTD or FE methods, because the dielectric is accounted for automatically by the Green's function and only the conductors must be modeled by the basis functions. This difference can be especially significant for arrays, but becomes less critical with the increasing power of personal workstations. Recent hybrid techniques combining analytically known results into FDTD and FE methods show promise of reducing the computational effort of these methods.

II.3 Commercially Available Microstrip Antenna CAD Software

Below we list a table of some commercially available software packages that can be used for microstrip antenna and array design. Of the four, ENSEMBLE and em use full-wave moment method solutions for microstrip antennas and small arrays, including the effect of coplanar feed networks. PCAAD uses cavity models for several microstrip elements and arrays, and MICROPATCH uses the multiport segmentation method for elements and arrays. A photograph of the ENSEMBLE package in use is shown in Figure 2.

Fig. 2. **Ensemble**, a Windows-compatible full-wave moment method-based CAD package for microstrip antenna design, can be used to model elements and small arrays. Photo courtesy of Doris Wu, Boulder Microwave Technologies, Inc.

III. Microstrip Antenna Design with and without CAD

It may come as a surprise to the newcomer to practical antenna development, but it is important to realize that many microstrip antenna designs have been successfully completed with little or no CAD support. But there are, of course, many situations where antennas and arrays can be designed more effectively, with better performance and less experimental iteration, when the proper CAD software tools are available. There are other situations, involving larger arrays of microstrip elements, which critically rely on the use of CAD software for design and the evaluation of tolerance effects. Thus the point we wish to emphasize here is that CAD software is not absolutely necessary for all facets of microstrip antenna design work, but good software tools can be very useful for dealing with the more complicated microstrip geometries. Another point that seems to be especially true for antenna design in general is that CAD software, no matter how versatile or accurate, cannot substitute for experience and understanding of the fundamentals of antenna operation. In this section we will examine the above premises in more detail under the following specific headings.

III.1 Design of Simple Microstrip Elements

The most common microstrip antenna element is the single-layer, linearly polarized, rectangular microstrip patch having a probe or microstrip line feed. If the substrate parameters are specified, there remain effectively only two design parameters: the patch length, which controls the resonant frequency, and the feed point, which controls the resonant resistance. In this case the antenna can be designed quite easily, without the need for CAD. The approximation that the resonant frequency is

TABLE I. SOME COMMERCIALLY AVAILABLE MICROSTRIP ANTENNA CAD SOFTWARE.

Software Package	Company	Theoretical Model	Approximate Cost (USD)
ENSEMBLE	Boulder Microwave Technologies, Inc Boulder, CO	Full-wave moment method	$10,000.
MICROPATCH	Microstrip Designs, Inc Boulder, CO	Segmentation model	$ 300.
PCAAD	Antenna Design Associates, Inc Leverett, MA	Cavity model	$ 200.
em and PATVU	Sonnet Software, Inc Liverpool, NY	Full-wave moment method	$40,000.

given by $c/(2L\sqrt{\epsilon_r})$ can be used to estimate the length of the patch, and the fact that resistance varies as $\cos^2(\pi x/L)$ from the edge of the patch can be used to estimate the resistance. One or two experimental iterations are generally all that are needed to converge to the desired operating frequency and impedance level. A similar procedure can be used for linearly polarized circular patch elements, or elements of other shapes (although other shapes rarely offer any advantage over rectangular or circular elements, and often have poor polarization performance). In fact, even if an accurate CAD tool is available for modeling such elements, variations in substrate dielectric constant and fabricational tolerances often require one or more experimental iterations.

If one wishes to consider the effect of different substrates, in order to obtain maximum impedance bandwidth or to explore the use of different materials for cost or sizing considerations, the availability of a CAD tool can be very useful. In this case even a simple CAD program based on the cavity or transmission line model can give helpful information on trends in bandwidth, efficiency, and element size, and thus reduce the need for experimental trials. A more sophisticated solution could be used, but high accuracy and rigor are seldom necessary in this case. The assumption that tolerances in the substrate relative permittivity and thickness can be accommodated either by experimental iteration or with the assistance of CAD software, relies on the substrate being homogeneous and of uniform thickness. Whilst this is likely to be the case for high-quality (expensive) substrates, other materials may have very variable geometry and composition not amenable to iterative or CAD procedures. The use of ill-defined substrates in microstrip antenna manufacture is thus to be avoided even for simple patch elements.

The inclusion of a cover or radome layer is a relatively simple and common variation of the basic microstrip element geometry, but its analysis is generally beyond the capabilities of a cavity or transmission line model. Sometimes an effective dielectric constant can be used to approximately account for the cover layer; otherwise a more sophisticated exact full-wave Green's function solution [3] can be used. In practice, however, the downward shift in element resonant frequency can be simply corrected by scaling the element length from measurements of one or two prototypes. The above-mentioned comments about poor quality substrates also apply to cover layers.

III.2 Design of More Complicated Microstrip Elements

The design of a simple microstrip element consists of only a few parameters, offering little control of the radiation pattern, gain, or bandwidth of the antenna. More complex elements provide more degrees of freedom for design optimization, and some of the many variants include proximity-coupled elements, stacked patches, and aperture-coupled patches, as well as elements designed for circular polarization. For these cases, because of the large number of design variables, it is often very helpful to have a CAD tool to study the effect of varying a few critical parameters. Design graphs are also available for several forms of these geometries (several of which can be found in the papers in this volume).

In a proximity-coupled patch geometry there are two substrates to specify in terms of thickness and dielectric constant, as well as the patch length and width, and the feed line width and offset. The presence of eight design variables makes a purely empirical development very difficult, but design curves [4] can be used to estimate initial values for the most critical variables in order to obtain a particular bandwidth and operating frequency. Accurate modeling of this geometry requires a full-wave solution, which has been accomplished [4], [5], but in fact the complexity of these solutions and the tight interaction between the design parameters makes computer optimization over more than one variable very costly in terms of runtime on even the fastest workstations. An effective design strategy may then be to use design curves as a starting point, and iterate the design using a CAD model, if available, followed by experimental iteration. These remarks apply equally well to the stacked patch antenna geometry.

The aperture-coupled patch geometry also has a large number of design variables, but the ground plane that separates the two substrates simplifies the design procedure by largely isolating the antenna and feed effects. Thus, both cavity models [6] and full-wave solutions [7] have been implemented for this geometry, and both can be used effectively to design an aperture coupled element accurately, with minimal need for experimental iteration. Design data is also available for certain sets of parameters [7].

Circularly polarized microstrip elements using a square or circular element with two feed points and a separate network for quadrature phasing can generally be designed by treating the element as a linearly polarized element in each direction. Circular polarized elements using a single feed point are more problematic, because of the wide variety of geometries that have been suggested for this purpose, as well as the fact that the very narrow axial ratio bandwidth makes for a very sensitive design [8]. Such elements were originally designed by experimental tuning, and this is still a perfectly viable option. Design graphs are available for several of the more popular single-feed circular polarized geometries [9], and some of the more sophisticated commercially available CAD packages can be used in some cases.

III.3 Design of Fixed-Beam Microstrip Arrays

The majority of arrays are designed as fixed-beam broadside antennas; these may be linear ($N \times 1$ elements), or planar ($N \times M$) arrays. Consider first moderate-sized arrays when N and M do not exceed about five. In such cases the full advantage of planar technology is exploited by using a microstrip feed network, often coplanar with the radiating elements. The design procedure can then be divided into three parts: first determining an appropriate element design, then finding the size and element spacing of the array, and finally designing the feed network.

Mutual coupling can be ignored for the majority of arrays of this type, a fact that greatly simplifies the design, as well as making irrelevant much of the academic analysis of mutual coupling effects. Thus the design of the array elements can proceed as if they were isolated, using the procedures discussed above. The array size, element spacing, and excitation can be found using basic array theory to meet the directivity, sidelobe level, and grating lobe specifications. Then a series or corporate feed network can be designed to produce the necessary amplitude and phase distributions. Often the initial feed design can be done simply with impedance matching and power divider circuits; the designer may then exploit the power of modern microwave circuit CAD packages (e.g., Touchstone, SuperCompact, Sonnet em, among others) to account for loss effects, bandwidth considerations, microstrip discontinuity effects, and other aspects of the feed network.

The above situation can quickly become more complicated, however, making the need for complete array CAD more critical. For example, while the effect of a dielectric cover layer on the radiating element can be treated as discussed above, the cover layer will also affect the impedance and guide wavelength of a coplanar feed network. A similar problem arises with arrays using two-layer stacked patches or proximity-coupled elements. Unfortunately, most microwave circuit CAD packages do not treat the presence of a cover layer.

For larger arrays with N and M much greater than five, the effect of the totality of tolerances becomes progressively worse and experimental trimming is not a practical proposition. The figure of N and M of about five is intended to be a very rough indication of the threshold where design without CAD software is not tenable, and even with it difficulties are still present that relate to the specified sidelobe and cross-polarization levels. Feed radiation, surface wave scattering, and related effects may be partially abated by the array factor, but ultimately they set a spurious signal floor that dictates sidelobe and cross-polarization levels. Next, mechanical and electrical tolerances in the feed network have a randomization effect on the element excitations that affects both patterns and feed impedances. All this assumes good quality dielectric materials throughout, but even then the likelihood of encountering substrate thickness and permittivity changes (sometimes related to temperature variation) increases with the size of the array. For low sidelobe levels and cross-polarization the success of an array design depends critically on CAD software for modeling these tolerances to the required accuracy, which is determined by the design specifications. A perhaps surprising factor is that some mathematical estimates of these tolerances are in themselves insufficiently accurate to create CAD software with the necessary precision. In some cases reliable empirical data can help, but this approach is obviously a last resort in the absence of adequate modeling software.

III.4 Design of Microstrip Phased Arrays

Scanning phased array antennas probably constitute another definitive need for rigorous and versatile CAD software. The primary driver for this requirement is the fact that such antennas are very expensive, so it is critical that the antenna performance be analyzed and optimized in a complete and thorough manner. Experimental trials using small arrays and active element patterns are an important part of the process, but accurate CAD models can provide much more information about important effects such as scan blindness, impedance mismatch, losses, random errors, sidelobe levels, and cross-polarization as a function of any design parameter. Unlike the case for fixed-beam arrays, scanning arrays usually require the consideration of mutual coupling effects, and the effect of the feed network, for a complete analysis of the array. This is a difficult problem in general, and producing CAD software for microstrip arrays is further complicated by the wide variety of element geometries and feeding methods that are characteristic of microstrip antenna technology.

As discussed in Chapter 7 of this book, the infinite-array approximation can usually be used for arrays having a hundred or more elements. This was done for the large microstrip phased array discussed in [10].

III.5 The Roles of CAD Software, Experience, and Fundamental Understanding

One of the premises in this paper has been that while CAD software can be an invaluable design tool, it is not a substitute for antenna design experience or a thorough understanding of the principles of operation of microstrip antennas and arrays. While microstrip antenna design is based on solid science, it also retains a strong component of intuitive understanding and a creative problem-solving approach that can only come from experience. As discussed above, the design methodology used for a particular microstrip antenna is highly dependent on the inherent complexity of the element, as well as the stringency of the design specifications. While it is probably true that the beginner will have little problem with a simple microstrip element or array, the number and range of design variables involved with more complex elements or larger arrays quickly become formidable enough to make a successful conclusion unlikely unless the designer is able to use CAD software tools in conjunction with his or her own experience and understanding of the problem.

IV. Conclusions

We have highlighted many of the important features related to microstrip antenna CAD, and have endorsed the premises stated in the Introduction. The variety and complexity of microstrip radiating structures determine the need for accurate and versatile CAD software. In the simplest cases CAD tools may not be essential for success, while the design and manufacture of microstrip arrays (particularly those with beam scanning) critically relies on CAD software tools. We have emphasized that present CAD software falls short of what is adequate for the confident, low-cost manufacture of most microstrip arrays. What then are prospects for the future?

There is unabated pressure to exploit the low-profile, integrated circuit-compatible, and rugged geometry of microstrip radiators and the perceived low manufacturing costs for large printed arrays, but the cost benefits are usually found to be concomitant with the availability of improved CAD software. A future implementation of the so-called Maxwell solver [1], giving precise parameterization of the complex electromagnetic configurations without reliance on independent validations for each problem of interest, is indeed a daunting but welcome prospect, but the economic viability of such a software package for a diverse and relatively small market remains an open question. It is concluded that, at least for the near future, CAD software will continue to aid, rather than actually replace, the experienced designer.

References

[1] J. R. James, "Printed antennas—new research frontiers," *Asia-Pacific Microwave Conf. Proceedings,* pp. 21–26, Mar. 1992.

[2] D. M. Pozar, "Microstrip antennas," *IEEE Proceedings,* vol. 80, pp. 79–91, Jan. 1992.

[3] N. K. Das and D.M. Pozar, "Multiport scattering analysis of general multilayered printed antennas fed by multiple feed ports: Part I—Theory, Part II—Applications," *IEEE Trans. Antennas and Prop.,* vol. 40, pp. 469–491, May 1992.

[4] G. Splitt and M. Davidovitz, "Guidelines for design of electromagnetically coupled microstrip patch antennas on two-layer substrates," *IEEE Trans. Antennas and Prop.,* vol. AP-38, pp.1136–1140, July 1990.

[5] D. M. Pozar and S. M. Voda, "A rigorous analysis of a microstripline fed patch antenna," *IEEE Trans. Antennas and Prop.,* vol. AP-35, pp. 1343–1350, Dec. 1987.

[6] M. Himdi, J. P. Daniel, and C. Terret, "Analysis of aperture coupled microstrip antenna using cavity method," *Electronics Letters,* vol. 25, pp. 391–391, Mar. 1989.

[7] P. L. Sullivan and D. H. Schaubert, "Analysis of an aperture coupled patch antenna," *IEEE Trans. Antennas and Prop.,* vol. AP-34, pp. 977–984, Aug. 1986.

[8] P. Hall, "Review of techniques for dual and circularly polarized microstrip antennas," Chapter 3, No. 1, this volume.

[9] P. C. Sharma and K. C. Gupta, "Analysis and optimized design of single feed circularly polarized microstrip antennas," *IEEE Trans. Antennas and Prop.,* vol. AP-31, pp. 949–955, Nov. 1983.

[10] J. J. Schuss, J. D. Hanfling, and R. L. Bauer, "Design of wideband patch radiator phased arrays," *IEEE Antennas and Propagation Symp. Digest,* pp. 1220–1223, 1989.

Chapter 2
Basic Microstrip Antenna Elements and Feeding Techniques

THE basic microstrip antenna element is comprised of a metal patch supported above a larger ground plane. The patch is usually printed on a microwave substrate material with relative permittivity in the range 2 to 10, but a variety of materials may be used, depending on the application. Air or low-density foam usually offer the lowest loss and highest radiation efficiency, but higher permittivity substrates result in smaller elements with broader radiation patterns. Microstrip antennas have been successfully operated on GaAs ($\epsilon_r = 12.8$) as well as $LaAlO_3$ and other substrates commonly used for high Tc superconductors ($\epsilon_r = 20-25$). Although rectangular and circular patches are most common, any shape that possesses a reasonably well defined resonant mode can be used, including ellipses, annular rings, and triangles [1], [2], [3]. The shape of the patch can be used to obtain special effects, such as circular polarization [1], [4], [5].

Power can be coupled into or out of the antenna by a variety of methods that can be broadly classified into contacting and noncontacting. Contacting feeds involve the direct connection of a transmission line, typically coax or microstripline, to the patch antenna. The input impedance depends on the location of the connection within the patch boundaries, which provides a commonly used means of impedance matching. Noncontacting feeds use electromagnetic field coupling to transfer power between the feedline and the radiating patch. Noncontacting feeds typically have more degrees of freedom than contacting feeds, which makes them harder to design but provides greater flexibility in mechanical form and electrical performance.

This chapter begins with a review article by Schaubert that covers some of the topics that could not be included in their original form. It includes (1) input impedance, radiation characteristics (especially cross-polarization), and surface wave excitation; (2) feed techniques, with emphasis on aperture coupling; (3) frequency tuning and multifrequency operation; and (4) operation in higher order modes. The article by Munson (1974) represents the beginning of the explosion of published works on microstrip antennas. It illustrates the use of the transmission line model for first-order design of rectangular antennas, and contains examples of very wide antennas that are fed at several points to insure excitation of the desired mode. This technique was used successfully in lieu of an array of patches to obtain omnidirectional coverage from wrap-around antennas on a variety of missiles.

The simple transmission line model, along with Hammerstad's [6] length extension for an open-circuited microstripline, can be used to give the approximate operating frequency of a rectangular patch antenna as $c/[2\,(L + t)\,\epsilon_r^{1/2}]$, where c is the velocity of light, ϵ_r is the relative permittivity of the substrate, L is the length of the patch, and t is the substrate thickness. This value is reasonably close for electrically thin substrates, $t/\lambda o < 0.02$. The paper by Chang, Long, and Richards discusses the behavior of antennas on thicker substrates, and compares expressions based on Hammerstad's work and from James, Hall, and Wood [7] to experimental results. They also present measured results of impedance bandwidth that are useful for estimating the performance of simple patch antennas.

The next two papers relate to circular patch antennas, with particular emphasis on radiation patterns and cross polarization. The paper by Kishk and Shafai describes the level of excitation of the dominant TM_{11} mode and some of the higher order modes of the circular structure. The subsequent effects on principal plane radiation patterns are shown for various substrate thicknesses and permittivities, and for various sizes of the ground plane. The paper by Lee, Luk, and Tam emphasizes the cross-polarization effects occurring in circular antennas, showing the peak cross-polarization levels for probe-fed antennas on various substrates. They also show that cross-polarization can, in some cases, increase from -30 dB in the principal planes to -11 dB in the diagonal plane.

The paper by Splitt and Davidovitz describes proximity coupling from a microstrip feed structure on one layer of the antenna to the radiating patch on a second layer. The authors present design curves for square and circular patches fed by a microstripline on a lower level than the radiating patch.

Patch antennas often are protected from the environment by some form of radome. The thin, conformal nature of the antenna can be preserved if the radome is a thin layer of dielectric laid directly over the patch element. This works quite well, except that the resonant frequency of the antenna must be adjusted to account for the loading effects of the cover layer. The paper by Bahl, Bhartia, and Stuchly presents data on changes in resonant frequency caused by a cover layer. Some additional information on cover layer effects is discussed in [8].

The final paper in this chapter presents some results on the effects of ground plane size. Diffraction from the edges of the ground has a profound effect on typical patch antenna patterns near the horizon and below it. In the main beam region, the ideal patch pattern is modulated by diffracted fields. The cross-polarization is also affected by diffraction. Huang's paper shows that the fields in the E-plane are predicted quite well by considering the geometrical optics field plus single- and double-edge diffractions. In the H-plane, the fields consist of the geometrical optics fields, the slope diffraction from the H-plane edges of the ground plane, and the E-plane edge equivalent currents.

The volume of printed material on microstrip patch elements is staggering, but some of the following additional references may be of particular interest. Schaubert, Adrian, and Pozar present measured data illustrating the effects of microstrip and probe feeds, and of substrate thickness and permittivity [9]. Annular ring microstrip antennas are described in two articles, one related to single layer antennas [10] and one related to stacked, dual-frequency antennas [11]. Two antennas that could be used as primary feeds for reflectors are described in [12] and [13].

References

[1] L. C. Shen, "The elliptical microstrip antenna with circular polarization," *IEEE Trans. Antennas and Prop.,* vol. AP-29, pp. 90–94, Jan. 1981.

[2] W. C. Chew, "Broadband annular ring microstrip antenna," *IEEE Trans. Antennas and Prop.,* vol. AP-30, pp. 918–922, Sept. 1982.

[3] M. Cuhaci and D. S. James, "Radiation from triangular and circular resonators in microstrip," *IEEE Int'l Microwave Symp. Digest,* pp. 438–441, June 1977.

[4] H. D. Weinschel, "Cylindrical array of circularly polarized microstrip antennas," *IEEE Int'l Antennas and Propagation Symp. Digest,* pp. 177–180, 1975.

[5] G. G. Sanford and R. E. Munson, "Conformal VHF antenna for the Apollo-Soyuz test project," *IEEE Int'l Antennas and Propagation Symp. Digest,* pp. 113–116, 1976.

[6] E. O. Hammerstad, "Equations for microstrip circuit design," *Proc. 5th European Microwave Conf.,* pp. 268–272, Sept. 1975.

[7] J. R. James, P. S. Hall, and C. Wood, *Microstrip Antenna Theory and Design,* Peter Peregrinus, Stevenage, UK, 1980.

[8] N. G. Alexopoulos and D. R. Jackson, "Fundamental superstrate (cover) effects on printed circuit antennas," *IEEE Trans. Antennas and Prop.,* vol. AP-32, pp. 807–816, Aug. 1984.

[9] D. H. Schaubert, D. M. Pozar, and A. Adrian, "Effect of microstrip antenna substrate thickness and permittivity: comparison of theories with experiment," *IEEE Trans. Antennas and Prop.,* vol. AP-37, pp. 677–682, June 1989.

[10] S. E. El-Khamy, R. M. El-Awadi, and E-B. A. El-Sharrawy, "Simple analysis of annular ring microstrip antennas," *IEE Proc.,* part H, vol. 133, pp. 198–202, June 1986.

[11] J. S. Dahele, K. F. Lee, and D. P. Wong, "Dual-frequency stacked annular-ring microstrip antenna," *IEEE Trans. Antennas and Prop.,* vol. AP-35, pp. 1281–1285, Nov. 1987.

[12] C. J. Prior and P. S. Hall, "Microstrip disc antenna with short-circuited annular ring," *Electronics Letters,* vol. 21, pp. 719–721, Aug. 15, 1985.

[13] A. A. Kishk and L. Shafai, "Optimization of microstrip feed geometry for prime focus reflector antennas," *IEEE Trans. Antennas and Prop.,* vol. AP-37, pp. 445–451, Apr. 1989.

A Review of Some Microstrip Antenna Characteristics

DANIEL H. SCHAUBERT
Electrical and Computer Engineering
University of Massachusetts
Amherst, Massachusetts 01003

Abstract—The basic microstrip antenna is a resonant patch of metal on the surface of a grounded dielectric slab. It radiates power in a beam broadside to the plane of the antenna and displays an input impedance similar to a parallel resonant circuit near its operating frequency. Impedance bandwidths of 1 to 3% are typical for antennas fabricated on $\epsilon_r = 2.5$ substrates that are 0.01 to 0.02 λ_o thick. Patch antennas are inherently linearly polarized, although they usually can be made to radiate circular or elliptical polarization by exciting two orthogonal linear modes in phase quadrature. The level of cross-polarization is typically −20 dB or better, but the use of thick substrates to increase bandwidth increases cross-polarization. Monolithic construction of patches and feed lines on the top surface of a substrate is a strong motivation for the use of microstrip antennas, but several other methods of feeding are described. The paper concludes with a brief discussion of frequency tuning and operation in higher order modes to achieve special radiation properties.

1. Introduction

This paper reviews some characteristics of microstrip antennas. It is intended to supplement the other papers reprinted in this book by presenting material that could not be included in its original form because of space limitations, and so it is not a comprehensive treatment of these antennas or of the specific topics. Readers may wish to refer to other papers in this book, and to referenced works not included in this book, for more information on these topics. Microstrip patch antennas can take a variety of forms, but the basic element consists of a single patch of conductor on the upper surface of a grounded dielectric slab (substrate). The patch radiates efficiently when it is "resonant," which generally means that some characteristic dimension of the patch is nearly equal to one-half wavelength in the substrate medium. The shape of the patch can be rather arbitrary, but rectangular and circular patches have several desirable characteristics and are most often used in practice. Most of the characteristics discussed here apply qualitatively, if not quantitatively, to both rectangular and circular microstrip antennas. For example, the impedance bandwidth of rectangular and circular microstrip antennas increases with increasing substrate thickness. Some of the characteristics described here are easily interpreted in terms of the simple transmission-line and cavity models.

Four topics are included in this review: (1) basic characteristics, (2) feed techniques, (3) frequency tuning and multifrequency operation, and (4) operation in higher order modes.

2. Basic Characteristics

All antennas have at least two basic characteristics important for any application; radiation pattern(co- and cross-polarized, axial ratio, gain, beamwidth, sidelobe level, etc.) and input impedance(resonant resistance, bandwidth, etc.). These two characteristics will be treated in turn.

Radiation Characteristics

The co-polarized radiation patterns of a typical patch antenna (rectangular or circular) are depicted in Figure 1. The E-plane pattern is broad and smooth, having a beamwidth of the order of 100 degrees. The H-plane pattern is similar, except it goes to zero at the horizon because of the conducting ground plane. The beamwidth of the E-plane pattern and its value at angles far from broadside can be controlled by changing the length of the patch antenna, but this change necessitates a change in the substrate permittivity to maintain the same resonant frequency. Nevertheless, this procedure can be useful because a patch fabricated with air dielectric, or a low-density foam, is approximately one-half wavelength long and its E-plane radiation pattern goes to zero at the horizon. This feature can be useful for applications that seek reception of signals from the zenith and rejection of signals at the horizon. Conversely, when using a high permittivity substrate, the patch is much shorter than one-half wavelength and the E-plane radiation pattern is broadened. It will be noted below that the use of higher permittivity substrates generally leads to reduced bandwidth for the antenna.

Purists of electromagnetic theory will note that both the E-plane and the H-plane space wave radiation patterns of Figure 1 will go to zero at the horizon if the grounded substrate is of infinite extent. This result follows from the electromagnetic boundary condition at the air-dielectric interface, which precludes power flow parallel to the interface between two dissimilar media except in guided modes. In practice, however, antenna substrates are not infinite and the radiation pattern obtained from the two-slot model is close to what one observes.

Microstrip patch antennas can be designed to produce reasonably pure linear polarization. However, several mechanisms exist to create cross-polarization, one of which is higher order modes of the antenna cavity (see [1]-[4] for information on the cavity model). Higher order modes usually radiate less power than the dominant mode, but they also usually radiate in a

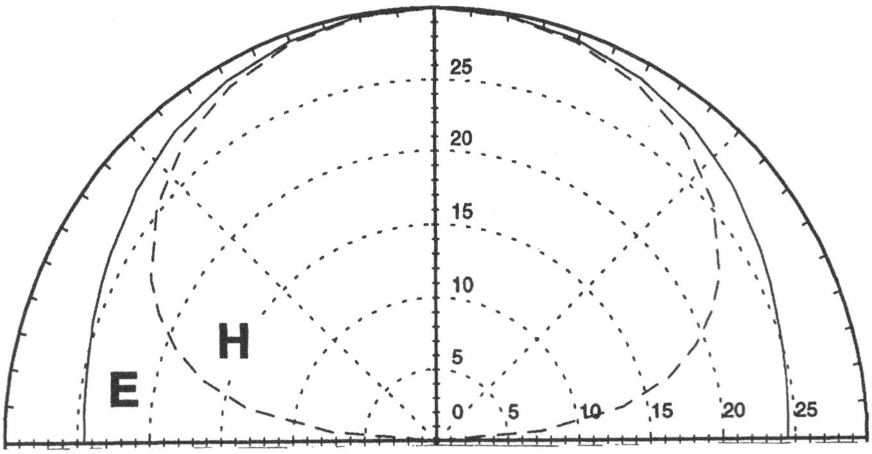

Fig. 1. Typical principal-plane radiation patterns of microstrip patch antenna.

different pattern and polarization. Generally, the lower the Q of the antenna (i.e., the wider the bandwidth), the more likely it is to radiate power from higher order modes and thus it is more prone to high levels of cross-polarization. A second source of cross-polarization is present in patch antennas operating only in the dominant mode. This feature can be most easily observed by considering the fringing fields of a rectangular, which give rise to the slot model for patch antenna radiation. The fringing fields around the four sides of a patch antenna are depicted in Figure 2. The fields at the top and bottom represent the preferred sources of radiation for the antenna, which is vertically polarized. The fields along the side walls radiate horizontally polarized power and are undesirable, but they cannot be avoided. In the H-plane, the symmetry of the side-wall fields results in cancellation and, hence, purely vertical polarization can be obtained in the ideal case. However, in the intercardinal planes, even the ideal, single-mode patch will radiate some cross-polarized power. It should be apparent that a one-quarter wavelength patch, which would be obtained from the antenna in Figure 2 by removing the upper half of the patch and shorting the remaining patch to the ground plane along the dashed line, will have moderately high levels of cross-polarization in the H-plane as well as the intercardinal planes.

Input Impedance

The input impedance of patch antennas can be estimated by using either transmission-line [5] or cavity models[3], [4]. The antennas generally resemble a parallel resonant circuit with the resonant resistance controlled primarily by the feed location, width of the radiating element, and substrate material and thickness. Figure 1 of [6] shows that the bandwidth of patch antennas increases as substrate thickness increases and decreases as substrate permittivity increases. Further aspects of patch antenna input impedance are discussed below in relation to methods of feeding.

Surface Wave Excitation

The grounded dielectric slab on which a microstrip patch is etched can support a TM_0 surface wave that has zero cutoff frequency. Therefore, at any frequency of operation, a single microstrip can launch power into the space wave and into the surface wave. The surface wave power decays at a rate of $1/\rho$, slower than the decay of the space wave, so the surface wave can have a significant impact on mutual coupling between antennas and on diffraction from the edges of the substrate. For most substrates that have been used to date, the velocity of

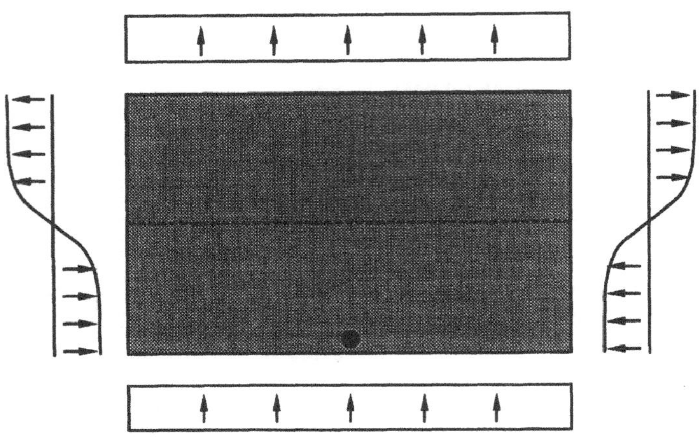

Fig. 2. Fringing fields associated with rectangular patch antenna.

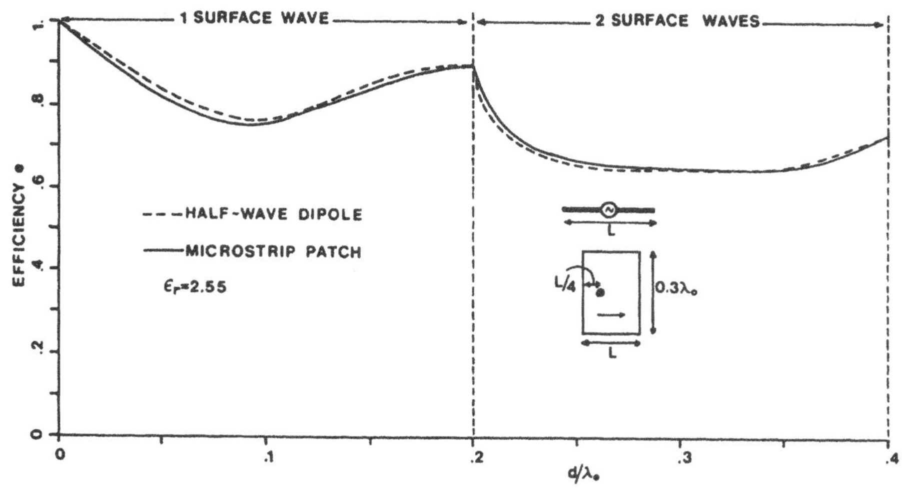

Fig. 3. Antenna efficiency based on power coupled to surface wave for printed dipole and patch antennas versus substrate thickness, d. $\epsilon_r = 2.55$. $W = 0.3\lambda_o$ for the patch. (Reprinted from [7].)

propagation of the surface wave is close to that of free space, but substrates that are very thick or that have a very high permittivity may support surface waves with a much slower velocity of propagation.

Pozar [7] has presented several results related to surface wave excitation by microstrip patch antennas. Two of the most important results from the point of view of the antenna designer who wishes to avoid detrimental performance effects are (i) reduced radiation efficiency caused by power "lost" to the surface wave and (ii) mutual coupling between antennas. Figure 3 shows the radiation efficiency, defined as the ratio of the power in the space wave to the sum of the powers in the space and surface waves, for patch and dipole antennas on a typical substrate. For substrate thicknesses less than $0.2\lambda_o$, only the TM_o surface wave propagates. The radiation efficiency (due to surface wave excitation) is only about 75% for a substrate thickness of $0.1\lambda_o$. In most practical applications, the power coupled into the surface wave reemerges as a space wave after diffraction from a substrate boundary or scattering from another obstacle, such as another antenna, a connector, or a mounting bracket. In [7], mutual coupling in the E-plane, where the surface wave is most strongly excited, is also shown to increase for substrate thicknesses corresponding to strong excitation of the surface wave.

3. Feed Techniques

One of the initial and continuing motivations for using microstrip patches is the ability to construct array antennas with the feed network and the radiating elements on one surface (monolithic). This arrangement implies that the antennas are fed by a microstripline connected directly to the patch. There are, however, many other ways to feed patch antennas. Four methods are summarized in Table 1.

When fed by a microstripline or a probe, the input impedance of the patch antenna exhibits some dependence on the substrate thickness and permittivity, but it is strongly dependent on the location of the connection between the feedline and the patch. A simple first-order theory that gives a reasonable approximation to the resonant resistance of a rectangular patch fed at an arbitrary point predicts the input resistance to be $R_o\cos^2\theta_f$, where R_o is the input resistance when the patch is fed at a radiating edge and θ_f is the electrical distance of the feed point from the radiating edge. This effect is very useful in matching antennas fed by coaxial probes or microstriplines, which can be inset from the radiating edge or attached along a nonradiating edge (parts a–c of Figure 4). The E-plane asymmetry of the structures a, b, and c in Figure 4 can lead to increased levels of cross-polarized radiation, especially from antennas constructed on thick substrates to achieve wider bandwidth. Chiba, et al. [8], and Hanfling and Schuss [9] have suggested using two symmetrically placed probes as indicated in Figure 4d. They find that this configuration improves the cross-polarization performance of the antenna, but they have also observed that undesirable impedance anomalies can occur in scanning arrays if the circuit that provides the 0–180 degree excitation of the two probes does not have high isolation.

Proximity coupling [10],[11] offers some opportunity to reduce feedline radiation while maintaining a relatively thick substrate for the radiating patch. The input impedance of the antenna is affected by the overlap of the patch and the feedline, and by the substrates. This feature adds degrees of freedom in the design, but may complicate the task of selecting an optimum design.

Aperture coupling [12] is becoming increasing popular as a means of producing patch arrays with enhanced performance. Because the feedlines are behind the ground plane, no spurious radiation escapes to corrupt the sidelobes or polarization of the antenna. The coupling aperture is usually centered under the patch and low levels of cross-polarization can be achieved. As in the proximity-coupled patch, additional degrees of freedom are available to the designer. Sullivan and Schaubert [13] have

TABLE 1. MICROSTRIP ANTENNA FEED TECHNIQUES.

Technique	Advantages	Disadvantages
Microstripline • Radiating Edge 	Monolithic. Good Polarization.	Spurious radiation. Must be inset or use transformer to match impedance.
• Nonradiating Edge	Impedance matching is easier.	Excites cross-pol.
Coaxial Probe  or 	Impedance matching by probe location. Probe location can selectively excite additional modes. Can be used with plated vias for multilayer circuits.	Impedance is highly inductive when thick substrates are used.
Proximity Coupling • Monolithic 	No dc contact between feed and radiating patch.	Direct radiation from coupling region. Dimensional tolerance.
• Multilayer 	Can have large effective thickness for patch substrate and much thinner feed substrate. Several degrees of freedom available for matching/tuning.	Multilayer fabrication is required. Difficult to optimize.
Aperture Coupling 	Independent choice of substrates for feed and radiators. No spurious radiation from feed. No via connectors.	Multilayer fabrication required.

demonstrated the effects of the various design parameters. They have developed a full-wave, method of moments analysis for these antennas and have conducted experimental studies. Their explanations of the basic characteristics follow.

The geometry of the antenna is shown in Figure 5 and the impedance loci are plotted in the figures that follow as a function of frequency in Smith chart form. Some of the figures contain both measured and calculated values and they contain loci for several values of a particular antenna parameter. The numbers identifying data points are frequencies in MHz. The effect of the feedline stub's length is shown in Figure 6, along with typical comparisons between calculated and measured results. If the input impedance at a single frequency (e.g., 2225 MHz) is plotted for various stub lengths, the locus approximately follows a constant resistance contour, implying that the aperture and antenna appear as a series load along an open circuited transmission line. An equivalent circuit of this type has been found to represent the antenna quite well near resonance.

The long dimension of the aperture was varied to obtain the curves given in Figure 7. The antenna dimensions are given in the figure legend and are very similar to the dimensions of the antenna of Figure 6. As the aperture length is reduced the radius of the impedance circle decreases and the center of the circle moves toward the short circuit location. This might be thought of as decreasing the coupling factor between the feedline and the patch antenna. The resonant frequency (where Z_{in} is real) of the antenna is determined primarily by the patch length, but it is affected slightly by the aperture length. The resonant frequency versus aperture length is plotted in Figure 8. The resonant frequency, which in this case is also the minimum voltage standing-wave ratio (VSWR) frequency, decreases with increasing slot length. Also plotted in Figure 8 is the input impedance at resonance versus slot length, which can be used to approximately determine the slot length required to achieve a perfect match and the corresponding resonance frequency. In this case, a perfect match would be obtained for an aperture length of 1.09 cm at a resonant frequency of 2.233 GHz. For comparison the resonant frequency of this antenna based on the cavity model is 2.306 GHz [1].

The results in Figures 6 and 7 illustrate how the antenna can be designed to have a specified input impedance. The aperture length can be adjusted to obtain the desired resistive part of the impedance and the open-circuited stub length can be adjusted to obtain the desired reactance.

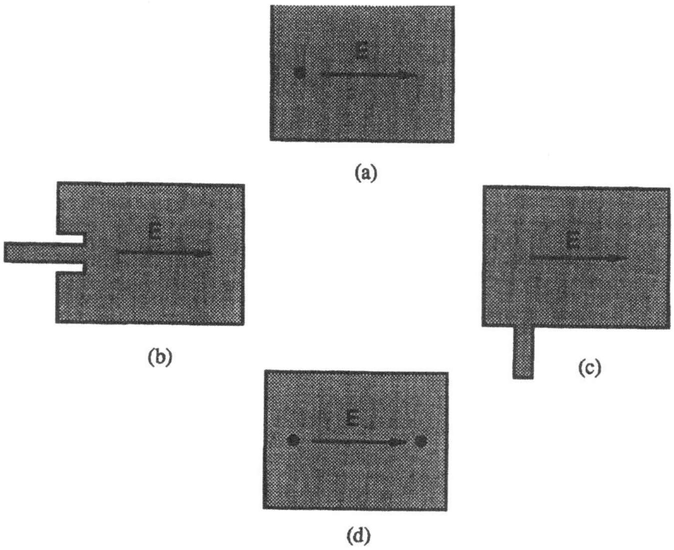

Fig. 4. Feed point position used to control input impedance. (a) Probe feed inset from radiating edge. (b) Microstripline feed inset from radiating edge. (c) Microstripline attached at nonradiating edge. (d) Balanced feed to reduce cross polarization.

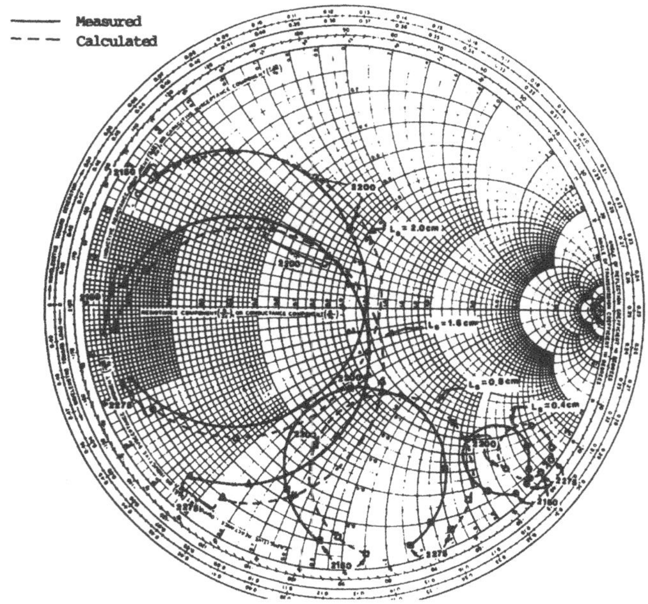

Fig. 6. Measured versus calculated input impedance as a function of stub length. $\epsilon_r^b = 2.54$, $d_b = 0.16$ cm, $L_p = 4.0$ cm, $W_p = 3.0$ cm, $x_{0s} = 0.0$ cm, $y_{0s} = 0.0$ cm, $L_{ap} = 1.12$ cm, $W_{ap} = 0.155$ cm, $\epsilon_r^a = 2.54$, $d_a = 0.16$ cm, $W_f = 0.442$ cm.

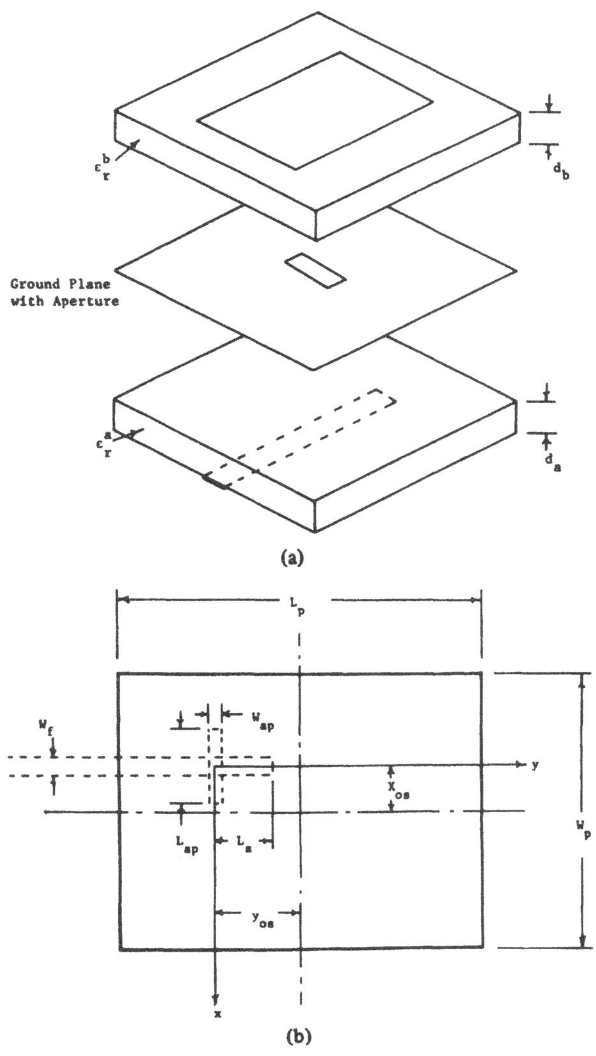

Fig. 5. Aperture-coupled patch antenna.

The input impedance is relatively insensitive to small variations in patch position over the aperture, but changes significantly for larger patch offsets. Measured and calculated plots are given in Figure 9 corresponding to movement of the patch in the y-direction, that is, along the resonant dimension (see Figure 5). The zero offset case is shown in Figure 6. The coupling factor, as defined by the radius of the impedance circle, is greatest when the patch is centered over the aperture and decreases significantly as the patch is moved in the y-direction. This is in accordance with Pozar's [12] simple model for this antenna based on Bethe hole theory and the cavity model. In addition, as the patch is offset in the y-direction the centers of the resonant loops move approximately in a straight line toward the edge of the Smith chart just to the inductive side of the short position, probably because, when the patch is offset by a large amount, the structure looks like a stub with a slightly capacitive input impedance in series with a small aperture in a ground plane, which is inductive.

In contrast to movement of the patch in the y-direction, lateral movement of the patch in the x-direction causes little change in the coupling factor, provided the entire slot remains under the patch. From the measured data in Figure 10 it can be seen that the coupling factor actually increases as the edge of the patch aligns with the edge of the slot and then monotonically decreases as the slot emerges from under the patch. The calculated impedance does not show an increase as the aperture moves to the edge of the patch. This disagreement is not surprising since the model utilizes only one mode in the aperture. A single aperture mode makes the analysis numerically more tractable but cannot account for skewing of the aperture electric field distribution as the patch is offset in a direction parallel to the long dimension of the slot. In addition, the patch current is assumed uniform in the x-direction, which may not be adequate for large

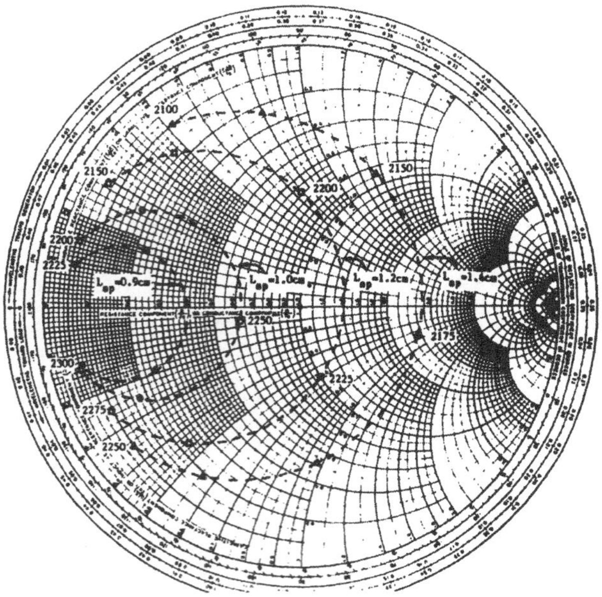

Fig. 7. Calculated input impedance as a function of aperture length. Other antenna parameters are: $\epsilon_r^b = 2.54$, $d_b = 0.16$ cm, $L_p = 4.0$ cm, $W_p = 3.0$ cm, $x_{0s} = 0.0$ cm, $y_{0s} = 0.0$ cm, $W_{ap} = 0.11$ cm, $\epsilon_r^a = 2.54$, $d_a = .16$ cm, $W_f = 0.495$ cm, $L_s = 2.0$ cm.

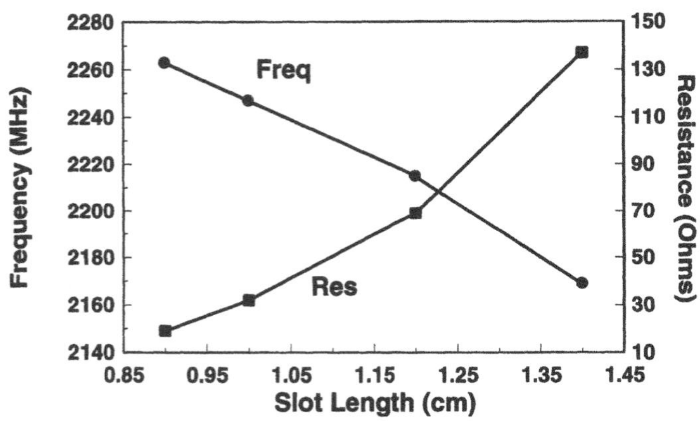

Fig. 8. Resonant frequency and input resistance at resonance versus slot length (data from Figure 7).

offsets in that direction. The calculated curve for case 3 lies midway between the measured curves for 3 and 4, and the calculated curve for case 4 lies midway between measured curves 4 and 5.

It is also of interest to examine the influence of feed substrate dielectric constant and thickness on the input impedance. As dielectric constant and thickness are varied in these studies the feedline width and stub length are modified to maintain a characteristic impedance of 50 Ω and a stub length of 0.22 λ_f. All other antenna parameters were held constant and are given in the figure legends.

The variation with dielectric constant is shown in Figure 11. The key features are the increase in the coupling factor and the invariance of the resonance frequency with increasing dielectric constant. The increase in the coupling factor is probably due to the slot appearing electrically longer as the dielectric constant of the feed increases.

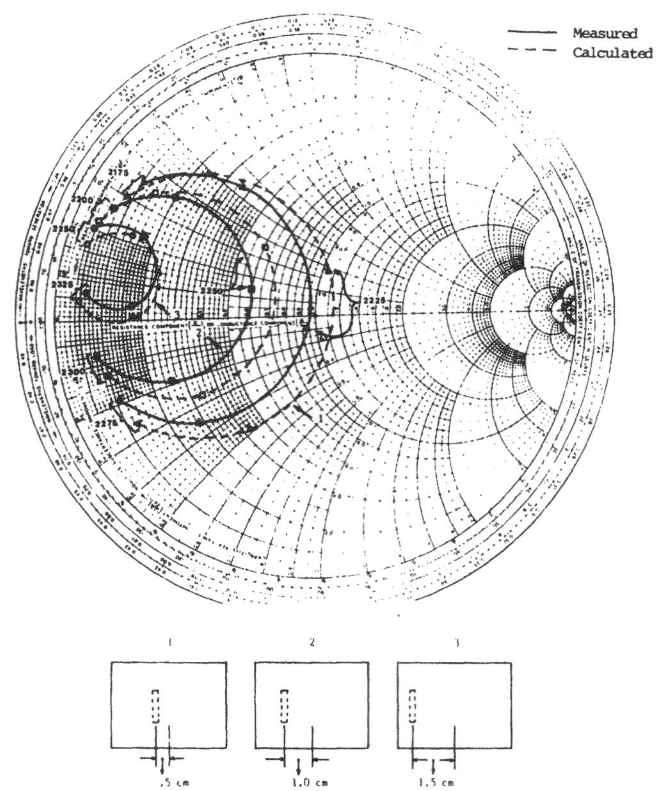

Fig. 9. Measured and calculated input impedance as a function of patch offset in the direction of resonance. $L_s = 2.0$ cm and other antenna parameters are the same as Figure 5.

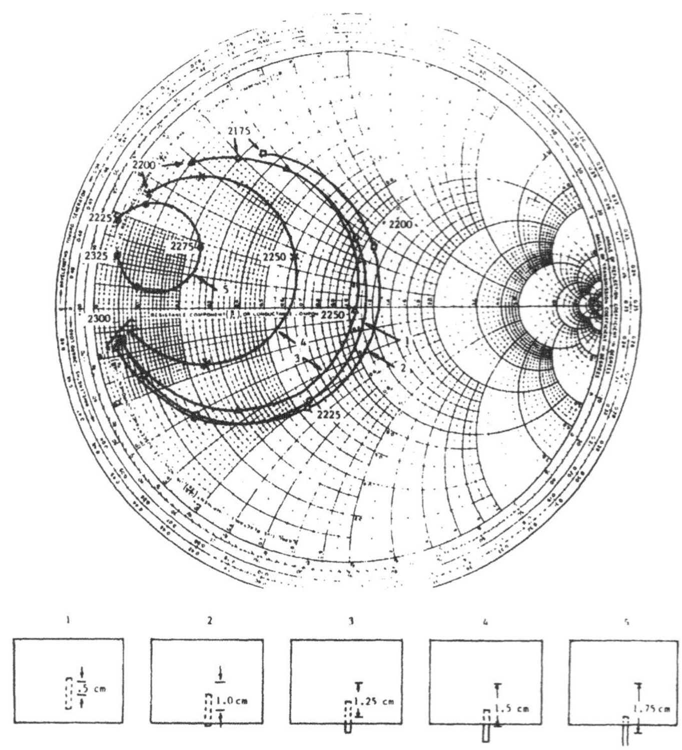

Fig. 10. Measured input impedance as a function of patch offset in the direction orthogonal to resonance. $L_s = 2.0$ cm and other antenna parameters are the same as Figure 5.

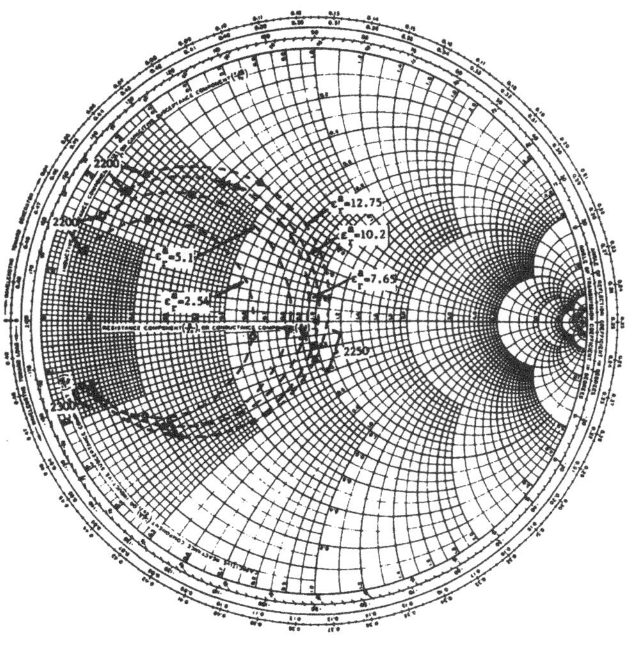

ε_r^a	W_f	L_s
2.54	.495 cm	2.000 cm
5.10	.310 cm	1.493 cm
7.65	.225 cm	1.255 cm
10.20	.173 cm	1.108 cm
12.75	.139 cm	1.004 cm

Fig. 11. Calculated input impedance for various feed substrate dielectric constants. Tabular data give feedline width and stub length used to maintain 50 Ω characteristic impedance and stub length of 0.22 λ_f for each value of ϵ_r^a. Other antenna parameters are: $\epsilon_r^b = 2.54$, $d_b = 0.16$ cm, $L_p = 4.0$ cm, $W_p = 3.0$ cm, $x_{0s} = 0.0$ cm, $y_{0s} = 0.0$ cm, $L_{ap} = 1.0$ cm, $W_{ap} = 0.11$ cm, $d_a = 0.16$ cm.

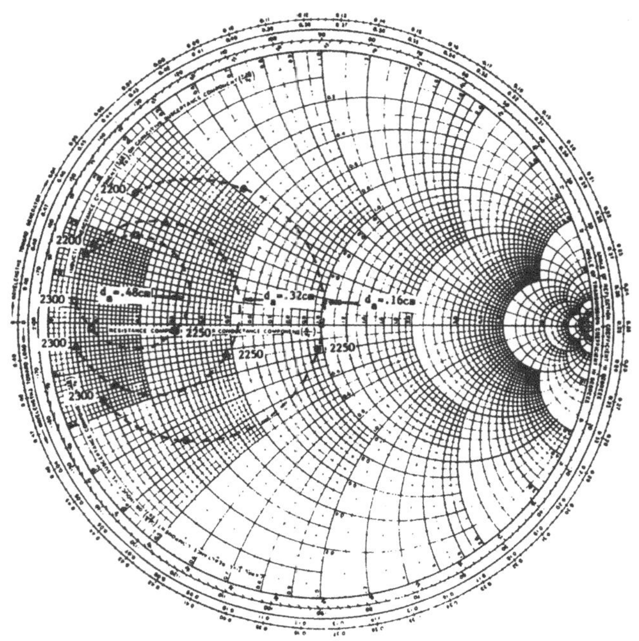

d_a	W_f	L_s
.16 cm	.173 cm	1.108 cm
.32 cm	.375 cm	1.083 cm
.48 cm	.613 cm	1.056 cm

Fig. 12. Calculated input impedance as a function of feed substrate thickness. Tabular data give feedline width and stub length used to maintain 50 Ω characteristic impedance and stub length of 0.22 λ_f for each value of d_a. Other antenna parameters are: $\epsilon_r^b = 2.54$, $d_b = 0.16$ cm, $L_p = 4.0$ cm, $W_p = 3.0$ cm, $x_{0s} = 0.0$ cm, $y_{0s} = 0.0$ cm, $L_{ap} = 1.0$ cm, $W_{ap} = 0.11$ cm, $\epsilon_r^a = 10.2$.

The last set of impedance data to be presented here involves substrate thickness. The thickness of the feed substrate of the antenna of Figure 11 in the case of $\epsilon_r^a = 10.2$ was increased. As the distance between the feedline and aperture increases, the coupling factor decreases as can be seen in Figure 12. As with the dielectric constant variations, the resonant frequency is unchanged with changes in substrate thickness over the range studied. Other computations involving increased thickness of the antenna substrate showed effects similar to increasing the feed substrate thickness. These effects are summarized in Figure 13, which shows the relations between the resonant resistance, the antenna substrate thickness, and the aperture length. The impedance loci resemble those in Figure 12. No significant inductive shift was noted as the substrate thicknesses were increased to 0.48 cm. Increasing the aperture length can increase coupling to help compensate for thicker substrates, but larger apertures can radiate more power on the feedline side of the ground plane, which is an undesirable effect. Antennas on substrates that are 0.01–0.03 λ_o have yielded front-to-back ratios on the order of 20 dB.

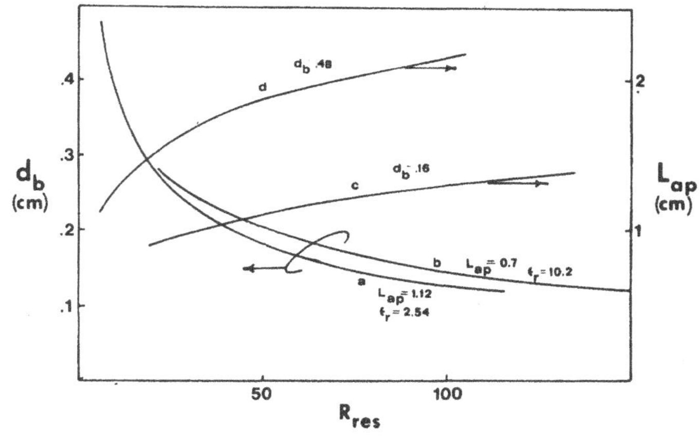

Fig. 13. Relationship of resonant resistance, antenna substrate thickness, and aperture length. Aperture offsets are zero and feedlines are 50 Ω, a: $\epsilon_r^a = \epsilon_r^b = 2.54$, $d_a = 0.16$ cm, $L_p = 4$ cm, $W_p = 3$ cm, $W_{ap} = 0.1545$ cm, $L_s = 2$ cm, b: $\epsilon_r^a = \epsilon_r^b = 10.2$, $d_a = 0.16$ cm, $L_p = 2$ cm, $W_p = 1.5$ cm, $W_{ap} = .1$ cm, c, d: $\epsilon_r^a = \epsilon_r^b = 2.54$, $d_a = 0.16$ cm, $L_p = 4$ cm, $W_p = 3$ cm, $L_s = 2$ cm.

Aperture coupling to a patch antenna can also be implemented with stripline feed circuits. Coupling through the aperture from the stripline to the patch radiator is not as strong as for a microstripline feed because only one of the ground planes (one-half of the total current) is interrupted by the coupling aperture. Also, care must be taken to suppress the parallel plate mode that can be excited in the stripline structure. Nevertheless, this technique can be very useful for arrays that require multilayer feed circuits.

4. Frequency Tuning and Multifrequency Operation

The frequency of operation of a microstrip antenna is controlled primarily by its size and the permittivity of the substrate. However, it is possible to tune the operating frequency over a modest range by means of reactive loading and this can be useful. Schaubert, et al. [14] found that shorting posts located at various positions within the patch cavity can raise the operating frequency in a predictable manner. The input impedance and radiation pattern remain well behaved over a 30–40% tuning range. Fixed or variable capacitors also can be used to alter the resonant frequency of a patch antenna. In particular, adding capacitance lowers the operating frequency of an antenna. Kerr [15],[16] noted that frequency tuning could be accomplished by varying the length of a printed or coaxial transmission line stub attached to the antenna. He also noted that the operating frequency of a patch decreases when some of the metal that comprises the patch and/or the ground plane is removed. Figure 14 depicts four methods of frequency tuning.

Patch antennas can sometimes be operated at more than one frequency. An obvious example is operation at the dominant-mode frequency and a frequency that corresponds to a higher order mode of the cavity, as discussed in the next section. Another example is a rectangular patch that is resonant at different frequencies in the horizontal and vertical directions. This will, of course, radiate different polarizations at the two frequencies. If patches of different sizes are stacked and properly fed, radiation in the same polarization can be obtained at two or three frequencies [17], [18]. The scheme is shown in Figure 15. A single probe passes through a small opening in the lower patch and attaches to the upper patch. The upper patch is usually smaller than the lower patch and operates nearly as if the lower patch were extended to form a large ground plane. The lower patch operates similar to a simple patch with a dielectric cover layer. As long as the two resonant frequencies are separated by a few percent (i.e., somewhat more than the bandwidth), the antennas function almost independently and provide dual-frequency operation from the same aperture and feed port. Montgomery demonstrates an antenna comprised of three stacked patches that operate at three discrete frequencies and can also be operated in two orthogonal polarizations by using two feed probes[18].

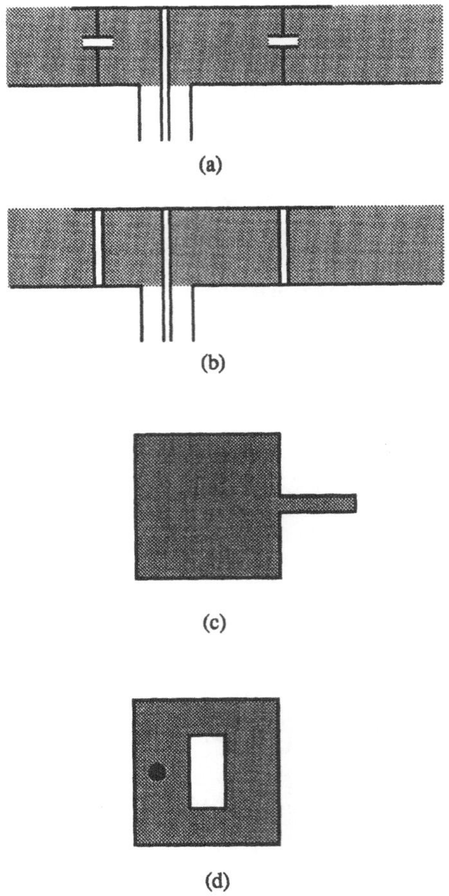

Fig. 14. Methods for tuning operating frequency of a patch of fixed dimensions. (a) Capacitive loading. (b) Inductive loading accomplished by using shorting posts. (c) Reactive loading by using transmission line stub. (d) Removal of metal from patch and/or ground plane.

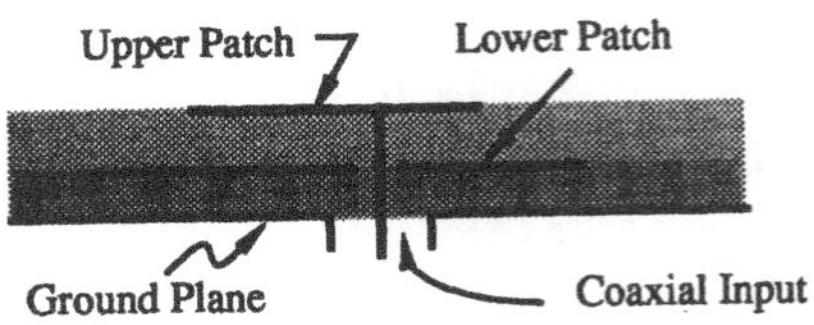

Fig. 15. Stacked patches fed by coaxial probe for dual-frequency operation.

5. Higher Order Mode Operation

The microstrip antenna is usually operated in its "dominant" mode, which is the (1,0) mode for rectangular patches, and has a maximum of its radiation pattern broadside to the plane of the antenna. However, some applications can benefit from the use of higher order modes of the structure. Farrar and Schaubert [19] describe the use of shorting posts placed at various locations to obtain a variety of radiation patterns from one patch antenna. Vaughan [20] has proposed the use of a two-port circular patch to obtain two different higher order radiation patterns that might be useful for mobile communications. The radiation properties of higher order modes are often different from the dominant mode and this must be taken into account when considering the use of these modes. The polarization of some of the higher order modes is different from the dominant mode, especially in the intercardinal planes. Also, the radiation pattern shape can be quite different. Zhong and Lo [21] demonstrate the use of the the TM_{10} and the TM_{30} modes. By using the post-tuning effect [14], the ratio of the two operating frequencies can be controlled. A combination of the tuning posts and a matching stub yields good impedance at both operating frequencies. The beamwidth of the TM_{30} mode radiation is much less than that of the dominant mode.

6. Future Trends

The basic operating principles of microstrip antennas are understood and a wide variety of antenna configurations have been developed. Recent developments in antenna elements have been driven by three major motivations: (1) greater bandwidth, which increases the applicability of patch antennas and the robustness of designs (making them less susceptible to material variations and fabrication tolerances); (2) wide-angle circular polarization to accommodate communication requirements; and (3) dual-polarization with good isolation between the polarizations. Although much of the early development of microstrip antennas was motivated by military and aerospace applications, much of the present work reflects commercial applications where the thin, conformal structure of the antennas is more important for marketing a product than for the product's overall performance. Other attractive features of microstrip antennas for commercial products are their ruggedness and manufacturability. When electronic products already use etched circuit boards, it is easy to include a patch antenna in the fabrication process. Low-cost construction using inexpensive materials such as foam and stamped metal parts have motivated variations of the traditional microstrip antenna design that eliminate connectors and other expensive parts. These kinds of developments are expected to continue, as are efforts to increase the operating bandwidth of the antennas (see Chapter 4 for information on increasing element bandwidth). One of the chief benefits of increased bandwidth is robustness of the design, important for low-cost manufacturing using alternative materials. Another area of development that is driven by both commercial and military applications is multifrequency antennas. There will continue to be advances in materials, and microstrip antennas will be developed to take advantage of the special properties, or to work in spite of the properties, of these materials. Examples that have already been demonstrated include antennas on ferrite substrates and on high-permittivity substrates needed for high-temperature superconductors. Chiral materials and new polymers may lead to other design variations. Finally, methods for feeding microstrip antennas from a variety of transmission media will continue to be developed. Connectors and coaxial cables are expensive to purchase and install, so novel ways of coupling the antennas to the electronic circuits will remain an area of fruitful investigation.

7. Summary

Some of the key characteristics of microstrip antennas were reviewed and results from several references that could not be reprinted in this book were described. Particular attention was given to cross-polarized radiation, feeding techniques, frequency tuning, and higher order mode operation.

References

[1] K. R. Carver and J. W. Mink, "Microstrip antenna technology," *IEEE Trans. Ant. and Propagat.*, AP-29, pp. 2–24, Jan. 1981.

[2] Y. T. Lo, D. Solomon, and W. F. Richards, "Theory and experiment on microstrip antennas," *IEEE Trans. Ant. and Propagat.*, AP-27, pp. 137–145, Mar. 1979.

[3] W. F. Richards, Y. T. Lo, and D. D. Harrison, "An improved theory for microstrip antennas and applications," *IEEE Trans. Ant. and Propagat.*, AP-29, pp. 38–46, Jan. 1981.

[4] K. R. Carver, "A modal expansion theory for the microstrip antenna," *Dig. IEEE Int'l Symp. Ant. and Propagat.*, pp. 101–104, 1979.

[5] R. E. Munson, "Conformal microstrip antennas and microstrip phased arrays," *IEEE Trans. Ant. and Propagat.*, AP-22, pp. 74–78, Jan. 1974.

[6] D. M. Pozar, "A review of bandwidth enhancement techniques for microstrip antennas," Chapter 4, No. 1 of this book.

[7] D. M. Pozar, "Considerations for millimeter wave printed antennas," *IEEE Trans. Ant. and Propagat.*, AP-31, pp. 740–747, Sept. 1983.

[8] T. Chiba, Y. Suzuki, and N. Miyano, "Suppression of higher modes and cross polarized component for microstrip antennas," *Dig. IEEE Int'l. Symp. Ant. and Propagat.*, pp. 285–288, 1982.

[9] J. D. Hanfling and J. J. Schuss, "Experimental results illustrating performance limitations and design tradeoffs in probe-fed microstrip-patch element phased arrays," *Dig. IEEE Int'l Symp. Ant. and Propagat.*, pp. 11–14, 1986.

[10] H. G. Oltman and D. A. Huebner, "Electromagnetically coupled microstrip dipoles," *IEEE Trans. Ant. and Propagat.*, AP-29, pp. 151–157, Jan. 1981.

[11] P. B. Katehi and N. G. Alexopoulos, "On the modeling of electromagnetically coupled microstrip antennas—the printed strip dipole," *IEEE Trans. Ant. and Propagat.*, AP-32, pp. 1179–1186, Nov. 1984.

[12] D. M. Pozar, "Microstrip antenna aperture-coupled to a microstripline," *Electronics Letters*, vol. 21, pp. 49–50, Jan. 1985.

[13] P. L. Sullivan and D. H. Schaubert, "Analysis of an aperture coupled microstrip antenna," *IEEE Trans. Ant. and Propagat.*, AP-34, pp. 977–984, Aug. 1986.

[14] D. H. Schaubert, F. G. Farrar, A. R. Sindoris, and S. T. Hayes, "Microstrip antennas with frequency agility and polarization diversity," *IEEE Trans. Ant. and Propagat.*, AP-29, pp. 118–123, Jan. 1981.

[15] J. L. Kerr, "Terminated microstrip antenna," *Proc. Antenna Applications Symp.*, Sept. 1978.

[16] J. L Kerr, "Microstrip antenna developments," *Proc. Workshop on Printed Circuit Antenna Technology*, pp. 3–1 to 3–20, Oct. 1979.

[17] S. A. Long and M. D. Walton, "A dual-frequency stacked circular-disc antenna," *IEEE Trans. Ant. and Propagat.*, AP-27, pp. 270–273, Mar. 1979.

[18] N. W. Montgomery, "Triple-frequency stacked microstrip element," *Dig. IEEE Int'l Antennas and Propagation Symp.*, pp. 255–258, 1984.

[19] F. G. Farrar and D. H. Schaubert, "Selectable-mode microstrip antenna and selectable-mode microstrip antenna arrays," US Patent No. 4,379,296, 5 April 1983.

[20] R. G. Vaughan, "Two-port higher mode circular microstrip antennas," *IEEE Trans. Ant. and Propagat.*, AP-36, pp. 309–321, Mar. 1988.

[21] S. S. Zhong and Y. T. Lo, "Single-element rectangular microstrip antenna for dual-frequency operation," *Electronics Letters*, vol. 19, pp. 298–300, 1983.

Conformal Microstrip Antennas and Microstrip Phased Arrays

ROBERT E. MUNSON

Abstract—**A new class of antennas using microstrips to form the feed networks and radiators is presented in this communication. These antennas have four distinct advantages: 1) cost, 2) performance, 3) ease of installation, and 4) the low profile conformal design. The application of these antennas is limited to small bandwidths. Phased arrays using these techniques are also discussed.**

I. Introduction

High-velocity aircraft, missiles, and rockets require conformal, thin antennas. Ideally, an antenna "paper thin" would best suit the aerodynamic and mechanical engineer. This antenna would neither disturb the aerodynamic flow, nor would it protrude inwardly to disrupt the mechanical structure.

With a microstrip (a single side etched) printed circuit board antenna, the two aforementioned goals are nearly attained. In addition, the desire for a lower cost antenna can be met because the single printed circuit (PC) board (microstrip) antenna is manufactured with the same low cost photo-etch processes used to make electronic printed circuit boards. The single board is photo etched on one side only (no front-to-back registration is required); no board alignments are required.

The microstrip phased array to be discussed is an antenna incorporating the basic radiating aperture with its associated microwave feed system all printed on the outside of a printed circuit board. It is a new microstrip device[1] that includes an efficient electrically thin microstrip radiator and integrated feed network, matching network, phasing network, switching network, and filter network, if required.

Currently, solid-state components are also added directly to this board to provide oscillators, amplifiers, phase shifters, switches, and receivers. It would appear that the feed lines would interfere with the radiation but they do not because they are electrically close to the ground plane which is the back of the antenna, and because the feed lines are perpendicular to the electric field being emitted by the radiator, i.e., a metal septum perpendicular to the electric field.

This communication will discuss microstrip arrays of three general types: wraparound microstrip antennas that wrap around missiles, rockets, and satellites to provide omnidirectional coverage; flat thin microstrip antennas that provide a high gain fan beam or a pencil beam; a phased array that consists of flat (or curved) thin microstrip antennas with pin diodes added to the microstrip substrate to provide an electronic beam steering capability.

II. Microstrip Wraparound Antennas

The wraparound antennas which provide omnidirectional coverage are similar in performance (coverage and bandwidth) to the stripline (two layer PC board) antennas discussed by Waterman and Henry [1], Campbell [2], and Johnson [3]. In general, stripline and microstrip antennas will produce bandwidths (VSWR $< 2{:}1$) of 30 MHz to 100 MHz in the L band and S band regions with a 1- to 2-dB variation in the roll plane. The microstrip wraparound antenna consists of two parts: 1) microstrip feed network and 2) microstrip radiator.

[1] Patent #3 713 162 "Single slot cavity antenna assembly," dated Jan. 23, 1973.

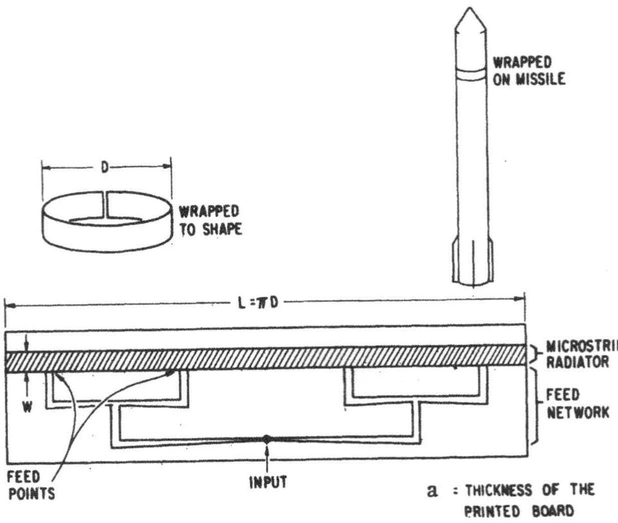

Fig. 1. Microstrip wraparound antenna.

III. Microstrip Feed Network

The microstrip feed network (Fig. 1) is a parallel (corporate) feed network where two-way power splits and equal line lengths result in equal power and equal phase to all of the feed points. The number of power divisions can be 2, 4, 8, 16, etc. The number of feeds, power divisions, required is dictated by the microstrip radiator. The number of feed points N_F must exceed the number of wavelengths in the dielectric in the L direction: $N_F > L_D$; L_D is the number of wavelengths in the dielectric $= L(\epsilon_r)^{1/2}/\lambda_0$; ϵ_r is the relative dielectric constant of the board material being used: $\epsilon_r = 2.45$ is typical; if only the TEM mode is to be excited. This mode will in turn excite only TM_{OM} modes in free space (no roll pattern variation). If $N_F < L_D$, then higher order modes will be excited on the microstrip radiator. These modes will excite TM_{NM} modes in free space [4, p. 276]. The excitation of higher order modes on the microstrip radiator will result in breakup of the roll (ϕ) plane patterns. As an example, the number of feeds required for an S band 2290 MHz ($\lambda_0 = 12.7$ cm) wraparound for a 25.4-cm missile would be

$$L = \pi D = 79.756 \text{ cm}$$

$$L_D = \frac{L(\epsilon_r)^{1/2}}{\lambda_0} = \frac{79.756(2.45)^{1/2}}{12.7} = \frac{79.657 \cdot 1.6}{12.7} = 10.05$$

$N_F > 10.05$ and N_F can be 2, 4, 8, 16, 32, 64, etc.

Thus N_F must be 16.

Two types of feed network are used to acomplish a 2, 4, 8, 16, etc., power split. Most often tapered lines, Fig. 2(a), are used to transfer a 50-Ω impedance to 100 Ω, so that it can be combined in parallel with another 100-Ω line. The same procedure is shown in

Reprinted from *IEEE Trans. Antennas Propaga.*, vol. AP-22, pp. 74–78, Jan. 1974.

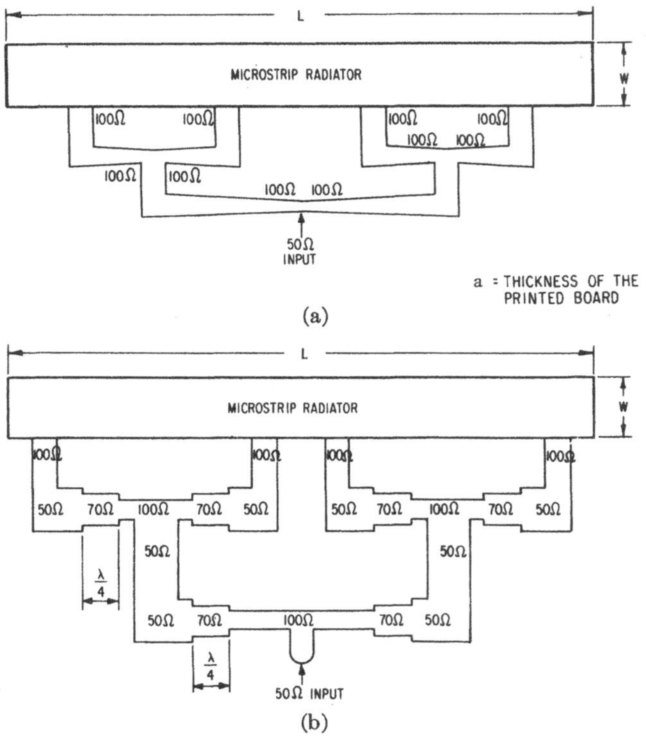

Fig. 2. (a) Tapered line parallel feed network. (b) Quarter-wave transformer parallel feed network.

Fig. 2(b) for a quarter-wave transformer technique. The impedance of the quarter-wave transformer is given by

$$Z_{\text{transformer}} = (Z_{\text{in}} \times Z_{\text{out}})^{1/2} = (100 \cdot 50)^{1/2} \approx 70\Omega.$$

The number of feed points possible for a very long radiator is limited only by the allowable system losses that can be allocated to the feed network. However, it is desirable to use the minimum N_F satisfying the condition $N_F > L_D$. If 32 feeds were used instead of 16 the preceding example would result in input impedances exceeding 300 Ω which would be impossible to match efficiently with microstrip feed lines.

IV. Microstrip Radiator

Two types of microstrip radiators are generally used: the long microstrip radiator and the patch radiator. The long microstrip radiator shown in Figs. 2(a) and (b) is shown in top and side view in Figs. 3(a) and (b), respectively. Gap A is an infinitesimal slot (in 0.79 mm microstrip $a/\lambda \approx 1/150$ at S band). The admittance of a slot radiator is given in Harrington [4, p. 183] for small $ka(a/\lambda < 0.1)$ which is always the case in microstrip antenna practice

$$G_a \approx \frac{\pi}{\lambda\eta}\left[1 - \frac{(ka)^2}{24}\right]$$

$$B_a \approx \frac{3.135 - 2\log ka}{\lambda\eta}.$$

In most microstrip applications $ka/24 \ll 1$ and the conductance simplifies to $G_a = \pi/\lambda\eta = 1/\lambda(120)$ mho/m or $R_a = 120\lambda \cdot \Omega \cdot \text{m}$. The conductance is expressed in per unit length so that the resistance of the Slot A in Figs. 3(a) and (b) is obtained by dividing R_a by the length

$$r_a = \frac{R_a}{L} = \frac{120\lambda}{2\lambda} = 60\Omega.$$

The dielectric under the microstrip radiator can be treated as a transmission line approximately $\lambda/2$ long. The problem with the microstrip transmission line is its very low impedance, typically 1 to 10 Ω. This section of parallel-plate transmission line does transform the Slot A impedance from 60 Ω through small imedances near the center and back to 60 Ω at Slot B [see Fig. 3(c)]. At this

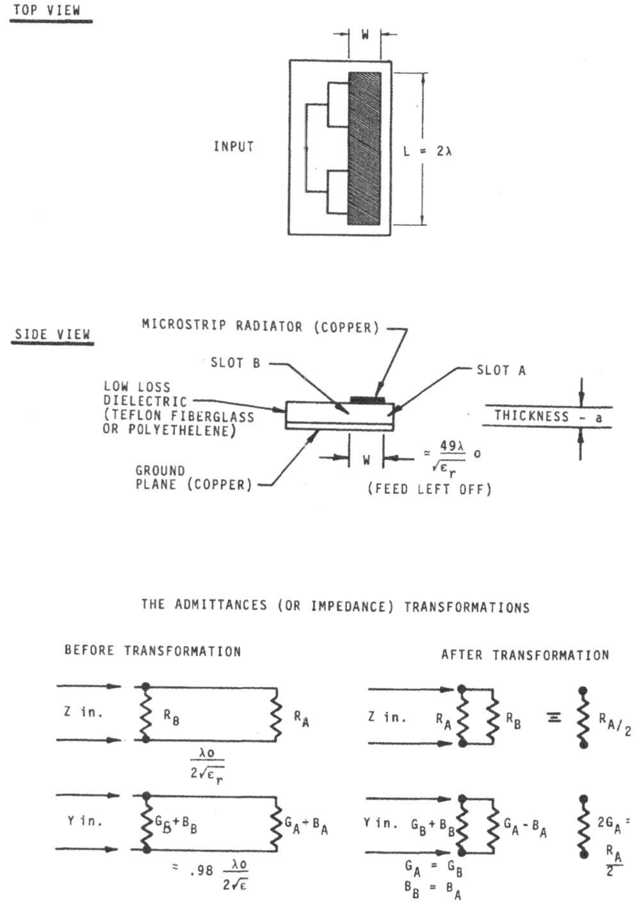

Fig. 3. Microstrip radiator.

point the two impedances combine in parallel to give

$$\frac{1}{r_{\text{in}}} = \frac{1}{r_a} + \frac{1}{r_b} = \frac{1}{60} + \frac{1}{60}$$

$$r_{\text{in}} = 30\ \Omega.$$

In the example shown in Fig. 3(a) this impedance is split between four feed points with each feed theoretically seeing 120 Ω. In practice, this is the measured impedance. This theory is very accurate in predicting the input impedances for many designs each with different frequencies, thicknesses, feed point separations, and number of feed points. The previous discussion did not treat the implications of the reactive component of the admittance B_A because it does not affect the conductance component of admittance G_A. The effect of the reactance B_A is to produce a resonance slightly short of a half-wavelength. For example, we can consider the admittance of Slot A to be

$$Y_A = G_A + B_A.$$

At a distance of 0.5λ on the parallel-plate transmission line, the admittance has been transformed to $Y_A = G_A + B_A$ and these admittances combine directly in parallel with Y_B to produce $Y_{\text{in}} = 2G_A + 2B$ which is not resonance. At a distance just short (usually 0.49λ to 0.48λ) of a half-wavelength in the parallel-plate transmission line transformer the transformed admittance of Slot A is

$$\tilde{Y}_A = G_A - B_A$$

and at this length slightly short of a half-wavelength $[\lambda_0/2(\epsilon_r)^{1/2}]$ resonance is established with no susceptance

$$Y_{\text{in}} = G_A + G_B = 2G_A$$

$$Z_{\text{in}} = R_A/2$$

and for the example

$$Z_{\text{in}} = R_{\text{in}} = 30\ \Omega \text{ (total resistance)}$$

$$R_{\text{in}} = 120\ \Omega \text{ (per feed)}.$$

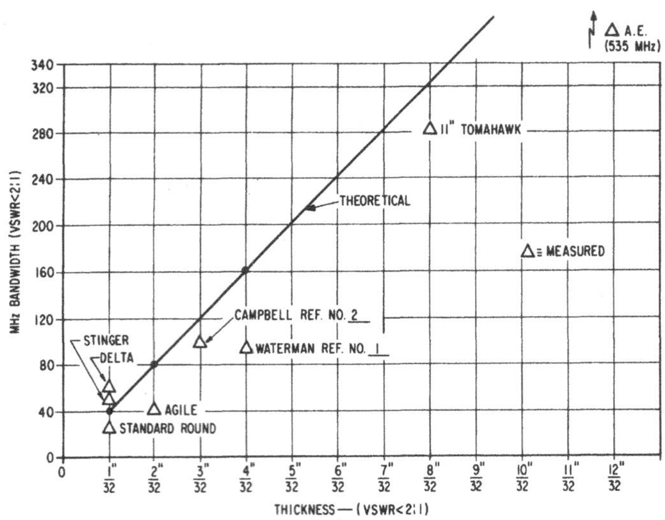

Fig. 4. *S* band bandwidth (VSWR 2–1) as function of antenna thickness.

The bandwidth of a microstrip antenna is dominated by the microstrip parallel-plate transmission line between Slot *A* and *B*. Since the transmission line usually has an impedance close to 1 Ω and the two slots have impedances close to 100 Ω, the transformation exists usually for 1-percent bandwidth for VSWR < 2:1. The bandwidth can be easily calculated by adding

$$Y_{\text{in}} = \tilde{Y}_A + Y_B$$

(where the amount that Y_A is transformed depends upon frequency.), and then evaluating the two frequency points at which the reactances cause the VSWR to equal 2:1. Several measured bandwidths of microstrip phased arrays are shown in Fig. 4 in conjunction with the theoretical bandwidth as calculated earlier.

The major limitation of the microstrip antenna is the bandwidth. To substantially increase the bandwidth of microstrip antennas requires an increase of the thickness of the parallel plate transformer which increases the characteristic impedance of the transformer. This increase in thickness is undesirable if the antenna is to remain low profile and conformal. In most applications the advantages of a low profile antenna outweighs the disadvantage of its narrow bandwidth because present applications require less than 1 percent. Three other methods of increasing the bandwidth are currently being investigated: 1) use of a high (ϵ_r) dielectirc constant to decrease the cavity length; 2) increasing the inductance of the microstrip radiator by cutting holes or slots into it. Experiments show increased bandwidth but at the cost of efficiency, in fact the same increase could have been attained by using a more lossy substrate; 3) broadbanding by addition of reactive components as discussed in Jasik [5] to reduce VSWR across a limited bandwidth. This technique is very limited usually to 50 percent of $\Delta f_0/f_0$.

V. Microstrip Antenna Pattern Coverage for Omniapplications

The pattern coverage for the omniantenna shown in Fig. 1 depends on the diameter of the missile. The limiting factor in omnidirectional pattern coverage is a singular hole at the tip and tail of the missile which gets *narrower* as the diameter of the missile *increases*. For instance, a 15-in diameter antenna produces a null along the missile axis of radius 1° at the −8-dB gain level. The fraction area with gain below −8 dB is given by

$$F_N = \left(\int_{0^\circ}^{360^\circ}\int_{0^\circ}^{1^\circ} \sin\theta\, d\theta\, d\phi + \int_{0^\circ}^{360^\circ}\int_{179^\circ}^{180^\circ} \sin\theta\, d\theta\, d\phi\right) \Big/ \int_{0^\circ}^{360^\circ}\int_{0^\circ}^{180^\circ} \sin\theta\, d\theta\, d\phi$$

$$\approx 0.0002.$$

Conversely, the fraction of the area with gain above −8 dB is 0.9998, or 99.98 percent coverage with gain greater than −8 dB. The percent coverage increases without limit for larger diameters until a nearly perfect coverage is attained for a single linear polarization.

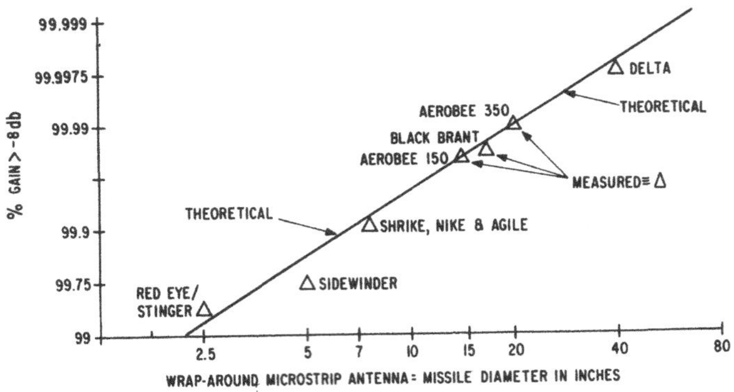

Fig. 5. Pattern coverage versus diameter for microstrip wraparound antennas on smooth cylinders.

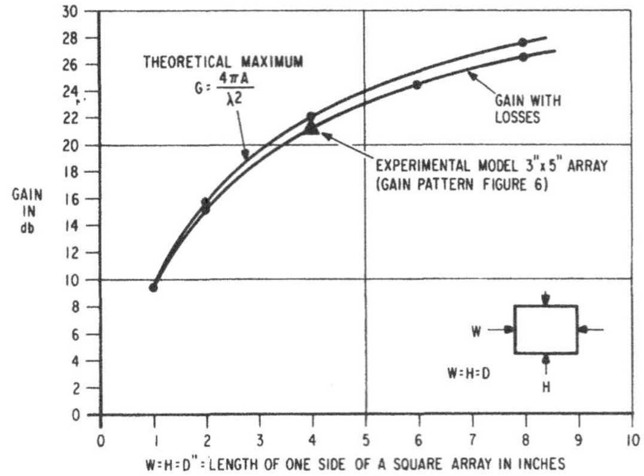

Fig. 6. Gain versus size for flat microstrip arrays (frequency is *X* band or 10 GHz and λ_0 = 1.18").

The percent coverage is only a function of diameter and is independent of antenna thickness. The theoretical and experimental pattern coverages for microstrip antennas on a smooth cylinder are given in Fig. 5 for gain greater than −8 dB.

IV. Flat-Plate Microstrip Antennas

Unwrapping omniwraparound antennas and mounting them flat on a metal surface or in free space produces a high gain fan beam antenna pattern. By arraying several antennas side by side, a pencil beam is produced. Theoretically, the microstrip radiators produce a uniform illumination of the aperture and the gain of a uniformly illuminated aperture is given by Silver [6] as

$$G_0 = \frac{4\pi A}{\lambda^2}.$$

In practice, the microstrip feed line attenuation subtracts from this gain

$$G_{\text{actual}} = 10 \log\left(\frac{4\pi A}{\lambda^2}\right) - \alpha$$

$$\alpha_{\text{line}} = \alpha L.$$

The attenuation is dependent on frequency and line length. At *X* band a microstrip line on 0.79-mm board has an attenuation α = 0.047 dB/cm. The length of the microstrip feed line for a given array is half of the height plus half of the width of the array

$$L = \frac{W}{2} + \frac{H}{2}$$

therefore

$$\alpha = (\alpha/2)(W + H)$$

at *X* band for a 12.7-cm × 7.62-cm antenna α_{line} = 0.48 dB. Gain as a function of size for a square microstrip array is shown in Fig. 6.

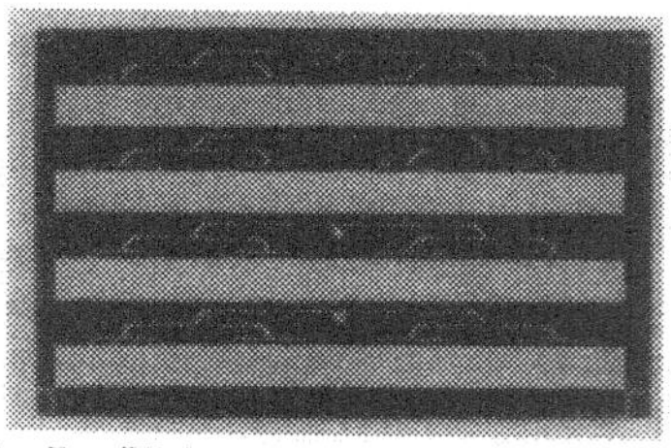

Fig. 7. High gain flat microstrip antenna.

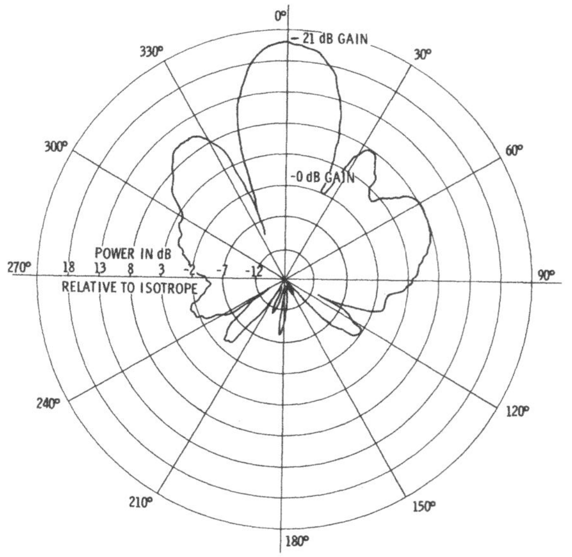

Fig. 8. Gain and pattern of 7.62 cm x 12.7 cm x 0.79 cm microstrip array at 9.92 GHz.

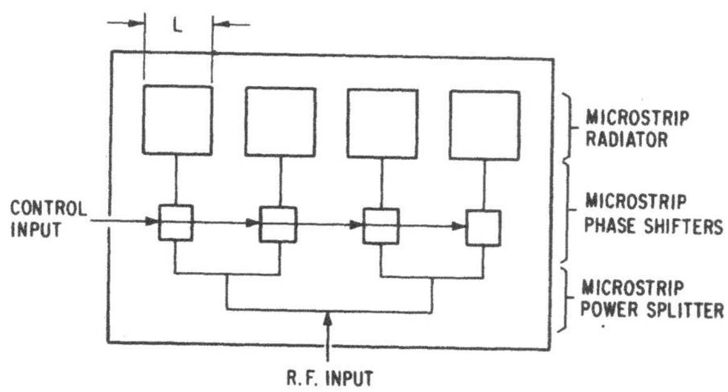

Fig. 9. Electrically scanned microstrip phased array (low cost and low profile).

An experimental model 7.62 cm × 12.7 cm × 0.79 mm (Fig. 7) was built and tested and confirms a gain (Fig. 8) in agreement with the theoretical predictions (Fig. 6). The measured gain of 21 dB is also plotted on the predicted gain curve (Fig. 6). The microstrip antenna offers high gain for a low cost. It also offers a low profile antenna that can operate flush mounted to a metal surface.

VII. Microstrip Phased Arrays

By adding "pin diodes" for digital phase shifting, Fig. 9, to the microstrip substrate an integrated electrically scanned antenna is attained. The process of phasing the radiators to scan the beam requires breaking up the microstrip radiators into individual elements. The individual microstrip elements (a sample is shown in Fig. 10) work just like the long microstrip radiator described in the previous section. By using L the length of the individual microstrip radiators we can calculate the resonant length, input impedance, and bandwidth of the microstrip radiator just as was done in the previous section.

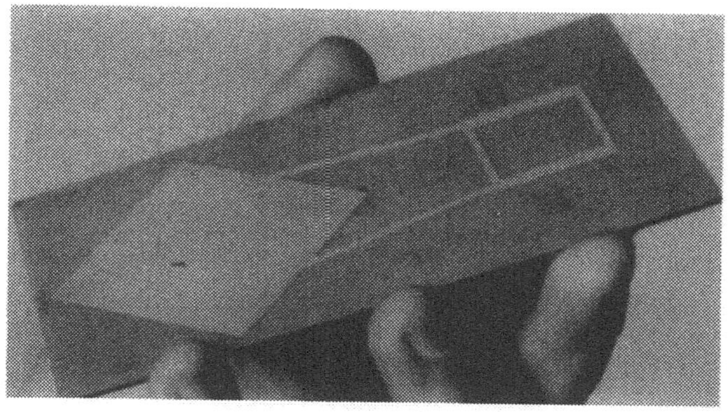

Fig. 10. Microstrip radiator.

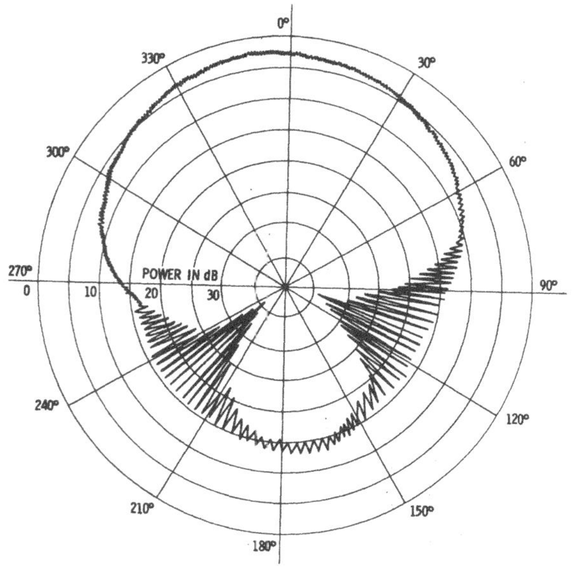

Fig. 11. Radiation pattern of microstrip patch. Patterns were measured with spinning dipole to demonstrate low axial ratios to wide angle.

This works quite well except when the L of the individual radiator is not reduced below 0.25 λ_0. For $L < 0.25\ \lambda_0$ the radiation resistance of the microstrip radiator rapidly disappears, i.e., the slots A and B are not long enough to match free-space efficiently because their size has been reduced below cutoff for the modes that must be matched to free space as described by Harrington [4, p. 278].

Each of these microstrip radiators are rectangular microstrip elements and each one produces a hemispherical coverage pattern, Fig. 11. A conceptual model of the phased array shown in Fig. 9 was built and tested to demonstrate a complete microstrip electrically scanned phased array. The patterns scanned to the angles predicted with a gain within 1 dB of the expected gain, Fig. 12. The phase shifters used were microstrip 90° hybrid phase shifters with diodes in the two output legs. Driving two diodes in the two output legs of the hybrid changes the phase of the reflected power in the reflected port of the hybrid. The phase shift attained is twice the distance the short reference is moved in the two output legs. Three phase shifters were used in series for each element to produce 0°, 45°, 90°, 135°, 180°, 225°, 270°, or 315° phasing of each element. The phase shifters along with all of their dc feed lines, dc blocks, RF blocks, the RF corporate feed network, the matching network, and the microstrip radiators were all photo etched on one side of one microstrip board.

VIII. Conclusions

Microstrip antennas constitute a new class of onmidirectional antenna for missiles and satellites. These antennas are capable of producing a predictable and nearly perfect onmidirectional coverage. A new low cost low profile flat microstrip array is shown to have 90-percent aperture efficiency. In addition, the flat microstrip

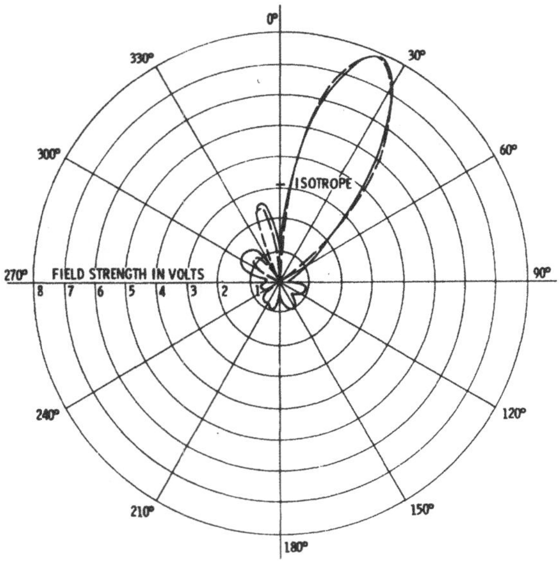

Fig. 12. Electronically scanned 4 element array. — measured pattern. --- predicted pattern.

arrays can be electronically scanned with the addition of phase shifters.

These antennas are inexpensive to fabricate because of the photo etch process used in their manufacture, and inexpensive to install because they are conformal. Electronically scanned microstrip arrays make possible an ultra low profile (conformal), low cost design for phased arrays. It may be possible to entirely cover the outer surface of a missile or aircraft with these antennas without large cost or weight penalties.

Acknowledgment

The author wishes to thank G. Sanford for his support and advice in the preparation of this paper, and M. Perdue for her assistance in editing and typing.

References

[1] A. Waterman and D. Henry, "Stripline strap-on antenna array," presented at the 21st USAF Antenna Symp.

[2] T. G. Campbell, "An extremely thin omnidirectional microwave antenna array for spacecraft applications," NASA Tech. Note D-5539, Nov. 1969.

[3] H. P. Johnson, "An extremely thin flush mounted slotted linear array," presented at the 16th USAF Antenna Symp.

[4] R. F. Harrington, *Tme Harmonic Electromagnetic Fields.* New York: McGraw-Hill, p. 276.

[5] Jasik, *Antenna Engineering Handbook*, p. 3125.

[6] S. Silver, *Microwave Antenna Theory and Design* (M.I.T. Rad. Lab. Series, vol. 12). New York: McGraw-Hill, 1949, p. 178.

An Experimental Investigation of Electrically Thick Rectangular Microstrip Antennas

ESIN CHANG, MEMBER, IEEE, STUART A. LONG, SENIOR MEMBER, IEEE, AND WILLIAM F. RICHARDS, MEMBER, IEEE

Abstract—**The electromagnetic properties of electrically thick rectangular microstrip antennas were investigated experimentally. Antennas were fabricated with different patch sizes and with electrical thicknesses ranging from 0.03 to 0.23 wavelengths in the dielectric substrate. The resonant frequencies were measured and compared to existing formulas. The bandwidth was calculated as a function of electrical thickness and the antenna radiation patterns were measured.**

I. Introduction

During the past ten years, microstrip antennas experienced a great gain in popularity and have become a major research topic in both theoretical and applied electromagnetics. They are well known for their highly desirable physical characteristics such as low profile, light weight, low cost, ruggedness, and conformability. Numerous researchers have investigated their basic characteristics and recently extensive efforts have also been devoted to the design of "frequency agile," "polarization agile," or dual-band microstrip antennas [1], [2], [3]. Most of the previous theoretical and experimental work has been carried out only with electrically thin microstrip antennas. Recent interest has developed in radiators etched on electrically thick substrates. This interest is primarily due to two major reasons. First, as these antennas are used for applications with increasingly higher operating frequencies, and consequently shorter wavelengths, even antennas with physically thin substrates become thick when compared to a wavelength. Second, microstrip antennas have inherently narrow bandwidths and are normally not suitable for broad bandwidth applications. Increasing the bandwidth is possible, but the methods used [4], [5], [6] invariably increase the volume of the antenna by either extending the radiating surface or by increasing the overall antenna thickness. To aid in the design of broader band microstrip antennas, a careful experimental study of the resonant frequency, bandwidth, and radiation patterns of rectangular microstrip antennas as a function of electrical thickness of the substrate was undertaken. The measured resonant frequencies were compared to formulas previously developed for predicting the resonant frequency of electrically thin rectangular microstrip antennas.

II. Experimental Procedures

The microstrip antennas investigated are rectangular patches with geometry as illustrated in Fig. 1. They are fabricated on 3M CuClad 233 and on Rogers RT/duroid 5870 microwave substrates. The CuClad material is made of a polytetrafluoroethylene (PTFE) woven glass laminate material while the RT/duroid material is made of a glass microfiber reinforced PTFE composite. Both substrates have a nominal dielectric constant (ϵ_r) of 2.33, and all the antennas are fed using an SMA coaxial feed. In this investigation the feed is located at the midpoint of the longer side ($x' = a/2$) and at a distance from the edge ($y' = 0.15$ cm). In each case the dimension "a" has been chosen to be approximately one and one-half times the dimension "b" with a 10 cm $\times$ 10 cm ground plane. Two sets of regular microstrip antennas have been fabricated. The ones in the first set have the same substrate thickness "h" but have nine different patch sizes; the ones in the second set have the same patch size but have three different substrate thicknesses. In addition a so-called "air-dielectric" model radiator has been fabricated to allow an even more detailed study of the resonant frequency. Its geometry models that of the regular rectangular microstrip antenna shown in Fig. 1 but with a substrate whose height can be changed by placing sheets of styrofoam ($\epsilon_r \approx 1.05$) with varying thicknesses between the ground plane and the radiating patch. The 1.78 cm $\times$ 2.67 cm aluminum radiating patch has a thickness of $t = 0.16$ cm and is coaxial fed over a 14 cm $\times$ 14 cm aluminum ground plane. This fixture allows the resonant frequency to be measured for a wide range of electrical thicknesses using exactly the same rectangular radiating patch.

The resonant frequency, impedance and radiation pattern measurements were all performed at the University of Houston-University Park, Applied Electromagnetics Laboratories, using an automated network analyzer system and dedicated computer programs. Accuracy enhancement techniques [7] have been used to partially correct for effective directivity, effective source match, and frequency tracking errors when taking impedance measurements. The radiation pattern measurements were taken with the antenna under test placed inside an anechoic chamber and mounted on a one meter diameter circular aluminum ground plane.

Reprinted from *IEEE Trans. Antennas Propaga.*, vol. AP-34, no. 6, pp. 767–772, June 1986.

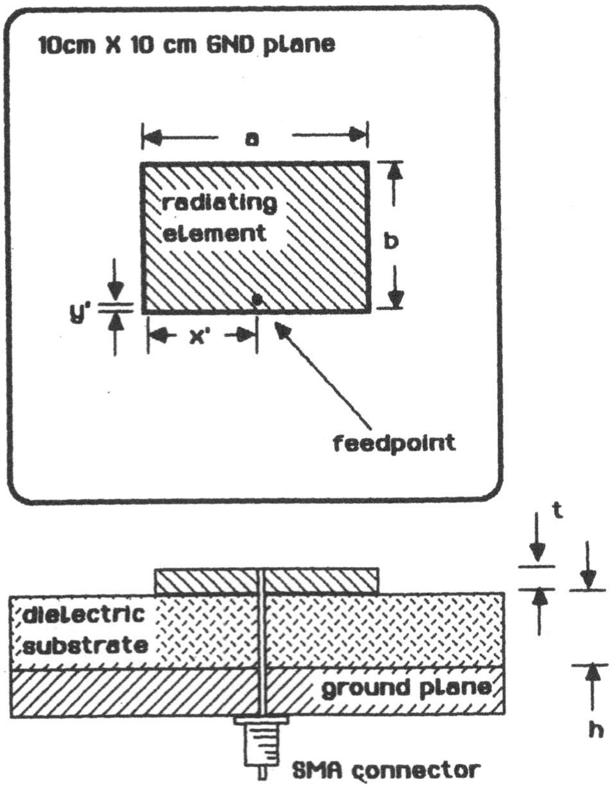

Fig. 1. Rectangular microstrip antenna geometry.

III. Experimental Results

A. Impedance

During the course of this research, the input impedances ($Z = R + jX$) and the radiation patterns of each of the antennas have been measured. Before the impedance data were used to determine the resonant frequencies (f_r) and the bandwidths (BW), they were smoothed in order to take out any residual ripples or oscillations that are due to reflections internal to the measurement equipment and have not been corrected by the accuracy enhancement routines. The values of admittance, $Y = G + jB$, as a function of frequency (f) were computed from the measured impedance versus frequency data through the relation $Y = Z^{-1}$. Depending on the values of admittance, either a cubic or quadratic least squares regression polynomial is fitted through the values of the admittance versus frequency curve. From this fitted polynomial, the smoothed admittance at each measured value of frequency is computed and then the reciprocal is taken to obtain the smoothed impedance. The actual smoothing operation is carried out with the admittance data since both the real and imaginary parts of the admittance are monotonic functions in the neighborhood of resonance and thus result in a better polynomial fit. The smoothed curves follow the general form of the measured traces very closely and allow the true peak of the resistance curve to be determined more accurately for resonant frequency measurements. Fig. 2 shows a comparison of typical smoothed and measured impedance versus frequency curves with data points taken every 10 MHz for an antenna with $h/\lambda_d = 0.110$.

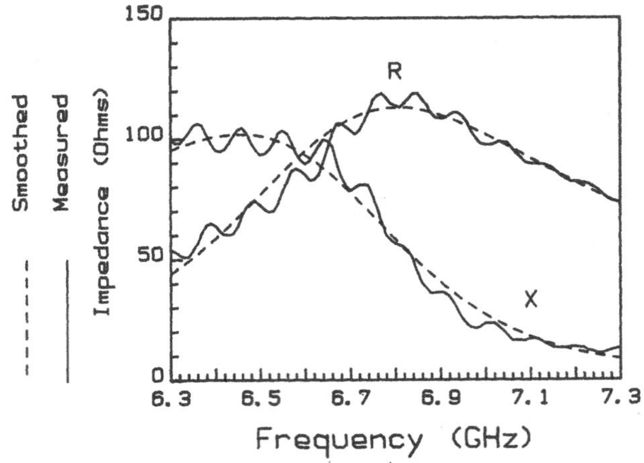

Fig. 2. Comparison of measured and smoothed impedance (1.1 cm × 1.7 cm radiating patch, 0.3175 cm substrate, ϵ_r = 2.33).

B. Resonant Frequency

Generally, the resonant frequency of a microstrip antenna is defined as the frequency at which the reactance is equal to zero. For electrically thin antennas, this point is also very close to the frequency where the resistance reaches a maximum. However, in this investigation many of the reactance curves exhibit an inductive shift due to the coaxial feed passing through the electrically thick substrate [8], [9]. In fact, for the thicker antennas, the reactance curve never passes through zero at all (see Fig. 2). For this reason, the resonant frequency has been redefined as the point at which the resistance reaches a maximum, independent of the value of reactance. Furthermore, since the bandwidth of an electrically thin microstrip antenna is commonly defined in terms of the impedance (and thus is dependent on the reactance), an alternate definition of bandwidth that is not affected by the inductive shift of the reactance is used to obtain the antenna bandwidths of this paper. This last point will be discussed in detail in a later section.

Since the main concern in the measurement of the resonant frequency is the effect of the changing electrical thickness of the substrate, a normalized resonant frequency is defined where $f_{\text{norm}} = f_r/f_{r0}$, and f_{r0} is the zeroth-order prediction of the resonant frequency. This approximation for f_{r0} assumes that the antenna thickness is infinitesimally thin and that b is equal to $\lambda_d/2$. Then knowing that $\lambda_d = c/(f_{r0}\sqrt{\epsilon_r}) = 2b$, f_{r0} can be computed. For the units of c in m/s and those of b in cm,

$$f_{r0} = 15/(b\sqrt{\epsilon_r}) \text{ GHz.} \tag{1}$$

Table I shows the measured resonant frequency, zeroth order prediction, physical dimensions, and electrical thickness of each antenna. Fig. 3 shows a plot of the normalized resonant frequency plotted as a function of electrical thickness for the nine antennas etched on the same thickness of substrate and for the three identically sized antennas on different substrate thicknesses.

Table II shows the measured resonant frequencies and the

TABLE I
MEASURED AND PREDICTED ANTENNA RESONANT FREQUENCIES

a (cm)	b (cm)	h (cm)	Meas'd (GHz)	James (GHz)	Hammerstad (GHz)	h/λ_d
5.7	3.8	0.3175	2.31	2.30	2.38	0.037
4.55	3.05	0.3175	2.89	2.79	2.90	0.047
2.95	1.95	0.3175	4.24	4.11	4.34	0.068
1.95	1.3	0.3175	5.84	5.70	6.12	0.094
1.7	1.1	0.3175	6.80	6.47	7.01	0.110*
1.4	0.9	0.3175	7.70	7.46	8.19	0.125
1.2	0.8	0.3175	8.27	8.13	9.01	0.141
1.05	0.7	0.3175	9.14	8.89	9.97	0.148
0.9	0.6	0.3175	10.25	9.82	11.18	0.166
1.7	1.1	0.1524	7.87	7.46	7.84	0.061
1.7	1.1	0.3175	6.80	6.47	7.01	0.110*
1.7	1.1	0.9525	4.73	4.32	5.27	0.229

*These two are the same antennas.

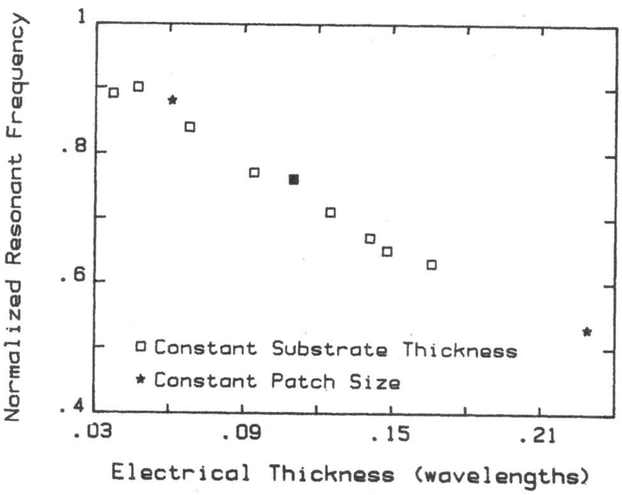

Fig. 3. Normalized antenna resonant frequency versus electrical thickness.

TABLE II
RESONANT FREQUENCY VERSUS h FOR AIR-DIELECTRIC FIXTURE (1.78 cm × 2.67 cm RADIATING PATCH)

h (cm)	Meas'd (GHz)	James (GHz)	Hammerstad (GHz)	h/λ_d
0.64	5.14	4.54	5.75	0.112
0.79	5.12	4.18	5.42	0.138
0.99	4.33	3.78	5.07	0.146
1.19	4.27	3.46	4.77	0.174
1.44	3.32	3.13	4.46	0.163
1.64	3.06	2.91	4.25	0.171
2.04	2.56	2.55	3.91	0.178
2.34	2.29	2.34	3.70	0.183

corresponding electrical thicknesses of the air dielectric fixture, while the circles in Fig. 4 represent the same data in graphical form. It is evident from Figs. 3 and 4 that the resonant frequency indeed decreases as the antennas become electrically thicker as has been shown in previously published results [10], [11]. It is perhaps unexpected, however, that this trend continues even to thicknesses approaching one quarter wavelength.

C. Bandwidth

The percent bandwidth of the antennas was determined from the impedance data. For ease of notation the term bandwidth refers to percent bandwidth unless otherwise specified. Bandwidth is normally defined as

$$\text{percent BW} = [(f_{r2} - f_{r1})/f_r]\ 100 \text{ percent} \qquad (2)$$

where f_r is the resonant frequency, while f_{r2} and f_{r1} are the frequencies between which the magnitude of the reflection coefficient of the antenna is less than or equal to 1/3 (which corresponds to a voltage standing-wave ratio (VSWR) ≤ 2.0). However, this definition is found not to be directly applicable

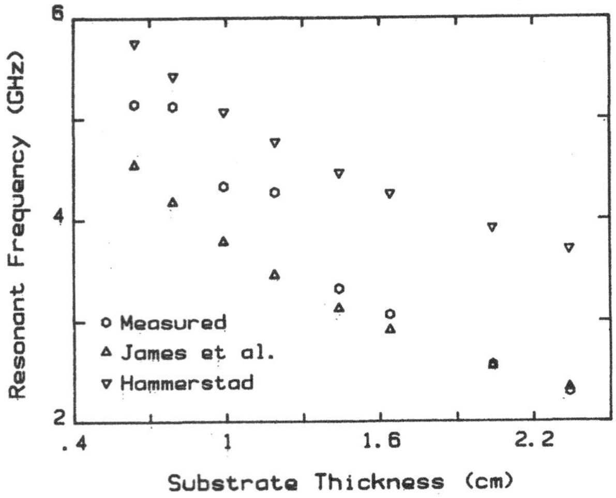

Fig. 4. Air-dielectric fixture resonant frequencies versus substrate thickness (1.78 cm × 2.67 cm radiating patch, $\epsilon_r \approx 1.05$).

to the experimental data because of the inductive shift. Thus, an alternate definition is found in order to determine the bandwidth of the test antennas.

The case where the impedance at resonance is purely resistive ($Z_{res} = R_{max}$) may be represented by a parallel RLC circuit, and an analytical expression for the input impedance in terms of the antenna Q-factor, R_{max} and f_r may be written as

$$Z(f)=\frac{\dfrac{1}{R_{max}}+j\left[\dfrac{Qf_r}{R_{max}f_r}-\dfrac{fQ}{R_{max}f_r}\right]}{\dfrac{1}{R^2_{max}}+\left[\dfrac{Qf}{R_{max}f_r}-\dfrac{Qf_r}{R_{max}f_r}\right]^2}. \quad (3)$$

Using (3), it was found that R and the magnitude of Z had a definite relationship to R_{max} that is independent of the other parameters. Namely,

$$|Z(f_{r1})|=|Z(f_{r2})|=0.818R_{max}, \quad (4)$$

and

$$R(f_{r1})=R(f_{r2})=0.670R_{max}. \quad (5)$$

For the special case of a prescribed VSWR = 2.0, this method is equivalent to the previously derived analytical expression [12]

$$\mathrm{BW}=(\mathrm{VSWR}-1)/(Q\sqrt{\mathrm{VSWR}}). \quad (6)$$

For electrically thick radiators with the associated large inductive shift in the impedance, the VSWR may not be below 2.0 for any range of frequencies. Using only the resistance data a projected bandwidth can be calculated, however, following this resonant circuit model by locating the frequencies where the resistance is equal to 0.670 times the value of R_{max}. Table III shows the bandwidths so obtained arranged in order of increasing electrical thickness along with the corresponding f_{r1} and f_{r2}, while Fig. 5 shows the same bandwidth data in graphical form. It is seen that values of bandwidth on the order of 20 percent may be achieved using electrically thick antennas.

It should be noted, however, that no attempt has been made in this investigation to actually match the antennas to a standard 50 Ω transmission line by a technique such as moving the feed point away from the edge. Thus the bandwidths reported in Table III and Fig. 5 are, in effect, projected ones that might be obtainable under the more usual definition of bandwidth. These values are most useful for comparison purposes to characterize the dependence of the bandwidth on the various antenna parameters.

D. Radiation Pattern

The radiation pattern for each element was measured at its resonant frequency in both the E-plane and the H-plane. The experimental results show that the radiation patterns of electrically thick antennas are very similar to those of electrically thin antennas. The H-plane patterns remain virtually unchanged while the E-plane ones begin to show some asymmetries only for the larger substrate thicknesses. Fig. 6 shows the radiation patterns of a representative thicker antenna ($h/\lambda_d = 0.110$, taken at 6.8 GHz in 0.5° steps) and it is typical of the remaining antennas.

IV. Comparison of Measured and Predicted Resonant Frequencies

The measured resonant frequencies can be compared to the predicted resonant frequencies as a mutual check of the experimental data and of the validity of the theories. The equations for the resonant frequency proposed by Hammerstad [13] and by James, Hall, and Wood [11] are used for these comparisons. Both methods share the concept of an effective dielectric constant (ϵ_{eff}) given by [14]

$$\epsilon_{eff}(w)=\frac{(\epsilon_r+1)}{2}+\frac{(\epsilon_r-1)(1+10h/w)^{-1/2}}{2} \quad (7)$$

where w is a variable and can be either the patch dimension "a" or "b." Hammerstad gives a predicted

$$f_r=\frac{c}{2(b+2\Delta b)\sqrt{\epsilon_r}}, \quad (8)$$

$$\Delta b=\frac{0.412h(\epsilon_{eff}(a)+0.3)(a/h+0.264)}{(\epsilon_{eff}(a)-0.258)(a/h+0.8)}. \quad (9)$$

James *et al.* give a predicted

$$f_r=\frac{f_{r0}\epsilon_r}{\sqrt{\epsilon_{eff}(a)\epsilon_{eff}(b)(1+\delta)}}, \quad (10)$$

where

$$\delta=(h/b)0.882+\left[\frac{0.164(\epsilon_r-1)}{\epsilon_r^2}\right]+\left[\frac{(\epsilon_r+1)[0.758+\ln(b/h+1.88)]}{\pi\epsilon_r}\right]. \quad (11)$$

The predicted resonant frequencies obtained using these two methods are shown in Tables I and II for the test antennas and for the air-dielectric fixture.

TABLE III
PROJECTED VALUES OF BANDWIDTH AND CORRESPONDING F_{r1} AND F_{r2}

h/λ_d	f_{r1} (GHz)	f_{r2} (GHz)	BW (%)
0.037	2.280	2.352	3.117
0.047	2.834	2.952	4.083
0.061	7.632	8.152	6.607
0.068	4.120	4.396	6.509
0.094	5.632	6.140	8.699
0.110	6.494	7.272	11.441
0.125	7.314	8.468	14.987
0.141	7.848	9.024	14.220
0.148	8.380	10.560	23.850
0.166	9.442	11.560	20.660
0.229	4.320	5.180	18.180

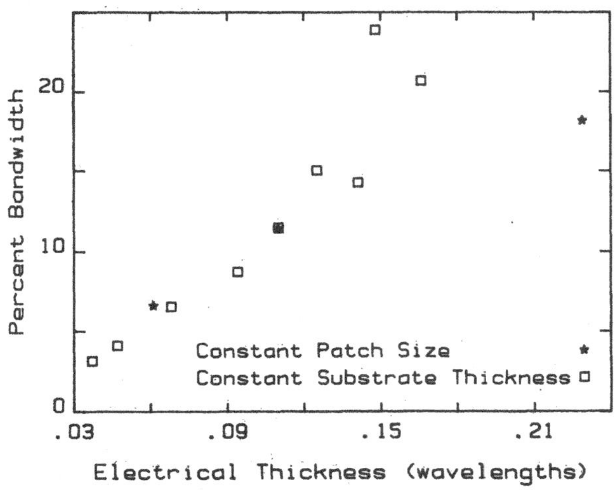

Fig. 5. Projected antenna bandwidth versus electrical thickness.

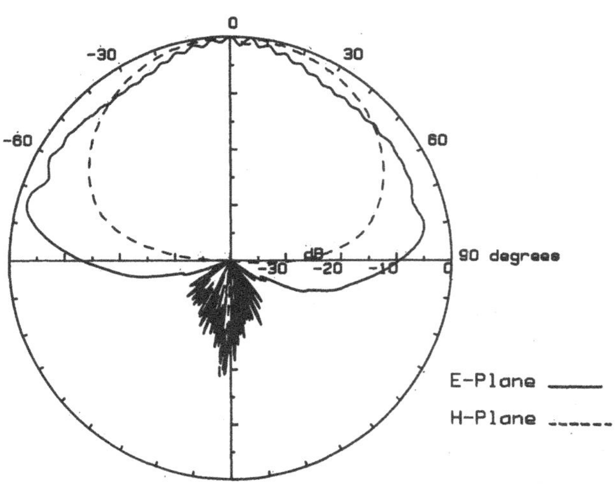

Fig. 6. Antenna radiation pattern (1.1 cm × 1.7 cm radiating patch, 0.3175 cm substrate, $\epsilon_r = 2.33$).

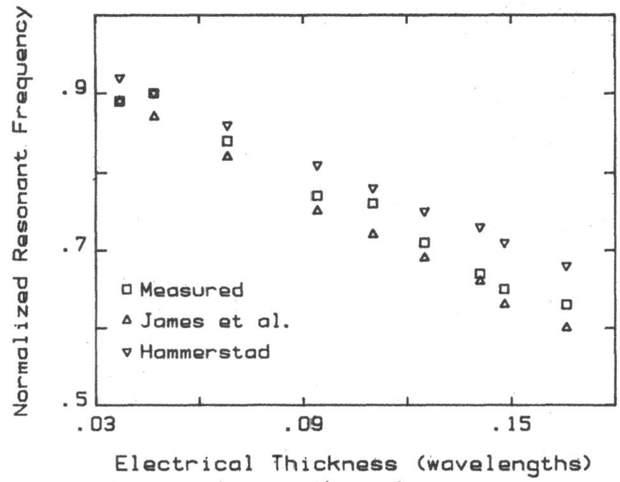

Fig. 7. Comparison of normalized antenna resonant frequencies versus electrical thickness (h = 0.3175 cm, ϵ_r = 2.33).

The predicted resonant frequencies are normalized to the zeroth order predictions and are compared to the measured, normalized resonant frequencies. Fig. 7 shows this comparison for the nine different patches on the substrates with identical physical thickness, while Fig. 4 shows the predicted values of resonant frequency for the air-dielectric fixture compared to the actual measured values. It should be noted that the theoretical data in Fig. 7 are presented as discrete points rather than a continuous curve so that the theory can correspond to the exact cases of the experimental cases, some of which have slightly varying values of a/b ratio and of dielectric constant. This use of individual points causes the data to no longer be aligned along smooth curves. In both cases it is clear that the theories follow the trend of the experimental data quite well even for electrically thick substrates. In fact, based on the information presented here and on additional research data [15], some general observations may be made concerning these two methods. Both predict the resonant frequency very closely for electrically thin rectangular microstrip antennas, but as h becomes greater than 0.1 λ_d, James *et al.* give consistently better predictions than the method by Hammerstad. Specifically, James *et al.* usually predict values approximately 4 percent lower than the measured resonant frequency while Hammerstad predicts values around 8 percent higher than the measured resonant frequency. It should be noted, however, that these two theories are not intended for use with electrically thick substrates. For $h \leq 0.1\ \lambda_d$, the measured resonant frequency is very nearly the mean of the predicted resonant frequencies from the two different formulas. A third algebraic formula for the resonant frequency has been proposed by Sengupta [16], but as the author states, it only applies to electrically thin structures and does not predict the proper behavior for the thicker substrates measured in this investigation.

V. CONCLUSION

The effect of varying the electrical thickness for rectangular microstrip antennas has been investigated experimentally during the course of this research. In addition, an air-dielectric model radiator has been fabricated with a single patch size and a variable substrate thickness. The resonant frequency, bandwidth, and radiation pattern have been measured over a range of substrate thickness from 0.03 to 0.23 of a wavelength in the dielectric.

The resonant frequency of a rectangular microstrip antenna was found to decrease as a function of electrical thickness. The validity of this finding is confirmed by previously published results and by the existing theories of Hammerstad and of James *et al.* The predicted resonant frequencies are all very close to the measured resonant frequency for electrically thin substrates. As the electrical thickness is increased, the theories due to Hammerstad and to James *et al.* generally predict values that are approximately 8 percent high and 4 percent low respectively when compared to the actual values. The impedance of the thicker antennas is characterized by an inductive shift in the reactance away from zero at resonance.

The projected bandwidth of the unmatched antennas was calculated to determine the effect of substrate thickness on this characteristic as well. The resulting data show that bandwidths as high as 20 percent could be achieved by simply using electrically thick substrates. Finally, the radiation patterns of electrically thick antennas were seen to be very similar to those of the more usual thin ones. Overall it has been shown that electrically thick rectangular microstrip antennas retain most of the desirable electrical characteristics of thinner ones and may be utilized for broad-band applications assuming a reactive network is used for impedance matching.

References

[1] D. H. Schaubert, F. G. Farrar, A. R. Sindoris and S. T. Hayes, "Microstrip antennas with frequency agility and polarization diversity," *IEEE Trans. Antennas Propagat.*, vol. AP-29, pp. 118–123, 1981.

[2] S. A. Long and M. D. Walton, "A dual-frequency stacked circular-disk antenna," *IEEE Trans. Antennas Propagat.*, vol. AP-27, pp. 270–273, 1979.

[3] W. F. Richards, S. E. Davidson, and S. A. Long, "Dual-band reactively loaded microstrip antenna," *IEEE Trans. Antennas Propagat.*, vol. AP-33, pp. 556–561, 1985.

[4] A. Sabban, "A new broadband stacked two-layer microstrip antenna," in *IEEE Antenna Propagat. Soc. Int. Symp. Digest,* 1983 pp. 63–66.

[5] K. F. Lee, K. Y. Ho, and J. S. Dahele, "Circular-disk microstrip antenna with an air gap," *IEEE Trans. Antennas Propagat.*, vol. AP-32, pp. 80–884, 1984.

[6] G. Kumar and K. C. Gupta, "Broadband microstrip antennas using coupled resonators," in *IEEE Antennas Propagat. Soc. Int. Symp. Dig.*, 1983, pp. 67–70.

[7] Hewlett Packard Appl. Note AP-221A, pp. 5-8, June 1980.

[8] W. F. Richards, J. R. Zinecker, R. D. Clark and S. A. Long, "Experimental and theoretical investigation of the inductance associated with a microstrip antenna feed," *Electromagn.*, vol. 3, pp. 327–346, 1983.

[9] D. M. Pozar, "Considerations for millimeter wave printed antennas," *IEEE Trans. Antennas Propagat.*, vol. AP-31, pp. 740–747, 1983.

[10] K. R. Carver and J. W. Mink, "Microstrip antenna technology," *IEEE Trans. Antennas Propagat.*, vol. AP-29, pp. 2–24, 1981.

[11] J. R. James, P. S. Hall, and C. Wood, *Microstrip Antennas-Theory and Design.* Stevenage, U.K.: Peter Peregrinus Ltd., 1981.

[12] A. G. Derneryd, "The circular microstrip antenna element," in *Proc. Inst. Elec. Eng. Int. Conf. Antennas Propagat.*, Oct. 1978, pp. 307–310.

[13] E. O. Hammerstad, "Equations for microstrip circuit design," in *Proc. 5th European Micro. Conf.*, Hamburg, Sept. 1975, pp. 268–272.

[14] M. V. Schneider, "Microstrip dispersion," *Proc. IEEE,* pp. 144–146, Jan. 1972.

[15] E. Chang, "An experimental study of electrically thick rectangular microstrip antennas," M.S. thesis, Dept. Elec. Eng., Univ. Houston, University Park, 1985.

[16] D. L. Sengupta, "Approximate expressions for the resonant frequency of a rectangular patch antenna," *Electron. Lett.*, pp. 834–835, July 29, 1983.

The Effect of Various Parameters of Circular Microstrip Antennas on Their Radiation Efficiency and the Mode Excitation

A. A. KISHK, STUDENT MEMBER, IEEE, AND LOTFOLLAH SHAFAI, SENIOR MEMBER, IEEE

Abstract—**The numerical solution of circular microstrip antenna is carried out using the method of moment. The effect of the probe position, The dielectric permittivity of the substrate, and the substrate thickness on the radiation pattern and the mode excitation efficiency are studied. It is found that the probe position and the patch size can be used to control the mode excitation efficiency, and heigher order modes can be generated using only one feed location. Also, the finite ground plane can be used to improve the symmetry of the radiation patterns. The technique is general and can be used to investigate other scattering and antenna problems involving axisymmetric geometries.**

INTRODUCTION

MICROSTRIP ANTENNAS are one of the most popular antenna types, since they are lightweight, have simple geometries, are inexpensive to fabricate and can be easily made conformal to the host body. These attractive features have increased their application recently and stimulated an ever increasing effort to investigate their performance. The analysis of the microstrip antenna is normally difficult to handle, which is primarily due to the existence of a dielectric substrate to support the conductor. Early studies have therefore been focused on developing approximate methods, such as the transmission line model [1], [2], cavity model [3], [4], modal analysis [5], [6] and a full wave analysis method for the rectangular [7] and for circular geometries [8]. A numerical method was developed by Newman [9] using the moment method with the image theory to calculate the input impedance of a rectangular patch. Shortly after Baily and Deshpande [10], [11] used the same technique to calculate the input impedance using an exact dyadic Green's function. Each method has made a certain approximation to simplify the problem and in particular has obtained solutions for an infinite substrate and ground plane geometry, which is not the case practically.

In this paper, a rigorous treatment of the problem is carried out using a numerical method applicable to circular microstrip patch geometries. Integral equations are developed which are valid for a multiple region problem consisting of dielectrics and conductors. These integral equations are then applied to rotationally symmetric objects and reduced to a matrix equation using the procedure common in solving the problem of bodies of revolution [12], [13]. When the method is applied to a circular microstrip antenna, it provides a convenient approach to solve for the contribution of various modes that are present in the structure and correspond to those of the modal expansion method. The difference, however, is that here the excitation efficiency of each mode can be determined accurately, and each mode's radiation patterns in combined or individual forms can be calculated. Also, the effect of the geometrical and physical parameters of the the antenna such as the patch size, substrate permittivity and height, and the size of the ground plane on the mode resonances and their relative magnitude can be determined. The number and location of the excitation probes are also significant to the relative intensity of the modes and their effects can be studied by the present method.

FORMULATION OF THE PROBLEM

The formulation of the problem is in terms of the surface integral equations. These equations are exact. Consequently, the accuracy of the solution depends on the nature of the numerical technique selected to solve them. The derivation of the equations governing the problem may be based on the equivalence principle [11], [12]. Fig. 1(a) shows the general electromagnetic problem under consideration, where a dielectric object is partially coated with a conductor. The surfaces S_{ce}, S_{cd}, and S_{de} refer, respectively to the boundaries between the conductor and the exterior region, conductor and dielectric, and dielectric and the exterior region. Also, $\vec{E}^d$, $\vec{H}^d$ and $\vec{E}^e$, $\vec{H}^e$ refer to the field vectors within the dielectric and the exterior regions, respectively.

In Fig. 1(a), V^d is a finite volume filled with a homogeneous material of permittivity ϵ_d and permeability μ_d and bounded by two surfaces S_{de} and S_{cd}. The surface S_{cd} may consist of several subsurfaces, to represent multiple dielectric and conducting interfaces. V^e represents the external region, with a permittivity of ϵ_e and permeability of μ_e, and is bounded by two surfaces S_{de} and S_{ce}. Again, the surface S_{ce} may consist of several subsurfaces. In the present work, all these surfaces are assumed to be rotationally symmetric, to represent bodies of revolution. The sources of the electromagnetic excitation are provided by the impressed electric and magnetic currents $\vec{J}^{id}$ and $\vec{M}^{id}$ in V^d.

The equivalence principle [14] is used to obtain the auxiliary problems shown in Figs. 1(b) and 1(c). In Fig. 1(b), the equivalent currents $\vec{J}_{ce}$, $\vec{J}_{de}$, and $\vec{M}$ radiate in the presence of the homogeneous medium (μ_e, ϵ_e) to produce $(\vec{E}^e, \vec{H}^e)$ in V^e

Reprinted from *IEEE Trans. Antennas Propaga.*, vol. AP-34, no. 8, pp. 969–976, Aug. 1986.

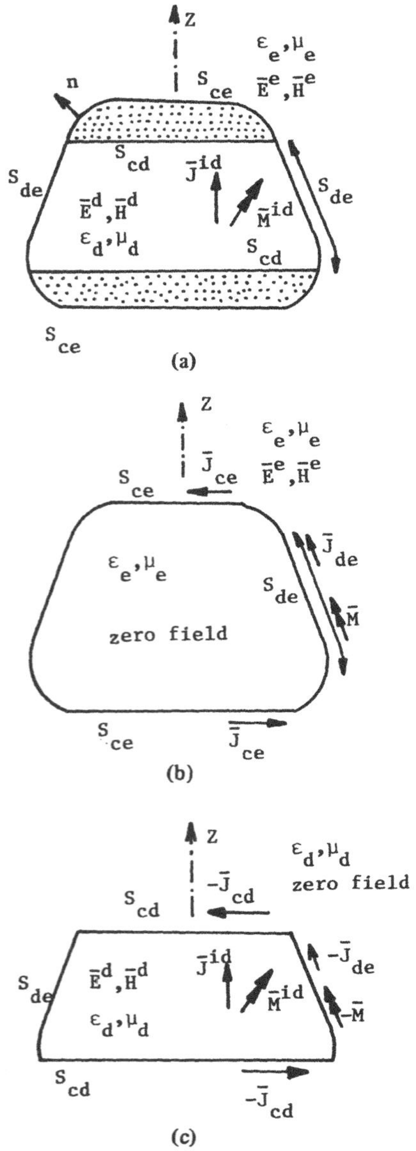

Fig. 1. Problem representation by equivalence principle. (a) Original problem. (b) External equivalence. (c) Internal equivalence.

and zero field elsewhere. Here, $\vec{J}_{ce}$ is an electric current on S_{ce}, $\vec{J}_{de}$ is an electric current on S_{de}, and $\vec{M}$ is a magnetic current on S_{de}. In Fig. 1(c), $(\vec{J}^{id}, \vec{M}^{id})$ and equivalent currents $-\vec{J}_{cd}$, $-\vec{J}_{de}$, and $-\vec{M}$ radiate in the presence of the homogeneous medium (μ_d, ϵ_d) to produce $(\vec{E}^d, \vec{H}^d)$ in V^d and zero field elsewhere. Here, $-\vec{J}_{cd}$ is the electric current on S_{cd}, $-\vec{J}_{de}$ is an electric current on S_{de}, and $-\vec{M}$ is a magnetic current on S_{de}. Since the surfaces S_{ce} and S_{cd} are perfectly conducting in the original problem of Fig. 1(a), only equivalent electric currents are needed on them in Figs. 1(b) and 1(c). The choice of $-\vec{J}_{cd}$ rather than $\vec{J}_{cd}$ on S_{cd} in Fig. 1(c) depends on a personal preference. However, the minus sign relationship between the aperture currents in Figs. 1(b) and 1(c) is dictated by the zero field stipulations in Figs. 1(b) and 1(c) and the continuity of the tangential components of the electric and magnetic fields across the aperture in Fig. 1(a). If the zero field stipulations in Figs. 1(b) and 1(c) are enforced, then the minus sign relationship between the aperture currents in these figures ensures the continuity of the tangential fields across the aperture.

The equivalence principle states that the equivalent currents in Figs. 1(b) and 1(c) are unique, but it does not indicate their evaluation method. The expressions in [14] for the equivalent currents in terms of the tangential components of the fields can not be used, since the fields are not known. The equivalent currents can be determined by enforcing the boundary conditions for the fields in Fig. 1(a) as in [15]. These surface equivalent currents are

$$\vec{J}_{cd} = \hat{n} \times \vec{H}^d, \qquad \text{on } S_{cd} \tag{1}$$

$$\vec{J}_{ce} = \hat{n} \times \vec{H}^e, \qquad \text{on } S_{ce} \tag{2}$$

$$\left.\begin{aligned} \vec{J}_{de} &= \hat{n} \times \vec{H}^e, \\ \vec{M} &= -\hat{n} \times \vec{E}^e, \end{aligned}\right. \qquad \text{on } S_{de}. \tag{3}$$

Again, the currents $\vec{J}_{cd}$, $\vec{J}_{ce}$, and $\vec{J}_{de}$ are the equivalent electric currents on each respective surface, and $\vec{M}$ is the magnetic current on the interface surface between the dielectric and the exterior region. The boundary conditions can be written as

$$-\frac{1}{\eta_e} \vec{E}^d_{\tan}(\vec{J}_{cd} + \vec{J}_{de}, \vec{M}) = -\frac{1}{\eta_e} \vec{E}^d_{\tan}(\vec{J}^{id}, \vec{M}^{id}), \qquad \text{on } S_{cd} \tag{4}$$

$$-\frac{1}{\eta_e} \vec{E}^e_{\tan}(\vec{J}_{ce} + \vec{J}_{de}, \vec{M}) = 0, \qquad \text{on } S_{ce} \tag{5}$$

$$-\frac{1}{\eta_e} \vec{E}^e_{\tan}(\vec{J}_{ce} + \vec{J}_{de}, \vec{M}) - \frac{1}{\eta_e} \vec{E}^d_{\tan}(\vec{J}_{cd} + \vec{J}_{de}, \vec{M}) = -\frac{1}{\eta_e} \vec{E}^d_{\tan}(\vec{J}^{id}, \vec{M}^{id}), \qquad \text{on } S_{de} \tag{6}$$

$$-\vec{H}^e_{\tan}(\vec{J}_{ce} + \vec{J}_{de}, \vec{M}) - \vec{H}^d_{\tan}(\vec{J}_{cd} + \vec{J}_{de}, \vec{M}) = -\vec{H}^d_{\tan}(\vec{J}^{id}, \vec{M}^{id}), \qquad \text{on } S_{de} \tag{7}$$

where $\vec{E}^e_{\tan}(\vec{J}, \vec{M})$ and $\vec{E}^d_{\tan}(\vec{J}, \vec{M})$ are the tangential components of the electric fields due to currents $\vec{J}$ and $\vec{M}$, radiating in media characterized by ϵ_e, μ_e and ϵ_d, μ_d, respectively. $\vec{H}^e_{\tan}(\vec{J}, \vec{M})$ and $\vec{H}^d_{\tan}(\vec{J}, \vec{M})$ denote the tangential components of the corresponding magnetic fields. These equations are dependent on the scalar Green's function G^q, which is given by

$$G^q = \frac{\exp(-jk_q R)}{4\pi R} \tag{8}$$

where $R = |\vec{r} - \vec{r}'|$ is the distance between the field point $\vec{r}$ and the source point $\vec{r}'$ on the surface, $k_q = \omega(\epsilon_q \mu_q)^{1/2}$ is the propagation constant and q represents e or d.

Matrix Formulations

The reduction of integral equations to matrix equations involving unknown surface currents follows the procedure well known for bodies of revolution. Here, both electric and magnetic surface currents exist and are represented by Mautz and Harrington [12], [13] and Iskander *et al.* [16] as

$$\vec{J}(\vec{r}') = \hat{u}_t J^t(t, \phi) + \hat{u}_\phi J^\phi(t, \phi) \tag{9}$$

$$\vec{M}(\vec{r}') = \hat{u}_t M^t(t, \phi) + \hat{u}_\phi M^\phi(t, \phi) \tag{10}$$

where $\hat{u}_t$ and $\hat{u}_\phi$ are the unit tangents to the body as in [9] and J^t, J^ϕ and M^t, M^ϕ are the current components. The electric current $\vec{J}$ exists on both conducting and dielectric surfaces, but $\vec{M}$ exists only on the dielectrics. If the electric and the magnetic currents are expanded into Nc and Nd expansion functions, respectively, the surface currents can be represented as

$$\vec{J}(\vec{r})=\sum_{n=-M0}^{M0}\sum_{j=1}^{Nc} I^t_{nj}J^t_{nj}(t,\ \phi)\hat{u}_t+I^\phi_{nj}J^\phi_{nj}(t,\ \phi)\hat{u}_\phi \quad (11)$$

$$\vec{M}(\vec{r})=\eta_e \sum_{n=-M0}^{M0}\sum_{j=Nc+1}^{Nc+Nd} M^t_{nj}K^t_{nj}(t,\ \phi)\hat{u}_t+M^\phi_{nj}K^\phi_{nj}(t,\ \phi)\hat{u}_\phi \quad (12)$$

where J^t_{nj}, J^ϕ_{nj}, K^t_{nj}, K^ϕ_{nj} are expansion functions defined as

$$J^t_{nj}=J^\phi_{nj}=K^t_{nj}=K^\phi_{nj}=f_j(t)e^{jn\phi} \quad (13)$$

The range $-M0$ to $+M0$ gives the total number of azimuthal modes. The coefficients I^t_{nj}, I^ϕ_{nj}, M^t_{nj}, M^ϕ_{nj} are the current coefficients to be determined by solving the matrix equation which results when (11) and (12) are substituted into (4) to (7) and the inner product, integrated over the surface, of the resulting equation with testing functions $\vec{W}^t_{li}$ and $\vec{W}^\phi_{li}$ are carried out. The testing functions are

$$\vec{W}^t_{li}=\hat{u}_t f_i(t)e^{-jl\phi} \quad (14)$$

$$\vec{W}^\phi_{li}=\hat{u}_\phi f_i(t)e^{-jl\phi} \quad (15)$$

and the details of above steps are provided in [12], [13].

The general matrix form takes the form

$$[\bar{T}_n][\bar{I}_n]=[\bar{V}_n],\ n=0,\ \pm 1,\ \pm 2,\ \cdots \quad (16)$$

where $\bar{T}_n$ is a square matrix, representing the impedance and the admittance submatrices, $\bar{I}_n$ is a column matrix for the unknown expansion coefficients of $\vec{J}$ and $\vec{M}$, and $\bar{V}_n$ is the excitation column matrix. Each mode has a matrix equations of the form

$$\begin{bmatrix} Z^e_{ce,ce} & 0 & Z^e_{ce,de} & Y^e_{ce,de} \\ 0 & \eta_r Z^d_{cd,cd} & \eta_r Z^d_{cd,de} & Y^d_{cd,de} \\ Z^e_{de,ce} & \eta_r Z^d_{de,cd} & Z^e_{de,de}+\eta_r Z^d_{de,de} & Y^e_{de,de}+Y^d_{de,de} \\ Y^e_{de,ce} & Y^d_{de,cd} & Y^e_{de,de}+Y^d_{de,de} & -Z^e_{de,de}-\dfrac{1}{\eta_r}Z^d_{de,de} \end{bmatrix} \begin{bmatrix} I_{ce,n} \\ I_{cd,n} \\ I_{de,n} \\ M_{de,n} \end{bmatrix} = \begin{bmatrix} 0 \\ -V^d_{cd,n} \\ -V^d_{de,n} \\ -I^d_{de,n} \end{bmatrix} \quad (17)$$

where $\eta_r = \eta_d/\eta_e$ and $V^d_{cd,n}$, $V^d_{de,n}$ and $I^d_{de,n}$ are the excitation submatrices, due to the electric and magnetic field sources on the surfaces S_{cd}, S_{de} from the interior region, respectively. The submatrices Z and Y with superscripts e and d denote the impedance and admittance matrices for the exterior or interior media, respectively, the first and second pairs of suffixes identify the field and source surfaces, and the index n implies azimuthal mode number. $I_{ce,n}$, $I_{cd,n}$, $I_{de,n}$ and $M_{de,n}$ are the unknown expansion coefficients of the electric and magnetic currents on S_{ce}, S_{cd}, and S_{de} respectively. In the above equations, each submatrix Y^q or Z^q consists of four submatrices, which are obtained by the procedure used in [9].

Excitation Matrix

For the microstrip problem the coaxial feed probe is simulated by an electric dipole in the dielectric substrate. The electric and magnetic fields due to an electric dipole are

$$\vec{E}^{\mathrm{inc},\,q}=-j\omega\vec{A}^q-\nabla\Phi^q \quad (18)$$

$$\vec{H}^{\mathrm{inc},\,q}=-\frac{1}{\mu_q}\nabla\times\vec{A}^q \quad (19)$$

where

$$\vec{A}^q=\frac{-jk_q\mu_q}{4\pi r}\vec{I}l\ h^{(2)}_0(k_q|\vec{r}-\vec{r}'|) \quad (20)$$

$$\Phi^q=\frac{\eta_q}{4\pi r}\vec{I}l\cdot\nabla h^{(2)}_0(k_q|\vec{r}-\vec{r}'|) \quad (21)$$

and $h^{(2)}_0$ is the spherical Hankel function of the second kind and zero order and $\vec{I}l$ is the dipole moment, in the z-direction. If the Hankel function is represented by

$$h^{(2)}_0(k_q|\vec{r}-\vec{r}'|)=\sum_{-\infty}^{\infty} G_m e^{jm(\phi-\phi)} \quad (22)$$

with

$$G_m=G_{-m}=\begin{cases}\displaystyle\sum_{n=m}^{\infty} a_n h^{(2)}_n(k_q r')j_n(K_q r)P^m_n(\cos\theta)P^m_n(\cos\theta'), & r'>r \\ \displaystyle\sum_{n=m}^{\infty} a_n h^{(2)}_n(k_q r)j_n(k_q r')P^m_n(\cos\theta)P^m_n(\cos\theta'), & r'<r\end{cases} \quad (23)$$

where $a_n = (2n + 1)(n - m)!/(n + m)!$, the inner products of $\vec{E}^{\mathrm{inc}}$ and $\vec{H}^{\mathrm{inc}}$ with testing functions $\vec{W}_{li}$ provide the elements of the excitation matrix. This column matrix has the elements in the form

$$(V^q_m)^t_{iz}=(\eta_q|Il|_z/2)\left[-k^2_q\int_0^{tu}dt\ \rho f_i(t)\cdot\cos\nu G_m-\int_0^{tu}dt\frac{d}{dt}(\rho f_i(t))\frac{1}{\rho}\frac{\partial G_m}{\partial\theta}\right] \quad (24)$$

$$(V^q_m)^\phi_{iz}=(\eta_q|Il|_z/2)\left(jm\int_0^{tu}dt\frac{1}{\rho}f_i(t)\frac{\partial G_m}{\partial\theta}\right) \quad (25)$$

$$(I^q_m)_{iz_t} = (-jk_q|Il|_z/2) \int_0^{tu} dt\, f_i(t) \sin \nu G_m \quad (26)$$

$$(I^q_m)^\phi_{iz} = (-jk_q|Il|_z/2) - \int_0^{tu} dt\, \rho f_i(t) \frac{\partial G_m}{\partial r} \quad (27)$$

where ρ is the distance from the field point on the surface to the Z-axis and v is the angle between the Z-axis and the unit tangent t at the field point.

The Measurement Coefficients

Once the induced currents $\vec{J}$ and $\vec{M}$ on the surface are determined after the solution of the matrix equations, the far-field components E_θ and E_ϕ at a far-field point (r_0, θ_0, ϕ_0) can be determined [17] as

$$E_\theta = \frac{-j\omega\mu_e}{4\pi r_0} e^{-jkr_0} F_1(\theta_0, \phi_0) \quad (28)$$

$$E_\phi = \frac{-j\omega\mu_e}{4\pi r_0} e^{-jkr_0} F_2(\theta_0, \phi_0) \quad (29)$$

where F_1 and F_2 are the measurement coefficients in this form

$$F_1(\theta_0, \phi_0) = \int_s \left(\vec{J} \cdot \hat{u}_\theta + \frac{1}{\eta_e} \vec{M} \cdot \hat{u}_\phi\right) e^{-jk_e \hat{r}_0 \cdot \vec{r}} \, ds \quad (30)$$

$$F_2(\theta_0, \phi_0) = \int_s \left(\vec{J} \cdot \hat{u}_\phi - \frac{1}{\eta_e} \vec{M} \cdot \hat{u}_\theta\right) e^{-jk_e \hat{r}_0 \cdot \vec{r}} \, ds \quad (31)$$

where S is the total outer surface of the body, $\hat{r}_0$ is a unit vector in the direction from the origin of the coordinates to the field point, $\vec{r}'$ is the positional vector of the source point (x', y', z') on the body, and $\hat{u}_\theta$ and $\hat{u}_\phi$ are unit vectors in the direction of increasing θ and ϕ, respectively. Note that E_θ and E_ϕ are the total fields in the exterior region.

Results and Discussions

The antenna geometry for a circular microstrip antenna with a finite ground plane is shown in Fig. 2, where the excitation is simulated by an electric dipole immersed in the dielectric substrate under the conducting patch. The patch radius, for all data in this paper, is selected as [4]

$$a_e = \left[1 + \frac{2h}{\pi a \epsilon_r}\left(\ln \frac{\pi a}{2h} + 1.7726\right)\right]^{1/2} \quad (32)$$

where a is the actual radius of the conducting patch, a_e is the effective radius due to the spread of the fringing field from the patch edge to the ground plane, h is the dielectric thickness and ϵ_r is the relative permittivity of the dielectric substrate. The effective radius is calculated from

$$a_e = \frac{K_{nm}}{2\pi\sqrt{\epsilon_r}} \quad (33)$$

where K_{nm} is the mth zero of the derivative of the Bessel function of order n.

In this paper the patch size is selected according to (32) together with (33) which are functions of the substrate height, the dielectric permittivity, and the desired mode of excitation. For each selected patch size the effects of other parameters on the antenna performance are studied and summarized in the following sections.

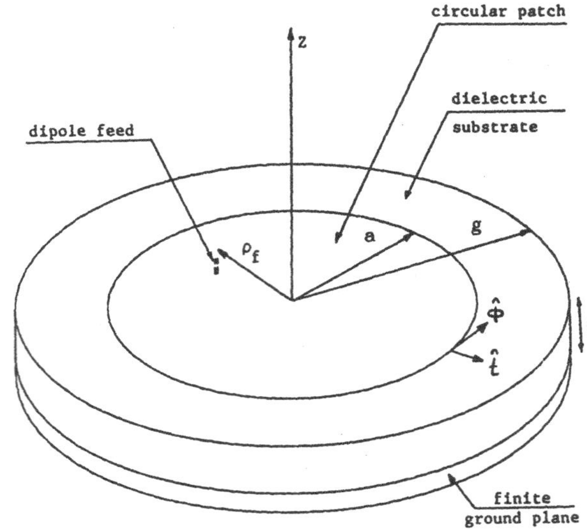

Fig. 2. Microstrip antenna geometry.

A. Feed Location

For coaxial feeds, location is usually selected to provide a good impedance match. Here, the dipole location is selected to ensure the proper excitation of the required mode. Fig. 3 shows the effect of the feed position ρ_f on the excitation efficiency for the TM_{11} mode. In this figure the peaks of the radiation patterns of TM_{01}, TM_{11}, TM_{21} modes are plotted to indicate their relative excitation level. The dominant mode is the strongest for all feed locations and the influence of the TM_{01} and TM_{21} modes increases by moving the feed toward the patch edge. It is evident that increasing the substrate thickness increases the influence of the TM_{01} and TM_{21} modes and decreases the relative excitation of the dominant mode. The excitation of the TM_{21} mode is shown in Fig. 4. Moving the feed away from the patch center increases the excitation of the TM_{21} mode initially, but decreases its excitation for $\rho_f >$ 0.68 a. The excitation of the other modes oscillate around a certain range, which depends on the order of each mode. Although the TM_{21} mode has the highest excitation at $\rho_f =$ 0.68 a, its relative excitation, with respect to the other modes, is strongest around $\rho_f \approx$ 0.75 a. The effect of the substrate permittivity was found to be insignificant. These results indicate that the excitation of TM_{11} or TM_{21} modes can be controlled by the feed location alone and in principle multiple feed locations are not necessary to excite higher order modes.

The radiation patterns of circular patchs for the dominant TM_{11} and TM_{21} modes excitation are shown in Fig. 5. The patterns for TM_{11} mode are calculated by including the contributions of the first four TM_{01}, TM_{11}, TM_{21}, and TM_{31} modes and the patterns for the TM_{21} mode are calculated by including the first five modes, TM_{01}, TM_{11}, TM_{21}, TM_{31} and TM_{41} modes. In each case the feed locations are selected to optimize the dominant mode excitation and the TM_{21} mode excitation, respectively. One example is selected to compare

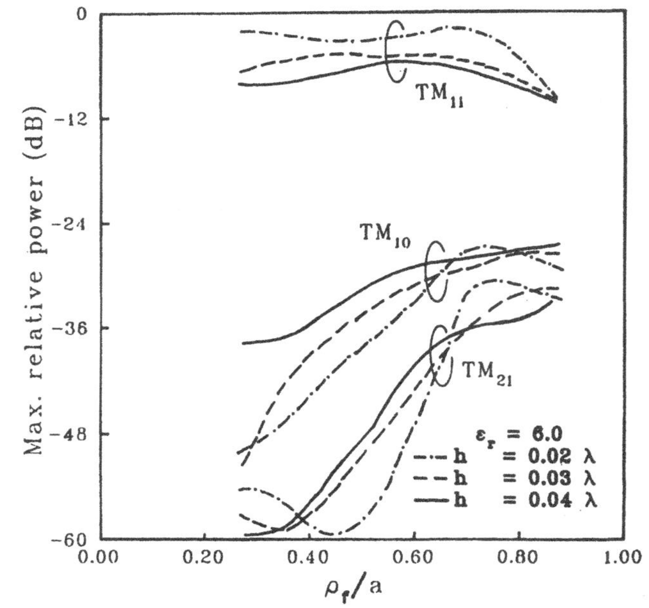

Fig. 3. The effect of the feed position on the excitation efficiency of TM_{11} mode.

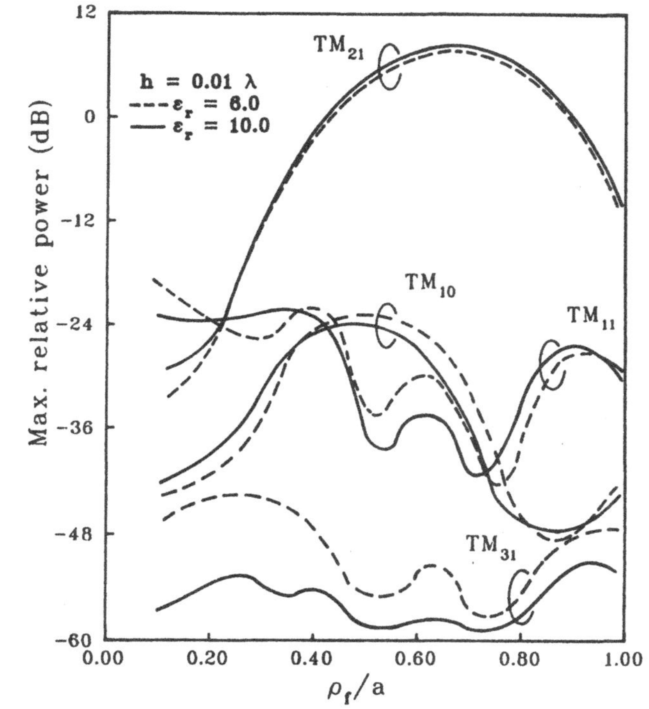

Fig. 4. The effect of the feed position on the excitation efficiency of TM_{21} mode.

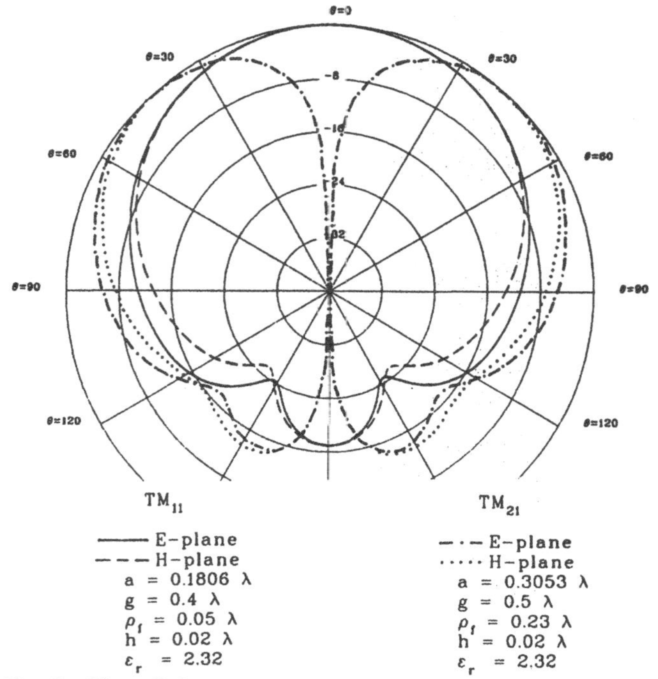

Fig. 5. The radiation patterns of a circular patch for the TM_{11} and TM_{21} mode excitations.

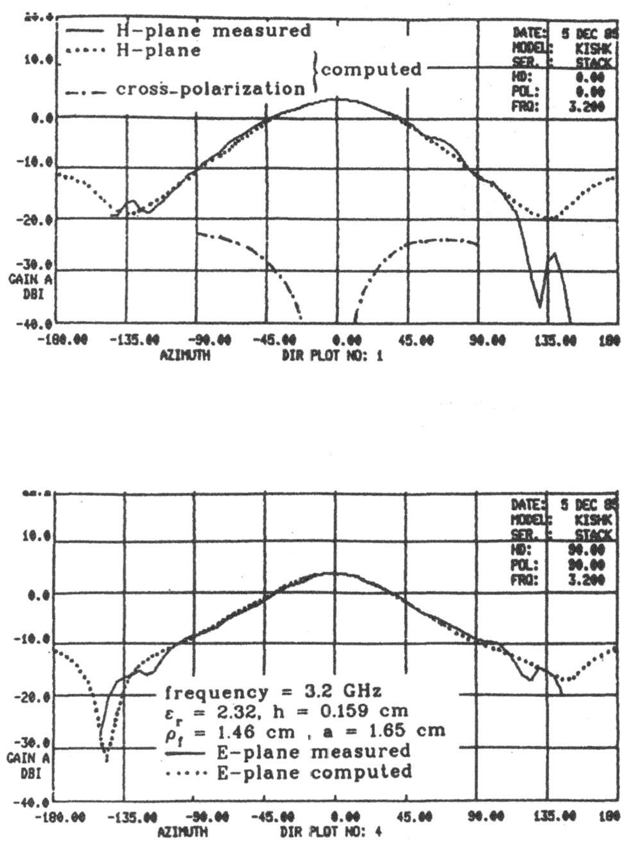

Fig. 6. Experimental and computed data of a circular patch excited with a coaxial probe.

the computed and the measured data as shown in Fig. 6 with the ground plane $g = 0.5\ \lambda$. Excellent agreement between the measured and computed data was found.

In practice, it is also desirable to understand the nature of radiation from a microstrip patch antenna. In the past, it was assumed that the fringing field near the patch edge is responsible for the radiations. To investigate this, the surface currents are also determined and plotted for both TM_{11} and TM_{21} modes. Fig. 7 shows the computed electric and magnetic surface currents for the TM_{11} mode on the outer boundary of

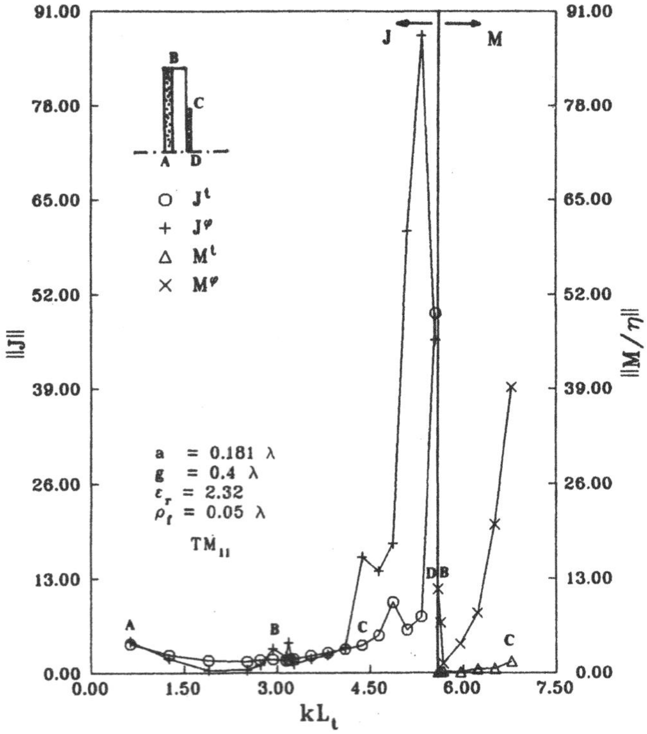

Fig. 7. The computed electric and magnetic surface currents of the TM_{11} mode on the outside boundary.

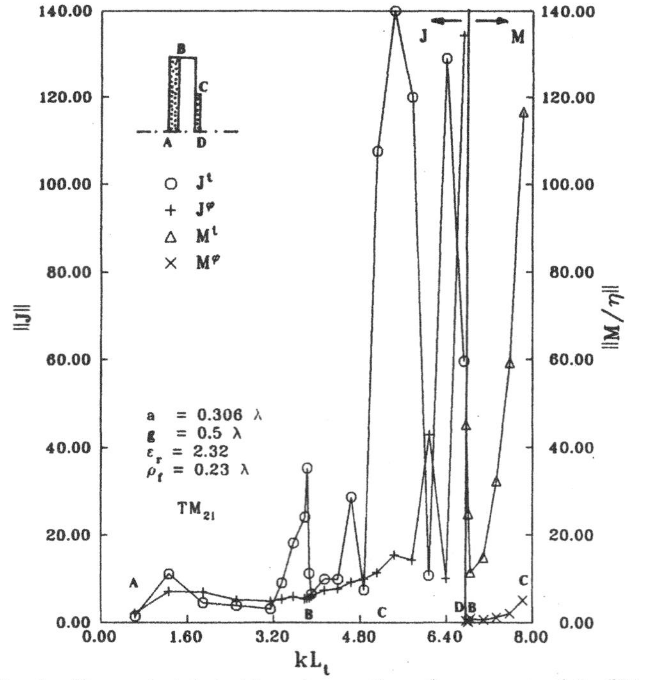

Fig. 8. The computed electric and magnetic surface currents of the TM_{21} mode on the outside boundary.

the microstrip surface. Due to the structure symmetry, only half of the geometry is considered as shown in Fig. 7. In this figure the currents are plotted with respect to its locations of the surface and the points A to B correspond to the ground plane, the points B to C correspond to the dielectric substrate which has electric and magnetic currents and C to D correspond to the patch surface. L_t represent the contour length. The magnetic current M^ϕ is considerably stronger than M^t and is maximum near the patch edge. It reduces towards the dielectric edge, but rises again on its end surface. This indicates that the main radiation comes from the dielectric surface near the patch edge, its truncated end face, and the electric current on the upper patch surface. The corresponding results for the TM_{21} mode are shown in Fig. 8. The main radiation zones are the same as the dominant mode case.

B. The Effect of the Substrate Permittivity

The effect of the substrate permittivity is shown in Fig. 9, which presents the radiation patterns of a circular patch with the dominant mode in resonance. The upper part shows the H-plane and cross-polar patterns and the lower one the E-plane patterns. It can be seen that the beamwidth increases with ϵ_r, but the effect of ϵ_r is stronger on the E-plane than on the H-plane. As a result, increasing the permittivity of the substrates deteriorates the symmetry of the radiation patterns and consequently increases the cross-polar level.

C. Effect of the Substrate Thickness

The bandwidth of microstrip antennas normally increases by increasing the substrate thickness. It is therefore desirable

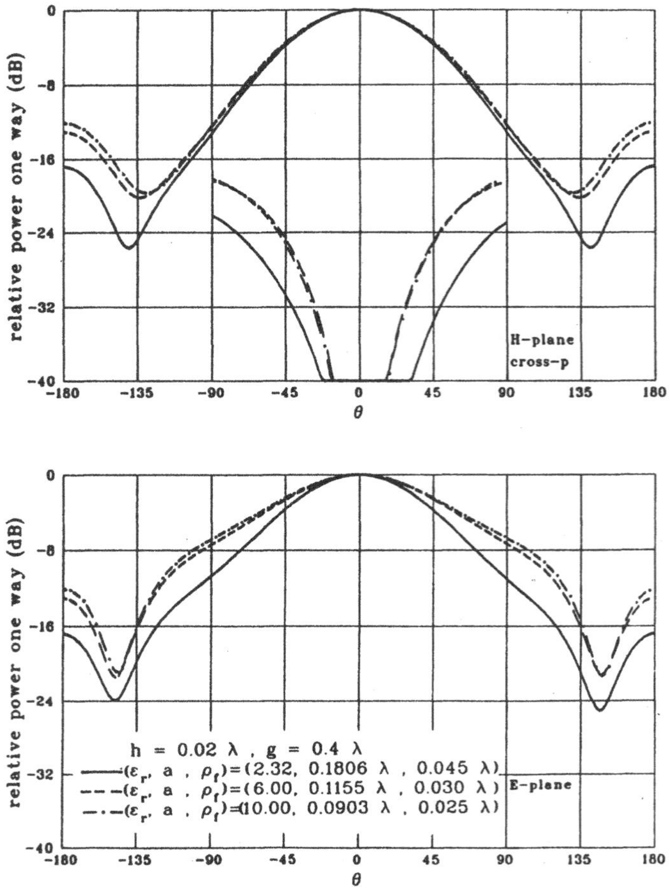

Fig. 9. The radiation pattern of a circular patch with different substrate permittivity.

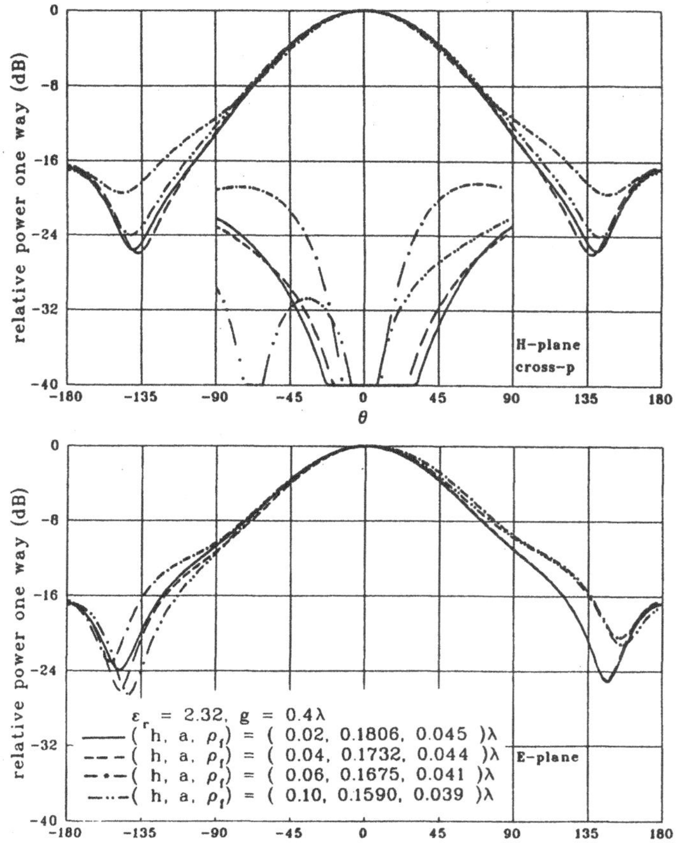

Fig. 10. The radiation patterns of a circular patch with different substrate height.

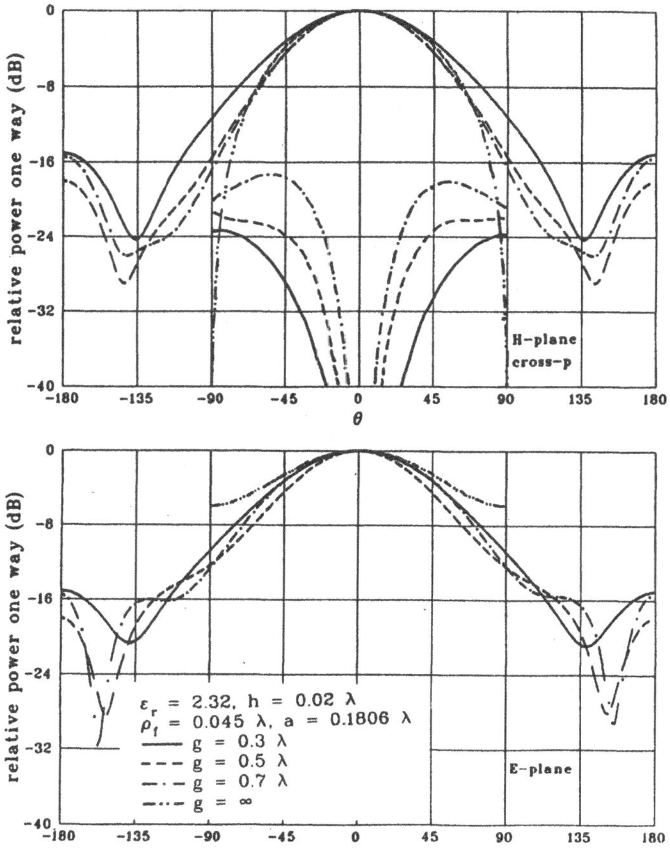

Fig. 11. The radiation patterns of a circular patch with different ground plane diameter for the dominant TM_{11} mode.

to study its effect on the radiation patterns. Fig. 10 shows the radiation patterns for a circular patch when TM_{11} is at resonance. It shows that increasing *h* increases the beamwidth in the *E*-plane, but reduces it in the *H*-plane, until *h* reaches 0.06 λ, after which the relationship reverses. The substrate thickness, generally, has a small effect on the radiation patterns.

D. Effect of the Ground Plane Size

The size of the ground plane has a pronounced effect on the far-field patterns. Fig. 11 shows its effect on the radiation patterns of a dominant TM_{11} mode patch, where the patterns for an infinite ground plane are calculated up to $\theta = 90°$. The results for an infinite ground plane are calculated using the image theory and truncating the dielectric at a radius of 0.4 λ. It seems that increasing the ground plane size increases the beamwidth of the *E*-plane patterns and decreases it for the *H*-plane patterns. This means the pattern symmetry can be improved by modifying the ground plane radius. The corresponding results for the TM_{21} mode are shown in Fig. 12, which are similar to the dominant mode case of Fig. 11. The radiation pattern are broader in the *E*-plane and tend to become narrow in the *H*-plane. Also, the difference between the peaks of the patterns in the *E*- and *H*-planes increases, and, the peaks move in opposite directions. For an infinite ground plane, the peaks move to $\theta = 52°$ and 42° for the *E*- and *H*-planes, respectively.

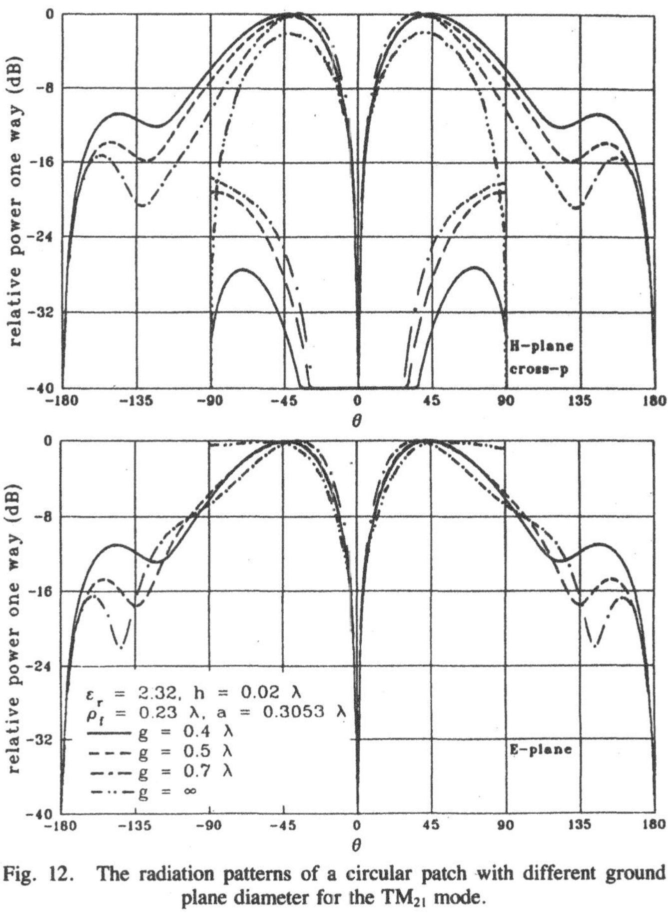

Fig. 12. The radiation patterns of a circular patch with different ground plane diameter for the TM_{21} mode.

Conclusion

The radiation characteristics of a circular microstrip patch antenna were studied numerically using the method of moment. The study included the effects of the finite ground plane, the substrate thickness, the feed locations and the material permittivity. The antenna geometry was considered as a multiple region problem, and the solution was obtained by applying the exact boundary conditions.

It was found that the feed location affects the excitation of each mode, and by its proper selection the resonant mode could be made dominant. In this manner the excitation of the higher order modes could be achieved using only a single feed location rather than multiple feed excitations. It was also shown that the ground plane size can be used to improve the pattern symmetry which is desirable for low cross polarization and circularly polarized applications. The permittivity of the substrate, also, could be used to reduce the physical dimension of the patch and to control the symmetry of the radiation patterns.

The method allowed us to study the effect of each mode separately and to determine their excitation efficiency. The technique can also be used to study the annular ring microstrip patch or annular slot antennas on a finite ground plane, as well as covered microstrip antennas and stacked multiple band configurations.

References

[1] R. E. Munson, "Conformal microstrip antennas and microstrip phased arrays," *IEEE Trans. Antennas Propagat.*, vol. AP-22, pp. 74–78, 1974.

[2] A. G. Derneryd, "Linearly polarized microstrip antennas," *IEEE Trans. Antennas Propagat.*, vol. AP-24, pp. 267–270, 1976.

[3] A. G. Derneryd and A. G. Lind, "Extended analysis of rectangular microstrip antennas," *IEEE Trans. Antennas Propagat.*, vol. AP-27, pp. 846–849, 1979.

[4] I. J. Bahl and R. Bhartia, *Microstrip Antennas.* Dedham, MA: Artech House, 1980.

[5] W. F. Richards, Y. T. Lo, and D. D. Harrison, "An improved theory for microstrip antennas and applications," *IEEE Trans. Antennas Propagat.*, vol. AP-29, pp. 38–46, 1981.

[6] S. Yano and A. Ishimaru, "A theoretical study of the input impedance of a circular disk antenna," *IEEE Trans. Antennas Propagat.*, vol. AP-29, pp. 77–83, 1981.

[7] T. Itoh and W. Menzel, "A full-wave analysis method for open microstrip structures," *IEEE Trans. Antennas Propagat.*, vol. AP-29, pp. 63–68, 1981.

[8] K. Araki and T. Itoh, "Hankel transform domain analysis of open circular microstrip radiating structures," *IEEE Trans. Antennas Propagat.*, vol. AP-29, pp. 84–89, 1981.

[9] E. D. Newman and P. Tulyathan, "Analysis of microstrip antennas using moment methods," *IEEE Trans. Antennas Propagat.*, vol. AP-29, pp. 47–53, 1981.

[10] M. C. Bailey and M. D. Deshpande, "Integral equation formulation of the microstrip antennas," *IEEE Trans. Antennas Propagat.*, vol. AP-30, pp. 651–656, 1982.

[11] M. D. Deshpande and M. C. Bailey, "Input impedance of microstrip antennas," *IEEE Trans. Antennas Propagat.*, vol. AP-30, pp. 645–650, 1982.

[12] J. R. Mautz and R. F. Harrington, "H-field, E-field, and combined field solutions for conducting bodies of revolution," *AEU* 32, 157–164.

[13] ——, "Electromagnetic scattering from a homogeneous material body of revolution," *AEU,* vol. 33, 71–80, 1979.

[14] R. F. Harrington, *Time-Harmonics Electromagnetic Fields.* New York: McGraw-Hill, 1961.

[15] J. R. Mautz and R. F. Harrington, "Boundary formulation for aperture coupling problem," *AEU,* vol. 34, 377–384, 1980.

[16] K. A. Iskander, L. Shafai, A. Frandsen and J. E. Hansen, "Application of impedance boundary conditions to numerical solution of corrugated circular horns," *IEEE Trans. Antennas Propagat.*, vol. AP-30, pp. 366–372, 1982.

[17] S. Silver, *Microwave Antenna Theory and Design.* New York: McGraw-Hill, 1949, pp. 87–89.

Crosspolarisation Characteristics of Circular Patch Antennas

K. F. Lee, K. M. Luk and P. Y. Tam

Indexing terms: Antennas, Polarisation

The crosspolarisation characteristics of coaxially fed microstrip patch antennas are studied using the cavity model. Numerical results showing the variation of crosspolarisation level for different feed positions, substrate thicknesses, substrate permittivities and resonance frequencies are given when the antenna is excited in the TM_{11} mode.

Introduction: The crosspolarisation level due to the excitation of modes with orthogonal polarisation is an important consideration in the design of microstrip antennas. For the rectangular patch, Oberhart *et al.* [1] showed that the quantity $|E_{copol}/E_{xpol}|$ is dependent on the aspect ratio a/b. More detailed results showing the variation of this quantity for different feed positions, substrate thicknesses, substrate permittivities and resonance frequencies are given by Huynh *et al.* [2]. We present similar results for the circular patch microstrip antenna which do not appear to be available in the literature. The calculations are based on the cavity model.

Theory: The geometry of the circular microstrip antenna is shown in Fig. 1. The coaxial feed is located at $(\rho_0, 0)$ and is

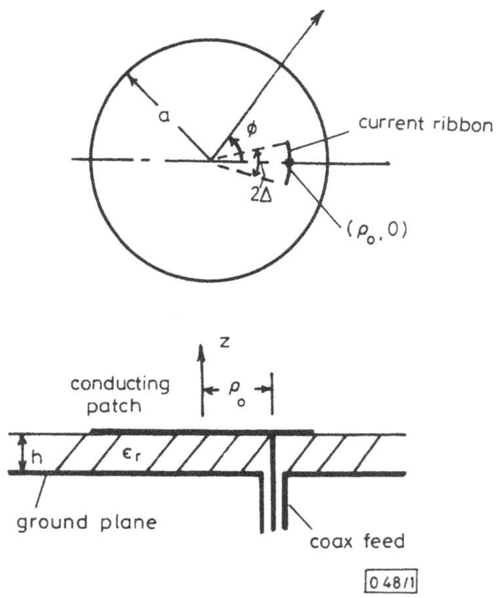

Fig. 1 *Geometry of circular microstrip patch antenna*

modelled by a current ribbon of width $d \simeq 2\rho_0\Delta$ where d is approximately 2·24 times the feed pin diameter. The components of the far-zone electric field for a feed current of 1 A are given by [3]

$$E_\theta = -\frac{k_0^2 a e^{-jk_0 r}\mu_0 h}{\pi r}\sum_{m=1}^{\infty}\sum_{n=0}^{\infty}\frac{j^{n+1}k_{nm}^2 J_n(k_{nm}\rho_0)\,\mathrm{sinc}\,(n\Delta)}{(k^2-k_{nm}^2)J_n(k_{nm}a)(k_{nm}^2a^2-n^2)} \times \cos(n\phi)[J_{n-1}(k_0 a\sin\theta) - J_{n+1}(k_0 a\sin\theta)] \tag{1a}$$

$$E_\phi = \frac{k_0^2 a e^{-jk_0 r}\mu_0 h}{\pi r}\cos\theta\sum_{m=1}^{\infty}\sum_{n=0}^{\infty}\frac{j^{n+1}k_{nm}^2 J_n(k_{nm}\rho_0)\,\mathrm{sinc}\,(n\Delta)}{(k^2-k_{nm}^2)J_n(k_{nm}a)(k_{nm}^2a^2-n^2)} \times \sin(n\phi)[J_{n-1}(k_0 a\sin\theta) + J_{n+1}(k_0 a\sin\theta)] \tag{1b}$$

where $J'_n(k_{nm}a) = 0$, $k = k_0\sqrt{[\varepsilon_r(1 - j\delta_{eff})]}$ and δ_{eff} is the effective loss tangent. The other symbols have the usual meanings.

The resonance frequencies of TM_{nm} modes are given by

$$f_{nm} = \frac{\alpha_{nm}c}{2\pi a_{eff}\sqrt{(\varepsilon_r)}} \tag{2}$$

where α_{nm} is the mth zero of the derivative of the Bessel function $J_n(x)$ and a_{eff} is the effective radius [3].

Following Oberhart [1], the crosspolarisation level is defined as the ratio of the maximum magnitude of E_{copol} to the maximum magnitude of E_{xpol} in a specified plane. The third definition of Ludwig [4] is used to define the principal polarisation E_{copol} and the crosspolarisation E_{xpol}:

$$E_{copol} = E_x = E_\theta\cos\phi - E_\phi\sin\phi \tag{3a}$$

$$E_{xpol} = E_y = E_\theta\sin\phi + E_\phi\cos\phi \tag{3b}$$

It follows from eqn. 1 that, in the E plane ($\phi = 0$), $E = 0$ and there are no crosspolarised fields due to orthogonal modes. In the following Section, numerical results will be given illustrating the crosspolarisation level in the H plane ($\phi = 90°$) and in other planes.

Numerical results: Radiation patterns in the H plane ($\phi = 90°$) for both copolar and crosspolar components are given in Fig. 2 for two feed positions, when the antenna is excited at the resonance frequency of the TM_{11} mode. It is noted that the copolar components are maximum in the broadside direction and nearly zero in the endfire direction. On the other hand, the crosspolar components are maximum in the endfire direction and nearly zero in the broadside direction. Varying the

Reprinted with permission from *Elect. Lett.*, K. F. Lee, K. M. Luk and P. Y. Tam, "Crosspolarisation Characteristics of Circular Patch Antennas," vol. 28, no. 6, pp. 587–589, March 1992. © Institution of Electrical Engineers.

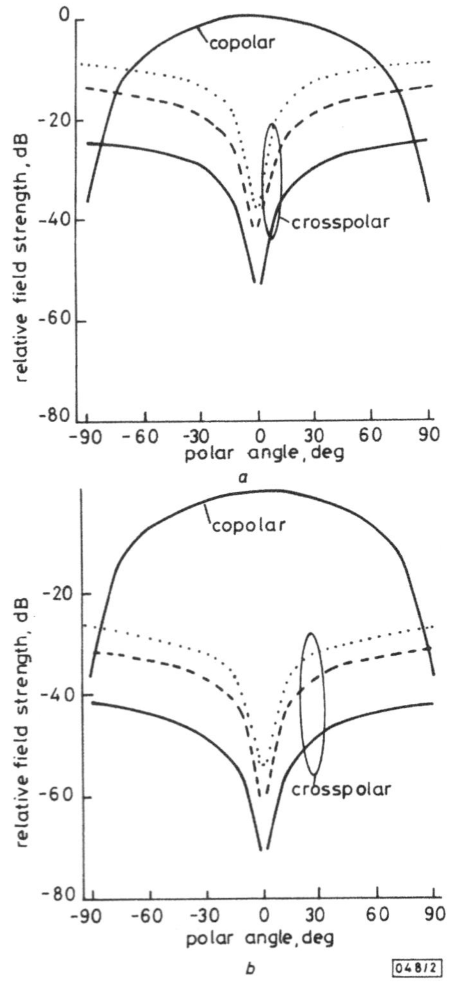

Fig. 2 *Radiation patterns in H plane*

$\varepsilon_r = 2{\cdot}32$, $h = 1{\cdot}59$ mm
a Feed at $\rho_0/a = 0{\cdot}1$
b Feed at $\rho_0/a = 0{\cdot}9$
——— copolar, $f_{11} = 1, 5, 9$ GHz
······ crosspolar, $f_{11} = 9$ GHz
– – – – crosspolar, $f_{11} = 5$ GHz
——— crosspolar, $f_{11} = 1$ GHz

feed position alters the relative sizes of the two components. This is illustrated in Fig. 3, in which it is observed that, as the

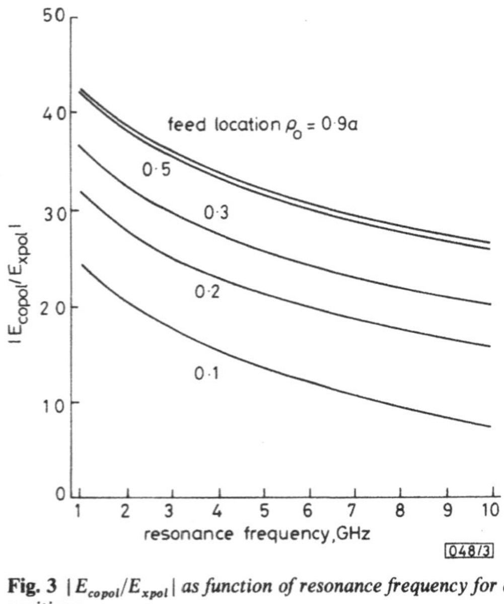

Fig. 3 $|E_{copol}/E_{xpol}|$ *as function of resonance frequency for different feed positions*
$\varepsilon_r = 2{\cdot}32$, $h = 1{\cdot}9$ mm, $\phi = 90°$

feed moves toward the centre of the patch, the cross-polarisation level increases. Also, as the resonance frequency increases, the quantity $|E_{copol}/E_{xpol}|$ decreases. The rate of decrease of $|E_{copol}/E_{xpol}|$ is approximately the same for various feed positions. This feature resembles the results for a rectangular patch [2].

Fig. 4 shows the result for three different substrate thicknesses. It is noted that $|E_{copol}/E_{xpol}|$ increases with decreasing substrate thickness. Fig. 5 is basically the same as Fig. 4 except that $\varepsilon_r = 9{\cdot}8$ instead of $2{\cdot}32$. It is observed that a high ε_r can improve the ratio $|E_{copol}/E_{xpol}|$.

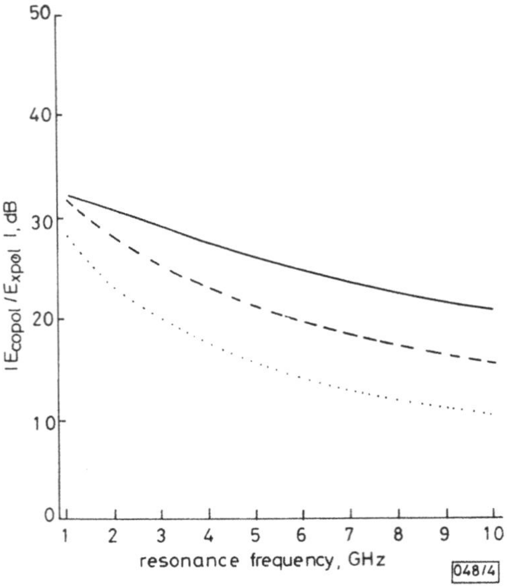

Fig. 4 $|E_{copol}/E_{xpol}|$ *as function of resonance frequency for different substrate thickness;* $\varepsilon_r = 2{\cdot}32$, $\rho_0/a = 0{\cdot}2$, $\phi = 90°$

——— $h = 0{\cdot}795$ mm
– – – – $h = 1{\cdot}59$ mm
······ $h = 3{\cdot}18$ mm

The above results are for the H plane. The ratio $|E_{copol}/E_{xpol}|$ for a number of planes defined by $\phi = 15, 30, 45,$

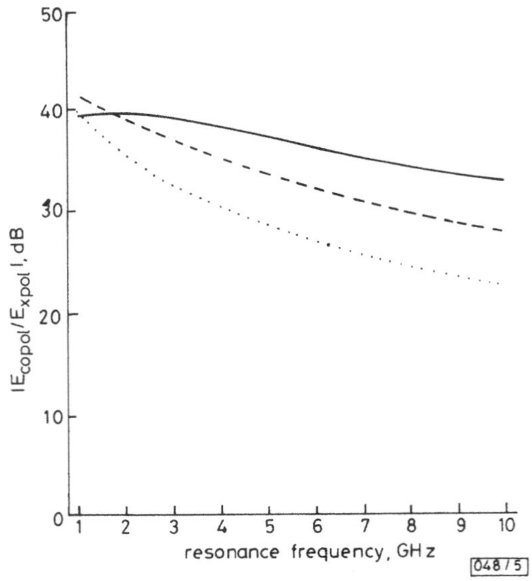

Fig. 5 $|E_{copol}/E_{xpol}|$ *as function of resonance frequency for different substrate thickness;* $\varepsilon_r = 9{\cdot}8$, $\rho_0/a = 0{\cdot}2$, $\phi = 90°$

——— $h = 0{\cdot}795$ mm
– – – – $h = 1{\cdot}59$ mm
······ $h = 3{\cdot}18$ mm

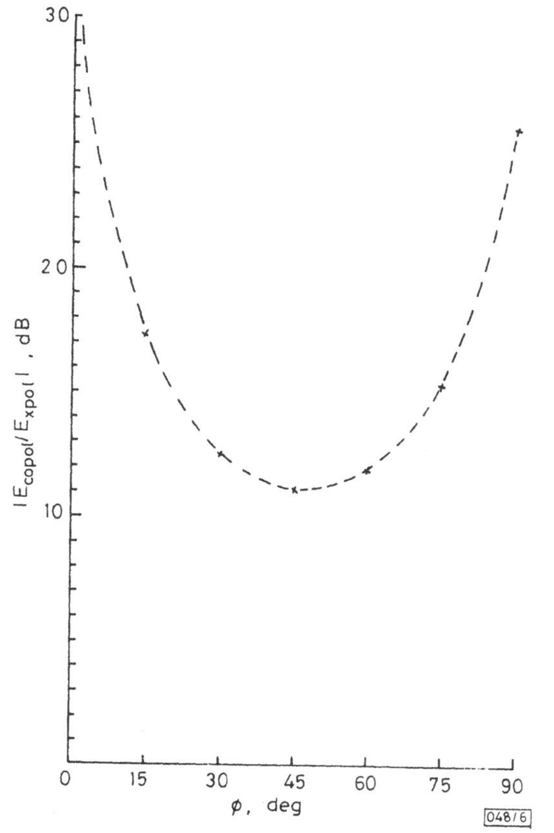

Fig. 6 *$|E_{copol}/E_{xpol}|$ in planes defined by various values of ϕ*
$\varepsilon_r = 2{\cdot}32$, $h = 1{\cdot}59$ mm, $f_{11} = 3$ GHz, $\rho_0/a = 0{\cdot}2$

60, and 75° are indicated by crosses in Fig. 6. It is seen that the crosspolarisation level is maximum in the plane $\phi = 45°$.

References

1 OBERHART, M. L., LO, Y. T., and LEE, R. Q. H.: 'New simple feed network for an array module of four microstrip elements', *Electron. Lett.*, 1987, **23**, (9), pp. 436–437

2 HUYNH, T., LEE, K. F., and LEE, R. Q.: 'Cross polarisation characteristics of rectangular patch antennas', *Electron. Lett.*, 1988, **24**, (8), pp. 463–464

3 DERNERYD, A. G.: 'Analysis of the microstrip disk antenna element', *IEEE Trans.*, 1979, **AP-27**, (5), pp. 660–664

4 LUDWIG, A. C.: 'The definition of cross polarization', *IEEE Trans.*, 1973, **AP-21**, (1), pp. 116–119

Guidelines for Design of Electromagnetically Coupled Microstrip Patch Antennas on Two-Layer Substrates

GEORG SPLITT and MARAT DAVIDOVITZ, Member, IEEE

Abstract—**Graphical guidelines for design of electromagnetically coupled square and circular microstrip antennas are given. Substrates composed of two different dielectric layers are considered. The analysis is extended to electrically thick substrates. Given the required resonant frequency and the bandwidth, material parameters are selected. Patch dimension and the optimal position of the feedline are obtained thereafter from the provided graphs. The design data were computed by applying the method of moments in the spectral domain to solve the integral equation for the currents on the patch and portion of the microstrip feed line. The integral equation was formulated using the appropriate dyadic Green function for the grounded multilayered slab.**

I. Introduction

Electromagnetically coupled microstrip dipoles and patches have been investigated [2], [3], [7], [9]–[11], [13] and found suitable as single antenna elements or for arrays applications. Several distinct advantages of this type of feed over the direct edge feed and probe feed have been noted. Among them is the possibility of placing the feed network closer to the ground plane, resulting in reduction of radiation from various transmission line discontinuities. At the same time the patch-ground plane spacing can be increased to obtain greater bandwidth. Match of the patch to the feed line is simply achieved by selecting an appropriate line-patch overlap. The absence of physical connections between resonator and feed line facilitates fabrication of the antenna. The results presented here were computed by applying the method of moments to a rigorously derived integral equation for the currents on the patch and the microstrip feed line. Sophisticated basis functions were employed to approximate the patch current. Circular and square antennas (see Fig. 1) with substrates composed of two distinct dielectric layers were considered. Resonant frequency and bandwidth of the fundamental resonant current mode were computed for a wide range of substrate parameters. The patch-to-feed line coupling was studied, for various dielectrics, as a function of the increasing overall substrate thickness. The results are presented as a set of curves intended for use as guidelines for a first-order design.

II. Theory

Detailed presentation of the theory, used to compute the results for the electromagnetically coupled circular and rectangular patch antennas, is given in [2], [3], [11], [13]. A brief outline of the theory is given below.

The dyadic Green function for the grounded multilayered dielectric slab is derived by applying the two-dimensional Fourier transform to the Maxwell equations and decomposing the field into transverse electric and transverse magnetic components. The problem is thereby reduced to a set of one-dimensional transmission line equations, which can be solved by known procedures [4], [5]. The integral equation for the currents on the patch and the feed line is then formulated using the dyadic Green function and forcing the total tangential electric field to vanish on the patch and the feed line. The Galerkin method of moments is applied in the spectral domain to solve the integral equation numerically. Expansion functions for the circular and the square patches consist of combinations of Chebyshev polynomials, with additional factors to incorporate the edge condition. The current on the microstripline is approximated by a subsectional basis set consisting of triangle functions in the direction of current flow and rectangular pulse function across the width of the line. It has been verified [8], [11] that for moderate width of the feed line, fulfilment of the edge condition was of secondary importance. The moment method matrix elements are represented by improper spectral single and double integrals. Accuracy and efficiency of the numerical integration have been significantly improved by acceleration techniques discribed in [2], [11]. Resonant frequencies and quality factors of the fundamental resonant modes of the circular and square patches were found by searching for the complex zero of the generalized impedance matrix determinant [2], [5]. Bandwidth of the antennas was estimated from the quality factor by the formula given in [1]. Input impedance was calculated by performing a standing wave analysis on the line. The line was made several wavelengths long and excited by a voltage gap generator near the unloaded end. It was assumed that in the region away from the voltage source and the patch the line supports only the quasi-transverse electromagnetic mode. The reflection coefficient was derived by finding the minima and maxima of the current standing-wave distribution in this region. The radiation efficiency of the patch antenna was estimated by assuming that only the fundamental resonant mode is excited on the patch, and computing the ratio of the radiated to the total power loss.

Reprinted from *IEEE Trans. Antennas Propaga.*, vol. AP-38, no. 7, pp. 1136–1140, July 1990.

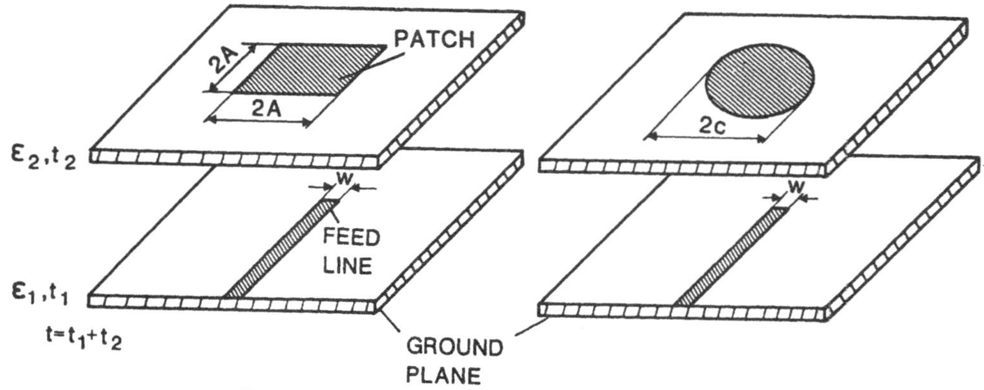

Fig. 1. Electromagnetically coupled microstrip antennas.

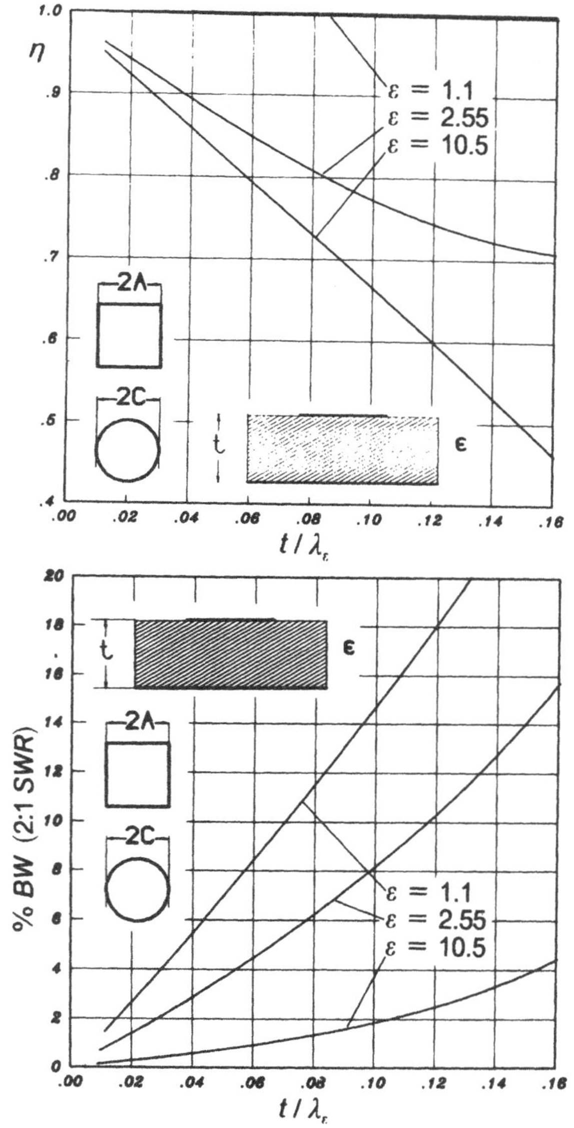

Fig. 2. Bandwidth and efficiency in one-layer dielectric.

The latter consisted of the radiated, surface wave, and dielectric power loss components. Copper loss was neglected, since this communication is concerned primarily with thicker patches, for which it constitutes a small fraction of the total power dissipated.

III. Design Procedures

The radiation efficiency and the impedance bandwidth are two particularly important and critical parameters in microstrip antenna design. A comparison of the data computed for the circular and square patches reveals that the efficiency and bandwidth are almost independent of the patch shape, being determined primarily by the substrate properties, particularly the thickness and permittivity. Therefore, the bandwidth and efficiency plots shown in Fig. 2 for single-layer, and in Fig. 3 for double-layer, substrates apply to both the square and circular patches. The data are presented as functions of the dimensionless parameter $t/\lambda_{\epsilon_1} = t\sqrt{\epsilon_1}/\lambda$, where t is the overall

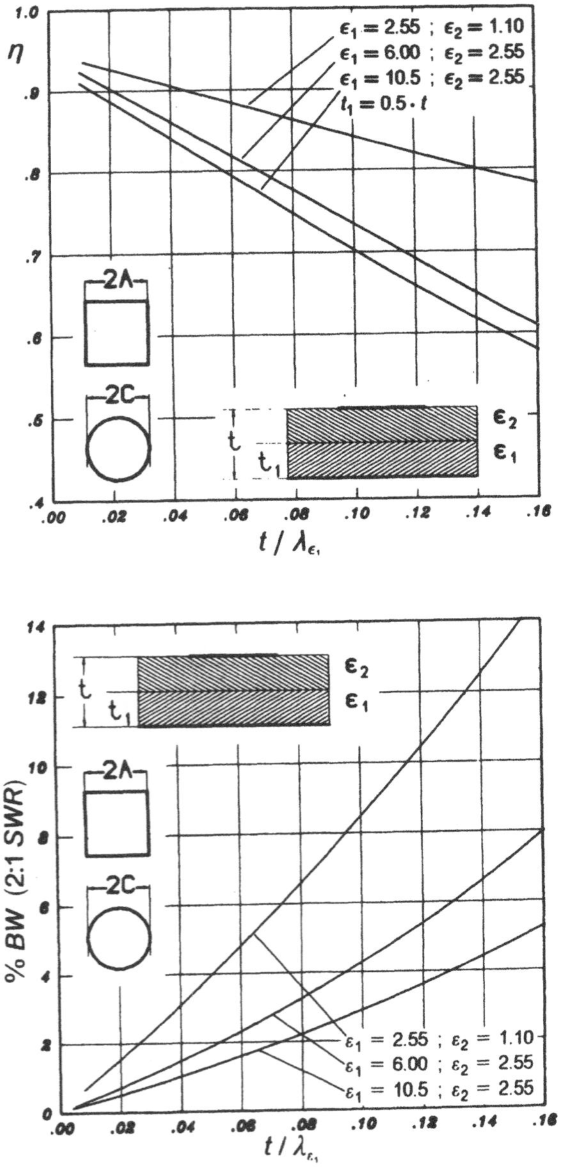

Fig. 3. Bandwidth and efficiency in two-layer dielectric.

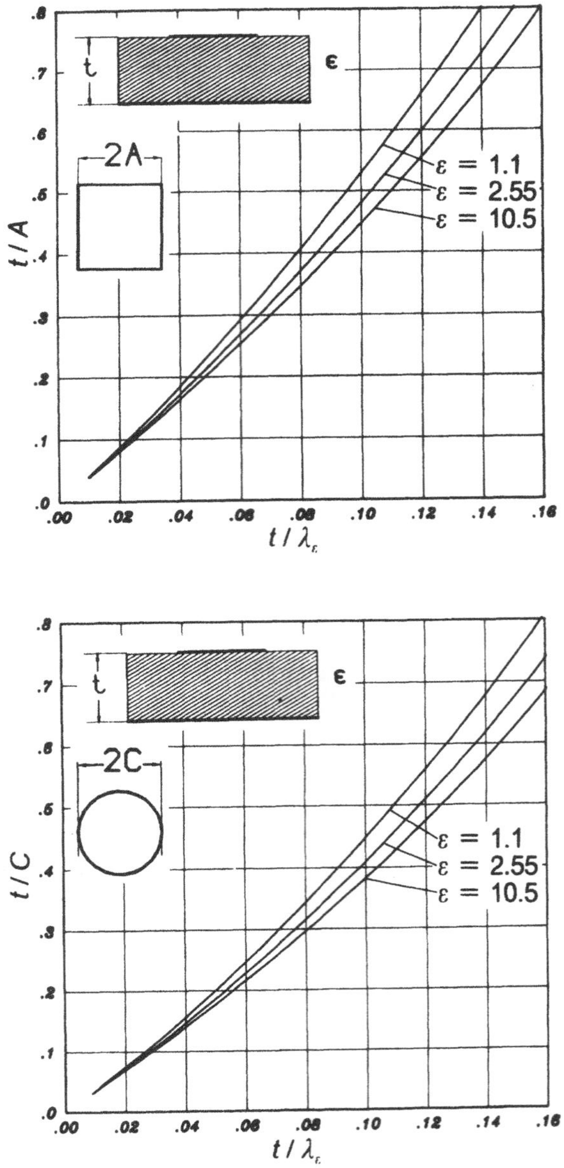

Fig. 4. Resonant frequency and patch size in one-layer dielectric.

substrate thickness, λ is the free-space wavelength corresponding to the resonant frequency, and ϵ_1 is the relative dielectric constant of the lower substrate layer (in the case of a single-dielectric substrate $\epsilon_1 = \epsilon$). These graphs are intended to guide the selection of the substrate parameters, given such design specifications as the resonant frequency, bandwidth, and efficiency.

Having chosen the substrate, the square and circular patch dimensions corresponding to the specified resonant frequency can be determined from Fig. 4 and 5, respectively. The quantities t/A and t/C are presented as functions of $t/\lambda_{\epsilon 1}$, where A,C are the dimensions of the square and circular patches, respectively (see Fig. 1). Note, these graphs also permit the resonant frequency to be computed when the patch dimensions are given.

In all cases of composite substrates considered here, the permittivity of the grounded lower layer is greater than that of top layer, i.e., $\epsilon_1 > \epsilon_2$, with $\epsilon_{1,2}$ defined in Fig. 1. In certain situations it is advantageous to use high permittivity dielectric in order to reduce the size of the feed network, or as in the case of semiconductor substrates to enable integration of active and passive circuit components. On the other hand, surface wave excitation increases with the dielectric constant of the substrate material. This undesirable effect can be reduced by using a lower permittivity dielectric in the feed-to-patch layer, thereby reducing the effective dielectric constant of the composite substrate supporting the patch antenna.

As already mentioned, the spurious radiation from the feed network can, to some measure, be controlled by placing the feed as close as possible to the ground plane. For a given overall substrate thickness (t), the smallest feed-to-ground plane spacing (t_1) for which the patch can still be matched to the feed line, is considered optimal. Fig. 6 shows the dependence of the optimal ratio $t_{1\min}/t$ on the normalized thickness $t/\lambda_{\epsilon 1}$. The lowest ratio is obtained when $\epsilon_1 = \epsilon_2$, i.e., the permittivities of both layers are equal. When $\epsilon_1 > \epsilon_2$, the fields tend to concentrate more in the lower layer. Consequently the feed-to-patch coupling is reduced and the optimal ratio $t_{1\min}/t$ increases. The curves in Fig. 6 represent a straight line fit of points obtained through manual analysis of numerous input impedance calculations.

To demonstrate the utility of the presented graphs, a possible design procedure is outlined:

1) Given the bandwidth and resonant frequency specifications, the substrate parameters can be selected from Figs. 2 and 3,

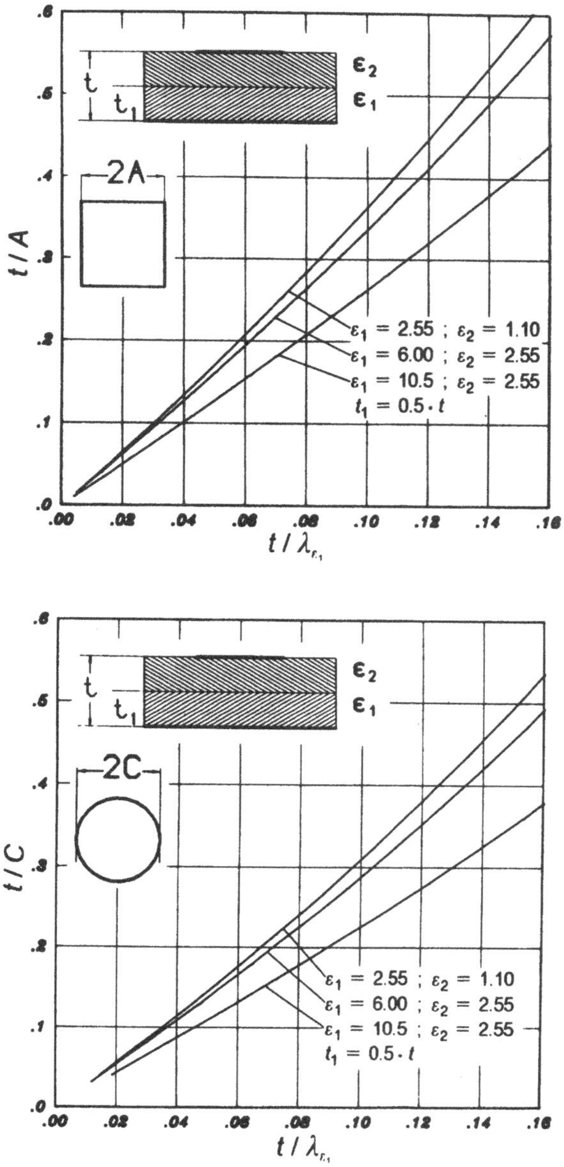

Fig. 5. Resonant frequency and patch size in two-layer dielectric.

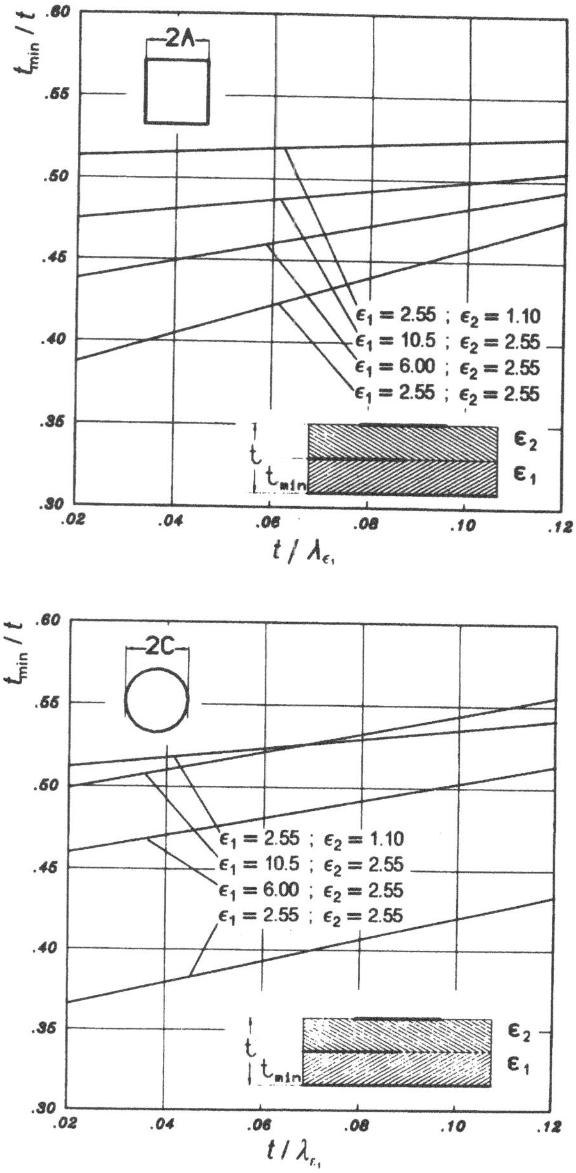

Fig. 6. Optimal choice of the ratio t_{min}/t for characteristic line impedance of 50 Ω.

subject to any efficiency constraints. The trade-off between the efficiency and bandwidth is facilitated by the efficiency plots provided in Figs. 2 and 3. Note, in Fig. 3 the two layers of the composite substrates are of equal thickness, or t_1/t = 0.5. This ratio was chosen because it is sufficiently close to the optimal for most cases of two-layer substrate considered and is easily realizable in practice.

2) Having determined the substrate parameters, the patch dimensions for the specified resonant frequency can be found from Figs. 4 and 5.
3) The optimal ratio $t_{1\min}/t$ for the given t/λ_{ϵ_1} can be obtained from Fig. 6.

Note that patch-to-feed coupling depends upon the extent to which the feed line and the patch overlap. For the optimal thickness ratio, a perfect match is generally possible for only one value of the overlap. Although this value is, to a small degree, thickness dependent, for most cases the match is obtained when the end of the transmission line is located under the center of the patch. Fig. 7 shows the minimum attainable reflection coefficient as a function of the patch-feed overlap. The presence of the feed under the patch also perturbs the resonant frequency from the values given in Figs. 4 and 5. Depending of the value of the overlap, frequency shifts of up to 2% are observed in Fig. 7. This effect must be accounted for in the design process.

IV. Conclusion

Design guidelines for square and circular electromagnetically coupled antennas were presented. The results were computed by using a rigorous moment method formulation, employing the Green function for the double-layered dielectric slab. Therefore, all the significant effects, such as surface waves and radiation, were included. The possibility and advantages of using composite substrates were demonstrated.

For a specified bandwidth and resonant frequency all the necessary antenna parameters, such as the patch size and efficiency, the substrate thickness, as well as the optimal feeding configuration, can be determined from the presented graphs. The designer is provided

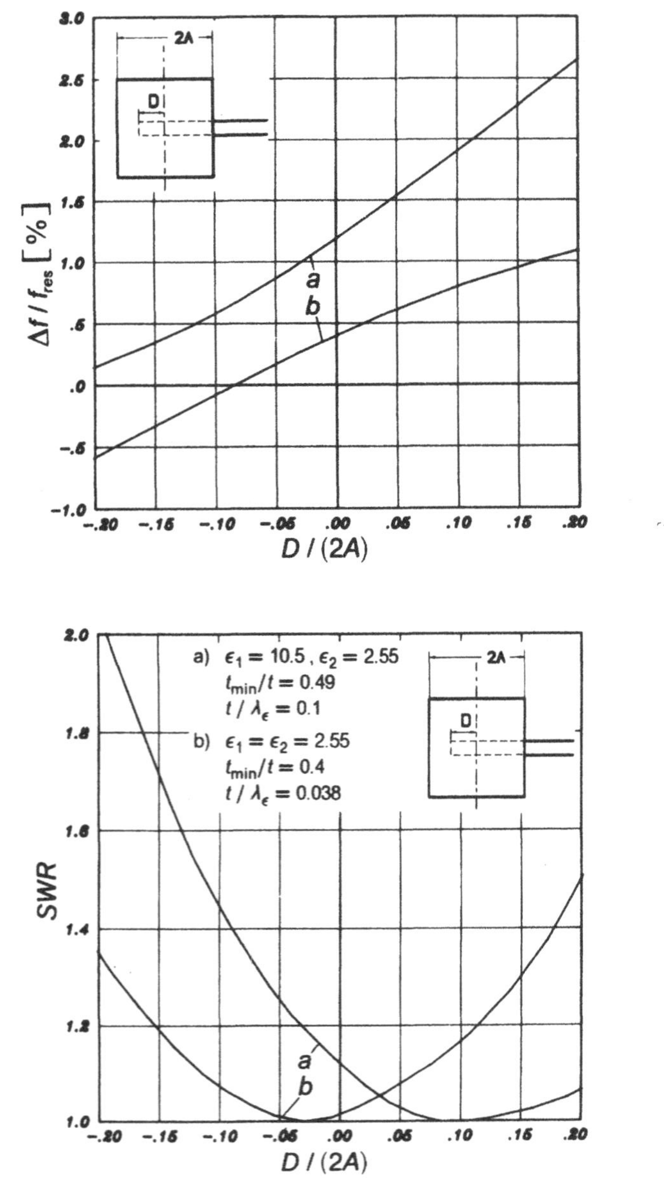

Fig. 7. Frequency shift and VSWR in dependency of line-patch overlap.

with a very good starting point for the final optimization of the antenna.

REFERENCES

[1] K. R. Carver, "Practical analytical techniques for the microstrip antenna," in *Proc. Workshop Printed Antenna Tech.*, New Mexico State Univ., Las Cruces, pp. 1–19, Jan. 1981.

[2] M. Davidovitz, "Feed analysis for microstrip antennas," Ph.D. dissertation, Dept. Elec. Eng. Univ. Illinois, Urbana-Champaign, 1985.

[3] M. Davidovitz and Y. T. Lo, "Rigorous analysis of a circular patch antenna excited by a microstrip transmission line," *IEEE Trans. Antennas Propagat.*, vol. 37, pp. 949–958, Aug. 1989.

[4] L. B. Felsen and N. Marcuvitz, *Radiation and Scattering of Waves.* Englewood Cliffs, NJ: Prentice-Hall, 1973.

[5] T. Itoh and W. Menzel, "A full-wave analysis method for open microstrip structure," *IEEE Trans. Antennas Propagat.*, vol. AP-29, pp. 63–67, Jan. 1981.

[6] D. R. Jackson and N. G. Alexapoulos, "Analysis of planar strip geometries in a substrate-superstrate configuration," *IEEE Trans. Antennas Propagat.*, vol. AP-34, pp. 1430-1438, Dec. 1986.

[7] P. B. Katehi and N. G. Alexopoulos, "On the modeling of electromagnetically coupled microstrip antennas—the printed strip dipole," *IEEE Trans. Antennas Propagat.*, vol. AP-32, no. 11, pp. 1179–1186, Nov. 1984.

[8] Y. T. Lo, S. M. Wright, W. F. Richards and B. F. Wang, "A study of microstrip antennas for multiple band and high frequency operations," Univ. Illinois, RADC-TR-86-8, Mar. 1986.

[9] H. G. Oltman and D. A. Huebner, "Electromagnetically coupled microstrip dipoles," *IEEE Trans. Antennas Propagat.*, vol. AP-29, pp. 151–157, Jan. 1981.

[10] D. M. Pozar and S. M. Voda, "A rigorous analysis of a microstrip fed patch antenna," *IEEE Trans. Antennas Propagat.*, vol. AP-35, pp. 1343–1150, Dec. 1987.

[11] G. Splitt, "Rectangular electromagnetically coupled microstrip antennas in multilayered structures," in *Proc. European Microwave Conf.*, Stockholm, Sweden, 1988, pp. 1043–1048.

[12] ——, "Die Modalanalyse für Microstripantennen unter Einbeziehung der Green'schen Funktion geschichteter Dielektrika," *ITG Fachtagung Antennen in Würzburg,* band 99, pp. 233–239, 1987.

[13] ——, "Moment method for electromagnetically and edge-feed coupled microstrip antennas," presented at SMBO Int. Microwave Symp., Brazil, July 1989.

[14] N. K. Uzunoglu, N. Alexopoulos, and J. K. Fikioris, "Radiation properties of microstrip dipoles," *IEEE Trans. Antennas Propagat.*, vol. AP-27, pp. 853–858, 1979.

[15] J. R. James, P. S. Hall, and C. Wood, *Microstrip Antennas: Theory and Design.* New York: Peregrinus, 1981.

Design of Microstrip Antennas Covered with a Dielectric Layer

INDER J. BAHL, Member, IEEE, PRAKASH BHARTIA, Senior Member, IEEE, and STANISLAW S. STUCHLY, Senior Member, IEEE.

Abstract—**The design of a microstrip antenna covered with a dielectric layer is presented. Due to loading, the resonant frequency of the antenna changes. The absolute value of the change increases with the operating frequency, the relative permittivity (except plasma), and the thickness of the dielectric layer. This change may cause degradation in performance due to the inherent narrow bandwidth of microstrip antennas if the effect of loading is not considered in the design. The curves presented here may be used to design microstrip antennas that may be subjected to icing or a plasma environment or coated with protective layers. Numerical and experimental results for the fractional change in the resonant frequency are found to be in good agreement.**

INTRODUCTION

Microstrip antennas have been employed in airborne and spacecraft systems because of their low profile and conformal nature [1]-[6]. Many of these applications require a dielectric cover over the radiating element to provide protection against heat, physical damage, and the environment. In addition, a dielectric cover increases the peak power-handling capability of microstrip antennas [7].

When microstrip antennas are coated with protective layers, are subjected to icing conditions, or come into contact with plasma, the resonant frequency is altered, causing detuning which may seriously degrade system performance. As the bandwidth of microstrip antennas is inherently low, typically of the order of 1-2 percent [1], it is important to determine the effect of a dielectric layer on the resonant frequency of microstrip antennas in order to introduce appropriate corrections in the design of the antenna.

RESONANT FREQUENCY

For the rectangular (or square) geometry (patch antenna) shown in Fig. 1, the lowest resonant frequency f_r can be accurately predicted from [8]

$$f_r = \frac{c}{2(L + 2\Delta l)\sqrt{\epsilon_e}}, \qquad c = 3 \times 10^8 \text{ m/s} \tag{1}$$

$$\Delta l = 0.412\, h \, \frac{(\epsilon_e + 0.3)}{(\epsilon_e - 0.258)} \, \frac{(W/h + 0.264)}{(W/h + 0.8)} \tag{2}$$

$$\epsilon_e = \frac{\epsilon_r + 1}{2} + \frac{\epsilon_r - 1}{2} \, \frac{1}{(1 + 12h/W)^{1/2}} \quad \text{for } W/h \geqslant 1 \tag{3}$$

where all the dimensional parameters used above are defined in Fig. 1, and ϵ_r and ϵ_e are the relative and effective dielectric constants, respectively.

For microstrip antennas the choice of the width of the patch radiator is very important. Small values of W result in low antenna efficiencies while large W values lead to higher order modes. The optimum value of W is given by [4]

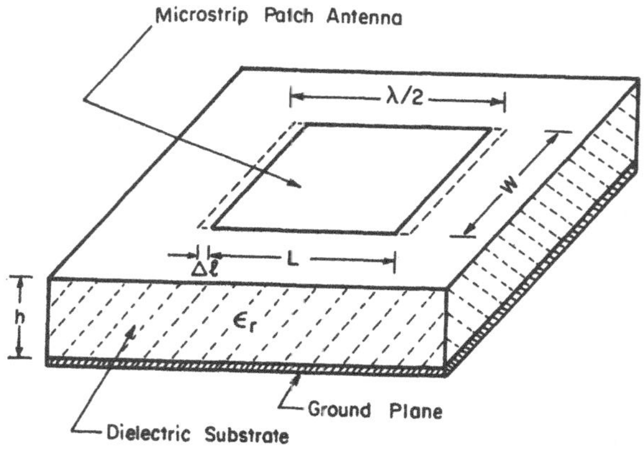

Fig. 1. Microstrip antenna geometry.

$$W = \frac{\lambda_0}{2} \left(\frac{\epsilon_{r+1}}{2}\right)^{-1/2} \tag{4}$$

where λ_0 is the free space wavelength.

MICROSTRIP ANTENNAS COVERED WITH A DIELECTRIC LAYER

When the microstrip line is covered by a dielectric layer the characteristic impedance, phase velocity, losses, and Q factor of the line change as a function of the dielectric constant, loss tangent and thickness of the layer. The configuration under investigation is shown in Fig. 2 (inset). The properties of a microstrip covered by a dielectric layer have been studied by the variational technique [9], [10]. The resonant frequency of a microstrip antenna covered with a dielectric layer can be determined when the effective dielectric constant of the structure is known [8]. An example of the variation of the effective dielectric constant as a function of d/h for various values of W/h is shown in Fig. 2.

For a matched antenna, the change in the fractional resonant frequency relative to the unloaded case can be calculated using the following expression:

$$\frac{\Delta f_r}{f_r} = \frac{f_r(d = 0) - f_r(d)}{f_r(d = 0)}. \tag{5}$$

The first-order change in the resonant frequency may be expressed as

$$\frac{\Delta f_r}{f_r} = \frac{\sqrt{\epsilon_e} - \sqrt{\epsilon_{eo}}}{\sqrt{\epsilon_o}} \tag{6}$$

Reprinted from *IEEE Trans. Antennas Propaga.*, vol. AP-30, no. 2, pp. 314–318, March 1982.

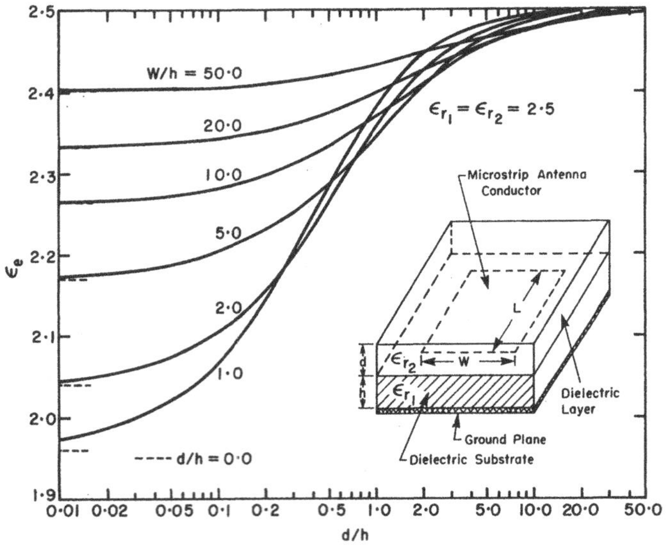

Fig. 2. The effective dielectric constant of a microstrip line covered with a dielectric layer as a function of dielectric cover thickness.

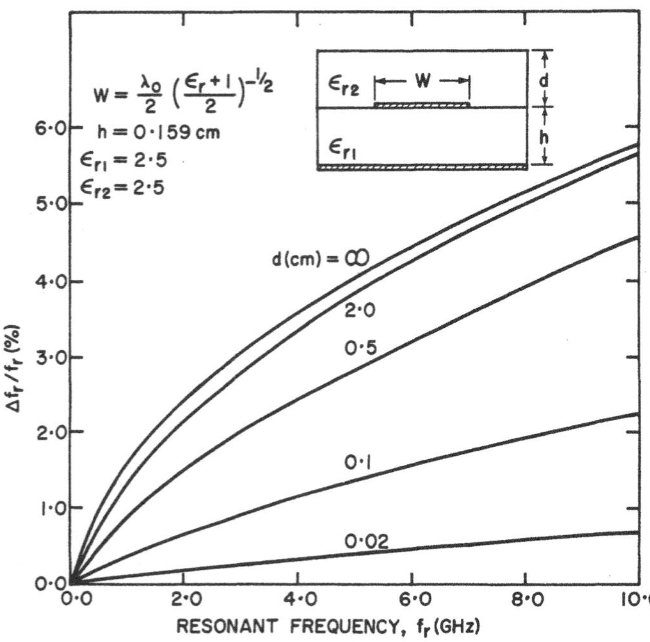

Fig. 3. The fractional resonant frequency of a microstrip antenna covered with a dielectric layer as a function of resonant frequency for $\epsilon_{r2} = 2.5$ (polystyrene).

where ϵ_{eo} is the effective dielectric constant without cover. If $\epsilon_e = \epsilon_{eo} + \Delta\epsilon_e$ and $\Delta\epsilon_e \leqslant 0.1\ \epsilon_{eo}$, then

$$\frac{\Delta f_r}{f_r} = \frac{1}{2}\,\frac{\Delta\epsilon_e/\epsilon_{eo}}{1 + \frac{1}{2}\,\Delta\epsilon_e/\epsilon_{eo}}. \tag{7}$$

The fractional resonant frequency of a microstrip antenna covered with a dielectric layer is plotted as a function of resonant frequency in Figs. 3–6.

Fig. 3 depicts the fractional resonant frequency for $\epsilon_{r1} = \epsilon_{r2} = 2.5$. The decrease in the resonant frequency for thin dielectric layers ($d \leqslant 1$ mm) is less than 1 percent for frequencies below 3 GHz. The maximum change in the resonant frequency for antennas operating below 10 GHz is less than 5.8 percent.

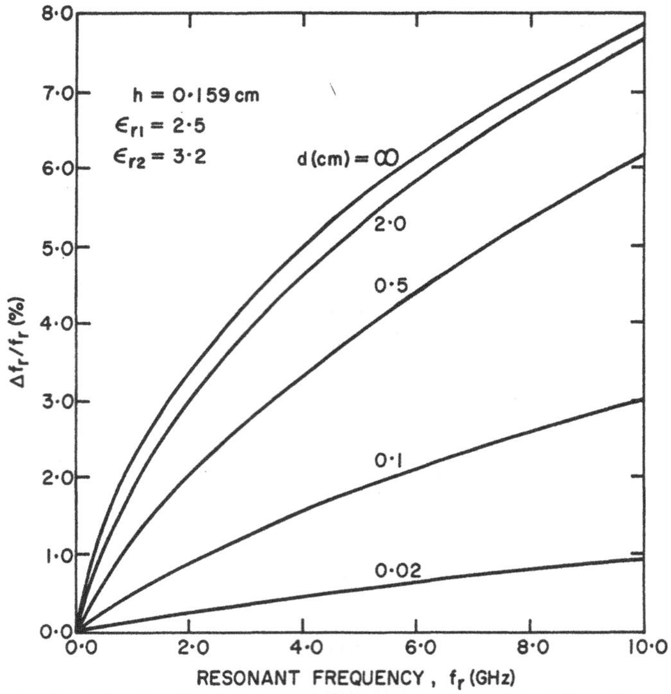

Fig. 4. The fractional resonant frequency of a microstrip antenna covered with a dielectric layer as a function of resonant frequency for $\epsilon_{r2} = 3.2$ (ice).

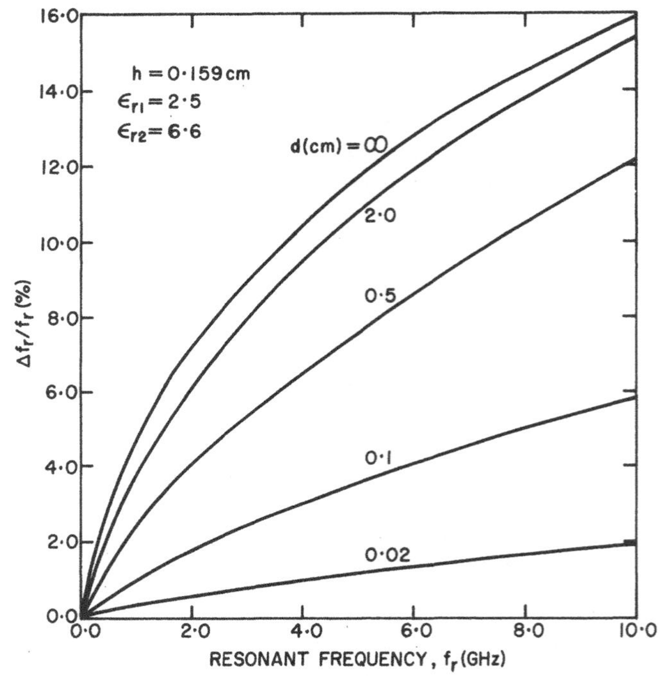

Fig. 5. The fractional resonant frequency of a microstrip antenna covered with a dielectric layer as a function of resonant frequency for $\epsilon_{r2} = 6.6$ (beryllium oxide).

The fractional change of the resonant frequency of an antenna versus operating frequency for various ice layers is shown in Fig. 4. The dielectric constant ϵ_{r2} represents the absolute value of the relative permittivity of ice. The resonant frequency of a rectangular microstrip antenna operating at 10 GHz covered with a semi-infinite ice layer decreases by 7.8 percent as compared to an unloaded antenna. This figure can be used to calculate the change in the resonant frequency of a microstrip antenna subjected to icing conditions.

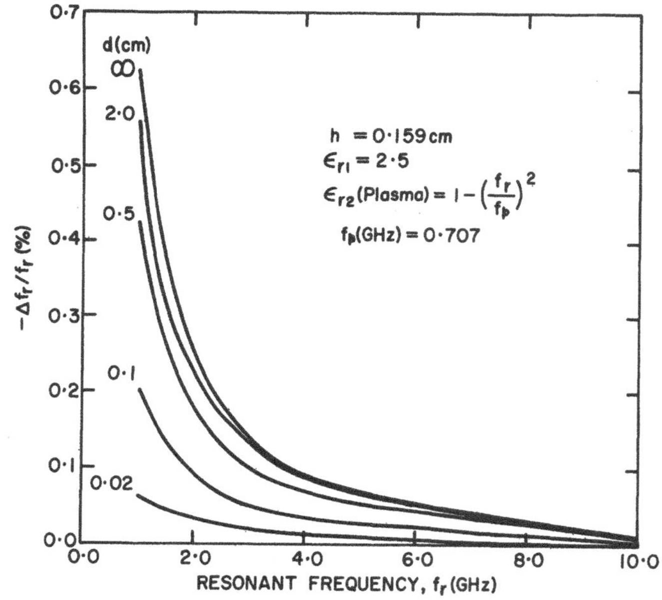

Fig. 6. The fractional resonant frequency of a microstrip antenna covered with a dielectric layer as a function of resonant frequency for plasma.

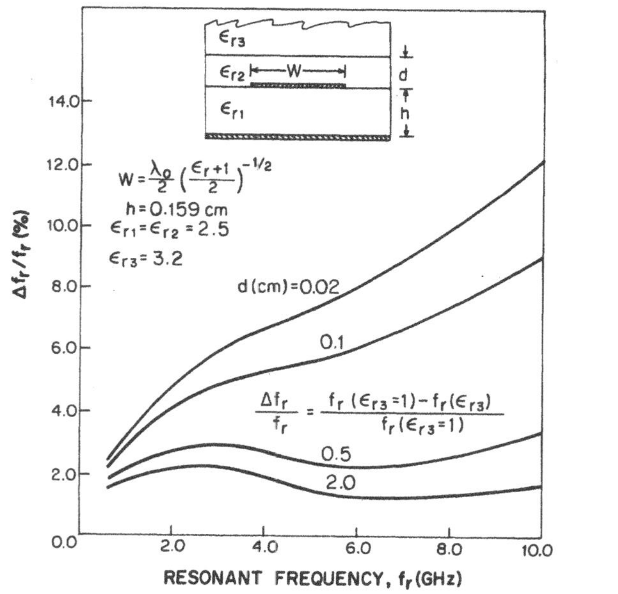

Fig. 7. The fractional resonant frequency of a microstrip antenna with a dielectric cover and loaded with ice, as a function of frequency for various thicknesses of dielectric cover.

A thin layer of beryllium oxide (BeO) over microstrip antennas may be used to increase their average power-handling capability [11]. Fig. 5 shows the fractional resonant frequency of a microstrip antenna covered with BeO layers of various thicknesses as a function of frequency. The maximum change in the resonant frequency of an antenna operating at 10 GHz is about 16 percent. The decrease in the resonant frequency for thin BeO layers ($d \leqslant 1$ mm) is less than 2 percent at frequencies below 2.3 GHz.

Finally, Fig. 6 depicts the fractional resonant frequency of a microstrip antenna in contact with plasma layers of different thickness. In the calculations, the collisions in the plasma were neglected and the plasma frequency was assumed to be 0.707 GHz. It may be noted that the increase in the resonant frequency which is larger at lower frequencies is less than 0.7 percent for frequencies above 1 GHz. This shows that for a microstrip antenna in a plasma medium the detuning is not as serious as that for an antenna subjected to icing.

DESIGN OF A MICROSTRIP ANTENNA WITH TWO DIELECTRIC LAYERS

If the relative permittivity and the thickness of the dielectric protective layer are known *a priori*, the antenna element may be designed using (1)–(3). For example, when the dielectric substrate ($h = 0.159$ cm) and the protective layer for the microstrip antennas are polystyrene ($\epsilon_{r1} = \epsilon_{r2} = 2.5$), a 1-mm thick dielectric cover lowers the resonant frequency of a 10-GHz antenna by 2.25 percent (Fig. 3). Thus a microstrip antenna designed at 10.225 GHz using (1)–(3) will be resonant at 10 GHz when covered with a dielectric layer having dielectric constant of 2.5 and thickness of 1 mm.

However, in the case of microstrip antennas subjected to icing or similar conditions, it is not possible to use the above design procedure since the thickness of the ice layer is not known beforehand. One can, however, cover the antenna with a thick dielectric layer of the same permittivity as that of the dielectric substrate as shown in Fig. 7. In this case the thickness of the dielectric cover (d) is selected such that the interaction of fringe fields with the external dielectric medium (viz. ice) above the dielectric cover becomes insignificant.

Fig. 7 shows the fractional change of the resonant frequency of a microstrip antenna with a dielectric cover loaded with a dielectric having relative permittivity ϵ_{r3}, for various values of dielectric cover thickness. It may be noted that for dielectric cover thicknesses larger than 2 cm, the effect of the external dielectric loading on the resonant frequency of the microstrip antenna is small.

The analysis presented above is a first-order solution to the problem and second-order effects such as change in Δl due to loading and reflections due to mismatch have not been considered in Figs. 3–6. However, these results are accurate provided only the microstrip length is covered with the dielectric. In Fig. 7, the second-order effects such as change in Δl due to loading have been measured and incorporated.

MEASURED RESULTS AND DISCUSSION

In order to verify the theoretical results for the fractional change in the resonant frequency, experiments were carried out on a rectangular patch resonator. A microstrip antenna using duroid substrate ($\epsilon_r = 2.32$ and 0.159-cm thick) was fabricated. The dimensions of the microstrip patch were 2.29×1.9 cm^2. The dielectric sheets were placed on the microstrip antenna and pressed with the help of styrofoam block ($\epsilon_r \cong 1.05$) such that the dielectric sheets were in good contact with the surface of microstrip resonator.

A comparison of theoretical and experimental results for $\Delta f_r/f_r$ is presented in Table I. The data refer to a microstrip line of length L covered with a dielectric sheet. The results agree fairly well with the calculated values. The effect of dielectric loading on the characteristics of microstrip antennas is shown in Table II. Here, the microstrip antenna substrate (10×10 cm^2) is completely covered with the dielectric sheets. It may be noted from Table II that the return loss first increases with increasing thickness of dielectric sheet and then decreases as observed previously [12]. The bandwidth of the microstrip antennas also increases with increasing thickness of dielectric sheet for low dielectric constant materials, and decreases for high dielectric constant materials. For example, when a microstrip duroid-substrate antenna is designed for $f = 4.1$ GHz and loaded with a 0.318-cm thick

TABLE I
COMPARISON BETWEEN THEORETICAL AND EXPERIMENTAL RESULTS OF $\Delta f_r/f_r$

Dielectric Cover	ϵ_{r1}	d(cm)	$\Delta f_r/f_r$ (%)	
			Experimental	Theoretical
Styrofoam	~1.05	20	0.02	0.015
Duroid	2.32	0.08	1.36	1.29
	(±0.01)	0.159	2.19	2.11
		0.318	2.73	2.61
Custom High-K	10.0	0.154	6.19	7.08
	(±0.2)	0.312	8.65	9.70

(W = 1.9 cm, L = 2.29 cm, h = 0.159 cm and ϵ_r = 2.32)

TABLE II
EXPERIMENTAL DATA FOR THE EFFECT OF DIELECTRIC LOADING ON THE CHARACTERISTICS OF MICROSTRIP ANTENNAS

Dielectric Cover	ϵ_{r1}	d(cm)	f_r (GHz)	$\frac{\Delta f_r}{f_r}$ (%)	Ret. Loss (-dB)	Bandwidth (%)
Air	1.0	∞	4.104	0	32	2.17
Duroid	2.32	0.08	4.008	2.34	35	2.18
		0.159	3.934	4.14	26	2.22
		0.318	3.895	5.09	22	2.31
Plexiglass	2.6	0.112	3.952	3.70	33	2.18
		0.159	3.912	4.68	25	2.18
		0.318	3.874	5.60	22	2.20
		0.636	3.806	7.26	16	2.32
Mylar	3.0	0.0064	4.070	0.83	37	2.18
		0.0128	4.058	1.21	39	2.18
		0.0384	4.010	2.29	40	2.20
Epsilam-10	10.2	0.0635	3.640	11.30	36	2.0
Custom High-K	10	0.154	3.482	15.16	24	1.75
		0.312	3.26	20.56	18	1.90

(W = 1.9 cm, L = 2.29 cm, h = 0.159 cm and ϵ_r = 2.32)

duroid sheet, the bandwidth increases by 7 percent, whereas, for small thicknesses of loading dielectric sheets (≤0.1 cm) the bandwidth is almost unchanged.

CONCLUSION

The design principles of microstrip antennas covered with dielectric layers have been presented. The resonant frequency of a microstrip antenna covered with a dielectric layer can be predicted accurately if the effective dielectric constant of the composite structure is known. The effective dielectric constant can be calculated using the variational technique. Numerical results show that the effective dielectric constant of a microstrip line covered with a thick sheet of high dielectric constant is drastically affected by the cover. The effect is more pronounced for small values of W/h ratio.

The fractional change of the resonant frequency for optimum width of the rectangular patch radiator was calculated. The calculations show that the maximum changes in the resonant frequency at 10 GHz are 5.8, 7.8, and 16 percent for infinitely thick dielectric covers of polystyrene, ice, and beryllium oxide, respectively. The curves presented here may be used to account for the detuning of microstrip antennas subjected to icing, a plasma environment, or coated with protective layers.

Numerical and experimental results for the fractional change in the resonant frequency have been found to be in good agreement. Measured results showed that the return loss increases for thin loading while the bandwidth increases when the thickness of the low dielectric constant sheets increases.

REFERENCES

[1] R. E. Munson, "Conformal microstrip antennas and microstrip phased arrays," *IEEE Trans. Antennas Propagat.*, vol. AP-22, pp. 74–78, Jan. 1974.

[2] G. G. Sanford and R. E. Munson, "Conformal VHF antenna for the Appollosoyuz test project," presented at the Inst. Elec. Eng. *Int. Conf. Antennas for Aircraft and Spacecraft*, pp. 130–135, 1975.

[3] H. D. Weinschel and K. R. Carver, "A medium-gain circularly polarized microstrip UHF antenna for marine DCP communication to the GOES satellite systems," in *IEEE Antennas Propagat. Soc. Int. Symp. Digest*, pp. 391–394, 1976.

[4] C. W. Garvin *et al.*, "Missile base mounted microstrip antennas," *IEEE Trans. Antennas Propagat.*, vol. AP-25, pp. 604–610, Sept. 1977.

[5] F. W. Cipolla, " A 7.5 GHz microstrip phased array for aircraft to satellite communications," in *Proc. Workshop on Printed Circuit Antennas Technol.*, New Mexico State Univ., Las Cruces, pp. 19.1–19.18, Oct. 1979.

[6] I. J. Bahl and P. Bhartia, *Microstrip Antennas.* Dedham, MA: Artech House, 1980.

[7] K. C. Gupta, R. Garg, and I. J. Bahl, *Microstrip Lines and Slotlines.* Dedham, MA: Artech House, 1979, p. 79.

[8] I. J. Bahl, "Build microstrip antennas with paper-thin dimensions," *Microwaves*, vol. 18, pp. 50–63, Oct. 1979.

[9] E. Yamashita and R. Mittra, "Variational method for the analysis of microstrip lines," *IEEE Trans. Microwave Theory Tech.*, vol. MTT-16, pp. 251–256, Apr. 1968.

[10] I. J. Bahl and S. S. Stuchly, "Analysis of a microstrip covered with a lossy dielectric," *IEEE Trans. Microwave Theory Tech.*, vol. MTT-28, pp. 104–109, Feb. 1980.

[11] I. J. Bahl and K. C. Gupta, "Average Power-Handling Capability of Microstrip Lines," *Inst. Elec. Eng. J. Microwaves, Opt. Acoust.*, vol. 3, pp. 1–4, Jan. 1979.

[12] P. K. Agrawal and M. C. Bailey, "An analysis technique for microstrip antennas," *IEEE Trans. Antennas Propagat.*, vol. AP-25, pp. 756–759, Nov. 1977.

The Finite Ground Plane Effect on the Microstrip Antenna Radiation Patterns

JOHN HUANG, Member, IEEE.

***Abstract*—The uniform geometrical theory of diffraction (GTD) is employed for calculating the edge diffracted fields from the finite ground plane of a microstrip antenna. The source field from the radiating patch is calculated by two different methods: the slot theory and the modal expansion theory. Many numerical and measured results are presented to demonstrate the accuracy of the calculations and the finite ground plane edge effect.**

I. INTRODUCTION

A microstrip patch antenna is a thin conducting strip radiator separated from its ground plane by a layer of dielectric substrate as described in Fig. 1. This communication presents the approach of combining the slot theory [1], [2] and the method of uniform geometrical theory of diffraction (GTD) [3] to account for the finite ground plane edge diffractions. In doing so, the radiation in the backlobe and wide angle regions can be accurately predicted while the other theories fail to do so. Even though the slot theory can only be employed for rectangular patch and copolar calculation, the GTD, however, can be combined with other theories, such as the modal expansion theory [4], to compute the patterns (include cross-polar information) for many different shaped-microstrip radiators. A discussion of the modal expansion theory is also included in this communication. The author wishes to point out that the method described here should not be applied without modifications when the product of the substrate thickness (in wave length) and dielectric constant is much greater than 0.1; otherwise accuracy degrades. This is due to the fact that the surface wave effect of the dielectric substrate and the dielectric wedge diffraction have not been taken into consideration. Since GTD is a high frequency technique, the rule of thumb is that the distance between the ground plane edge and the edge of radiating patch should not be less than a quarter wavelength. Fortunately, most of the applications that have been encountered to date are in the valid region of the formulations to be described in the following section.

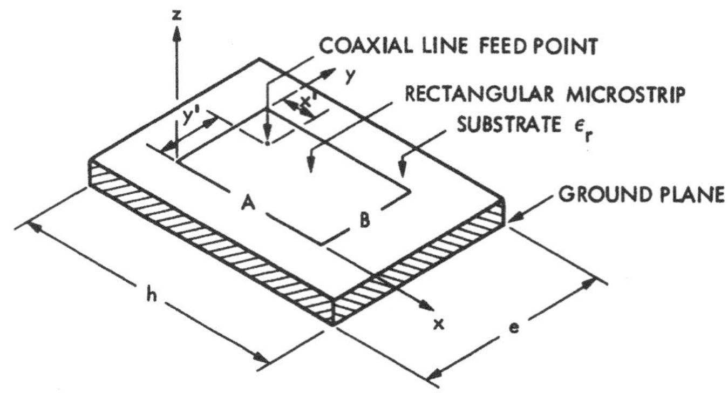

Fig. 1. Microstrip antenna configuration.

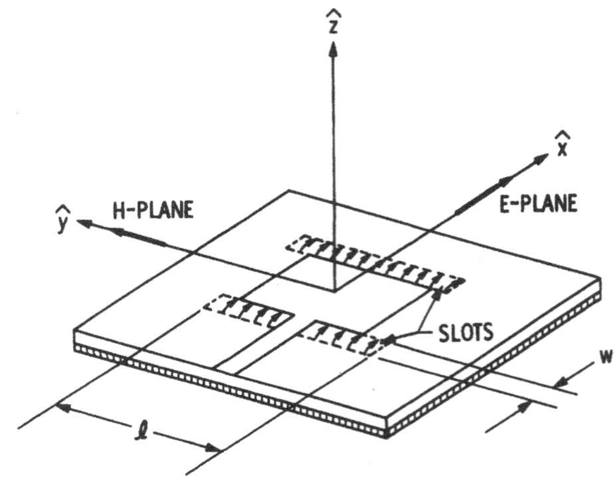

Fig. 2. Slot model configuration of a microstrip patch.

II. RADIATION PATTERN FORMULATIONS

A. Slot Theory and GTD

The slot theory is presented here because its combination with the GTD is much easier to be understood by the readers. The slot theory considers that the radiation from a rectangular microstrip patch is equivalent to that from two parallel slots adjacent to the metallic patch as shown in Fig. 2. The width (W) of each slot is approximated by the thickness of the substrate, and the length (l) is equal to the length of the patch (A) plus the substrate thickness [5] (due to fringing effect). The E-plane pattern can be calculated by summing three rays from each of the two slots as illustrated in Fig. 3(a). The direct geometrical optics (GO) field from each slot is given by [2]

$$\vec{E}_{GO} = \hat{\mu} \frac{\sin(\pi W\sqrt{\epsilon_r} \cos\mu)}{\pi W\sqrt{\epsilon_r}\cos\mu} \frac{e^{-jkS}}{\sqrt{S}}, \tag{1}$$

where W is the slot width in terms of wave length, ϵ_r is relative dielectric constant of substrate, and S is the distance from slot center to the observation point. The singly diffracted GTD field from each edge generated from the same slot is given by

$$\begin{aligned}\vec{E}_d &= \hat{\mu}_i E_{\text{inc}} D_h \frac{e^{-jkS_i}}{\sqrt{S_i}} \\ &= \hat{\mu}_i \frac{\sin(\pi W\sqrt{\epsilon_r})}{\pi W\sqrt{\epsilon_r}} \frac{e^{-jkd_i}}{\sqrt{d_i}} D_h \frac{e^{-jkS_i}}{\sqrt{S_i}},\end{aligned} \tag{2}$$

Reprinted from *IEEE Trans. Antennas Propaga.*, vol. AP-31, no. 4, pp. 649–653, July 1983.

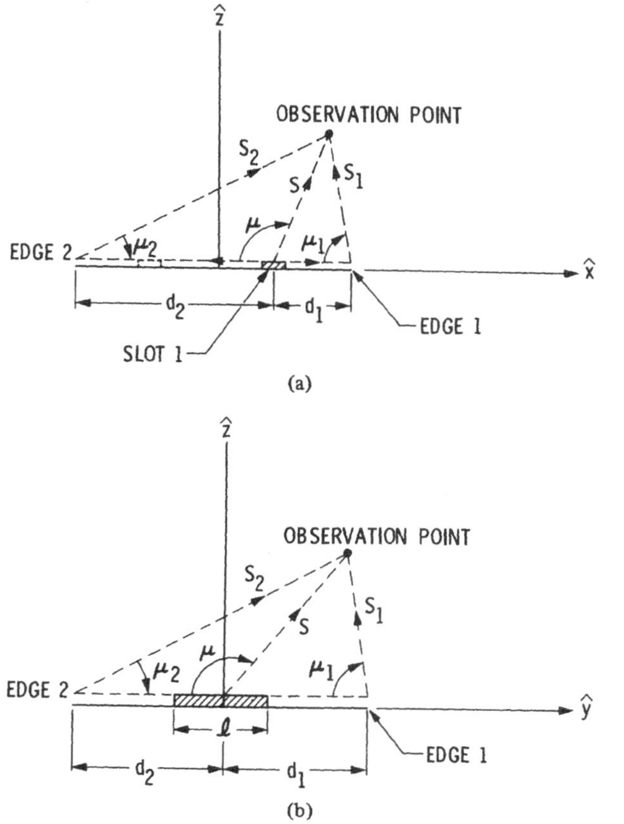

Fig. 3. (a) E-plane radiation and diffraction mechanisms. (b) H-plane radiation and diffraction mechanisms.

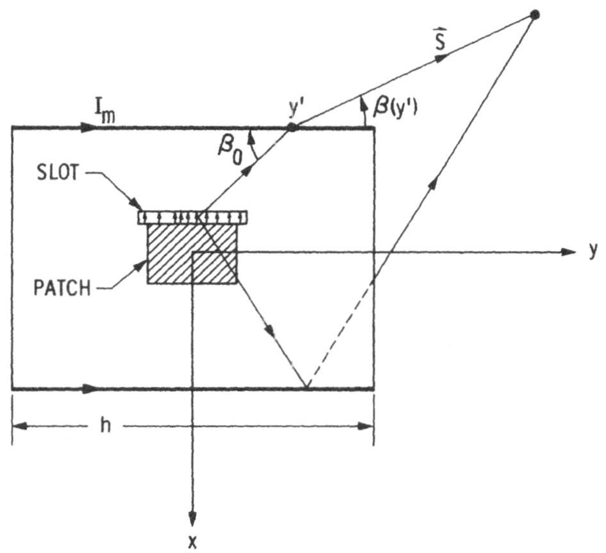

Fig. 4. Equivalent magnetic line current for the H-plane pattern calculation.

where D_h is the hard-boundary diffraction coefficient without the dielectric effect and has been given in [3]. In addition to the GO field and the singly diffracted fields, the doubly diffracted fields need to be included if a continuous pattern is required in the regions of the two shadow boundaries ($\mu = 0°$ and $180°$).

The H-plane pattern in the forward region can also be calculated by summing three rays as illustrated in Fig. 3(b). The direct GO field from the slot is given by [2]

$$\vec{E}_{GO} = \hat{x}\, \frac{\sin(\pi l \cos\mu)}{\pi l \cos\mu} \sin\mu\, \frac{e^{-jkS}}{\sqrt{S}}. \tag{3}$$

Because the electric field on the surface of a conductor wedge vanishes for polarization of the grazing incident wave being parallel to the surface (soft boundary condition), the first-order diffracted field from each edge is zero. However, a second-order diffracted field derived from the Maxwell's equations is nonzero and can be viewed as a result of the rapid change of GO field. This diffracted field, known as slope diffraction [6], is given by

$$\vec{E}_{\text{slope}} = \hat{x}\, \frac{1}{jk}\, \frac{1}{d_i}\, \frac{\partial E_{GO}}{\partial \mu}\bigg|_{\mu = 0° \text{ or } 180°} D_{sp}\, \frac{e^{-jkS_i}}{\sqrt{S_i}} \tag{4}$$

where D_{sp} is the slope diffraction coefficient and has been given in [3]. In the backlobe region of the H-plane pattern, one needs to include the contributions from the E-plane edge diffractions simply because the E-plane edge diffraction has a much larger magnitude than that of the H-plane edge slope diffraction. This E-plane edge contribution can best be calculated by an equivalent current technique [7] as described by the following equation (see Fig. 4):

$$\vec{E}_{eq} = \frac{jk}{4\pi} \int_{-h/2}^{h/2} \hat{S} \times \vec{I}_m(y')\, \frac{e^{-jkS}}{S}\, dy' \tag{5}$$

where S, with $\hat{S}$ being its unit vector, is the distance between instantaneous diffraction point y' and the observation point. h is the length of E-plane edge and $I_m(y')$ is the equivalent magnetic line current given by

$$I_m(y') = -\hat{y} \cdot \frac{\vec{H}_i(y')}{Y_0} D_h \sqrt{8\pi/k}\, e^{-j\pi/4} \tag{6}$$

with $H^i(y')$ being the incident field at y', D_h the diffraction coefficient for hard-boundary condition, and Y_0 the free space admittance. To summarize, the H-plane field is the vectorial summation of the GO field (E_{GO}), the slope diffracted field (E_{slope}) and the integrated equivalent current field (E_{eq}).

B. Modal Expansion Technique

The modal expansion technique, in the past, has been extensively applied to copolar pattern and input impedance calculations [4] for the microstrip radiators on an infinite ground plane. It is employed here, in conjunction with the GTD, not only for the copolar prediction but also to have a closer look at the cross-polar behavior on a finite ground plane. The fields under the patch can be determined by modeling the patch as a cavity bounded by perfect magnetic walls [4]. Once the fields within the cavity region are known, the induced magnetic current in the magnetic wall at the perimeter can be determined and in turn the radiated field can be calculated by integrating this magnetic current. This radiated field can then be used as the incident field to calculate the finite ground plane edge diffracted fields in the same fashion as that shown in (2), (4), and (5).

For a rectangular microstrip as shown in Fig. 1, the z-directed electrical field in the cavity (underneath the patch) can be separated into different modes and can be written as [4]

$$E_{mn}(x, y) = \sum_{m=0}^{\infty} \sum_{n=0}^{\infty} C_{mn} \phi_{mn}(x, y) \phi_{mn}(x', y') \tag{7}$$

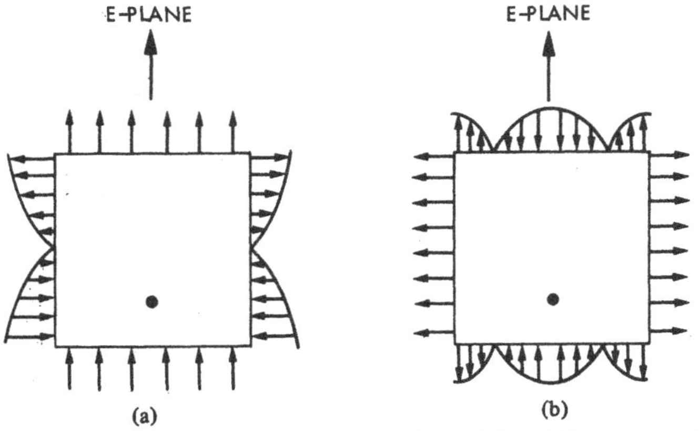

Fig. 5. Perimeter fields of a square microstrip patch for (a) E_{10} mode and (b) E_{02} mode. The heavy dot indicates the feed probe location.

where C_{mn} are the coefficients that depend on m, n, A, and B dimensions of the patch, dielectric constant, and feed size. Their details have been shown previously [4] and need not be repeated here. The modal function ϕ_{mn} is composed of two cosine functions and is shown in the following equation:

$$\phi_{mn}(x, y) = \cos(m\pi x/A) \cos(n\pi y/B), \tag{8}$$

where (x, y) is an arbitrary point under the patch, and (x', y') is the feed location. It is found that the series in (7) only needs to be summed to the fourth term and still preserve the accuracy. For linear polarization and fundamental mode operation, the dominant term, $E_{mn} = E_{10}$, generates the copolar field, while the term E_{02} generates the cross-polar field. The other terms contribute to either copolar or cross-polar fields with less significant effect. As an example, the perimeter fields of E_{10} and E_{02} modes are illustrated in Fig. 5 where E_{02} mode has a smaller magnitude than the E_{10} mode. The vertical arrows in the E_{10} mode indicate the copolar edge field, and the horizontal arrows in the E_{02} mode indicate the cross-polar edge field. The sinusoidally varied edge fields in both modes contribute very little in the far field because its net effect cancels itself. Notice that the cross-polar arrows in Fig. 5(b) are pointed in opposite directions. This is why that the cross-polar field of a rectangular or square patch always yields a null at the broadside direction.

III. RESULTS

Both the E- and H-plane patterns of a single microstrip patch have been calculated and compared with the measured results as shown in Fig. 6. The antenna dimensions in inches are (Fig. 1):

$$A = 2.126 \qquad e = 10.5$$
$$B = 1.488 \qquad h = 14.0 \tag{9}$$

substrate thickness = 0.125.

The relative dielectric constant of substrate is 2.55, and the operating frequency is 2.295 GHz. For practical purpose, the overall comparison between the measurement and the prediction is quite good. The effect of finite ground plane and different edge diffractions are demonstrated in Figs. 7 and 8. The double diffraction has been included in all the E-plane pattern calculations. Fig. 7 illustrates the difference between the E-plane patterns when the patch radiation is calculated on an infinite ground

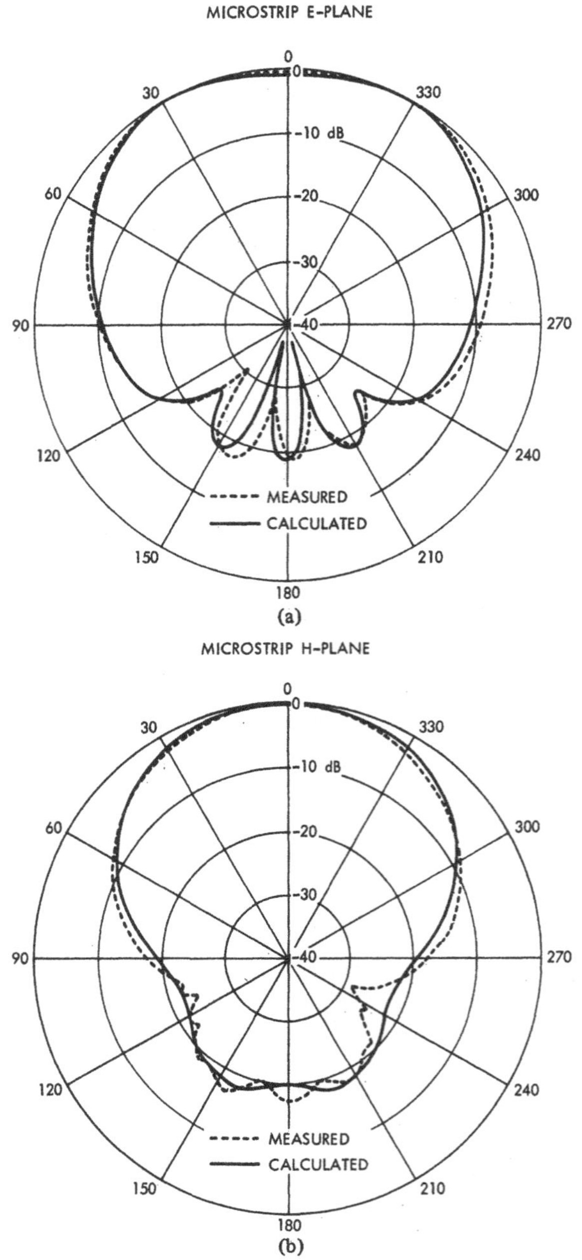

Fig. 6. (a) Microstrip E-plane. (b) Microstrip H-plane. Radiation patterns of a rectangular microstrip antenna. Antenna dimensions are (see Fig. 1): $A = 2.126$ in, $B = 1.488$ in $e = 10.5$ in, $h = 14.0$ in, substrate thickness = 0.125 in, $\epsilon_r = 2.55$, frequency = 2.295 GHz.

plane and on a two-wavelength ground plane. This comparison shows that the amount of error can be introduced when the pattern is calculated for an infinite ground plane while the measurement is performed on a finite ground plane. Fig. 8 shows the difference of the H-plane patterns when the radiation is calculated without the slope diffraction and without the E-plane edge equivalent current contribution. The importance of the edge diffractions is again clearly demonstrated here. The calculation in Figs. 6–8 are based on the slot theory which does not yield any cross-polar information. In order to demonstrate the accuracy in predicting both the copolar and cross-polar fields by the modal expansion theory, a microstrip antenna is constructed and meas-

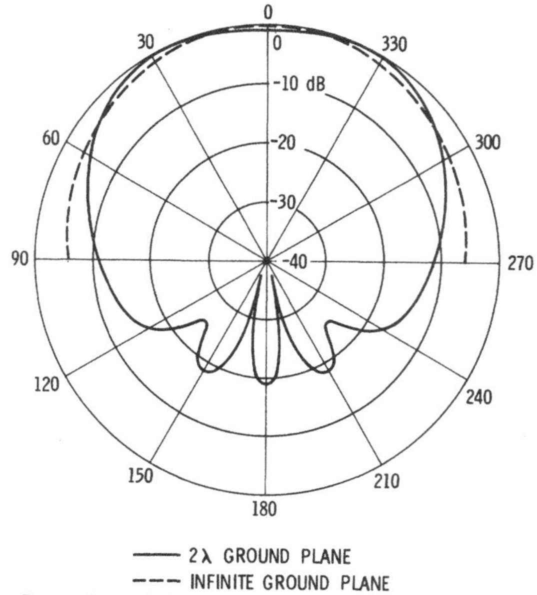

Fig. 7. Comparison of the E-plane calculated patterns when the patch is on a two-wavelength ground plane (see (9)) and that on an infinite ground plane.

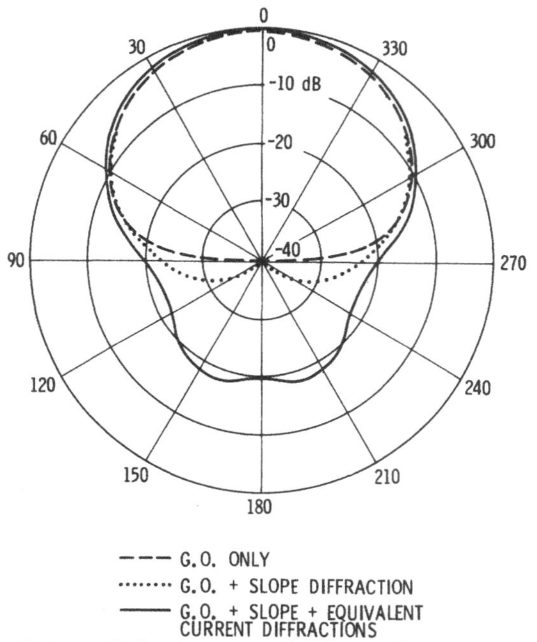

Fig. 8. H-plane calculated patterns of different edge contributions. The ground plane sizes are shown in (9).

ured with results compared with calculations as shown in Fig. 9. The dimensions of the antenna (see Fig. 1) are $A = B = 1.8$ in, $h = e = 38.7$ in, substrate thickness $= 0.125$ in, $\epsilon_r = 2.17$, and frequency $= 2.115$ GHz. Excellent agreements are observed in both the copolar and the cross-polar patterns. The ripples in the copolar of the E-plane pattern and in the cross polar of the H-plane pattern are due to ground plane edge diffractions. These diffractions, especially in the forward region, are very well predicted by the GTD technique. In the backlobe region, however, the prediction is not quite as well as that predicted in Fig. 6. This is due to the fact that a larger ground plane ($7\lambda \times 7\lambda$) is being used here. This larger ground plane in turn requires a larger back mounting structure which has a pronounced scattering effect to the field in the back lobe direction.

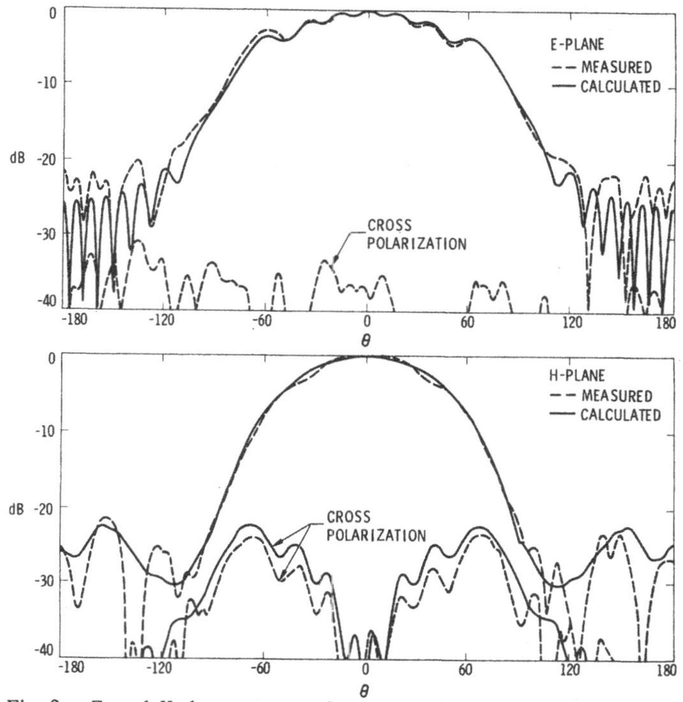

Fig. 9. E- and H-plane patterns of a square microstrip antenna. Calculation is done by model expansion theory and GTD. Antenna dimension are (see Fig. 1): $A = B = 1.8$ in, $e = h = 38.7$ in, substrate thickness = 0.125 in, $\epsilon_r = 2.17$ and frequency = 2.115 GHz.

IV. CONCLUSION

The slot theory and the modal expansion theory augmented by the uniform GTD diffraction solution for the prediction of microstrip antenna radiation have been presented. The GTD edge diffractions are included for the finite ground plane effect in both E- and H-plane calculations. In the E-plane, single and double edge diffractions plus the direct GO field contribute to the total field. In the H-plane, the total field consists of the direct GO field, the slope diffracted field and the E-plane edge equivalent current field. The measured results indicate that the theoretical predictions for both large ground plane ($7\lambda \times 7\lambda$) and small ground plane ($2\lambda \times 2.7\lambda$) are quite good despite the exclusion of the dielectric effect in the diffraction calculations. Numerical examples demonstrates that the finite edge calculation is essential if accurate pattern levels at wide angles and backlobe information are required. The pattern cuts other than at the principal planes, such as diagonal cuts, can be predicted by GTD with its well-established corner diffraction solution. GTD's creeping wave solution can also be employed to calculate microstrip radiation on a curved surface.

ACKNOWLEDGMENT

The author would like to thank Mr. H. Marlin for performing the measurements, and Dr. K. Woo and Dr. Y. Rahmat-Samii for their comments and suggestions.

REFERENCES

[1] A. G. Derneryd, "Linearly polarized microstrip antennas," *IEEE Trans. Antennas Propagat.*, vol. AP-24, no. 6, pp. 846–850, Nov. 1976.

[2] A. G. Derneryd and A. G. Lind, "Extended analysis of rectangular microstrip resonator antennas," *IEEE Trans. Antennas Propagat.*, vol. AP-27, no. 6, pp. 846–849, Nov. 1979.

[3] R. G. Kouyoumjian and P. H. Pathak, "A uniform geometrical theory of diffraction for an edge in a perfectly conducting surface," *Proc. IEEE*, vol. 62, pp. 1448–1461, Nov. 1974.

[4] W. F. Richards, Y. T. Lo, and D. D. Harrison, "An improved theory for microstrip antennas and applications," *IEEE Trans. Antennas Propagat.*, vol. AP-29, no. 1, pp. 38–46, Jan. 1981.

[5] P. Hammer, D. Van Bouchaute, D. Vershraeven, and A. Van DeCapelle, "A Model for Calculating the Radiation Field of Microstrip Antennas," *IEEE Trans. Antennas Propagat.*, vol. AP-27, no. 2, pp. 267–270, Mar. 1979.

[6] C. A. Mentzer, L. Peters, and R. C. Rudduck, "Slope diffraction and its application to horns," *IEEE Trans. Antennas Propagat.*, vol. AP-23, pp. 153–159, Mar. 1979.

[7] C. E. Ryan and L. Peters, "Evaluation of edge-diffraction fields including equivalent currents for the caustic regions," *IEEE Trans. Antennas Propagat.*,vol. AP-27, pp. 292–299, May 1969.

Chapter 3
Dual and Circularly Polarized Elements

MANY applications in communications and radar require circular or dual linear polarization, and the flexibility afforded by microstrip antenna technology has led to a wide variety of designs and techniques to fill this need. Again by necessity, we cannot include papers on all of these areas, but instead have tried to select papers that represent some of the most successful approaches to the problems of circular polarization and dual or switched linear polarization. The list of additional references at the end of this introduction can be consulted for further work in this area.

This chapter begins with a review article by Hall, written for this reprint book. This article surveys many of the proposed designs for circular and dual linear polarization, and contains an extensive list of references, many of them to the European and Japanese literature.

The operation of virtually all circularly polarized microstrip antennas can be viewed as the superposition of two (or more) linearly polarized modes with equal amplitude excitation and the proper phasing to generate a rotating field (some exceptions are a spiral element, and a patch antenna on a biased ferrite substrate). The multitude of circularly polarized patch designs differ primarily in how these linearly polarized modes are excited, but can be classified into three main types: those using a single feed point, those using two feeds in phase quadrature, and those using sequential rotation. Designs using a single feed point rely on coupling two orthogonal linearly polarized modes such that their amplitudes are equal and their phases are in quadrature. This is a simple and economical approach, but with the serious disadvantage that the resulting axial ratio is rather narrowband, often being less than 1%. Implementations of such elements, using slightly off-square patches, patches with notches, and patches with slots, among others, are discussed in the review article by Hall. More design details for single-point fed CP patches can be found in the article by Sharma and Gupta, who provide practical design-oriented data for three patch geometries, with an emphasis on optimizing the axial ratio bandwidth. This article treats only probe-fed patches, but it should be noted that any of the feeding techniques discussed in Chapter 2 (e.g., probe feed, edge feed, aperture coupled, or proximity feed) can be applied to this scheme. For example, the article by Iwasaki, Sawada, and Kawabata discusses a circularly polarized element with a single proximity coupled feed line; the impedance bandwidth was about 3.5%, while the axial ratio bandwidth was about 0.55%. Further design details for singly fed circularly polarized elements can be found in [1], [2].

The use of two feed points generally gives much better axial ratio bandwidths than singly fed CP elements, since the amplitude and phase of the linearly polarized field components is determined by a relatively broadband power divider circuit. Such designs are also more robust in terms of degradation due to manufacturing and material tolerances, but have the drawback of requiring a separate power divider network, which adds complexity, takes up space, and increases loss. Quadrature hybrids and reactive dividers are commonly used for this application, although Lange couplers are often used in MMIC circuits. In addition, the feeds may be probes, microstrip lines, apertures, or proximity coupled lines. The paper by Adrian and Schaubert presents results for a circularly polarized element using two separate apertures, while the paper by Targonski and Pozar shows design details and results for a circularly polarized element using a crossed slot feed. The crossed slot feed is inherently symmetrical and balanced, important conditions for axial ratio purity. Such symmetry can only be obtained with probe feeds if two pairs of balanced feed probes are used [3].

One logical extension of the two-feed circularly polarized patch is the sequentially rotated array, where (typically) four linearly polarized elements are fed in phase rotation to achieve circular polarization. This concept has a long history, but one of the first papers to present this idea in the context of microstrip antennas is the one by Teshirogi, Tanaka, and Chujo. They point out that this technique not only has very good axial ratio bandwidth, but may also have an enhanced impedance bandwidth over that of a single patch element. An interesting question arose with this approach, whereby large diagonal plane grating lobes were found in some sequentially rotated subarray designs; this issue was resolved in the paper by Hall, Huang, Rammos, and Roederer, with further details relative to array design discussed in [4].

Conceptually, dual polarized elements can be simply considered as a superposition of two linearly polarized modes, or separate elements. Thus, dual polarized elements can be made from square or circular patches using any of the usual feeding methods; the dual aperture coupled patch design in the paper by Adrian and Schaubert, for example, can readily be used for dual polarization. In practice, besides the usual problem of element bandwidth, isolation between polarizations is often an important parameter. Dual polarized designs using unbalanced feeds (such as two probe feeds or two offset slot feeds) have an inherent asymmetry that typically limits isolation to 20 dB or less; one way to improve this figure, at the expense of more complexity, is to use balanced feed points (two pairs of balanced probe feeds or a crossed slot feed). A comparison of two very practical dual polarized crossed-slot-fed elements are discussed in the paper by Edimo, Sharaiha, and Terret, where it is shown that a bandwidth of about 30% and a polarization isolation of about 25 dB can be obtained. Other examples of dual polarized and polarization-agile elements are discussed in [5] and [6]. In an array environment additional problems arise because of the complexity of two feed networks and a constrained interelement spacing [7]. If simultaneous dual linear polarization is not required, it is possible to use a single feed line with a patch element having switchable polarization; this approach is discussed in the paper by Schaubert, Farrar, Sindoris, and Hayes. In a similar vein, many communication applications that require different polarizations or frequencies for transmit and receive can benefit from a self-diplexing antenna design, whereby separate feed points are an advantage. The paper by Nakano, Arai, Chujo, Fujise, and Goto describes a very successful self-diplexing circularly polarized element operating at L-band, with an isolation in excess of 40 dB.

References

[1] Y. T. Lo and W. F. Richards, "Perturbation approach to design circularly polarized microstrip antennas," *Electronics Letters,* vol. 17, pp. 383–385, 1981.

[2] M. Haneishi and S. Yoshida, "A design method of circularly polarized rectangular microstrip antenna by one-point feed," *Electronics and Commun. in Japan,* vol. 64-B, pp. 46–54, 1981.

[3] J. J. Schuss and R. L. Bauer, "Axial ratio of balanced and unbalanced fed circularly polarized patch radiator arrays," *IEEE Int'l Symp. on Antennas and Propagation Digest,* pp. 286–289, June 1987.

[4] P. S. Hall, J. S. Dahele, and J. R. James, "Design principles of sequentially fed, wide bandwidth, circularly polarised microstrip antennas," *IEE Proceedings,* vol. 136, pt. H, pp. 381–389, Oct. 1989.

[5] R. I. Wolfson and W. G. Sterns, "A high-performance microstrip dual-polarized radiating element," *IEEE Int'l Symp. on Antennas and Propagation Digest,* pp. 555–558, June 1984.

[6] A. J. Sangster, "Polarization agile microstrip array antenna element," *Microwave and Optical Technology Letters,* vol. 4, pp. 419–421, Sept. 1991.

[7] J. Huang, "Dual-polarized microstrip array with high isolation and low cross-polarization," *Microwave and Optical Technology Letters,* vol. 4, pp. 99–103, Feb. 1991.

Review of Techniques for Dual and Circularly Polarised Microstrip Antennas

P. S. HALL
School of Electronic and Electrical Engineering
The University of Birmingham
Edgbaston, Birmingham
B15 2TT UK

1. Introduction

Many current communication and sensor systems require a high degree of polarisation control to optimise system performance. For microstrip antennas to be fully exploited in such systems high polarisation purity and isolation between orthogonal polarisations, be they linear or circular, are needed. This review paper examines the polarisation control capabilities of microstrip antennas and in particular relates these capabilities to the current demands for circularly polarised and dual polarised planar antennas.

Historically, single linearly polarised microstrip patch antennas were the first to be developed. Soon after, techniques for circular polarisation were demonstrated, but again involving only a single hand of polarisation. The quality of polarisation control in either linear or circular systems is linked to how well the two orthogonal modes in the antenna can be controlled, which is to some extent related to the inherent isolation between them. This isolation is in turn dependent on the patch quality factor and the excitation geometry. Thus the likely cross-polarisation or axial ratio is determined early on in the design process and may in fact be determined by other parameters such as bandwidth or the desired mechanical construction, unless specific measures are taken. The more recent desire for dual polarised antennas has put further emphasis on these difficult issues. A new class of microstrip types, known as self-diplexing antennas, have arisen, which aim to maximise the isolation between polarisations in such dual systems. It is the progress in two areas of polarisation control in circularly polarised and dual polarised microstrip antennas that this paper aims to review; the area of polarisation control in linearly polarised antennas is also touched on. After specifying some of the parameters important when discussing polarisation in antenna systems, the paper examines the basic action of a patch antenna supporting two orthogonal modes to clarify the likely degree of polarisation control that can be obtained from currently used microstrip antennas. Techniques are then reviewed in the following order: patches for circular and dual polarisation, microstrip spirals and special types including ferrite substrates, and finally circular and dual polarised array configurations.

2. Polarisation in Antenna Systems

Several excellent references describe polarised electromagnetic waves [1],[2], and the interested reader is referred to these for details. For the purpose of this overview it is noted that in general waves are elliptically polarised and are defined by three variables, namely, axial ratio, tilt angle, and sense. The IEEE standard definition of sense states that for an approaching wave, counter-clockwise vector rotation corresponds to a right-handed wave and vice versa. For an infinite or zero axial ratio, linear polarisation results and the tilt angle defines the orientation of the electric vector; sense is not applicable. For waves with close to linear polarisation, axial ratio is not used but rather the level of cross-polarisation in an orthogonal plane is quoted. Sense is usually not quoted in this case, although for calculation of coupling between slightly-off linearly polarised antennas (whether similarly or orthogonally polarised) it is necessary to know all three parameters. For unity axial ratio, circular polarisation results, and the tilt angle is not applicable. The quality of slightly off circularly polarised waves is specified by the axial ratio and tilt angle is not usually quoted, although again for coupling between antennas all three parameters are needed.

For single polarisation systems the antenna can be considered as a two-port device with one port comprising the interface with the transmitter or receiver, and the other port as free space (Figure 1a). The S parameters involve the usual antenna characteristics. For dual polarisation a four-port representation must be used and additional parameters become evident. S_{21} is the isolation between the two input ports and represents that part of the signal to be transmitted on polarisation 1 that is coupled into

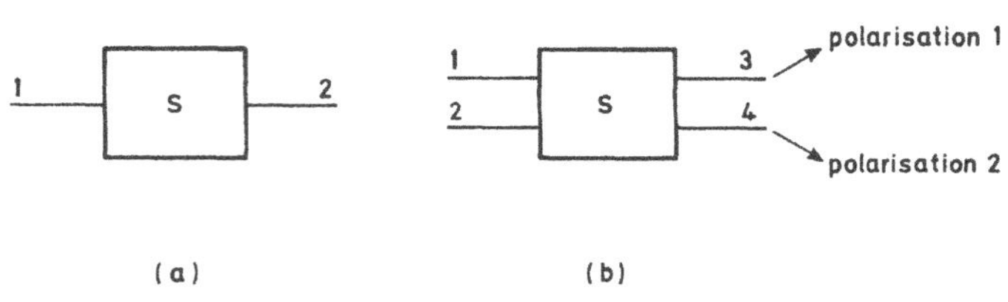

Fig. 1. S parameter representation of antennas.
(a) single polarisation
(b) dual polarisation

port 2, assuming both polarisations are being transmitted. S_{41} represents the amount of signal that was to be transmitted on polarisation 1 but appears as polarisation 2, and similarly for S_{32}. The isolation usually quoted for dual polarisation antennas is S_{21} or S_{12}. S_{41} or S_{32} are usually specified by the cross-polarisation or axial ratio of the radiated wave.

3. Generation of Orthogonal Polarisations

Design of dual and circular polarisation microstrip antennas demands precise control of the individual orthogonal radiated polarisations. In some microstrip antennas the structure favourably supports a given polarisation. For example a high aspect ratio rectangular patch will give a relatively pure linearly polarised wave. Similarly, the microstrip spiral or patch on a biased ferrite substrate readily gives circular polarised waves. In general, however, the wanted polarisations are synthesised from a pair of orthogonal linear polarisations and the coupling (S_{21} in Figure 1b) is a critical guide to the quality of the antenna. Figure 2a [3] illustrates the coupling between orthogonal ports in a dual linearly polarised circular microstrip patch. At resonance, the high Q patch with $t = 3.2$ mm substrate thickness has better than 50 dB isolation. For the low Q patch on a 12.3 mm substrate higher order modes are generated that degrade the isolation to about -28 dB. Feed geometry is also critical here with, in general, increasing feed port size increasing coupling. This limits the upper frequency range of, for instance, probe feeds that mate to coaxial cable. However, isolation can be improved by optimising the feed position [4]. Use of notched or slotted patches with two feeds to give dual circular polarisation further increases this undesirable mode coupling. Figure 2b [5] compares coupling in a dual linear and dual circular patch, with isolation degrading from about 20 dB for dual linear to less than 10 dB for dual circular.

The radiation pattern shape is also significant in this discussion of the fundamentals of polarisation control. From the gross features of the pattern several important points emerge. The beamwidth in the two principal planes of a patch are unequal, which will give rise to unequal radiation amplitudes off-broadside in dual linearly polarised antennas, or increasing axial ratio off-broadside in circularly polarised ones. In scanned arrays this means that polarisation control degrades with scan angle until at very low angles only vertical polarisation with respect to the ground plane can be radiated. Furthermore, it is obvious that in most patches some azimuthal variation of the polarisation for dual linear or circular will take place. However the circular symmetry of the circular patch radiating circular polarisation gives no azimuthal variation [6] and is thus widely used in circularly polarised arrays.

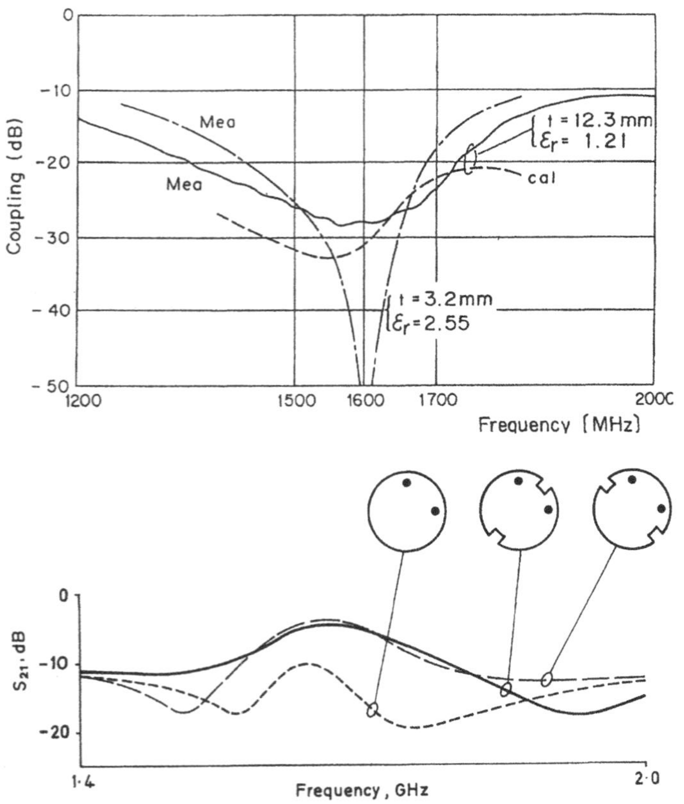

Fig. 2. Coupling between orthogonal feed ports in circular microstrip patches.
(a) Effect of patch Q [3]
(b) Comparison of dual linear and dual circular polarisation [5]; patch dia = 40 mm, substrate height = 0.79 mm, $\epsilon_r = 2.3$.

4. Circularly Polarised Patches

Circularly polarised radiation can be generated by exciting two orthogonal patch modes in phase quadrature with the sign of the relative phase determining polarisation hand. These modes may be excited in a number of ways, described below. Before reviewing the methods it is instructive to compare the relative performance of the feed arrangements. Figure 3 shows three methods of excitation applied to a square patch fed by microstrip lines. The comments that follow apply to all patch shapes and feed connection geometries. Figure 3a and Figure 3b show the two orthogonal modes excited by orthogonal feed lines. In Figure 3a the quadrature phasing is achieved by the difference in the line lengths to the patch feeds. In Figure 3b a hybrid provides the phase offset and in addition gives isolation between the two feed points. In both cases the patch input mismatch

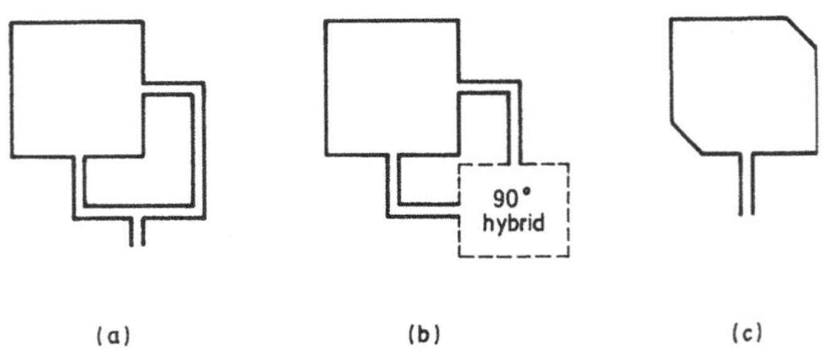

Fig. 3. Excitation methods for circular polarisation.
(a) Orthogonal feeds, reactive splitter
(b) Orthogonal feeds, isolating splitter
(c) Single feed degenerate mode patch

determines the overall performance [7]. In the non-isolated case the 90 degree phase shift between feeds means that the mismatch reflections tend to cancel at the input port, and the input match to the element remains acceptable over the bandwidth of a single mode. However, the reflections coupled to the splitter output ports result in radiation of the opposite hand of polarisation. Calculations show about 3 dB axial ratio when the element VSWR exceeds about 1.4. The isolated feed gives good axial ratio and input VSWR over the band as the reflected power is absorbed in the matched load on the fourth port of the hybrid coupler. This absorbed power is equal to the power radiated in the unwanted hand using a non-isolated feed; thus in both cases the gain is identical. The degenerate mode patch fed by a single line, Figure 3c, has been examined in the same way. The patch asymmetry excites the orthogonal mode. It is found that the performance is very similar to the reactive splitter-fed patch, Figure 3a, with the axial ratio degrading rapidly with frequency away from resonance while the input VSWR remains acceptable. Figure 3c is a more compact structure and is adopted in many practical antennas.

4.1 Orthogonal Patches

The two orthogonal modes noted above may in fact be supported by two orthogonal patches. In general the radiation pattern of such a pair will be asymmetric but this can be improved by the use of quarter wavelength shorted patches [8], [9] to give useful low-angle performance (Figure 4a). The technique, using conventional patches, has recently been applied to a circularly polarised phased array [10] to reduce the number of splitters (Figure 4b). Useful scanning and beam control was obtained but gain loss due to grating lobes occurs unless small element spacing is used [11].

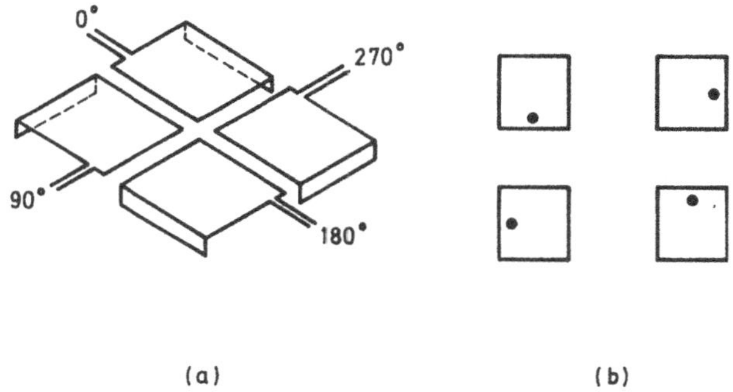

Fig. 4. Generation of circular polarisation using orthogonal linearly polarised patches.
(a) Using rotated quarter wavelength shorted patches, closely spaced [8], [9]
(b) Using conventional square patches [10]

4.2 Multi-point Feeds

Examples of circularly polarised patches fed by a pair of feeds are numerous and the technique has been used for over two decades [12], [13]. Directly-attached microstrip line, through-the-substrate pin, and, more recently, aperture coupled [4], [14] feeds have been used on square and circular shaped patches. Figure 5 shows axial ratio for thick circular patches [15]. In the thick patch, higher order modes are excited, which give rise to the coupling noted in Section 2. Use of more than two feeds [16] will reduce this coupling, as the figure shows, with both three- and four-feed systems having beneficial effects. The three-feed

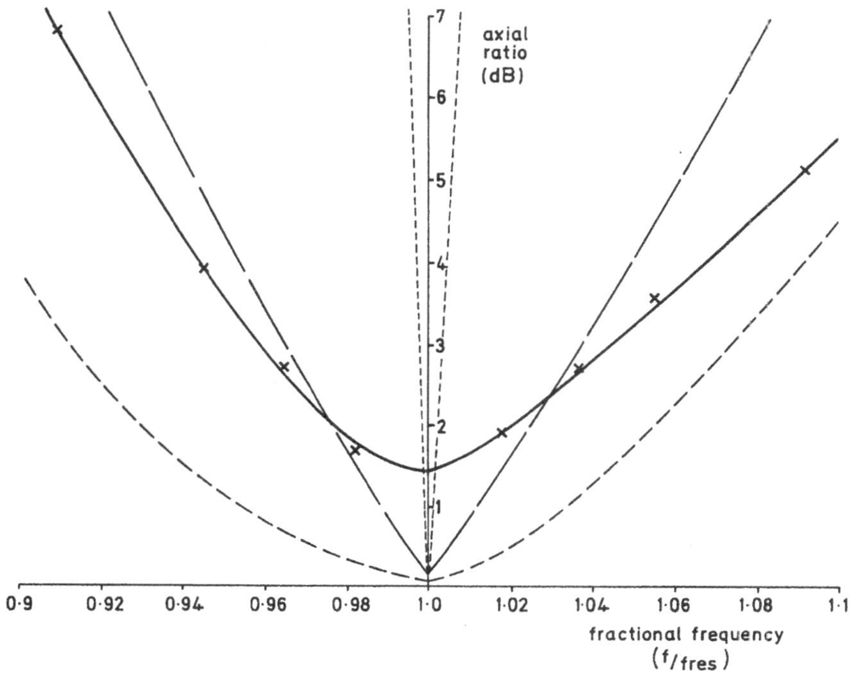

Fig. 5. Computed axial ratio of multiple feed microstrip circular patches (disc radius = 39mm, feed pin radius = 21mm, substrate ϵ_r = 1.06, h = 25mm,——2-feed, – – – 4-feed - - - - - - 3-feed; substrate ϵ_r = 2.32, h = 3.2mm . . . . . 2-feed)

patch has lower axial ratio than the four-feed type as the reflections from the non-isolating splitter radiate in the wanted polarisation. In the thin patch, coupling is low and good axial ratio on resonance results. It is noted in [17] that the coupling between the orthogonal feed ports in a 0.05 wavelength thick circular patch on $\epsilon_r = 1.21$ material can be reduced from −28 dB on resonance for two probes to about −60 dB for four probes. These values are close to those deduced from the computations of Figure 5. The use of notches on dual-fed patches [18] has also been shown to reduce axial ratio by about 50%.

Bandwidth can be increased in several ways. The use of parasitic elements [19] has been shown to achieve about 7% axial ratio bandwidth and 10 dBi gain, with two parasitic patches giving an overall element height of 0.56 wavelength. A coplanar reactive splitter feed was used. A crossed slot [20] used to aperture couple series and parallel feeds to a square patch of approximate thickness 0.1 wavelength produced, for the series arrangement, a 12% axial ratio bandwidth and, for the parallel feed using isolating splitters, a 22% bandwidth. These configurations are similar in principle to the four-point feeding discussed above, and give similar bandwidths.

4.3 Single Point Feeds

Figure 3c shows a square patch with single point feed where circular polarisation is induced by a so-called perturbation segment, in this case a pair of truncated corners. Figure 6 [21] illustrates how such perturbation can generate circular polarisation. Modes 1 and 2, in the diagonal planes, are equal amplitude and in phase quadrature at f_0. It is clear that off f_0 phase and amplitude errors will rapidly degrade the axial ratio. Figure 7 [22] shows typical axial ratio and input VSWR, which confirm the

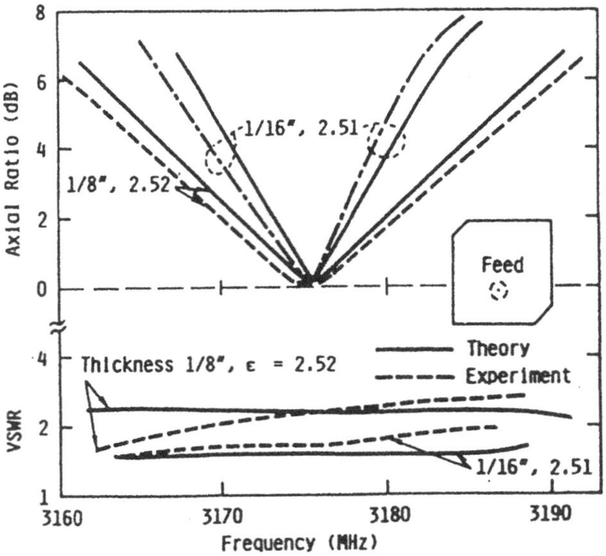

Fig. 7. Axial ratio and input VSWR of truncated corner square microstrip antenna [22].

qualitative behaviour noted earlier in this section. In addition, it is clear that axial ratio bandwidth is determined by the Q of the individual modes with a thicker, lower-Q patch giving better axial ratio bandwidth.

The shape of the perturbation segment or patch can vary widely; rectangular patch [23], patch with tabs [24], patch with notches [25], patch with centre slot [26], patch with truncated corner [22], elliptical patch [27], pentagonal patch [28], triangular patch [29], ring with notches [30], and loop [31] have been studied. There are some comparative studies available, although no exhaustive work appears to have been done. For instance, it is shown in [32] that significant differences in design procedure result for the feed point on the patch principal planes or on the diagonal, although for two particular cases chosen, the truncated corner square patch and the rectangular patch, almost identical axial ratio bandwidth is noted. Similarly, results given in [22] indicate that the square patch with diagonal slot has the largest axial ratio bandwidth, whereas minimum VSWR is obtained with the diagonal-fed nearly square patch. The truncated corner patch has best axial ratio but least axial ratio bandwidth. It is clear the optimisation of patch type will depend on system requirements.

Aperture coupling is advantageous for many reasons and patches fed from a single line using a cross-shaped slot [33] or single slot with a second parasitic slot [34] have been described. The latter is claimed to have higher isolation when used in simultaneous transmit and receive systems. Proximity coupling to a feed line with an overlaid patch [35] is also possible.

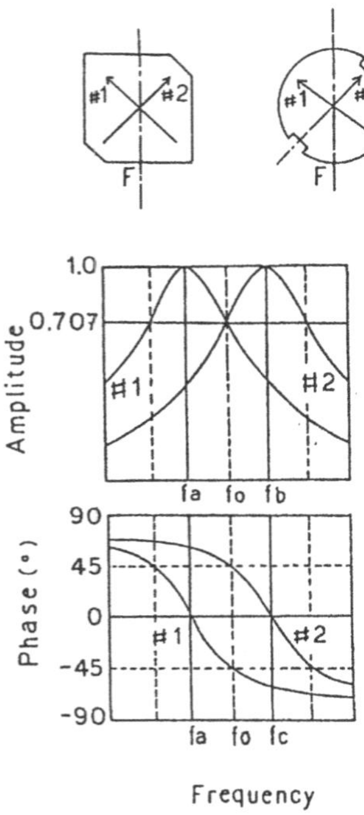

Fig. 6. Amplitude and phase of orthogonal modes in single point feed circularly polarised microstrip patch [21].

5. Dual Polarised Patches

There is currently much interest in patches that can produce either simultaneous orthogonal linear or circular polarisations at the same or at two close frequencies to reduce the size of equipment operating with diplexed signals.

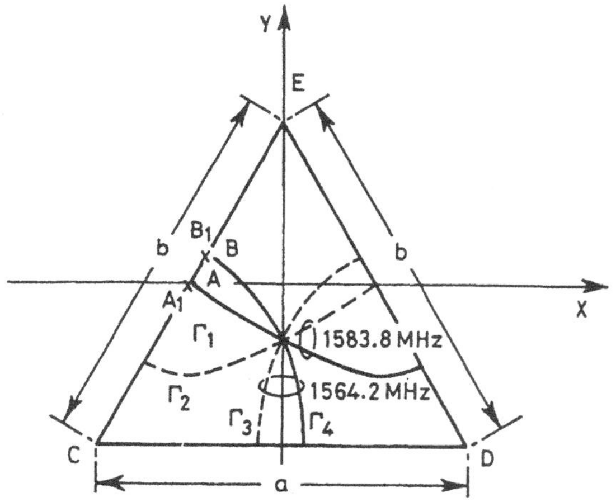

Fig. 8. Feed location loci for circular polarisation in equilateral triangle microstrip antenna [29]
(b/a = 0.98; a = 76 mm, substrate height = 3.2 mm, $\epsilon_r = 2.55$)
——RHCP, - - - - - L H C P
Γ_1 and Γ_2 are contours for 1583.8 MHz operation and Γ_3 and Γ_4 are for 1564.2 MHz.

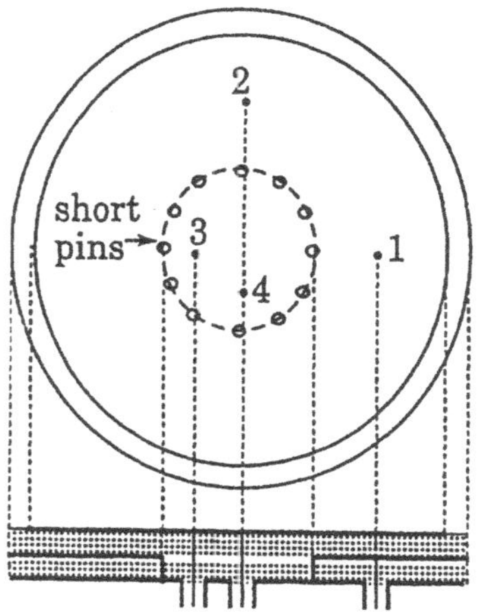

Fig. 9. Dual circularly polarised ring and patch microstrip antenna [40], [41].

If two orthogonal linear polarisations at separate frequencies are required, then a rectangular patch with two feed points exciting the orthogonal modes [36] can be used. A multilayer construction, with each pin feeding separate patches, gives further flexibility. If the frequency separation is greater than the individual patch bandwidth, then isolation is primarily determined by frequency separation to patch bandwidth ratio. For closely spaced frequencies, such as those used at L band for satellite communication systems, square or circular patches can be used and isolation is dependent on geometry, as noted in Section 2.

Isolations greater than −35 dB for dual linear polarisations have been achieved [4], [37] using optimised aperture coupling. The position and size of the orthogonal slot apertures is adjusted to minimise the higher order mode excitation. Isolation of about −40 dB has been achieved for an etched cross patch [38] in which a gridded structure is used to polarise the surface currents in the direction of resonance.

Dual circular polarisation has been achieved in two ways. Analysis of the triangular patch [29] reveals that there are a variety of feed points that will give either hand of circular polarisation at different frequencies, as Figure 8 shows. The frequency spacing can be controlled to some degree by the aspect ratio a/b. Similar characteristics can be obtained for a rectangular patch loaded with stubs [39]. Isolation between the dual polarisations is not reported in either case. Alternately, a multilayer structure using a short-circuited ring and patch (Figure 9) can be used [40], [41]. By rotation of the patch with respect to the ring, isolation of about −50 dB can be achieved over a narrow range of frequencies.

6. Microstrip Spirals

The application of the spiral concept to microstrip was first investigated by Wood [42], who analysed the radiation from curved microstrip lines and fabricated a number of single arm spirals. He concluded that, due to the tight wave trapping action of microstrip, the amount of power radiated per turn was significantly less than that from a conventional cavity-backed two or more turn spiral. This meant that radiation from the outer turns perturbed the pattern and led him to produce single turn spirals having bandwidths up to 40% and radiation efficiencies of about 50% with well behaved radiation paterns. Similar elements have been suggested recently for L-band Land Mobile Communications applications [43]. One advantage of the one-arm centre-fed spiral is that a wideband balun feed is not needed. An alternative arrangement is to feed a one-arm spiral with a small number of turns from the outside by a microstrip line, thus allowing use in a corporately fed array. An optimised open circuited spiral with 1.5 turns [44] gives less than 3 dB axial ratio over a 2.6% bandwidth on a 0.08λ thick substrate; the measured gain of a four-element array is 13.7 dBi. Resistive loading [45] reduces the axial ratio to less than 1 dB and the gain by about 0.7 dB.

Centre-fed two-arm microstrip spirals [46] are now being examined as alternatives to the cavity-backed type where multi-octave bandwidths are required. Wideband baluns are still required. By careful resistive loading at the outer edge, good performance over a 2–18 GHz range can be obtained, although the axial ratio is not as small as the best cavity-backed type [47]. Square-shaped spirals [48] and multi-mode types [49] with some beam scanning are also being investigated. Use of two dielectric layers has been shown to give a conical circularly polarised beam [50].

7. Special Substrates and Active Antennas

Both ferrite and chiral substrates have been examined recently in the search for improved polarisation control of microstrip antennas. Das [51] reports an early example of a patch on a ferrite substrate. The unique features of such an antenna [52] are, firstly, that a square patch with a single feed probe will give circular polarisation switchable between right hand and left hand, and frequency-tunable by adjusting the magnetic bias field [53]. Secondly, a phased array of such elements can be wide-angle impedance matched, again by bias field control. Finally the radar cross-section can be reduced in its "off" state by 20 to 40 dB [54]. For a patch on a 0.03 wavelength thick, $\epsilon_r = 15$ substrate, the impedance and axial ratio bandwidths are 1% and 13% respectively. The wideband axial ratio behaviour is attributed to the generation of an inherently circularly polarised mode within the ferrite, and is seen to be an important and advantageous feature of such antennas.

Chiral substrate, although having an inherent handedness, has been found to possess some disadvantages [55] when used for microstrip patches. In particular there are increased losses due to surface wave excitation and high cross-polarisation. As yet, good system advantages have not been identified for the use of chiral substrates, although this may well happen in the future. Little work appears to have been done on the circular polarisation on conformally shaped substrates, although reference [56] derives the circular polarisation conditions for a rectangular patch on a cylinder.

Several active antennas with polarisation control have been reported. Circular polarisation has been generated using four quarter-wavelength active dielectric resonator antennas [57]. The use of external locking of two orthogonally polarised patch oscillators [58] has been shown to allow selectable polarisation, both linear and circular. Selectable polarisation has also been demonstrated with switched lines located beneath the ground plane [59].

8. Dual and Circularly Polarised Arrays

In general, dual and circularly polarised arrays can be formed from the elements described in the preceding sections. This section reviews progress in array design or special array techniques that either simplify design or enhance performance of such arrays.

8.1 Patch Arrays

Improvements in the performance of two-dimensional patch arrays continue to be made. A four-element array of electromagnetically coupled patches with parasitic patches above [60] has been shown to have over an 85% efficiency and less than 3 dB axial ratio across a 13% bandwidth, using honeycomb substrates. Dual linear polarisation at 12.6 GHz and 14.3 GHz respectively, with about 35 dB isolation, has been achieved with a multiple-layer 16-element array with two separate corporate feeds sandwiched between perforated ground planes [61].

8.2 Microstrip Line Arrays

The rampart line [62], chain antenna [63], square loop line [64], crank line [65], herringbone line [66], and strip/dipole array [67] are microstrip line arrays that give circular polarisation (Figure 10). Many of these travelling wave arrays have similar characteristics. As an example, a rampart array having ten periods [62] was found to give a peak axial ratio of less than 1 dB and an input return loss of -10 dB. The beam direction and axial ratio are, in general, frequency dependent. They should be operated with off-broadside beam to ensure good input VSWR. Feeding at opposite ends will produce circular polarisation of the opposite hand and, although they can be considered dual polarisation, the two hands will be radiated in beams oppositely displaced from broadside. Such arrays can be used to form simple two-dimensional arrays, but the frequency-dependent beam scan renders them suitable only for narrow bandwidth applications. The beam scan problem is overcome by forming such line arrays into cross structures [68] (Figure 11). Here, cross-polarisation is achieved over more than 10% bandwidth, with efficiencies greater than 80%.

8.3 Sequentially Rotated Arrays

Sequential rotation [69],[70] is a technique that improves the axial ratio of circularly polarised arrays. Figure 12 shows the method and Figure 13 shows two implementations. Each element in the subarray is rotated with respect to its neighbour and the phase change generated by the rotation of the circularly polarised element is offset by an appropriate phase change in the excitation, which is usually created by a line length change in the corporate feed. In Figure 13a two pairs having 0 degrees, 90 degrees rotations are shown and in Figure 13b 0 degrees, 90 degrees, 180 degrees, 270 degrees rotations are used. The principle of the technique is that the cross-circularly polarised components of the elliptically polarised elements are cancelled, as the feeding phase changes are correct for the desired sense of polarisation only. These changes are calculated for the main beam peak only, so that in some cases cross-polarisation sidelobes may be higher than in a conventional array [15]. An additional benefit arises due to the fact that reflections from mismatched elements cancel out in the feed. In the case of microstrip patches, axial ratio and input match both degrade off resonance, and sequential rotation hence serves to widen the apparent bandwidth. Figure 14 [69] clearly shows the improvement in both axial ratio and input VSWR of eight-element arrays with sequential rotation applied to groups of four. Analysis of various configurations [15], [71] indicate that the dominant factor determining the performance of most sequentially rotated patch arrays is multiple reflections between the patches and the non-isolating power splitters. Furthermore, the gain over a range of frequencies of a sequentially rotated array is similar to that of a conventional array using the same elements, because the power lost in the conventional array in main beam cross-polarisation and input VSWR is spread into grating lobes in the far-out radiation pattern in the sequentially rotated array. The position of the grating lobes is determined by the geometry of the rotations [72], with the array of Figure 13a having lobes in the vertical principal plane and Figure 13b in the diagonal planes. The lobe height is determined by four factors: the element axial ratio, the excitation imbalance caused by multiple feed reflections, the element pattern, and the element spacing. For example, a circular patch array on 0.06 wavelength thick substrate, at a frequency of 1.03 times the resonant frequency, has its axial ratio reduced from 4.7 dB to 0.4 dB by the

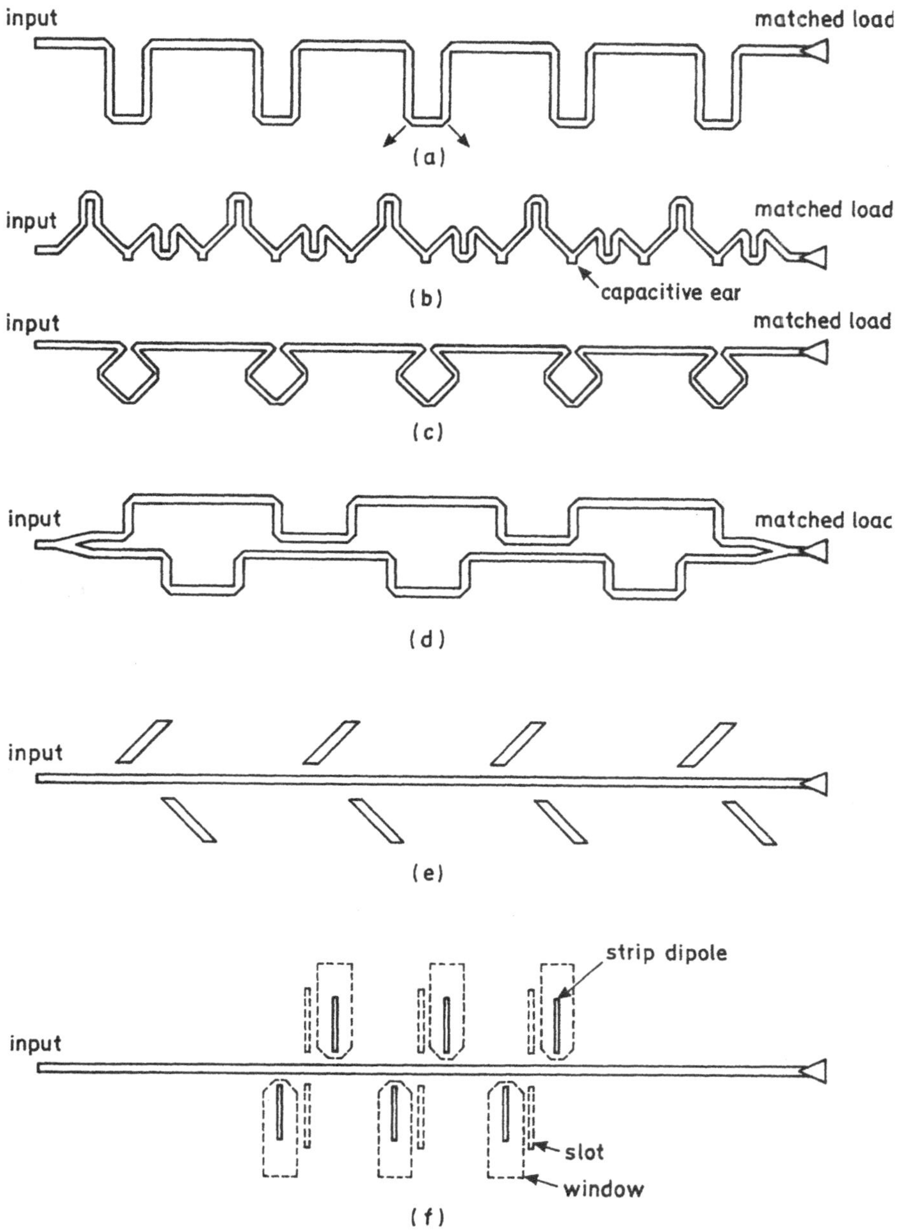

Fig. 10. Microstrip line, circularly polarised arrays. (a) rampart line, (b) chain antenna, (c) square loop line, (d) crank line, (e) herringbone, (f) slot-dipole array

Fig. 11. Cross-antenna for circular polarisation [68]. (Single arm, two-turn cross with log-periodic expansion)

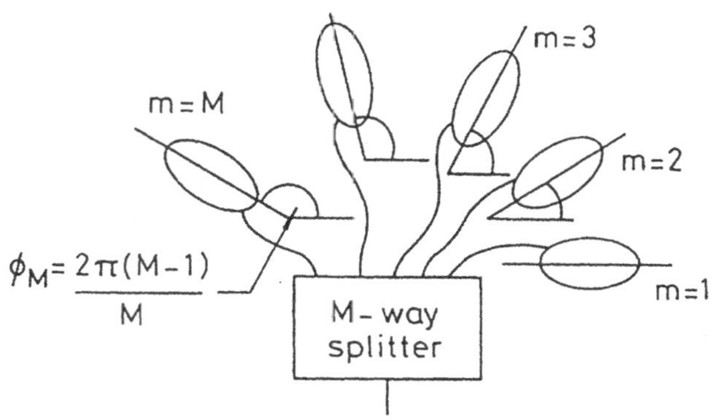

Fig. 12. Sequentially rotated feeding of circularly polarised array elements [69], [70].

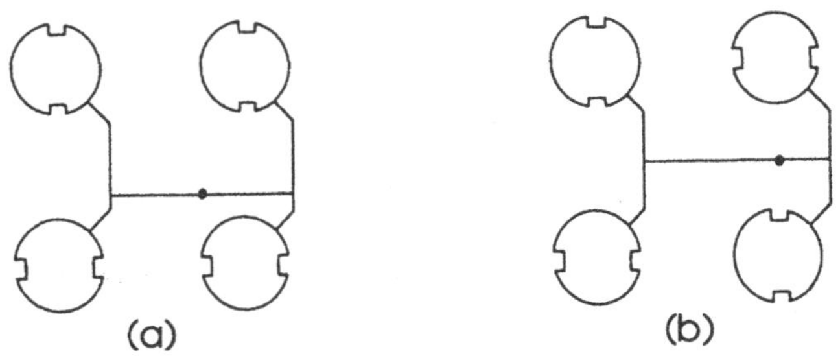

Fig. 13. Sequentially rotated feeding of notched circular patches. (a) Pairs rotation (b) Rotation of group of four

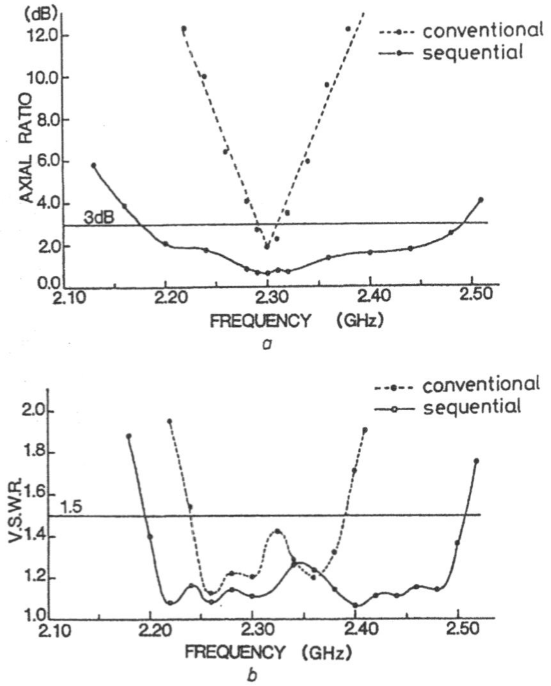

Fig. 14. Measured bandwidth characteristics of sequentially rotated microstrip patch arrays [69]. (a) axial ratio, (b) VSWR (eight-element arrays, substrate height = 4mm, $\epsilon_r = 2.6$)

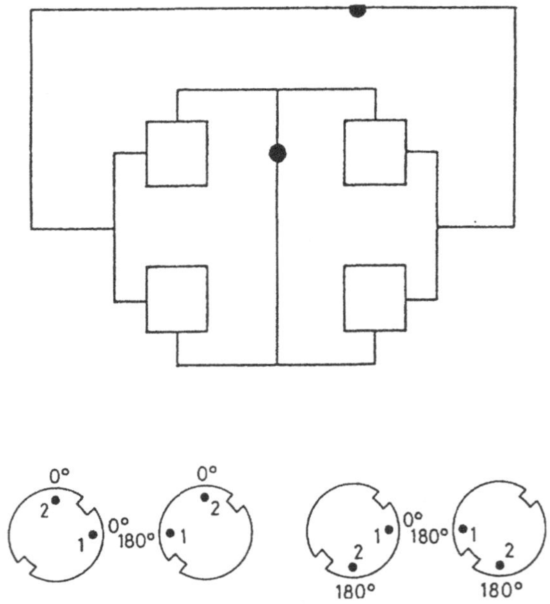

Fig. 15. Dual polarised sequentially rotated arrays. (a) Dual linear [75] (b) Dual circular polarisation [5]

application of Figure 13b-type subarraying, but diagonal plane grating lobes appear at a level of about −20 dB. Further rotation of the subarrays or application of rotation to larger groups has been shown [73] to reduce this grating lobe level to about −30 dB. The use of sequential rotation in a circularly polarised phased array [74] has emphasised the importance of such grating lobes for large scan angles.

Sequential rotation can also be used in dual polarisation arrays. Figure 15 shows two examples. Figure 15a is a dual linear array with planar feed [75] that gives reduced cross-polarisation and isolation. The reduction is attributed to feeding on opposite patch sides and the 180 degree phase offsets in the feed. However, the application of this method to dual circularly polarised arrays is more problematic. In the array shown in Figure 15b [5], cross-polarisation and input VSWR are reduced as expected. However, due to the opposite polarity coupling between feeds in patches 1 and 3 compared to 2 and 4, no improvement in isolation occurs. Nevertheless, good isolation of the order of −40 dB for a 0.023 wavelength thick substrate has been achieved by operating the array of Figure 15b with a slightly squinted beam [5]. The phase changes in the feed required to scan the beam now provide the cancellation of the coupled signals. Further dual polarised arrays have been reported [76], [77] that use concepts based on the generation of circular polarisation from rotated linearly polarised patches [10] and a variety of different feed networks. Isolations similar to those above are obtained.

9. Conclusions

The field of dual and circularly polarised microstrip antennas is a rich and diverse one, with the freedom offered by the medium giving rise to many configurations. This paper has attempted to give examples of the various types, and to highlight some of the underlying principles and limitations. The concepts necessary for the generation of circular polarisation have been known for some time, and place constraints on the performance and in particular the bandwidth if simple non-isolating feeds or single point feed patches are to be used. Reduction of the coupling between the required orthogonal modes is identified as important in obtaining both good quality circular polarisation and good isolation between dual polarisations and this reduction is related to patch Q and feeding geometry. Thick or multiple layer parasitic patches help to increase circular polarisation bandwidth, as does sequential rotation in arrays.

Dual polarisation is an increasingly important requirement, and optimised patch and array geometry now allows isolation of the order of −40 dB to be achieved for both linear and circular polarisation. System requirements then dictate if further diplexing components are needed. Other types such as microstrip spirals, ferrite substrates, and active antennas are also noted as being significant, and will no doubt have advantages in specific applications. In spite of the large number of types available, innovation and development continue apace in the quest for improved polarisation control for current and future applications.

References

[1] W. L. Stutzman, *Polarisation in Electromagnetic Systems,* Canton, Mass.: Artech House, 1993.

[2] H. Mott, *Polarisation in Antennas and Radar,* New York: John Wiley Inc, 1986.

[3] J. R. James and P. S. Hall, *Handbook of Microstrip Antennas,* IEE Electromagnetic Wave Series 28, London: Peter Perigrinus, 1989.

[4] Y. Murakami, W. Chujo, M. Fujise, "Mutual coupling between two ports of dual slot coupled circular patch antennas," *IEEE Int'l. Ant. and Prop. Symp.,* pp. 1469–1472, June 1993.

[5] P. S. Hall, "Dual polarisation antenna arrays with sequentially rotated feeding," *IEE Proc.,* vol. 139, pt H, no. 5, pp. 465–471, Oct. 1992.

[6] J. R. James, P. S. Hall, and C. Wood, *Microstrip Antenna Theory and Design,* IEE Electromagnetic Wave Series 12, London: Peter Perigrinus, 1981.

[7] Op cit, p. 176ff.

[8] G. G. Sanford and L. Klein, "Recent developments in the design of conformal microstrip phased arrays," *IEE Conf. on Maritime and Aeronautical Satellites for Communications and Navigation,* IEE Conf. Publ. No 160, pp. 105–108, 1978.

[9] P. S. Hall, C. Wood, and J. R. James, "Recent examples of conformal microstrip antenna arrays for aerospace applications," *2nd IEE Conf. on Ant. and Prop.,* pp. 397–401, 1981.

[10] J. Huang, "Technique for an array to generate circular polarisation with linearly polarised elements," *IEEE Trans. on Ant. and Prop.,* AP-34, (9), pp. 1113–1123, 1986.

[11] P. S. Hall, J. Huang, E. Rammos, and A. Roederer, "Gain of circularly polarised arrays with linearly polarised elements," *Electronics Letters, 25,* pp. 124–125, Jan. 1989.

[12] D. J. Brain and J. R. Mark, "The disc antenna—a possible L-band aircraft antenna," *IEE Conf. Publ. 95,* pp. 14–16, 1973.

[13] J. Q. Howell, "Microstrip antennas," *IEEE Trans. on Ant. and Prop.,* AP-23, pp. 90–93, 1975.

[14] A. Adrian and D. H. Schaubert, "Dual aperture coupled microstrip antenna for dual or circular polarisation," *Electronics Letters,* vol. 23, no. 23, pp. 1226–1228, Nov. 1987.

[15] P. S. Hall, J. S. Dahele, and J. R. James, "Design principles of sequentially fed, wide bandwidth, circularly polarised microstrip antennas," *IEE Proc.,* vol. 136, pt H, no. 5, pp. 381–389, Oct. 1989.

[16] T. Chiba, Y. Suzuki, N. Miyano, S. Muira, and S. Ohmori, "A phased array antenna using microstrip patch antennas," *12th European Microwave Conf.,* pp. 472–477, Sept. 1982.

[17] T. Chiba, Y. Suzuki, and N. Miyano, "Suppression of higher modes and cross polarised component for microstrip antennas," *IEEE Ant. and Prop. Symp.,* pp. 285–288, 1982.

[18] T. Teshirogi and N. Goto, "Recent phased array work in Japan," *ESA/COST 204 Phased Array Antenna Workshop,* pp. 37–44, 1983.

[19] E. Nishiyama, S. Egashira, and A. Sakitani, "Stacked circular polarised microstrip antenna with wideband and high gain," *IEEE Ant. & Prop. Symp.,* pp. 1923–1926, July 1992.

[20] S. D. Targonski and D. M. Pozar, "Design of wideband circularly polarised aperture coupled microstrip antennas," *IEEE Trans. on Ant. and Prop.,* AP-41, no. 2, pp. 214–220, Feb. 1993.

[21] Reference 3, p. 222.

[22] P. C. Sharma and K. C. Gupta, "Analysis and optimised design of single point feed circularly polarised microstrip antennas," *IEEE Trans. on Ant. and Prop.,* AP-31, pp. 949–955, 1983.

[23] G. G. Sanford and R. E. Munson, "Conformal VHF antenna for the Apollo-Soyuz Test Project," *IEE Int'l Conf. on Antennas for Aircraft and Spacecraft,* pp. 130–135, 1975.

[24] L. T. Ostwald and C. W. Garvin, "Microstrip command and telemetry antennas for communications and technology satellites," *IEE Int'l Conf. on Antennas for Aircraft and Spacecraft,* pp. 217–222, 1975.

[25] M. Haneishi et al, "Broadband microstrip array composed of single feed type circularly polarised microstrip elements," *IEEE Ant. and Prop. Symp.,* pp. 160–163, May 1982.

[26] J. Kerr, "Microstrip antenna developments," *Workshop on Printed Antenna Technology,* pp. 3.1–3.20, 1979.

[27] L. C. Shen, "Elliptical microstrip antenna with circular polarisation, *IEEE Trans. on Ant. and Prop.,* AP-29, pp. 90–94, 1981.

[28] H. D. Wienschel, "Cylindrical array of circularly polarised microstrip antennas," *IEEE Ant. and Prop. Symp.,* pp. 177–180, 1975.

[29] Y. Suzuki, N. Miyano, and T. Chiba, "Circularly polarised radiation from singly fed equilateral-triangular microstrip antenna," *IEE Proc.,* vol. 134, pt H, pp. 194–198, 1987.

[30] T. Tsutsumi, S. Tsuda, A. Matsui, and M. Haneishi, "Radiation properties of circularly polarised microstrip ring antenna excited by dominant mode," *Int'l Symp. on Ant. and Prop.,* ISAP92, pp. 805–808, 1992.

[31] K. Hirose and H. Nakano, "Circularly polarised printed loop antenna proximity coupled to a vertical probe," *IEEE Ant. and Prop. Symp.,* pp. 1919–1922, July 1992.

[32] M. Haneishi and S. Yoshida, "Design method of circularly polarised microstrip antenna by one point feed," *Electronics and Communications,* vol. 64-B, no. 4, pp. 46–54, 1981.

[33] H. Inasaki and K. Kawabata, "Circular microstrip antenna with a cross slot for circular polarisation," *IEICE Trans.,* vol. E74, no. 10, pp. 3274–3279, Oct. 1991.

[34] H. Shoki, K. Kawabata, and H. Iwasaki, "Circularly polarised slot coupled microstrip antenna using a parasitically excited slot," *IEICE Trans.,* vol. E74, no. 10, pp. 3268–3273, Oct. 1991.

[35] H. Iwasaki, H. Sawada, and K. Kawabata, "Circularly polarised microstrip antenna using singly fed proximity coupled feed," *ISAP92,* pp. 797–798, 1992.

[36] Reference 3, pp. 318–320.

[37] Y. Murakami, W. Chujo, I. Chiba, and M. Fujuse, "Dual slot coupled microstrip antenna for dual frequency operation," *Electronics Letters,* vol. 29, no. 22, pp. 1906–1907, Oct. 1993.

[38] L. Habib, G. Kossiavas, and A. Papiernik, "Cross shaped patch with etched bars for dual polarisation," *Electronics Letters,* vol. 29, no. 10, pp. 916–918, May 1993.

[39] Y. Murakami, K. Ieda, T. Yasuda, Y. Kawamura, and T. Nakamura, "Dual band circularly polarised stub loaded microstrip antenna," *Int'l Symp. on Ant. and Prop.,* ISAP92, pp. 793–796, 1992.

[40] M. Nakano, H. Arai, W. Chujo, M. Fujise, and N. Goto, "Double layered self diplexing antenna using ring patch antenna and its feed circuits," *IEEE Ant. and Prop. Symp.,* pp. 483–486, July 1992.

[41] W. Chujo, M. Fujise, H. Arai, and N. Goto, "Improvement of the isolation characteristics of a two layer self diplexing array antenna using a circularly polarised ring patch antenna," *IEICE Trans.,* vol. E76-B, no. 7, pp. 755–758, July 1993.

[42] C. Wood, "Curved microstrip line as compact wideband circularly polarised antennas," *IEE Proc. on MOA,* vol. 3, 1979, pp. 5–13, 1979.

[43] A. Huhaneu and S. Tallquist, "One turn antenna for L- band land mobile communications," *Proc. Progress in Electromagnetics Research Symp.,* p. 883, July 1993.

[44] H. Nakano, G. Hirokawa, J. Yamauchi, and K. Hirosi, "Spiral antenna with coplanar strip line feed," *Electronics Letters,* vol. 28, no. 23 pp. 2130–2131, Nov. 1992.

[45] H. Nakano, H. Mimaki, J. Yamauchi, and K. Hirose, "Numerical analysis of a low profile spiral antenna," *XXIV URSI General Assembly,* p. B3–10, 1993.

[46] J. J. H. Wang and V. K. Tripp, "Design of multioctave spiral mode microstrip antennas," *IEEE Trans. on Ant. and Prop.,* AP-39, pp. 332–335, March 1991.

[47] T. R. Holsheimer and J. C. Holloway, "Investigation of spiral performance over tightly spaced ground planes," *IEEE Ant. and Prop. Symp.,* pp. 454–457, June 1993.

[48] D. Shirley, "Spectral domain analysis of square spiral microstrip antennas," *IEEE Ant. and Prop. Symp.*, pp. 1466–1468, June 1993.

[49] J. J. H. Wang, G. D. Hopkins, and V. K. Tripp, "Beam switching and steering of spiral mode microstrip antennas," *Proc. ISAP92,* pp. 789–792, 1992.

[50] I. G. Tiglis et al, "Radiation of a planar spiral antenna above a double dielectric ground plane," *IEEE AP-S Symp.*, pp. 152–155, June 1993.

[51] N. Das and S. K. Chowdhury, "Rectangular microstrip antenna on a ferrite substrate," *IEEE Trans. on Ant. and Prop.*, AP-30, pp. 499–502, 1982.

[52] D. M. Pozar, "Radiation and scattering characteristics of microstrip antennas on normally biased ferrite substrates," *IEEE Trans. on Ant. and Prop.*, AP-40, no. 9, pp. 1084–1092, Sept. 1992.

[53] D. M. Pozar and V. Sanchez, "Magnetic tuning of a microstrip antenna on a ferrite substrate," *Electronics Letters 24,* pp. 729–731, 1988.

[54] D. M. Pozar, "Radar cross section of microstrip antenna on a normally biased ferrite substrate," *Electronics Letters 25,* pp. 1079–1080, 1989.

[55] D. M. Pozar, "Microstrip antennas and arrays on chiral substrates," *IEEE Trans. on Ant. and Prop.*, AP-40, no. 10, pp. 1260–1263, Oct. 1992

[56] K. L. Wong and S. Y. Ke, "Cylindrical rectangular microstrip patch antenna for circular polarisation," *IEEE Trans. on Ant. and Prop.*, AP-41, no. 2, pp. 246–249, Feb. 1993.

[57] T. B. Ng, Y. O. Yam, and L. M. Lam, "Active quarter-wavelength dielectric radiator with circularly polarised radiation pattern," *Electronics Letters,* vol. 27, no. 19, pp. 1758–1759, 1991.

[58] P. M. Haskins, P. S. Hall, and J. S. Dahele, "Polarisation-agile active patch antenna," *Electronics Letters,* vol. 30, pp. 98–99, 1994.

[59] M. L. Charles, "Active antennas with polarisation switching," *L'Onde Electrique,* vol. 73, no. 1, pp. 5–9, Jan.–Feb. 1993.

[60] S. Borchi, P. Capece, M. Di Fausto, V. Mazzieri, and M. Votta, "Electrical design of a wideband, high efficiency circularly polarised 4-patch subarray," *IEEE AP-S Symp.*, pp. 1931–1934, July 1992.

[61] T. Murata, M. Ohta, and H. Ishizaka, "Dual frequency and dual polarisation low sidelobe microstrip arrays antenna for satellite communications," *Proc. ISAP Conf.*, pp. 1113–1116, 1992.

[62] P. S. Hall, "Microstrip linear array with polarisation control," *IEE Proc.*, vol. 103H, pp. 215–224, 1983.

[63] J. Henriksson, K. Markus, and M. Turi, "Circularly polarised travelling wave chain antenna," *Proc. 9th European Microwave Conf.*, pp. 174–178, 1979.

[64] T. Makimoto and S. Nishimura, "Circularly polarised microstrip line antenna," *US Patent 4 398 199,* 1983.

[65] S. Nishimura, Y. Sugio, and T. Makimoto, "Crank type circularly polarised microstrip line antenna," *IEEE Ant. and Prop. Symp.*, pp. 162–165, 1983.

[66] R. P. Owens and J. Thraves, "Microstrip antenna with dual polarisation capability," *Proc. Military Microwaves Conf.*, pp. 250–254, Oct. 1994.

[67] K. Ito, K. Itoh, and H. Kogo, "Improved design of series fed circularly polarised printed linear arrays," *IEE Proc.*, vol. 133H pp. 462–466, 1986.

[68] A. G. Roederer, "The cross antenna: a new low profile circularly polarised radiator," *IEEE Trans.*, AP-38, pp. 704–710, 1990.

[69] T. Teshirogi, M. Tanaka, and W. Chujo, "Wideband circularly polarised array antenna with sequential rotations and phase shift of elements," *Int'l Symp. on Ant. and Prop.*, ISAP85, pp. 117–120, 1985.

[70] M. Haneishi, "Circularly polarised SHF planar array composed of microstrip pairs element," *Int'l Symp. on Ant. and Prop.*, ISAP85, pp. 125–128, 1985.

[71] P. S. Hall, "Application of sequential feeding to wide bandwidth circularly polarised microstrip patch arrays," *IEE Proc.*, vol. 136H, no. 5, pp. 390–398, Oct. 1989.

[72] M. S. Smith and P. S. Hall, "Analysis of radiation pattern effects in sequentially rotated arrays," *IEE Proc.*, vol. 141, pp. 313–320, Aug. 1994.

[73] P. S. Hall and M. S. Smith, "Sequentially rotated arrays with reduced sidelobe levels," *IEE Proc.*, vol. 141, pp. 321–325, Aug. 1994.

[74] J. Notter and C. G. Panni, "Moments method analysis of a finite array of arbitrary shaped microstrip patch radiating elements," *Int'l J. of Microwave and Millimetre Wave Computer Aided Design,* vol. 4, no. 1, pp. 18–30, Jan. 1994.

[75] J. Huang, "Dual polarised microstrip array with high isolation and low cross polarisation," *Microwave and Opt. Tech. no. 2. Letts.*, vol. 41, no. 3, pp. 99–103, 1991.

[76] E. Rammos and A. Roederer, "Self diplexing CP antenna," *IEEE Ant. and Prop. Symp. Digest,* pp. 803–806, 1990.

[77] Q. Garcia, C. Martin-Pascual, N. Merce, and P. Tejedor, "Self diplexing and active radiating elements for active antenna systems," *COST 223 Workshop on Active Antennas,* pp. 3.2.1–3.2.18, June 1992.

Analysis and Optimized Design of Single Feed Circularly Polarized Microstrip Antennas

P. C. SHARMA, MEMBER, IEEE, AND KULDIP C. GUPTA, SENIOR MEMBER, IEEE

Abstract—**Analysis and optimized designs are presented of three types of single feed circularly polarized microstrip antennas, namely, a diagonal fed nearly square, a truncated-corners square, and a square with a diagonal slot. The Green's function approach and the desegmentation methods are used. The resonant frequencies are calculated for two orthogonal modes which together yield circular polarization. Optimum feed locations are determined for the best impedance match to a 50 Ω coaxial feed line. Axial-ratio bandwidths, voltage standing-wave ratio (VSWR) bandwidths and radiation patterns are evaluated and verified experimentally.**

I. INTRODUCTION

SEVERAL CIRCULARLY polarized microstrip antenna configurations have been reported during the last decade [1]-[10]. In this paper, three types of single feed circularly polarized microstrip patch antennas (diagonal-fed nearly square, truncated-corners square, and square with a diagonal slot) have been studied analytically as well as experimentally. For a diagonal-fed nearly square patch antenna ($\epsilon_r = 2.62$, substrate thickness is 1.588 mm), an experimental value of axial ratio of 1.4 dB has been reported earlier by Carver and Coffey [11] for a ratio of length of sides of the rectangle equal to 1.029. A successful attempt has been made in the present work to improve the axial ratio to 0.17 dB. The truncated-corners square patch antenna and the square patch antenna with a diagonal slot had been studied experimentally by Kerr [7], but no theoretical analysis and design procedure for these structures have been available so far. In the present investigation, the optimum dimensions and the feed locations for these antennas have been determined. Axial ratio and input voltage standing-wave ratio (VSWR) are evaluated as functions of frequency. Radiation patterns are also evaluated. The analysis is based on Green's functions for rectangular and triangular segments [12]-[14] and recently reported segmentation and the desegmentation methods [14]-[16].

II. METHOD OF ANALYSIS

In this method, the antenna is modeled as multiport network. The procedure is illustrated in Fig. 1. The physical periphery of the antenna is extended outward to obtain a planar model with a magnetic wall boundary. This planar model is treated as a lossless resonator during the initial steps of the analysis. The periphery of the planar model, with effective dimensions, is divided into several sections of small widths so that the field variation over the width of each of these sections is negligibly small. Each one of these sections is considered as a port (Fig. 1(c)). The radiated power is taken into account by terminating the ports of the multiport network by resistors corresponding to the radiation resistances [17]. The values of radiation resist-

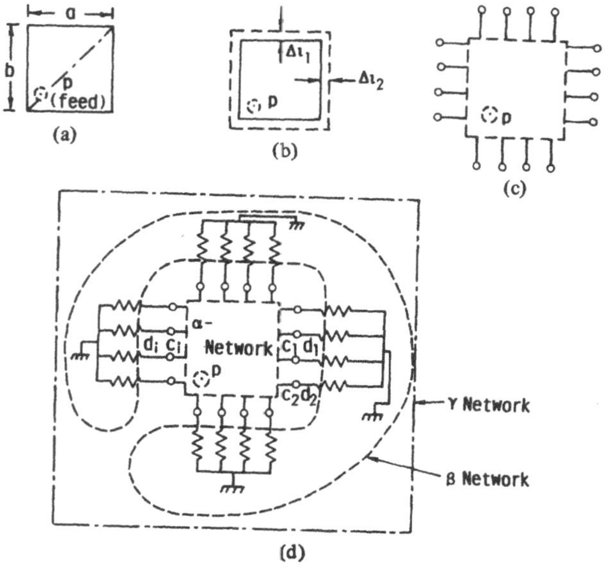

Fig. 1. Various steps in the analysis of microstrip patch antennas. (a) Antenna shape. (b) Planar model. (c) Equivalent multiport network. (d) Loading by radiation conductance.

ances are calculated as follows: 1) the radiation conductance [17] for each of the straight edges of the radiating patch is evaluated; and 2) the conductance so calculated is distributed amongst all the ports on the corresponding edge in proportion to the port widths. The entire radiation resistance network is treated as one multiport network β as shown in Fig. 1(d). The entries in the Z-matrix of the β-network are all zeros except the diagonal elements Z_{ii} which are equal to resistances connected to various ports. The multiport network that represents the lossless planar model is treated as another network α.

The input impedance and voltage around the periphery of the antenna are evaluated using the segmentation method [15] and/or desegmentation method [17] as discussed in the following subsections.

B. Application of Segmentation Method [15]

When a two-dimensional configuration γ can be considered as combination of several segments, the Z-matrix for the combination γ can be expressed in terms of the Z-matrices of the constituent segments. The Z-matrices of the various segments are grouped together as [16],

$$\begin{bmatrix} V_p \\ V_c \\ V_d \end{bmatrix} = \begin{bmatrix} \tilde{Z}_{pp} & \tilde{Z}_{pc} & \tilde{Z}_{pd} \\ \tilde{Z}_{cp} & \tilde{Z}_{cc} & \tilde{Z}_{cd} \\ \tilde{Z}_{dp} & \tilde{Z}_{dc} & \tilde{Z}_{dd} \end{bmatrix} \begin{bmatrix} I_p \\ I_c \\ I_d \end{bmatrix} \quad (1)$$

Reprinted from *IEEE Trans. Antennas Propaga.*, vol. AP-31, no. 6, pp. 949–955, Nov. 1983.

where p refers to the unconnected ports of the various segments of γ (i.e., the external ports of the circuit γ). Subscripts c and d represent the interconnected ports which are numbered in such a way that the port c_i is connected to the port d_i as illustrated in Fig. 1(d). The submatrices in (1) are obtained from the Z-matrices of the individual segments as [15],

$$\tilde{Z}_\gamma = \tilde{Z}_{pp} + [\tilde{Z}_{pc} - \tilde{Z}_{pd}]\tilde{Z}'_{cp} \tag{2}$$

where

$$\tilde{Z}'_{cp} = [\tilde{Z}_{cc} - \tilde{Z}_{cd} - \tilde{Z}_{dc} + \tilde{Z}_{dd}]^{-1}[\tilde{Z}_{dp} - \tilde{Z}_{cp}].$$

For a electric current I_p fed into the pth port, the voltages at the interconnected c and d ports are given by

$$\mathbf{V}_c = \mathbf{V}_d = [\tilde{Z}_{cp} + [\tilde{Z}_{cc} - \tilde{Z}_{cd}]\tilde{Z}'_{cp}]\mathbf{I}_p. \tag{3}$$

C. Application of Desegmentation Method [16]

Consider the configuration (α) of a truncated-corners square patch antenna shown in Fig. 2(a). This configuration can be considered as obtained by removal of two isosceles triangular segments β_1 and β_2 from the two opposite corners of a square patch (γ-segment) as illustrated in Fig. 2(b). The interfaces between α- and β-segments are divided into discrete number of ports. These interconnected ports are named as c-ports on α-segment and d-ports on β-segments (Fig. 2(b)). The unconnected ports on the α-segment are named as p-ports and those on the β-segments are named as q-ports. It has been shown [16] that when the number of q-ports equals that of d-ports (equals that of c-ports), the Z-matrix of the α-segment is given in terms of the Z-matrices of β- and γ-segments as

$$\tilde{Z}_\alpha = \begin{bmatrix} \tilde{Z}_{pp\alpha} & \tilde{Z}_{pc} \\ \tilde{Z}_{cp} & \tilde{Z}_{cc} \end{bmatrix} = \begin{bmatrix} \tilde{Z}_{pp\gamma} - \tilde{Z}_{pq}\tilde{Z}'_{qp}, & -\tilde{Z}_{pq}\tilde{Z}'_{qd} \\ -\tilde{Z}_{dq}\tilde{Z}'_{qp} - \tilde{Z}_{dd}, & -\tilde{Z}_{dq}\tilde{Z}'_{qd} \end{bmatrix} \tag{4}$$

where

$$\tilde{Z}'_{qp} = [\tilde{Z}_{qq\gamma} - \tilde{Z}_{qq\beta}]^{-1}\tilde{Z}_{qp}$$
$$\tilde{Z}'_{qd} = [\tilde{Z}_{qq\gamma} - \tilde{Z}_{qq\beta}]^{-1}\tilde{Z}_{qd}. \tag{5}$$

Z_{dd}, Z_{dq}, Z_{qd}, $Z_{qq\beta}$ are submatrices of Z_β for β-segments, and $Z_{pp\gamma}$, Z_{pq}, Z_{qp}, $Z_{qq\gamma}$ are submatrices of Z_γ for γ-segment. Z_β and Z_γ are evaluated by using the Green's functions for an isosceles triangle [13] and for a rectangle [12].

For the square patch antenna with a diagonal slot (Fig. 2(c)), the γ-segment of Fig. 2(d) is obtained when a rectangular patch (β-segment) is added to the α-segment of Fig. 2(c). In this case both Z_β and Z_γ are evaluated from the Green's function for a rectangle, and Z_α is obtained from (4) and (5).

As shown in Fig. 1(d), the ports of the α-segment are terminated into radiation resistances. The input impedance and the voltage along the radiating edges are evaluated employing (2) and (3).

D. Radiation Characteristics

The radiation characteristics are evaluated in terms of the equivalent magnetic current distribution along the periphery of the antenna. The far field at a distance r is given by [11]

$$\bar{E}_\theta = \eta\bar{H}_\phi = jk_0\bar{F}_\phi = jk_0(-\bar{F}_x \sin\phi + \bar{F}_y \cos\phi)$$

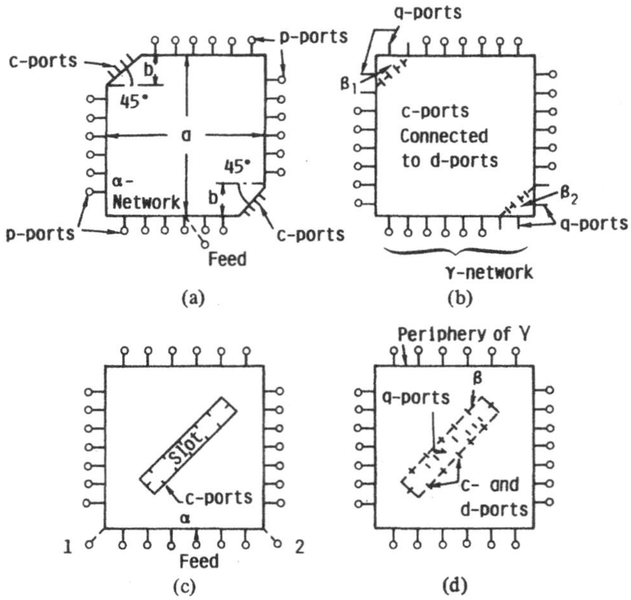

Fig. 2. (a), (b) Desegmentation method applied to a corners-truncated antenna. (c), (d) Desegmentation method applied to a square antenna with a diagonal slot.

$$\bar{E}_\phi = -\eta\bar{H}_\theta = -jk_0\bar{F}_\theta = -jk_0(\bar{F}_x \cos\phi + \bar{F}_y \sin\phi)\cos\theta \tag{6}$$

where $\bar{F}_x$ and $\bar{F}_y$ are the x- and y-components of electric vector potential $F(r)$ defined as

$$\bar{F}(\bar{r}) = \epsilon_0 \int_c \frac{\bar{K}(\bar{r})}{4\pi|\bar{r}-\bar{r}'|} e^{-jk_0|\bar{r}-\bar{r}'|}\, dl(\bar{r}') \tag{7}$$

and $\bar{K} = -2(\hat{n} \times \hat{z})E_z$ is the equivalent magnetic current, E_z is the electric field along the periphery of the antenna directed along the thickness of the substrate, and k_0 is the free-space wavenumber $(\omega\sqrt{\mu_0\epsilon_0})$.

III. NEARLY SQUARE DIAGONAL FED ANTENNA

The antenna configuration is shown in Fig. 1(a). In this case the circular polarization is obtained because the two modes of resonance (corresponding to the adjacent sides of the rectangle) are spatially orthogonal [6], [11]. The antenna is excited at a frequency in between the resonant frequencies of these two modes in order to obtain the phase quadrature relationship between the voltages (and therefore magnetic currents) of two modes. Corner or diagonal feeding is required to allow both the modes to be excited with a single feed.

A. Optimum Dimensions

For analysis, each of the sides of the rectangle is divided into seven ports. The impedance matrix for the multiport planar model is evaluated by employing the Green's function for a rectangle [12]. For the chosen value of the width of the rectangle a (Fig. 1(a)), the length b has been varied and the axial ratio has been evaluated at several frequencies. For a 1/8 in thick polystyrene substrate (ϵ_r = 2.52) and a = 2.66 cm, the best axial ratio (0.45 dB) is found to occur at 3.101 GHz when the ratio b/a = 1.0526 and the feed is located at one of the corners (A in Fig. 3 inset) of the rectangle. The theoretical values of the resonant frequencies of the two orthogonal modes corresponding to the length b and the width a are 3.035 GHz and 3.175 GHz. The corresponding measured values are 3.032 GHZ and 3.169

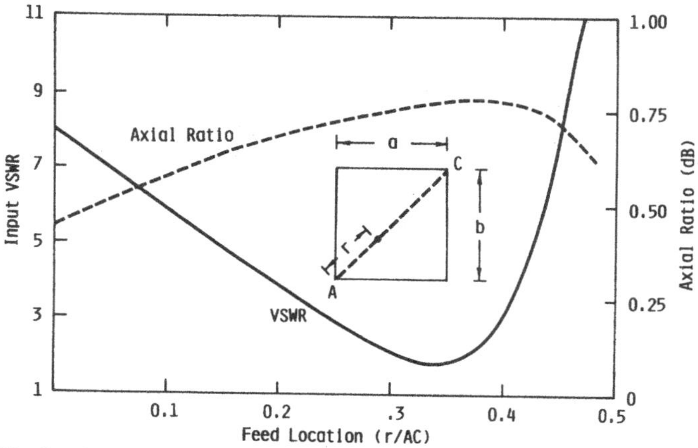

Fig. 3. Variations of input VSWR and axial ratio with feed location for a diagonal-fed antenna (thickness = 1/8 in, ϵ_r = 2.52, frequency = 3.101 GHz).

GHz, respectively. The best axial ratio (0.45 dB) is obtained at 3.101 GHz.

B. Optimum Feed Point Location

Although initial experiments were reported with a corner fed antenna, it is found that the circular polarization can be obtained even when the feed is located elsewhere on the diagonal AC. The variation of the input VSWR and the axial ratio with feed location on the diagonal AC is shown in Fig. 3. The axial ratio degrades from 0.45 dB to 0.79 dB and the input VSWR decreases from 8.1 to 1.73 as the feed is moved from corner A to a point 0.3441 times AC away from the corner. For feed locations at a distance r greater than 0.3441 AC, the input VSWR increases again. At the optimum feed location where the input VSWR is minimum (= 1.73), the value of axial ratio is 0.77 dB. Further calculations showed that, for this optimum feed location, an axial ratio equal to 0.45 dB is obtained again if the frequency is shifted to 3.103 GHz (2 MHz higher than the previous value for excitation at the corner A).

The input impedance (Z_{in}) at the optimum feed location is $(62.42 + j28.4)\ \Omega$ which is higher than the feed line impedance of 50 Ω. Thus input VSWR could be improved by decreasing Z_{in}. It may be recalled from the scaling principle of two dimensional components [18] that, for the same effective dimensions of a planar component, the impedance level (reactive component) reduces to half the original value when the thickness of the substrate is reduced to half. Therefore, another antenna on 1/16 in thick substrate with $\epsilon_r = 2.52$ as before was designed. The width of the rectangle (a in Fig. 3) was chosen such that the effective width equals the effective width of the 1/8 in thick antenna. The optimum ratio of lengths of sides of the rectangle, for obtaining circular polarization with axial ratio equal to 0.17 dB, is found to be 1.026 which is different than that for 1/8 in thick antenna. Thus the ratio b/a is found to depend upon the thickness of the substrate. The minimum input VSWR, for the antenna on 1/16 in thick substrate, is found to be 1.33, the corresponding input impedance being $(53.3 + j14.2)\ \Omega$. The reactive component of the input impedance is thus reduced to half as compared to that for the 1/8 in thick substrate and the resistive component, representing the radiated power, is reduced from 62.42 Ω to 53.2 Ω. The optimum feed location for minimum input VSWR is found to be on the diagonal at a distance 0.3522 times AC. For this optimum feed location, the minimum axial ratio (equal to 0.17 dB) is obtained at 3.1372 GHz. As the dielectric constant for the available substrate with thickness equal to 1/16 in was different (2.49), the antenna designed was optimized again and these results are summarized in Table I.

The input VSWR can be reduced further if a 1/32 in thick substrate is used. Extrapolating the results, the input impedance is expected to be around $(45.3 + j7)\ \Omega$ and the input VSWR is likely to be about 1.19. Although the input VSWR improves with reduction in the thickness of the substrate, it has been observed that the axial ratio limited bandwidth also decreases. Thus a design trade-off is involved in the selection of the substrate thickness.

C. Bandwidth and Radiation Patterns

Theoretical and measured values of input VSWR and axial ratio as functions of frequency for one of the antennas is shown in Fig. 4. The measured values of bandwidth (for axial ratio less than 6 dB) is 34.8 MHz (1.12 percent), the corresponding theoretical value being 33.7 MHz (1.086 percent). The VSWR variation over this band of frequencies is small (Fig. 4). The bandwidth of the antenna is therefore limited by the axial ratio and not by the input impedance. Similar results have also been observed for the antenna on 1/16 in thick substrate, but the axial ratio bandwidth is lower by nearly 40 percent (Table I).

The experimental and theoretical radiation patterns for the antenna (thickness = 1/8 in, $\epsilon_r = 2.52$) in the $\theta = 90°$ plane are shown in Fig. 5. The beamwidth is calculated from the radiation pattern. Table I gives the summary of the result for the diagonal-fed nearly square patch antennas investigated.

IV. TRUNCATED-CORNERS SQUARE PATCH ANTENNA

In this case (Fig. 2(a)), the two orthogonal modes of resonance are diagonal modes which would individually yield linear polarization along the directions of the two diagonals. Chopping of the two corners makes the resonant frequency of the mode along this diagonal to be higher than that for the mode along the unchopped diagonal. The frequency of operation and the feed point are chosen such that the two modes are excited in phase quadrature.

A. Optimum Configuration

The periphery of the truncated-corners antenna (Fig. 2(a)) is divided into 32 ports which include four c-ports at each of the truncated corners. An additional port is considered to represent the feed point. Thus, there are 25 p-ports and four d-ports for each of the β-segments. The desegmentation method is used to evaluate the Z-matrix of the multiport planar model. The antenna characteristics are evaluated as discussed in Section II. It has been found that for the chosen dimensions (2.73 cm × 2.73 cm) of the square patch, and 1/8 in thick polystyrene substrate ($\epsilon_r = 2.52$), the best value of axial ratio (= 0.12 dB) is obtained at 3.175 GHz when $b/a = 0.04578$ where b is the amount of truncation in cm and a is the length of sides of the square patch. The theoretical values of the resonant frequencies of the two orthogonal modes, which can be excited independentally by locating the feed point at the corners, are 3.1340 GHz and 3.212 GHz, respectively. The frequency for the best circular polarization (axial ratio equal to 0.12 dB) is 3.1750 GHz.

TABLE I
PERFORMANCE OF DIAGONAL FED NEARLY SQUARE PATCH ANTENNAS

	ANTENNA I		ANTENNA II	
I. Parameters				
1. Thickness, ϵ_r	1/8", 2.52		1/16", 2.49	
2. Width 'a' (cm)	2.66		2.80	
3. Length to width ratio (b/a)	1.0526		1.0296	
II. Performance	Theoretical	Experimental	Theoretical	Experimental
1. Best axial ratio (dB)	0.45	0.5	0.17	0.25
2. Center frequency f_c (GHz)	3.1030	3.101	3.1658	3.1664
3. Resonant frequencies of orthogonal modes (GHz)	3.035 3.175	3.032 3.169	3.122 3.210	3.116 3.203
4. Input VSWR at f_c	1.73	1.72	1.33	1.55
5. Bandwidth (MHz) for axial ratio < 6 dB	33.70 (1.086%)	34.80 (1.12%)	20.00 (0.632%)	21.20 (0.670%)
6. Beamwidth for 3 dB difference between $\lvert E_\theta \rvert$ and $\lvert E_\phi \rvert$	110°	140°	116°	140°

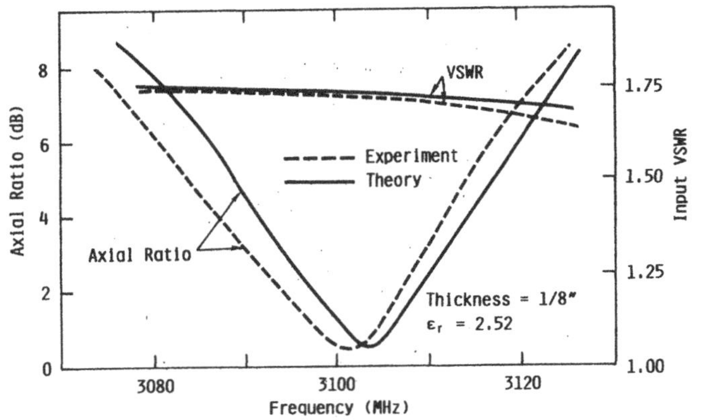

Fig. 4. Theoretical and experimental results for axial ratio and input VSWR for a diagonal fed antenna on 1/8 in thick substrate ($\epsilon_r = 2.52$).

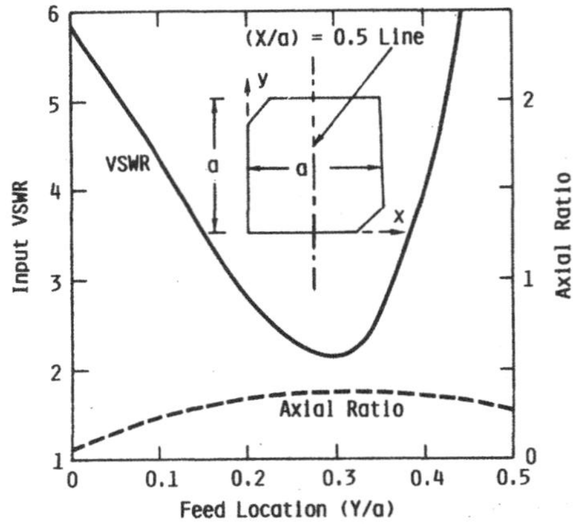

Fig. 6. Variations of input VSWR and axial ratio with feed location for truncated-corners square antenna (thickness = 1/8 in, ϵ_r = 2.52, frequency = 3.175 GHz).

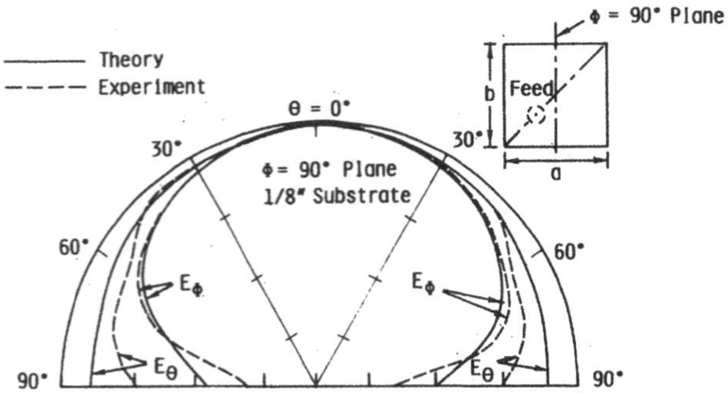

Fig. 5. Radiation pattern for diagonal fed antenna; thickness = 1/8 in, ϵ_r = 2.52, frequency = 3.103 GHz.

B. Feed Point Location

Variations of the axial ratio and the input VSWR with the feed location, on the line joining the midpoints of two opposite sides, are illustrated in Fig. 6. The input VSWR improves from 5.8 for feed location at $(x/a, y/a) = (0.5, 0.0)$ to 2.26 for feed at $(x/a, y/a) = (0.5, 0.326)$ and increases again for feed $y/a > 0.326$ on the line $x/a = 0.5$. At the location of the feed where the input VSWR is minimum the axial ratio is 0.36 dB at 3.175 GHz. The axial ratio improves to 0.02 dB when the operating frequency is changed to 3.1758 GHz for feed at the location of minimum input VSWR. For another antenna designed on 1/16 in thick substrate ($\epsilon_r = 2.51$), the minimum input VSWR is found to be 1.6 and occurs at feed location $(x/a, y/a) = (0.5, 0.3204)$. The input VSWR thus improves when the thickness is reduced. Details of these two antennas are summarized in Table II.

C. Bandwidth and Radiation Pattern

The calculated and measured values of axial ratio and input VSWR for one of the truncated-corners square antennas (ϵ_r = 2.52), substrate thickness is 1/8 in) are shown in Fig. 7. The theoretical and experimental values of bandwidth defined for an axial ratio less than 6 dB are 26.4 MHz (0.831 percent) and 29.4 MHz (0.925 percent). The corresponding values for the antenna on the 1/16 in thick substrate ($\epsilon_r = 2.51$) are found to be 14.0 MHz (0.44 percent) and 14.4 MHz (0.4535 percent). The reduction in the substrate thickness to half reduces the theoretical axial ratio bandwidth by 47 percent and the measured value by nearly 51 percent. The radiation patterns, at center frequencies, have been evaluated and verified experimentally. These are shown in Fig. 8.

TABLE II
PERFORMANCE OF CORNERS CHOPPED SQUARE PATCH ANTENNAS

I. Parameters	ANTENNA I		ANTENNA II	
1. Thickness, ε_r	1/8'', 2.52		1/16'', 2.51	
2. Dimensions axa cm^2	2.73 x 2.73		2.86 x 2.86	
3. Truncation b/a	0.04578		0.0573	
II. Performance	**Theoretical**	**Experimental**	**Theoretical**	**Experimental**
1. Center frequency f_c (GHz)	3.1758	3.1750	3.1756	3.1753
2. Resonant frequencies of orthogonal modes (GHz)	3.1340 3.2155	3.1325 3.2125	3.1370 3.2340	3.1343 3.2298
3. Axial ratio at center frequency (dB)	0.02	0.0	0.12	0.15
4. Bandwidth (MHz) for axial ratio < 6 dB	26.4 (0.831%)	29.4 (0.925%)	14.0 (0.44%)	14.4 (0.4535%)
5. Input VSWR at center frequency	2.26	2.26	1.6	1.8
6. Beamwidth for 3 dB difference between $\lvert E_\theta \rvert$ and $\lvert E_\phi \rvert$	129°	152°	129°	138°

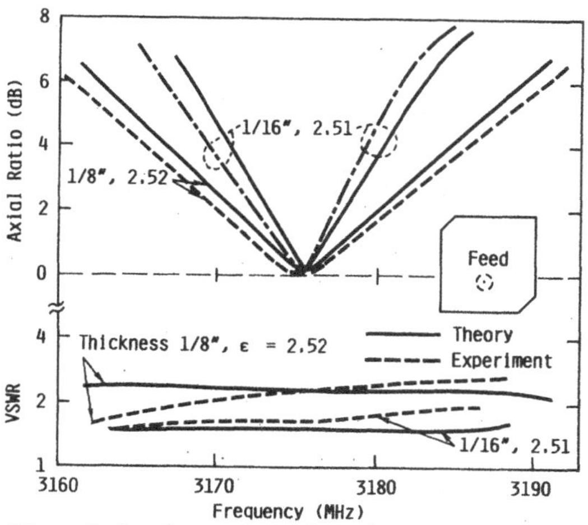

Fig. 7. Theoretical and experimental results for axial ratio and input VSWR for truncated-corners square antenna.

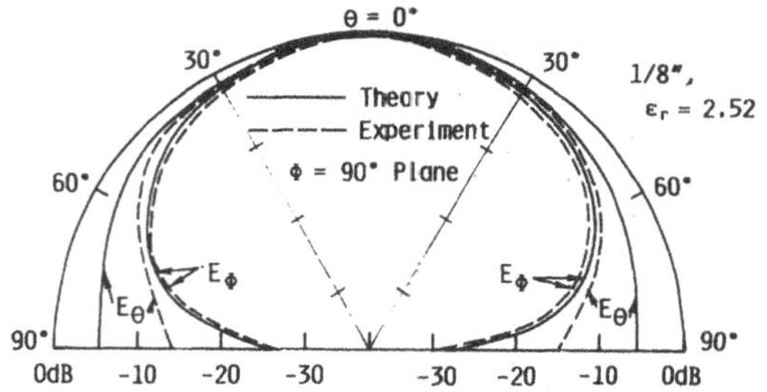

Fig. 8. Radiation pattern for truncated-corners square antenna; (thickness = 1/8 in, $\epsilon_r = 2.52$, frequency = 3.176 GHz).

V. SQUARE PATCH ANTENNA WITH A DIAGONAL SLOT

For a square patch antenna with a diagonal slot (Fig. 2(c)) also, the two orthogonal modes are diagonal modes. The difference in the resonant frequencies is caused by the rectangular slot which disturbs one mode more than the other. The desegmentation method is used to evaluate the Z-matrix of the multiport planar model as explained in Section II-C. The outer periphery has been divided into 24 ports which constitute p-ports. The number of q-ports needed (and hence that of e-ports and d-ports also) is 27. Equation (4) is used to evaluate the Z-matrix of the multiport model.

A. Optimum Configuration

The thickness and dielectric constant of the substrate are 1/8 in and 2.52, respectively. The outer dimensions of the square are 2.602 cm X 2.602 cm. The optimum dimensions of the slot are 2.89 cm X 0.47 cm. These yield an axial ratio of 0.198 dB at 3.130 GHz. The two orthogonally spaced modes of the antenna structure can be excited independently by feeding at points 1 and 2 (Fig. 2(c)) respectively. The measured values of resonant frequencies of the two orthogonal modes are 3.060 GHz and 3.210 GHz. The respective theoretical values are 3.063 GHz and 3.212 GHz. As before the operating frequency for circular polarization lies in between the two values.

The variation of axial ratio and input VSWR with feed location on the line joining the midpoints of two opposite sides ($x/a = 0.5$) is shown in Fig. 9. The optimum feed location is found to be at $(x/a, y/a) = (0.5, 0.1636)$ where input VSWR is 2.3 and axial ratio equals 0.2 dB. For $y/a > 0.1636$ on the line $x/a = 0.5$, the input VSWR improves further but axial ratio starts degrading.

B. Bandwidth and Radiation Pattern

The axial ratio and input VSWR as functions of frequency are plotted in Fig. 10. Input VSWR and axial ratio have been calculated for two feed locations. The theoretical values of axial ratio bandwidths are same for the two feed locations. VSWR variations in the two cases are shown in Fig. 10. Experiments have been conducted for feed location at $(x/a, y/a) = (0.5, 0.064)$. The theoretical and experimental values of bandwidths (for axial ratio to be less than 6 dB) are 35.5 (1.134 percent) and 38.0 MHz (1.214 percent), respectively. The variation in input VSWR over this frequency range is small as compared to the variations in axial ratio values. The theoretical and experimental radiation patterns are illustrated in Fig. 11. Table III summarizes the performance of this antenna.

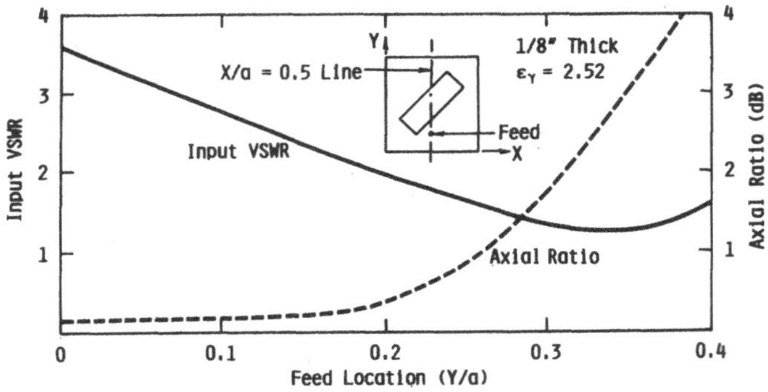

Fig. 9. Variations of input VSWR and axial ratio with feed locations for square antenna with a diagonal slot (thickness = 1/8 in, ϵ_r = 2.52, frequency = 3.130 GHz).

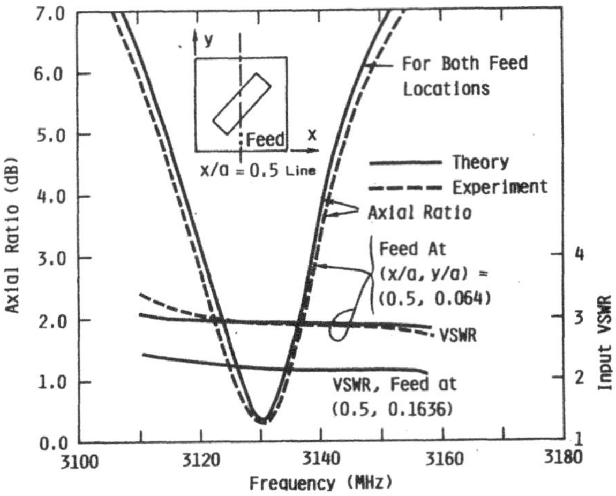

Fig. 10. Theoretical and experimental results for square antenna with a diagonal slot (thickness = 1/8 in, ϵ_r = 2.52, frequency = 3.130 GHz).

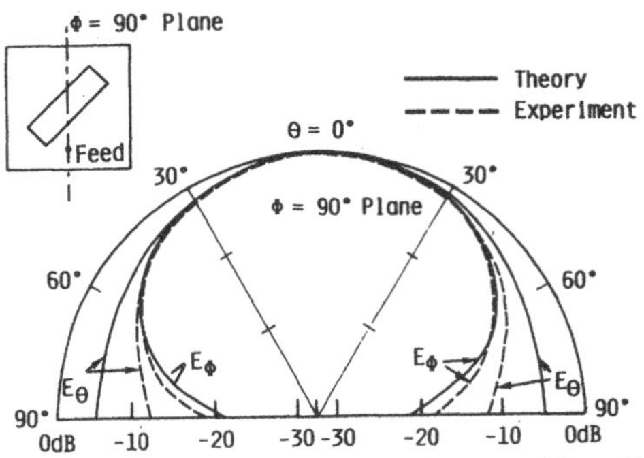

Fig. 11. Radiation pattern for a square antenna with a diagonal slot (thickness = 1/8 in, ϵ_r = 2.52, frequency = 3.130 GHz).

TABLE III

PERFORMANCE OF SQUARE PATCH ANTENNA WITH A DIAGONAL SLOT

		Theoretical	Experimental
1.	Center frequency f_c (GHz)	3.130	3.130
2.	Resonance frequency of orthogonal modes (GHz)	3.063 3.212	3.060 3.210
3.	Axial ratio at f_c	0.198	0.2
4.	Bandwidth for axial ratio less than 6 dB	35.5 MHz (1.134%)	38.0 MHz (1.214%)
5.	Input VSWR at chosen feed location	2.9	2.9
6.	Beamwidth for 3 dB difference between $\lvert E_\theta \rvert$ and $\lvert E_\phi \rvert$	116°	124°

Substrate thickness = 1/8 in, ϵ_r = 2.52. Dimensions of square patch = 2.602 × 2.602 cm. Dimensions of the slot = 2.89 × 0.47 cm.

VI. CONCLUDING REMARKS

A technique employing impedance Green's functions for segments with magnetic wall boundary is used for analysis and design of three types of single feed circularly polarized microstrip patch antennas. The dimensions of the three types of antennas are optimized for obtaining the best axial ratios. The input VSWR and axial ratio variations with feed locations are investigated in an attempt to achieve a better input VSWR without using an external impedance matching network. It has been observed that for the three types of antennas investigated, perfect matching with a 50 Ω feed line is not practical unless an impedance matching network is used. Better input VSWR can be realized by using a thinner substrate, but the axial-ratio bandwidth is reduced by nearly 40 to 50 percent when the thickness of the substrate is halved. Among the three types of antennas reported, the square patch antenna with a diagonal slot has the largest axial ratio bandwidth, whereas the minimum VSWR is obtained with diagonal-fed nearly square patch antenna. The truncated-corners antenna exhibits the best axial ratio (0.02 dB) but has the least axial-ratio bandwidth. The input VSWR values of the same order as the square antenna with a diagonal slot. The theoretical and experimental results are found to be in a reasonable agreement.

REFERENCES

[1] K. R. Carver and J. W. Mink, "Microstrip antenna technology," *IEEE Trans. Antennas Propagat.*, vol. AP-29, pp. 2–24, Jan. 1981.

[2] J. R. James *et al.*, *Microstrip Antenna Theory and Design*. Stevenage, UK: Peter Peregrinus, 1981.

[3] I. J. Bahl and P. Bhartia, *Microstrip Antennas*. Dedham, MA: Artech House, 1981.

[4] J. Q. Howell, "Microstrip Antennas," *IEEE Trans. Antennas Propagat.*, vol. AP-23, pp. 90–93, Jan. 1975.

[5] H. D. Weinschel, "A cylindrical array of circularly polarized microstrip antennas," in 1975 *Antennas Propagat. Soc. Int. Symp. Digest*, pp. 177–180.

[6] C. M. Coloi, "Corner fed electric microstrip dipole," Naval Missile Center, Ft. Mugu, CA, Mar. 1978.

[7] J. L. Kerr, "Microstrip antenna developments," in *Proc. Workshop Printed Circuit Antennas*, New Mexico State Univ., pp. 3.1–3.20, Oct. 1979.

[8] L. C. Shen, "Elliptical microstrip antenna with circular-polarization," *IEEE Trans. Antennas Propagat.*, vol. AP-29, pp. 90–94, Jan. 1981.

[9] S. A. Long *et al.*, "An experimental study of the circularly-polarized elliptical printed circuit antenna," *IEEE Trans. Antennas Propagat.*, vol. AP-29, pp. 95–99, Jan. 1981.

[10] R. E. Munson, "Conformal microstrip antennas and arrays," *IEEE Trans. Antennas Propagat.*, vol. AP-22, pp. 74–78, Jan. 1974.

[11] K. R. Carver and E. L. Coffey, "Theoretical Investigations of Microstrip Antennas," Mexico State Univ., Tech. Rep. PT-00929, Jan. 1979.

[12] T. Okoshi and T. Miyoshi, "The planar circuit—An approach to microwave integrated circuitry," *IEEE Trans. Microwave Theory Tech.*, vol. MTT-20, pp. 245–252, Apr. 1972.

[13] R. Chadha and K. C. Gupta, "Green's functions for triangular segments in planar microwave circuits," *IEEE Trans. Microwave Theory Tech.*, vol. MTT-28, pp. 1139–1143, Oct. 1980.

[14] K. C. Gupta and P. C. Sharma, "Segmentation and desegmentation techniques for analysis of planar microstrip antennas," in 1980 *Antennas Propagat. Soc. Int. Symp. Digest*, pp. 19–22.

[15] R. Chadha and K. C. Gupta, "Segmentation method using impedance matrices for analysis of planar microwave circuits," *IEEE Trans. Microwave Theory Tech.*, vol. MTT-29, pp, 71–74, Jan. 1981.

[16] P. C. Sharma and K. C. Gupta, "Desegmentation method for analysis of two-dimensional microwave circuits," *IEEE Trans. Microwave Theory Tech.*, vol. MTT-29, pp. 1094–1098, Oct. 1981.

[17] H. Pues and A. Van de Capelle, "A simple accurate formula for the radiation conductance of a rectangular microstrip antenna," *Int. Antennas Propagat. Soc. Symp. Digest*, pp. 23–26, June 1981.

[18] K. C. Gupta *et al.*, *Computer Aided Design of Microwave Circuits*. Dedham, MA: Artech House, Dec. 1981, pp. 256–258.

A CIRCULARLY POLARIZED MICROSTRIP ANTENNA USING SINGLY-FED PROXIMITY COUPLED FEED

Hisao IWASAKI, Hisashi SAWADA, and Kazuaki KAWABATA

Toshiba Research and Development Center
1,Komukai Toshiba-cho,Saiwai-ku,Kawasaki 210,Japan

1. Introduction

Many types of antenna have been proposed and investigated to enable airplanes, ships, and cars to maintain mobile satellite communications[1],[2]. Mobile satellite communications require circularly polarized patch antennas.

A proximity coupled microstrip antenna excited by a microstrip line is suitable for constructing a thin, light, and multi-layered feed network. This antenna has several well-known advantages compared with an edge-fed patch antenna. An optimal feed-patch configuration has been proposed for linear polarization[3]. Moreover, for circularly polarized operation, a 90 degree hybrid coupler is commonly used.

The purpose of this paper is to propose a simple new antenna configuration using a rectangular patch antenna with an offset microstrip line from center for circular polarization without the need for a 90 degree hybrid coupler. The results of experiments are described. Good impedance and axial ratio characteristics have been obtained.

2. Antenna configuration for circular polarization

An antenna configuration which gives circular polarization is shown in Fig.1. The rectangular patch antenna and microstrip line are formed of substrates with a dielectric constant εr, and thicknesses h and t, respectively. The length of the patch antenna is Lp and the width is Wp. Lo is the offset length from the center of the rectangular patch antenna to the microstrip line. The characteristic impedance of the microstrip line is 50 Ω. S is the distance between the edge of the patch antenna and the edge of the microstrip line.

The operation of this antenna is based on the fact that the generated mode, which is excited in the electrically thin cavity of the microstrip antenna, can be separated into two orthogonal modes due to the effects of the offset microstrip line as shown in Fig.2. The generated modes are separated into the orthogonal modes #1 and #2, which are excited in equal amplitude and 90 degree out of phase at frequency fo by adjusting the aspect ratio Lp/Wp and offset length Lo. Hence, a circularly polarized antenna can be obtained with this configuration.

Reprinted with permission from *Proc. ISAP,* H. Iwasaki, H. Sawada, and K. Kawabata, "A Circularly Polarized Microstrip Antenna Using Singly-fed Proximity Coupled Feed," pp. 797–800, Sept. 1992.

3.Experimental results

Antennas were designed and tested to verify the circularly polarizing operation of the proposed configuration. The experimental models were made of copper-clad thick Teflon Fiberglass with a εr=2.6. The aspect ratio, Lp/Wp, was 0.966.

Figure 3 shows the measured impedance and return loss for a proximity coupled rectangular patch antenna with Lo = 14.4 mm and S = 13.8 mm, respectively. Good impedance matching was achieved and the bandwidth for VSWR<2 was 3.5 %. Figure 4 shows the measured radiation pattern in the y-z plane at 1.575GHz. A 0.3 dB boresight axial ratio was obtained and an axial ratio of less than 2 dB was obtained in the 60 degree range. Figure 5 shows the measured axial ratios. The axial ratios are given as a parameter of the offset length Lo. The bandwidth of the 2dB axial ratio was 0.55 %.

Table 1 shows the measured axial ratios as parameters of the offset length Lo and the microstrip line length S. Boresight axial ratio < 2 dB was obtained over a wide range from the center to the edge of the rectangular patch antenna.

In the case of Lp<Wp and 0<Lo<Wp/2, left-hand circularly polarized waves are radiated by this antenna configuration. On the other hand, right-hand circularly polarized waves are radiated when -Wp/2<Lo<0.

4.Array antenna

Sequential array antennas consisting of 3, 4, and 6 of the proposed antenna elements were designed to increase the bandwidth of return loss and axial ratio, respectively.

Figure 6 shows the measured return loss of the 6-element array antenna. Figure 7 shows the measured axial ratios. From Fig.7, the bandwidth of the 2 dB axial ratio was determined to be about 6 %. These bandwidths satisfy the required values for L-band mobile satellite communications.

5.Conclusion

This paper describes the results of measurements made on a simple circularly polarized microstrip antenna. A patch antenna with an axial ratio of less than 0.3 dB was obtained using the proposed antenna configuration. The feed network can be made simplified, because only a microstrip line, offset from the center of the rectangular patch antenna, is used to generate the circularly polarized waves.

The proposed antenna is suitable for a phased array antenna with a multi-layered feed network in mobile satellite communications.

References

[1] J.L.Keer, "Microstrip Polarization Techniques," Proc.1978 Antenna and Propagat., Symp., Sept., 1978.

[2] H.Iwasaki and K.Kawabata,"A Circularly Polarized Microstrip Antenna with a Cross Slot," The 3rd Asia and Pacific Microwave Conference, Sept., 1990.

[3] M.Davidovitz and Y.T. Lo, "Rigorous Analysis of a Circular Patch Antenna Excited by a Microstrip Transmission Line," IEEE Trans., Antenna and Propagt., vol.AP-37, no.8, Aug., 1989.

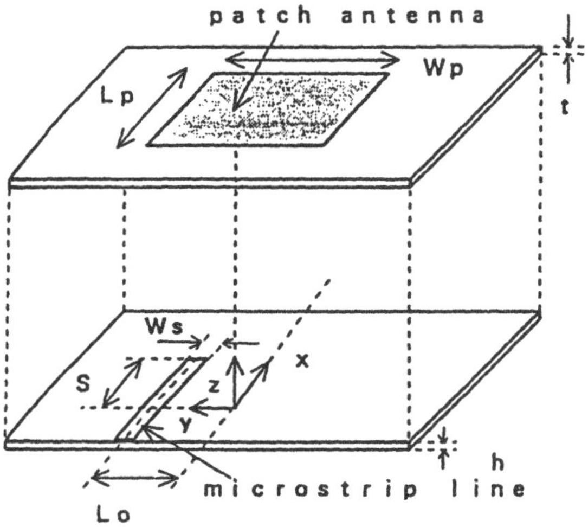

Fig.1 Configuration of a rectangular patch antenna.

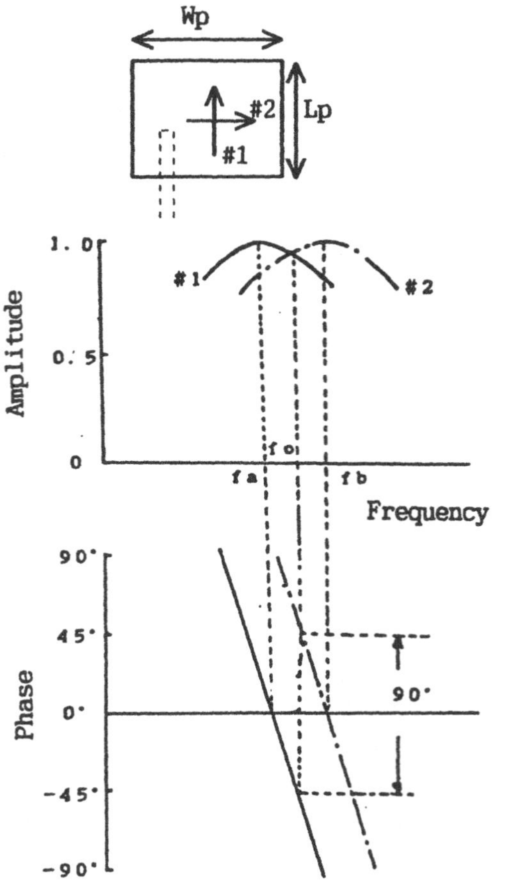

Fig.2 Schematic explanation of circularly polarized operation.

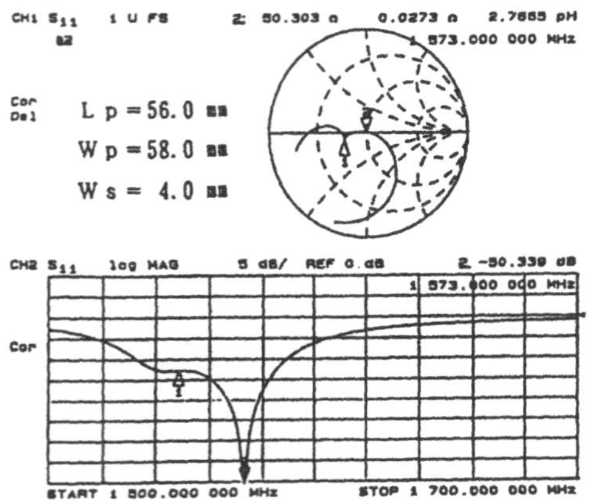

Fig.3 Measured input impedance and return loss.

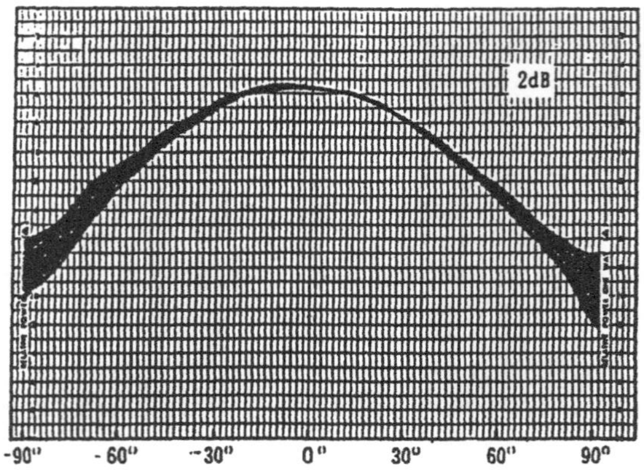

Fig.4 Measured radiation pattern. (y-z plane)

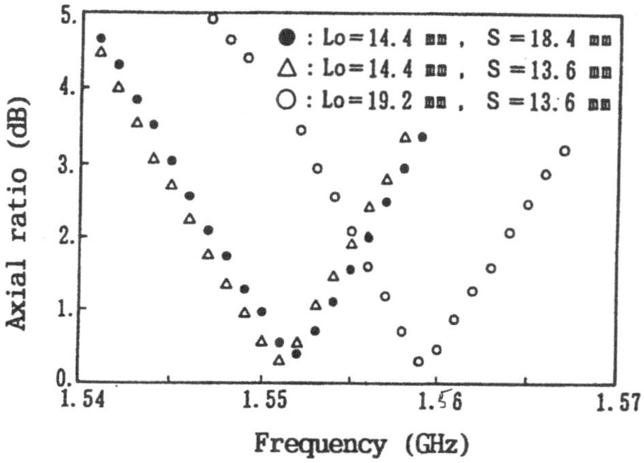

Fig.5 Measured axial ratio.

Table1 Measured axial ratios as a function of offset length Lo and microstrip line length S.

S \ Lo	4.8	9.6	14.4	19.2	24.0	28.8 (mm)
52.0	2.0	3.52	1.20	2.32	3.60	3.24
47.2	2.43	3.68	1.44	1.76	3.28	2.43
42.4	3.45	3.45	3.36	1.45	2.77	3.20
37.6	2.80	2.88	1.36	1.60	3.20	2.64
32.8	3.39	3.36	1.55	1.74	2.16	2.24
28.0	3.23	2.88	0.86	1.31	2.22	1.47
23.2	3.37	3.17	0.56	0.72	1.66	1.68
18.4	3.93	2.60	0.40	0.77	2.56	1.45
13.6	3.60	2.83	0.32	0.32	1.92	1.26
8.8	4.24	2.48	0.32	0.32	1.56	1.07
4.0	5.15	2.08	1.42	0.48	1.36	0.88 (dB)
(mm)						

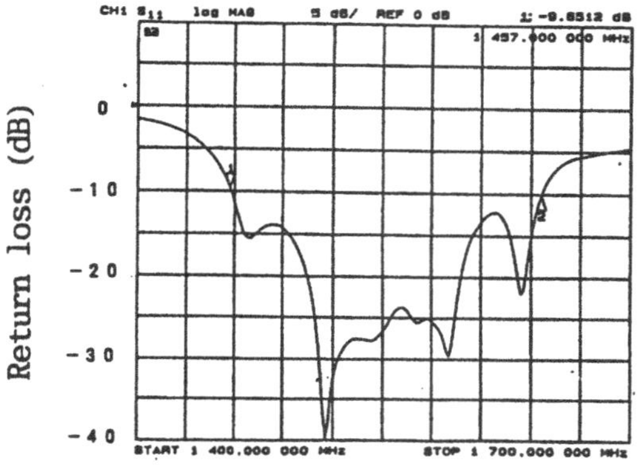

Fig.6 Measured return loss of 6-element array antenna.

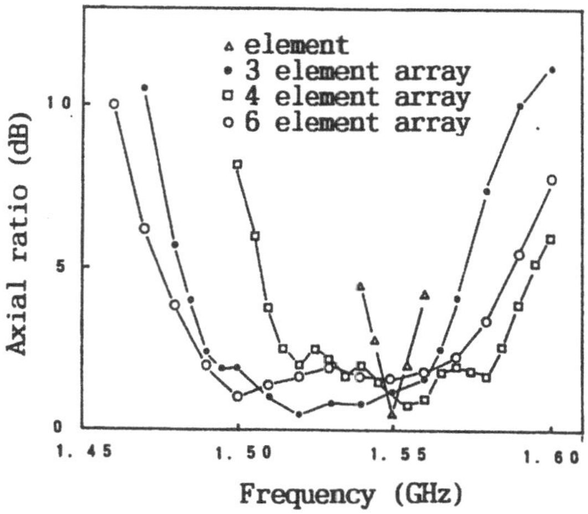

Fig.7 Measured axial ratios of array antennas.

Dual Aperture-Coupled Microstrip Antenna for Dual or Circular Polarisation

A. Adrian and D. H. Schaubert

Indexing terms: Antennas, Microstrip, Polarisation

A new method for radiating dual or circular polarisation with a printed circuit antenna element is described. A square microstrip patch on one substrate is coupled to a pair of microstriplines on another substrate via two orthogonal, rectangular apertures in a common ground plane. Quadrature excitation of the system results in circularly polarised radiation.

Introduction: The dual aperture coupled circularly polarised patch (DACCPP) described here utilises the technique of feeding a microstrip patch with a microstripline through an aperture.[1] However, in the case of the DACCPP, there are two orthogonally placed rectangular slots beneath a square patch (Fig. 1). Each aperture couples the radiator to one of two microstrip feed lines, each of which is tuned with an open-circuited stub. Aperture 1 resonates the patch in the *x*-direction while aperture 2 resonates the patch in the *y*-direction (Fig. 1*b*). Electrically, this structure forms two identical antennas that are collocated, but which are isolated and orthogonally polarised. Quadrature excitation at the apertures results in circularly polarised radiation.

The resonant frequency is mainly determined by the size of the square patch. Aperture lengths dictate the amount of coupling, whereas the lengths of the stubs are adjusted to achieve the desired reactance. In addition, lateral movement of a slot in a patch nonresonant direction causes little change in the input impedance, provided the entire aperture remains under the radiator.[2] This property allows for the design extension from single-feed, linearly polarised to double-feed, dual-polarised aperture-coupled patches.

In this letter experimental characteristics of a DACCPP are presented; *S*-parameter, pattern, axial ratio and bandwidth data are included.

Antenna performance: A DACCPP with the dimensions shown in Fig. 1*b* has been developed for operation at 2·215 GHz. The patch and feed lines were fabricated on 0·062 in- (1·575 mm)-thick Oak 601 substrate with $\varepsilon_r = 2{\cdot}55$. As illustrated in Fig. 1*b*, each DACCPP layer is symmetric about a line drawn from corners P_1 to P_2.

The best return loss for port 1 occurs at 2·215 GHz, while for port 2, 2·220 GHz is the frequency with the best return loss. However, both ports have almost the same return loss at 2·215 GHz. The isolation in the vicinity of resonance is greater than 18 dB, and probably could be increased by more careful fabrication that maintains perfect symmetry. Thus dual-polarisation operation is possible. Also, by feeding the ports from a quadrature hybrid, circular polarisation can be obtained over the entire operating band of the antenna.

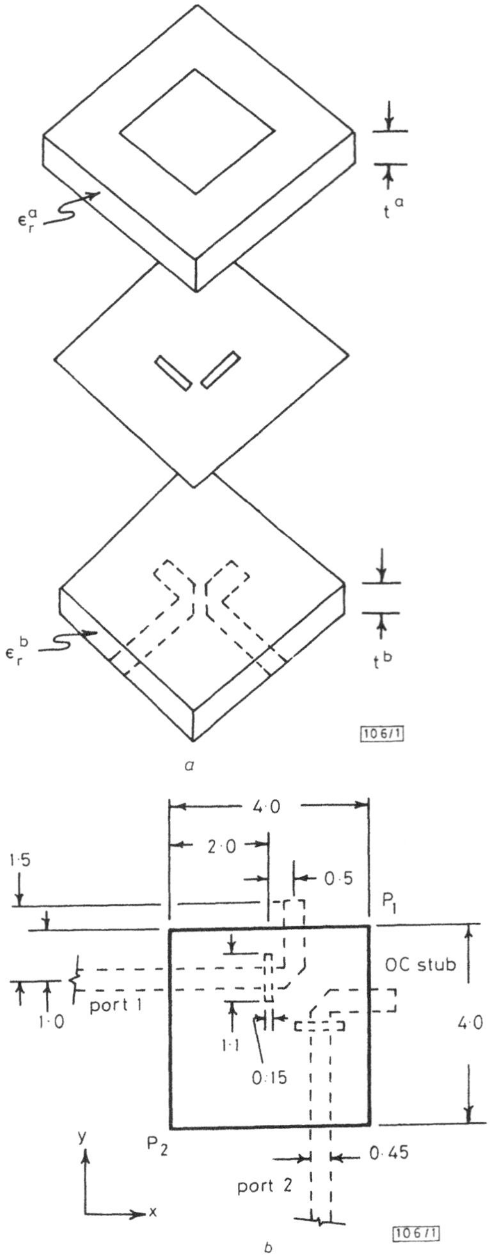

Fig. 1 *Dual aperture-coupled circularly polarised patch (DACCPP)*
Dimensions are in centimetres

Reprinted with permission from *Elect. Lett.*, A. Adrian and D. H. Schaubert, "Dual Aperture-Coupled Microstrip Antenna for Dual or Circular Polarization," vol. 23, no. 23, pp. 1226–1228, Nov. 1987. © Institution of Electrical Engineers.

A spin-linear pattern of the DACCPP at the centre frequency is illustrated in Fig. 2. The patterns are essentially the same throughout the operating band defined by the 3 dB gain bandwidth, which is 3·5% (Fig. 3). Axial ratios vary from 1·3 to 2·0 dB across the 3 dB gain bandwidth (Fig. 3).

Bandwidth considerations: In this structure the gain bandwidth is a function of the patch radiator and its aperture-coupled excitation. In general, the bandwidth is the same as for the patch when fed by a probe or microstripline. The axial ratio bandwidth is a direct result of the antenna symmetry and the hybrid bandwidth. If the hybrid can maintain an equal power split and a 90° phase shift between the coupled ports over the entire gain bandwidth, then the axial ratio will not degrade between the 3 dB gain points. However, small errors in amplitude and/or phase shift in the coupled hybrid ports or slight asymmetries of the antenna can greatly reduce the axial ratio bandwidth of the DACCPP system.

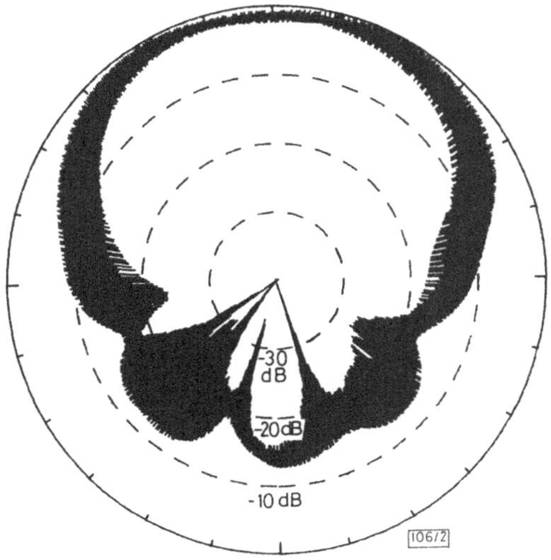

Fig. 2 *Far-field, spin-linear pattern of DACCPP at 2·215 GHz*

Conclusion: A new method for radiating a dual- or circularly polarised wave from a single element has been presented. The circular polarisation sense is determined by the lead-lag phase relationship between the aperture excitations. Either polarisation sense can be obtained.

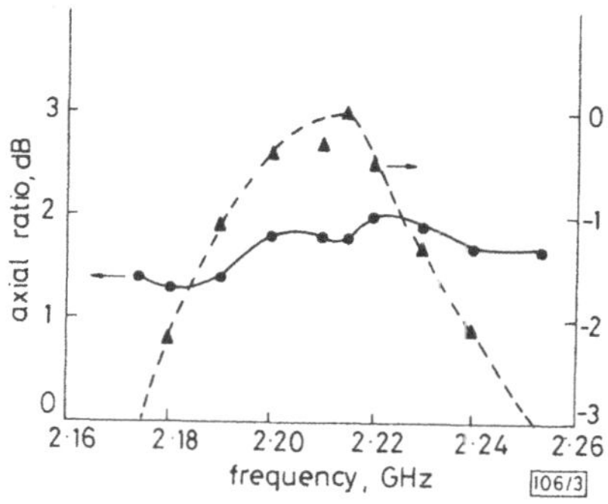

Fig. 3 *Axial ratio and relative gain of DACCPP over frequency*

In addition to circular polarisation, this antenna element lends itself to monolithic array integration, with all the advantages of aperture-coupled microstrip patches. The feed network and the radiator are separated by a ground plane, thereby reducing interference in the element pattern from stray feed line radiation. This layered design also allows for the fabrication of the feed network on a high dielectric substrate while concurrently allowing for a low dielectric patch substrate. The high dielectric feed substrate allows for size reduction in the feed circuit, while the lower dielectric patch substrate is preferable because it has a better radiation efficiency, broader bandwidth and an increased angle off broadside at which scan blindness occurs.[3] In addition, via connections are completely eliminated, along with the problems of their consistent fabrication and self-reactances.

Acknowledgment: This work was supported by General Electric Company Electronics Laboratory, Syracuse, NY, USA.

References

1 POZAR, D. M.: 'Microstrip antenna aperture-coupled to a microstripline', *Electron. Lett.*, 1985, **21**, pp. 49–50

2 SULLIVAN, P. L., and SCHAUBERT, D. H.: 'Analysis of an aperture coupled microstrip antenna', *IEEE Trans.*, 1986, **AP-34**, pp. 977–984

3 POZAR, D. M., and SCHAUBERT, D. H.: 'Scan blindness in infinite arrays of printed dipoles', *ibid.*, 1984, **AP-32**, pp. 602–610

Design of Wideband Circularly Polarized Aperture-Coupled Microstrip Antennas

Stephen D. Targonski and David M. Pozar, *Fellow, IEEE*

***Abstract*— Two variations of a novel feeding technique for a wideband circularly polarized aperture-coupled microstrip antenna are described. Prototype designs for wideband linearly polarized elements are first presented, and then used for circularly polarized designs. Techniques used for design of the feed network are detailed, for both series feed and parallel feed versions. Experimental results are shown for each antenna, and results for the two designs are compared. The impedance and axial ratio bandwidths for these antennas are among the best yet achieved for microstrip antenna elements. Several design variations are also discussed.**

I. Introduction

The aperture-coupled microstrip antenna [1] has several advantages over transmission line or probe fed patch antennas. Separate substrates can be used for the feed circuit and the antenna element to isolate spurious feed radiation from the antenna element by use of a common ground plane, and to allow more space for the feed network. The input impedance is easily controlled by the size and position of the aperture, and any excess reactance caused by the coupling aperture can be removed through the use of a tuning stub. The aperture-coupled configuration also exhibits very low cross-polarization levels, making it well suited to circularly polarized antenna designs.

A common technique for producing circular polarization is to excite two orthogonal linearly polarized elements with a 90° phase difference. This method can be utilized in the aperture-coupled microstrip antenna by using either two off-center coupling apertures [2] or a crossed slot [3]. The use of two off-center apertures results in an inherent asymmetry in the antenna which produces higher cross-polarization levels, thereby diminishing the level of circular polarization purity that can be achieved. This effect is especially apparent when a thicker antenna substrate is used to increase bandwidth. In addition, the offset slot configuration limits the slot length to less than half the patch dimension, which in turn limits the substrate thickness over which impedance matching can be obtained and, hence, the maximum bandwidth. The crossed slot retains symmetry and therefore is capable of producing circularly polarized radiation with very good polarization purity. It also permits the use of slot lengths greater than half the patch width, which is critical to achieving adequate coupling when thick antenna substrates are used for wide bandwidths.

The aperture-coupled microstrip antenna can generally be impedance matched with an aperture that is well below resonant size, thus limiting the level of back radiation to about −20 dB relative to the main lobe. But a microstrip antenna coupled in this fashion is capable of only about 5% bandwidth, owing to the fact that the small coupling aperture limits the antenna substrate thickness that may be used. By using a thick antenna substrate with a low dielectric constant, a bandwidth of 20% to 25% can be achieved [4], [5]. However, because of the thick antenna substrate, a larger slot size is needed to obtain the necessary coupling to impedance match the antenna, resulting in a higher level of back radiation. As an alternative, the required coupling can also be achieved through the use of a "dogbone" aperture [6], which can provide increased coupling compared with a rectangular slot of the same length.

In this paper the steps taken in the design of three separate wideband circularly polarized aperture-coupled microstrip antennas are discussed, and the relative advantage and disadvantages of each design are compared. Details of a series feed design and of two parallel feed designs will be presented. The series design is capable of 15% bandwidth for an axial ratio of less than 3 dB and return loss of better than −10 dB, with a gain greater than 7 dB over this range. The parallel feed designs display even better performance, exhibiting impedance and axial ratio bandwidths of 22% and 25%, respectively, and an axial ratio of better than 2 dB.

II. Brief Discussion of Aperture Coupling Using Crossed Slots and Dual Feedlines

The geometry of the aperture-coupled microstrip antenna is shown in Fig. 1. Notice that the patch is square, thereby retaining the symmetry needed for good circular polarization purity. Also notice that the coupling aperture may be a single rectangular slot for linear polarization or a crossed slot for circular polarization. The single rectangular slot will be used in the preliminary design of wideband linearly polarized elements, which will serve as the basis for a circularly polarized element using a crossed slot.

The two orthogonal apertures of the crossed slot can be modeled as short-circuited transmission lines, giving rise to an equivalent transmission line circuit. Consider only one slot being fed. Then the two slotline stubs of the unfed slot appear as series inductances to the fed slotline. With a single unbalanced offset feed line, a voltage will be induced across the unfed orthogonal slot, resulting in a high level of aperture cross-coupling, which will cause amplitude and phase errors in the network and degrade the purity of circular polarization

Reprinted from *IEEE Trans. Antennas Propaga.*, vol. 41, no. 2, pp. 214–220, Feb. 1993.

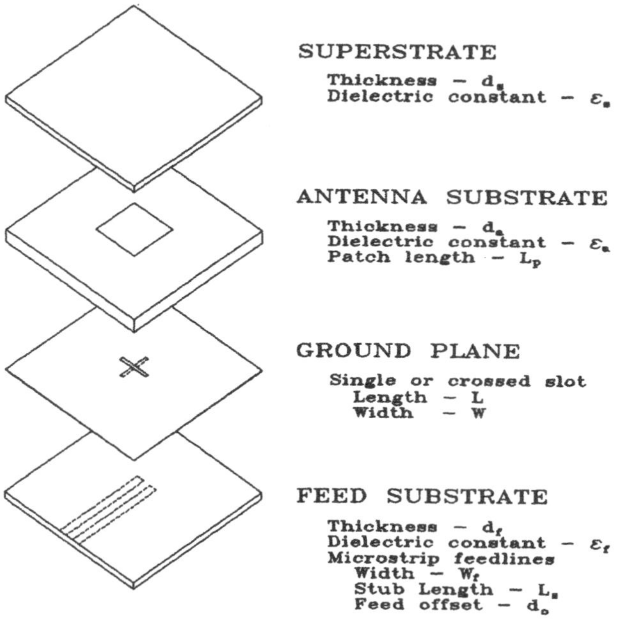

Fig. 1. Geometry of the aperture-coupled microstrip antenna. The center-to-center spacing of the microstrip feedlines is equal to $2d_0$. The stub length, L_s, is measured from the center of the aperture to the edge of the feed line.

that can be achieved. By feeding the slot in a balanced configuration with two symmetric feed lines, no voltage is induced across the unfed slot, and the two crossed apertures are perfectly decoupled. This is essential for good circular polarization purity.

Since a balanced feed is needed for the crossed slot to operate properly, the design of the linearly polarized prototype antennas should also have this balanced feed network. This consists of two feed lines offset from the center of the coupling aperture by a distance d_0, as shown in Figure 1. The addition of a second feed line to the coupling aperture causes a change in the input impedance of the antenna. An effective series impedance, Z_{eff}, is present to each feed line at the aperture when it is fed in a balanced manner. This effective impedance cannot be directly measured, but it can be easily derived from measurements by considering the dual-fed aperture-coupled microstrip antenna as a two-port network. If the S parameters of this two-port network are measured, it can be shown that the effective reflection coefficient, Γ_{eff}, seen by one feed line when both feed lines are excited in phase with equal amplitudes can be expressed as

$$\Gamma_{\text{eff}} = S_{11} + S_{12}. \tag{1}$$

Then the effective series impedance, Z_{eff}, can be found as

$$Z_{\text{eff}} = Z_0 \frac{1+\Gamma_{\text{eff}}}{1-\Gamma_{\text{eff}}}. \tag{2}$$

This computation can be easily performed using the trace math option on the HP8510B network analyzer.

The effective impedance may also be computed theoretically. A full-wave solution using exact Green's functions for the multilayered structure and allowing the introduction of multiple feed ports can be employed [7]. As an alternative, a modified cavity model allowing the superstrate and multiple feed ports could be used to obtain an approximate solution [8]. A computer program utilizing the cavity model can be run much faster than the full-wave solution, making it useful for initial designs.

III. Design of the Wideband Linearly Polarized Prototype Element

A dual-fed linearly polarized element was constructed on a feed substrate of 0.031-in.-thick Rogers 5880 ($\epsilon_r = 2.2$), with an antenna substrate of 0.5-cm-thick ROHACELL foam ($\epsilon_r = 1.07$). The thick foam antenna substrate provides wide bandwidth. Fig. 2(a) shows the measured effective impedance for this element. The large bandwidth of this antenna can be explained by the fact that the aperture acts as a second resonator in combination with the patch element. This creates an effect similar to the stacked patch antenna, which uses two patches of slightly different resonant frequency to achieve a larger bandwidth. The antenna uses a large aperture with a resonant frequency that is reasonably close to that of the patch element, creating a bandwidth (SWR$< 2:1$) of 34% centered at 5.23 GHz. However, the fact that the resonant frequency of the aperture is close to that of the patch causes it to radiate a high level of back radiation. For this large bandwidth configuration, the level of back radiation is approximately -10 dB.

In order to lower the level of back radiation, a second element was designed on a 0.025-in.-thick Rogers 6010 feed substrate ($\epsilon_r = 10.2$). The original design of this low back lobe configuration used the same antenna substrate as in the previous case. However, the high dielectric constant of the feed substrate caused the fields to be bound tightly to the feed substrate and the aperture did not couple strongly to the patch. Therefore, the antenna substrate thickness was decreased, resulting in a reduction of bandwidth and an increase in the resonant frequency of the element. These effects can be seen in the measured effective impedance locus of Figure 2(b). Also, an aperture with a higher resonant frequency was used. This lowered the level of back radiation to -15 dB but also lowered the bandwidth. These two factors combined to lower the bandwidth to 22% centered at 5.84 GHz. Since the aperture acts as a second resonator, there is a trade-off between the bandwidth and the level of back radiation of the element; in order to increase the bandwidth the level of back radiation must also be increased, and vice versa.

A balanced feed configuration as used in the linearly polarized elements described above is difficult to implement on each arm of the crossed slot. In [3], a microstrip crossover was used in the design, but this technique creates potential fabrication problems. This problem can be overcome by implementing a 180° phase shift in the feed design, and using a series or parallel feed for the two arms of the crossed slot, as discussed below.

IV. The Series Feed Configuration

The series feed configuration is shown in Fig. 3(a) and its equivalent circuit in Fig. 3(b). Each port is fed 180° out of

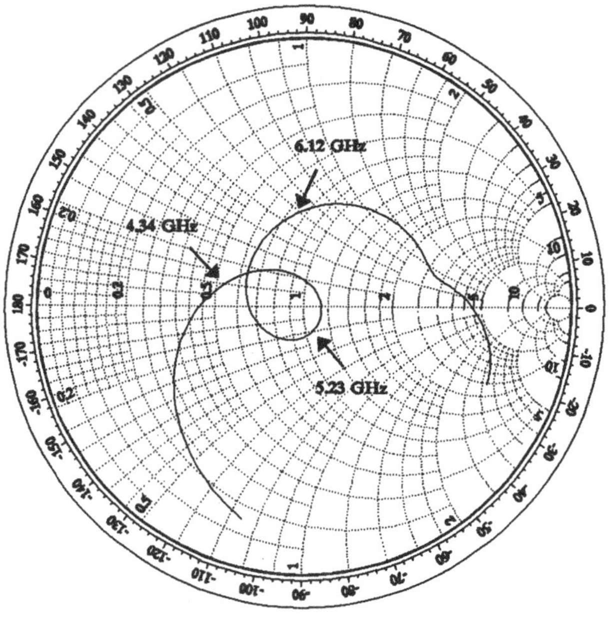

(a)

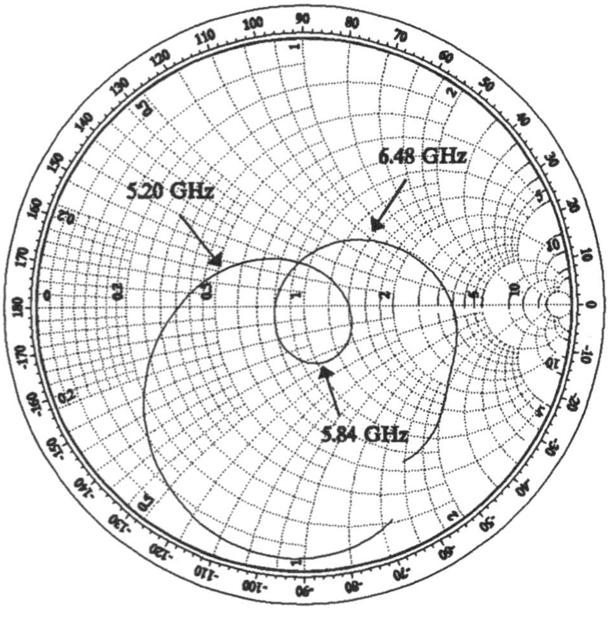

(b)

Fig. 2. (a) Effective impedance for large bandwidth linearly polarized prototype element. $d_s = 0.16$ cm, $\epsilon_s = 2.2$, $d_a = 0.5$ cm, $\epsilon_a = 1.07$, $L_p = 1.7$ cm, $L = 1.7$ cm, $W = 0.08$ cm, $d_f = 0.08$ cm, $\epsilon_f = 2,2$, $W_f = 0.232$ cm, $L_s = 0.4$ cm, $d_O = 0.52$ cm. The frequency range is from 4 to 8 GHz. (b) Effective impedance for low back lobe linearly polarized prototype element. $d_s = 0.16$ cm, $\epsilon_s = 2.2$, $d_a = 0.32$ cm, $\epsilon_a = 1.07$, $L_p = 1.7$ cm, $L = 1.01$ cm, $W = 0.07$ cm, $d_f = 0.0635$ cm, $\epsilon_f = 10.2$, $W_f = 0.06$ cm, $L_s = 0.145$ cm, $d_O = 0.278$ cm. The frequency range is from 4 to 8 GHz.

phase in order to account for the phase reversal caused by the oppositely directed feed lines at the two feed points for each aperture. A quarter-wavelength section of transmission line is placed between each aperture to create the 90° phase difference required for circular polarization. In order to achieve an excitation of the two orthogonal apertures that is equal in amplitude to a 90° phase difference, the effective impedance

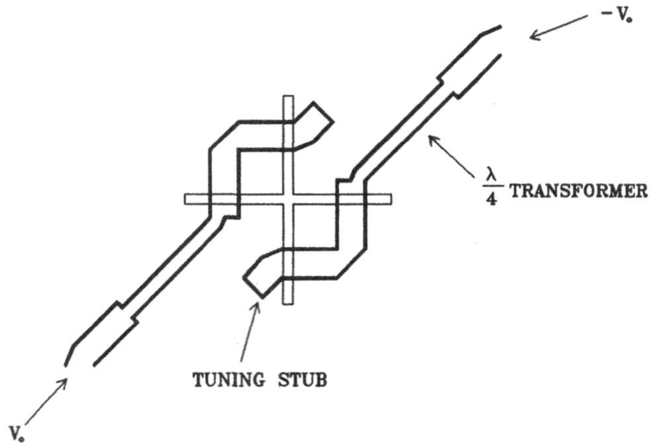

(a)

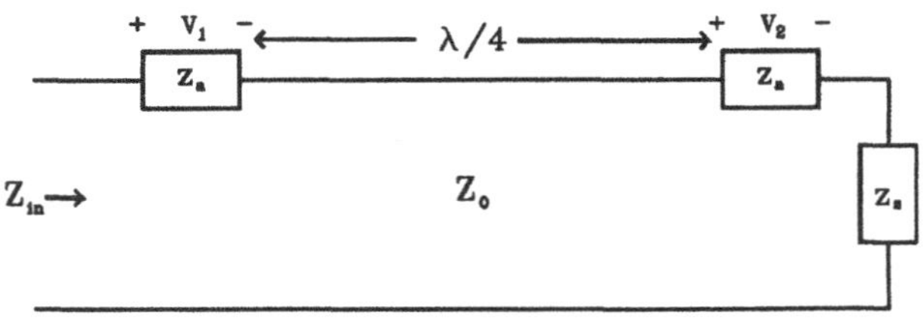

(b)

Fig. 3. (a) Series feed configuration. (b) Equivalent circuit for series feed configuration.

must be matched to 50 Ω. Any impedance mismatch will create a standing wave on the quarter-wave section of line between the orthogonal apertures and produce phase and amplitude errors, resulting in an increased axial ratio.

In the equivalent circuit, the series impedance, Z_a, is equal to the effective impedance minus the impedance of the tuning stub. Because of the lack of a tuning stub on the first aperture, a large reactance will be present in the input impedance referenced at the first aperture. This reactance may be removed by implementing a short section of transmission line before the first aperture. The resulting input impedance can then be easily matched to 50 Ω with a quarter-wave transformer.

A series fed antenna was constructed from the first linearly polarized prototype design. The effective impedance, Z_{eff}, was nominally matched to 50 Ω, thereby creating a VSWR of less than 2 : 1 on the quarter-wave section of line between the apertures. Since the VSWR was small a good axial ratio was obtained.

The measured gain and axial ratio of the series fed element are plotted versus frequency in Fig. 4(a), and the return loss is plotted in Fig. 4(b). Also, a measured spinning linear far-field pattern for this element is shown in Fig. 5. The axial ratio at 5.1 GHz is less than 1 dB over a wide angle; the size of the ground plane in this measurement was 61×45.75 cm. An axial ratio of less than 3 dB was obtained over a 12% bandwidth. The bandwidth for 10 dB return loss was in excess of 30%, leaving the axial ratio as the limiting factor.

It should be noted that the series feed configuration is extremely sensitive to the length of transmission line present between the two orthogonal apertures, and also to the effective impedance seen at each aperture. Therefore, small errors in

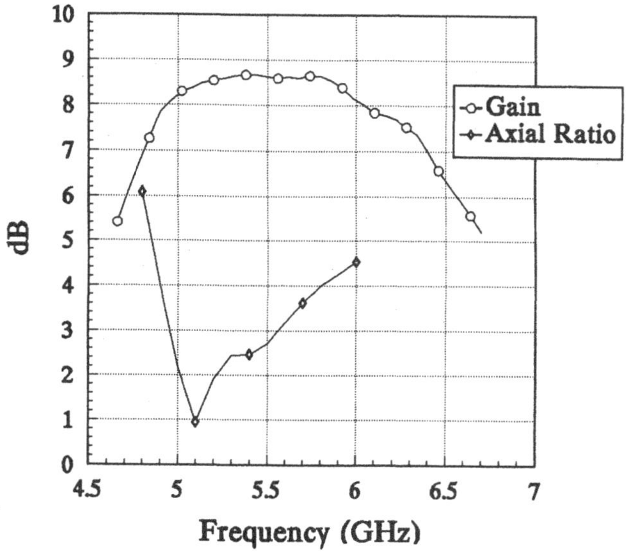

(a)

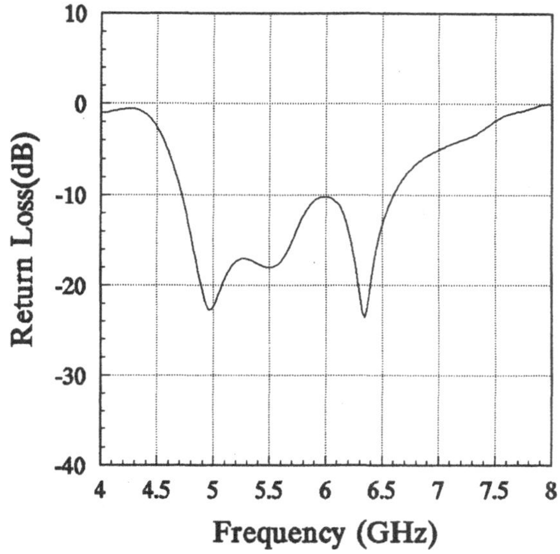

(b)

Fig. 4. (a) Axial ratio and gain plot versus frequency for large bandwidth series fed antenna. (b) Return loss plot versus frequency for large bandwidth series fed antenna.

Fig. 5. Spinning linear pattern for series fed element taken at 5.1 GHz.

fabrication, especially errors in alignment between the crossed slot and feed network, will cause the element to not operate properly.

V. The Parallel Feed Configuration

The series feed configuration has an advantage in that it is easy to design and fabricate, but it also has several drawbacks. One disadvantage is that the feed offset from the center of the aperture must be at least $\lambda/8$ in order to accommodate a $\lambda/4$ section of transmission line between the two arms of the crossed slot. This places a severe restriction on the design of the feed network and the antenna element. Also, the axial ratio degrades fairly rapidly with frequency because of increased amplitude error in the excitation of the two apertures. This amplitude error has a much greater effect on the axial ratio than does the phase error [9]. A parallel feed structure (shown in Fig. 6(a) along with its equivalent circuit in Fig. 6(b) overcomes these disadvantages. Here the two orthogonal apertures are fed through a power divider, with one arm of the output feedlines a quarter-wavelength longer than the other in order to produce a 90° phase shift.

The parallel feed structure allows for a variable feed offset, the only restriction being the separation of the feed lines near the crossed slot. Using the generally accepted microwave design procedures, this distance should be at least two substrate thicknesses from edge to edge to avoid coupling between the two arms of the parallel feed. A variable offset allows the feed points of each aperture to be moved closer to the center, thereby producing a greater amount of coupling to the antenna. This increased coupling allows a shorter aperture to be used in the design with the beneficial effect of a decreased back lobe level.

The original parallel feed configuration employed a reactive divider to feed the two slots. However, since the reactive divider does not provide any isolation between the two output ports, reflected power from one aperture can couple through to the other aperture. This coupling, albeit small, can cause enough phase and amplitude error to degrade the axial ratio beyond an acceptable level. This problem was alleviated by replacing the reactive power dividers with Wilkinson power dividers, which provide isolation between the output ports. With the two output ports isolated, the amplitude error between the two slot excitations is zero because of symmetry. An increased axial ratio is then due only to phase error.

Two antennas were constructed using the parallel feed, one using the large bandwidth configuration and the other using the low back lobe configuration of Section III. The higher dielectric constant of the feed substrate in the second element allowed thinner feed lines to be used, which in turn allowed a smaller feed offset from the center of the slots. A spinning linear pattern for the low back lobe design is shown in Fig. 7. The pattern for the large bandwidth design is the same as

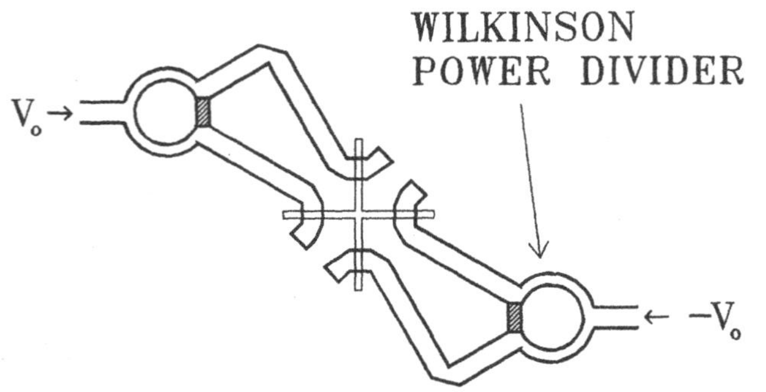

(a)

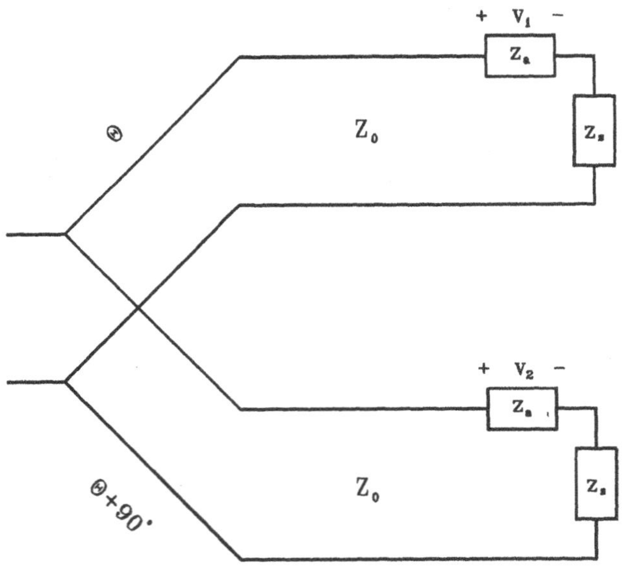

(b)

Fig. 6. (a) Parallel feed configuration. (b) Equivalent circuit for parallel feed configuration.

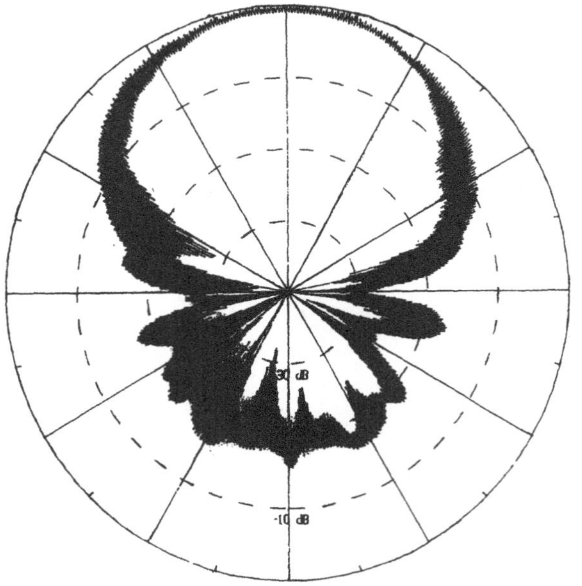

Fig. 7. Spinning linear pattern for low back lobe parallel fed antenna taken at 5.84 GHz.

that for the series fed antenna, since the same linearly polarized element was used in the design. Notice again that the axial ratio is quite good over a wide angle. Parts (a) and (b) of Fig. 8 show the axial ratio and gain of each parallel feed design versus frequency. Note that the axial ratio characteristics versus frequency are much improved over the series feed, owing to the fact that amplitude error between the two orthogonal polarizations has been eliminated. The gain of the antenna then becomes the limiting factor in the bandwidth. The bandwidths for the gain, defined as being within 1 dB of the maximum gain over the band, are 29% and 22% for the larger bandwidth and low back lobe designs, respectively. As mentioned in Section III, the second antenna element was designed to have a smaller back lobe by sacrificing some bandwidth, which is evident by comparing Figs. 5 and 7, and also parts (a) and (b) of Fig. 8.

The return loss measurement for the large bandwidth parallel design (Fig. 9) shows an excellent match over a very wide band, more than 50%. This is due to the fact that some of the reflected power is absorbed in the chip resistor used in the Wilkinson power divider. The actual bandwidth for return loss less than -10 dB is the same as that of the linearly polarized element used in the design. The return loss measurement for the low back lobe parallel design is similar to Fig. 9, and is not included here.

VI. Discussion

Other interesting possibilities exist for the parallel feed configuration. The Wilkinson power divider may be replaced by a quadrature hybrid, which still provides isolation between the two output lines. With the quadrature hybrid a 90° phase difference is already present at the output, thereby removing the need for an extra quarter-wavelength section of transmission line in the feed network. The quadrature hybrid will also create less phase error in the excitation of the two orthogonal polarizations over a wider frequency range, thereby establishing better axial ratio characteristics versus frequency. It should be noted here, however, that the hybrid will take up a large amount of space on the feed substrate. This could be a drawback if there are spatial limitations, such as in the design of an array. One way to reduce the size of the feed network is to switch to a higher dielectric constant for the feed substrate; however, this is not always feasible.

For a simple uniform linear or planar array, a single 180° hybrid can be used for the input to an entire feed network for several elements. For a phased array, however, each element must be fed separately. This would require a ring hybrid for each element in the array, which most likely would not be feasible because of space limitations. As an alternative in this case, a half-wavelength section of transmission line can be used to produce the 180° phase shift needed to drive both ports.

The half-wavelength section of transmission line will produce a 180° phase difference only at the center frequency, and a phase error in the excitation of each aperture will result over the band. Following the discussion presented in Section II, a phase error in the excitation of each aperture excites a mode in the orthogonal aperture. Owing to the symmetry of the crossed slot, however, the effects of this excitation on the axial ratio will cancel and no decrease in circular polarization purity will result. The input impedance match will degrade, however, as

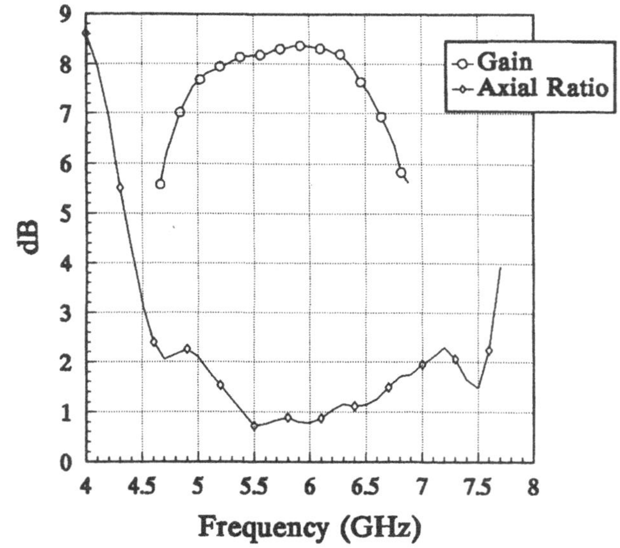

(a)

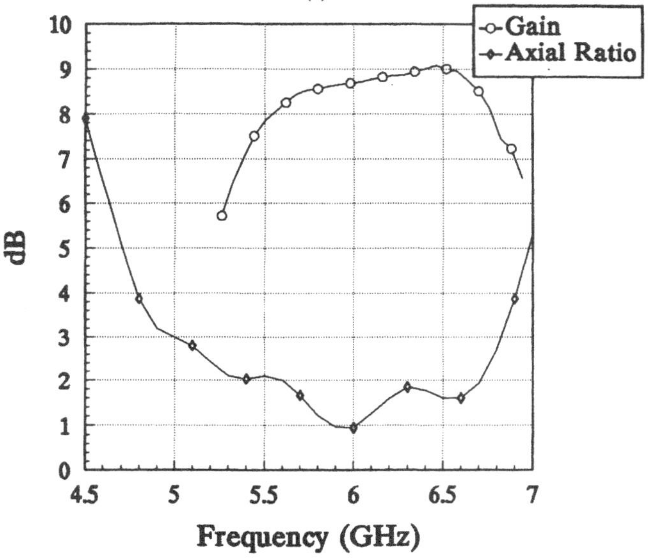

(b)

Fig. 8. (a) Axial ratio and gain plot versus frequency for large bandwidth parallel fed antenna. (b) Axial ratio and gain plot versus frequency for low back lobe parallel fed antenna.

can be easily seen by modifying (1), which becomes

$$\Gamma_{\text{eff}} = S_{11} + S_{12}e^{j\theta} \qquad (3)$$

where θ is the phase error in the excitation of the two ports. Equation (3) then can be inserted into (2) to find the effective impedance. Experimental results performed on the wide bandwidth parallel fed CP antenna presented in Section V have shown that a phase error of as much as 25° produced no noticeable effects on the axial ratio, pattern, or gain of the element over the entire bandwidth, thereby showing that a half-wavelength section of transmission line used between the two ports can operate over a wide bandwidth. This type of arrangement will not work on the series feed configuration, however, since the feed network lacks the symmetry of the parallel configuration.

Two novel geometries for feeding a wideband circularly polarized aperture-coupled microstrip element have been presented, using either a series or a parallel feed. The parallel

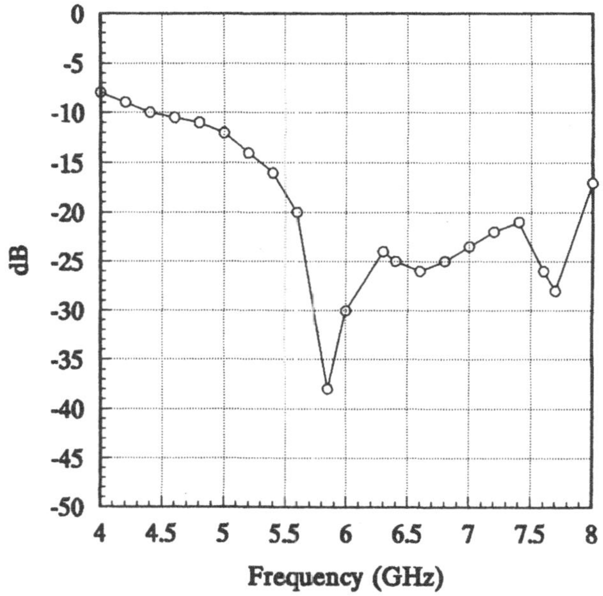

Fig. 9. Return loss for large bandwidth parallel fed antenna. Frequency range is 4 to 8 GHz. Vertical scale is 10 dB per division.

feed configuration provides better axial ratio and bandwidth characteristics, but is more complicated to design and fabricate because of the Wilkinson power dividers needed for isolation between the output ports. If a very large array of these elements is being constructed and a bandwidth in excess of 12% is not needed, the series feed design will greatly improve the ease of fabrication.

ACKNOWLEDGMENT

The authors would like to thank Prof. R. W. Jackson for his input and help in the design of the parallel feed structure. They also would like to thank B. Pelin at RohmTech for the donation of several samples of ROHACELL foam which were used in the antenna.

REFERENCES

[1] D. M. Pozar, "Microstrip antenna aperture-coupled to a microstripline," *Electron. Lett.*, vol. 21, pp. 49–50, Jan. 1985.
[2] A. Adrian and D. H. Schaubert, "Dual aperture-coupled microstrip antenna for dual or circular polarization," *Electron. Lett.*, vol. 23, pp. 1226–1228, Nov. 1987.
[3] C. H. Tsao, Y. M. Hwang, F. Killburg, and F. Dietrich, "Aperture-coupled patch antennas with wide-bandwidth and dual polarization capabilities," *IEEE Antennas and Propagation Symp. Dig.*, 1988, pp. 936–939.
[4] J.-F. Zurcher, "The SSFIP: A global concept for high-performance broadband planar antennas," *Electron. Lett.*, vol. 24, pp. 1433–1435, Nov. 1988.
[5] F. Croq, A. Papiernik, and P. Brachat, "Wideband aperture coupled microstrip subarray," in *IEEE Antennas and Propagation Symp. Dig.*, 1990, pp. 1128–1131.
[6] D. M. Pozar and S. D. Targonski, "Improved coupling for aperture coupled microstrip antennas," *Electron. Lett.*, vol. 27, pp. 1129–1131, June 1991.
[7] N. K. Das and D. M. Pozar, "Multiport scattering analysis of general multilayered printed antennas fed by multiple feed ports" (parts I and II), *IEEE Trans. Antennas Propagat.*, vol. 40, pp. 469–491, May 1992.
[8] S. D. Targonski, "Analysis and design of circularly polarized aperture-coupled microstrip antennas," M. Eng. Thesis, University of Massachusetts at Amherst.
[9] D. M. Pozar and S. D. Targonski, "Axial ratio of circularly polarized antennas with amplitude and phase errors," Antenna Designer's Notebook, *IEEE Antennas and Propagation Magazine*, vol. 32, pp. 45–46, Oct. 1990.

WIDEBAND CIRCULARLY POLARIZED ARRAY ANTENNA WITH SEQUENTIAL ROTATIONS AND PHASE SHIFT OF ELEMENTS

Tasuku TESHIROGI, Masato TANAKA, Wataru CHUJO
Radio Research Laboratories, MOPT
Nukui-Kitamachi 4-2-1, Koganei-shi, Tokyo, 184 JAPAN

1. INTRODUCTION

Circularly polarized microstrip array antennas have been widely used as phased arrays, mobile antennas, satellite antennas, and receiving antennas for direct satellite broadcasting, due to their thin and compact structures.

The problem, however, is that frequency characteristics of polarization and impedance of microstrip arrays are narrow. Several broadband techniques for circularly polarized microstrip antenna or array are reported(1),(2),(3).

This paper proposes a new composition of array antenna which has good circular polarization and low VSWR over the wide frequency band in spite of use of narrow band elements and describe the results of the verification experiment.

2. GENERATION OF WIDEBAND CIRCULAR POLARIZATION BY PROVIDING SEQUENTIAL ROTATIONS AND PHASE SHIFTS TO ELEMENTS

Let's consider a N-element planar array antenna as shown in Fig.1. In the following analysis, it is assumed that all elements are the same, and the mutual coupling can be neglected.

The n-th element is located at an arbitrary position but with orientation angle of

$$\phi_n = p(n-1)\pi/N \text{ (rad.)}, \quad (1)$$

where p is an integer and $1 \leq p \leq N-1$, with respect to the first element, say #1, and is fed with a differential phase shift of ϕ_n. In other words, each element is provided sequential rotation and phase shift. Therefore, for convenience, we call it a sequential array. We assume that the polarization of electric field radiated by #1 element in the boresight direction is elliptical and expressed as

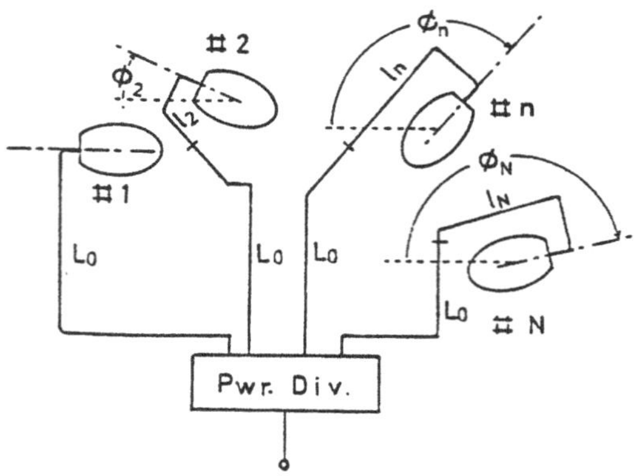

Fig.1. Configuration of a sequential array

$$\mathbf{E}_1 = a\mathbf{U}_1 + jb\mathbf{V}_1, \quad (2)$$

where $\mathbf{U}_1$ and $\mathbf{V}_1$ are orthogonal unit vectors, and a, b are the amplitude of the components. Then, the boresight field radiated by the n-th element $\mathbf{E}_n$ becomes

$$\mathbf{E}_n = \left[(a\cos\phi_n - jb\sin\phi_n)\mathbf{U}_1 + (a\sin\phi_n + jb\cos\phi_n)\mathbf{V}_1\right]\exp(j\phi_n), \quad (3)$$

Using the following relations:

$$\sum_{n=1}^{N}\cos^2\phi_n = \sum_{n=1}^{N}\sin^2\phi_n = N/2, \quad (4)$$

$$\sum_{n=1}^{N}\cos 2\phi_n = \sum_{n=1}^{N}\sin 2\phi_n = 0, \quad (5)$$

we obtain the total boresight field E radiated by the array, such as

$$E = \sum_{n=1}^{N} E_n = \frac{(a+b)}{2} N(U_1+jV_1). \quad (6)$$

This means that the sequential array radiates perfect circular polarized wave in the boresight direction independently of the polarization of the element. In general, because microstrip antenna is narrowband, polarization becomes rapidly as frequency changes from its center. But, the sequential array can greatly reduce the cross polarization, even at off-center frequency. Consequently, we can get a wideband circularly polarized microstrip array.

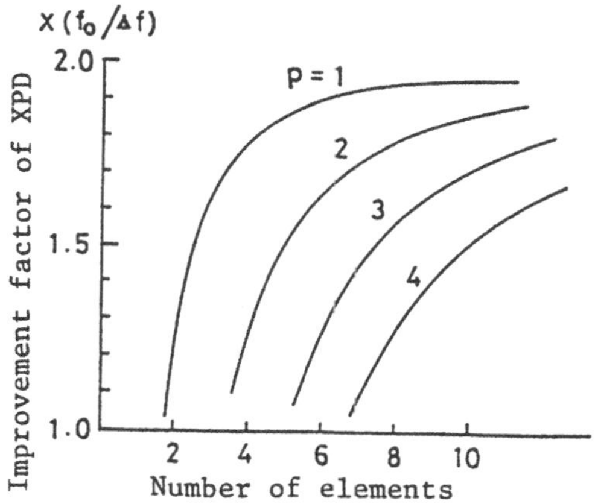

Fig.2. Improvement factor of XPD

Fig. 2 shows the improvement factor of XPD, η, the ratio of the XPD of the array to the XPD of the element. From this figure, it is clear that XPD is improved as N increases, and the case of p=1 is the best.

Next, we consider VSWR at the input terminal of a sequential array. In the Fig.1, we assume that the input power is equally divided to each element, so let the amplitude of the input voltage to each element be V_0. Owing to the differential path length of each feed line, the reflected waves from the n-th element have a differential phase shift of $2\phi_n$. Therefore, if all reflection coefficients of the elements are the same, the sum of the total reflected wave V_r returning to the input terminal of the array becomes

$$V_r = V_0\Gamma\sum_{n=1}^{N}\exp(j2\phi_n) = 0. \quad (7)$$

From the above discussion, it can be seen that the sequential array provides not only perfect circular polarization in the boresight direction but also, no reflection at the input terminal.

3. OFF-AXIS RADIATION OF PLANAR SEQUENTIAL ARRAY

In this section we consider a M x N planar sequential array in which the elements are arranged on rectangular lattices as shown in Fig.3. In this array, each column and each row are the sequential linear arrays. Let the differential phase shift and rotation to be given to the (m,n)th element be $\phi_{m,n}$. The radiation pattern of the element in the E-plane and H-plane are expressed by $E_p(\theta)$ and $H_p(\theta)$, respectively, and let the

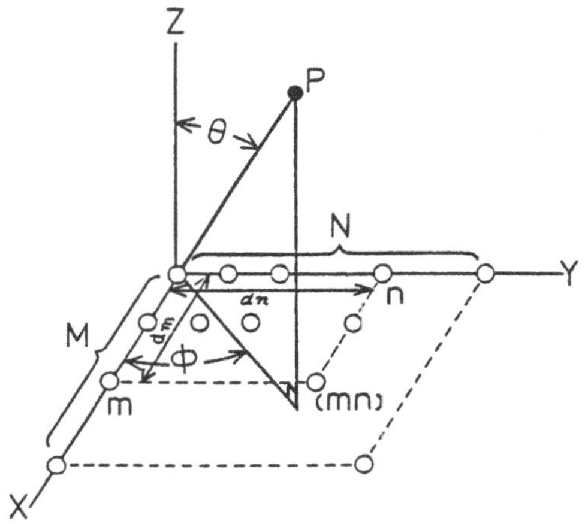

Fig.3. Geometry of a planar sequential array

complex excitaion factor for the quadrature components be α ($\alpha=\pm 1$ means perfect circular polarization). We can obtain the expression of the radiation pattern for arbitrary angular cut plane. Particularly, in the $\phi=0$ (X-Z plane), using the sequential conditions, one has

$$E(\theta,0) = \frac{(1+\alpha)}{2}N(\sum_{m=1}^{M}e^{jkd_m\sin\theta})(E_p(\theta)\mathbf{e}_\theta+jH_p(\theta)\mathbf{e}_\phi), \quad (8)$$

where $\mathbf{e}_\theta$ and $\mathbf{e}_\phi$ are unit vectors. Similarly, in the plane $\phi=\pi/2$,

$$E(\theta,\pi/2) = j\frac{1+\alpha}{2}M(\sum_{n=1}^{N}e^{jkd_n\sin\theta})(E_p(\theta)\mathbf{e}_\theta+jH_p(\theta)\mathbf{e}_\phi). \quad (9)$$

Therefore, in the two principal planes, within the angular region where E (θ)=H (θ) holds, the array radiates circular polarized wave independently of the polarization of the element. In ordinary microstrip antennas, the patterns in the E- and H-planes almost agree up to considerable angular region, so the sequential array composed of these element has excellent polarization isolation over the wide angle.

4. EXPERIMENTAL RESULTS

In order to verify the principle of the sequential array, some basic experiments have been carried out.

The test array used are two 8-element arrays. One is a sequemtial, and the other is a conventional arrays as shown in Fig.4. All elements are the same microstrip circular patches which have small notches and is excited by one-point-backside feeding(3). The material of the substrate is glass cloth PTFE and its is 2.6 and the thickness is 4 mm.

Fig. 5 and 6 show the axial ratio and VSWR of these arrays. From these figures, it is clear that the sequential array has much more wideband characteristics of polarization and impedance than the conventional array. For example, 3 dB axial ratio bandwidth of the sequential array exceeds 14% and this is about 15 times of the conventioal array, while

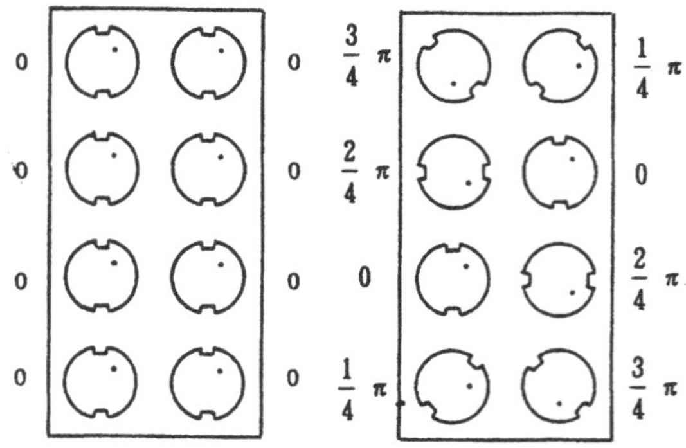

(1) Conventional array (2) Sequential array

Fig.4. Arrangement of elements for two test arrays

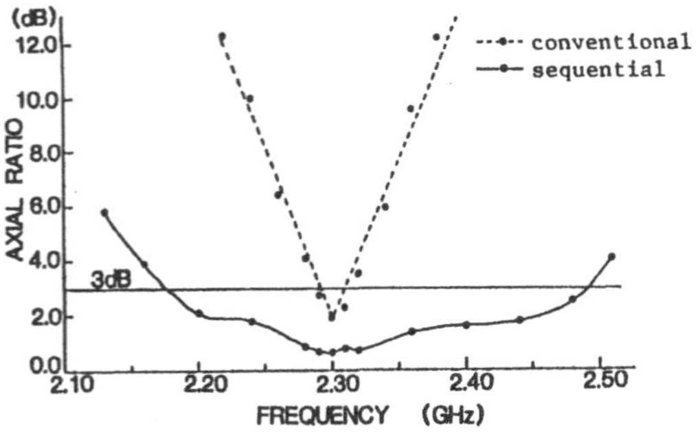

Fig.5. Measured axial ratio vs. frequency

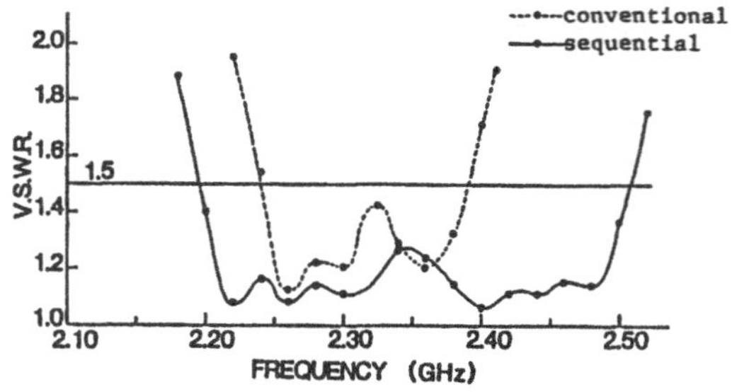

Fig.6. Measured V.S.W.R. vs. frequency

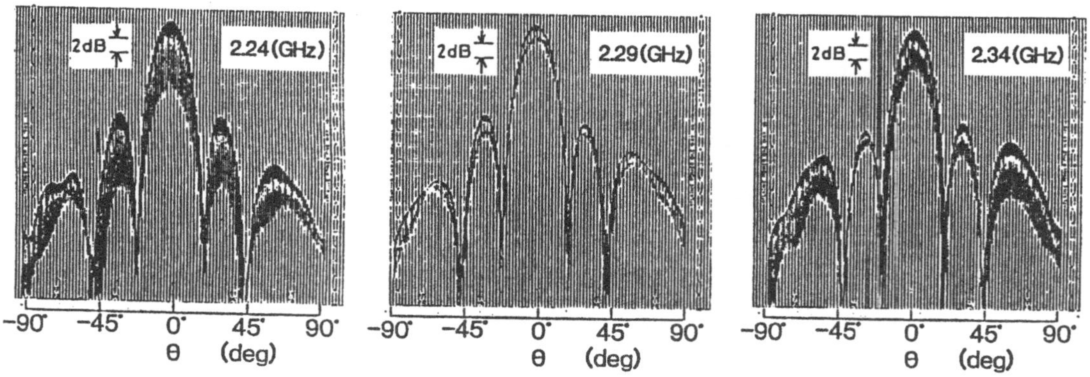

Fig.7. Radiation patterns of the conventional array

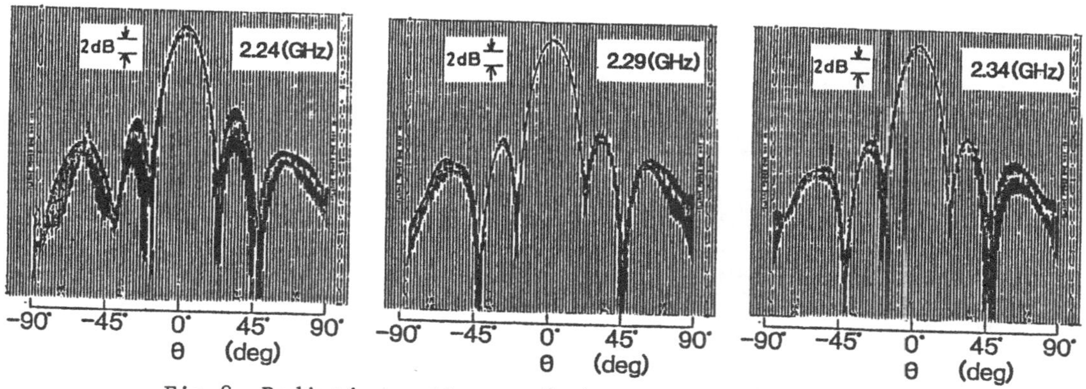

Fig.8. Radiation patterns of the sequential array

1.5 VSWR width is 13.7% and about twice as that of the conventioal array.

Finally, the comparison of the radiation patterns is shown. Fig. 7 and Fig. 8 show the radiation patterns of the conventional array and the sequential array, respectively. It can be seen that the polarization of the conventional array deteriorates execpt at the center frequency, while the sequential array maintains good circular polarization over the range from 2.24 GHz to 2.34 GHz.

REFERENCES

(1) K. R. Carver et al."Microstrip antenna technology", IEEE Trans. AP-29, No.1 (Jan. 1981)

(2) Y. Suzuki et al."Expanding the frequency bandwidth of a microstrip antenna",IEEE AP-S Int. Symp. Digest vol.1 p 336 (June 1981)

(3) M. Haneishi et al,"A broadband microstrip array composed of single feed type circularly polarized microstrip antennas", IEEE AP-S Int. Symp. Digest pp.160-163 (May 1982)

Gain of Circularly Polarized Arrays Composed of Linearly Polarised Elements

P. S. Hall, J. Huang, E. Rammos, and A. Roederer

Indexing terms: Antennas, Antenna arrays, Microstrip

The gain of circularly polarised (CP) array antennas realised by proper phasing of sequentially rotated linearly polarised (LP) elements is compared to that of arrays using CP elements and demonstrated by calculations for microstrip patch elements. When element spacing is large and array size is small, the advantages of LP elements are offset by the significant reduction in gain due to high cross polarised lobes in the diagonal planes. For large arrays of closely spaced elements, this gain loss reduces to a negligible amount. However, for spacings above a critical value of about 0·7 wavelengths, unacceptably high gain losses will be incurred.

Introduction: Sequential rotation with proper phasing of elliptically polarised elements improves the on axis circular polarisation purity, radiation pattern symmetry and input impedance match of array antennas. The technique has been proposed for microstrip patch arrays by Teshirogi[1] and implemented in several prototype arrays.[2,3] A four element example is shown in Fig. 1*a* where each element is a notched elliptically polarised microstrip disc.[4] The disc can also be fed by two pins as shown on Fig. 1*b*. Recently Huang[5] has extended the technique to generate circular polarisation from arrays of linearly polarised elements, as shown in Fig. 1*c*. In this case, mutual coupling between the elements is reduced and pattern symmetry is improved. In addition we note here that this configuration allows dual polarisation to be simply obtained. Fig. 1*c* indicates that transposition of the phases of a diagonal pair of elements switches the hand of polarisation, a property not possessed by the sequentially rotated array of notched or CP elements.

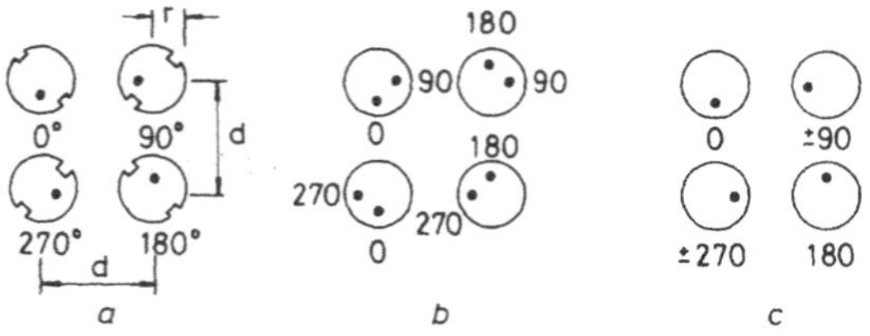

Fig. 1 *Circularly polarised sequentially rotated array configurations*

a Notched elements *b* CP elements

c LP elements showing alternate phase arrangements to switch between hands of polarisation

Huang[5] has demonstrated that in a 2 × 2 array of LP elements phased for CP operation high cross circularly polarised lobes appeared in the diagonal planes. It has been suggested† that these lobes are critically dependent on element spacing and that the consequent array gain loss can be offset by closer spacing at the expense of increased mutual coupling. This paper further quantifies the gain loss and confirms this point. Preliminary results for larger arrays are also given that indicate constraints on suitable element spacing.

† ROEDERER, A., and RAMMOS, E.: Private communication, March 1988

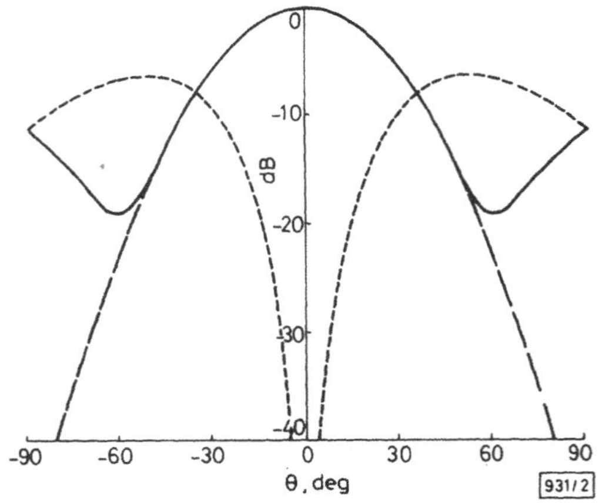

Fig. 2 *Computed diagonal plane radiation patterns*

——— copolarised LP discs

— — copolarised notched discs

- - - - crosspolarised LP discs

Array as Fig. 1*a* and $r = 6{\cdot}5$ mm, $d = 23{\cdot}0$ mm ($0{\cdot}61\,\lambda_0$), substrate height = 1·59 mm, $\varepsilon_r = 2{\cdot}32$, frequency = 8·0 GHz

Arrays of 2 × 2 elements: Fig. 2 shows computed radiation patterns in the diagonal plane that are confirmed by measurements in Reference 5. The high off-axis lobes occurring with LP elements phased for CP operation are due to incomplete cancellation of the crosspolarised components in this plane. Fig. 2 is deduced using a cavity model of the patch including higher order modes and a transmission line model of the feed network which uses a nonisolating four way reactive power splitter. Integration over the complete radiation patterns and estimates of array ohmic losses provides array gain. Fig. 3 shows array normalised gain over the patch bandwidth. It can

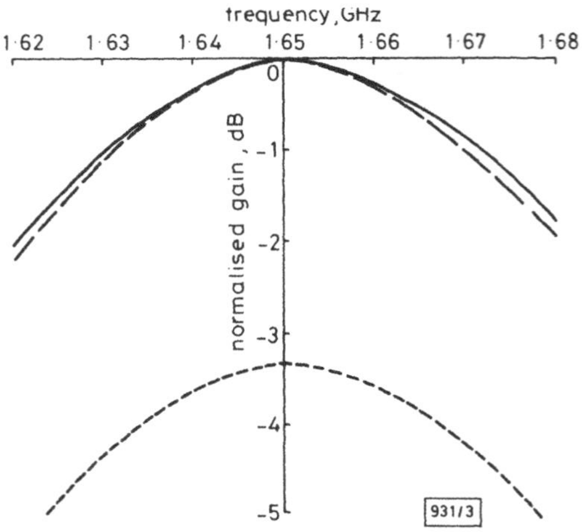

Fig. 3 *Computed normalised gain of sequentially rotated disc arrays*

——— conventionally fed array of notched discs

— — sequentially rotated array of notched discs

- - - - sequentially rotated array of LP discs

Array gain normalised to gain of conventionally fed array of notched discs at centre frequency $r = 32$ mm, $d = 120$ mm ($0{\cdot}66\,\lambda_0$), substrate height = 3·2 mm, $\varepsilon_r = 2{\cdot}32$

be seen that the use of LP elements results in a gain loss exceeding 3 dB across the whole band compared to that for notched elements with the same spacing. The effect of patch spacing is shown in Fig. 4. For a 2 × 2 array, the loss is seen to decrease rapidly with reduced spacing confirming the previous supposition.[6] Mutual coupling effects, which are not included in the analysis, will of course increase. As the minimum likely spacing is of the order of $0{\cdot}4\,\lambda_0$, a minimum loss of about 0·7 dB is indicated. As gain is normalised to that

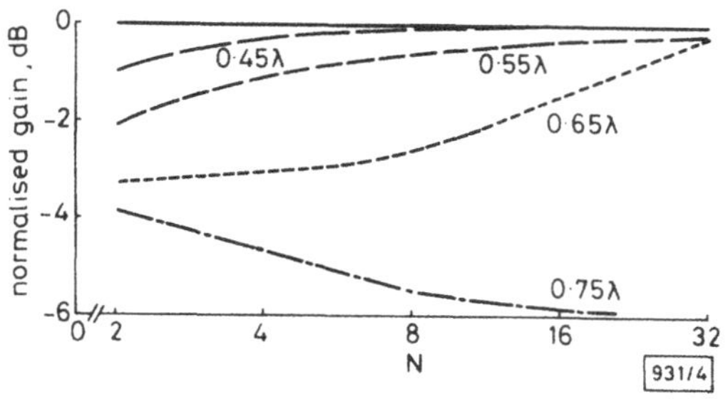

Fig. 4 *Computed normalised gain of N × N element sequentially rotated disc arrays*

——— notched discs $d = 0{\cdot}45\,\lambda_0$ to $0{\cdot}75\,\lambda_0$;
LP discs — — $d = 0{\cdot}45\,\lambda_0$ ------ $d = 0{\cdot}55\,\lambda_0$
- - - - $d = 0{\cdot}65\,\lambda_0$ —·—· $d = 0{\cdot}75\,\lambda_0$

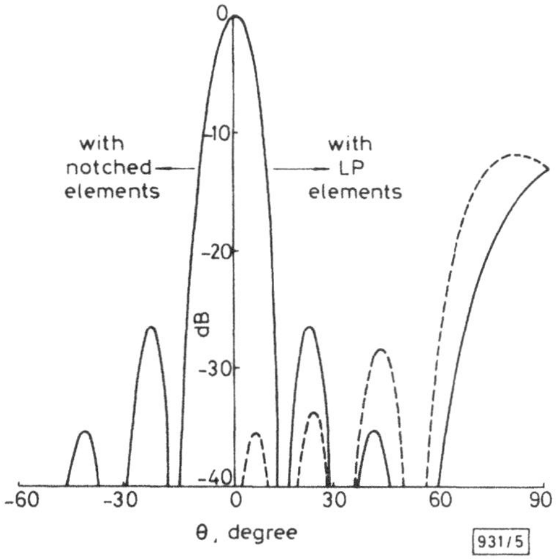

Fig. 5 *Computed diagonal plane radiation patterns of 8 × 8 element sequentially rotated array*

——— copolarised - - - - crosspolarised
Array details as Fig. 3, $d = 0{\cdot}65\,\lambda_0$, patterns are symmetric about $\theta = 0$

of a conventionally fed array, this is in addition to ohmic and surface wave losses.

Large arrays: It can be seen from Fig. 4 that the gain loss for small spacing decreases rapidly with array size, due to the increased suppression of the crosspolarised lobes by the array factor in the diagonal planes. For $0{\cdot}55\,\lambda_0$ spacing, this loss is less than a few tenths of a dB for 16 × 16 arrays and larger. However, for larger element spacing this suppression is insufficient to reduce the loss and indeed for spacings greater than about $0{\cdot}7\,\lambda_0$, which corresponds to a diagonal spacing of about $1{\cdot}0\,\lambda_0$, high crosspolarised lobes occur. Fig. 5 compares the pattern with LP elements to that with notched ones for an 8 × 8 array and illustrates the unwanted lobe effect.

Conclusion: The loss in gain in small arrays of linearly polarised elements sequentially rotated to produce circular polarisation has been quantified and shown to be dependent on element spacing. Reduced spacing is indicated to minimise the loss although this will to some extent offset the reduction in mutual coupling that comes from such arrangements. This gain loss disappears in large arrays with small element spacing. However above a critical spacing of about 0·7 wavelengths high gain losses are likely to occur, due to unsuppressed crosspolarised lobes in the diagonal planes of the radiation pattern. The above results are preliminary and further work is on-going to consolidate the analysis.

References

1 TESHIROGI, T., TANAKA, M., and CHUJO, W.: 'Wideband circularly polarised array antenna with sequential rotations and phase shift of elements'. Proceedings of International Symposium on Antennas and Propagation, Japan 1985, pp. 117–120
2 HANEISHI, M., HAKURA, Y., SAITO, S., and HASEGAWA, T.: 'A low profile antenna for DBS reception'. IEEE AP-S International Symposium Digest, June 1987, pp. 914–917
3 HALL, P. S.: 'Feed radiation effects in sequentially rotated microstrip patch arrays', *Electron. Lett.*, 1987, **23**, pp. 877–878
4 TESHIROGI, T., and GOTO, N.: 'Recent phased array work in Japan'. ESA/CST204 Phased Array Workshop, ESTEC, Noordwijk, The Netherlands, 13 June 1983
5 HUANG, J.: 'A technique for an array to generate circular polarisation with linearly polarised elements', *IEEE Trans.*, 1986, **AP-34**, pp. 113–1124

Optimised Feeding of Dual Polarized Broadband Aperture-Coupled Printed Antenna

E. Edimo, A. Sharaiha and C. Terret

Indexing terms: Antennas, Microstrip, Antenna feeders

A new feeding technique is proposed for the dual polarised broadband aperture-coupled printed antenna, to optimise the port decoupling and crosspolarisation level. Experimental results are compared with those of a more classical feeding technique.

Introduction: The technique of feeding a microstrip patch with a microstrip line through an aperture is attracting an increasing amount of interest [1]. This is mainly because, in the aperture-coupled patch antenna (ACPA), there is physical isolation between the feeding port and the radiating element.

For dual polarisation radiation, a square patch is coupled to a pair of microstrip lines through either two separated orthogonal slots (off-centre with respect to the antenna axis) [2, 3], or two centred crossed slots fed offset to match the antenna [4]. Generally, these antennas exhibit, in the entire 3dB-gain narrow bandwidth ($<5\%$), a crosspolarisation level in both E and H planes of ~ -20 dB and input isolation between -18 and -25 dB in the vicinity of resonance. Wideband operation can be obtained, using the aperture-coupled stacked patch antenna (ACSPA), with a nonresonating slot [5]. To achieve dual polarisation, it is better to use two orthogonal crossed slots located beneath the centre of the patches, as the symmetry of the antenna elements is advantageous in improving the radiation characteristics.

The main purpose of this Letter is to present an optimum method of feeding a dual polarisation aperture-coupled stacked patch antenna (DPACSPA) which leads to excellent performance over a very large frequency range.

Antenna performance: A DPACSPA with the dimensions shown in Fig. 1*a* and *b* has been developed to operate at C-band. The antennas are fabricated on Cu-clad dielectric slabs. The square stacked patches are double fed by two crossed and stacked microstrip lines through two crossed and nonresonating slots formed in the ground plane ($150 \times 150\,mm^2$). A thin substrate ($0{\cdot}01\lambda_g$) is placed between the microstrip lines to enhance the port decoupling. These lines of characteristic impedances $Z_c = 50\,\Omega$, are terminated in a series stub ($\sim\lambda_g/4$) to match the DPACSPA. The slot axis

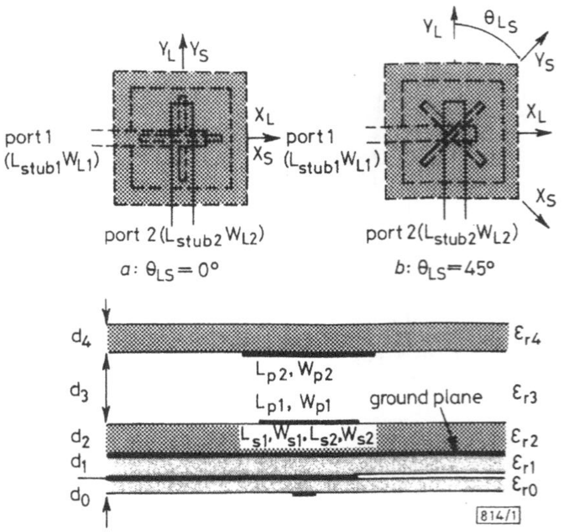

Fig. 1 *Dual polarisation aperture coupled stacked patch antennas (DPACSPAs)*

$d_0 = 0{\cdot}38$ mm, $d_1 = 0{\cdot}76$ mm, $d_2 = d_4 = 1{\cdot}52$ mm, $d_3 = 4{\cdot}5$ mm, $\varepsilon_{r0} = 2{\cdot}2$, $\varepsilon_{r1} = \varepsilon_{r2} = \varepsilon_{r4} = 2{\cdot}55$, $\varepsilon_{r3} = 1$, $L_{P1} = W_{P1} = 14$ mm, $L_{P2} = W_{P2} = 17$ mm, $L_{S1} = L_{S2} = 13$ mm, $W_{S1} = W_{S2} = 0{\cdot}8$ mm $L_{stub1} = L_{stub2} = 9$ mm, $W_{L1} = 2{\cdot}12$ mm, $W_{L2} = 3{\cdot}25$ mm

(X_S, Y_S) are shown for two different orientations (*a*) and (*b*) relative to the feeding microstrip lines. The classical slot orientation (*b*) has already been investigated with triplate feeding lines [6], whereas in this Letter, case *a* is the proposed new feeding technique.

The spectral domain approach method [1] applied to single fed ACSPA is used to design the double-fed antennas presented here.

The 2:1 VSWR measurements (Fig. 2), exhibit $\sim 30\%$ bandwidth around the centre frequency $f_0 = 5{\cdot}45$ GHz, in both cases *a* and *b*.

The coupling coefficient (Fig. 3) between the inputs for case *a*, is less than -26 dB over 80% of the bandwidth, which is 7dB better (at 5·45 GHz) than case *b*. The coupling enhancement observed at the lower bandwidth edge (-20 dB) can be

Reprinted with permission from *Elect. Lett.*, M. Edimo, A. Sharaiha and C. Terret, "Optimised Feeding of Dual Polarised Broadband Aperture-Coupled Printed Antenna," vol. 28, no. 19, pp. 1785–1787, Sept. 1992.

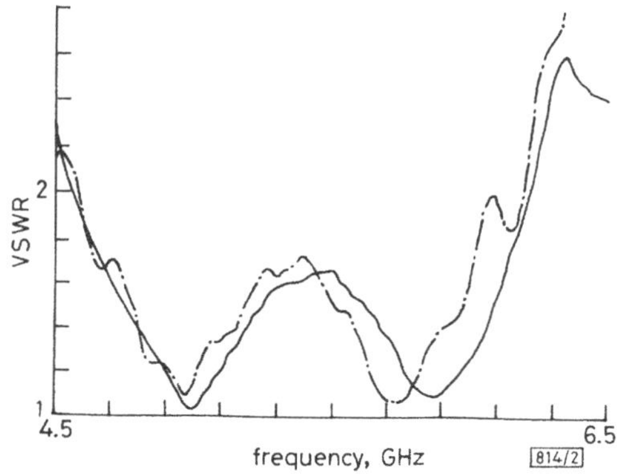

Fig. 2 *Comparison of measured VSWR at port 1*

——— case *a*
— · — case *b*

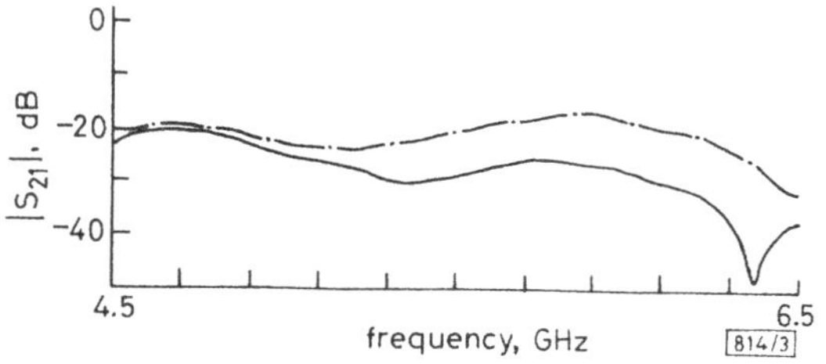

Fig. 3 *Measured* $|S_{21}|$ *between feeding ports*

——— case *a*
— · — case *b*

attributed to the antenna fabrication. Globally, the poorer decoupling of the DPACSPA (case *b*) is caused by each of the feeding lines strongly exciting both slots, whereas in case *a*, each input line excites only the corresponding orthogonal aperture.

From the radiation patterns plotted in the E-plane (Fig. 4) and H-plane (Fig. 5), it appears that the crosspolar level (for −3 dB beamwidth) of the classical technique is greater than that of the latter technique (< −25 dB). Once more, according to how the crossed slots and feed lines are placed, the crosspolar level and the symmetry in the copolar patterns are attractive (DPACSPA case *a*).

The electrical characteristics of the two DPACSPA structures are summarised in Table 1. For comparison, experimental performances of a dual polarisation aperture-coupled patch antenna (DPACPA) with the same feeding technique as Fig. 1*a*, are also given. For all these antenna structures, a

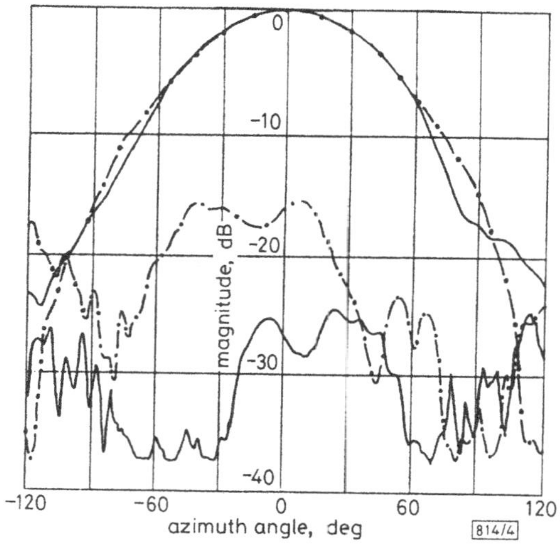

Fig. 4 *Measured E-plane far field patterns at centre frequency* f_0 = *5·45 GHz*

——— case *a*
— · — case *b*

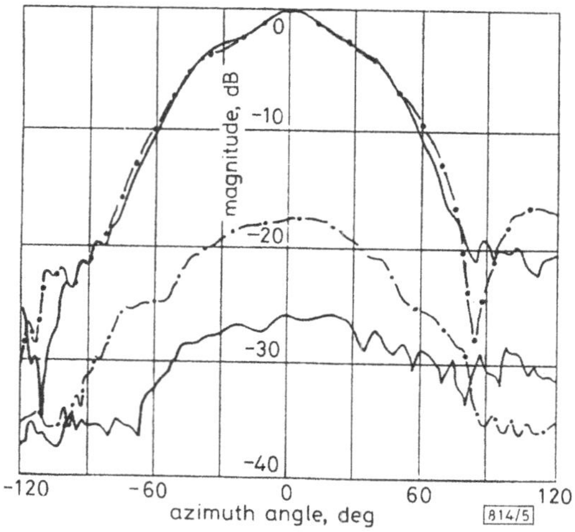

Fig. 5 *Measured H-plane far field patterns at centre frequency* f_0 = *5·45 GHz*

——— case *a*
— · — case *b*

Table 1 MEASURED CHARACTERISTICS OF VARIOUS DUAL POLARISED APERTURE-COUPLED PRINTED ANTENNAS

		DPACSPA case *a*		DPACSPA case *b*		DPACPA case *a*	
		Port 1	Port 2	Port 1	Port 2	Port 1	Port 2
Bandwidth (VSWR 2 : 1)		31%	33%	30%	31%	18%	15%
Maximum crosspolar within the beamwidth	E plane	−25 dB	−22 dB	−16 dB	−17 dB	−23 dB	−18 dB
	H plane	−25 dB	−25 dB	−17 dB	−16 dB	−22 dB	−20 dB
Beam width (−3 dB)	E plane	76°		76°		91°	
	H plane	66°		66°		80°	
S_{21} at f_0 = 5·45 GHz		−28 dB		−21 dB		−22 dB	

slight difference in performance is observed between the feeding ports. This is mainly due to the influence of the thin dielectric layer separating the two microstrip lines. However, this spacer improves the port decoupling.

Conclusion: Measured performances of dual polarised and wideband printed antennas have been presented. With proper choice of symmetrical feed line and aperture geometries, the characteristics (coupling, crosspolar) of the DPACSPA (case *a*) are improved.

References

1 POZAR, D. M.: 'A reciprocity method of analysis for printed slot and slot-coupled microstrip antennas', *IEEE Trans.*, 1986, **AP-34,** pp. 1439–1446

2 ADRIAN, A., and SCHAUBERT, D. H.: 'Dual aperture-coupled microstrip antenna for dual or circular polarisation', *Electron. Lett.*, 1987, **23,** pp. 1226–1228

3 CHARES, M. L., PENARD, E., MOULINARD, M. L., HIMDI, M., and DANIEL, J. P.: 'Technology and design of an active antenna with dual polarisation'. Proc. COST 223-ESA Workshop on Active Antennas, Noordwijk, 1992, pp. 231–238

4 HERSCOVICI, N., and POZAR, D. M.: 'Application of aperture coupled microstrip lines'. Progress in Electromagnetics Research Symp., Massachusetts, 1991, p. 362

5 CROQ, F., and PAPIERNIK, A.: 'Large bandwidth aperture-coupled microstrip antenna', *Electron. Lett.*, 1990, **26,** pp. 1293–1294

6 DUBOST, G., and FRIN, R.: 'Dual polarized microstrip arrays in S, C, X, and Ku bands'. Progress in Electromagnetics Research Symp., Massachusetts, 1991, p. 286

Feed Circuits of Double-Layered Self-Diplexing Antenna for Mobile Satellite Communications

MASAYUKI NAKANO, HIROYUKI ARAI, WATARU CHUJO, MASAYUKI FUJISE, and NAOHISA GOTO

***Abstract*—This communication presents an analysis of feed circuits for a double-layered self-diplexing antenna (SDA) for mobile satellite communications. It is possible to reduce the weight of the diplexer for a self-diplexing antenna by getting large internal isolation between transmitting and receiving. This communication presents that the internal isolation is heavily dependent on the configuration and errors of the feed circuits of the antenna. We discuss the feed circuits of two- and four-point feeds for the antenna by using the cavity model of microstrip elements, and also present error analysis of the feed circuit.**

I. Introduction

The diplexer is very heavy and large to get internal isolation greater than 90 dB between transmitting and receiving. To solve this problem, it has already been shown that some kinds of SDA can act as a part of a diplexer [1]. Among them, a double-layered circularly polarized SDA composed of a circular microstrip patch and a ring patch is suitable for minimizing the size of the array [2]. The same circular polarization for transmitting and receiving provides this SDA with a peculiar polarization isolation in addition to isolations by the frequency and the antenna spacing. In this communication, the isolation of the double-layered SDA is analyzed to present the effects of feed circuits and manufacturing errors.

II. Configuration and Analysis

A configuration of a SDA is shown in Fig. 1. A circular microstrip antenna (CMA) lies on the upper layer for transmitting at 1.635 GHz, and a ring patch antenna (RPA) stays on the lower layer for receiving at 1.535 GHz. The disk of the RPA operates as a ground plane for the CMA that provides the double-layered structure. The key feature is that each layer of the SDA can be independently fed, because the CMA is fed through an electrical shorted center conductor of the RPA. To obtain circular polarizations, each level of the SDA fed at two points with a phase difference of 90°, at four points with a sequential phase difference of 90°. The feed point locations are shown in Fig. 2, where the phase sequence of the CMA is the reverse direction.

The analysis is based on an electromotive force method considering a boundary admittance at the cavity edge to calculate the mutual impedance between feed pins of an antenna element [3]. The mutual impedance between the CMA and the RPA is evaluated by assuming magnetic currents at the cavity edge [4]. The validity of the analysis was confirmed by the two-point feed SDA [5]. The purpose of this communication is to present the effect of the feed circuit configuration and manufacturing errors.

We present the analysis of the SDA including the feed circuit for the four-point feed, because the two-point feed antenna element is treated as a part of the four-point feed. The impedance of the four-feed SDA is expressed by 8 × 8 matrix. The impedance matrix is converted to the scattering matrix $\mathbf{S}_a$, according to the following relation:

$$[\mathbf{S}_a] = \{[\mathbf{Z}_i] + [\mathbf{E}]\}^{-1}\{[\mathbf{Z}_i] - [\mathbf{E}]\} \tag{1}$$

where $[\mathbf{E}]$ is a unit matrix, $[\mathbf{Z}_i] = [\mathbf{Z}]/\mathbf{Z}_0$, and the Z_0 is chosen as 50 Ω for the characteristic impedance of the feed line. The feed circuit of four-point feed consisting of one 180° hybrid and two 90° hybrids, is expressed as an 8 × 8 scattering matrix $\mathbf{S}_{f4}$. Finally, the output ports of $\mathbf{S}_{f4}$ are connected to $\mathbf{S}_a$, and the 8 × 8 matrix $\mathbf{S}_h$ is obtained for the four-point feed SDA including its feed circuit, and provides the isolation characteristics of the SDA as mutual coupling.

We have already presented that the SDA fed at two points gives 35-dB isolation by using a feed circuit of 90° hybrid experimentally, and its analysis in the previous paper [5]. This communication presents a simple feed circuit for two-point feed SDA using a T-junction, and the analysis of the four-point feed SDA and error effects of the feed circuits. It indicates theoretical limitation to reduction of the internal isolation of the SDA.

III. Isolation Characteristics of SDA

In the experiments, a shorted center conductor of the RPA is made by short pins, and the region inside the pins is filled with the substrate dielectric. This difference from the analysis model

Reprinted from *IEEE Trans. Antennas Propaga.*, vol. 40, no. 10, pp. 1269–1271, Oct. 1992.

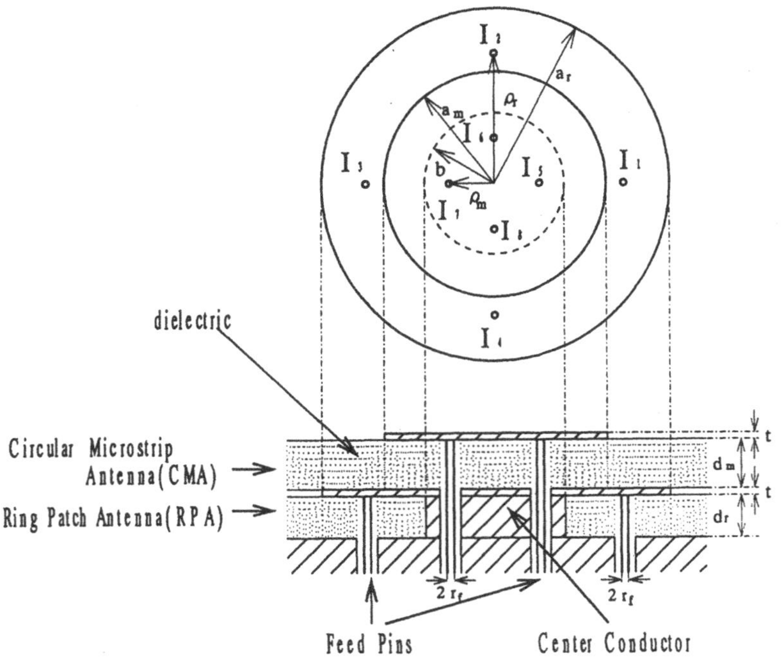

Fig. 1. Configuration of a double-layered self-diplexing antenna.

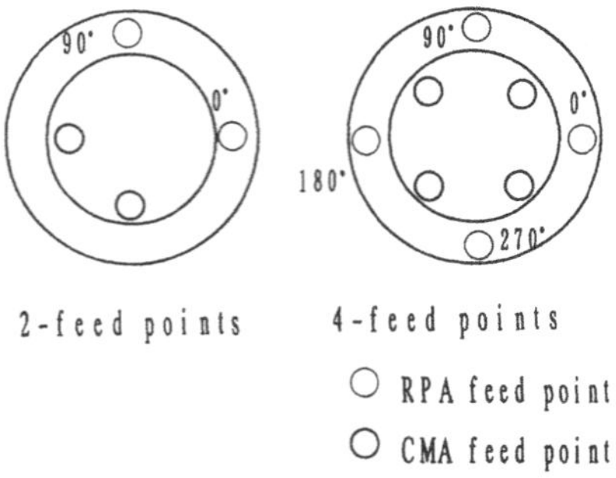

Fig. 2. Location and numbers of feed points of SDA.

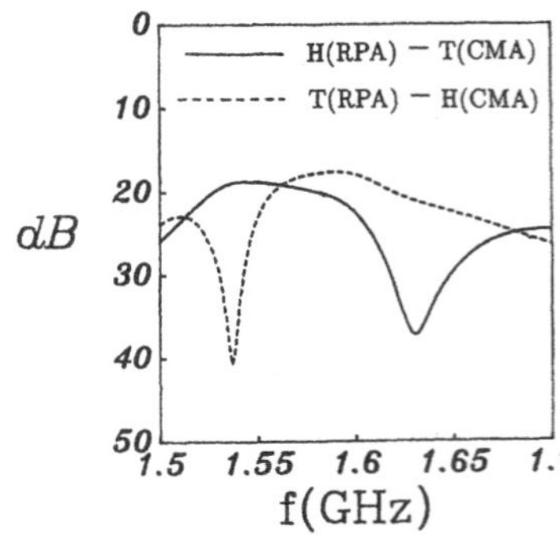

Fig. 3. Isolation of SDA fed at two points by combination of *T*-junction and hybrid. *H* and *T* denote hybrid feed circuit and *T*-junction feed, respectively. $a_r = 38.5$, $a_m = 30.2$, $b = 12.2$, $\rho_r = 17.2$, $\rho_m = 9.0$ (8.47), $d_r = 3.15$, $d_m = 3.15(5.20)$, $t = 0.05$, $r_f = 0.5$ (mm), $\epsilon_r = 2.60(2.80)$, $\epsilon_m = 2.60(2.68)$ [( ) = effective value for calculation].

is compensated by a small change of antenna parameters [5]. The two-point feed SDA is excited by 3-dB hybrid couplers for the circular polarization, however, it is also excited by *T*-junctions with delay lines. We may use the combination of a *T*-junction feed and a 3-dB hybrid for feed circuits, where delay lines are matched at the transmitting or receiving frequency. The isolations by these feed circuits are shown in Fig. 3. The combination of the *T*-junction and the 3-dB hybrid feed has the maximum point of the isolation. The internal isolation should be large at only the transmitting frequency, and the combination of the 3-dB hybrid coupler for the RPA and *T*-junction feed for the CMA is useful to simplify the feed circuit.

To increase the polarization isolation, a four-point feed was proposed for the SDA [6]. The ideal amplitudes of this feed circuit are $1/4$ ($= -6.0$ dB) for each output port, however, the actual feed circuit has an imbalance of amplitude and phase. The ideal feed circuits eliminate the coupling, because higher order modes are not excited and a pure circular polarization is obtained. However, output imbalances of the feed circuit decrease the isolation. Amplitude imbalance is about $-6.2 \sim -6.9$ dB, and phase difference are not exactly 90°. The isolations shown in Fig. 4, taking into account these imbalances, agree well with the measured results, where the measured output imbalance of the feed circuit is used for the *S*-parameter matrix element. It should be noted that the coupling is zero for ideal feed circuits. Therefore, the isolation of the four-point feed greatly depends on errors of the feed circuits.

IV. Error Analysis of Feed Circuits

The errors discussed in the preceding section are the amplitude and the phase imbalance at output ports of the feed circuits. Fig. 5 shows how the isolation depends on the errors for the four-point feed of the SDA at the transmitting frequency (1.635 GHz). The maximum error in the figure means each output port of the feed circuit has random errors within δa and δp.

The four-point feed SDA is able to increase the isolation to more than 45 dB. These results indicate that $\delta a \leq 0.6$ dB and $\delta p \leq 0.5°$ are required to suppress the coupling to less than -50 dB. We also estimate the isolation of the four-point feed SDA for the manufacturing error in addition to the above feed circuit errors. There are many types of manufacturing errors, but we only consider errors in feed-point locations in the radial direction. We define that the maximum location error δr is 0.1 mm, because the manufacturing precision is less than 0.1 mm at the present stage. The calculated examples are shown in Fig. 6

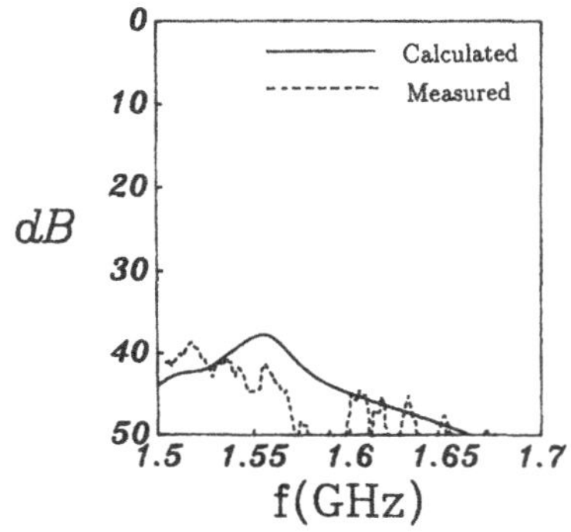

Fig. 4. Isolation between transmitting and receiving of SDA fed at four points. $a_r = 40.7$, $a_m = 30.7$, $b = 15.1$, $\rho_r = 20.1$, $\rho_m = 13.0(11.8)$, $d_r = 3.15$, $d_m = 3.15(5.20)$, $t = 0.05$, $r_f = 0.5$ (mm), $\epsilon_r = 2.60(2.97)$, $\epsilon_m = 2.60(2.74)$ [( ) = effective value for calculation].

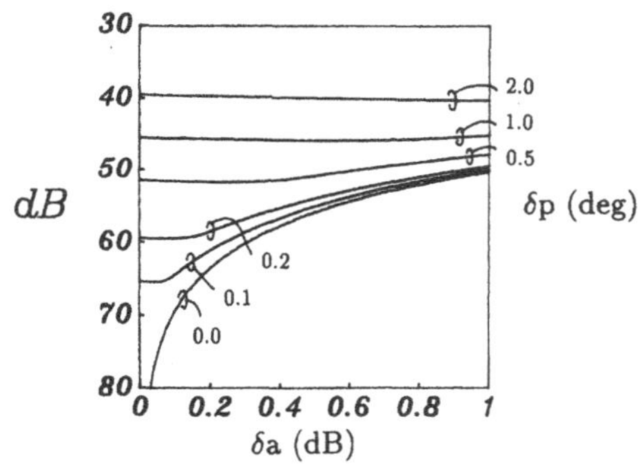

Fig. 5. Isolation of the four-feed SDA for maximum amplitude error δa and phase error δp. Antenna parameters are the same with those of Fig. 4.

for $\delta p = 0.0°, 1.0°$. The manufacturing error decreases the isolation of the maximum amplitude error less than 0.4 dB for $\delta p = 0.0$, however its effect is not observed for $\delta p = 1.0°$. These results indicate that δp decreases the isolation more than δr for $\delta a \geq 0.4$ dB. As a result, the phase and amplitude errors are dominant factors in determining the isolation characteristics.

V. Conclusion

In this paper, a circularly polarized SDA for mobile satellite communications has been analyzed from the view point of the internal isolation between transmitting and receiving. The isolation of the SDA fed at four points is ideally infinite, however, it greatly depends on the output amplitude and phase imbalance of the feed circuits. The isolation of the actual feed circuits is more than 40 dB for the four-point feed of the SDA, and the calculated values are in agreement with measured values. The simulation of the output amplitude and phase errors for the four-point feed circuit shows the theoretical limit for the isolation.

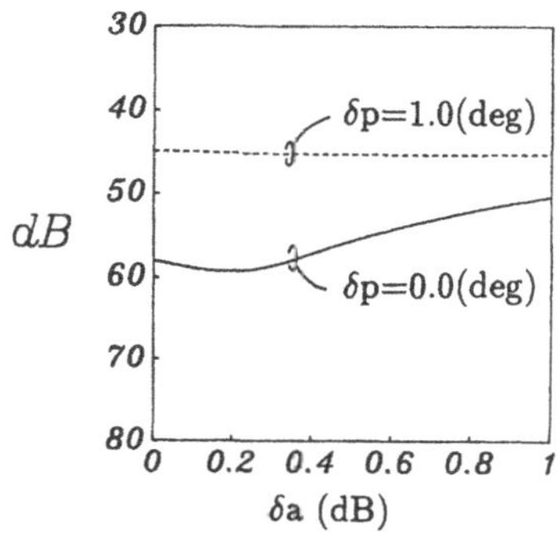

Fig. 6. Isolation of four types of the four-feed SDA with manufacturing error for maximum amplitude error δa ($\delta l = 0.1$ mm).

References

[1] E. Rammos and A. Roederer, "Self-diplexing circularly polarized antenna," in *1990 Int. IEEE/AP-S Symp. Digest*, pp. 803–806, May 1990.

[2] M. Yasunaga, F. Watanabe, T. Shiokawa, and M. Yamada, "Phased array antennas for aeronautical satellite communications," in Fifth Int. Conf. Antennas Propagat. 87, (Apr. 1987), pp. 47–50.

[3] S. Yano and A. Ishimaru, "A theoretical study of the input impedance of a circular microstrip disk antenna," *IEEE Trans. Antennas Propagat.*, vol. AP-29, no. 1, pp. 77–83, Jan. 1981.

[4] Haneishi, M., "Studies on circularly polarized microstrip antennas," Ph.D. dissertation, Tokyo Institute of Technology, Tokyo, Japan, 1982.

[5] W. Chujo, M. Fujise, M. Nakano, H. Arai, and N. Goto, "A two-layer self-diplexing antenna using a circularly polarized ring patch antenna," *IEICE Trans.*, vol. E-74, no. 10, pp. 3261–3267, Oct. 1991.

[6] W. Chujo, K. Yasukawa, H. Arai, and N. Goto, "Two-layer self-diplexing antenna composed of microstrip and ring patches fed at four points," *The 3rd Asia-Pacific Microwave Conf. Proc. 90*, Sept. 1990, pp. 273–276.

Microstrip Antennas with Frequency Agility and Polarization Diversity

DANIEL H. SCHAUBERT, SENIOR MEMBER, IEEE, FREDERICK G. FARRAR, MEMBER, IEEE, ARTHUR SINDORIS, SENIOR MEMBER, IEEE, AND SCOTT T. HAYES

***Abstract*—A technique is investigated for controlling the operating frequency and polarization of microstrip antennas. The control is achieved by placing shorting posts at appropriate locations within the antenna's boundaries. By changing the number and locations of the posts, the operating frequency can be tuned over a 1.5-to-1 range, and the polarization can be changed from horizontal to vertical, right-hand circular, or left-hand circular. All of these changes are obtained without significantly altering the input impedance or radiation patterns of the antena and without increasing the complexity of the external microwave feed network. The frequency and polarization can be electronically controlled by using microwave switching diodes for the shorting posts. Antennas that have two feeds and operate simultaneously in two orthogonal polarizations have been constructed with the capability to switch between linear and circular polarization. Also, a thin frequency-scanned array has been built with the frequency-agile microstrip elements.**

I. INTRODUCTION

THE MICROSTRIP antenna has been shown to be an excellent radiator for many applications that require only a narrow bandwidth [1]-[5]. It is rugged and can be fabricated by using standard printed-circuit techniques. A single microstrip radiator has a moderately broad radiation pattern, but high-gain arrays suitable for space applications have been built [6].

In its simplest form, the microstrip antenna radiates linearly polarized signals over a bandwidth of one or two percent. However, by modifying the geometry of the basic antenna it is possible to obtain circularly polarized radiation [7], [8] or a shift in the operating frequency [9]. Since these techniques require permanent physical changes to the antenna they cannot be used to electronically modify or control the antenna's performance. Electronic control of the antenna's performance can be accomplished by means of varactors [10] or variable-length transmission lines [11]. However, varactors require a precise dc bias voltage, and switched-length transmission lines require space outside the basic microstrip antenna's boundaries. Both of these disadvantages can be overcome by using shorting posts (e.g., switching diodes) at appropriate points within the antenna's boundaries. By changing the number and locations of the shorting posts both the operating frequency and polarization of the microstrip antenna can be controlled. Also, Malagisi has shown that a phase-shifting reflector can be built by using circular microstrip elements with shorting posts [12]. Of course, all of the modifications that use diodes, capacitors, inductors, or shorting posts sacrifice the monolithic construction of the basic microstrip antenna. However, the additional capabilities of these modified antennas should offset the increased complexity of their fabrication.

The post-loading technique has been experimentally investigated by using machine screws as removable shorting posts. A single microstrip element has been made to radiate fields that are polarized horizontal linear, vertical linear, right-hand circular, or left-hand circular. Also, single elements and an eight-element array have been tuned to operate over a 1.5-to-1 range of frequencies. This frequency tuning is accomplished without the serious degradation of input impedance that was observed by Kernweis and McIlvenna [13]. A simple analytical model has been developed and used to generate useful design data for the post-loaded microstrip antennas.

II. FREQUENCY-AGILE ANTENNA

The operating characteristics of a typical rectangular-patch microstrip antenna are determined by the antenna's size and feed location and by the substrate permittivity. The antenna in Fig. 1, without the shorting posts, is a typical configuration designed for x-oriented linearly polarized radiation. The antenna operates at a fundamental frequency f_0,

$$f_0 \approx \frac{c}{2a\sqrt{\epsilon_r}}, \tag{1}$$

where the patch length a is approximately one-half wavelength in the dielectric. At this frequency the voltage and current distributions on the patch resemble those of an open-circuited microstrip transmission line with propagation in the $\pm x$ directions. The input impedance of the antenna is determined primarily by the patch width b and the feed location f [1], [14].

The addition of shorting posts along the centerline $y = b/2$ increases the operating frequency of the antenna. This frequency increase may be explained by considering the transmission-line model for the microstrip antenna [1], [15]. This model is depicted in Fig. 2, where Z_0 is the characteristic impedance of a microstrip line of width b on the substrate material. The length extensions Δl account for fringe-field reactance at the open-circuit ends, and the conductance G accounts for radiation from the ends. The formulas of Hammerstad [16] were used to calculate Z_0 and Δl, and Harrington's formula [17] for slot conductance was used for G:

$$Z_0 = \frac{377}{\sqrt{\epsilon_e}}\{b/t + 1.393 + 0.667 \ln (b/t + 1.444)\}^{-1} \tag{2}$$

$$\Delta l = 0.412t \frac{(\epsilon_e + 0.3)(b/t + 0.262)}{(\epsilon_e - 0.258)(b/t + 0.813)} \tag{3}$$

$$\epsilon_e = \frac{\epsilon_r + 1}{2} + \frac{\epsilon_r - 1}{2(1 + 10t/b)^{1/2}} \tag{4}$$

Reprinted from *IEEE Trans. Antennas Propaga.*, vol. AP-29, no. 1, pp. 118–123, Jan. 1981.

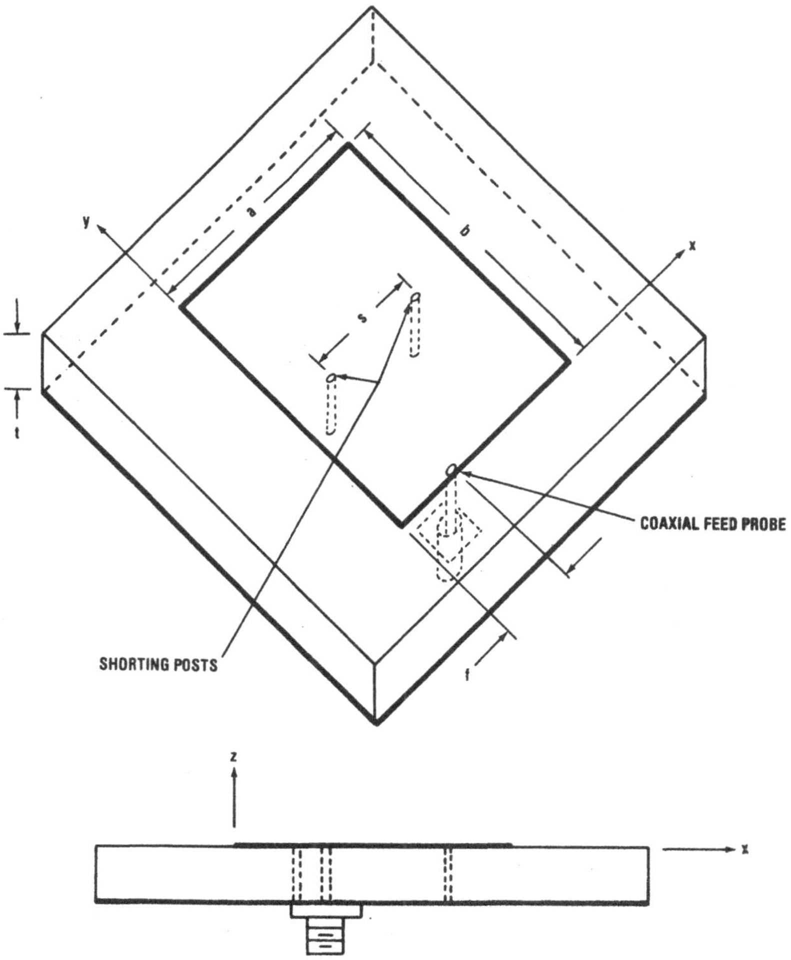

Fig. 1. Typical microstrip antenna with shorting posts for changing operating frequency.

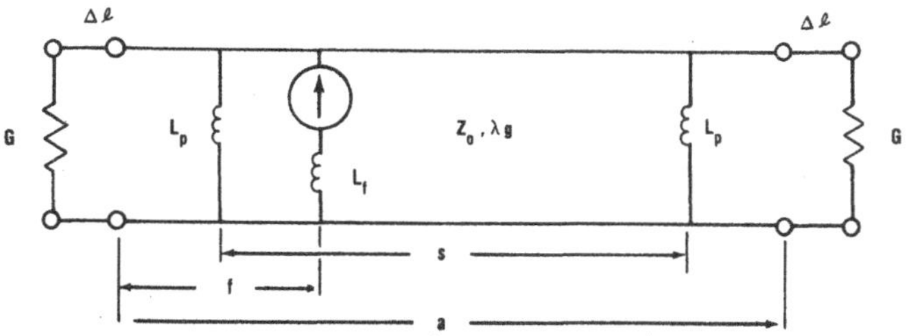

Fig. 2. Transmission-line model for calculating operating frequency and input impedance of frequency-tuned antenna.

$$G = \frac{\pi b}{377\lambda_0}\left[1 - \frac{(kt)^2}{24}\right] \qquad (5)$$

$$\approx 0.0083\, b/\lambda_0, \qquad t/\lambda_0 \ll 1.$$

The RF feed is represented by a current source with a series inductor to represent the feed probe inductance [18]. The shorting posts are represented as shunt inductances at the locations of the posts. The inductive reactance of the posts and the feed probe are calculated from the formula

$$X_L = \frac{377}{\sqrt{\epsilon_r}} \tan \frac{2\pi t}{\lambda_0}. \qquad (6)$$

The input impedance and radiation loss are calculated from this model. Plots of the operating frequency and voltage standing-wave ratio (VSWR) of a 6.2 X 9.0-cm antenna on 1.6-mm (1/16 in) Teflon fiberglass are shown in Fig. 3. The agreement between the calculated and measured frequencies is quite good (within five percent) and demonstrates that this simple transmission-line model is useful for predicting the performance of the post-tuned microstrip antenna.

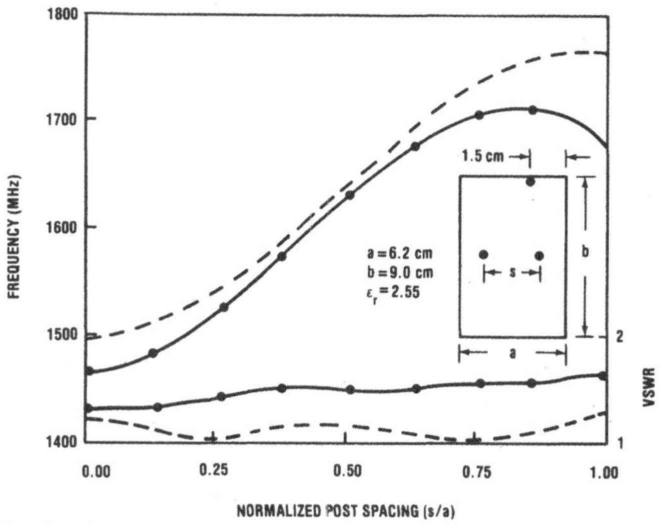

Fig. 3. Operating frequency (upper curves) and VSWR (lower curves) of frequency-tuned antenna. 1.6-mm Teflon fiberglass substrate. --- Calculated. —— Measured.

The range of frequency tuning achieved by a pair of posts along the center line is about 20 percent, but tuning ranges in excess of 50 percent have been achieved by adding more posts. These additional posts may be placed along the centerline $y = b/2$ (Fig. 1) or offset from the centerline along $y = b/2 \pm c$. (When placing posts away from the centerline it is preferred that they be added in pairs symmetric about the centerline. This avoids introducing cross-polarized signals.) The radiation patterns of the antenna are not significantly changed by the shorting posts. The bandwidth (VSWR < 2) is approximately one percent at each operating frequency.

The simple transmission-line model is not adequate to describe the antenna when several pairs of posts are used. In that case, a more detailed model (e.g., a leaky cavity with posts) will be required to obtain accurate performance predictions.

The major disadvantage of using many diodes to actively control the antenna's operating frequency appears to be the need to individually bias the diodes on or off. The bias circuit may require many components to properly distribute the dc bias signals while isolating the RF. However, this requirement is no more complicated than that already performed by solid-state RF switches, and it is not expected to greatly limit the usefulness of the post-tuned antenna.

The use of post tuning does not prevent the use of other antenna-tuning techniques. In particular, the inductive shorting posts may be combined with capacitive varactors to obtain a very wide tuning range extending above and below f_0. This combination will possess the benefits and problems of both techniques.

III. POLARIZATION DIVERSITY

The polarization of the microstrip antenna also can be selectively altered by proper location of the shorting posts [19]. A case of particular interest is the square patch fed along a diagonal with shorting posts located along the centerlines (Fig. 4). This antenna will radiate x- or y-oriented linear polarization, or right-hand or left-hand circular polarization, depending upon the locations of the posts. Typical radiation patterns obtained by using a spinning linearly polarized receive antenna are shown in Fig. 5. The pattern shapes of the two linearly polarized antennas are different because Fig. 5(a) is an E-plane pattern cut and Fig. 5(b) is an H-plane pattern cut. In Fig. 5(c) and (d) the antenna is configured for circular polarization. The axial ratios of the circularly polarized antennas are less than 3 dB over a wide sector around the zenith. (The axial ratio does not become infinite at the horizon because the ground plane is relatively small.)

The polarization changes can be explained by considering the frequency-tuning effects described above. The square patch without shorting posts supports both x-oriented and y-oriented modes, which have the same resonant frequency. Since the feed probe is located on the diagonal of the patch, both the x- and y-oriented modes are excited with equal amplitude and phase. By adding shorting posts along the centerline $x = a/2$ (Fig. 4), the resonant frequency of the y-oriented mode can be raised without affecting the x-oriented mode. Similarly, by adding posts along $y = a/2$, the resonant frequenty of the x-oriented mode can be raised without affecting the y-oriented mode. Therefore a single mode (x or y oriented) may be selected by shifting the resonant frequency of the undesired mode far above that of the desired mode. This large frequency shift is obtained by placing shorting posts at or near the edges of the patch. The result is linear polarization.

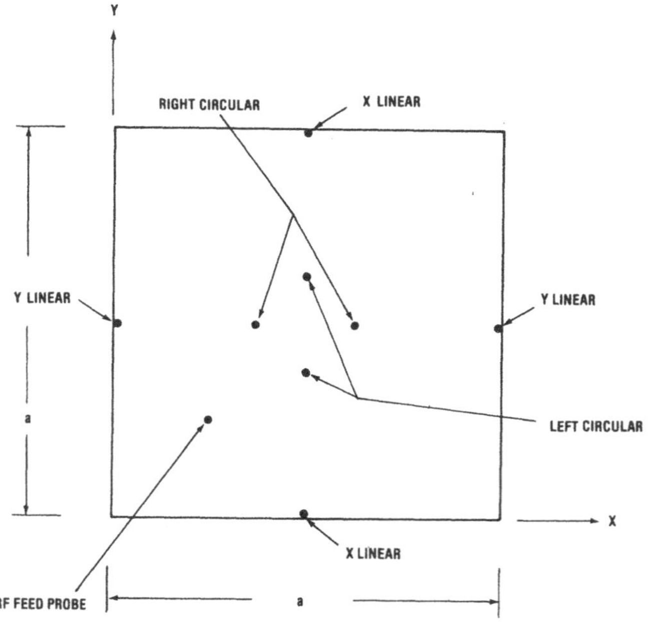

Fig. 4. Square patch antenna with four pairs of posts for obtaining four different polarizations.

Circular polarization may be obtained by exciting both the x- and y-oriented modes with equal amplitudes, but with 90° phase difference. This can be accomplished by raising the resonant frequency of one mode slightly above the other and operating at a frequency between the two resonances. Then the input impedance of one mode is inductive and the other mode is capacitive. By adjusting the difference between the resonant frequencies, both modes can be excited with equal amplitudes and 90° phase difference.

Fig. 6 shows the measured axial ratio of a typical antenna as the separation between a pair of shorting posts is changed. When $s/a = 0$, the posts are at the center and they do not affect either mode. In this case the antenna is linearly polarized along the diagonal with the feed. When $s/a = 0.09$, the resonant frequencies of the two modes are offset enough to obtain a phase difference of approximately 90° and the antenna is circularly polarized. As the posts are moved further apart, the resonant frequency of the vertical mode is further increased and the antenna's polarization becomes horizontal linear (see the drawing in Fig. 6). The input impedance of the antenna changes as the posts are moved, but the VSWR remains very good for all senses of polarization. (Although the best circular polarization occurs over a narrow band of frequencies slightly above the resonance for the linear polarization, the bandwidth of the linearly polarized antenna is adequate to permit it to operate at the same frequency as the circularly polarized antenna.)

Experimental versions of the circularly polarized antenna have been built by using microwave switching diodes instead of machine screws for shorting posts. These antennas verified that diode tuning can be used for precise control of the frequency and polarization. However, it may be necessary to slightly adjust the diode locations to account for the parasitic effects inherent in the diodes.

IV. APPLICATIONS AND OTHER CONFIGURATIONS

The frequency-agile microstrip antenna has many potential applications. By using microwave switching diodes for shorting

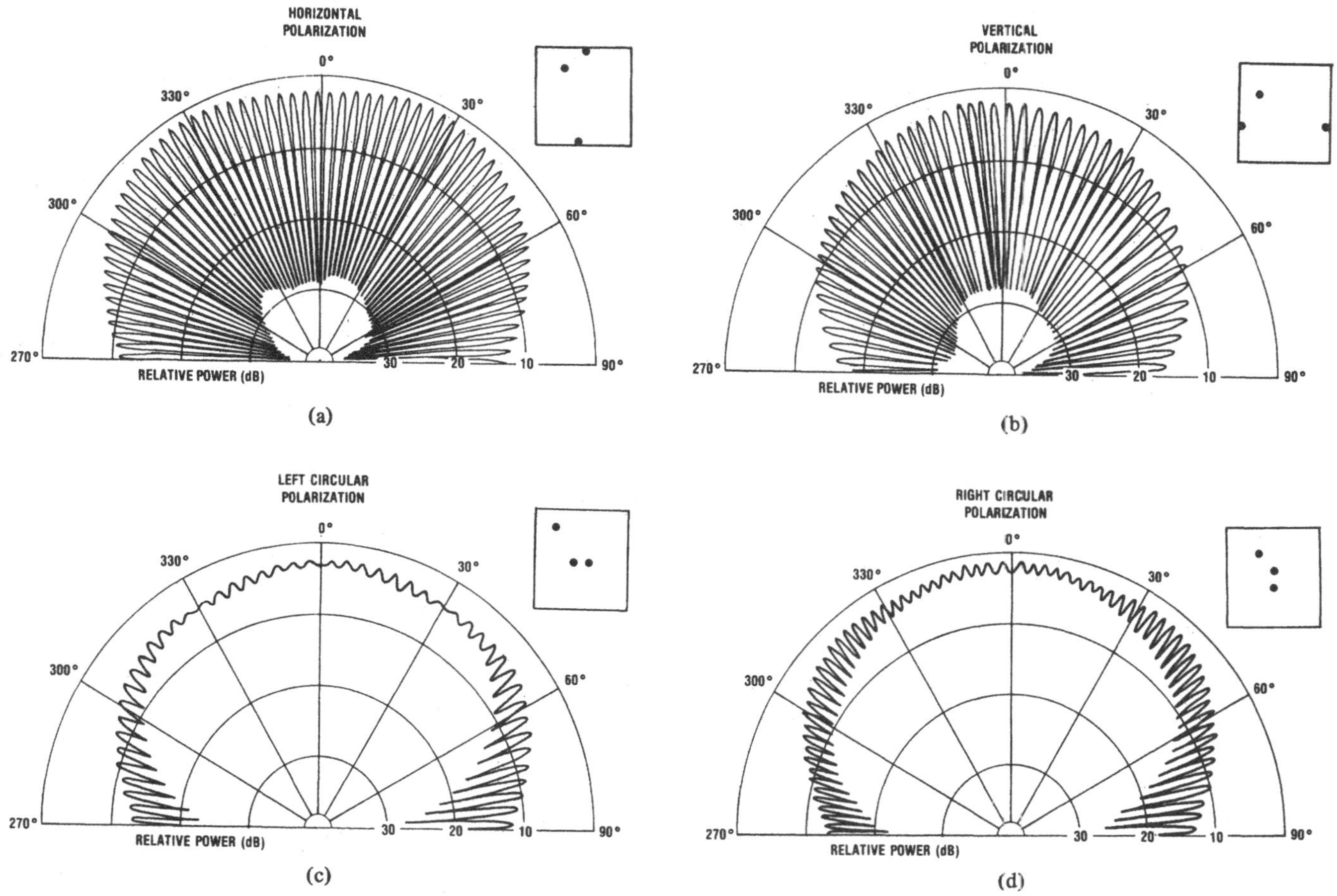

Fig. 5. Typical spin-linear radiation patterns of 6.2-cm square patch antenna at 1470 MHz. Substrate and ground plane are 22-cm square.

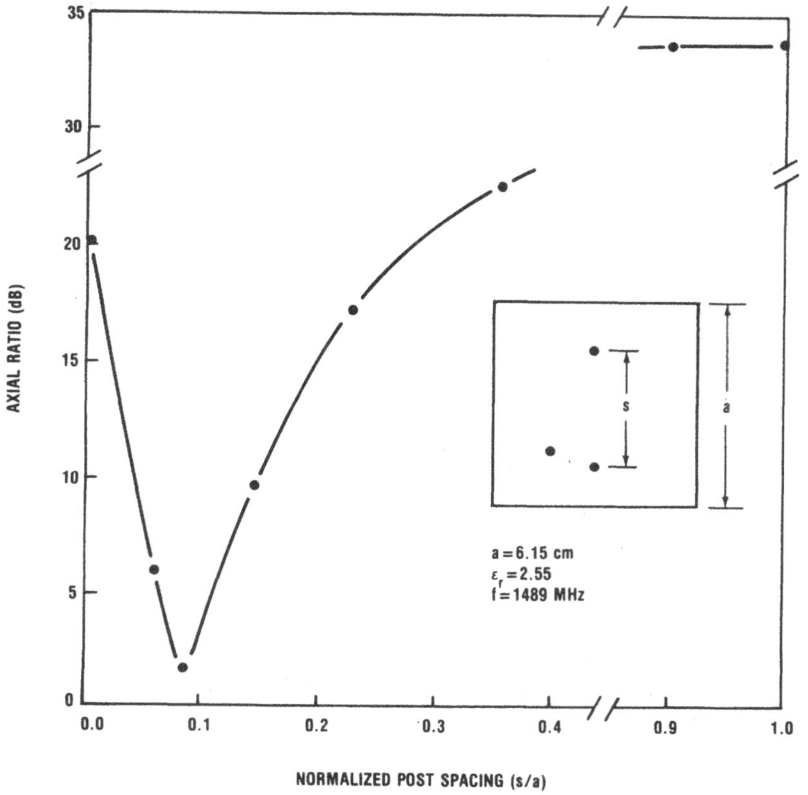

Fig. 6. Measured axial ratio of 6.15-cm square patch antenna with a pair of symmetrically located posts. 1.6-mm Teflon fiberglass substrate.

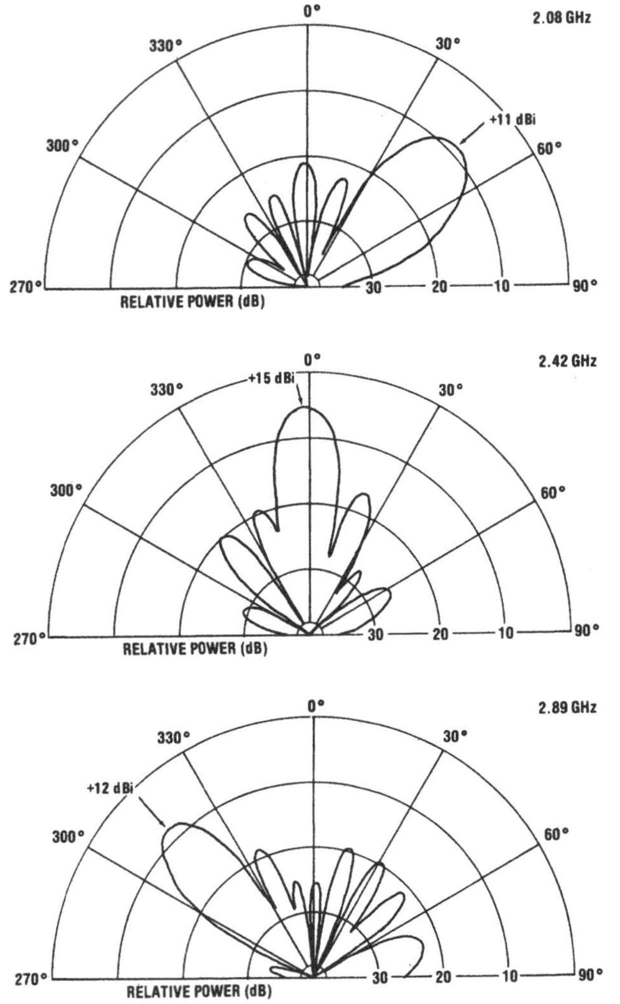

Fig. 7. Typical *E*-plane radiation patterns of linear eight-element frequency-scanned array with uniform amplitude distribution. Array gain is 11 dBi at 2.08 GHz, 15 dBi at 2.42 GHz, and 12 dBi at 2.89 GHz.

posts, a thin conformal communication or radar antenna can be fabricated with the ability to track the tuning of the transmitter or receiver. The antenna remains well-matched at each operating frequency and provides bandpass filtering of the transmitted and received signals. This type of frequency agility provides added flexibility in avoiding interfering signals.

Thin conformal frequency-scanned arrays can also be built with the frequency-agile microstrip antenna. An eight-element linear array consisting of 4.32 × 6.27-cm patches has been fabricated on a 1.6-mm (1/16-in) thick Teflon fiberglass substrate. This experimental array was tuned with small machine screws inserted into holes in the antennas. A corporate feed network that provides progressive phase shift was used to create an antenna that scans ±45° from broadside as the frequency varies from 2.08 to 2.89 GHz (Fig. 7). This type of array with diode tuning posts would perform well in a computer-controlled system that simultaneously increments the operating frequency of the antenna and the transmitter/receiver.

Simple low-power polarization-diverse antennas have been fabricated as shown in Fig. 8(a). A positive or negative bias voltage may be inserted through the RF feed line to select one of the two polarizations.

Other configurations of the microstrip antenna that provide polarization diversity include circular patches and square patches fed along a center line. The centerline-fed square patches (Fig. 8(b)) provide one linear polarization (no posts) and both circular polarizations by means of shorting posts located along the diagonals of the patch. The addition of a second feed (Fig. 8(c)) permits simultaneous operation in two polarizations. The isolation between the two linear polarizations is greater than 30 dB, and the isolation between the two circular polarizations is greater than 20 dB.

V. CONCLUSION

The operating frequency and polarization of microstrip antennas can be conveniently controlled by inserting shorting posts at appropriate locations within the antenna's boundary. By using microwave switching diodes, an electronically controlled frequency-agile or polarization-diverse antenna can be obtained.

The operating frequency of a rectangular microstrip antenna can be tuned over a 1.5-to-1 range without changing its size or the feed location. The tuning is accomplished by varying the number and locations of the shorting posts. The radiation patterns of the microstrip elements do not change significantly as the operating frequency is varied.

Most of the temperature drift and bias control problems that are encountered when varactors are used to electronically tune the antennas are eliminated by using the microwave switching diodes. Also, since the frequency depends on post locations and not on variable reactances, it should be easier to insure that all elements of an array are resonant at the same frequency. The switching diodes are also capable of operating in the high-power environment encountered in transmitting antennas.

The polarization of square and circular microstrip antennas can be varied by changing the locations of shorting posts. Very narrow bandwidth circular polarization of either sense can be obtained, as well as horizontal or vertical linear polarization. The axial ratio of the circularly polarized antenna is less than 3 dB over a wide portion of the beam.

In addition to providing active control of the frequency and polarization of a microstrip antenna, these post-tuning techniques can ease the burdens on designing and manufacturing the antennas. In order to obtain the desired performance, the antenna's precise operating frequency and polarization can be altered by inserting shorting posts during manufacturing or prior to use.

The frequency agility and polarization diversity provide added versatility to the microstrip antenna. Furthermore, these features are obtained without sacrificing the thin conformal structure of the microstrip antenna and without increasing the complexity of the external microwave feed network.

REFERENCES

[1] R. E. Munson, "Conformal microstrip antennas and microstrip phased arrays," *IEEE Trans. Antennas Propagat.*, vol. AP-22, pp. 74–78, Jan. 1974.

[2] J. S. Yee and W. J. Furlong, "An extremely lightweight elec-

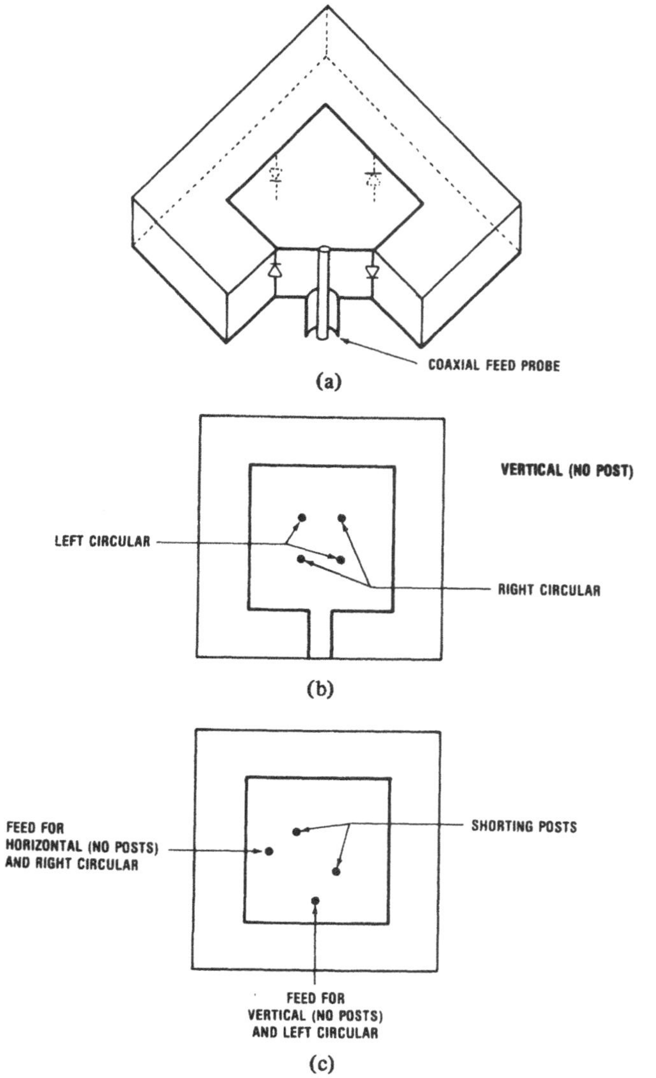

Fig. 8. Some configurations for polarization diversity. (a) Diagonal-fed two-polarization antenna with bias voltage inserted through RF feed. (b) Centerline-fed patch for three polarizations. (c) Dual-feed antenna for horizontal and vertical or left and right circular polarizations.

tronically steerable microstrip phased array antenna," in *IEEE Antennas Propagat. Soc. Int. Symp. Digest*, pp. 170–173, May 1978.

[3] R. E. Munson and G. G. Sanford, "Conformal microstrip antenna arrays," in *Proc. 1977 Antenna Applications Symp.*, Univ. Illinois, Apr. 1977.

[4] J. L. Kerr, "Microstrip antenna developments," in *Proc. Workshop on Printed Circuit Antenna Technol.*, New Mexico State Univ., Las Cruces, Oct. 1979, pp. 3-1 to 3-20.

[5] S. W. Bartley and D. A. Heubner, "A dual beam low sidelobe microstrip array," in *IEEE Antennas Propagat. Soc. Int. Symp. Digest*, pp. 130–133, June 1979.

[6] L. R. Murphy, "SEASAT and SIR-A microstrip antennas," in *Proc. Workshop on Microstrip Antenna Technol.*, New Mexico State Univ., Las Cruces, Oct. 1979, pp. 18-1 to 18-20.

[7] H. D. Weinschel, "A cylindrical array of circularly polarized microstrip antennas," in *IEEE Antennas Propagat. Soc. Int. Symp. Digest*, June 1975, pp. 177–180.

[8] J. L. Kerr, "Microstrip polarization techniques," in *Proc. 1978 Antenna Applications Symp.*, Univ. Illinois, Sept. 1978.

[9] ——, "Other microstrip antenna applications," in *Proc. 1977 Antenna Applications Symp.*, Univ. Illinois, Apr. 1977.

[10] W. F. Richards, Y. T. Lo, P. Simon, and D. D. Harrison, "Theory and applications for microstrip antennas," in *Proc. Workshop on Printed Circuit Antenna Technol.*, New Mexico State Univ., Las Cruces, Oct. 1979, pp. 8-1 to 8-23.

[11] J. L. Kerr, "Terminated microstrip antenna," in *Proc. 1978 Antenna Applications Symp.*, Univ. Illinois, Sept. 1978.

[12] C. S. Malagisi, "Electronically scanned microstrip antenna array," U.S. Patent No. 4 045 895, Oct. 11, 1977.

[13] N. P. Kernweis and J. McIlvenna, "Microstrip antenna elements for hemispherically scanned arrays," Rome Air Development Center Rep. RADC-TR-79-43, Feb. 1979.

[14] Y. T. Lo, D. Solomon, and W. F. Richards, "Theory and experiment on microstrip antennas," *IEEE Trans. Antennas Propagat.*, vol. AP-27, pp. 137–145, Mar. 1979.

[15] A. G. Derneryd, "Linearly polarized microstrip antennas," *IEEE Trans. Antennas Propagat.*, vol. AP-24, pp. 846–851, Nov. 1976.

[16] E. O. Hammerstad, "Equations for microstrip circuit design," in *Proc. 5th European Microwave Conf.*, Sept. 1975, pp. 268–272.

[17] R. F. Harrington, *Time-Harmonic Electromagnetic Fields*. New York: McGraw-Hill, 1961, p. 183.

[18] K. R. Carver and E. L. Coffey, "Theoretical investigation of the microstrip antenna," New Mexico State Univ., Las Cruces, Phys. Sci. Lab. Tech. Rep. PT-00929, Jan. 1979, prepared for US Army Research Office under Grant DAAG29-78-G-0082.

[19] D. H. Schaubert and F. G. Farrar, "Microstrip antenna with polarization diversity," in *Program and Abstracts of Nat. Radio Sci. Meeting*, USNC/URSI, Nov. 1979, p. 139.

Chapter 4
Techniques for Improving Element Bandwidth

PROBABLY the most serious limitation of the basic microstrip antenna element is its narrow impedance bandwidth, so it is not surprising that bandwidth enhancement techniques are so prevalent in the literature [1]. In this chapter we give an overview of some of the most practical methods for bandwidth improvement, with an assessment of the performance and costs that can be expected from these methods. We begin with the review paper by Pozar, written specially for this reprint volume. This paper first discusses several general issues related to bandwidth, including definitions of impedance bandwidth, pattern bandwidth, and gain bandwidth, as well as fundamental limitations on the bandwidth of the microstrip antenna element. The paper then categorizes bandwidth enhancement methods into three basic types: impedance matching, the use of multiple resonances, and the use of lossy materials. The remaining papers in this chapter can be grouped into these categories as well.

Although the impedance bandwidth of the microstrip antenna element can be a serious problem, application in an array environment can adversely affect pattern and gain bandwidth, as well as the impedance bandwidth. This is especially true in the case of series feeds, which are often used due to their simplicity and compactness. There are very few papers [2] that specifically address these issues, but the review article by Pozar provides some material on this topic, as do some of the array design articles in Chapter 6. Similarly, axial ratio bandwidth for circularly polarized elements is discussed in Pozar's review article, but the reader can find further papers on this topic in Chapter 3.

The narrow impedance bandwidth of the basic microstrip element is ultimately a consequence of its electrically thin ground-plane-backed dielectric substrate, which leads to a high-Q resonance behavior. Bandwidth improves as the substrate thickness is increased, or as the dielectric constant is reduced, but these trends are limited by an inductive impedance offset that increases with thickness. A logical approach, therefore, is to use a thick substrate with some type of additional impedance matching to cancel this inductance. For example, the paper by Pues and Van de Capelle takes a fairly general view of the microstrip element by treating it in terms of the Bode-Fano criteria, with an example showing that a properly designed matching network can increase impedance bandwidth to about 10–12%; the paper by Hall directly addresses the problem of inductive shift with increasing substrate thickness by introducing a series capacitor in the probe feed circuit. Several variations on this idea have been published [1], [3], with reported bandwidths of up to 30%. A useful extension of this method can be achieved with a microstrip patch proximity coupled to a microstrip line, as this geometry has an intrinsic series capacitance that can provide a built-in matching mechanism [4]. The article by Pozar and Kaufman describes a simple and practical version of such an element, having a bandwidth of 13%.

Besides impedance matching, another very popular bandwidth extension technique involves the use of two or more stagger-tuned resonators, implemented with stacked patches, parasitic patches, or a combination of dissimilar elements (as in the case of an aperture coupled element). The stacked patch arrangement is very popular, with reported bandwidths ranging from 10 to 20%. Early results for this geometry were reported in [5] and [6], but since then a considerable amount of design and analysis work has been done by K.F. Lee and his colleagues. The paper by Lee, Lee, and Bobinchak in this chapter is representative, and contains some useful design data for stacked patches. It is also possible to use parasitically coupled coplanar patches for stagger tuning [7], but it should be realized that this approach does not have a fixed phase center with frequency, and the effectively increased element size can be problematic for array applications.

Aperture coupling introduces a new degree of freedom—the slot size—that can be used for enhanced bandwidth. One of the most successful examples of this idea is described in the paper by Zurcher, where a foam antenna substrate was used at X-band to achieve a bandwidth in excess of 20%. In this design, the slot is close to resonance, which has the effect of increasing the back radiation level. It is also possible to use aperture coupling with stacked patches [8], as described here in the paper by Croq and Pozar, where bandwidths in excess of 20% at K-band were demonstrated. Parasitically coupled coplanar resonators can also be implemented with aperture coupling [9].

An extension of the notion of using multiply tuned elements to increase bandwidth is the log-periodic array. In its microstrip form, the patches of a log periodic array are fed from a main feed line, and vary in size so that different sections of the array become resonant at different frequencies. Since only a few patches of the array radiate significant power at any frequency, the gain is usually limited to about 8–10 dB. Strictly speaking, a log-periodic array should be an endfire array, but this is not possible with elements printed on a ground plane, so log-periodic microstrip arrays usually have their beam scanned somewhere between endfire and broadside. This type of antenna can be considered to be a broadband element, and may be useful where electrical size is not critical. A very thorough design study of various types of log-periodic microstrip arrays is presented in this chapter's final paper by Hall.

References

[1] A. Henderson, J. R. James, and C. M. Hall, "Bandwidth extension techniques in printed conformal antennas," *Military Microwaves*, MM 86, Brighton, England, pp. 329–334, June 1986.

[2] D. M. Pozar and D. H. Schaubert, "Comparison of three series-fed microstrip array geometries," *IEEE Int'l Symp. on Antennas and Propagation Digest*, pp. 728–731, June 1993.

[3] F. S. Fong, H. F. Pues, and M. J. Withers, "Wideband multilayer coaxial-fed microstrip antenna element," *Electronics Letters*, vol. 21, pp. 497–499, 1985.

[4] P. S. Hall, C. Wood, and C. Garrett, "Wide bandwidth microstrip antennas for circuit integration," *Electronics Letters*, vol. 15, pp. 458–460, 1979.

[5] A. Sabban, "A new broadband stacked two-layer microstrip antenna," *IEEE Antenna and Propagation Symp. Digest*, pp. 63–66, May 1983.

[6] C. H. Chen, A. Tulintseff, and R. M. Sorbello, "Broadband microstrip antenna," *IEEE Antennas and Propagation Symp. Digest*, pp. 251–254, June 1984.

[7] G. Kumar and K. C. Gupta, "Non-radiating edges and four-edges gap-coupled with multiple resonator, broadband microstrip antennas," *IEEE Trans. Antennas and Propagation*, vol. AP-33, pp. 173–178, 1985.

[8] F. Croq and A. Papiernik, "Stacked slot-coupled printed antenna," *IEEE Microwave and Guided Wave Letters*, vol. 1, pp. 288–290, Oct. 1991.

[9] P. A. Miller, J. C. Mackichan, M. R. Staker, and J. S. Dahele, "A wide bandwidth low sidelobe low profile microstrip array antenna for communication applications," *ISAP Proceedings Digest*, pp. 525–528, Aug. 1989.

A Review of Bandwidth Enhancement Techniques for Microstrip Antennas

DAVID M. POZAR
ECE DEPARTMENT
UNIVERSITY OF MASSACHUSETTS AT AMHERST
AMHERST, MA 01003

***Abstract*—Narrow bandwidth has been one of the most serious limitations hindering the wider application of microstrip antenna technology. In the last fifteen years this subject has thus received considerable attention from workers throughout the world, and as a consequence there now exist many practical innovative extensions of the basic element that increase impedance bandwidths to as much as 30%. This paper will first discuss bandwidth issues in general, and then review some of the more useful and practical techniques for microstrip antenna bandwidth enhancement. This discussion will include impedance bandwidth, pattern and gain bandwidth, and axial ratio bandwidth.**

INTRODUCTION

Microstrip antenna elements have a number of useful and interesting features, but probably the most serious limitation of this technology is the narrow bandwidth of the basic element. While competing antenna elements such as dipoles, slots, and waveguide horns have operating bandwidths of 15–50%, the traditional microstrip patch element typically has an impedance bandwidth of only a few percent. For this reason much of the large volume of research and development in the area of microstrip antennas in the last fifteen years has been devoted to various techniques for the enhancement of microstrip antenna bandwidth [1], [2]. Although bandwidth is a dominant topic in the microstrip antenna literature, there are sometimes confusing and misleading conclusions presented due to a lack of clear definitions of bandwidth, and the failure to consider all the relevant electrical characteristics.

This paper will begin by discussing several different types of bandwidth that are relevant to microstrip antenna elements and arrays, and present results for the impedance bandwidth of the basic microstrip element. The inherent limits on impedance bandwidth will be discussed in relation to the Chu-Harrington criteria. Most of the work on microstrip antenna bandwidth enhancement has dealt with impedance bandwidth improvement, and we will categorize this wide variety of work along the lines of three basic techniques, with examples of typical results. We will also discuss the issue of pattern and gain bandwidth in the context of microstrip arrays, and conclude with a study of axial ratio bandwidth degradation for circularly polarized elements.

As a result of the attention and creative energies of both university and industry workers, the bandwidth performance of microstrip antennas has been substantially improved in recent years, and it is now possible to design microstrip elements with impedance bandwidths ranging from 10–30%, or more. This paper will review some of the most significant contributions in this area, and present an evaluation of the costs and benefits of these different techniques. The discussion here will concentrate on microstrip patch elements, and will not cover related printed antenna configurations that may have substantially better bandwidths, such as tapered slot elements and planar spiral antennas.

Bandwidth Definitions

There is no unique definition of antenna bandwidth, since the operating specifications of an antenna may involve a variety of parameters, so it is important to specify the criteria being used when antenna bandwidth is quoted. Several definitions of interest for microstrip antenna work are listed below:

- Impedance Bandwidth: The impedance variation with frequency of the antenna element results in a limitation of the frequency range over which the element can be matched to its feed line. Impedance bandwidth is usually specified in terms of a return loss or maximum SWR (typically less than 2.0 or 1.5) over a frequency range. Conversion of bandwidth from one SWR level to another can be accomplished by using the relation between bandwidth B, and Q:

$$B = \frac{SWR - 1}{Q\sqrt{SWR}} \tag{1}$$

- Pattern Bandwidth: The beamwidths, sidelobe level, and gain of an antenna all vary with frequency. If any of these quantities is specified as a minimum or maximum, the operating frequency range can be determined.
- Polarization or Axial Ratio Bandwidth: The polarization properties (linear or circular) of an antenna are usually preferred to be fixed with frequency. Specifying a maximum cross-pol or axial ratio level can be used to find this bandwidth.

The above bandwidths can be applied to both single elements and arrays of microstrip elements, but the bandwidths for the two cases are typically quite different. For single microstrip antenna elements, the impedance bandwidth is generally the limiting factor; the patterns and directivity of a microstrip element generally vary little with frequency. The same is true for the polarization properties of a linearly polarized element, although the axial ratio of circularly polarized elements may be very narrow-band in some cases. In the case of arrays, the type of feeding network used may either increase or reduce the impedance bandwidth, while the pattern bandwidth is usually less than that of an isolated element due to amplitude and phase errors, and in some cases (such as series-fed resonant arrays) may be quite narrow.

In view of the above discussion it should be apparent that specifying the bandwidth of a microstrip element or array may involve a single consideration, such as impedance, but other characteristics may have to be considered as well. An example of the misrepresentation that can occur when this notion is ignored is the case of an unusually-shaped microstrip antenna element design that was reported several years ago to have an impedance bandwidth in excess of 20%. Closer examination showed that the polarization from this element was linear, but varied in direction over its operating frequency range.

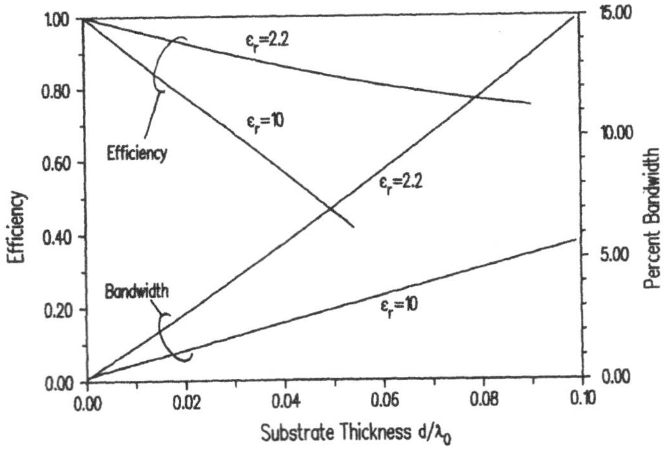

Fig. 1. Impedance bandwidth (SWR<2) and efficiency for a square microstrip antenna element versus substrate thickness.

Impedance Bandwidth of Basic Microstrip Element

The operating range of the basic microstrip antenna element on a thin substrate is generally limited by its narrow impedance bandwidth. The equivalent circuit of a probe- or edge-fed microstrip antenna appears as a parallel RLC resonator with a fairly high Q, primarily because of the electrically thin substrate. Figure 1 shows the bandwidth that is typically obtained with a square element, versus substrate thickness, as computed using a cavity model. Since substrates are usually on the order of 0.01λ to 0.02λ thick, the bandwidth is limited to a few percent. Bandwidth increases monotonically with thickness, but the problem with using a substrate thicker than the above range is that the impedance locus of the element becomes increasingly inductive [3], [4], making impedance matching increasingly difficult. In addition, a thick substrate does not lend itself to the effective use of coplanar microstrip lines for feeding purposes, since spurious radiation from microstripline bends and other discontinuities is unacceptably high for substrates thicker than a few hundredths of a wavelength. Also note from Figure 1 that bandwidth decreases with an increase in substrate dielectric constant. This effect can be explained by the fact that element size decreases with an increase in dielectric constant, which raises the Q of the resonator. This data leads us to the important conclusion that thick substrates with low dielectric constants are preferred for good bandwidth, but this bandwidth will still be relatively narrow for a single microstrip element.

Another disadvantage of using thick high-dielectric-constant substrates is that surface wave excitation will be higher, which will lower efficiency, and may lead to spurious radiation and pattern degradation. The efficiency of a microstrip element versus substrate thickness is also shown in Figure 1, where it is seen that efficiency drops rapidly with increasing substrate thickness and dielectric constant. This efficiency only includes power lost to surface wave generation, and neglects dielectric and conductor losses. (Dielectric loss is generally negligible for all but the thinnest substrates, but conductor loss may become significant at millimeter wave frequencies.) Thick substrates also have the disadvantages of greater weight and higher cost.

Element bandwidth can also be affected by the shape of the element, as illustrated in Table I below, where it is seen that wide rectangular patch elements have slightly higher bandwidths than very narrow rectangular elements. This table shows the approximate impedance bandwidths for three rectangular microstrip elements of varying aspect ratios, and a circular patch element, all on the same substrate with the same resonant frequency. The wider elements have better bandwidths primarily because their radiation resistance is lower, since the radiating edges are larger. The circular element has a bandwidth on the same order as a square patch.

Fundamental Bandwidth Limits

The Chu-Harrington theory concerning the minimum Q of an antenna of a given size [5] is often invoked as an explanation for the narrow bandwidth of the basic microstrip element. This theory gives an explanation for the trends of bandwidth versus size, but in fact does not quantitatively account for the narrow bandwidth of the microstrip element. The Chu-Harrington theory states that the minimum Q of an antenna with 100% radiation efficiency that can be enclosed within a spherical surface of radius R is given approximately by,

TABLE I. BANDWIDTH OF MICROSTRIP ANTENNAS WITH VARIOUS SHAPES

Element Shape	Element Size	Bandwidth (SWR<2)
Narrow Rectangular	L = 4.924 cm W = 2.0 cm	0.7%
Square	L = 4.82 cm W = 4.82 cm	1.3%
Wide Rectangular	L = 4.79 cm W = 7.2 cm	1.6%
Circular	R = 2.78 cm	1.3%

ϵ_r = 2.32, d = 0.159 cm, f = 2.0 GHz

$$Q = \frac{1 + 3(k_0R)^2}{(k_0R)^3\,[1 + (k_0R)^2]} \qquad (2)$$

It is important to consider the volume of the enclosing sphere, and not simply the volume contained below the surface of the patch. As an example, for the circular patch listed in Table I, (2) gives a Q of less than 2, for a bandwidth of more than 50%; the actual bandwidth of this element, however, is less than 2%. The Chu-Harrington criteria gives the *minimum* Q of an antenna enclosed by the spherical volume, but does not preclude this antenna from having a much higher Q (narrower bandwidth), as is the case with microstrip antennas. The narrow bandwidth of the basic microstrip patch is ultimately caused by the electrically thin grounded substrate, not the size of its enclosing volume. There are other antenna elements that can be enclosed by the same volume as a microstrip patch, but with much better bandwidths (e.g., dielectrically loaded dipoles or open-ended waveguides).

Of course, the basic microstrip element does not make very effective use of the entire volume of its enclosing sphere, because of its low-profile geometry. Several of the enhanced bandwidth designs discussed below can be interpreted as either employing the enclosing spherical volume more effectively (e.g., stacked patches), or as increasing the enclosing volume (e.g., using adjacent parasitic elements). Thus, if the grounded substrate feature is to be retained, improved microstrip antenna impedance bandwidth can only come with greater complexity (to make more effective use of the enclosing volume), larger size (which leads to a larger enclosing sphere), or the introduction of loss (which reduces the radiation efficiency).

Methods for Impedance Bandwidth Enhancement

Most of the work in the area of bandwidth enhancement has been directed to improving the impedance bandwidth of the microstrip antenna element, since the narrow bandwidth of the basic element is usually the dominant characteristic that limits its application. There have been dozens of different element designs and variations proposed for bandwidth improvement, and we will review some of those that are most useful. In spite of the wide variety of approaches proposed as a solution to the bandwidth problem, it is possible to categorize them according to three canonical approaches. First, the element can be viewed as a high-Q circuit element and matched over the desired operating band, using a matching network. Next, the general technique of introducing dual (or multiple) resonances can be applied by adding one or more resonant elements that are tuned to slightly different frequencies. Finally, it is always possible to increase impedance bandwidth at the expense of efficiency by introducing loss to the system. Note that these three methods are fairly general, and have previously been applied to many other types of antennas and RF systems; uniqueness is apparent only when these techniques are implemented for specific microstrip antenna geometries.

By Impedance Matching

Probably the most direct way of improving the impedance bandwidth of a microstrip antenna is to attach a separate lossless matching network, without altering the antenna element itself. As shown in Figure 2, this can be done conveniently in microstrip form using a coplanar matching network, or with an off-board network. Tuning stubs, quarter-wave transformers, and active devices can easily be implemented in microstrip form, with little added expense, as long as there is room on the substrate (this may be difficult in planar arrays). Some consideration may also have to be given to spurious radiation from the tuning network, especially if the substrate is not very thin. The overall efficiency and bandwidth performance of the matched antenna is generally best when the network is mounted as close as possible to the radiating element. The amount of bandwidth improvement that can be obtained with this method is ultimately governed by the Bode-Fano criteria [6], but in practice size, complexity, and loss effects generally limit the achievable bandwidth to about 10–30%.

A comprehensive application of this technique has been reported by Pues and Van de Capelle [7], who obtained bandwidths of 10–12% using a passive coplanar matching network. Similar techniques have been applied by Paschen [8] for coverage of both GPS bands with a single microstrip element, for a bandwidth of more than 25%. It is also possible to include transistors in the matching network to combine amplification with the matching function. One example of this is the work by An, et al. [9], where an impedance bandwidth of about 24% was obtained, with an added gain of about 10 dB. Another is the work by Svitak, et al. [10], where a patch radiator and an FET amplifier were combined with an optically fed photodiode; the resulting bandwidth was about 7%.

As noted above, the input impedance of a probe- or microstripline-fed microstrip antenna becomes increasingly

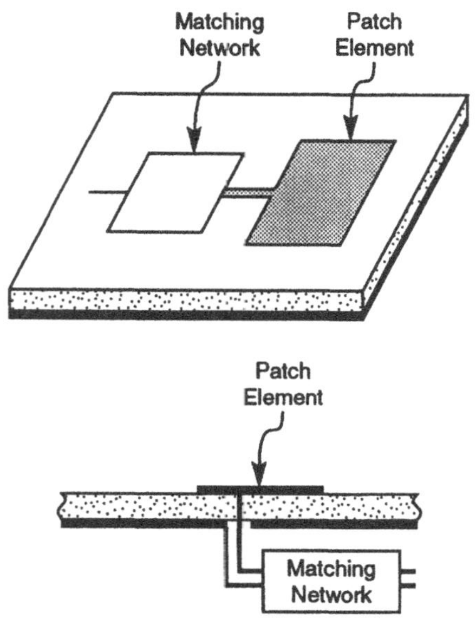

Fig. 2. Bandwidth improvement by impedance matching: (a) using an on-board matching network, (b) using an off-board matching network.

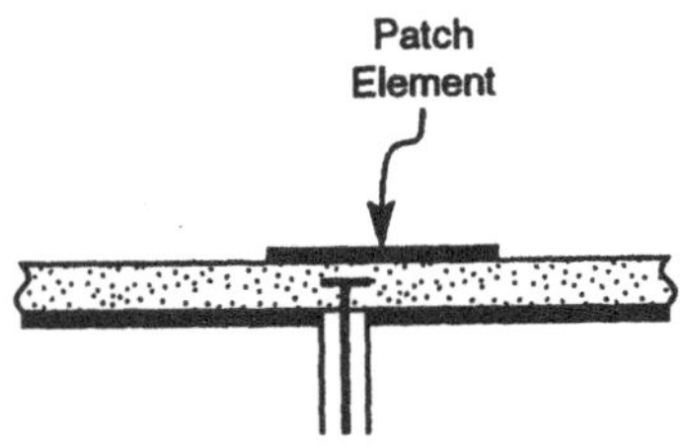

Fig. 3. Bandwidth improvement by tuning out probe inductance with a series capacitor in the feed probe.

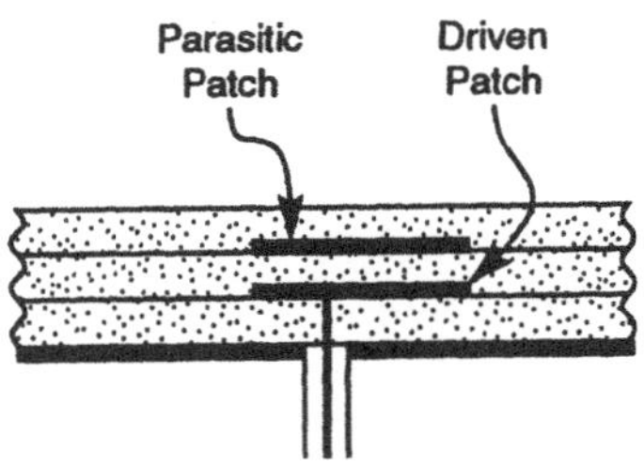

Fig. 5. Bandwidth improvement using dual resonances obtained with stacked patches.

inductive as the substrate thickness increases [3],[4], so an obvious approach to bandwidth improvement is to tune out this inductance with a series capacitor. One way of implementing this technique is shown in Figure 3, where the end of the coax feed probe is formed into a tab that does not directly contact the patch element [11]. This arrangement forms a series capacitor that can be controlled by the size of the tab and the spacing from the patch. In practice, this gap can be fixed by using a separate thin dielectric layer, but the design remains very sensitive to fabrication tolerances. Another approach, suggested by Hall [12], is to feed the patch with a coax probe in the usual manner, but with a circular or linear gap in the patch conductor around the feed point. This gap, however, must be very narrow to obtain sufficient coupling to the patch, so fabrication is again problematic. Bandwidths of up to 30% have been obtained with these techniques, however.

A related method is to use a two-layer proximity-coupled design [13], as shown in Figure 4. In this case a microstrip feed line is placed on the lower substrate, and terminated in an open-end at a point approximately below the midpoint of the patch element, which is fabricated on a superstrate layer. This forms a fairly tight capacitive coupling to the patch, resulting in an equivalent circuit with a capacitor in series with the parallel RLC resonator. Performance can be further improved by including a small tuning stub on the feed line, with a resulting bandwidth of 13%. This type of element has been integrated with an MMIC module for a high-volume smart munitions application.

Using Multiple Resonances

A proven bandwidth broadening technique, borrowed from tuned electronic amplifier design, is to stagger-tune two or more resonators to cover the frequency range of interest. This approach has been applied to several types of antennas over the years (e.g., the sleeve dipole), and it is possible to employ this basic concept in a variety of ways to increase the impedance bandwidth of microstrip antennas. The basic idea is to introduce additional resonant patches to provide two or more closely spaced resonances. Usually only one element is fed directly, with the other patches being coupled by proximity effects. The impedance locus of such multiply tuned antennas has the characteristic feature of two or more loops enclosing the center of the Smith chart, while a single microstrip resonator typically has one loop enclosing the center of the chart, in the vicinity of its operating range.

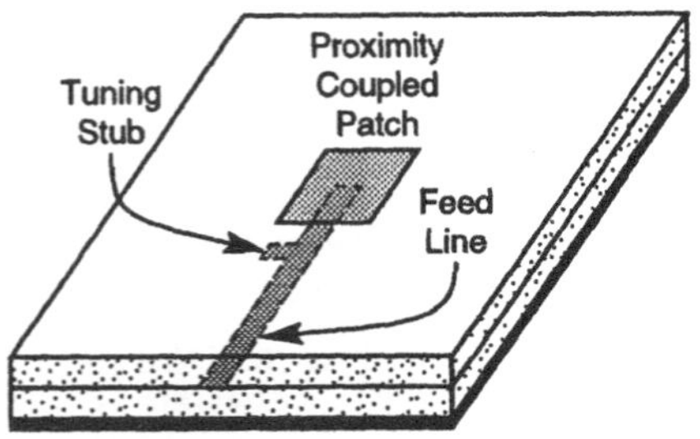

Fig. 4. Bandwidth improvement using a proximity-coupled patch in conjunction with a simple stub tuner.

One of the most practical ways of implementing a doubly tuned microstrip element is to use the stacked patch configuration, as shown in Figure 5. The bottom patch may be fed by coaxial probe or microstrip feed line [14], [19], or by aperture coupling [20], [22]. Impedance bandwidths typically range from 10% to 20% with this approach. In practice, the two patches are usually very close in size, with the top element being slightly larger than the bottom (driven) element. Square, rectangular, or circular patches can be used. If the patches are not very close in size, two distinct resonances will result, which may be useful for dual band performance. Practical implementation often uses two patches with several dielectric layers, with air or other low-dielectric-constant materials, sometimes with a built-in cover layer. It is possible to use three stacked elements, but performance may not be much better than that obtained from a properly optimized two-element design.

The stacked patch design is attractive for several reasons. Since it does not increase the surface area occupied by the element (as does a coplanar matching network, or the parasitic multiple resonator design discussed below), a stacked patch can be used in array configurations without the need for increased element spacings and the concomitant danger of grating lobes. The close proximity of the stacked patch element ensures tight coupling to the fed element, which simplifies design. And the pattern and phase center of the stacked element remains symmetric over its operating band, which is an important consideration for reflector feed or array applications.

The large number of parameters associated with the stacked patch geometry (two patch sizes, two or more substrate thicknesses and dielectric constants, feed position), implies design freedom for optimization, but also makes such optimization very difficult unless a computer solution or model is available. In recent years there has been a considerable amount of work by

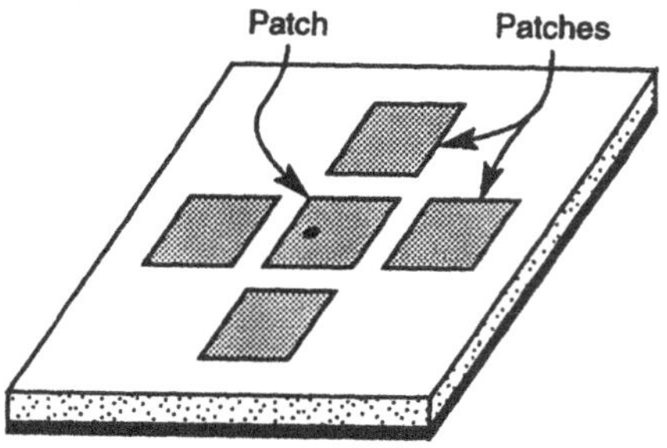

Fig. 6. Bandwidth improvement using multiple resonances obtained with edge-coupled parasitic patches.

several different authors who have applied full-wave Green's function moment method techniques to this problem [22], [26]. Another possible disadvantage of the stacked element is that the multilayer construction may complicate or prevent the use of discrete components, or MICs, in a coplanar arrangement with the feed network.

An alternative to the stacked patch geometry is the coplanar arrangement of a fed element with one or more parasitic patches [27], [28], as shown in Figure 6. Tight coupling requires small gaps between elements, which can cause fabricational problems, but bandwidths up to 25% have been demonstrated with one central-fed patch and four surrounding parasitic patches. This design may have some drawbacks when compared with the stacked patch approach, however. First, because different parts of the configuration radiate with different relative amplitudes and phases at different frequencies, the patterns and phase center usually change markedly over the frequency band of operation, especially for wider bandwidth designs. In addition, even though the geometry is coplanar, the presence of the parasitic elements restricts the placement of coplanar feed lines and integrated components. For this reason, the driven patch is usually fed by coaxial probe, although aperture coupling can be used very conveniently, as demonstrated by the 1280-element planar array reported in [29]. This array used 256 subarrays consisting of a central aperture-coupled element with four surrounding parasitic elements; grating lobes were avoided by using a triangular array grid.

Another aperture-coupled coplanar multiple resonator antenna is shown in Figure 7, where it is seen that the patches have been replaced with several pairs of thin printed-dipole elements [30]. These resonators have varying lengths to create a stagger-tuned effect over the operating band, but each pair is actually excited to some degree by the coupling aperture as well as neighboring dipoles. The work reported in [30] illustrates multifrequency operation at three different frequencies, but broadband operation is also possible. Since this geometry is compact and retains a high degree of symmetry, its patterns and phase center are generally better behaved than designs that used parasitically coupled coplanar patches.

With Reduced Efficiency

It is always possible to increase impedance bandwidth by introducing loss into the antenna system, but at the cost of radiation efficiency. Thus, adding a 6 dB attenuator in series with a microstrip antenna will lead to a minimum of 12 dB return loss over a very broad band, but the antenna gain will be reduced by 6 dB. Besides adding loss external to the antenna element, there are several ways of fabricating a microstrip antenna with integral loss. Lossy substrate materials can be used, or lossy films can be added underneath, or on top of, the conducting patch element. It is also possible to connect discrete chip resistors or similar loads to the patch element, or its feed lines.

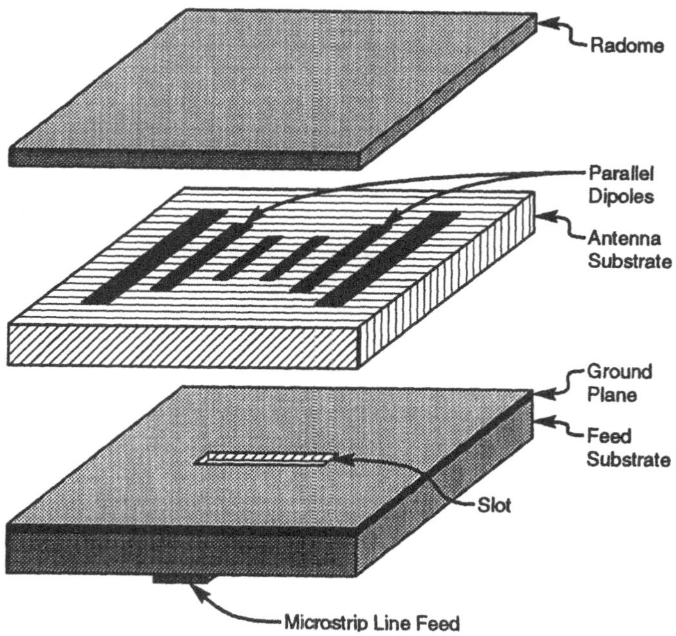

Fig. 7. Bandwidth improvement using multiple resonances obtained with aperture-coupled dipoles of variable size.

In practice, good radiation efficiency is usually desirable, so lossless bandwidth enhancement methods are generally preferable to adding loss. Thus there is very little work to report on the tradeoffs between loss and bandwidth for microstrip antennas, but it is not difficult to estimate the loss in gain.

Considerations for Gain and Pattern Bandwidth

As discussed above, the primary bandwidth limitation of the microstrip element is its impedance bandwidth. Over a wide frequency range the pattern of a microstrip antenna element is relatively constant, and its gain is in the range of 6–7 dB (depending on aspect ratio and substrate dielectric constant). Beyond the operating range of a microstrip element, patterns and gain vary with frequency according to the electrical size of the patch element, in the same manner as any radiating aperture.

Placing elements in an array environment introduces a new frequency dependence, according to the electrical spacing between elements, and the frequency variation of the amplitude and phase of the excitation applied to each element by the feed network. Since there are a multitude of different array feeding schemes, some array configurations may have characteristics that are very frequency sensitive, while others may have bandwidth properties that improve on that of the isolated constituent microstrip elements that make up the array.

Log-periodic and Yagi Arrays

The stagger-tuning concept discussed above meets its culmination in the log-periodic array, where a set of progressively sized elements allow efficient radiation over a very wide bandwidth. The log-periodic array has been used in dipole form for many years, and recently this principle has been applied to broadband linear microstrip arrays [31], [32]. The log-periodic dipole array is an endfire array, but because the microstrip element is a broadside radiator a microstrip log-periodic array is usually designed to radiate at or near broadside. Another point to note about the log-periodic microstrip array is that, at any given frequency, only a portion of array elements actually radiate efficiently. This "active region" usually consists of the two or three elements that are closest to resonance, so that the gain of a log-periodic array is usually much less than that of a conventional array of the same aperture size.

Probably the most complete treatment of log-periodic microstrip arrays has been carried out by Hall [31]. This work illustrates several practical implementations of such arrays, including proximity-coupled elements, direct coupled elements, and equivalent circuits for the feed networks and radiating elements. One example presented in [31] is a 36-element log-periodic array consisting of patch elements proximity-coupled to a microstrip feed line. This array has an operating bandwidth of more than two octaves, with a beam that is about 10 degrees off broadside, and a gain of 10 dB over the band.

Parallel effort in this area has been carried out by Mayes and colleagues [32]. An example from [32] is a ten-element array of patch elements fed with a compound feed consisting of a probe and an aperture. This array was designed to have its main beam about 30 degrees from broadside, and demonstrated a bandwidth of about an octave, with a gain of 2 to 6 dB over this band.

The dipole Yagi array is also an endfire array, and is similar in several respects to the log-periodic array, although its bandwidth is not as large. In microstrip form, the Yagi array must have its main beam scanned up from endfire for efficient radiation, similar to the log-periodic microstrip array. The microstrip Yagi has not received as much development as the log-periodic microstrip array, but Huang has developed a very practical design for mobile satellite applications [33]. This array uses one fed element, with two director elements and one reflector element, and operates over two L-band frequency bands with circular polarization.

Corporate-fed and Series-fed Arrays

A corporate feed network supplies excitation individually to each array element. Most corporate feed networks use equal line lengths and power dividers for each element, so the amplitude and phase tracking with frequency is usually quite good. The commonly used binary microstrip feed network is an example of such a corporate feed. The result is that a corporate-fed array will usually have good pattern and gain bandwidth, but the impedance bandwidth of the overall array will be approximately limited to that of a single patch element; if losses are significant, the bandwidth may appear larger, but of course this is at the expense of efficiency. The above element impedance bandwidth enhancement methods may be applicable if broader array bandwidth is needed.

Series-fed arrays have the advantages of being less complex and requiring less substrate real estate, and often have lower loss, compared to corporate feeds, but amplitude and phase tracking with frequency can be more problematic. Multiple reflections and interactions between elements can also cause both impedance and pattern bandwidth limitations. There are, however, many different types of series-array feeds, so it is not possible to generalize about the achievable characteristics. In fact, series-fed microstrip arrays offer the designer many degrees of freedom that can be exploited to obtain practical designs having very good performance [34]. The literature contains more than one instance of erroneously concluding that the well-known bandwidth limitations of series-fed waveguide slot arrays also apply to microstrip series-fed arrays, but this has been shown to be false. In particular, it has been stated that the impedance bandwidth of any series-fed array decreases with an increase in size, but in fact it is sometimes possible to actually have the bandwidth improve with an increase in size.

Figure 8 shows several possible series-fed linear microstrip array designs. Most of these configurations can be used for either traveling-wave designs, or for standing-wave (resonant) designs. In a traveling-wave array the main beam usually is scanned off broadside, and will have a beam position that changes with frequency, but it is possible to use two back-to-back traveling-wave arrays to achieve a broadside beam over the range of operation. Standing-wave arrays are usually broadside, and often made in two sections with a center feed point. We will direct most of our discussion here to the case of standing-wave arrays.

The main problem with standing-wave arrays is that correct phase and amplitude excitation at each element requires that the branch tap points on the main feed line be λ_g apart. For an array with a large number of elements, the phase at the end of the main feed line changes very rapidly with frequency, relative to the phase at a point near the beginning of the line. This is true for either the two-port patch design of Figure 8a, or the standard series-fed linear array using one-port patches shown in Figure 8b. The result is that the impedance, pattern, and gain bandwidths of such arrays may decrease with array size, and the array bandwidth can be much less than that of a single element for even small arrays. For example, a 16-element array consisting of two eight-element sections of the type shown in Figure 8b was reported in [34] to have an impedance bandwidth of less than 2%, and a pattern bandwidth (defined as the band over which sidelobes remained below 13 dB) of 2.3%. It is possible, however, to vary parameters such as the characteristic impedances of the main or branch lines, and to use quarter-wave transformers, to improve at least the impedance bandwidth. But the complexity of this type of optimization requires CAD modeling for the

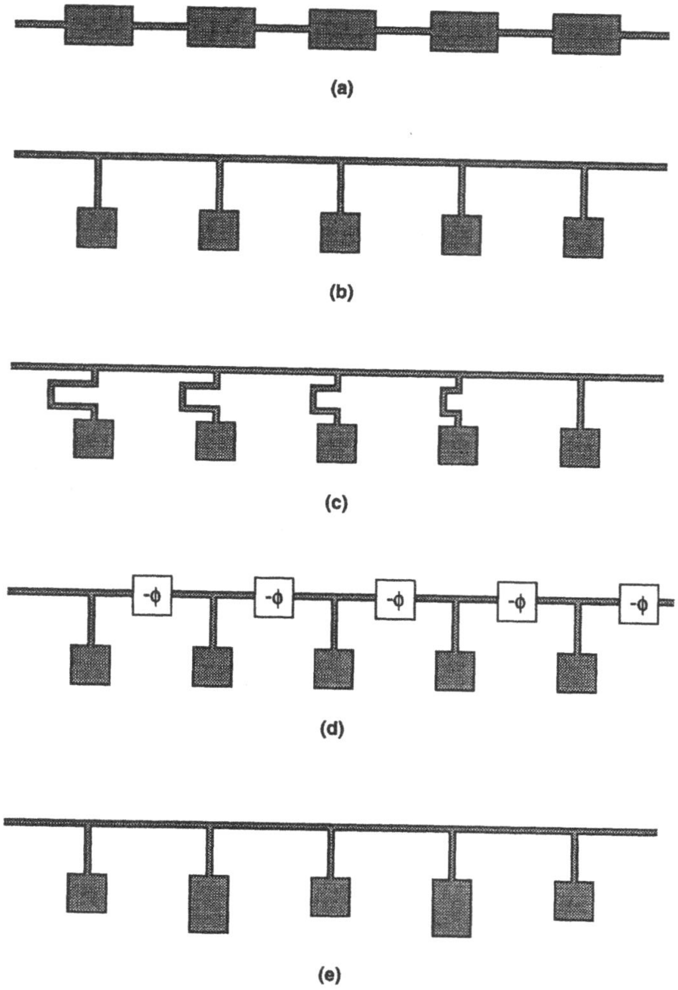

Fig. 8. Series-fed linear array designs: (a) using two-port patches, (b) standard feed network with one-port patches, (c) using phase delay compensation, (d) using phase equalization networks, (e) using alternating patch sizes.

particular array, and it is therefore difficult to make any sort of generalizations about the performance of arrays of this type.

One obvious way to design around this problem is to incorporate integer multiples of λ_g in the branch lines from the main feeder to the patches, so the phase delay from the input point to all the elements is equal. Such phase delay compensation is schematically shown in Figure 8c, but in practice it is difficult to run the necessary line lengths in the available space, especially for planar arrays. The longer lines also increase losses, especially at millimeter wave frequencies. As a compromise, some combination of subarraying, corporate feeding, and phase delay compensation can be used to improve performance over that of the standard array design, but the advantages of simplicity of design and layout, and low loss, are forfeited with this scheme.

Another approach is to employ some type of passive network to equalize the phase response of the main feed line, as conceptually shown in Figure 8d. Such a network can be designed to have a negative phase response over a narrow frequency band using microstrip line sections and stubs, and can improve both impedance and pattern bandwidths. Again, CAD modeling and optimization is required for each particular array design. An example reported in [34] for a 16-element array of two eight-element sections demonstrated an impedance bandwidth of about 4%, and a pattern bandwidth (sidelobes less than 13 dB) of about 12%.

One final design idea is shown in Figure 8e, where the layout is similar to a standard series-fed array, but the patches are made with alternating sizes, to introduce a staggered-tuning effect around the nominal center frequency. This type of array differs from a log-periodic design in that the amount of detuning is only on the order of 5% or less, so that the entire array radiates effectively over the operating band. There are many parameters and variations associated with this type of design, but recent work indicates that this technique offers an easy way to increase the bandwidth of a series-fed array by approximately a factor of two. More work in this area may lead to improved performance for series-fed microstrip arrays.

Considerations for Axial Ratio Bandwidth

The bandwidth over which good circular polarization can be obtained might be considered under the heading of pattern bandwidth, but there are enough special considerations that occur with this issue that a separate discussion is justified. We will begin by listing four distinct causes of axial ratio degradation that can occur for any type of circularly polarized antenna. Then we will discuss two specific microstrip designs that are commonly used for circular polarization, and offer some general remarks about their axial ratio performance. More discussion of circularly polarized microstrip antenna design can be found in the review paper by Hall in Chapter 3 of this book.

Causes of Axial Ratio Degradation

Many (but not all) circularly polarized radiators can be viewed in terms of a superposition of two orthogonal linearly polarized radiation components in phase quadrature [35]. These two components may be radiated by two separate antenna elements (e.g., two crossed dipoles), or by two modes of a single element (e.g., orthogonal modes in a square waveguide). In either case, four distinct sources of error that degrade the circular polarization axial ratio have been identified [35], [36]:

- *Amplitude error* Perfect circular polarization requires that both linearly polarized components have the same amplitude.
- *Phase error* Perfect circular polarization requires the phase shift between the two linearly polarized components to be ±90 degrees.
- *Orthogonality error* Perfect circular polarization requires the two linearly polarized components to be orthogonal.
- *Polarization error* Perfect circular polarization requires the two orthogonal components to be linearly polarized, with no cross-polarization.

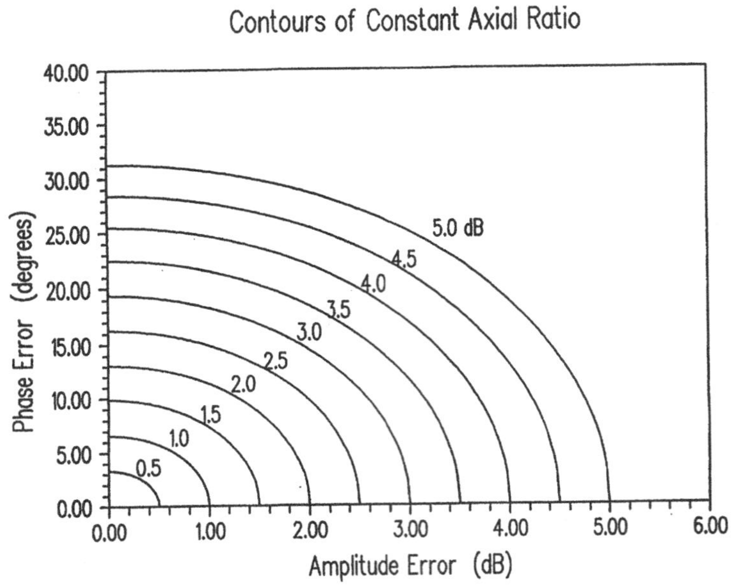

Fig. 9. Contours of constant axial ratio for a circularly polarized antenna, versus amplitude and phase errors.

The first two errors listed above usually depend on the electrical design and characteristics of the antenna, while the last two generally depend on the physical design of the antenna. An approximate equation for the axial ratio in the presence of the first three errors listed above has been derived by Parekh [36]:

$$\text{AR (dB)} = \sqrt{A_e^2 + 0.0225(\phi_e + \beta_e)^2} \tag{3}$$

where A_e is the amplitude error in dB, ϕ_e is the phase error in degrees, and β_e is the orthogonality error in degrees. This assumes both components are linearly polarized; the interested reader should see [36] for the more general case when the two components are elliptically polarized. Figure 9 shows contours of constant axial ratio in the presence of amplitude and phase errors, assuming no orthogonality or polarization errors.

Bandwidth of Single-fed CP Elements

The most conventional and robust way of obtaining circular polarization from a microstrip patch is to use a square or circular element fed on two edges to excite two orthogonal modes, with a power divider or hybrid to obtain equal amplitude excitations in phase quadrature [1]. The axial ratio bandwidth of this approach is generally much better than the impedance bandwidth of the patch itself, being limited primarily by the amplitude and phase tracking of the power divider. The patch geometry ensures that the orthogonality and polarization errors, as defined above, will be negligible (a square patch has a slight advantage in this regard).

It is also possible to excite two orthogonal modes in a patch antenna using a single feed point, and thus obtain circular polarization. This can be done in a wide variety of ways, including the use of a slightly non-square patch, a patch with notched edges, a patch with trimmed corners, and a patch containing a diagonal slot. The principle of operation in all cases, however, is that the two orthogonal modes are made to have resonant frequencies slightly above, and slightly below, the nominal center frequency in order to achieve phase quadrature. Such elements are advantageous because of their simplicity, but the resulting axial ratio bandwidth performance is very poor, typically being 1/6 to 1/12 of the impedance bandwidth of the patch element [2]. The problem is that the stagger-tuned mode effect used to obtain the necessary phase shift is very sensitive to frequency, due to the large input reactance slope of the element. The amplitude error can be made fairly small in most cases by ensuring symmetry with the feed point, and the orthogonality and polarization errors are usually negligible due to the modal characteristics of the element. But the phase error seriously limits the performance of this type of design, especially when fabrication tolerances are considered.

Many workers have addressed the issue of optimizing the performance of single-fed circularly polarized microstrip antennas, of which [37], [38] represent comprehensive studies together with experimental results.

Sequential Rotation

The sequential rotation technique is a method of obtaining circular polarization using a subarray of linearly polarized patches [39], [40]. Typically, four patches are used with single feed points to radiate linear polarizations oriented at 0 degrees, 90 degrees, 180 degrees, and 270 degrees in space, and excited with equal amplitudes and phased at 0 degrees, 90 degrees, 180 degrees, and 270 degrees. Orthogonality and polarization errors can be made very small by the physical placement of the elements, and by cancellation of errors due to symmetry. Amplitude and phase errors are affected mainly by the feeding network. In one of its implementations [39], it is possible for an array of this type to have impedance and axial ratio bandwidths well in excess of the impedance bandwidth of the single element, because of cancellation of reflections between pairs of elements. As an example, a four-element sequentially rotated subarray reported in [39] had an impedance bandwidth (SWR<1.5) of about 14% and a 3 dB axial ratio bandwidth of more than 14%. It is also possible to use circularly polarized elements with this technique.

Conclusions

This paper has reviewed the current state of the art in the area of bandwidth enhancement for microstrip antennas. Methods for impedance bandwidth improvement were reviewed, along with special considerations for the bandwidth of microstrip arrays and circularly polarized microstrip antennas. We see that element bandwidths as high as 20 to 30% can now be achieved in a variety of ways, thus removing one of the most serious limitations of the microstrip antenna element, and promoting the application of microstrip antenna technology to a wider variety of applications.

The problem of bandwidth has been the subject of a large volume of research and development carried out by many workers throughout the world, demonstrating an enormous amount of creativity. The author regrets that he was not able to include references to the entire body of this work.

References

[1] D. M. Pozar, "Microstrip antennas," *IEEE Proceedings*, vol. 80, pp. 79–91, Jan. 1992.

[2] A. Henderson, J. R. James and C. M. Hall, "Bandwidth extension techniques in printed conformal antennas," *Military Microwaves*, MM 86, pp. 329–334, June 1986.

[3] E. Chang, S. A. Long and W. F. Richards, "Experimental investigation of electrically thick rectangular microstrip antennas," *IEEE Trans. Antennas and Propagation*, vol. AP-34, pp. 767–772, June 1986.

[4] D. H. Schaubert, D. M. Pozar, and A. Adrian, "Effect of microstrip antenna substrate thickness and permittivity: comparison of theories and experiment," *IEEE Trans. Antennas and Propagation*, vol. AP-37, pp. 677–682, June 1989.

[5] R. C. Hansen, "Fundamental limitations in antennas," *IEEE Proceedings*, vol. 69, pp.170–182, Feb. 1981.

[6] R. M. Fano, "Theoretical limitations on the broadband matching of arbitrary impedances," *J. of the Franklin Institute*, vol. 249, pp. 57–83 and 139–154, Jan.–Feb. 1950.

[7] H. F. Pues and A. R. Van de Capelle, "An impedance matching technique for increasing the bandwidth of microstrip antennas," *IEEE Trans. Antennas and Propagation*, vol. AP-37, pp. 1345–1354, Nov. 1989.

[8] D. A. Paschen, "Practical examples of integral broadband matching of microstrip antenna elements," *Proceedings of the 1986 Antenna Applications Symp.*, pp. 199–217, 1986.

[9] H. An, B. Nauwelaers, and A. Van de Capelle, "Broadband active microstrip array elements," *Electronics Letters*, vol. 27, pp. 2378–2379, Dec. 5, 1991.

[10] A. J. Svitak, D. M. Pozar, and R. W. Jackson, "Optically fed aperture-coupled microstrip patch antennas," *IEEE Trans. Antennas and Propagation*, vol. 40, pp. 85–90, Jan. 1992.

[11] F. S. Fong, H. F. Pues and M. J. Withers, "Wideband multilayer coaxial-fed microstrip antenna element," *Electronics Letters*, vol. 21, pp. 497–499, 1985.

[12] P. S. Hall, "Probe compensation in thick microstrip patches," *Electronics Letters*, vol. 21, pp. 606–607, May 1987.

[13] D. M. Pozar and B. Kaufman, "Increasing the bandwidth of a microstrip antenna by proximity coupling," *Electronics Letters*, vol. 23, pp. 368–369, Apr. 1987.

[14] P. S. Hall, C. Wood, and C. Garrett, "Wide bandwidth microstrip antennas for circuit integration," *Electronics Letters*, vol. 15, pp. 458–460, 1979.

[15] A. Sabban, "A new broadband stacked two-layer microstrip antenna," *IEEE Antennas and Propagation Symp. Digest*, pp. 63–66, May 1983.

[16] C. H. Chen, A. Tulintseff, and R. M. Sorbello, "Broadband microstrip antenna," *IEEE Antennas and Propagation Symp. Digest*, pp. 251–254, June 1984.

[17] J. S. Dahele, S. H. Tung, and K. F. Lee, "Normal and inverted configurations of the broadband electromagnetic-coupled microstrip antenna," *IEEE Antennas and Propagation Symp. Digest*, pp. 841–844, June 1986.

[18] R. Q. Lee, K. F. Lee and J. Bobinchak, "Characteristics of a two-layer electromagnetically coupled rectangular patch antenna," *Electronics Letters*, vol. 23, pp. 1070–1072, Sept. 1987.

[19] G. Kossiavas and A. Papiernik, "A circularly or linearly polarized broadband microstrip antenna operating in L-band," *Microwave J.*, pp. 266–272, May 1992.

[20] J.-F. Zurcher, "The SSFIP: a global concept for high performance broadband planar antennas," *Electronics Letters*, vol. 24, pp. 1433–1435, Nov. 1988.

[21] F. Croq and A. Papiernik, "Stacked slot-coupled printed antenna," *IEEE Microwave and Guided Wave Letters*, vol. 1, pp. 288–290, Oct. 1991.

[22] F. Croq and D. M. Pozar, "Millimeter wave design of wide-band aperture-coupled stacked microstrip antennas," *IEEE Trans. Antennas and Propagation*, vol. 39, pp. 1770–1776, Dec. 1991.

[23] L. Barlatey, J. R. Mosig, and T. Sphicopoulos, "Analysis of stacked micro-strip patches with a mixed potential integral equation," *IEEE Trans. Antenna and Propagation*, vol. 38, pp. 608–615, May 1990.

[24] A. N. Tulintseff, S. M. Ali, and J. A. Kong, "Input impedance of a probe-fed stacked circular microstrip antenna," *IEEE Trans. Antennas and Propagation*, vol. 39, pp. 381–390, Mar. 1991.

[25] C. Wu, J. Wang, R. Fralich, and J. Litva, "A rigorous analysis of an aperture-coupled stacked microstrip antenna," *Microwave and Optical Technology Letters*, vol. 3, pp. 400–404, Nov. 1990.

[26] J. S. Herd, "Modeling of wideband proximity coupled microstrip array elements," *Electronics Letters*, vol. 26, pp. 1282–1284, Aug. 1990.

[27] C. Wood, "Improved bandwidth of microstrip antennas using parasitic elements," *Proc. IEE*, vol. 127, Part H, pp. 231–234, 1980.

[28] G. Kumar and K. C. Gupta, "Non-radiating edges and four-edges gap-coupled with multiple resonator, broadband microstrip antennas," *IEEE Trans. Antennas and Propagation*, vol. AP-33, pp. 173–178, 1985.

[29] P. A. Miller, J. C. MacKichan, M. R. Staker, and J. S. Dahele, "A wide bandwidth low sidelobe low profile microstrip array antenna for communication applications," *ISAP Proceedings Digest*, pp. 525–528, Aug. 1989.

[30] F. Croq and D. M. Pozar, "Multifrequency operation of microstrip antennas using aperture coupled parallel resonators," *IEEE Trans. Antennas and Propagation*, vol. 40, pp. 1367–1374, Nov. 1992.

[31] P. S. Hall, "Multioctave bandwidth log-periodic microstrip antenna array," *IEE Proc.*, vol. 133, Part H, pp. 127–136, April 1986.

[32] H. K. Smith and P. E. Mayes, "Log-periodic array of dual-feed microstrip patch antennas," *IEEE Trans. Antennas and Propagation*, vol. 39, pp. 1659–1664, Dec. 1991.

[33] J. Huang and A. C. Densmore, "Microstrip Yagi antenna for mobile satellite vehicle application," *IEEE Trans. Antennas and Propagation*, vol. 39, pp. 1024–1030, July 1991.

[34] D. M. Pozar and D. H. Schaubert, "Comparison of three series-fed microstrip array geometries," *IEEE Int'l Symp. on Antennas and Propagation Digest*, pp. 728–731, June 1993.

[35] D. M. Pozar and S. Targonski, "Axial ratio of circularly polarized antennas with amplitude and phase errors," *IEEE Antennas and Propagation Magazine,* vol. 32, pp. 45–46, Oct. 1990.

[36] S. V. Parekh, "Simple formulae for circular-polarization axial-ratio calculations," *IEEE Antennas and Propagation Magazine,* vol. 33, pp. 30–32, Feb. 1991.

[37] P. C. Sharma and K. C. Gupta, "Analysis and optimized design of single feed circularly polarized microstrip antennas," *IEEE Trans. Antennas and Propagation,* vol. AP-31, pp. 949–955, Nov. 1983.

[38] M. Haneishi and S. Yoshida, "A design method of circularly polarized microstrip antennas," *Electronics and Communications in Japan,* vol. 64-B, pp. 46–54, 1981.

[39] T. Teshirogi, M. Tanaka, and W. Chujo, "Wideband circularly polarized array with sequential rotation," *Proc. ISAP,* pp. 117–120, Aug. 1985.

[40] P. S. Hall, J. Huang, E. Rammos and A. Roederer, "Gain of circularly polarized arrays composed of linearly polarized elements," *Electronics Letters,* vol. 25, pp. 124–125, Jan. 1989.

An Impedance-Matching Technique for Increasing the Bandwidth of Microstrip Antennas

HUGO F. PUES, MEMBER, IEEE, AND ANTOINE R. VAN DE CAPELLE, MEMBER, IEEE

***Abstract*—The nature of the inherent narrow bandwidth of conventional microstrip patch antennas is considered. It is observed that, except for single-feed circularly polarized elements, their bandwidth is limited only by the resonant behavior of the input impedance and not by radiation pattern or gain variations, which usually are negligible over a moderate 10 to 20 percent bandwidth. Therefore, broad-band impedance-matching is proposed as a natural solution to increase the bandwidth. The maximum obtainable bandwidth is calculated using Fano's broad-band matching theory. It is found that by using an optimally designed impedance-matching network, the bandwidth can be increased by a factor of at least 3.9, the exact value depending on the degree of matching required. In view of practical realizations, a transmission-line prototype for a proper matching network is developed. Attention is paid to the translation of this prototype network into a practical structure (e.g. a microstrip or stripline circuit). Practical design examples along with experimental results are given which clearly show the validity of the technique.**

I. Introduction

MICROSTRIP ANTENNAS have many interesting properties (e.g., low profile, light weight, cheapness), but their application in many systems is impeded by their inherent narrow bandwidth [1]. Many elements with enhanced bandwidth have already been investigated; e.g., electrically thick elements [2], stacked multipatch, multilayer elements [3], multiple-resonator elements [4], [5]. All these wider band elements, however, are characterized by increased complexity and/or enlarged size of the radiating structure. Mostly, their increased impedance bandwidth is also paid for by poorer radiation characteristics. For example, multiple-resonator elements [4], [5] exhibit frequency-dependent array effects that disturb, more or less, the radiation pattern. Increasing the substrate thickness [2], [3], causes increased excitation of substrate waves [6]. Apart from lowering the radiation efficiency, these substrate waves diffract at the substrate edges and deteriorate the quality of the radiation pattern. Although the excitation of substrate waves can be largely avoided by using substrate materials with very low dielectric constant (i.e., $\epsilon_r \approx 1$), the application of electrically thick antennas only becomes feasible if proper feeding techniques can be developed [1], [3], [7].

In this paper, broad-band impedance-matching [8] is proposed as a method for bandwidth enhancement of microstrip antennas [9], [10]. The method is unique in that it does not alter the radiating element itself. Instead, a reactive matching network is added to compensate for the rapid frequency variations of the input impedance. The validity of the technique is based upon the relative frequency insensitivity of the radiation pattern and gain characteristics as compared to the resonant behavior of the input impedance. This is explained in Section II. In Section III, both the normally obtained bandwidth and the bandwidth that can be obtained using broad-band matching, are calculated. Dividing the latter quantity by the former one, a bandwidth-enlargement factor is found which depends only on the bandwidth criterion expressed as a maximum allowable voltage standing-wave ratio (VSWR). In Section IV, a transmission-line matching-network prototype is derived that can serve as a basis for practical designs. A complete design procedure for an impedance-matched microstrip antenna is outlined in Section V. It is indicated that because of approximations in both the derivation of the prototype and the translation of this prototype to a practical structure, good final designs can be obtained only if proper use is made of computer simulation and optimization. Finally, in Section VI, two practical design examples are described. Both concern S-band microstrip antenna elements: a single substrate rectangular element with a coplanar microstrip matching network, and a square multilayer element with a stripline matching network.

II. Frequency Dependence of Antenna Parameters

An experimental investigation of the frequency dependence of the operational characteristics of common microstrip patch antennas clearly shows that the impedance variations are the dominant bandwidth-limiting factor, whereas the gain (=directivity × radiation efficiency) and radiation pattern variations are almost negligible over a moderate 10 to 20 percent bandwidth. This behavior can be explained easily using the theory of modal expansion in cavities [11] as applied in microstrip antenna cavity analysis models [12]. According to these models, the total input impedance can be written as a sum of modal impedances where each modal impedance behaves as a parallel-resonant circuit. In the same way, the total radiation field can be written as a vector sum of modal radiation fields where each modal field is given as the product of a nearly frequency independent normalized pattern and a frequency dependent scalar excitation coefficient. Thus, it follows that in all cases where only one dominant mode is excited, the input impedance will behave as a parallel-resonant circuit, whereas the (relative) radiation pattern will show almost no frequency variation. Because the operation of single-

Reprinted from *IEEE Trans. Antennas Propaga.*, vol. AP-37, no. 11, pp. 1345–1354, Nov. 1989.

feed circularly polarized (SFCP) microstrip antennas [12], [13] is based upon the simultaneous excitation of two orthogonal modes, the above does not apply for SFCP elements. But in nearly all other cases, there will exist a band of some 10 to 20 percent, where the excitation level of higher order modes is negligible, and the impedance is the only bandwidth-limiting factor. This even applies to microwave scanning arrays [14].

III. Bandwidth-Enlargement Factor

In the vicinity of its fundamental resonant frequency, the input impedance of a microstrip antenna can be modeled by either a series-resonant or a parallel-resonant RLC circuit. Indeed, it suffices to choose a proper reference plane on the feed line (preferably as close as possible to the element) or to devise some very simple impedance-transforming circuit, for such a behavior to occur in a more or less approximate fashion. So, assuming an $\exp(j\omega t)$ time dependence, the input impedance can be written as either

$$Z_{\text{in}} = R_0(1 + jQv) \tag{1}$$

in the series-resonant case, or as

$$Z_{\text{in}} = \frac{R_0}{1 + jQv} \tag{2}$$

in the parallel-resonant case. In these equations R_0 is the resonant resistance, Q is the quality factor and

$$v = \frac{f}{f_r} - \frac{f_r}{f} \tag{3}$$

where f is the frequency variable and f_r the resonant frequency. If the feed line has a characteristic impedance Z_0, the input VSWR is given by

$$\left|\frac{Z_{\text{in}}(f) - Z_0}{Z_{\text{in}}(f) + Z_0}\right| = \frac{\text{VSWR}(f) - 1}{\text{VSWR}(f) + 1}. \tag{4}$$

If the bandwidth criterion is taken to be VSWR $\leq S$, and f_1 and f_2 are the lower and upper band edge frequencies, respectively, so that VSWR(f_1) = VSWR$(f_2) = S$, the bandwidth is given by

$$B = \frac{f_2 - f_1}{f_r}. \tag{5}$$

It follows from (1)–(5) that

$$B = \frac{1}{Q}\sqrt{\frac{(TS-1)(S-T)}{S}} \tag{6}$$

where $T = Z_0/R_0$ in the series-resonant case, and $T = R_0/Z_0$ in the parallel-resonant case. Because, normally, an antenna is designed to be perfectly matched at its resonant frequency (e.g., by properly locating the position of a coxial feed probe or by using a quarter-wavelength transformer), T normally equals unity. Equation (6) then reduces to the well-known expression [12]

$$B|_{T=1} = \frac{1}{Q}\frac{S-1}{\sqrt{S}}. \tag{7}$$

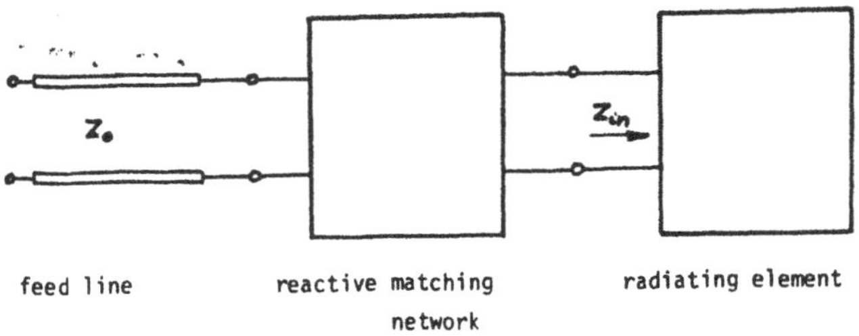

Fig. 1. Principle of broad-band matching.

Note, however, that, in order to maximize B, it would be best to take $T = T_{\text{opt}} \neq 1$ where

$$T_{\text{opt}} = \frac{1}{2}\left(S + \frac{1}{S}\right). \tag{8}$$

The application of (8) turns out to be the most elementary form of broad-band impedance-matching (case $n = 1$ as explained below).

It is evident that the above-calculated bandwidth (7) can be increased, at least in principle, by using an impedance-matching network, as shown in Fig. 1. Ideally, this network would transform the frequency-dependent complex antenna impedance Z_{in} to a pure real resistance Z_0 over as large a bandwidth as required. However, there appear to exist some theoretical limitations on such a transformation which are imposed by nature itself [8]. Indeed, it is impossible to realize a perfect match over a continuous band of frequencies by means of a purely reactive (i.e., linear, passive and lossless) network. The best one can do is to realize a constant (but not perfect) match within the band of operation and a total mismatch outside this band. In that way, one can either optimize the degree of matching if the bandwidth is given *a priori*, or maximize the bandwidth if the degree of matching (e.g., VSWR $\leq S$) is given. The maximum VSWR $= S$ bandwidth obtainable for a series- or parallel-resonant circuit, can be calculated in a straightforward manner using Fano's theory [8], [15]. The result is given by

$$B_m = \frac{1}{Q}\frac{\pi}{\ln\{(S+1)/(S-1)\}}. \tag{9}$$

This equation expresses that the maximum realizable bandwidth is inversely proportional to both the element quality factor and the specified return loss (expressed in dB).

Because (9) represents the optimum that is theoretically achievable using broad-band matching and (7) gives the normally obtained bandwidth, the maximum bandwidth-enlargement factor is found by dividing both quantities:

$$F = \frac{\pi\sqrt{S}}{(S-1)\ln\{(S+1)/(S-1)\}}. \tag{10}$$

Fig. 2 shows this factor which only depends on S and has a minimum value of 3.90 for $S = 2.64$.

IV. Transmission-Line Matching-Network Prototype

For increasing the bandwidth by impedance matching, a proper matching network has to be designed. In this section, a transmission-line matching-network consisting of half-

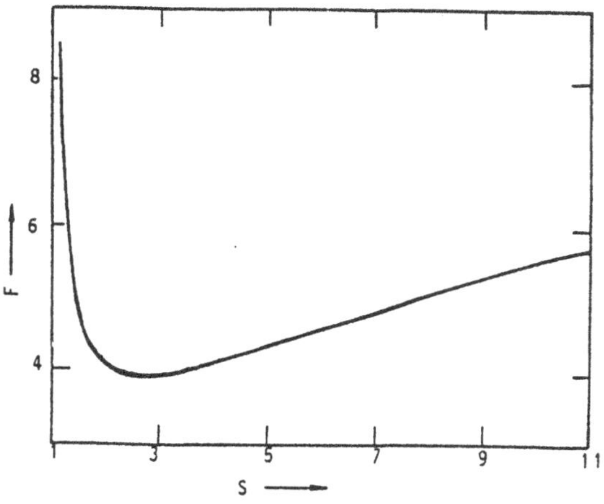

Fig. 2. Bandwidth-enlargement factor versus specified VSWR.

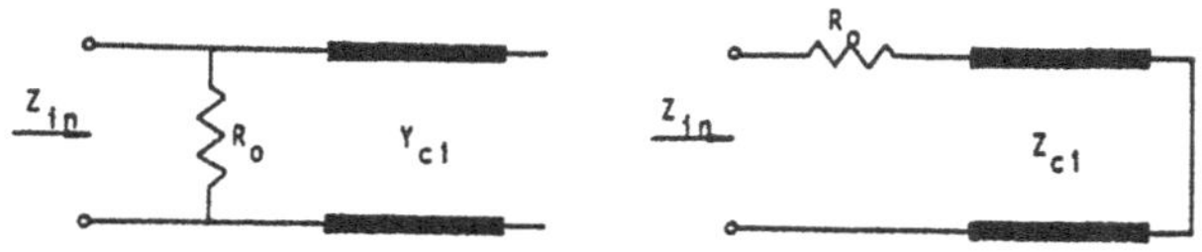

(a) (b)

Fig. 3. Transmission-line models for antenna impedance. (a) Parallel-resonant case. (b) Series-resonant case.

wavelength open-circuited stubs and quarter-wavelength interconnecting lines, is derived that is useful as a prototype for practical realizations at microwave frequencies. This prototype has enough degrees of freedom to ensure practical realizability in microstrip or stripline, if the design bandwidth is not less than about 4 percent. It is clear that other prototypes could be devised depending on the desired practical realization form of the matching network (e.g., quasi-lumped-element prototypes for MMIC realizations or coupled-transmission-line prorotypes for compact interdigital realizations), but such other prototypes are not considered in this paper (except for some short references to lumped-element approaches in this and the following section).

In general, the design of a broad-band matching network is a difficult network synthesis problem. Therefore, published results are used as much as possible in the present derivation. Particularly, the modified Chebyshev equal-ripple characteristic as proposed by Fano [8], is adopted. In [16], normalized low-pass prototype element values for an LC-ladder circuit having this characteristic, are given for the case of either a parallel-RC or a series-RL load. These normalized design parameters (called g_i-parameters) are used below to synthesize the present prototype.

The parallel-RC or series-RL loads of the low-pass prototype are transformed to the band-pass resonant models of Fig. 3 by setting

$$f_{\mathrm{LP}} = \frac{\tan(\pi f/f_r)}{2\pi\Lambda} \tag{11}$$

where

$$\Lambda = \tan\left(\frac{\pi}{2}B\right) \tag{12}$$

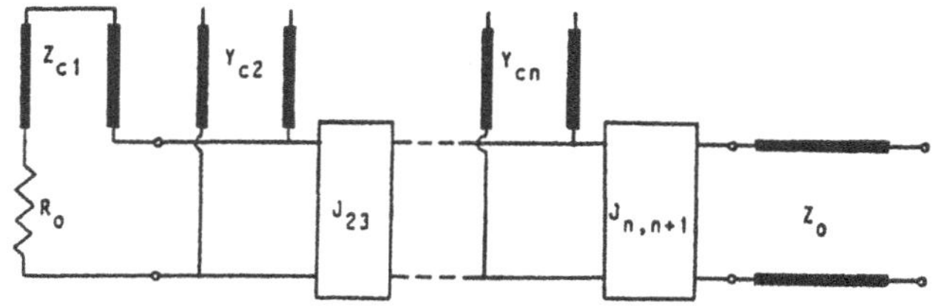

Fig. 4. Intermediate matching-network prototype consisting of open-circuited transmission-line stubs and admittance inverters (series-resonant case).

and f_{LP} is the low-pass frequency variable. By this frequency transformation, parallel-C elements are transformed into parallel open-circuited half-wavelength stubs and series-L elements into series short-circuited half-wavelength stubs. Because the latter are not physically realizable, they are removed from the matching network by using admittance inverters J characterized by their Y-matrix

$$Y = \begin{bmatrix} 0 & jJ \\ jJ & 0 \end{bmatrix}. \tag{13}$$

In this way, the intermediate matching-network of Fig. 4 is obtained in the series-resonant case, and a similar one (containing an additional inverter J_{12}) in the parallel-resonant case.

The transmission-line resonant models of Fig. 3 are almost equivalent (at least over a moderate bandwidth) to the lumped-element RLC-circuits used in Section III (using $f_{\mathrm{LP}} = V/B$ instead of (11) would have yielded these). Their quality factor is given by

$$Q = \frac{\pi}{2}\frac{R_0}{Z_{c1}} \tag{14}$$

in the parallel-resonant case (Fig. 3(a)), and

$$Q = \frac{\pi}{2}\frac{Z_{c1}}{R_0} \tag{15}$$

in the series-resonant case (Fig. 3(b)).

With respect to Figs. 3 and 4, it can be observed that all line sections are a half-wavelength long at the resonant frequency f_r, R_0 is the resonant antenna resistance, $Y_{ci}(Z_{ci})$ is the characteristic admittance (impedance) of the ith transmission-line resonator, $J_{i,i+1}$ is the admittance inverter between resonators i and $i+1$, $J_{n,n+1}$ is a final impedance-scaling admittance inverter, and Z_0 is the (real) system impedance (usually 50 Ω). It can be seen that the first resonator ($i = 1$) belongs to the antenna model itself, whereas the following ones ($i = 2, 3, \cdots, n$) belong to the matching network. If one restricts the antenna model to the patch element proper so it does not include a possible feed probe inductance, the latter can be included in the $i = 2$ resonator [7], [17], [18], as discussed in Section V.

The different network parameters Y_{ci} and $J_{i,i+1}$ must satisfy the following:

$$J_{12} = \sqrt{\frac{\Lambda Y_{c2}}{R_0 g_2}} \quad \text{(parallel-resonant case)} \tag{16}$$

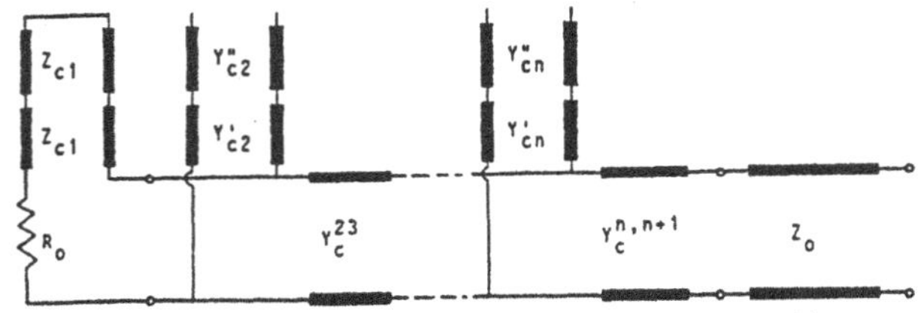

Fig. 5. Final transmission-line prototype for broad-band matching network (series-resonant case).

$$Y_{c2} = \frac{g_2}{\Lambda R_0} \quad \text{(series-resonant case)} \tag{17}$$

$$J_{i,i+1} = \Lambda \sqrt{\frac{Y_{ci} Y_{c,i+1}}{g_i g_{i+1}}}, \qquad i = 2, 3, \cdots, n-1 \tag{18}$$

$$J_{n,n+1} = \sqrt{\frac{\Lambda Y_{cn}}{g_n g_{n+1} Z_0}}. \tag{19}$$

The g_i-parameters are found from [16], and are a function of the order of the network n (to be chosen by the designer) and the decrement

$$\delta = \frac{\pi}{2\Lambda Q}. \tag{20}$$

Observe that, by definition, $g_0 = 1$ and $g_1 = 1/\delta$.

To obtain a prototype that is practically realizable, the admittance inverters are replaced by quarter-wavelength lines. Furthermore, to increase the number of degrees of freedom, the half-wavelength stubs are splitted up in two quarter-wavelength sections with different characteristic impedances. In this way, the final prototype is obtained which is depicted in Fig. 5 for the series-resonant case. For the networks of Fig. 4 and Fig. 5 (series-resonant case) and their corresponding ones (parallel-resonant case) to be approximately equivalent, the following equations have to be satisfied for $i = 2, 3, \cdots, n$ [15]:

$$Y_c^{i,i+1} = J_{i,i+1} \cos\left(\frac{\pi}{4} B\right) \tag{21}$$

$$Y'_{ci} = [Y_{ci}\Lambda - (Y_c^{i-1,i} + Y_c^{i,i+1})\Gamma] \frac{a_i - \Gamma^2}{(1 + a_i)\Gamma} \tag{22}$$

$$Y''_{ci} = a_i Y'_{ci} \tag{23}$$

where

$$\Gamma = \tan\left(\frac{\pi}{4} B\right) \tag{24}$$

and the a_i-parameters can be freely chosen. In the parallel-resonant case, (21) also applies for $i = 1$, and in the series-resonant case, the first term between the inner parentheses in (22) vanishes for $i = 2$.

V. Design Procedure for an Impedance-Matched Antenna

This section summarizes the complete procedure for designing a broad-band impedance-matching network for a given microstrip antenna element. First, the antenna impedance is made to be resonant at the center frequency of the band, as explained in Section III. Then, the antenna model parameters f_r, R_0, and Q are determined. This has to be done very carefully, by preference trough accurate measurements, because most analysis models are not accurate enough for this purpose [15].

Once the antenna parameters are known, the order n and the bandwidth B (if not given *a priori*) are to be determined. Using (20) and [16], a deliberate choice can be made. The choice of n typically reflects a trade-off between increased bandwidth and/or degree of matching (the larger n, the nearer the optimum (9) will be approached) on the one hand and increased complexity (the network will become larger and lossier) on the other. Typical values for n are 2, 3, or 4. The case $n = 1$ is trivial and has been dealt with in Section III (8). The approaches of [7] and [17] could be described as $n = 1.5$ (feed probe inductance resonated by series capacitor at center frequency without first optimizing the inductance value) whereas [18] gives a real $n = 2$ lumped-element approach.

Knowing n and δ, the g_i-parameters ($i = 2, 3, \cdots, n$) are found from [16]. The parameters of the intermediate prototype (Fig. 4) then follow from (16) or (17), (18) and (19). Subsequently, the parameters of the final prototype are derived from (21)–(23). In this process, there are $2n - 3$ degrees of freedom in the series-resonant case and $2n - 2$ in the parallel-resonant case. One could, for example, choose freely the Y_{ci}-parameters (except Y_{c2} in the series-resonant case) and the a_i-parameters. By making these choices in a deliberate fashion, it is normally possible to obtain a practically realizable prototype, i.e., a prototype that, when translated to a physical lay-out, yields line widths that are neither too wide nor too narrow.

The final step of translating the prototype to a practical circuit is a very critical one. Indeed, for getting good results, it is absolutely essential that the effects of discontinuities (such as open ends, steps and T-junctions) are compensated. Therefore, to avoid lengthy trial-and-error tuning procedures, the application of computer simulation and optimization techniques is highly desirable. This also allows to compensate for the different approximations in the design of the prototype itself, i.e., the use of approximate transmission-line models for the antenna impedance (Fig. 3) and the approximation of the intermediate prototype (Fig. 4) by the final prototype (Fig. 5).

VI. Applications

A. Single-Substrate Impedance-Matched Rectangular Antenna

The first design example concerns an integrated impedance-matched antenna consisting of a rectangular microstrip antenna and a coplanar microstrip impedance-matching network. The whole structure is laid out on top of a 20 cm × 15 cm × 1.6 mm RT/duroid 5880 substrate ($\epsilon_r = 2.20$), as shown in Fig. 6. A similar antenna with a shielded-microstrip matching network (where the shield height was tuned to optimize the network response), has been described elswehere [10], [19].

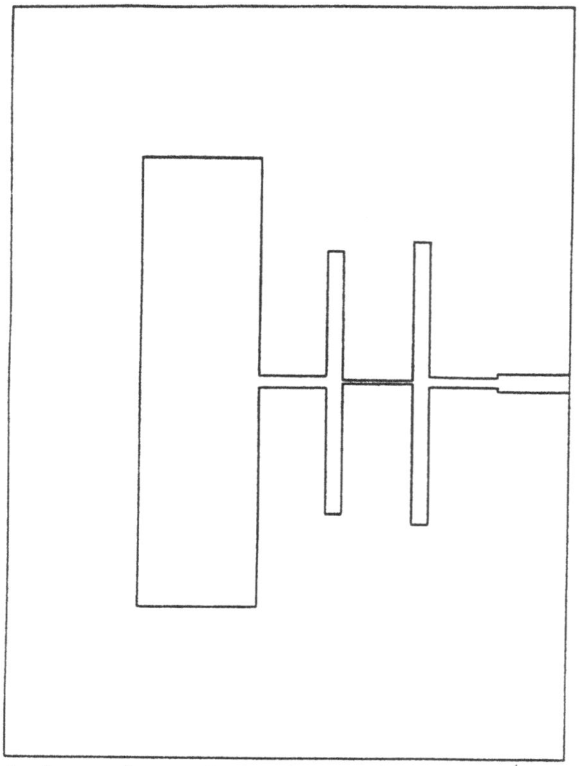

Fig. 6. Layout of rectangular impedance-matched antenna (antenna #1).

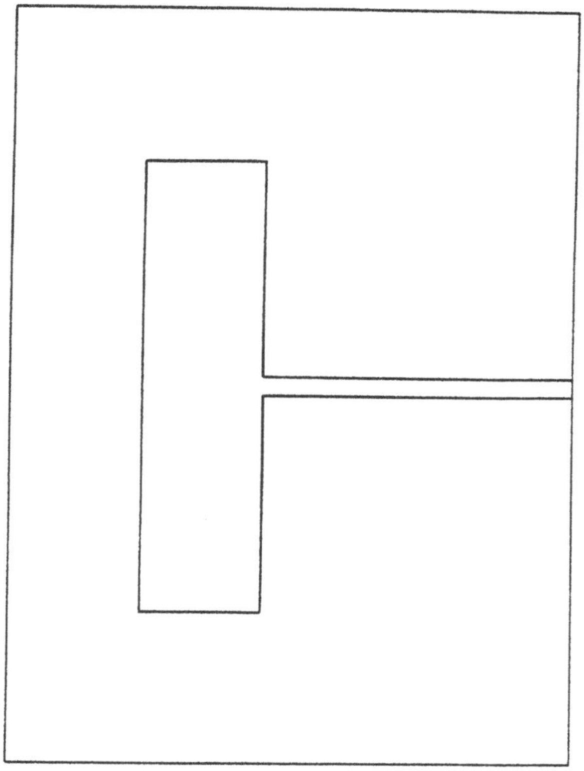

Fig. 7. Layout of reference antenna (antenna #2).

The following antenna parameters, calculated from an improved transmission-line model [20], were used in the present design: $f_r = 3.027$ GHz, $R_0 = 48.88\ \Omega$ and $Q = 22.64$ (parallel-resonant case). The design of the circuit was based on the following choices: $n = 3$, $B = 10$ percent, $Z_c^{23} = 130$ Ω, $Y_{c2} = Y_{c3}$, and $a_2 = a_3 = 1$. With $Z_0 = 50\ \Omega$, this yielded: $Z_c^{12} = 65.72\ \Omega$, $Z_c^{34} = 72.28\ \Omega$, $Z'_{c2} = Z''_{c2} = 25.78$ Ω and $Z'_{c3} = Z''_{c3} = 25.33\ \Omega$. When translating these values to the microstrip circuit shown in Fig. 6, both the $i = 2$ and $i = 3$ resonators were realized as two parallel identical stubs in order to reduce their line width.

To be able to judge the performance of this impedance-matched antenna properly, a reference antenna (Fig. 7) has been built in the same process (a piece of substrate cut from the same sheet was used). This reference antenna is completely identical to the impedance-matched antenna except that the matching network is replaced by a simple 50 Ω microstrip line. Note that the calculated edge-fed impedance of the antenna element (i.e., 48.88 Ω) is very nearly equal to 50 Ω. Hence, the reference antenna should be well matched at $f = f_r$. Fig. 8 shows the return loss of both antennas. The reference antenna has its best match at 3.025 GHz (−21.5 dB) and has a higher order mode dip at 3.424 GHz. This higher order mode dip is very much suppressed by the matching network as shown by the other curve. Within the band of operation, the impedance-matched antenna has its worst match at 3.035 GHz (−8.8 dB). It can be seen, that the bandwidth at this level ($S = 2.14$) has been increased by a factor of 3.2 to a value of 275 MHz or 9.1 percent, whereas the theoretical maximum bandwidth-enlargement factor for this degree of matching equals 4.0 (Fig. 2).

It is clear from Fig. 8 that, except in a small band around

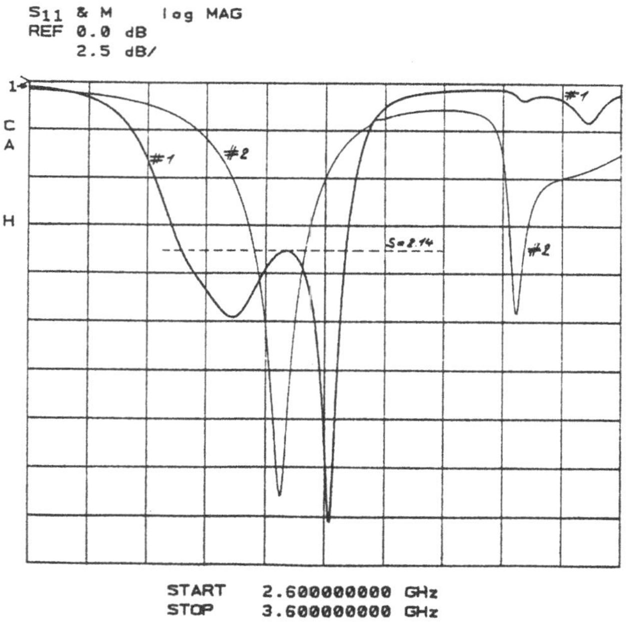

Fig. 8. Return loss versus frequency of antennas #1 and #2.

f_r, the mismatch loss of antenna #1 (impedance-matched antenna) within its band of operation is less than that of antenna #2 (reference antenna). However, because the matching network will inevitably be somewhat lossy, one could ask if the decrease of the mismatch loss is not annihilated by the increase of the dissipation loss. That this is not the case, is demonstrated by Fig. 9 which shows the transmission performance of both antennas. Particularly, a radiation link was established between a standard gain horn on the one side and antenna #1 or #2 on the other. The figure shows the transmission co-

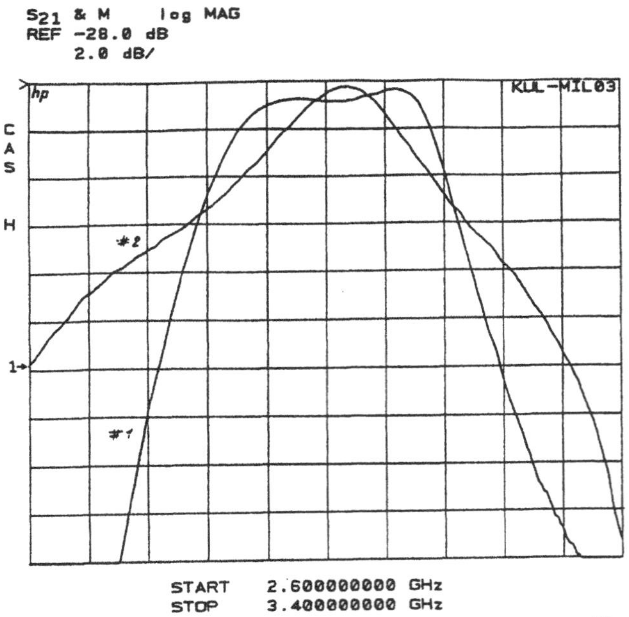

Fig. 9. Transmission characteristic versus frequency of antennas #1 and #2.

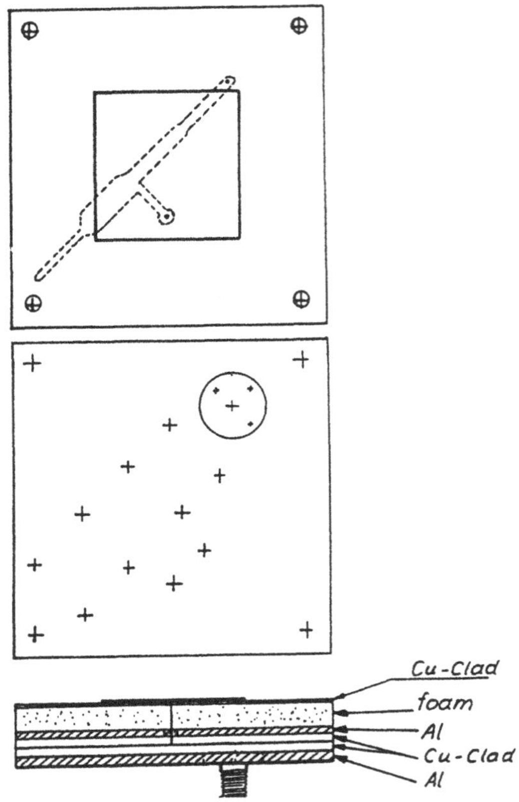

Fig. 10. Multilayer impedance-matched antenna (antenna #3).

efficient measured in these two cases. This characteristic is almost proportional to the realized gain. It follows that antenna #1 is a more efficient radiator over the 2.832 – 2.988 GHz band and the 3.055 – 3.174 GHz band, whereas antenna #2 is more efficient in between. The maximum difference in this center band equals 0.61 dB and occurs at 3.026 GHz (i.e., the frequency of best match of antenna #2).

Concerning radiation patterns, *E*- and *H*-plane cuts for both antennas have been measured at 2.9, 3.0, and 3.1 GHz [15]. They do not show any appreciable difference, which proves that the matching network, although it is coplanar with the patch, does not affect the radiation characteristics. It is to be observed, however, that only copolar patterns were measured.

B. Multilayer Impedance-Matched Square Antenna

The second design example concerns a multilayer square microstrip antenna with a stripline matching network situated underneath the antenna ground plane. A similar antenna

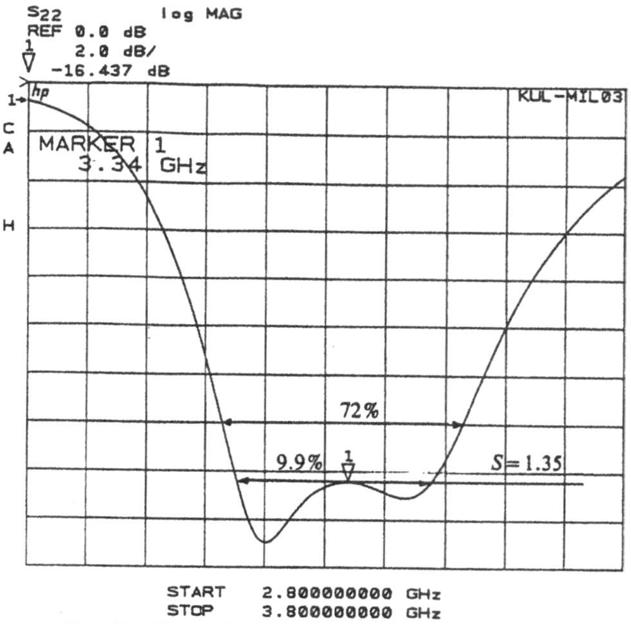

Fig. 11. Return loss versus frequency of antenna #3.

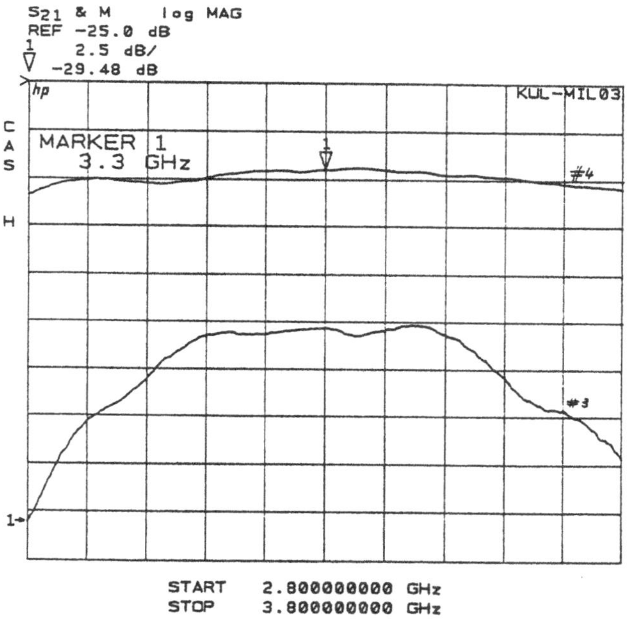

Fig. 12. Transmission characteristic of antenna #3 and standard gain horn (antenna #4).

with an underneath microstrip matching network has been described elsewhere [21].

The present antenna is shown in Fig. 10. It is a sandwich structure consisting of (from top to bottom) a 0.5 mm Cu-Clad 217 substrate bearing the antenna patch, a 6.4 mm Eccofoam PP-2 layer, a first metal ground plate (the antenna ground plane), two 1.6 mm Cu-Clad 217 substrate layers bearing the stripline matching network, and a second bottom ground plate onto which an OSM 203-1 stripline connector is attached. The overall dimensions (apart from the connector and four connecting screws) are 70 mm × 70 mm × 14 mm.

The antenna model parameters were $f_r = 3.28$ GHz, $R_0 = 33.3\ \Omega$ and $Z_{c1} = 151.5\ \Omega$ (series-resonant case). Choosing $n = 2$, $b = 12$ percent, $a_2 = 0.3$ and $Z_0 = 50\ \Omega$, the design was carried out straightforwardly. Using computer-aided simulation and optimization, adjustments were made to compensate for the different approximations. The measured return loss diagram is shown in Fig. 11. Considering the -16.44 dB (or $S = 1.35$) level, which is the maximum level in the band of operation, a bandwidth of 324 MHz or 9.9 percent is obtained. Using (7) and (15), the unmatched antenna is found to have a bandwidth of only 4.2 percent at this level. Observe also that a better match than -14 dB is obtained within the design bandwidth of 12 percent.

The transmission performance is illustrated in Fig. 12. This figure shows the transmission coefficient between a log-periodic dipole array antenna on the one side and the impedance-matched antenna or a standard gain horn (Narda Model 644) on the other side. It follows that the realized gain is about 8 dB over a bandwidth of 12 percent. This high gain

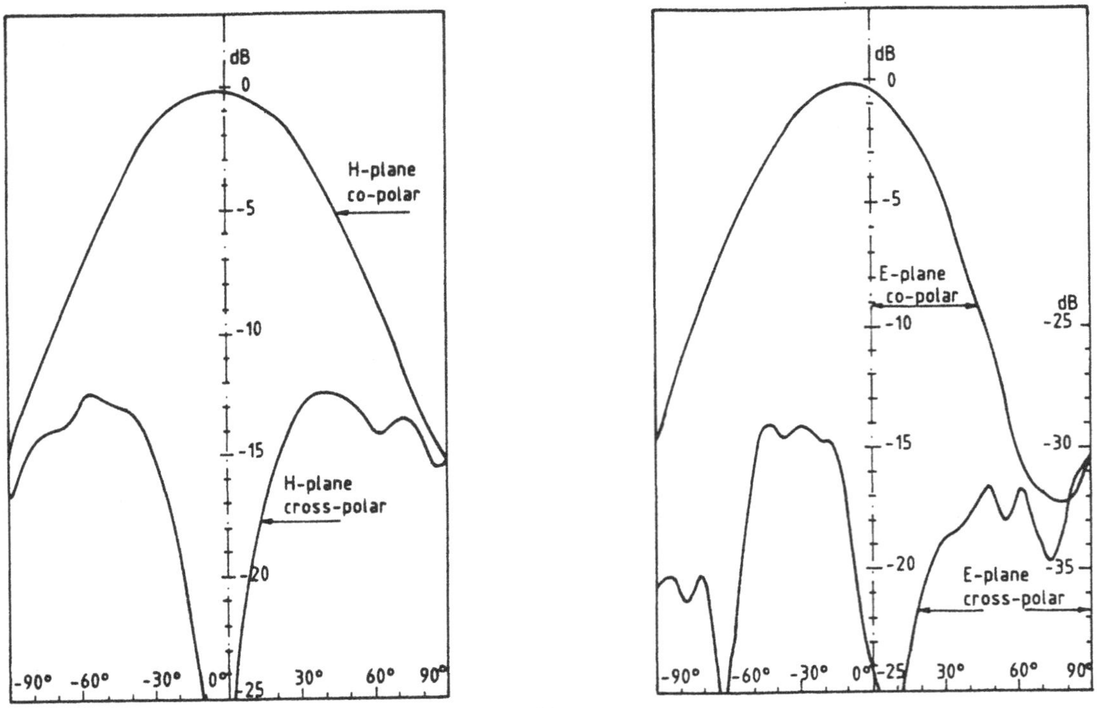

(a)

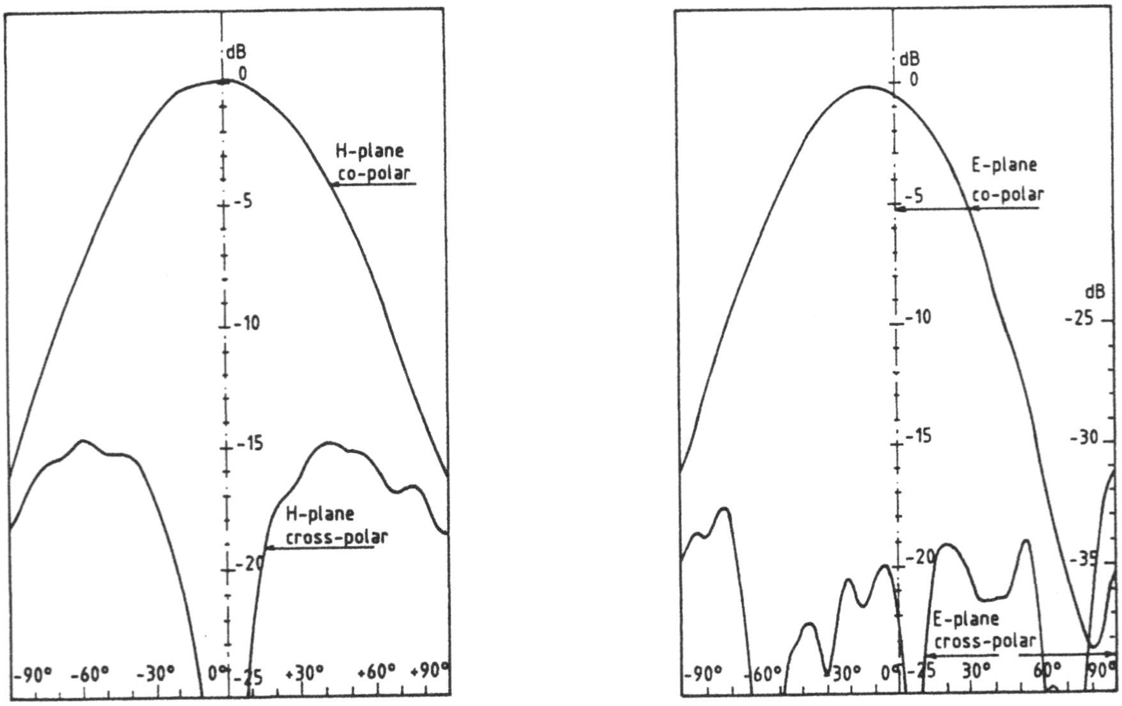

(b)

Fig. 13. (a) Measured radiation patterns at 3.100 GHz of antenna #3. (b) Measured radiation patterns at 3.300 GHz of antenna #3. (c) Measured radiation patterns at 3.500 GHz of antenna #3.

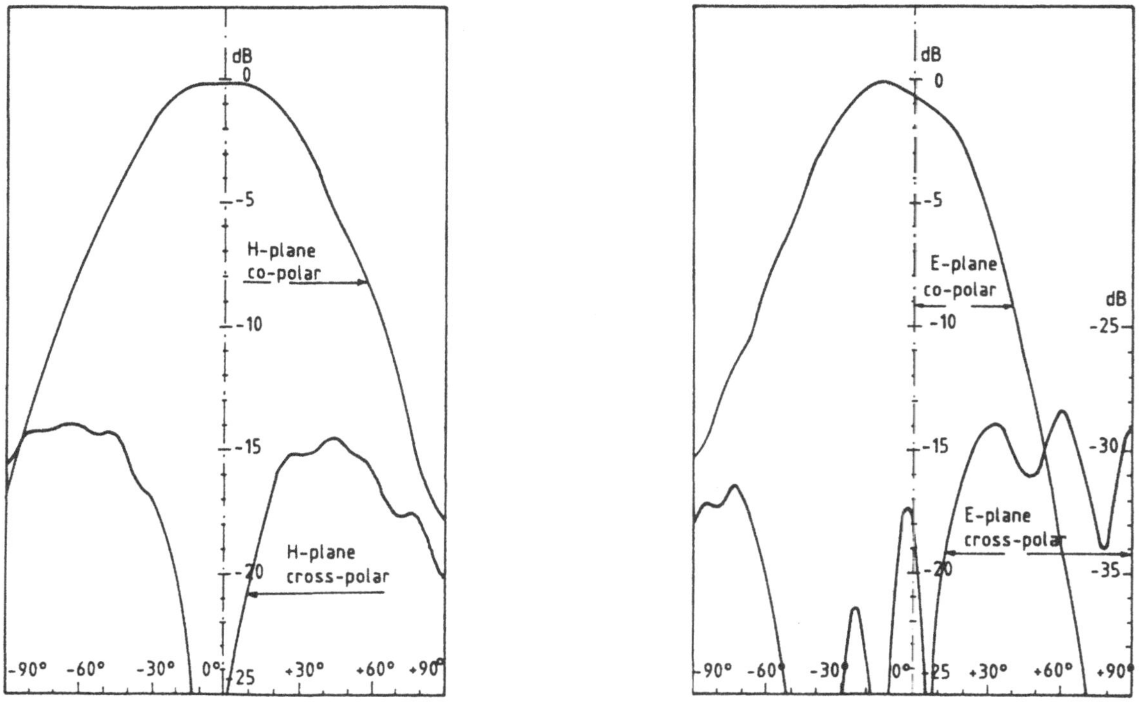

(c)

Fig. 13. *(Continued.)*

value for a single square element is partly due to the deliberate choice of the horizontal dimensions (70 mm × 70 mm). Mounted on a large ground plane, the gain would be somewhat less.

Finally, Fig. 13 shows the *E*- and *H*-plane co- and cross-polar patterns at 3.1, 3.3, and 3.5 GHz. These patterns do not show any significant change within the band of operation.

VII. Conclusion

In this paper, broad-band impedance matching has been proposed as a powerful technique to increase the bandwidth of microstrip antennas. The theoretical limitations have been described and a practical design method for the required matching networks has been outlined. The validity of this design procedure has been illustrated by two representative design examples. However, it should be stressed that impedance-matching is a very general technique and that many other design procedures and realization forms could be devised, which possibly could yield better results.

References

[1] D. M. Pozar, "An update on microstrip antenna theory and design including some novel feeding techniques," *IEEE Antennas Propagat. Soc. Newsletter*, vol. 28, no. 5, pp. 5–9, Oct. 1986.

[2] E. Chang, S. A. Long, and W. F. Richards, "An experimental investigation of electrically thick rectangular microstrip antennas," *IEEE Trans. Antennas Propagat.*, vol. AP-34, pp. 767–772, June 1986.

[3] C. H. Chen, A. Tulintseff, and R. M. Sorbello, "Broadband two-layer microstrip antenna," in *IEEE Antennas Propagat. Soc. Int. Symp. Dig.*, 1984, pp. 251–254.

[4] G. Kumar and K. C. Gupta, "Directly coupled multiple resonator wideband microstrip antennas," *IEEE Trans. Antennas Propagat.*, vol. AP-33, pp. 588–593, June 1985.

[5] H. Pues, J. Bogaers, R. Pieck, and A. Van de Capelle, "Wideband quasi-log-periodic microstrip antenna," *Inst. Elec. Eng. Proc.*, vol. 128, pt. H, pp. 159–163, June 1981.

[6] A. K. Bhattacharyya and R. Garg, "Effect of substrate on the efficiency of an arbitrarily shaped microstrip patch antenna," *IEEE Trans. Antennas Propagat.*, vol. AP-34, pp. 1181–1188, Oct. 1986.

[7] K. S. Fong, H. F. Pues, and M. J. Withers, "Wideband multilayer coaxial-fed microstrip antenna element," *Electron. Lett.*, vol. 21, pp. 497–499, May 23, 1985.

[8] R. M. Fano, "Theoretical limitations on the broadband matching of arbitrary impedances," *J. Franklin Inst.*, vol. 249, nos. 1-2, pp. 57–83 and 139–154, Jan.–Feb. 1950.

[9] H. F. Pues and A. R. Van de Capelle, "Impedance-matching of microstrip resonator antennas," in *URSI North Amer. Radio Sci. Meet. Dig.*, Québec, June 1980, p. 189.

[10] "Broad-band microstrip antenna," U.S. Patent 4445122, Apr. 24, 1984.

[11] R. F. Harrington, *Time-Harmonic Electromagnetic Fields*. New York: McGraw-Hill, 1961, pp. 431–440.

[12] K. R. Carver and J. W. Mink, "Microstrip antenna technology," *IEEE Trans. Antennas Propagat.*, vol. AP-29, pp. 2–24, Jan. 1981.

[13] P. C. Sharma and K. C. Gupta, "Analysis and optimized design of single feed circularly polarized microstrip antennas," *IEEE Trans. Antennas Propagat.*, vol. AP-31, pp. 949–955, Nov. 1983.

[14] J. S. Lee and W. J. Furlong, "An extremely lightweight fuselage-integrated phased array for airborne applications," *IEEE Trans. Antennas Propagat.*, vol. AP-29, pp. 178–182, Jan. 1981.

[15] H. F. Pues, "Study of the bandwidth of microwave integrated antennas: Development of design models for wideband microstrip antennas" (in Dutch), Ph.D. dissertation, Microwaves and Lasers Div., Catholic Univ. Louvain, Louvain, Belgium, 1983.

[16] G. L. Matthaei, L. Young, and E. M. T. Jones, *Microwave Filters, Impedance-Matching Networks, and Coupling Structures*. New York: McGraw-Hill, 1964, sec. 4.09–4.10.

[17] J. M. Griffin and J. R. Forrest, "Broadband circular disc microstrip antenna," *Electron. Lett.*, vol. 18, pp. 266–269, Mar. 18, 1982.

[18] D. A. Paschen, "Practical examples of integral broadband matching of microstrip antenna elements," in *Proc. 1986 Antenna Appl. Symp.*, Monticello, IL, Sept. 17–19, 1986.

[19] H. F. Pues and A. R. Van de Capelle, "Wideband impedance-matched microstrip resonator antennas," in *Inst. Elec. Eng. Conf. Pub. 195* (Antennas and Propagation), pt. 1, pp. 402–405, Apr. 1981.

[20] ——, "Accurate transmission-line model for the rectangular microstrip antenna," *Inst. Elec. Eng. Proc.*, vol. 131, pt. H, pp. 334-340, Dec. 1984.

[21] H. Pues, A. Van Kauteren, J. Vercruysse, and A. Van de Capelle, "Broadband microstrip radar antenna element," in *Proc. Int. Conf. Radar*, Paris, May 1984, pp. 298–303.

Probe Compensation in Thick Microstrip Patches

P. S. Hall

Indexing terms: Antennas, Microstrip

In thick microstrip patches, probe inductance prevents matching of the patch impedance to the input connector. The probe inductance can be tuned out with a capacitive gap. To maintain simplified construction the gap is here etched on the patch surface. Bandwidths equal to or greater than that theoretically predicted are realised. Use of a single probe-compensated feed results in radiation pattern distortion, high crosspolarisation and low efficiency due both to higher-order modes and surface-wave generation. Two-probe feeding is used here to overcome these problems and to give a wide-band antenna with good radiation pattern control and high efficiency.

Introduction: There is a continuing need for wider bandwidth in antenna systems. The bandwidth of microstrip patch antennas can be widened by increasing the substrate thickness. However, the use of very thick substrates leads to an increase in the inductive component of the input impedance which ultimately prevents a match being obtained.[1] This limit has been given as for example $h = 0{\cdot}11\lambda_0$ for $\varepsilon_r = 2{\cdot}2$.[2] The use of a capacitive gap to offset the inductance and match the device has been suggested both in the feed network behind the patch[3] and in the feed probe itself.[4] To maximise constructional simplicity, the gap is here located on the patch surface and takes the form of an annular gap around the feed probe as shown in Fig. 1. Patches have been made on $\varepsilon_r = 2{\cdot}32$ and $\varepsilon_r \simeq 1{\cdot}06$ substrates for TM_{11}- and TM_{21}-mode operation, respectively. The efficiency and crosspolarisation performance of probe-compensated discs has not previously been given and results are presented here. The high levels of crosspolarisation expected from such thick patches are confirmed in these results. Significant suppression can, however, be obtained by the use of a second feed point symmetrically located with respect to the disc centre.[5] This has been implemented with the probe-compensated feeds and results are given to illustrate the degree of suppression available.

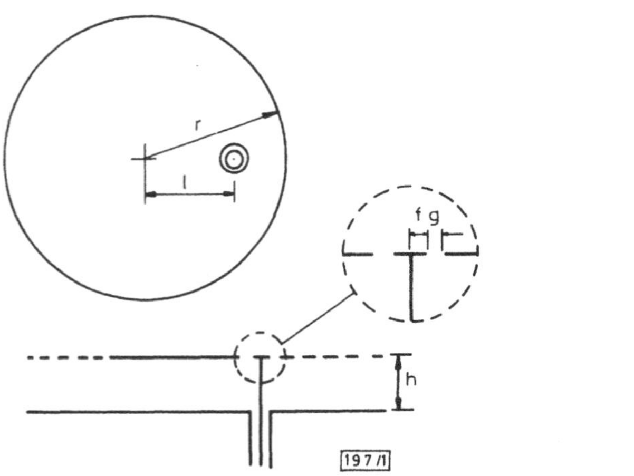

Fig. 1 *Microstrip disc with annular gap probe compensation*

Description of antenna: The circular disc antenna with probe compensation is shown in Fig. 1. The coaxial inner conductor is used to form the feed probe, which is connected to the disc, as in a conventional patch. The annular gap is concentric with this probe. The feed position and gap dimensions, which were determined empirically to match the input impedance, are given in the footnote to Table 1 for patches on PTFE ($\varepsilon_r = 2{\cdot}32$) and foam ($\varepsilon_r = 1{\cdot}06$) substrates. Both patches have $h/\lambda_0 > 0{\cdot}1$ and hence cannot be matched without the probe-compensation gap. The foam patch was designed for the TM_{21}-mode while that on PTFE was for the fundamental TM_{11}-mode. The gap dimensions in wavelengths for both patches are comparable, which suggests that the probe inductance is relatively insensitive to substrate dielectric constant.

Performance: Fig. 2 shows the return loss of the two patches with probe-compensated feeds. The 10 dB return loss bandwidth of each is given in Table 1 compared to that estimated using a single-mode cavity model.[6] It can be seen in the case of the TM_{21} patch that the measured bandwidth is considerably in excess of the calculated figure, while for the TM_{11} the measurement and calculation agree well.

Patch gain has been measured, and by comparison to calculated patch directivity[6] efficiency has been estimated; results are also shown in Table 1. Measurement uncertainty and

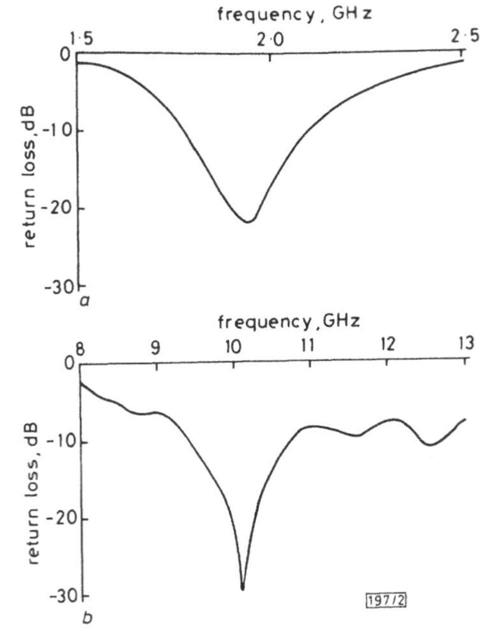

Fig. 2 *Input return loss of probe-compensated patches*
a TM_{21}, $\varepsilon_r = 1{\cdot}06$
b TM_{11}, $\varepsilon_r = 2{\cdot}32$
Patch details in Table 1

Reprinted with permission from *Elect. Lett.*, P. S. Hall, "Probe Compensation in Thick Microstrip Patches," vol. 23, no. 11, pp. 606–607, May 1987.

errors due to diffraction by the substrate edges are estimated to be about ±10%. It can be seen that significant losses occur for the single-probe antennas. In the case of the $\varepsilon_r = 2.32$ patch, 1·5 dB loss is attributed to dielectric, metal and surface-wave losses.[7] Measured crosspolarisation levels are of the order of −10 dB, representing 0·5 dB loss, leaving greater than 0·5 dB unaccounted for, possibly representing losses in the matching circuit. For the $\varepsilon_r = 1.06$ TM_{21} patch crosspolarisation levels are very high indeed, as shown in Fig. 3*a*, and this offsets the lower expected loss to surface waves.

Fig. 3*b* shows the same pattern cut for the TM_{21} patch as Fig. 3*a* but with two probe-compensated feeds located as shown in the inset. The two inputs are fed in phase. Substantial suppression of the crosspolarisation is achieved, and the distortion in the copolarised pattern is reduced. In addition,

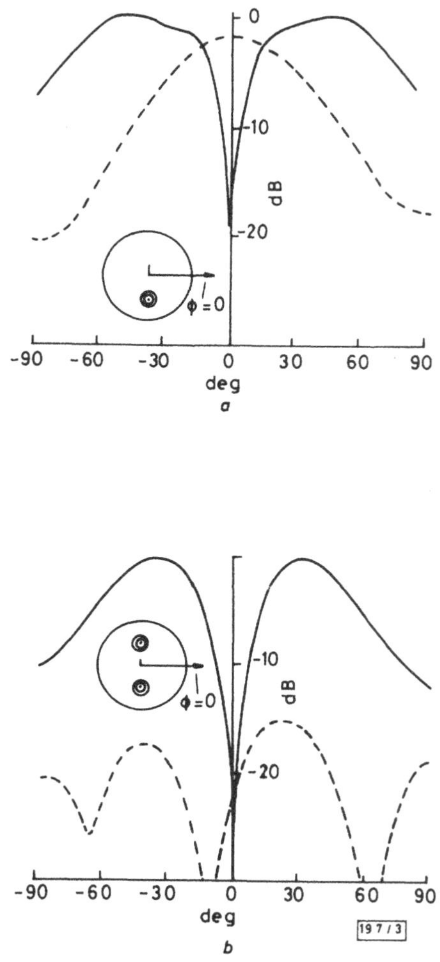

Fig. 3 *Measured radiation pattern in $\phi = 0$ plane of TM_{21}, $\varepsilon_r = 1.06$ patch*

a Single feed point
b Two-probe feeding

as noted in Table 1, efficiency is greatly increased, indicating that much of the loss in the one-probe antenna is due to crosspolarisation.

Table 1 BANDWIDTH AND EFFICIENCY OF PROBE-COMPENSATED MICROSTRIP DISCS

Mode	TM_{21}	TM_{11}
ε_r	1·06	2·32
Substrate thickness (mm)	25	3·18
(h/λ_0)	0·15	0·12
Bandwidth, %		
(−10 dB return loss)		
—measured	15·8	13·2
—theory[6]	10·9	12·3
Efficiency, %		
(at band centre)		
—single probe	22 (±10)	44 (±10)
—two probe	92 (+8, −10)	

For $TM_{21}(TM_{11})$ patch, radius = 70·0 (4·0) mm, L = 38·0 (2·5) mm, f = 5 (1·0) mm = 0·027 (0·033)λ_0, g = 2·5 (0·4) mm = 0·013 (0·013)λ_0

Conclusion: A simplified form of probe compensation for thick microstrip patches has been demonstrated in which an annular gap located on the patch surface is used to tune out the probe inductance and hence allow matching. Bandwidths equal to or greater than those predicted theoretically have been achieved. For single-probe feeding in both $\varepsilon_r = 1.06$ TM_{21} and $\varepsilon_r = 2.32$ TM_{11} patches, losses are considerable with crosspolarisation loss dominant in the low dielectric constant case and surface waves in the other. The use of two-probe feeding has been shown to reduce these problems significantly for an $\varepsilon_r = 1.06$ substrate, suggesting that the thick patch with two-probe compensated feeding can form a good wide-bandwidth antenna for many applications.

Acknowledgments: The author would like to acknowledge the support and advice of members of the Wolfson RF Engineering Centre, RMCS.

References

1 CHANG, E., LONG, S. A., and RICHARDS, W. F.: 'Experimental investigation of electrically thick rectangular microstrip antennas', *IEEE Trans.*, 1986, **AP-34**, pp. 767–772
2 POZAR, D. M.: 'Considerations for millimeter wave printed antennas', *ibid.*, 1983, **AP-31**, pp. 740–741
3 GRIFFIN, J. M., and FORREST, J. R.: 'Broadband circular disc microstrip antenna', *Electron. Lett.*, 1982, **18**, pp. 266–269
4 FONG, K. S., PUES, H. F., and WITHERS, M. J.: 'Wideband multi-layer coaxial-fed microstrip antenna element', *ibid.*, 1985, **21**, pp. 497–499
5 CHIBA, T., SUZUKI, Y., MIYANO, N., MIURA, S., and OHMORI, S.: 'Phased array antenna using microstrip patch antennas'. Proc. 12th European microwave conference, Helsinki, Sept. 1982, pp. 472–477
6 JAMES, J. R., HALL, P. S., and WOOD, C.: 'Microstrip antenna theory and design'. IEE Electromagnetic Waves Series No. 12 (Peter Peregrinus, London, 1981), pp. 84–86
7 FONSECA, S. B. A., and GARIOLA, A. J.: 'Microstrip disc antennas, Part 1: efficiency of space wave launching', *IEEE Trans.*, 1984, **AP-32**, pp. 561–567

Increasing the Bandwidth of a Microstrip Antenna by Proximity Coupling

D. M. Pozar and B. Kaufman

Indexing terms: Antennas, Microstrip

The letter presents experimental results for a proximity-coupled microstrip patch antenna capable of 13% bandwidth. The impedance match (VSWR ≤ 2), copolarised radiation patterns and crosspolarised radiation were measured over this bandwidth to confirm operation. The construction is quite simple, consisting of a microstrip feedline on a substrate proximity-coupled to a rectangular microstrip patch on a covering superstrate; a small open-circuit tuning stub is connected in shunt with the feed line.

Introduction: Small bandwidth is probably the most serious disadvantage of microstrip antennas and, as discussed in a recent review paper,[1] a large number of attempts have been made to increase the bandwidth of printed antennas beyond the typical values of a few per cent. The most straightforward way of doing this is to use a thicker substrate, but it is difficult to achieve more than about 4–5% bandwidth in this manner before the impedance locus becomes very inductive, causing matching problems.[2,3] Other methods have included the use of parasitic elements to obtain a double resonance response, or the use of a matching network to obtain improved bandwidth from a patch on a thick substrate.[4]

This letter describes an alternative method of obtaining enhanced bandwidth from a microstrip antenna, using a microstrip feed line proximity-coupled to a patch antenna printed on a superstrate above the feedline. A small shunt tuning stub is also required, with the result that about 13% bandwidth (VSWR ≤ 2) is obtained. The copolarised and crosspolarised radiation patterns were measured over this band, with good results. The geometry of the antenna is described below, and an equivalent circuit is developed to gain an understanding of how the antenna operates and is matched. The measured impedance loci are shown before and after matching.

When discussing antenna bandwidth, one must realise that two separate criteria should be satisfied over the frequency range of interest: impedance bandwidth (the bandwidth over which the antenna remains matched to the feedline to some specified level, such as VSWR ≤ 2), and the pattern bandwidth (the bandwidth over which the pattern remains, in some sense, constant). The ideal broadband element will satisfy both of these criteria. In addition, it is often necessary in array applications to restrict the element size so that elements may be spaced in the order of $\lambda_0/2$ apart. The present element meets these conditions.

Description of antenna and measured results: The geometry of the proximity-coupled patch is shown in Fig. 1. The microstrip feedline is of width W_f and is printed on the bottom substrate. The microstrip patch is of length L (resonant length) and width W, and is printed on a substrate (superstrate) bonded to the feed substrate. The feedline is centred with respect to the patch width, and is inset a distance S from the edge of the patch. In general, the two substrates may be of different thicknesses and permittivity, but in the case at hand both substrates were 60 mil ($d_1 = d_2 = 0{\cdot}158$ cm) Duroid, with $\varepsilon_r = 2{\cdot}20$. A short tuning stub of length l_s is connected in shunt with the feedline a distance d_s from the edge of the patch. For the example discussed here, the dimensions were $L = 2{\cdot}5$ cm, $W = 4{\cdot}0$ cm, $W_f = 0{\cdot}5$ cm, $l_s = 0{\cdot}65$ cm, $d_s = 3{\cdot}3$ cm and $S = 1{\cdot}25$ cm.

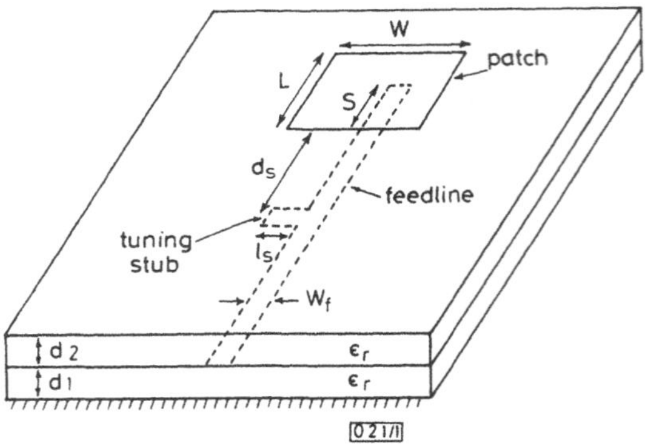

Fig. 1 *Geometry of proximity-coupled microstrip antenna*

The proximity-coupled (also referred to in the literature as 'electromagnetically coupled') patch antenna is similar to a geometry used by Huebner.[5] The present design differs from that of Reference 5, however, in that a short tuning stub is used in conjunction with the proximity-coupled element to provide broad bandwidth without an excessively thick substrate. Note that multiple-tuned circuits can always be devised for improving the impedance bandwidth of an antenna, but when implemented in printed form may involve $\lambda/4$ or $\lambda/2$ stubs, which will lead to undesirable radiation. The present design, using a relatively short tuning stub, does not suffer from this drawback.

We first discuss the impedance of the antenna without a tuning stub. If the phase reference is set at a point along the feedline just below the front edge of the patch, the measured impedance locus shown in Fig. 2 is obtained. For low frequencies this impedance is capacitive, but it becomes inductive, and then capacitive again, for frequencies above resonance. The equivalent circuit is shown in Fig. 3, consisting of a capacitor in series with a parallel *RLC* resonator, and is valid through the first resonance. The component values of this circuit were obtained by fitting to the measured data using the optimisation feature of SUPERCOMPACT. This equivalent circuit shows that the choice of phase reference at the patch edge is probably the most appealing one, since intuitively it seems that the proximity coupling mechanism involves a series capacitor, while the patch is usually modelled as a parallel *RLC* circuit when fed near an edge.[6]

Reprinted with permission from *Elect. Lett.*, D. M. Pozar and B. Kaufman, "Increasing the Bandwidth of a Microstrip Antenna by Proximity Coupling," vol. 23, no. 8, pp. 368–369, April 1987.

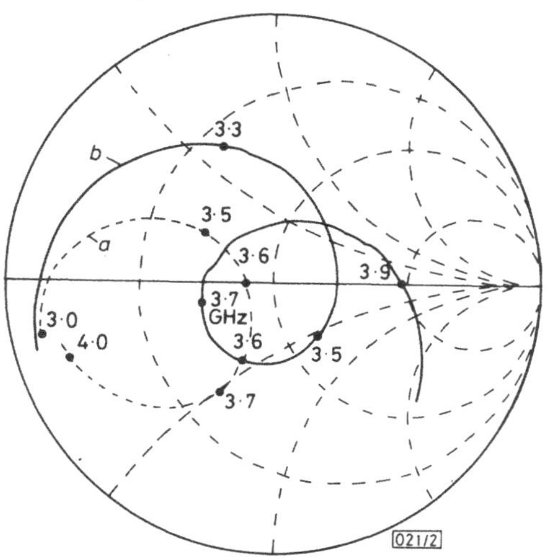

Fig. 2 *Smith chart plot of measured impedance locus*

a With a phase reference at edge of patch, without tuning stub
b With a phase reference at stub location, with tuning stub (matched antenna)

The size of the impedance locus of Fig. 2 is controlled by the amount of feedline inset S and the patch width W. The maximum locus diameter occurs when $S = L/2$; for an inset smaller or greater than this, the locus will be smaller. Thus, it appears that maximum coupling of the feedline to the patch occurs for $S = L/2$. The locus diameter (or coupling) will also increase as the patch width is made smaller; a smaller patch width also has the effect of slightly raising the resonant frequency.

Moving the phase reference down the line to a point 3·55 cm (about $\lambda_g/2$) from the patch edge gives an impedance locus which consists of a loop indicating a double-tuned resonance, but is not centred on the centre of the chart. Thus, the impedance bandwidth can be improved by using a short (capacitive) tuning stub connected in shunt with the feedline at this point. The resulting matched locus is shown in Fig. 2, and is seen to be nearly optimum for bandwidth. The measured VSWR ≤ 2 bandwidth extends from 3·375 GHz to 3·855 GHz for a fractional bandwidth of 13%.

The patterns of the matched proximity-coupled antenna were measured from 3·3 to 3·8 GHz in 1 GHz steps. The E-plane patterns show a slight hump of a few decibels on the main beam at some frequencies; this is believed to be due to the standing wave on the feedline between the patch and the tuning stub. If the tuning stub is moved towards the patch by

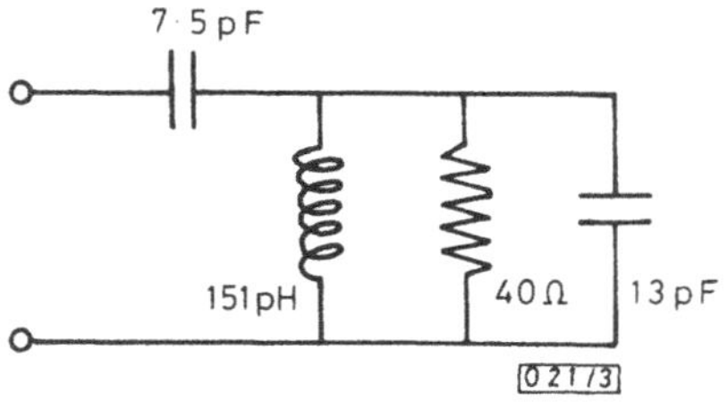

Fig. 3 *Equivalent circuit of impedance locus of Fig. 2 without tuning stub (curve a in Fig. 2)*

Phase reference at edge of patch

about $\lambda_g/2$, this effect will be eliminated, without much change in the bandwidth. This was verified experimentally, but because the tuning stub then lies below the edge of the patch and couples to the fringing fields of the patch, the tuning must be done by trial and error (which is cumbersome when the two substrates must be bonded together).

The H-plane patterns were well formed at all frequencies. Crosspolarisation is mainly due to the tuning stub; since the present stub is relatively short, this was not a serious problem. An earlier model, however, used a longer stub (at a different location) and had relatively high crosspolarisation levels (−15 dB). Thus, it is desirable to use as short a tuning stub as possible, not only to reduce the crosspolarisation, but also to reduce the VSWR on the line between the stub and the patch. (If a longer stub must be used, one way to reduce the effect of crosspolarisation from the stub is to bend the stub around so that it is mostly parallel with the feedline.) Fig. 4 shows the measured crosspolarisation levels of the matched proximity-coupled patch antenna, where it is seen that the crosspolarisation is generally 20 dB or more below the copolarisation pattern.

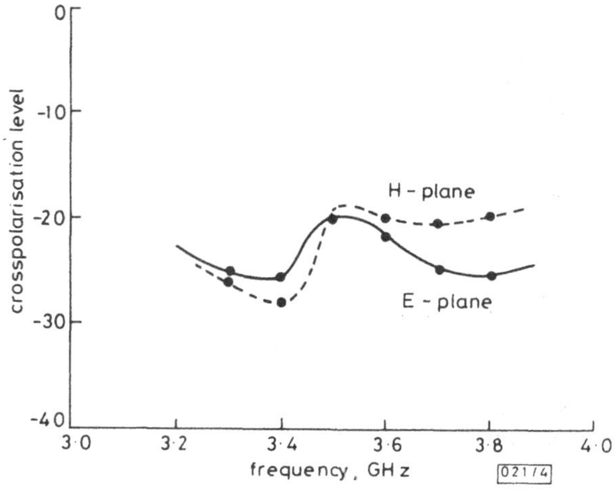

Fig. 4 *Measured crosspolarisation levels of matched proximity-coupled patch antenna*

Conclusion: A simple printed antenna having 13% bandwidth (impedance and pattern) has been described. The antenna consists of a microstrip patch proximity-coupled to a microstrip feedline below the patch. A small tuning stub is connected in shunt with the feedline, and may be located either near the patch edge, or about $\lambda_g/2$ away, resulting in an element that is small enough to be used in array applications. A rigorous full-wave moment method solution to analyse the proximity-coupled patch, and related geometries, has been completed and will be described in a future publication.

References

1 HENDERSON, A., JAMES, J. R., and HALL, C. M.: Bandwidth extension techniques in printed conformal antennas'. Conference proceedings, Military Microwaves, June 1986, Brighton, UK, pp. 329–334

2 CHANG, E., ZONG, S. A., and RICHARDS, W. F.: 'An experimental investigation of electrically thick rectangular microstrip antennas', *IEEE Trans.*, 1986, **AP-34**, pp. 767–772

3 POZAR, D. M.: 'Considerations for millimetre wave printed antennas', *ibid.*, 1983, **AP-31**, pp. 740–747

4 FONG, K. S., PUES, H. F., and WITHERS, M. J.: 'Wideband multilayer coaxial-fed microstrip antenna element', *Electron. Lett.*, 1985, **21**, pp. 497–499

5 OLTMAN, H. G., and HUEBNER, D. A.: 'Electromagnetically coupled microstrip dipoles', *IEEE Trans.*, 1981, **AP-29**, pp. 151–157

6 ZHANG, Q., FUKUOKA, Y., and ITOH, T.: 'Analysis of a suspended patch antenna excited by an electromagnetically coupled inverted microstrip feed', *ibid.*, 1985, **AP-33**, pp. 895–899

Characteristics of a Two-Layer Electromagnetically Coupled Rectangular Patch Antenna

R. Q. Lee, K. F. Lee, and J. Bobinchak

Indexing term: Antennas

Experimental results on the characteristics of a two-layer electromagnetically coupled rectangular patch antenna are presented. The variations of pattern shape, 3 dB beamwidth and impedance bandwidth with spacing s between the two layers are studied for s between 0 and $0{\cdot}37\,\lambda_0$. A relatively high-gain region is found for s between $0{\cdot}31\,\lambda_0$ and $0{\cdot}37\,\lambda_0$.

Introduction: There has recently been considerable interest in the two-layer electromagnetically coupled patch (EMCP) antenna consisting of a driven patch in the bottom and a parasitic patch on the top.[1–4] The two layers are separated by an air region of thickness s, as shown in Fig. 1. Experiments using circular and equitriangular patches showed that this antenna can provide an impedance bandwidth of between 10–20% compared to about 2% for a single patch. Somewhat surprisingly, detailed results using rectangular patches do not appear to have been reported in the literature.

The emphasis of previous studies on this antenna was concerned mainly with the improvement in bandwidth. Little attention was paid to the variation of the pattern shape, 3 dB beamwidth (and hence the gain) with the spacing between the layers s. In this letter we report an experimental study of a rectangular EMCP antenna excited in the TM_{01} mode. The major finding is that, while a relatively large bandwidth is obtained for s below $0{\cdot}1\,\lambda_0$, the 3 dB beamwidth in the

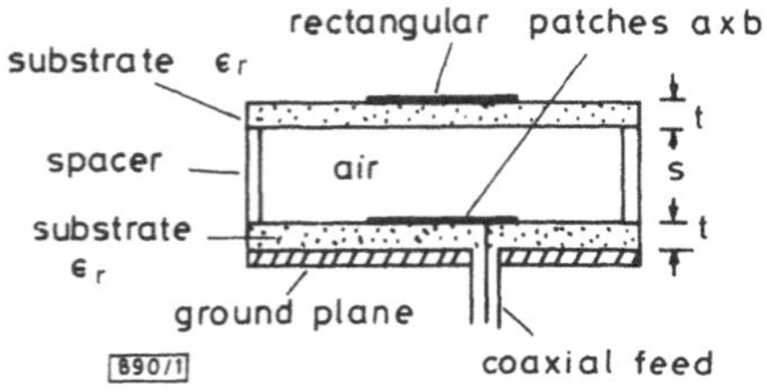

Fig. 1 *Geometry of rectangular electromagnetically coupled patch antenna*

H-plane is considerably narrower and the gain significantly higher, when s is about $0{\cdot}31\,\lambda_0$. The gain at $s = 0{\cdot}31\,\lambda_0$ is estimated to be 8·9 dB compared to about 5·3 dB for the single patch. This high-gain region, however, is associated with an impedance bandwidth of about 1·3%.

Experimental results: In our experiment, two rectangular patches of dimensions $a = 1{\cdot}5$ cm, $b = 1$ cm are fabricated in Cuflon substrate with $\varepsilon_r = 2{\cdot}17$ and thickness 0·0254 cm (10 mil). They are stacked in the manner shown in Fig. 1. The upper patch is parasitic and the lower patch is fed with a coaxial probe excited at the resonant frequency of the TM_{01} mode ($f_{01} \simeq 10{\cdot}2$ GHz). Cardboard pieces with large holes in the centre are used as spacers separating the two layers. The thickness of each piece is 0·0508 cm (20 mil) and s can be made to increase in multiples of 0·0508 cm. For each spacing, measurements were made on E- and H-plane patterns, sweep frequency responses, and input impedance as a function of frequency. The results are summarised in Table 1. Explanations and observations follow:

(i) Our data cover the range $0 \leq s \leq 0{\cdot}37\,\lambda_0$ while previous studies of EMCPs did not go beyond $0{\cdot}1\,\lambda_0$.

(ii) The gain in column 5 of Table 1 is estimated from the formula[5]

$$\text{gain (dB)} \simeq 10 \log \left(\frac{26\,000}{HP_E^0\, HP_H^0}\right)$$

(iii) The percentage bandwidth is obtained from $1/Q \times 100\%$, where $Q = f_{01}/\Delta f$ and Δf is the width between the frequencies at which the input resistance drops to one-half its value at resonance.

(iv) The pattern shapes are described as normal when the patterns in both planes are maximum at broadside and are symmetrical. Deviations from this shape, such as a dip at broadside, are referred to as abnormal. Data for beamwidth and gain are given only for 'normal' patterns.

(v) The bandwidth exceeds 10% for $s = 0{\cdot}0508$ cm ($\simeq 0{\cdot}017\,\lambda_0$) and $s = 0{\cdot}102$ cm ($\simeq 0{\cdot}034\,\lambda_0$). The gain associated with these spacings is estimated to be about 7 dB, compared to about 5·3 dB for the single patch.

(vi) In the range 0·457 cm ($\simeq 0{\cdot}15\,\lambda_0$) $\leq s \leq$ 0·864 cm ($0{\cdot}29\,\lambda_0$), the E-plane patterns show a dip at broadside. These patterns are described as 'abnormal' and no data for beamwidth and gain are given.

(vii) Beginning at $s = 0{\cdot}914$ cm ($\simeq 0{\cdot}31\,\lambda_0$) and continuing to $s = 1{\cdot}118$ cm ($\simeq 0{\cdot}37\,\lambda_0$), the patterns become 'normal' again. Moreover, the 3 dB beamwidth in the H-plane is only 37°, which is considerably narrower than those in the range

Table 1 CHARACTERISTICS OF A RECTANGULAR ELECTROMAGNETICALLY COUPLED PATCH ANTENNA

Spacing s	f_{01}	Pattern shape	3 dB beamwidth	Estimated gain	Bandwidth
cm	GHz			dB	%
0	9·9	normal	95° × 73°	5·7	9·0
0·0508	9·95	normal	75° × 65°	7·3	13·0
0·102	10·10	normal	75° × 70°	7·0	10·5
0·152	10·45	normal	75° × 70°	7·0	6·2
0·204	10·46	normal	75° × 70°	7·0	4·8
0·254	10·48	normal	70° × 70°	7·2	3·4
0·305	10·46	normal	73° × 78°	6·6	2·9
0·356	10·46	normal	75° × 85°	6·1	2·9
0·406	10·40	normal	85° × 90°	5·3	2·6
0·457	10·37	abnormal	—	—	1·5
0·508	10·37	abnormal	—	—	1·5
0·610	10·34	abnormal	—	—	1·4
0·762	10·30	abnormal	—	—	1·3
0·864	10·30	abnormal	—	—	1·3
0·914	10·28	normal	90° × 37°	8·9	1·3
0·965	10·28	normal	90° × 37°	8·9	1·3
1·016	10·28	normal	85° × 37°	9·2	1·3
1·118	10·30	normal	70° × 37°	10·0	1·2
Single patch	10·20	normal	110° × 70°	5·3	2·3

$a = 1·5$ cm, $b = 1$ cm, $\varepsilon_r = 2·17$, $t = 0·254$ mm

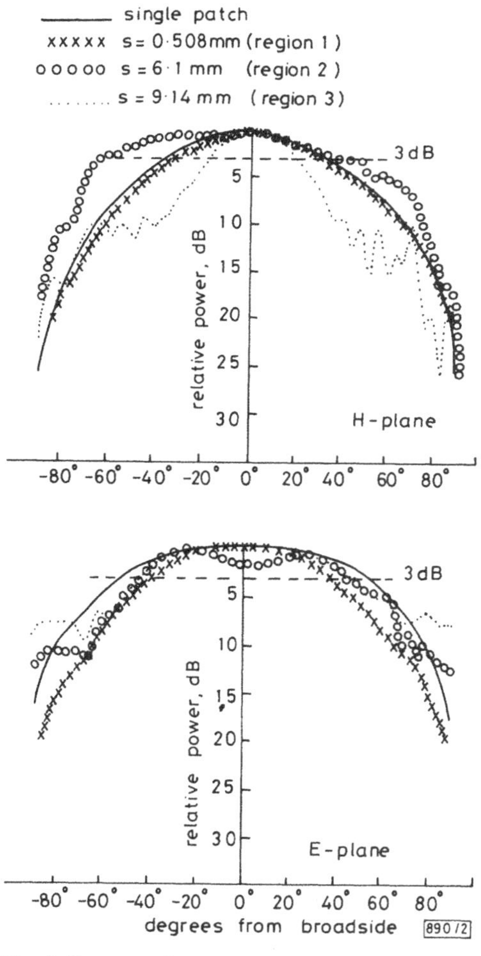

Fig. 2 *Patterns of a rectangular electromagnetically coupled patch antenna*

$a = 1·5$ cm, $b = 1$ cm, $\varepsilon_r = 2·17$, $t = 0·0254$ cm
$s = 0·0508$ cm (region 1), 0·61 cm (region 2) and 0·9 cm (region 3)
Patterns of a single patch are also shown (solid curves)

$0 \leq s \leq 0·406$ cm ($\simeq 0·14\lambda_0$). The gain is estimated to be about 8·9 dB. However, the bandwidth is only about 1·3%.

In conclusion, detailed experimental results for a rectangular EMCP antenna have been presented. It is found that, depending on the spacing s, the characteristics of the antenna can be separated into three regions. In region 1, occurring when s is between 0 and 0·406 cm ($\simeq 0·14\lambda_0$), the patterns show good broadside features. The bandwidth rises to 13% at $s = 0·0508$ cm ($\simeq 0·017\lambda_0$) and the gain is about 7 dB. At the upper boundary of this region ($s = 0·406$ cm), the bandwidth and the gain are about the same as the single patch. In region 2, occurring when s is between 0·457 cm and 0·864 cm, the E-plane patterns show a dip at broadside and the bandwidth is less than 2%. Little advantage is gained in operating the antenna in this region. In region 3, which begins at 0·914 cm ($\simeq 0·31\lambda_0$), the patterns return to the 'normal' shape and the gain increases to 8·9 dB. This high-gain region may be utilised in applications where narrow bandwidth is not a disadvantage. Representative patterns in the three regions are shown in Fig. 2. Those of the single patch are also included for the purpose of comparison.

References

1 SABBIN, A.: 'A new broadband stacked two-layer microstrip antenna'. IEEE AP-S int. symp. digest, 1983, pp. 63–66
2 CHEN, C. H., TULINTSEFF, A., and SORBELLO, R. M.: 'Broadband two-layer microstrip antenna'. Ibid., 1984, pp. 251–254
3 BHATNAGAR, P. S., DANIEL, J.-P., MAHDJOUBI, K., and TERRET, C.: 'Experimental study on stacked triangular microstrip antennas', *Electron. Lett.*, 1986, **22**, pp. 864–865
4 DAHELE, J. S., TUNG, S. H., and LEE, K. F.: 'Normal and inverted configurations of the broadband electromagnetic coupled microstrip antenna'. IEEE AP-S int. symp. digest, 1986, pp. 841–844
5 STUTZMAN, W. L., and THIELE, G. A.: 'Antenna theory and design' (John Wiley & Sons, 1981), p. 397

The SSFIP: A Global Concept for High Performance Broadband Planar Antennas

J. F. Zürcher

Indexing terms: Antennas, Planar antennas, Microstrip

The SSFIP (strip–slot–foam–inverted patch) antenna presents significant advantages over standard microstrip antennas: very broad bandwidth, high efficiency, low crosspolarisation level, integrated radome, lightweight and rigid construction and low cost. A 16-element array with more than 16dB gain and 21·1% bandwidth for SWR ≤ 2 shows what can be achieved.

Introduction: Microstrip patch antennas present significant advantages in terms of size, ease of fabrication and compatibility with printed circuits, but also a number of drawbacks, ranging from narrow bandwidth to low efficiency. Surface waves produce diffraction at the dielectric's edges and coupling between array elements, contributing to higher sidelobes and crosspolarisation levels. The use of a single substrate for both the radiating elements and the feed network is a poor compromise, since two contradictory functions are expected from the same structure. These problems and others were discussed recently by Hall *et al.*[1] Last but not least, patch antennas should be protected by a cover against environmental effects.

Taking all these factors into consideration a new global concept, the strip–slot–foam–inverted patch antenna, has been developed. Low-cost materials are used, the substrate being a low loss, low-permittivity foam to prevent surface wave propagation and to increase the bandwidth. The radiating patches are deposited on the underside of a thin plastic sheet (standard epoxy fibre-glass substrate), that also serves as protective cover. They are fed via wide coupling slots by a microstrip line located on a high-quality dielectric substrate underneath the ground plane (Fig. 1).

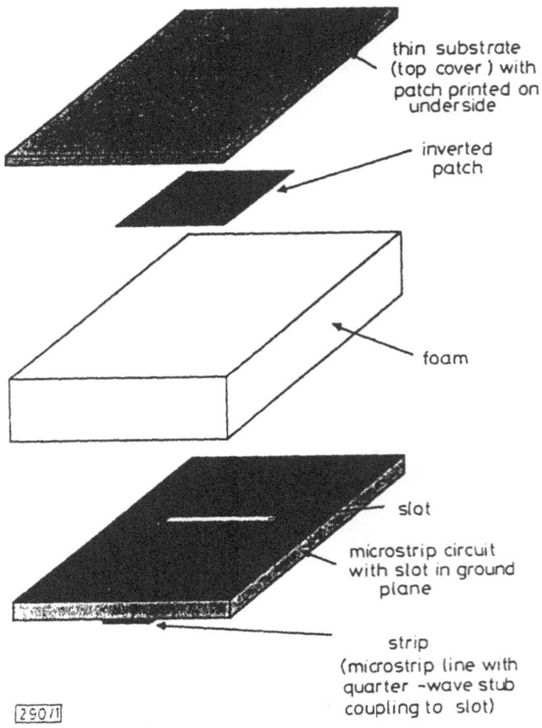

Fig. 1 *Exploded view of SSFIP antenna structure*

16-element SSFIP array with 21·1% bandwidth: A single-element antenna was first realised with the following materials:

(*a*) Microstrip substrate: RT/Duroïd 5870, $\varepsilon = 2{\cdot}33$, $h = 0{\cdot}254$ mm, with a 100Ω microstrip line exciting a slot in the ground plane.

(*b*) Foam: polymethacrylamid hard foam, $\varepsilon = 1{\cdot}07$, $\tan\delta = 8 \times 10^{-4}$, $h = 2$ mm.

(*c*) Patch support: epoxy fibre-glass, $\varepsilon = 4{\cdot}32$, $h = 0{\cdot}1$ mm, patch on underside.

The relative bandwidth of the final design, determined by experiment, was 11·7% for a SWR ≤ 2. The measured maximum gain was 5·8dB.

Using the optimised dimensions of this single-element antenna, a 16-element broadside antenna array (4 × 4) with a corporate feed network was designed. The feed network was carefully designed to reduce as well as possible the frequency sensitivity. The CAD/CAM program MICROS 6[2] was used to design and realise this array. Fig. 2*a* shows the mask of the

Reprinted with permission from *Elect. Lett.*, J.-F. Zurcher, "The SSFIP: A Global Concept for High Performance Broadband Planar Antennas," vol. 24, no. 23, pp. 1433–1435, Nov. 1988. © Institution of Electrical Engineers.

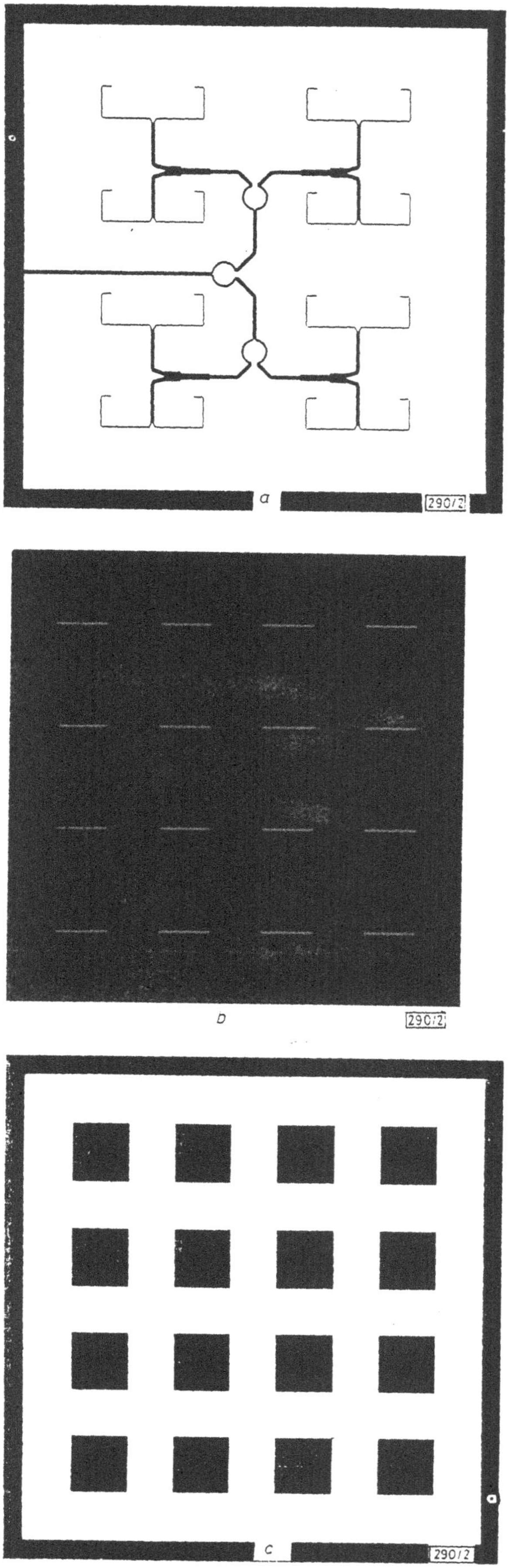

Fig. 2 *Masks of 16-element 9 GHz SSFIP array antenna*
a Feed network *b* Slot array *c* Patch array

feed network and Fig. 2*b* one of the slot array, both etched on an RT/Duroïd substrate (after aligning the masks). The mask of the 16 square patches, spaced 0·66λ apart, is shown in Fig. 2*c*.

The antenna was assembled with two-component epoxy cement and maintained under pressure while curing the cement. The sandwich structure provides high mechanical rigidity and flatness, together with a very low weight. Without the connector, the antenna only weighs 10·3 g with dimensions of 9 × 9 cm: this corresponds to a specific weight of 0·127 g/cm^2. With an SMA connector, the total weight is 11·7 g.

Fig. 3 shows the SWR between 7 and 11 GHz. The bandwidth for a SWR ≤ 2 is 1·9 GHz, yielding a relative bandwidth of 21·1%. The antenna can be used between 8 and 10 GHZ (markers 2 and 3 respectively on Fig. 3) with a maximum SWR of 2·5.

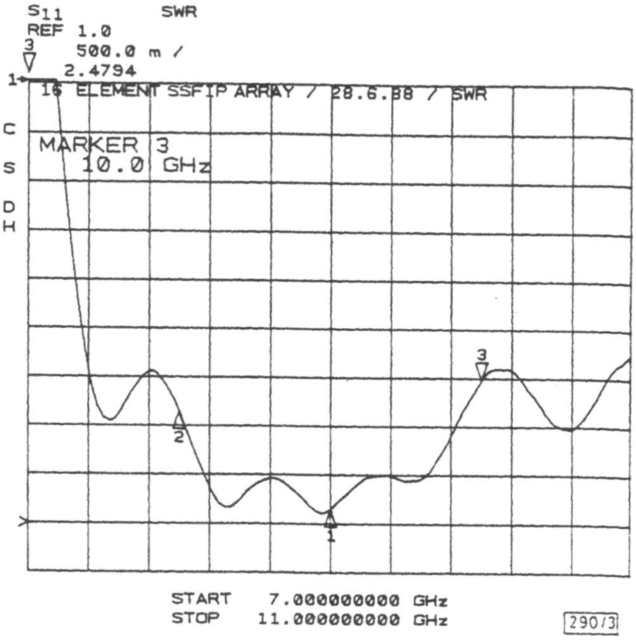

Fig. 3 *16-element SSFIP array antenna: measured SWR*

The radiation patterns, measured in our automated anechoïc chamber, are shown in Figs. 4*a* and *b* for the *H*- and *E*-plane patterns, respectively, at the centre frequency of 9 GHz. The patterns are quite symmetrical, with low crosspolar levels (worst case = −30 dB). The relatively high level of the first sidelobes results from the uniform feeding of all elements. The radiation patterns were measured from 7·5 to 11 GHz, and show practically no degradation when compared to the patterns at 9 GHz. Also, the crosspolar levels remain always below −25 dB. Altogether, if one can tolerate a maximum SWR of about 2·6, this antenna exhibits outstanding performance from 7·5 to 11 GHz.

The −3 dB aperture angles of the main lobe are 19·5° in the *H*-plane and 22·5° in the *E*-plane at 9 GHz, decreasing regularly with increasing frequency. The maximum gain measured using a substitution technique is:

Frequency	Gain
GHz	dB
8	16·1
9	16·8
10	16·6

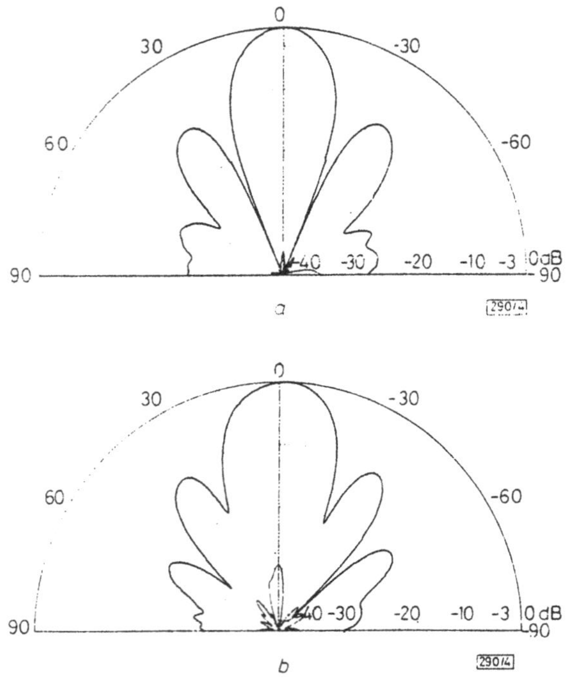

Fig. 4 *16-element SSFIP array antenna radiation patterns*
a 9 GHz, *H*-plane *b* 9 GHz, *E*-plane

At all three frequencies, the gain of the SSFIP array surpassed the gain of our Sivers standard X-band horn antenna (size = 77 × 53 × 135 mm, weight 166 g). Over the 7·5–11 GHz frequency range, the lowest measured gain was 15·7 dB.

Since the patches were excited by relatively large slots, a backward radiation is observed (front to back ratio of −17 dB), which could be suppressed by enclosing the microstrip feed circuit or by using a triplate technology.

Conclusion: Using a global concept (SSFIP = strip–slot–foam–inverted patch) which optimises both the electrical and the mechanical characteristics and includes a radome, a high-performance 16-element planar array antenna has been constructed, that could replace horn antennas in many applications. It has high gain, very wide bandwidth for a printed antenna, strong mechanical characteristics, very low weight and low price.

Future developments are planned to further increase the bandwidth and to provide accurate design tools.

I thank Prof. D. Pozar of the University of Massachusetts, Amherst for his valuable suggestions and encouragement.

References

1 HALL, P. S., and HALL, C. M.: 'Coplanar corporate feed effects in microstrip patch array design', *IEE Proc. H, Microwaves, Antennas & Propag.*, 1988, **135**, pp. 180–186

2 ZÜRCHER, J.-F.: 'MICROS3—a CAD/CAM program for fast realization of microstrip masks, description and typical applications'. IEE coll. on computer-aided design of microwave circuits, London, Nov. 1985, pp. 11/1–5

Millimeter-Wave Design of Wide-Band Aperture-Coupled Stacked Microstrip Antennas

Frederic Croq and David M. Pozar, *Fellow, IEEE*

Abstract—**K-band aperture-coupled stacked microstrip antennas are studied and numerical results, based on the solution of integral equations solved in the spectral domain are presented. The effects of varying physical parameters of the structure are investigated with a goal of designing millimeter-wave wideband microstrip antennas. Antennas with different characteristics are then analyzed and compared with experimental data. Bandwidths in excess of 20% are obtained and applications to phased array antennas are discussed.**

I. Introduction

THE design of microstrip antennas at millimeter wave frequencies is closely related to the feeding technique. At these frequencies, there are several problems associated with classical feeding techniques, such as coaxial probe or edge feeds: 1) performance can be severely degraded by the size of the feed, which can be comparable to the size of the patch itself, and 2) soldering of probe-feeds is prone to repeatability problems, while edge-feeding provides very little room for the feed network and associated devices. These considerations are even more important for wideband applications, which require thicker substrates. On the other hand, the aperture coupled feeding technique [1] has intrinsic properties which make it an attractive feature for millimeter wave applications [2]. Wide-band operation of this type of microstrip antenna has been demonstrated at microwave frequencies (1–10 GHz) using either single [3], [4] or stacked patch configurations [5]–[7]. Although both of them have been shown to give excellent bandwidth characteristics, the former structure gives rise to a high back-radiation level because the slot is near resonance, while this does not occur in the latter configuration. In this paper, the objective is to present a set of data for an understanding of the coupling phenomena which occur in the stacked patch structure shown in Fig. 1. These phenomena will be shown to determine the relative excitation of the resonators as well as the bandwidth and frequency characteristics. Based on this study, the feasibility of wideband stacked aperture coupled microstrip antennas is demonstrated at K-band both theoretically and experimentally through two examples. The comparison of these two examples demonstrates the necessary compromise between bandwidth and potential scanning performance when designing a wide-band radiating element for phased array antenna applications.

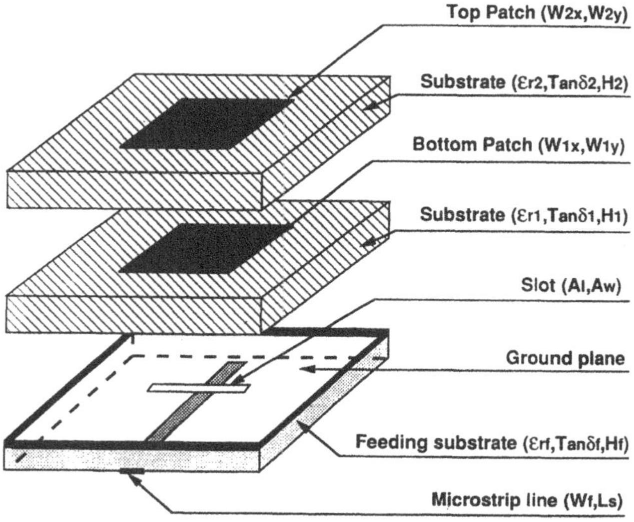

Fig. 1. Stacked patches aperture coupled to a microstrip line.

II. Method of Analysis

The theoretical analysis was based on the solution of integral equations solved in the spectral domain by the method of moments. Since this analysis is an extension of the method proposed for the single patch configuration [8], the details will not be included here. The analysis differs mainly in that a supplementary electric field integral equation is imposed on the second patch and that the exact Green's functions for the multilayered structure are necessary for the evaluation of reaction terms between the different sources. The reciprocity formulation permits the calculation of the equivalent series impedance introduced by the antenna on the microstrip line without calculating the current on the microstrip line itself. The Galerkin testing procedure was used on the patches and the slot. The unknown resonant currents on the patches were expanded in a set of entire domain basis functions, while the electric field in the nonresonant aperture was expanded with a single piecewise sinusoidal basis function.

III. Parameter Study

In this section the sensitivity of the geometric parameters will be studied for five parameters which are most critical in

Reprinted from *IEEE Trans. Antennas Propaga.*, vol. 39, no. 12, pp. 1770–1776, Dec. 1991.

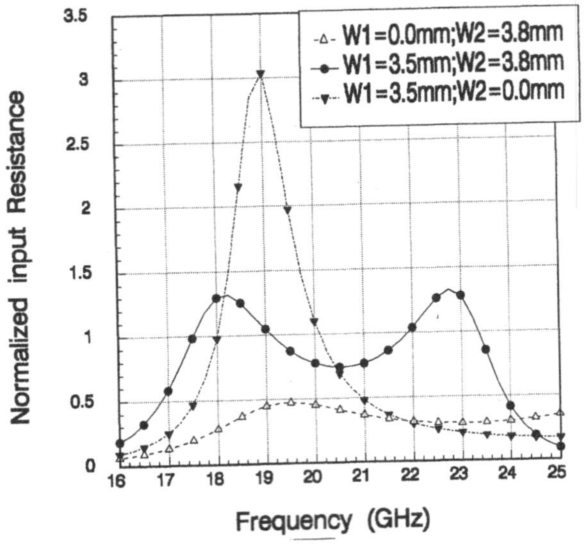

(a)

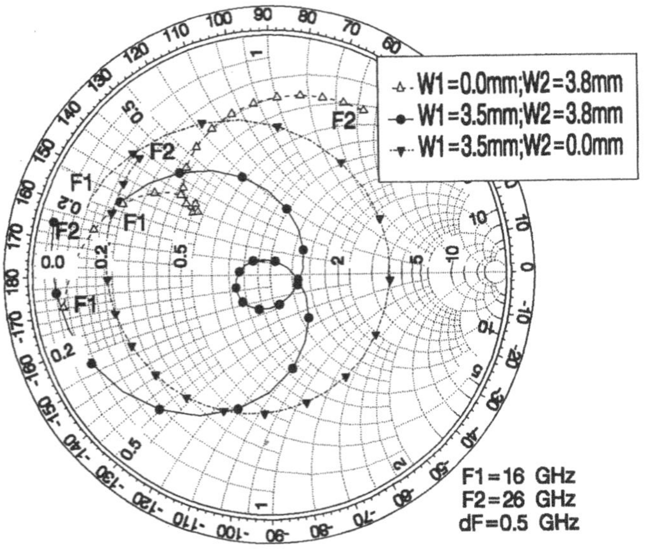

(b)

Fig. 2. (a) Real part of the input impedance of coupled and uncoupled patches. Other parameters: $\epsilon_{r2} = 2.20$; $H_2 = 1.0$ mm; $\tan \delta 2 = 0.0009$; $\epsilon_{r1} = 2.20$; $H_1 = 0.50$ mm; $\tan \delta 1 = 0.0009$; $A_1 = 3.2$ mm; $A_w = 0.4$ mm; $\epsilon_{rf} = 2.2$; $H_f = 0.508$ mm; $\tan \delta f = 0.0009$; $W_f = 1.55$ mm; $L_s = 1.8$ mm. (b) Impedance loci of the input impedance of coupled and uncoupled patches. Other parameters are similar to (a).

the design of aperture coupled stacked patch. These parameters are listed below:

- W_1 and W_2 the dimensions of the lower and upper square patches
- H_1 and H_2 the thicknesses of the two substrates supporting the patches
- A_1 the length of the coupling slot.

Data will be presented showing the individual effect of each of these key parameters. The other parameters (mainly the parameters of the feeding line) behave similarly to those presented for the single patch configuration [9], and thus are not discussed here. To aid in the understanding the operation of the aperture coupled patch antenna, much of the data are presented in two forms: a rectangular plot of input resistance versus frequency, and a Smith chart plot of input impedance versus frequency. It may at first appear that this is an unnecessary duplication, but in fact the former plot offers the best way of determining the coupling levels of the dual resonance system, while the latter plot includes the effect of reactance, and thus shows the overall impedance matching of the antenna.

As a basis for this parametric study, an antenna with wide-band characteristics was calculated such that the two resonances (defined as a maximum of the real part) resulting from the coupled patches are of equal amplitude. Fig. 2(a) shows the real part of the input impedance of such an antenna, as well as the input impedance of each patch when taken in isolation. The bottom patch alone was overexcited, while the top patch alone was underexcited, with a low Q. The input resistance and resonant frequency for these isolated patches are referenced as (R_{b0}, f_{b0}) for the bottom patch and (R_{t0}, f_{t0}) for the top patch, and are used for comparison with the parametric study of the stacked configurations. It appears that the coupling of these two resonators generates two new resonances, one below and one above the former independent ones. These new resonant frequencies do not have a simple or direct relation to the former ones, and it is very difficult to make general statements about their characteristics. The Smith chart of Fig. 2(b) shows the impedance loci of the coupled and uncoupled configuration. The top patch impedance locus contains a loop which implies its coupling to the resonance of the slot.

Fig. 3(a) shows the evolution of the real part of the stacked patch antenna impedance as a function of the bottom patch size ($W_{x1} = W_{y1} = W_1$). These results should at least in some ways be consistent with the known behavior [9] of single patches. For example, it is known [9] that as the ratio W_1 / A_1 (patch size to slot size) is increased, the coupling to the patch is decreased. From another point of view, as W_1 is increased, the top patch becomes isolated from the excitation field of the slot, and the coupling to the fringing field is reduced. Both actions contribute to reducing the coupling of the top patch. Thus, it seems that both resonances should be lowered as W_1 is increased. In fact, the excitation of the lower resonance (R_{1w}, f_{1w}) decreased when the size of the patch decreased and converges to the behavior of the top patch alone (R_{t0}, f_{t0}) as W_1 tends to zero. But as W_1 was decreased, the upper resonance $(R_{\mathrm{up}}, f_{\mathrm{up}})$ was strongly excited at a higher frequency. With regard to the frequencies of these resonances, an increase of 10% in W_1 reduced the lower resonance by 4% in frequency, while the upper resonance only decreased by 2.6%; Conversely, when W_1 was reduced by 10%, the increase in the lower resonance was 1.1%, while the upper resonance increased by 4.3%. This behavior suggests that each resonator affects both resonances. Starting with resonances equivalently excited, when the dimensions of the bottom patch were increased it became the dominant factor for the lower resonance, and inversely when its dimensions were decreased it tended to dominate the higher resonance. The corresponding Smith chart representation of Fig. 3(b) shows that the impedance locus is more

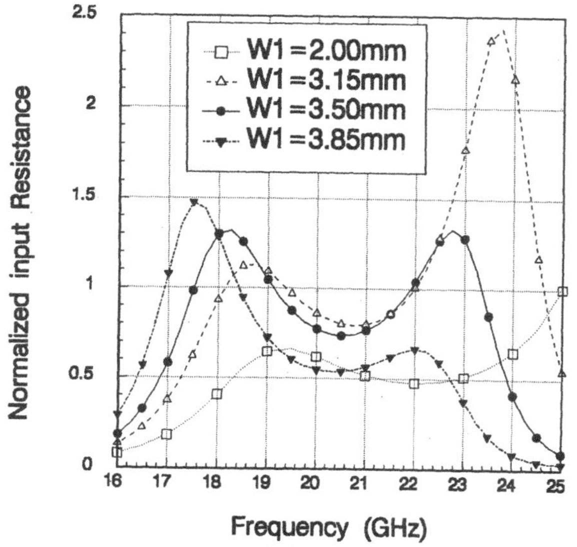

(a)

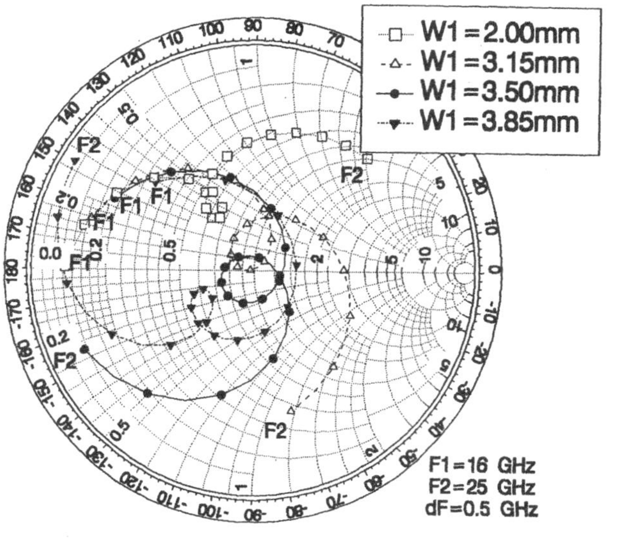

(b)

Fig. 3. Real part of the input impedance as a function of the bottom patch dimensions $W_1 = W_{x1} = W_{y1}$. Other parameters: $W_{x2} = W_{y2} = 3.8$ mm; $\epsilon_{r2} = 2.20$; $H_2 = 1.0$ mm; tan $\delta 2 = 0.0009$; $\epsilon_{r1} = 2.20$; $H_1 = 0.50$ mm; tan $\delta 1 = 0.0009$; $A_1 = 3.2$ mm; $A_w = 0.4$ mm; $\epsilon_{rf} = 2.2$; $H_f = 0.508$ mm; tan $\delta f = 0.0009$; $W_f = 1.55$ mm; $L_s = 1.8$ mm. (b) Impedance loci of the input impedance as a function of the bottom patch dimensions $W_1 = W_{x1} = W_{y1}$. Other parameters are similar to (a).

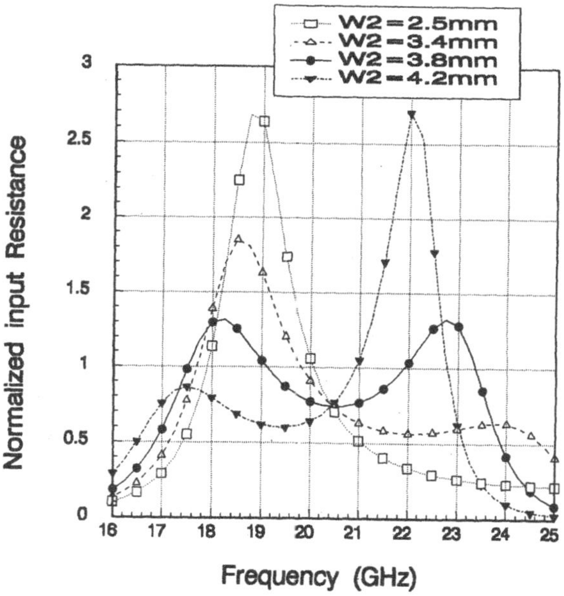

(a)

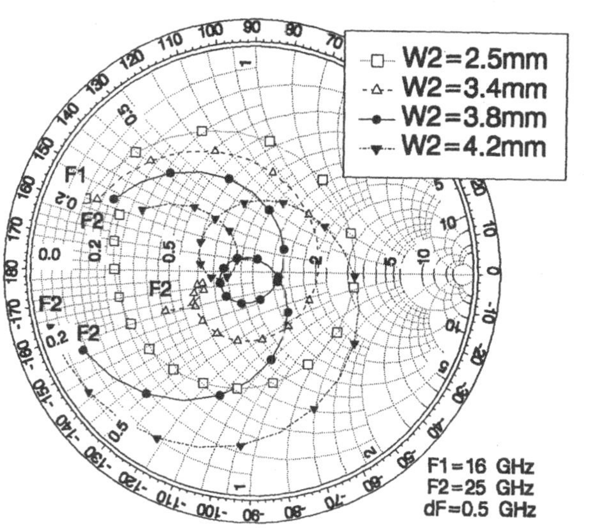

(b)

Fig. 4. (a) Real part of the input impedance as a function of the top patch dimensions $W_2 = W_{x2} = W_{y2}$. Other parameters: $\epsilon_{r2} = 2.20$; $H_2 = 1.0$ mm; tan $\delta 2 = 0.0009$; $W_{x1} = W_{y1} = 3.5$ mm; $\epsilon_{r1} = 2.20$; $H_1 = 0.50$ mm; tan $\delta 1 = 0.0009$; $A_1 = 3.2$ mm; $A_w = 0.4$ mm; $\epsilon_{rf} = 2.2$; $H_f = 0.508$ mm; tan $\delta f = 0.0009$; $W_f = 1.55$ mm; $L_s = 1.8$ mm. (b) Impedance loci of the input impedance as a function of the top patch dimensions $W_2 = W_{x2} = W_{y2}$. Other parameters are similar to (a).

capacitive as W_1 increases. The size of the loop of the impedance locus had a maximum when the resonant peaks of the input resistance are equal. This criteria was found to be a good measure of the coupling between the two resonators, as will be seen in the following curves. For the case where $W_1 = 2$ mm, we can see that the upper resonance tends to behave like the case $W_1 = 0$ mm in Fig. 2(b) where the upper resonance is the slot resonance. This suggests that the coupling slot also contributes to the resonant behavior, and becomes dominant in the upper resonance as W_1 was reduced to zero.

Fig. 4(a) shows the influence of the size of the top square patch ($W_{x2} = W_{y2} = W_2$) on the real part of the impedance of the antenna. As W_2 is decreased, the coupling to the lower resonance increases and tends, as W_2 becomes small, to behave like the resonance of the bottom patch alone (R_{b0}, f_{b0}). The fringing fields of the bottom patch explain this behavior since, as the size of the top patch is reduced, its coupling to the fringing field of the bottom patch becomes negligible. This behavior is opposite to what was noted for the variation of W_1. This seems logical since increasing W_1 and reducing W_2 have the same effect of decoupling the top patch from the fringing field of the bottom patch. Resonant frequencies were again affected by 6 to 7% for a 20% variation of W_2, which also implies that the overall structure will be more stable and less sensitive to fabrication tolerances than a single patch structure. The Smith chart representation of Fig. 4(b) shows behavior opposite to that observed with varying W_1, since the impedance locus in the useful frequency range becomes more inductive with increasing W_2. Again a maximum in the size of the coupling loop was observed, which quickly disappeared as W_2 increased.

Fig. 5(a) presents the resonance characteristics as a function of the thickness (H_1) of the substrate supporting the bottom patch. As H_1 is increased, both patches are moved away from the coupling slot. Relative to the single patch behavior, it seems that in the stacked patch configuration both resonances should decrease with increasing H_1. However, it is seen from Fig. 5(a) that this did not occur. Even though coupling to the lower resonance was strongly reduced with increasing H_1, the upper resonance increased to a maximum before decreasing. The resonant frequencies were only weakly affected by changing H_1. The imaginary part shown in the Smith chart plot in Fig. 5(b) became capacitive as H_1 decreased.

Fig. 6(a) shows the effect of distance between the two patches (H_2) on the resonances of the structure. The lower resonance for small value of H_2, and the upper resonance for large value of H_2, both tend to similar behavior. This implies that when the patches are very close they appear to the slot as a single patch, though the coupling between them generates another resonance at a high frequency. Conversely, for large values of H_2 the behavior of the overall structure was close to the behavior of the bottom patch alone (R_{b0}, f_{b0}) since the coupling to the top patch was very weak.

It was also seen that the coupling between the two patches affected the frequency of the upper resonance, which increased as H_2 decreased. In the Smith chart of Fig. 6(b), the degree of coupling is implied by the size of the loop of the impedance locus, which became bigger as the gap between the patches was reduced. The size of this loop will obviously have a primary effect on the bandwidth characteristics of the antenna, as well as on the scan blindness (which is known to be related to the total thickness of the antenna).

Lastly, the influence of the length of the slot is represented in Fig. 7(a). Contrary to the single patch configuration, where A_1 had an opposite effect to that of W_1 on the excitation of the patch, in the double patch structure these two parameters have the same effect, although to a different degree. It is interesting to note that the slot length affects the resonant frequency of the antenna as much as a variation of patch dimensions would have (i.e., Figs. 3 and 5).

The above effects imply that the three resonators of this structure are intimately coupled and interact strongly. For the two resonances that make up the useful bandwidth of the antenna, it is seen that any variation of the geometric parameters results in a modification of the relative excitation of the resonances. These effects are interdependent, and sometimes have countering trends, allowing design flexibility. Thus, for example, the length of the slot and the back radiation associated with it can be reduced by reducing the thickness of the bottom patch substrate. This would also increase the Q of the antenna, and reduce the bandwidth. Other examples of interactive effects can be used for dual frequency operation, or high gain capability.

The set of data presented in this section should be useful for the design of wide-band microstrip antennas, and for the evaluation of their sensitivity to fabrication errors.

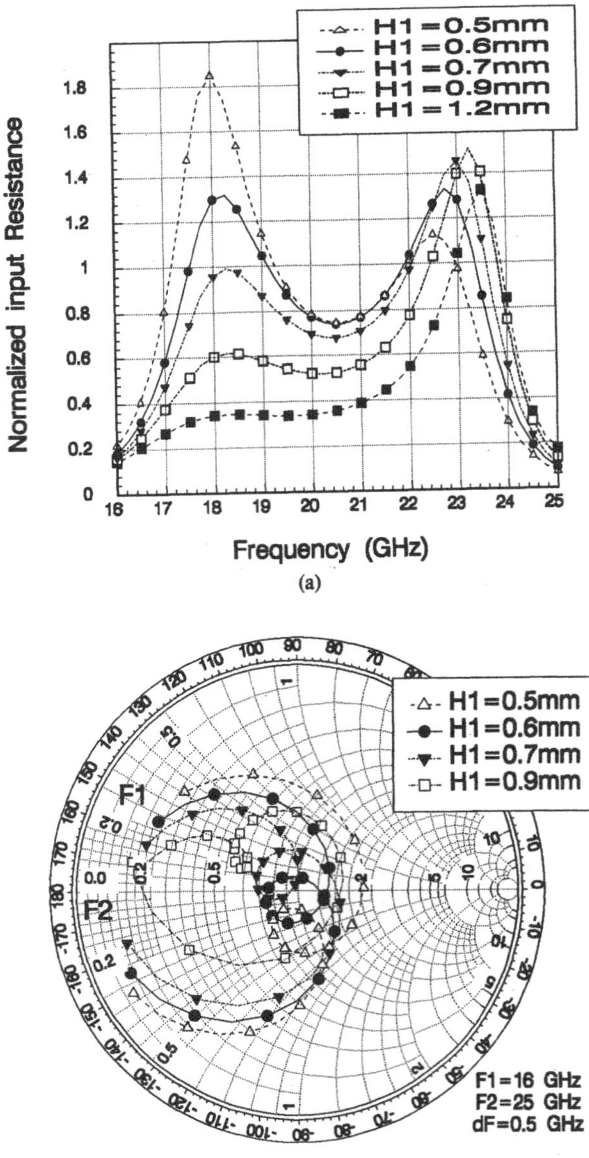

Fig. 5. (a) Real part of the input impedance as a function of the bottom substrate thickness H_1. Other parameters: $W_{x2} = W_{y2} = 3.8$ mm; $\epsilon_{r2} = 2.20$; $H_2 = 1.0$ mm; $\tan\delta 2 = 0.0009$; $W_{x1} = W_{y1} = 3.5$ mm; $\epsilon_{r1} = 2.20$; $\tan\delta 1 = 0.0009$; $A_1 = 3.2$ mm; $A_w = 0.4$ mm; $\epsilon_{rf} = 2.2$; $H_f = 0.508$ mm; $\tan\delta f = 0.0009$; $W_f = 1.55$ mm; $L_s = 1.8$ mm. (b) Impedance loci of the input impedance as a function of the bottom substrate thickness H_1. Other parameters are similar to (a).

IV. Design Considerations, Results, and Discussion

Two wide-band antennas, with two different design objectives, were fabricated and tested. The first antenna, antenna #1 was designed for phased array antenna applications. Consequently, for scan blindness considerations [10], efforts were made to limit the thickness of the dielectric slabs within the design bandwidth goal which was BW > 20% for S11 < − 10 dB. The second antenna, antenna #2, was designed for better matching over a wide bandwidth (BW > 20% for S11 < − 15 dB), without concern for scanning mismatch.

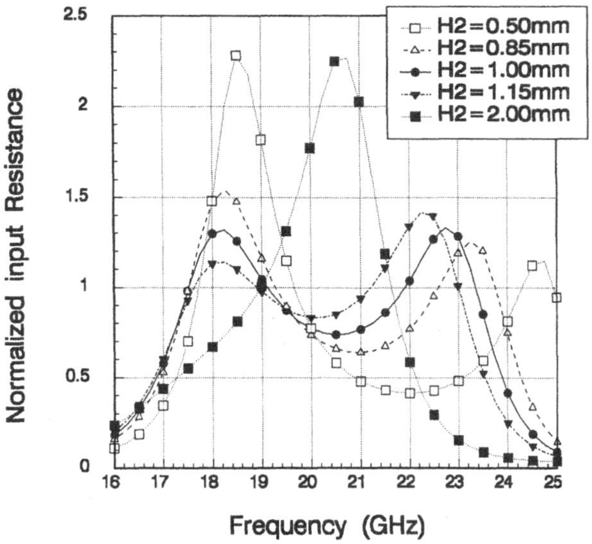

(a)

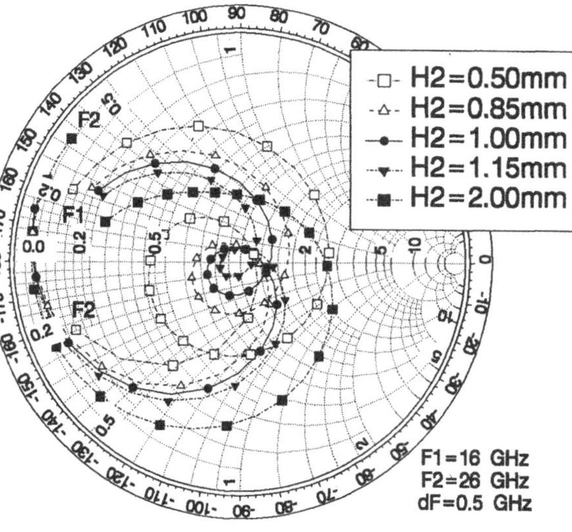

(b)

Fig. 6. (a) Real part of the input impedance as a function of the top substrate thickness H_2. Other parameters: $W_{x2} = W_{y2} = 3.8$ mm; $\epsilon_{r2} = 2.20$; $\tan \delta 2 = 0.0009$; $W_{x1} = W_{y1} = 3.5$ mm; $\epsilon_{r1} = 2.20$; $H_1 = 0.50$ mm; $\tan \delta 1 = 0.0009$; $A_1 = 3.2$ mm; $A_w = 0.4$ mm; $\epsilon_{rf} = 2.2$; $H_f = 0.508$ mm; $\tan \delta f = 0.0009$; $W_f = 1.55$ mm; $L_s = 1.8$ mm. (b) Impedance loci of the input impedance as a function of the top substrate thickness H_2. Other parameters are similar to (a).

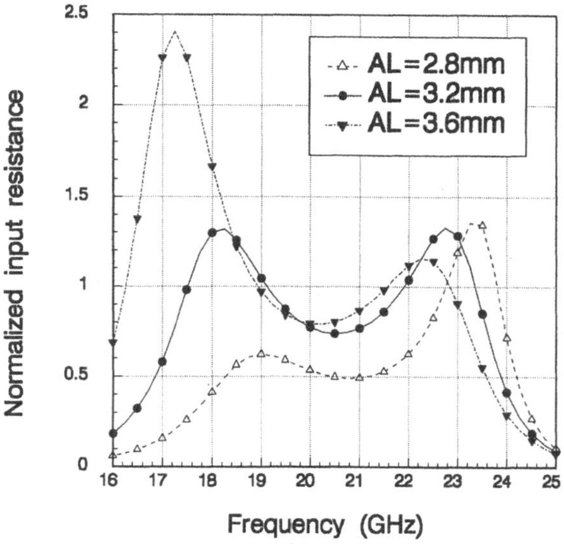

(a)

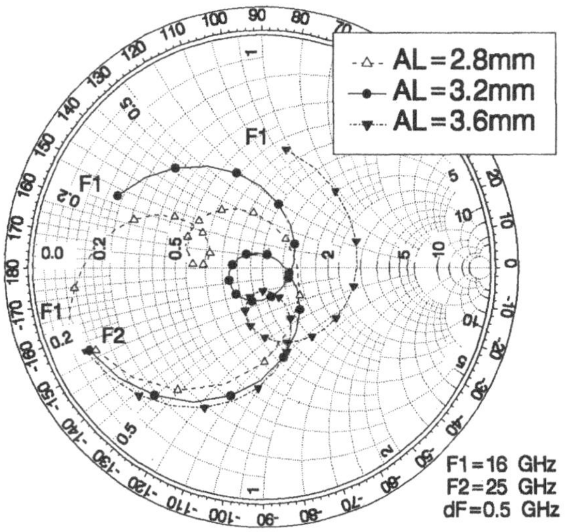

(b)

Fig. 7. (a) Real part of the input impedance as a function of the length of the coupling aperture A_1. Other parameters: $W_{x2} = W_{y2} = 3.8$ mm; $\epsilon_{r2} = 2.20$; $H_2 = 1.0$ mm; $\tan \delta 2 = 0.0009$; $W_{x1} = W_{y1} = 3.5$ mm; $\epsilon_{r1} = 2.20$; $H_1 = 0.50$ mm; $\tan \delta 1 = 0.0009$; $A_w = 0.4$ mm; $\epsilon_{rf} = 2.2$; $H_f = 0.508$ mm; $\tan \delta f = 0.0009$; $W_f = 1.55$ mm; $L_s = 1.8$ mm. (b) Impedance loci of the input impedance as a function of the length of the coupling aperture. Other parameters are similar to (a).

The dimensions of these antennas are given below with reference to Fig. 1.

Antenna #1: $W_{x2} = W_{y2} = 3.8$ mm; $\epsilon_{r2} = 2.33$; $H_2 = 0.7874$ mm; $\tan \delta 2 = 0.0012$; $W_{x1} = W_{y1} = 3.5$ mm; $\epsilon_{r1} = 2.2$; $H_1 = 0.508$ mm; $\tan \delta 1 = 0.0009$; $A_1 = 3.2$ mm; $A_w = 0.4$ mm; $\epsilon_{rf} = 2.2$; $H_f = 0.508$ mm; $\tan \delta f = 0.0009$; $W_f = 1.55$ mm; $L_s = 1.8$ mm.

Antenna #2: $W_{x2} = W_{y2} = 3.5$ mm; $\epsilon_{r2} = 2.33$; $H_2 = 1.15$ mm; $\tan \delta 2 = 0.0012$; $W_{x1} = W_{y1} = 3.3$ mm; $\epsilon_{r1} = 2.20$; $H_1 = 0.508$ mm; $\tan \delta 1 = 0.0009$; $A_1 = 3.1$ mm; $A_w = 0.4$ mm; $\epsilon_{rf} = 2.2$; $H_f = 0.508$ mm; $\tan \delta f = 0.0009$; $W_f = 1.55$ mm; $L_s = 1.9$ mm.

The $\tan \delta i$ are the loss tangents of the dielectric materials.

According to Fig. 6(a), the limitation of the substrate thicknesses in the first design leads to a configuration where the patches are strongly coupled. This coupling tends to separate the resonant frequencies of the resonators with a high Q factor. Conversely, a good match for the design of antenna #2 was obtained by minimizing the coupling between two resonant patches to that their resonant frequencies are close, and the loop resulting from the coupling is small and centered on the Smith chart. This result was obtained by increasing the inter-resonator distance, H_2.

Fig. 8 gives the theoretical and experimental impedance loci of antenna #1. The coupling between the patches was well described, as shown by the size of the loop on the Smith chart. The comparison of theoretical and measured results was reasonably good. The center frequencies and bandwidths are compared below:

Experimental: f_0 = 20.42 GHz, BW = 32.5% (17.1–23.75 GHz) S11 < −10 dB)

Theoretical: f_0 = 20.375 GHz, BW = 27.2% (17.6–23.15 GHz) (S11 < −10 dB)

Fig. 9 shows the theoretical and experimental impedance loci of antenna #2. In this case, even though the coupling between the two resonators was still well-formed (as shown by the size of the loop), the experimental loop was off-centered from the Smith chart. This emphasizes the fact that when the patches are less coupled, they were also more sensitive to fabrication tolerances. These errors were probably increased by the use in this case of two dielectric slabs to achieve the necessary inter-resonator thickness. Figs. 5(a) and 6(a) have shown how a small change in H_1 and H_2 could affect the excitation of the lower resonance, which was the problem here. For this reason, the matching goal for this antenna was not achieved. However, the size of the loop shows the feasibility of such characteristics and an acceptable comparison was still available for the criteria S11 < −12 dB:

Experimental: f_0 = 20.90 GHz, BW = 24.4% (18.35–23.45 GHz)

Theoretical: f_0 = 20.75 GHz, BW = 24.6% (18.20–23.3 GHz).

It must be mentioned that the measurements have been corrected for loss effects and connector mismatch, and that part of the differences observed between theoretical modeling and measurements may have been introduced by the imperfect nature of these corrections, which are difficult at these frequencies. Additionally, an electrical delay was applied to shift the reference plane to the slot, but this shift did not account for dispersion in the line, which may not be negligible in a microstrip line at these frequencies. The effect of the connector was removed using the HP8510B time domain gating technique, although some error could have been introduced due to the limited bandwidth of the measurement. Losses in the feeding line were evaluated on a separate board, and then removed from the measurements of the antennas by normalizing the measurements to the losses using trace math. Lastly, in the calculations, the glue used to bond the various substrates was simulated by a dielectric slab 0.05 mm thick with a 2.2 dielectric constant.

For phased array applications, an important criteria is the total thickness of the antenna, which greatly influences the scanning properties [10]. It is then interesting to compare the normalized total thicknesses, h_t, of the slabs used for these antennas at two extreme frequencies in the bandwidth (f_1 = 17 GHz, f_h = 23 GHz):

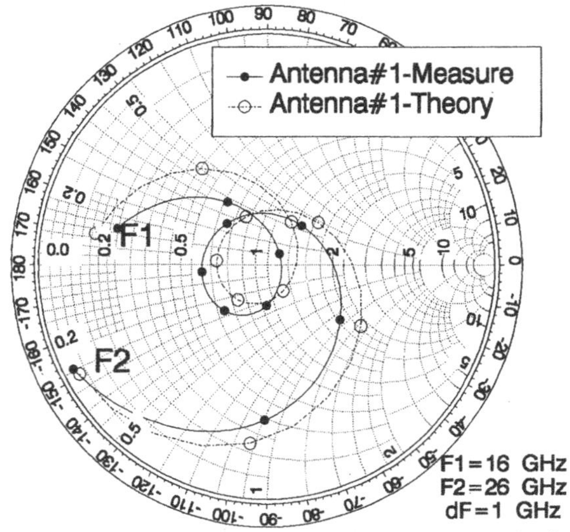

Fig. 8. Antenna #1 impedance locus, reference plane at the slot.

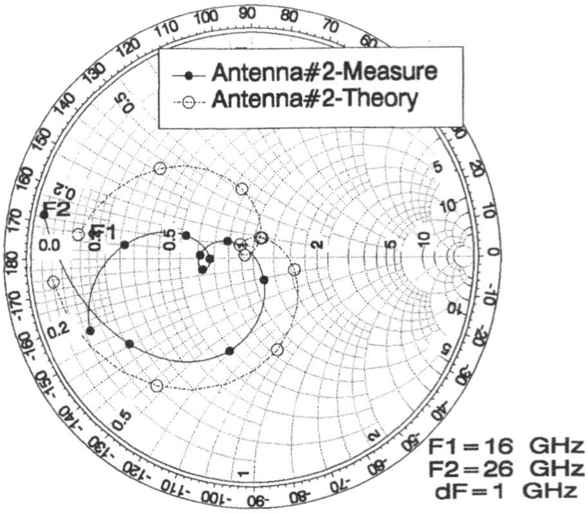

Fig. 9. Antenna #2 impedance locus, reference plane at the slot.

V. Conclusion

Wide-band operation of aperture coupled stacked microstrip patches was studied at K-band. Results based on the solution of integral equations solved in the spectral domain by a moment method have been presented. A study of the principal parameters of the structure has given a more complete understanding of the coupling mechanism between the different resonators. Based on this study, two different antennas were designed and tested. Good agreement between theory and experiment confirm the high frequency capabilities of this aperture feeding technique, and the validity of the model. Due to the slab thicknesses necessary to achieve the desired bandwidth, these antennas will generate scan blindnesses problems when integrated in a phased array antenna, but it should be possible to locate these blindnesses beyond the desired scanning range.

	$h_t/\lambda_0\ (f_1)$	θ_b ($\lambda_0/2$ spacing)	$h_t/\lambda_0\ (f_h)$	θ_b ($\lambda_0/2$ spacing)
Antenna #1	0.074	75°	0.099	69°
Antenna #2	0.094	70°	0.127	63°

This table also shows the location of the scan blindness angle, assuming a $\lambda_0/2$ element spacing. From these data, it is clear that the thicknesses involved in both antennas are fairly large and that this might be the source of potential scan blindness problems, especially at higher frequencies.

References

[1] D. M. Pozar, "A microstrip antenna aperture coupled to a microstripline," *Electron. Lett.*, vol. 21, no. 2, pp. 49-50, Jan. 17, 1985.

[2] D. M. Pozar and D. H. Schaubert, "Comparison of architectures for monolithic phased array antennas," *Microwave J.*, pp. 93-103, Mar. 1986.

[3] J. F. Zürcher, "The SSFIP: A global concept for high performance broadband planar antennas," *Electron. Lett.*, vol. 24, no. 23, pp. 1433-1435, Nov. 10, 1988.

[4] F. Croq and A. Papiernik, "Wideband aperture coupled microstrip antenna," *Electron. Lett.*, vol. 26, no. 16, pp. 1293-1294, Aug. 2, 1990.

[5] C. H. Tsao, Y. M. Hwang, F. Killburg, and F. Dietrich, "Aperture coupled patch antenna with wide bandwidth and dual polarization capabilities," in *IEEE Antennas Propagat. Soc. Symp. Dig.*, Syracuse, NY, 1988, pp. 936-939.

[6] F. Croq, "Antenne microruban multicouches a large bande passante et haute purete de polarisation," 3rd cycle thesis, Univ. Nice-Sophia Antipolis, France, Oct. 1990.

[7] J. Wang, R. Fralich, C. Wu, and J. Litva, "Multifunctional aperture coupled stack antenna," *Electron. Lett.*, vol. 26, no. 25, Dec. 1990.

[8] D. M. Pozar, "A reciprocity method of analysis of printed slots and slot coupled microstrip antennas," *IEEE Trans. Antennas Propagat.*, vol. AP-34, pp. 1439-1446, Dec. 1986.

[9] P. L. Sullivan and D. H. Schaubert, "Analysis of an aperture coupled microstrip patch," *IEEE Trans. Antennas Propagat.*, vol. AP-34, pp. 977-984, 1986.

[10] D. M. Pozar, "Analysis of an infinite phased array of aperture coupled microstrip patches," *IEEE Trans. Antennas Propagat.*, vol. 37, pp. 418-425, Apr. 1989.

Multioctave bandwidth log-periodic microstrip antenna array

P. S. Hall, M.Eng., Ph.D., C.Eng., M.I.E.E.

Indexing term: *Antennas (microstrip)*

Abstract: The application of the log-periodic technique to the series-fed electromagnetically coupled overlaid-patch array allows antennas with flat conformal characteristics and wide bandwidths to be obtained. A k-β analysis of this and other microstrip array types indicate that the microstrip patch is not an optimum element for log-periodic arrays and that direct connection will result in arrays having a limited bandwidth. The addition of series capacitance to the patch equivalent circuit, implemented by electromagnetic coupling, allows an optimum to be approached. Log-periodic overlaid patch array design and measured results for an array with a 4 : 1 bandwidth are presented. These, together with a transmission-line analysis, indicate the array-design trade offs available and that the ultimate bandwidth is limited primarily by changes in the input return loss and radiation pattern due to the use of uniform thickness substrates.

List of principal symbols

d = element spacing
d_p, d_l = reference plane extensions
f = frequency
g_r, b_r = conductance and susceptance of microstrip line open end
h_p, h_f = patch and feed substrate heights
k, k_0 = feed-line and free-space wavenumbers
l, l_q = patch and quarter-wave feed-line lengths
m = radiating element and patch number
n = transformer turns ratio of overlaid patch coupling
p = overlaid patch displacement
t_1, t_2 = patch-line lengths
w, w_f, w_q, w_{ep} = patch, feed-line, quarter-wave feed-line and patch equivalent width, respectively
(x, y) = Cartesian co-ordinates
$(ABCD)$ = transmission-line matrix elements
I_1, I_2 = patch feed currents
H_1, M_2 = patch magnetic field and magnetic current source
L = array length
M = number of array elements
S_{11}, S_{12} = scattering parameters
V = voltage
W = array width
X_c = reactance of overlaid patch coupling
Y, Y', Y_r, Y_{11}, Y_{1m} = admittances
Z_0, Z_p = feed ($= 1/Y_0$) and patch-line ($= 1/Y_p$) impedance
Z_1, Z_{12}, Z_{21} = patch terminal and patch mutual impedances, respectively
α = array attenuation constant
β', β, β_0 = array complex, array and free-space propagation constant
γ_p, γ_0 = patch-line and feed-line complex propagation constant
$\tan\delta$ = loss tangent
ε_0, ε_r, ε_e = free-space, relative and effective permittivities
θ = angular variable in radiation pattern
λ_m = microstrip wavelength
μ_0 = free-space permeability
τ, τ_0, τ_p = scale factors for generalised array, eqn. (5), patch width, length and spacing and patch displacement, respectively
ω = angular frequency = $2\pi f$

1 Introduction

The need for antennas to cover very wide bandwidths is of continuing importance, particularly in the field of electronic warfare [1] and wideband radar and measuring systems [2]. The application of microstrip antenna techniques to these areas is limited by the inherently narrow bandwidth capability of conventional microstrip radiating elements. Examples of these, such as the patch with single [3] and multiple layers [4], the spiral [5] and travelling-wave arrays [6, 7], show that, in general, bandwidths of the order of 40% cannot be readily exceeded. Arrays of stagger tuned patches have been fed corporately [8] in an attempt to overcome this limitation, although, due to the problem of impedance control in the feed, the extension to wide bandwidth is limited.

The application of log-periodic techniques to microstrip series fed arrays suggest that wideband action can, however, be obtained. Indeed, log-periodic electromagnetically coupled overlaid patch arrays [9, 10] and quarter-wavelength line-coupled patch arrays [11] have been made, although to date the maximum bandwidth obtained in either case is less than 50%. Although many various forms of narrowband microstrip series fed array exist that have similar array action, the choice of array configuration for very wideband log periodic action is less clear cut, due to the particular behaviour required of the radiating elements over the whole array bandwidth. The suitability of the various configurations can be assessed by examining their k-β characteristics in the light of criteria established for other log-periodic arrays. This is carried out in Section 2 for representative array forms, where, together with design recommendations, some factors that limit the ultimate bandwidth are identified. These recommendations are implemented in Section 3, where the design procedure and measured results for an overlaid patch array having a 4 : 1 bandwidth are presented. Trade offs in the design process have been investigated analytically using a transmission-line model, and the results are given where they relate to design parameter selection. In addition, limitations in the array performance are noted and results for antennas designed to overcome them are presented. Overall conclusions and an Appendix containing additional technical detail then follow.

Reprinted with permission from *Proc. IEE,* P. S. Hall, "Multioctave Bandwidth Log-Periodic Microstrip Antenna Array," vol. 133, pt. H, no. 2, pp. 127–136, April 1986. © Institution of Electrical Engineers.

2 Application of the log-periodic technique to microstrip series arrays

2.1 k-β analysis

The propagation characteristic of an equivalent uniform array has been shown to predict well the first-order behaviour of a log-periodic array [12], particularly when the change of image impedance from period to period is small [13]. The characteristic equation of an infinite uniform structure composed of loaded periodic cells Fig. 1*a* is [12]

$$\cos \beta' d = \cos kd + j \times \left\{ Y_{11} + 2 \sum_{m=1}^{M} Y_{1(m+1)} \cos m\beta' d \right\} \frac{\sin kd}{2Y_0} \quad (1)$$

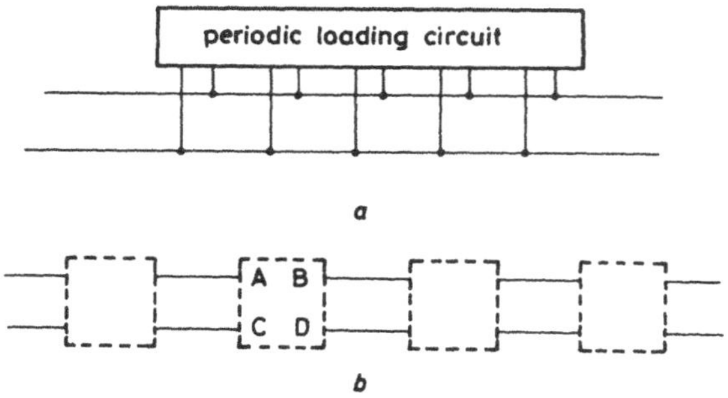

Fig. 1 *Sections of infinite uniform array*
a Feed-line with periodic loading circuit
b Cascaded two-port networks

where $\beta' = \beta + j\alpha$ is the complex propagation constant, d is the array period length, and Y_0 and k are the feed-line admittance and wavenumber, respectively. For a microstrip line $k = 2\pi f\sqrt{(\varepsilon_e)}/c$, where f is the frequency, $c = 3 \times 10^8$ m/s and ε_e is the effective dielectric constant. Y_{11} and Y_{1m} are the self and mutual admittance of the loading. If only adjacent cell coupling is included, then eqn. 1 simplifies to

$$\cos \beta' d = \frac{\cos kd + jY_{11} \dfrac{\sin kd}{2Y_0}}{1 - jY_{12} \dfrac{\sin kd}{Y_0}} \quad (2)$$

If no mutual coupling is included, then eqn. 2 reduces to the result given in Reference [13].

$$\cos \beta' d = A \quad (3)$$

where A is a parameter of the $ABCD$ matrix representing a single cell of the structure as a two-port network, as shown in Fig. 1*b*. In application to log-periodic arrays, the propagation constant β' at low frequencies represents the behaviour of the transmission-line region between the input and the active region and the characteristics around resonance give insight into the structure of the active region.

2.2 Criteria for wideband array action

Such analysis, together with empirical deductions, has allowed the following recommendations to be evolved [14] for good wideband array action:

(*a*) To prevent excitation of the higher order resonances in the low-frequency elements beyond the active region:

(i) the array should be fed from the high frequency end

(ii) the array should have high attenuation within and beyond the active region

(iii) the array radiation pattern should have a null in the direction of the wave propagating along the array.

(*b*) To ensure wideband action, the propagation characteristic should have no stopbands below the frequency of the active region.

2.3 Propagation characteristics of microstrip arrays

Propagation characteristics have been calculated for the overlaid patch array, the quarter-wavelength line-coupled array, the comb-line array [15] and the series-connected patch array [7]. These types are representative of many series-fed array configurations [16], although further types may well be developed to provide optimised performance. Series arrays formed by bends in the feed line, such as the rampart line [6], are not considered. Although similar shapes have been used to form wire-skeleton log-periodic arrays [17], the high wave trapping action of microstrip means that radiation is confined to the vicinity of the bends, unlike wire types where strong radiation occurs from the resonant wire lengths. Such non-resonant action is inappropriate for good log-periodic action.

2.4 Overlaid patch array propagation characteristics

The overlaid array is shown in Fig. 7 and the equivalent circuit of a single period in Fig. 8*a*. The propagation characteristic Fig. 2 is derived from eqn. 2. Calculation of the

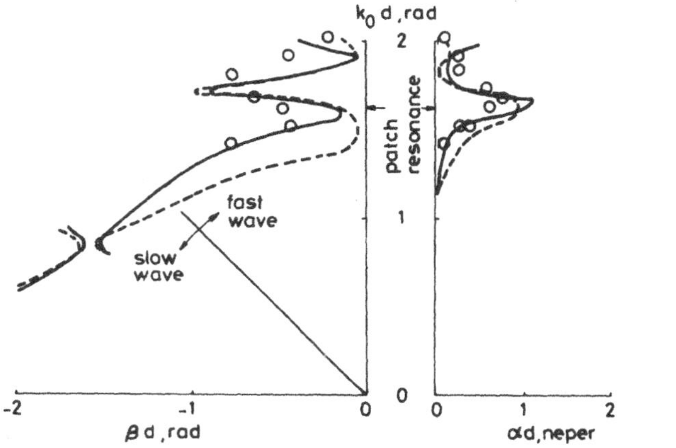

Fig. 2 *Propagation characteristic of electromagnetically coupled overlaid patch array*
Equivalent circuit of single period shown in Fig. 8
——— calculated with mutual coupling
- - - - calculated without mutual coupling
0 measured points
$l = 10$ mm, $w = 8$ mm, $p = 1.25$ mm, $d = 9.82$ mm, $w_f = 2.5$ mm, $h_p = 1.586$ mm, $h_f = 0.793$ mm, $\varepsilon_r = 2.32$

mutual admittance Y_{12} is described in Appendix 7.1. $k_0 = 2\pi f/c$ is the free-space wave number. The normalised propagation constant βd is $-\pi$ at zero frequency due to the alternation of the patch feeding. βd rises smoothly from $-\pi$ to the resonant region at $k_0 d \simeq 1.7$, where heavy attenuation takes place due to strong radiation. The measured results are obtained from a 20-element uniform array, where β is deduced by measuring the radiated beam angle from broadside θ and using

$$\sin \theta = \frac{\beta}{k} \quad (4)$$

The normalised attenuation constant αd is deduced from the array transmission loss, where the exponential power falloff down the array is accounted for. The measured results confirm the trends shown by the computation, although the small differences can be attributed to the approximations in the equivalent circuit. The attenuation constant on resonance is considerably less than that found in switched dipole arrays [12] where 8 nepers/unit length occurs in a typical case. α for the overlaid array can be increased by tighter coupling, but this leads to the onset of stopbands. Apart from this proviso, the overlaid patch array is seen largely to satisfy the criteria for wideband action of Section 2.2. The radiation null on end fire satisfies criteria 1c, and, although mutual coupling and surface waves will contribute to higher-order mode excitation of the larger patches, this is considered to be a second-order effect.

It is believed that the parasitically coupled patch array [18], an array constructed on a single substrate, will have a similar equivalent circuit to the overlaid patch and hence will also give good log-periodic action. However, the coupling to such patches is considerably less than can be achieved for overlaid types, unless very narrow patch feed-line gaps are used. This may, however, give rise to difficulty in accurate manufacture, thus offsetting any advantage in the use of a single substrate.

2.5 Quarter-wavelength line-coupled patch array

The array configuration and equivalent circuit of this element is shown inset in Fig. 3. The parameters are calculated using expressions similar to those used in Reference

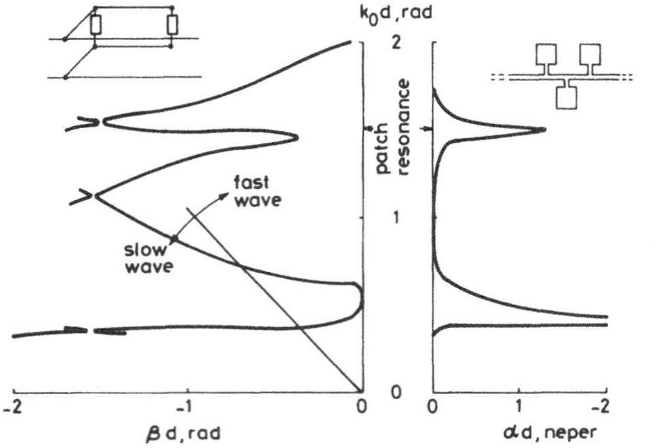

Fig. 3 *Calculated propagation characteristic of quarter-wavelength line-coupled patch array*

Array configuration and equivalent circuit of single period shown inset top right and left respectively
l = 10 mm, w = 8 mm, d = 9.82 mm, w_f = 2.5 mm, h = 1.586 mm, ε_r = 2.32, w_q = 0.5 mm, l_q = 7.0 mm

11, but mutual coupling is not included. The propagation characteristic [19] is derived from eqn. 3.

Resonance occurs at $k_0 d \simeq 1.5$. It can be seen that a stopband occurs at $k_0 d \simeq 0.5$. This is not associated with high radiation, as it occurs at a frequency well below patch resonance. It will result in a poor input return loss at these frequencies. This action is due to the high input admittance of the patch at the frequency when $l \simeq \lambda_m/4$, where λ_m is the microstrip wavelength, and this will prevent power being transmitted past these patches to the resonant elements in a log-periodic array. It can be seen that there is a region between $k_0 d = 1.1$ and 2.0 where a limited-bandwidth log-periodic action may be obtained. This represents close to a 2 : 1 frequency band, and it is concluded that the directly coupled patch is unsuitable for wideband log-periodic action. Although it appears possible to introduce a gap in the quarter-wavelength line to produce a series capacitance to prevent this stopband action, computer synthesis of such a structure indicates that only by the use of a very short connecting line was the stopband signficantly reduced, and this was accompanied by reduced coupling and hence αd. This then leads to a structure that is similar to the overlaid patch or the parasitically coupled patch [18].

2.6 Comb-line array

Fig. 4 shows array configuration, element equivalent circuit and the propagation characteristic deduced from

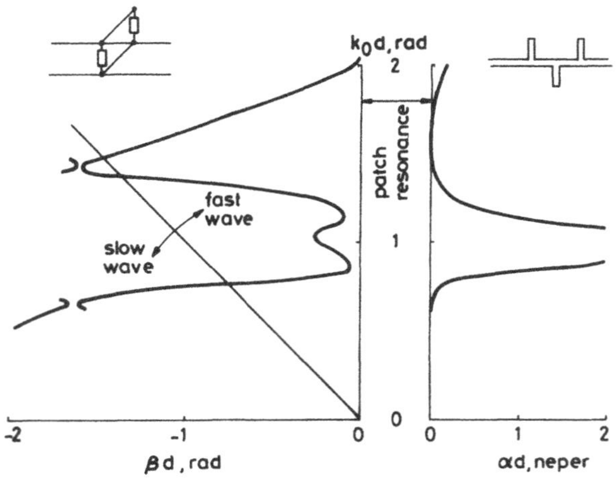

Fig. 4 *Calculated propagation characteristic of comb-line array*

Array configuration and equivalent circuit of single period shown inset top right and left, respectively
l = 10 mm, w = 8 mm, d = 9.82 mm, w_f = 2.5 mm, h = 1.586 mm, ε_r = 2.32

eqn. 3. Resonance occurs at $k_0 d \simeq 1.8$. Again low frequency stopband action is noted when the stub length $l \approx \lambda_m/4$, indicating that only limited-bandwidth log-periodic action can be obtained.

2.7 Series connected patch array

The array configuration, element equivalent circuit and propagation characteristic are shown in Fig. 5, where a

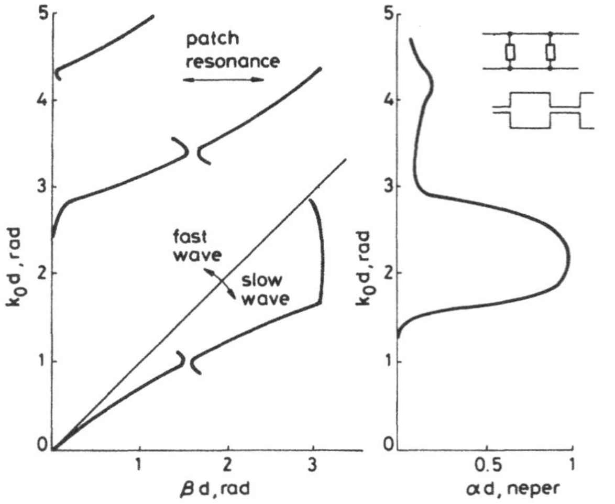

Fig. 5 *Calculated propagation characteristic of series connected patch array*

Array configuration and equivalent circuit of single period shown inset top right
w = 15.0 mm, w_f = 1.0 mm, l = 11.0 mm, d = 11.0 mm, h = 1.586 mm, ε_r = 2.32

low frequency stopband is seen below the resonant frequency ($k_0 d \simeq 4.2$), again indicating limited-bandwidth log-periodic action. As the patch feeding is not alternated, the propagation constant rises from zero at low frequency.

2.8 Deductions on optimum array form

The requirement for no low-frequency stopbands indicate that the series-connected patches, comb-lines and quarter-wavelength coupled patches are inappropriate choices, except where limited bandwidths are required. The use of a patch element with series capacitance coupling is then indicated. This conclusion is illustrated in Fig. 6, which shows the Smith chart characteristic of a typical patch with a feeder matched to the resonant impedance. At low frequency, point A, the patch input impedance is capacitive, which is appropriate for the transmission-line region

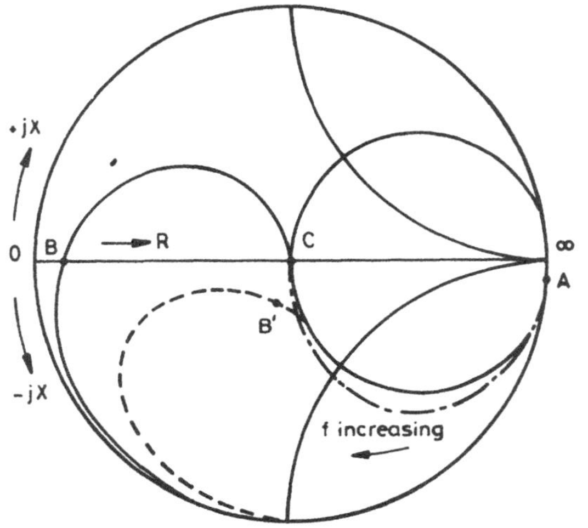

Fig. 6 *Calculated input impedance loci*
——— directly fed microstrip patch
– – – – series capacitance fed microstrip patch
– · – · – directly fed dipole

[13]. When $l \simeq \lambda_m/4$ for the directly coupled case, point B, the impedance is real and low, and stopbands occur with little radiation. On resonance, point C, the impedance is matched. When a series capacitance is added, the characteristic is lowered, resulting in points B and C merging so that both stopband action and radiation will occur simultaneously around point B′, where the patch is best matched. These characteristics can be compared to the dipole impedance plot, where the impedance moves smoothly to resonance as the frequency is increased. The microstrip patch thus appears not to be an optimum element for log-periodic action, but, by the use of a capacitive coupling circuit, this optimum can be approached. Such coupling is implemented in the electromagnetically coupled patch which satisfies the conditions for good wideband action.

It is also noted that nonlog-periodic scaling, inherent in the use of uniform thickness substrates, will give rise to deviations from frequency independent action, thus limiting the bandwidth.

3 Wide-bandwidth electromagnetically coupled patch array

3.1 Basic array design

Design of the wideband array is based on frequency-independent antenna principles [20], which, when applied to a periodic structure, result in scaling of the dimensions from period to period so that the performance is periodic with the logarithm of frequency [14]. If this principle is applied to the electromagnetically coupled overlaid patch array [21], then the structure of Fig. 7*a* results. To make construction simpler, uniform substrate thickness is used as shown in Fig. 7*b*. The patch length l, width w and spacing d of the mth and $(m+1)$th elements are related to the scale factor τ by

$$\tau = \frac{l_{m+1}}{l_m} = \frac{w_{m+1}}{w_m} = \frac{d_{m+1}}{d_m} \tag{5}$$

In the initial array designs, the patch displacement p is constant and this, together with the use of constant substrate thicknesses h_f and h_p, and feed-line width w_f, means that the array deviates from a completely log-periodic design. Although the log-periodic principle is derived from frequency-independent considerations, it is not clear

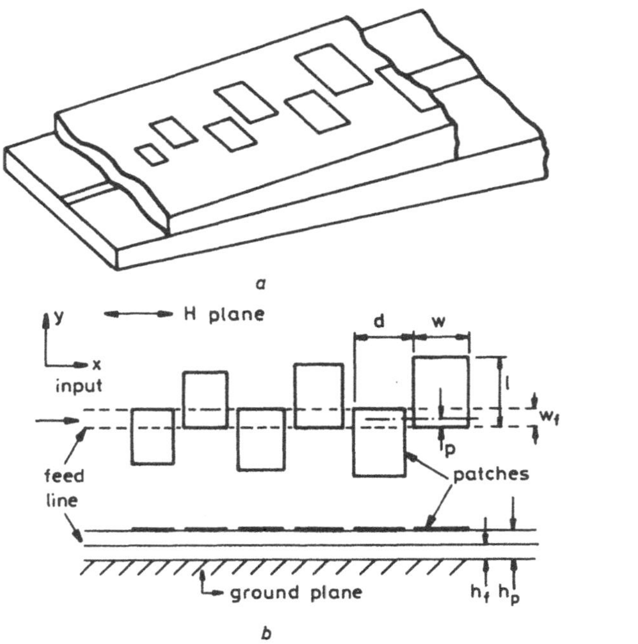

Fig. 7 *Log-periodic electromagnetically coupled overlaid patch array*
a With scaled feed line and substrate
b With uniform patch displacement p, substrate thicknesses h_p and h_f, and feed-line width w_f

whether the technique will here produce an optimum wideband structure due to these significant deviations from periodic scaling. However, this principle is taken as the fundamental design rule, and it is believed that the results will indicate the order of performance that can be expected for such antennas. Other design parameters, not given by eqn. 5, are either deduced empirically or derived from the results of the parametric study given in Section 3.3, which is based on the analysis in the following Section.

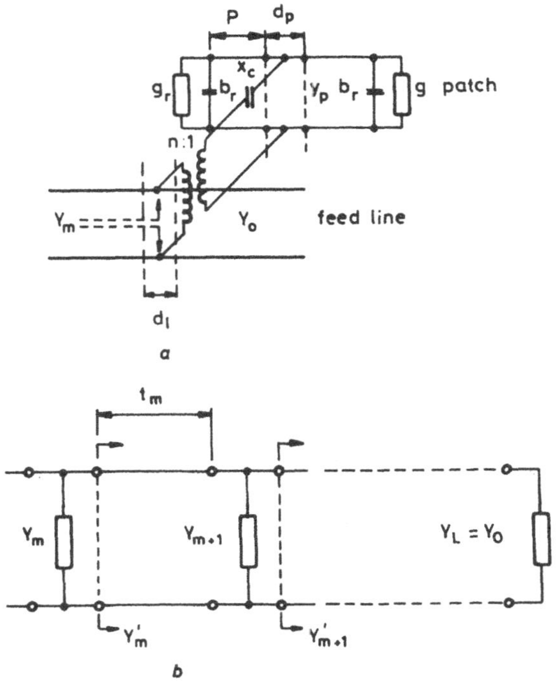

Fig. 8 *Equivalent circuits of overlaid patch array*
a Overlaid patch-element equivalent circuit
b Loaded feed-line equivalent circuit

3.2 *Array analysis*

Analysis of the log-periodic overlaid patch array is performed using a transmission-line model. The equivalent circuit of Fig. 8*a* is used to represent the overlaid patch and is derived by physical reasoning from equivalent circuits of other microstrip discontinuities [22]. The reactance of the capacitor, X_c, is that of an equivalent parallel-plate capacitor at zero frequency, formed by the overlapping circuits, and is given by

$$X_c = (h_p - h_f)/(2\pi f w_f w_{ep} \varepsilon_0 \varepsilon_r) \qquad (6)$$

where w_f, h_p and h_f are defined in Fig. 7, and ε_r and ε_0 are the relative dielectric and free-space permittivities, respectively. f is the operating frequency and w_{ep} is the patch effective width given by [23]

$$w_{ep} = \frac{120\pi h_p}{Z_m \sqrt{\varepsilon_e}} \qquad (7)$$

where Z_m and ε_e are the impedance and effective-dielectric constant of the transmission line forming the patch. g_r and b_r are the patch end radiation conductance and susceptance, respectively. The transformer turns ratio n, and the reference plane extensions d_p and d_l are deduced from scattering parameter measurements on isolated patches which have $h_f = 0.793$ mm, $h_p = 1.586$ mm, $\varepsilon_r = 2.32$, $l/w \simeq 0.8$ and which operate around 10 GHz. n, d_p and d_l are given by

$$n = 9\left(\frac{l}{h_p}\right)^{-0.9} \cosh 2\left\{0.5 - \left|0.5 - \frac{p}{l}\right|\right\} \qquad (8)$$

$$\frac{d_p}{l} = -0.1\left[1 + 0.85 \cos\left\{2.38\pi\left(\frac{p}{l} - 0.08\right)\right\}\right] \qquad (9)$$

$$\frac{d_l}{w_{ep}} = 0.101 \qquad (10)$$

The admittance across the feed-line due to the mth patch Y_m is derived using successive transmission-line matrix transformations [24] and is

$$Y_m = \{(Z_1 - jX_c)n^2\}^{-1} \qquad (11)$$

where

$$Z_1 = \left\{Y_p \sum_{i=1}^{2} \frac{Y_r Z_p \cosh \gamma_p t_i + \sinh \gamma_p t_i}{\cosh \gamma_p t_i + Z_p Y_r \sinh \gamma_p t_i}\right\} \qquad (12)$$

The complex propagation constant γ_p is given by $\gamma_p = (0.0115 + j2\pi)\lambda_{mp}$, where λ_{mp} is the wavelength on the transmission line formed by the patch, and where the line attenuation is approximately 0.1 dB/λ_m. $Z_p (= Y_p^{-1})$ is the patch-line impedance and $Y_r = g_r + jb_r$. $i = 1$ and 2 refer to the left- and right-hand side of the patch, respectively, and

$$t_1 = p\left(1 + \frac{d_p}{l}\right) \qquad (13)$$

and

$$t_2 = (1 - p)\left(1 + \frac{d_p}{l}\right) \qquad (14)$$

The array input impedance is then deduced by successively transforming the matched load admittance and patch admittances down the feed-line to the input, using the recurrence relationship

$$Y_m = Y_0\left\{\frac{(Y'_{m+1} + Y_{m+1})Z_0 \cosh \gamma_0 t_m + \sinh \gamma_0 t_m}{\cosh \gamma_0 t_m + (Y'_{m+1} + Y_{m+1})Z_0 \sinh \gamma_0 t_m}\right\} \qquad (15)$$

where Y'_m is the admittance, at the position of Y_m, due to patches immediately to the right, as in Fig. 8*b*. Here $\gamma_0 = (0.0115 + j2\pi)/\lambda_{m0}$ where λ_{m0} is the feed-line microstrip wavelength, $Z_0 (= Y_0^{-1})$ is the feed-line impedance and

$$t_m = d_m + d_l \qquad (16)$$

where d_l is the reference-plane extension in the feed-line, eqn. 10, due to the mth patch.

The voltage on the radiation conductances, and hence the radiation pattern, is found by successive transformation from the array input up the feed-line and patches. The excitation voltage on the ith patch end V_{ei} is given by

$$V_{ei} = \frac{V_m}{n}\left(1 - \frac{jX_c}{Z_1}\right)\{\cosh \gamma_p t_i + Y_r Z_p \sinh \gamma_p t_i\}^{-1} \qquad (17)$$

where n, X_c and Z_1 are given by eqns. 8, 6 and 12, respectively, for the nth patch, and V_m is the voltage on the patch-line admittance Y_m.

V_m is deduced by placing a 1 V source on the input and using the recurrence relationship

$$V_{m+1} = V_m\{\cosh \gamma_0 t_m + (Y'_{m+1} + Y_{m+1})Z_0 \sinh \gamma_0 t_m\}^{-1} \qquad (18)$$

Hence, the array scattering parameters S_{11} and S_{12} are found from the input impedance and voltage on the terminating matched load [24].

The array efficiency is determined by comparing the power dissipated in the radiation conductances to the input power. Directivity is found by integrating over the E- and H-plane patterns; hence array gain is obtained.

3.3 *Performance trade offs relating to array design*

The primary performance trade off relates to input return loss and radiation pattern coverage in the H-plane. This relationship is controlled largely by the coupling between the feed-line and patches. Fig. 9*a* shows the computed

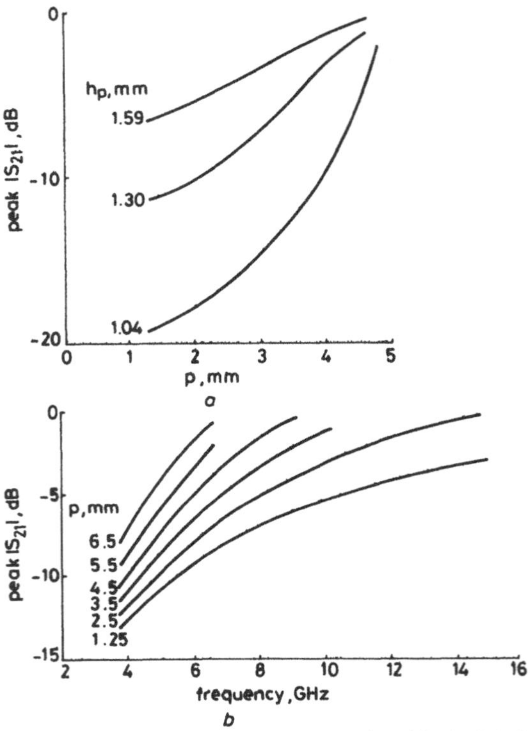

Fig. 9 *Computed resonance insertion loss* $|S_{21}|$ *of isolated overlaid patch*

$h_f = 0.793$ mm, $\varepsilon_r = 2.32$, $w_f = 2.5$ mm

a $l = 10$ mm, $w = 8$ mm, curves are annotated with h_p (mm)

b $h_p = 1.59$ mm, curves are annotated with p (mm)

variation of insertion loss $|S_{21}|$ on resonance due to a single patch with parameters h_p and p, for patches designed to operate at 10 GHz. Maximum coupling was found to occur when the patch edge was above the feed-line edge, which, for $w_f = 2.5$ mm, corresponds to $p = 1.25$ mm. Coupling values for $p < 1.25$ mm are not given, but measurements indicate that coupling reduces smoothly as p is decreased. A coupling range of about 20 dB is noted, suggesting that a wide range of array designs is possible. However, as $|S_{11}| = |S_{21}| - 1$, high values of coupling will lead to a high reflected waves and poor input return loss [25], limiting the useable coupling range. Fig. 9*b* shows the resonance insertion loss against frequency and indicates that, for constant h_p and p, coupling will change significantly over the length of the log periodic array, resulting in changes both in the input match and radiation characteristics. This can be offset by scaling p, but the use of a uniform substrate in this case will still result in a significant change. For example, over a 4 to 15 GHz frequency range the patch insertion loss changes by 6.4 dB. To offset fully this change, p should be calculated using eqn. 5 with a scale factor higher than that used for the other patch dimensions, as described in Section 3.5.

The amount of coupling from the feed-line to the patch will determine the input match, H-plane beamwidth and power lost in the load. For long arrays the power lost in the load will be small, as indicated in Fig. 10*b* for a 36 element array, although power lost in the feed-lines may be significant. Heavy coupling will lead to a poor array input return loss $|S_{11}|$ and short active regions resulting in wide H-plane beamwidths, as indicated in Fig. 10*a*, where the coupling is varied by changing h_p. Light coupling will lead to long active regions but better input match. Truncation effects in which the active region length is similar to the array length lead to minimum beamwidths of about 30° and 25° for the 9 and 18 element arrays, respectively. These results indicate that very wide beamwidths are not possible, due to poor input match, and very narrow beamwidths will be limited by array length. For very wide bandwidth action, the active region must be kept as short as possible to minimise the truncation effects, implying heavy coupling, and this means that the maximum bandwidth per array unit length will be determined primarily by the specified input return loss. The input return loss can be improved by altering the patch spacing d to scan the beam away from the broadside direction, as indicated in Fig. 10*b*. This behaviour is analogous to periodic travelling-wave arrays [26]. The use of $\lambda_m/4$ element spacing with feeding alternation of each pair of elements has been suggested for periodic arrays [26, 15] and some improvement may be obtained, although this will be limited by the high mutual coupling between each element in the pair.

Results in Section 3.4 indicate that active region lengths of the order of 10 elements long are typically achieved, and thus, although the results presented here are for relatively short log-periodic arrays they will serve to indicate the performance obtained at various frequencies within the overall bandwidth of a long array.

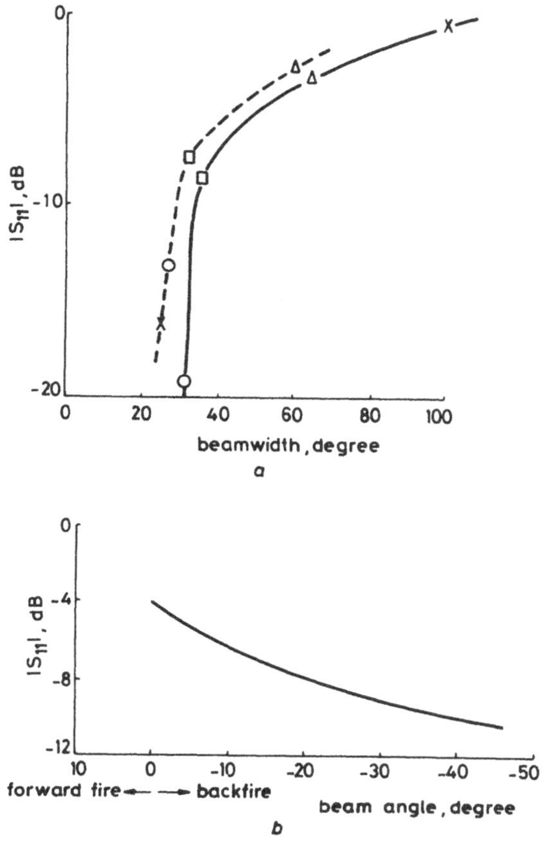

Fig. 10 *Computed input return loss $|S_{11}|$ of overlaid patch log-periodic arrays*

$\tau_0 = 1.05$, $l/w = 0.8$, $p = 1.25$ mm, $h_f = 0.793$ mm, $w_f = 2.5$ mm, $l_1 = 6.8$ mm
——— $M = 9$
– – – – $M = 18$ (M = number of array elements)
a $d_1 = 6.97$ mm, ×, $h_p = 0.794$ mm, △, $h_p = 1.191$ mm, □, $h_p = 1.586$ mm, ○, $h_p = 2.379$ mm, +, $h_p = 3.172$ mm
b $h_p = 1.586$ mm

3.4 Typical array design and performance

Array design proceeds by calculating the smallest patch length for operation at the upper band edge using conventional patch design expressions, together with the reference plane extensions d_p given by eqn. 9. Patch width is chosen to prevent excitation of the orthogonal patch mode; $w = 0.8\,l$ is used here. Patch spacing is chosen to ensure a backfire beam so that acceptable input return loss results, Fig. 10*b*; in this case $|S_{11}| < -8$ dB was used as the criterion. Substrate height and patch displacement are chosen to produce the desired H-plane beamwidth consistent with the same return loss criterion, Fig. 10*a*; $h_p = 1.586$ mm and $p = 1.25$ mm were chosen. The latter is not scaled which ensures maximum patch coupling at all frequencies for this substrate height, Fig. 9*b*. A feed substrate height of $h_f = 0.793$ mm was used with a 50 Ω feed-line of width $w_f = 2.5$ mm. The number of patches required, M, is determined by the ratio of the required bandwidth to the average patch bandwidth. The scale factor τ is thus found. Specification of M, τ, l_1 and d_1 fix the overall size of the array. The width of the array, W, is determined by

$$W = l_1 \tau^{(M-1)} \tag{19}$$

and the overall length L by

$$L = d_1 \left\{ \frac{1 - \tau^{(M-1)}}{1 - \tau} \right\} \tag{20}$$

Variations in M and τ will effect the input return loss and overall array size. Table 1 shows the computed input return loss and effective bandwidth deduced from the overall bandwidth for arrays of nine elements. To maximise patch coupling, $p = 1.25$ mm is used for all patches, and hence the scale factor for patch displacement is $\tau_p = 1.0$. Bandwidth variation is achieved by altering the scale factor τ_0 for patch length, width and spacing. The -10 dB input return loss bandwidth for an equivalent isolated

Table 1: Computed input return loss and equivalent patch bandwidth for nine element log-periodic arrays

τ_0	Peak input return loss within array bandwidth, dB	% bandwidth per patch
1.02	−10.9	1.0
1.03	− 9.9	1.9
1.05	− 8.6	3.5
1.07	− 7.2	4.8
1.11	− 5.2	7.1

$l_1 = 6.8$ mm, $l/w = 0.8$, $d_1 = 6.97$ mm, $h_f = 0.793$ mm, $h_p = 1.586$ mm, $p = 1.25$ mm, $\varepsilon_r = 2.32$, $w_f = 2.5$ mm, $\tau_p = 1.0$

patch with a directly connected feed is about 4%. It can be seen that if this is used in array design, $\tau_0 = 1.05$ is indicated. Improved return loss can be obtained by lower values of τ_0, but this will result, in the case of $\tau_0 = 1.02$, in an approximately sixfold increase in array length.

Figs. 11, 12, 13 and 14 show measured and computed results for a 36 element array designed for a 4 to 16 GHz bandwidth and having $\tau_0 = 1.05$ and $\tau_p = 1.0$. Fig. 11 shows the array silhouette, scattering parameters and gain. Calculated input return loss $|S_{11}|$ shows good agreement with measurement. The calculated transmission loss $|S_{21}|$ is much larger than measured, but both indicate low levels of power lost in the load. $|S_{11}|$ peaks at the low frequencies and this is consistent with the interpretation of Fig. 9*b*, where use of constant displacement p results in heavier coupling and hence worse $|S_{11}|$ at low frequencies. The measured gain is better than 8 dB over the bandwidth 4 to 16 GHz. Agreement between measured and calculated gain is good. The calculated efficiency ranges from 85% to 70% across the frequency band.

The *E*-plane radiation patterns, Fig. 12 show beamwidths similar to isolated patches, although considerable distortion occurs at higher frequencies. Poorer agreement

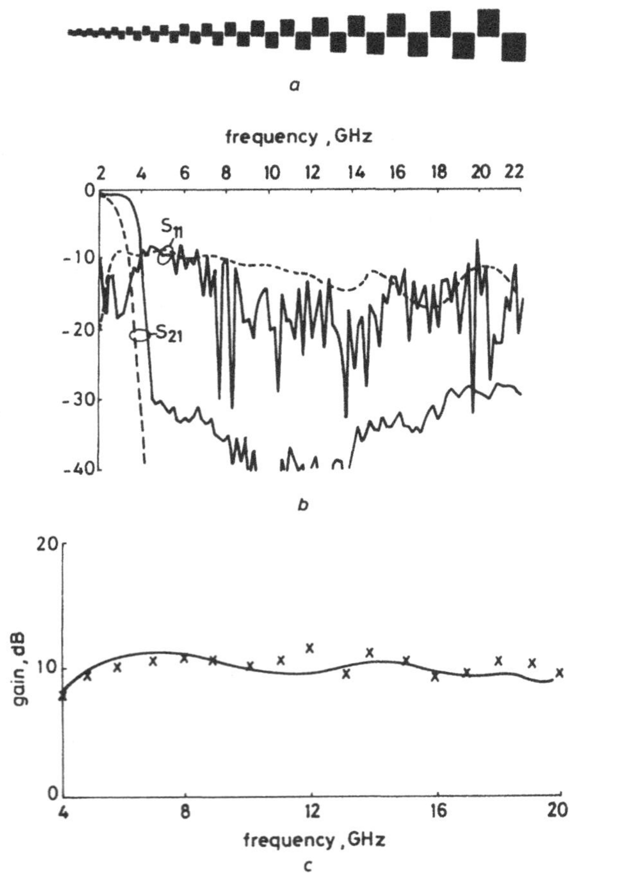

Fig. 11 *36-element overlaid patch log-periodic array*
a array silhouette
b Measured input return loss $|S_{11}|$ and transmission loss $|S_{21}|$
——— measured
- - - - calculated
c array gain
——— measured
× calculated
$l = 3.67$ mm, $w_1 = 2.92$ mm, $d_1 = 3.67$ mm, $\tau_0 = 1.05$, $h_f = 0.793$ mm, $h_p = 1.586$ mm, $p = 1.25$ mm, $\varepsilon_r = 2.32$, $\tau_p = 1.0$
Overall array size = 340 mm × 50 mm

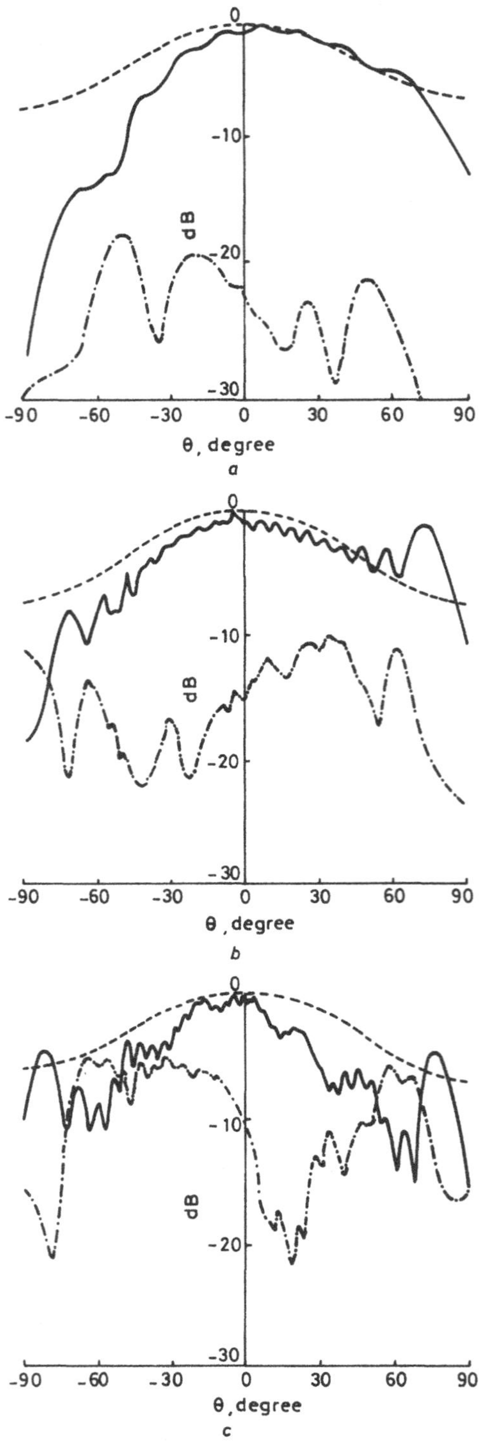

Fig. 12 *E-plane radiation patterns of array of Fig. 11*
a 4 GHz
b 10 GHz
c 16 GHz
——— measured copolar
- - - - calculated copolar
-·-·- measured cross-polar

between measured and calculated patterns is noted, due, primarily, to diffraction effects at the ground-plane edges which is not accounted for in theory. These effects will mean that the E-plane coverage will be very dependent on the mounting-body shape. The H-plane patterns, Fig. 13, show a transition from smooth, wide beamwidths to narrower beams with significant sidelobes as frequency is increased. This behaviour is due to the use of a uniform thickness substrate and patch displacement as indicated in Section 3.3. Significant errors in the theoretical H-plane beam-pointing angle can be seen. It is believed that these

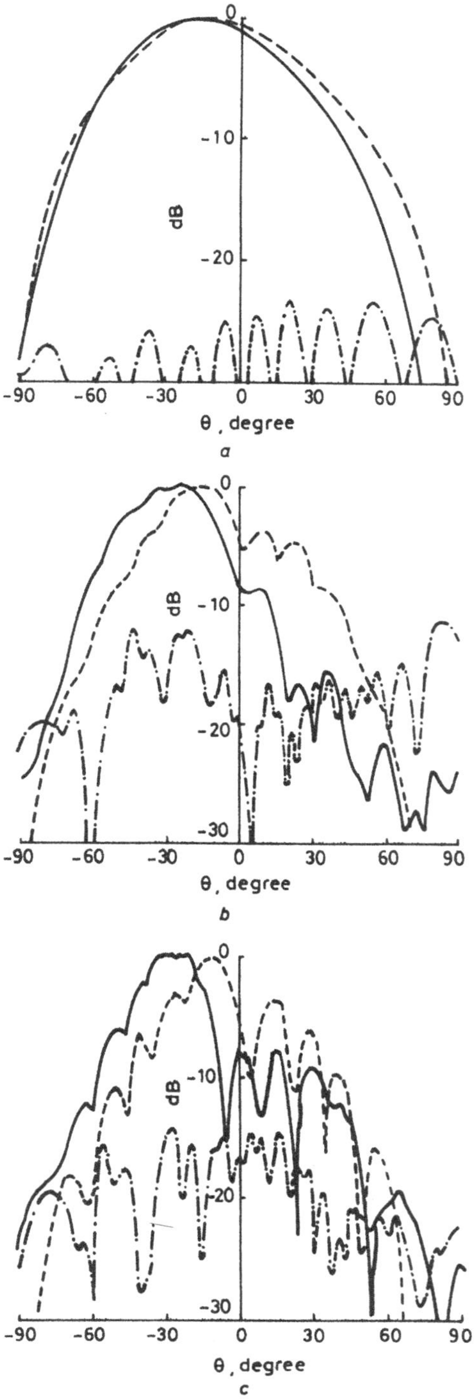

Fig. 13 *H-plane radiation patterns of array of Fig. 11*

a 4 GHz
b 10 GHz
c 16 GHz
——— measured copolar
– – – – calculated copolar
–·–·– measured cross-polar

are due both to lack of mutual coupling in the transmission-line analysis and to further approximations in the equivalent circuit. Fig. 2 indicates that βd around patch resonance is significantly changed when mutual coupling is included in the $k-\beta$ analysis. To include mutual effects in the transmission-line analysis, corrections to t_m, eqn. 16, need to be established across the whole frequency band. It is also noted that the other parameters in the equivalent circuit were empirically modelled only at 10 GHz. Cross-polarisation is not predicted in the simple patch model used in the analysis. Measured levels are high, particularly at high frequencies where the patches are electrically thick. Unwanted radiation off the input transition was also found to be significant at the higher frequencies.

The poor H-plane pattern shape at high frequency is associated with the change in patch feed coupling with frequency and this is illustrated in Fig. 14. This shows the

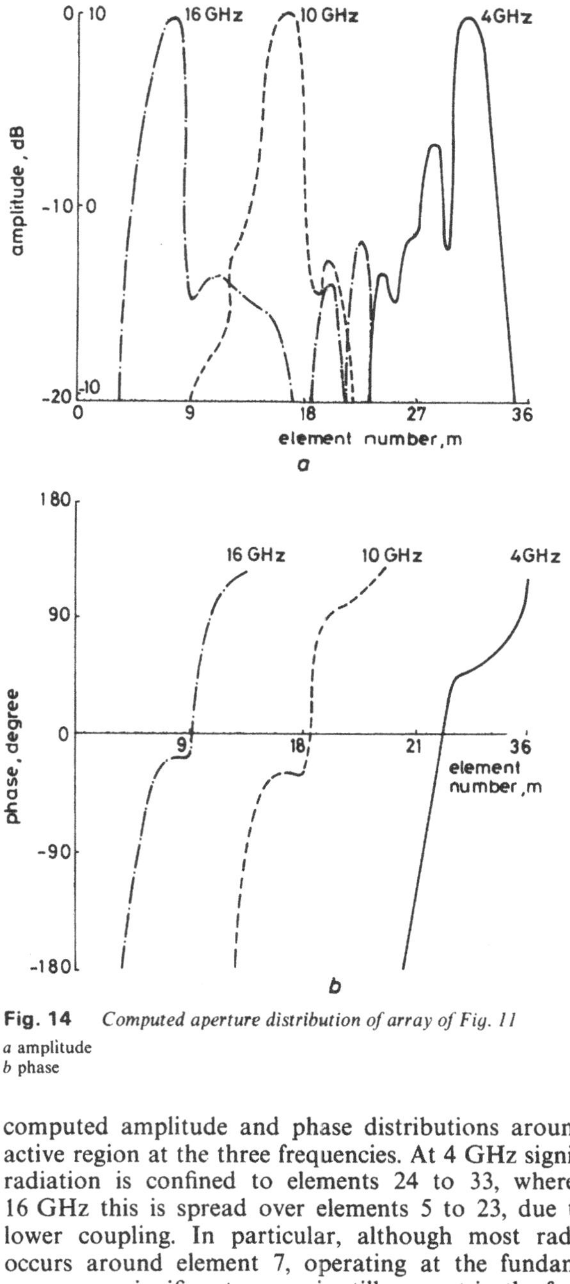

Fig. 14 *Computed aperture distribution of array of Fig. 11*

a amplitude
b phase

computed amplitude and phase distributions around the active region at the three frequencies. At 4 GHz significant radiation is confined to elements 24 to 33, whereas at 16 GHz this is spread over elements 5 to 23, due to the lower coupling. In particular, although most radiation occurs around element 7, operating at the fundamental resonance, significant power is still present in the feed-line to excite those around element 21 which are operating near the first harmonic. On this basis, patterns will begin to degrade for bandwidths greater than 2 : 1 unless tighter coupling is used for the high frequency patches; such an array is described in the following Section.

3.5 Measures to improve bandwidth and radiation pattern performance

The change of H-plane beamwidth with frequency, and the radiation pattern quality at high frequencies, can, in principle, be improved by designing the array for equal feed-line to patch coupling across its bandwidth. Coupling is controlled for a given substrate type by patch displacement, p, and patch substrate height h_p. Fig. 15 shows the

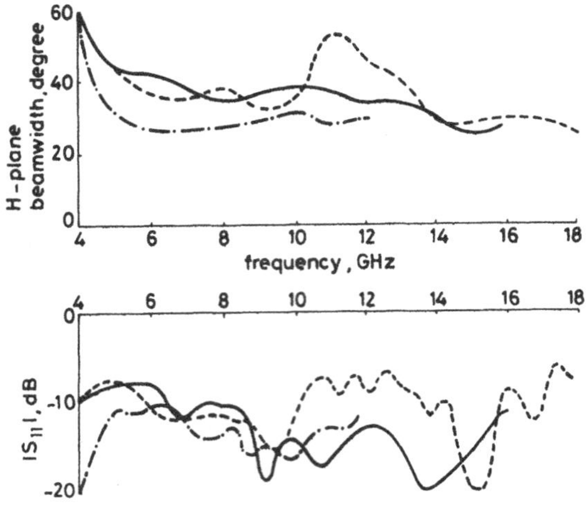

Fig. 15 *H-plane beamwidth and input return loss of 36 element overlaid patch log-periodic arrays*

——— array of Fig. 11
–·–·– array of Fig. 11 with $\tau_p = 1.056$
– – – – array of Fig. 11 with $h_p = 1.04$ mm for patches $1 < m < 14$ and $h_p = 1.59$ mm for patches $15 < m < 36$

H-plane beamwidth and input return loss for arrays similar to that of Fig. 11 but with (i) scaled patch displacement and (ii) two patch substrate thicknesses. These are compared to the results for the array of Fig. 11 where $\tau_p = 1$ and $h_p = 1.59$ mm throughout. The results are truncated when the significant radiation pattern sidelobes in the region $0 < \theta < 90$ are greater than about -10 dB. It can be seen that the array with scaled p exhibits narrower beamwidths, a flatter beamwidth response with frequency and an improved return loss, but has a reduced bandwidth. The array with two patch substrate heights gives a sharp discontinuity in beamwidth at 10 GHz corresponding to the change in substrate heights, but has better radiation pattern control at higher frequencies and exhibits the widest bandwidth. Its input return loss is, however, slightly worse. It is concluded then, that constant beamwidth can be obtained, but at the expense of bandwidth, and that wider bandwidths can be obtained, but at the expense of beamwidth control and additional construction complexity.

The maximum bandwidth that can be obtained from a microstrip log-periodic array is seen thus to be dependent primarily on radiation pattern constraints and also on the input return loss specification. In the case of both uniform thickness feed and patch substrates, the change in pattern beamwidth and the onset of low-frequency stopbands suggest that it may well not be possible to obtain bandwidths far in excess of the two octaves already achieved. Some bandwidth extension may be possible by adjusting the array parameters in a nonlog-periodic way to optimise the performance characteristics. In addition, the array presented can also be scaled to cover other frequency bands, and it may then be possible to connect such arrays in series to operate over bandwidths of many octaves. The use of various patch substrate thicknesses on a uniform feed substrate will extend the bandwidth of a single array, although this will be limited at the high frequency end when the patch length is of the same order as the line width and at the low frequency end by the feed-line fringing fields being insufficient to excite the patch.

4 Conclusions

Operation of microstrip antennas over multioctave bandwidths has been shown to be possible using electromagnetically coupled patches in a log-periodic series-fed array configuration. Although k-β analysis of uniform patch arrays indicates that the microstrip patch is not an optimum choice for log-periodic applications, the addition of electromagnetic coupling to the feed-line produces an element closer to the optimum choice, and such arrays fulfil the criteria for wideband operation deduced from previous work. Directly coupled patch arrays are only suitable for bandwidths limited to about 2 : 1.

Measured results for overlaid patch arrays indicate that a 4 : 1 bandwidth can be obtained with an input return loss of 8 dB, a gain of 8 dB and a 30 degree backfire beam whose beamwidth varies from 63 to 32 degrees across the band. These results, together with those from a transmission-line analysis, also indicate that:

(*a*) The constraints imposed on the element and the feed arrangement by the use of uniform thickness substrates will limit the maximum achievable bandwidth.

(*b*) Arrays with bandwidths significantly greater than two octaves will incur considerable degradation in the radiation pattern and input return loss. However, the use of nonlog-periodic parameter optimisation, multiple patch substrates or series connection of several arrays covering various bands may alleviate this.

(*c*) Radiation pattern control and input return loss are critical performance parameters. Some improvement in the currently obtained levels can be obtained at the expense of bandwidth or overall length.

However, in spite of limitations to the ultimate bandwidth obtainable, the log-periodic microstrip array substantially extends the useful application area of microstrip antennas and provides the designer of wide bandwidth systems with a versatile, low profile, lightweight antenna with conformal mounting capabilities.

5 Acknowledgments

The author would like to acknowledge Capt. K.P. Barrett and Lt. G.J.T. Rafferty for many array measurements and help in developing the overlaid patch equivalent circuit, and the help and advice of members of the Electromagnetic Systems Group at RMCS.

6 References

1 HARDIE, G.S., and SEFTON, H.B.: 'Fixed beam and mechanically steerable antennas', *Microwave J.*, 1984, **27**, pp. 143–156
2 BENNETT, C.L., and ROSS, G.F.: 'Time domain electromagnetics and its applications', *Proc IEEE*, 1978, **66**, (3), pp. 299–318
3 POZAR, D.M.: 'Considerations for millimetre wave printed antennas', *IEEE Trans.*, 1983, **AP-31**, (5), pp. 740–747
4 HALL, P.S., WOOD, C., and GARRETT, C.: 'Wide bandwidth microstrip antennas for circuit integration', *Electron. Lett.*, 1979, **15**, (15), pp. 458–460
5 WOOD, C.: 'Curved microstrip lines as compact wideband circularly polarised antennas', *IEE, Microwaves, Opt. & Antennas*, 1979, **3**, pp. 5–13
6 HALL, P.S.: 'Microstrip linear array with polarisation control', *IEE Proc. H, Microwaves, Opt. & Antennas*, 1983, **130**, (3), pp. 215–224

7 DONG, W.R., and SENGUPTA, L.L.: 'A class of broad-band patch microstrip travelling wave antennas', *IEEE Trans.*, 1984, **AP-32,** (1), pp. 98–100

8 PUES, H., VANDESANDE, J., and VAN DE CAPELLE, A.: 'Broadband microstrip resonator antennas', Proceedings IEEE International Symposium on Antennas and Propagation, Washington DC, 1978, pp. 268–271

9 HALL, P.S.: 'New wideband microstrip antenna using log-periodic technique', *Electron. Lett.*, 1980, **16,** pp. 127–128

10 HALL, P.S.: 'Log-periodic microstrip patch array', UK Patent GB 2064877B

11 PUES, H., BOGAERS, J., PIECK, R., and VAN DE CAPELLE, A.: 'Wideband quasilog-periodic microstrip antenna', *IEE Proc. H, Microwaves, Opt. & Antennas*, 1981, **128,** (3), pp. 159–163

12 MITTRA, R., and JONES, K.E.: 'Theoretical brillouin (k-β) diagrams for monopole and dipole arrays and their application to log-periodic arrays', *IEEE Trans.*, 1962, **AP-12,** pp. 533–540

13 INGERSOL, P.G., and MAYES, P.E.: 'Log periodic antennas with modulated impedance feeders', *ibid.*, 1968, **AP-16,** (6), pp. 633–642

14 JORDAN, E.C., and BALMAIN, K.G.: 'Electromagnetic waves and radiating systems' (Prentice Hall, New Jersey, 1968) Chap. 15

15 JAMES, J.R., and HALL, P.S.: 'Microstrip antennas and arrays, Pt 2 — new array design technique', *IEE J. Microwaves, Opt. & Antennas*, 1977, **1,** (5), pp. 175–181

16 HALL, P.S., and JAMES, J.R.: 'Cross polarisation behaviour of series-fed microstrip linear arrays', *IEE Proc. H, Microwaves, Opt. & Antennas*, 1984, **131,** (4), pp. 247–257

17 DUHAMEL, R.H., and BERRY, D.G.: 'Logarithmically periodic antenna arrays', *IRE Wescon. Conv. Record, Pt 1*, 1958, pp. 161–174

18 OWENS, R.P.: 'Design and manufacture of serpent arrays and parasitic patch arrays', *in* 'Advances in printed antenna design and manufacture', IEE Colloquium Digest 1982/19 February 1982, pp. 4-1 to 4-3

19 HALL, P.S.: 'Bandwidth limitations of log-periodic microstrip patch antenna arrays', *Electron. Lett.* 1984, **20,** pp. 437–438

20 RUMSEY, V.H.: 'Frequency independent antennas' (Academic Press, London, 1966) Chaps. 5 and 6

21 OLTMAN, G.H.: 'Microstrip dipole antenna elements and arrays', US Patent No 4,054,874, 18 October 1977

22 HAMMERSTAD, E.O., and BEKKADAL, F.: 'Microstrip handbook', ELAB report STF 44 A74169, University of Trondheim, Norwegian Institute of Technology, 1975

23 JAMES, J.R., HALL, P.S., and WOOD, C.: 'Microstrip antenna theory and design', IEE Electromagnetic Wave Series No 12, (Peter Perigrinus, London, 1981) p. 35

24 RAGAN, G.L.: 'Microwave Transmission Circuits', MIT Radiation Lab Series, No. 9, (McGraw Hill, London, 1951) Chap. 9

25 HALL, P.S.: 'Multi-octave bandwidth microstrip antenna arrays', 4th International Conference on Antennas and Propagation, University of Warwick, UK, April 1985

26 STARR, A.T.: 'Radio and radar technique', (Pitman, 1953) pp. 266–267

27 PENARD, E., and DANIEL, J.P.: 'Mutual coupling between microstrip antennas, *Electron. Lett.*, 1982, **18,** (14), pp. 605–607

7 Appendix

7.1 Patch mutual coupling

The patch mutual coupling is calculated using the reaction theorem with a cavity model of the patch fields and is based on Reference 27, but here expanded to deal with patches in echelon.

The mutual impedance Z_{21} is given by

$$Z_{21} \frac{1}{I_1 I_2} \oint H_1 M_2 \, dc \tag{21}$$

where I_1 and I_2 are the feed currents on the patches, H_1 is the magnetic field set up by patch 1 on patch 2, which has magnetic source M_2, and c is a contour around the patch edge. The effect of the feed-line under the patch is ignored, and if I_1 and I_2 are the patch currents at the radiating edges, then Z_{21} is referred to the position of g_r and b_r in Fig. 8*a*. Assuming thin patches supporting TM_{01} modes, then

$$Z_{21} = -\frac{j h_p^2 \omega^2 \mu_0^2}{240\pi^2 k_0^2 \beta_0^2 w^2 \tan^2(\beta_0 l)} \times (R_1 + R_2 + R_3) \tag{22}$$

where μ_0 and k_0 are the permeability of and wavenumber in free space, respectively, $\beta_0^2 = k_0^2(1 - j \tan \delta)$ where $\tan \delta$ is a coefficient that takes into account radiation, dielectric and copper loss and $\omega = 2\pi f$. R_1, R_2 and R_3 are given by

$$R_1 = \left(\frac{\pi^2}{l} - k_0^2\right) \int_0^l \int_e^{e+l} \cos\frac{\pi}{l}(y_2 - e) \times \cos\frac{\pi}{l} y_1 \left\{ \frac{2e^{-jk_0R}}{R}\bigg|_{x_2 - x_1 = d} - \frac{e^{-jk_0R}}{R}\bigg|_{x_2 - x_1 = d - w} - \frac{e^{-jk_0R}}{R}\bigg|_{x_2 - x_1 = d + w} \right\} dy_2 \, dy_1 \tag{23}$$

$$R_2 = 2\left(\frac{\pi}{l}\right) \int_0^l \sin\frac{\pi}{l} y_2 \times \left\{ \frac{2e^{-jk_0R}}{R}\bigg|_{x_2 - x_1 = d} - \frac{e^{-jk_0R}}{R}\bigg|_{\substack{x_2 - x_1 = d - w \\ y_1 = e}} - \frac{e^{-jk_0R}}{R}\bigg|_{\substack{x_2 - x_1 = d + w \\ y_1 = e}} \right\} dy_2 \tag{24}$$

$$R_3 = -k_0^2 \int_0^w \int_d^{d+w} \times \left\{ \frac{2e^{-jk_0R}}{R}\bigg|_{y_2 - y_1 = e} + \frac{e^{-jk_0R}}{R}\bigg|_{y_2 - y_1 = e - l} + \frac{e^{-jk_0R}}{R}\bigg|_{y_2 - y_1 = e + l} \right\} dx_2 \, dx_1 \tag{25}$$

where the co-ordinate system is given in Fig. 7*b* and the subscripts 1 and 2 refer to patch 1 and 2 respectively. R and e are given by

$$R^2 = (x_2 - x_1)^2 + (y_2 - y_1)^2 \tag{26}$$

$$e = l - 2p \tag{27}$$

Transmission-line matrix transformations through the equivalent circuit are then used to find the mutual admittance Y_{12} at the feed-line terminals. It is assumed that $Y_{12} = Y_{21}$.

Fig. 2 shows the propagation characteristic for uniform overlaid patch array with and without mutual coupling, where it can be seen that significant changes in βd occur. The effect of mutual coupling is to move the phase constant characteristic away from $\beta d = 0$, just below resonance, thus reducing the amount of attenuation due to the stopband effect. It is assumed that in this case power is being transmitted past the stopband by mutual coupling. A similar phenomenon is noted in dipole arrays [12]. Mutual coupling is included in the dispersion analysis of Section 2, but not in the transmission line analysis of Section 3.

Chapter 5
Modeling Techniques for Microstrip Antenna Elements

A microstrip antenna or array commonly consists of a single thin dielectric sheet with an etched copper pattern, but this structural simplicity belies a very difficult analysis problem in electromagnetics. While antenna analysis in general often involves formidable difficulties, the nature of the microstrip element introduces further complexity. This complexity is partly due to the high-Q nature of the microstrip antenna, which makes accurate impedance determination difficult, and the presence of an inhomogeneous dielectric, whose loading, loss, and surface wave effects are often critical. Another factor is the wide variety of microstrip geometries that have been found to be of practical interest, including different patch shapes, different feeding methods, the use of parasitic or stacked elements, and the integration with coplanar feed networks and active circuitry. These considerations, coupled with the fact that microstrip antenna technology is relatively new, perhaps explains why we still do not have microstrip antenna models or CAD codes that are capable of treating adequately even a fraction of the large number of problems that are of interest today. (Further discussion of CAD software for microstrip antennas can be found in the review article by Pozar and James in Chapter 1.)

One of the main reasons for developing an accurate model for a particular antenna, or class of antennas, (and one which is often overlooked by academic researchers) is to provide a tool to allow design of an antenna without costly and tedious experimental iteration. For this purpose the designer needs to be able to predict quantities such as input impedance, loss effects, patterns, gain, and cross-polarization. Of these, input impedance is usually the most difficult to predict accurately, primarily because of the narrow bandwidth of most microstrip elements, but also because of dielectric material tolerances.

The above difficulties may explain the profusion of published articles on microstrip analysis. This subject has in fact proved to be very fertile ground for antenna and electromagnetics researchers worldwide (largely from academia). Most of these analysis techniques can be separated into two broad categories: approximate methods based on simplifying assumptions, and solutions that are full-wave. Examples of the former include the transmission line model, the cavity model, and the segmentation model. These models generally treat the element as a transmission line or cavity resonator, thus simplifying the analysis considerably. Such models should not be dismissed, however, since they often are accurate enough for first-cut designs. They also have a clear advantage in terms of computational simplicity and speed, and in providing a physical insight that is usually missing in more numerical solutions. Solutions such as the transmission line and cavity models were among the first to be proposed for the microstrip antenna element [1]–[3] (also see the review articles in Chapter 1), but these models have since undergone many revisions and extensions to their original form. At the present time, one of the most satisfactory transmission line models for rectangular patches is described in the paper by Pues and Van de Capelle. This model has been validated with a large amount of experimental data, with good results for resonant frequency and input impedance for elements on thin, low-dielectric constant substrates. There are several variations of the cavity model, and it is probably not possible to identify any single one as being substantially better than the others, but the model described in the paper by Thouroude, Himdi, and Daniel represents a recent version of this popular solution. The cavity model has also been applied to aperture-coupled microstrip antenna elements, as described in the paper by Himdi, Daniel, and Terret, and in [4], [5]. To date, the transmission line and cavity models have been applied to rectangular, circular, and triangular patches, with probe feeds, microstrip line feeds, and aperture feeds. A generalization of the transmission line model is capable of treating more complex shapes, such as annular rings [6]. In practice, these models generally work well for thin, low dielectric constant substrates, but exhibit less accuracy as the substrate thickness and/or the dielectric constant increases. This trend is quantified by comparison with experimental data in [7]. The segmentation technique described in the paper by Palanisamy and Garg is an extension of the cavity model, whereby a multiport network is formed by segmenting the antenna patch into sections. This technique was originally developed by Gupta [8] for planar circuit analysis, and has the advantage of being able to treat patches of arbitrary shape, and connecting feedlines.

The term **full-wave** generally refers to an electromagnetic solution that includes all relevant wave mechanisms, allowing the enforcement of boundary conditions to an accuracy limited only by the numerical implementation of the solution. Cavity and transmission line models, for example, do not enforce boundary conditions on the dielectric-air interface of a microstrip antenna substrate, and thus cannot rigorously include surface wave effects, mutual coupling, or even radiation. Full-wave solutions may take various forms, but most that have been applied to microstrip antennas and arrays employ the moment method with the exact Green's function for the dielectric substrate. There are many good papers on this topic, with the ones by Alexopoulos and Jackson, Mosig and Gardiol, and Pozar being representative. The full-wave moment method has been applied to probe-fed rectangular patches [9], [10], circular patches, mutual coupling between patches [11], aperture-coupled patches [12], [13], stacked patches [14]–[16], and many other problems of practical importance. While such solutions have demonstrated versatility and accuracy, they suffer from the drawback of requiring extensive computer time, and from the fact that any change in the geometry (patch shape, feeding method, addition of a cover layer, etc.) requires the development of a new solution. This may be a good way to generate journal articles, but not a very desirable situation from the point of view of the design engineer looking for a versatile CAD tool. In addition, the complexity of these solutions often requires a considerable investment to implement a new code.

Recently there has been some progress toward implementing more general solutions for the treatment of a class of printed antenna problems. One example is described in the paper by Das and Pozar, where patch antenna or printed dipole problems having multiple dielectric layers and feeding methods have been treated in a general purpose computer code. We have included only Part II (Applications) of this article, to show some of the problems that can be treated with this technique; see [17] for Part I on the theory of the solution. Other techniques, such as finite-difference time-domain solutions, also offer the promise of such flexibility, with the paper by Wu, Wu, Bi, and Litva being a recent example. There is much more work to be done, however, to exploit the increasing computational power of personal computers and workstations to provide accurate and flexible design tools to the microstrip antenna designer.

References

[1] R. E. Munson, "Conformal microstrip antennas and microstrip phased arrays," *IEEE Trans. Antennas and Prop.,* vol. AP-22, pp. 74–78, Jan. 1974.

[2] A. G. Derneryd, "A theoretical investigation of the rectangular microstrip antenna," *IEEE Trans. Antennas and Prop.,* vol. AP-26, pp. 532–535, July 1978.

[3] W. F. Richards, Y. T. Lo, and D. Harrison, "An improved theory for microstrip antennas and applications," *IEEE Trans. Antennas and Prop.,* vol. AP-29, pp. 38–46, Jan. 1981.

[4] A. Ittipiboon, R. Oostlander, Y. M. Antar, and M. Cuhaci, "A modal expansion method of analysis and measurement on aperture-coupled microstrip antenna," *IEEE Trans. Antennas and Prop.,* vol. 39, pp. 1567–1573, Nov. 1991.

[5] M. El Yazdi, M. Himdi, and J. P. Daniel, "Analysis of aperture-coupled circular microstrip antenna," *Electronics Letters,* vol. 29, pp. 1021–1022, May 1993.

[6] A. K. Bhattacharyya and R. Garg, "Generalized transmission line model for microstrip patches," *Proc. IEE,* Part H, vol. 132, pp. 93–98, 1985.

[7] D. H. Schaubert, D. M. Pozar, and A. Adrian, "Effect of microstrip antenna thickness and permittivity: comparison of theories and experiment," *IEEE Trans. Antennas and Prop.,* vol. 37, pp. 677–682, June 1989.

[8] K. C. Gupta and P. C. Sharma, "Segmentation and desegmentation techniques for the analysis of planar microstrip antennas," *IEEE Int'l Symp. on Antennas and Propagation Digest,* pp. 19–22, June 1981.

[9] M. D. Deshpande and M. C. Bailey, "Input impedance of microstrip antennas," *IEEE Trans. Antennas and Prop.,* vol. AP-31, pp. 740–747, Sept. 1983.

[10] J. T. Aberle, D. M. Pozar, and C. R. Birtcher, "Radiation and scattering from probe-fed microstrip patch antennas," *IEEE Trans. Antenna and Prop.,* vol. 39, pp. 1691–1696, Dec. 1991.

[11] D. M. Pozar, "Input impedance and mutual coupling of rectangular microstrip antennas," *IEEE Trans. Antennas and Prop.,* vol. AP-30, pp. 1191–1196, Nov. 1982.

[12] P. L. Sullivan and D. H. Schaubert, "Analysis of an aperture coupled micro-strip antenna," *IEEE Trans. Antennas and Prop.,* vol. AP-34, pp. 977–984, Aug. 1986.

[13] A. K. Bhattacharyya, Y. M. Antar, A. Ittipiboon, "Full wave analysis of an aperture coupled patch antenna," *Electronics Letters,* vol. 27, pp. 153–155, Jan. 17, 1991.

[14] L. Barlatey, J. R. Mosig, and T. Sphicopoulos, "Analysis of stacked micro-strip patches with a mixed potential integral equation," *IEEE Trans. Antennas and Prop.,* vol. 38, pp. 608–615, May 1990.

[15] A. N. Tulintseff, S. M. Ali, and J. A. Kong, "Input impedance of a probe-fed stacked circular microstrip antenna," *IEEE Trans. Antennas and Prop.,* vol. 39, pp. 381–390, Mar. 1991

[16] C. Wu, J. Wang, R. Fralich, and J. Litva, "A rigorous analysis of an aperture-coupled stacked microstrip antenna," *Microwave and Optical Technology Letters,* vol. 3, pp. 400–404, Nov. 1990.

[17] N. K. Das and D. M. Pozar, "Multiport scattering analysis of general multi-layered printed antenna fed by multiple feed ports: Part I—Theory," *IEEE Trans. Antennas and Prop.,* vol. 40, pp. 469–481, May 1992.

Accurate transmission-line model for the rectangular microstrip antenna

H. Pues, Mem. I.E.E.E., and A. Van de Capelle, Mem. I.E.E.E.

Indexing terms: *Antennas, Microwave components*

Abstract: An accurate and numerically efficient model for the rectangular microstrip antenna is presented. It concerns a transmission-line model which features the following three major improvements with respect to earlier such models: the mutual radiative coupling (both real and imaginary parts) between the equivalent slots is fully taken into account; the influence of the side slots on the radiation conductance is taken into account implicitly; simple analytic expressions are introduced for all relevant model parameters. By way of illustration, the new model is applied to antennas with a single microstrip feed line. Excellent agreement is shown with available experimental and theoretical results for the input impedance of a rectangular antenna. The improvements with respect to previous transmission-line models are illustrated for a square antenna.

1 Introduction

Microstrip antennas are becoming more and more popular as both feeds and array elements. Therefore the need for accurate models is growing. Not only accuracy is required, but also numerical efficiency, in order that the models are suited for computer-aided-design (CAD) procedures involving optimisation. For rectangular elements, transmission-line models [1, 2] are very attractive, especially for arrays fed by a coplanar microstrip network [16].

Fig. 1 shows a rectangular microstrip antenna of patch

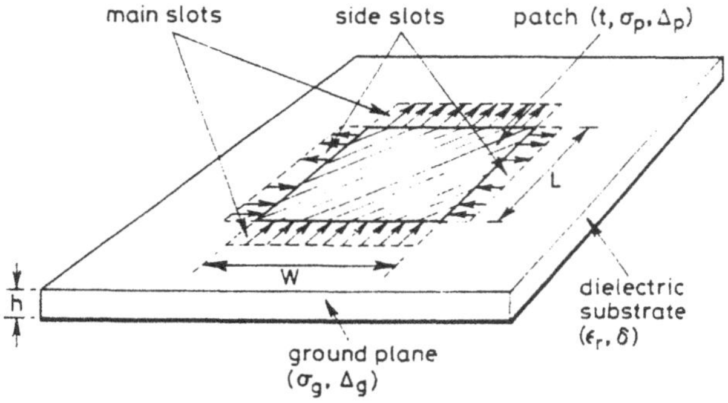

Fig. 1 *Rectangular microstrip antenna: geometry, parameters and equivalent radiating slots*

width W and length L, where L is the resonant dimension of the fundamental radiating mode. The Figure also shows four imaginary radiating slots which form a useful model to calculate the radiation field of the antenna [3]. These so-called equivalent slots consist of two main slots with a uniform distribution and two side slots with a sinusoidal distribution.

The transmission-line models available up to now represent the antenna by a line section terminated at both ends by a radiation admittance Y_T. A general three-port representation is shown in Fig. 2. Depending on the particular feed structure, the model shown has to be extended at the relevant input ports. In the case of a probe-fed antenna, for example, where $I_1 = I_2 = 0$, a series inductance has to be added to port 3 [4].

Munson [1] published the first transmission-line model. He proposes $Y_T = Wy_s$, where y_s is the admittance per unit length of a TE-excited slot having an infinite length and a width equal to the substrate thickness h. The model has the advantage of yielding very simple expressions for

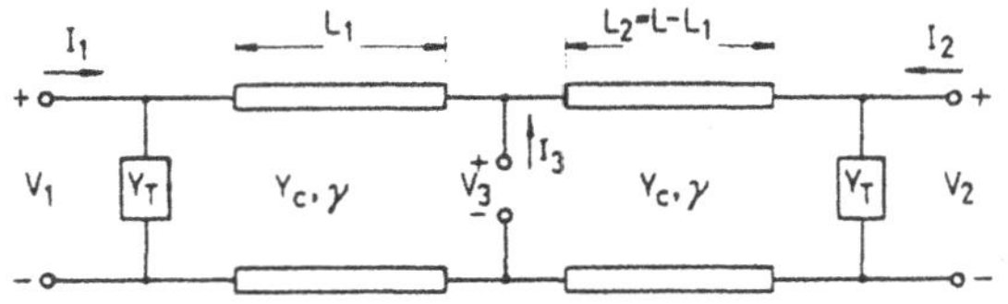

Fig. 2 *Conventional transmission-line model represented as a three-port*

both the real and imaginary part of Y_T, but it has three important shortcomings:

(*a*) The expressions used for Y_T are inaccurate for narrow patches (i.e. for $W \leqslant \lambda_0$; λ_0 = free-space wavelength).

(*b*) The mutual coupling between the main radiating slots is neglected.

(*c*) The influence of the side slots on the radiation conductance is neglected.

Derneryd [2, 5] has partly corrected the first two of these shortcomings:

(*a*) To determine the conductance $G_T = \text{Re}\,(Y_T)$, he considers the two main slots with an identical excitation and a negligible width. He finds an integral expression for G_T, for which an approximate analytical solution has been derived by Lier [4]. Derneryd's model corrects the first two shortcomings of Reference 1 for the real part of Y_T, but it still neglects the influence of the side slots on G_T.

(*b*) To determine the susceptance $B_T = \text{Im}\,(Y_T)$, Derneryd equals this parameter to the open-end self susceptance of the microstrip line formed by the patch. This corrects only the first shortcoming of Reference 1 for the imaginary part of Y_T.

An alternative for transmission-line models are cavity models [15, 18], which take into account higher-order modes but suffer from similar problems. As indicated in Reference 18 the good correspondence between predictions and experiments is obtained by introducing an experimentally determined equivalent width of the feed.

It is obvious that the shortcomings of the transmission-line model can be avoided by applying more sophisticated techniques, such as rigorous moment-method treatments [7], but then the desired numerical efficiency is sacrificed.

In this paper an improved transmission-line model is presented which corrects the three shortcomings of Reference 1 and has a broad range of validity.

Reprinted with permission from *Proc. IEE*, H. Pues and A. Van de Capelle, "Accurate Transmission-line Model for the Rectangular Microstrip Antenna," vol. 131, pt. H, no. 6, pp. 334–340, Dec. 1984. © Institution of Electrical Engineers.

2 General description of the model

The circuit representation of the present model is shown in Fig. 3. In this network Y_s is the self admittance of the main

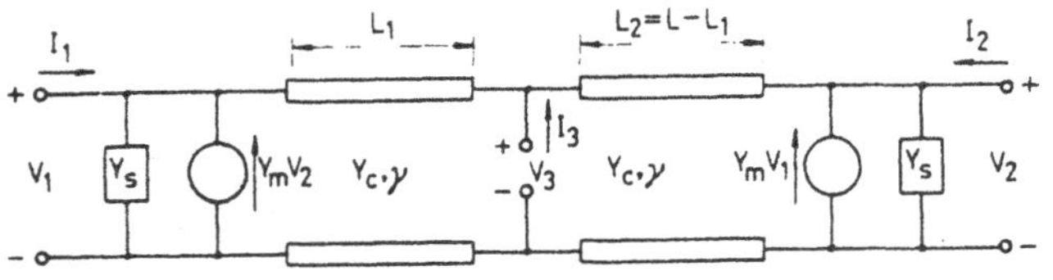

Fig. 3 *Improved transmission-line model represented as a three-port*

radiating slots and Y_m is their mutual (radiative) admittance. It can be seen that the mutual coupling is formally taken into account by voltage-dependent current generators.

The admittance matrix of this three-port model is given by

$$[Y] = \begin{bmatrix} Y_s + Y_c \coth(\gamma L_1) & -Y_m & -Y_c \operatorname{csch}(\gamma L_1) \\ -Y_m & Y_s + Y_c \coth(\gamma L_2) & -Y_c \operatorname{csch}(\gamma L_2) \\ -Y_c \operatorname{csch}(\gamma L_1) & -Y_c \operatorname{csch}(\gamma L_2) & Y_c \coth(\gamma L_1) + Y_c \coth(\gamma L_2) \end{bmatrix} \tag{1}$$

where coth (z) and csch (z) are the complex hyberbolic cotangent and cosecant functions of argument z, respectively; Y_c is the characteristic admittance of the microstrip line formed by the patch; and $\gamma = \alpha + j\beta$ is the complex propagation constant of this line. Hence, the copper and dielectric losses of the antenna are taken into account by the attenuation constant α of the transmission-line of the model.

If there is only one feed point, an input admittance can be defined. Assuming $I_1 = I_2 = 0$, it follows from eqn. 1 that

$$Y_{in} = 2Y_c\left[\frac{Y_c^2 + Y_s^2 - Y_m^2 + 2Y_sY_c \coth(\gamma L) - 2Y_mY_c \operatorname{csch}(\gamma L)}{(Y_c^2 + Y_s^2 - Y_m^2)\coth(\gamma L) + (Y_c^2 - Y_s^2 + Y_m^2)\cosh(2\gamma\Delta)\operatorname{csch}(\gamma L) + 2Y_sY_c}\right] \tag{2}$$

where

$$\Delta = \left|\frac{L}{2} - L_1\right| = \left|\frac{L}{2} - L_2\right| \tag{3}$$

and L_1 and L_2 are defined in Figs. 2 and 3.

3 Determination of the model parameters

The model of Fig. 3 contains the following unknowns: the line parameters (Y_c, γ), the self admittance of the slots (Y_s), and the mutual admittance (Y_m).

3.1 Line parameters

Using the planar waveguide model [8], the line parameters can be expressed as follows:

$$Z_c = \frac{\eta_0}{\sqrt{\varepsilon_{re}}}\frac{h}{W_e} \tag{4}$$

$$\beta = k_0\sqrt{\varepsilon_{re}} \tag{5}$$

$$\alpha = 0.5\beta \tan \delta_e \tag{6}$$

where

$\eta_0 = \sqrt{\mu_0/\varepsilon_0}$ = wave impedance in free space
$k_0 = 2\pi/\lambda_0$ = free-space wave number
W_e = effective width
ε_{re} = effective dielectric constant
$\tan \delta_e$ = effective loss tangent.

Accurate formulas and techniques to determine W_e, ε_{re}, $\tan \delta_e$ can be found in the literature [9].

3.2 Self admittance

To determine $Y_s = G_s + jB_s$, Derneryd's model [2] has been refined to improve accuracy and efficiency.

For the self susceptance B_s, the correct transmission-line formula

$$B_s = Y_c \tan(\beta\Delta l) \tag{7}$$

is used, where Δl is the open-end extension of the patch considered as a (semi-infinite) microstrip line of width W. A careful comparative study revealed that the expression given in Reference 6 is to be recommended to compute Δl.

To determine the self conductance G_s, the radiation conductance of a uniformly TE-excited slot radiator of length W_e and width Δl is computed. It is given by the following integral expression:

$$G_s = \frac{1}{\pi^2\eta_0}\int_0^{\pi}\int_0^{\pi}\frac{\sin^2(w/2\cos\alpha)}{\cos^2\alpha} \times \frac{\sin^2(s/2\sin\alpha\cos\beta)}{(s/2\sin\alpha\cos\beta)^2}\sin^3\alpha\, d\alpha\, d\beta \tag{8}$$

where

$$w = k_0 W_e \tag{9}$$

and

$$s = k_0 \Delta l \tag{10}$$

are the normalised slot dimensions. A Maclaurin series expansion in the normalised slot width s has been derived for this integral [17]. Retaining only the first two terms, the result is given by

$$G_s = \frac{1}{\pi\eta_0}\left\{\left(wSi(w) + \frac{\sin w}{w} + \cos w - 2\right)\left(1 - \frac{s^2}{24}\right) + \frac{s^2}{12}\left(\frac{1}{3} + \frac{\cos w}{w^2} - \frac{\sin w}{w^3}\right)\right\} \tag{11}$$

The truncation error of eqn. 11 has been found to be less than 0.1% for $s \leqslant 1$, irrespective of w.

It has to be remarked that surface waves were neglected in the calculation of G_s, which is acceptable [10] if

$$k_0\sqrt{\varepsilon_r}h \leqslant 0.3 \tag{12}$$

This also seems to be a useful limit for neglecting the frequency dependence of Δl [10].

3.3 Mutual admittance

Accurate closed-form expressions have been derived for both the real and imaginary parts of the mutual admittance $Y_m = G_m + jB_m$ [17]. The result can be written as

$$G_m = G_sF_gK_g \tag{13}$$

$$B_m = B_sF_bK_b \tag{14}$$

where

$$F_g = g_m/g_s \tag{15}$$

$$F_b = b_m/b_s \tag{16}$$

K_g, K_b = correction functions to be determined further.

The functions F_g and F_b are coupling functions expressing the ratio between the per-unit-length mutual admittance ($y_m = g_m + jb_m$) and the per-unit-length self admittance ($y_s = g_s + jb_s$) of two infinite-length TE-excited slot radiators. The following expressions, which are accurate for $s \leqslant 1$, have been derived [17]:

$$F_g = J_0(l) + \frac{s^2}{24 - s^2} J_2(l) \tag{17}$$

$$F_b = \frac{\pi}{2} \frac{Y_0(l) + \dfrac{s^2}{24 - s^2} Y_2(l)}{\ln\left(\dfrac{s}{2}\right) + C^e - \dfrac{3}{2} + \dfrac{s^2/12}{24 - s^2}} \tag{18}$$

where
$l = k_0 L_e$ = normalised centre distance between the slots
$\ln(x)$ = natural logarithm
$J_i(x)$ and $Y_i(x)$ are the ith-order Bessel functions of the first and second kind, respectively
$C^e = 0.577216\ldots$ = Euler's constant.

The functions K_g and K_b are correction functions introduced to take into account the finite length of the main slots and the influence of the side slots. The derivation of these correction functions is rather lengthy [17], so that only the fundamentals are given here.

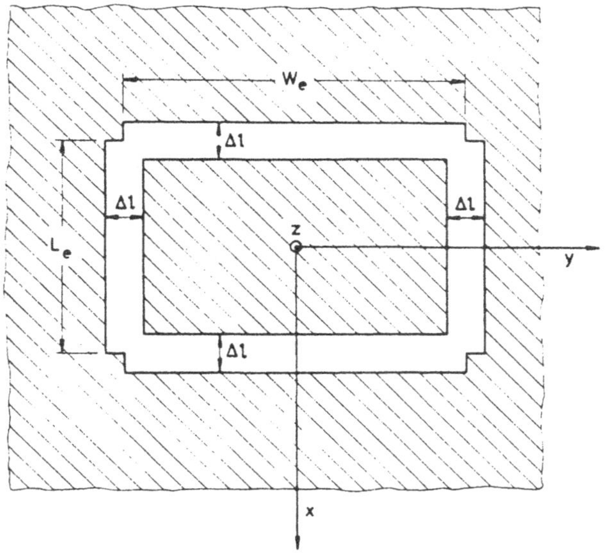

Fig. 4 *Four-slot aperture radiation model: geometry, dimensions and co-ordinates*

Table 1: Radiation conductance for l = 2 and s = 0

w	G_r^1, mS	G_r^2, mS	G_r^4, mS	G_r^{mod}, mS
1	0.55	0.75	0.69	0.68
2	2.11	2.84	2.63	2.58
3	4.40	5.86	5.48	5.38
4	7.08	9.33	8.79	8.67
5	9.87	12.84	12.23	12.08
6	12.61	16.19	15.58	15.43
7	15.26	19.39	18.82	18.68
8	17.86	22.54	22.00	21.86

First, the determination of K_g is discussed. As a reference, the radiation conductance of the four-slot structure shown in Fig. 4 was calculated. The tangential electric field in the aperture plane $z = 0$ was taken to be

$$E_a = \begin{cases} \hat{x}\dfrac{V_s}{\Delta l}\,; & |y| \leqslant \dfrac{W_e}{2},\ \dfrac{L_e - \Delta l}{2} \leqslant |x| \leqslant \dfrac{L_e + \Delta l}{2} \\ \hat{y}\dfrac{V_s}{\Delta l}\sin\left(\pi\dfrac{x}{L_e}\right); & |x| \leqslant \dfrac{L_e}{2},\ \dfrac{W_e - \Delta l}{2} \leqslant y \leqslant \dfrac{W_e + \Delta l}{2} \\ -\hat{y}\dfrac{V_s}{\Delta l}\sin\left(\pi\dfrac{x}{L_e}\right); & |x| \leqslant \dfrac{L_e}{2},\ \dfrac{W_e - \Delta l}{2} \leqslant -y \leqslant \dfrac{W_e + \Delta l}{2} \\ 0; & \text{otherwise} \end{cases} \tag{19}$$

where
V_s = excitation voltage
$\hat{x}, \hat{y}$ = unit vectors in the positive x- and y-directions.
This aperture field is an acceptable approximation of the true aperture field corresponding to the fundamental mode, and enables an accurate computation of the far-field and radiation conductance [3]. This computation is straightforward using the plane-wave spectral method [3, 11]. The resulting integral expression for the radiation conductance is too complicated for analytic integration, but can be evaluated numerically without any difficulty. This numerical quantity has been used as a reference to determine K_g, and is further indicated as G_r^{ref}. The corresponding radiation conductance given by the model (Fig. 3) is indicated as G_r^{mod}:

$$G_r^{mod} = 2(G_s + G_m) = 2G_s(1 + F_g K_g) \tag{20}$$

As a good correspondence between G_r^{mod} and G_r^{ref} is required, K_g has to be a good approximation of the numerical quantity K_g^{ref}:

$$K_g^{ref} = \left(\frac{G_r^{ref}}{2G_s} - 1\right)\Big/ F_g \tag{21}$$

This quantity has been evaluated for a large number of parameter values in the ranges $w \geqslant 0.1$; $1 \leqslant 3.2$ and $s \leqslant 1$. This evaluation showed the totally unexpected result that K_g^{ref} can be approximated by the expression

$$K_g = 1 \tag{22}$$

It has to be interpreted that this result is caused by a compensating effect: the influence of the side slots on the radiation conductance and the influence of the finite length of the main slots cancel each other nearly perfectly.

This effect is illustrated in Table 1, which shows the following quantities:

$G_r^1 = 2G_s$ = twice the radiation self conductance of one main slot

G_r^2 = radiation conductance of the two-slot system consisting of the main slots

$G_r^4 = G_r^{ref}$ = radiation conductance of the reference system consisting of four slots: two main slots and two side slots

G_r^{mod} = radiation conductance as given by eqns. 20 and 22.

The results of our model (G_r^{mod}) follow the reference values (G_r^{ref}) much better than the two-slot-system values (G_r^2).

The error introduced by taking $K_g = 1$ has been studied systematically by calculating the relative error

$$E_p = \frac{G_r^{ref} - G_r^{mod}}{G_r^{ref}} \qquad (23)$$

This led to the following conclusions:

(a) E_p is always positive

(b) E_p is a decreasing function of w if l and s are constant.

(c) E_p is an increasing function of l if w and s are constant.

(d) E_p is a decreasing function of s if w and l are constant.

Hence, the maximum error for a fixed value of l occurs for $w \to 0$ and $s \to 0$. The maximum error for $w \geqslant 0.1$ and $s \leqslant 1.0$ is shown in Fig. 5. For the fundamental mode, l is given by

$$l \simeq \pi/\sqrt{\varepsilon_e} \simeq \pi/\sqrt{\varepsilon_r} \qquad (24)$$

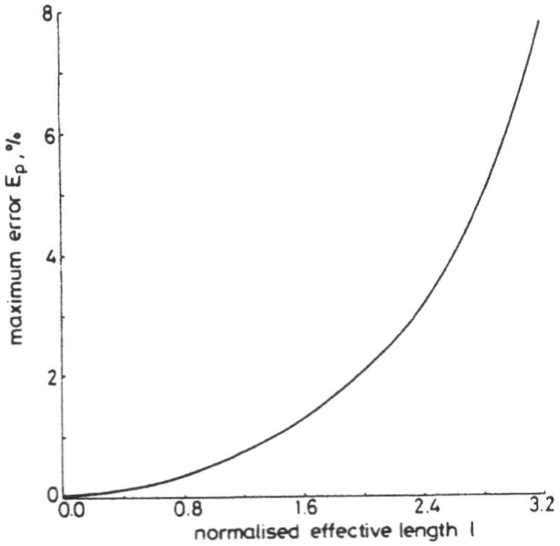

Fig. 5 *Maximum relative error of model radiation conductance with respect to reference radiation conductance*
$w \geqslant 0.1, \quad s \leqslant 1.0$

For commonly used substrate materials with a dielectric constant $\varepsilon_r > 2$, the error E_p remains smaller than 2.5%.

In Reference 4 Lier argues that the influence of the side slots on the radiation conductance can be neglected. However, the present authors' investigations show that a distinctly better correspondence with experiments is obtained if the side slots are taken into account.

Secondly, the determination of K_b will be discussed. It was assumed that the influence of the side slots on the value of the mutual susceptance is negligible. Therefore K_b was determined with reference to a system consisting of only the two main slots. In Reference 12, Rhodes shows that the susceptance of a radiating aperture is much more sensitive to the precise form of the aperture field distribution than the conductance. With the conditions of Reference 12 in mind, a linear tapering of the aperture field near the ends of the slots has been assumed:

$$E_a = \begin{cases} \hat{x}\dfrac{V_s}{\Delta l} \;;\; |y| \leqslant \dfrac{W_e}{2} - T, \quad \dfrac{L_e - \Delta l}{2} \leqslant |x| \leqslant \dfrac{L_e + \Delta l}{2} \\ \hat{x}\dfrac{V_s}{\Delta l}\dfrac{W_e - 2|y|}{2T}; \\ \dfrac{W_e}{2} - T \leqslant |y| \leqslant \dfrac{W_e}{2}, \quad \dfrac{L_e - \Delta l}{2} \leqslant |x| \leqslant \dfrac{L_e + \Delta l}{2} \\ 0; \quad \text{otherwise} \end{cases} \qquad (25)$$

where T (assumed to be very much less than W_e) is the taper distance. For the determination of K_b the mutual susceptance of this two-slot system was taken as a reference and is indicated here by B_m^{ref}. This quantity is given by the integral expression:

$$B_m^{ref} = \frac{4}{\pi^2 \eta_0} \int_0^{\pi/2} \int_0^{\pi/2} (1 - \sec^2 u \sin^2 v) \frac{\sin^2\left(\frac{s}{2}\sec u \cos v\right)}{\left(\frac{s}{2}\sec u \cos v\right)^2} \times \frac{\sin^2\left(\frac{t}{2}\sec u \sin v\right)}{\left(\frac{t}{2}\sec u \sin v\right)^2} \frac{\sin^2\left(\frac{w-t}{2}\sec u \sin v\right)}{\sin^2 v} \times \cos(l \sec u \cos v)\, dv\, du \qquad (26)$$

where $t = k_0 T$ = the normalised taper distance.

This quantity is only slightly influenced by the precise values of s and t, at least if $s \ll l$ and $t \ll w$, because of the averaging out effect of the cosine function in the integrand.

This mutual susceptance of two finite-length slots has been compared with the mutual susceptance of an equivalent section of length W_e of two infinite-length slots with the same width (Δl) and the same separation distance ($L_e = L + \Delta l$). Hence, the reference correction function can be written as

$$K_b^{ref} = \frac{B_m^{ref}}{b_m W_e} \qquad (27)$$

where b_m = the mutual susceptance per unit of length between two infinite-length slots for which an analytical expression has been derived [17].

A numerical investigation proved that K_b^{ref} is nearly independent of s and t, at least for small values of these parameters. Table 2 shows some results for K_b^{ref} as a function of w and l if $s = t = 0.2$. The influence of l is rather unimportant in the range $1.5 \leqslant l \leqslant 2.25$, which corresponds to $2 \leqslant \varepsilon_r \leqslant 4$. Hence, it can be assumed that K_b only depends on w, and the following expression for K_b has been obtained by curve fitting:

$$K_b = 1 - \exp(-0.21w) \qquad (28)$$

Table 2: Correction function for the mutual susceptance

K_b	Reference eqn. 27 with $s = t = 0.2$				Model eqn. 28
w \ l	1.50	1.75	2.00	2.25	
1	0.20	0.16	0.13	0.14	0.19
2	0.34	0.33	0.32	0.33	0.34
3	0.36	0.44	0.46	0.49	0.47
4	0.52	0.55	0.55	0.60	0.57
5	0.63	0.64	0.64	0.68	0.65
6	0.67	0.71	0.74	0.75	0.72
7	0.77	0.76	0.79	0.79	0.77

Eqn. 28 is also tabulated in Table 2. There is a good correspondence with K_b^{ref}, at least for $w \geqslant 2$. The not so good correspondence for $w < 2$ can be ascribed to the condition $T \ll W_e$ not being met in this range. Also, eqn. 28 has the correct asymptotic behaviour for $w \to \infty$. Indeed, for $w \to \infty$, the influence of the presence of the side slots and of the finite length of the main slots is negligible, so that K_b has to approach unity.

4 Microstrip-line-fed antenna

Consider a microstrip antenna fed by a coplanar microstrip line (see Figs. 6 or 7). Referring to Fig. 3, this corresponds to the case $I_2 = I_3 = 0$ or, alternatively, $I_1 = I_2 = 0$ and $L_1 = 0$. It follows from eqns. 2 and 3 that

$$Y_{in} = \frac{Y_c^2 + Y_s^2 - Y_m^2 + 2Y_s Y_c \coth(\gamma L) - 2Y_m Y_c \operatorname{csch}(\gamma L)}{Y_s + Y_c \coth(\gamma L)} \tag{29}$$

To model the parasitic effects of the feed line on the antenna behaviour, the self admittance of the main slot facing the feed line can be reduced by a factor

$$r = 1 - \frac{W_m}{W_e} \tag{30}$$

where W_m is the width of the microstrip feed line. This

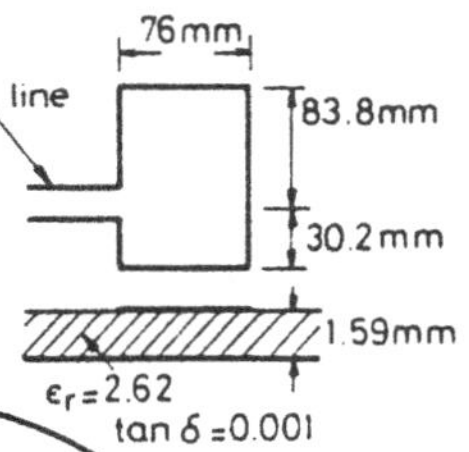

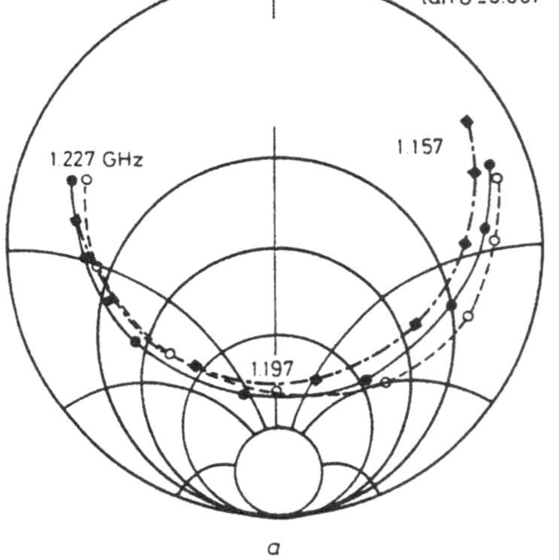

a

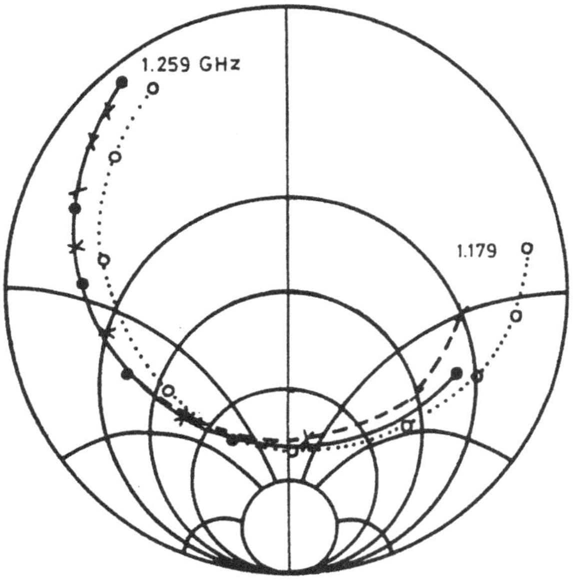

b

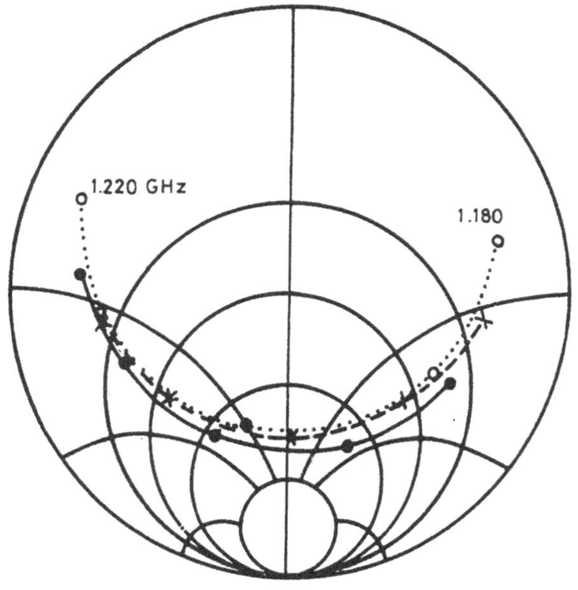

c

Fig. 6 *Input impedance of a rectangular ($W/L = 1.5$) microstrip antenna fed by a 50 Ω microstrip line*

a O---O measured [13]
●—● moment method [7]
◆---◆ present model
b O· · ·O calculated [13]
●—● [7]
×---× present model
c O· · ·O [14]
●—● [7]
×---× present model

reduction takes into account the partial covering of the equivalent slot by the feed line.

The reduction of the self admittance at terminal 1 can be considered as an addition of a parallel admittance:

$$Y_F = (r - 1)Y_s \tag{31}$$

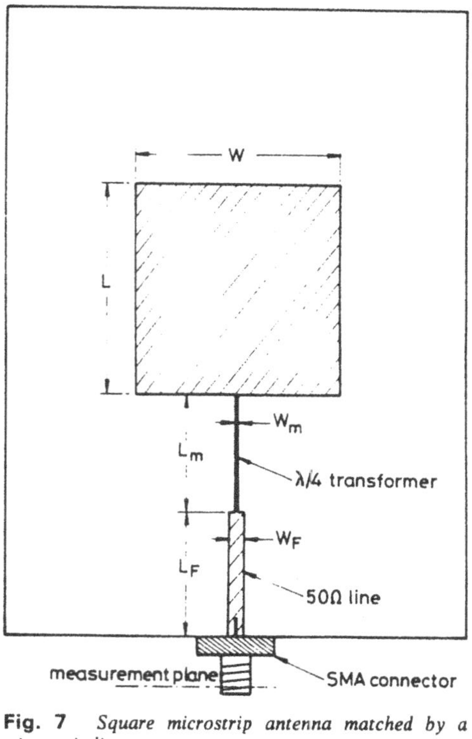

Fig. 7 *Square microstrip antenna matched by a quarter-wavelength microstrip line*

$W = 33.147$ mm, $L = 33.165$ mm, $W_m = 0.473$ mm
$L_m = 18.713$ mm, $W_F = 2.403$ mm, $L_F = 20$ mm
$h = 0.7874$ mm, $\varepsilon_r = 2.20$, $\tan\delta = 0.0009$
$t = 0.018$ mm, $\sigma_p = \sigma_g = 0.556 \times 10^5$ S/mm
$\Delta_p = \Delta_g = 0.0005$ mm

The antenna input admittance is thus given by

$$Y'_{in} = Y_{in} + Y_F = rY_s + \frac{Y_c^2 - Y_m^2 + Y_s Y_c \coth(\gamma L) - 2Y_m Y_c \operatorname{csch}(\gamma L)}{Y_s + Y_c \coth(\gamma L)} \quad (32)$$

Apart from an input impedance computation, the model enables an efficient computation of radiation conductance, antenna efficiency, resonant frequency, Q-factor and bandwidth [17]. Combined with an equivalent-slot radiation model, the directivity and gain can also be computed easily [17].

5 Experimental verification

As a first verification, the present transmission-line model was compared with the measured results of Lo *et al.* [13] and the moment-method results of Deshpande and Bailey [7] (see Fig. 6*a*). The Figure shows a rectangular microstrip antenna fed by a 50 Ω microstrip line. The following parameter values were used in evaluating eqn. 32: $W = 114$ mm, $L = 76$ mm, $W_m = 4.3$ mm, $h = 1.59$ mm, $\varepsilon_r = 2.62$, $\tan\delta = 0.001$ (loss tangent), $t = 0.035$ mm (copper thickness), $\sigma_p = \sigma_g = 0.556 \times 10^5$ S/mm (conductivity of patch and ground plane), $\Delta_p = \Delta_g = 0.0015$ mm (rms surface roughness of patch and ground plane).

The moment-method solution [7] is more in agreement with the experimental results [13] than the present results. However, it has been verified that the error of the present results remains within the limits caused by the tolerances on ε_r. Moreover, a comparison with other theoretical results quoted in Reference 7 shows that the present transmission-line model results are more accurate than the cavity-model results of Lo *et al.* [13] and the moment-method results of Newman and Tulyathan [14] (see Figs. 6*b* and *c*).

A second varification is based on our own experimental results for the square microstrip antenna shown in Fig. 7. This antenna was manufactured by photoetching a double-clad 'RT/duroid 5880' substrate of thickness 0.031 in. (0.7874 mm), and is fed by an 'OSM 215-3' connector. The precise (measured) dimensions of the conductive pattern and the parameters of the substrate and the copper cladding are indicated. Fig. 8 shows the measured reflection diagram and three calculated curves. To produce these calculated diagrams, the particular model for the antenna element (different in the three cases) was completed with appropriate models (identical in the three cases) for the two microstrip-line sections, the step discontinuity and the microstrip-coax transition. To represent the antenna element, eqn. 32 was evaluated three times with different values for Y_s and Y_m.

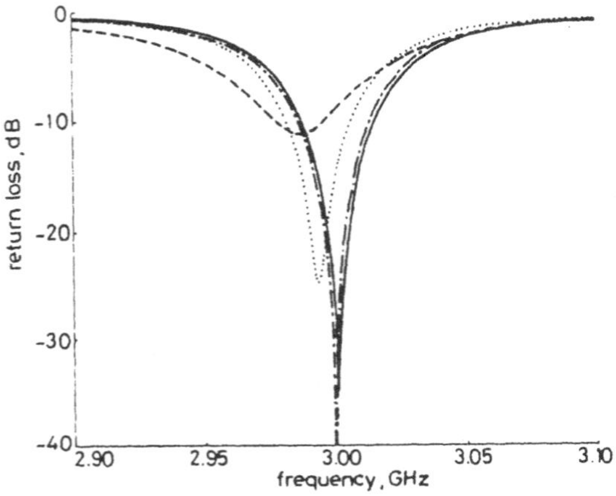

Fig. 8 *Return loss of antenna shown in Fig. 7*
——— measured
- - - - Munson's model [1]
········ Derneryd's model [2]
-·-·-· present model

First, the present transmission-line model was used as described in this paper. Secondly, the mutual admittance ($Y_m = 0$) was neglected to simulate Derneryd's model [2]. Thirdly, in addition to $Y_m = 0$, the expression $Y_s = W_e y_s$ was used to stimulate Munson's model [1].

This comparison shows clearly the influence of the mutual coupling between the slots (comparing Derneryd's results with the present ones) and the influence of the finite length of the slots (comparing Munson's model with the results of Derneryd).

6 Discussion

An improved transmission-line model has been described that forms an accurate tool for the analysis of rectangular microstrip antennas. Due to its numerical efficiency, this model is extremely well suited for design purposes. The model has a very broad range of validity in terms of patch aspect ratio (W/L), substrate dielectric constant (ε_r) and substrate electrical thickness (h/λ_0). However, some physical effects are still neglected by the model, such as the excitation of substrate waves, the mutual coupling with neighbouring elements, and the diffraction at the substrate and ground plane edges. Further, because it is a transmission-line model, the model does not take into account the effects of the higher-order modes. However, it is believed that, in many cases in the present situation, there is more to be gained with an improvement of the description of the fundamental mode than with the inclusion in the analysis model of the higher-order modes without further caring about the fundamental mode. In this context, it can be observed that the present results for Y_s and Y_m could be used profitably in cavity models such as that of Carver and Mink [15], in order to model accurately the cavity-wall admittances for the dominant mode.

7 Acknowledgments

Dr. Van de Capelle is, and Dr. Pues was formerly, supported by the National Fund for Scientific Research of Belgium, as a Senior Research Associate and a Research Assistant, respectively.

The authors thank Ir. B. Nauwelaers for his very useful comments and discussions.

8 References

1 MUNSON, R.E.: 'Conformal microstrip antennas and microstrip phased arrays', *IEEE Trans.*, 1974, **AP-22**, pp. 74–78
2 DERNERYD, A.G.: 'Linearly polarised microstrip antennas', *ibid.*, 1976, **AP-24**, pp. 846–851
3 HAMMER, P., VAN BOUCHAUTE, D., VERSCHRAEVEN, D., and VAN DE CAPELLE, A.: 'A model for calculating the radiation field of microstrip antennas', *ibid.*, 1979, **AP-27**, pp. 267–270
4 LIER, E.: 'Improved formulas for input impedance of coax-fed microstrip patch antennas', *IEE Proc. H, Microwaves, Opt. & Antennas*, 1982, **129**, (4), pp. 161–164
5 DERNERYD, A.G.: 'A theoretical investigation of the rectangular microstrip antenna element', *IEEE Trans.*, 1978, **AP-26**, pp. 532–535
6 KIRSCHNING, M., JANSEN, R.H., and KOSTER, N.H.L.: 'Accurate model for open end effect of microstrip lines', *Electron. Lett.*, 1981, **17**, (3), pp. 123–125
7 DESHPANDE, M.D., and BAILEY, M.C.: 'Input impedance of microstrip antennas', *IEEE Trans.*, 1982, **AP-30**, pp. 645–650

8 KOMPA, G., and MEHRAN, R.: 'Planar waveguide model for calculating microstrip components', *Electron. Lett.*, 1975, **11**, (19), pp. 459–460

9 HAMMERSTAD, E., and JENSEN, Ø.: 'Accurate models for microstrip computer-aided design'. Dig. 1980 IEEE MTT-S International Microwave Symposium, Washington, 1980, pp. 407–409

10 JAMES, J.R., and HENDERSON, A.: 'High-frequency behaviour of microstrip open-circuit terminations', *IEE J. Microwaves, Opt. & Acoust.*, 1979, **3**, (5), pp. 205–218

11 COLLIN, R.E., and ZUCKER, F.J. (Eds.): 'Antenna theory, Part 1' (McGraw-Hill, New York, 1969), pp. 61–68

12 RHODES, D.R.: 'On a new condition for physical realisability of planar antennas', *IEEE Trans.*, 1971, **AP-19**, pp. 162–166

13 LO, Y.T., SOLOMON, D., and RICHARDS, W.F.: 'Theory and experiment on microstrip antennas', *ibid.*, 1979, **AP-27**, pp. 137–145

14 NEWMAN, E.H., and TULYATHAN, P.: 'Analysis of microstrip antennas using moment methods', *ibid.*, 1981, **AP-29**, pp. 47–53

15 CARVER, K.R., and MINK, J.W.: 'Microstrip antenna technology', *ibid.*, 1981, **AP-29**, pp. 2–24

16 VAN DE CAPELLE, A., NAUWELAERS, B., LEPLA, R., and KISSEMBEEK, F.: 'Analysis of linear microstrip resonator arrays'. Proc. International URSI Symposium on Electromagnetic theory, Santiago de Compostella, 1983, pp. 433–437

17 PUES, H., and VAN DE CAPELLE, A.: 'Accurate transmission-line model for the rectangular microstrip antenna', Catholic University of Louvain Internal Report, 1983

18 RICHARDS, W.F., LO, Y.T., and HARRISON, D.D.: 'An improved theory for microstrip antennas and applications', *IEEE Trans.*, 1981, **AP-29**, pp. 38–46

CAD-Oriented Cavity Model for Rectangular Patches

D. Thouroude, M. Himdi, and J. P. Daniel

Indexing terms: Antennas, Modelling

A cavity model well suited for computed-aided design is presented. The patch antenna is described by geometrical and electrical parameters. Using a cavity model, input impedance as a function of frequency is then calculated with a fast computer program implemented on a PC. Resonant resistance and resonant frequency are deduced.

Introduction: The cavity model is a classical method for patch antenna analysis.[1–5] The rectangular microstrip patch antenna is treated as a cavity bounded by four magnetic walls. When the field-matching technique is used, the fields are expressed as a series of waveguide modes. All the losses in the antenna are represented by means of an effective loss tangent. The purpose of this letter is to shorten the iterative procedure using a proper value of δ_{eff}.

Cavity model: The electric field is expressed as a series using a mode-matching technique (a similar expression for magnetic field can be obtained)

$$E_z^i(x, y) = \sum_{p=0}^{\infty} A_p \cos\left(\frac{p\pi x}{a}\right) \frac{\cos \beta_p(y - v)}{\cos \beta_p v} \tag{1}$$

where $v = b - y_1$ for $y > 0$, $v = -y_1$ for $y < 0$.

$$A_p = \frac{4j\omega\mu_0}{\pi} \frac{J \cos \dfrac{p\pi x_0}{a} \sin \dfrac{p\pi d}{2a}}{p\beta_p[tg\beta_p(b - y_1) + tg\beta_p y_1]} \tag{2}$$

$$A_0 = \frac{j\omega\mu_0}{a} \frac{Jd}{\beta_0[tg\beta_0(b - y_1) + tg\beta_0 y_1]}$$

$$\beta_p^2 = k_0^2 \varepsilon_r(1 - j\delta_{eff}) - \left(\frac{p\pi}{a}\right)^2 = k^2 - \left(\frac{p\pi}{a}\right)^2 \tag{3}$$

Fig. 1*a* shows the geometrical dimensions of the patch, feeding being represented by a sheet of current parallel to Oz, located between $x_0 - d/2$ and $x_0 + d/2$, with $J = I/d$ (I is the intensity of current).

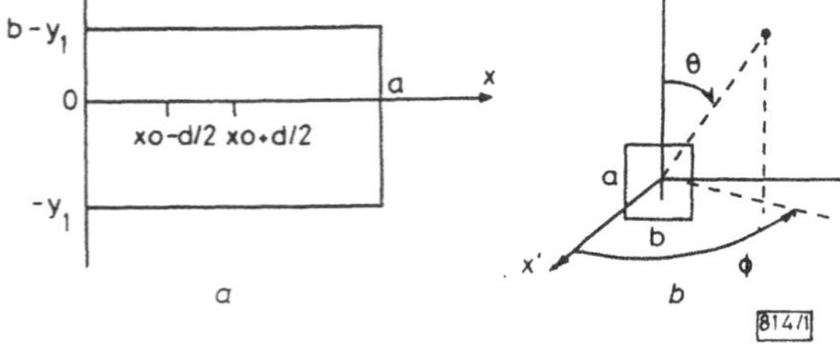

Fig. 1 *Structure of rectangular microstrip antenna*
a Geometry
b Co-ordinate system

Conductor, dielectric and radiation losses are represented by means of an effective loss tangent δ_{eff} in eqn. 3

$$\delta_{eff} = \frac{P_r + P_d + P_c}{\omega(w_e + w_h)} = \frac{1}{Q} \tag{4}$$

where Q is the quality factor, w_e and w_h are electric and magnetic energy stored in the cavity, P_r is the power radiated, P_d represents dielectric loss and P_c the conductor losses.

The input impedance is

$$Z = -\frac{t}{Id} \int_{x_0 - d/2}^{x_0 + d/2} E_z^i(x, 0)\, dx \tag{5}$$

where t is the substrate thickness.

The iterative procedure is then

Input: characteristic parameters (electrical and geometrical)

Step 1: $\delta_{eff} = \delta$

Step 2: calculation of fields (with eqn. 1)

Step 3: calculation of $(\delta_{eff})_{i+1}$ with eqn. 4

Step 4: test if $\dfrac{(\delta_{eff})_{i+1} - (\delta_{eff})_i}{(\delta_{eff})_i} > P\%$ then step 2

P is chosen arbitrarily (usually lower than 5)

Output: Impedance Z

It has been shown[4] that δ_{eff} does not depend strongly on k or δ. It is recommended an initial value of δ_{eff} very close to the real value is used to shorten computation time.

Effective loss tangent determination: The effective loss tangent may be written, near the resonant frequency, according to eqn. 4, as

$$\delta_{eff} = \delta + \frac{\Delta}{t} + \frac{P_r}{2\omega_0 w_e} \tag{6}$$

where Δ is the skin depth.

In order to find a proper value of δ_{eff0}, w_e and P_r must be calculated.

Stored electric energy: The electric energy stored, at resonance, for the dominant mode, can be shown to be

$$w_e = \frac{\varepsilon_0 \varepsilon_r abV_0^2}{8t} \tag{7}$$

where V_0 is the input voltage.

Radiated power: The far fields may be calculated by modelling the radiator as four radiating slots. These fields are expressed in $x'y'z'$ co-ordinates (Fig. 1*b*)

$$E_\theta = -jk_0 F_{y'} \cos\phi$$
$$E_\phi = jk_0(F_{y'} \cos\theta \sin\phi - F_{z'} \sin\theta) \tag{8}$$

with

$$F_{z'} = \frac{V_0 a}{\pi r} e^{-jk_0 r} \cos\left(\frac{k_0 b}{2} \sin\theta \sin\phi\right) \times \frac{\sin\left(\frac{k_0 a}{2}\cos\theta\right)}{\frac{k_0 a}{2}\cos\theta} \tag{9a}$$

$$F_{y'} = \frac{V_0 b^2 k_0}{2\pi r} e^{-jk_0 r} \cos\left(\frac{k_0 b}{2} \sin\theta \sin\phi\right) \times \frac{\sin\left(\frac{k_0 a}{2}\cos\theta\right) \sin\theta \sin\phi}{\left(\frac{\pi}{2}\right)^2 - \left(\frac{k_0 b}{2}\sin\theta\sin\phi\right)^2} \tag{9b}$$

To calculate the radiated power these expressions may be approximated. In eqn. 9*a* and eqn. 9*b* we substitute $\cos(u)$ by $(\pi/8)[(\pi^2/4) - u^2]$ and $\sin(u)$ by $u - u^3/6$. The variations of the exact functions and approximate polynomials are plotted on Fig. 2*a*; there is a good agreement between 0 and $\pi/2$. Using these approximations, analytical integration for the radiated power, P_r, can be performed

$$P_r = \frac{V_0^2 A\pi^4}{23040} \times \left[(1-B)\left(1 - \frac{A}{15} + \frac{A^2}{420}\right) + \frac{B^2}{5}\left(2 - \frac{A}{7} + \frac{A^2}{189}\right)\right] \tag{10}$$

with $A = (\pi a/\lambda_0)^2$ and $B = (2b/\lambda_0)^2$.

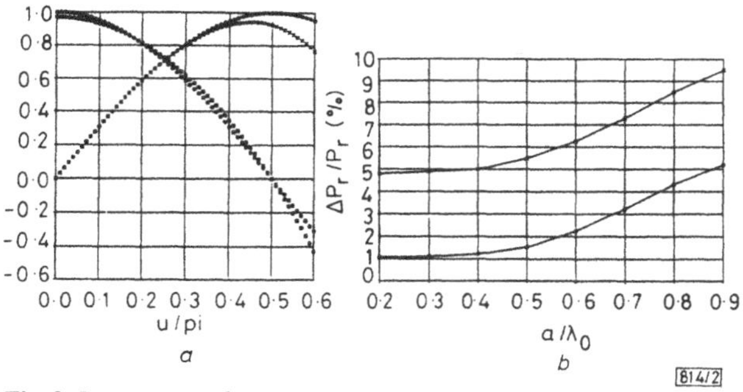

Fig. 2 *Integration aids*

a Approximate polynomials
b Relative difference between analytical and numerical methods
—●— $b/\lambda_0 = 0{\cdot}30$
—×— $b/\lambda_0 = 0{\cdot}15$

Comparison was made between eqn. 10 and numerical integration. Eqn. 10 gives radiated power with an accuracy better than 2·5% for $b/\lambda_0 = 0{\cdot}3$ (with a typical limit of $a/\lambda_0 = 0{\cdot}6$) and 4% for $b/\lambda_0 = 0{\cdot}15$ (and a similar typical limit $a/\lambda_0 = 0{\cdot}3$). (See Fig. 2*b*).

New iterative procedure and results: The electric energy stored is calculated using analytical formulas. The iterative procedure is similar to the previous one, except for the initial value of the first step where $\delta_{eff} = \delta_{effo}$.

In Table 1 we present measured and computed values of the resonant frequency and resonant resistance. Measured values are those given by Schaubert *et al.*[6] using microstrip feeding line ($y_1 = 0$ in Fig. 1). The present method gives reasonable results for the resonant frequency and the resistance. Resonant frequencies are obtained with about 2% or less error. When ε_r is high or when the substrate is thick, there is some discrepancy between measured and calculated resonant resistance. Two reasons can explain these differences

(*a*) The surface wave effect has been assumed to be negligible.

(*b*) The width of the feeding line is considered to be small enough to keep identical radiating slot lengths both on the edge connected to the microstrip feeding line and on the opposite edge.

A single matching network often uses a quarter wave transformer with a high impedance characteristic section. Two other examples of microstrip feeding line excitation have been tested.[7,8] In each case the previous approximations lead to good results as shown in Figs. 3 and 4. The computation time

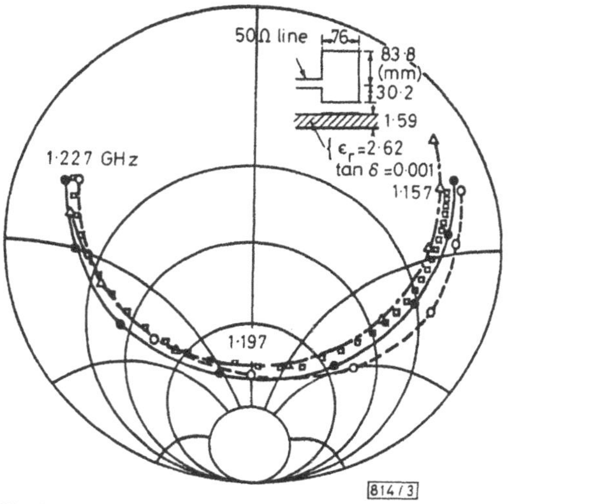

Fig. 3 *Return loss of antenna*

$a = 114$ mm; $b = 76$ mm; $d = 4{\cdot}3$ mm; $t = 1{\cdot}59$ mm
$\varepsilon_r = 2{\cdot}62$; $\delta = 0{\cdot}001$
O- -O measured[3]
●—● moment method[7]
△- -△ theory[10]
□□□ this method

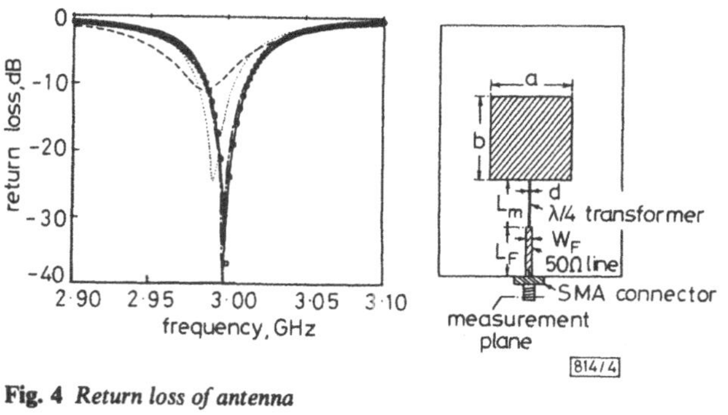

Fig. 4 *Return loss of antenna*

$a = 33{\cdot}147$ mm; $b = 33{\cdot}165$ mm; $d = 0{\cdot}473$ mm; $t = 0{\cdot}7874$ mm
$\varepsilon_r = 2{\cdot}2$; $\delta = 0{\cdot}0009$; $L_m = 18{\cdot}713$ mm; $W_F = 2403$ mm;
$L_F = 20$ mm
——— measured[10]
- - - - model[8]
······· model[9]
—·— model[10]
OOO this method

is less than 20 s for a range of 60 frequencies on a PC-AT with an 8087 arithmetic coprocessor.

Table 1 COMPARISON OF MEASURED AND CALCULATED RESULTS

						Measured[6]		Calculated		Moment method[6]	
ε_r	t	b	a	x_0	d	f_r	R_r	f_r	R_r	f_r	R_r
	mm	mm	mm	mm	mm	GHz	Ω	GHz	Ω	GHz	Ω
10·2	1·27	20	30	6·5	1·19	2·26	335	2·31	343	2·25	350
10·2	1·27	9·5	15	3·2	1·19	4·43	339	4·49	389	4·5	350
10·2	2·54	19	30	6·5	2·38	2·18	363	2·29	394	2·33	420
2·22	0·79	25	40	4	2·42	3·92	136	3·92	136	3·92	130
2·22	0·79	12·5	20	2	2·42	7·56	152	7·61	153	7·6	160
2·22	1,52	25	40	4	4·66	3·82	119	3·82	153	3·8	143
2·22	1·52	12	20	2	4·66	7·72	69	7·55	147	7·75	145

References

1 BAHL, I. J., and BHARTIA, P.: 'Microstrip antennas' (Artech House, Dedham, 1980)

2 JAMES, J. R., HALL, P. S., and WOOD, C.: 'Microstrip antenna—theory and design', *in* 'IEE Electromagnetic wave series 12' (Peter Peregrinus, 1981)

3 LO, Y. T., SOLOMON, D., and RICHARDS, W. F.: 'Theory and experiment on microstrip antennas', *IEEE Trans.*, 1979, **AP-27**, pp. 137–145

4 CARVER, K. R., and COFFEY, E. L.: 'Theoretical investigation of the microstrip antenna'. Technical report 00929, Physical Science Laboratory, New Mexico State University, Las Cruces, New Mexico, 1979

5 PENARD, E.: 'Etude d'antennes imprimées par la méthode de la cavité'. Thesis, Rennes, 1982

6 SCHAUBERT, D., POZAR, D., and ADRIAN, A.: 'Effect of microstrip antenna substrate thickness and permittivity: Comparison of theories with experiment', *IEEE Trans.*, 1989, **AP-37**, pp. 677–682

7 DESHPANDE, M., and BAILEY, M.: 'Input impedance of microstrip antennas', *IEEE Trans.*, 1982, **AP-30**, pp. 645–650

8 MUNSON, R.: 'Conformal microstrip antennas and microstrip phased arrays', *IEEE Trans.*, 1974, **AP-22**, pp. 74–78

9 DERNERYD, A.: 'Linearly polarised microstrip antennas', *IEEE Trans.*, 1976, **AP-24**, pp. 846–851

10 VAN DE CAPELLE, A.: 'Transmission-line model for rectangular microstrip antennas'. Handbook of microstrip antennas in 'IEE Electromagnetic wave series 28' (Peter Peregrinus, 1989), pp. 527–578

Analysis of Aperture-Coupled Microstrip Antenna Using Cavity Method

M. Himdi, J. P. Daniel and C. Terret

Indexing terms: Antennas, Microstrip

The letter presents an original analysis of aperture coupling of a microstrip antenna. The theory is based on the cavity model, and results are compared with the moment method and measurement.

Introduction: The structure has been previously described (Fig. 1) and analysed by the moment method in References 1 and 2. This letter proposes an application of the cavity model

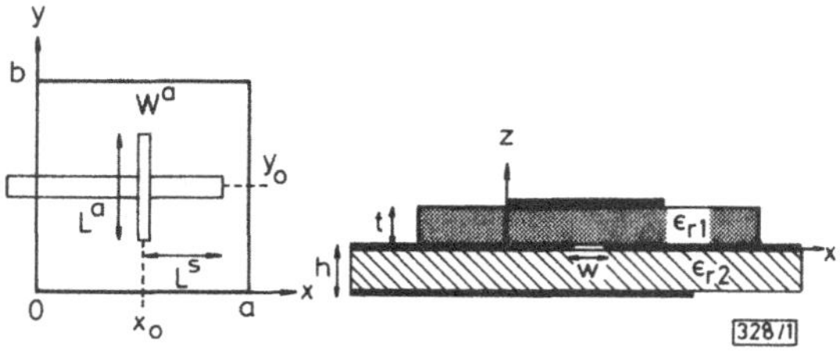

Fig. 1 *Aperture coupled microstrip antenna*

with a magnetic current excitation. The main objective is to obtain a simple calculation, keeping the physical comprehension of the phenomenon. In this context, only the case of a thin cavity is considered in a first approximation; therefore the dominant mode TM_{10} is sufficient.

Analysis: The first step is to consider the microstrip antenna as a cavity bounded by four perfect magnetic walls and two electric walls in $z = 0$ and t (Fig. 1). A magnetic current source $\boldsymbol{M}$ located in the aperture, is determined using the principle of equivalence: $\boldsymbol{M} = 2\boldsymbol{E}^a \wedge \boldsymbol{z}$. Here $\boldsymbol{E}^a$ is the aperture electric field, expressed by[1]

$$\boldsymbol{E}^a = \frac{V_0}{W^a} \frac{\sin k^a(L^a/2 - |y - y_0|)}{\sin(k^a L^a/2)} \boldsymbol{x}$$

$$x_0 - w^a/2 \leq x \leq x_0 + w^a/2 \qquad y_0 - L^a/2 \leq y \leq y_0 + L^a/2 \quad (1)$$

where k^a has been determined by Cohn's method,[4] taking into account the two dielectric constants and the near-metallic top plane (patch).

To obtain the z-component of the electric field into the overall volume cavity, the magnetic current source is presumed to be uniformly distributed in the volume above the slot. The other electric field components near the slot aperture are considered later with the slotline reactive power evaluation. The equivalent magnetic current density (Fig. 2) can be written as

$$\boldsymbol{J}_m = \frac{2V_0}{tW^a} \frac{\sin k^a(L^a/2 - |y - y_0|)}{\sin(k^a L^a/2)} \boldsymbol{y}$$

$$x_0 - w^a/2 \leq x \leq x_0 + w^a/2$$

$$y_0 - L^a/2 \leq y \leq y_0 + L^a/2 \qquad 0 \leq z \leq t \quad (2)$$

The time harmonic form of Maxwell's equations with magnetic source is

$$\nabla \wedge \boldsymbol{E}^i = -j\omega\mu_0 \boldsymbol{H}^i - \boldsymbol{J}_m \quad (3)$$

$$\nabla \wedge \boldsymbol{H}^i = j\omega\varepsilon \boldsymbol{E}^i \quad (4)$$

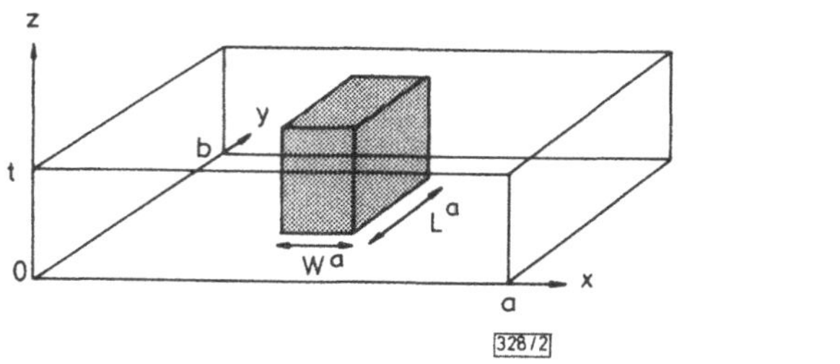

Fig. 2 *Model of magnetic current density in cavity*

where $\boldsymbol{E}^i$ and $\boldsymbol{H}^i$ represent electromagnetic fields in the cavity, and are solutions of the propagation equation with perfect magnetic walls.

Simple expressions for $\boldsymbol{E}^i$ and $\boldsymbol{H}^i$ can be obtained for the dominant mode TM_{10}:

$$\boldsymbol{E}^i = A \cos\left(\frac{\pi x}{a}\right)\boldsymbol{z} \quad (5)$$

$$\boldsymbol{H}^i = B \sin\left(\frac{\pi x}{a}\right)\boldsymbol{y} \quad (6)$$

where A and B are expressed by

$$A = \frac{8\pi V_0}{k^2 - (\pi/a)^2} \frac{\sin(\pi x_0/a)}{ba^2} \frac{\operatorname{sinc}(\pi w^a/2a)}{t k^a \sin(k^a L^a/2)} \times [1 - \cos(k^a L^a/2)] \quad (7)$$

$$B = j\left(\frac{a}{\pi}\right)\omega\varepsilon_{r1}\varepsilon_0 A \quad (8)$$

The second step deals with the radiation of the magnetic current source $\boldsymbol{K}(x, y) = t\boldsymbol{E}^i(x, y) \wedge \boldsymbol{I}$ at the edges of the cavity. This source is allowed to radiate into space and the radiative losses P_r may be computed in the usual manner. The stored electric and magnetic energy can be also obtained. Therefore one can define

$$\delta_{eff} = \frac{1}{Q} = \frac{P_{rad} + P_{cu} + P_d}{2\omega W_{elec}} \quad (9)$$

and replace k by k_{eff}, where $k_{eff} = k_0\sqrt{[\varepsilon_{r1}(1 - \delta_{eff})]}$, where k_0 is the free space wavenumber. The admittance of the antenna at the aperture is given by

Reprinted with permission from *Elect. Lett.*, M. Himdi, J. P. Daniel and C. Terret, "Analysis of Aperture Coupled Microstrip Antenna Using Cavity Method," vol. 25, no. 6, pp. 391–392, March 1989. © Institution of Electrical Engineers.

$$Y_{ant} = [P_{rad} + P_{cu} + P_d + 2j\omega(W_{elec} - W_{mag})]\,|V_0|^{-2} \tag{10}$$

The susceptance component due to the stored energy of the local field near the slot can be simply obtained from the two short-circuit slot lines (with proper characteristic impedance Z_{ca} and wavenumber k^a)

$$Y_{ap} = -\frac{2j}{Z_{ca}} \cot\left(k^a \frac{L^a}{2}\right) \tag{11}$$

Then the total admittance at the aperture is

$$Y_{at} = Y_{ap} + Y_{ant} \tag{12}$$

The last step is to transform the impedance along the microstrip line (Fig. 3). The discontinuity ΔV in modal voltage in

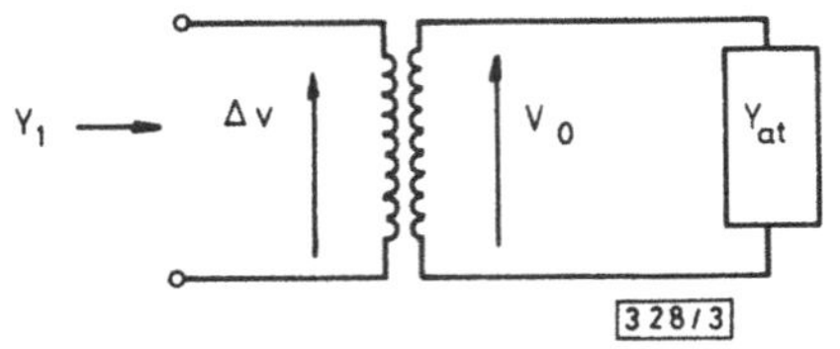

Fig. 3 *Equivalent transformer of transition microstrip line/slotline*

the microstrip line due to the slot cut on its ground plane may be determined[3] by

$$\Delta V = \int_{slot} E^a \wedge h^l \,.\, ds \tag{13}$$

where h^l is the normalised magnetic field for a microstrip line.

Then $Y_l = Z_l = Y_{at}\,\Delta V^{-2}$. Finally, the normalised input impedance is expressed by

$$Z_{in} = Z_l - jZ_c \cot(k^l L_s) \tag{14}$$

where L_s is the length of open-circuited stub of and k^l the wavenumber of microstrip line.

Results and conclusion: Calculations using this method have been compared with theoretical and experimental results from References 1 and 2; both the impedance curves and the resonant frequencies are similarly displayed on the Smith chart (Fig. 4), and only a very small shift of the resonant frequency remains. The agreement between measured and computed results supports the validity of this method. The effects of all parameters of the antenna were compared successfully with theory.[1] This method can be stretched to antenna substrates of larger thickness, but higher-order modes have to be considered. All these results will be published later.

References

1 SULLIVAN, P. L., and SCHAUBERT, D. H.: 'Analysis of an aperture coupled microstrip antenna', *IEEE Trans. Antennas & Propag.*, 1986, **34**, pp. 977–984

2 POZAR, D. M.: 'A reciprocity method of analysis for printed slot and slot-coupled microstrip antenna', *IEEE Trans. Antennas & Propag.*, 1986, **34**, (12)

3 RAO, I. S., JOSHI, K. K., and DAS, B. N.: 'Analysis of small aperture coupling between rectangular waveguide and microstripline', *IEEE Trans. Microwave Theory & Tech.*, 1981, **29**, (2)

4 COHN, S. B.: 'Slot-line on a dielectrique substrate', *IEEE Trans. Microwave Theory & Tech.*, 1969, **17**, (10), pp. 768–778

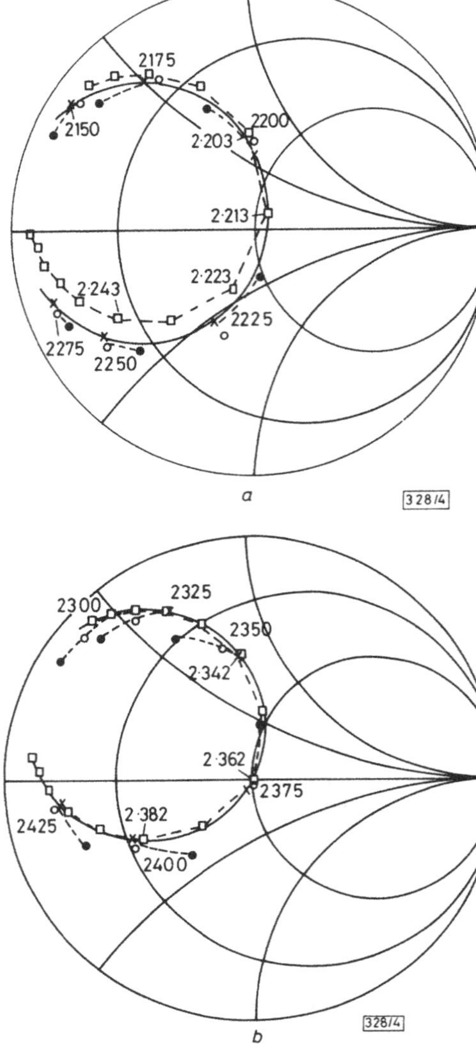

Fig. 4

(*a*) $a = 4{\cdot}0\,$cm, $b = 3{\cdot}0\,$cm, $\varepsilon_{r1} = 2{\cdot}54$, $t = 0{\cdot}16\,$cm, $W = 0{\cdot}442\,$cm, $\varepsilon_{r2} = 2{\cdot}54$, $h = 0{\cdot}16\,$cm, $L_s = 2{\cdot}0\,$cm, $x_0 = a/2$, $y_0 = b/2$, $L^a = 1{\cdot}12\,$cm, $W^a = 0{\cdot}155\,$cm

(*b*) $a = 4{\cdot}0\,$cm, $b = 3{\cdot}0\,$cm, $\varepsilon_{r1} = 2{\cdot}22$, $t = 0{\cdot}16\,$cm, $W = 0{\cdot}116\,$cm, $\varepsilon_{r2} = 10{\cdot}2$, $h = 0{\cdot}127\,$cm, $L_s = 1{\cdot}1\,$cm, $x_0 = a/2$, $y_0 = b/2$, $L^a = 1{\cdot}0\,$cm, $W^a = 0{\cdot}11\,$cm

—●— theory of Reference 1
—○— theory of Reference 2
—×— measured in Reference 1
—□— cavity method

Analysis of Arbitrarily Shaped Microstrip Patch Antennas Using Segmentation Technique and Cavity Model

V. PALANISAMY AND RAMESH GARG, MEMBER, IEEE

Abstract—**Arbitrarily shaped microstrip patch antennas have been analyzed for resonant frequency, input impedance, and radiation patterns. The segmentation technique and the cavity model have been used for this purpose. The usefulness and the accuracy of the method are shown through comparison with experimental results for a rectangular ring antenna. The agreement is seen to be very good. The method appears to be more efficient compared to those reported so far for arbitrary shapes. Moreover, feed reactance is built into the analysis. The method presented here can also be used to analyze microstrip antennas with various types of loadings, e.g., shorting pins, matched loads, etc.**

I. Introduction

THERE ARE A number of techniques available for analyzing microstrip patch antennas. The analytical techniques include transmission line model [1] and cavity model [2]. The most common numerical techniques are moment method [3] and the finite element method [4]. The later technique is time consuming while the former method and the analytical techniques have been applied to regular shapes only like rectangular, circular, and elliptical shapes. These techniques are dealt with comprehensively in a special issue [5]. For an arbitrary shape, the finite element method can be used; however, this is computationally expensive. Recently, Suzuki and Chiba [6] have proposed a technique based on variational method and the modal expansion technique. The expansion functions used are quite general and thus time consuming. In the segmentation technique, on the other hand, proposed by Okoshi *et al.* [7], the expansion functions are the eigenfunctions of the segments into which the given shape is segmented. This technique appears to be efficient. Moreover, it is amenable to computer-aided design based on gradient optimization methods.

Segmentation/desegmentation method has been utilized earlier for the analysis of patch antennas [8], [9]. For this, the antenna is segmented into regular shapes for which Green's function can be determined. The effect of radiation losses has been taken into account by dividing the radiating aperture into discrete number of ports and loading these ports by lumped resistors [9]. In this paper, we avoid dividing the radiating aperture into a number of ports. This makes the method efficient. Secondly, the error introduced by assuming the radiation losses for a given segment will not be present. The input impedance of the antenna has also been determined, which was not attempted in [8], [9]. Finally the efficacy of the improved technique is proved by analyzing a rectangular ring microstrip antenna.

II. Analysis

The geometry of a microstrip patch antenna along with the coordinate system employed is shown in Fig. 1. It consists of an arbitrarily shaped patch located on the surface of a grounded dielectric substrate of thickness h and dielectric constant ϵ_r.

The various steps in the analysis of the patch are: 1) development of magnetic wall model, 2) determination of electric field distribution in the patch using the segmentation technique and the multiport connection method, and 3) evaluation of antenna characteristics using cavity model.

The magnetic wall model of the geometry is developed first by replacing the fringing fields at the peripheries by equivalent outward extensions. These extensions depend on the planar dimensions of the patch, relative dielectric constant, dielectric thickness and the field distribution at the peripheries. Equivalent extensions for rectangular and circular patches are well known [10]–[12]. It is difficult to estimate the exact extensions for patches with arbitrary shapes. However, for a given shape, the extension available for the shape closest to the given shape can be used. For example, for a pentagonal patch, Suzuki and Chiba [6] have used the same extension as applicable to a circular disk of same planar area as that of given pentagonal patch. In the absence of any such information for the given structure, the edge extension applicable to various segments of the patch can be used. It is desirable that the edge extension be the same all around the periphery because the full patch and not the individual segments acts as a resonator. It has been proved qualitatively [13], [14] that this extension should be almost equal to the dielectric thickness unless for a specific geometry like circular disk, the extension is determined to be different [12]. The modified geometry with magnetic walls at the periphery will have the same shape as the original geometry but with different dimensions.

Next, the electric field distribution in the patch will be determined.

Reprinted from *IEEE Trans. Antennas Propaga.*, vol. AP-34, no. 10, pp. 1208–1213, Oct. 1986.

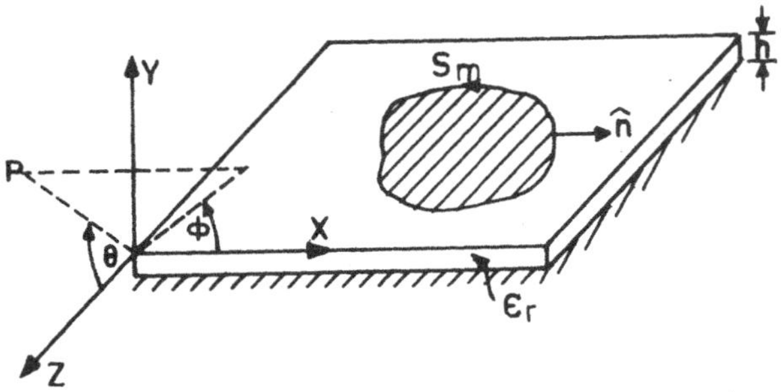

Fig. 1. Configuration of an arbitrarily shaped microstrip antenna along with the coordinate system.

A. Field Distribution

In practice, the substrate is electrically thin ($h \ll \lambda_0$). Therefore, only the y-component of electric field and the x- and z-components of the magnetic field exist in the region bounded by the patch and the ground plane. Assuming $e^{j\omega t}$ time variation, the electric field E_y due to a current source J_y located at (x_0, z_0) in the patch must satisfy the following:

$$(\nabla_t^2 + k^2)E_y = -j\omega\mu_0 J_y(x_0, z_0) \tag{1}$$

where $\bar{\nabla}_t = \hat{x}(\partial/\partial x) + \hat{z}(\partial/\partial z)$, ω is the angular frequency and $k^2 = k_0^2\epsilon_{r\,\text{eff}}$ with k_0 the free space wavenumber and $\epsilon_{r\,\text{eff}}$ the effective dielectric constant of the patch. The boundary condition satisfied by E_y will be

$$\frac{\partial E_y}{\partial n} = 0 \quad \text{on} \quad S_m. \tag{2}$$

S_m is the magnetic wall boundary of the patch and is shown in Fig. 1. It coincides with the outward extensions of the patch.

The above problem can be solved using the Green's function $G(x, z/x_0, z_0)$. The solution for E_y then becomes

$$E_y(x, z) = \iint G(x, z|x_0, z_0) J_y(x_0, z_0)\, dx_0\, dz_0. \tag{3}$$

The Green's function, in general, can be expressed in terms of the set of eigenfunctions for the patch shape. It is given as [15]

$$G(x, z|x_0, z_0) = \frac{j\omega\mu_0 h}{ab} \sum_m \sum_n \frac{\psi_{mn}(x, z)\psi_{mn}^*(x_0, z_0)}{k_{mn}^2 - k^2} \tag{4}$$

where ψ_{mn} and k_{mn} are the eigenfunctions and the corresponding eigenvalues for the mnth mode of the path. These must satisfy the wave equation

$$(\nabla_t^2 + k_{mn}^2)\psi_{mn} = 0 \tag{5}$$

and $(\partial/\partial n)\psi_{mn} = 0$ on S_m.

The eigenfunctions for some regular shapes such as rectangular, circular, and triangular shapes are available in the literature [16]. Only numerical solutions are available for an arbitrary shape. Pang *et al.* [17] and Suzuki and Chiba [6] have used the Rayleigh–Ritz method, which is a variational technique, to determine the eigenfunctions and the eigenvalues. In their analysis of pentagonal patch [6], the basis functions used for expanding the eigenfunctions are $x^m z^n$ (m, $n = 0, 1, 2, \cdots$), which are very general. If the basis functions are chosen suitably, the convergence will be faster.

In the approach given here, we determine E_y by segmenting the given patch shape into a number of regular shapes for which Green's functions can be determined. The Green's functions for rectangles, circles, triangles, circular sectors, annular rings and annular sectors are available [15]. Most of the useful practical patch shapes will have symmetry planes and can be decomposed into regular shapes; the available Green's function can be used. A completely arbitrary shape with no plane of symmetry, in general, will give higher level of cross polarization in the radiation patterns.

1) Segmentation: It is possible to determine the fields in an arbitrary geometry by expanding fields in various segments in terms of their natural modes and then matching fields along the interconnection lengths. Gruner [18] has used this technique to determine the cutoff wavelengths of rectangular coaxial waveguides. However, we shall use circuit theory to make the analysis computer oriented. For this, the continuous interconnections between the segments are discretized by interconnections only at a finite number of points. With each interconnection point, we associate a port. While approximating the continuous interconnections by a finite number of ports, each port width is kept less than or equal to $\lambda_g/20$ to optimize the discretization error and efficiency. Here, λ_g denotes the intrinsic wavelength in the patch. The small size of the port width allows us to make the assumption that the current density is uniform over the width of the port.

The individual segments are now treated as multiport planar networks, and the z-matrices for the same are evaluated using the corresponding impedance Green's function. It is given as [15],

$$Z_{ij}^s = \frac{1}{W_i W_j} \int_{\text{PW}_i} \int_{\text{PW}_j} G^s(s|s_0)\, ds\, ds_0 \tag{6}$$

where Z_{ij}^s is the ijth element of the Z-matrix of the segment, W_i, W_j and PW_i, PW_j are the effective and the physical width of the ith and the jth ports, respectively. The effective widths of the ports includes the fringing field extension, when the ports span the patch metalization and the extension. For ports entirely within the patch metalization physical widths and effective widths will be equal. The Green's function G^s is of the following form [15]:

$$G^s(x, z|x_0, z_0) = \frac{j\omega\mu_0 h}{ab} \cdot \sum_{m=-\infty}^{\infty} \sum_{n=-\infty}^{\infty} \frac{\psi_{mn}^s(x, z)\psi_{mn}^{s*}(x_0, z_0)}{k_{mn}^2 - k^2}. \tag{7}$$

Here $\psi_{mn}^s(x, z)$ is the eigenfunction for the mnth mode of the segment, k_{mn} is the corresponding eigenvalue, h is the dielectric thickness and k^2 has been defined earlier.

It may be pointed out that for calculating the impedance matrix of a segment, the local coordinate system can be oriented independent of the coordinate systems chosen for other segments. It should be oriented in such a way that the maximum number of ports lie on the coordinate axis. The integrals involved in (6) are simple and can be obtained in closed form.

The microstrip antenna can be fed either by a coax probe through the substrate or by a microstrip/stripline. In any case, the input feed is also considered a port (or several ports if the width is more than $\lambda_g/20$) and can be treated like other port/ports of the segment in the Z-matrix evaluations. For a microstrip feed, the effective width of the feed [19] is used for this purpose.

2) Multiport Connection Method: The multiport Z-matrices corresponding to the various segments are now combined one by one, by using the multiport connection method [15], to obtain the overall Z-matrix of the given structure. For this, the ports of the segments (to be combined) are separated into external (p) ports and connected (c) ports. The connected ports are equally divided into two groups labeled q and r ports such that q ports are the connected ports of one segment and r ports are the corresponding connected ports of the other segment, to be combined. Based on this labeling, the Z-matrix of the combination can now be written as

$$\begin{bmatrix} \vec{V}_p \\ \vec{V}_q \\ \vec{V}_r \end{bmatrix} = \begin{bmatrix} Z_{pp} & Z_{pq} & Z_{pr} \\ Z_{qp} & Z_{qq} & Z_{qr} \\ Z_{rp} & Z_{rq} & Z_{rr} \end{bmatrix} \begin{bmatrix} \vec{i}_p \\ \vec{i}_q \\ \vec{i}_r \end{bmatrix} \tag{8}$$

where $\vec{V}_p$, $\vec{V}_q$, $\vec{V}_r$, and $\vec{i}_p$, $\vec{i}_q$, $\vec{i}_r$ are the vectors corresponding to radio frequency (RF) port voltages and port currents, respectively, and Z_{pp}, etc. are the impedance submatrices. Since ports q and ports r are respective ports of two physically separate segments (that are being connected together), submatrices Z_{qr} and Z_{rq} are identically zero. The boundary conditions, i.e., the continuity of the tangential components of the electric and magnetic fields at the boundary plane between the two combining segments is expressed in terms of the continuity of port voltages and port currents. These are known as interconnection constraints [15] and are expressed as

$$\vec{V}_q = \vec{V}_r \tag{9}$$

and

$$\vec{i}_q + \vec{i}_r = 0. \tag{10}$$

Upon substituting (9) and (10) in (8), one obtains the RF currents at the interconnecting q ports $\vec{i}_q$ and the impedance matrix of the combination as

$$\vec{i}_q = [Z_{qq} + Z_{rr}]^{-1}[Z_{rp} - Z_{qp}]\vec{i}_p \tag{11}$$

$$[Z_p] = Z_{pp} + (Z_{pq} - Z_{pr})(Z_{qq} + Z_{rr})^{-1}(Z_{rp} - Z_{qp}). \tag{12}$$

It is to be noted that the process of combining the Z-matrices of various segments one by one results in faster computation compared to when all the segments are combined simultaneously. The increase in speed occurs because the size of the matrix to be inverted in implementing (11) or (12) becomes smaller. This procedure is consistent mathematically also because the constraints can be applied either in groups or all of them simultaneously.

III. Resonant Frequency

The input impedance evaluated from (12) gives the input reactance of the lossless cavity since the radiation, conductor, dielectric and surface wave losses have not yet been accounted for. The variation of input reactance with frequency is now determined. A very large value of the input reactance indicates resonance because of the antiresonant nature of the patch.

While evaluating the Z-matrices of various segments using (6), the various combinations of m and n in the evaluation of the Green's function represent the contribution of higher order modes. Since the series is converging, the values of m and n can be limited to $m = M$ and $n = N$, as discussed below. The values of M and N depend on the dimensions of the segment, frequency of operation and the permittivity of the substrate through k_{MN} and k. They (M and N) should be selected such that the contribution of the $[(M + 1), (N + 1)]$th mode to the Z-matrix is insignificant. Pang *et al.* [17] observe that k_{MN} should be greater than $4k$ to determine M and N. The values of M and N can be different for different segments depending upon their shape and size.

Next we determine the electric field distribution in the patch. This is required to determine the radiation characteristics and the input impedance.

The electric field distribution E_y^s in a segment can be expressed as

$$E_y^s = \sum_m \sum_n e_{mn} \psi_{mn}^s(x, z) \tag{13}$$

$$\bar{H} = -\frac{1}{j\omega\mu_0} \hat{y} \times \bar{\nabla}_t E_y^s. \tag{14}$$

The expansion coefficients e_{mn} depend on the excitation and are given as

$$e_{mn} = \frac{-j\omega\mu_0}{k^2 - k_{mn}^2} \frac{\langle J_s \psi_{mn}^s \rangle}{\langle \psi_{mn}^s \psi_{mn}^s \rangle} \tag{15}$$

where

$$\langle J_s \psi_{mn}^s \rangle = \iint_S J_s \psi_{mn}^{s*} \, dy \, ds$$

and

$$\langle \psi_{mn}^s \psi_{mn}^s \rangle = \iiint \psi_{mn}^s \psi_{mn}^{s*} \, dx \, dy \, dz.$$

The element of arc length along the periphery of the segment is ds. The periphery of the segment over which the integration is to be carried out is denoted by s. It corresponds to the interconnection interfaces of the segment over which the tangential component of magnetic field is nonzero.

The excitation current density J_s for a segment is related to the tangential component of magnetic field on its peripheries through the relation $\bar{J}_s = \hat{n} \times \bar{H}$. In the present case, J_s can be determined from the port currents. For this, we expand the current density along the interconnections of the segment in terms of P modes of the segment [20]. For a rectangular segment, it is given by

$$J_s(s_0) = -\frac{1}{W} \sum_{k=1}^{P} a_k \cos\left\{\frac{(k-1)\pi}{L} s_0\right\} \tag{16}$$

where W is the port width, L is the length of the segment along the interconnection, a_k is the expansion coefficient for the kth mode, and s_0 is the running coordinate on the interconnection. The subscript $k = 1$ corresponds to the transverse electromagnetic (TEM) mode and P represents the number of modes in the segment which should be taken equal to the number of interconnecting ports r or q. The coefficients a_k are determined by equating the port current i_k determined in (11) at the kth port to $i(s_0) = WJ_s(s_0)$ evaluated at the middle of kth port, and solving the resultant set of simultaneous equations.

Using the definition of Green's function $G^s(s/s_0)$ yields the expression for E^s_y. It can be written as

$$\bar{E}^s_y = \hat{y}\frac{1}{h}\int_{S_m} G^s(s/s_0)J_s(s_0)\, ds_0. \tag{17}$$

Using either (13) or (17) (both lead to the same result) gives the expression for the electric field distribution in the segment. Similarly, the electric field distribution in other segments can be evaluated.

The electric field for the antenna structure is evaluated and plotted at the resonant frequency. From the nature of variation, the mode of operation can be identified.

IV. Input Impedance

Knowing the electric field distribution at the periphery of antenna structure, one can calculate the radiation patterns using the magnetic current model [11]. From the power radiated and the electric energy stored, the cavity Q can be evaluated. This is then used to determine the input impedance of the antenna by expressing the losses as an effective loss tangent [2] as follows. Write

$$\delta_{\text{eff}} = \frac{1}{Q}. \tag{18}$$

Then replace k^2 by

$$k^2_{\text{eff}} = k^2_0\epsilon_{r\,\text{eff}}(1 - j\delta_{\text{eff}}). \tag{19}$$

Introducing the effective wavenumber changes the Green's function (7) for a segment to

$$G^s(x, z|x_0, z_0) = \frac{j\omega\mu_0 h}{ab}\sum_m\sum_n \frac{\psi^s_{mn}(x, z)\psi^{s*}_{mn}(x_0, z_0)}{k^2_{mn} - k^2_0\epsilon_{r\,\text{eff}}(1 - j\delta_{\text{eff}})}. \tag{20}$$

The Z-matrices for the individual segments are again evaluated using the modified Green's function (20) and combined as detailed earlier to give the complex input impedance of the antenna. By varying the frequency, the input impedance can be evaluated at and near the resonant frequency. It is corrected for feed reactance. Resonance is indicated by real input impedance.

Application of the above method to the analysis of a rectangular ring microstrip antenna is described next.

V. Rectangular Ring Microstrip Antenna

The geometry of a rectangular ring microstrip antenna is shown in Fig. 2. The outer dimensions of the ring are $a \times b$ and the inner dimensions $s \times d$. The outward extensions are taken to be the same all around and equal to h, the dielectric thickness. The magnetic wall model of the antenna is shown in Fig. 3(a).

The dimensions of the ring chosen are $a = 6.9$ cm, $b = 5.5$ cm, $s = 1.8$ cm and $d = 2.6$ cm. The dielectric constant of the substrate is 2.50 and the dielectric thickness $h = 0.159$ cm. For these dimensions the effective interconnection length between the segments is 1.767 cm. The antenna is expected to resonate at 1080 MHz for the dominant TM_{10}-mode. Therefore, interconnection lengths are approximated by two ports each only, each port width approximately equal to $\lambda_g/20$. The various segments and the interconnecting ports are illustrated in Fig. 3(b). The total number of ports are 17 and port 1 is the feed port. The width of the feed port is taken equal to the diameter of the feed probe.

Each segment is now considered a multiport component and the impedance matrix is evaluated using the following Green's function:

$$G(x, z|x_0, z_0) = \frac{1j\omega\mu_0 h}{a'b'}\sum_{m=0}^{\infty}\sum_{n=0}^{\infty}\frac{\sigma_m\sigma_n}{k^2_x + k^2_z - k^2} \cdot \{\cos(k_x x)\cos(k_z z)\cos(k_x x_0)\cos(k_z z_0)\} \tag{21}$$

where

$$\sigma_i = \begin{cases} 2, & \text{if } i = 0 \\ 1, & \text{if } i \neq 0 \end{cases}$$

$$k_x = \frac{m\pi}{a'},\quad k_z = \frac{n\pi}{b'}\quad \text{and}\quad k^2 = \omega^2\mu_0\epsilon_0\epsilon_{r\,\text{eff}}.$$

a' and b' are the dimensions of the segment. The maximum values of m and n used were five and ten, respectively.

Impedance matrices for the various segments of the ring are now combined one by one using multiport connection method. The maximum size of the matrix to be inverted in implementing (12) for this case was 2×2. Next, the resonance frequency was determined from the behavior of input reactance as a function of frequency. The currents at the various interconnected ports of the ring were determined at this frequency using (11) and (10). These port currents were converted into a continuous current density along the interconnection lengths using (16), and the electric field distribution was determined for the TM_{10}-mode. The electric field distribution is shown in Fig. 4 at the inner and the outer peripheries of the ring antenna. It is found to be almost constant along the radiating edges BC and DA, similar to that in a rectangular patch operated in the TM_{10}-mode. The radiation patterns obtained from these distributions are shown in Fig. 5. Comparison with measurements shown there indicates good agreement.

Input impedance for the antenna is compared in Fig. 6. Again the agreement is found to be very good, except for a slight shift in the resonance frequency which could be due to tolerances in dielectric constant, fabrication tolerances and/or inaccurate modeling of fringing field extension.

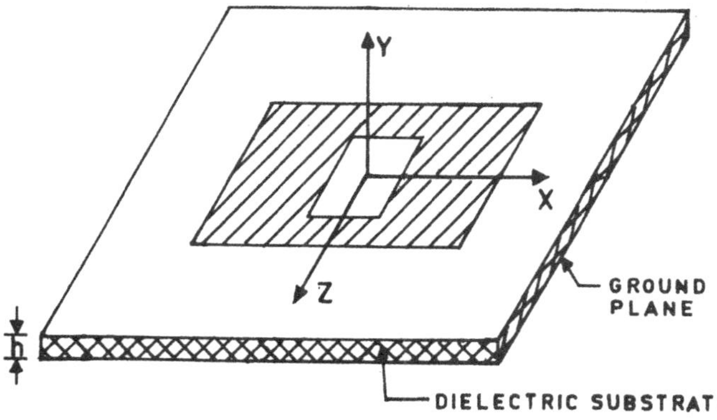

Fig. 2. Geometry of a rectangular ring microstrip antenna.

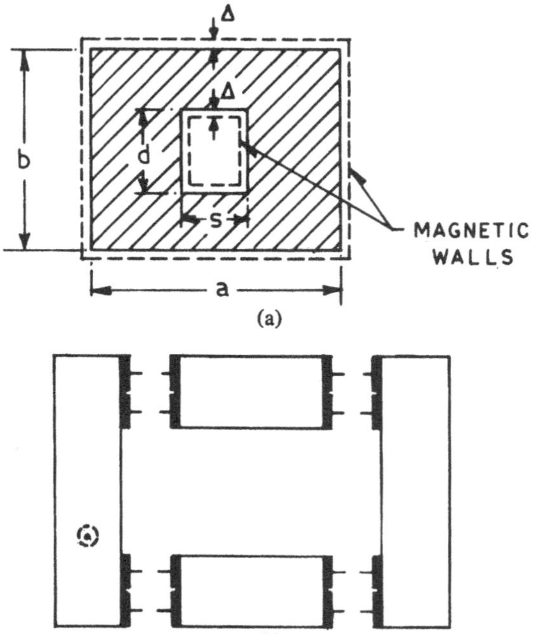

Fig. 3. (a) Configuration showing the magnetic wall model of a rectangular ring microstrip antenna. (b) Segmentation in terms of rectangular shapes of the geometry shown in (a).

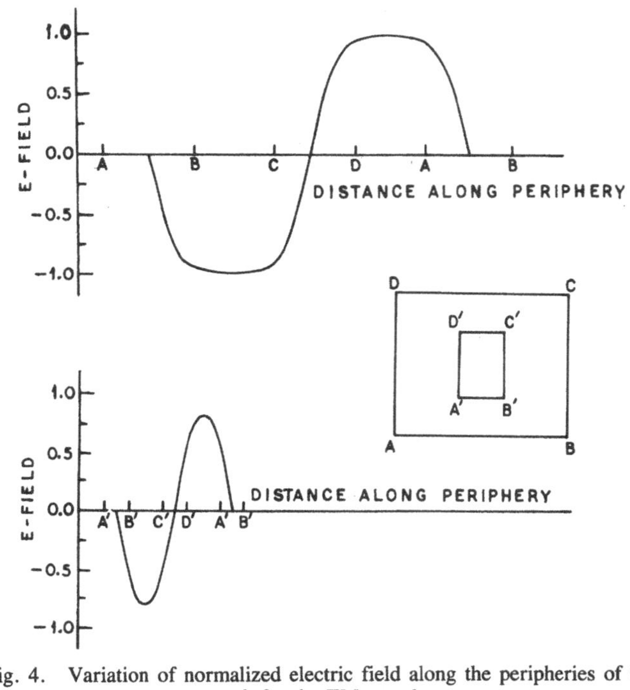

Fig. 4. Variation of normalized electric field along the peripheries of the patch for the TM_{10} mode.

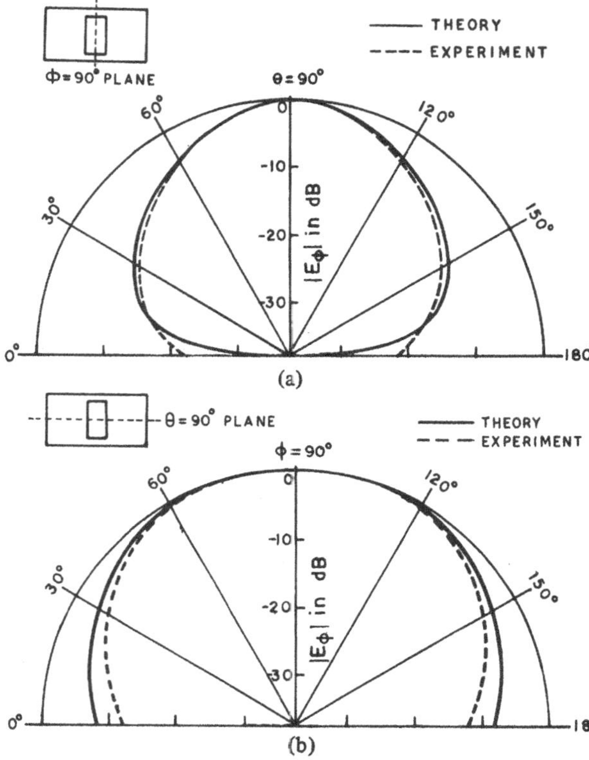

Fig. 5. Radiation patterns for the TM_{10} mode of a rectangular ring microstrip antenna. (a) $\phi = 90°$ plane. (b) $\theta = 90°$ plane.

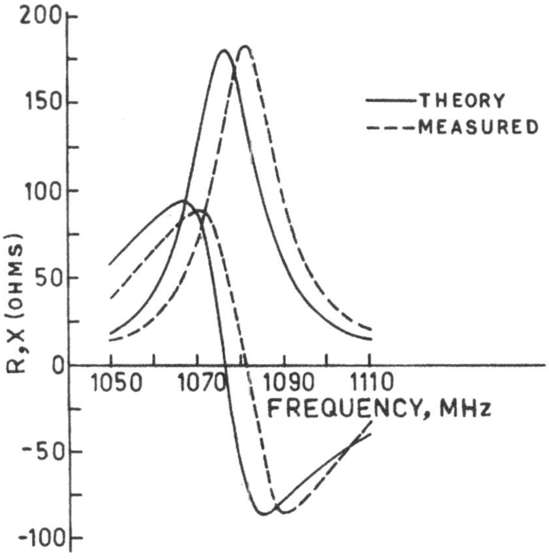

Fig. 6. Comparison of the theoretical and measured values of the input impedance for the TM_{10} mode of rectangular ring microstrip antenna (a = 6.9 cm, b = 5.5 cm, s = 1.8 cm, d = 2.6 cm, ϵ_r = 2.50, h = 0.159 cm, x_0 = −1.05 cm, z_0 = 1.65 cm).

VI. Conclusion

A generalized cavity method has been developed, using segmentation technique, for analyzing arbitrary shaped microstrip patch antennas. Resonance frequency, field distribution, radiation patterns and input impedance can be obtained using this technique. A rectangular ring microstrip antenna was analyzed based on the technique developed. Comparison with measurements shows good agreement. Unlike other techniques, the feed reactance is built into the theory. It does not have to be calculated separately and then added later on.

The proposed method can be applied to patch antennas with various types of loadings like shorting pins for frequency agility [21], traveling wave microstrip antennas [11], etc. The cutoff frequencies for various modes of coaxial lines of arbitrary cross section can be obtained because of the Babinet's equivalence between microstrip rings and coaxial waveguide cross sections. Microstrip resonators and circuit elements of arbitrary shape can also be analyzed. The proposed technique can be easily extended to the computer-aided design of microstrip patch antennas utilizing gradient optimization methods.

The main limitation of the above method is that the effect of dielectric substrate on the input impedance has not been considered. Although this effect is small for the dominant mode, it can be evaluated using the method given in [22], once the electric field distribution is known.

References

[1] A. K. Bhattacharyya and R. Garg, "A generalized transmission line model for microstrip patches," *Proc. Inst. Elec. Eng.*, vol. 132, pt. H, no. 132, pp. 93–98, Apr. 1985.

[2] W. F. Richards *et al.*, "An improved theory for microstrip antennas and applications," *IEEE Trans. Antennas Propagat.*, vol. AP-29, pp. 38–46, Jan. 1981.

[3] D. M. Pozar, "Input impedance and mutual coupling of rectangular microstrip antennas," *IEEE Trans. Antennas Propagat.*, vol. AP-30, pp. 1191–1196, Nov. 1982.

[4] P. Silvester, "Finite element analysis of planar microwave networks," *IEEE Trans. Microwave Theory Tech.*, vol. MTT-21, pp. 104–108, Feb. 1973.

[5] *IEEE Trans. Antennas Propagat.*, vol. AP-29, special issue on microstrip antennas, Jan. 1981.

[6] Y. Suzuki and T. Chiba, "Computer analysis method for arbitrarily shaped microstrip antenna with multiterminals," *IEEE Trans. Antennas Propagat.*, vol. AP-32, pp. 585–590, June 1984.

[7] T. Okoshi *et al.*, "The segmentation method—An approach to the analysis of microwave planar circuits," *IEEE Trans. Microwave Theory Tech.*, vol. MTT-24, pp. 662–668, Oct. 1976.

[8] K. C. Gupta and P. C. Sharma, "Segmentation and desegmentation techniques for the analysis of planar microstrip antennas," in *Proc. IEEE Int. Symp. Antennas Propagat.*, 1981, pp. 19–22.

[9] P. C. Sharma and K. C. Gupta, "Analysis and optimized design of single feed circularly polarized microstrip antennas," *IEEE Trans. Antennas Propagat.*, vol. AP-31, pp. 949–955, Nov. 1983.

[10] J. R. James *et al.*, *Microstrip Antenna—Theory and Design.* London, U.K.: IEE, Peter Peregrinus Ltd., 1981.

[11] I. J. Bahl and P. Bhartia, *Microstrip Antenna.* Dedham, MA: Artech House, 1980.

[12] W. C. Chew and J. A. Kong, "Effect of fringing fields on the capacitance of circular microstrip disc," *IEEE Trans. Microwave Theory Tech.*, vol. MTT-28, pp. 98–104, Feb. 1980.

[13] P. Hammer *et al.*, "A model for calculating the radiation field of microstrip antennas," *IEEE Trans. Antennas Propagat.*, vol. AP-27, pp. 267–270, Mar. 1979.

[14] E. Lier and J. R. Jakobson, "Rectangular microstrip patch antennas' with infinite and finite ground plane dimensions," *IEEE Trans. Antennas Propagat.*, vol. AP-31, pp. 978–984, Nov. 1983.

[15] K. C. Gupta *et al.*, *Computer-Aided-Design of Microwave Circuits.* Dedham, MA: Artech House, 1981.

[16] Y. T. Lo *et al.*, "Theory and experiment on microstrip antennas," *IEEE Trans. Antennas Propagat.*, vol. AP-27, pp. 137–145, Mar. 1979.

[17] H. J. Pang *et al.*, "Computer analysis of microwave planar circuit with impedance matrix," *Electron. Commn.* (Japan), vol. 64-B, no. 9, pp. 55–63, 1981.

[18] L. Gruner, "Higher order modes in rectangular coaxial waveguides," *IEEE Trans. Microwave Theory Tech.*, vol. MTT-15, pp. 483–485, Aug. 1967.

[19] K. C. Gupta *et al.*, *Microstrip Lines and Slot Lines.* Dedham, MA: Artech House, 1979.

[20] T. Miyoshi and S. Miyauchi, "The design of planar circulators for wideband operation," *IEEE Trans. Microwave Theory Tech.*, vol. MTT-28, pp. 210–214, Mar. 1980.

[21] D. H. Schaubert *et al.*, "Microstrip antennas with frequency agility and polarization diversity," *IEEE Trans. Antennas Propagat.*, vol. AP-29, pp. 118–123, Jan. 1981.

[22] A. K. Bhattacharyya and R. Garg, "Effect of substrate on the efficiency of an arbitrarily shaped microstrip patch antenna," *IEEE Trans. Antennas Propagat.*, pp. 1181–1188, this issue.

Fundamental Superstrate (Cover) Effects on Printed Circuit Antennas

NICÓLAOS G. ALEXÓPOULOS, SENIOR MEMBER, AND DAVID R. JACKSON

***Abstract*—The fundamental effects of superstrate (cover) materials on printed circuits antennas are investigated. Substrate–superstrate resonance conditions are established which maximize antenna gain, radiation resistance, and radiation efficiency. Criteria are determined for material properties and dimensions for which surface waves are eliminated and a radiation efficiency due to substrate–superstrate effects of $e_s = 100$ percent is obtained. Criteria for nearly omnidirectional $\bar{H}$-plane patterns and nearly omnidirctional $\bar{E}$-plane patterns are presented. Finally, a general criterion is given for choosing a superstrate to optimize efficiency for the important case of nonmagnetic layers with the antenna at the interface.**

I. INTRODUCTION

SUPERSTRATE (cover) dielectric layers are often used to protect printed circuit antennas (PCA's) from environmental hazards, or may be naturally formed (e.g. ice layers) during flight or severe weather conditions [1], [3]. Whether a cover layer is naturally formed or imposed by design, it may affect adversely the antenna basic performance characteristics, such as gain, radiation resistance and efficiency. For this reason, it is important to analyze superstrate effects from a fundamental point of view, so that the PCA performance may be understood better or a proper choice of cover parameters may be implemented to advantage in the enhancement of gain and radiation efficiency.

Furthermore, in the case of PCA integration in millimeter and submillimeter wave integrated circuits, where substrates such as GaAs or Si are used [4], [5], the antenna radiation efficiency is quite low due to high substrate dielectric constant and associated surface wave effects [6]-[8]. Since GaAs and Si are natural materials for hybrid or monolithic integrated circuit technologies, it is important to investigate radiation efficiency and gain optimization by incorporating them with covers or as covers.

In order to analyze the basic properties of microstrip antennas with a superstrate, the problem of the Hertzian dipole is solved exactly. Firstly, the Green's function for the infinitesimal dipole is derived using Sommerfeld's method [9]-[12]. After obtaining the Green's function a contour integration is used in order to compute the surface wave power in the transverse electric (TE) and transverse magnetic (TM) surface waves. Subsequently, the gain and radiation resistance are obtained by application of the reciprocity theorem. Once the radiation resistance and surface wave power have been found, the radiation efficiency is computed.

The results show that a proper choice of substrate and superstrate thicknesses generates resonance conditions in the composite layers, which greatly enhance the PCA gain, radiation resistance, and efficiency.

It is also demonstrated that by employing the proper combination of materials and dimensions, surface waves can be eliminated in the substrate–cover structure. This implies then that the radiation efficiency (component pertaining to substrate-superstrate effects) is optimized to a value of $e_s = 100$ percent.

Finally, a set of criteria is established for nearly omnidirectional $\bar{H}$-plane patterns and $\bar{E}$-plane patterns, and optimum cover thickness.

II. GREEN'S FUNCTION FOR THE HERTZIAN DIPOLE

The problem of the infinitesimal dipole embedded in the top layer of a two-layer dielectric is shown in Fig. 1. Following the Sommerfeld method [9], the incident Hertzian magnetic vector potential due to an $\hat{x}$-directed source is written in region 2 as as follows.

$$\Pi_x^{\{2\}} = \int_0^\infty J_0(\lambda r) \frac{e^{-u_2|z-z_0|}}{u_2} \lambda\, d\lambda \tag{1}$$

(a multiplicative factor $-j\omega\mu_0\mu_2/4\pi k_2^2$ and the time dependence $e^{+j\omega t}$ are being suppressed throughout this paper). In order to satisfy all the required boundary conditions, x and z components of scattered potentials are considered in each of the regions. The total potentials in each of the three regions, after simplifying and combining terms are

Region 1 $(0 \leqslant z \leqslant B)$

$$\Pi_x^{(1)} = \int_0^\infty \frac{h_1(\lambda)}{D_e(\lambda)} \sinh(u_1 z) J_0(\lambda r)\, d\lambda \tag{2}$$

$$\Pi_z^{(1)} = \cos\phi \int_0^\infty \frac{I_1(\lambda)}{D_e(\lambda) D_m(\lambda)} \cosh(u_1 z) J_1(\lambda r)\, d\lambda. \tag{3}$$

Region 2a $(B \leqslant z \leqslant z_0)$

$$\Pi_x^{(2a)} = \int_0^\infty \left(\frac{h_{2a}^{(1)}(\lambda)\sinh(u_2 z) + h_{2a}^{(2)}(\lambda)\cosh(u_2 z)}{D_e(\lambda)} \right) J_0(\lambda r)\, d\lambda. \tag{4}$$

Region 2b $(z_0 \leqslant z \leqslant H)$

$$\Pi_x^{(2b)} = \int_0^\infty \left(\frac{h_{2b}^{(1)} \sinh(u_2 z) + h_{2b}^{(1)}(\lambda)\cosh(u_2 z)}{D_e(\lambda)} \right) J_0(\lambda r)\, d\lambda. \tag{5}$$

Region 2 $(B \leqslant z \leqslant H)$

$$\Pi_z^{(2)} = \cos\phi \int_0^\infty \left(\frac{I_2^{(1)}(\lambda)\sinh(u_2 z) + I_2^{(2)}(\lambda)\cosh(u_2 z)}{D_e(\lambda) D_m(\lambda)} \right) \cdot J_1(\lambda r)\, d\lambda \tag{6}$$

Reprinted from *IEEE Trans. Antennas Propaga.*, vol. AP-32, no. 8, pp. 807–816, Aug. 1984.

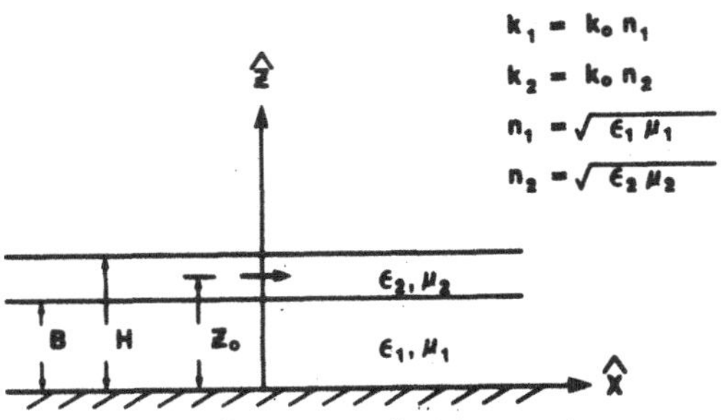

Fig. 1. Superstrate–substrate geometry.

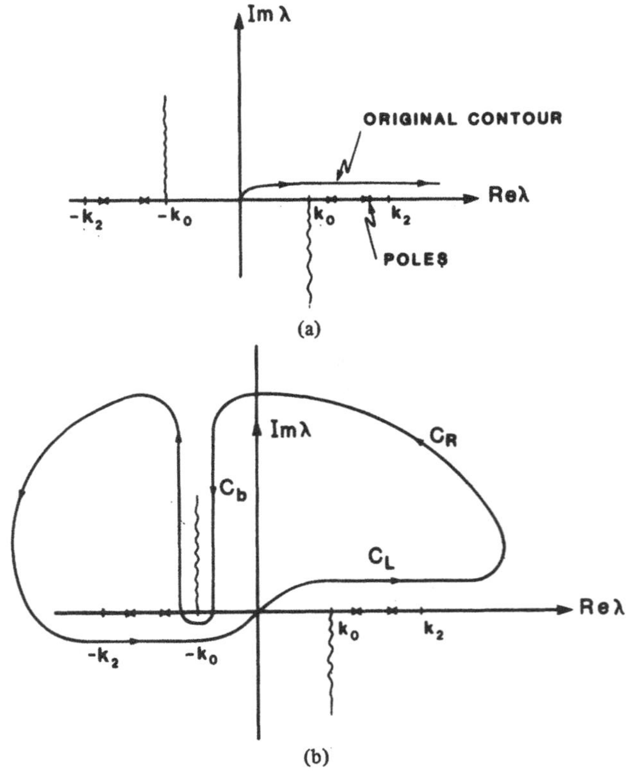

Fig. 2. Contour of integration in complex λ-plane.

and

Region 3

$$\Pi_x^{(3)} = \int_0^\infty \frac{h_3(\lambda)}{D_3(\lambda)} e^{-u(z-H)} J_0(\lambda r)\, d\lambda \tag{7}$$

$$\Pi_z^{(3)} = \cos\phi \int_0^\infty \frac{I_3(\lambda)}{D_3(\lambda) D_m(\lambda)} e^{-u(z-H)} J_1(\lambda r)\, d\lambda \tag{8}$$

where

$$u_1 = (\lambda^2 - k_1^2)^{1/2}, \quad u_2 = (\lambda^2 - k_2^2)^{1/2}, \quad u = (\lambda^2 - k_0^2)^{1/2}$$

with k_1, k_2, k_0 being wavenumbers in regions 1, 2, and 3, respectively, and they are defined as $k_1 = k_0 n_1$, $k_2 = k_0 n_2$, $k_0 = \omega\sqrt{\mu_0 \epsilon_0}$, where $n_1 = \sqrt{\mu_1 \epsilon_1}$, $n_2 = \sqrt{\mu_2 \epsilon_2}$ with $\mu_{1,2}$ and $\epsilon_{1,2}$ being the relative permeability and permittivity of the corresponding layer.

The Sommerfeld condition at ∞ determines the branch interpretation of $u = (\lambda^2 - k_0^2)^{1/2}$ as

$$u = |(\lambda^2 - k_0^2)^{1/2}|, \qquad |\lambda| \geqslant k_0$$

and

$$u = j|(\lambda^2 - k_0^2)^{1/2}|, \qquad |\lambda| \leqslant k_0.$$

The zeros of

$$D_e(\lambda) = u_2 \epsilon_2 \mu_2 \cosh[u_2(H-B)]\{u + u_1 \mu_1^{-1} \coth(u_1 B)\} + u_2 \epsilon_2 \mu_2 \sinh[u_2(H-B)] \left\{ u_2 \mu_2^{-1} + \left(\frac{u u_1}{u_2}\right)\left(\frac{\mu_2}{\mu_1}\right) \coth(u_1 B) \right\} \tag{9}$$

and

$$D_m(\lambda) = -\sinh[u_2(H-B)]\{u_2^2 \epsilon_1 \mu_1 + \mu_1 \epsilon_2^2 u u_1 \tanh(u_1 B)\} - u_2 \epsilon_2 \mu_2 \cosh[u_2(H-B)] \cdot \left\{ \left(\frac{\mu_1}{\mu_2}\right) u_1 \tanh(u_1 B) + \epsilon_1 \left(\frac{\mu_1}{\mu_2}\right) u \right\} \tag{10}$$

define the surface wave poles in the composite layer. The remaining pertinent parameters in (2)–(8) are given in [14]. This result has been checked to agree in the limiting cases where no cover is present [6], [7].

Fig. 2(a) shows the path of integration. The integration contour is shifted above the x-axis to avoid the poles of $D_e(\lambda)$ and $D_m(\lambda)$. Furthermore, by using

$$J_0(x) = \tfrac{1}{2}(H_0^{(1)}(x) + H_0^{(2)}(x)) \tag{11}$$

$$J_1(x) = \tfrac{1}{2}(H_1^{(1)}(x) + H_1^{(2)}(x)) \tag{12}$$

and

$$H_0^{(2)}(x) = -H_0^{(1)}(-x) \tag{13}$$

$$H_1^{(2)}(x) = H_1^{(1)}(-x) \tag{14}$$

$$-\pi < \arg(x) < \pi,$$

the path of integration is extended from $-\infty$ to $+\infty$ and the contour is closed as shown in Fig. 2(b) [9]. Although the radiated field can be obtained from the contour integration, in order to reduce the tedious algebraic manipulations, the reciprocity theorem is invoked. The surface wave fields, however, are determined by computing the residues at the poles. The residue contributions from the roots of $D_e(\lambda)$ give rise to TE surface waves, while those from $D_m(\lambda)$ give rise to TM surface waves.

If the transformations

$$\Pi_\phi = -\Pi_x \sin\phi \tag{15}$$

$$\Pi_r = \Pi_x \cos\phi \tag{16}$$

$$\Pi_z = \Pi_z \tag{17}$$

are incorporated into the relations

$$\bar{H} = j\omega\epsilon\, \nabla \times \bar{\Pi} \tag{18}$$

$$\bar{E} = \frac{1}{j\omega\epsilon} \nabla \times \bar{H} \tag{19}$$

and $\epsilon = \epsilon_0 \epsilon_i$, ϵ_i being the relative permittivity in the ith layer ($i = 1, 2$), then the Poynting vector can be integrated over a large cylinder of radius $r \to \infty$ to obtain the surface wave power

$$P_{sw} = \int_0^\infty \int_0^{2\pi} \frac{1}{2} \operatorname{Re}[\bar{E} \times \bar{H}^*] \cdot \hat{r} r\, d\phi\, dz. \tag{20}$$

The surface wave power is the sum of the power in each TE and TM wave, since these are orthogonal [13]. Although the expressions are excessively lengthy, they are simply algebraic, not involving integrals, since integration in ϕ and z can be performed explicitly.

Although in the derivation of the Green's function the dipole was assumed to be in the top layer, it is easy, by a reciprocity argument, to determine the surface wave power for the dipole in the lower layer. To achieve this, the surface wave power that would be excited if the dipole were at the interface, $P_{sw}^{(0)}$, is determined. The actual power in a given surface wave for the dipole at $z = z_0 (0 \leqslant z_0 \leqslant B)$ is given then by

$$P_{sw} = P_{sw}^{(0)} \left(\frac{\sin(\alpha z_0)}{\sin(\alpha B)} \right)^2 \tag{21}$$

where $\alpha = (k_1^2 - \beta^2)^{1/2}$, β being the propagation constant for the surface wave mode.

III. RADIATION FIELD

By reciprocity, the E_θ and E_ϕ fields at a point $P(\theta, \phi)$ are the same as the E_x field at the dipole, due to an infinitesimal dipole source at P, in the $\hat{\theta}$ and $\hat{\phi}$ direction.

The following field expressions can be written (spherical coordinates)

$$E_\phi = \sin\phi \left(\frac{j\omega\mu_0}{4\pi R} \right) e^{-jk_0 R} F(\theta) \tag{22}$$

$$E_\theta = -\cos\phi \left(\frac{j\omega\mu_0}{4\pi R} \right) e^{-jk_0 R} G(\theta) \tag{23}$$

where $F(\theta)$ and $G(\theta)$ are functions of θ which represent the field inside of the dielectric structure due to an incident unit strength plane wave. For $F(\theta)$, the incident E-field is perpendicular to the plane of incidence, and for $G(\theta)$, it is in the plane of incidence. $F(\theta)$ and $G(\theta)$ can be determined in a straightforward manner from transmission-line analogy [14], [15].

The power density in space is given as $(\eta_0 = \sqrt{\mu_0/\epsilon_0})$

$$P_0 = \frac{1}{2\eta_0} \left[|E_\theta|^2 + |E_\phi|^2 \right] \tag{24}$$

while the total radiated power is

$$P_{rad} = \int_0^{\pi/2} \int_0^{2\pi} P_0(\theta, \phi) R^2 \sin\theta \, d\phi \, d\theta. \tag{25}$$

The ϕ-integration is simple to perform explicitly, involving $\sin\phi$ and $\cos\phi$. The integration in θ is performed numerically.

The gain of the dipole is given by

$$G = \frac{P_0(0,0)(4\pi R^2)}{P_{rad}}, \qquad k_0 R \gg 1 \tag{26}$$

which simplifies to

$$G = \frac{4|F(0)|^2}{\displaystyle\int_0^{\pi/2} \sin\theta (|F(\theta)|^2 + |G(\theta)|^2) \, d\theta} \tag{27}$$

and in dB, $G_{dB} = 10 \log_{10} G$.

The radiation resistance can be defined from

$$P_{rad} = \frac{1}{2} I^2 R_{ro} = \frac{1}{2} \left(\frac{1}{l^2} \right) (Il)^2 R_{ro} \tag{28}$$

where l is the length of dipole and I is the (constant) current in the infinitesimal dipole. The field strength is proportional to the moment (Il). After simplification,

$$R_{ro} = 30\pi^2 \left(\frac{l}{\lambda_0} \right)^2 \int_0^{\pi/2} \sin\theta (|F(\theta)|^2 + |G(\theta)|^2) \, d\theta \tag{29}$$

results, where λ_0 is the free space wavelength.

The radiation efficiency of the dipole is the ratio of radiated power to total (radiated plus surface wave) power,

$$e_s = \frac{P_{rad}}{P_{rad} + P_{sw}}. \tag{30}$$

Both P_{rad} and P_{sw} are proportional to $(Il)^2$, and hence the result is independent of dipole length (as long as the dipole is short, $l \ll \lambda_0$).

IV. SURFACE-WAVE STRUCTURE

The overall behavior of surface waves is similar to that for the single dielectric layer [6], [7]. The dominant lowest order TM_1 mode is always above cut-off regardless of slab thicknesses. The next mode excited is the TE_1 mode, then the TM_2 mode, etc.

The transcendental equations for the TE mode roots and the TM mode roots may be written as $(t = H - B)$ [14], [15]

TE Mode

$$\cos(t\delta) \left[\left(\frac{\mu_1}{\mu_2} \right) (\gamma B) + \left(\frac{1}{\mu_2} \right) (\alpha B) \cot(\alpha B) \right] = \frac{\sin(t\delta)}{t\delta} \left[\left(\frac{1}{\mu_2} \right) \left(\frac{\mu_1}{\mu_2} \right) (\delta t)(\delta B) - (\alpha B)(\gamma t) \cot(\alpha B) \right]. \tag{31}$$

TM Mode

$$\cos(t\delta) \left[\left(\frac{\epsilon_1}{\epsilon_2} \right) (\gamma B) \cot(\alpha B) - \left(\frac{1}{\epsilon_2} \right) (\alpha B) \right] = \frac{\sin(t\delta)}{t\delta} \left[\left(\frac{1}{\epsilon_2} \right) \left(\frac{\epsilon_1}{\epsilon_2} \right) (\delta t)(\delta B) \cot(\alpha B) + (\alpha B)(\gamma t) \right] \tag{32}$$

where

$$\delta^2 = k_2^2 - \beta^2 \qquad \delta = \pm j u_2(\lambda_{root})$$

$$\gamma^2 = \beta^2 - k_0^2 \quad \text{or} \quad \gamma = u(\lambda_{root})$$

$$\alpha^2 = k_1^2 - \beta^2 \qquad \alpha = \pm j u_1(\lambda_{root})$$

$$\beta = \lambda_{root} = \text{propagation constant for mode.}$$

It is easy to prove that $k_0 \leqslant \beta \leqslant \max(k_1, k_2)$. Thus γ, which determines the decay constant in the air region, is always real. It can be seen that if $k_2 > k_1$, then δ is real, and if $k_2 < k_1$, then α is real. In most practical cases, $k_2 > k_1$ so δ will always be real. α may be real or pure imaginary in this case, however, depending upon whether $\beta < k_1$, or $\beta > k_1$. When a given mode turns on, we have $\beta = k_0$.

For a proper choice of substrate-superstrate parameters, we can cause the condition $\beta = k_1$ to be met for a given mode. At this point, we will have $\alpha = 0$ for that mode. This will be seen to have important implications for the case of only one propagating mode (the dominant TM_1 mode).

V. SUPERSTRATE EFFECT ON PCA PROPERTIES

The superstrate layer may prove beneficial or detrimental to printed antenna radiation characteristics, depending on the thicknesses of the substrate and cover, as well as relative dielectric and permeability constants.

As a first example, the effect of an ice layer on an elementary dipole on a duroid substrate ($\epsilon_1 = 2.45$) has been considered. The radiation efficiency e_s, radiation resistance R_{r0}, and antenna gain are shown in Fig. 3. The variation in directive gain for the case of thin ($n_1B/\lambda_0 = 0.100$) substrates typically used in practice, indicates that the formation of the ice layer will reduce gain by at most 2 dB for $n_2t/\lambda_0 \cong 0.20$; for thick substrate layers such as $n_1B/\lambda_0 = 0.400$, the ice cover improves gain up to a thickness of $n_2t/\lambda_0 \cong 0.33$ while beyond that it reduces it substantially, up to the shown range of $n_2t/\lambda_0 = 0.5$. Of more interest are the observations which may be drawn from Figs. 3(a) and 3(b). The e_s curves indicate that for a given substrate thickness, the radiation efficiency can be optimized by the presence of the proper superstrate layer thickness ($\epsilon_2 > \epsilon_1$). The maxima of the e_s curves coincide with those of the corresponding R_{r0} curves of Fig. 3(b).

This result, aside from the fact that it shows how a cover layer affects the basic properties of a PCA, indicates the direction of analysis to improve the radiation efficiency of printed antennas. The radiation efficiency of an elementary PCA on a GaAs substrate (no cover) is shown in Fig. 4(a). The efficiency e_s achieved is small unless $n_1B/\lambda_0 \ll 1$. The total radiated power tends to zero as $B \to 0$ due to image cancellation of the PCA current with the ground plane. As $B \to 0$, however, it can be shown analytically that for a single layer ($\epsilon_1 \gg 1$)

$$P_{\text{rad}} \sim I^2 \left(\frac{l}{\lambda_0}\right)^2 (80\pi^2)(k_0B)^2 \mu_1^2 \tag{33}$$

while

$$P_{\text{sw}} \sim I^2 \left(\frac{l}{\lambda_0}\right)^2 (60\pi^3)(k_0B)^3 \mu_1^3 \tag{34}$$

and therefore,

$$e_s = \frac{P_{\text{rad}}}{P_{\text{rad}} + P_{\text{sw}}} \cong 1 - \frac{P_{\text{sw}}}{P_{\text{rad}}} \tag{35}$$

or

$$e_s \cong 1 - \frac{3\pi}{4} \mu_1(k_0B). \tag{36}$$

It follows therefore that $P_{\text{sw}} \to 0$ faster than P_{rad} and therefore $e_s \to 1.0$ for a single layer as $B \to 0$. However, $R_{r0} \to 0$ quickly as $n_1B/\lambda_0 \to 0$, as seen in Fig. 4(b). To improve the radiation efficiency of a PCA on a GaAs layer, the PCA is integrated on the lower side of a GaAs superstrate which in turn is supported by some low permittivity material ($\epsilon_2 > \epsilon_1$) such as e.g. a Teflon substrate ($\epsilon_1 \cong 2.1$). The radiation efficiency graphs for this arrangement are shown in Fig. 5(a), where a significant (for moderate thickness of n_1B/λ_0) improvement in e_s is observed by comparison with the case of Fig. 4(a). Thus, for a Teflon substrate with $n_1B/\lambda_0 = 0.1$ and for $n_2t/\lambda_0 \cong 0.09$

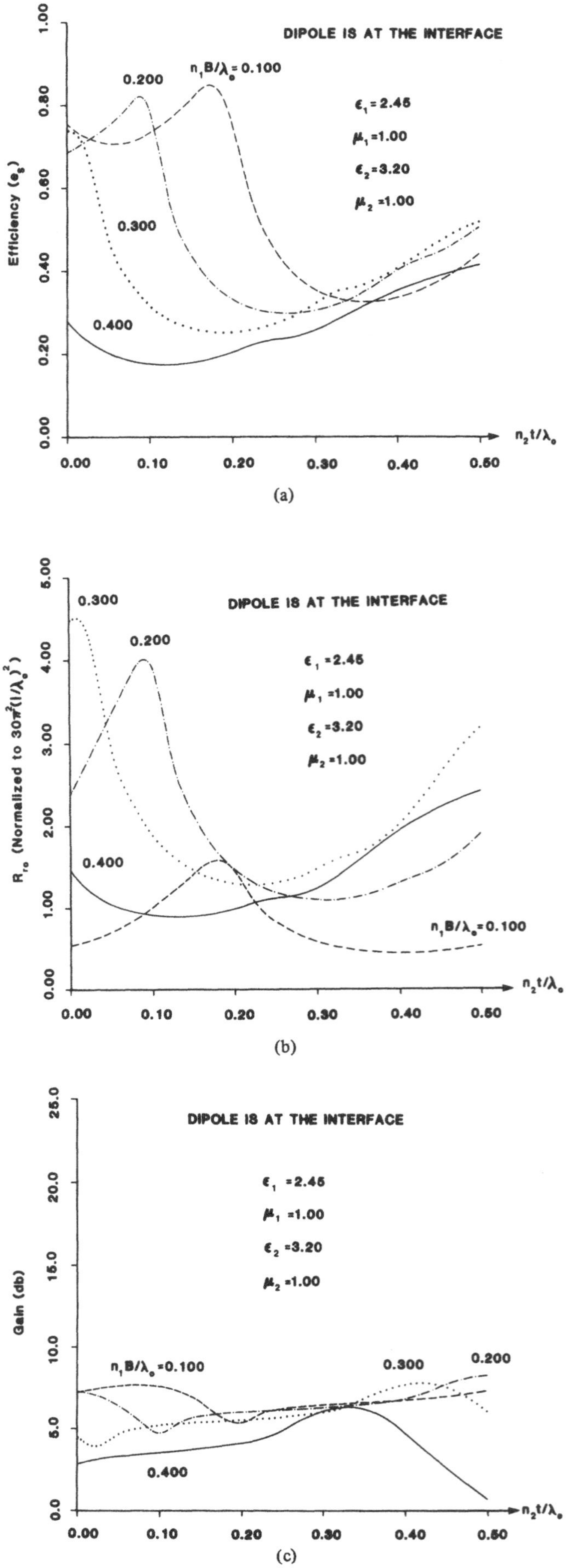

Fig. 3. (a) e_s versus n_2t/λ_0. (b) R_{r0} versus n_2t/λ_0. (c) Gain versus n_2t/λ_0.

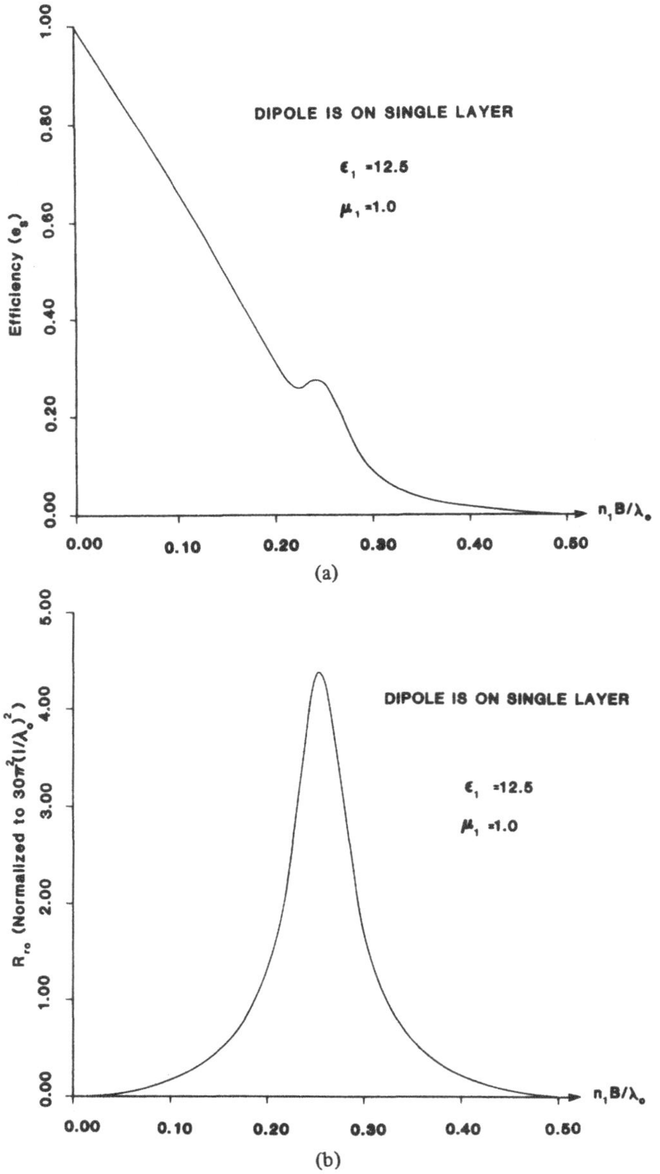

Fig. 4. (a) e_s versus n_1B/λ_0. (b) R_{ro} versus n_1B/λ_0.

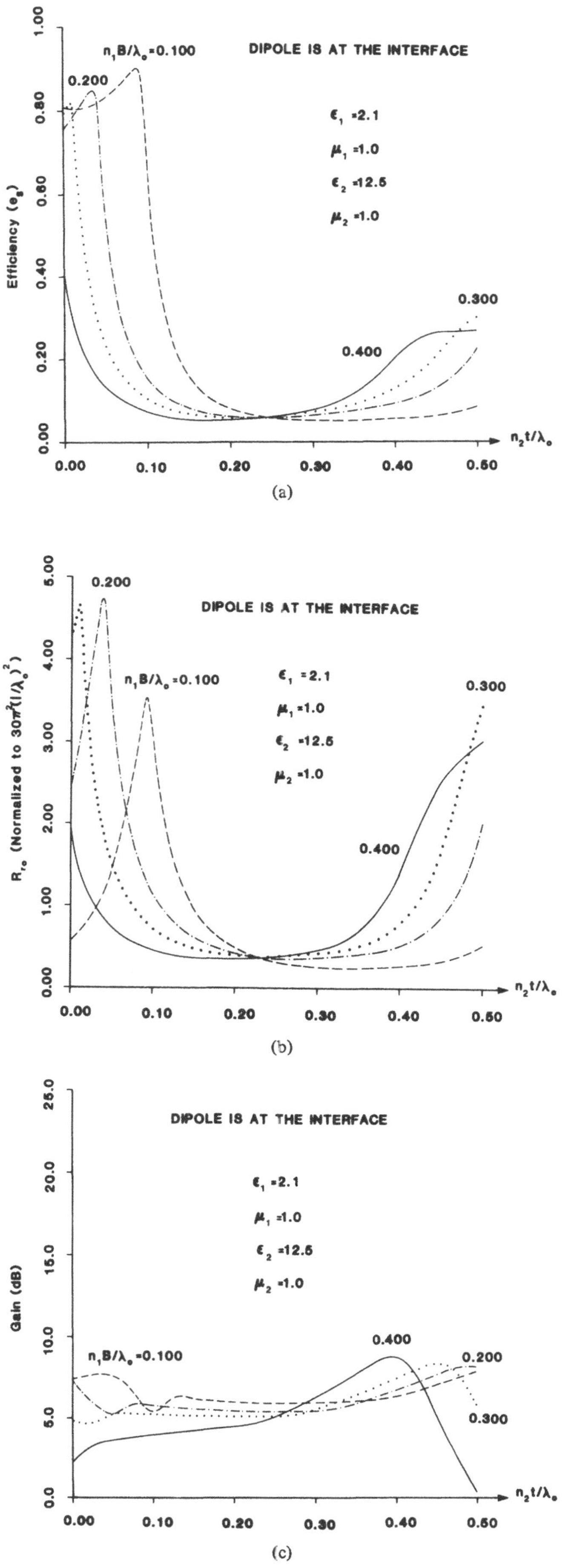

Fig. 5. (a) e_s versus n_2t/λ_0. (b) R_{ro} versus n_2t/λ_0. (c) Gain versus n_2t/λ_0.

a maximum efficiency of $e_s \cong 90$ percent is obtained. Figs. 5(b) and 5(c) show R_{r0} and gain for these cases. It is observed that the general trend of these curves is very similar to the case of ice on duroid (Fig. 3). Similar improvement of e_s can be obtained for other materials used in imaging arrays such as, e.g., quartz. Finally, it is important to note that the e_s maxima coincide with R_{r0} maxima and gain minima when the dipole is at the interface.

VI. RESONANCE CONDITIONS

In order to improve gain, radiation resistance, and e_s by optimizing substrate-cover effects, it has been determined that the various parameters may be chosen so that a substrate-cover resonance condition [16] is obtained. There are two such resonance conditions. The first condition requires a lower layer thickness of $n_1B/\lambda_0 \cong 0.50$ while the cover thickness must be $n_2t/\lambda_0 \cong 0.25$. In this case, the antenna must be located in the middle of the substrate and the condition $\epsilon_2 \gg \epsilon_1$ must be satisfied. Under these conditions, a resonance is created in a transmission line analogy, which causes a large voltage at the location

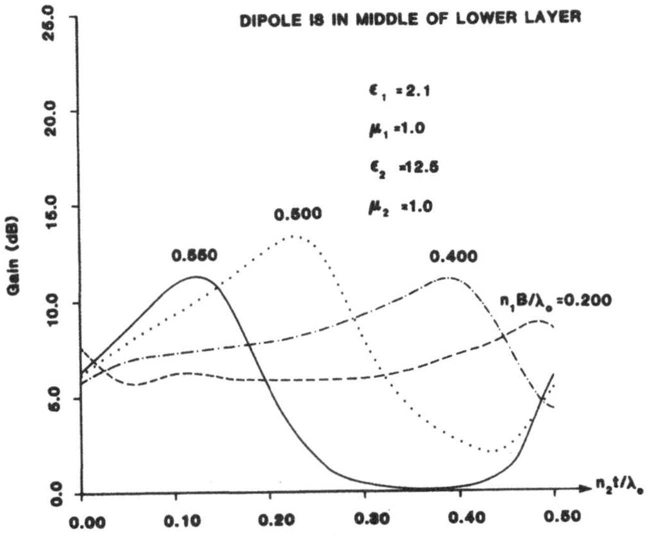

(a)

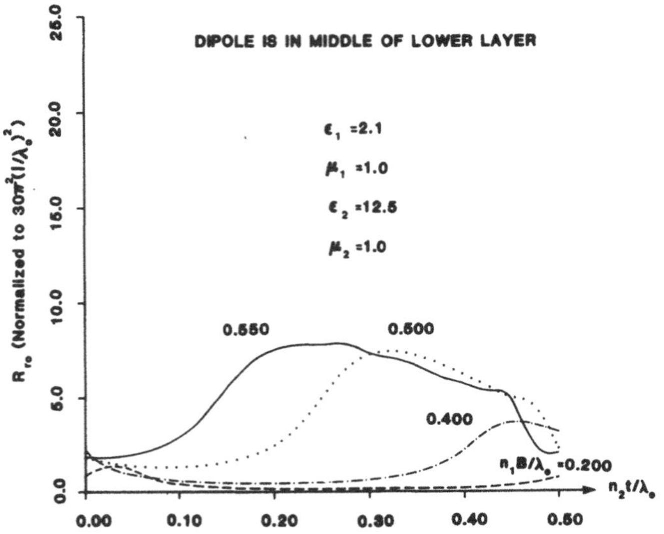

(b)

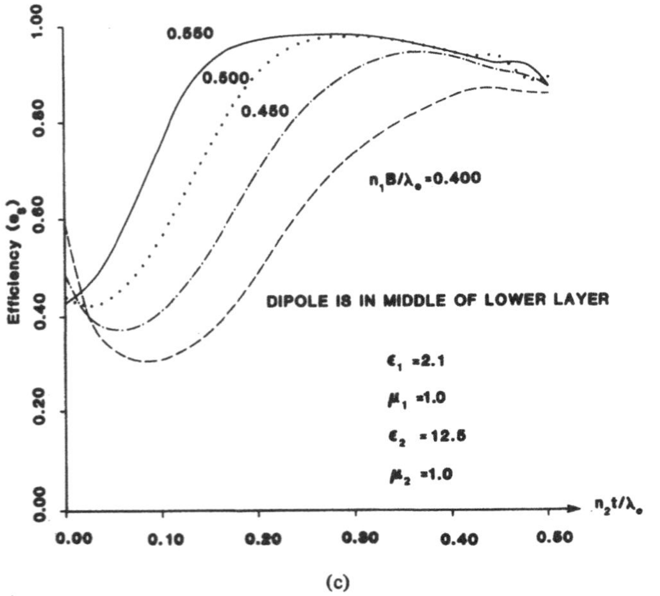

(c)

Fig. 6. (a) Gain versus n_2t/λ_0. (b) R_{ro} versus n_2t/λ_0. (c) e_s versus n_2t/λ_0.

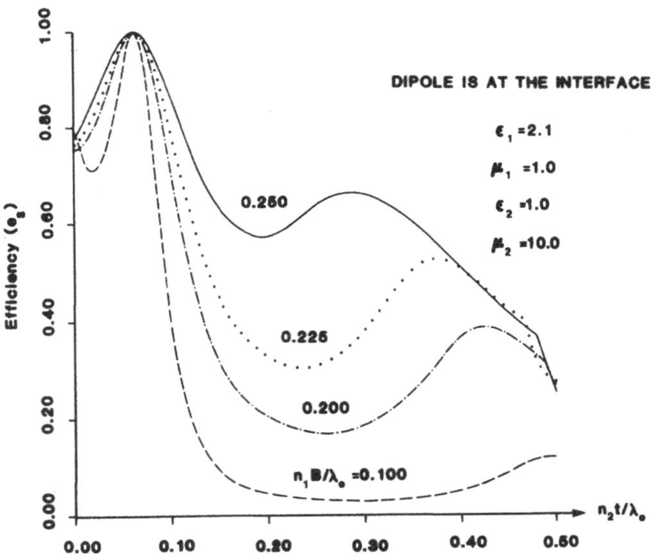

Fig. 7. e_s versus n_2t/λ_0.

of the antenna at $z = z_0$. This leads to greatly enhanced gain and R_{r0} is also improved. The results for GaAs over Teflon are presented in Fig. 6, where it is observed that all pertinent antenna parameters are enhanced substantially. The second substrate-superstrate resonance condition involves a configuration wherein the PCA is on the interface and the superstrate is a high permeability magnetic material ($\mu_2 \gg \mu_1$). This resonance condition is a kind of dual to the first one. This resonance is achieved for smaller composite thickness, since now $n_1B/\lambda_0 \cong 0.25$ and $n_2t/\lambda_0 \cong 0.25$, but it may be less practical due to the requirement of low loss, high permeability materials.

VII. ELIMINATION OF SURFACE WAVES

It is significant to note that for lossless materials, it is possible to design a composite substrate-superstrate structure to yield $e_s = 100$ percent. This is clearly demonstrated in Fig. 7 where for $n_2t/\lambda_0 \cong 0.06$, $e_s = 100$ percent. For a very thin substrate layer this is impractical since R_{r0} is exceedingly small. However, for a thicker substrate such as e.g. $n_1B/\lambda_0 \cong 0.250$, R_{r0} obtains reasonably useful values for $e_s = 100$ percent design. The $e_s = 100$ percent requirement necessitates that $k_2 \gg k_1$ and the proper value of n_2t/λ_0 must be found so as to enforce the condition $\beta = k_1$ or $\alpha = 0$ for the dominant mode. This condition implies that there is no variation of the surface wave fields with z in the substrate. The tangential surface wave $\bar{E}$ field must then be zero everywhere in the substrate. By reciprocity, this implies that this mode is not excited.

If the superstrate is a nonmagnetic material, such as GaAs, the $e_s = 100$ percent condition requires a very thin substrate as shown in Fig. 8 which renders this particular example impractical since R_{r0} is very small. As Figs. 7 and 8 indicate, an additional disadvantage of achieving $e_s = 100$ percent is the narrow bandwidth imposed by the substrate-superstrate structure. It can be shown that the thickness t at which the TE_1 mode is turned on is given by the equation

$$\tan\left\{2\pi\left(\frac{n_2 t_{E1}}{\lambda_0}\right)\sqrt{1-1/n_2^2}\right\} = \frac{\mu_2}{\mu_1}\frac{\sqrt{n_1^2-1}}{\sqrt{n_2^2-1}}\cot\left[2\pi\left(\frac{n_1 B}{\lambda_0}\right)\sqrt{1-1/n_1^2}\right] \tag{37}$$

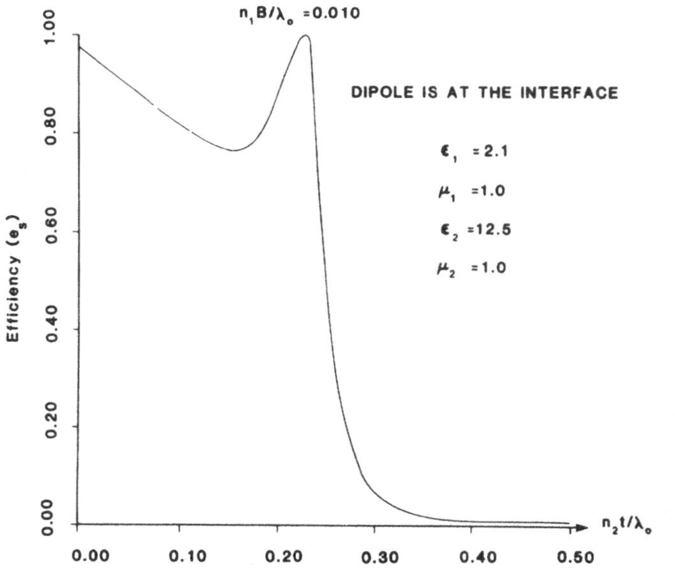

Fig. 8. e_s versus n_2t/λ_0.

while the value of t for which $\alpha = 0$ is derived from

$$\cot\left\{2\pi\left(\frac{n_2t_c}{\lambda_0}\right)\sqrt{1-\left(\frac{n_1}{n_2}\right)^2}\right\} = \frac{\sqrt{n_2^2-n_1^2}}{\epsilon_2\sqrt{n_1^2-1}}. \tag{38}$$

It follows that in order to have $\alpha = 0$ and simultaneously only the dominant TM_1 mode then $t_c < t_{E1}$. We have that t_c is independent of B. However, a complete elimination of all surface wave fields is possible only if $t_c \leqslant t_{E1}$, since we are only eliminating ($\alpha = 0$) the dominant mode. This implies that the substrate thickness must not exceed some value n_1B_{max}/λ_0, for a given superstrate material. The maximum substrate thickness allowable results from the condition

$$t_c = t_{E1} \tag{39}$$

and it is derived from

$$\frac{1}{\sqrt{n_2^2-1}}\tan^{-1}\left\{\frac{\mu_2}{\mu_1}\frac{\sqrt{n_1^2-1}}{\sqrt{n_2^2-1}}\cot\left[2\pi\left(\frac{n_1B_{max}}{\lambda_0}\right)\sqrt{1-\frac{1}{n_1^2}}\right]\right\}$$

$$=\frac{1}{\sqrt{n_2^2-n_1^2}}\tan^{-1}\left[\frac{\epsilon_2\sqrt{n_1^2-1}}{\sqrt{n_2^2-n_1^2}}\right]. \tag{40}$$

The normalized critical superstrate thickness n_2t_c/λ_0 for e_s = 100 percent is shown in Fig. 9(a) (nonmagnetic layers) as a function of ϵ_2 for various cases of ϵ_1. The maximum normalized substrate thickness corresponding to the data of Fig. 9(a) is depicted in Fig. 9(b). These figures demonstrate very clearly that superstrate materials such as GaAs and Si which are basic to integrated circuit technology require very thin substrate thickness (Fig. 9(b)) which yields exceedingly small R_{r0} values for the antenna to be practical. However, as Fig. 5 indicates addition of a top layer may be beneficial, even if e_s = 100 percent cannot be obtained. A magnetic superstrate allows for much thicker substrates for achieving e_s = 100 percent. Design curves for the antenna to be practical. However, as Fig. 5 indicates, Fig. 10. The use of magnetic superstrates to achieve e_s = 100 percent is discussed in [15].

VIII. RADIATION INTO THE HORIZON

It is important to note the fact that in general the radiation patterns for printed antennas with a cover tend to zero along

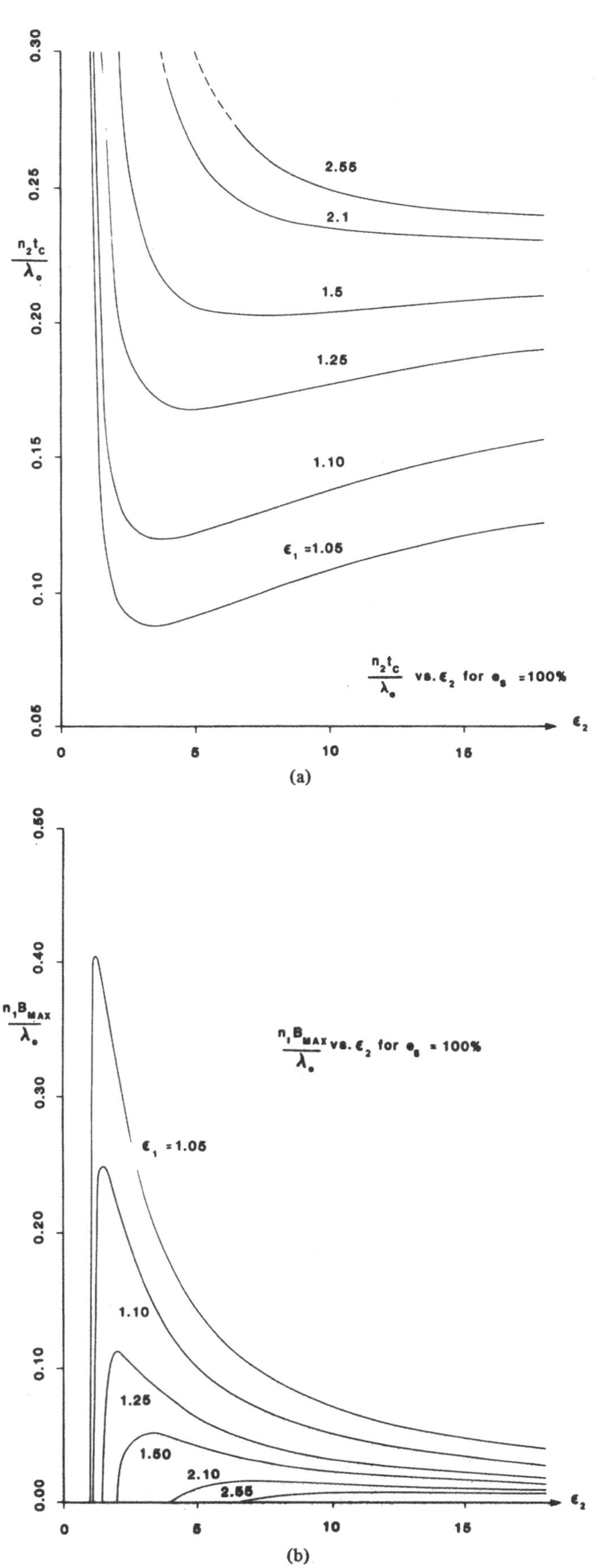

Fig. 9. (a) n_2t_c/λ_o versus ϵ_2 for e_s = 100 percent. (b) n_1B_{max}/λ_0 versus ϵ_2 for e_s = 100 percent.

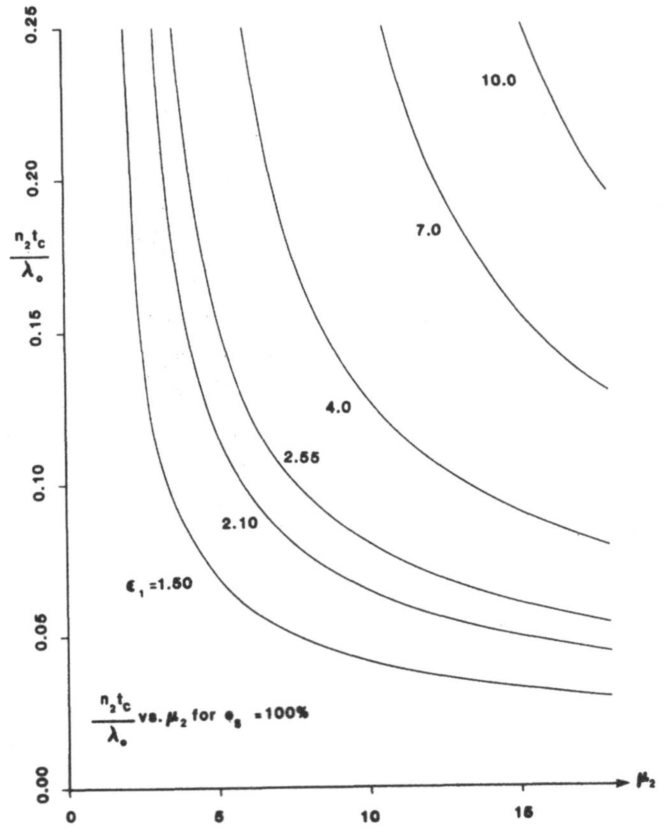

(a)

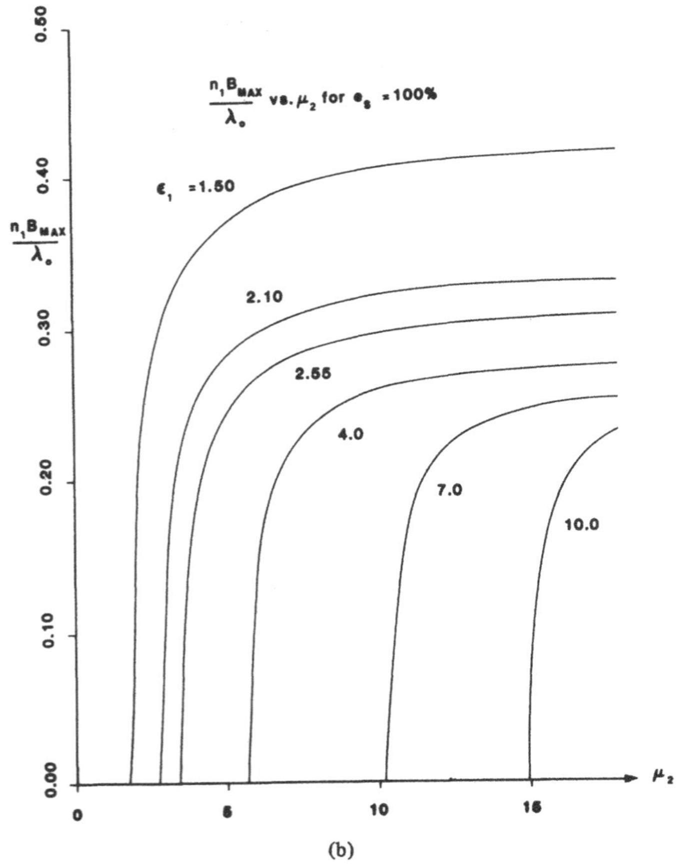

(b)

Fig. 10. (a) $n_2 t_c/\lambda_0$ versus μ_2 for e_s = 100 percent. (b) $n_1 B_{max}/\lambda_0$ versus μ_2 for e_s = 100 percent.

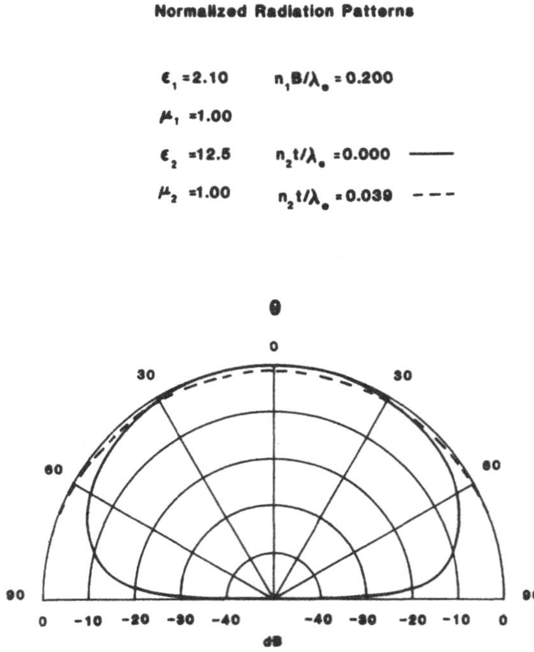

(a)

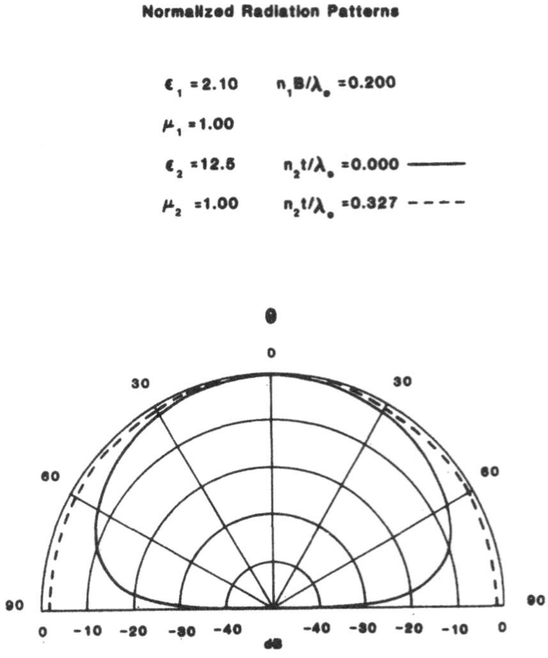

(b)

Fig. 11. (a) $\bar{H}$-plane pattern. (b) $\bar{E}$-plane pattern.

the horizon ($\theta \to \pi/2$). This in fact is always true, unless the superstrate thickness is chosen to have certain critical values. These critical values are those for which the TE or TM modes turn on ($\beta = k_0$). When a TE mode turns on, the $F(\theta)$ function in (22) remains nonzero as $\theta \to \pi/2$. When a TM mode turns on, the $G(\theta)$ function in (23) remains nonzero as $\theta \to \pi/2$. This causes the $\bar{H}$-plane pattern to become nearly omnidirectional when the TE_1 mode turns on, and the $\bar{E}$-plane pattern to become nearly omnidirectional when the TM_2 mode turns on.

An example of each case is shown in Figs. 11(a) and 11(b). In these figures we see the pattern for a dipole on a Teflon substrate superimposed with the patttern obtained by addition

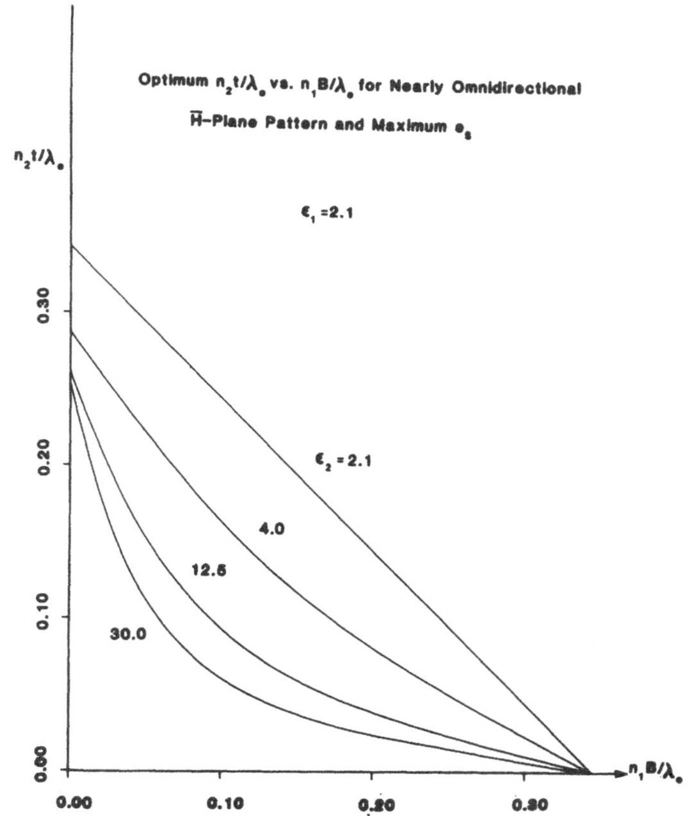

Fig. 12. Optimum n_2t/λ_0 versus n_1B/λ_0 for nearly omnidirectional $\overline{H}$-plane pattern and maximum e_s.

of the appropriate thickness GaAs superstrate. Although both $\overline{H}$-plane and $\overline{E}$-plane radiation into the horizon can be achieved with a cover, the case of $\overline{H}$-plane radiation is more important because it has a significant implication, which is now discussed.

IX. OPTIMUM SUPERSTRATE THICKNESS

As mentioned previously, the addition of a superstrate can yield $e_s = 100$ percent if the substrate is thin enough. With nonmagnetic layers having typical dielectric constants used in practice, the substrate may have to be prohibitively thin. However, the efficiency can still be optimized by the addition of a superstrate, even if $e_s = 100$ percent cannot be obtained. For nonmagnetic materials with the dipole at the interface of the substrate and superstrate, the efficiency is maximized when the TE_1 mode turns on, and we have $\overline{H}$-plane radiation into the horizon. At this point, the radiation resistance also peaks, while the gain dips. This is illustrated in Figs. 5(a)-5(c). The cause of this behavior is the radiation near the horizon caused by the TE_1 mode turning on. When the TM_2 mode turns on, the layer thickness is generally too large to have a similar effect. When $n_1B/\lambda_0 = n_1B_{max}/\lambda_0$, it can be seen that $n_2t_c/\lambda_0 = n_2t_{E1}/\lambda_0$. Thus, we have a general rule for optimizing the efficiency: For $n_1B/\lambda_0 \leqslant n_1B_{max}/\lambda_0$, choose $n_2t/\lambda_0 = n_2t_c/\lambda_0$. For $n_1B/\lambda_0 \geqslant n_1B_{max}/\lambda_0$, choose $n_2t/\lambda_0 = n_2t_{E1}/\lambda_0$. This will generally give the optimum superstrate thickness in terms of maximizing the efficiency. Fig. 12 shows superstrate thicknesses required for $\overline{H}$-plane radiation into the horizon for a Teflon substrate. For a magnetic superstrate or for a dipole not close to the interface with dielectric layers, resonance conditions tend to dominate the radiation into the horizon effect, and no such simple rule holds for choosing the superstrate thickness to optimize the efficiency.

X. CONCLUSION

Care must be taken when using a superstrate cover for a microstrip antenna so as not to adversely affect performance. By properly choosing superstrate parameters, a significant increase in gain, radiation resistance, and efficiency may be achieved, enabling the cover to act as a desirable part of the antenna as well as a protective layer. By choosing the thicknesses properly, a resonance condition may be created, whereby gain and radiation resistance are substantially improved over a significant bandwidth. This requires fairly thick layers, about a half-wavelenth in the media. Alternatively, it may be possible to achieve 100 percent efficiency, with no surface wave power being excited. This usually requires the substrate to be thin, unless a magnetic superstrate is used. Also, this scheme tends to be narrow band. However, even for cases where the substrate is too thick to achieve a 100 percent efficiency, a significant improvement may still be obtained by using a superstrate with an optimum thickness determined by the turning on of the TE_1 mode, for the important case of nonmagnetic layers with the dipole at the interface.

ACKNOWLEDGMENT

The authors wish to thank Dr. P. Bargeliotes and Dr. J. F. Cashen of Northrop Corporation for encouraging initiation of this research. Also appreciation is due to Ms. I. Andreadis and Ms. P. Parris for typing the manuscript and to Mr. K. Abolhassani for drawing the figures.

REFERENCES

[1] K. R. Carver and J. W. Mink, "Microstrip antenna technology," *IEEE Trans. Antennas Propagat.*, vol. AP-29, no. 1, pp. 2-24, Jan. 1981.

[2] I. J. Bahl, P. Bhartia, and S. S. Stuchly, "Design of microstrip antennas covered with a dielectric layer," *IEEE Trans. Antennas Propagat.*, vol. AP-30, pp. 314-318, Mar. 1982.

[3] I. J. Bahl and P. Bhartia, *Microstrip Antennas.* Dedham, MA: Artech House, 1980.

[4] D. B. Rutledge, S. E. Schwartz, and A. T. Adams, "Infrared and submillimeter antennas," *Infrared Phys.*, vol. 18, pp. 713-729, Dec. 1978.

[5] P. T. Parrish, T. C. L. G. Sollner, R. H. Mathews, H. R. Fetterman, C. D. Parker, P. E. Tannenwald, and A. G. Cardiasmenos, "Printed dipole-Schottky diode millimeter wave antenna array," *SPIE Proc.*, vol. 337, Millimeter Wave Technology, May, 1982.

[6] P. B. Katehi and N. G. Alexopoulos, "On the effect of substrate thickness and permittivity on printed circuit antennas," *IEEE Trans. Antennas Propagat.*, vol. AP-31, no. 1, pp. 34-39, Jan. 1983.

[7] N. G. Alexopoulos, P. B. Katehi, and D. B. Rutledge, "Substrate optimization for integrated circuit antennas," *IEEE Trans. Microwave Theory Tech.*, vol. MTT-31, pp. 550-557, July 1983.

[8] D. M. Pozar, "Considerations for millimeter wave printed antennas," *IEEE Trans. Antennas Propagat.*, vol. AP-31, pp. 740-747, Sept. 1983.

[9] A. Sommerfeld, *Partial Differential Equations.* New York: Academic, 1941, vol. VI.

[10] J. R. Wait, "Characteristics of antennas over lossy earth," in *Antenna Theory*, Pt. 2, R. E. Collin and F. J. Zucker, Eds. New York: McGraw-Hill, 1969.

[11] I. E. Rana and N. G. Alexopoulos, "Current distribution and input impedance of printed dipoles," *IEEE Trans. Antennas Propagat.*, vol. AP-29, no. 1, pp. 99-105, Jan. 1981.

[12] N. G. Alexopoulos and I. E. Rana, "Current distribution and input impedance of printed dipoles," *IEEE Trans. Antennas Propagat.*, vol. AP-30, pp. 822, July 1982.

[13] D. Marcuse, *Light Transmission Optics.* Van Nostrand Reinhold, 1972.

[14] N. G. Alexopoulos and D. R. Jackson, "Fundamental superstrate (cover) effects on printed circuit antennas," UCLA Rep. ENG-83-50, Oct. 1983.

[15] N. G. Alexopoulos and D. R. Jackson, "Radiation efficiency optimization for printed circuit antennas using magnetic superstrates," UCLA Rep. ENG-84-01, Dec. 1983.

[16] Y. Sugio, T. Makimoto, S. Nishimura, and H. Nakanishi, "Analysis for gain enchancement of multiple-reflection line antenna with dielectric plates," *Trans. IECE*, pp. 80-112, Jan. 1981.

General integral equation formulation for microstrip antennas and scatterers

J.R. Mosig, D.Appl.Sc., and Prof. F.E. Gardiol, M.Sc., D.Appl.Sc., Sen. Mem. IEEE

Indexing terms: *Antennas (Microstrip), Numerical Analysis*

Abstract: The paper deals with the dynamic analysis of microstrip structures. It is shown that the mixed-potential integral equation for stratified media, which was introduced in a previous publication, provides a rigorous and powerful approach. The Green's functions belonging to the kernel of the integral equation are expressed as Sommerfeld integrals, in which surface wave effects are automatically included. A two-dimensional moment's method using subsectional basis functions has been chosen. Thus, microstrip patches of any shape can be analysed at any frequency and for any substrate. Practical numerical aspects are carefully discussed, and special numerical devices are introduced to reduce computation time without loss of accuracy. Complete results for a rectangular patch and for a slotted patch are given and compared with measured values. Radiation patterns corresponding to the ideal situation of a substrate with infinite transverse dimensions are presented for a rectangular patch.

1 Introduction

Over the past decade, the range of application of microstrip structures has broadened considerably. In particular, microstrip antennas are used in an increasing numer of applications, ranging from biomedical diagnosis to satellite communications. Such a wide range of applications, coupled with the fact that microstrip structures are relatively simple to produce with good reproducibility, has turned microstrip analysis into a cornerstone problem, to which almost all the mathematical models developed in the field of electromagnetics have been applied. This is witnessed by the huge amount of technical literature and several monograph books published in recent years [1–3].

Models used to study microstrip patch antennas range from very simplified ones, such as the transmission-line model [4], through cavity models [5], segmentation techniques [6], full-wave analysis [7] and up to quite sophisticated approaches based on an integral formulation and numerical resolution on a computer [8–10]. Whereas simple approximations yield directly usable simplified formulas, the more complex approaches require increasingly lengthy calculations. Many models are directly linked to simple patch shapes (rectangular or circular), and in some instances an approximate distribution of the current is introduced, determined by an educated guess. Unfortunately, the range of validity of many of the assumptions made was not defined; as a matter of fact, this would be difficult to make in the absence of a rigorous solution as a basis for comparison. Also, some effects such as the presence of surface waves are lost in the approximation process (but not in the actual device!). Detailed surveys of the previously developed methods are available [11–12].

The purpose of the present study is to provide a rigorous treatment of the general problem, free from oversimplifying assumptions and applicable to arbitrarily shaped patches. The mixed-potential integral equation [13] was found to be better suited for numerical analysis than the previously used electric-field integral formulation [8–10]. Of particular interest is the fact that the numerical techniques presented by the authors in a previous publication can be taken full advantage of [14].

The integral equation is solved by means by a moment's method using rooftop subsectional basis functions [15], which are much more flexible to use than sine waves defined over the entire domain. These choices, and a thorough treatment of the antenna's excitation, lead to a very general technique, suitable for the analysis of complex shaped patches with any combination of thickness and permittivity, taking dielectric and conducting losses into account. The frequency behaviour of a given structure can be determined from the quasistatic range up to its higher-order resonances.

2 Mixed-potential integral equation (MPIE)

This formulation was extensively used in the analysis of wire antennas by the moment's method [16]. Here, it will be applied to lossy microstrip structures. With reference to Fig. 1, the boundary conditions for the electric field on the

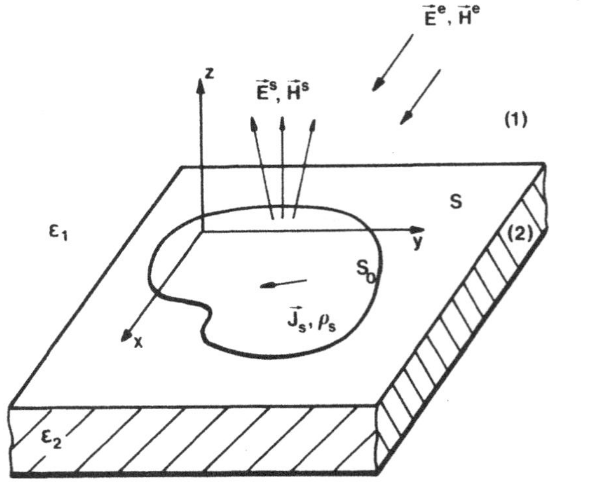

Fig. 1 *Arbitrarily shaped microstrip structure with dynamical excitation*

$E^{(e)}$: Excitation field (source)
$E^{(s)}$: Scattered field
J_s, ρ_s: Induced current and charge densities
S_0: Conducting patch (upper conductor)
S: air-substrate interface

surface of a patch of real conductor (non ideal) is

$$e_z \times [E^{(s)}(r) + E^{(e)}(r)] = Z_s[e_z \times J_s(r)] \qquad r \in S_0 \tag{1}$$

This equation simply expresses that the total electric field, sum of the excitation field $E^{(e)}$ and of the scattered field $E^{(s)}$ must be proportional to the electric surface current J_s. The proportionality factor Z_s has the dimensions of an impedance and depends on the metal conductivity σ, the thickness of the upper conductor t and the frequency f. In a perfect conductor, Z_s vanishes, whereas in most practical situations the metal skindepth is much smaller than the conductor thickness, so that Z_s becomes the classical plane-wave surface impedance $Z_s = (1 + j)\sqrt{\mu_0 f/\pi\sigma}$. As the upper conductor is always much thinner than the dielectric substrate, it can be replaced by a current sheet at all frequencies. The surface current density J_s in eqn. 1 is thus a total value, the sum of the actual surface currents flowing over both sides of the patch.

The scattered field derives from a scalar and a vector potential, which in turn are expressed in terms of superposition integrals of the corresponding Green's functions, weighted by the unknown distributions of surface electric charge and current [14]

$$A(r) = \int_{S_0} \bar{G}_A(r \mid r') \cdot J_s(r')\, dS';$$

$$V(r) = \int_{S_0} G_V(r \mid r') \rho_s(r')\, dS' \qquad (2)$$

The Green's functions $\bar{G}_A$ and G_V can be expressed in terms of Sommerfeld integrals. They are related to those introduced in Reference 10 to solve Pocklington's equation for printed wires. Their analytical and numerical properties have been extensively studied in a previous paper [14] for the lossless case. The modifications needed to account for a complex relative permittivity $\varepsilon_r = \varepsilon_r'(1 - j \tan \delta)$ have been outlined in Reference 17.

When the observer is very close to the source, the dominant term in the Green's functions is given by the static Green's functions corresponding to an homogeneous medium of permittivity $\varepsilon_0(\varepsilon_r + 1)/2$ and permeability μ_0

$$\frac{4\pi}{\mu_0} G_A^{xx}(r \mid r') = 2\pi(\varepsilon_r + 1)\varepsilon_0 G_V(r \mid r') = 1/|r - r'| \qquad (3)$$

This fact can be taken advantage of to expand any component of the Green's functions G in the following way:

$$G = G_H + (G - G_H) \qquad (4)$$

where H stands for homogeneous.

The term G_H, given by eqns. 3, exhibits at the origin a weak singularity of the r^{-1} type. Its integral over a rectangular domain can be performed analytically. Therefore, numerical techniques are only needed to evaluate the difference term, which is a regular well-behaved function at the origin.

3 Moment's method

To obtain an exact solution, one would have to satisfy the boundary condition, eqn. 1, at every point within the patch. This is clearly not feasible, as it would require the resolution of an infinite set of equations. Some kind of truncation of the set is an absolute requirement: the boundary condition, eqn. 1, will then be satisfied over a limited number of points, carefully chosen over the patch, using a method of moments.

3.1 Charge and current cells

The basis and test functions best suited to the study of arbitrary shaped patches at any frequency are selected. No *a priori* assumptions will be made for the distribution of currents on the patch: this actually eliminates the use of basis functions defined over the entire domain [8]. A comparison of available possibilities [11] led to the selection of rooftop functions for the surface current J_s, which were successfully used in similar problems [15]. To implement these functions, the patch's boundary is replaced by a Manhattan-type polygonal line (Fig. 2A). As most commonly used antennas exhibit this kind of geometry anyway, this requirement is easily satisfied.

The patch's surface is then divided into rectangular cells, called charge cells, which are all chosen of equal size, with dimensions $a \times b$ (Fig. 2A). This is not a basic requirement, but the use of different cell sizes would considerably increase the length of the computations.

Two adjacent charge cells, sharing a common border perpendicular to the x-direction (y-direction), will form an x-directed (y-directed) current cell (Fig. 2B). An automatic overlapping of current cells is obtained in this manner, in which a charge cell may belong to up to four different current cells. The number of charge cells is thus related to the number of current cells, although the relationship is not a simple one, depending as it does on the shape of the patch. For rectangular patches with $m \times n$ charge cells, the number of x-directed current cells is $M = (m - 1)n$, and that of y-directed current cells $N = m(n - 1)$.

Every current cell supports one rooftop basis function, to which is associated one test segment joining the centres of the two charge cells belonging to the current cell. The centre of the segment C_{xj} associated to the j-th x-directed current will be denoted by the vector r_{xj}, its ends by r_{xj}^{-}

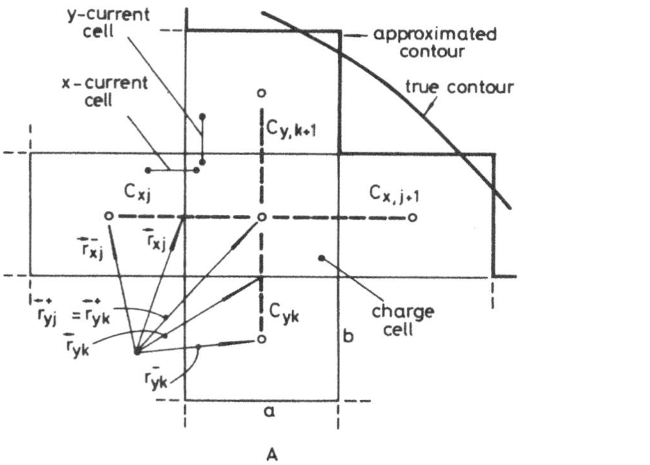

Fig. 2A *Segmentation of the patch in elementary charge and current cells, showing the network of test segments*

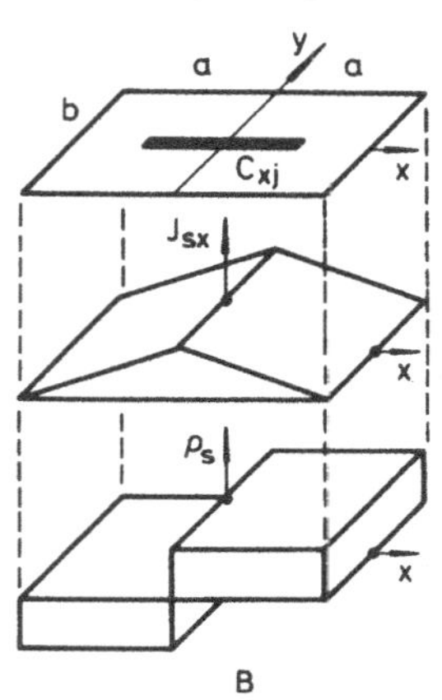

Fig. 2B *x-directed current cell centred at r = 0 and its associate surface current density $J_{sx} = T_x(r)$, and surface charge density $\rho_s = \Pi(r - e_x a/2) - \Pi(r + e_x a/2)$*

and r_{xj}^+ (Fig. 2), with these three vectors related through

$$r_{xj}^{\pm} = r_{xj} \pm e_x(a/2) \qquad j = 1, 2, \ldots M \tag{5}$$

A similar relationship is written for y-directed segments C_{yj} ($j = 1, 2 \ldots, N$).

3.2 Basis functions

The Cartesian components of the surface current are expanded over a set of basis functions T_x, T_y

$$J_{sx} = \frac{1}{b}\sum_{j=1}^{M} I_{xj}\, T_x(r - r_{xj})$$
$$J_{sy} = \frac{1}{a}\sum_{j=1}^{N} I_{yj}\, T_y(r - r_{yj}) \tag{6}$$

where the basis functions are of rooftop type defined as (Fig. 2)

$$T_x(r) = \begin{cases} 1 - |x|/a & |x| < a, |y| < b/2 \\ 0 & \text{elsewhere} \end{cases} \tag{7}$$

A similar expression is obtained for T_y by interchanging $a \leftrightarrow b$, $x \leftrightarrow y$ in eqn. 7.

The introduction of factors $1/a$ and $1/b$ in eqn. 6 yields unknown coefficients I_{xj} and I_{yj} having dimensions of a current. Moreover, every coefficient gives the total current flowing across the common boundary of two charge cells.

The associated surface charge density is obtained from eqn. 6 by using the continuity equation, yielding

$$\rho_s = \frac{1}{j\omega ab}\left\{\sum_{j=1}^{M} I_{xj}[\Pi(r - r_{xj}^+) - \Pi(r - r_{xj}^-)] + \sum_{j=1}^{N} I_{yj}[\Pi(r - r_{yj}^+) - \Pi(r - r_{yj}^-)]\right\} \tag{8}$$

where $\Pi(r)$ is a two-dimensional unit pulse function defined over a rectangle of dimensions $a \times b$, centred at $r = 0$ (Fig. 2A).

The charge density within every elementary cell remains constant, justifying the appellation of charge cell. For the charge cell of Fig. 2B, with four test segments ending at its centre, the surface charge density is simply given by

$$\rho_s = \frac{1}{j\omega ab}[I_{x,j+1} - I_{x,j} + I_{y,k+1} - I_{y,k}] \tag{9}$$

The charge density is discontinuous on the borders between charge cells. The scalar potential remains bounded, while the electric field becomes singular, as ρ_s does not satisfy a Hölder condition [11]. This means that the test functions must be selected carefully, avoiding the locations where the electric field is singular.

3.3 Discrete Green's functions

The notation and the computational task can be simplified by introducing discrete Green's functions, which have as source a complete basis function, instead of the traditional elementary point source.

The vector potential Γ_A is created by a rooftop distribution of surface current, whereas Γ_V is the scalar potential resulting from a rectangular distribution of unit surface charge. It is convenient in practice to deal with dimensionless quantities, in a normalised space where physical lengths are replaced by electrical lengths. The following adimensional expressions are therefore introduced, defining the discrete Green's functions

$$\Gamma_A^{xx}(r \mid r_{xj}) = \int_{S_{xj}} \frac{1}{\mu_0 k_0} G_A^{xx}(r \mid r')T_x(r' - r_{xj})(k_0^2\, dS') \tag{10a}$$

$$\Gamma_V(r \mid r_{0j}) = \int_{S_{0j}} \frac{\varepsilon_0}{k_0} G_V(r \mid r')\Pi(r' - r_{0j})(k_0^2\, dS') \tag{10b}$$

A similar expression holds for Γ_A^{yy}. In these formulas, $r_{xj}(r_{0j})$ denotes the centre and $S_{xj}(S_{0j})$ the surface of a current (charge) cell.

The discrete Green's functions exhibit the same properties of translational invariance and of symmetry as the conventional ones do [14]. In the general case, the surface integrals in eqn. 10 must be evaluated numerically. When the observation point r belongs to the source cell, some difficulties arise in the integration process. It is then recommended to separate the Green's functions into their singular and regular parts, as indicated in eqn. 4, where the singular part can be integrated analytically. For an observation point at the centre of a charge cell, replacement of eqn. 3 in eqn 10b yields the singular part of the discrete Green's function as

$$2\pi(\varepsilon_r + 1)\Gamma_V(0 \mid 0) \simeq 2k_0 a \ln \tan\left(\frac{\alpha}{2} + \frac{\pi}{4}\right) - 2k_0 b \ln \tan(\alpha/2) \tag{11}$$

with $\tan \alpha = b/a$

When the observer is located many cells away from the sources, the latter can be concentrated at the centre of the cell. The following approximations may then be used:

$$\Gamma_A^{xx}(r \mid r_{xi}) \simeq \frac{1}{\mu_0 k_0} G_A^{xx}(r \mid r_{xi})(k_0 a)(k_0 b) \tag{12}$$

$$\Gamma_V(r \mid r_{0i}) \simeq \frac{\varepsilon_0}{k_0} G_V(r \mid r_{0i})(k_0 a)(k_0 b)$$

Discrete Green's functions provide a very compact notation for the potentials created by the whole structure. Introducing eqns. 6 and 8 in the definitions 2, and making use of eqn. 10, yields

$$A(r) = e_x \frac{\mu_0}{k_0 b}\sum_{j=1}^{M} I_{xj}\Gamma_A^{xx}(r \mid r_{xj}) + e_y \frac{\mu_0}{k_0 b}\sum_{j=1}^{N} I_{yj}\Gamma_A^{yy}(r \mid r_{yj}) \tag{13a}$$

$$V(r) = \frac{Z_0}{j(k_0 a)(k_0 b)}\left\{\sum_{j=1}^{M} I_{xj}[\Gamma_V(r \mid r_{xj}^+) - \Gamma_V(r \mid r_{xj}^-)] + \sum_{j=1}^{N} I_{yj}[\Gamma_V(r \mid r_{yj}^+) - \Gamma_V(r \mid r_{yj}^-)]\right\} \tag{13b}$$

where Z_0 is the characteristic free-space impedance.

3.4 Test functions

The last step of the resolution with a moment's method is the selection of a suitable test function. Previous work [11] has shown that the most adequate choice, compatible with the basis functions selected, is the use of unidimensional rectangular pulses. This actually means that the boundary condition, eqn. 1, is integrated along all the test segments, yielding

$$j\omega \int_{C_{xi}} A_x\, dx + V(r_{xi}^+) - V(r_{xi}^-) + Z_s \int_{C_{xi}} J_{sx}\, dx = \int_{C_{xi}} E_x^{(e)}\, dx = -V_{xi}^{(e)} \tag{14}$$

where C_{xi} is the x-directed test segment extending from r_{xi}^- to r_{xi}^+ and $V_{xi}^{(e)}$ is the excitation (impressed) voltage along the segment. A similar relationship is obtained for y-directed test segments. It is worth mentioning that this choice eliminates the need for computing field values near the edges, where field singularities can negatively affect the performances of the moment's method.

Eqns. 14 are well suited to a numerical treatment, as all derivatives have been removed. The integration of J_{sx} can be done easily by using expansion 6 with the result

$$\int_{C_{xi}} J_{sx}\, dx = \frac{a}{4b}[2I_{xi} + I_{xi+1} + I_{xi-1}] \simeq \frac{a}{b} I_{xi} \qquad (15)$$

The last approximation is valid for a reasonably smooth current distribution.

3.5 Matrix equation

Introducing expansions 13 into eqn. 14 yields the following matrix equation:

$$\begin{pmatrix} C^{xx} & C^{xy} \\ C^{yx} & C^{yy} \end{pmatrix} \begin{pmatrix} I_x \\ I_y \end{pmatrix} = \frac{1}{jZ_0} \begin{pmatrix} V_x^{(e)} \\ V_y^{(e)} \end{pmatrix} \qquad (16)$$

The elements in the submatrices are given by

$$C_{ij}^{xx} = \frac{1}{k_0 a k_0 b}[-\Gamma_V(r_{xi}^+ | r_{xj}^-) - \Gamma_V(r_{xi}^- | r_{xj}^+) + \Gamma_V(r_{xi}^+ | r_{xj}^+) + \Gamma_V(r_{xi}^- | r_{xj}^-)] - \frac{1}{k_0 b}\int_{C_{xi}} \Gamma_A^{xx}(r | r_{xj}) k_0\, d_x + j\frac{Z_s}{Z_0}\frac{a}{b}\delta_{ij}$$

$$i = 1 \ldots M, j = 1 \ldots M \qquad (17a)$$

$$C_{ij}^{xy} = \frac{1}{k_0 a k_0 b}[-\Gamma_V(r_{xi}^+ | r_{yj}^-) - \Gamma_V(r_{xi}^- | r_{yj}^+) + \Gamma_V(r_{xi}^+ | r_{yj}^+) + \Gamma_V(r_{xi}^- | r_{yj}^-)]$$

$$i = 1 \ldots M, j = 1 \ldots N \qquad (17b)$$

where δ_{ij} is the Kronecker delta. The expression for C_{ij}^{yy} is obtained by interchanging the couples (x, y), (a, b) and (M, N) within eqn. 17a. Finally, it is easily shown that $C_{ij}^{yx} = C_{ji}^{xy}$.

For distances $|r_{xi} - r_{xj}|$ much greater than the dimensions of a cell, the integrals in eqn. 17a can be replaced by

$$\int_{C_{xi}} \Gamma_A^{xx}(r | r_{xj}) k_0\, dx \simeq k_0 a \Gamma_A^{xx}(r_{xi} | r_{xj}) \qquad (18)$$

In principle, this approximation is not valid for short distances between cells. For these situations, however, the contribution of the vector potential to eqn. 17a is overshadowed by the one of the scalar potential, so that the approximation of eqn. 18 still suffices. As a matter of fact, eqn. 18 may be used everywhere but in the diagonal terms. This assumption was confirmed by extensive numerical tests.

A last point worth mentioning concerns the number of discrete Green's functions which must be calculated. For a rectangular patch with $m \times n$ charge cells, the number of matrix elements is $(M + N)^2$, with $M = (m - 1)n$ and $N = (n - 1)m$. When all the cells have identical sizes, only $m \times n$ values of Γ_V, M values of Γ_A^{xx} and N values of Γ_A^{yy} are needed in order to completely fill the matrix. This is the great advantage of using cells of equal size. It is generally more convenient to use a larger number of identical cells, rather than fewer cells of different sizes.

4 Numerical details

4.1 Interpolation among Green's functions

The evaluation of the matrix in eqn. 21 requires a large amount of computation. For a rectangular patch divided into 10×10 cells, the order of the matrix is 180, hence the number of elements in it is $180^2 = 32\,400$. Even when a simple 4×4 Gaussian quadrature is used to evaluate the discrete Green's functions, eqns. 16, the number of Sommerfeld integrals which should be evaluated would exceed half a million.

Fortunately, for a given structure these integrals only depend upon the distance from source to observer. It is thus possible to tabulate the integrals for a small number of distances, and then to interpolate between the tabulated values. The distances to be considered range from zero to the maximum linear dimension of the patch. Several interpolation schemes have been tried [11]. The best solution was obtained by separating the Green's functions according to eqn. 4, and then using a simple parabolic Lagrange interpolation for the regular part.

For a square patch with 10×10 cells, at frequencies for which the patch's length is less than a free-space wavelength, the error obtained when interpolating from 25 tabulated values is hardly noticeable: less than 0.5%, even though the computation time was reduced by a factor of 100!

4.2 Resolution of the linear system

The system of linear equations 16 is solved by standard Gaussian elimination. The $[C_{ij}]$ matrix is ill-conditioned, so that a careful evaluation of its elements is needed. This matrix is diagonal dominated, so that the accuracy requirements may be relaxed for the off-diagonal terms. The following approximation was therefore considered: the double numerical integration, eqn. 10, is replaced by its analytical approximation, eqn. 12, whenever the distance between cells i and j of a given element C_{ij} exceeds a certain critical distance D.

As the physical characteristics of the substrate itself do not significantly affect the numerical problem, the values $\varepsilon_r = 1$, $h \to \infty$ (isolated patch in free space) have been selected to determine the effect of the parameter D. A rectangular patch with 10×10 cells is again considered. When D becomes larger than the patch's diagonal ($D/a > 10\sqrt{2}$), no approximation is introduced, and the solution of the rigorous computation is obtained. When $D/a = 0$, the approximation is used everywhere but for the diagonal terms. The resulting relative error for the RMS value of the currents is about 25%. For $D/a = 4$, the error is reduced to 4%, and for $D/a = 8$, it further drops down to 0.1%. Hence, the use of this approximation can reduce considerably the length of computations, without significantly affecting the accuracy.

4.3 Relevance of surface waves and losses

The diagonal terms of eqn. 17a, which dominate the behaviour of the matrix, can be written as

$$C_{ii}^{xx} = \frac{2}{k_0 a k_0 b}[\Gamma_V(\mathbf{0} | \mathbf{0}) - \Gamma_V(e_x a | \mathbf{0})] - \frac{1}{k_0 b}\int_{-a/2}^{+a/2} \Gamma_A^{xx}(r | \mathbf{0}) k_0\, dx + j\frac{Z_s}{Z_0}\frac{a}{b} \qquad (19)$$

For electrically small cells, the main contribution to C_{ii}^{xx} comes from the self term $\Gamma_V(\mathbf{0} | \mathbf{0})$, which gives the scalar potential produced by a cell on its own centre. In a homogeneous lossless case, this self term has a negative imaginary part. But it has been shown in previous works [11, 14] that in the microstrip case the imaginary part of $\Gamma_V(\mathbf{0}|\mathbf{0})$ is positive, due to the presence of a surface wave. The dielectric losses will also contribute a small positive imaginary term, due to the presence of the complex permittivity in the denominator of eqn. 11. Finally, it is apparent from eqn. 19 that conductor losses further add a positive imaginary contribution.

Summarising the above, surface waves and losses considerably affect the imaginary part of the diagonal terms. Still, their effect may go unnoticed, as the real part, which is not affected, is larger by several orders of magnitude. Nevertheless, at resonance, the currents are in quadrature with the excitation, in which case they are mainly determined by the imaginary part of the moment's matrix. This means that losses and surface waves play a significant role at resonance.

4.4 Resonant frequencies and matrix condition

The roots of the complex determinant of the moment's matrix yield the resonant frequencies, which are, in general, complex (they correspond to an open radiating structure). On the real frequency axis, the determinant does not vanish, but goes through sharp minima at the points closest to the complex roots. These minima, detected with standard numerical techniques, provide the real resonant frequencies of the antenna. The condition number of the matrix, which is often obtained as a by-product of resolution techniques for linear systems may also be used to locate the resonances.

As was shown in the previous Section, in electrically small cells the scalar potential is predominant in the matrix elements. As a result, when four test segments form a square loop, a test along one of them is practically equivalent to a test along the remainder of the loop. In other words, some rows of the moment's matrix are almost linear combinations of three other rows. It is for this reason that the matrix is severely ill-conditioned. In some extreme situations, some resonances may actually be missed, the numerical value of the determinant being masked by numerical noise and round-off errors. A considerable improvement can be obtained by systematically replacing every test segment closing a loop by the complete loop. Eqn. 14 for a test segment is now replaced by the loop's equation

$$+j\omega \oint_\Gamma A_x \, dx + Z_s \oint_\Gamma J_{sx} \, dx = + \oint_\Gamma E_x^{(e)} \, dx \qquad (20)$$

which is independent of the scalar potential [18].

4.5 Excitation and input impedance

The column vector $V^{(e)}$ in eqn. 16 is obtained by integrating the tangential excitation electric field. The simplest kind of excitation is provided by a plane wave impinging on the patch. In the previous numerical tests, this excitation was used.

For transmitting antennas, more complex excitations have to be considered. The numerical problems encountered with a coaxial probe excitation will be described here. A rigorous model would require the introduction of a frill of magnetic currents $\boldsymbol{M}_s$ within the ground plane (Fig. 3A). In practice, however, simpler models may be used. The inner coaxial conductor carries a total current $I = 1A.$, supposed to be uniformly distributed on the surface, so that

$$\boldsymbol{J}_s = \boldsymbol{e}_z \frac{I}{2\pi r_c} \qquad (21)$$

where r_c is the radius of the inner conductor. The current then spreads radially across the patch, and several sophisticated attachment models have been devised to describe it [19]. A new, simpler model is introduced here, in which the excitation current spreads over a charge cell (Fig. 3A). The postulated current distribution is (Fig. 3B)

$$\boldsymbol{J}_s = \boldsymbol{e}_x \frac{I}{4b} \operatorname{sgn}(x)(1 - 2|x|/a) + \boldsymbol{e}_y \frac{I}{4a} \operatorname{sgn}(y)(1 - 2|y|/b) \qquad (22)$$

A

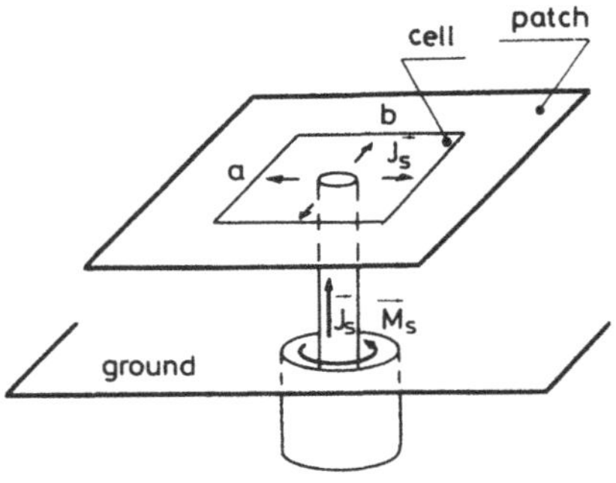

Fig. 3A *Coaxial-fed microstrip patch*

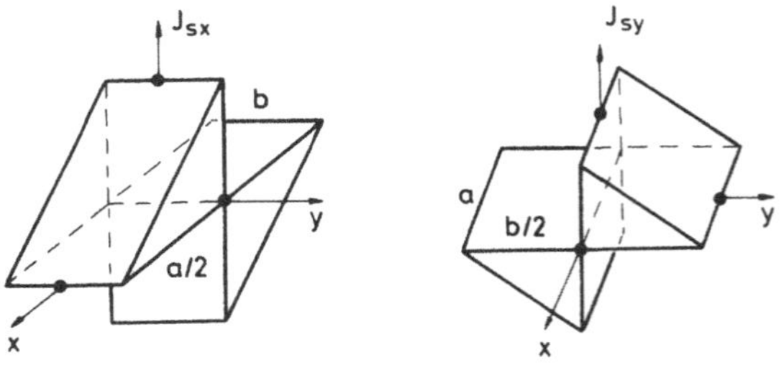

B

Fig. 3B *Electric surface current distribution associated with the coaxial excitation*

The surface current $\boldsymbol{J}_s$ thus defined is not a continuous function at the junction between coaxial line and patch. Still, the continuity equation is globally satisfied. The associated surface charge is constant over the spreading rectangle, having the value $\rho_s = I/j\omega ab$. Consequently, the total charge is $I/j\omega$, as required.

This model was developed to be compatible with the basis functions. For a coaxial excitation located at the centre of a charge cell, the excitation vector may be obtained from the matrix elements with little additional computation. When the surface currents have been determined over the whole patch, the antenna's input impedance is easily calculated as

$$Z_{IN} = -\frac{1}{I} \int_0^h (\boldsymbol{E}^{(e)} + \boldsymbol{E}^{(s)}) \cdot \boldsymbol{e}_z \, dz \qquad (23)$$

The field $\boldsymbol{E}^{(s)}$ results from the currents on the antenna J_{sx} and J_{sy}, whereas $\boldsymbol{E}^{(e)}$ is the excitation field, produced by the excitation currents (Fig. 3). Previous works [8] sometimes neglected $\boldsymbol{E}^{(e)}$ in the calculation of the input impedance. A somewhat artificial correction was then added to account for the 'inductive effect of the coaxial probe'. Even when $\boldsymbol{E}^{(e)}$ only represents a second-order contribution at resonances, it must be retained in the general theory, to obtain the correct input impedance at low frequencies, given by $Z_{in} = 1/(j\omega C_{stat})$ plus a small positive real term.

At low frequencies, in fact, the mixed potential integral

equation is formally identical to the scalar potential equation used to calculate microstrip capacitances [20]. This provides a useful way to check the computer implementation.

5 Results

5.1 Numerical convergence

As the number of charge cells m, n increases, the calculated values converge towards the true solution. The process was investigated for a rectangular patch having an aspect ratio of $B/A = \sqrt{2}$, excited by a normally incident plane wave with an x-polarised electric field. Relative errors were determined from comparison with extrapolated values (Fig. 4), for both the resonant frequency and the RMS current at the resonance. With 10×10 cells, the relative error on these two quantities is, respectively, 0.7% and 1.3% (Fig. 4). In this example, the number of cells n taken in the y-direction has little effect, as the transverse currents (along y) at resonance are quite small.

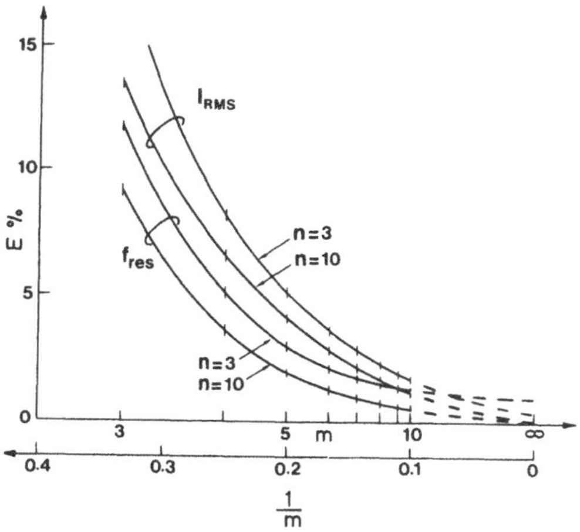

Fig. 4 *Relative error in the resonant frequency and the RMS value of the current at resonance as a function of the number of cells ($m \times n$)*

5.2 Rectangular patch

A rectangular patch of 60×40 mm was analysed theoretically and then measured. The substrate parameters are $\varepsilon_r = 4.34$, $\tan\,\delta = 2\ 10^{-3}$ and $h = 0.8$ mm. An effective conductivity $\sigma_e = \sigma_{Cu}/4$ was assumed, taking into account the surface roughness of the conductors. The number of cells taken is 9×6. To obtain adequate impedance levels at the first four resonances, the coaxial probe (a standard APC connector) was located at the centre of cell (2, 2). Computer-generated plots of the calculated surface currents are given in Fig. 5, at a low frequency (one half of the first resonant frequency), and at the first four resonances. The numerical values given correspond to the largest current value (longest arrow in every plot). The real part of $\boldsymbol{J}_s$ clearly depends on the position of the coaxial excitation, corresponding roughly to a total unit current spreading radially from the injection point. On the other hand, the imaginary part (in quadrature with the excitation current) is practically independent of the position of the excitation. This component is negligible out of the resonances, but becomes the dominant one at resonant frequencies. The current patterns are thus easily recognisable as those of the TM_{ij0} modes in the microstrip cavity. Fig. 6 provides the computed and the measured values of the input impedance near the two first resonances. A good agreement is observed in every case. Similar agreement was observed for the next two resonances and in the study of a more complicated structure, an L-shaped patch.

5.3 Slotted patch

To establish the accuracy limitations of the present approach, the slotted rectangular patch of Fig. 7 has been considered. This geometry provides a rather severe test, because a significant part of the structure is modelled with only one row of charge cells, over which transverse effects cannot be accounted for. For the slot dimensions of Fig. 7, two close resonant modes have been found at the frequencies of 1.28 and 1.32 GHz. The surface-current plots show that one mode is just the dominant mode, slightly perturbed by the slot, whereas the second one is an annular-like mode having dominant currents perpendicular to those of the first mode. Slotted patches may be used to generate circularly polarised radiation at some intermediate frequency, a fact which was recognised experimentally [21]. This work provides the first theoretical justification for these phenomena. Computed and measured values for the input impedance are presented in Fig. 7. The theoretical predictions are still qualitatively valid, but a shift in frequency and in impedance level is observed. A larger number of cells would be required to accurately study this structure. Also, the excitation point is here very close to the edge of the patch, so that a more sophisticated modelling of the excitation would be needed to describe the current distribution.

6 Radiation pattern

6.1 Asymptotic expressions for the radiated field

The dyadic Green's function formulation derived in Section 2 can still be used to determine the far field. As the fields are to be calculated far from the sources it is possible to define in a unambiguous way a dyadic Green's function associated with the electric field as

$$\bar{G}_E = -j\omega \bar{G}_A + \frac{1}{j\omega}\nabla\nabla G_V \tag{24}$$

Thus, G_E^{st} gives the s-component of the electric field created by a t-directed unit electric dipole located on the substrate.

The radiated field can be obtained by asymptotic evaluation of the Sommerfeld integrals appearing in eqn. 24. It is found [10, 11] that the far field is composed of a spherical and a cylindrical wave, currently termed spatial and surface wave. The surface wave is only relevant at grazing angles ($\theta \simeq \pi/2$). For a horizontal electric dipole, on a electrically thin substrate, the ratio between the power carried by the surface wave and the spatial wave is [22]

$$\eta = \frac{\text{Power surface wave}}{\text{Power spatial wave}} = \pi^2 \frac{(\varepsilon_r - 1)^3 h/\lambda_0}{\frac{2}{3}\varepsilon_r^2(\varepsilon_r - 1) + \frac{4}{15}\varepsilon_r} \tag{25}$$

The spatial wave is described by the spherical components G_E^{st} ($s = \theta, \phi, t = x, y$). Explicit expressions for these components can be found elsewhere [10, 11].

6.2 Radiation from a patch

In the previous Section the current density was obtained numerically as a set of discrete currents I_{xi} ($i = 1 \ldots M$) and I_{yj} ($j = 1 \ldots N$), each current related to a charge cell of dimensions a, b, with its centre defined by the vector

$$\boldsymbol{\rho}'_k = \boldsymbol{e}_x x_k + \boldsymbol{e}_y y_k \quad \text{with} \quad k = i, j \tag{26}$$

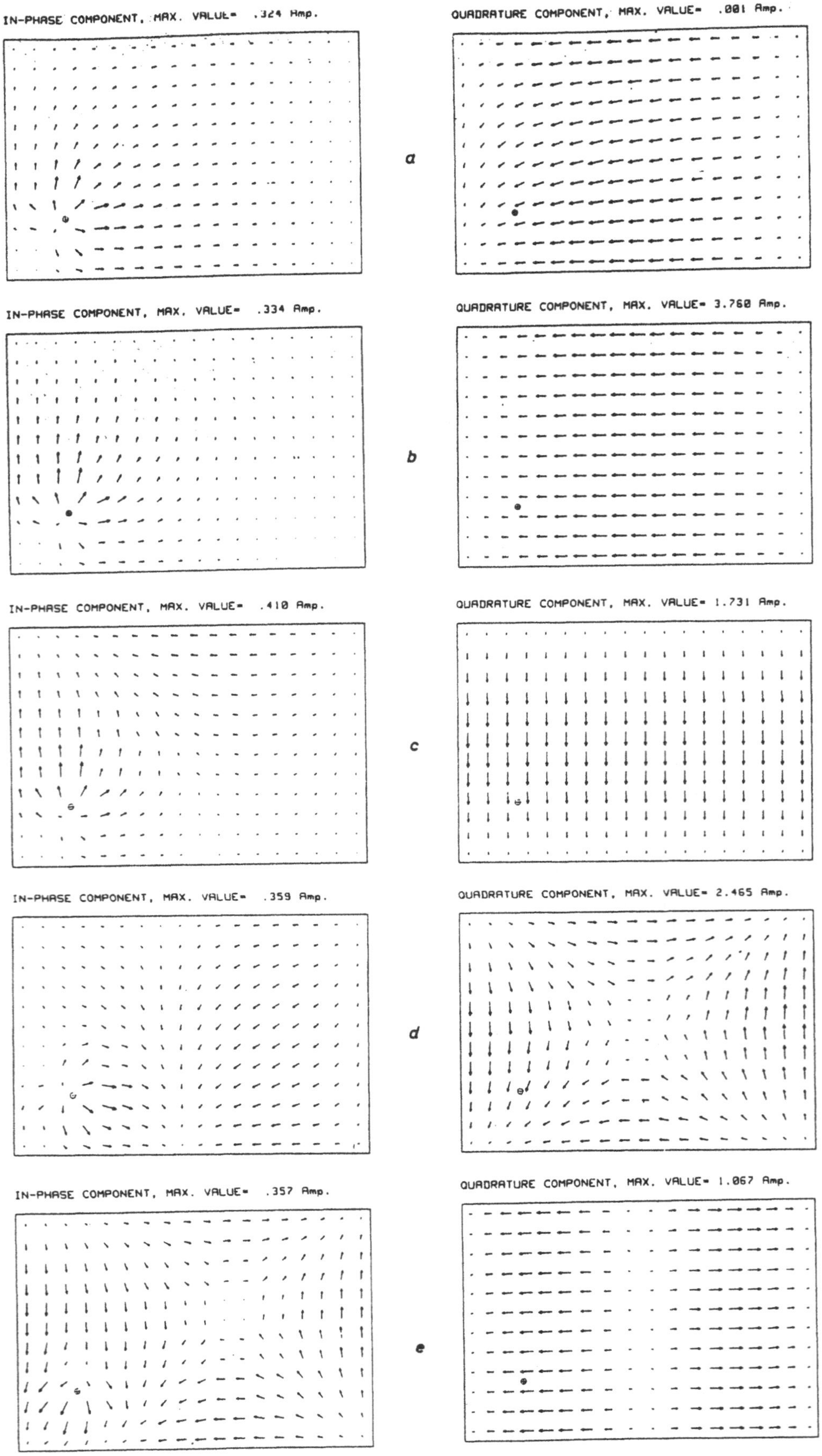

Fig. 5 *Real and imaginary parts of the surface current for a rectangular coaxial-fed microstrip antenna*

a Below resonance $f = f_{10}/2 = 0.603$ GHz
b TM_{10} resonance $f = f_{10} = 1.206$ GHz
c TM_{01} resonance $f = f_{01} = 1.783$ GHz
d TM_{11} resonance $f = f_{11} = 2.177$ GHz
e TM_{20} resonance $f = f_{20} = 2.405$ GHz

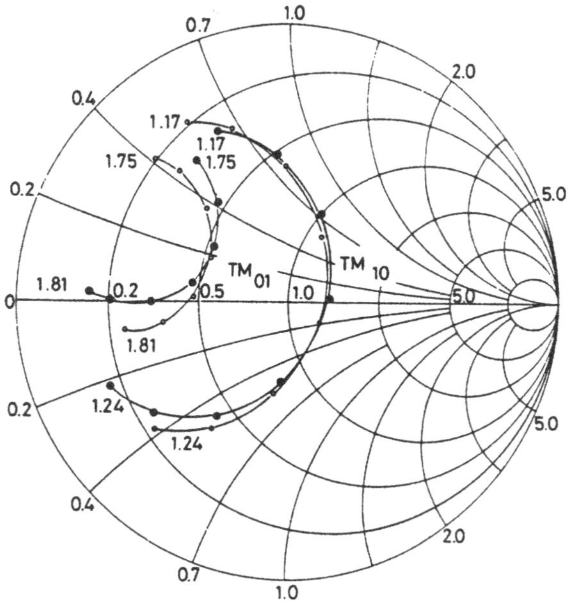

Fig. 6 *Input impedances near the TM_{10} and TM_{01} resonances*
Frequency increases clockwise by 0.01 GHz steps
○—○ theory
●—● measured

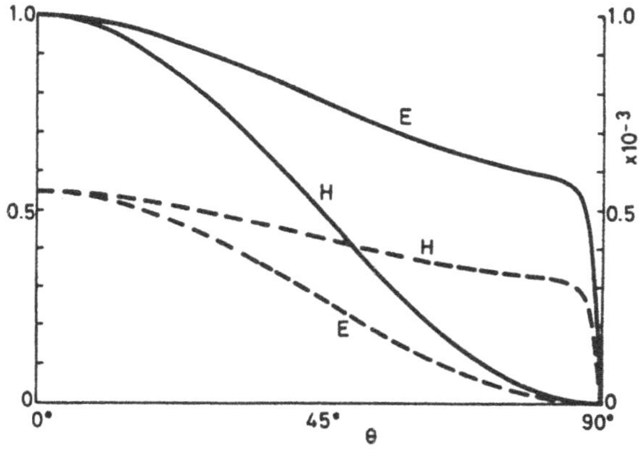

Fig. 8 *Theoretical predictions for the radiation pattern of a rectangular microstrip antenna*
E: E-plane ($\phi = 0°$)
H: H-plane ($\phi = 90°$)
——— copolar radiation (left vertical scale)
- - - - crosspolar radiation (right vertical scale)

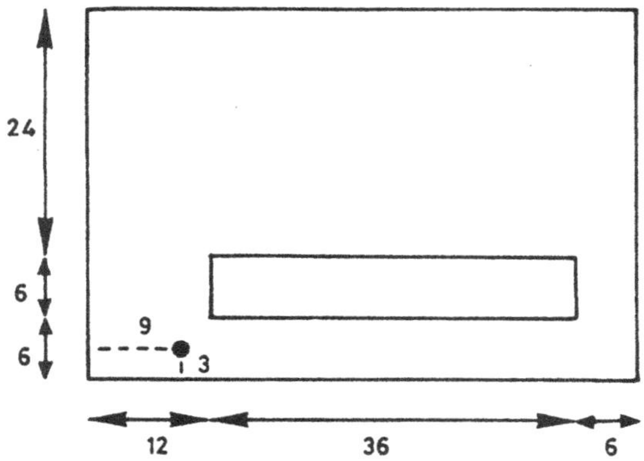

a

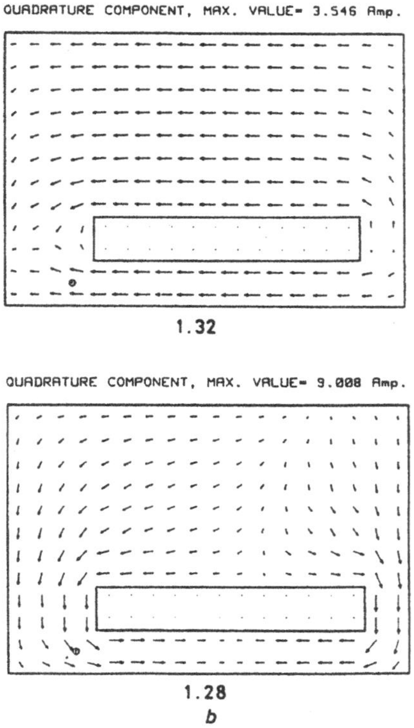

b

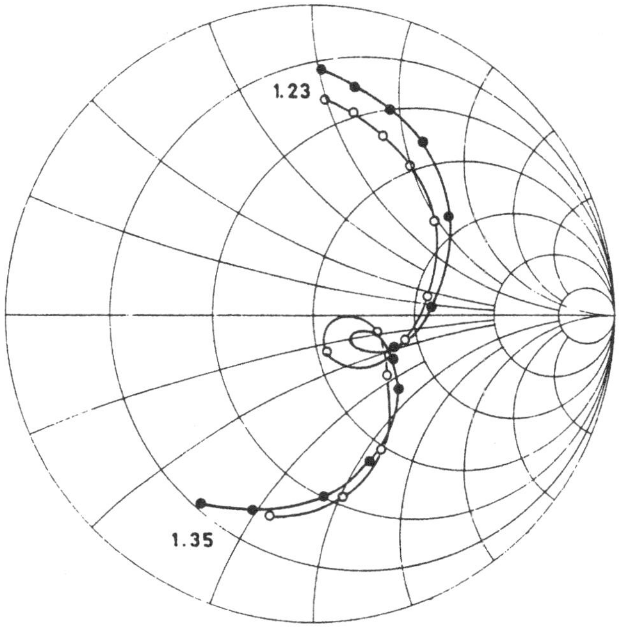

c

Fig. 7 *Slotted microstrip rectangular antenna $\varepsilon_r = 4.34$, $h = 0.8$ mm*
a Dimensions in mm. and coaxial location
b Surface current distribution at two nearby resonant frequencies (1.28 and 1.32 GHz)
c Input impedance in the 1.23–1.35 GHz band. Frequency increases clockwise by 0.01 GHz steps
○—○ theory
●—● measured

The patch is then replaced by an array of Hertz dipoles, for which the radiated field is given by

$$E_\alpha = G_E^{\alpha x}(\mathbf{r}\,|\,\mathbf{0}) \sum_{i=1}^{M} aI_{xi} \exp(jk_0 \mathbf{e}_r \cdot \boldsymbol{\rho}_i')$$

$$+ G_E^{\alpha y}(\mathbf{r}\,|\,\mathbf{0}) \sum_{i=1}^{N} bI_{yj} \exp(jk_0 \mathbf{e}_r \cdot \boldsymbol{\rho}_j'), \; \alpha = \theta, \phi \quad (27)$$

The radiated power density can then be determined, yielding the radiation pattern and related antenna properties.

As a typical example, the rectangular antenna of Section 5 was considered, at its first resonant frequency (1.206 GHz). The radiation patterns obtained, respectively, in the E-plane ($\phi = 0°$) and the H-plane ($\phi = 90°$) are represented in Fig. 8. In the E-plane, the radiation pattern is strongly affected by the substrate: radiation remains large even close to the substrate ($\theta \simeq \pi/2$). On the other hand, the H-plane pattern resembles the one of a half-wavelength dipole in free space. The polarisation is practically linear, with an electric field directed along x. The cross-polarisation component (dotted lines) is mostly due to currents in phase with the coaxial excitation.

7 Conclusion

The integral equation technique is a powerful tool for the analysis of planar microstrip antennas. Combined with the Green's function treatment introduced in a previous work [14], it provides a flexible and accurate numerical algorithm able to handle arbitrary microstrip shapes at any frequency and for any substrate parameters.

Standard feeds like coaxial probes and microstrip lines can be easily included in the model. In addition to the examples presented, other practical devices like coupled and short-circuited patches, parasitic elements and multiple-fed antennas can be studied without added complexity. The proposed model provides a good quantitative description of the electric surface currents on the patch. Hence, accurate theoretical predictions can be made for related quantities such as input impedance, near field values and polarisation purity.

In the theoretical developments, substrate and ground plane are assumed to be infinite. This departure from the real situation is not a drawback as far as near-field quantities (resonant frequencies, input impedances) are considered. On the other hand, the theoretical predictions for the radiation pattern can be considerably modified by a substrate having finite dimensions. As a surface wave reaches the antenna's edge, it is scattered, producing both a reflected surface wave and a radiated wave. The presence of secondary sources of radiation on the dielectric edges proved most troublesome in practice, as it contributes to secondary lobes and to cross-polarised radiation. Moreover, the spatial wave itself can no longer be considered separately from the surface wave.

The present technique allows one to determine the amplitude of the incident surface wave. Further study should look at the scattering of surface waves by edges and the resulting effect on the spatial radiation pattern.

8 References

1 BAHL, I.J., and BHARTIA, P.: 'Microstrip antennas' (Artech House, Dedham, MA, USA, 1980)

2 JAMES, J.R., HALL, P.S., and WOOD, C.: 'Microstrip antenna theory and design' (Peter Peregrinus, London, 1981)

3 DUBOST, G.: 'Flat radiating dipoles and application to arrays'. Research Studies Press, (John Wiley), New York, 1981)

4 LIER, L.: 'Improved formulas for input impedance of coax-fed microstrip patch antennas', *IEE Proc. H, Microwaves, Opt. & Antennas*, 1982, **129**, pp. 161–164

5 RICHARDS, W.F., LO, Y.T., and HARRISON, D.D.: 'An improved theory for microstrip antennas and applications', *IEEE Trans.*, 1981, **AP-29**, pp. 38–46

6 GUPTA, K.C., and SHARMA, P.C.: 'Segmentation and desegmentation techniques for the analysis of planar microstrip antennas', IEEE AP-S International Symposium, Los Angeles, 1981, pp. 19

7 ARAKI, K., and ITOH, T.: 'Hankel transform domain analysis of open circular microstrip radiating structures', *IEEE Trans.*, 1981, **AP-29**, pp. 84–89

8 POZAR, D.M.: 'Input impedance and mutual coupling of rectangular microstrip antennas', *ibid.*, 1982, **AP-30**, pp. 1191–1196

9 BAYLEY, M.C., and DESHPANDE, M.D.: 'Integral equations formulation of microstrip antennas', *ibid.*, 1982, **AP-30**, pp. 651–655

10 UZUNOGLU, N.K., ALEXOPOULOS, N.G., and FIKIORIS, J.G.: 'Radiation properties of microstrip dipoles'. *ibid.*, 1979, **AP-27**, pp. 853–858

11 MOSIG, J.R., and GARDIOL, F.E.: 'A dynamical radiation model for microstrip structures', *In* HAWKES, P. (Ed): 'Advances in electronics and electron physics' (Academic Press, New York, 1982), pp. 139–237

12 CARVER, K.R., and MINK, J.W.: 'Microstrip antenna technology', *IEEE Trans.*, 1981, **AP-29**, pp. 2–24

13 MILLER, E.K., and DEADRICK, F.: 'Some computational aspects of thin wire modeling'. *In* MITTRA, R. (Ed.): 'Numerical and asymptotical techniques in electromagnetics' (Springer Verlag, New York, 1975)

14 MOSIG, J.R., and GARDIOL, F.E.: 'Analytical and numerical techniques in the Green's function treatment of microstrip antennas and scatterers', *IEE Proc. Microwaves, Opt. & Antennas*, 1983, **130**, pp. 175–182

15 GLISSON, A.W., and WILTON, D.R.: 'Simple and efficient numerical methods for problems of electromagnetic radiation and scattering from surfaces', *IEEE Trans.*, 1980, **AP-28**, pp. 593–603

16 HARRINGTON, R.F.: 'Field computation by moment methods' (McMillan, New York, 1968)

17 MOSIG, J.R., and GARDIOL, F.E.: 'Dielectric losses, ohmic losses and surface wave effects in microstrip antennas', Int. U.R.S.I. Symposium, Santiago de Compostela, August 1983, pp. 425–428

18 WILTON, D.R., and GLISSON, A.W.: 'On improving the stability of the electric field integral equation at low frequency'. IEEE AP-S International Symposium., Los Angeles, June 1981

19 NEWMAN, E.M., and POZAR, D.M.: 'Electromagnetic modeling of composite wire and surface geometry'. *IEEE Trans.*, 1978, **AP-26**, pp. 784–789

20 SILVESTER, P., and BENEDEK, P.: 'Electrostatics of the microstrip revisited'. *ibid.*, 1972, **MTT-20**, pp. 756–758

21 KERR, J.L.: 'Microstrip polarization techniques'. Proceedings Antenna Application Symposium, Allerton Park, Illinois, USA, April 1977

22 MOSIG, J.R., and GARDIOL, F.E.: 'Radiation of an arbitrarily shaped microstrip antenna', *Ann. Telecommun.*, 1985, **40**, pp. 181–189

A Reciprocity Method of Analysis for Printed Slot and Slot-Coupled Microstrip Antennas

DAVID M. POZAR, MEMBER, IEEE

Abstract—**A method is presented for the analysis of slot-type discontinuities in microstripline. The approach is based on the reciprocity theorem and uses the exact Green's functions for the grounded dielectric slab in a moment method solution for the unknown antenna currents. The method is applied to two specific geometries: a radiating slot in the ground plane of a microstripline, and an aperture coupled microstrip patch antenna. Results for antenna impedance are compared with measurements, and far-zone patterns are calculated. The method is shown to be quite versatile, and should find application to related problems.**

I. Introduction

MILLIMETER WAVE printed antennas can take many forms, including microstrip patch elements, slot elements, and a variety of proximity coupled printed radiators [1]–[4]. The microstripline-fed printed slot [3], [5], and the aperture-coupled microstrip patch [2] are examples of this latter type, and may be useful in certain planar array applications. The present paper describes a method of analysis that can be applied to these geometries, as well as related configurations. The theory is described, and impedance results for the microstripline-fed printed slot antenna and the aperture-coupled microstrip patch antenna are given and compared with measured data.

The method is derived in Section II using the reciprocity theorem in a manner similar to the analysis of waveguide slot elements [6]. The exact Green's functions in spectral domain form are used to find the necessary field components from electric and magnetic currents in the presence of a dielectric slab. Expressions are derived for the amplitudes of reflected and transmitted waves on the microstrip line, and an equivalent circuit representing the slot discontinuity is found. In Sections III and IV, the basic method is extended to moment method solutions for a slot antenna with a number of expansion modes in the slot, and for an aperture coupled patch antenna with a number of expansion modes on the patch. Results are compared with impedance measurements, and far-zone patterns are calculated.

The combination of the reciprocity analysis and a moment method solution using the exact Green's functions for the planar structure results in a very versatile technique that should find application in a number of printed antenna and planar circuit problems. The method is very similar to that used in [4], for coupling of printed dipoles to a microstripline. The method avoids the more "brute-force" approach of modeling the actual (nonuniform) currents on a feed line, as was done in [7], [8], although it could be argued that the present method is less rigorous than that of [7], [8], as it does not include the existence of higher order modes on the microstrip feed line. The utility of any solution, however, is determined by the accuracy of the results, as well as its simplicity.

II. Derivation of the Basic Method

The basic method will be derived here for the problem of a microstripline-fed slot with one piecewise sinusoidal mode representing the aperture field. The following section will generalize this method to a full moment method solution using a number of expansion modes for the slot field, and Section IV will extend it to the aperture coupled patch geometry.

The geometry of a microstripline-fed printed slot is shown in Fig. 1. The microstripline is assumed to be infinitely long, and propagating a quasi-transverse electromagnetic (TEM) mode, with transverse modal fields given by

$$\bar{e}(y, z)=e_y\hat{y}+e_z\hat{z}, \tag{1}$$

$$\bar{h}(y, z)=h_y\hat{y}+h_z\hat{z}. \tag{2}$$

These fields are assumed to be normalized so that

$$\int_{y=-\infty}^{\infty}\int_{z=0}^{\infty} \bar{e}\times\bar{h}\cdot\hat{x}\, dy\, dz=1. \tag{3}$$

The propagating microstripline fields are then

$$\bar{E}^{\pm}=\bar{e}e^{\pm j\beta x}, \tag{4}$$

$$\bar{H}^{\pm}=\pm\bar{h}e^{\pm j\beta x}, \tag{5}$$

where β is the propagation constant of the line. The fields ($\bar{e}$, $\bar{h}$) and the propagation constant β are found from the Green's function for an electric current element on a grounded dielectric slab [7] (as given in the Appendix).

Now if the slot discontinuity is centered at $x = 0$, the total microstrip line fields can be written as

$$\bar{E}=\begin{cases}\bar{E}^{+}+R\bar{E}^{-}, & \text{for } x<0\\ T\bar{E}^{+}, & \text{for } x>0\end{cases} \tag{6}$$

$$\bar{H}=\begin{cases}\bar{H}^{+}+R\bar{H}^{-}, & \text{for } x<0\\ T\bar{H}^{+}, & \text{for } x>0\end{cases} \tag{7}$$

Reprinted from *IEEE Trans. Antennas Propaga.*, vol. AP-34, no. 12, pp. 1439–1446, Dec. 1986.

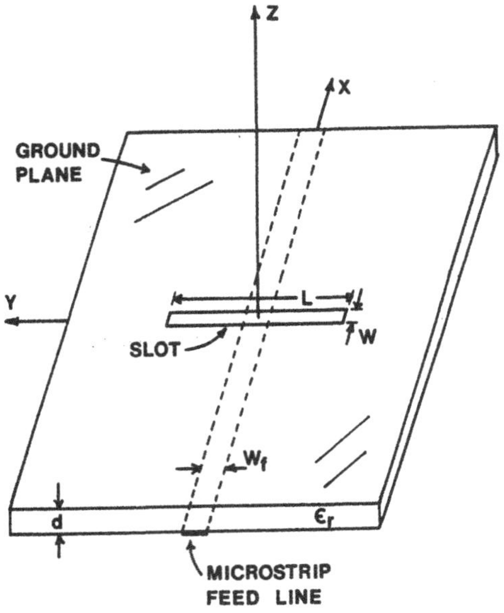

Fig. 1. Geometry of a printed slot in the ground plane of an infinite microstrip line.

where R and T are the voltage reflection and transmission coefficients on the line, respectively. Applying the reciprocity theorem to the total fields $\bar{E}$, $\bar{H}$ and the positive traveling wave fields $\bar{E}^+$, $\bar{H}^+$ gives

$$\int_S \bar{E}\times\bar{H}^+ \cdot d\bar{s}=\int_S \bar{E}^+\times\bar{H}\cdot d\bar{s}, \tag{8}$$

where S is a closed surface consisting of three pieces, as shown in Fig. 2:

$$S=S_0+S_a+S_w,$$

where

S_0 = the effective cross section of the microstripline ($-\infty < y < \infty;\ 0 \leqq z < \infty$)
S_a = the aperture surface
S_w = the "walls" of the microstripline ($y\rightarrow \pm\infty$, $z\rightarrow \infty$, $z = 0$).

On S_w, $\hat{n}\times\bar{E}=\hat{n}\times\bar{E}^+=0$, so the contributions to the integrals in (8) from this portion of the surface S are zero. On S_a, $\hat{n}\times\bar{E}^+=0$, but $\hat{n}\times\bar{E}=V_0\hat{n}\times\hat{x}e_x^a$, where $V_0e_x^a$ is the unknown aperture field which, for example, may be taken as a piecewise sinusoidal (PWS) mode of the form,

$$V_0e_x^a=V_0\frac{\sin k_e(L/2-|y|)}{W\sin k_eL/2},\qquad \text{for } |x|<w/2,\quad |y|<L/2. \tag{9}$$

In (9), k_e is the effective wavenumber of the PWS mode; a good choice is the average of the wavenumbers in the two regions adjacent to the slot, i.e., $k_e=k_0\sqrt{(\epsilon_r+1)/2}$. Equation (8) can then be evaluated to give

$$R=\frac{V_0}{2}\int_{S_a} e_x^a(x,\ y)h_y(x,\ y)\ ds=\frac{V_0}{2}\,\Delta v, \tag{10}$$

where

$$\Delta v=\int_{S_a} e_x^a(x,\ y)h_y(x,\ y)\ ds. \tag{11}$$

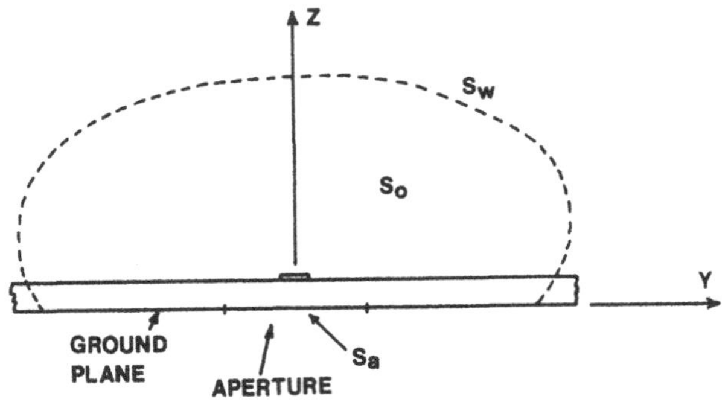

Fig. 2. The closed surface $S=S_0+S_a+S_w$ used in the reciprocity analysis.

(Note: an $e^{\pm j\beta x}$ term could be included in the integrals of (10), (11) to account for propagation phase shift across the width of the slot, but for narrow slots this effect is negligible.)

Another application of the reciprocity theorem of the form,

$$\int_s \bar{E}\times\bar{H}^- \cdot d\bar{s}=\int_s \bar{E}^-\times\bar{H}\cdot d\bar{s}, \tag{12}$$

results in the following expression for T:

$$T=1-\frac{V_0}{2}\int_{S_a} e_x^a(x,\ y)h_y(x,\ y)\ ds=1-R. \tag{13}$$

At this point the two applications of the reciprocity theorem (similar to conserving reaction) have led to two equations for the three unknowns R, T, and V_0 (the unknown amplitude of the aperture field). The required third equation comes from enforcing the continuity of H_y across the aperture:

$$H_y^e=H_y^f+H_y^i, \tag{14}$$

where

H_y^e = exterior field ($z<0$) due to $V_0e_x^a$
H_y^i = interior field ($z>0$) due to $V_0e_x^a$
H_y^f = interior field ($z>0$) due to feed line modes.

Now, at $x=0^-$ (or $x=0^+$, since $1-R=T$),

$$H_y^f=(1-R)h_y. \tag{15}$$

Now define a Green's function G_{yy}^{HM} to account for the H_y fields on both sides of the aperture ($z=0$) due to a $\hat{y}$ magnetic aperture current:

$$H_y^e-H_y^i=V_0\int_{S_a} G_{yy}^{HM}(x,\ y;\ x_0,\ y_0)\cdot e_x^a(x_0,\ y_0)\ ds_0. \tag{16}$$

(The spectral domain expression for G_{yy}^{HM} is given in the Appendix.) Combining (14), (15), and (16) gives

$$V_0\int_{S_a} G_{yy}^{HM}(x,\ y;\ x_0,\ y_0)e_x^a(x_0,\ y_0)\ ds_0=(1-R)h_y(x,\ y),\qquad \text{for } x,\ y\in S_a. \tag{17}$$

This equation can be enforced in a least-mean-square sense

over the aperture by multiplying by e_x^a and integrating over S_a:

$$V_0 \int_{S_a} \int_{S_a} e_x^a(x, y) G_{yy}^{HM}(x, y; x_0, y_0) e_x^a(x_0, y_0) \, ds \, ds_0$$
$$= (1-R) \int_{S_a} e_x^a(x, y) h_y(x, y) \, ds, \quad (18)$$

or,

$$V_0 Y^e = \Delta v(1-R), \quad (19)$$

where Y^e is the external admittance of the slot defined as

$$Y^e = \int_{S_a} \int_{S_a} e_x^a(x, y) G_{yy}^{HM}(x, y; x_0, y_0) \cdot e_x^a(x_0, y_0) \, ds \, ds_0. \quad (20)$$

Expressions for the three unknowns can now be written in terms of Y^e and Δv as follows:

$$V_0 = \frac{2\Delta v}{\Delta v^2 + 2Y^e} \quad (21)$$

$$R = \frac{\Delta v^2}{\Delta v^2 + 2Y^e} \quad (22)$$

$$T = \frac{2Y^e}{\Delta v^2 + 2Y^e} = 1 - R. \quad (23)$$

The above result that $T = 1 - R$ implies that the slot discontinuity appears as a simple series impedance Z to the microstripline, as shown in Fig. 3. (Note that this equivalence is not assumed *a priori,* but is a consequence of the analysis.) This series impedance Z can be found as

$$Z = Z_c \frac{2R}{1-R} = Z_c \frac{\Delta v^2}{Y^e}, \quad (24)$$

where Z_c is the characteristic impedance of the microstripline. The equivalent circuit provides a very convenient way to apply the results of this analysis, as transmission line theory can be used to account for the presence of tuning stubs and other external circuitry.

III. Moment Method Solution for the Microstripline-Fed Slot

The geometry of a slot in the ground plane of a microstripline is shown in Fig. 1. In practice such a slot is usually resonant, so a one-mode approximation to the slot fields of the form of (9) may not be a sufficient approximation (although convergence checks using many modes show that the one mode approximation is actually quite good). This section, then, generalizes the basic method of Section II to employ a full moment method solution for the aperture distribution, and compares the results with measured data.

Let the aperture field e_x^a be expanded in a set of N PWS modes:

$$e_x^a(x, y) = \sum_{n=1}^{N} V_n e_{xn}^a(x, y) = \sum_{n=1}^{N} V_n \frac{1}{W} f_p(y - y_n), \quad (25)$$

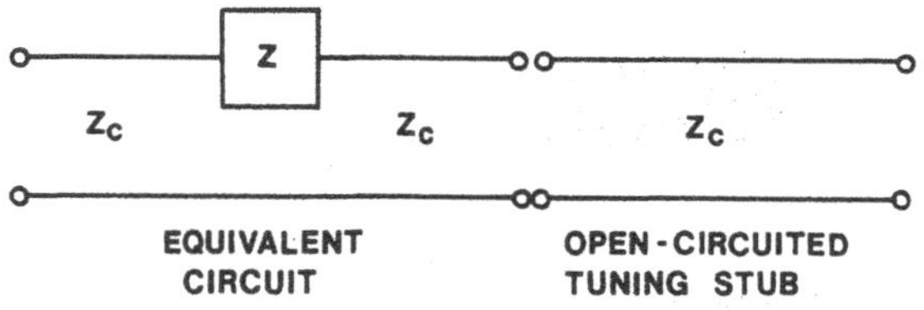

Fig. 3. The equivalent circuit of a slot in the ground plane of a microstrip line.

where V_n is the unknown mode coefficient, f_p is a PWS mode as defined in (57) of the Appendix, and y_n is the center point of the nth expansion mode:

$$y_n = -L/2 + (n+1)h, \quad (26)$$

where $h = L/(N+1)$ is the PWS mode half-length. An admittance matrix $[Y^e]$ can be defined for the aperture with elements

$$Y_{mn}^e = \int_{S_a} \int_{S_a} e_{xm}^a(x, y) G_{yy}^{HM}(x, y; x_0, y_0) \cdot e_{xn}^a(x_0, y_0) \, ds \, ds_0. \quad (27)$$

Using the results listed in the Appendix, (27) can be written in spectral form as

$$Y_{mn}^e = \frac{1}{4\pi^2} \iint_{-\infty}^{\infty} F_u^2(k_x) \tilde{G}_{yy}^{HM}(k_x, k_y) F_p^2(k_y) \cdot \cos k_y(y_m - y_n) \, dk_x \, dk_y, \quad (28)$$

and the integrations done numerically as described in [9]. By extension of (10), the reflection coefficient on the microstripline can be expressed as

$$R = \frac{1}{2} \sum_{n=1}^{N} V_n \int_{S_a} e_{xn}^a(x, y) h_y(x, y) \, ds, \quad (29)$$

where h_y is the normalized magnetic field of the microstripline. A discontinuity voltage vector $[\Delta v]$ can then be defined (similar to (11)) with elements given by

$$\Delta v_n = \int_{S_a} e_{xn}^a(x, y) h_y(x, y) \, ds$$
$$= \frac{1}{2\pi\sqrt{Z_c}} \int_{-\infty}^{\infty} F_u(k_y) \tilde{G}_{yx}^{HJ}(k_x = -\beta, k_y) F_p(k_y) \cdot \cos k_y y_n \, dk_y. \quad (30)$$

The width to be used in F_u in (30) is W_f, the width of the feed line. At this point an offset-fed slot (feed line not centered in aperture) can be easily treated by replacing y_n in (30) by $y_n - y_{0s}$, where y_{0s} is the offset distance of the feed line from the center of the slot.

The boundary condition that corresponds to enforcing continuity of H_y through the aperture can then be written in matrix form as

$$[Y^e][V] = (1-R)[\Delta v], \quad (31)$$

which corresponds to (19). The vector of expansion mode amplitudes can be found as

$$[V]=\left\{[Y^e]+\frac{1}{2}[\Delta v][\Delta v]^t\right\}^{-1}[\Delta v], \tag{32}$$

and the reflection coefficient computed from

$$R=\frac{1}{2}[V]^t[\Delta v]. \tag{33}$$

The equivalent series impedance Z of the slot can then be found using R in (24).

The above solution was carried out for a number of slot geometries. A necessary (but not sufficient) check is to compute the admittance matrix elements of (28) for a slot in a substrate with $\epsilon_r = 1$, and compare with the results of a PWS mode expansion on a free-space dipole of appropriate equivalent radius [10]. Babinet's principle relates the dipole impedance to the slot admittance.

Fig. 4 shows a comparison with the measured series impedance Z of a printed slot antenna. Initially, the method described in [3] of measuring S_{12} and calculating the series impedance Z was tried, but this was found to be very sensitive to errors. A more reliable measurement was obtained by terminating one port of the microstrip feed line with a matched load and measuring the input impedance. The normalized series impedance $\bar{Z} = \bar{R} + j\bar{X}$ is then $\bar{Z} = Z_{in}/50 - 1$, for a 50 Ω characteristic line impedance. The measurements were made with an HP8510 network analyzer.

The calculations were made with three PWS modes in the aperture, and the edge condition was applied to both the y-distribution of current on the feed line and the x-distribution of the aperture field. The solution was quite stable, in terms of convergence. Results using one PWS mode were very close to those using three or more modes, and the edge condition had only a small effect. The calculations are in reasonable agreement with measurements.

It is interesting to see the effect of a tuning stub on this slot antenna, as a very strong loading effect occurs. As can be seen from the data of Fig. 4, the resonant frequency of the slot when fed by an infinite microstrip line is about 3.0 GHz. There is, however, a very severe impedance mismatch at this frequency. A stub-tuned slot of the same dimensions was measured [11] to have a resonant frequency of about 2.5 GHz, and a perfect impedance match was obtained at this frequency. This shift in resonant frequency of 17 percent is explained by the data of Fig. 4, where it is seen that a normalized input resistance of about 1 is obtained at 2.5 GHz. The normalized reactance at this frequency is about $j4$, and is cancelled by the open-circuited tuning stub, which is about 0.04 λ_g long at this frequency.

Fig. 5 shows the calculated E- and H-plane far-field patterns of the slot antenna, at a frequency of 2.5 GHz (stub tuned). The far-fields of the slot are calculated from the stationary phase evaluation of the Green's functions for the field components of the slot. It is seen that the presence of the thin dielectric layer has no significant effect on the radiation patterns of the slot, and that the radiation is bidirectional. This

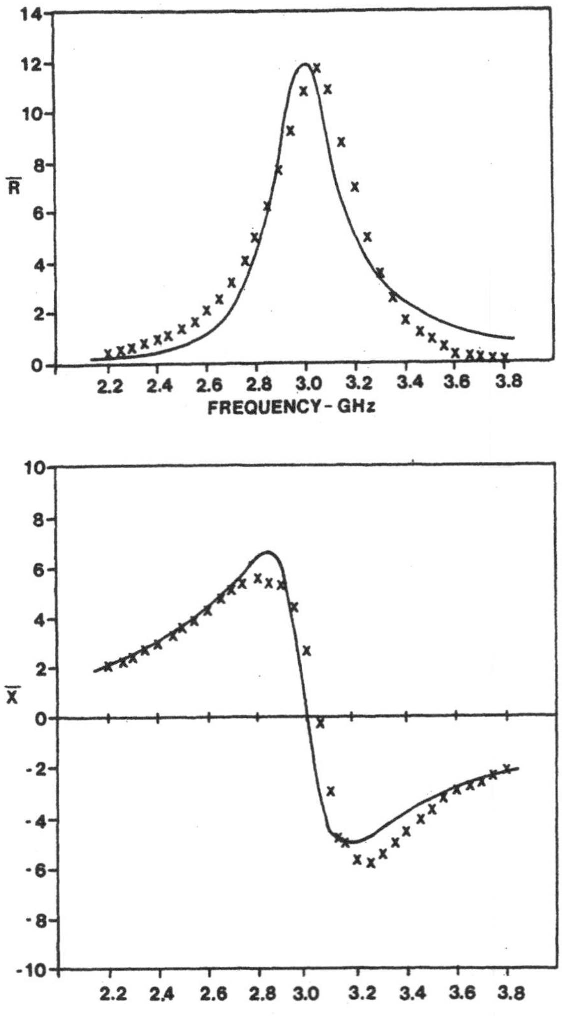

Fig. 4. Measured (XXX) and calculated (——) normalized equivalent series impedance of a slot in the ground plane of a microstripline. $\epsilon_r = 2.20$, $d = 0.16$ cm, $L = 4.02$ cm, $W = 0.07$ cm, $W_f = 0.50$ cm.

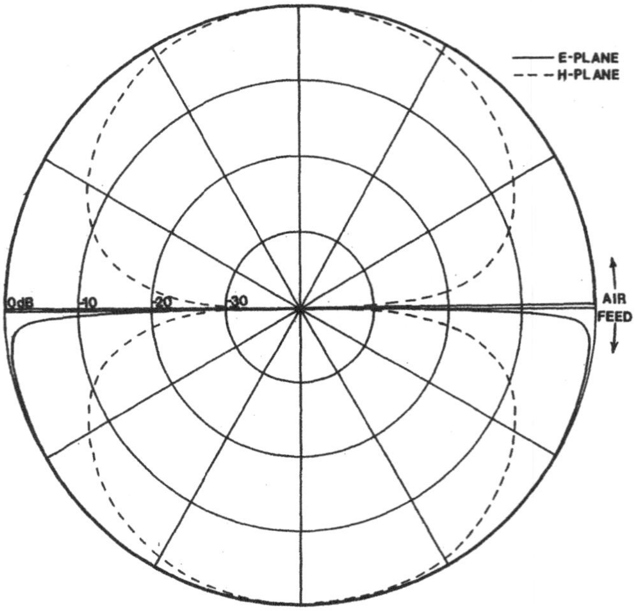

Fig. 5. E- and H-plane far-field plots of the microstrip-fed slot antenna (stub-tuned) of Fig. 4.

bidirectional property is not desirable in most array applications, so a ground plane spaced $\lambda_0/4$ away from the array plane is often used to achieve a unidirectional pattern [5]. This procedure works well for broadside arrays, but may lead to excitation of parallel plate waveguide modes in a scanning array.

IV. Moment Method Solution for the Aperture Coupled Patch Antenna

The geometry of the aperture coupled microstrip patch antenna is shown in Fig. 6; this configuration is similar to the printed slot antenna shown in Fig. 1, except that an additional substrate with a microstrip patch element is placed over the slot on the ground plane side. The slot is smaller than resonant size, so most radiation occurs from the resonant patch element. As discussed in [2], this configuration has some interesting features when used in a monolithic phased array application.

Since the coupling aperture is electrically small, a single PWS mode is assumed to be adequate to represent the field. The analysis of Section II is then modified to account for the presence of the patch as seen by the aperture. The unknown currents on the patch are expanded in a set of entire domain functions, and the additional boundary condition that $E_{\tan} = 0$ on the patch is enforced. The coupling of the slot to the feed line is the same as in Section II, with the self-admittance Y^e of the slot replaced with $Y^e + Y^a$, where Y^a is the admittance of the slot looking at the patch antenna. It is assumed that the Q of the patch is high enough so that only x-directed currents are necessary—an assumption validated in [9].

Let E_x^i be the "incident" field at the patch due to the equivalent magnetic current source $\bar{M}_s = \hat{y}e_x^a$ on the aperture, and let the patch current J_{sx} be expanded in a set of N entire domain modes for the x-variation and uniform (pulse) modes for the y-variation:

$$J_{sx} = \sum_{n=1}^{N} I_n f_s^n(x) f_u(y). \tag{34}$$

Enforcing the boundary condition that E_x must vanish on the patch surface leads to the following:

$$E_x^i = -\int_{S_p} \sum_{n=1}^{N} I_n f_s^n(x_0) f_u(y_0)\, ds, \tag{35}$$

where S_p denotes the patch surface. Weighting of this equation with the same functions as the expansion modes gives, in matrix form,

$$[V] = [Z][I], \tag{36}$$

where $[I]$ is a column vector of current coefficients I_n, $[Z]$ is the patch impedance matrix with elements given by

$$Z_{mn} = -\int_{S_p}\int_{S_p} f_s^m(x) f_s^n(x_0) f_u(y) \cdot f_u(y_0) G_{xx}^{EJ}(x, y; x_0, y_0)\, ds\, ds_0, \tag{37}$$

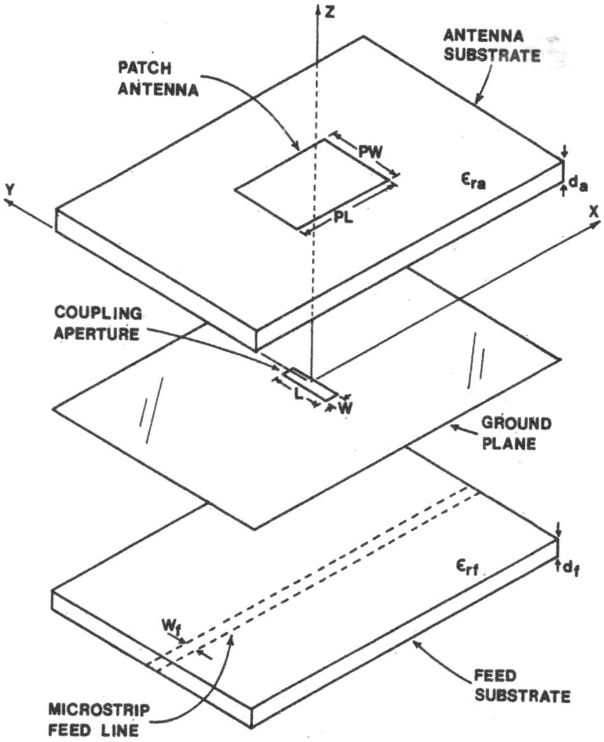

Fig. 6. Geometry of an aperture coupled microstrip patch antenna.

and $[V]$ is a voltage vector with elements given by

$$V_n = \int_{S_p} E_x^i f_s^n(x) f_u(y)\, ds = \int_{S_p}\int_{S_a} e_x^a(x_0, y_0) G_{xy}^{EM}(x, y; x_0, y_0) f_s^n(x) f_u(y)\, ds\, ds_0. \tag{38}$$

These expressions can be written in the following spectral domain form:

$$Z_{mn} = \frac{-1}{4\pi^2} \iint_{-\infty}^{\infty} F_s^m(k_x) F_s^n(k_x) F_u^2(k_y) \cdot \tilde{G}_{xx}^{EJ}(k_x, k_y)\, dk_x\, dk_y \tag{39}$$

$$V_n = \frac{1}{4\pi^2} \iint_{-\infty}^{\infty} F_s^n(k_x) F_p(k_x) F_u^2(k_y) \cdot \tilde{G}_{xy}^{EM}(k_x, k_y)\, dk_x\, dk_y. \tag{40}$$

The patch contribution to the aperture admittance seen by the slot is then

$$Y^a = [V]^t [Z]^{-1} [V], \tag{41}$$

and the equivalent series impedance seen by the microstrip feed line is, from (24),

$$Z = Z_c \frac{\Delta v^2}{Y^e + Y^a}, \tag{42}$$

where Δv is given by (30).

The above analysis is general enough to accommodate different substrate permittivities and thicknesses for the separate antenna and feed substrates. A coupling aperture that is offset from either the feed line or the patch, or both, can also be treated (by adjusting y_n in (30), or y in (38)), but from a practical point of view such offsets are not generally desirable (unless two feed points are needed, as, for example, when circular polarization is required), as the coupling would be reduced [2], [8]. The above analysis assumes the slot is centered with respect to the patch and the feed line.

The aperture coupled patch antenna is usually tuned with an open-circuited stub of microstrip line, approximately $\lambda_g/4$ long. If the stub length is L_s, the input impedance seen looking into the microstrip feed line referenced to the aperture is

$$Z_{\text{in}} = Z - jZ_c \cot \beta L_s. \tag{43}$$

Slightly more accurate results can be obtained by adding a length extension to L_s to account for fringing fields at the end of the open stub; for the cases considered here the length extension is approximately 0.4 d_f, where d_f is the feed substrate thickness.

Figs. 7 and 8 show Smith chart plots of the input impedance of two stub-tuned aperture coupled patch antennas. The case in Fig. 7 has the same substrate parameters for both the feed and antenna substrates. Measurements are compared with the present theory, as well as the theory of [8]; it is seen that the present theory is actually a bit better than that of [8], which shows a small shift in the resonant frequency. Part of the reason for this difference may be that [8] used PWS modes on the patch, while the present theory uses entire domain modes on the patch.

Fig. 8 shows results for a low dielectric constant (2.22) substrate for the antenna and a high dielectric constant (10.2) substrate for the feed line. This configuration simulates the monolithic phased array application, where the feed substrate would be Gallium Arsenide for phase shifters and other active circuitry. Again the measurements are compared with the present theory and the theory of [8], and good results are obtained.

Fig. 9 shows the calculated far-zone E- and H-plane patterns of the stub-tuned aperture coupled patch of Fig. 8. The far fields are calculated from the stationary phase evaluations of the electric current on the patch [12], in addition to the far-fields of the slot. A front-to-back ratio of about 23 dB is realized, showing that the microstrip patch element is radiating much more effectively than the slot element. Measured patterns were found to be in reasonable agreement with these calculations, with some distortions being due to finite ground plane and feed line diffractions.

V. Conclusion

A method has been presented for analyzing slot-type discontinuities in microstripline. The method is based on the reciprocity theorem and uses the exact Green's functions for the dielectric slab, and a moment method solution for the unknown antenna currents. The method has been applied to the microstrip-fed slot antenna, and to the aperture coupled patch antenna, with quite good results when compared with mea-

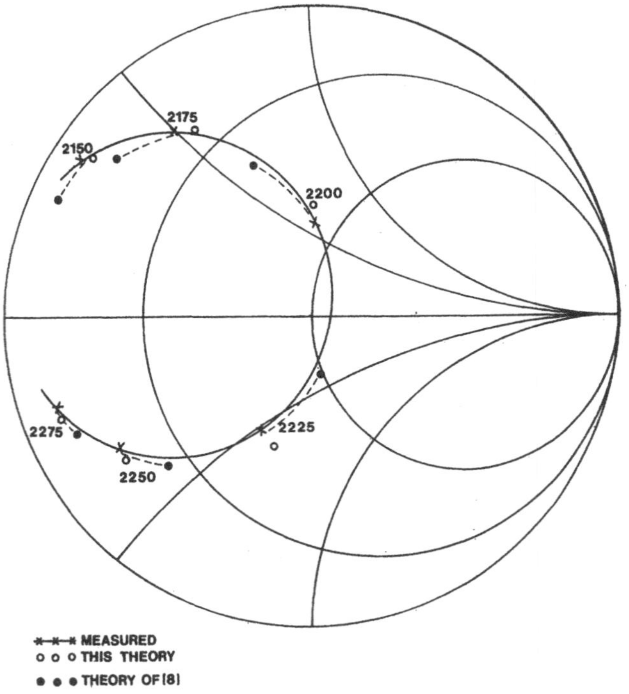

Fig. 7. Smith chart plot of the input impedance of a stub-tuned aperture coupled patch antenna. ϵ_{ra} = 2.54, d_a = 0.16 cm, PL = 4.0 cm, PW = 3.0 cm, ϵ_{rf} = 2.54, d_f = 0.16 cm, L = 1.12 cm, W = 0.155 cm, W_f = 0.442 cm, L_s = 2.0 cm.

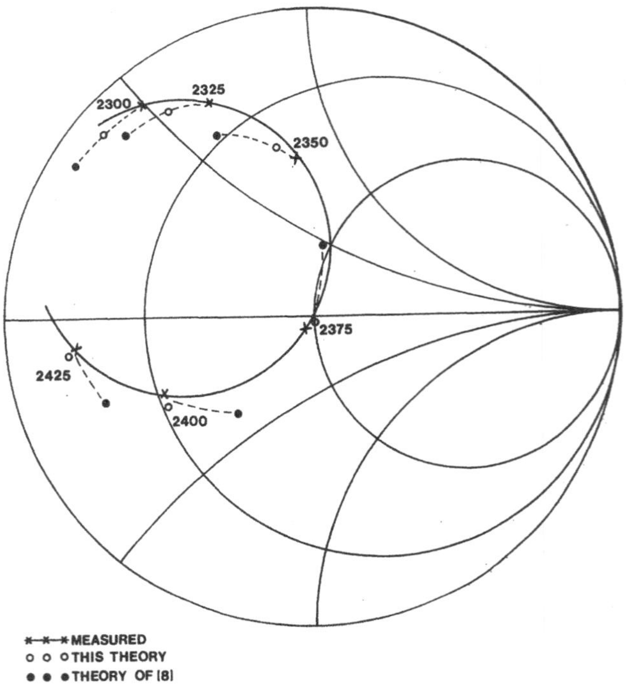

Fig. 8. Smith chart plot of the input impedance of a stub-tuned aperture coupled patch antenna. ϵ_{ra} = 2.22, d_a = 0.16 cm, PL = 4.0 cm, PW = 3.0 cm, ϵ_{rf} = 10.2, d_f = 0.127 cm, L = 1.0 cm, W = 0.11 cm, W_f = 0.116 cm, L_s = 1.1 cm.

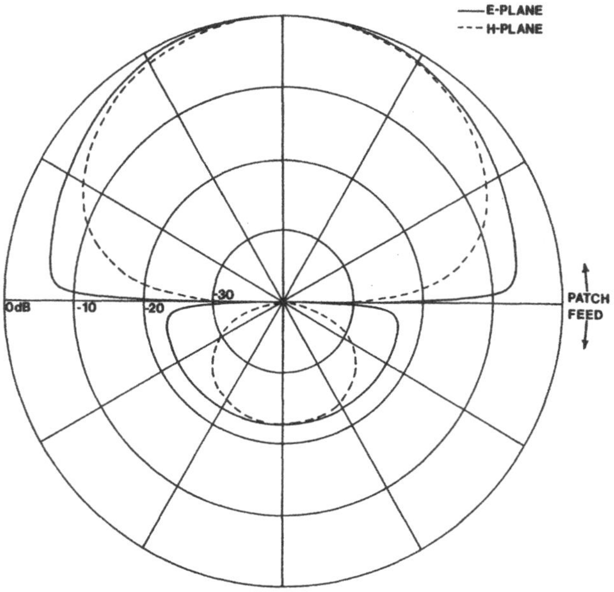

Fig. 9. E- and H-plane far-field plots of the stub-tuned aperture coupled patch antenna of Fig. 8.

sured data. This method should prove to be useful for related problems.

Appendix

This Appendix lists the required Green's function components:

G_{xx}^{EJ} = E_x at (x, y, d) due to a unit $\hat{x}$ electric current element at (x_0, y_0, d)

G_{yx}^{HJ} = H_y at $(x, y, 0)$ due to a unit $\hat{x}$ electric current element at (x_0, y_0, d)

G_{yy}^{HM} = H_y at $(x, y, 0)$ due to a unit $\hat{y}$ magnetic current (or e_x^a slot field) element at $(x_0, y_0, 0)$

G_{xy}^{EM} = E_x at (x, y, d) due to a unit $\hat{y}$ magnetic current (or e_x^a slot field) element at $(x_0, y_0, 0)$.

Define the Fourier transform $\tilde{G}$ of G as

$$G(x, y)=\frac{1}{4\pi^2}\iint_{-\infty}^{\infty} \tilde{G}(k_x, k_y)e^{jk_x(x-x_0)}e^{jk_y(y-y_0)}\, dk_x\, dk_y. \tag{44}$$

Then,

$$\tilde{G}_{xx}^{EJ}=\frac{jZ_0}{k_0} \cdot \frac{(\epsilon_r k_0^2-k_x^2)k_z \cos k_1 d+jk_1(k_0^2-k_x^2) \sin k_1 d}{T_e T_m} \sin k_1 d \tag{45}$$

$$\tilde{G}_{yx}^{HJ}=\frac{-jk_x^2(\epsilon_r-1) \sin k_1 d}{T_e T_m}+\frac{k_1}{T_e} \tag{46}$$

$$\tilde{G}_{yy}^{HM}=\frac{-j}{k_0 Z_0}\left[\frac{j(k_1 \cos k_1 d+jk_2\epsilon_r \sin k_1 d)(\epsilon_r k_0^2-k_y^2)}{k_1 T_m} - \frac{jk_y^2 k_1(\epsilon_r-1)}{T_e T_m}-\frac{(k_0^2-k_y^2)}{jk_2}\right] \tag{47}$$

$$\tilde{G}_{xy}^{EM}=-\tilde{G}_{yx}^{HJ}. \tag{48}$$

In the above,

$$T_e=k_1 \cos k_1 d+jk_2 \sin k_1 d \tag{49}$$

$$T_m=\epsilon_r k_2 \cos k_1 d+jk_1 \sin k_1 d \tag{50}$$

$$k_1^2=\epsilon_r k_0^2-\beta^2, \qquad \text{Im } k_1<0 \tag{51}$$

$$k_2^2=k_0^2-\beta^2, \qquad \text{Im } k_2<0 \tag{52}$$

$$\beta^2=k_x^2+k_y^2 \tag{53}$$

$$k_0^2=\omega^2\mu_0\epsilon_0=(2\pi/\lambda_0)^2 \tag{54}$$

$$Z_0=\sqrt{\mu_0/\epsilon_0}. \tag{55}$$

The one-dimensional Fourier transforms of the following expansion/test modes are also required:

$$\text{pulse function: } f_u(y)=\begin{cases}1/W, & \text{for } |y|<w/2\\ 0, & \text{for } |y|>w/2\end{cases} \tag{56}$$

piecewise sinusoidal mode:
(h = mode half-length)

$$f_p(y)=\begin{cases}\dfrac{\sin k_e(h-|y|)}{\sin k_e h}, & \text{for } |y|<h\\ 0, & \text{for } |y|>h\end{cases} \tag{57}$$

entire domain (sine) mode:
($m=1, 3, 5, \cdots$)

$$f_s^m(x)=\begin{cases}\sin \dfrac{m\pi}{2a}(x+a), & \text{for } |x|<a\\ 0, & \text{for } |x|>a\end{cases} \tag{58}$$

edge condition mode:

$$f_e(y)=\begin{cases}\dfrac{1}{\pi\sqrt{(w/2)^2-y^2}}, & \text{for } |y|<w/2\\ 0, & \text{for } |y|>w/2.\end{cases} \tag{59}$$

The Fourier transform is defined as

$$F(k_x)=\int_{-\infty}^{\infty} f(x)e^{jk_x x}\, dx, \tag{60}$$

and the transforms of the above functions are

$$F_u(k_y)=\frac{\sin k_y W/2}{k_y W/2} \tag{61}$$

$$F_p(k_y)=\frac{2k_e[\cos k_y h-\cos k_e h]}{\sin k_e h(k_e^2-k_y^2)} \quad (62)$$

$$F_s^m(k_x)=\frac{m\pi}{a}\frac{\cos k_x a}{(m\pi/2a)^2-k_x^2} \quad (63)$$

$$F_e(k_y)=J_0(k_y W/2). \quad (64)$$

The fields and the propagation constant β of the infinite microstrip line can be found from the above expressions, as discussed in [7]. The H_y field at $(x, y, 0)$ due to the microstrip line is, for example,

$$H_y(x, y)=\frac{1}{2\pi}\int_{-\infty}^{\infty} F_u(k_y) \cdot \tilde{G}_{yx}^{HJ}(k_x=-\beta, k_y)e^{-j\beta x}e^{jk_y y}\, dk_y, \quad (65)$$

where the feed line width, W_f, is the proper width to use in $F_u(k_y)$ in (65). This model of the microstrip line assumes uniform current (56) across the width of the line. The edge condition (59) could be used just as easily, but it has been found that this has negligible effect. To normalize the microstrip line fields in accordance with relation (3), it is noted that the modal fields under the quasi-TEM approximation are real, so that the power flow down the infinite line is

$$P=\int_{-\infty}^{\infty}\int_{0}^{\infty} \bar{E}x\bar{H}^* \cdot \hat{x}\, dz\, dy=VI=Z_c I^2=Z_c, \quad (66)$$

where $\bar{E}$, $\bar{H}$ are the fields from the microstripline with a total current of I = 1 A. The appropriate normalization constant is thus $\sqrt{Z_c}$, where Z_c is the characteristic impedance of the line. Then,

$$h_y=H_y/\sqrt{Z_c}. \quad (67)$$

Acknowledgment

The author would like to thank Allan C. Buck for fabricating the experimental models, and plotting the far-field pattern of the aperture-coupled patch antennas.

References

[1] J. R. James and A. Henderson, "Planar millimeter-wave antenna arrays," in *Infrared and Millimeter Waves,* vol. 14, pt. V, K. Button, Ed. London: Academic, 1985.

[2] D. M. Pozar, "A microstrip antenna aperture coupled to a microstrip line," *Electron. Lett.*, vol. 21, pp. 49–50, Jan. 17, 1985.

[3] B. N. Das and K. K. Joshi, "Impedance of a radiating slot in the ground plane of a microstripline," *IEEE Trans. Antennas Propagat.*, pp. 922–926, Sept. 1982. (Also see, D. M. Pozar, N. K. Das, B. N. Das, and K. K. Joshi, "Comments on 'Impedance of a radiating slot in the ground plane of a microstripline'," *IEEE Trans. Antennas Propagat.*, vol. AP-34, pp. 958–959, July 1986.)

[4] M. Kominami, T. Takei, and K. Rokushima, "A printed dipole electromagnetically coupled to a microstrip feed line," in *1985 ISAP Symp. Proc.*, Kyoto, Japan, pp. 93–96.

[5] Y. Yoshimura, "A microstripline slot antenna," *IEEE Trans. Microwave Theory Tech.*, vol. MTT-20, pp. 760–762, Nov. 1972.

[6] R. E. Collin and F. J. Zucker, *Antenna Theory, Part I.* New York: McGraw-Hill, 1969, ch. 14.

[7] R. W. Jackson and D. M. Pozar, "Full-wave analysis of microstrip open-end and gap discontinuities," *IEEE Trans. Microwave Theory Tech.*, vol. MTT-33, pp. 1036–1042, Oct. 1985.

[8] P. L. Sullivan and D. H. Schaubert, "Analysis of an aperture coupled microstrip antenna," *IEEE Trans. Antennas Propagat.*, vol. AP-34, pp. 977–984, Aug. 1986.

[9] D. M. Pozar, "Input impedance and mutual coupling of rectangular microstrip antennas," *IEEE Trans. Antennas Propagat.*, vol. AP-30, pp. 1191–1196, Nov. 1982.

[10] ——, *Antenna Design Using Personal Computers.* Dedham, MA: Artech House, 1985.

[11] A. C. Buck, "Investigation of printed circuit antennas," M.S. thesis, Elec. Comput. Eng. Dept., Univ. Massachusetts, Amherst, MA 1986.

[12] D. M. Pozar, "Analysis of finite phased arrays of printed dipoles," *IEEE Trans. Antennas Propagat.*, vol. AP-33, pp. 1045–1053, Oct. 1985.

Multiport Scattering Analysis of General Multilayered Printed Antennas Fed by Multiple Feed Ports: Part II—Applications

Nirod K. Das, *Member, IEEE*, and David M. Pozar, *Fellow, IEEE*

***Abstract*—This part of the paper describes the application of the general analysis of Part I to several practical geometries of multilayer/multifeed printed antennas. These examples include a dual-feed circularly polarized geometry; a stacked patch geometry; a stripline-aperture coupled geometry with a radome; an open-end proximity coupled patch; and dipole and slot geometries inclined or perpendicularly coupled to different feedlines. Features of the selected geometries cover many practical aspects of multilayer integrated phased arrays. Experimental results for several geometries are compared with the analytical results to demonstrate the accuracy and versatility of the analysis used. Various design considerations for the use of these multilayered printed antenna geometries in integrated phased array applications are discussed.**

I. Introduction

PRINTED antennas monolithically integrated with active circuits and a feed network may be a cost-effective way of realizing phased arrays for electronic beam steering [1], [2]. Early printed antennas and arrays usually consisted of patch elements with microstrip feed lines on a single dielectric layer, a geometry which is simple and easy to fabricate, but less than ideal in terms of electrical performance. Besides a notoriously small bandwidth, the performance of this type of design is often limited by spurious radiation from the feed network which affects sidelobe and cross-polarization levels. In a monolithic phased array application, the thin high dielectric constant substrate usually preferred for the integration of active devices may not be optimum for the antenna elements (a thick, low dielectric constant substrate is preferred for the microstrip antenna)—scan blindness is hence a potential problem for the single layer design. There are also mechanical problems with the simple single layer design, in terms of available substrate area, and the topological constraints of laying out the feed networks and active circuitry. Also, it is often desirable to add a protective radome layer to the antenna or array. Further, in addition to the multilayer architectures, multiple feed geometries can provide design flexibility and/or performance enhancement for circular or dual polarization designs, cross-polarization balancing, or dual-frequency operations.

Thus, in recent years, increasingly complex multilayer/multifeed printed antenna geometries have been proposed to improve electrical and mechanical characteristics [3]–[7]. Using multilayer substrates, feed networks and circuitry can be isolated from radiating elements, while also providing more surface area. Coupling between the layers can be accomplished with proximity feeds [3], or by using coupling apertures [4]. These types of noncontacting proximity or aperture coupled feeds are also advantageous from a fabricational point of view, when compared with coaxial probe feeds. Bandwidth can be enhanced by using stacked parasitic patches in two or more layers, and a radome layer or a wide angle impedance matching (WAIM) layer [8] can be used to cover the entire antenna or array for environmental protection and/or increasing the array scan volume. Due to complexity of analytical characterization and design, application of multifeed geometries in multilayered configurations have been generally limited to dual-aperture feeds for dual- or circular-polarizations [9], [10], but carry significant promise and potential for diverse design requirements.

Fig. 1 depicts a possible multilayer geometry. It shows a subarray module consisting of two distinct layers of primary and secondary feed networks (for topological and space considerations), the primary feed network for circular polarization, a two-layer stacked configuration for bandwidth enhancement, and a protective radome layer. Similar complex multilayer/multifeed geometries can be rigorously analyzed using the general analysis of Part I. Here we discuss the application of the analysis described in Part I to several specific geometries of practical interest. The analytical considerations as well as various performance and design considerations for the selected geometries are separately discussed in the following sections.

A. Dual-Aperture Coupled Microstrip Antenna for Circular Polarization

Fig. 2(a) shows the geometry of a circularly polarized microstrip antenna configuration with two orthogonal coupling slots and a reactive power divider to excite the two slots with necessary amplitudes and phases. Due to mutual cou-

Reprinted from *IEEE Trans. Antennas Propaga.*, vol. 40, no. 5, pp. 482–491, May 1992.

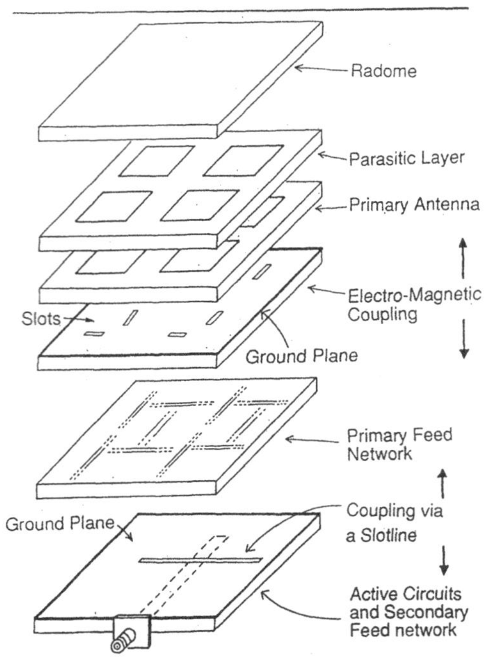

Fig. 1. An integrated multilayer architecture of printed antennas and transmission feed lines.

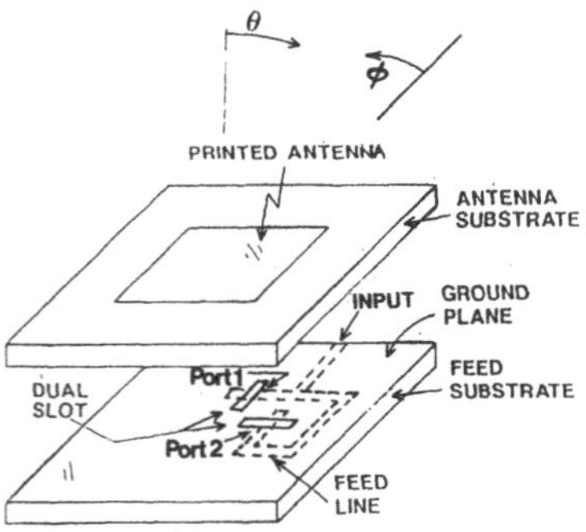

(a)

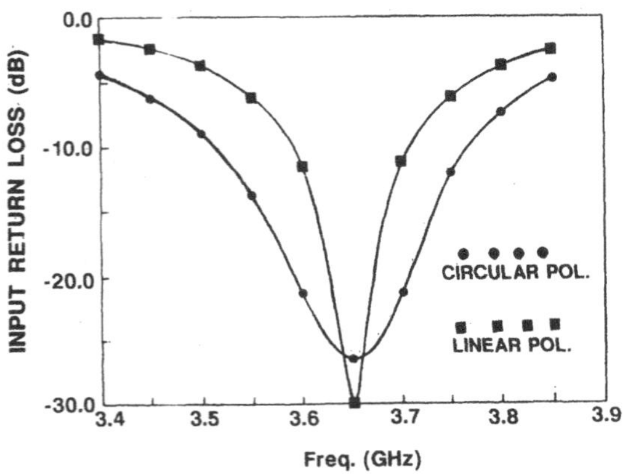

(b)

Fig. 2. (a) Geometry of a two slot coupled, dual feed microstrip antenna for circular polarization. (b) return loss (dB) at the input of reactive power divider, as compared with that of a single slot fed linearly polarized antenna. Antenna and feed substrates: $\epsilon_r = 2.2$, 0.16 cm; feed lines: $w = 0.5$ cm, 50 Ω, $\epsilon_{\text{eff}} = 1.9$; antenna: 2.5 × 2.5 cm; Slot: 1.1 × 0.15 cm, offset 0.75 from patch center in nonresonant dimension; array unit cell: 4.11 × 4.11 cm; broadside scan.

pling between the two offset-slot ports, exciting the two slots with exactly equal amplitudes and a 90° phase difference does not create perfect circularly polarized radiation, and so suitable compensation is necessary to account for the mutual coupling effects. The two slots can also be fed in series using a single microstrip line, instead of the parallel feeding of Fig. 2(a), to obtain the required excitations for a circular polarization.

As discussed in Part I of this paper, the geometry of Fig. 2(a) is first analyzed as a four-port circuit with two through transmission lines feeding the two slots. Then the transmission line stubs (quarter wavelength at 3.65 GHz) and the power divider circuitry are incorporated via the four-port scattering parameters. This procedure completely characterizes the impedance properties of the geometry of Fig. 2(a). In order to obtain the radiation characteristics, the two orthogonal radiation field components, E_θ and E_ϕ, are treated as two additional ports. Given the current distributions on the patch and slots obtained from the general analysis, computation of the far-field components using suitable Green's functions and a stationary phase evaluation method is relatively straightforward [11], [12].

Fig. 2(b) shows the computed return loss of the geometry of Fig. 2(a), as compared with that of a simple linearly polarized single-slot fed microstrip antenna, with all physical parameters remaining the same. As can be seen, the return loss bandwidth of the circularly polarized geometry is about twice that of the linearly polarized geometry. This is due to the fact that the almost identical values of reflection coefficients (S_{11} and S_{22}) at the planes of the slots transform through the two transmission lines with a $\lambda g/4$ length difference to give out-of-phase reflections at the input plane of the power divider. The almost out-of-phase reflections tend to cancel with each other resulting in the predicted bandwidth improvement.

The scan variation of input reflection coefficient of the dual slot coupled antenna is shown in Fig. 3, along with that of the corresponding single-slot coupled linearly polarized geometry. As is clear from Fig. 3, the reflection characteristics of the circularly polarized structure are more or less constant over a wider range of scan angles as compared with that of the linearly polarized structure. As in the case of the improvement of return-loss bandwidth discussed earlier, the effect of the reactive feed circuitry of the dual feed structure is also responsible for the wider angle scan behavior of Fig. 3. This type of improvement of scan performance of a circularly polarized geometry over the corresponding linearly polarized structure will, however, not be observed with all types of antennas. The appropriate scan characteristics of the particular type of antenna element (here an aperture-coupled microstrip antenna), together with the mutual interaction of reflections in the feed circuit, are responsible for such wide-angle performance. As presented in Part I, the S_{11} and S_{22} scan characteristics of our dual-aperture coupled microstrip antenna geometry, with its two stub-tuned ($\lambda g/4$) transmission feed line ports, and the port references at the respective coupling slots, follow in phase with each other as the array is scanned off broadside direction. The almost in-phase port

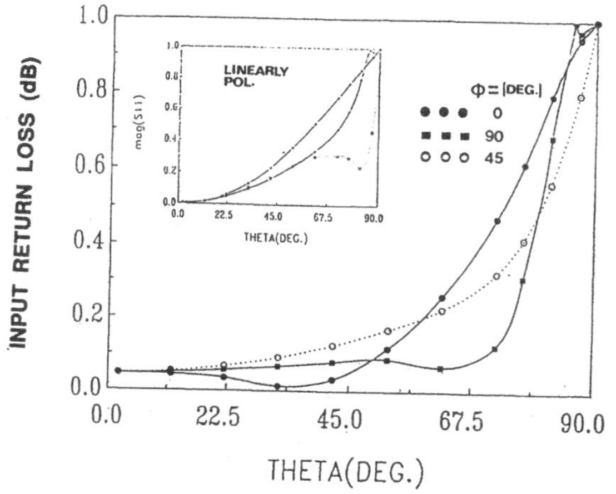

Fig. 3. Scan variation of input return loss of the dual slot coupled parallel-fed circularly polarized antenna of Fig. 2.

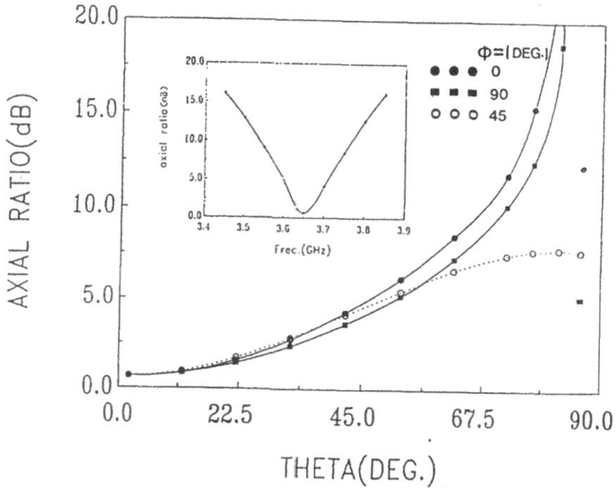

Fig. 4. Axial ratio performance of the dual slot coupled circularly polarized antenna of Fig. 2.

reflections (port reflections are approximately equal to S_{11} and S_{22}, neglecting S_{12}) at the plane of the slot transform into approximately out-of-phase reflections at the input plane of the feed circuitry due to the $\lambda g/4$ line-length difference. Consequently, it results in mutual cancellation of the out-of-phase reflections, and hence improvement of the scan dependence of the input return loss. As discussed in Part I, because the scan characteristics of other types of antenna elements (a pair of crossed free-space dipoles for circular polarization, for example) do not necessarily exhibit the required in-phase variations of the orthogonal port reflections, overall improvement in the scan variation of return loss of a circularly polarized geometry over that of the linearly polarized geometry may not always occur for other types of antennas. Such desirable characteristics seem to be exhibited by microstrip antenna geometries on thin substrates.

The scan as well as frequency variation of axial ratio of the circularly polarized antenna are shown in Fig. 4, and were computed using the orthogonal radiation fields. For the results of Fig. 4, the feed network was designed for perfect circular polarization at 3.65 GHz without accounting for mutual coupling effects. However, as Fig. 4 shows, perfect circular polarization could not be obtained at broadside ($\sim$ 1.0 dB axial ratio is typical) as designed. As mentioned previously, this is a result of mutual coupling effects, and is due to the slot offsets. Ideally, if the two slots could be placed orthogonally to each other at the center of the patch, such mutual coupling would be strictly absent at broadside scan. Using the computed scattering parameters of the multiport antenna, including the scattering parameters for the two additional ports for the two orthogonal far field components, suitable compensation for the amplitudes and phases of incident port excitations at the planes of the slots were computed. These results are presented in Fig. 5. Simple modifications in the feed circuitry can be made to obtain the necessary compensated incident excitations.

Axial ratio characteristics of an infinite array of dual-aperture coupled microstrip antennas cannot be verified in a waveguide simulator. Thus, an isolated geometry was designed with a parallel feed network, neglecting the mutual

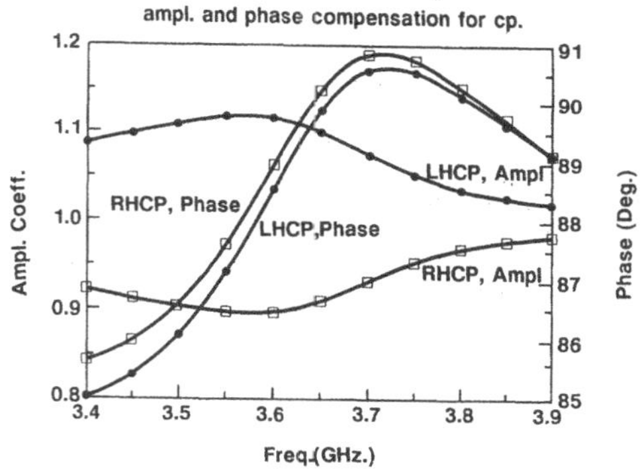

Fig. 5. Phase compensation ($|\phi_2 - \phi_1|$, ideally 90°) and amplitude compensation [$|v_{+2}/v_{+1}|$, ideally 1.0] required to obtain perfect circular polarization. Resonant frequency = 3.65 GHz.

coupling between the slots as a first-order approximation. The measured results for axial ratio are compared in Fig. 6 with the rigorous theoretical results including mutual coupling, and show a reasonable agreement of the frequency characteristics with a frequency shift of about 1%. As the theoretical and experimental results show, due to mutual coupling effects perfect circular polarization could not be obtained.

B. Aperture-Coupled Microstrip Antenna with a Parasitic Patch

As demonstrated in recent publications, [10], [13], [14], it is possible to improve the bandwidth of a microstrip antenna by using a two-layer stacked structure with two or more patches properly spaced with dielectric layers in between. In addition, an aperture-coupled stacked geometry might be more useful for integrated phased array applications from fabrication and feed isolation considerations.

Fig. 7 shows a stacked patch geometry aperture-coupled to a microstrip feed line. This two-port geometry is characterized for an infinite array as well as an isolated element using the analysis of Part I. Five entire basis sinusoids (EBS) modes for each current direction (x and y) on both patches (total 20 modes) were used for the patch current expansion,

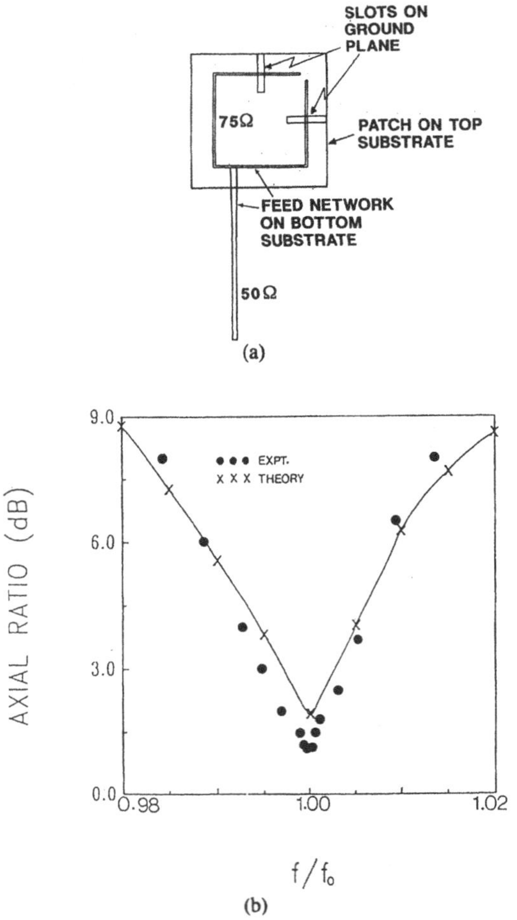

Fig. 6. (1) Geometry of a dual slot coupled circularly polarized microstrip antenna designed, built and tested. Antenna substrate $\epsilon_r = 2.2$, 0.16 cm, feed substrate: $\epsilon_r = 10.2$, 0.127 cm, patch: 3.9 × 3.9 cm, slots: 1.2 × 0.17 cm, patch offset with respect to slots (center to center): 0.7 cm in the nonresonant dimension, 75 Ω feed lines: 0.4 mm width, 90° phase difference between the two lines, 50 Ω feed line: width 0.12 cm with a 37.5–50 Ω tapered transition. (b) Comparison of theoretical and experimental results of normalized frequency performance of axial ratio of the isolated dual-slot fed circularly polarized antenna: f_0 (theory) = 2.34 GHz, f_0 (experiment) = 2.3875 GHz.

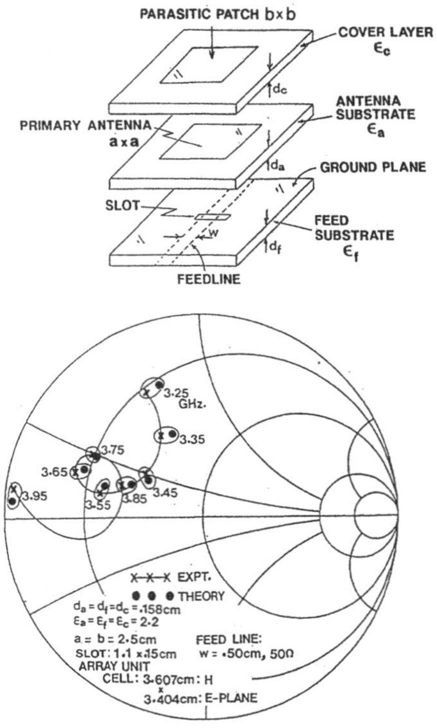

Fig. 7. Waveguide simulator measurements and calculation for a stacked microstrip antenna configuration with $a = 2.5$ cm $= b$.

whereas only one piecewise sinusoidal (PWS) mode was used on the slot. Nontraveling-wave (subsectional) modes were included on the feed along with the traveling-wave modes, but the results were not significantly different from those with traveling-wave modes alone. The results of the normalized equivalent series impedance $(1 - S_{11} = S_{12})$ as seen by the microstrip line are compared in Fig. 7 with a waveguide simulator experiment [15]. The theory and experiment compare reasonably well, and show a loop on the impedance locii due to overlapping of two resonant modes corresponding to the two patches. The relative levels of excitation of the two resonant modes, as well as the separation between the resonant frequencies of the two modes, control the available bandwidth of the antenna, and are determined by the relative dimensions of the two patches and the separation between them.

The bandwidth (VSWR < 2) of the aperture-coupled stacked patch geometry of Fig. 7 computed by suitably tuning the reactance, and matching it to a transmission line of characteristic impedance close to the resistance level of the center of the double-tuned loop, is about 12%. In contrast, only about 3% bandwidth is obtained without the top patch and substrate. However, it should be noted that the impedances of Fig. 7(b) are as seen by the simulator, and correspond to different scan angles at different frequencies. Fig. 8 shows the broadside bandwidth of an infinite array with unit cell dimensions 4.11 × 4.11 cm, as a function of the length of the top square patch, keeping the bottom patch dimensions fixed at 2.5 × 2.5 cm. As seen from Fig. 8, better bandwidth can be obtained by using a cover patch of larger size than the primary antenna. However, there is a limit to how large the cover patch can be relative to the bottom patch, beyond which the two resonant modes radiate independently resulting in a dual frequency operation of the antenna, rather than the broad-band operation. When the cover patch is smaller than the primary patch, the top patch is more isolated from the feeding slot, and therefore is not strongly excited. Hence, the weakly excited resonant mode of the top patch fails to enhance the total bandwidth of the antenna. On the other hand, a larger cover patch effectively couples to the fringing fields of the bottom patch to result in excitation of two distinct and overlapping resonant modes that ensures improvement of the overall bandwidth. The above reasoning is true only for cases with thin substrates, and may not be valid if the substrates are electrically thick. This is because the significant fringing field of the bottom patch can now effectively couple to the top patch even when the top patch is smaller than the bottom patch.

Fig. 9 shows the normalized E-plane scan behavior of the infinite array of Fig. 8 for $a = b = 2.5$ cm, and frequency

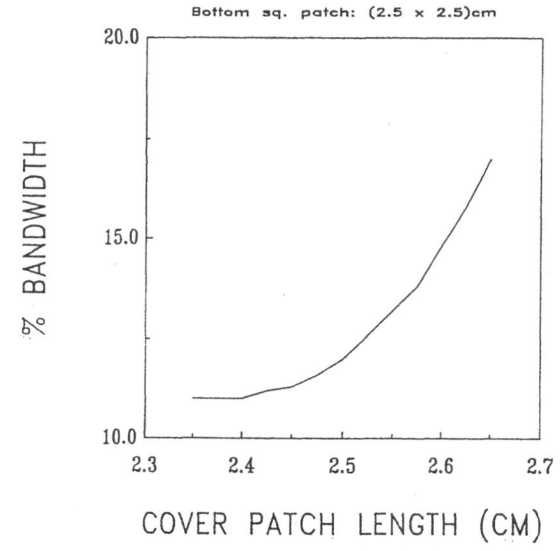

Fig. 8. Bandwidth of the stacked aperture-coupled configuration of Fig. 7 with array unit cell 4.11 × 4.11 cm for different values of cover patch dimensions ($b \times b$).

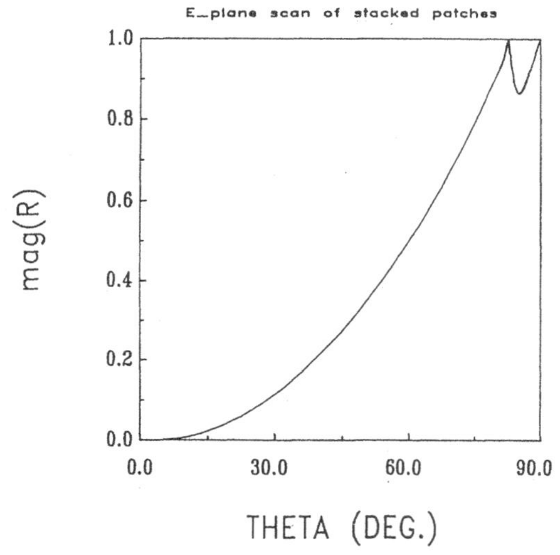

Fig. 9. *E*-plane scan performance of the broadside conjugate matched input reflection coefficient, R, for the stacked patch of Fig. 7 with array unit cell 4.11 × 4.11 cm.

= 3.75 GHz, and shows a blind spot close to the horizon. Qualitatively, the general trend of the *E*-plane scan behavior of a stacked microstrip array is similar to that of a standard single-layered microstrip array [15]. The possibility of a scan-blindness angle is an important factor limiting the bandwidth optimization with substrate thickness and dielectric constant. The larger the substrate thickness or dielectric constant, the closer the blindness angle is to the broadside.

C. Stripline Aperture-Coupled Microstrip Antenna with a Radome

A stripline feed, instead of a microstrip feed, would be useful to avoid the back radiation from an aperture coupled geometry. Also, use of striplines to feed the antenna elements can be unavoidable in a multilayer phased array as a conse-

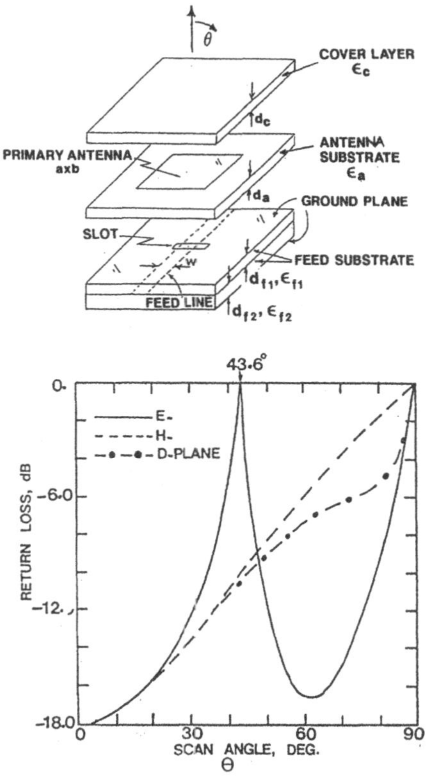

Fig. 10. Scan performance of an infinite array of stripline-fed aperture coupled covered microstrip patches. Array unit cell: 0.4 × 0.4 λ_0, patch: 2.5 × 2.5 cm; slot 1.1 × 0.15 cm; feedline: 0.1 cm width, 50 Ω; antenna substrate: $\epsilon_r = 2.2$, 0.158 cm; cover substrate: $\epsilon_r = 2.55$, 0.158 cm; feed substrate: top, $\epsilon_r = 10.2$, 0.127 cm, bottom, $\epsilon_r = 2.2$, 0.158 cm; frequency = 3.45 GHz.

quence of the isolating ground planes necessary to electrically separate the different feed layers (see Fig. 1). However, potential problems of scan blindness due to excitation of the parallel plate mode, and possible leakage of power from a stripline with two different dielectric layers above and below the center strip (inhomogeneous stripline) [16] require careful consideration.

The geometry of a representative stripline-aperture coupled antenna with a radome layer is shown in Fig. 10 along with its scan performance of the input return loss for a specific set of physical parameters; one port of the feedline is $\lambda_g/4$ stub tunes at 3.45 GHz. The results were obtained by modeling the geometry as a two-port circuit using the multiport analysis of Part I, following essentially a similar procedure as for the microstrip-aperture coupled geometry, resulting in an equivalent series impedance across the stripline. Characteristics of the two-layer stripline such as the characteristics impedance, propagation constant and the transverse fields, were obtained using the general multilayer transmission line analysis of [16]–[18], that must be used with the general analysis to electrically describe the feedline. The mode selection was similar to the stacked patch case (10 modes for currents (x and y) on the patch), but for scan characteristics exact cavity currents discussed in Part I were used for better convergence. As Fig. 10 shows, for the chosen set of parameters, the scan performance exhibits a prominent *E*-plane scan blindness at 43.6° that corresponds to excitation of the parallel plate mode of the feed structure. The scan blindness here is not due to the resonance of the characteristic source-free modes of the substrate layers associated with the radiating face of the antenna, but due to the resonance of the characteristic mode of the layer configuration associated with the stripline feed, and is forced by the non-radiating slots. In fact, in reference to the general structure of Fig. 1, such a potential blindness can result due to excitation of any characteristic mode of the entire multilayer configuration, and can be excited by the currents in the associated feed circuitry that undergo identical phase variation as that of the scanning antennas on the radiating face. However, this scan blindness effect due to levels of feed circuitry below the radiating surface and primary feed network would probably be much less prominent, and possibly result in a narrow band or high "Q" scan blindness of the infinite array. Specific measures to help avoid or suppress such unwanted resonance effects would be desirable for the safe operation of a phased array.

Possible solutions for avoiding the potential dangers of blindness due to parallel plate feed structures are to use low dielectric constant substrates for the feed and/or reduce the size of the array unit cell. In Fig. 11 the resulting blindness angles due to forced slot resonance of the dominant parallel plate mode, for different layered geometries of the stripline feed, are compared for two values of element spacings. The corresponding values due to the surface resonance of the radiating patch are also shown (curve IV) in Fig. 11 for comparison, that clearly demonstrates the dominant effect of the parallel plate resonance excited due to the coupling slot.

As mentioned, the other potential danger of a stripline-aperture-coupled structure is due to the stripline feed alone. There is a possibility of excessive surface radiation loss due to leakage of power from a stripline with two dielectrics to the characteristic source free mode of the parallel plate structure. As discussed in [16] and [19] such parallel plate leakage loss does not, however, occur when the two substrates on both sides of the center strip are of the same thickness, but could occur when the thinner substrate has a lower dielectric constant. Carefully choosing the physical parameters of the substrate layers would, accordingly, avoid such problem of power leakage.

When using the stripline feed structure with more than one level of feed networks (see Fig. 1), in order to establish effective direct coupling to subsequent layers below, the stripline should not use highly different dielectric constants on the two sides of its center strip. Otherwise, due to an unbalanced field strength, a weaker coupling to the side with lower dielectric constant will result. Figure 12 quantitatively demonstrates such magnetic coupling unbalance to a slot in the groundplane, and as expected, clearly shows an equal coupling strength to both sides when the two substrates are the same. Also as a reference, Fig. 13 shows the broadside frequency variation of impedance of a stripline coupled structure for two similar geometries with different parameters for the stripline structure. The levels of coupling of the two cases, (a) and (b), are significantly different, dominantly due to the different stripline feed configurations. As expected, case (a) with a larger dielectric constant substrate in the antenna side has a larger effective coupling as compared with case (b). The results of Figs. 12 and 13 clearly indicate a trade-off between the aperture coupling of the feedline to the radiating antenna above and to the secondary feed network below. Lower coupling to the antenna results in a lower level of equivalent impedance of the antenna, whereas lower coupling to the feed network below results in a higher insertion loss of the transitions.

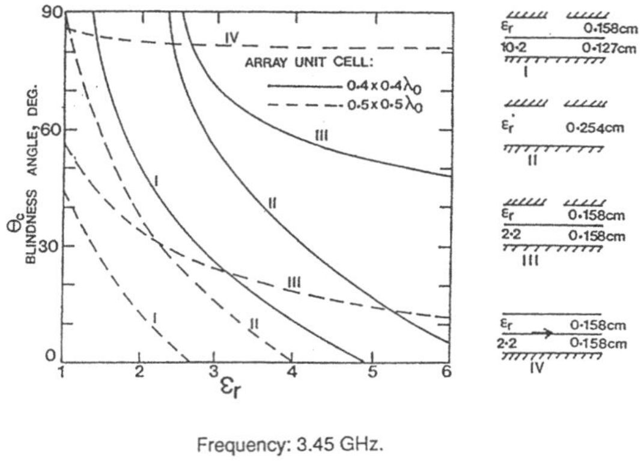

Fig. 11. Approximate scan blindness angles of a stripline fed antenna at 3.45 GHz forced by characteristic modes of various multilayer structures excited by slots (I–III), or an electric current source (IV). Note no blindness of (IV) for 0.4 × 0.4 λ_0 array unit cell.

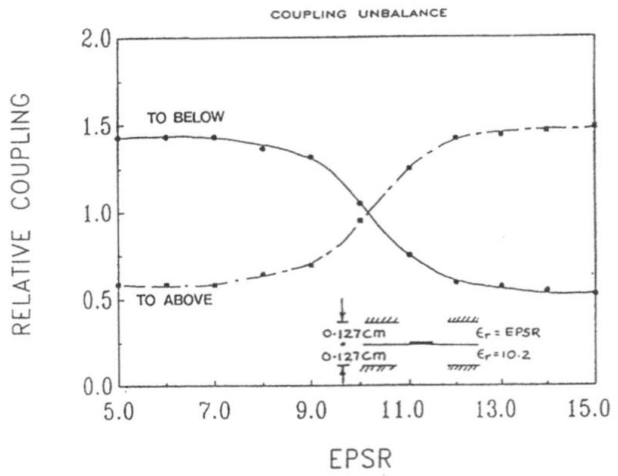

Fig. 12. Unbalanced coupling from a stripline with different dielectrics on two sides of the center conductor at 3.45 GHz.

Following the discussion above, coupling from a stripline feed to the aperture coupled antenna is inherently lower than that from a microstrip feedline, for the same physical parameters of the antenna and the slot. The fields of a microstripline concentrate on the side of its ground plane, and thus result in a stronger coupling to the slot than that due to a stripline, where the transmission line fields are rather split into two sides of the center conductor, out of which only one part is responsible for coupling to the slot. As has been verified, using the analysis of Part I, for the case when the two substrates of the stripline feed are the same, equivalent impedances of the order of one fourth should be expected in comparison with that of a microstripline coupled structure. Coupling can, however, be improved by increasing the size of the slot, or using inhomogeneous stripline.

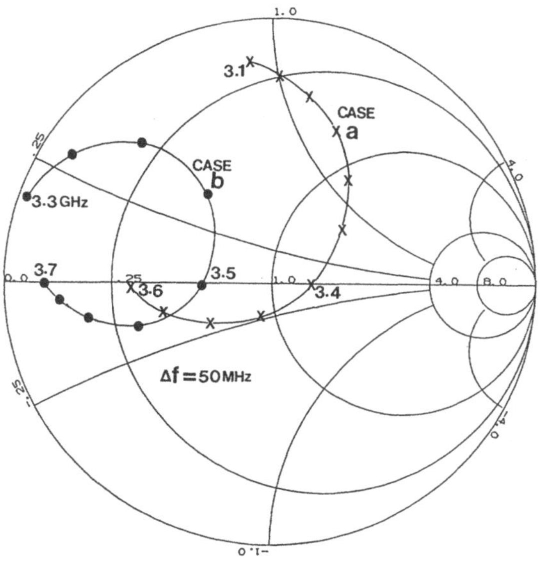

Fig. 13. Broadside frequency variation of equivalent impedance of the stripline-aperture coupled structure of Fig. 10, for two sets of parameters: (a) same as of Fig. 10. (b) cover substrate $\epsilon_r = 2.2$, 0.16 cm, stripline feed: $d_{f1} = d_{f2} = 0.127$ cm, $\epsilon_{f1} = \epsilon_{f2} = 10.2$ $w = 0.046$ cm, 50 Ω; array unit cell: 4.11 × 4.11 cm.

D. A Covered Microstrip Antenna Proximity Coupled to the Open End of a Covered Microstrip Line

Unlike the previous examples, where the radiating elements were coupled to the dominant transverse fields of a through transmission line, in this case the antenna is coupled to the fringing fields of the open end. As discussed in Part I, the assumption of no direct interaction of the transmission line terminations on the antenna elements, that simplified the generalized scattering analysis, do not apply for this case. However, as discussed in Part I, the general multiport scattering principle is still rigorously applicable here, but may not be more efficient than an alternative approach of using a non-Galerkin testing procedure to solve for the traveling-wave currents [20]. Both methods were used with suitable modifications of the general analysis providing comparable results. Extra PWS current modes are used to model the nontraveling-wave current components near the open end, that play the dominant role in establishing effective coupling with the antenna. In contrast, for cases of through transmission line coupled antennas, the coupling from the feed line to the antenna is dominantly due to the fields of the traveling wave current components on the feed line; and any extra nontraveling wave current expansion on the feed line in the vicinity of the antenna only describes a refinement of the solution. In fact, for the through transmission line feeding of an aperture coupled microstrip antenna [21], for example, fairly accurate results can be obtained without including the nontraveling wave current expansion modes on the microstrip feed line, whereas for the present open-end proximity coupled geometry a sufficiently large number of finite length expansion functions are required to obtain reasonable results. Thus, the solution of such proximity coupled geometries is computationally more involved, and also requires careful attention to the convergence of the solution with respect to the number and density of the nontraveling-wave modes near the open end.

Fig. 14 shows the comparison of the experimental results with the results of the present analysis for a covered proximity coupled patch antenna. The theoretical results were obtained with eight PWS modes of longitudinal currents on the feed line over a length of about 4.0 cm near the open end, and five entire domain sinusoid modes (even and odd, with uniform transverse variation) on the patch with currents in the same direction as on the transmission line. However, other possible current expansion modes could be chosen from the large set of possibilities that the solution can handle, but these did not drastically affect the final results. Fig. 14 shows the equivalent impedance as seen by the feed line with the phase reference at the open end. Clearly, the agreement between the results is good, with a shift in resonant frequency on the order of approximately 1.0%, which probably can be attributed to tolerances in the value of the dielectric constant and other fabrication tolerances. Similar agreement was also obtained without the cover substrate as well as with other patch dimensions.

As observed theoretically as well as experimentally, the coupling is not sensitive to offset, δ, of the feed line along the transverse dimension (y) of the patch. On the other hand, the coupling to the antenna significantly changes with offset in the longitudinal direction (x). This is expected, because the field due to the dominant radiating current on the patch that strongly couples to the feed line is associated with a fairly uniform variation in the transverse dimension, but a cosinusoidal variation longitudinally. Fig. 15 shows the theoretical variation of the resonant equivalent admittance G of the open-end coupled microstrip antenna of Fig. 14 with offset, δ, both in the transverse and longitudinal directions, and it clearly demonstrates the trend discussed above. It should be noted, that the resonant frequency of the antenna also changes with the feed line offset.

E. Microstrip Dipole Coupled to an Inclined Covered Microstrip Line

The general analysis of Part I is now applied to analyze an infinite array of printed dipoles coupled to inclined covered microstrip lines. The S_{11}, S_{22}, and S_{12} of the two-port circuit are plotted in Fig. 16 as a function of the angle of inclination, θ. Three PWS modes were used on the dipole, and the nontraveling wave currents on the feedline are included. Unlike the perpendicular case [22], for a general inclined dipole $S_{11} \neq S_{22}$ due to the physical and electrical asymmetry with respect to the two input ports [23]. Also, unlike the perpendicular dipole [22], we now have $S_{21} \neq 1 + S_{11}$, which implies that for the general inclined dipole the shunt equivalent impedance model is not valid. In order to be able to obtain an equivalent circuit model for such inclined dipoles, it is required to use a general T or π network with three independent impedance parameters. Using the general multiport analysis described in Part I of this paper, the appropriate asymmetries are incorporated via the generalized scattering conditions. The equivalent impedance parameters of a T or

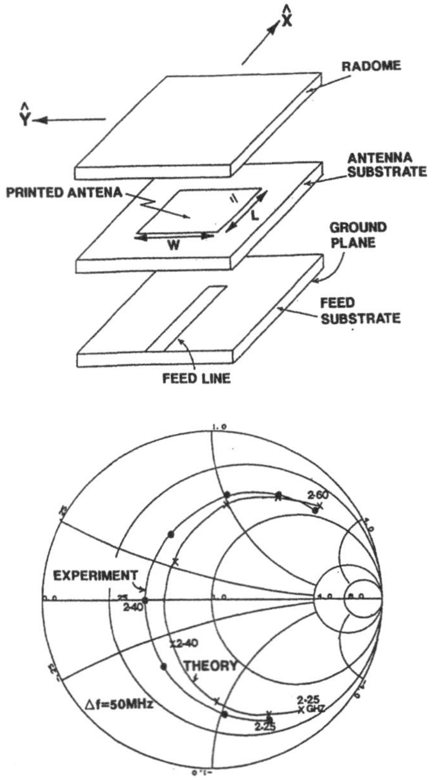

Fig. 14. Comparison of theoretical and experimental results for an isolated microstrip antenna covered by a dielectric sheet proximity coupled to the open end of a covered microstrip line. Patch: 3.9 × 3.9 cm, feed line: $w = 0.5$ cm, $Z_c = 50\ \Omega$, $\epsilon_{eff} = 2.144$, all substrates: $\epsilon_r = 2.2$, 0.16 cm; open end at the patch center.

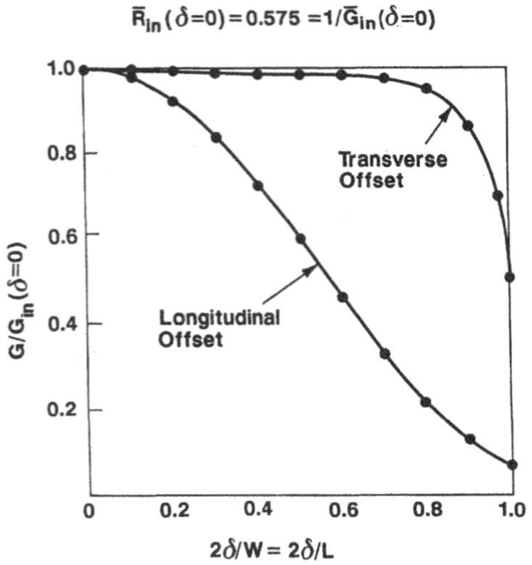

Fig. 15. Variation of equivalent resonant admittance of the open-end coupled proximity fed microstrip antenna of Fig. 14 as a function of transverse and longitudinal feedline offset, δ.

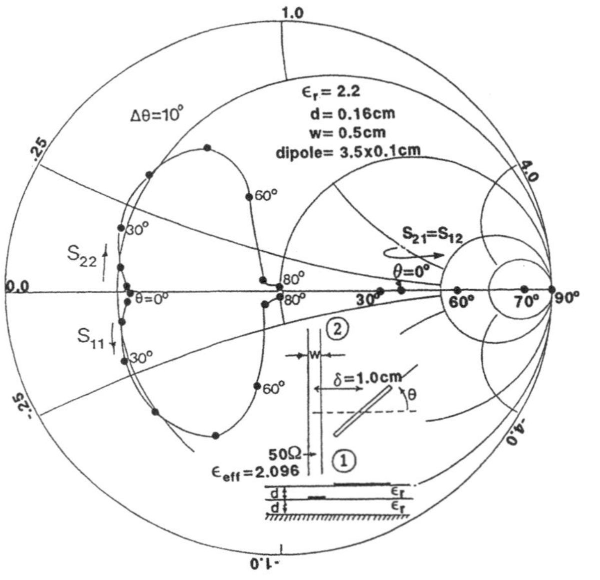

Fig. 16. Locus of S_{11}, S_{22}, and S_{12} of an infinite array of printed inclined dipoles coupled to covered microstrip feed lines, as a function of the inclination angle, θ. Substrate: $\epsilon_r = 2.2$, 0.16 cm, superstrate: $\epsilon_r = 2.2$, 0.16 cm, feed line: 0.5 cm width, 50 Ω, $\epsilon_{eff} = 2.096$, dipole: 3.5 × 0.1 cm, array unit cell: 5.0 × 5.0 cm, offset, $\delta = 1.0$ cm, freq. = 3.143 GHz, scan angle: broadside.

π network can always be derived from the scattering parameters.

The dipoles of Fig. 16 can also be fed by slotlines on the ground plane, instead of the covered microstrip lines. As discussed in the analysis of Part I, like a covered microstrip line coupled perpendicular dipole geometry [22], the equivalent circuit for a slotline-coupled perpendicular dipole geometry is also a shunt impedance across the slotline. In fact, the same shunt equivalent impedance model is applicable to any perpendicular dipole geometry, irrespective of the type of the feed line used. The values of the equivalent shunt resonant impedances normalized to the feed line characteristic impedance are plotted in Fig. 17 as a function of the transverse offset, δ, of the center of the dipole from the centerline of the feed line. The results for the covered microstrip line case are compared with the case of the slotline feed. The coupling levels of the slotline feed are stronger than those for the covered microstrip feed line, resulting in values of equivalent resonant impedances for the former case smaller than those of the latter by a factor of about 10. This is because of the tighter coupling between the dominant transverse fields of the slotline and the dipole, in contrast to a relatively weak coupling to the transverse fringing fields of the covered microstrip line. Also, the slotline feed has maximum coupling to the dipole when the dipole is symmetrically across the slotline, whereas the corresponding symmetric positioning of the dipole results in zero coupling for the microstrip feed case. This is expected considering the even and odd symmetries of the electric fields of the slotline and the microstrip feed line, respectively, which exhibit maximum or no effective coupling to the dominant radiating current mode of the dipole with an even distribution about the center point.

F. Odd-Mode Coupled Microstrip Line Fed Printed Slot Antenna

An odd-mode microstrip line, sometimes also referred to as a coplanar microstrip line, can be used instead of a regular microstrip line to feed printed slot radiators or slot-coupled patch antennas [21]. This geometry is analyzed using the multiport analysis of Part I, with the coupled microstrip line

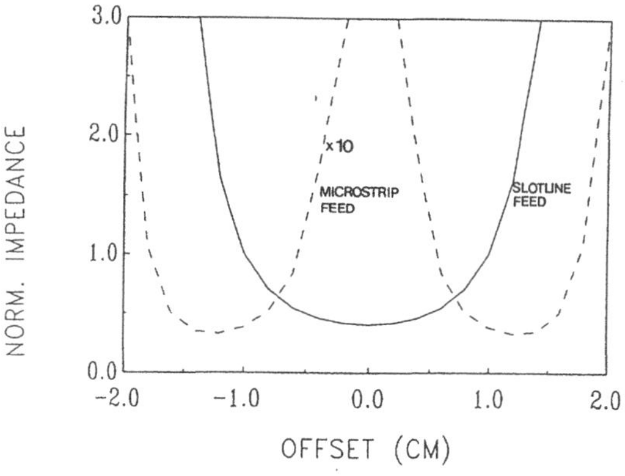

Fig. 17. Equivalent shunt resonant resistances of an infinite array of printed dipoles of Fig. 16 coupled to a perpendicular covered microstrip feed line as a function of center-to-center offset, compared to that with a slotline feed on the ground plane. Slotline: $w = 0.1$ cm, 117 Ω, $\epsilon_{eff} = 1.38$ cm. The impedances are normalized to respective feedline impedances.

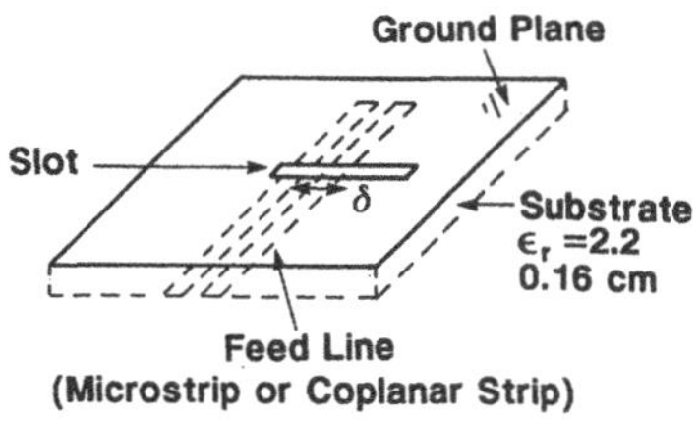

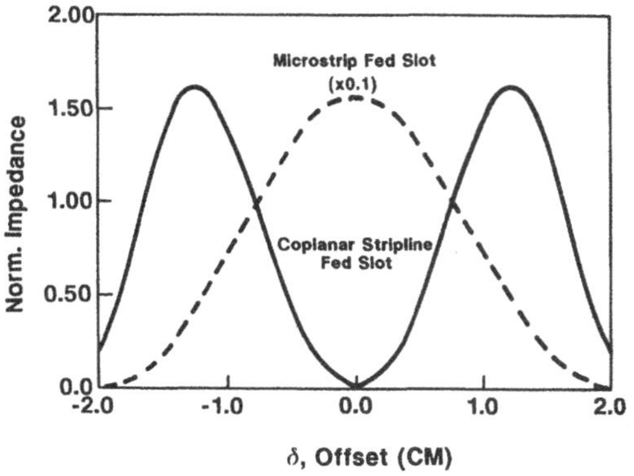

Fig. 18. Equivalent series resonant resistances normalized to the feed line impedances of an infinite array of slot radiators. Coupled line: $w = 0.5$ cm, $Z_c = 41$ Ω, $\epsilon_{eff} = 1.76$, center-to-center line separation = 0.6 cm; slot: 3.5 × 0.1 cm; array unit cell: 5.0 × 5.0 cm; frequency = 3.54 GHz; microstrip line: $w = 0.5$ cm, $Z_c = 50$ Ω, broadside scan.

characterized by the general transmission line analysis of [17], [18]. The equivalent series resonant resistance values of an infinite array of coplanar microstrip line coupled slot radiators are compared in Fig. 18 with the corresponding values for a microstrip-feed case, all other physical parameters remaining the same. As expected, due to the odd symmetry of the magnetic fields on the ground plane of a coplanar microstrip feed line, the slot radiator is not coupled to such a feed line when placed symmetrically about the center ($\delta = 0$). In contrast, the microstrip-feed case exhibits the strongest coupling when $\delta = 0$. Also, due to mutual cancellation of the magnetic fields produced by the oppositely directed currents on the two strips of a coplanar microstrip line, the relative level of the coupling is weaker than that for a microstrip line feed. As Fig. 18 shows, the maximum level of normalized equivalent impedance for the former case is about 10 times smaller than that for the latter case. In contrast to the high impedance loading of the microstrip line coupled geometry [21], such lower levels of impedance with coplanar microstrip feeding can make slot radiators more practical for series fed array designs.

II. Conclusion

The main purpose of this paper has been to apply the unified solution presented in Part I to a wide range of practical multilayer and/or multifeed printed antenna geometries. As the results of the work demonstrate, a multilayer/multifeed integrated antenna architecture has the potential of solving various design problems for future integrated phased array systems with improved performance and added versatility. The specific geometries studied in this paper constitute only a representative class. Many other interesting multilayer/multifeed geometries can be similarly investigated using the general analysis of Part I.

References

[1] R. J. Mailloux, "Phased array architecture for millimeter wave active arrays," *IEEE Antennas Propagat. Soc. Newsletter*, vol. 28, pp. 5–7, Feb. 1986.

[2] J. A. Kinzel, "GaAs technology for millimeter wave active arrays," *IEEE Antennas Propagat. Soc. Newsletter*, vol. 29, pp. 12–14, Feb. 1987.

[3] H. G. Oltman and D. A. Huebner, "Electromagnetically coupled microstrip dipoles," *IEEE Trans. Antennas Propagat.*, vol. AP-29, pp. 151–157, Jan. 1981.

[4] D. M. Pozar, "A microstrip antenna aperture coupled to a microstrip line," *Electron. Lett.*, vol. 21, pp. 49–50, Jan. 1985.

[5] D. R. Jackson and N. G. Alexopoulos, "Analysis of planar strip geometries in a substrate-superstrate configuration," *IEEE Trans. Antennas Propagat.*, vol. AP-34, pp. 1430–1438, Dec. 1986.

[6] P. B. Katehi, N. G. Alexopoulos, and I. Y. Hsia, "A bandwidth enhancement method for microstrip antennas," *IEEE Trans. Antennas Propagat.*, vol. AP-35, pp. 5–12, Jan. 1987.

[7] J. Herd, "Scanning impedance of electromagnetically coupled rectangular microstrip patch arrays," *IEEE Antennas Propagat. Soc. Symp. Dig.*, 1989, pp. 1150–1153.

[8] E. G. Magill and H. A. Wheeler, "Wide angle impedance matching of a planar array antenna by a dielectric sheet," *IEEE Trans. Antennas Propagat.*, vol. AP-14, pp. 49–53, Jan. 1966.

[9] A. Adrian and D. H. Schaubert, "Dual aperture-coupled microstrip antenna for dual or circular polarization," *Electron. Lett.*, vol. 23, no. 23, pp. 1226–1227, Nov. 1987.

[10] C. H. Tsao *et al.*, "Aperture-coupled patch antenna with wideband-width and dual-polarization capabilities," in *IEEE Antennas Propagat. Soc. Symp. Dig.*, vol. 3, Syracuse, NY, 1988, pp. 836–839.

[11] N. K. Das, "Study of multilayer printed antennas," Ph.D. dissertation, Dept. Elect. Comput. Eng., Univ. Massachusetts, Amherst, MA, Sept. 1989.

[12] D. M. Pozar, "Radiation and scattering from a microstrip patch on a uniaxial substrate," *IEEE Trans. Antennas Propagat.*, vol. AP-35, June 1987.

[13] R. Q. Lee *et al.*, "Characteristics of a two-layer electromagnetically coupled rectangular patch antenna," *Electron. Lett.*, vol. 23, no. 20, pp. 1070–1072, Sept. 1987.

[14] H. J. Stalzer, Jr., A. Hessel, and J. Shmoys, "Microstrip stacked strip element phased arrays," *IEEE Trans. Antennas Propagat.*, vol. 38, pp. 770–773, May 1990.

[15] D. M. Pozar, "Analysis of an infinite phased array of aperture

coupled microstrip patches," *IEEE Trans. Antennas Propagat.*, vol. 37, pp. 418-428, Apr. 1989.

[16] N. K. Das and D. M. Pozar, "Full-wave analysis of material, surface wave and radiation losses in multilayered printed transmission lines," *IEEE Trans. Microwave Theory Tech.*, vol. 39, pp. 54-63, Jan. 1991.

[17] N. K. Das and D. M. Pozar, "Generalized spectral-domain Green's function for multilayer dielectric substrates with applications to multilayer transmission lines," *IEEE Trans. Microwave Theory Tech.*, vol. MTT-35, pp. 326-335, Mar. 1987.

[18] —, "PCAAMT, personal computer aided analysis of multilayer transmission lines," Version 1.0, User's Manual, Antenna Design Associates, Inc., Leverett, MA, June 1990. (Also, N. K. Das and D. M. Pozar, "Perform full-wave multilayer analysis on a PC," *Mirowaves and RF Mag.*, pp. 125-132, Feb.1992.

[19] —, "Printed antennas in multiple layers: General considerations and infinite array analysis using a unified method," in *Proc. Inst. Elec. Eng. Int. Conf. Antennas Propagat.*, ICAP, Univ. Warwick, UK, pt. I, Apr. 1989, pp. 364-368.

[20] P. L. Sullivan and D. H. Schaubert, "Analysis of an aperture coupled microstrip antenna," *IEEE Trans. Antennas Propagat.*, vol. AP-34, pp. 977-984, Aug. 1986.

[21] D. M. Pozar, "A reciprocity method of analysis of printed slot and slot coupled microstrip antennas," *IEEE Trans. Antennas Propagat.*, vol. AP-34, pp. 1439-1446, Dec. 1986.

[22] N. K. Das and D. M. Pozar, "Analysis and design of series-fed arrays of printed-dipoles proximity-coupled to a perpendicular microstripline," *IEEE Trans. Antennas Propagat.*, vol. 37, pp. 435-444, Apr. 1989.

[23] M. Kominami, T. Takei, and K. Rokushima, "A printed dipole electromagnetically coupled to a microstrip feed line," *Proc. Int. Symp. Antennas Propaga.* (ISAP), 1985, pp. 93-95.

Accurate Characterization of Planar Printed Antennas Using Finite-Difference Time-Domain Method

Chen Wu, *Member, IEEE,* Ke-Li Wu, *Member, IEEE,* Zhi-Qiang Bi, *Student Member, IEEE,* and John Litva, *Member, IEEE*

Abstract—**The finite-difference time-domain method (FDTD) is used to accurately characterize complex planar printed antennas with various feed structures, which include coaxial probe feed, microstrip line feed, and aperture coupled feed structures. A new coaxial probe model is developed by using a three-dimensional FDTD technique. This model is shown to be an efficient and accurate tool for modeling coaxial-line fed structures. A novel use of a dispersive absorbing boundary condition is presented for a printed antenna with a high dielectric constant. All the numerical results obtained by the FDTD method are compared with experimental results, and the comparison shows excellent agreement over a wide frequency band.**

I. Introduction

The popularity of planar printed antennas has steadily increased over the past decade, or so, due to a number of advantages such as low cost, low weight, low profile, conformability with existing structures, and ease of fabrication and integration with active devices. During this time they have become an important area of activity within the antenna community and have led to a major innovation in antenna theory. Usually, printed antennas are fabricated on a substrate, or on a number of substrates backed by a metallic sheet (the ground plane). The radiating elements, consisting of thin metallic patches or slots in a metallic sheet, are located at an interface, commonly consisting of a dielectric and air. Multilayered or stacked structures are often used to increase antenna bandwidth. This can be achieved, for example, by simply introducing an air gap between the dielectric layers. Usually, the bandwidth can be increased to more than 10%. Practically, there are three common structures that are used to feed planar printed antennas. These are coaxial probe feeds, microstrip line feeds, and aperture-coupled feeds. The coaxial-fed structure is often used in a single element or a small array because of the ease of matching its characteristic impedance to that of the antenna; and, as well as, the parasitic radiation from the feed network tends to be insignificant. Furthermore, it can also be used as the transition from a printed circuit located on one side of a substrate to the printed antenna on the other side. Compared to probe feeds, microstrip line-fed structures are more suitable for larger arrays due to the ease of fabrication and lower costs, but the serious drawback of this feed structure is the strong parasitic radiation [1]. The aperture-coupled structure has all of the advantages of the former two structures, and isolates the radiation from the feed network, thereby leaving the main antenna radiation uncontaminated. All three of these practical feed structures will be discussed in this paper.

To date, many numerical techniques [1]–[6] have been developed to analyze planar printed antennas in the spectral domain. For coaxial-fed patch antennas, the earliest model to be adopted for full wave analysis is the delta current source model [2]. The model is based on the use of sinusoidal expansion modes and the assumption that the current on the probe is constant. The assumption restricts the model to the point where reasonable results can be obtained only near the resonant frequency of the patch antenna. Another popular model is based on sophisticated attachment models [3], in which the excitation current was spread over a charge cell. This model was developed to be compatible with the rooftop basis functions. Unfortunately, the resulting matrix needs to be carefully treated because it is severely ill conditioned in the vicinity of the resonant frequency. Recently, a more accurate spectral domain model was developed [4], in which the fringing field is replaced by a frill of magnetic current. However, the discontinuity between the coaxial line and the patch substrate, as well as the higher mode near the connector region, cannot be easily accounted for, even though a primary transverse electromagnetic (TEM) mode excitation concept is incorporated in the model. It is found that the spectral domain methods can provide a more accurate model for microstrip line-fed antennas than that for coaxial-fed antennas, even though some non-practical assumptions must be imposed in the line-fed model. A number of assumptions, such as the transverse directed currents [5] are not being taken into account and little consideration being given to contributions from higher modes propagating down the feed line, will cause the numerical results to diverge as the frequency increase. Furthermore, when a antenna consists of a multilayered structure, the spectral domain methods become more difficult to use because of the complexity of the Sommerfeld-type integral treatment.

The finite-difference time-domain (FDTD) method has been

Reprinted from *IEEE Trans. Antennas Propaga.*, vol. 40, no. 5, pp. 526–534, May 1992.

widely used to solve electromagnetic problems since 1966. Because Maxwell's equation are discretized directly, using central difference in both space and time, the FDTD method is more flexible for modeling complex structures. In the last few years, a number of investigators have used the FDTD method to analyze microstrip problems [7]–[9], but in the case of the coaxial-line feed problem the analysis is based on assumptions that deviate from practice. For example, the discontinuity between the coaxial line and patch region is replaced by an equivalent lump resistance, and as well, the characteristic impedance of the coaxial line is not included in the model [7]. Obviously, it is very difficult to obtain an accurate equivalent resistance to incorporate all of the effects of the discontinuity near the connector, especially if the modeling is being carried out over a wide frequency range. On the other hand, although a number of researchers have given attention to modeling line-fed printed antennas using the FDTD method, as of yet, none has addressed the problem of strong dispersion when the dielectric constant is high. This situation will be addressed here using a dispersive absorbing boundary condition.

In this paper it will be shown that the FDTD method provides a technique for accurate modeling of planar printed antennas. There are three features of this full-wave analysis technique that will be highlighted. First, rather than being limited to a treatment of simple printed antenna structures, this study focuses on various complex printed antennas, such as coaxial-fed stacked microstrip antennas, microstrip line-fed aperture coupled stacked microstrip antennas, and printed slot antennas. Second, a new coaxial feed model is presented, which provides a robust description of probe feeds, as well as allowing for modeling of complex printed antennas. The model takes into account contributions from the higher order modes at the junction between the probe and the antenna. The validity of the model is demonstrated by a comparison of simulated and experimental results. The example, which will be discussed in detail, is the coax-to-microstrip transition. This problem often occurs in practical printed antenna designs. The third feature of this paper is the novel use of a dispersive absorbing condition. Its implementation will be shown to be quite straightforward. This boundary condition is useful in analyzing printed antenna structures which contain microstrip lines, where the dielectric constant of the substrate is high.

The antenna structures that are analyzed in this paper can be considered to be representative of printed antenna structures. Also, the results of the sophisticated numerical treatment will be shown to be in excellent agreement with the experimental results over a very wide frequency range, the experimental results that are used to validate the numerical modeling were obtained using the HP8510B network analyzer. Details with regard to calibration and measurement error will be provided in the following sections.

II. Numerical Implementations and Experimental Considerations

The FDTD method is formulated using a central difference discretization of Maxwell's curl equations in both time and space. Yee's original algorithm [10] solving Maxwell's equations in three dimension is adopted. The field values on the nodal points of the discretized finite volume are calculated in a leapfrog fashion. In order to enhance the capabilities of the FDTD method with planar antennas, two major developments are used in conjunction with the algorithm. The first is a coaxial feed model and the second is the dispersive boundary condition.

A. Leapfrog Algorithm

Since the FDTD algorithm is well known, only the fundamentals of its operation will be described here. For simplicity, the antenna substrates will be assumed to be isotropic, homogeneous and lossless. With these assumptions, Maxwell's curl equations can be expressed as

$$\mu \frac{\partial \mathbf{H}}{\partial t} = -\nabla \times \mathbf{E} \tag{1}$$

$$\epsilon \frac{\partial \mathbf{E}}{\partial t} = \nabla \times \mathbf{H} \tag{2}$$

and may be discretized by using the central difference scheme. The central difference technique reduces the round-off error for accuracy to the second-order. With time and space discretized, the **E**- and **H**-fields are interlaced within the spatial 3-D grid. All of these points are brought to light by the leapfrog formula; a representative sample of which is given by

$$\begin{aligned} E_x^{n+1}(i,j,k) &= E_x^n(i,j,k) + \frac{\Delta t}{\epsilon} \\ &\cdot \left[\frac{H_z^{n+1/2}(i,j+1,k) - H_z^{n-1/2}(i,j,k)}{\Delta y} \right. \\ &\left. - \frac{H_y^{n+1/2}(i,j,k+1) - H_y^{n-1/2}(i,j,k)}{\Delta z} \right]. \end{aligned} \tag{3}$$

The time step in (3) must be limited by the stability criterion

$$\Delta t \leq \frac{1}{v_{\max}\sqrt{\dfrac{1}{\Delta x^2} + \dfrac{1}{\Delta y^2} + \dfrac{1}{\Delta z^2}}} \tag{4}$$

where Δx, Δy, and Δz are the space steps in the x-, y-, and z-directions. The quantity, Δt, is the time step and $v_{\max}$ is the maximum velocity in the computational domain.

At this point, Maxwell's equations have been replaced by a system of computer recognizable finite-difference equations. The leapfrog algorithm is able to start working as soon as the boundary conditions are set up. The excitation plane is a special component of the boundary plane, which needs to be treated carefully when setting up the problem.

B. Excitation Treatment

For planar printed antenna problems, microstrip lines and coaxial probes are the basic structures used as feeds. It is assumed that the fields in the computational domain are identically zero at time $t = 0$. The Gaussian pulse is used as the source of excitation because its smooth Gaussian shaped spectrum can provide information from dc to the desired frequency simply by adjusting the width of the pulse.

In the case of microstrip line or microstrip line-fed problems, the electric or magnetic wall condition is used at the front plane of the device, i.e., at the point at which the wave is launched. An impulse of vertical electric field is applied underneath the microstrip line as the excitation. It is a plane in the spatial domain and has a Gaussian shape in the time domain. Although a fictitious source is used, the boundary conditions will force the field to take on a realistic distribution after the wave propagates a distance of a few lattices. Once the Gaussian pulse is well clear of the front plane, the front plane is shifted forward a few lattices and is transformed into an absorbing boundary. Because the dominant mode for the microstrip line is the *quasi*-TEM mode, which is known to be dispersive, the dispersive characteristics of the waves propagating on the line must be taken into account by using dispersive absorbing condition. This becomes more important when the dielectric constant of the substrate is very high, for example $\epsilon_r = 10.2$.

From a knowledge of the modes that exist on a coaxial line, a simple field distribution can be specified at the excitation plane, i.e., the plane between the feed and the antenna, in such a way that the field components in the rectangular coordinate system take on the projected values of the analytically derived radius-field-distribution. The non-TEM modes that are excited by the nonphysical excitation will decay after propagating at most a few lattices. The only mode which is able to propagate down the coaxial line is the TEM mode. Because the TEM mode is a nondispersive wave, the first-order absorbing boundary will absorb almost all the wave reflected from the antenna to excitation plane of the coaxial line.

C. Coaxial Feed Modeling

The coaxial line-fed connection is a critical part of coaxial-fed patch antennas and needs a special treatment. The curved boundary of the inner and outer conductors of a coaxial line is approximated by staircasing, and the tangential component of the **E**-field is forced to zero at the conductor surface. For the purpose of fitting the numerical coaxial line with the lattice, the numerical characteristic impedance of an SMA connector and its coaxial line is chosen to be about 47.0 Ω over a broad frequency range.

As shown in Fig. 1, the inner conductor of the coaxial line is attached on the patch antenna going through the dielectric substrate, and the outer conductor is connected to the ground plane. In this model, the antenna is divided into two computational regions. One is the coaxial line region and the other is the microstrip components region. The advantage of using two regions is that the electromagnetic field in the coaxial line can be defined by a small matrix, so that the computational space as well as CPU time expended in the coaxial line region is less than 2% of that expended in a single patch antenna region. Although the boundary of the coaxial line is approximated by using staircasing, the extent to which waves are scattered into the coaxial line is largely determined by the

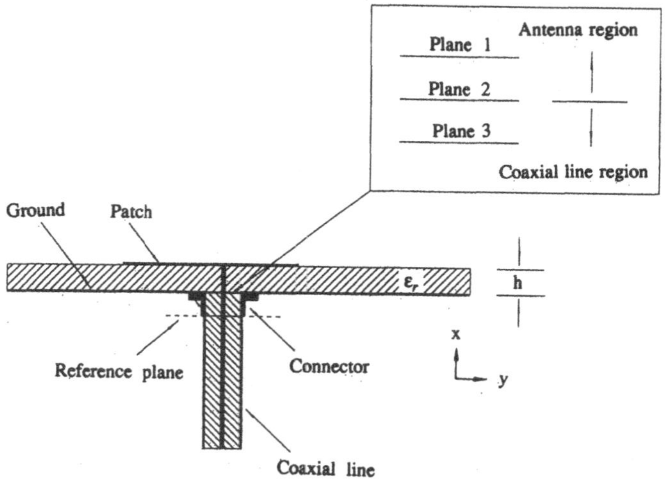

Fig. 1. Side view of a coaxial probe-fed printed antenna.

characteristic impedance of the coaxial line, i.e., its electric characteristics, but seldom upon the specific shape, i.e., its physical characteristics [11]. It is interesting to observe that a very good numerical result can be obtained provided that the numerical characterization impedance of the coaxial line is almost the same as that of the coaxial line used in the measurement.

The two computational regions must be carefully merged near the ground plane. Fig. 1 shows how the two regions are connected. The subscript a and c refer to the field in the antenna and coaxial line regions, respectively. The lattices are the same near the interface plane (plane 2), which is always located on the $E_y - E_z - H_x$ plane in Yee's lattice. Planes 1 and 3 are located at half a lattice immediately above and below plane 2. The E_x, H_y, and H_z components are located on these two planes. The fields in these two computational domains are calculated separately during each time iteration. In the interface region, the **H**-field components can be calculated by the following:

$$\begin{aligned}
h_{xx} &= \left(H_x^{(c)}(i,j,k)_c + H_x^{(a)}(i,j,k)_a\right)/2 \\
H_x^{(c)}(i,j,k)_c &= h_{xx} \\
H_y^{(c)}(i+1,j,k)_c &= H_y^{(a)}(i+1,j,k)_a \\
H_z^{(c)}(i+1,j,k)_c &= H_z^{(a)}(i+1,j,k)_a \\
H_x^{(a)}(i,j,k)_a &= h_{xx} \\
H_y^{(a)}(i,j,k)_a &= H_y^{(c)}(i,j,k)_c \\
H_z^{(a)}(i,j,k)_a &= H_z^{(c)}(i,j,k)_c
\end{aligned} \tag{5}$$

and the **E**-field components by

$$\begin{aligned}
e_{yy} &= \left(E_y^{(c)}(i,j,k)_c + E_y^{(a)}(i,j,k)_a\right)/2 \\
e_{zz} &= \left(E_z^{(c)}(i,j,k)_c + E_z^{(a)}(i,j,k)_a\right)/2 \\
E_x^{(c)}(i+1,j,k)_c &= E_x^{(a)}(i+1,j,k)_a \\
E_y^{(c)}(i,j,k)_c &= e_{yy} \\
E_z^{(c)}(i,j,k)_c &= e_{zz} \\
E_x^{(a)}(i,j,k)_a &= E_x^{(c)}(i,j,k)_c \\
E_y^{(a)}(i,j,k)_a &= e_{yy} \\
E_z^{(a)}(i,j,k)_a &= e_{zz}.
\end{aligned} \tag{6}$$

The two regions are stacked together after the above treatment.

D. Dispersive Boundary Condition

As is well known, a wave traveling down a microstrip line propagates in the waveguide direction. The side wave leakage and radiation are relatively small due to the guiding nature of the metal strip. This is quite similar to a one dimensional propagation problem. Based on the above observations, the first-order boundary condition, i.e.,

$$\left(\frac{\partial}{\partial z} - \frac{1}{v_i}\frac{\partial}{\partial t}\right)E = 0 \tag{7}$$

is usually used, where E represents the tangential electric field components relative to the boundary wall and v_i represents the velocity of propagation of the fields. If this condition is used solely, it is found that the reflections from the boundary can be quite large because the boundary condition only acts as a good absorber at the velocity v_i. Therefore, a dispersive boundary condition which can absorb fields in a wide frequency band needs to be used.

In fact, many wide-angle absorbing boundary condition can be adopted for dispersive problems. For example, it can be seen that the following boundary condition, which was originally developed for wide-angle absorption by Higden [12], can absorb plane waves traveling with velocity v_1 and v_2. The condition is given by

$$\left(\frac{\partial}{\partial z} - \frac{1}{v_1}\frac{\partial}{\partial t}\right)\left(\frac{\partial}{\partial z} - \frac{1}{v_2}\frac{\partial}{\partial t}\right)E = 0. \tag{8}$$

It is easily seen that the above boundary condition is fairly absorptive for any linear combination of plane waves propagating with velocity v_1 and v_2. By concatenating several absorbing boundary conditions, as given by (7), the number of velocities at which absorption is optimized can be increased.

E. Frequency Parameters of Interest

To describe the frequency parameters of planar printed antenna or the properties of the coax-to-microstrip transition, the frequency dependent generalized scattering matrix can be used, which is defined as

$$S_{ij}(f) = \left[\frac{V_i^- \sqrt{Z_{oj}}}{V_j^+ \sqrt{Z_{oi}}}\right]_{V_{\text{others}}^+ = 0}, \qquad i, j = 1 \text{ or } 2. \tag{9}$$

Ports 1 and 2 represent the coaxial line port of microstrip line port, respectively, and Z_{oi} is the characteristic impedance of the ith port. $V_i^+(f)$ and $V_i^-(f)$ are the incident and reflected voltage waves at the ith port, which are given from the Fourier transform of the voltages in the time domain,

$$V_i^{\pm}(f) = F\{V_i^{\pm}(t)\}. \tag{10}$$

Fig. 2(a) shows a microstrip line-fed rectangular patch antenna with a coax-to-microstrip transition. This problem can be solved by considering the equivalent circuit in Fig. 2(b), where the transition from coax-to-microstrip line is given by a reciprocal lossy two-port network.

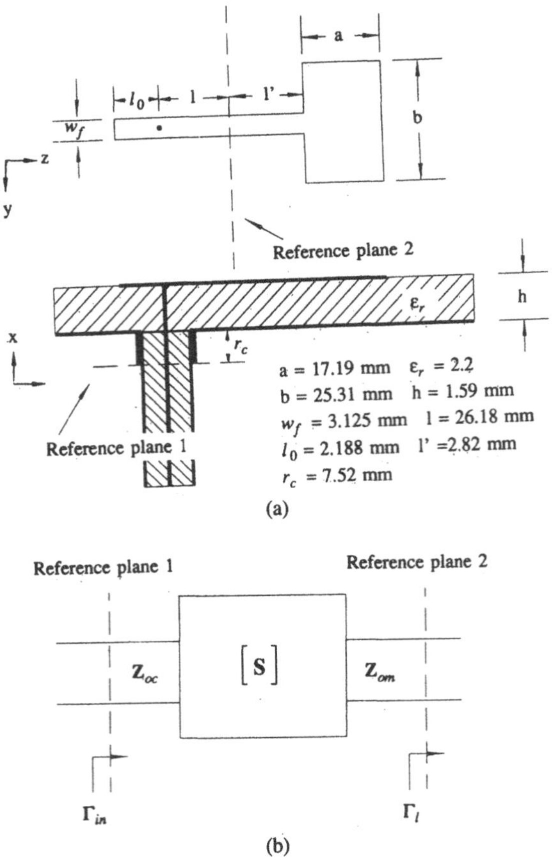

Fig. 2. (a) Microstrip line-fed rectangular patch antenna with coax-to-microstrip transition. (b) Equivalent two-port network of coax-to-microstrip transition.

Reference plane 1 is located in the coaxial line, where the characteristic impedance is Z_{oc}, and Γ_{in} is the reflection coefficient looking into the antenna from plane 1. Once Γ_{in} and $[S]$ for the transition are given, the reflection coefficient Γ_l, which is defined somewhere on the microstrip line can be written as

$$\Gamma_l = \left(S_{22} + \frac{S_{12}S_{21}}{\Gamma_{in} - S_{11}}\right)^{-1}. \tag{11}$$

Simple transmission line theory can then be used if the reference planes need to be shifted along the transmission lines.

Many techniques can be used for deriving the antenna radiation pattern. For example, one can take direct advantage of the FDTD method, because the field at any time step in the computational domain is known during the simulation process. Using an equivalent principle and assuming that the substrate is infinitely large, the air–dielectric interface can be replaced by a conducting sheet on which is superposed a magnetic current. By applying image theory, the surface magnetic current $\mathbf{M}_s$ can be written as

$$\mathbf{M}_s(f) = 2\mathbf{E}(f) \times \mathbf{n} \tag{12}$$

where $\mathbf{E}(f)$ is the electric field on the air–dielectric interface at a particular frequency and $\mathbf{n}$ is the outward unit vector perpendicular to the interface. After obtaining $\mathbf{M}_s$ and using the free space Green's function of magnetic current, the radiation pattern can be easily obtained.

F. Experimental Considerations

The measurements of the input characteristics of the planar printed antennas under discussion are carried out on an HP8510B network analyzer. To set the reference plane at a specific location, two kinds of calibration techniques are used; one is the standard coaxial line calibration and the other is the TRL calibration. The former can only be used to set the reference plane at the interface between the coaxial-cable and SMA connector (see Fig. 2(a) reference plane 1). The latter can be used to set the reference plane to any place on a line, so that the effect of the coax-to-microstrip transition can be eliminated from the measured results. In the TRL calibration, three calibration kits were required: a Thru line of length l_{thru}, an open-circuit reflect line of length $l_{\text{open}} = l_{\text{thru}}/2$, and a delay line of length $l_{\text{line}} = l_{\text{thru}} + \Delta l$. The resulting reference planes are defined at a distance $l_{\text{thru}}/2$ from the connector to the patch antenna. The characteristic impedance and propagation constant of the three lines must be known for the center frequency and must be the same as those for the line, on which the reference plane is located. Usually $\Delta l = \lambda_g/4$, where λ_g is the waveguide wavelength corresponding to the center frequency in the frequency range of interest. A limitation of the TRL calibration is the fact that only the center-frequency characteristic impedance and propagation constant for the line are used in the calibration. As is well known, the characteristic impedance and effective dielectric constant for a line vary with frequency. The effects of dispersion on the microstrip line can not be taken into account by means of experimental techniques. The limitation brought about by dispersion restricts the bandwith of the measurements, as well as causing measurement errors, especially when the dispersion is serious.

III. Numerical Results and Discussions

To validate the proposed coaxial feed model and to show the improvement that is brought about by using the dispersive boundary condition, four typical complex structures of planar printed antennas are analyzed. These are: a microstrip line-fed rectangular patch antenna with coax-to-microstrip transition, a coaxial-fed stacked patch antenna with an air-gap between two layers, a slot antenna that lies on the ground of a microstrip line, and a microstrip line-fed aperture coupled stacked rectangular patch antenna. Both simulated and measured results will be provided in each case.

With one exception, the first-order absorbing boundary condition (ABC) is used in this study for the top plane, as well as the side walls of the computational domain. The one exception is the strongly dispersive case. In comparison with the higher order ABC, the first-order ABC is considerably simpler to implement. Although a little more computational overhead would be expected. Furthermore, when the lattice size is small compared to the wavelengths of interest, the fields at a boundary are strongly correlated with the fields one lattice inside the boundary. Thus in this case, which is general to the problems being analyzed in this paper, the first-order ABC gives highly accurate and therefore acceptable results.

A. Microstrip Line-Fed Rectangular Patch Antenna with Coax-to-Microstrip Transition

The example shown in Fig. 2(a) consists of two parts. One is a simple patch antenna, which has been studied extensively. The other is a coax-to-microstrip transition. To the best of the authors' knowledge, this transition problem has never been adequately solved over a wide frequency band by any numerical technique, certainly not using analytic analysis. In practice, the transition is widely used in various printed antenna structures, as well as printed circuits. The de-embedding of the effects of the transition is urgently needed for carrying out accurate practical design. In the numerical analysis to follow, reference plane 1 is located at $19\Delta x$ away from the ground plane, and plane 2 is located at $84\Delta z$ from the connector, where $\Delta x = 1.272\Delta h$, $\Delta z = \Delta y = \Delta h = 0.315$ mm. In this example, the microstrip line has a characteristic impedance of 63 Ω at 6 GHz. The first order absorbing boundary condition is applied at a distance of $64\Delta h$ away from the patch. The numerical coaxial line length is $100\,\Delta x$, and the Gaussian pulse is applied at the second grid with respect to the bottom of the coaxial line. The 5% pulse width of the pulse corresponds to 15 space steps with the pulse maximum at $100\,\Delta t$.

The transition from a coaxial line to a microstrip line can be represented by the two-port network shown in Fig. 2(b), and described by *S*-parameters. Due to the existence of surface and radiation waves, and the fact that only isotropic substrates are considered, the network is lossy, as well as being reciprocal. The *S*-parameters for this example, calculated by the FDTD method, are shown in Fig. 3. It can be clearly seen that at low frequencies electromagnetic energy is easily transmitted between a coaxial line and a microstrip line. However, at a higher frequency the transmission characteristics degenerate due to the higher order modes at the discontinuity and radiation loss in the microstrip line. It is interesting to note that in this example the energy is seriously blocked at a frequency of around 18 GHz. This blockage is caused mainly by the open end stub, which shorts the circuit at a length of about one quarter of the waveguide wavelength.

The reflection coefficient Γ_{in} of the entire antenna is measured at the reference plane 1, which is located on the coaxial line. Fig. 4 shows the magnitude and phase of Γ_{in} from 3 to 9 GHz (group 1). The measured results are in very good agreement with the calculated results. The reflection coefficient tells us that the antenna is resonant at frequency of 5.53 GHz. The equivalent magnetic current for the air–dielectric interface and at the resonant frequency is given in Fig. 5. In the diagram, the direction and length of each arrow indicates the orientation, phase, and magnitude of the magnetic current at that point. It presents a very clear picture of how the antenna works. At the resonant frequency, the dominant mode on the patch is the fundamental (1, 0) mode. On the two wider edges of the patch, the magnetic current,

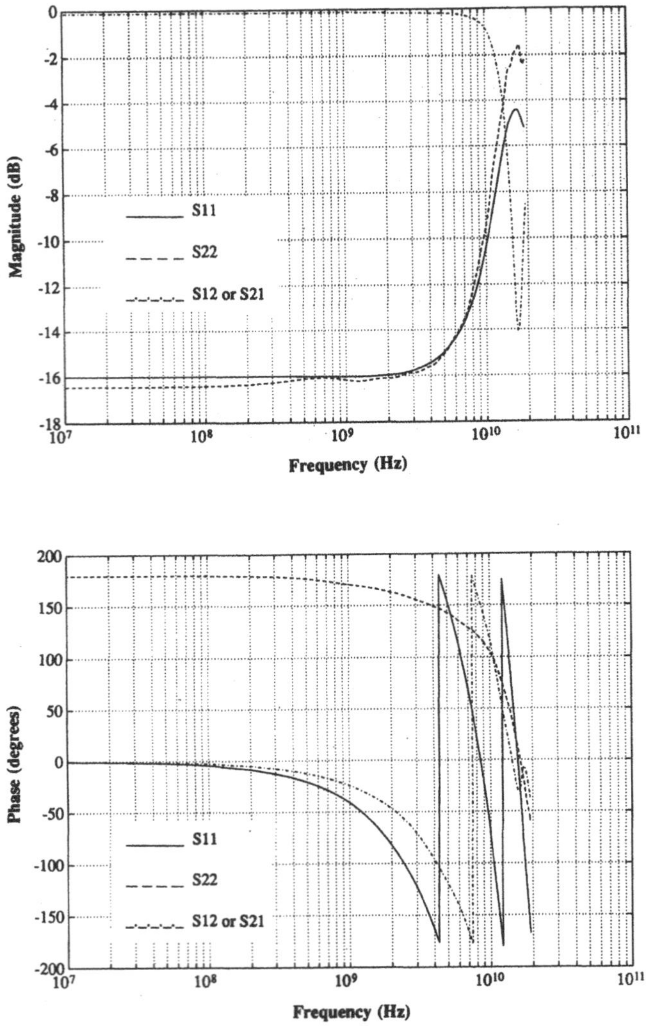

Fig. 3. *S* parameters of coax-to-microstrip transition.

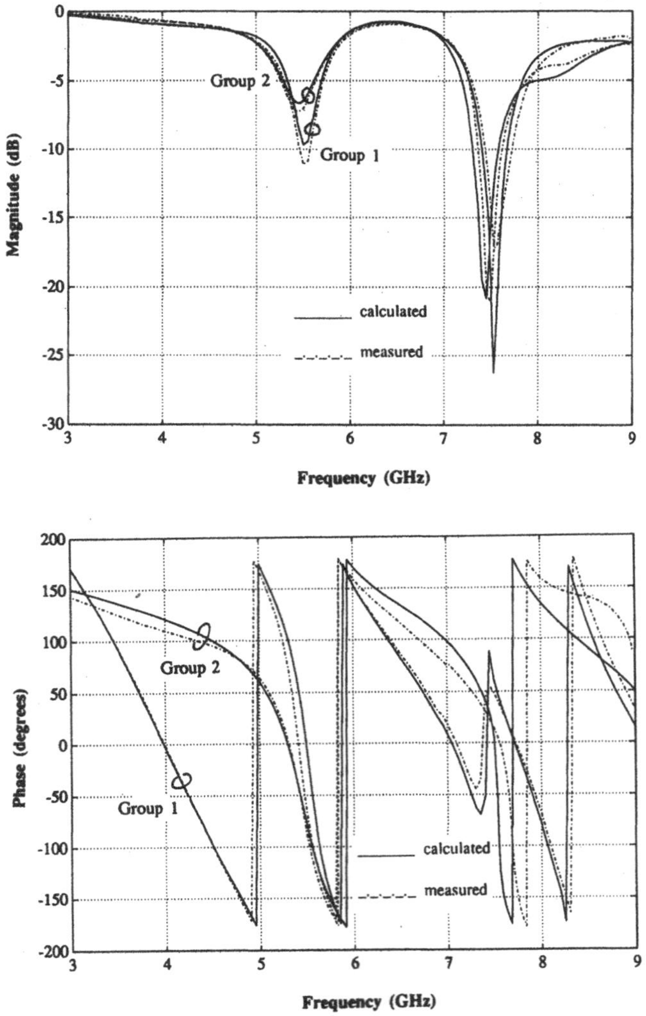

Fig. 4. Reflection coefficients of rectangular patch antenna with coax-to-microstrip transition. Group 1 and Group 2 is defined at reference plane 1 and 2, respectively.

which has almost the same magnitude and phase, contributes mainly to the far field. On the two narrow edges, the current phase changes. The radiation from these currents will almost be cancelled in the far field range.

By using the TRL calibration, the reflection coefficient Γ_l of the patch antenna defined at reference plane 2 (see Fig. 2(a)) is measured and is shown by the dashed line in group 2 of Fig. 4. With the help of (11), the numerical value for Γ_l is obtained by converting Γ_{in}, which is calculated by the FDTD method (see solid line of group 1 in Fig. 4), from reference plane 1 to reference plane 2, using the previously calculated *S*-parameters. It follows from the close agreement that the *S*-parameters of the transition obtained from the FDTD method are correct.

B. Coaxial Probe-Fed Stacked Rectangular Patch Antenna

To show the applicability of the coaxial feed model to more complicated printed antenna structure, the coaxial probe-fed stacked patch antenna is investigated. As shown in Fig. 6, the antenna consists of two patches. The coaxial probe is connected to the lower patch. An air gap is introduced between the two patches in order to increase the

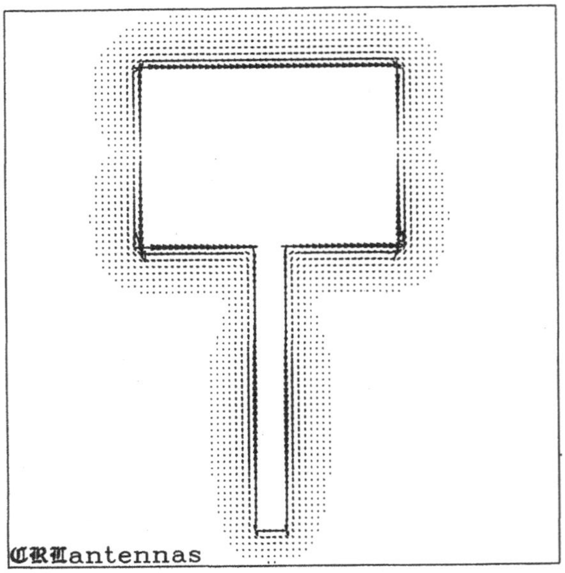

Fig. 5. Magnetic current distribution of the rectangular patch antenna at the interface between air and substrate.

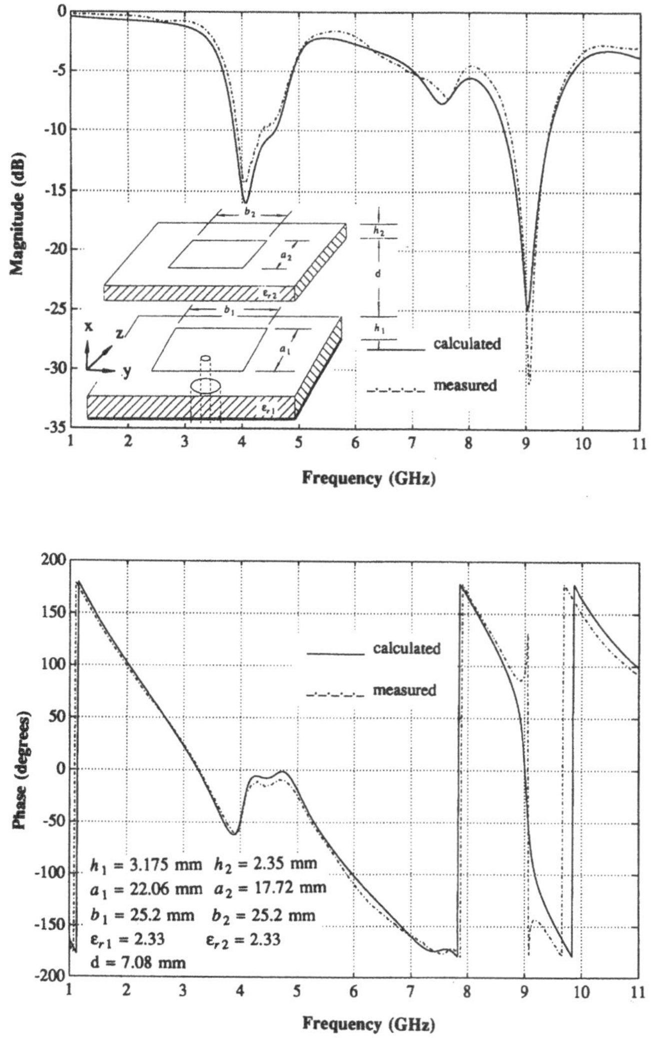

Fig. 6. Reflection coefficient of a coaxial probe-fed stacked patch antenna.

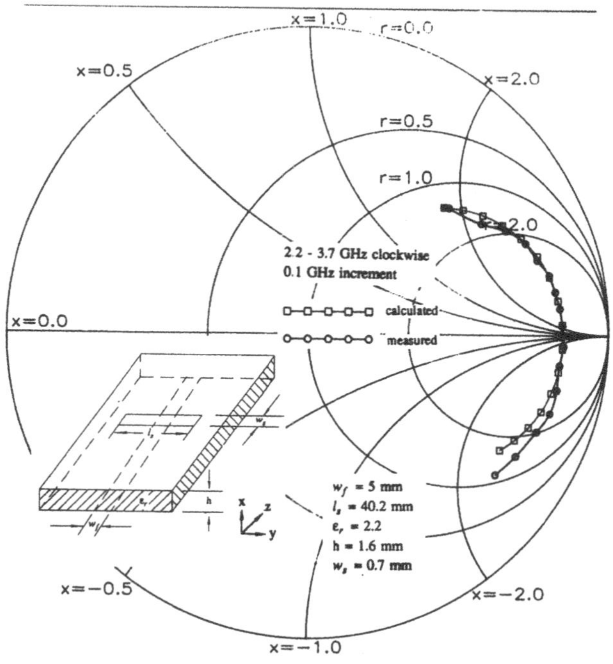

Fig. 7. Input impedance of a microstrip line-fed slot antenna.

bandwidth of the antenna. In this example, the feed probe is located at a point which is (12.5, 4) mm from the low left corner of the lower patch.

Fig. 6 gives the measured and calculated results for the reflection coefficient of the stacked patch antenna. It is obvious that the comparison is excellent both in magnitude and phase within a wide frequency range. From the reflection coefficient we discover that the antenna has a bandwidth that exceed 16% at the first resonate frequency, within which the return loss is less than -10 dB.

C. Microstrip Line-Fed Slot Antenna

Fig. 7 shows a microstrip line-fed slot antenna, which was analyzed previously in [13] using the spectral domain technique. The computational parameters used in the FD-TD analysis are

$$\Delta h = 0.4 \text{ mm}$$

$$\Delta x = \Delta h, \quad \Delta y = 1.5\Delta h, \quad \Delta z = 1.75\Delta h$$

$$\Delta t = 0.515\Delta h / c$$

and the first-order absorbing boundary is applied. It is assumed that the field across the slot is almost constant; therefore, only one lattice is used.

The input impedance of the slot antenna in the frequency range from 2.2 to 3.7 GHz is also given in Fig. 7. The reference plane is defined at the center of the slot. The square marks represent the FDTD simulated results and the circular marks give the measured results from [13].

D. Aperture-Coupled Stacked Microstrip Rectangular Patch Antenna

The treatment of the aperture-coupled patch antenna [14] is similar to that of the traditional microstrip antenna except that the microstrip patch antenna is located on one substrate with a relative dielectric constant ϵ_{pr} and a feed network on another substrate with relative dielectric constant ϵ_{fr}. Usually, ϵ_{fr} is higher than ϵ_{pr} in order to reduce the dimensions of the feed network. These two substrates are separated by a common ground plane. In order to couple electromagnetic power from the feed network to the patch antenna, an electrically small opening or aperture is made in the ground plane, as shown in Fig. 8. Since the radiator and the feeder are separated by the common ground plane, the radiation from the feed network can be eliminated from the far-field pattern. As well, the feed network will be decoupled from the antenna. Because ϵ_{fr} usually has a large value, the microstrip line will be strongly dispersive, thereby degrading the performance of the first-order absorbing boundary condition. From numerical experiments in the time domain, it is observed that the reflected wave for a first order boundary is about ten times greater than that from the dispersive absorbing boundary condition discussed in Section II [15]. Therefore, the dispersive boundary condition is used in the analysis to be carried out. In this example, the distance between the open end of microstrip line and the center of the aperture is 3.8 mm. The two velocities that are selected for designing the

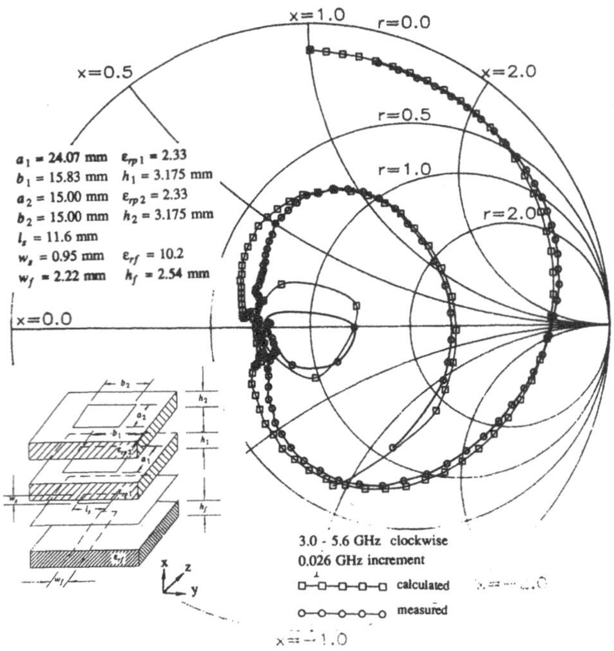

Fig. 8. Input impedance of a aperture-coupled stacked microstrip antenna.

absorbing boundary condition are $v_1 = C/\sqrt{7.12}$ and $v_2 = C/\sqrt{8.5}$. These correspond to frequency 1 and 8 GHz, respectively, where C is the speed of light.

Fig. 8 shows a Smith chart for the input impedance of the aperture coupled stacked patch antenna. Fairly good agreement is observed between calculated and measured results over the frequency band from 3 to 5.6 GHz. This is the band in which the antenna operates most efficiently. Because of the serious dispersion in the microstrip line, it is difficult to design a TRL calibration which is accurate over a wide frequency band. The measurement repeatability of the return loss is about ± 0.05 dB, and phase is about $\pm 8°$. The observed experimental error is primarily due to the uncertainties inherent in the calibration kits that were used for the TRL calibration.

IV. Conclusion

By carrying out a numerical analysis of a number of complex printed antennas, it has been shown that the FDTD method is a very powerful tool for analyzing planar printed antennas. The method can be used to accurately predict all the antenna parameters of interest over a wide frequency range, based on one time-domain simulation. It can provide not only input information for the antennas, but also very detailed field distributions, including the near and far fields. The proposed three-dimensional FDTD coaxial feed model provides a means to address more complicated, but practical printed antenna problems. The validity of the model is demonstrated by comparing the numerical and experimental results for four representative complex antenna structures. A one-dimensional simple dispersive absorbing boundary condition was used when analyzing components of printed antennas with large dielectric constant substrates. In these cases, the wave propagating on the structure suffers serious dispersion. In our experience, the dispersive absorbing boundary can be applied as easily as that of the first order absorbing boundary. The advantage of the dispersive absorbing boundary is that it is defined by the known dispersive characteristics of the transmission line, and it gives second-order performance when the wave propagates in a direction which is normal to the boundary. Finally, it should be noted that advances in the application of the FDTD method to printed antennas require the development of 1) lattices that provide greater numerical efficiency for the analysis of these structures, and 2) the adoption of appropriate signal processing techniques.

Acknowledgment

The authors wish to thank their colleagues, Russ Fralich, for his helpful discussions on this research, and Paul Chung, for help in preparing this manuscript. The authors also wish to thank Sharon R. Aspden of Rogers Corporation for providing the dielectric material used in this research under the auspices of their university program.

References

[1] K. L. Wu, M. Spenuk, J. Litva, and D. G. Fang, "Theoretical and experimental study of feed network effects on the radiation pattern of series-fed microstrip antenna arrays," *Inst. Elec. Eng. Proc.*, pt. H, vol. 138, pp. 238–242, 1991.

[2] D. M. Pozar, "Input impedance and mutual coupling of rectangular microstrip antennas," *IEEE Trans. Antennas Propagat.*, vol. AP-30, pp. 1191–1196, 1982.

[3] J. R. Mosig and F. E. Gardiol, "General integral equation formulation for microstrip antennas and scatterers," *Inst. Elec. Eng. Proc.*, pt. H, vol. 132, pp. 424–432, 1985.

[4] W. C. Chew, Z. Nie, H. Liu, and Y. T. Lo, "Analysis of a probe-fed microstrip disk antenna," *Inst. Elec. Eng. Proc.*, pt. H, vol. 138, pp. 185–191, 1991.

[5] D. M. Pozar and S. M. Voda, "A rigorous analysis of a microstrip line fed patch antenna," *IEEE Trans. Antennas Propagat.*, vol. AP-35, pp. 1343–1349, 1987.

[6] K. L. Wu, J. Litva, R. Fralich, and C. Wu, "Full wave analysis of arbitrarily-shaped line-fed microstrip antennas using triangular finite element method," *Inst. Elec. Eng. Proc.*, pt. H, vol. 138, no. 5, pp. 412–428, 1991.

[7] A. Reineix and B. Jecko, "Analysis of microstrip patch antennas using finite difference time domain method," *IEEE Trans. Antennas Propagat.*, vol. 37, pp. 1361–1368, 1989.

[8] D. M. Sheen, S. M. Ali, M. D. Abouzahra, and J. A. Kong, "Application of the three-dimensional finite-difference time-domain method to the analysis of planar microstrip circuits," *IEEE Trans. Microwave Theory Tech.*, vol. 38, pp. 849–857, 1990.

[9] X. Zhang and K. K. Mei, "Time-domain finite-difference approach to the calculation of the frequency-dependent characteristics of microstrip discontinuities," *IEEE Trans. Microwave Theory Tech.*, vol. 36, pp. 1775–1787, 1988.

[10] K. S. Yee, "Numerical solution of initial boundary value problems involving Maxwell's equations in isotropic media," *IEEE Trans. Antennas Propagat.*, vol. AP-14, pp. 302–307, 1966.

[11] C. Wu, K. L. Wu, Z. Q. Bi, and J. Litva, "Modeling of coaxial-fed microstrip patch antenna by finite difference time domain method," *Electron. Lett.*, vol. 27, no. 19, pp. 1691–1692, 1991.

[12] R. L. Higdon, "Numerical absorbing boundary conditions for the wave equation," *Math. Comput.*, vol. 49, 65–91, 1987.

[13] D. M. Pozar, "A reciprocity method of analysis for printed slot and slot-coupled microstrip antenna," *IEEE Trans. Antennas Propagat.*, vol. AP-34, pp. 1439–1446, 1986.

[14] C. Wu, J. Wang, R. Fralich, and J. Litva, "A rigorous analysis of an aperture-coupled stacked microstrip antenna," *Microwave Opt. Tech. Lett.*, vol. 3, pp. 400–404, 1990.

[15] Z. Q. Bi, K. L. Wu, C. Wu, and J. Litva, "A dispersive boundary condition for microstrip components analysis using FD-TD method," *IEEE Trans. Microwave Theory Tech.*, vol. 40, pp. 774–777, 1992.

Chapter 6
Microstrip Antenna Array Design

MICROSTRIP antenna arrays comprised of printed patches and printed lines for the feed network represent the goal of much of the research-and-development activities over the past two decades, and many successful examples of this type exist in the literature and in operational systems. The design of microstrip antenna arrays is fundamentally the same as the design of other types of arrays, so ultimately performance is dependent upon achieving the desired amplitude and phase distribution of currents on the elements of the array for all frequencies and scan angles of interest. The effects of mutual coupling can be more significant in microstrip arrays than in some other arrays, leading to scan blindness in severe cases [1], [2]. However, nonscanning arrays with a broadside beam are often required in practice, and these arrays frequently can be designed without considering mutual coupling effects.

The configurations of arrays to meet specific needs are nearly as varied as the applications that inspire them; therefore, it is difficult to select representative papers that will be generally useful to designers. The papers selected here provide insights into some important design considerations, and the effects of feedline radiation and loss on the performance of monolithic arrays. The first paper is a review by Schaubert of microstrip array design. The paper by Jones, Chow, and Seeto describes the use of the transmission line model to design series-fed linear arrays. Present workstations and CAD programs allow for more exact analysis of each element of the array and, in some cases, analysis of entire arrays, but the underlying design methodology is similar to that in this paper and another by Metzler [3], so it may be useful for designers who lack experience in this area.

The last four papers in this chapter deal with array performance. The papers by Hall and Hall, and by Levine, Malamud, Strikman, and Treves, provide extremely useful insights into the effects of loss and radiation from a corporate feed network that is printed on the same substrate surface as the patches. Limitations on gain, sidelobe level, and cross-polarization are described. The paper by Pozar and Kaufman describes a low-sidelobe array and presents several important considerations related to achieving low side lobes from microstrip arrays. The final paper, by Huang, presents a practical approach to improving the performance of fixed-beam arrays of moderate gain.

The review paper by Schaubert contains several additional references that may be useful for the design of arrays for specific applications. In addition, [4] and [5] contain further information about arrays with corporate feed networks printed on the surface of the substrate. Several millimeter wave arrays are described in [6] and a series-feeding scheme for multiple beam applications is described in [7]. Rampart lines are a simple form of the series-fed array that work through constructive addition of small amounts of radiation from several discontinuities along a microstripline. Design information for these types of arrays can be found in [8] and [9]. In [10], an undesirable surface-wave resonance on a moderate size substrate is identified as the cause of serious pattern degradation of a small array.

References

[1] D. M. Pozar and D. H. Schaubert, "Scan blindness in infinite phased arrays of printed dipoles," *IEEE Trans. Antennas and Prop.*, vol. AP-32, pp. 602–610, June 1984.

[2] D. M. Pozar and D. H. Schaubert, "Analysis of an infinite array of rectangular microstrip patches with idealized probe feeds," *IEEE Trans. Antennas and Prop.*, vol. AP-32, pp. 1101–1107, Oct. 1984.

[3] T. Metzler, "Microstrip series arrays," *IEEE Trans. Antennas and Prop.*, vol. AP-29, pp. 174–178, Jan. 1981.

[4] P. S. Hall and C. J. Prior, "Radiation control in corporately fed microstrip patch arrays," *Dig. 1986 Journees Internationales de Nice sur les Antennes,* JINA 86, pp. 271–275, 1986.

[5] J. Ashkenazy, P. Perlmutter, and D. Treves, "A modular approach for the design of microstrip array antennas," *IEEE Trans. Antennas and Prop.*, vol. AP-31, pp. 190–193, Jan. 1983.

[6] F. Lalezari and C. D. Massey, "mm-wave microstrip antennas," *Microwave J.*, vol. 30, no. 4, pp. 87–96, Apr. 1987.

[7] S. J. Vetterlein and P. S. Hall, "Novel multiple beam microstrip patch array with integrated beamformer," *Electronics Letters,* vol. 25, pp. 1149–1150, Aug. 1989.

[8] P. S. Hall, "Microstrip linear array with polarisation control," *IEE Proc.*, part H, vol. 130, pp. 215–224, Apr. 1983.

[9] L. Shafai and A. A. Sebak, "Radiation characteristics and polarisation of undulated microstrip line antennas," *IEE Proc,* part H, vol. 132, pp. 433–439, Dec. 1985.

[10] D. H. Schaubert and K. S. Yngvesson, "Experimental study of a microstrip array on high permittivity substrate," *IEEE Trans. Antennas and Prop.*, vol. AP-34, pp. 92–97, Jan. 1986.

Review of Microstrip Antenna Array Techniques

DANIEL H. SCHAUBERT
ELECTRICAL AND COMPUTER ENGINEERING
UNIVERSITY OF MASSACHUSETTS AT AMHERST
AMHERST, MASSACHUSETTS 01003

Abstract—**Microstrip antenna array techniques are reviewed with attention to the basic considerations important to a user or designer. Microstrip arrays are often chosen because of their ruggedness, ease of manufacturing by printed circuit techniques, compatibility with MMICs, and thin, conformal geometry. Parallel and series feeds are often used for microstrip arrays, and space feeds are also possible. Monolithic arrays with the feedlines and radiating patches all on one surface are attractive for some applications, but suffer from spurious feedline radiation and insufficient space for phase shifters and amplifiers. Mutual coupling is always a consideration in antenna arrays, but fixed, broadside beam microstrip arrays can often be designed successfully without regard to mutual coupling. Scanning arrays can suffer from impedance anomalies (blindness), but decreasing the element spacing will alleviate this problem. The paper concludes with several arrays that have successfully overcome particular design problems.**

1. INTRODUCTION

Microstrip antennas are important as single radiating elements, but their major advantages are realized in applications that require moderate size arrays. They can be produced by simple photolithographic techniques as thin, conformal, rugged antennas that are monolithic or integrable, without cables, connectors, or other additional components. Recent advances have increased the bandwidth of microstrip antennas, and the complete integration of active and passive components into an antenna system promises high performance in a reliable and reproducible package. Removal of heat from high-power transmitter arrays is still a challenge, but receive-only arrays with LNAs and phase shifters are reasonably straightforward to design and fabricate.

This paper reviews some of the technology that is useful for evaluating the potential benefits of microstrip antenna arrays in a particular application and for successfully designing ar-rays to meet typical system specifications. Two general topics are covered, array architectures and feed radiation/surface-wave effects, and several particular microstrip arrays are described.

2. ARRAY ARCHITECTURES

Array antennas can be designed to provide a fixed beam of specified shape or a beam that scans in response to a system stimulus. Scanning arrays typically use phase shifters or time-delay devices to provide beam scanning that is relatively independent of the instantaneous frequency or tapped feed lines to produce scanning as the excitation frequency changes. The limited bandwidth of microstrip elements makes them less desirable for use in frequency-scanned arrays, but some successes have been reported [1].

The choice of architecture for an antenna array is dependent on many factors: electrical performance, heat removal, power and logic distribution, operating environment, etc. Microstrip antenna arrays are often chosen when the thin, conformal nature of these antennas is valued, when the ability to expand the size of the array by adding additional "tiles" is important, or when the antenna includes MMIC components that must be mounted onto a microstrip circuit. Within each tile or subarray of a large array, the signal may be distributed via a parallel feed network, a series feed, or a space feed. Parallel and series feeds are most commonly used in microstrip arrays. The use of microstrip as the medium for the feed network as well as the radiating elements allows for easy control of the characteristic impedance, which provides a degree of freedom not so easily obtained in waveguide feed networks. The arrays in Figure 1 illustrate parallel and series feeds and a combination of these incorporated in a dual-polarized array. Parallel feed networks with equal path lengths offer wider instantaneous bandwidth than series feeds, but they also incur higher losses and this contributes to a limitation on the realizable gain achievable by microstrip arrays [2], [3, pp. 186–189].

One of the difficulties encountered in designing planar microstrip arrays is the limitation on space available in the unit cell. Unit cells are typically 0.5–0.8 wavelengths on a side and the radiating patch is typically 0.2–0.4 wavelengths on a side. As is evident from the array in Figure 1a, maintaining sufficient clearance between the feed network and the radiating patches can be difficult in a corporate-fed array with equal line lengths. Including phase shifters and amplifiers makes the problem worse. To alleviate the problems associated with crowding the feed network and the radiators onto a single surface (monolithic construction), the feed network can be placed behind the ground plane. The signal is coupled to the radiating patches by means of plated through vias or aperture coupling [4], [5]. Microstrip or stripline can be used for the feed when it is placed behind the ground plane, and some complicated networks may involve multiple layers with many interconnections.

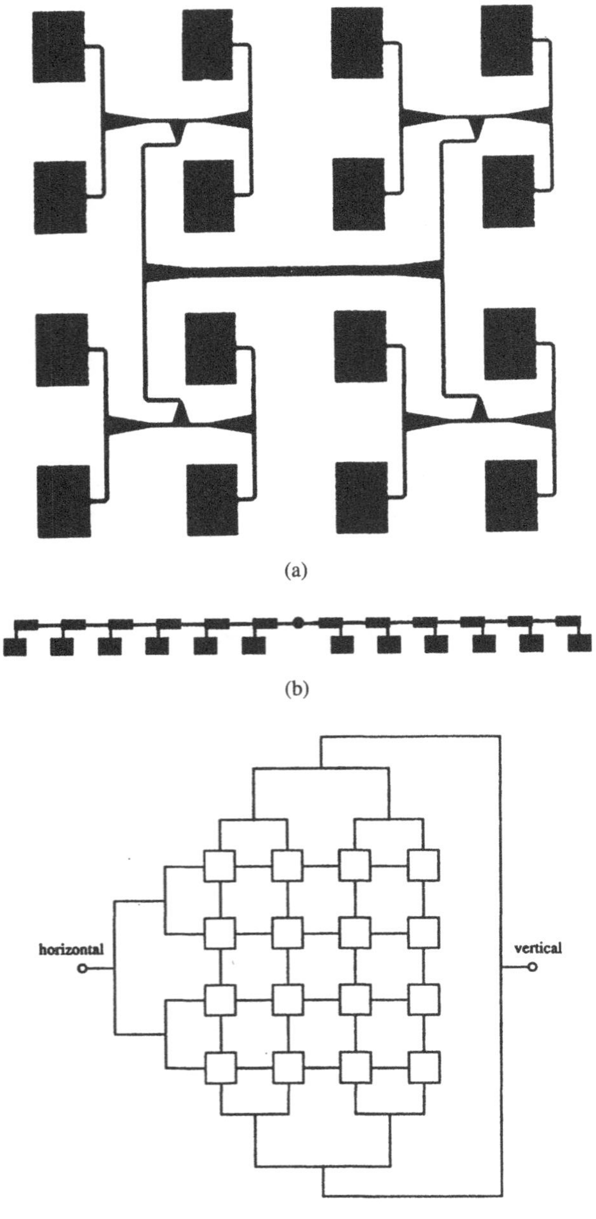

Fig. 1. Microstrip arrays with coplanar feedlines. (a) Parallel feed network (reprinted from [6]). (b) Series feed with impedance changes along feedline. (c) Parallel networks connected to series-fed lines and columns forming a dual-polarized array.

3. Feed Radiation and Surface Wave Effects

Monolithic construction is probably the least expensive way to fabricate the patch arrays and it provides the most reliability because the antenna is a single structure. However, the surface area available within the unit cell is limited and the exposed feed lines will radiate or receive signals, which can degrade the sidelobes and contribute to cross-polarization [6], [7]. Hall and Hall [6] have estimated that feedline radiation in a 16×16 corporate-fed array can degrade the sidelobe level by 10 dB. This is illustrated by the curves in Figure 2, which show the envelope of the sidelobes for an ideal array and the expected radiation from the corporate feed network. The radiated power is plotted in dB relative to the main beam. Chiba, et al.[8] and Huang [9] have shown that cross-polarization generated by higher order modes of a patch antenna can be cancelled by proper use of symmetry. Proper use of symmetry also reduces the effects of feedline radiation, especially in the symmetry planes. Hall and Hall [6] note a significant improvement for circularly polarized arrays when using sequential rotation, which has the appropriate symmetry.

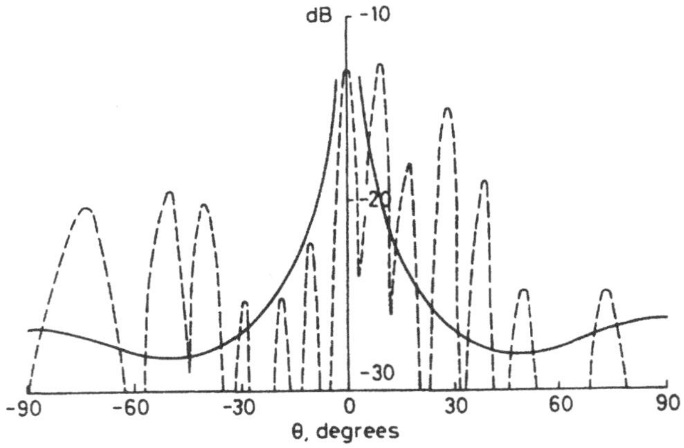

Fig. 2. Predicted radiation from 16 × 16 corporate feed network (- - -) compared to envelop of ideal patch array (__). Reprinted from [6].

Microstrip patch antennas couple power into surface waves that are guided on the substrate[10]. In arrays, these surface waves can increase coupling between elements and diffract from the boundaries of the substrate. Both of these effects are detrimental to array performance, but it has been found that patch arrays with a fixed, broadside beam are not greatly affected by mutual coupling unless extremely low sidelobes are required. However, scanning arrays can be adversely affected by mutual coupling. Impedance and pattern anomalies associated with scan blindness are potential hazards for array designers [11],[12]. The plots in Figure 3 show that the angle at which a scan blindness occurs moves closer to broadside as the

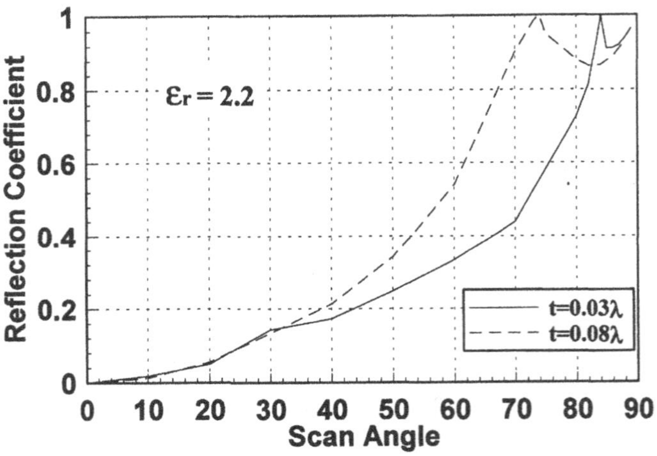

Fig. 3. Active reflection coefficient magnitude of infinite patch array for substrate thicknesses of $0.03\lambda_o$ and $0.08\lambda_o$. Patches are passively matched at broadside. Unit cell size: $0.5\lambda_o \times 0.5\lambda_o$. Patch size: $0.322\lambda_o \times 0.322\lambda_o$ for $t = 0.03\lambda_o$, $0.28\lambda_o \times 0.28\lambda_o$ for $t = 0.08\lambda_o$.

thickness of the substrate increases. The results in [11] and [12] show that scan blindness occurs along circular arcs in u-v scan space and that the blindness also moves closer to broadside as the permittivity of the substrate increases. Reducing the element spacing moves the blindness further from broadside and often can move it completely outside the region of visible radiation, but the increased element density increases the cost of active arrays because each element requires a phase shifter and, perhaps, an amplifier.

4. Other Types of Arrays

There are many variations of microstrip antenna arrays. Some are, like those discussed above, a collection of identical elements simultaneously fed to produce a desired radiation pattern. Others are a combination of two or more different elements that operate together to perform a desired function or produce a desired radiation pattern.

Some arrays resemble a series-fed array, except that there are no special radiating elements. Figure 4 illustrates some of the types that have been demonstrated [3, pp 123–129], [13]. The antennas consist of long microstriplines that radiate because they have bends. The radiation from each bend is small, but when the antenna is properly designed the radiation from all of the discontinuities combines to produce a well-formed beam, the polarization of which is controllable by the antenna's design. Another array that includes a series feed line supporting either traveling or standing waves, and that produces circular polarization, is the strip and slot array of Ito, et al. [14]. A log-periodic version employing overlaid patches capacitively coupled to the feedline has provided a 4:1 bandwidth, which is limited primarily by the lack of scaling caused by the use of uniform thickness dielectric substrates [15].

McGrath has demonstrated a thin lens antenna comprised of back-to-back patch arrays connected by various lengths of microstripline to produce the desired focussing [16]. The lens array is illustrated in Figure 5.

The feed network is one of the most troublesome aspects of a microstrip array. It occupies valuable space, radiates spurious signals, and consumes power through ohmic losses. By parasitically coupling several patches to each driven patch, it is possible to transfer some of the power division tasks of the feed network to the radiating elements. Entschladen and Nagel have demonstrated such arrays with up to fifteen elements, only one of which is driven [17]. A 1280-element array was constructed from 256 clusters of five elements [18]. The center element of each cluster was aperture-fed and the surrounding four elements were parasitically coupled.

A commonly occurring problem in array design is the desire for a single antenna aperture to operate at two widely separated frequencies. Sometimes the elements of a high-frequency array can be interleaved with those of a low-frequency array, but three problems occur when using microstrip antennas. First, it is often difficult to route the feed lines of both arrays on the surface of a single substrate that also contains the ele-

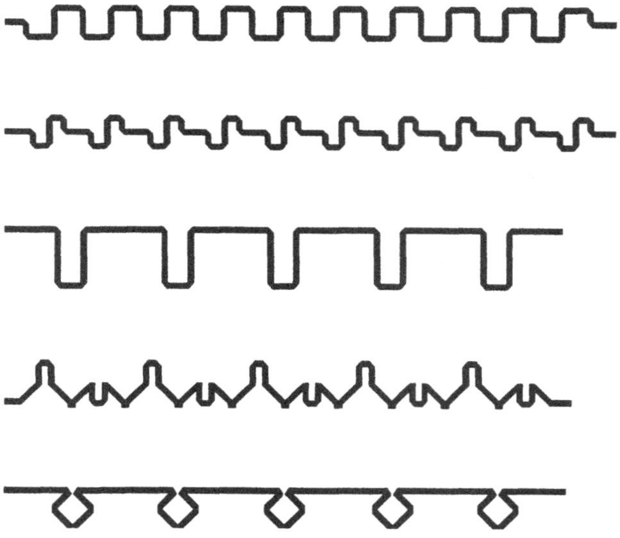

Fig. 4. Linear arrays that radiate because of discontinuities in microstrip lines.

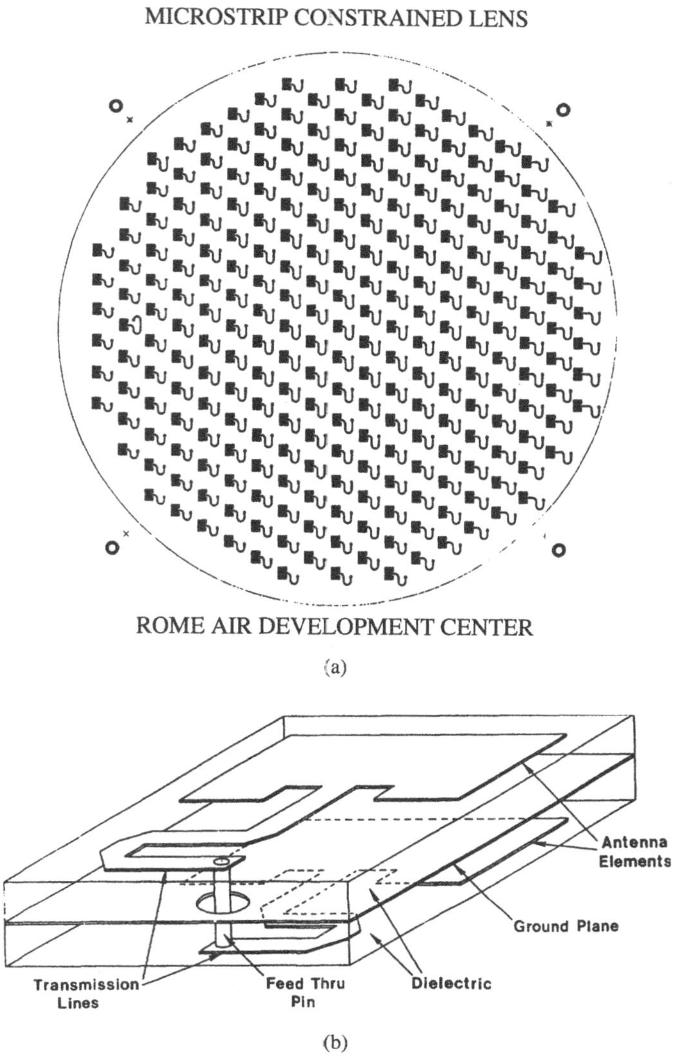

Fig. 5. Thin lens formed by back-to-back patch arrays. (a) Front face. (b) Detail of interconnection. Reprinted from [16].

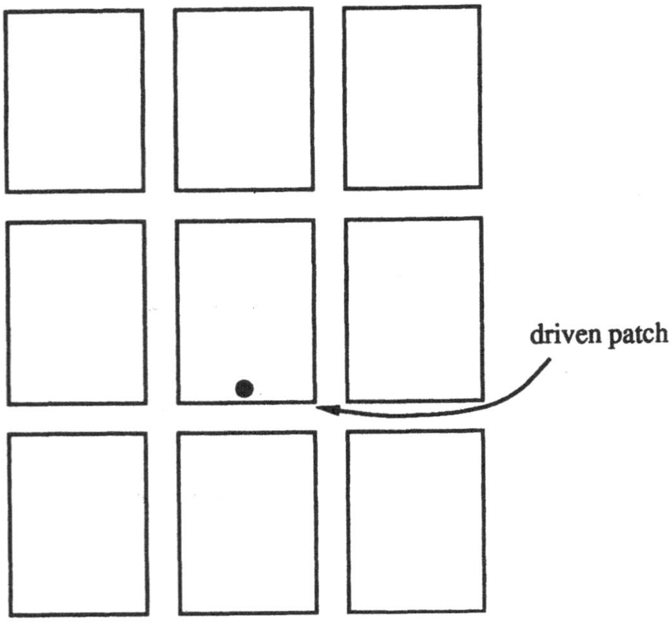

Fig. 6. Nine-element array with one driven patch and eight parasitically coupled patches.

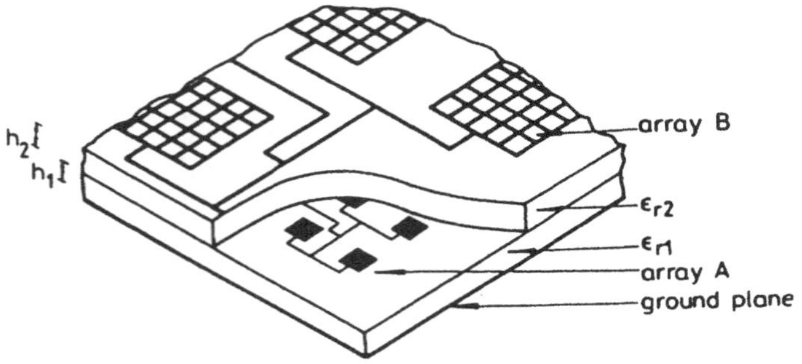

Fig. 7. Superimposed arrays using dichroic surface for two-frequency operation. Reprinted from [19].

ments of both arrays. Second, because the low-frequency elements may be larger than the wavelength of the higher operating frequency, the interleaved elements of the high-frequency array may be so widely spaced that grating lobes appear in the pattern. Third, in order to obtain a reasonable bandwidth at the lower frequency, the substrate often must be so thick that the high-frequency array does not work very well. James and Andrasic [19] have demonstrated a solution to this problem by using a two-layer structure with the closely spaced elements of the high-frequency array on the lower surface and the low-frequency array covering it on a second surface. The low-frequency patches are formed from dichroic surfaces that are transparent to the higher frequency but behave as conductors at the lower frequency. The scheme is illustrated schematically in Figure 7.

5. Future Trends

Antenna array development is more closely tied to particular applications than is element development because arrays are used to obtain specific characteristics, such as beamwidth, beam shape, gain, or selective rejection of signals from particular directions. Therefore, a comprehensive prediction of trends in microstrip array development requires a comprehensive evaluation of potential applications of microstrip antennas. However, there are some overarching issues that appear in a variety of applications and are likely to motivate microstrip antenna array development over the next few years.

Losses in array feed networks and compatibility of the feed architecture with the electronic circuits of the system often lead to array designs that do not use the "traditional" monolithic feed and antenna. It is likely that future developments will include hybrid structures comprised of feed networks fabricated in media such as stripline, coplanar waveguide, or metallic waveguide. Although it violates the "purely printed" notion of microstrip antennas, metallic waveguide is particularly attractive for millimeter wave applications where narrow-beam, high-gain patterns often are desired and losses in printed circuit transmission lines can be particularly troublesome. A combination of waveguide and printed microstrip power division often provides a useful compromise for high-gain arrays, and future developments may seek to optimize the trade-offs of these and other types of hybrid structures. Coplanar waveguide and microstripline are more easily integrated with MMICs than waveguide or stripline, but heat removal in transmitting circuits is a problem if there are no massive metal parts in the structure.

Accomplishing multiple functions with a single antenna aperture is another common motivation for array development. The multiple function requirement could be as simple as switched or multiple beams at a single frequency, or it could include multiple frequencies and/or polarizations operating within a single aperture and beams of differing shapes and gains. These extensive requirements are frequently placed on advanced military systems, but they also offer advantages in commercial applications, provided they can be accomplished in an affordable and reliable antenna. Therefore, robustness of the design and the cost of manufacturing and maintaining the system will stimulate development of microstrip antenna arrays.

6. Summary

Microstrip antenna arrays afford a unique design alternative for applications that benefit from thin, conformal, rugged antennas with low to moderately high gain or shaped radiation patterns. They can be produced in large quantities at reasonable cost by photolithographic techniques and they are compatible with modern integrated circuit fabrication. They also may be fabricated from stamped sheet metal parts and inexpensive dielectrics for low-cost systems. This paper has reviewed some of the basic considerations for microstrip antenna arrays. Among the key design considerations are feedline radiation and ohmic losses, coupling between antennas and between feedlines and antennas, and the overall architecture that the system requires. Several successful designs have been described here and others are presented in the papers reprinted in this chapter.

References

[1] M. Danielson and R. Jorgensen, "Frequency scanning microstrip antennas," *IEEE Trans. Ant. Propagat.*, AP-27, pp. 146–150, Mar. 1979.

[2] M. Collier, "Microstrip antenna array for 12 GHz TV," *Microwave J.*, vol. 20, no. 9, pp. 67–71, Sept. 1977.

[3] J. R. James, P. S. Hall, and C. Wood, *Microstrip Antenna Theory and Design,* Stevenage England: Peter Peregrinus, 1981.

[4] D. M. Pozar, "A microstrip antenna aperture coupled to a microstrip line," *Electronics Letters.*, vol. 21, pp. 49–50, Jan. 1985.

[5] D. H. Schaubert and D. M. Pozar, "Aperture coupled patch antennas and arrays," *Proc. Antenna Applications Symp.*, Sept. 1986.

[6] P. S. Hall and C. M. Hall, "Coplanar corporate feed effects in microstrip patch array design," *IEE Proc.*, part H, vol. 135, pp. 180–186, June 1988.

[7] E. Levine, G. Malamud, S. Shtrikman, and D. Treves," A study of microstrip antennas with the feed network," *IEEE Trans. Ant. Propagat.*, AP-37, pp. 426–434, Apr. 1989.

[8] T. Chiba, Y. Suzuki and N. Miyano, "Suppression of higher modes and cross polarized component for microstrip antennas," *Dig. of IEEE Int'l Ant. Propagat. Symp.*, pp. 285–288, 1982.

[9] J. Huang, "Dual-polarized microstrip array with high isolation and low cross-polarization," *Microwave and Optical Tech. Lett.*, vol. 4, pp. 99–103, Feb. 1991.

[10] D. M. Pozar, "Considerations for millimeter wave printed antennas," *IEEE Trans. Ant. Propagat.*, AP-31, pp. 740–747, Sept. 1983.

[11] D. M. Pozar and D. H. Schaubert, "Scan blindness in infinite arrays of printed dipoles," *IEEE Trans. Ant. Propagat.*, AP-32, pp. 602–610, June 1984.

[12] D. M. Pozar and D. H. Schaubert, "Analysis of an infinite array of rectangular patches with idealized probe feeds," *IEEE Trans. Ant. Propagat.*, AP-32, pp. 1101–1107, Oct. 1984.

[13] K. Ito, T. Teshirogi, and S. Nishimura, "Circularly polarised antenna arrays," Chap. 13, pp. 763–764, in *Handbook of Microstrip Antennas,* J. R. James and P. S. Hall, ed., London: Peter Peregrinus, 1989.

[14] K. Ito, K. Itoh, and H. Kogo, "Improved design of series-fed circularly polarised printed linear arrays," *IEE Proc.*, part H, vol. 133, pp. 462–466, Dec. 1986.

[15] P. S. Hall, "Multioctave bandwidth log-periodic microstrip antenna array," *IEE Proc.*, part H, vol. 133, pp. 127–136, Apr. 1986.

[16] D. T. McGrath, "A lightweight constrained lens for wide angle scan in two planes," *Proc. Antenna Applications Symp.*, Sept. 1986. Also available as RADC-TR-87-10, vol. I, Feb. 1987.

[17] H. Entschladen and U. Nagel, "Microstrip patch antenna array," *Electronics Letters.*, vol. 20, pp. 931–933, Oct. 1984.

[18] P. A. Miller, J. C. MacKichan, M. R. Staker, and J. S. Dahele, "A wide bandwidth low sidelobe low profile microstrip array antenna for communication applications," *Proc. ISAP,* pp. 525–528, 1989.

[19] J. R. James and G. Andrasic, "Superimposed dichroic microstrip antenna arrays," *IEE Proc.*, part H, vol. 135, pp. 304–312, Oct. 1988.

The Synthesis of Shaped Patterns with Series-Fed Microstrip Patch Arrays

BEVAN B. JONES, MEMBER, IEEE, FRANCIS Y. M. CHOW, AND ANTHONY W. SEETO, MEMBER, IEEE

***Abstract*—A method of designing series-fed microstrip patch arrays to produce a shaped pattern is described. The technique is based on a circuit representation of the patches in the array environment with experimentally determined parameters. The positions and widths of the patches are derived from the amplitudes and phases of the elements of a uniform array, which produces the desired pattern and which has the same extent as the patch array. For the array to have high efficiency, the amplitude distribution must not be strongly peaked. An algorithm for obtaining an approximation to the desired far-field amplitude distribution while retaining control over the amplitude of excitation is presented. Very close agreement has been obtained between calculated and measured performance of such arrays.**

I. INTRODUCTION

THE MECHANISM of radiation from rectangular microstrip patches has been described by several authors [1]-[3], and in particular the use of series fed arrays of microstrip patches to achieve pencil beams has been described by Metzler [4]. A method of accurately synthesizing shaped beams with series fed microstrip patch array is described in this paper.

The form of the arrays considered is shown in Fig. 1. The radiating patches themselves are resonant so that the input line to a patch is matched if the output line is terminated. In the simple transmission line model of a patch this corresponds to a length of $\lambda g/2$. Excitation at one end of the array with a termination at the other results in a traveling wave on the microstrip line, from which power is radiated at the edges of each patch which behave as a pair of in-phase slots [5]. The phases of the radiation sources can be controlled by selection of the positions of the patches on the line and their amplitudes by choice of the widths of the patches. Because the amplitude and phase of the radiation at each patch is determined by the cumulative transmission characteristics of the preceding patches on the line, the transmission characteristics of the patches must be determined with some accuracy in order to achieve a desired amplitude and phase distribution of radiating sources along the array. A circuit representation of the patches based on the transmission line model of Derneryd [6], but with empirically determined parameters, is used to represent the patch array.

The properties of series fed microstrip patch arrays are similar to those of edge slotted waveguide antennas as far as bandwidth, polarization, and efficiency are concerned, however, microstrip arrays appear to be capable of cheaper fabrication.

II. CIRCUIT REPRESENTATION OF PATCHES AND ARRAYS

The circuit representation used here for a single patch is shown in Fig. 2. The patch itself is represented by a length of microstrip line of width equal to that of the patch. The characteristic impedance and complex propagation constant are taken from the formulas for microstrip line properties given by Bahl and Trivedi [7]. The effects of fringing fields and of the junction between the wide and narrow lines are represented by the extensions Δ and δ in the wide and narrow lines, respectively, as shown. This representation of the junctions was chosen because Δ and δ are not strongly dependent on frequency. Radiation at the discontinuities is accounted for by shunt conductances g, which are referred to the impedance of the narrow line. Experimental characterization of the patches requires a determination of the parameters g, Δ, and δ as a function of patch width W and, since the parameters are intended to characterize the patch in the presence of mutual effects from other patches, it is expected that g and Δ should also be functions of the patch spacing S. δ which principally represents local junction effects is assumed to be independent of S. It has, however, been found experimentally that the dependence of Δ on S is negligible.

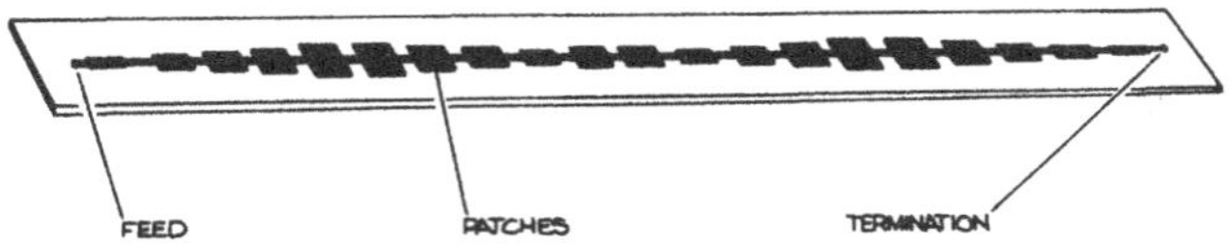

Fig. 1. Series fed microstrip patch array.

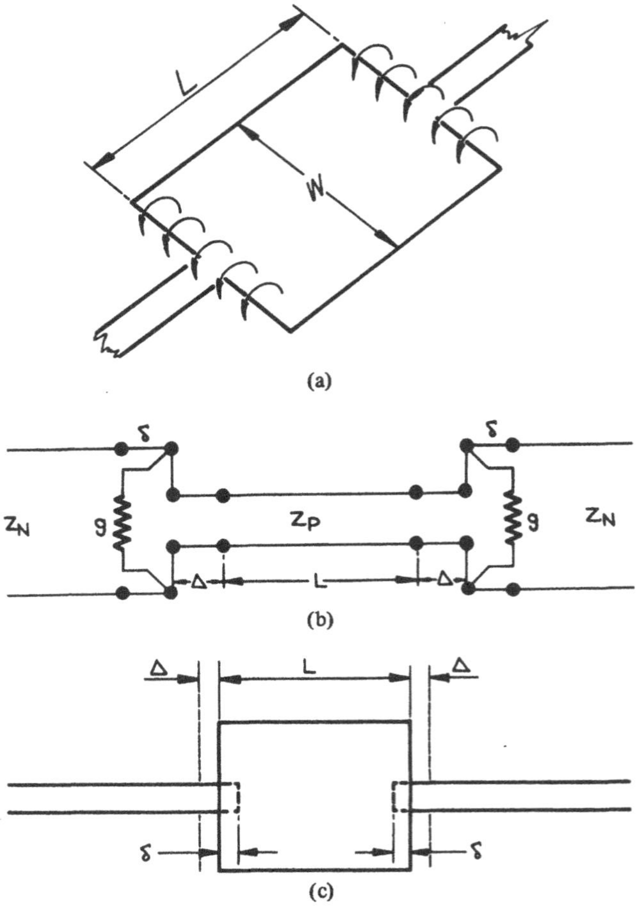

Fig. 2. Single patch and its equivalent circuit respresentation.

Reprinted from *IEEE Trans. Antennas Propaga.*, vol. AP-30, no. 6, pp. 1206–1212, Nov. 1982.

The assumption is made that the parameters for a patch in the array environment depend only on the patch dimensions and the local spacing of patches.

When the dependence of Δ, δ, and g on patch width and of Δ and g on patch spacing has been ascertained, analysis of the performance of a patch array can be carried out using the circuit representation of the elements. This is conveniently done using a T-matrix representation of line sections, shunt conductances, and extensions. Assuming unit forward and zero backward propagating wave at the terminating load, the forward and backward wave components at any other point in the array can be found by multiplication of the T-matrices. Insertion loss, return loss, and radiation efficiency (ratio of power dissipation in the radiation conductances to the input power), can be readily found. The radiation pattern in the principal plane (E-plane) is assumed to be that of an array of point sources located at the patch ends with amplitude $v\sqrt{g}$, where v is the complex voltage at the radiation conductance of the patch in the circuit representation.

III. EXPERIMENTAL CHARACTERIZATION OF PATCHES

In order that an approximate design can be carried out to establish the range of patch widths and spacings required, the parameters g, Δ, and δ can be determined for single patches, ignoring the dependence of g and Δ on S. When the approximate range of widths and spacings has been established, several uniform arrays spanning the range of spacings and widths can be made and the dependence on S of these quantities can be determined.

The parameters g, Δ, and δ can be determined from measurements on several patches of different widths and of length near $\lambda g/2$ where λg is the wavelength of the quasi-transverse electromagnetic (TEM) mode in microstrip of the width of the patch given by the formulas of [7].

Δ may be found by terminating the output line from the patch and finding the resonant frequency defined by a minimum standing-wave ratio (SWR) on the input line. Δ is then defined by

$$L + 2\Delta = \lambda g/2 \qquad (1)$$

at the resonant frequency, where g is evaluated from the formulas of [7].

δ is found from a measurement of the transmission phase

$$2\delta = \frac{\lambda_N}{2\pi}\,\phi_{\mathrm{exc}} \qquad (2)$$

where ϕ_{exc} is the phase angle in excess of π between the patch ends at resonance and λ_N is the wavelength in narrow microstrip line at resonance.

The conductance g can be measured by connecting an adjustable short circuit to the output line, finding the ratio or the input reflection coefficients when the short is positioned so that radiation from the patch is minimized, ρ_1 and at a quarter wavelength from this position ρ_2. The conductance is given by

$$2g = \frac{\rho_1 + \rho_2}{\rho_1 - \rho_2}. \qquad (3)$$

This technique largely separates the distributed line loss from the localized radiation loss. However, localized radiation loss from connectors still causes some inaccuracy. Graphs of the three parameters are shown in Fig. 3. Expressions for g for isolated patches without the feedline are given by Derneryd [6]. These are shown in the diagram for comparison.

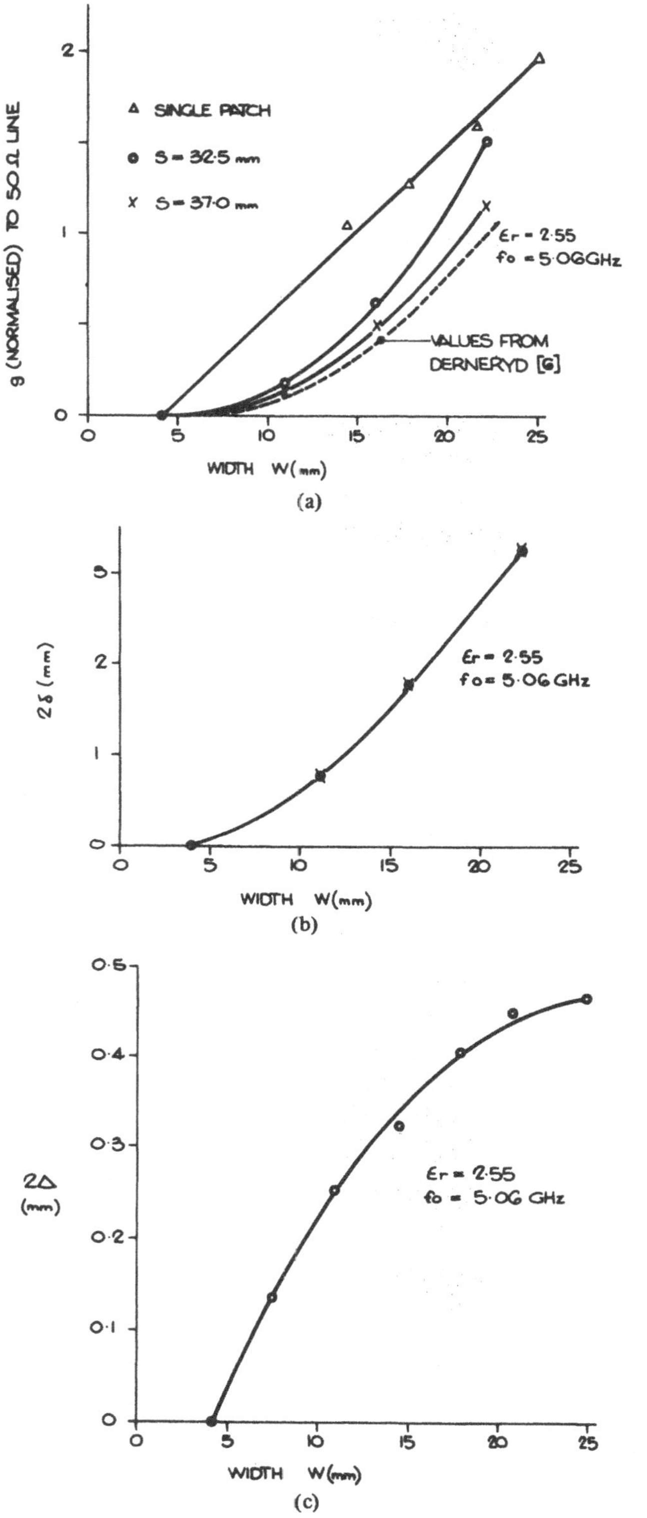

Fig. 3. Measured radiation conductance g and extensions Δ and δ as a function of patch width, W, and spacing S.

An initial design of the patch array can be carried out based on the circuit parameters determined from measurements on single patches using the techniques described in the following sections. From this initial design, the approximate ranges of patch widths and spacings which will occur in the final design are found. A more accurate determination of the patch characteristics that apply to patches in the array environment over the appropriate ranges of widths and spacings is made from measurements on several uniformly spaced arrays of constant width patches.

The insertion loss and insertion phase of the uniform array is first calculated using the T-matrix circuit analysis and the values of Δ, δ, and g found from the single patch measurements. The analysis is repeated using slightly perturbed values of Δ and g until the measured values of insertion loss and phase are obtained. It is found that the value of Δ so obtained is not significantly different from that obtained from the single patch method, although a higher accuracy is obtainable in this way. The values of g, however, differ significantly from those for a single patch. These results are also shown in Fig. 3.

IV. PATCH ARRAY GEOMETRY

The design of an array consists of finding the location of the center of each patch and its width. Its length follows from the width of the patch since it is chosen to make the patch resonant, i.e.,

$$l(w) = \frac{\lambda(w)}{2} - 2\Delta. \tag{4}$$

The approach which we have adopted to the design is firstly to design a uniform array of point source elements which produces a good approximation to the required pattern and which extends over approximately the same length as the required microstrip patch array. The spacing of the elements should be similar to the average element spacing in the final array. A suitable spacing may be found by equating the phase gradient of the sources formed by a uniform patch array to that appropriate for the direction of maximum radiation in the desired pattern. This leads to

$$S(1/\lambda_N - 1/\lambda_0 \cos \zeta) = 1/2 + \frac{1/2\lambda g - 2\Delta - 2\delta}{\lambda_N} \tag{5}$$

where ζ is the angle between the traveling wave and the direction of maximum radiation. From the amplitudes and phases of the elements of this uniform array, a set of patch widths and locations which will produce approximately the same pattern as the uniform array may be found by a process of interpolation.

Following is a technique that has been found to give satisfactory results.

1) A smooth curve of phase as a function of position $\Phi(x)$ is fitted to the phases φ_n of the elements of the uniform array, i.e., $\Phi(x_n) = \varphi_n$ where x_n is the location of the nth element.
2) A smooth curve is fitted to the cumulative amplitude distribution of the uniform array, as a function of position. The "cumulative amplitude" function $C(x_n)$ at element n located at x_n refers to

$$C(x_n) = \sum_1^n |a_i| \tag{6}$$

where a_i is the amplitude of excitation of the ith element.
3) The patches of the nonuniform patch array are located so that the phases of the sources which they form lie on the curve of condition 1). The widths of the patches are selected so that the cumulative amplitudes at each of the sources

$$C(z_n) = \sum_1^n |a_i| \tag{7}$$

lie on the curve of condition 2).

For this purpose, a patch is represented by a single source located at z_n, the center of the patch, with amplitude.

$$\alpha_n = (v_1 + v_2) \cdot \sqrt{g} \tag{8}$$

where v_1 and v_2 are the voltages at the ends of the patch and g is the radiation conductance at each end of the nth patch. The index of the patches is taken as increasing from the termination to the feed.

To do this in practice it is assumed that unit power is incident on the termination and a value for the total radiated power γ is assumed. The amplitudes of the elements of the uniform array are therefore normalized so that

$$\sum_1^N |a_i|^2 = \gamma \tag{9}$$

where N is the number of elements. The first patch (nearest the termination) is colocated with the first element of the uniform array. By condition (3) we have therefore $C(x_1) = \alpha_1$. The voltage is known and hence the conductance. The experimental data give the patch width and length necessary to achieve this conductance. If it is assumed that the second patch is identical to the first, using the T-matrix circuit analysis, the phase $\varphi(z_2)$ of the source representing this patch can be found as a function of z_2, its location. The patch is located so that $\varphi(z_2) = \Phi(z_2)$. The solution to this equation nearest to $z_2 = z_1$ is selected.

The amplitude of this patch is then given by $\alpha_2 = C(z_2) - C(z_1)$. The conductance of the patch is calculated and appropriate values of the patch width and length are found (leaving the phase slightly in error). This procedure is continued until all the patches have been located and assigned a width and length.

The entire procedure is iterated several times using the lengths, widths, and element spacing of each preceding iteration until the conditions in (3) are realized. If at the end of this process it is found that the widths of some patches are unacceptably high, the entire process is repeated with a reduced value of γ.

The radiation pattern of the final configuration can be calculated by the method outlined in Section II. In practice the patterns obtained agree closely with those of the uniform arrays from which they were derived. The radiation efficiency, insertion loss, and return loss are also given by the analysis.

V. OPTIMIZATION OF EXCITATION

It is desirable that the distribution of amplitude along the length of the array should not be highly peaked. A peak amplitude distribution requires high values of radiation conductance if good efficiency is to be achieved. A limit to the achievable conductance is set by the increasing excitation of the first transverse mode as the width of the patch approaches one wavelength (in dielectric).

If a limit (of, e.g., 0.6 λg) is set to the width of the patches then a peaked amplitude distribution will lead to a design with low total radiating power and consequently low efficiency. Maximum efficiency is achieved if all the patches have the maximum allowable width. The amplitude distribution would then drop exponentially away from the feed point.

A method of synthesis of the desired far-field amplitude pattern with an array is therefore required in which some control is maintained over the amplitudes of excitation of the array. Good approximations to the desired far-field amplitude pattern are possible with a variety of amplitude distributions on the array because only the far-field amplitude pattern is prescribed while the phase may be chosen arbitrarily.

If the phase of the far field is assigned, the amplitudes and phases of excitation which give the best approximation to the prescribed pattern in the least squares sense can be found directly. It suffices to find suitable amplitudes and phases of excitation for a uniform array since the interpolation procedure described in Section IV can be used to derive the geometry of the actual patch array from this. The complex excitations $\{a_n\}$ of the uniform array are found by seeking to match the pattern to the prescribed one $F(\theta_\nu)$, at a large number M (more than the number of elements) of angles θ_ν.

$$F(\theta_\nu) = \sum_{n=1}^{N} a_n b_{\nu n} \tag{11}$$

where

$$b_{\nu n} = \exp\{-jk_n \sin\theta_\nu\} \tag{12}$$

$$z_n = \left\{n - \frac{(N+1)}{2}\right\} d$$

where d is the element spacing.

In matrix notation:

$$F = Ba \tag{13}$$

where

$$F = \begin{bmatrix} F(\theta_1) \\ F(\theta_2) \\ \vdots \\ F(\theta_m) \end{bmatrix} \quad B = [b_{\nu n}] \quad a = \begin{bmatrix} a_1 \\ a_2 \\ \vdots \\ a_N \end{bmatrix}$$

The least squares solution to this overdetermined set of equations gives the complex excitation of the elements.

$$a = (B^*B)^{-1}B^*F \tag{14}$$

where the asterisk denotes transpose conjugate.

The application of this method to the ideal far-field amplitude pattern shown in Fig. 4(a) with the phase arbitrarily set constant leads to the array amplitude distribution and far-field amplitude distribution given in Fig. 4. This peaked distribution is not well suited to implementation of a patch array. An efficiency of 32 percent is obtained with a maximum patch width of 0.4 λg.

To obtain a more favorable amplitude distribution, we must remove the constraint on the far-field phase distribution and impose a constraint on the excitation amplitude distribution. An iterative algorithm which has been found to yield excellent results is the following. An initial coarse estimate is made of the far-field phase which is compatible with the desired far-field amplitude pattern and the desired distribution of excitation amplitude. A ray-optics method based on Chu's synthesis technique [8] has been found to be satisfactory.

In this method it is assumed that the power radiated by the nth element of the array is confined to the range of angle θ_{n-1} to θ_n. Referring to Fig. 5, we have

$$\sum_{1}^{n} |a_i|^2 = \int_{\theta_0}^{\theta_n} p(\theta)\cos\theta\, d\theta \tag{15}$$

where $p(\theta)$ is the prescribed power pattern, normalized so that

$$\sum_{1}^{N} |a_i|^2 = \int_{\theta_0}^{\theta_N} p(\theta)\cos\theta\, d\theta. \tag{16}$$

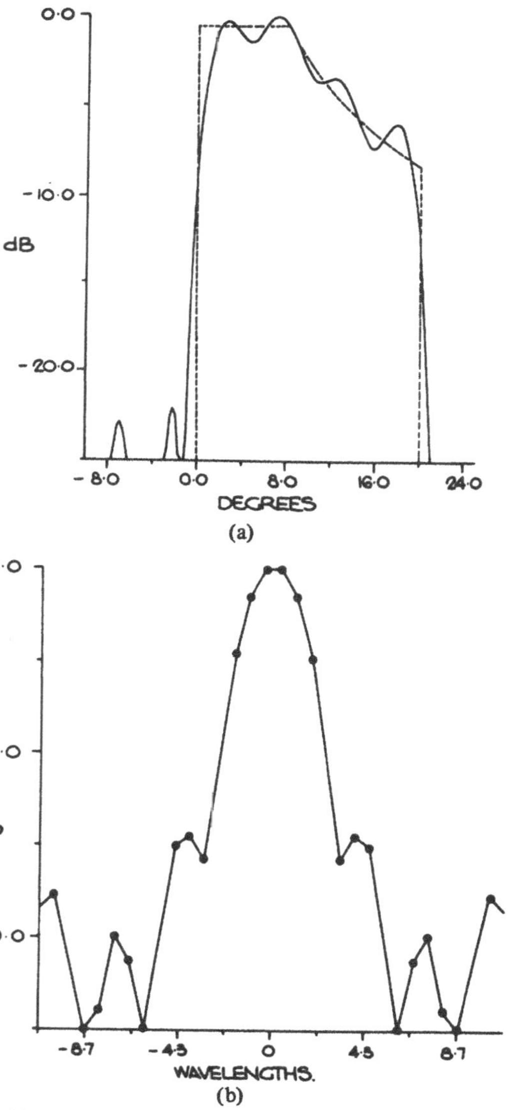

Fig. 4. (a) Ideal amplitude pattern and pattern obtained by least squares fitting assuming phase of far field to be zero. (b) Amplitudes of excitation of patch array obtained by this method.

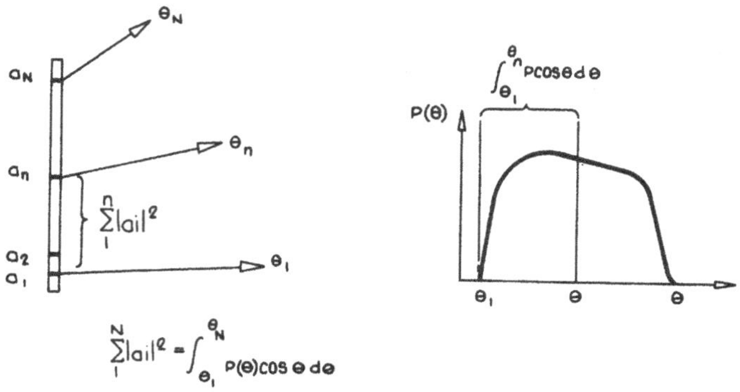

Fig. 5. Approximation used in Chu's optical method of pattern synthesis.

Using the equation, given the required amplitudes of excitation $|a_i|$, and the prescribed power distribution $p(\theta)$, the angles θ_n can be found.

Knowing the angles at which the radiation from each element is centered, a smooth phase distribution can be set up so that the phase gradient at each element is that appropriate to radiation in the required direction.

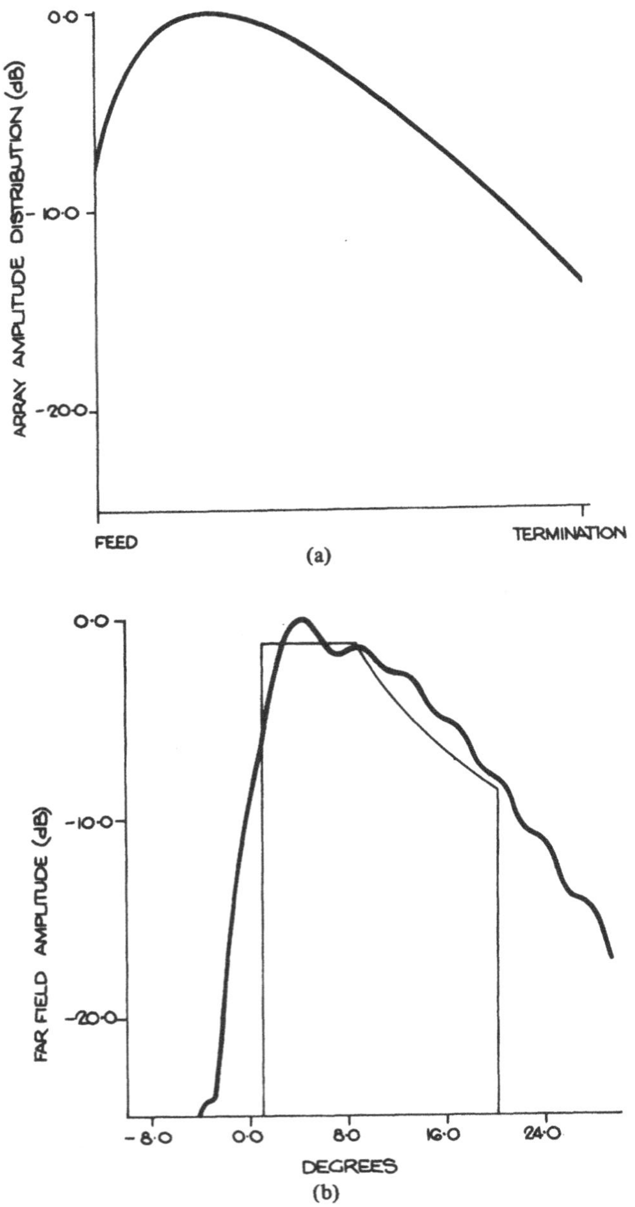

Fig. 6. (a) Assigned amplitude of excitation. (b) Amplitude pattern obtained by Chu's optical method of pattern synthesis.

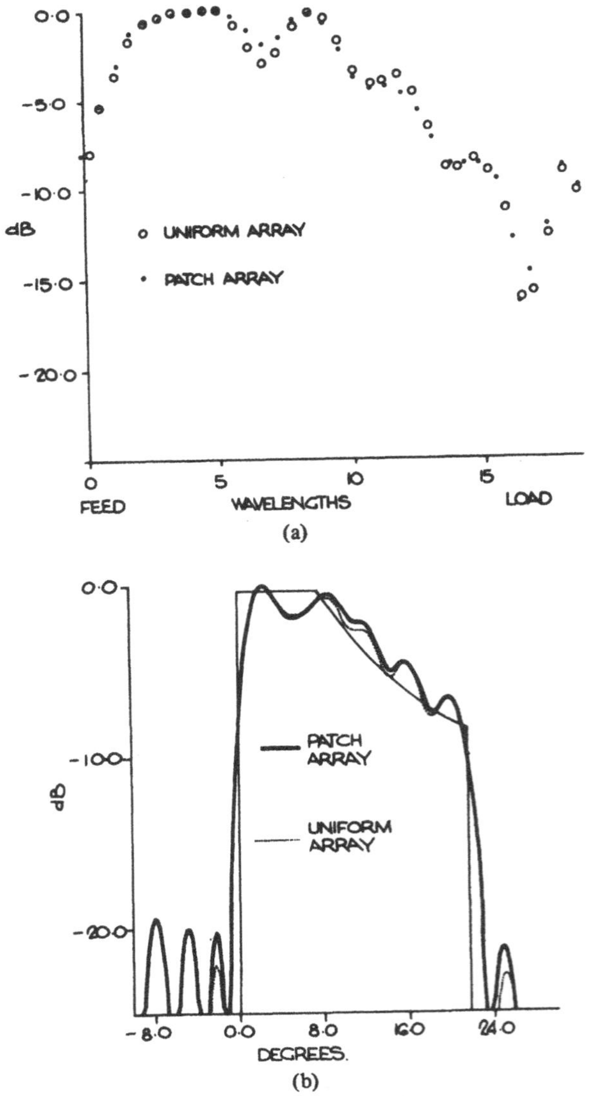

Fig. 7. (a) Amplitudes of excitation for uniform and patch array obtained by algorithm of text. (b) Far-field amplitude patterns of uniform and patch arrays.

The method becomes accurate if the θ_n are closely spaced so that adequate length of the array is available to confine the radiation to the required range of directions. A desired amplitude distribution and the far-field amplitude and phase pattern obtained by this method is shown in Fig. 6.

This coarse estimate of the far-field phase and the ideal far-field amplitude distribution are now used to calculate the required amplitude and phase of excitation of the array elements using the least squares method described above. The amplitude of excitation is discarded and replaced by the desired amplitude of excitation before calculating the far-field pattern with an improved estimate of the required far-field phase distribution.

The process is continued until convergence has occurred. The result is illustrated in Fig. 7. The amplitude of excitation obtained from the final least squares synthesis resembles closely the desired one. For comparison the amplitudes and pattern of the patch array derived from this by the method of Section VI are also given. It is seen that application of the method results in little degradation of the pattern.

TABLE I
PARAMETERS OF PATCH ARRAY

Substrate	PTFE Fiberglass, thickness 1.5 mm $\epsilon_r = 2.55$.
Frequency	5 GHz
Number of patches	32
Length of array	1100 mm
Characteristic impedance of connecting line	50 Ω
Radiation efficiency	71 percent
Maximum patch width	21 mm
Power dissipated in microstrip	22 percent
Power dissipated in termination	7 percent
Return loss	−25 dB

The pattern obtained with this distribution is also shown. The radiation efficiency obtained with this distribution and the parameters of Table I is 71 percent which may be compared with the figure of 32 percent obtained for the array of Fig. 4.

This iterative technique is similar to one described by Mautz and Harrington [9] for optimizing the excitation of an array for synthesizing a far-field amplitude pattern when the far-field phase is unconstrained.

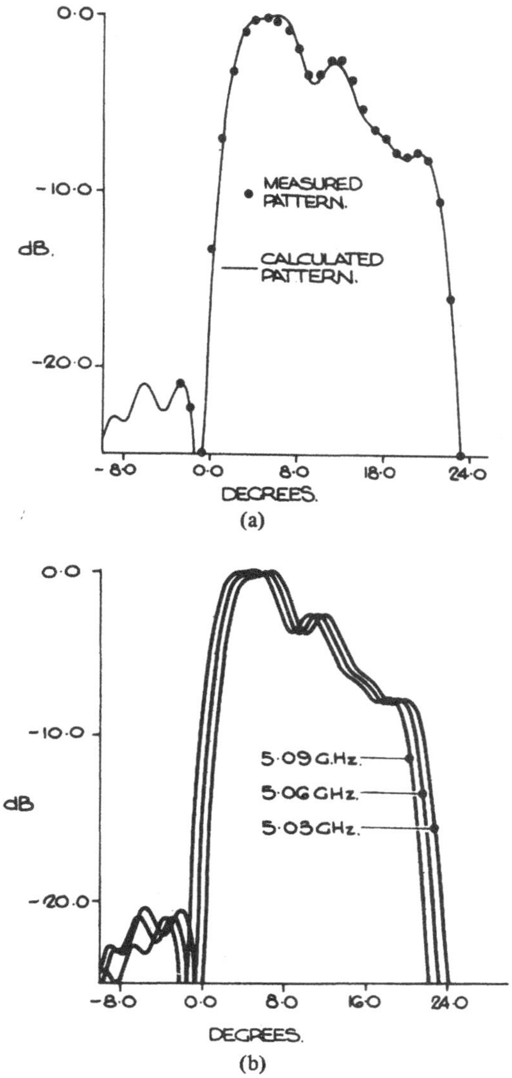

Fig. 8. (a) Comparison of measured and calculated amplitude patterns of a patch array. (b) Effect on pattern of a 1 percent change in frequency.

VI. RESULTS

An array was designed using the techniques described to give the pattern of Fig. 4(a) with the characteristics given in Table I. The measured pattern at the design frequency is given in Fig. 8(a), together with the pattern calculated from the circuit analysis. It is seen that the agreement is excellent. Difficulty was initially experienced with disturbance to the pattern caused by radiation from the coaxial to microstrip transitions. Small metal shields lined with absorber were later placed over these, and the pattern of Fig. 8 was obtained. The effect on the pattern of a 1 percent change in frequency is also shown. The effect is almost entirely a translation of the pattern resulting from a net change of phase gradient along the array. This dispersion is accurately predicted by the circuit model and is common to all antennas, such as slotted waveguide antennas, consisting of radiating elements loosely coupled to a traveling wave. The return loss of the antenna was better than −20 dB over a 3 percent frequency range.

VII. CONCLUSION

Using the design techniques described here it is possible to synthesize accurately shaped patterns with low sidelobe levels with series fed microstrip patch arrays and with high efficiencies. The arrays themselves are an attractive alternative to edge-slotted waveguide antennas.

REFERENCES

[1] J. Q. Howell, "Microstrip antennas," *IEEE Trans. Antennas Propagat.*, vol. AP-23, pp. 90–93, Jan. 1975.

[2] Y. T. Lo, D. Solomon, and W. R. Richards, "Theory and experiment on microstrip antennas," *IEEE Trans. Antennas Propagat.*, vol. AP-27, pp. 137–145, Mar. 1979.

[3] K .R. Carver and J. W. Mink, "Microstrip antenna technology," *IEEE Trans. Antennas Propagat.*, vol. AP-29, pp. 2–25, Jan. 1981.

[4] T. Metzler, "Microstrip series arrays," *IEEE Trans. Antennas Propagat.*, vol. AP-29, pp. 174–178, Jan. 1981.

[5] R. E. Munson, "Conformal microstrip antennas and microstrip phased arrays," *IEEE Trans. Antennas Propagat.*, vol. AP-22, pp. 74–77, Jan. 1974.

[6] A. G. Derneryd, "Linearly polarized microstrip antennas," *IEEE Trans. Antennas Propagat.*, vol. AP-24, pp. 846–851, Nov. 1976.

[7] I. J. Bahl and D. K. Trivedi, "A designer's guide to microstrip," *Microwaves*, pp. 174–182, May 1977.

[8] C. H. Walter, *Travelling Wave Antennas*. New York: McGraw-Hill, 1965.

[9] J. R. Mautz and R. F. Harrington, "Computational methods for antenna pattern synthesis," Dept. Elect. Comput. Eng., Syracuse Univ., Tech. Rep. TR-73-9, Aug. 1973.

Coplanar corporate feed effects in microstrip patch array design

P.S. Hall, MEng, PhD, CEng, MIEE
C.M. Hall, BSc, PhD

Indexing terms: Antennas (microstrip), Antennas (stripline), Microstrip and stripline

Abstract: The use of coplanar corporate feeds for microstrip patch arrays leads to constructional simplicity, but also to performance degradations due to feed radiation, in addition to limitations due to feed resistive loss, surface waves, mutual coupling and tolerances. These effects are quantified, and this allows specification of array performance limitations in addition to the recommendation of the use of smooth feed discontinuities, high line impedance, and thin substrates. Improvements due to the use of alternate feed geometries, such as sequentially rotated feeding and subarraying, are also quantified and are shown to be substantial.

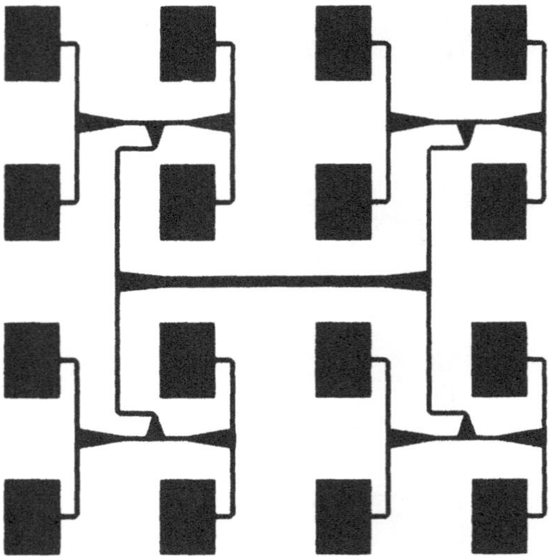

Fig. 1 *Silhouette of corporately fed patch array for linear polarisation with uniform aperture distribution*
$\varepsilon_r = 2.32$; $h = 1.59$ mm; $h/\lambda_0 = 0.06$; frequency = 12.0 GHz; $d/\lambda_0 = 0.7$

1 Introduction

Microstrip array antennas are now being actively considered for applications, such as satellite communication systems [1, 2], where their thin profile and light weight are important considerations. The array thickness, weight, and cost can be optimised by the use of a corporate feed structure etched on the same surface as the patch array, as shown in Fig. 1. However, feed resistive losses and radiation lead to gain and radiation pattern limitations, making feed effects a key issue in the design of such arrays. Experimental investigations [3] have highlighted these problems, although to the authors knowledge no detailed results allowing design optimisation have been published. In this paper these effects, together with other design perturbations such as surface waves, mutual coupling, and tolerances, are quantified to allow a more complete understanding of performance limitations and of methods [4] by which these limitations can be improved upon with a view to their implementation in computer-aided design (CAD).

2 Corporate feed loss mechanisms

Losses in microstrip corporate feeds are due to copper and to dielectric losses in straight lengths compounded with radiation and surface-wave losses from the overall structure, the complete quantification of which requires detailed knowledge of the current distribution on the feed network. First order estimates have been obtained by isolating and assessing these various components and in following Sections these estimates are used to deduce their effects on efficiency and on the array radiation pattern.

2.1 Line loss

I^2R and dielectric losses have been estimated by many authors. Based on comparative studies of applicability and accuracy we recommend estimates of attenuation coefficient due to dielectric loss α_d and copper loss α_c by Pucel [5], based on an analysis of the effective dielectric constant ε_e and the microstrip wavelength λ_m by Wheeler [6] and Kirschning *et al.* [7]. α_c must then be corrected for roughness effects [8] to give α_{cr}. Fig. 2*a* shows total line loss ($\alpha_d + \alpha_{cr}$) for two types of substrate. It can be seen that line loss increases with dielectric constant and line impedance, and decreases with increasing substrate thickness.

2.2 Radiation loss

Estimations of radiation from microstrip discontinuities are based on analyses of equivalent electric and polarisation currents [9, 10]. The ratio of power radiated to power incident on the discontinuity is given by Reference 9 as

$$\frac{P_r}{P_{inc}} = \frac{60(\beta h)^2}{Z} \cdot F \tag{1}$$

where $\beta = 2\pi/\lambda_0$, h is the substrate thickness, and Z is the impedance of the input line. For the coaxial-to-microstrip transition of Fig. 1, F is found by combining that owing to a conventional transition [9] and that owing to a continuous line, to give

$$F = \frac{1}{4}\left\{3 - \frac{1}{\varepsilon_e} - \left(3 + \frac{1}{\varepsilon_e}\right) \cdot \frac{\varepsilon_e - 1}{2\sqrt{\varepsilon_e}} \cdot \ln \frac{\sqrt{(\varepsilon_e)} + 1}{\sqrt{(\varepsilon_e)} - 1}\right\} \quad (2)$$

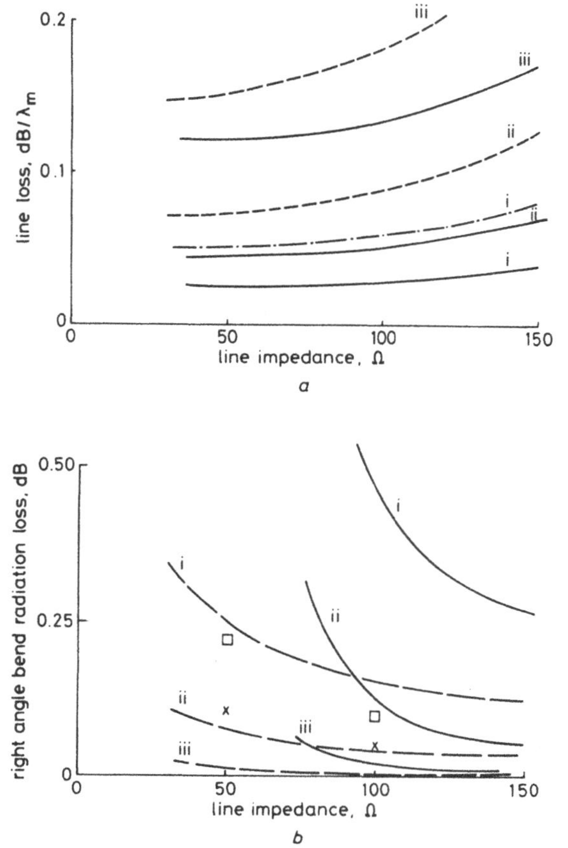

Fig. 2 *Calculated microstrip line loss and discontinuity radiation loss*
a Line loss $\alpha_d + \alpha_{cr}$
b Right angle bend radiation loss P_r/P_{inc}
——— $\varepsilon_r = 1.0$
- - - - $\varepsilon_r = 2.32$
i $h/\lambda_0 = 0.06$
ii $h/\lambda_0 = 0.03$
iii $h/\lambda_0 = 0.01$

Smooth 90° bend, $\varepsilon_r = 2.32$; $h/\lambda_0 = 0.06$; bend radius □ $\rho = 0.2\lambda_0$; × $\rho = 0.4\lambda_0$

where ε_e is the effective dielectric constant of the microstrip line. Radiation loss decreases with decreasing substrate thickness and increasing dielectric constant, being less than 0.1 dB for $h/\lambda_0 = 0.03$ with $\varepsilon_r = 1.0$ and for $h/\lambda_0 = 0.06$ with $\varepsilon_r = 2.32$. For the right angle bend [9], whether mitred or not, F is given by

$$F = \frac{\varepsilon_e + 1}{\sqrt{\varepsilon_e}} \ln \frac{\sqrt{(\varepsilon_e)} + 1}{\sqrt{(\varepsilon_e)} - 1} - \frac{2\varepsilon_e}{\sqrt{2\varepsilon_e - 1}} \ln \frac{\sqrt{2(\varepsilon_e - 1)} + 1}{\sqrt{2(\varepsilon_e - 1)} - 1} \quad (3)$$

while for the T junction [11]

$$F = \frac{(3\varepsilon_e + 1)^2}{8\varepsilon_e^{3/2}} \ln \frac{\sqrt{(\varepsilon_e)} + 1}{\sqrt{(\varepsilon_e)} - 1} - \frac{\varepsilon_e}{2\varepsilon_e - 1} \times \ln \frac{\varepsilon_e + \sqrt{(2\varepsilon_e - 1)}}{\varepsilon_e - \sqrt{(2\varepsilon_e - 1)}} - \frac{\varepsilon_e + 1}{4\varepsilon_e} \quad (4)$$

Fig. 2*b* shows radiation loss for the right angle bend. Similar values are obtained for the T junction. For a smooth-radiused bend a constant equivalent magnetic current $\boldsymbol{M}_e$ located at the microstrip line centre is assumed, where $\boldsymbol{M}_e$ is given by [12],

$$\boldsymbol{M}_e = \hat{\boldsymbol{\phi}} V k w \left\{\frac{1}{k\rho} + j \cos\phi \sin\theta\right\} \exp(-j\beta\phi\rho) \quad (5)$$

where V is the line voltage, ρ is the line radius, $k = 2\pi/\lambda_m$ and (θ, ϕ) are far-field spherical coordinates. Integration of the radiation vector due to this current $\boldsymbol{M}_e$ over the bend allows the radiated power to be obtained, and from this the radiation loss as shown in Fig. 2*b*. It is seen that considerable reduction occurs for increasing bend radius. Variational analysis of the radiation from a quarter-wavelength matching section [13] from 50 to 100 Ω shows that radiation loss is typically below 0.02 dB for low dielectric constant substrates. These results, therefore, indicate that the most significant radiation loss occurs in T junctions and right angle bends. Reductions can be obtained by the use of thin high dielectric constant substrates, high line impedances, and radiused bends.

3 Radiation pattern effects

3.1 Feed radiation

The effect of feed radiation on the array patterns has been analysed using a source distribution of magnetic currents [14, 15]. The patch-source distribution $\boldsymbol{M}$ is a continuous current around the patch periphery, whose value is determined from a patch cavity model including the fundamental and higher-order modes. The feed-source distribution M_q is made up of discrete Hertzian currents located at the feed discontinuities. Thus the radiated field $\boldsymbol{E}_{rad}$ is given by

$$\boldsymbol{E}_{rad} = jhk\left[\hat{\boldsymbol{R}} \times \int_c \boldsymbol{M} \exp(jk_0 r \cos\psi)\, dc\right] + jhK \sum_{q=1}^{Q} M_q \exp(jk_0 r_q \cos\psi_q)\{\sin(\phi - \gamma_q)\hat{\boldsymbol{\theta}} - \cos\theta \cos(\phi - \gamma_q)\hat{\boldsymbol{\phi}}\} \quad (6)$$

where $K = \exp(-jk_0 R)/\lambda_0 R$, $\boldsymbol{R}$ is a unit vector in the far field direction, and r, ψ, (r_q, ψ_q) define the magnetic-current source position [14]. γ_q defines the Hertzian current orientation [14]. The discontinuity magnetic current value M_q is given by

$$M_q = I_q Z_0 \sqrt{\left(\frac{3}{2} F_q\right)} \quad (7)$$

where Z_0 is the free space impedance: F_q is given by eqns. 2, 3 or 4.

The current on the qth discontinuity I_q is given by

$$I_q = \left\{\frac{P_{in} \lambda_m}{Z_q \alpha q l_q}\right\}^{1/2} \exp\{-j(k_m l_q + e_q)\} \quad (8)$$

P_{in} is the array input power; Z_q is the line impedance at the qth transition; $\alpha = \alpha_d + \alpha_{cr}$, l_q is the line length between the q and $(q-1)$th transitions; k_m is the microstrip line wavenumber; e_q is the reference plane extension for the qth discontinuity [8]. The discontinuity numbering is arranged so that q is the number of input lines to a particular stage of power division in the corporate feed. $q = 1$ for the coax to microstrip transition and for subsequent stages $q = 2, 4, 8$, etc.

Fig. 3 shows computed results compared to measurements for an 8 × 8 element array of the form of Fig. 1. Agreement is good down to a level of −15 dB but below this the analytical approximations, and in particular the difficulties in accurate calculation of the relative phase of the feed contributions, lead to errors. Examination of the measured patterns also indicates that the high sensitivity

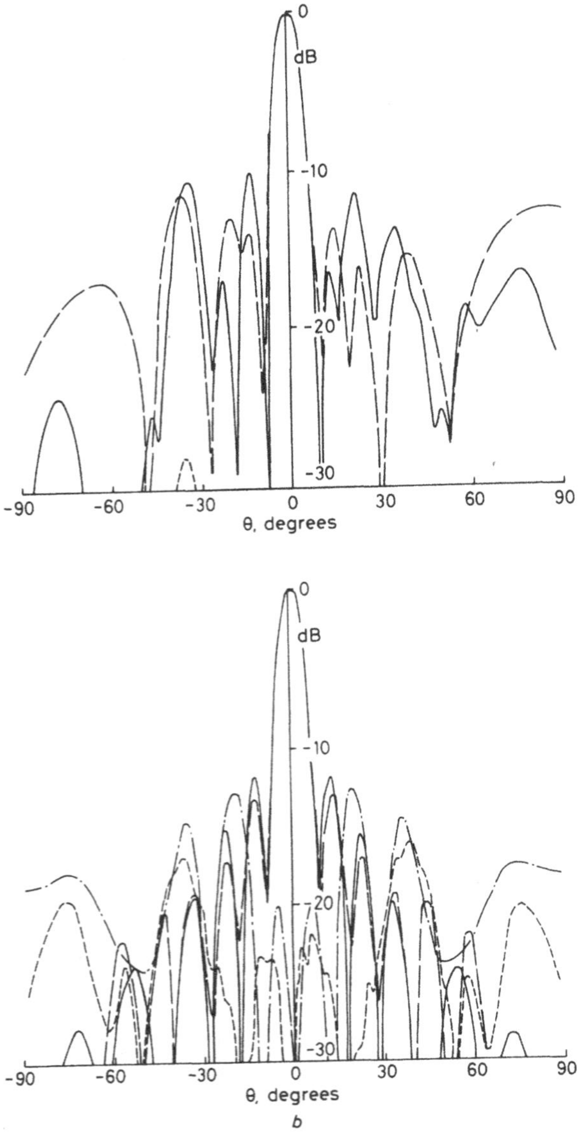

Fig. 3 *Radiation patterns of 8 × 8 element array as Fig. 1*

a E-plane
b H-plane
$h/\lambda_0 = 0.06$; $\varepsilon_r = 2.32$; frequency = 12 GHz
——— measured } copolarised
— — theoretical } copolarised
- - - - measured } crosspolarised
—·— theoretical } crosspolarised

of the feed-radiation pattern to frequency may reduce the overall array bandwidth, if this is determined by pattern quality in addition to input voltage standing-wave ratio (VSWR). Fig. 4 shows the computed patterns of a 16 × 16 element array. Here the pattern of the feed alone, that is with the patch radiation supressed, is compared to the envelope of the sidelobe pattern of the patches alone. It can be seen that feed sidelobes are in some cases of the order of 10 dB higher than patch radiation. A symmetrical grating structure is noted in the *H*-plane of the feed radiation.

The relative contributions of the various feed discontinuities can be estimated by multiplying the radiation levels given in Section 2 by a gain associated with the array effect for the particular discontinuity considered. Thus the discontinuity gain, *G*, is given by

$$G = \frac{Z_q I_q^2}{P_{in}} \cdot \frac{P_r}{P_{inc}} \cdot A_d g_d \qquad (9)$$

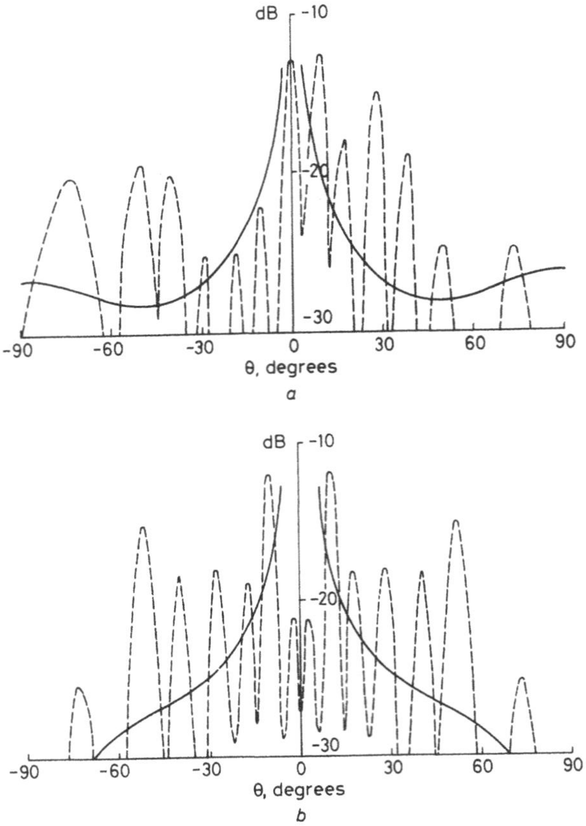

Fig. 4 *Calculated radiation pattern of 16 × 16 element array as Fig. 1*

——— sidelobe envelope of patches
a E-plane
- - - - copolarised feed pattern
b H-plane
- - - - crosspolarised feed pattern

$Z_q I_q^2/P_{in}$ is the ratio of power incident on the discontinuity to the array input power deduced from eqn. 8; P_r/P_{inc} is the fraction of incident power radiated by the discontinuity as deduced from eqn. 1; A_d is the discontinuity array factor involving the total number of each type of discontinuity and their spacing; g_d is the gain of a Hertzian magnetic current source. For a 16 × 16 element array the following values are found. For the *T*-junctions next to the patches $G = 5.0$ dBi, which is about 23 dB down on the peak array gain; for the input connector $G = -8.0$ dBi, about 36 dB down. *G* for the other discontinuities is between these values, and decreases monotonically towards the centre of the array. This result indicates that the discontinuities nearest the patches have the greatest effect on the array radiation pattern.

3.2 Substrate surface wave scattering

This is characterised by S_L, the ratio of surface-wave generation to radiation [16] representing a peak of likely sidelobe perturbation, and by $S_A = S_L/G_A$ where G_A is the array gain which represents the mean sidelobe level. S_L is approximately −13 dB, −18 dB and −22 dB for $h/\lambda_0 = 0.06$, 0.03 and 0.01, respectively, with $\varepsilon_r = 2.32$. The mean sidelobe level S_A will be significantly lower than this and will decrease with increasing array size. This trend agrees with results given by analysis of finite microstrip arrays using the moments method [17], although actual surface-waves levels are somewhat lower here.

3.3 Mutual coupling effects

The effect of mutual coupling has been calculated by incorporating mutual admittance values given by the magnetic-current method [18] into a forced excitation model of the feed-array interaction. Hence

$$[V] = [Z][I] \tag{10}$$

where $[V]$ is the element voltage matrix, $[Z] = [Y]^{-1}$ is the array self and mutual impedances matrix, and $[I]$ is the constant current excitation matrix. Table 1 shows the peak increase in the sidelobe level for linear patch arrays with uniform excitations. Increases in the E-plane are considerably higher than in the H-plane. Such increases may well account for the higher measured sidelobes noted in Fig. 3. For tapered aperture distributions larger increases are only noted in the smaller arrays. Mutual coupling effects also decrease with decreasing dielectric constant and substrate thickness.

Table 1: Computed peak increase in sidelobe level in linear patch array due to mutual coupling

Number of elements	Peak increase in sidelobe level, dB	
	E-plane	H-plane
8	2.6	0.5
16	4.0	0.5

$\varepsilon_r = 2.32$; $h/\lambda_0 = 0.06$; $d/\lambda_0 = 0.7$

3.4 Design and production tolerances

The maximum rise in sidelobe level L_{max} due to tolerance effects is given [19] by

$$L_{max} = 10 \log_{10}\{1 + 10^{s/10} K(\sigma_1^2 + 2\sigma_2^2 + 2f^2)\} \tag{11}$$

s is the design sidelobe level (in dB); K is a constant relating to array size; σ_1^2 and σ_2^2 are the variances of positional and angular element errors; f relates to errors in element excitation. Estimates have been based on experience of production tolerances achieved at RMCS, manufacturers quoted material tolerances in height and dielectric constant, and previous corporate feed analysis [20]. Table 2 gives typical values of L_{max}. It can be seen from the caption that excitation error f^2 dominate, owing primarily to material and production tolerance effects in the feed nctwork. However, overall sidelobe increases are well below those due to feed radiation. Excitation errors and hence sidelobe errors will increase with increasing substrate dielectric constant owing to the increased electrical line lengths involved for a given array size. The effect of substrate height is less clear.

Table 2: Maximum rise in sidelobe level L_{max} due to design and production tolerances in a two dimensional array

Design sidelobe level D, dB	L_{max}, dB	
	$N = 16$	$N = 64$
13 (uniform distribution)	0.04	0.00
20	0.20	0.04
30	2.10	0.42

Array size = N elements; $\varepsilon_e = 2.32$; $h/\lambda_0 = 0.06$; $d/\lambda_0 = 0.7$; $\sigma_1^2 = 0.0025$; $\sigma_2^2 = 0.01$; $f^2 = 0.09$

4 Array gain and efficiency

Gain reduction in microstrip patch arrays is due to losses in the feed line and patches and to losses due to feed radiation, surface wave generation, mutual coupling, and design and manufacturing tolerance errors. The total feed loss L_f in a two dimensional patch array is given by Reference 21 as follows:

$$L_f \approx \frac{\alpha d}{\lambda_m}(\sqrt{(N)} - 1) + n_t l_t + n_b l_b + l_c \tag{12}$$

$\alpha = \alpha_d + \alpha_{cr}$; N is the total number of elements; d is the element spacing; l_t, l_b and l_c are the radiation losses in dB of a T-junction, bend and coax to microstrip transition, respectively, deduced from eqns. 4, 3 and 2, respectively, with eqn. 1. n_t is the number of T-junctions; n_b is the number of bends. For the configuration of Fig. 1 $n_b = 3$. Gain loss L_t due to array design and production tolerances is given by Reference 22 as follows:

$$L_t = 1 + \sigma^2 \tag{13}$$

$\sigma^2 = \sigma_1^2 + \sigma_2^2 + f^2$ is the total error variance, eqn. 11; for values given in Table 2 $L_t = 0.4$ dB and is independent of array size; gain loss due to mutual coupling is given by the analysis of Section 3.3 and, assuming that the feed is matched to the active impedance of the array central element, is less than 0.1 dB for arrays of less than 16 × 16 elements. Patch resistive and surface wave losses are given by Reference 23.

Fig. 5 shows that overall array loss increases with increasing substrate height, decreasing dielectric constant and feed impedance. For arrays smaller than 16 × 16 elements total feed radiation is greater than resistive losses. Above this the high line losses make such arrays impractical. The rapid increase in loss with size gives rise to a maximum gain value for coplaner fed arrays which is illustrated in Fig. 6. A peak gain of about 35 dB is indicated, corresponding to an efficiency of about 10%.

5 Design implications and array optimisation

5.1 Substrate choice

Table 3 summarises the deductions relating to substrate choice. The overall requirements are contradictory, with low dielectric constant leading to wider bandwidths but in particular increased array losses and thin substrates leading to all round array benefits at the cost of narrow bandwidth. These trade-offs are further illustrated in Table 4, where the effect of substrate thickness on array efficiency and bandwidth is given. It can be seen that substantial increases in efficiency are obtained by use of thin substrates but at the expense of patch bandwidth. Decreases in substrate thickness are accompanied by reduced pattern perturbation due to feed radiation as shown in Fig. 7, and the perturbation may be reduced to

Table 3: Optimum substrate parameter choice for array performance requirements

Requirement	Dielectric constant	Substrate thickness
1 Low feed radiation	high	thin
2 Low surface waves	low	thin
3 Good tolerance control	low	thin
4 Low mutual coupling	low	thin
5 Low array losses	high	thin
6 Wide bandwidth	low	thick

For substrates $1 < \varepsilon_r < 2.5$; $0.01 < h/\lambda_0 < 0.1$

a relative level of -25 dB if the thickness is decreased to the order of $h/\lambda_0 = 0.01$. This perturbation level is comparable to surface-wave levels, and is less than those due to mutual coupling. The bandwidth constraint of thin substrates can be overcome to some extent by the use of electromagnetic coupling to overlaid patches [24] or the use of a separate feed layer, although the penalty of increased constructional complexity is incurred.

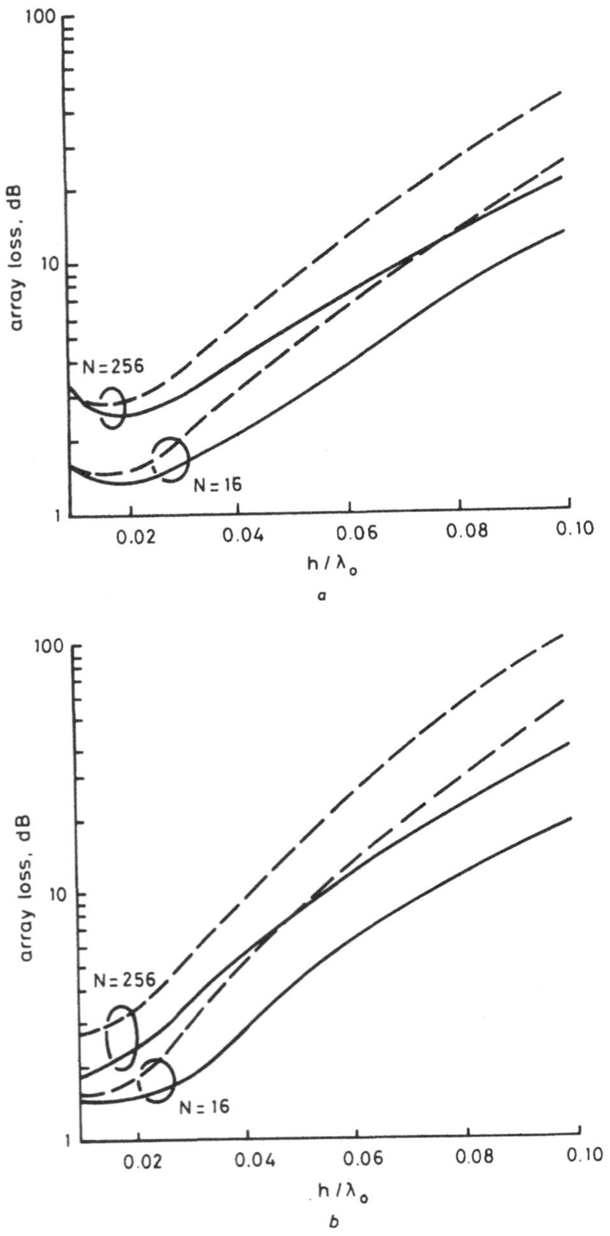

Fig. 5 *Overall loss of corporately fed microstrip array*

a $\varepsilon_r = 2.32$
b $\varepsilon_r = 1.06$
Assumes feed form of Fig. 1; element spacing $= 0.8\lambda_0$
—— feed impedance = 100 Ω
- - - - feed impedance = 50 Ω

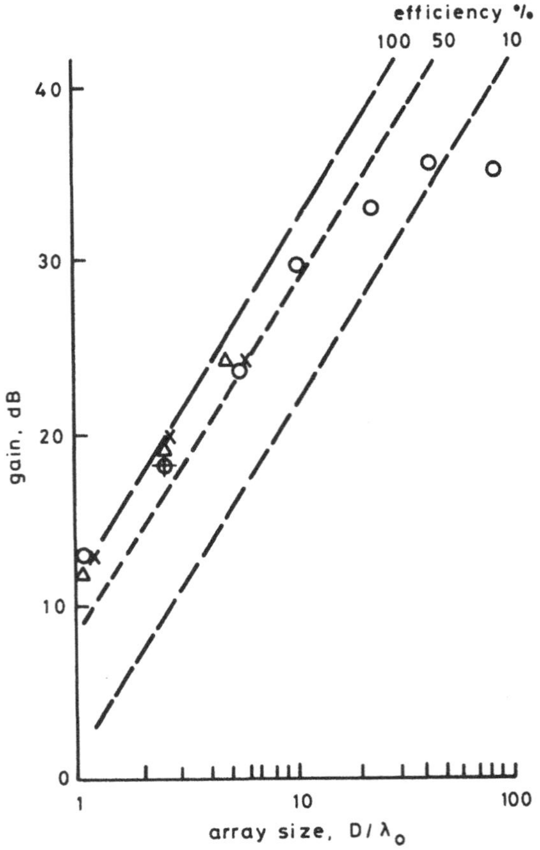

Fig. 6 *Array gain*

○ estimates based on results of Fig. 5
× measured, feed impedance 120 Ω, as Fig. 1
△ measured, Reference 3, feed impedance 200 Ω; $h/\lambda_0 = 0.06$; $\varepsilon_r = 2.1$
+ measured, Reference 21, feed impedance = 100 Ω; $h/\lambda_0 = 0.0036$; $\varepsilon_r = 2.32$

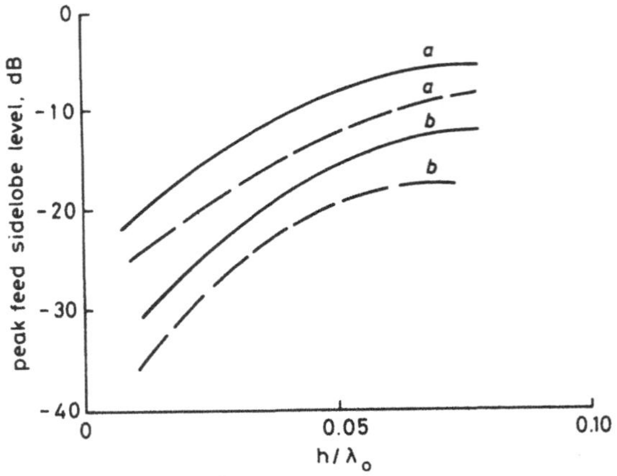

Fig. 7 *Calculated peak sidelobe level of feed radiation of 4 × 4 element array*

—— $\varepsilon_r = 1.06$
— — $\varepsilon_r = 2.32$
a Array as Fig. 1
b Array as Fig. 8

Table 4: Computed efficiency and bandwidth of 16 × 16 patch array

Feed type	Feed height, h/λ_0	Efficiency, %	Patch bandwidth, % (to -10 dB return loss points)
As Fig. 1	0.06	39	5
As Fig. 1 but with smoothed bends and splitters ($\rho/\lambda_0 = 0.5$)	0.03	57	2
	0.01	64	1
As Fig. 8	0.06	55	5

5.2 Feed geometry

The use of smooth bends to reduce feed radiation has been noted in Table 4, and is cited as an example of optimisation of the feed geometry. A further example is in the use of sparse arrays with spherical dielectric overlays [25], where at the expense of more constructional complexity the number of feed discontinuities is significantly reduced, thus decreasing feed loss and pattern disturbance. However, an example combining optimised performance with manufacturing simplicity is that of sequentially rotated feeding for circularly polarised arrays [26]. This technique involves providing to the mth array element both a physical rotation and a feeding phase shift Φ_m given by

$$\Phi_m = \frac{2\pi(m-1)}{N} \quad 1 < m < N \tag{14}$$

where N is the array size. N may constitute the complete array or be a sub-array of a larger one. The principle has been applied to $N = 2$ arrays [1] and here to $N = 2 \times 2$ arrays [4]. These subarrays are used to form the complete array by applying sequential rotation to their feeding, as shown in Fig. 8. The overall geometry has less

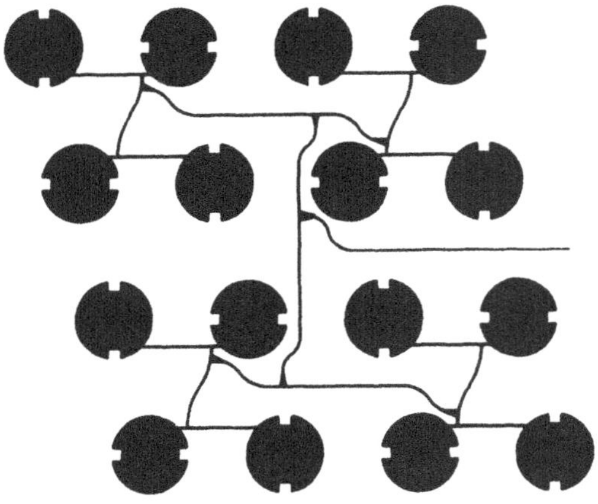

Fig. 8 *Silhouette of disc array for circular polarisation using sequentially rotated feeding*

feed discontinuities than the conventional feed of Fig. 1. Additionally, the feeding phase shifts cause radiation from adjacent T-junction pairs to be co-phased, and hence add into the main beam, reducing gain loss and sidelobe levels. The slanted patch configuration is used to reduce far out grating lobes in the 45° planes, owing to feed discontinuities with spacings greater than the free space wavelength.

Table 4 shows the increase in efficiency to be expected from this method, deduced from first-order consideration of the grating lobe structure of each pair of feed discontinuities. Figs. 9*a* and 9*b* show the computed radiation patterns of 16 × 16 element versions of the arrays of Fig. 8 and Fig. 1, respectively. Significant reduction of the overall sidelobe levels is obtained, particularly at wide angles. Fig. 7 shows that peak feed sidelobe levels for 4 × 4 element arrays fall by up to 10 dB for equivalent substrate thickness. Ultimately, the realisable sidelode levels for practical arrays using this feed method will of course depend on the relative levels of feed radiation, surface-wave, mutual coupling, and tolerance effects; however, significant reduction is expected for small arrays where feed radiation effects dominate.

5.3 Subarraying

Improved radiation pattern control and increased efficiency can be obtained by splitting the array into subarray sections, as indicated in the inset to Fig. 10. Each subarray is then fed by a low-loss corporate feed located behind the microstrip groundplane (as for example in Reference 1). Fig. 10 quantifies the advantage of subarraying within a 32 × 32 element patch array. It can be

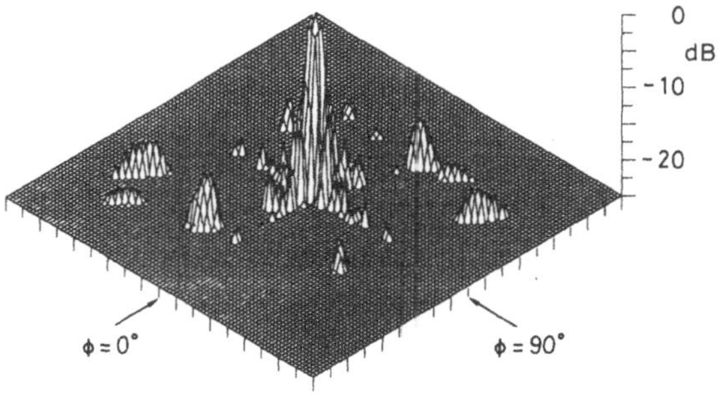

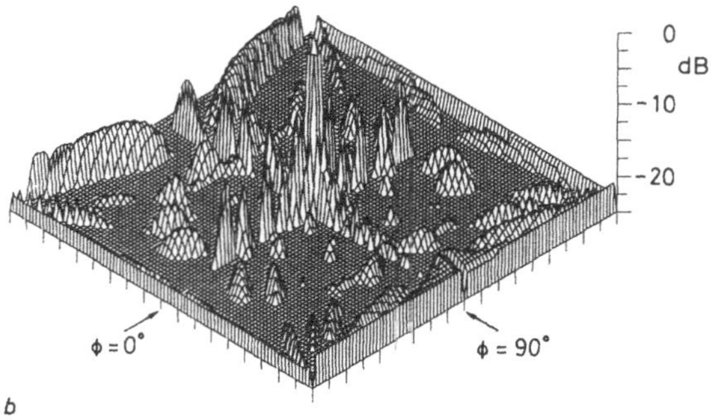

Fig. 9 *Computed radiation patterns of 16 × 16 patch array*
a Array as Fig. 8
b Array as Fig. 1
level shown is peak radiation whether co- or crosspolarised

seen from Fig. 10*a* that significant suppression of array sidelobes occurs for subarray sizes of 8 × 8 elements and less, particularly for the conventional corporate feed (Fig. 1), although sidelobe levels of the order of −25 dB are still expected. The use of sequential rotation for circular polarisation (Fig. 8), is seen to be advantageous for larger subarray sizes. The efficiency (Fig. 10*b*), for subarray size less than 4 × 4 elements, is determined primarily by feed-radiation loss and patch-dissipative and surface-wave losses. The advantages of subarray construction become greater for larger array size and indeed may be forced on designers by limited commercially available substrate sizes.

6 Conclusions

The use of coplanar microstrip feed networks for microstrip patch arrays allows simple construction, but incurs both gain loss and degradation in sidelobe and cross polarisation due to resistive loss and feed radiation. In comparison to these effects substrate surface wave and tolerance problems are small, although mutual coupling can be significant. For example in an 8 × 8 elemnt array with uniform distribution and 5% bandwidth, less than 55% efficiency is achieved with a sidelobe level of −10 dB. These effects are minimised by the use of smooth feed discontinuities, high feed line impedance, and thin substrates with a dielectric constant between 2 and 3, although the substrate recommendation will compromise bandwidth. Electromagnetic coupling to overlaid patches, shielded feeds or sparse arrays with overlaid lenses will overcome the bandwidth problem at the expense of constructional simplicity. The use of sequentially-rotated circularly-polarised arrays maintains simplicity, while reducing peak sidelobes by up to 10 dB.

Subarrays fed by low-loss shielded feeds will further improve performance.

The use of patch arrays with coplanar feed networks will thus give poorer performance than arrays with fully-shielded feeds or other antenna types such as reflectors. However, some degree of perfomance optimisation is possible and this, together with their potential low cost and volume, makes them attractive for a wide variety of applications. In addition such arrays lend themselves readily to computer aided design and production, and the first order analytical expressions and performance trade offs given here are likely to be an important element in such techniques.

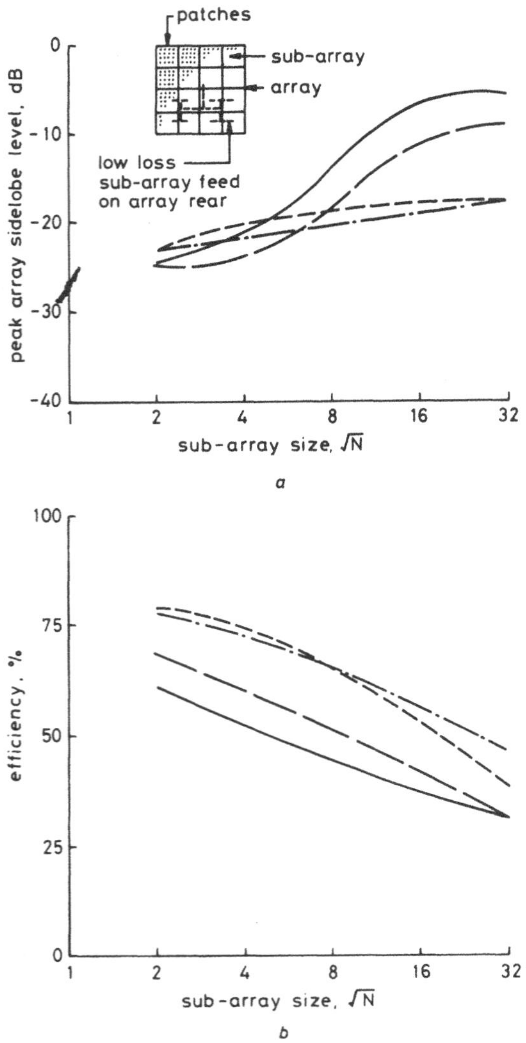

Fig. 10 *Performance of array of coplanar microstrip subarrays fed by low loss network*

a Peak array sidelobe level
b Array efficiency
Number of subarray elements = N; number of array elements = 32 × 32; inset shows example for N = 16 (i.e. subarray = 4 × 4 elements) microstrip coplanar feed within subarray; low loss (0.1 dB/λ_d, $\lambda_d = \lambda_0/\sqrt{2.5}$) feed connects subarrays
array of form of Fig. 1:
——— $\varepsilon_r = 1.06$
— — $\varepsilon_r = 2.32$
array of form of Fig. 8:
—·—· $\varepsilon_r = 1.06$
- - - - $\varepsilon_r = 2.32$

7 Acknowledgments

C.M. Hall was partly supported by the US Army, European Research Office. The authors would like to thank members of the Wolfson RF Engineering centre, RMCS, for useful discussions. The assistance of Mr L. Petterson of the Swedish Defence Research Institute, in the derivation of eqn. 2 is also acknowledged.

8 References

1 HANEISHI, M.: 'A circularly polarised SHF planar array composed of microstrip pairs elements', *Proc. ISAP 85*, Tokyo, Japan, August 1985, pp. 125–128
2 SHEEHAN, P.G., and FORREST, J.R.: 'Satellite-borne active phased array techniques for mobile communications' *IEE Proc. F, Commun., Radar & Signal Process.*, 1986, **133**, (4), pp. 339–344
3 ASHKENAZY, J., PERLMUTTER, P., and TREVES, D.: 'A modular approach for the design of microstrip array antennas', *IEEE Trans.*, 1983, **AP-31**, pp. 190–193
4 HALL, P.S.: 'Feed radiation effects in sequentially rotated microstrip patch arrays', *Electron. Lett.*, 1987, **123**, pp. 877–878
5 PUCEL, R.A., MASSE, D., and HARTWIG, C.P.: 'Losses in microstrip', *IEEE Trans.*, 1968, **MTT-16**, pp. 342–350 and p. 1064
6 WHEELER, H.A.: 'Transmission line properties of parallel strips separated by a dielectric sheet', *IEEE Trans.*, 1965, **MTT-13**, pp. 172–185
7 KIRSCHNING, M., and JANSEN, R.H.: 'Accurate model for effective dielectric constant of microstrip with validity up to millimetre-wave frequencies', *Electron. Lett.*, 1982, **18**, (6), pp. 272–273
8 HAMMERSTAD, E.O., and BEKKADAL, F.: 'Microstrip Handbook' ELAB report STF 44 A74169, University of Trondheim, Norwegian Institute of Technology
9 LEWIN, L.: 'Radiation from discontinuities in stripline', *IEE Proc. C*, 1960, **107**, pp. 163–170
10 HALL, P.S.: 'Microstrip linear array with polarisation control', *IEE Proc. H, Microwaves, Opt. & Antennas*, 1983, **130**, (3), pp. 215–224
11 LEWIN, L.: 'Spurious radiation from microstrip', *Proc. IEE*, 1978, **125**, (7), pp. 633–642
12 WOOD, C.: 'Curved microstrip lines as compact wideband circularly polarised antennas', *IEE J. Microwaves Opt. & Acoust.*, 1979, **3**, (1), pp. 5–13
13 HENDERSON, A., and JAMES, J.R.: 'Design of microstrip antenna feeds, Part 1: estimation of radiation loss and design implications', *IEE Proc. H, Microwaves, Opt. & Antennas*, 1981, **128**, (1), pp. 19–25
14 HALL, P.S., and JAMES, J.R.: 'Crosspolarisation behaviour of series-fed microstrip linear arrays', *IEE Proc. H, Microwaves, Antennas & Propag.*, 1983, **131**, (4), pp. 247–257
15 HALL, P.S., and PRIOR, C.J.: 'Microstrip array for reflector feed applications', 14th European Microwave Conference, Liege, Belgium, September 1984, pp. 631-636
16 JAMES, J.R., and HENDERSON, A.: 'High frequency behaviour of microstrip open-circuit terminations', *IEE J. Microwaves, Opt. & Acoust.*, 1979, **3**, pp. 205–211
17 POZAR, D.M.: 'Finite phased arrays of rectangular microstrip patches', *IEEE Trans.*, 1966, **AP-34**, pp. 658–665
18 PENARD, E., and DANIEL, J.P.: 'Mutual coupling between microstrip antennas', *Electron. Lett.*, 1982, **18**, pp. 605–607
19 ELLIOT, R.S.: 'Mechanical and electrical tolerances for two dimensional scanning arrays', *IRE Trans.*, 1958, **AP-6**, pp. 114–120
20 HALL, P.S., and JAMES, J.R.: 'Design of microstrip antenna feeds: Part 2', *IEE Proc. H, Microwaves, Opt. & Antennas*, 1981, **128**, (1), pp. 26–34
21 HALL, C.M.: 'Millimetre-wave microstrip antennas and hybrid types', PhD thesis, Royal Military College of Science, Shrivenham, February 1987
22 RUDGE, A.W., MILNE, K., OLVER, A.D., and KNIGHT, P.: 'Handbook of antenna design' (Peter Perigrinus, London, 1983) Vol 2, p. 76
23 JAMES, J.R., HENDERSON, A., and HALL, P.S.: 'Microstrip antenna performance is determined by substrate constraints', *Microwave Systems News*, 1982, **2**, pp. 73–84
24 HALL, P.S., and PRIOR, C.J.: 'Microwave feeds for prime focus fed reflector antennas, *IEE Proc. H, Microwaves, Antenna & Propag.*, 1987, **134**, (2), pp. 185–193
25 JAMES, J.R., HALL, C.M., and ANDRASIC, G.: 'Microstrip elements and arrays with spherical dielectric overlays', *IEE Proc. H, Microwaves, Antenna & Propag.*, 1986, **133**, (6), pp. 474–482
26 TESHIROGI, T., TANAKA, M., and CHUJO, W.: 'Wideband circularly polarised array with sequential rotation', Proc. ISAP, Tokyo, Japan, August 1985, pp. 117–120

A Study of Microstrip Array Antennas with the Feed Network

ELY LEVINE, MEMBER, IEEE, GABI MALAMUD, SHMUEL SHTRIKMAN, FELLOW, IEEE, AND DAVID TREVES, FELLOW, IEEE

Abstract—**The radiation and losses in microstrip antennas with a corporate feed network are studied. First, we apply a surface current approach in which the electrical currents in the feed lines are modeled as in ideal transmission lines. The free space radiation and the surface wave excitation of typical segments in printed feed networks are studied. A four-element array antenna with its printed feed network is analyzed and predicted radiation patterns, directivity, and gain are presented and compared with experimental results. The gain and directivity of large arrays of 16, 64, 256 and 1024 elements are calculated and measurements in the frequency range of 10 to 35 GHz are reported.**

I. Introduction

PRINTED ANTENNAS ARE promising candidates for microwave and millimeter wave applications, where the weight and the volume of the antenna should be kept to a minimum, or when conformal arrays are needed. However, printed arrays show low efficiency due to ohmic and dielectric losses in the feed network, due to parasitic radiation in the feed network, and due to the excitation of surface waves in the dielectric substrate [1]–[3]. The efficiency limitations are most severe in large arrays where the feed network is long and complicated, and at high frequencies (K or Ka bands) where the dielectric and ohmic losses are high. In recent years, several high-gain printed arrays were investigated [4], [5], especially for direct broadcast satellite (DBS) applications [6]–[8].

The purpose of this work is to present an analysis of the radiation and the losses of microstrip arrays including the feed network effects. The discussion is restricted to single-layer microstrip arrays built on commercial substrates with dielectric constants close to two. This analysis can give useful estimations of the available gains in various arrays, and also provides a theoretical prediction of the radiation patterns in the presence of feed lines. There are only a very few publications on such network effects [9], [10], mainly because solutions based on moment methods or on conjugate gradient methods require a vast amount of computation, and therefore are not very practical for general design purposes. A recent contribution [11] based on a magnetic current model gives a detailed assessment of the feed network effects.

Our approach is based on a surface current model [12] in which the printed radiators and the feed lines are replaced by assumed current distributions. The currents in the radiating elements are derived from cavity or equivalent transmission-line models and the currents in the feed lines are taken as traveling waves. The radiation fields and the surface-wave excitation are then found from the assumed currents using the appropriate Green's function in the Fourier domain. The ohmic and dielectric losses in the feed lines are calculated using known formulas from the literature [13]–[15].

In Section II, the mechanism of radiation and surface wave excitation emerging from typical microstrip segments is briefly reviewed. In Section III, an analysis of four element arrays, including the feed network, is presented and compared to results without the network. Experimental results of several four element arrays are also given. The directivity and the gain of modular designed arrays [16], [17] are studied in Section IV, and compared to experimental results of 16, 64, 256 and 1024 element arrays, built for frequencies of 10, 20, 30 and 35 GHz, respectively. These results can be used as a general estimation of the gain limitations in large printed arrays.

II. Radiation and Losses in Microstrip Lines

We consider a planar microstrip configuration with an infinite metallic ground plane at $z = 0$ and a planar dielectric substrate of height h and dielectric constant ϵ_r. We are interested in the electromagnetic fields due to any arbitrary current distribution $\bar{J}(x, y)$. Following the surface current analysis in the spectral domain described in [12] one can represent the current distribution by its Fourier decomposition:

$$\bar{J}(x, y) = \frac{1}{4\pi^2} \iint_{-\infty}^{+\infty} \tilde{\bar{J}}(k_x, k_y) e^{-j(k_x x + k_y y)} \, dk_x dk_y \quad (1)$$

where the tilde over a variable denotes its Fourier transform, and $j = \sqrt{-1}$. The same transformation is done for $\bar{E}$ and $\bar{H}$. The electric fields and the surface currents are related by

$$\tilde{E}_i(k_x, k_y) = \tilde{G}_{ij}(k_x, k_y) \tilde{J}_j(k_x, k_y), \qquad i, j = x, y \quad (2)$$

where the matrix $\tilde{G}$ is the dyadic Green's function for an elemental surface current source in the Fourier domain.

The complex input power into the antenna is

$$P_{\text{in}} = -\frac{1}{2} \iint_{-\infty}^{+\infty} \tilde{\bar{E}}(k_x, k_y) \cdot \tilde{\bar{J}}^*(k_x, k_y) \, dk_x dk_y. \quad (3)$$

The contribution due to radiation into free space comes from the "visible range" in which $k_x^2 + k_y^2 < k_0^2$. Transforming to spherical coordinates by $k_x = k_0 \sin\theta \cos\phi$ and $k_y = k_0 \sin\theta \sin\phi$ and integrating over a sphere of radius k_0, the

Reprinted from *IEEE Trans. Antennas Propaga.*, vol. 37, no. 4, pp. 426–434, April 1989.

power radiated into free space is given by

$$P_r = \int_0^{\pi/2} \int_0^{2\pi} f(\theta, \phi) \sin \theta d\theta d\phi \tag{4}$$

where

$$f(\theta, \phi) = \frac{15k_0^2}{\pi} \cdot \left(\frac{|-\tilde{J}_x \sin \phi + \tilde{J}_y \cos \phi|^2 \cos^2 \theta}{(\epsilon_r - \sin^2 \theta) \cot^2 (hk_0\sqrt{\epsilon_r - \sin^2 \theta}) + \cos^2 \theta} + \frac{|\tilde{J}_x \cos \phi + \tilde{J}_y \sin \phi|^2 \cos^2 \theta (\epsilon_r - \sin^2 \theta)}{(\epsilon_r - \sin^2 \theta) + \epsilon_r^2 \cos^2 \theta \cot^2 (hk_0\sqrt{\epsilon_r - \sin^2 \theta})} \right). \tag{5}$$

The expression for $f(\theta, \phi)$ is the free space radiation power pattern.

When we examine values of $k_t^2 = k_x^2 + k_y^2$ which are greater than k_0^2 (the "invisible range"), we find that the power propagated by surface waves in the first transverse magnetic (TM) mode is given by

$$P_s = \int_0^{2\pi} f_s(k_{tp}, \phi) \, d\phi \tag{6}$$

where

$$f_s(k_{tp}, \phi) = A_s 15k_0^2 |\tilde{J}_x(k_{tp}, \phi) \cos \phi + \tilde{J}_y(k_{tp}, \phi) \sin \phi|^2 \tag{7}$$

and

$$A_s = \frac{\epsilon_r(x_p^2 - 1)x_p}{\epsilon_r x_p \left(\frac{1}{\sqrt{x_p^2 - 1}} - \frac{\sqrt{x_p^2 - 1}}{x_p^2 - \epsilon_r} \right) + x_p k_0 h \left(1 + \frac{\epsilon_r^2(x_p^2 - 1)}{\epsilon_r - x_p^2} \right)}$$

$$x_p \equiv \frac{k_{tp}}{k_0}$$

$$k_{tp}^2 = k_x^2 + k_y^2$$

is the first solution of

$$j\epsilon_r\sqrt{k_0^2 - k_{tp}^2} \cot(\sqrt{\epsilon_r k_0^2 - k_{tp}^2} h) - \sqrt{\epsilon_r k_0^2 - k_{tp}^2} = 0$$

and k_x, k_y have been transformed to cylindrical coordinates by

$$k_x = k_{tp} \cos \phi; \quad k_y = k_{tp} \sin \phi.$$

The integrand in (6) represents a one-dimensional "radiation pattern," where ϕ is the "angle of radiation" with respect to the $\hat{x}$ axis. Additional aspects of this model, more details about the calculations, and numerical results can be found in [12].

Now consider a microstrip line printed on a grounded dielectric substrate of thickness h and dielectric constant ϵ_r where the width and length of the line are W_l and L, respectively (Fig. 1). The two cases of end-fed and center-fed lines shown in this figure are the most typical building blocks in antenna feed networks. At this stage the printed lines are regarded as lossless transmission lines, e.g., the current along the line is not influenced by the dielectric, ohmic, and radiative losses. The effective dielectric constant and the characteristic impedance of the line are [18]

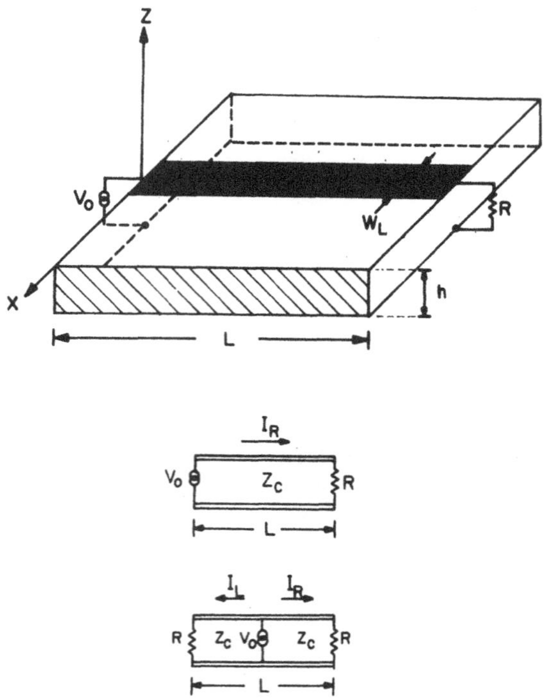

Fig. 1. A microstripline geometry. (a) The structure and the dimensions of an end-fed microstripline. (b) Two schematic feeds: end-fed and center-fed. In both cases the lines are matched in their ends.

$$\epsilon_{\text{eff}} = \frac{\epsilon_r + 1}{2} + \frac{\epsilon_r - 1}{2\sqrt{1 + 10\frac{h}{W_l}}} \tag{8}$$

$$Z_c = \frac{120\pi h}{\sqrt{\epsilon_{\text{eff}}}} \cdot \begin{cases} [W_l + 2.42h - 0.44h^2/W_l + h(1 - h/W_l)^6]^{-1}, & \text{if } h \le W_l \\ \ln (8h/W_l + W_l/4h)/2\pi h, & \text{if } h \ge W_l. \end{cases} \tag{9}$$

The current and voltage in a transmission line which is fed at its end by the voltage $V_0 e^{j\omega t}$ and loaded by the resistance R at its other side are [19]

$$V(y) = V_0(e^{-j\beta y} + \Gamma e^{+j\beta y}), \quad \text{for } 0 \le y \le L$$

and (10)

$$I(y) = I_0(e^{-j\beta y} - \Gamma e^{+j\beta y}), \quad \text{for } 0 \le y \le L$$

where $I_0 = V_0/Z_c$ and β is the constant of propagation in the transmission line

$$\beta = k_0 \sqrt{\epsilon_{\text{eff}}} \tag{11}$$

and Γ is the reflection coefficient at the end of the line:

$$\Gamma = \frac{R - Z_c}{R + Z_c}. \tag{12}$$

The time dependence $e^{j\omega t}$ is assumed but omitted everywhere in this work.

When $R = Z_c$ the line is matched and $\Gamma = 0$. Assuming that the current amplitude is constant across the line, one obtains that the surface current density in an end-fed line is

$$\bar{J}(x, y) = \hat{y} \frac{I_0}{W_l} e^{-j\beta y} \cdot \text{rect}\left(\frac{x}{W_l}\right), \quad \text{for } 0 \le y \le L \tag{13}$$

where

$$\text{rect}\left(\frac{x}{W_l}\right) \equiv \begin{cases} 1, & \text{for } -W_l/2 \le x \le +W_l/2 \\ 0, & \text{elsewhere.} \end{cases}$$

For a center-fed line, which is matched at both ends, the surface current density is

$$\bar{J}(x, y) = \begin{cases} \hat{y}\left(\frac{I_0}{W_l}\right) \cdot e^{-j\beta y} \ \text{rect}\left(\frac{x}{W_l}\right), \\ \quad \text{for } 0 \le y \le L/2 \\ -\hat{y}\left(\frac{I_0}{W_l}\right) \cdot e^{+j\beta y} \ \text{rect}\left(\frac{x}{W_l}\right), \\ \quad \text{for } -L/2 \le y \le 0. \end{cases} \tag{14}$$

In order to calculate the radiation from these microstriplines the Fourier transforms of the surface currents should be used. For an end-fed line one gets

$$\bar{\bar{J}}(k_x, k_y) = \hat{y} I_0 \left(\frac{e^{+j(k_y - \beta)L} - 1}{j(k_y - \beta)}\right) \text{sinc}\left(\frac{k_x W_l}{2}\right) \tag{15}$$

and for a center-fed line

$$\bar{\bar{J}}(k_x, k_y) = \hat{y} I_0 \left(\frac{e^{+j(k_y - \beta)L/2} - 1}{j(k_y - \beta)} - \frac{1 - e^{-j(k_y + \beta)L/2}}{j(k_y + \beta)}\right) \text{sinc}\left(\frac{k_x W_l}{2}\right) \tag{16}$$

where $\text{sinc}(x) \equiv \sin(x)/x$.

The radiated power is now calculated by substituting the relevant current distributions into (5). The surface waves are calculated by substituting these currents into (7). The input power to the microstripline is given by

$$P_{\text{in}} = \frac{1}{2} I_0^2 Z_c \tag{17}$$

so that for a current with an amplitude of 1 A the radiation *loss* and the surface wave *loss* are defined as

$$\frac{P_r}{P_{\text{in}}} = \frac{2P_r}{Z_c} \tag{18}$$

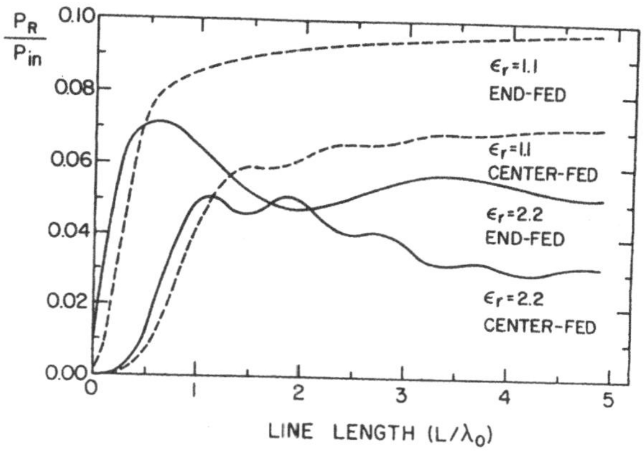

Fig. 2. Calculated free space radiation loss of a microstripline as a function of the line length. Four cases are studied: end-fed ($\epsilon_r = 1.1, 2.2$) and center-fed ($\epsilon_r = 1.1, 2.2$). Other parameters of the line are: characteristic impedance $Z_c = 100\ \Omega$ and thickness $h = 1.6$ mm. The frequency is $f = 10$ GHz.

and

$$\frac{P_s}{P_{\text{in}}} = \frac{2P_s}{Z_c}, \tag{19}$$

respectively.

A set of representative results has been derived using the above procedure. The radiation losses of typical end-fed and center-fed microstriplines with dielectric constants of $\epsilon_r = 1.1$ and 2.2 are shown in Fig. 2. The general conclusions concerning the losses in such lines are as follows.

- From (18) and (19) it follows that the losses are proportional to $1/Z_c$. The radiation power (P_r) and the surface wave excitation (P_s) are weakly dependent on Z_c so that the dependence of the losses on $1/Z_c$ is dominant. In general, the desire to minimize radiation and surface wave losses suggests that high characteristic impedance should be chosen.
- For a microstripline with given length and characteristic impedance, the radiation loss increases with $(h\sqrt{\epsilon_r}/\lambda_0)^2$ and the surface wave loss increases with $(h\sqrt{\epsilon_r}/\lambda_0)^3$. This is exact in the limits $h\sqrt{\epsilon_r}/\lambda_0 << 1$ and $\epsilon_r \rightarrow 1$, but is still a good approximation for standard dielectric substrates with $h\sqrt{\epsilon_r}/\lambda_0) \approx 0.1$ and $\epsilon_r \approx 2.2$. Thus, the desire to decrease radiation losses dictates low values of $h\sqrt{\epsilon_r}/\lambda_0$.
- For a microstripline with given thickness and characteristic impedance, the radiation loss depends on the line length as follows: in the range of $0 < L < \lambda_0$ the loss grows with $(L/\lambda_0)^2$. For the length greater than $L \gtrsim 3\lambda_0$ they are not sensitive to the length. The behavior of the surface wave loss as a function of the line length is oscillatory.
- The radiation losses are higher in end-fed microstrip lines than in center-fed lines. The reason is that a center-fed line carries two oppositely directed currents which tend to "cancel" each other. The surface wave losses are almost the same for end-fed and center-fed lines.
- Practical values for typical losses in microstriplines with $\epsilon_r = 2.2$, $h\sqrt{\epsilon_r}/\lambda_0 = 0.08$ and impedance of 200 Ω are about 3 percent (center-fed) to 5 percent (end-fed). These values are, as said, inversely proportional to the impedance, proportional to the square of the thickness, and not sensitive to the length.

The ohmic and dielectric losses are well known in the literature [13]–[15]. We consider only a nonmagnetic dielectric substrate in which the losses per unit length are small and can be calculated in terms of an attenuation factor α in the transmitted power $P(y)$:

$$P(y) = P_0 e^{-2\alpha y}. \tag{20}$$

Here y denotes a point along the direction of propagation (Fig. 1) and P_0 is the input power to the line at $y = 0$. The attenuation factor α is the sum of a dielectric factor α_d and an ohmic factor α_c:

$$\alpha = -\frac{dP/dy}{2P(y)} = \frac{P_c + P_d}{2p(y)} \quad \frac{\text{nepers}}{\text{unit length}} \tag{21}$$

or

$$\alpha_d = \frac{P_d}{2P(y)} \frac{\text{nepers}}{\text{unit length}}; \quad \alpha_c = \frac{P_c}{2P(y)} \frac{\text{nepers}}{\text{unit length}} \tag{22}$$

where P_d and P_c are the average dielectric and ohmic power losses per unit length, respectively. For the sake of convenience, the attenuation factor is given in dB/length units, thus the overall loss of the line is found by multiplying the attenuation factor by the length of the line. Several considerations for printed antennas should be pointed out. 1) The dielectric loss is not sensitive to the geometry of the line but it depends on the loss tangent, tan δ, of the material used. In common materials tan δ grows linearly with the frequency f. 2) The ohmic loss is high for narrow lines and for high impedances. It is also proportional to the skin resistance of the metal and thus goes up with $\sqrt{f}$. 3) The dissipation ohmic losses are therefore a technological factor, while the radiation and surface wave losses depend mainly on the line structure and can be minimized by appropriate design. 4) The dissipative losses increase linearly with the line length and thus play an important role in the efficiency of large arrays. It is interesting to notice that, once the substrate is chosen, the dissipation losses can be reduced by choosing feed lines with low impedances, but the radiation and surface wave losses would become higher. As a result, the total loss is not sensitive to the chosen impedances within the range of 100–200 Ω.

III. Four-Element Array Including Feed Network

In this section we describe the radiation and the surface wave excitation of a four-element array including its printed feed network. The printed layout and the relevant currents are shown in Fig. 3. The following results are presented: radiation patterns in the E- and H-planes, the radiation efficiency due to surface waves, the directivity, and the gain. Part of the results are shown also for the case where the feed network is omitted and some of the results are compared to experimental measurements.

The printed array includes four patches with distances D in both directions. Each patch has a length of L_p and a width of W_p. The patches are represented by four standing-wave currents I_p (put $\Gamma = 1$ in (12)), in a transmission-line with propagation constant β_p (derived by (8) and (11)) and characteristic impedance Z_c (9). The length of the patch is determined by the resonance condition: $\beta_p L_p = \pi/2$, and the width W_p is

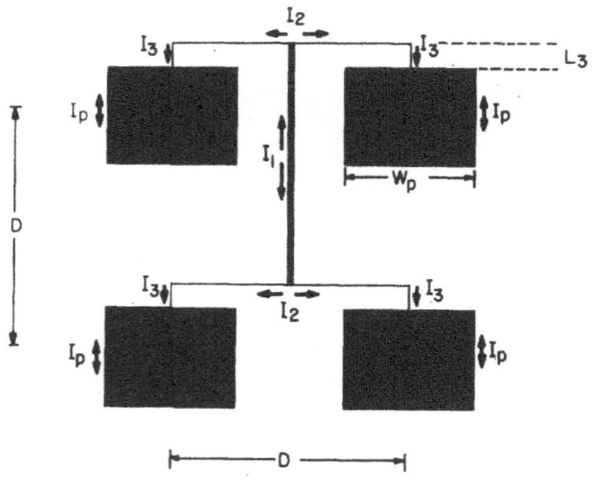

Fig. 3. Four element antenna geometry. The currents in the feed network are I_1, I_2, I_3. The currents in the patches are I_p. The distance between the patches is D in both directions. The length of the segments which enter the patches is L_3. Note that I_2 is x-directed and all the other currents are y-directed.

calculated iteratively as will be explained later on. The network includes a center line of 100 Ω fed at its center by an ideal 50 Ω source. In each one of the splitting points there are two 200 Ω lines going to the patches. The modeling of this network is done by traveling-wave currents I_1, I_2, I_3 following the approach described in the previous section. Notice that I_1 and I_2 are center-fed currents while I_3 is an end-fed current.

Fixing the relative amplitudes of the currents in the network and in the elements is an important point of this analysis. The basic assumption is that the supplied power is totally dissipated through free space radiation and surface waves. Hence, the resistance of each patch must be equal to the characteristic impedance Z_3. The procedure is the following: first, we normalize the amplitude of I_1 to be 1 A and the phase in the central feed point as zero. This choice determines all the amplitudes and phases in the feed network in such a way that the propagating power along the lines is conserved and the phase is continuous. Secondly we choose a value for W_p and use the condition of continuity of the voltages:

$$I_p Z_p = I_3 Z_3 e^{-j(\beta_3 L_3 + \pi/2)} \tag{23}$$

to find the amplitude of I_p. Then the radiation resistance R_r and surface wave resistance R_s are calculated and their sum is compared to Z_3. The width of the patch W_p is then changed iteratively until

$$R_r + R_s = Z_3. \tag{24}$$

The ratio between the current I_p and the current I_3 presents the quality factor Q of the radiating elements. It should be noted again that the above currents and the associated radiation properties are based on the approximation that the element is an ideal resonator or a section of an ideal transmission-line. From these currents one gets the radiated power, the surface wave power, and their equivalent real resistance. This resistance is then added to the feed network providing a match. The discussion on the validity ranges of this method is beyond the scope of this work. However, for high-Q microstrip elements it is a useful tool [20]. It is clear also that the approximation is good for the *feed lines,* whose radiation is much smaller than their input power (3–5 percent).

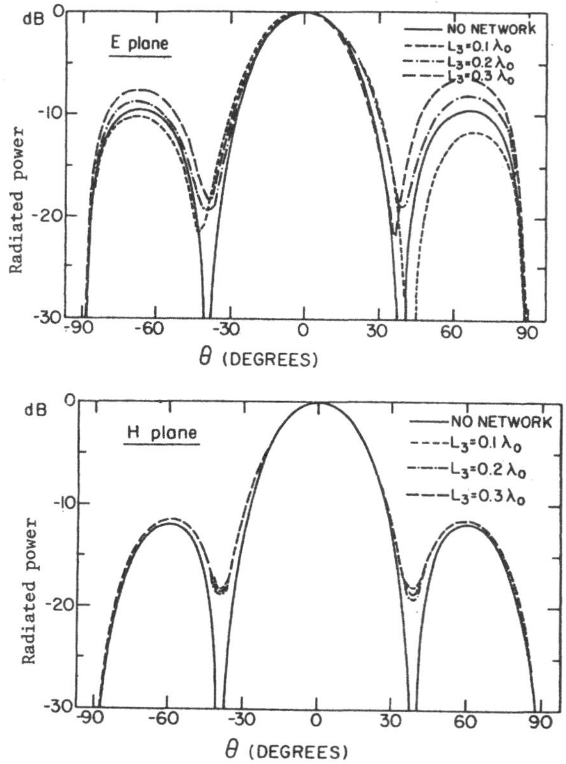

Fig. 4. Calculated radiation patterns of a four element array. The dielectric constant is $\epsilon_r = 2.2$, the thickness is $h\sqrt{\epsilon_r}/\lambda_0 = 0.08$, and the spacing is $D/\lambda_0 = 0.8$. Results are given for three values of L_3/λ_0 and for an array without the network. (a) For E-plane. (b) For H-plane.

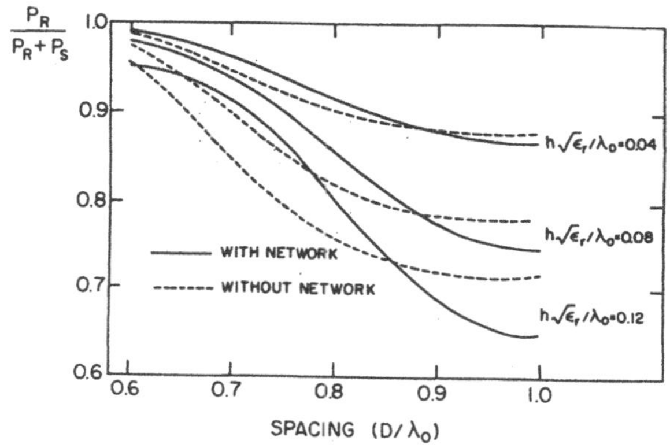

Fig. 5. Calculated efficiency due to surface wave losses of a four element array. P_r is the free space radiated power and P_s is the power radiated by surface waves. The segment $L_3/\lambda_0 = 0.1$, and the dielectric constant is $\epsilon_r = 2.2$. Results are given for three values of $h\sqrt{\epsilon_r}/\lambda_0$, with and without the network.

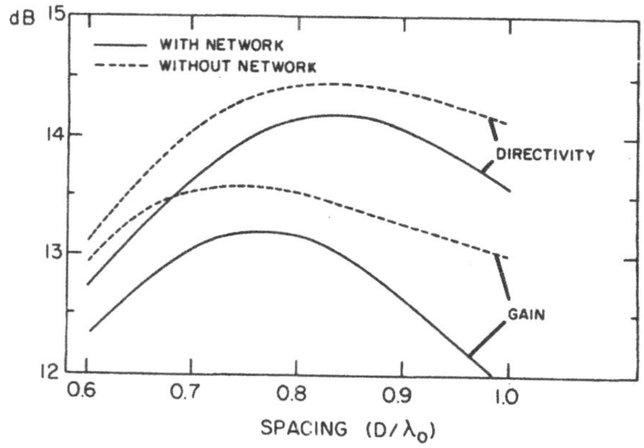

Fig. 6. Calculated directivity and gain of four-element arrays. The dielectric constant is $\epsilon_r = 2.2$, the thickness is $h\sqrt{\epsilon_r}/\lambda_0 = 0.08$ and $L_3/\lambda_0 = 0.1$.

A set of representative radiation patterns of this four element array is shown in Fig. 4. The results are shown for different values of the section length L_3. Although for practical design purposes one usually takes L_3/λ_0 to be small, it was found to be a suitable parameter for checking the analysis. The effects of the feed network on the sidelobes are significant in the E-plane pattern. The influence on the H-plane is less important due to the orthogonality of I_2 and the symmetry of all the other currents in this cut. Calculated results are shown also for an array without the feed network.

The radiation efficiency due to surface waves is defined as the ratio between the radiated power and the total power of radiation and surface waves:

$$\eta_s \equiv \frac{P_r}{P_r + P_s}. \tag{25}$$

Fig. 5 shows the calculated surface wave efficiency as a function of the spacing between close patches for three values of $h\sqrt{\epsilon_r}/\lambda_0$ and for a fixed value of L_3. Results without the feed network are also shown. The efficiency decreases as D/λ_0 increases and this effect is related to the strong interference between surface waves excited by co-linear elements on the $\hat{y}$ axes (7). The surface wave loss has a minimum when $D \approx \lambda_0/2$ and a maximum when $D \approx \lambda_0$. The gain is calculated by

$$G = D_{ir} - \eta, \quad \text{dB} \tag{26}$$

where η is the power efficiency of the antenna:

$$\eta = \eta_s + \eta_d + \eta_c + \eta_{cr}, \quad \text{dB} \tag{27}$$

and η_d and η_c are the efficiencies due to dielectric and ohmic losses. The efficiency due to connector losses η_{cr} is estimated theoretically [20] and checked experimentally [21] to be lower than 0.2 dB the mentioned substrate parameters. Fig. 6 shows the calculated directivity and gain as a function of D/λ_0 where $L_3/\lambda_0 = 0.1$ and $h\sqrt{\epsilon_r}/\lambda_0 = 0.08$. The directivity is calculated according to the definition:

$$D_{ir} = \frac{4\pi f(\theta, \phi)_{\max}}{P_r} \tag{28}$$

where $f(\theta, \phi)_{\max}$ denotes the maximum radiation intensity. Results without the feed network are again given for comparison. Maximum directivity and maximum gain occur for different values of D/λ_0 (0.83 and 0.76, respectively). The ohmic and dielectric losses are calculated to be 0.14 dB at 10 GHz. The difference between the directivity and the gain is caused mainly by surface wave losses.

A set of seven four-element arrays with different lengths L_3 had been built and tested. The antennas were designed to work at a resonance frequency of 10 GHz and the dielectric substrate was OAK-605, which has $\epsilon_r = 2.2$ and $h = 1.6$ mm ($h\sqrt{\epsilon_r}/\lambda_0 = 0.08$). The spacing between close elements was $D = 24$ mm ($0.8\,\lambda_0$) and the line widths were $W_1 = 1.5$ mm ($Z_1 = 98\,\Omega$) and $W_2 = W_3 = 0.2$ mm ($Z_2 = Z_3 = 193\,\Omega$). The width and length of the patches were chosen empirically

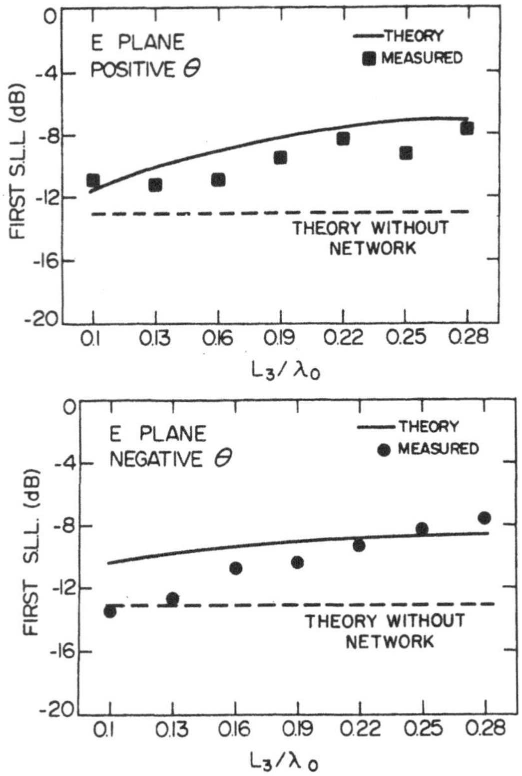

Fig. 7. The first sidelobe levels in the E plane for seven four-element arrays with different values of L_3/λ_0. (a) Positive Θ. (b) Negative Θ.

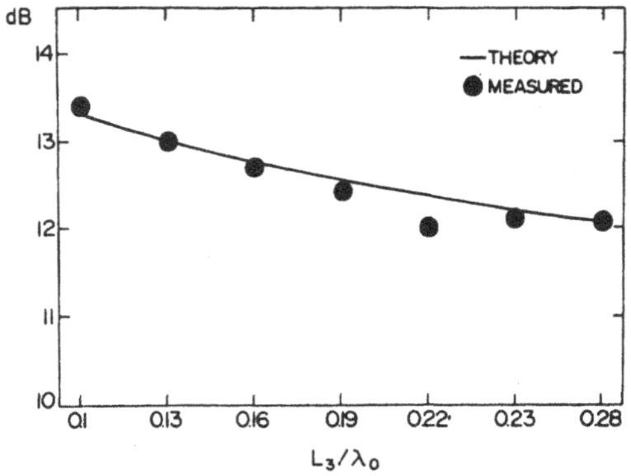

Fig. 8. The gains of four-element arrays with different values of the segment length L_3/λ_0. The spacing between close elements is $D/\lambda_0 = 0.8$.

TABLE I
CALCULATED DIRECTIVITY, LOSSES AND GAIN AND THE MEASURED GAIN (ALL IN dB) FOR A SET OF FOUR-ELEMENT ANTENNA ARRAYS WITH SEVEN VALUES OF THE SEGMENT L_3 (SEE FIG. 10)

L_3/λ_0	0.1	0.13	0.16	0.19	0.22	0.25	0.28
directivity without network	14.4	14.4	14.4	14.4	14.4	14.4	14.4
radiation loss	0.2	0.3	0.4	0.5	0.6	0.7	0.7
surface-wave loss	0.6	0.8	0.9	1.0	1.1	1.2	1.3
dielectric and ohmic losses	0.1	0.1	0.1	0.1	0.1	0.1	0.1
connector loss	0.2	0.2	0.2	0.2	0.2	0.2	0.2
calculated GAIN	13.3	13.0	12.8	12.6	12.4	12.2	12.1
measured GAIN	13.3	13.0	12.7	12.4	12.0	12.1	12.1

to match a central frequency of 10 GHz: $W_p = 8.9$ mm and $L_p = 11.2$ mm. It is interesting to compare these values with the values obtained by the iteration suggested before: $W_p = 9.4$ mm and $L_p = 10.2$ mm. The lengths of the last line segments were $L_3/\lambda_0 = 0.1, 0.13, 0.16, 0.19, 0.22, 0.25$, and 0.28. The measured resonance frequency was 9.9 GHz for all the measured antennas, but similar results were obtained at near-resonance frequencies of 9.8 and 10.0 GHz. Gain and pattern measurements were made outdoors at a distance of 25 m.

Fig. 7 shows measured and calculated sidelobe levels (SLL) in the E-plane as a function of L_3, both for positive and negative θ. An additional graph (the dashed line) shows the SLL of a four-element array without the feed network. The evaluation of the accuracy of the measurements is difficult. However, good reproducibility was obtained in many sets of measurements, including the inversion of the antenna direction in both the E- and H-planes. The SLL in the H-plane do not change with L_3, as expected from the theory, but in the E-plane we find an obvious tendency of the SLL to increase with L_3/λ_0. The measured and calculated gain as a function of L_3 are shown in Fig. 8. The accuracy of these measurements is better than 0.5 dB and the agreement of the measurement results with the theory is good. The dependence of the gain on L_3 is mainly caused by the increase in the surface wave loss, as can be seen from the loss budget in Table I.

The agreement between the measured and calculated results is better for the gain than for the sidelobes. We conclude that the main reason for the difference between the theory and the measurements is connected with the approximations made in the described model. The surface currents in the antenna are not self-consistently solved, and corrections due to the mutual interaction between the currents are not calculated. The effects of the vertical feed and the effects of the finite ground plane are also ignored. The accuracy of the current model is therefore limited, and seems to be better for integral quantities, like the gain. However, the overall behavior of the radiation patterns fits the measured results better than a simpler model that neglects the feed network.

IV. Directivity and Gain in Large Arrays

The effects of the feed network become important in high gain miscostrip array antennas with large numbers of radiating elements and complicated feed networks. It is the purpose of this section to present results of the gain limitations in large modular arrays. Consider $n \times n$ microstrip arrays where $n = 2, 4, 8, 16, \cdots$ with single layer power dividing network. Fig. 9 shows the layouts of 16, 64, and 256 element arrays built in such a way [16], [17]. The two basic building blocks are the four-element array (but with a center line of 200 Ω) and an "H" shaped feed network. The four-element subarray is changed a little in the layout shown in Fig. 3. The central line has been moved down and small diagonal sections have been inserted as can be seen in Fig. 9. This change was made in order to permit the connection of the "H" network to the center of the four-element subarray. The sensitivity of the directivity and the gain results to this change were checked and found negligible. The 16-element array is constructed from

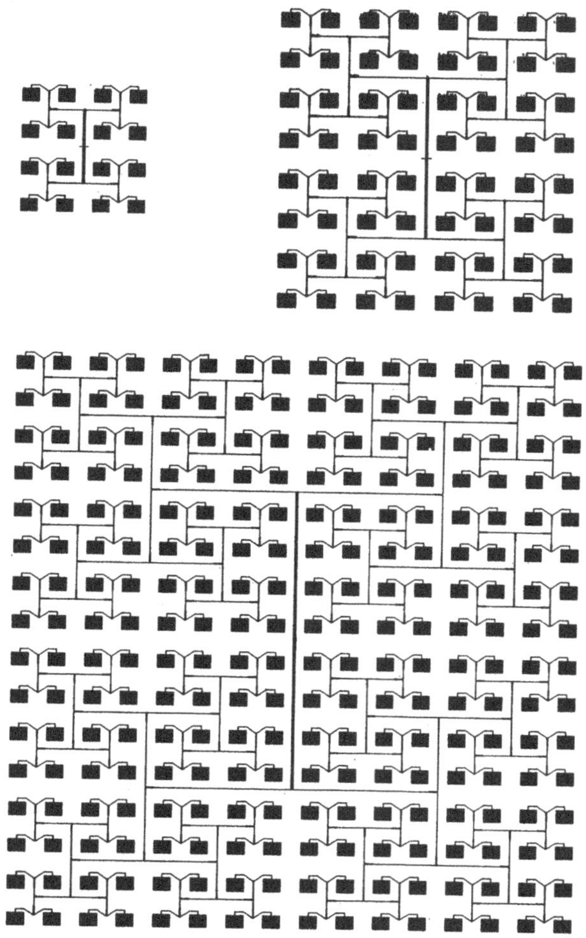

Fig. 9. Layouts of modular arrays of 16, 64, and 256 elements.

four subarrays connected and fed by the "H" network. The 64-element array is constructed again from four subarrays of 16 elements each and another "H" network, and so forth. In each one of the "H" networks there are several quarter wavelength transformers for matching. The current in each transformer is represented by two oppositely directed waves. Their amplitudes are determined by continuity of the voltages at the ends and conservation of the power.

The calculation of the radiation properties of these arrays is done easily by multiplying the building blocks with the appropriate array factors.

In addition to the radiation and the surface wave losses, the dissipative losses are derived by the multiplication of the electrical path length of the lines by the attenuation factor. The dissipative losses of the radiating elements are small but are also taken into account according to [15]. A detailed loss budget of a set of modular arrays with typical parameters is given in Table II. A comparison is made with the gain of a reflector antenna having the same area and an aperture efficiency of 50 percent. It can be seen that arrays of up to 1024 elements are predicted to have about the same gain as a reflector but larger printed arrays are not practical where reasonable efficiency is needed. It also can be seen that the dielectric and ohmic losses become dominant in these high gain arrays, and novel feeding techniques should be considered. Fig. 10 shows the calculated gain and the directivity as a function of the spacing between close elements D/λ_0. The dielectric, constant is 2.2, the thickness is 1.6 mm, and the frequency is 10 GHz. It can be seen that the directivity increases with D/λ_0 and a maximum is reached in the range of 0.8–0.9 λ_0. The gain is less sensitive to the spacing because the ohmic, dielectric and surface wave losses increase with the spacing. One can see that, for values of D greater than 0.9 λ_0, the surface wave loss increases rapidly. The results reported here agree quite well with the results of Hall and Prior [11]. For example, the estimated losses given there, at 12 GHz for a substrate with a thickness of 1.6 mm and a dielectric constant of 2.32, are higher than our estimated results by 0.3, 0.8, 1.3, 1.7, and 1.8 dB for the arrays of 16, 64, 256, 1024 and 4096 elements, respectively. The difference is mainly the result of the higher frequency and the higher dielectric constant. An adaptation of these parameters in our analysis shows that both estimates agree within 1 dB.

TABLE II
A DETAILED CALCULATED LOSS BUDGET OF MODULAR MICROSTRIP ARRAYS AT 10 GHz, DIELECTRIC CONSTANT IS $\epsilon_r = 2.2$, THICKNESS IS $h = 1.6$ mm, AND SPACING BETWEEN CLOSE ELEMENTS $0.8\lambda_0$; ALL NUMBERS ARE GIVEN IN dB

number of elements	16	64	256	1024	4096
directivity without network	20.9	27.0	33.0	39.2	45.1
radiation loss	0.8	1.0	1.3	1.9	2.6
surface wave loss	0.3	0.3	0.2	0.2	0.1
dielectric loss	0.1	0.3	0.5	1.0	2.1
ohmic loss	0.1	0.3	0.6	1.2	2.4
connetor loss	0.2	0.2	0.2	0.2	0.2
calculated gain	19.5	25	30	34.5	37.5
gain of a dish	18	24	30	36	42

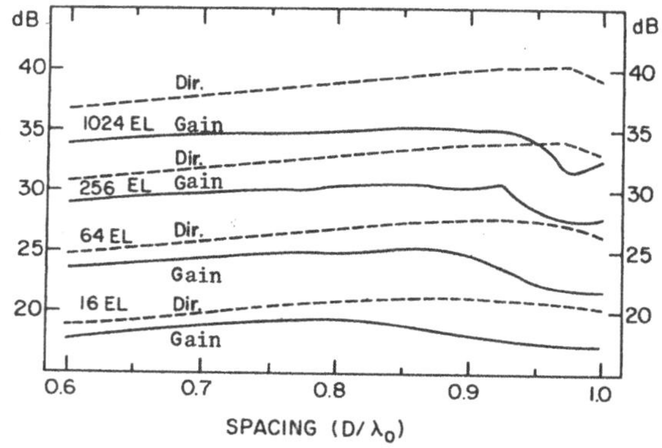

Fig. 10. Calculated directivity and gain of microstrip arrays with 16, 64, 256 and 1024 elements as a function of the spacing between close elements D/λ_0. The dielectric constant is 2.2, the thickness is $h\sqrt{\epsilon_r}/\lambda_0 = 0.08$, and the frequency is $f = 10$ GHz.

A set of 16, 64, 256, and 1024 element arrays was built and tested at frequencies of 10, 20, 30, and 35 GHz (except for the 1024 element array at 10 GHz whose dimensions are 80 × 80 cm). All the antennas were built on the same material (OAK-605 or Duroid 5880) with accurate scaling in all the relevant dimensions. The spacing was 0.8 λ_0 in all cases. The photoresist coating and the etching process were done in homemade facilities with excellent accuracies. The

TABLE III
CALCULATED AND MEASURED GAINS OF MICROSTRIP ARRAY ANTENNAS (IN dB); DIELECTRIC CONSTANT $\epsilon_r = 2.2$, THICKNESS $h\sqrt{\epsilon_r}/\lambda_0 = 0.08$ AND SPACING BETWEEN CLOSE ELEMENTS $D/\lambda_0 = 0.8$

No. of elements	16	64	256	1024
10 GHz				
calculated gain	19.2	24.8	30.3	34.9
measured gain	19.5	25.0	29.5	—
20 GHz				
calculated gain	19.1	24.7	29.9	34.2
measured gain	19.8	24.5	29.7	34.1
30 GHz				
calculated gain	19.0	24.5	29.6	33.5
measured gain	19.5	24.0	28.5	32.0
35 GHz				
calculated gain	19.0	24.4	29.4	33.2
measured gain	19.0	24.0	28.5	32.0

thin feed lines in the higher frequencies have a width of 70 ± 10 micron without any detected cuts. The comparison between calculated and measured results is given in Table III. The radiation and surface wave losses for a given number of elements are identical at all four frequencies and appear in Table II. The difference between the gains at different frequencies is the result of the growing dissipation losses. The agreement between calculated and measured results is in general very good. An independent efficiency measurement of these antennas which was made by a radiometric method [23] gave consistent results for the losses within an accuracy of 0.5 dB.

V. CONCLUSION

An analytical approach for the analysis of microstrip arrays including their feed network is introduced. This approach is based on an educated guess of the currents in the radiating elements and in the feed lines. It is a useful tool for high Q-radiators and for quasi-TEM transmission lines. Modular structures of feed networks are handled in a simple and efficient way using closed-form expressions.

Calculated results of radiation patterns, directivity, and gain of a four element array are shown and the experimental results show reasonable agreement with the theory.

The analysis has been extended to large modular arrays and a detailed loss budget for a typical set of parameters proves that single-layer arrays with up to 1024 elements have about the same gain as reflectors having the same area. The experimental results for a set of 16, 64, 256 and 1024 element arrays agree very well with the theoretical prediction.

REFERENCES

[1] J. R. James, P. S. Hall, and C. Wood, *Microstrip Antenna Theory and Design*. London: Peter Peregrinus, 1981, ch. 5, 6.

[2] R. J. Mailloux, J. F. McIlvenna, and N. P. Kernweis, "Microstrip array technology," *IEEE Trans. Antennas Propagat.*, vol. AP-29, pp. 25–37, Jan. 1981.

[3] M. A. Weiss, "Microstrip antennas for millimeter waves," *IEEE Trans. Antennas Propagat.*, vol. AP-29, pp. 171–174, Jan. 1981.

[4] S. Nishimura, Y. Sugio, and T. Makimoto, "Cranck-type circularly polarized microstrip line antenna," in *IEEE Antennas Propagat. Soc. Int. Symp. Dig.*, 1983, pp. 162–165.

[5] M. Ando, K. Sakurai, N. Goto, K. Arimura, and Y. Ito, "A radial line slot antenna for 12 GHz satellite TV reception," *IEEE Trans. Antennas Propagat.*, vol. AP-33, pp. 1347–1353, Dec. 1985.

[6] E. Rammos, "A new wideband high gain suspended substrate line planar array for 12 GHz satellite TV," in *Proc. 13th European Microwave Conf.* 1983, pp. 227–231.

[7] G. Dubost and C. Vinatier, "Large bandwidth and high gain array of flat folded dipoles acting at 12 GHz," in *Proc. IEE ICAP*, 1983, pp. 145–149.

[8] A. Henderson and J. R. James, "Improved microstrip flat plate array for domestic DBS reception," in *IEEE Antennas Propagat. Soc. Int. Symp. Dig.*, 1986, pp. 565–568.

[9] E. H. Newman and J. E. Tehan, "Analysis of a microstrip array and feed network," *IEEE Trans. Antennas Propagat.*, vol. AP-33, pp. 397–403, Apr. 1985.

[10] S. M. Voda and D. M. Pozar, "A rigorous analysis of a microstrip line fed patch antenna," in *IEEE Antenna Propagat. Soc. Int. Symp. Dig.*, 1986, pp. 825–828.

[11] P. S. Hall and C. J. Prior, "Radiation control in corporately fed microstrip patch arrays," in *JINA '86 Dig.*, Nice, 1986, pp. 271–275.

[12] P. Perlmutter, S. Shtrikman, and D. Treves, "Electric surface current model for the analysis of microstrip antennas with application to rectangular elements," *IEEE Trans. Antennas Propagat.*, vol. AP-33, pp. 301–311, Mar. 1985.

[13] R. A. Pucel, D. J. Masse, and C. P. Hartwig, "Losses in microstrip," *IEEE Trans. Microwave Theory Tech.*, vol. MTT-16, pp. 342–350, June 1968, p. 1064, Dec. 1968.

[14] E. J. Denlinger, "Losses of microstrip lines," *IEEE Trans. Microwave Theory Tech.*, vol. MTT-28, pp. 513–522, June 1980.

[15] W. F. Richards, Y. T. Lo, and J. Brewer, "A simple experimental method for separating loss parameters of a microstrip antenna," *IEEE Trans. Antennas Propagat.*, vol. AP-29, pp. 150–151, Jan. 1981.

[16] J. Ashkenazy, P. Perlmutter, and D. Treves, "A modular approach for the design of microstrip antennas," *IEEE Trans. Antennas Propagat.*, vol. AP-31, pp. 190–193, Jan. 1983.

[17] E. Levine, G. Malamud, and D. Treves, "High gain modular microstrip antennas," in *Proc. 16th European Microwave Conf.*, 1986, pp. 655–660.

[18] M. V. Schneider, "Microstrip lines for microwave integrated circuits," *Bell Syst. Tech. J.*, vol. 48, pp. 1422–1444, 1969.

[19] R. E. Collin, *Foundations for Microwave Engineering*. New York: McGraw-Hill, 1966, ch. 3.

[20] W. F. Richards, Y. T. Lo, and D. D. Harrison, "An improved theory for microstrip antennas and applications," *IEEE Trans. Antennas Propagat.*, vol. AP-29, pp. 38–46, Jan. 1981.

[21] S. Pinhas and S. Shtrikman, "Vertical currents in microstrip antennas," *IEEE Trans. Antennas Propagat.*, vol. AP-35, pp. 1285–1289, Nov. 1987.

[22] E. Levine and D. Treves, "Test technique improves coax-to-microstrip transmissions," *Microwaves*, vol. 25, pp. 99–102, July 1986.

[23] J. Ashkenazy, E. Levine, and D. Treves, "Radiometric measurement of antenna efficiency," *Electron. Lett.*, vol. 21, pp. 111–112, Jan. 1985.

Design Considerations for Low Sidelobe Microstrip Arrays

DAVID M. POZAR, FELLOW, IEEE, AND BARRY KAUFMAN

Abstract—**The factors affecting the realizable sidelobe performance of microstrip arrays are discussed and quantified. These include excitation amplitude and phase accuracies, mutual coupling, diffraction effects, positioning errors, and errors due to imperfect element matching and feed network isolation. Also, it is shown that low-sidelobe microstrip arrays require a very tight tolerance on the resonant frequencies of the elements, and the elimination of spurious radiation from the feed network. Cross-pol and surface wave effects are also discussed. An experimental 16-element microstrip array prototype incorporated these considerations into its design, and achieved a −35 dB relative sidelobe level.**

I. Introduction

THE PRACTICAL REALIZATION of low sidelobe arrays becomes increasingly difficult as the peak sidelobe level is reduced more than 20 or 30 dB below the main beam primarily because of random errors in the feed network and the array itself, although deterministic factors such as mutual coupling and diffraction effects can also be important. By very careful control of fabrication tolerances and taking into account mutual coupling effects, however, several slotted waveguide (AWACS, TPS-70, F-16 radar), open-ended waveguide (EAR), and dipole arrays have achieved peak sidelobe levels as low as −50 dB [1]. This impressive level of performance has not yet been obtained for microstrip arrays, although attempts have been made. The lowest sidelobe levels for microstrip arrays reported in the literature are generally on the order of −25 dB; besides tolerance effects, the reasons given for the lack of better sidelobe performance often include mutual coupling and surface wave effects. However, the relation of these effects to the sidelobe level of a microstrip array has not been rigorously established, except for some very rough estimates [2], [3]. In addition, there appears to be other factors that are more important but have gone unrecognized.

In this paper we will examine quantitatively the various factors that potentially affect the sidelobe level of microstrip arrays, with the interesting result that mutual coupling and surface waves often do not significantly degrade the sidelobe level of such arrays, but that the narrow bandwidth of the microstrip element does have a very substantial, although indirect, effect on sidelobe level, as does spurious radiation from the feed network. The results of this study were incorporated into the design of a 16-element linear microstrip array that achieved a measured relative sidelobe level of −35 dB, or −19 dBi; the average absolute sidelobe level was well below −20 dBi. Knowing the risks in making such a claim, we will nevertheless state that, to our knowledge, this is the lowest sidelobe level yet attained for a microstrip array.

This paper will consider the dominant factors that control the sidelobe level of an array, such as amplitude and phase accuracy of the excitation, mutual coupling, diffraction effects, flatness of the array face, and errors caused by nonperfect feed network isolation. In the context of microstrip arrays, several additional factors are of prime importance. One involves the spurious radiation from the feed network, which will imply that the feed network should not be on the same substrate face as the array elements for relative sidelobe levels of 20 dB or more. Another consideration is that significant phase errors can result from very small variations in the resonant frequencies of the microstrip elements, which can occur due to tolerances in the element dimensions or dielectric inhomogeneities. We will quantify this effect, show that it is caused by the typically narrow bandwidth of the patch element, and suggest a possible remedy. Yet another factor associated primarily with printed antennas is that of surface wave excitation and subsequent diffraction from substrate or ground plane edges. It is also possible that unacceptably high cross-polarization levels can result from small deviations in the position of the microstrip element feed.

Our goal, then, is not just to demonstrate an experimental low sidelobe microstrip array, but to describe in a systematic and quantitative way the various factors that should be considered in the design of such arrays. This may also explain why past attempts at achieving low sidelobe microstrip arrays have not been totally successful.

II. Factors Affecting Sidelobe Performance

Sidelobe levels can be measured in several ways [1]. The relative sidelobe level (or just sidelobe level) is the level of the highest sidelobe relative to the main beam. The absolute sidelobe level is the level of the highest sidelobe relative to isotropic. The average sidelobe level is a measure of the total power contained in the sidelobes, and is usually given relative to the isotropic level. As suggested in [1], average sidelobe levels from −5 to −20 dBi can be referred to as low sidelobes, while levels below −20 dBi can be referred to as ultralow sidelobes.

A. Amplitude and Phase Accuracies

Consider an N-element linear microstrip array with zero-mean Gaussian distributed excitation errors. Let σ_a and σ_ϕ be,

Reprinted from *IEEE Trans. Antennas Propaga.*, vol. 38, no. 8, pp. 1176–1185, Aug. 1990.

TABLE I
REQUIRED AMPLITUDE AND PHASE TOLERANCES FOR A GIVEN LINEAR BROADSIDE ARRAY SIZE AND AVERAGE RELATIVE SIDELOBE LEVEL (ASSUMING $\lambda_0/2$ ELEMENT SPACING, 6 dB ELEMENT SPACING, AND $\sigma_a = \sigma_\phi$)

$\overline{SLL}$ dB	$N = 16$	$N = 40$	$N = 100$
20	2.05 dB 15.2° ($D = 17.5$ dB)	2.96 dB 23.3° $D = 21.2$ dB)	3.98 dB 33.3° ($D = 24.3$ dB)
30	0.66 dB 4.5° ($D = 17.0$ dB)	1.03 dB 7.2° ($D = 21.0$ dB)	1.56 dB 11.3° ($D = 24.9$ dB)
40	0.20 dB 1.3° ($D = 16.4$ dB)	0.32 dB 2.2° ($D = 20.5$ dB)	0.50 dB 3.4° ($D = 24.4$ dB)
50	0.06 dB 0.4° ($D = 16.0$ dB)	0.10 dB 0.6° ($D = 20.0$ dB)	0.15 dB 1.0° ($D = 24.0$ dB)

respectively, the standard deviations of the normalized amplitude and phase errors. Then the array will have an average pattern given by

$$\bar{E}^2(\theta, \phi) = F_e^2(\theta, \phi)\left[E_0^2(\theta, \phi) + \frac{\sigma^2}{D_i}\right], \tag{1}$$

where $F_e(\theta, \phi)$ is the directivity pattern of a single element, $E_0(\theta, \phi)$ is the array pattern assuming isotropic elements without amplitude and phase errors, D_i is the directivity of the array assuming isotropic elements, and

$$\sigma^2 = \sigma_a^2 + \phi_\phi^2. \tag{2}$$

This result was originally given for arrays of isotropic elements ($F_e = 1$) [4], but the effect of element gain must be considered for meaningful results. If D_e is the element directivity then, for reasonably large arrays of low-gain elements, we can assume that

$$D = D_e D_i, \tag{3}$$

where D is the overall array directivity. Then (1) gives the average relative sidelobe level of the array as

$$\overline{\mathrm{SLL}} = \sigma^2 / D_i, \tag{4}$$

and the average sidelobe level relative to isotropic as

$$\overline{\mathrm{SLL}_i} = D \cdot \overline{\mathrm{SLL}} = D_e \sigma^2. \tag{5}$$

The average isotropic sidelobe level due to excitation errors is effectively raised by an amount equal to the element gain, and is independent of array size. The average relative sidelobe level, however, improves as array size (D_i) increases, for fixed amplitude and phase errors. Thus the sidelobe level relative to isotropic is the better indicator of array accuracy, although specifications are usually given in terms of the relative sidelobe level.

The data shown in Table I give the necessary amplitude and phase tolerances for the excitation of a given linear array size and average relative sidelobe level. A Chebyshev distribution is assumed, with $\lambda_0/2$ element spacing and an element gain

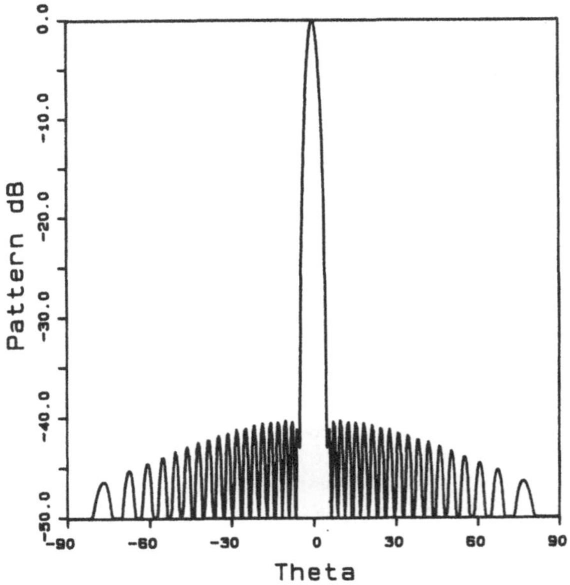

Fig. 1. Calculated pattern for a 40-element E-plane linear array of rectangular microstrip patches. 40 dB Chebyshev excitation, no errors. Patch size: 1.9 × 1.85 cm, substrate permittivity: $\epsilon_r = 2.22$, substrate thickness: 0.16 cm, element spacing: 3.0 cm, frequency: 5 GHz.

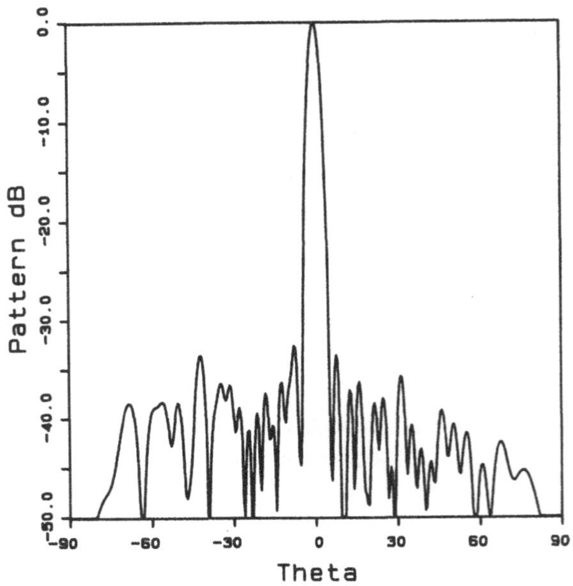

Fig. 2. Calculated pattern for array of Fig. 1 having Gaussian distributed excitation errors with $\sigma_a = 0.32$ dB and $\sigma_\phi = 2.2°$.

$D_e = 6$ dB. (The directivity of a single microstrip element is generally closer to 7 dB, but we allow about 1 dB for losses.) These results apply only to the average sidelobe levels, so some sidelobes will be above these levels. Statistical analysis suggests that more than 80% of the sidelobes should be lower than the average level plus 3 dB.

As an example, consider a 40-element E-plane linear array of rectangular microstrip patches, designed for a 40 dB sidelobe level. Fig. 1 shows the calculated pattern in the absence of excitation errors. If we apply Gaussian distributed random errors with $\sigma_a = 0.32$ dB and $\sigma_\phi = 2.2°$ to the element excitations, the pattern shown in Fig. 2 is obtained. Note that several sidelobes are higher than 35 dB below the main beam.

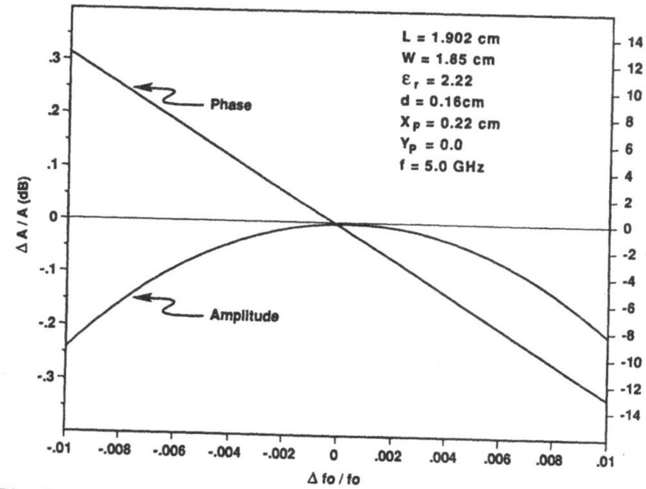

Fig. 3. Normalized amplitude and phase errors in the radiated field of a rectangular microstrip antenna caused by a shift in resonant frequency. Patch length: 1.902 cm, patch width: 1.85 cm, substrate permittivity: 2.2, substrate thickness: 0.16 cm, frequency: 5 GHz. The patch is probe-fed and driven with a $1\angle 0$ V incident wave.

B. Effect of Narrow Bandwidth on Phase Accuracy

An ideal feed network will deliver incident voltages with the correct amplitudes and phases to the array elements. In the absence of mutual coupling (which will be considered later), the amplitude and phase of the radiated field from each element will be proportional to the excitation of that element, and a function of the driving point impedance of the element. If all elements are identical, the amplitude and phase distribution of the feed network will be preserved across the component waves radiating from each element. But if there are variations in the elements, due to fabricational tolerances or other perturbations, which produce variations in the driving point impedances, then the amplitudes and phases of the radiated fields will be in error. The magnitude of these errors will be dependent on the sensitivity of the driving point impedance to element perturbations.

For elements that have moderate bandwidths, such as dipoles, open-ended waveguides, and waveguide slots, this sensitivity is not too severe. But for microstrip patches, which generally have bandwidths of a few percent or less, the driving point impedance can change rapidly with relatively small changes to the element. The important variables are primarily those that affect the resonant frequency of the patch element, such as element size (length and width), and substrate dielectric constant.

This effect is quantified in Fig. 3, which shows the normalized change in amplitude, A, and phase ϕ, of the field radiated from a microstrip patch versus a variation in the resonant frequency, f_0 (the operating frequency is fixed). This data was calculated using a full-wave moment method solution [5], [6], with the patch excited with a $1\angle 0$ V incident wave. The phase variation is seen to be linear over the range in Δf_0, since it is related to the essentially linear variation of the input reactance of the patch near resonance. The amplitude variation is due to the impedance mismatch of the patch off resonance, and is symmetric about the resonant frequency ($\Delta f_0 = 0$).

Consider, for example, a 1% shift in resonant frequency. Fig. 3 shows that this will lead to a phase change of about 12.5° in the radiated field and an amplitude drop of about 0.2 dB. Comparison with the data in Table I indicates that the phase error will always be the more critical quantity. Thus, for a given array size and average sidelobe level, the required element phase tolerance given in Table I can be used with Fig. 3 to obtain the necessary tolerance on the resonant frequencies of the microstrip elements. For example, an $N = 40$ array with an average sidelobe level of 40 dB will require a phase accuracy of 2.2° which, from Fig. 3, implies a tolerance of less than $\pm 0.17\%$ for the element resonant frequencies. This level of accuracy is generally not achieved in microstrip arrays unless special efforts are made, as discussed below.

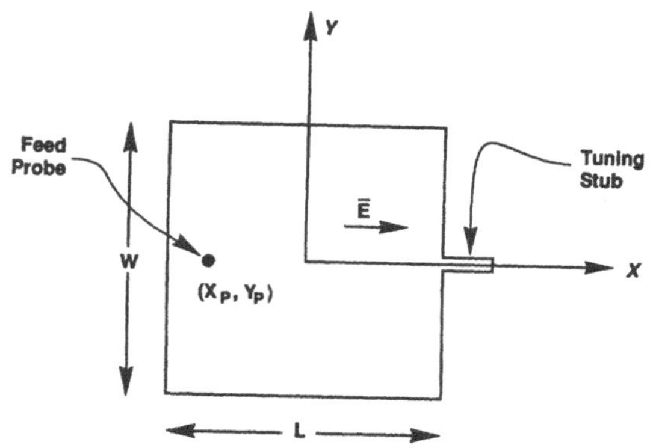

Fig. 4. Geometry of a rectangular probe-fed microstrip antenna tuned with a small stub.

Inaccuracies in the resonant frequency of a microstrip element can be caused by several factors. Since resonant frequency f_0 is proportional to the length L of a rectangular patch, a change of ΔL in the length will produce a change in resonant frequency given by

$$\Delta f_0 = \frac{-\Delta L}{L} f_0. \tag{6}$$

Thus, the required accuracy for resonant frequency translates into a corresponding accuracy for the resonant length of the patch element. The resonant frequency is also affected by a change in the permittivity of the substrate. Since resonant frequency varies as $1/\sqrt{\epsilon_r}$, a change of $\Delta\epsilon_r$ leads to a change in resonant frequency given by

$$\Delta f_0 = \frac{-\Delta \epsilon_r}{2\epsilon_r} f_0. \tag{7}$$

Good substrates have dielectric constant tolerances of a few percent or less, but the real problem is caused by variations in permittivity across the substrate, which can be as high as 0.5 to 1%, especially for high permittivity substrates. Other variables, such as patch width and substrate thickness, can also affect the resonant frequency of the element, but to a much lesser degree than either the length or permittivity.

In order to achieve the accuracy in element resonant frequencies required for low sidelobe designs in the presence of inevitable deviations in element size and substrate permittivity, it is useful to be able to individually trim each element. A practical way of doing this is to use a small tuning stub at the end of each patch, as shown in Fig. 4 and discussed in [7]. The resonant frequency can then be tuned with an accuracy better than 0.05% by manually trimming this stub. Of course, this technique requires access to the input port of each element, which is not possible if the feed network is printed on the same substrate. We will see in the next section, however, that this type of feed geometry is not suitable for low sidelobe arrays.

C. Radiation from the Feed Network

One of the often-stated advantages of microstrip antennas is that fabrication can be simplified by printing a microstrip feed network on the same substrate as the microstrip patches. The discontinuities, bends, power dividers, and other components in such a feed network, however, cause spurious radiation that limits the minimum sidelobe level. Such radiation increases with the substrate thickness, a trend which is counter to the desirability of using a thicker substrate for increased bandwidth. For typical substrate thicknesses of $0.01\lambda_0$ to $0.03\lambda_0$, the spurious radiation levels have been estimated theoretically to limit the relative sidelobe level to about 20–30 dB [2], [3]. It is difficult to be precise with such estimates, however, because of the complexity in accurately calculating the radiation from a complicated feed network. Our own experience with several microstrip array designs [8]–[10], and the experience of others [11], [12], suggests a slightly more conservative value of 15–25 dB for the relative sidelobe level, when the feed network is printed on the same substrate as the antenna elements.

The most practical resolution of the problem of spurious feed radiation is to place the feed network behind the antenna substrate [9], [11], and connect to the radiating elements with feedthrough pins or coaxial connectors. This increases the complexity of the array, but it seems that such a compromise must be made if low sidelobe performance is to be achieved. Separate coaxial connectors also offer the advantages of being able to access the feed network ports and patch elements for testing and tuning, and of providing modularity between feed and radiator substrates. The spurious radiation from a coaxial feed probe has been calculated to be less than 25 dB below the radiation from the microstrip patch, and is modified by the same array factor as the patches, so it should be negligible.

Alternatively, the feed network can be made on one substrate and then aperture coupled [13], [14] to the radiating elements, thus eliminating the need for feed pins or connectors. In either case, using separate substrates for the radiating elements and feed network allows the substrates to be individually matched to these two distinct electrical functions.

D. Mutual Coupling

In the presence of mutual coupling the amplitude and phase of the field radiated by an array element will not be directly proportional to the amplitude and phase of the excitation on that element. But if the mutual coupling between array elements can be measured or calculated, the feed excitations can be adjusted to compensate for this effect. Consider the equivalent circuit for the feed of a microstrip antenna element shown in Fig. 5. The current I is the amplitude of the current on the patch element; the radiated field is proportional to this current. Let $[I]$ be the vector of all the patch current amplitudes

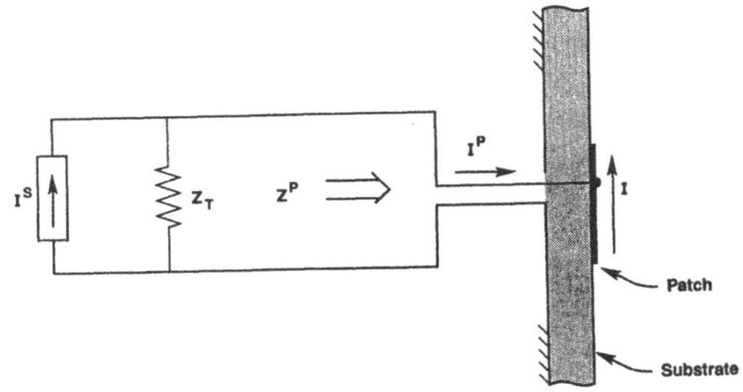

Fig. 5. Feeding circuit for a probe-fed microstrip patch antenna.

in an N-element array. To synthesize a particular low-sidelobe pattern, the required amplitude and phase distribution must be applied to $[I]$. Then the necessary feed (or source) currents, $[I^s]$, can be determined as [15]

$$[I^s] = -V_0\{[Z^P]^{-1} + [Z^T]^{-1}\}[I], \tag{8}$$

where V_0 is a modal voltage relating patch current to port voltage, $[Z^T]$ is a diagonal termination impedance matrix with elements Z_T, and $[Z^P]$ is the open-circuit port impedance matrix of the patch array. $[Z^P]$ can be calculated, as in [15], or determined from measurements of the scattering matrix of the array. Since V_0 is a constant, it does not affect the amplitude or phase distribution of $[I^s]$, and does not actually need to be determined. If no mutual coupling were present, $[Z^P]$ would be a diagonal matrix with elements equal to the input impedance of a single element; then the amplitude and phase distributions of $[I]$ and $[I^s]$ would be the same.

The above theory assumes that the current on each patch can be adequately represented with one expansion mode. This has been found to be a good assumption [15], because the patch antenna is a relatively high-Q component, although it is straightforward to extend the theory to use several expansion modes per patch [15]. It should be noted, though, that if more than one mode is used per patch, the field radiated by each element is a more complicated function of the patch currents (element pattern distortion), and a more sophisticated synthesis technique is required [16]. In this case, it would not be possible to synthesize correctly a low-sidelobe pattern with measurements alone. But this does not appear to be a significant concern for practical microstrip arrays.

To study the effect of mutual coupling on sidelobe level, the patterns for several microstrip arrays were calculated with and without coupling effects. A typical result is shown in Fig. 6, where it is seen that the sidelobes are perturbed by a few decibels at most. Similar or less deviations were found for H-plane arrays, for arrays with more elements, and for arrays printed on thicker substrates. Broadside arrays showed less deviations than scanned arrays. Only when the element spacing was greater than $\lambda_0/2$ and the beam was scanned off broadside did errors more than a few decibels occur, but in these cases a grating lobe was close to or in visible space. Similar results were reported in [11] and [17]; in [11] it was found that mutual coupling had to be accounted for in a reduced sidelobe array of four unequally spaced patches, where some spacings were greater than $\lambda_0/2$.

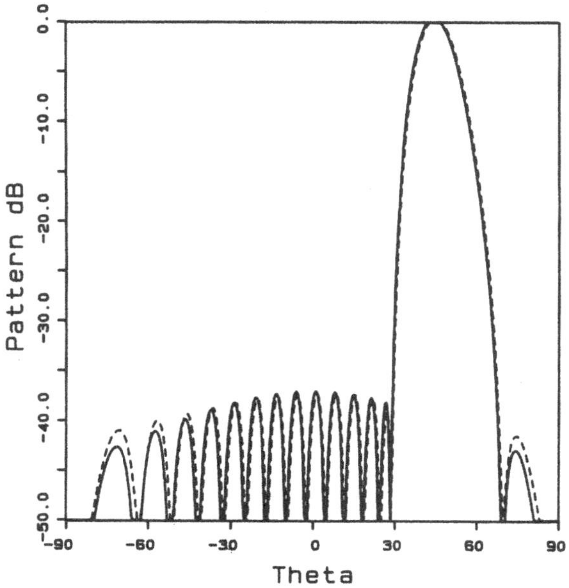

Fig. 6. Calculated patterns of a 16-element E-plane microstrip array showing the effect of mutual coupling. Patch length: 1.9 cm, patch width: 1.85 cm, substrate permittivity: 2.22, substrate thickness: 0.16 cm, element spacing: 3.0 cm, frequency: 5 GHz. — without mutual coupling, --- with mutual coupling.

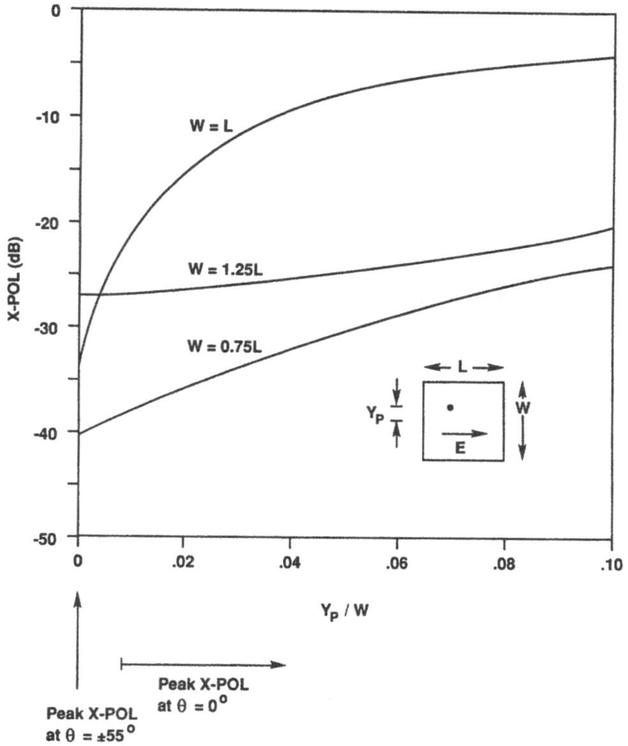

Fig. 7. Calculated cross-pol level for a probe-fed rectangular patch versus patch width w, and feed probe positioning error y_p/w. Patch length: 1.9 cm, substrate permittivity: 2.22, substrate thickness: 0.16 cm, frequency: 5 GHz.

Thus we conclude that, for microstrip arrays with uniform element spacings of about $\lambda_0/2$, mutual coupling should not cause significant degradation in the sidelobe level. This is in contrast with the situation for low-sidelobe waveguide or dipole arrays. But if necessary, mutual coupling effects can easily be incorporated into the design of the feed network.

E. Cross Polarization

The cross-pol level of a properly designed microstrip element is usually 20 dB or more below the co-pol level. If all the elements in the array are identical, then the cross-pol pattern of the array would be given by the product of the array factor and the cross-pol pattern of a single element, and so would be 20 dB or more below the co-pol pattern at all angles (ignoring diffraction effects, which will be discussed later). But if perturbations in the elements cause the cross-polarized element patterns to differ in magnitude and phase across the array, then the resulting cross-pol level may be much higher (possibly higher than the co-pol level of the array) because of the loss of phase and amplitude accuracy.

Such perturbations can be caused by small variations in element resonant frequency, as discussed in Section II-B. But tuning the elements to the same resonant frequency should correct such phase and amplitude errors for the cross-pol fields as well as the co-pol fields.

A more important effect occurs when high element cross-pol levels are generated by a square patch element with a slightly misplaced feed probe. Probe-fed rectangular patches typically have the feed positioned along the middle of the H-plane dimension ($y_p = 0$), as shown in Fig. 4, to excite the dominant TM_{10} mode. The x_p coordinate then controls the impedance match. The principal plane cross-pol patterns result from higher order TM_{20} and TM_{02} modes, and have nulls at broadside and maxima between 40° and 60° from broadside. But when fabricational tolerances cause the feed point to be positioned slightly off the midplane ($y_p \neq 0$), a cross-polarized TM_{01} mode is excited, with maxima at or near broadside. When the patch is square ($W = L$), this mode becomes resonant and the resulting cross-pol level can be quite large for even small errors in probe position. Fig. 7 shows an example of this effect, where the H-plane cross-pol level for a probe-fed rectangular patch is calculated (via a full-wave moment method solution) versus the feed positioning error y_p/W, for three different patch widths. Observe that for a square patch ($W = L$), the cross-pol level becomes higher than -20 dB (an increase of about 15 dB) for a positioning error greater than about 1% (about 0.25 mm). The best results occur for the narrow patch, $W = 0.75L$. Similar levels appear in the E-plane of the patch, except when $y_p = 0$, in which case the cross-polarization is zero.

If the random probe positioning errors are large enough, the cross-pol fields from the patches will add incoherently. Then the resulting average cross-pol level of the array, relative to isotropic, will be about equal to the isotropic cross-pol level of a single element, which is its relative cross-pol level (in negative decibels) plus the directivity of the element. Thus, to achieve a -20 dBi average cross-pol sidelobe level in an array with significant feed positioning errors, the cross-pol level of the element should be better than -26 dB (assuming $D_e = 6$ dB). This is the worst-case requirement, but it can be achieved with a narrow patch ($W < L$) and reasonable ($\sim 2\%$) probe positioning accuracies. The most important conclusion here is that a square patch should be avoided for low-sidelobe arrays.

If square patches are necessary (to produce circular polarization, for example), the balanced-feed technique [18], where two feed probes are positioned symmetrically at $\pm x_p$ and fed 180° out of phase, could be used to reduce element cross-pol. Such an arrangement is considerably more complicated, however.

F. Diffraction Effects

The ground plane or other structure on which the microstrip array is mounted can diffract surface and space waves radiated by the array to degrade the sidelobe level. For general structures, the analysis of such effects can be very complicated, although asymptotic methods can often be used because such structures are usually electrically large in practice. We will limit our discussion here to some estimates of the power diffracted by flat finite-sized ground planes.

First consider surface wave power. As shown in [19], the relative amount of surface wave power radiated by a single element increases with the electrical thickness of the substrate; for thin ($d \leq 0.03\lambda_0$) low-dielectric constant substrates ($\epsilon_r \leq 2.55$), the surface wave power is less than 15% of the total radiated power [19]. But in arrays of microstrip elements, destructive interference reduces the surface wave power as the number of array elements increases [15]; this occurs in both planar and E-plane linear arrays for all scan angles except near the blindness angle. If we assume we are not operating near blindness, the surface wave power will be less than 1% for $N \geq 10$, for thin low permittivity substrates. The surface wave power excited by an H-plane linear array is about equal to that of a single element.

Surface wave power is launched along the substrate primarily in the E-plane direction of the patch element, with a cos ϕ pattern factor. This power does not enter the radiation pattern (co- or cross-pol) until it diffracts from the substrate edges or other discontinuities, and will be partially attenuated by dielectric and conductor losses. If we assume isotropic radiation and a worst-case surface wave power of 1%, then we should have an average radiated surface wave power level of less than −20 dBi. In practice, radiated surface wave power appears to be even less than this, so it should not be an important factor for low-sidelobe arrays that are large and operated away from blindness angles. This conclusion is somewhat less pessimistic than that expressed in [2], because the cancellation effect noted in [15] was not considered. In the worst case of a linear H-plane array (or other geometries, if necessary), excessive surface wave power can be reduced by using thin absorber material along the substrate edges parallel to the array plane.

Diffraction of space wave power by the ground plane edges can also degrade sidelobe levels. It has been shown in [20] that geometrical theory of diffraction (GTD) techniques can be used to predict the effect of a finite ground plane on the radiation pattern for a single element, and such an analysis can easily be extended to arrays. Diffractions from edges perpendicular to the plane of a linear array appear to be negligible due to the very low level of incident space wave field at these edges (in the endfire directions of the array). The incident field is much stronger, however, at the edges parallel to a linear array. Calculations and measurements indicate that a ground plane extending about 5λ from the array on all edges is enough to keep diffraction effects below the −40 dB level. Diffraction effects should be less for planar arrays having reduced sidelobes in both planes.

TABLE II
MAXIMUM AMPLITUDE AND PHASE ERRORS DUE TO ELEMENT MISMATCH AND IMPERFECT FEED NETWORK ISOLATION

Return Loss + Isolation (dB)	Max. Amplitude Error, ΔA (dB)	Max. Phase Error, $\Delta\phi$ (deg)
10	0.414	17.5
20	0.043	5.7
30	0.004	1.8
40	0.000	0.6
50	0.000	0.2

G. Element Positioning Errors

Errors in the physical location of an array element leads to phase errors. If the face of the array is not perfectly flat, so that an element has a positioning error of Δz above or below the nominal array plane, a phase error

$$\Delta\phi = k_0 \, \Delta z \cos\theta, \tag{9}$$

occurs at the far-field angle θ. For a positioning error Δx in the plane of the array, the phase error is

$$\Delta\phi = k_0 \, \Delta x \sin\theta. \tag{10}$$

From these results we can calculate the root mean square (rms) phase error over visible space for the array to obtain

$$\sigma_\phi = k_0 \, \Delta R/\sqrt{2}, \tag{11}$$

where $\Delta R = \sqrt{(\Delta x)^2 + (\Delta z)^2}$ is the total positioning error for any element. Knowing σ_ϕ (from Table I) for a given array size and sidelobe level then allows the necessary element positioning tolerance to be calculated from (11).

H. Errors Due to Imperfect Element Match and Feed Network Isolation

If an array element is not perfectly matched, some of the incident power from the feed network will be reflected, causing an amplitude error given by

$$\Delta A = 10 \log(1 - |\Gamma|^2) \quad \text{dB}, \tag{12}$$

where $|\Gamma|$ is the reflection coefficient magnitude of the element This error is not the same as the amplitude error associated with resonant frequency variations as discussed in Section II-B (but it includes this effect), because in the latter case the element was not necessarily matched. If the element mismatches are identical across the array (which should be approximately true if they have been individually tuned, as discussed in Section II-B), then this effect should not be significant. Otherwise the elements should be matched well enough so that the resulting amplitude errors are within the acceptable tolerance limits for the desired sidelobe level. Table II can be used for this purpose.

A related problem can occur when reflections from one element return back through the feed network and are transmitted toward another element (or the same element). These "back-feed reflections" are possible because of a combination of element mismatch and imperfect feed network isolation or matching, and can cause phase and amplitude errors. The maximum amplitude error will occur when a phase difference between the desired excitation and the back feed reflection is either 0° or 180°, and is given by

$$\Delta A(\max) = 10 \log (1 + |\Gamma|^2 |I|^2) \quad \text{dB}, \tag{13}$$

where $|I|$ is the magnitude of isolation between the feed network ports. The maximum phase error will occur when the phase difference is $\pm 90°$, and is given by

$$\Delta \phi(\max) = \tan^{-1} |\Gamma|\,|I|. \tag{14}$$

Table II lists these maximum amplitude and phase errors for various values of element return loss plus feed network isolation. Comparison with the data of Table I suggests that the phase errors are more severe than the amplitude errors, but in any case it should not be very difficult to control these errors, since in practice element return losses and feed network isolations are typically 20 dB (each), or better. In addition, the amplitude and phase errors given in Table II are the maximum possible errors that result for specific phase relationships between the desired and undesired signals; in practice the almost-random nature of these phases across the array should serve to further reduce the effect of such errors.

Table II can also be used to determine the necessary element return loss for a given amplitude tolerance, by reading the "return loss + isolation" column as just return loss; this is very accurate for return losses greater than 10 dB.

I. Other Considerations

There are several other factors that often must be considered for low-sidelobe array design, but since these have been adequately treated elsewhere and are not especially different for the case of printed arrays, we make only a passing mention here.

If digital phase shifters are used, several additional sources of phase and amplitude errors can arise. These include differential amplitude errors, phase quantization errors, and quantization lobes [4]. The latter problem can always be minimized by using decorrelated line lengths between the phase shifters and radiating elements.

Additional factors include temperature effects, pattern degradation caused by a radome, and the effect of failed components or elements, which is especially important for large phased arrays. Finally, it should be realized that measuring low-sidelobe antenna patterns requires special considerations. Besides the obvious need for a measurement system with a dynamic range in excess of the relative sidelobe levels to be measured, a low-reflectivity anechoic chamber or test site is also required. And, as discussed in [21], [22], the range distance requirement for measuring low-sidelobe patterns is much greater than the usual $2D^2/\lambda$ criterion.

III. Design Example: 16-Element *H*-Plane Microstrip Array

The preceding considerations were incorporated into the design of a 16-element linear microstrip array test bed. A photograph of the hardware is shown in Fig. 8. The microstrip patch elements were 1.85-cm square, with a short tuning stub as shown in Fig. 4. Each element was fed with an SMA coax connector 0.65 cm from the edge opposite the tuning stub. The substrate was 0.16-cm thick Duroid 5880 with $\epsilon_r = 2.22$, and was 72.4 × 11.4 cm in size. This substrate was mounted on a larger aluminum ground plane of size 183 × 61 cm. A 16-way equal-split power divider was made with commercial coaxial components, and connected to the array elements through a set of 16 manually adjustable attenuators and phase trimmers. In this way, the amplitude and phase of the excitation for each element could be adjusted independently and with high accuracy.

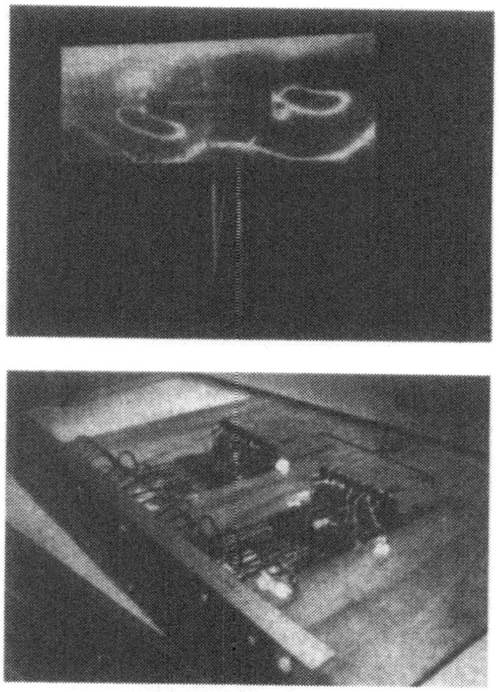

Fig. 8. Photographs of the 16-element low-sidelobe microstrip array prototype.

This array was tested using a 40 dB Chebyshev amplitude distribution for broadside and 30° scanned beams. Measurements were made before and after the microstrip elements were tuned; Table III summarizes the "error budgets," in terms of variance (σ^2), for these two cases. We will discuss each of the factors presented in Section II as they pertain to this array.

A. Amplitude and Phase Accuracies

With an HP8510 network analyzer, the feed network was adjusted to an amplitude tolerance of ±0.1 dB and a phase tolerance of ±1.5°. These lead to variances of $\sigma_a^2 = (10^{0.1/20} - 1)^2 = 0.00013$, and $\sigma_\phi^2 = (1.5\pi/180)^2 = 0.00069$, respectively. The relatively large phase tolerance was primarily due to cable flexing.

B. Phase Errors Caused by Variations in Element Resonant Frequency

Before tuning the patch elements, the element resonant frequencies were measured and found to vary by as much as

TABLE III
SUMMARY OF ERRORS FOR THE 16-ELEMENT MICROSTRIP ARRAY

Type of Error	Before Tuning		After Tuning	
	Error	Variance	Error	Variance
Excitation Amplitude	± 0.1 dB	0.00013	±0.1 dB	0.00013
Excitation Phase	±1.5°	0.00069	±1.5°	0.00069
Resonant Freq. Variation ($\Delta\phi$)	±3°	0.00274	±0.6°	0.00011
Positioning Errors (±0.2mm)	±0.8°	0.00022	±0.8°	0.00022
RL of Element (ΔA)	-0.19 dB	0.00016	-0.05 dB	0.00001
$RL + I$ ($\Delta\phi$)	0.7°	0.00016	0.4°	0.00004
TOTAL VARIANCE		0.0041		0.0012
RESULTING $\overline{SLL}_i$ (ISOTROPIC)		-18 dB		-23 dB
RESULTING $\overline{SLL}$ (RELATIVE)		-34 dB		-40 dB

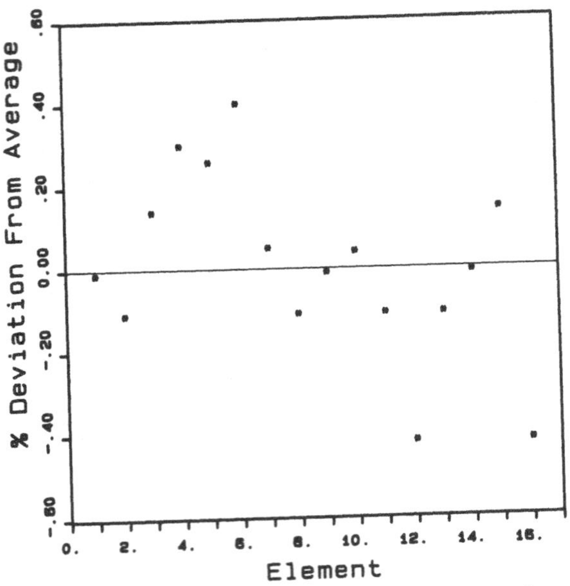

Fig. 9. Measured resonant frequency deviations (in percent) for the untuned 16-element array prototype. The average resonant frequency is 4.8505 GHz, and the rms deviation is 0.225%.

0.42% from the average resonant frequency of 4.8505 GHz; Fig. 9 shows a scatter plot of these deviations (in percent). The rms deviation in resonant frequency for the entire array was 0.225% which, from Fig. 3, is seen to cause an rms phase error of about ±3°. The elements were later tuned to 5.0 GHz by trimming the tuning stubs; the rms deviation in resonant frequency was then reduced to 0.05%, or a phase error of ±0.6°.

C. Feed Network Radiation

This problem was avoided by using a feed network mounted behind the array.

D. Mutual Coupling

As demonstrated in Section II-D, mutual coupling was not strong enough to degrade the sidelobe levels, and so its effect was not included in the feed network design.

E. Cross Polarization

By chance, square patch elements were used in the first few prototype arrays; the resulting cross-pol levels were unacceptably high (about −30 dB), which led to the reasoning developed in Section II-E. A later prototype array used rectangular elements, and achieved lower cross-pol levels, particularly in the sidelobe region.

F. Diffraction Effects

Without the use of the large aluminum ground plane, (e.g., using only the ground plane of the substrate itself), diffraction effects degraded the sidelobe level to about −31 dB, compared with −35 dB when the larger ground plane was used.

G. Element Positioning Errors

We estimated the accuracy of the element positions to be better than 0.1 mm, and perturbations in the flatness of the array to less than 0.2 mm. If we use $\Delta R = 0.2$ mm in (11), we obtain $\sigma_\phi = 0.85°$.

H. Errors Due to Element Match and Feed Network Isolation

Before tuning, the return loss of each element was measured at 4.8505 GHz; the resulting rms amplitude error was −0.19 dB. After tuning, the return loss of each element was measured at 5.0 GHz, with a resulting rms amplitude error of −0.05 dB. Phase errors are accounted for under Section III-B.

The feed network isolation and matching were better than 22 dB. With typical return losses of 16 dB (before tuning) to 22 dB (after tuning), the amplitude errors due to back feed reflections were negligible. The phase errors, however, were more significant: about ±0.7° before tuning, and about ±0.4° after tuning.

Table III summarizes the variances for the amplitude and phase errors before and after tuning. It is clear from this data that tuning the resonant frequency of the elements was the most important single factor in reducing the total error variance, as the associated phase errors were reduced from ± 3° to ± 0.6°; errors due to element mismatch and feed network isolation were also reduced. Such an "error budget" is invaluable when designing a low-sidelobe array, as it shows clearly where time and effort should best be spent in an attempt to reduce the total error to its achievable minimum. In the present case, for example, we see that after tuning the largest remaining variance is associated with the excitation phase errors; better phase trimmers would have helped here, although phase errors were also introduced by flexing of the semirigid cables between the feed network and array. Using (4), we estimate the average relative sidelobe levels at 34 dB before tuning and 40 dB after tuning; the directivity of the array was calculated to be 16.4 dB.

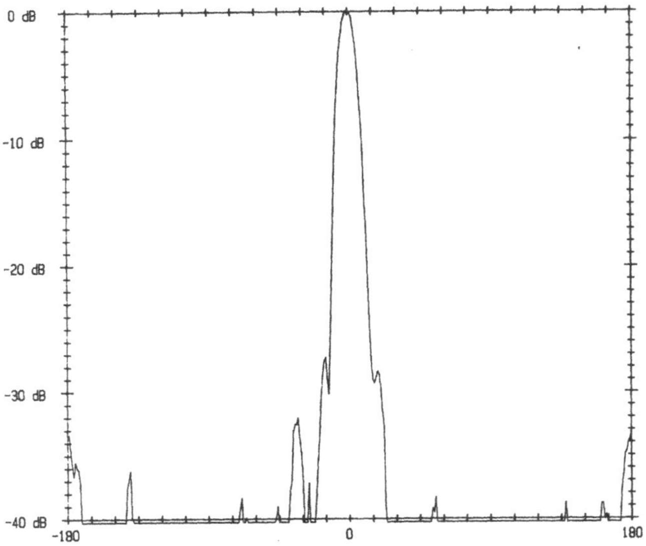

Fig. 10. Measured H-plane pattern of the 16-element low-sidelobe microstrip array before tuning. Broadside beam.

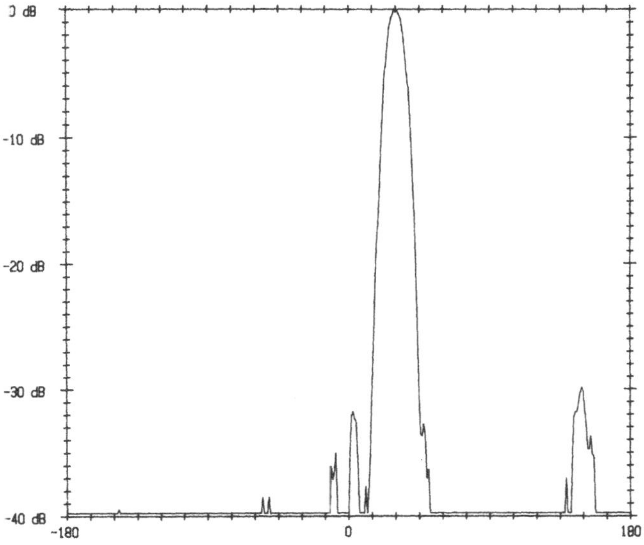

Fig. 12. Measured H-plane pattern of the 16-element low-sidelobe microstrip array after tuning. Beam scanned to 30°.

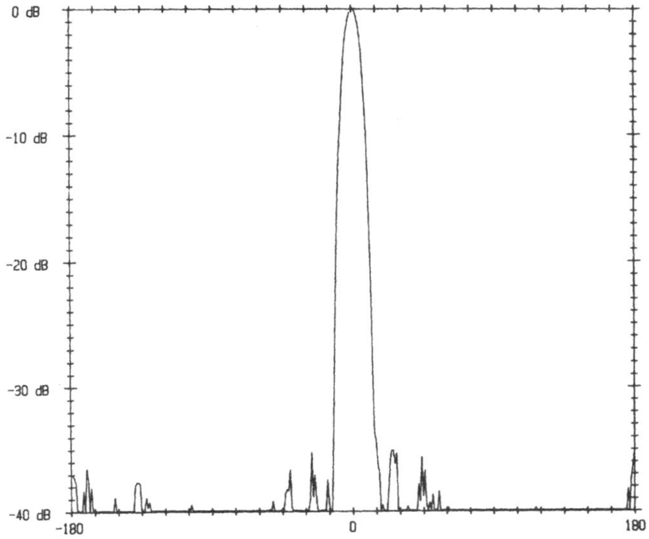

Fig. 11. Measured H-plane pattern of the 16-element low-sidelobe microstrip array after tuning. Broadside beam.

Measured patterns are shown in Figs. 10–12. Before tuning the resonant frequency of the elements, the pattern shown in Fig. 10 was measured. The relative sidelobe level was about 27 dB; it is difficult to estimate the average relative sidelobe level, but the value of 34 dB from Table III appears to be reasonable. The measured pattern after frequency tuning is shown in Fig. 11, with a relative sidelobe level of 35 dB. This is an improvement of 8 dB from the untuned case, and implies an isotropic sidelobe level of about −19 dBi. According to the definitions suggested in [1], this is within 1 dB of being categorized as an "ultralow" sidelobe array.

Fig. 12 shows the measured pattern when the phase trimmers were adjusted to scan the beam to 30°. The resulting sidelobe level was 32 dB relative to the main beam, but the main beam level was lowered by about 2.1 dB due to element pattern roll-off and the $\cos\theta$ scanning factor, so the isotropic sidelobe level of this array was about −18 dBi. The large lobe at 150° occurs behind the ground plane of the array, and was caused by leakage from the front of the array through small gaps between the large aluminum ground plane and the substrate ground plane; sealing these gaps with copper tape eliminated this lobe in later measurements.

Our tapered anechoic chamber was not well-suited for these measurements, as the wall reflectivity was on the order −30 dB. In addition, it was only large enough to obtain a range distance of about $2D^2/\lambda$, while [21], [22] suggest a range of more than $4D^2/\lambda$ for 1 dB accuracy in the first sidelobe level. Also, our receiver system had some nonlinearities over its dynamic range, so we suspect that the actual sidelobe level of these arrays may be a few decibels lower than what was measured. Using rectangular patches, we measured cross-pol levels at −34 dB in the main beam region, and below −40 dB in the sidelobe region; this measurement is in doubt, however, because the cross-pol level of the transmit horn (or the chamber itself) was only about −30 dB.

IV. Conclusion

This paper has examined the various factors that affect the successful design of low-sidelobe microstrip arrays, including several considerations that have generally gone unrecognized but may have contributed to the past lack of success in achieving such designs in practice. The utility of this work was demonstrated with a 16-element linear microstrip array that achieved a −19 dBi peak sidelobe level. During the course of this effort a computer-aided design program was developed for low-sidelobe linear microstrip arrays to predict the effect of various amplitude and phase distributions, mutual coupling, Gaussian-distributed excitation errors, and finite ground-plane size.

Finally, it might be noted that this work represents a good example of how theoretical analyses, on which most of the results in this paper are based, can be applied to a practical but difficult problem in antenna engineering that would probably remain unsolved if attacked with purely empirical techniques.

References

[1] H. E. Schrank, "Low sidelobe phased array antennas," *IEEE Antennas Propagat. Soc. Newsletter*, pp. 5–9, Apr. 1983.

[2] P. S. Hall and C. J. Prior, "Radiation control in corporately fed microstrip patch arrays," in *1985 Int. Antenna Symp. Dig.*, pp. 271-275. Nice, France.

[3] P. S. Hall and J. R. James, "Design of microstrip antenna feeds: Parts I and II," *Inst. Elec. Eng. Proc.*, Pt. H, vol. 128, pp. 19-34, Feb. 1981.

[4] A. W. Rudge, K. Milne, A. D. Olver, and P. Knight, Eds. *The Handbook of Antenna Design*, vol. 2. London: Peter Peregrinus, 1983, ch. 9.

[5] D. M. Pozar, "Input impedance and mutual coupling of rectangular microstrip antennas," *IEEE Trans. Antennas Propagat.*, vol. AP-30, pp. 1191-1196, Nov. 1982.

[6] —, "Radiation and scattering from a microstrip patch on a uniaxial substrate," *IEEE Trans. Antennas Propagat.*, vol. AP-35, pp. 613-621, June 1987.

[7] —, "Trimming stubs for microstrip feed networks and patch antennas," *IEEE Antennas Propagat. Soc. Newsletter*, pp. 26-28, Dec. 1987.

[8] J. S. Herd and D. M. Pozar, "Design of a microstrip antenna array fed by a Rotman lens," presented at *1984 Int. IEEE Antennas Propagat. Soc./URSI Symp. Antennas Propagat.*, Boston, MA.

[9] D. M. Pozar, *Antenna Design Using Personal Computers*. Dedham, MA: Artech House, 1985.

[10] N. K. Das and D. M. Pozar, "Analysis and design of series-fed arrays of printed-dipole proximity-coupled to a perpendicular microstripline," *IEEE Trans. Antennas Propagat.*, vol. 37, pp. 435-444, Apr. 1989.

[11] G. Gronau, H. Moschuring, and I. Wolff, "Microstrip antenna arrays fed from the backside of the substrate," in *Proc. 1985 Int. Symp. Antennas Propagat.*, Kyoto, Japan.

[12] J. H. Cloete and L. J. du Toit, "Linear patch array pattern degradation due to corporate feed radiation," *1988 IEEE Int. Symp. Antennas Propagat.*, Syracuse, NY, pp. 466-469, June 1988.

[13] D. M. Pozar, "A microstrip antenna aperture coupled to a microstrip line," *Electron. Lett.*, vol. 21, pp. 49-50, Jan. 17, 1985.

[14] D. M. Pozar and R. W. Jackson, "An aperture coupled microstrip antenna with a proximity feed on a perpendicular substrate," *IEEE Trans. Antennas Propagat.*, vol. AP-35, pp. 728-731, June 1981.

[15] D. M. Pozar, "Finite phased arrays of rectangular microstrip patches," *IEEE Trans. Antennas Propagat.*, vol. AP-34, pp. 658-665, May 1986.

[16] Y.-W. Kang and D. M. Pozar, "Correction of error in reduced sidelobe synthesis due to mutual coupling," *IEEE Trans. Antennas Propagat.*, vol. AP-33, pp. 1025-1028, Sept. 1985.

[17] C. E. Grove, D. J. Martin, and C. Pepe, "Active impedance effects in low sidelobe and ultra wideband phased arrays," in *Proc. 1985 Phased Arrays Symp.*, RADC-TR-85-171, pp. 187-206.

[18] J. J. Schuss and R. L. Bauer, "Axial ratio of balanced and unbalanced fed circularly polarized patch radiator arrays," *IEEE Antennas Propagat. Soc. Symp. Dig.*, June 1987, pp. 286-289.

[19] D. M. Pozar, "Considerations for millimeter wave printed antennas," *IEEE Trans. Antennas Propagat.*, vol. AP-31, pp. 740-747, Sept. 1983.

[20] J. Huang, "The finite ground plane effect on the microstrip antenna radiation patterns," *IEEE Trans. Antennas Propagat.*, vol. AP-37, pp. 649-653, July 1983.

[21] P. S. Hacker and H. E. Schrank, "Range distance requirements for measuring low and ultralow sidelobe antenna patterns," *IEEE Trans. Antennas Propagat.*, vol. AP-30, pp. 956-966, Sept. 1982.

[22] R. C. Hansen, "Measurement distance effects on low sidelobe patterns," *IEEE Trans. Antennas Propagat.*, vol. AP-32, pp. 591-594, June 1984.

A Parallel-Series-Fed Microstrip Array with High Efficiency and Low Cross-Polarization

John Huang

KEY TERMS

Antenna, parallel/series feed, microstrip

ABSTRACT

A linearly polarized traveling-wave microstrip array antenna is parallel and series fed by microstrip transmission lines. Excellent antenna efficiency is achieved by having proper impedance matching throughout the array and by properly utilizing the reflected power from the end of the array. Very low cross-polarization radiation is achieved by exciting the patches and transmission lines with antiphase technique.

BACKGROUND

A low-profile antenna with a vertically polarized fan beam (approximately 2° × 50°) is needed for the C-band aircraft interferometric SAR (synthetic aperture radar) application. The main beam of the antenna is required to be fixed at the broadside direction. The available physical area for the antenna is 1.7 m × 0.17 m. A microstrip array with thin substrate material is ideal to conformally mount the antenna outside the aircraft's surface. The simplest form of feed system for such a relatively long microstrip array is series feeding, which not only minimizes the dielectric insertion loss of the feed transmission lines, but also reduces the radiation leakage from the lines when compared to a complete corporate feed system. In addition, the space usage of the given aperture is significantly improved in a series-fed array architecture.

There are two types of series-feeding techniques [1, 2]: resonant feed and traveling-wave feed. In a resonant array, no impedance matching to the elements is necessary and the resulting multiply bounced waves in the transmission line will radiate into space through the elements with phases equal to the primary radiated waves due to proper element spacing. However, because of the multiple bounces, the insertion loss that occurs in the transmission line of a resonant array is generally higher than that in a traveling-wave array. In addition, because of the phase coherence requirement of the multiply bounced waves, the resonant array has extremely narrow bandwidth. Due to these drawbacks of the resonant array, a traveling-wave array technique is employed here. In the traveling-wave array designed here, as shown in Figure 1, the impedance is not only matched at the input location but also matched to all the elements and all the power division points. Generally, a small percentage of power is lost in a matched load at the end of a traveling array. In this application, however, a half-wavelength-long open-circuited stub is used at each end of the array so that the energy remaining after the last element is reflected from the stub and radiated into space through the patch elements. Because of the required broadside beam radiation and the consequent design of one-wavelength spacing (in dielectric) between elements, the reflected energy from this open-circuit stub is in phase with all the forward-traveling waves at all the element locations. As a result, very little energy is wasted. Another special feature of the array designed here is that, as shown in Figures 1 and 2, the two rows of series-fed arrays are excited with opposite feed locations and opposite phases [3]. In doing so, not only is the higher-order-mode radiation from the patches canceled, but so is the spurious leakage radiation from the transmission lines, which results in a very pure veritcally polarized radiation with very low cross-polarization.

In a complete series-feed array, the input power to the antenna should come from one end of the array. With this end feeding, the main beam angle will be very sensitive to frequency change due to the progressive phase change of the series-fed elements. To avoid this main beam squint as frequency changes, a combination of parallel- and series-feed techniques can be used. As an example, if a linear array is parallel fed at the center of the antenna while each half of the array is series fed, although the beam angle of each half array will squint away from broadside as frequency changes, the combined beam of the whole array will remain pointed in the broadside direction. Certainly, gain degradation will occur due to the combination of the two off-broadside pointed beams. Consequently, the gain bandwidth product of a parallel-series-fed array is generally small. This gain bandwidth performance, however, can be improved if the number of parallel-fed stages increases. The array design presented in this article, as illustrated in Figure 1, has a three-stage parallel-fed configuration. Good gain bandwidth performance has been achieved.

DESCRIPTION OF THE ARRAY

The array, as shown in Figure 1, consists of a total of 72 identical square microstrip patches that are arrayed in two rows of 36 elements; the array is designed to resonate at 5.30 GHz. The dielectric substrate of the microstrip array has a

Reprinted with permission from *Microwave and Optical Tech. Lett.*, J. Huang, "A Parallel-Series-Fed Microstrip Array with High Efficiency and Low Cross-Polarization," vol. 5, no. 5, pp. 230–233, May 1992.

Figure 1 Microstrip array with parallel/series feed

relative dielectric constant of 2.17 and a thickness of 0.16 cm. Element spacing in the horizontal direction is 1 dielectric wavelength or 0.74 free-space wavelength. This 1-dielectric-wavelength spacing is needed to achieve broadside radiation with equal phases from all the series-fed elements. Element spacing in the vertical direction is 0.56 free-space wavelength, which is designed to achieve the required elevation beamwidth. Overall length of the array, including mounting areas at both ends, is 1.68 m, and the width is 0.17 m. Because no manufacturer can supply a single low-loss dielectric board of such a length, the whole antenna is made of two identical halves that are combined electrically by a coaxial power divider (matched T) and two coaxial cables. Along each row of the array, the middle 12 elements are designed to have uniform power distribution, while the 12 elements at each end of the array have tapered power distribution which is computer designed for a −20-dB sidelobe performance. The power distribution of half the array is shown in Figure 3, where the relative power in ratio (referenced to the center elements) is plotted as a function of element number.

The right half of the complete array shown in Figure 1 is sketched in Figure 2 for more detailed presentation. It is clearly indicated in this figure that the coaxial probe is fed off center in the vertical direction by 90° in phase, so that the top-row and bottom-row elements are excited 180° out of phase. With this antiphase feeding and opposite feed locations for these two rows of elements, the undesirable cross-polarization radiations from the higher-order modes of the patches will cancel each other in the far field [3]. In addition, due to this antiphase feeding, most of the leakage radiations from the two rows of microstrip transmission lines will also cancel in the far field, which will further reduce the cross-polarization level. One reason that the array is coaxially fed in the horizontal direction between the sixth and seventh elements from the center of the array is to achieve proper amplitude taper with an appropriate amount of energy reflected from the end of the array. In this design, approximately 11% of input power went into and was reflected by the open-circuit stubs at the two ends. Another reason for the feed location is to avoid a design with too thin a microstrip line which may cause fabrication tolerance problems and be more prone to damage. In Figure 2, the array section to the right side of the probe feed has tapered amplitude distribution with all element sections having identical microstrip lines. In each element section, one sixth of the incoming power traveling to the right is radiated by the patch. To achieve such a power division, a very high impedance (≈250 Ω) and very thin (≈0.05 mm) line is generally needed to transform a 300-Ω high-impedance line to a 236-Ω high-input-impedance patch. This extremely thin line is avoided by using two quarter-wave transformers for impedance matching in each element section, as shown in Figure 4. In this figure, the highest impedance line has a impedance of 173 Ω with a linewidth of 0.3 mm, which is much more tolerable than 0.05 mm. For the array, if the probe feed location is moved to the left in Figure 2, the fraction of power radiated by the patch in each element section will be smaller in order to achieve a similar amplitude taper. This will result in lines thinner than 0.3 mm, which is not acceptable. On the other hand, if the feed probe is moved to the right of the array, not only will the reflected energy from the end of the array become significant and travel into the feed to cause a mismatched input impedance, but the length of the coaxial cables that combine the two half arrays will become longer and result in a higher loss. From the preceding dis-

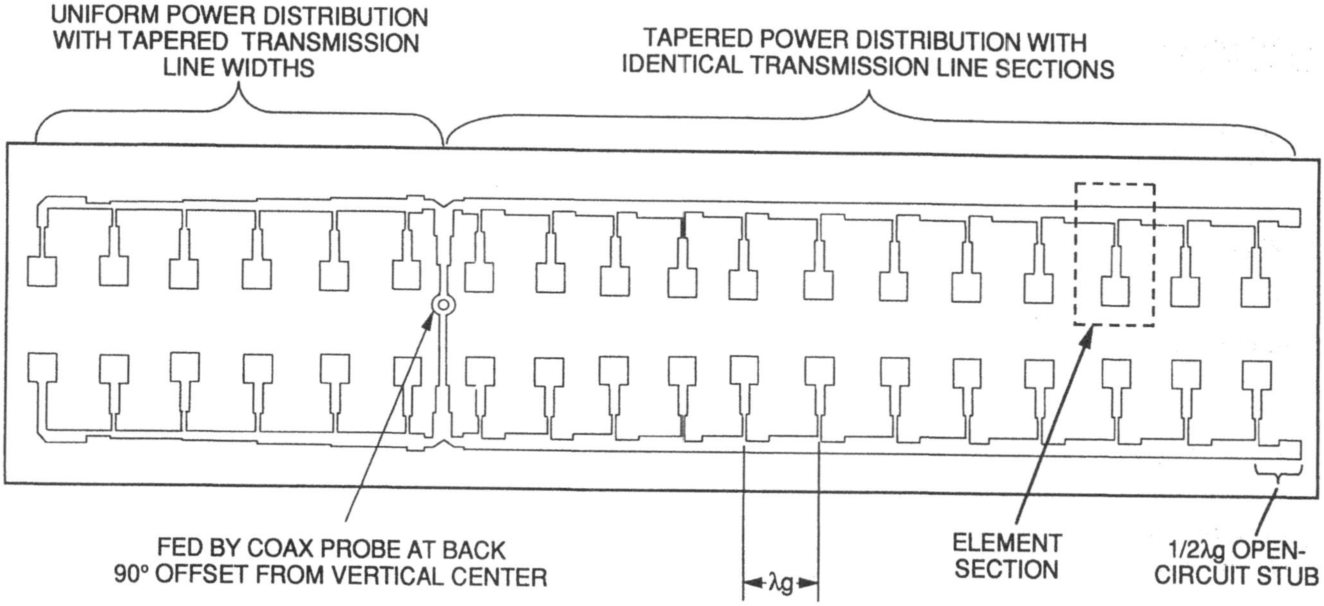

Figure 2 Right half-array of that shown in Figure 1. Dimensions not to scale and not to proportion

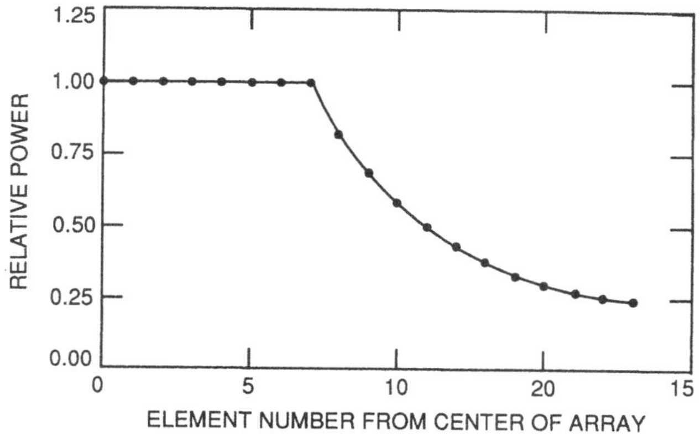

Figure 3 Relative power distribution of half the array

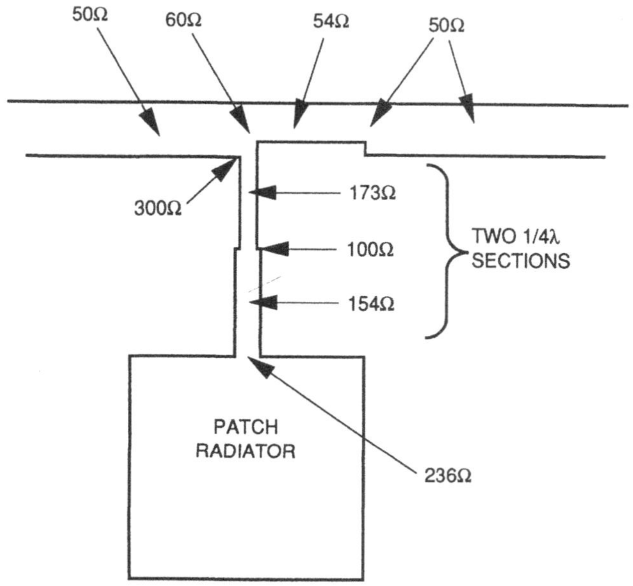

Figure 4 Impedance transformations of the element section shown in Figure 2

cussion, it is apparent that there are many factors that determine the probe feed location for this array. One other point is that if the whole array could have been made by a single dielectric board, microstrip lines would have been used to combine the two half arrays instead of the coaxial cables, which should make the overall array more efficient.

In this array, the transmission lines are impedance matched at every junction for all the waves that travel toward the two ends of the array (travel to the right in Figure 4). However, the junction impedances where 60 Ω meets 50 Ω and 300 Ω are slightly mismatched for the waves that reflected back from the open-circuit terminations (travel to the left in Figure 4). Fortunately, because only a small amount of power is reflected back from each end termination as compared to the total antenna input power and only a slight mismatch is encountered ($\frac{1}{6}$ power division is maintained), the performance of the array is good and agrees fairly well with that calculated.

ARRAY PERFORMANCE

The measured two principal-plane patterns of the complete assembled array are presented in Figure 5, where the narrow beamwidth is 2.1° and the broad beamwidth is 57.2°. Since the design of the amplitude taper is of some importance here, the measured narrow-beam pattern is compared with that of the calculated, as presented in Figure 6. Relatively good agreement between the two patterns indicates that the array is performing properly according to the design. Figure 7 gives the input return loss measured at the coaxial input to each half array. The 1.5:1 VSWR bandwidth is 58 MHz, while the 2:1 VSWR bandwidth is 120 MHz. The complete array suffered a 1-dB gain drop at about ± 30 MHz away from the center frequency of 5.30 GHz. At the center frequency, the measured antenna peak gain, referenced to the input of the coaxial power divider, is 23.80 dBi, while the calculated directivity is 25.26 dBi. The insertion loss of the coaxial power divider and coaxial cables is measured to be 1.10 dB, which implies that the loss in the microstrip array is only 0.36 dB (92% efficiency). It is estimated that 86%–88% efficiency can be achieved by the complete antenna if the two half-arrays are connected by microstrip lines instead of coaxial cables.

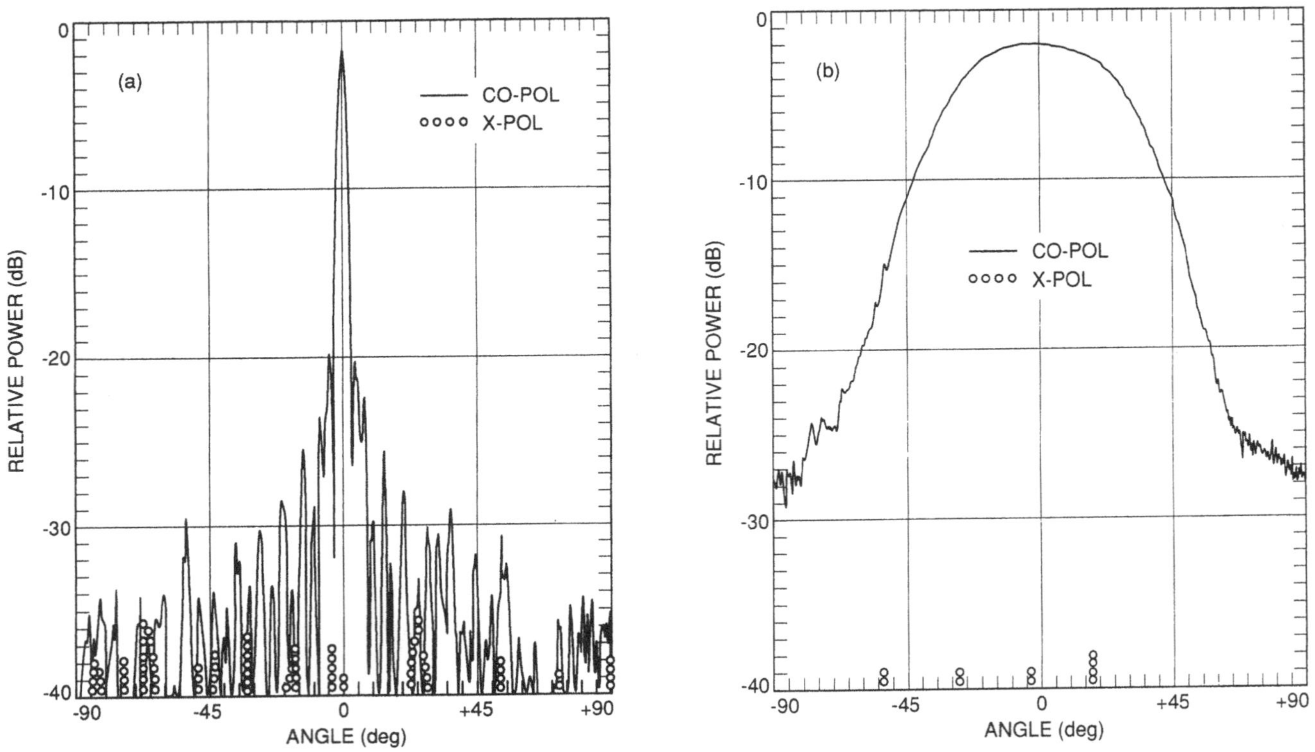

Figure 5 Measured principal-plane patterns. (a) H plane, (b) E-plane

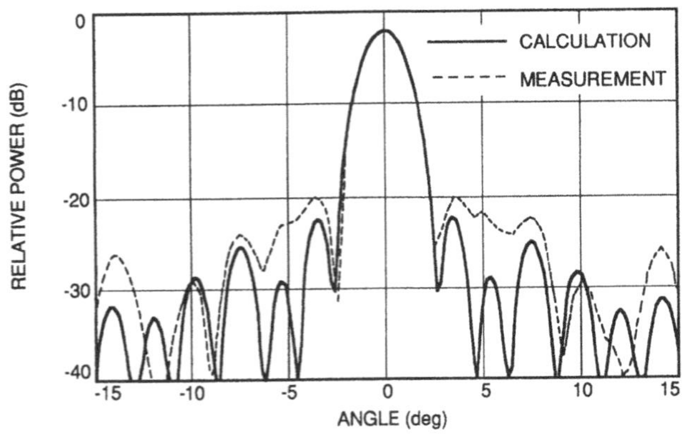

Figure 6 Comparison of measured and calculated narrow-beam patterns

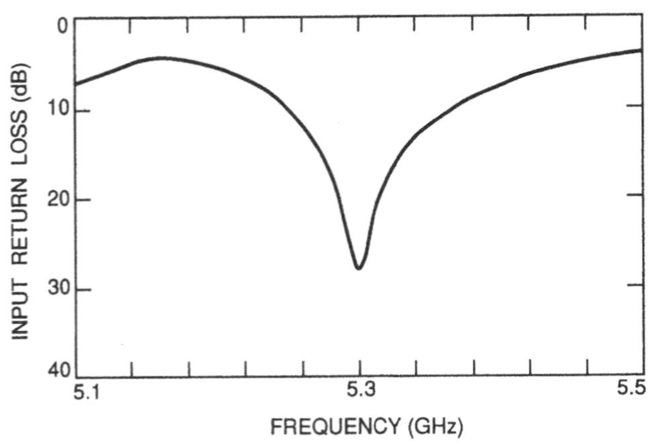

Figure 7 Input return loss versus frequency for each half-array

This good antenna efficiency is mainly attributed to the unique parallel- and series-feed configuration designed here and to the effective utilization of the reflected power from two ends of the array.The cross-polarization measured at all angular directions (within ±90° from array broadside) in the two principal planes, as shown in Figure 5, has a peak value of −33 dB from the peak of the co-polarization and an average value of about −45 dB. This low cross-polarization level is primarily the result of the antiphase feed technique being utilized here.

ACKNOWLEDGMENTS

This work was supported by the Jet Propulsion Laboratory, California Institute of Technology, under contract with the National Aeronautics and Space Administration. The assistance of Mr. Cosme Chavez in performing the experiment is greatly appreciated.

REFERENCES

1. J. R. James, P. S. Hall, and C. Wood, *Microstrip Antenna—Theory and Design*, Peter Peregrinus Ltd., Stevenage, UK, pp. 111–159.
2. J. R. James and P. S. Hall (Eds.), *Handbook of Microstrip Antennas*, Peter Peregrinus Ltd., London, UK, pp. 825–849.
3. J. Huang, "Dual-Polarized Microstrip Array with High Isolation and Low Cross-Polarization" *Microwave Opt. Technol. Lett.*, Feb. 1991, pp. 99–103.

Chapter 7
Analysis of Arrays and Mutual Coupling

IF mutual coupling were nonexistent, the input impedance of an element in an array would be the same as that of an isolated element, and the input impedance of any particular element in an array would not be a function of the excitations of other elements. Also, the pattern of the array could be predicted by simple pattern multiplication of the element factor and the array factor. In reality, however, all arrays do have mutual coupling between elements, and the effect of such coupling will be to make the input impedance a function of the excitation, and to perturb the excitation of the array elements, and thus the array pattern. So the real question for the array designer is whether or not mutual coupling effects will be significant enough to warrant inclusion in the design of the array.

Fortunately, many practical microstrip arrays can be designed by treating separately the radiating element and the feed network, without concern for mutual coupling effects. This is especially true for broadside or other fixed-beam microstrip arrays, and may come as a surprise to array designers familiar with dipole or slot arrays, where mutual coupling effects are often more significant. This is an instance where the electrically thin substrate of the typical microstrip antenna plays a beneficial role, by reducing the level of mutual coupling between adjacent elements. There are microstrip arrays, however, for which mutual coupling effects are significant, and may even be detrimental. Large phased arrays having a wide scanning range, for example, will generally exhibit impedance mismatch and gain loss versus scan angle due to mutual coupling effects. And if the array substrate is thicker than usual, has a thick cover layer, or a high dielectric constant, mutual coupling effects may be important even if the array is fixed-beam. The analysis and compensation of these effects are the subject of the papers in this chapter.

Although it is possible to get useful results for mutual coupling using simpler models [1], full-wave solutions ensure that both space wave and surface wave coupling effects are properly included. Thus we begin with the paper by Pozar, which describes a full-wave analysis of the mutual coupling between two probe-fed microstrip patch antennas. These results, and later results for elements having different shape and feeds [2], [3], indicate that typical coupling levels for elements spaced about a half wavelength apart (center-to-center) are in the range of -15 to -20 dB. Mutual coupling between elements can also be found by measurement, which may be easier than calculation if an appropriate computer code is not available. After mutual coupling data has been found for the array of interest, the next step is to use this data either to predict the performance of the array, or to design the feed network to compensate for the effects of mutual coupling. The paper by Solbach presents a good example of this procedure for a phased array of circular patches, using measured coupling data. The next paper, by Pozar, presents a complete analytical solution for the mutual coupling analysis of rectangular probe-fed patch arrays, including the radiation pattern, the active element pattern, and input impedance. Similar work has been described for finite arrays of aperture-coupled patches [3], and circular probe-fed patches [4]. The above work generally assumes implicitly a corporate-type feeding arrangement, but it is sometimes necessary to consider series-type feeds, which introduce a bit more complexity. This was done for a series-fed C-band array of aperture coupled patches in the paper by Wu, Wang, Fralich, and Litva. Similar work has been done for series-fed arrays of printed dipole elements [5]–[7].

As the size of an array increases, it becomes increasingly costly to calculate the effects of mutual coupling on an element-by-element basis. But if the array is uniform (uniform spacing, uniform amplitude and phase progression, and identical elements), and large enough that edge effects can be ignored, then the infinite array approximation can be used to great advantage. The periodicity of an infinite array simplifies the analysis to that of a single unit cell, while still including all mutual coupling effects. Infinite array analysis follows very nicely from full-wave moment method solutions that are formulated in the spectral domain, as described in [8] for infinite arrays of printed dipoles. Since then this technique has been applied to a wide variety of microstrip and other printed antenna geometries [9]–[11]. The paper by Liu, Hessel, and Shmoys gives a comprehensive study of probe-fed rectangular microstrip antennas using the infinite-array approach, and the paper by Aberle and Pozar presents an infinite-array analysis for circular patches having multiple probe feeds. The paper by Herd discusses an infinite-array solution for proximity-fed patches, and illustrates the usefulness of such a solution by using it for optimization of the bandwidth and scanning performance of the array. The final paper, by Pozar, shows how to extend infinite-array solutions to include subarrays, with application to arrays with parasitic elements, arrays of sequential subarrays, and arrays of hexagonal subarrays.

CAD software for microstrip arrays is in the same minimally developed state as CAD for single elements, as discussed in the introduction to Chapter 5 and the review article on CAD in Chapter 1. Infinite-array models have probably received the most attention in terms of developing general-purpose software, with the moment method solutions for multilayer microstrip elements described in [12], and the hybrid finite-element solution for infinite arrays of very arbitrary elements presented in [13], being typical examples. As can be seen from the papers in this chapter and the references below, the theoretical treatment for microstrip arrays is well developed, but much work remains to implement these models into user-oriented software that is both accurate and versatile.

References

[1] E. Penard and J. P. Daniel, "Mutual coupling between microstrip antennas," *Electronics Letters*, vol. 18, pp. 605–607, 1982.

[2] J. T. Aberle and D. M. Pozar, "A full-wave solution for the mutual coupling between circular microstrip patch antennas," *Microwave and Optical Technology Letters*, vol. 2, pp. 130–132, Apr. 1989.

[3] S. G. Pan and I. Wolff, "Computation of mutual coupling between slot-coupled microstrip patches in a finite array," *IEEE Trans. Antennas and Prop.*, vol. 40, pp. 1047–1053, Sept. 1992.

[4] M. D. Deshpande and M. C. Bailey, "Analysis of finite phased array of circular microstrip patches," *IEEE Trans. Antennas and Prop.*, vol. 37, pp. 1355–1360, Nov. 1989.

[5] P. B. Katehi, "Mutual coupling between microstrip dipoles in multi-element arrays," *IEEE Trans. Antennas and Prop.*, vol. 37, pp. 275–280, Mar. 1989.

[6] N. K. Das and D. M. Pozar, "Analysis and design of series-fed arrays of printed-dipoles proximity-coupled to a perpendicular microstripline," *IEEE Trans. Antennas and Prop.*, vol. AP- 37, pp. 435–444, Apr. 1989.

[7] D. R. Jackson and P. Manghnani, "Analysis and design of a linear array of electromagnetically coupled microstrip patches," *IEEE Trans. Antennas and Prop.*, vol. 38, pp. 754–759, May 1990.

[8] D. M. Pozar and D. H. Schaubert, "Scan blindness in infinite arrays of printed dipoles," *IEEE Trans. Antennas and Prop.*, vol. AP-32, pp. 602–610, June 1984.

[9] J. T. Aberle and D. M. Pozar, "Analysis of infinite arrays of probe-fed rectangular microstrip antennas using a rigorous feed model," *Proc. IEE*, pt. H. vol. 136, pp. 110–119, Apr. 1989.

[10] M. D. Deshpande and P. D. R. Prabhakar, "Analysis of dielectric covered infinite array of rectangular microstrip antennas," *IEEE Trans. Antennas and Prop.*, vol. AP-37, pp. 732–736, June 1987.

[11] A. K. Skrivervik and J. R. Mosig, "Finite phased array of microstrip patch antennas: the infinite array approach," *IEEE Trans. Antennas and Prop.*, vol. 40, pp. 579–582, May 1992.

[12] N. K. Das and D. M. Pozar, "Multiport scattering analysis of general multilayered printed antenna fed by multiple feed ports: Part I—Theory, and Part II—Applications," *IEEE Trans. Antennas and Prop.*, vol. 40, pp. 469–491, May 1992.

[13] D. T. McGrath, "A finite element code for phased array radiator analysis," *Antenna Applications Symp. Proceedings*, pp. 1–27, Sept. 1994.

Input Impedance and Mutual Coupling of Rectangular Microstrip Antennas

DAVID M. POZAR, MEMBER, IEEE

Abstract—**A moment method solution to the problem of input impedance and mutual coupling of rectangular microstrip antenna elements is presented. The formulation uses the grounded dielectric slab Green's function to account rigorously for the presence of the substrate and surface waves. Both entire basis (EB) and piecewise sinusoidal (PWS) expansion modes are used, and their relative advantages are noted. Calculations of input impedance and mutual coupling are compared with measured data and other calculations.**

I. INTRODUCTION

THE INCREASING USE of microstrip antenna technology [1], [2] requires analysis methods capable of accurately predicting the input impedance, mutual coupling, and radiation of these antennas. Two methods which have been somewhat successful for calculating input impedance and radiation are the transmission line model [3], and the cavity model [4], [5]. Although proven to be useful for predicting input impedance, the above two methods do not rigorously account for the presence of surface waves on the antenna substrate, and do not account for mutual coupling between closely spaced antenna elements. Also, they lack a convenient way of improving solution accuracy.

Recently, a moment method solution to the microstrip antenna problem was proposed [6]. In [6], image theory was used with an integral equation to solve for the antenna patch current; the dielectric slab was treated with equivalent currents. This method, too, gave good results for input impedance, but suffered from the necessity of extremely accurate numerical evaluations of the impedance matrix elements.

Presented here is a moment method solution for rectangular microstrip antenna elements, including mutual coupling. This approach uses the exact Green's function for the dielectric slab, and thus rigorously accounts for surface waves and coupling to adjacent antenna elements. Since the ground plane is accounted for analytically by the Green's function, no severe numerical problems are encountered as in [6], which relied upon image theory to treat the ground plane. The present method can handle edge (microstrip) or probe (coax) type feeds, and can be generalized to treat antenna elements of arbitrary shape. The solution is similar in principle to the previously treated case of printed dipoles [7], [8], and [9].

Two types of expansion modes are considered—entire basis (EB), and piecewise sinusoidal (PWS). The relative advantages and disadvantages of these modes are noted and discussed, and it is shown how accuracy can be improved by the appropriate choice and number of expansion modes.

Section II describes the theory of the method. The Green's function and its efficient evaluation are discussed, and definitions of the expansion modes and impedance matrix elements are given. Section III presents numerical results for input impedance and mutual coupling for various antenna geometries. These results are compared with measurements and with other solutions. Also presented is a design data curve for resonant resistance of a rectangular element with variable width and feed position.

II. THEORY

The Dielectric Slab Green's Function

For the calculation of impedance matrix elements and voltage vector elements, the electric field from a horizontal electric current element on a grounded dielectric slab is needed. No generality is lost by having the current element directed in the $\hat{x}$-direction, since all impedance matrix elements can be found by orienting the test mode for current flow in the $\hat{x}$-direction. So for the geometry shown in Fig. 1, the vector potential is [7]-[10] (assuming $e^{j\omega t}$ time dependence).

$$A_x = \frac{\mu Il}{4\pi^2} \iint_{-\infty}^{\infty} zG_1 e^{jk_x(x-x_0)+jk_y(y-y_0)}\, dk_x\, dk_y \tag{1}$$

$$A_y = 0$$

$$A_z = \frac{\mu Il}{4\pi^2} \iint_{-\infty}^{\infty} k_x G_2 e^{jk_x(x-x_0)+jk_y(y-y_0)}\, dk_x\, dk_y \tag{2}$$

where

$$zG_1 = \frac{\sin k_1 z}{T_e} \tag{3}$$

$$G_2 = \frac{(\epsilon_r - 1)\sin k_1 d \cos k_1 z}{T_e T_m} \tag{4}$$

$$T_e = k_1 \cos k_1 d + jk_2 \sin k_1 d \tag{5}$$

$$T_m = \epsilon_r k_2 \cos k_1 d + jk_1 \sin k_1 d \tag{6}$$

$$k_1{}^2 = \epsilon_r k_0{}^2 - \beta^2, \quad (\text{Im } k_1 < 0) \tag{7}$$

$$k_2{}^2 = k_0{}^2 - \beta^2, \quad (\text{Im } k_2 < 0) \tag{8}$$

$$\beta^2 = k_x{}^2 + k_y{}^2 \tag{9}$$

$$k_0{}^2 = \omega^2 \mu_0 \epsilon_0, \tag{10}$$

the field point is at (x, y, z); the source location is at (x_0, y_0, d). Note that dielectric loss is easily included by replacing ϵ_r by $\epsilon_r(1 - j \tan \delta)$ in (7), where $\tan \delta$ is the loss tangent of the substrate material. The electric fields are then found by

$$\bar{E} = \frac{-j\omega}{\epsilon_r k_0{}^2} (\epsilon_r k_0{}^2 \bar{A} + \nabla\nabla \cdot \bar{A}). \tag{11}$$

The integrals in (1) and (2) will be evaluated numerically, however this can be facilitated by changing to polar coordi-

Reprinted from *IEEE Trans. Antennas Propaga.*, vol. AP-30, no. 6, pp. 1191–1196, Nov. 1982.

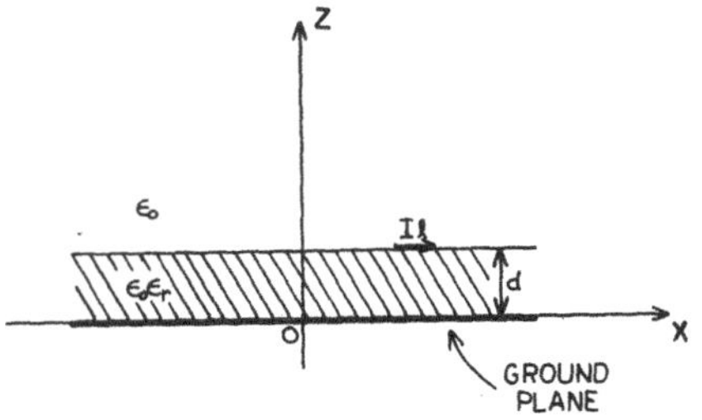

Fig. 1. Geometry for an infinitesimal dipole on a grounded dielectric slab.

nates β, α where

$$k_x = \beta \cos \alpha \tag{12}$$

$$k_y = \beta \sin \alpha. \tag{13}$$

The integration ranges are $\beta = 0 \to \infty$ and $\alpha = 0 \to 2\pi$. The terms T_e and T_m are functions of β (not of α), and represent transverse electric (TE) and Transverse magnetic (TM) surface wave poles, respectively. The first TM surface wave has a zero cutoff frequency, so there is always at least one surface wave pole (unless $\epsilon_r = 1$). These poles occur for real values of $\beta = \beta_0$ such that $k_0 \leqslant \beta_0 < \sqrt{\epsilon_r} k_0$, if $\tan \delta = 0$ (no loss). If dielectric loss is present, the poles move off the real β axis to $\beta = \beta_0 - j\gamma$, $\gamma > 0$. The exact pole location $\beta_0 - j\gamma$ can be determined by using the Newton-Rhapson procedure to find the zeros of (5) and (6). To avoid numerical difficulties when numerically integrating on β, the integration from $\beta_0 - \delta$ to $\beta_0 + \delta$, where $\delta \sim 0.001 k_0$, is done analytically [7]. For example, if only one surface wave mode is present,

$$\iint_{-\infty}^{\infty} (\;)\, dk_x\, dk_y = \int_0^{2\pi} \left\{ \int_0^{\beta_0-\delta} (\;)\, d\beta + \int_{\beta_0-\delta}^{\beta_0+\delta} (\;)\, d\beta + \int_{\beta_0+\delta}^{\infty} (\;)\, d\beta \right\} d\alpha \tag{14}$$

Then,

$$I_\delta = \int_{\beta_0-\delta}^{\beta_0+\delta} (\;)\, d\beta = \int_{\beta_0-\delta}^{\beta_0+\delta} \frac{f(\beta)}{T_m(\beta)}\, d\beta, \tag{15}$$

where $T_m(\beta)$ has a pole at $\beta = \beta_0 - j\gamma$ and $f(\beta)$ represents the remaining nonsingular portion of the integrand. Using two terms of a Taylor series expansion of $T_m(\beta)$ about $\beta = \beta_0 - j\gamma$ then yields

$$I_\delta = \frac{jf(\beta_0 - j\gamma)}{T_m'(\beta_0 - j\gamma)} \tan^{-1} \frac{-2\delta\gamma}{\gamma^2 - \delta^2}. \tag{16}$$

For the lossless case $\gamma = 0$, and (16) reduces to

$$I_\delta = \frac{-\pi j f(\beta_0)}{T_m'(\beta_0)},$$

which is the same as the result obtained using residue theory. Higher order surface wave modes can be treated in the same manner. The infinite integration in (14) can generally be terminated at $\beta \simeq 150\, k_0$.

The Moment Method Formulation

The moment method solution used here is a Galerkin solution of the electric field integral equation with the dielectric slab Green's function. The unknown surface current density $\bar{J}_s$ on the microstrip antenna elements is expanded in a set of N basis functions,

$$\bar{J}_s = \sum_{n=1}^{N} I_n \bar{J}_n, \tag{17}$$

where $\bar{J}_n$ is the nth basis function and I_n is its unknown amplitude. The basis functions $\bar{J}_n$ may represent currents in the x or y directions. Use of the same set of functions as testing functions leads to a set of linear algebraic equations to be solved for the unknown I_n:

$$[Z][I] = [V], \tag{18}$$

where

$$Z_{mn} = -\int_{S_n} \bar{E}_m \cdot \bar{J}_n\, ds, \tag{19}$$

$$V_m = \int_{v_i} \bar{E}_m \cdot \bar{J}_i\, dv. \tag{20}$$

In (19) and (20), $\bar{E}_m$ is the electric field due to current $\bar{J}_m$ in the presence of the dielectric slab, and $\bar{J}_i$ is the impressed (source) current. Since the same set of functions is used for testing and expansion (Galerkin), $Z_{mn} = Z_{nm}$.

Fig. 2(a) shows a possible layout of expansion modes for a microstrip antenna problem. The two expansion modes are shown on different antenna patches, but they could be on the same patch, and could have currents in the same direction or in orthogonal directions. Therefore, all impedance elements are of one of two possible types: Z^{xx} (test current in x, expansion current in x), or Z^{xy} (test current in x, expansion current in y). Using (1)–(4), (11), and (19) allows these impedance matrix elements to be written as

$$Z_{mn}{}^{xx} = \iint_{-\infty}^{\infty} Q_x(k_x, k_y) F_x(J_n) F_y(J_n) \cdot F_x{}^*(J_m) F_y{}^*(J_m)\, dk_x\, dk_y \tag{21}$$

$$Z_{mn}{}^{xy} = \iint_{-\infty}^{\infty} Q_y(k_x, k_y) F_x(J_n) F_y(J_n) \cdot F_x{}^*(J_m) F_y{}^*(J_m)\, dk_x\, dk_y \tag{22}$$

where

$$Q_x = \frac{jZ_0}{4\pi^2 k_0} \cdot \sin k_1 d \cdot \frac{(\epsilon_r k_0{}^2 - k_x{}^2) k_2 \cos k_1 d + jk_1 (k_0{}^2 - k_x{}^2) \sin k_1 d}{T_e T_m} \tag{23}$$

$$Q_y = \frac{-jZ_0}{4\pi^2 k_0} \frac{k_x k_y \sin k_1 d [k_2 \cos k_1 d + jk_1 \sin k_1 d]}{T_e T_m} \tag{24}$$

and F_x, F_y are the Fourier transforms of the x and y dependences of J:

$$F_x(J_n) = \int_{x_n} J_n{}^x(x) e^{jk_x x}\, dx, \tag{25a}$$

$$F_y(J_n) = \int_{y_n} J_n{}^y(y) e^{jk_y y}\, dy, \tag{25b}$$

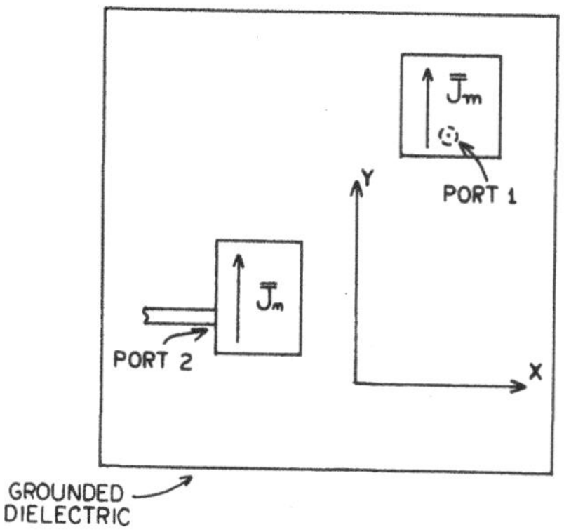

(a)

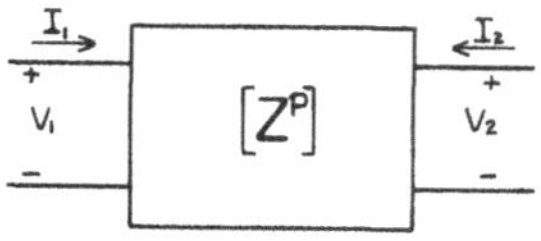

(b)

Fig. 2. (a) Example of two coupled microstrip antenna elements and two expansion modes, $\bar{J}_n$ and $\bar{J}_m$. (b) Definitions of two-port voltages and currents.

where $J_n(x, y)$ has been factored into the product of a function of x and a function of y:

$$J_n(x, y) = J_n{}^x(x) J_n{}^y(y). \tag{26}$$

If the impressed source current is expressed as

$$\bar{J}_i = \hat{z}\delta(x_0 - x_p)\delta(y_0 - y_p), \tag{27}$$

where the feed position is at (x_p, y_p), then the voltage vector elements can be found by using (2)–(4), (11), and (20):

$$V_m = \iint_{-\infty}^{\infty} Q_v(k_x, k_y) F_x{}^*(Jm) F_y{}^*(J_m) \cdot e^{jk_x x_p + jk_y y_p}\, dk_x\, dk_y, \tag{28}$$

where

$$Q_v = \frac{-jZ_0}{4\pi^2 k_0} \frac{\beta^2 k_x(\epsilon_r - 1)\sin k_1 d + jk_x k_1 T_m}{\epsilon_r T_e T_m} \frac{\sin k_1 d}{k_1}. \tag{29}$$

Note that (21) and (22) consist of six integrations; however, the integrals for F_x and F_y can be done in closed form leaving only two numerical integrations. Equation (28) similarly requires two numerical integrations. As mentioned previously, the infinite k_x, k_y integrations can be facilitated by converting to polar coordinates. Also, the even and odd properties of the integrand can be used to reduce the $\alpha = 0 \to 2\pi$ integration range to $\alpha = 0 \to \pi/2$. Equations (21), (22), and (28) can then be written as

$$Z_{mn}{}^{xx} = 4 \int_0^{\pi/2} \int_0^{\infty} Q_x \operatorname{Re}[F_x(J_n) F_x{}^*(J_m)] \cdot \operatorname{Re}[F_y(J_n) F_y{}^*(J_m)]\, \beta\, d\beta\, d\alpha \tag{30}$$

$$Z_{mn}{}^{xy} = -4 \int_0^{\pi/2} \int_0^{\infty} Q_y \operatorname{Im}[F_x(J_n) F_x{}^*(J_m)] \cdot \operatorname{Im}[F_y(J_n) F_y{}^*(J_m)]\, \beta\, d\beta\, d\alpha \tag{31}$$

$$V_m = 4j \int_0^{\pi/2} \int_0^{\infty} Q_v \operatorname{Im}[F_x{}^*(J_m) e^{jk_x x_p}] \cdot \operatorname{Re}[F_y{}^*(J_m) e^{jk_x y_p}]\, \beta\, d\beta\, d\alpha. \tag{32}$$

Equations (30)–(32) are the final forms used for the computation of the impedance matrix and voltage vector elements.

Feed Modeling

A coaxial or probe-type feed can be modeled by using (27) to represent a unit current source at the probe position (x_p, y_p), and accounting for the probe self-inductance by adding jX_p to the input impedance, where [1]

$$X_p = \frac{Z_0}{\sqrt{\epsilon_r}} \tan(\sqrt{\epsilon_r}\, k_0 d). \tag{33}$$

A microstrip-type feed can also be modeled by using (27) to represent an equivalent $\hat{z}$ current source at the point where the feed line joins the antenna patch. Strictly speaking, the width of the feed line should be incorporated by using a ribbon of current of width equal to that of the feed line, but it has been found that the line element of (27) gives an identical result for the narrow feed lines that are in common use. It has also been found [9] that the voltage term (32) should be modified by the factor $\sqrt{W_e/d}$ to account for edge effects of the microstrip line. If W is the width of the feed line, the effective width W_e is [1]

$$W_e = W + 0.412 \left(\frac{\epsilon_e + 0.30}{\epsilon_e - 0.258}\right)\left(\frac{W + 0.262d}{W + 0.813d}\right) d,$$

where ϵ_e is the effective dielectric constant [1],

$$\epsilon_e = \frac{\epsilon_r + 1}{2} + \frac{\epsilon_r - 1}{2}\left(1 + \frac{10d}{W}\right)^{-1/2}. \tag{34}$$

Port Impedances

Consider the two-element microstrip antenna geometry shown in Fig. 2(a). The relation between the port voltages and currents (defined in Fig. 2(b)) is

$$[V^p] = [Z^p][I^p], \tag{35}$$

where the superscript "p" is used to differentiate between these "port" quantities and the moment method quantities of (18). This distinction is an important one and necessary to avoid confusion and erroneous results. That these two sets of quantities are distinct can easily be seen by noting that $[Z]$ of (18) can be of order N if N expansion modes are used on the two antenna elements, while $[Z^p]$ of (35) is always a 2×2 matrix (for a two-port geometry). The "port" impedance matrix $[Z^p]$ is the one of interest for determining input impedance and mutual coupling, so its relation to (18) is now described.

$Z_{11}{}^p$ is the input impedance of element one with element two open-circuited, and can be written as

$$Z_{11}{}^p = \frac{-\int_v \bar{E}^{(1)} \cdot \bar{J}_i{}^{(1)}\, dv}{(I_i)^2}, \tag{36}$$

where $\bar{E}^{(1)}$ is the total electric field from the N expansion modes caused by the source current $\bar{J}_i{}^{(1)}$ at port one, and I_i is the terminal current of the source (1 A in this case). The use of (20) then gives

$$Z_{11}{}^p = -\sum_{n=1}^{N} I_n V_n{}^{(1)}, \tag{37}$$

where I_n are the expansion mode currents found from (18), and $V_n^{(1)}$ are the voltages due to source 1. Note that all I_n for expansion modes on element 2 should be set to zero, since element 2 is open-circuited. Equation (37) also gives the input impedance of a single antenna element, and can be used for Z_{22}^p by substituting $V_n^{(2)}$, the voltage due to source 2, for $V_n^{(1)}$ and setting all I_n on element 1 to zero.

The open-circuit transfer impedance between ports 1 and 2 can be written as

$$Z_{21}^p = \frac{-\int_v \bar{E}^{(2)} \cdot \bar{J}_i^{(1)}\, dv}{(I_i)^2}, \tag{38}$$

where $\bar{E}^{(2)}$ is the total electric field caused by source current $\bar{J}_i^{(2)}$ at port 2. Using (20) gives

$$Z_{21}^p = -\sum_{n=1}^{N} I_n V_n^{(2)}, \tag{39}$$

where $V_n^{(2)}$ is the voltage due to source 2, and all I_n on element 2 are set to zero. Of course, $Z_{12}^p = Z_{21}^p$.

Expansion Modes

Both EB and PWS expansion modes were used in this solution. The current density for the EB modes with current in the $\hat{x}$-direction can be written as $J(x, y) = J^x(x)J^y(y)$, where

$$J^x(x) = \sin \frac{m\pi}{2a} [x - (x' - a)], \quad \text{for } x' - a < x < x' + a, \tag{40a}$$

and

$$J^y(y) = \frac{1}{w}, \tag{40b}$$

for $y' - w/2 < y < y' + w/2$, and x', y' are the coordinates of the center of the mode and a and w are the half-length and width of the mode, respectively. This mode is constant in the y-direction and has a sinusoidal variation in the x-direction, with zero current at the ends of the mode and continuous derivatives over the range of the mode. The constant m in (40a) determines the order (number of variations in x) of the expansion mode, and assumes integer values starting at $m = 1$. Fig. 3(a) shows EB modes of order $m = 1$ and 2 arranged on a microstrip antenna element. Often symmetry arguments can be used to eliminate certain expansion modes, such as all odd modes or all even modes. As the number of expansion modes increases, the solution should converge and accuracy increase.

The EB modes have the useful property that many combinations of modes on the same antenna element are uncoupled ($Zmn = 0$), thus making evalution of the impedance matrix faster. For example, (30), (31), (25), and (40) can be used to show the coupling and uncoupling of impedance elements as shown in Table I.

In Table *I* m_t and m_e are the values of m for the test mode and expansion mode, respectively, and Z_{xx} is the impedance between two modes with current in the same direction, while Z_{xy} is the impedance between two modes with current in orthogonal directions. This data are only valid for modes on the same antenna element (identical x', y' for each mode); modes on different elements are not generally uncoupled.

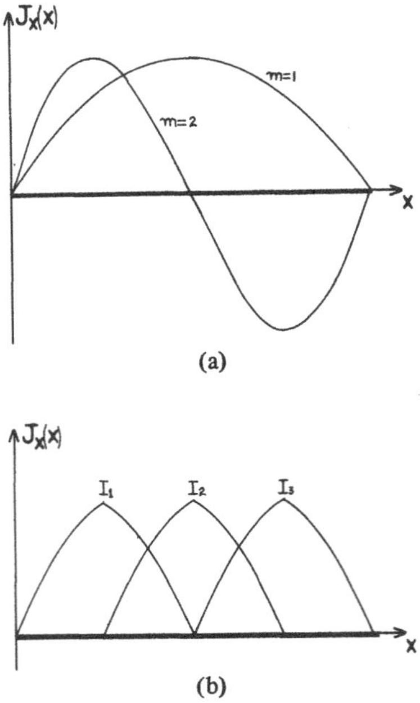

Fig. 3. (a) Example of layout of two EB expansion modes on an antenna patch. (b) Example of layout of three PWS modes on an antenna patch.

TABLE I
COUPLING OF EB IMPEDANCE MATRIX ELEMENTS

m_t	m_e	Z_{xx}	Z_{xy}
1	1	$\neq 0$	$= 0$
2	1	$= 0$	$= 0$
1	2	$= 0$	$= 0$
2	2	$\neq 0$	$\neq 0$

The current density for a PWS mode centered at x', y' is

$$J^x(x) = \frac{\sin k(a - |x - x'|)}{\sin ka} \tag{40c}$$

for $x' - a < x < x' + a$, and a y-dependence as given by (40b). This mode is constant in y, piecewise sinusoidal in x, with zero current at the ends of the mode and a discontinuous derivative at $x = x'$. Fig. 3(b) shows three PWS modes arranged on an antenna element. Accuracy is improved by increasing the number of overlapping PWS modes. In contrast to the EB modes, no uncoupled PWS modes exist, however symmetry between PWS modes often reduce the number of matrix elements which need to be calculated. For example, in Fig. 3(b) symmetry gives $Z_{11} = Z_{22} = Z_{23}$, and $Z_{12} = Z_{23}$. Another advantage comes from the fact that a voltage vector element V_m is negligibly small unless the feed position lies within the bounds of mode M. Table II summarizes the computational cost of using EB or PWS modes for some typical mode layouts for a calculation of input impedance of a single antenna element.

In this table, a unit "cost" is defined as the effort required for evaluation of a double integration like that in (30)–(32). Also, Table II assumes that the EB modes are used sequentially (no modes removed by symmetry considerations), and N_x, N_y refer to the number of expansion modes in the x and y-directions, respectively. The wavenumber k for the PWS mode in (40c) can be chosen arbitrarily, however, a judicious choice will improve convergence. It has been found that setting k

TABLE II
COMPARISON OF COMPUTATIONAL COST FOR EB AND PWS MODES

Mode Layout	Quantity Calculated	Cost Using EB Modes	Cost Using PWS Modes
$N_x = 1$, $N_y = 0$	Z	1	1
	V	1	1
	total	2	2
$N_x = 2$, $N_y = 0$	Z	2	2
	V	2	1
	total	4	3
$N_x = 3$, $N_y = 0$	Z	4	3
	V	3	1
	total	7	4
$N_x = 1$, $N_y = 1$	Z	2	3
	V	2	2
	total	4	5

equal to the "effective" wavenumber for the substrate is a good choice:

$$k = k_e = \omega\sqrt{\mu_0 \epsilon_e}, \tag{41}$$

where ϵ_e is defined by (34). This choice for k was also made in [6].

Neither the EB or PWS modes included an edge condition; this simplification was justified on the basis of previous surface patch modeling work [11], in which it was discovered that, if sophisticated expansion modes are used (e.g., PWS or EB), the edge condition does not need to be enforced for accurate results.

III. RESULTS

In this section numerical results using the above-described method are compared with measurements and other calculations of input impedance and mutual coupling for microstrip antennas.

Fig. 4 compares calculated input impedance of an edge-fed microstrip antenna with measurements from [12]. Four sets of calculations are shown here, using both one and two expansion modes for both EB and PWS modes. As can be seen, agreement with measurements is very good for all cases, and improves as the number of modes used increases from one to two. Also, it can be seen that the EB results are generally closer to the measured values than the PWS results, for the same number of modes. The tradeoff here is that, as Table II shows, the PWS calculation is faster for $N > 1$.

Fig. 5 shows calculated input impedance of a coax-fed antenna using one EB mode compared with calculations and measurements from [6]. The present calculations compare well with the measured data and, for frequencies above resonance, appear to be more accurate than the calculations from [6]. This is probably due to the fact that the method of [6] suffered from numerical sensitivities and an approximate treatment of the dielectric slab. (Note: The measured data point at $f = 639$ MHz appears to be incorrectly labeled as 640 MHz in [6].)

Fig. 6 shows calculated input impedance of a coax-fed antenna using one EB mode compared with measurements from Lo [13]. Again the agreement is good.

Fig. 7 gives design data for the resonant resistance of a coax-fed microstrip antenna versus feed position and width of the patch. These data show the radiation resistance decreasing

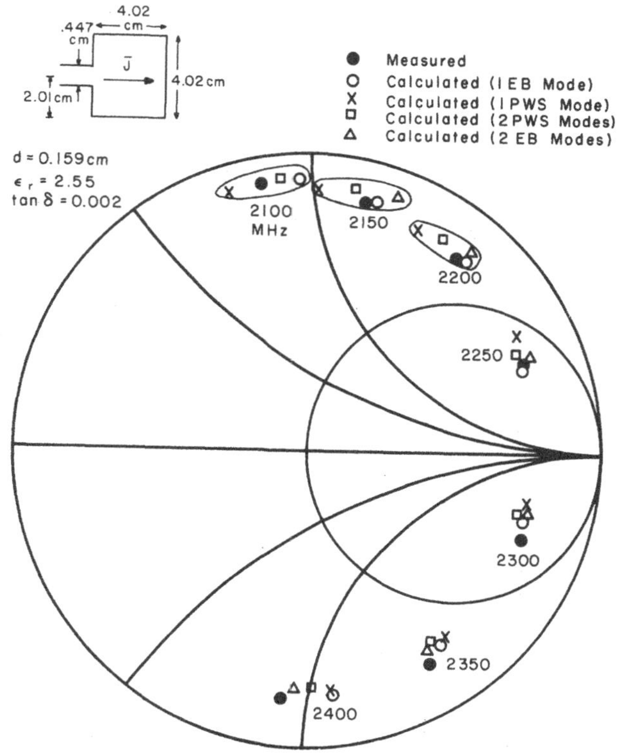

Fig. 4. Measured and calculated input impedance of an edge-fed microstrip antenna. Calculations made using one and two EB and PWS modes.

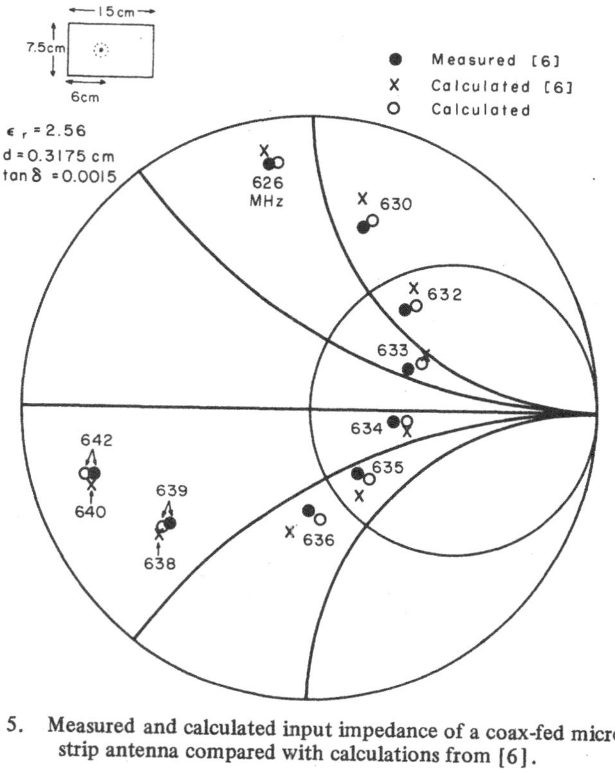

Fig. 5. Measured and calculated input impedance of a coax-fed microstrip antenna compared with calculations from [6].

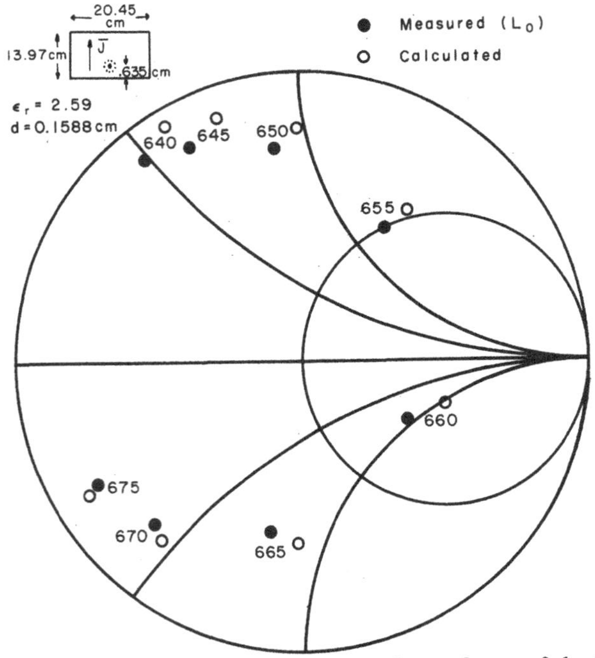

Fig. 6. Measured and calculated input impedance of a coax-fed microstrip antenna.

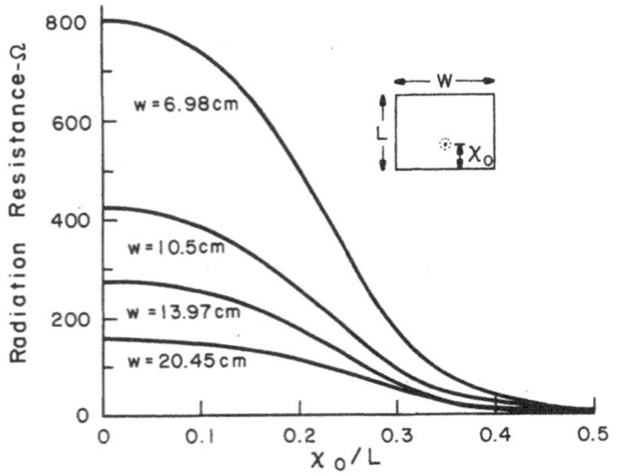

Fig. 7. Radiation resistance for a coax-fed microstrip antenna versus feed position and width. L = 13.97 cm, ϵ_r = 2.60, d = 0.1588 cm, tan δ = .002.

as the feed approahces the center of the patch, and increasing as the width is made larger. The resonant frequency is independent of feed position, but decreases from 666 MHz for $W = 6.98$ cm to 659 MHz for $W = 20.45$ cm.

Fig. 8 shows the calculation of mutual coupling (S_{12} parameter) between two coax-fed microstrip antenna compared with measurements by Carver [14]. Both $\bar{E}$-plane and $\bar{H}$-plane couplings are shown. The $\bar{E}$-plane coupling is larger due to a stronger surface wave being excited for this case. The good agreement demonstrates the accuracy and versatility of the method.

It should be noted that, due to the very narrow bandwidth of microstrip antennas, the good agreement between measured and calculated input impedance is somewhat fortuitous—the percentage tolerance on the substrate permittivity and patch size can be of the same order of magnitude as the antenna bandwidth (a few percent is typical). This error causes a shift in resonant frequency, while the impedance level is unaffected. Thus, in Figs. 4–6, the substrate permittivity is nominally 2.55, but values of 2.55 to 2.59 were used in the calculations to obtain best results.

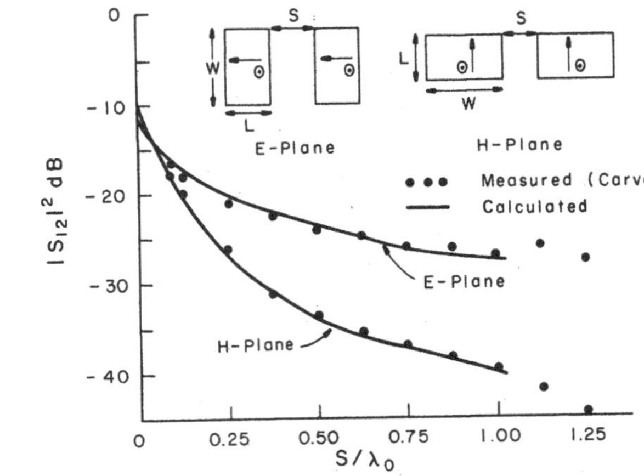

Fig. 8. Measured and calculated mutual coupling between two coax-fed microstrip antennas, for both $\bar{E}$-plane and $\bar{H}$-plane coupling. W = 10.57 cm, L = 6.55 cm, d = 0.1588 cm, ϵ_r = 2.55, f = 1410. MHz.

IV. CONCLUSION

This paper has presented a moment method solution for microstrip antennas using the rigorous grounded dielectric slab Green's function. Both entire base and piecewise sinusoidal modes have been used, and their relative advantages and disadvantages noted. Good agreement between calculated and measured values of input impedance and mutual coupling has been shown.

REFERENCES

[1] K. R. Carver and J. W. Mink, "Microstrip antenna technology," *IEEE Trans. Antennas Propagat.*, vol. AP-29, pp. 2–24, Jan. 1981.

[2] R. J. Mailloux, J. F. McIlvenna, and N. P. Kernweis, "Microstrip array technology," *IEEE Trans. Antennas Propagat.*, vol. AP-29, pp. 25–37, Jan. 1981.

[3] A. G. Derneryd, "Linearly polarized microstrip antennas," *IEEE Trans. Antennas Propagat.*, vol. AP-24, pp. 846–851, Nov. 1976.

[4] Y. T. Lo, D. Solomon, and W. F. Richards, "Theory and experiment on microstrip antennas," *IEEE Trans. Antennas Propagat.*, vol. AP-27, pp. 137–145, Mar. 1979.

[5] K. R. Carver and E. L. Coffey, "Theoretical investigations of the microstrip antenna," Univ. New Mexico Phys. Sci. Lab. Rep. PT-00929, Jan. 1979.

[6] E. H. Newman and P. Tulyathan, "Analysis of microstrip antennas using moment methods," *IEEE Trans. Antennas Propagat.*, vol. AP-29, pp. 47–53, Jan. 1981.

[7] N. K. Uzunoglu, N. G. Alexopoulos, and J. G. Fikioris, "Radiation properties of microstrip dipoles," *IEEE Trans. Antennas Propagat.*, vol. AP-27, pp. 853–858, Nov. 1979.

[8] I. E. Rana and N. G. Alexopoulos, "Current distribution and input impedance of printed dipoles," *IEEE Trans. Antennas Propagat.*, vol. AP-29, no. 1, pp. 99–105, Jan. 1981.

[9] N. G. Alexopoulos and I. E. Rana, "Mutual impedance computation between printed dipoles," *IEEE Trans. Antennas Propagat.*, vol. AP-29, no. 1, pp. 106–111, Jan. 1981.

[10] M. C. Bailey, "Analysis of the properties of microstrip antennas using strips embedded in a grounded dielectric slab," in *IEEE 1979 AP-S Int. Symp. Digest*, Seattle, WA, pp. 373–379.

[11] D. M. Pozar and E. H. Newman, "Analysis of a monopole mounted near or at the edge of a half-plane," *IEEE Trans. Antennas Propagat.*, vol. AP-29, pp. 488–495, May 1981.

[12] M. C. Bailey and M. D. Deshpande, "Resonant frequency of rectangular microstrip antennas," in *IEEE 1981 AP-S Int. Symp. Digest*, Los Angeles, CA, pp. 3–6.

[13] Y. T. Lo, D. D. Harrison, D. Solomon, G. A. Deschamps, and F. R. Ore, "Study of microstrip antennas, microstrip phased arrays, and microstrip feed networks," RADC Tech. Rep. TR-77-406, Oct. 21, 1977.

[14] R. P. Jedlicka, M. T. Poe, and K. R. Carver, "Measured mutual coupling between microstrip antennas," *IEEE Trans. Antennas Propagat.*, vol. AP-29, pp. 147–149, Jan. 1981.

Phased Array Simulation Using Circular Patch Radiators

KLAUS SOLBACH, SENIOR MEMBER, IEEE

Abstract—The circuit properties of a circular patch with center short-circuit pin is investigated in a rectangular grid array. It is found that feed probe direct radiation produces large variations in element-to-element coupling coefficients in the *E*-plane. For the investigated pairwise arrangement of feed probe positions the additional effect is unsymmetric *E*-plane coupling with respect to the direction toward the coupled element; this leads to unsymmetric *E*-plane scanning-induced impedance variation. The principal method of investigation is the small-array method using an 8 × 8 element array, but two waveguide simulators are used to check the results. It is concluded that the small array calculation provides reliable results for scanning angles greater than 60° off broadside. Active element match is observed to vary from −8.5 dB at broadside to about −20 dB on a 60° scanning cone. Finally, a direct measurement of the dissipation losses of the patch radiator is presented using a waveguide simulator as a two-port network. At 5.5 GHz a −0.3 dB dissipation loss is found.

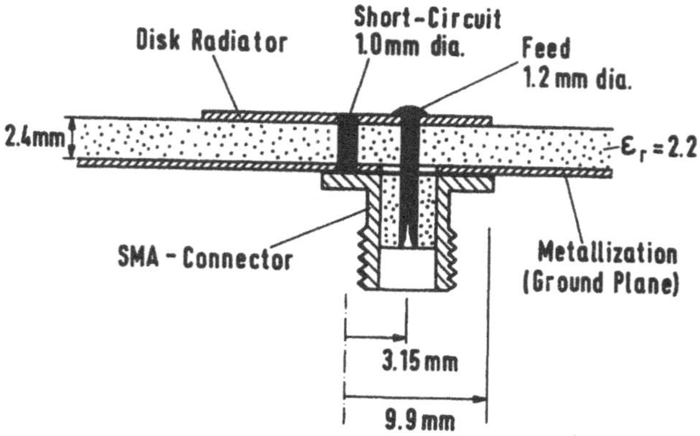

Fig. 1. Cross section of circular patch radiator.

I. Introduction

During the last few years, microstrip patch radiators have received increasing interest as candidates for planar or conformal phased array antennas. Already some complete antennas have been built [1], [2] but there is still some critical lack of theoretical/experimental knowledge concerning the fundamental properties of the patch radiators in the phased array environment. There have been some reports about experimental studies of the coupling properties of array elements [3], [4] and a very recent theoretical treatment of the infinite rectangular patch radiator array [5].

The present paper is intended to supplement the results of the cited papers and to enhance the understanding of the properties of the circular patch radiator in a large phased array.

II. Characteristics of Isolated Circular Patch Radiator

For a practical phased array development a circular patch radiator was chosen as shown in Fig. 1.

The patch has a through-metallization center short circuit mainly to provide a bias ground for the phase shifter connected to the radiator terminals at the back of the element. In the radiator design for 5.5 GHz the dielectric substrate (RT/Duroid 5880) was chosen to have a thickness of 2.4 mm. This choice initially was intended to be a safe compromise between large bandwidth (thick substrate) and high immunity to "blind spot" effects from surface modes on the dielectric substrate (thinn substrate). Later, by interpolating results from [5] (Figs. 3 and 4), and also from experiences with an experimental phased array on the basis of the present element, it was learned that the choice was sound in that blind spots do appear only far outside the intended 60° scanning cone.

With the remaining dimensions chosen as indicated in Fig. 1, the element is well matched to the coaxial line impedance at 5.5 GHz and the measured reflection coefficient exhibits a single pole resonator behavior as presented in Fig. 2. Bandwidth (voltage standing wave ratio VSWR = 2) was measured at about 5 percent slightly increasing with a smaller diameter short circuit. In the phased array the elements are fed via a reactance compensation transformer section increasing the resultant bandwidth to far above 10 percent. The transformer employs a near quarter-wave high-impedance microstrip line etched on a substrate bonded to the back of the radiator plate and separated from the front face by an aluminum interlayer.

III. Patch Radiator Array (Small Array)

Mainly for reasons of wide angle matching a rectangular grid of near half-wavelength spacing at 5.5 GHz center frequency ($A = 0.484\ \lambda$) was chosen for the patch array. As it was intended to combine at least two elements in one phase shifter module it was decided to group pairs of radiators with their feed probes face to face, Fig. 3. This is an unusual arrangement producing unsymmetric coupling of radiators in the *E*-plane as compared to symmetric behaviour in the usual arrangement of equal feed probe positions (e.g. [4]).

The basic method employed in our work for the simulation of a full-size phased array was the small-array method, rather than extensive use of waveguide simulators or use of the array element radiation pattern. The small-array method employs

Reprinted from *IEEE Trans. Antennas and Propaga.*, vol. AP-34, no. 8, pp. 1053–1058, Aug. 1986.

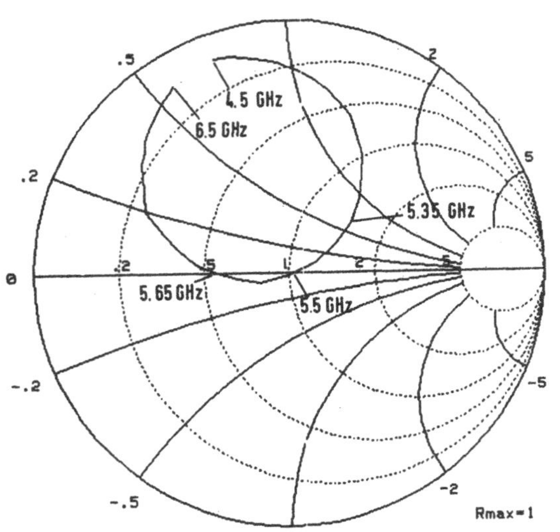

Fig. 2. Measured reflexion coefficient of isolated circular patch radiator.

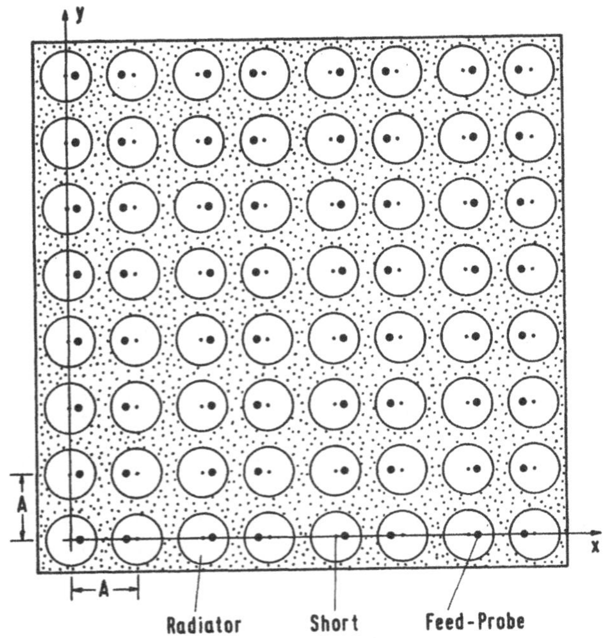

Fig. 3. Small array of circular patch radiators.

measured coupling coefficients of elements in a small (8 × 8 elements) experimental array to calculate the active element reflection coefficient or terminal impedance. Two simple waveguide simulators were also employed for the purpose of verification of the calculated results at two scan directions.

A. Measurements of Coupling

Coupling of elements in the array, Fig. 3, was determined by measuring the transmission scattering coefficient between a reference element and all remaining elements in the array. Choosing a reference element situated right in the center of the small array should yield the most realistic results for the simulation of a large array but in that case measurements could only cover three or four elements in x- and y-directions. When an element near the array edges is chosen as the reference element, six or seven elements can be covered. In a comparison of the first coupling coefficients measured with the reference element situated either in the center or on the outer column of the small array it was found that discrepancies in amplitude were well within some 2 to 3 dB and in phase were within 10° to 20° for coupling coefficients in the −20 to −30 dB range.

Thus for the experimental 8 × 8 element array, useful coupling data could be measured for spacings conforming to a much larger array than actually available. There may be, however, a cumulative effect of the measurement discrepancies on the calculated active-element match, but this has not been investigated. Due to the coupling environment the resonant frequency of the patch radiators in the small array exhibits slight deviations from the isolated radiator situation of less than 1 percent, depending on the position in the array. This variation is, however, not significant since repeatability of isolated element characteristics was found to vary at the same level. Element match, on the other hand, was changed significantly in the array environment with reflection coefficients (all other element match terminated) varying from −21 to −26 dB, depending on the position in the array.

Both effects are not considered to seriously degrade the usefulness of the measured coupling data. Results of the measurements are plotted in Figs. 4 and 5. H-plane coupling, Fig. 4, is found to decay steeply with element spacing and exhibits a high degree of symmetry with respect to the direction of element spacing. Coupling phase practically alternates from one element to the next, which is due to the half-wavelength spacing of elements. Due to the limited range of the network analyzer and errors introduced through reflections from absorbing walls surrounding the small array, measurement accuracy is decreasing rapidly for transmission coefficients below 40 dB. Thus H-plane coupling could only be evaluated for element spacings up to $l = 4 \cdots 5$. The situation is very much different in the E-plane coupling case. Here, coupling decays weaker and the coupling value over element spacing dependence exhibits large variations in amplitude and phase. It is very important to note that E-plane coupling is very unsymmetric with respect to the direction (left or right) of coupling. This effect can be attributed to the pair-wise arrangement of the element feed probe positions in the array. It is a strong indication of direct radiation from the feed probes interfering with the ordinary radiation from the element, effectively increasing or reducing coupling depending on the direction of the coupled element.

The coupling phase for far-field coupled elements (elements spaced more than one wavelength) tends to be constant from one element to the next. This also is due to the pair-wise alternating positions of the feed probes since every change in feed probe position is equivalent to a phase reversal of the element excitation, thus countering the alternating coupling phase due to the half-wavelength array spacing. The irregular phase and amplitude behavior exhibited by the first two elements to both sides of the reference element is an indication of strong reactive (near-field) coupling superposing the regular radiation (far-field) coupling.

B. Active Element Match

It is well known that a radiator (active element) in a large phased array can exhibit large variations of impedance match

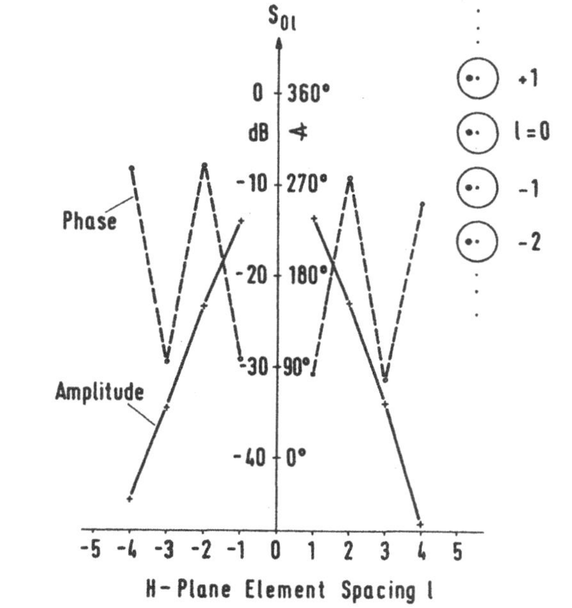

Fig. 4. *H*-plane coupling of elements in the small array.

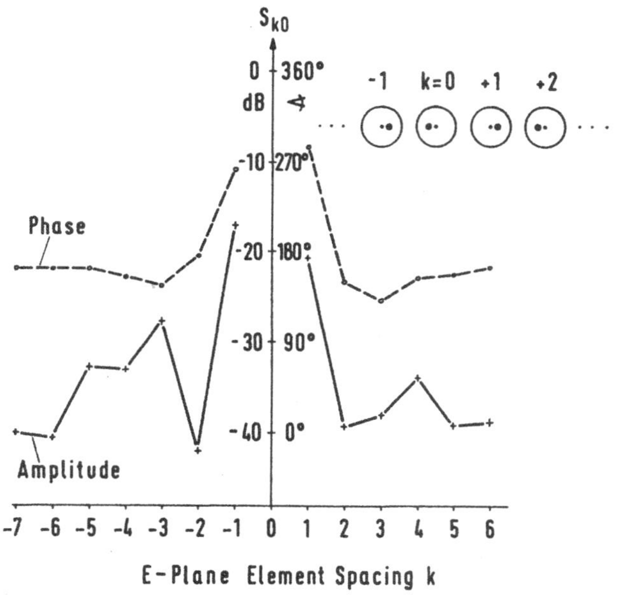

Fig. 5. *E*-plane coupling of elements in the small array.

with beam scan position due to coupling to the remaining radiators of the array. Theoretically, the transmitted (coupled) waves from all elements superimpose to result in the reflected wave from the active element. The reflection coefficient of the active element thus can be calculated as

$$r(\theta, \phi) = \sum_{k=-n}^{n} \sum_{l=-m}^{m} s_{kl} \cdot \exp(-k_0 A \cdot \{k \sin\theta \cos\phi + l \sin\theta \sin\phi\}) \quad (1)$$

where s_{kl} is the transmission scattering coefficient measured for an element at a distance $k \cdot A$ in x-direction and $l \cdot A$ in y-direction, s_{00} is the reflection coefficient of the radiator with all other elements matched terminated and k_0 is the free space wavenumber. For the angles θ and ϕ the conventional spherical coordinate system is employed with $\theta = 0°$ conforming to the z-axis direction and $\phi = 0°$ conforming to the x-axis direction (E-plane) in the Cartesian system of Fig. 3.

The summation limits n and m in full scale phased arrays usually take the order of several thousands but in our simulation can only go up to six or seven in the E-plane and four to five in the H-plane due to the limited range of measured coupling data. It is thus important to look first at the convergence behavior of the calculated active element reflection coefficient.

In Fig. 6, the calculated reflection coefficient is plotted for several scan directions and with the number of elements included in the summation varying from zero to $m = n = 5$ (quadratic array of coupled elements). It is seen that for a broadside beam ($\theta = 0°$) convergence is achieved with fewer elements than for a beam scanned to a $\theta = 60°$ position.

Furthermore, for scan angles $\theta > 60° \cdots 70°$ sufficient convergence can no longer be achieved with the available number of coupling terms as is demonstrated for $\theta = \phi = 90°$. For this particular case the reflection coefficient as calculated from a rigorous theoretical model [5] should approach the unit circle ($|r| \approx 1$) but in the limited small array summation the absolute value of the calculated reflection coefficient falls far below that value.

Thus, it can be concluded that in the present small array simulation method realistic results representing the properties of large phased array antennas are indeed achieved for scanning volumes usually required of such antennas (i.e., $\theta \leq 60°$).

In Figs. 7 and 8, the active element reflection coefficient resulting from (1) with $m = n = 5$ is plotted for H-plane ($\phi = \pm 90°$) and E-plane ($\phi = 0°, 180°$) scan. It is noted that for broadside scan ($\theta = 0°$) the active element is heavily mismatched. Only for $\theta = 50° - 60°$ does the element reflection coefficient nearly vanish in both planes, and for scan angles larger than $\theta = 60°$ increases rapidly. The good match of the active element at the outer limits of a 60° scanning cone had earlier been found for a similar radiator element in a triangular array [2].

In a phased array, however, the active element impedance may be matched at an arbitrary scan angle using, e.g., a quarter-wave transformer section between radiator element and phase shifter.

It is interesting to note that for the E-plane scan two different plots result for the two directions of scan, $\phi = 0°$ and $\phi = 180°$. This fact is due to the unsymmetric behavior of the E-plane element coupling terms versus the direction of coupling. In a phased array scanned in the E-plane this means that half the elements exhibit the impedance variation with scan angle as plotted and the other half of the elements has the corresponding scan directions interchanged. Thus, due to the alternating feed probe position in the array design, we find two different active element terminal impedances in the array for scan positions other than pure H-plane scan.

IV. Waveguide Simulator Measurements

Waveguide simulator measurements originally were performed to serve as a reference for the small array calculations.

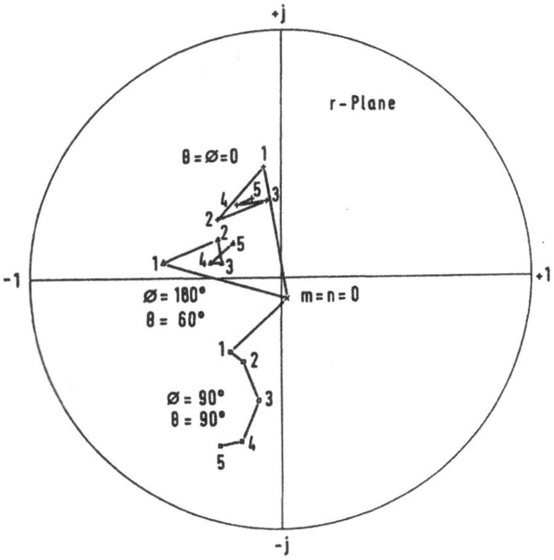

Fig. 6. Convergence behavior of calculated active element reflexion coefficient for various scan directions.

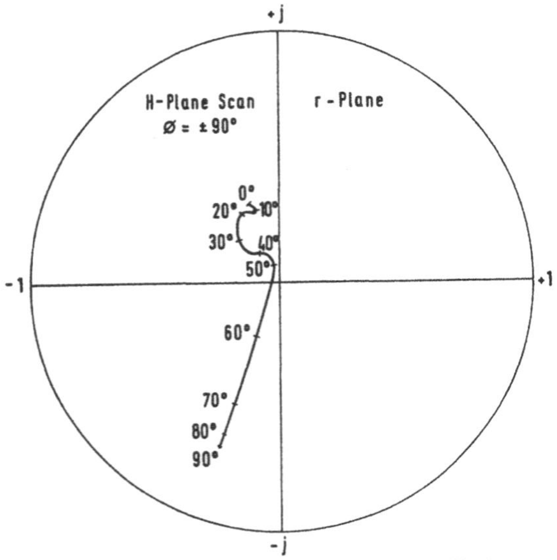

Fig. 7. Active element reflexion coefficient for *H*-plane scan.

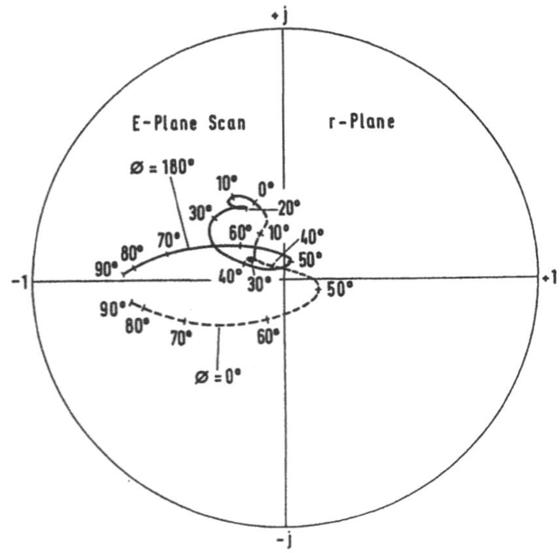

Fig. 8. Active element reflexion coefficient for *E*-plane scan.

As a byproduct, with one of the simulators a direct measurement of the radiator efficiency (dissipation losses) was performed.

A. Waveguide Simulator Impedance Measurements

Only two simulators were built, Fig. 9, both for *H*-plane scan, one simulator employing five complete elements, corresponding to $\theta = 12°$. The other simulator employs only one complete element and one "half-patch"-element carrying zero feed current, corresponding to $\theta = 44°$. With the five-element simulator, measurements were made only "looking in," i.e., at the waveguide port, while all radiator feed terminals were match terminated. With the one-and-a-half-element simulator, measurements were made at both the active element feed connector ("looking out") and at the waveguide port ("looking in"). Measured absolute values of the active element reflection coefficient for both simulators are plotted in Fig. 10 versus frequency together with the corresponding results from the small array calculation. It is seen that agreement is satisfactory in both cases, with a slight frequency shift of the calculated curves against the waveguide simulator results. Above 5.67 GHz the waveguide simulator height exceeded one half-wavelength and the TE_{01}-waveguide mode should be excited by the unsymmetric patch radiators. There was, however, no indication of this occuring in the simulators within the observed frequency range up to 5.7 GHz.

B. Measurement of Dissipation Loss of Active Element

In a large planar phased array the broadside gain of the array is ruled by the radiation aperture area rather than the element gain. Element gain, however, is strongly connected with the phased array scanning properties in that high element gain produces strong impedance mismatch for large scan angles.

On the other hand, for the array gain, dissipation losses in the radiators are of great importance, but such specifications are not readily available for many types of radiators including the present patch element. Losses in high-gain antennas can be measured to a good approximation by comparing the antenna gain (using a standard gain horn as a reference) with the antenna directivity using *E*-plane and *H*-plane half-power beamwidths. Directivity formulas employing only the beamwidth data are valid only for antennas exhibiting well defined radiation patterns, a condition usually true for large aperture antennas and arrays of high gain. On the other hand, the usual method of deriving the dissipation losses of an isolated low-gain radiator is to integrate the gain-pattern of the element across the complete enclosing sphere. Losses are then derived from the difference in the total radiated power (result of integration) and the input power to the radiator.

Since it is very difficult to perform this measurement with the required accuracy, to the knowledge of the author, no one has yet published such a measurement for the patch-radiator or a similar element. The dissipation losses of an element in an infinite phased array may be measured in a completely different manner, alleviating the need for gain-pattern measurements. In the proposed method a waveguide simulator is used to provide a two-port network with the radiator element

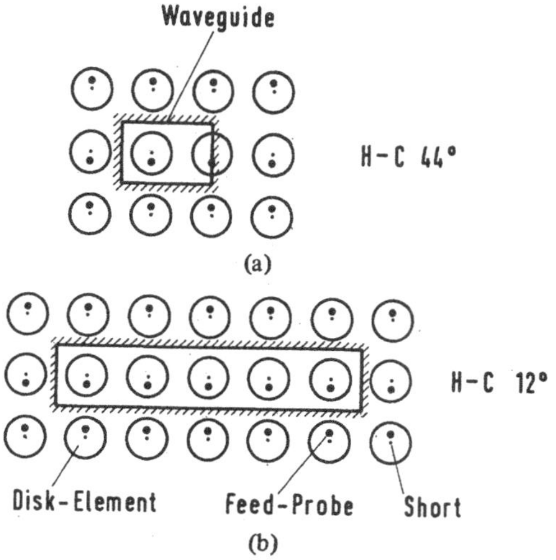

Fig. 9. Configuration of two waveguide simulators. (a) H-plane scan 44°, one element fed. (b) H-plane scan 12°, five elements fed.

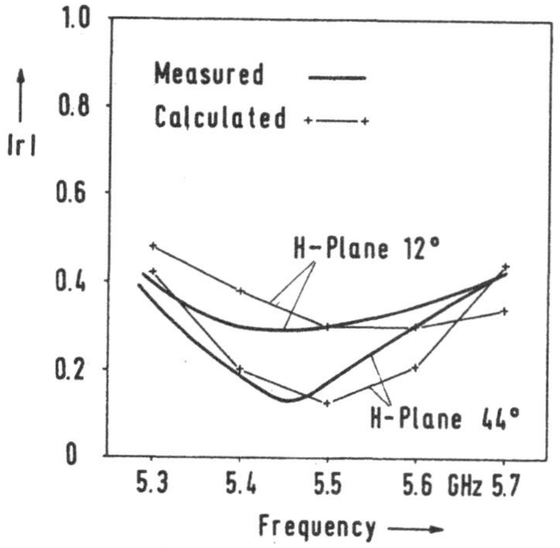

Fig. 10. Absolute values of the measured reflection coefficient for both waveguide simulators as a function of frequency together with calculated results from the small array method.

terminals as one port and the waveguide representing the radiation port of the element. In our specific problem the one-and-a-half-element simulator, Fig. 9(a), was employed and the complete scattering matrix was measured with an automatic network analyzer. From the measured reflection and transmission coefficients the element efficiency η (dissipated power over incident power) valid for the simulated scan angle is calculated from

$$\eta = 1 - |S_{11}|^2 - |S_{21}|^2, \tag{2}$$

where port 1 was the waveguide port and port 2 the radiator port.

As the measurement accuracy of the employed automatic network analyzer, due to several calibration inaccuracies, is limited to about ± 0.1 dB several frequency points were averaged to yield $\eta = -0.3$ dB. This result was checked in a further measurement where the reflection coefficient "looking in" was measured with the radiator coaxial output terminated in either a shielded open or a short circuit. From these two measurements two additional values for $|s_{12} \cdot s_{21}|$ were calculated and inserted in place of $|s_{21}|^2$ in (2). The element efficiency resulting was $\eta = -0.29$ dB.

Two remarks concerning the proposed method are in order. Firstly, the basic method is not restricted to a single-element simulator as used in the presented experiment simulating relatively large scanning angles. Rather, with an N-element simulator, as, e.g., the five-element ($N = 5$) simulator in Fig. 9(b), it would be necessary to measure the transmission scattering coefficients for each of the N-elements s_{i1}, where $i = 2, 3, 4, \cdots, N + 1$, and calculate the element efficiency from

$$\eta = 1 - |S_{11}|^2 - \sum_{i=2}^{N+1} |S_{i1}|^2. \tag{4}$$

Secondly, it has to be cautioned that the simulated situation for which the element efficiency is given correctly, is not exactly the same physical situation occuring in a phased array antenna. Rather, in a waveguide simulator two partial plane waves (H-plane scan) are incident to the array, creating a current distribution on the radiator patch different from the one obtained if only a single wave is incident. Concerning the reflexion coefficient there is no difference for the two partial waves (identical angle θ) in an H-plane scan simulator. In an E-plane (or intermediate plane) scan simulator the partial waves incident to the array experience different reflexion coefficients as shown in Fig. 8, thus the simulation only yields some average reflexion coefficient for the involved angles of incidence.

If losses in the radiator element are concerned, even in the H-plane scan simulator situation the result is only approximate to the result for a single incident wave. The difference in losses appears since losses depend on the squares of field intensities and currents, rather than on the field quantities directly as in the case of the reflection coefficient (the superposition theorem does not hold for power quantities). The resulting discrepancies for the effective element efficiency are supposed to be quite small, however, at least for angles of incidence not too far from broadside ($\theta = 0°$).

V. Conclusion

It has been shown that practical circular patch radiators may exhibit unusual scanning properties in a planar phased array configuration. In the investigated array with pairwise symmetric feed probe configuration considerable unsymmetric E-plane coupling is observed, which is due to the direct radiation contributions of feed probes. As a result, active elements (in a phased array) exhibit two distinct terminal impedances versus E-plane scan angle, depending on the direction of the scanned beam. The basic effect is present, although less distinct, in all scan directions other than pure H-plane directions. The effect will be a periodic error in the phase/amplitude taper of the phased array antenna inducing degradations of the array radiation pattern (increased sidelobe level).

As a byproduct of the waveguide simulator development a direct measurement of the radiator element dissipation losses is performed, yielding -0.3 dB efficiency for the circular patch element at 5.5 GHz. This type of measurement may be of increasing value as more and more printed, planar array elements come in use, since these have to be assumed to be lossy rather than practically lossless as in the case of conventional waveguide aperture elements.

References

[1] F. W. Cipola, "A 7.5 GHz microstrip phased array for aircraft-to-satellite communication," *IEEE Trans. Antennas Propagat.*, vol. AP-29, no. 1, pp. 166–170, Jan. 1981.

[2] C. A. Chuang, D. J. Martin, and N. Moldovan, "Miniaturization techniques benefit conformal arrays," *Microwaves and Radio Frequency*, pp. 87–96, Mar. 1984.

[3] F. G. Parks and M. C. Bailey, "Mutual coupling between microstrip disk antennas," in *Dig. Int. Symp. Antennas Propagat. Soc.*, Amherst, MA, Oct. 1976, pp. 399–402.

[4] R. P. Jedlicka, M. T. Poe, and K. R. Carver, "Measured mutual coupling between microstrip antennas," *IEEE Trans. Antennas Propagat.*, vol. AP-29, no. 1, pp. 147–149, Jan. 1981.

[5] D. M. Pozar and D. H. Schaubert, "Analysis of an infinite array of rectangular microstrip patches with idealized probe feeds," *IEEE Trans. Antennas Propagat.*, vol. AP-32, no. 10, pp. 1101–1107, Oct. 1984.

Finite Phased Arrays of Rectangular Microstrip Patches

DAVID M. POZAR, MEMBER, IEEE

Abstract—**Finite phased arrays of rectangular microstrip patch antennas are analyzed. Reflection coefficient magnitudes, element patterns and efficiency (based on power lost to surface waves) are calculated for various sized arrays on substrates of practical interest and are compared with previous infinite array solutions. Measured element patterns and mutual coupling data for a small array are presented and compared with calculations.**

I. Introduction

PHASED ARRAYS of printed antennas are of considerable current interest, particularly in the context of millimeter-wave monolithic arrays, where active devices for phase shifters and amplifiers are integrated on the same substrate as the radiating elements. Such systems potentially offer high reliability and low cost, as well as high performance. The microstrip patch antenna is well suited to such integrated arrays, since it has low gain, is compatible with monolithic fabrication, and can be fed in a variety of ways [1], [2].

Isolated microstrip patch and printed dipole elements have been studied by a number of authors [3], [4], [5], and infinite arrays of such elements have been considered more recently [6], [7], [8]. The infinite array approximation generally models the central elements of a large but finite array quite well and implicitly includes the effect of mutual coupling between elements. Infinite array solutions yield the active impedance variation of the array with scan angle, and are typically no more complex than the corresponding solution (spectral transform) for a single element. The infinite array solution, however, does not account for edge effects of a finite array, and it is not generally known how large a finite array must be before it can be adequately modeled as infinite. Thus, the study of finite phased arrays may yield more practical information than the study of infinite arrays.

Finite arrays, however, are considerably more difficult to analyze than either single elements or infinite arrays. Using the "element-by-element" approach, the mutual coupling between each pair of elements in the array must be calculated, and matrices of order equal to the number of elements in the array (or larger, if there is more than one expansion mode per element) must be inverted. The size of planar arrays that can be handled by this method is thus quite limited.

Finite arrays of printed dipoles have been recently treated [10], and the present solution for rectangular microstrip patches follows that work. In [10], it was shown that finite arrays of printed dipoles excite surface waves on the grounded dielectric substrate. This surface wave power is related to scan blindness, as will be seen, and is generally considered to be detrimental to the operation of a phased array (due to loss, diffraction, spurious coupling, etc.), and so it is desirable to predict surface wave excitation versus scan angle and array size. An infinite array analysis does not provide this information, as an infinite array cannot support surface waves, except perhaps at a scan blindness angle.

Section II presents the theory of the solution for finite rectangular patch arrays. The self- and mutual impedances of the patch elements are computed by the spectral domain moment method of [3]. Theory is developed for the active input impedance, as well as the active element pattern of the array. Consideration is given to computational efficiency and convergence of the current expansion on the patch.

Section III presents two sets of results for $\epsilon_r = 2.55$ and $\epsilon_r = 12.8$ substrates. Reflection coefficient magnitudes and efficiency based on power lost to surface waves are plotted versus scan angle; active element patterns are also plotted. These results are presented for various sized arrays and compared to infinite array results.

Active element patterns of a 7×7 patch array on a low dielectric constant substrate were measured and are compared with calculations. The S-parameters of some of the elements of this array were also measured and compared to calculated values.

II. Theory

The geometry of a finite microstrip patch array is shown in Fig. 1. The patches are assumed to be resonant in the x-direction and have a length L in this direction. The patch width is W, and the patches are uniformly spaced by distances a in the x-direction and b in the y-direction. Rectangular grids are assumed here, but triangular grids could be handled as well. Each patch is assumed to be probe fed; the feed is modeled as a current source as in [3], [7]. This idealized feed model is not rigorous (e.g., current continuity is not enforced), but has been found to give good results for thin substrates. In addition, as discussed in [7], [11], details of the feed do not greatly affect the dominant scan characteristics of printed phased arrays.

Reprinted from *IEEE Trans. Antennas Propaga.*, vol. AP-34, no. 5, pp. 658–665, May 1986.

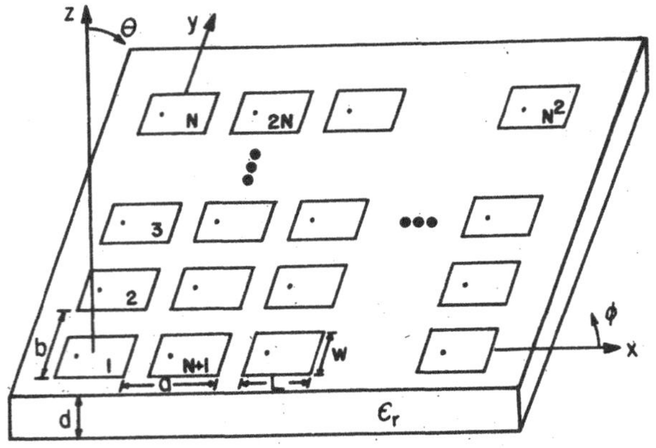

Fig. 1. Geometry of a finite array of rectangular microstrip patches.

A. Moment Method Solution

As discussed in [3] and [10], the mutual impedance between two piecewise sinusoidal (PWS) dipole expansion modes m, n on a grounded dielectric slab can be expressed as

$$Z_{mn}=\frac{jZ_0}{4\pi^2 k_0}\iint_{-\infty}^{\infty} Q_{xx}(k_x,\ k_y)F_x^2(k_x)F_y^2(k_y) \cdot e^{jk_xS_x}e^{jk_yS_y}\ dk_x\ dk_y, \quad (1)$$

where

$$Q_{xx}(k_x,\ k_y)=\frac{(\epsilon_r k_0^2-k_x^2)k_2\cos k_1d+jk_1(k_0^2-k_x^2)\sin k_1d}{T_eT_m}\sin k_1d; \quad (2)$$

$$T_e=k_1\cos k_1d+jk_2\sin k_1d \quad (3)$$

$$T_m=\epsilon_r k_2\cos k_1d+jk_1\sin k_1d \quad (4)$$

$$k_1^2=\epsilon_r k_0^2-\beta^2,\ \text{Im}\ (k_1)<0 \quad (5)$$

$$k_2^2=k_0^2-\beta^2,\ \text{Im}\ (k_2)<0 \quad (6)$$

$$\beta^2=k_x^2+k_y^2 \quad (7)$$

$$k_0=\omega\sqrt{\mu_0\epsilon_0}=2\pi/\lambda_0 \quad (8)$$

$$Z_0=\sqrt{\mu_0/\epsilon_0}. \quad (9)$$

In addition, F_x and F_y represent the Fourier transforms of the x and y dependencies of the PWS expansion current $J(x, y)$:

$$J(x,\ y)=\frac{\sin k_e(h-|x|)}{W\sin k_eh},\ \text{for}\ |x|<h,\ |y|<W/2, \quad (10)$$

$$F_x(k_x)=\frac{2k_e(\cos k_xh-\cos k_eh)}{\sin k_eh(k_e^2-k_x^2)}, \quad (11)$$

$$F_y(k_y)=\frac{\sin k_yW/2}{k_yW/2}. \quad (12)$$

In (10)–(12), h is the half-length of the PWS mode, W is the width of the mode, and k_e is the effective wavenumber of the PWS mode. In (1), S_x and S_y are the distances in the x and y directions, respectively, separating the two PWS modes in m and n.

It must be realized that the mutual impedances defined by (1) are not the same as those seen by the probe feeds of the patches. That is, unlike the dipole case [10], there is a distinction between the moment method impedance matrix [Z] and the "port" impedance matrix $[Z^p]$ defined at the probe terminals. The current flow on the patch is related to the voltage excited at the probe terminals by the mode voltage V_0 [3]:

$$V_0=\frac{-jZ_0}{4\pi^2k_0}\iint_{-\infty}^{\infty} Q_{zx}(k_x,\ k_y)F_x(k_x)F_y(k_y) \cdot e^{jk_xx_p}e^{jk_yy_p}\ dk_x\ dk_y, \quad (13)$$

where

$$Q_{zx}(k_x,\ k_y)=\frac{-jk_xk_2\sin k_1d}{k_1T_m}. \quad (14)$$

In (13), x_p, y_p represents the probe position relative to the patch center.

The expressions for Z_{mn} and V_0 in (1) and (13) constitute the bulk of the computational effort for the finite patch array solution, and so it is critical that these terms be calculated as efficiently as possible. As described in [3], the integration variables in (1) and (13) can be converted to polar coordinates, and one of the integrals reduced to a finite range. Most significant for large array problems, however, is to use the technique described in [10] to compute the entire $[Z]$ matrix at once to avoid recalculation of Q_{xx}, F_x, and F_y. This results in orders of magnitude time savings for large arrays. Toeplitz symmetries can also be exploited. Another useful technique, particularly for mutual impedances of widely separated elements, is to subtract the mutual impedance between the same two elements in a homogeneous media having the same asymptotic behavior of the integrand [12]. Techniques such as these are mandatory for the solution of practical array problems.

Up to this point the solution is general enough to handle an arbitrary number of patch elements with an arbitrary number of expansion modes on each patch. Because of limitations on matrix size, however, it is important to use as few expansion modes on each patch as possible. For example, a 17 × 17 finite patch array with three PWS expansion modes on each patch requires a matrix of size 867 × 867, while the same array with one mode per patch requires a more manageable matrix size of 289 × 289.

As in [10], the use of one expansion mode per patch can be justified by a number of arguments. First, results for single patch antennas [3] show that a judicious choice of an expansion mode can lead to good results when only one mode is used. A PWS mode with

$$k_e = k_0 \sqrt{\epsilon_e}, \tag{15}$$

$$\epsilon_e = \frac{\epsilon_r + 1}{2} + \frac{\epsilon_r - 1}{2}\left(1 + \frac{10d}{W}\right)^{-1/2}, \tag{16}$$

has been found to give good results. In addition, the microstrip patch is a highly resonant structure, and near resonance its current distribution can be well approximated by a single (properly chosen) mode. This distribution is not greatly affected by the proximity of nearby elements in an array (assuming the elements are not too closely spaced). The best argument for a single mode approximation, however, is to compare the single mode results with those computed using more than one mode per element. These results are shown in the next section for a specific case, but in general it has been found that reflection coefficients, efficiencies, and element patterns differed by 10 percent or less when the number of expansion modes was increased from one to three.

Thus, most of the results in this paper, particularly for large arrays, were calculated using one expansion mode per patch. In order to simplify the following theory, it is assumed that only one mode is used on each patch.

B. Active Input Impedance and Element Patterns

Let $[Z]$ be the moment method impedance matrix with $N^2 \times N^2$ elements as defined by (1), and let $[Y] = [Z]^{-1}$. Let the index i run through each of the N^2 array elements as ordered in Fig. 1, that is, counting across the H-plane of the array. The column k, and row l, in the array of the ith element can be broken out as

$$k = 1 + (i-1)/N = \text{column in array} \tag{17}$$

$$l = i - (k-1) \cdot N = \text{row in array}. \tag{18}$$

Let $[I^p]$ be the N^2-element column vector of port currents on the probes. For scanning at the angle θ, ϕ, these port currents should be driven as

$$I_i^p = e^{jk_0(kau + lbv)}, \tag{19}$$

for the ith patch, where k, l are defined in (17) and (18), and u, v are direction cosines defined as

$$u = \sin\theta \cos\phi, \tag{20}$$

$$v = \sin\theta \sin\phi. \tag{21}$$

The patch current amplitudes are then given by the column vector $[I]$ as [3],

$$[I] = V_0[Y][I^p]. \tag{22}$$

where V_0 is defined by (13). The active input impedance at the ith patch is then [3],

$$Z_{\text{in}}^i(\theta, \phi) = \frac{-V_0 I_i}{I_i^p}. \tag{23}$$

The reflection coefficient at the ith port can then be calculated as

$$R^i(\theta, \phi) = \frac{Z_{\text{in}}^i(\theta, \phi) - Z_{\text{in}}^i(0, 0)}{Z_{\text{in}}^i(\theta, \phi) + Z_{\text{in}}^{i*}(0, 0)}, \tag{24}$$

which assumes the array is conjugate matched for broadside scan. The radiation efficiency based on power lost to surface waves can be defined as [4]

$$e = 1 - \frac{P_{\text{sw}}}{P_{\text{in}}}, \tag{25}$$

where P_{in} is the total input power to the array, and P_{sw} is the surface wave power. These quantities can be calculated by

$$P_{\text{in}} = \text{Re}\left\{\sum_{m=1}^{N^2}\sum_{n=1}^{N^2} I_m^* Z_{mn} I_n\right\} \tag{26}$$

$$P_{\text{sw}} = \sum_{m=1}^{N^2}\sum_{n=1}^{N^2} I_m^* R_{mn}^{\text{sw}} I_n, \tag{27}$$

where R_{mn}^{sw} is the surface wave contribution to the impedance element Z_{mn} (from the residue of the surface wave pole [3]).

The active element pattern $\bar{E}^j(\theta, \phi)$ of the array is found by exciting the jth element of the array and terminating all other elements. For a finite array, the active element pattern depends on the position of the excited element.

To calculate the active element pattern, reference is made to the multiport network of Fig. 2, which shows the N^2 element array as an N^2-port network. All ports are terminated in an impedance Z_T, and port j is fed with a 1 A current source. The relation between the port currents and voltages is

$$[V^p] = [Z^p][I^p], \tag{28}$$

where the port impedance matrix $[Z^p]$ is found as [3]

$$[Z^p] = [Y^p]^{-1} = -V_0^2[Y]. \tag{29}$$

Using (22), (28), and (29), the patch currents $[I]$ due to a set of port voltages $[V^p]$ can be calculated as

$$[I] = V_0[Y][Y^p][V^p] = \frac{-1}{V_0}[V^p]. \tag{30}$$

Now define $[I^s]$ as the driving current source vector feeding the terminated ports of Fig. 2. Then all elements of $[I^s]$ are zero except for the jth element, which is unity and corresponds to the fed element. The port voltages due to $[I^s]$ are related as

$$\{[Y^p] + [Y^T]\}[V^p] = [I^s], \tag{31}$$

where $[Y^T]$ is a square diagonal matrix with elements $1/Z_T$. Finally, the patch currents due to the 1 A source at element j are, from (30) and (31),

$$[I] = \frac{-1}{V_0}\{[Y^p] + [Y^T]\}^{-1}[I^s]. \tag{32}$$

The element pattern can then be computed as

$$\bar{E}^j(\theta, \phi) = \bar{E}^0(\theta, \phi)\sum_{i=1}^{N^2} I_i e^{jk_0(kau + lbv)}, \tag{33}$$

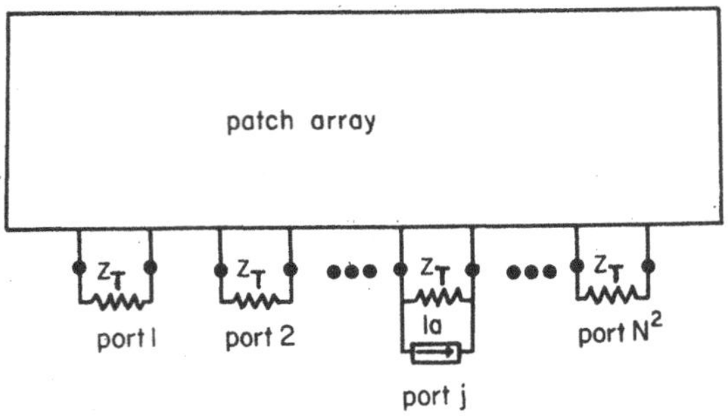

Fig. 2. The N^2-element finite patch array as an N^2-port network.

where k, l are the row, column of element i, as defined in (17), (18), and $\bar{E}^0(\theta, \phi)$ is the pattern of an isolated PWS mode [10]. The active element gain is derived as

$$G^j(\theta, \phi) = \frac{4\pi |\bar{E}^j(\theta, \phi)|^2}{Z_0 P_{in}}. \tag{34}$$

This definition does not include the power lost in the terminating impedance of the fed element.

Once the open-circuit port impedance matrix is found from (29), the S-parameters can be found from

$$[S] = \{[Z^p] - [Z_c]\}\{[Z^p] + [Z_c]\}^{-1}, \tag{35}$$

where $[Z_c]$ is a diagonal matrix with elements Z_c, the characteristic impedance of the connecting transmission lines.

The above theory, derived assuming one expansion mode per patch, remains essentially unchanged, except for indexing, when more expansion modes are used. Since all of the above square matrices are symmetric, storage can be saved by storing only half the matrix, and inversion time saved by using a Cholesky square-root method.

III. Results

Figs. 3–6 show calculated results for finite patch arrays on an $\epsilon_r = 12.8$, $d = 0.06\ \lambda_0$ substrate. The reflection coefficient magnitude versus scan angle is shown in Fig. 3 for the center element ($j = 41$) of a 9×9 planar array, and is compared with the results for an infinite array [7]. Note that even the relatively small 9×9 array results follow closely the infinite array theory, up to scan angles of 70° or 80°. At 90°, the infinite array, having infinite gain and no grating lobes, has unity reflection coefficient magnitude while the finite array can still transmit power, since it has a nonzero beamwidth. Also, note the presence of a scan blindness in the E-plane at 46.1° for the infinite array theory. The reflection coefficient magnitude of the finite array at this angle is actually greater than unity. This merely means that power is being delivered to port 41 from other array elements; total power is conserved.

Fig. 4 shows the efficiency computed from (25) for E-plane scan for various sized arrays. This result is of particular interest because it shows the transition in terms of surface wave power from a single element to a finite array to an infinite array. For an isolated dipole, the efficiency is about 0.283, for any scan angle. For arrays, this efficiency shows a variation with scan angle, and as array size increases the efficiency increases for all scan angles except near $\theta = 46°$,

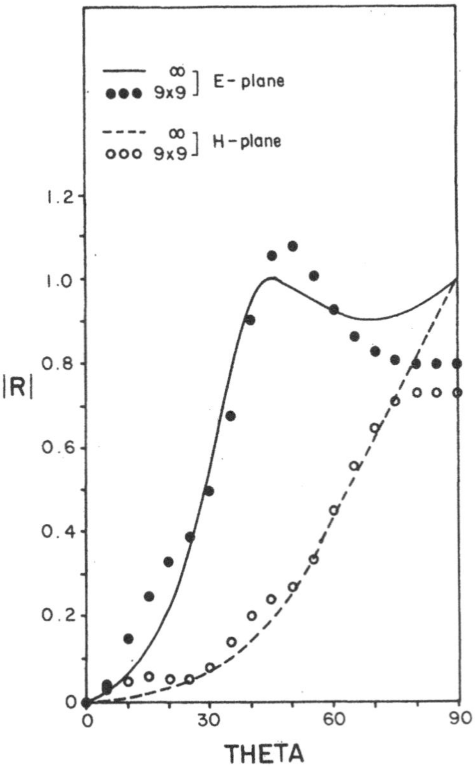

Fig. 3. Reflection coefficient magnitude versus scan angle (E- and H-plane) for a finite (9×9, center element) patch array, compared with infinite array results. $\epsilon_r = 12.8$, $d = 0.06\ \lambda_0$, $a = b = 0.5\ \lambda_0$, $L = 0.1074\ \lambda_0$, $W = 0.15\ \lambda_0$, $X_p = -L/2$, $Y_p = 0$.

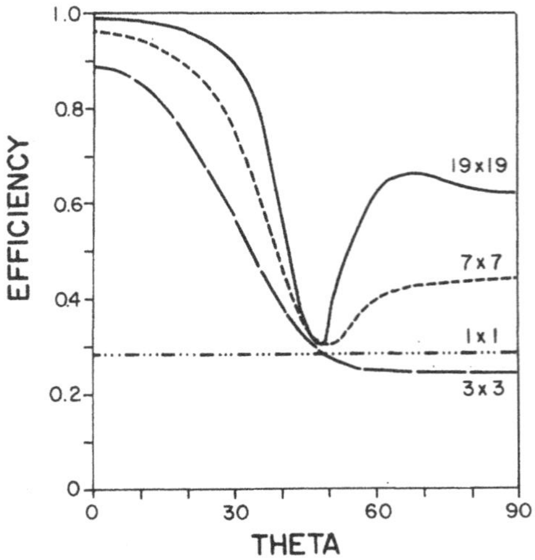

Fig. 4. Efficiency of a finite patch array versus E-plane scan angle for various array sizes. $\epsilon_r = 12.8$, $d = 0.06\ \lambda_0$, $a = b = 0.5\ \lambda_0$, $L = 0.1074\ \lambda_0$, $W = 0.15\ \lambda_0$, $X_p = -L/2$, $Y_p = 0$.

where it tends to decrease. This result shows that destructive interference of the surface wave occurs with increasing array size, except near the blindness angle. In the infinite array, the efficiency would be unity at all scan angles except 46.1°, where it would be zero. Since no blindness occurs in the H-plane scan case, the efficiency there quickly approaches unity with increasing array size for all scan angles. These results are qualitatively similar to the printed dipole case [10].

Figs. 5(a) and 5(b) show the active element patterns of the center element for finite patch arrays of various size, compared with the element pattern of the infinite array. The

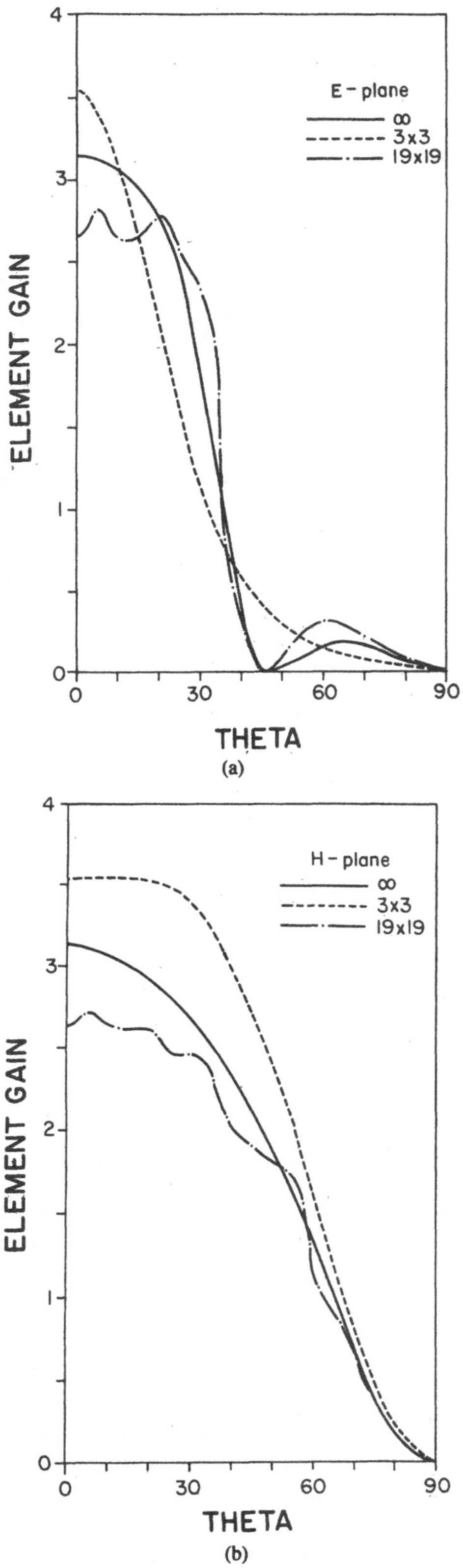

(a)

(b)

Fig. 5. (a) E-plane. (b) H-plane active center element gains for patch arrays of various sizes. $\epsilon_r = 12.8$, $d = 0.06\ \lambda_0$, $a = b = 0.5\ \lambda_0$, $L = 0.1074\ \lambda_0$, $W = 0.15\ \lambda_0$, $X_p = -L/2$, $Y_p = 0$.

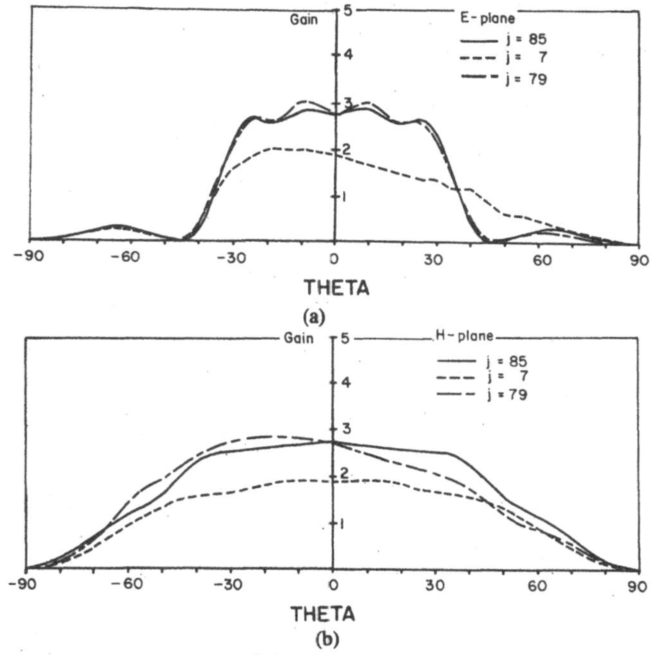

(a)

(b)

Fig. 6. (a) E-plane. (b) H-plane active element gain patterns for various elements of a 13 × 13 patch array. $\epsilon_r = 12.8$, $d = 0.06\ \lambda_0$, $a = b = 0.5\ \lambda_0$, $L = 0.1074\ \lambda_0$, $W = 0.15\ \lambda_0$, $X_p = -L/2$, $Y_p = 0$.

patterns are gain patterns, as computed from (34). These results show that finite array element patterns tend toward the infinite array element patterns in an oscillatory manner as array size increases. The existence of scan blindness at $\theta = 46.1°$ in the E-plane shows up as a null at this angle for the infinite array, and also for the 19 × 19 array. The 3 × 3 case, while not achieving a deep null, also drops considerably by 46°. These patterns and those that follow were made with $Z_T = Z_{11}^*$, where Z_{11} is the impedance of an isolated patch element; the element patterns are not a strong function of the terminating impedance Z_T, however.

Figs. 6(a) and 6(b) show the variation in element patterns for different elements in the array. E- and H-plane patterns are shown for three different fed elements in a 13 × 13 array: $j = 85$ (center element), $j = 7$ (element at edge of E-plane, centered along H-plane), and $j = 79$ (element at edge of H-plane, centered along E-plane). As can be seen, the element patterns can be quite asymmetric in the plane of the element offset. The input impedance and reflection coefficient versus scan angle also varies substantially as the array edge is approached, but space does not permit presentation of these results (see [10] for corresponding data for printed dipoles).

Figs. 7–9 show results for an $\epsilon_r = 2.55$, $d = 0.04\ \lambda_0$ substrate, which is significantly thinner electrically than the previous case, and has its scan blindness angle much closer to endfire. The reflection coefficient magnitude (at the central element) versus scan angle for a 7 × 7 array is shown in Fig. 7, and is compared with infinite array results [7]. Again the finite array result follows the infinite array result reasonably well up to about 70°. The infinite array has a scan blindness in the E-plane at 81.1°.

Fig. 8 shows the efficiency versus E-plane scan angle for finite arrays of various sizes. Since the substrate for this case is electrically thinner than the case of Fig. 4, the efficiency is

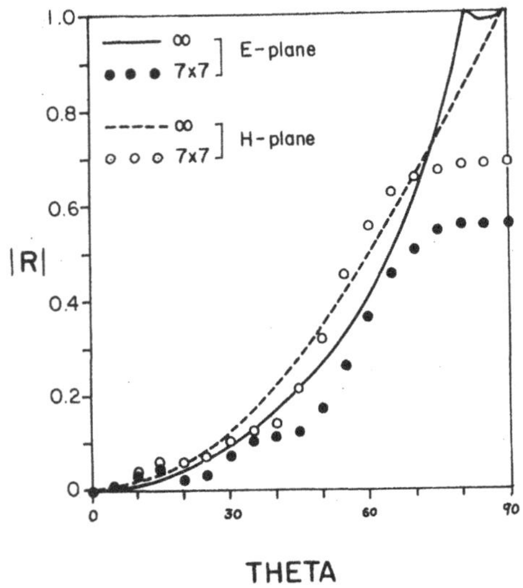

Fig. 7. Reflection coefficient magnitude versus scan angle (E-plane and H-planes) for a finite (7×7, center element) patch array, compared with infinite array results. $\epsilon_r = 2.55$, $d = 0.04\ \lambda_0$, $a = b = 0.5\ \lambda_0$, $L = W = 0.3\ \lambda_0$, $X_p = -L/2$, $Y_p = 0$.

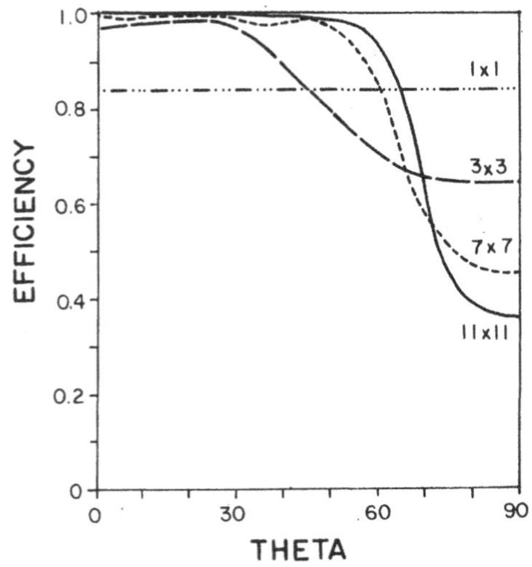

Fig. 8. Efficiency of a finite patch array versus E-plane scan angle for various array sizes. $\epsilon_r = 2.55$, $d = 0.04\ \lambda_0$, $a = b = 0.5\ \lambda_0$, $L = W = 0.3\ \lambda_0$, $X_p = -L/2$, $Y_p = 0$.

higher for a given array size. As before, the efficiency approaches unity with an increase in array size at all scan angles except in the vicinity of the blindness angle, where it approaches zero.

Figs. 9(a) and 9(b) show the E-plane and H-plane active element gain patterns for three different array sizes. Again the finite array patterns tend to converge to the infinite array results in an oscillatory manner. As noted for other phased arrays of printed antennas [6], [7], the use of thinner substrates leads to more desirable scan performance, since the blindness angle is closer to endfire.

Fig. 10 shows the convergence of the reflection coefficient magnitude with an increase in the number of PWS expansion modes on the patch, for a 7×7 array. Note that the results for one mode and three modes generally agree to within a few percent, except for wide scan angles in the H-plane where the disagreement is about 10 percent. Corresponding curves for

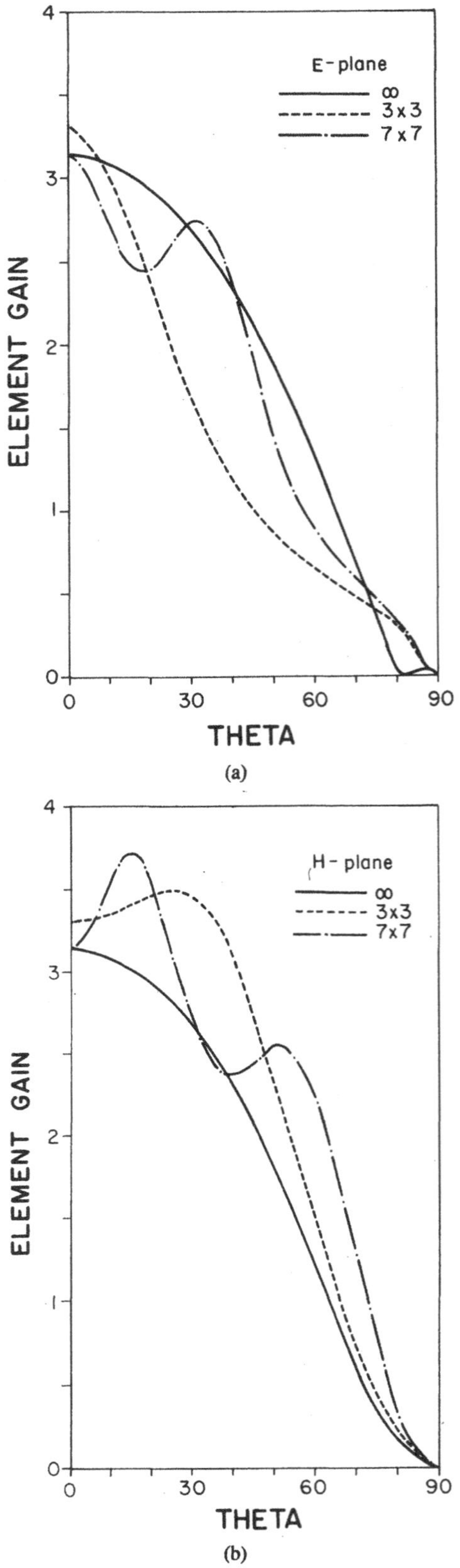

Fig. 9. (a) E-plane. (b) H-plane active center element gain patterns for patch arrays of various sizes. $\epsilon_r = 2.55$, $d = 0.04\ \lambda_0$, $a = b = 0.5\ \lambda_0$, $L = W = 0.3\ \lambda_0$, $X_p = -L/2$, $Y_p = 0$.

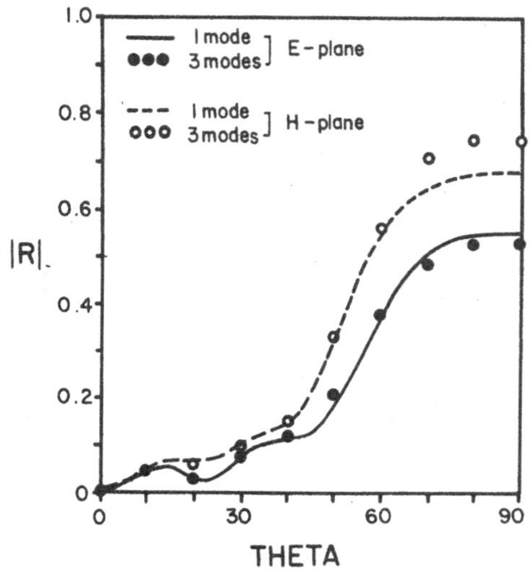

Fig. 10. Convergence of reflection coefficient magnitude with number of piecewise sinusoidal expansion modes. $\epsilon_r = 2.55$, $d = 0.04\ \lambda_0$, $a = b = 0.5\ \lambda_0$, $L = W = 0.3\ \lambda_0$, $X_p = -L/2$, $Y_p = 0$. 7 × 7 array.

TABLE I
MEASURED AND CALCULATED S-PARAMETERS

i,j	S_{ij} Measured	S_{ij} Calculated
25,24	-12.5 dB/-147°	-13.4 dB/-140°
25,26	-12.5 dB/-145°	-13.4 dB/-140°
25,23	-21.0 dB/44°	-21.5 dB/59°
25,27	-21.0 dB/49°	-21.5 dB/59°
24,26	-21.5 dB/45°	-21.8 dB/57°
25,18	-24.5 dB/113°	-26.0 dB/122°
25,32	-25.0 dB/112°	-26.0 dB/122°
25,11	-29.5 dB/-113°	-29.8 dB/-105°
25,39	-30.0 dB/-128°	-29.8 dB/-105°
18,32	-30.0 dB/-115°	-29.8 dB/-105°

7 × 7 element array; $\epsilon_r = 2.55$; $d = 0.16$ cm; $a = b = 3.45$ cm; $L = 2.0$ cm, $W = 3.0$ cm; $X_p = 0.55$ cm, $Y_p = 0$; $f = 4.35$ GHz.

efficiency and element patterns (being far-field quantities) are indistinguishable for the one- and three-mode cases. This data substantiates the single mode approximation.

To verify the above theory, measurements were made on a 7 × 7 patch array on a thin substrate. With all ports terminated in 50 Ω, S-parameters were measured and calculated for a variety of element pairs at different locations in the array; typical data are shown in Table I. The elements are numbered across the H-plane, as in Fig. 1. Because of the difficulty in obtaining an accurate phase reference, it is estimated that the measured phase data in the Table I is in error by about 10°.

Element patterns were measured for the above array by terminating all elements but the center element. Figs. 11(a) and 11(b) show the measured and calculated patterns. Agreement is generally within a dB or so, although it is clear that the measurements are not much better than this, since asymmetries of this order are evident in the measured patterns. Both E- and H-plane patterns are close to the ideal cos θ case, so such an array should have good scanning performance. The ground plane size was 35.7 cm (E) by 31.7 cm (H).

IV. Conclusion

This paper has presented an analysis of finite arrays of rectangular microstrip patch antennas, in terms of scan performance, surface wave excitation, and active element patterns. Data were presented for arrays on substrates of practical interest, and measurements were made on a 7 × 7 patch array. From the above results, it can be concluded that the scan performance of the center element of a finite array of even moderate size (7 × 7 or larger) can be well approximated by the infinite array solution, although this performance may be somewhat different for elements near the edge of the array. The analysis of finite arrays shows that even small arrays (3 × 3 or 7 × 7) can have dips in the element pattern where the infinite array has a scan blindness, and that this effect becomes more pronounced as array size increases.

The role of surface waves in finite and infinite arrays is clearly demonstrated by finite array analysis, where it is shown that surface wave power diminishes with increasing array size at all scan angles except the scan blindness angle, where it becomes predominant.

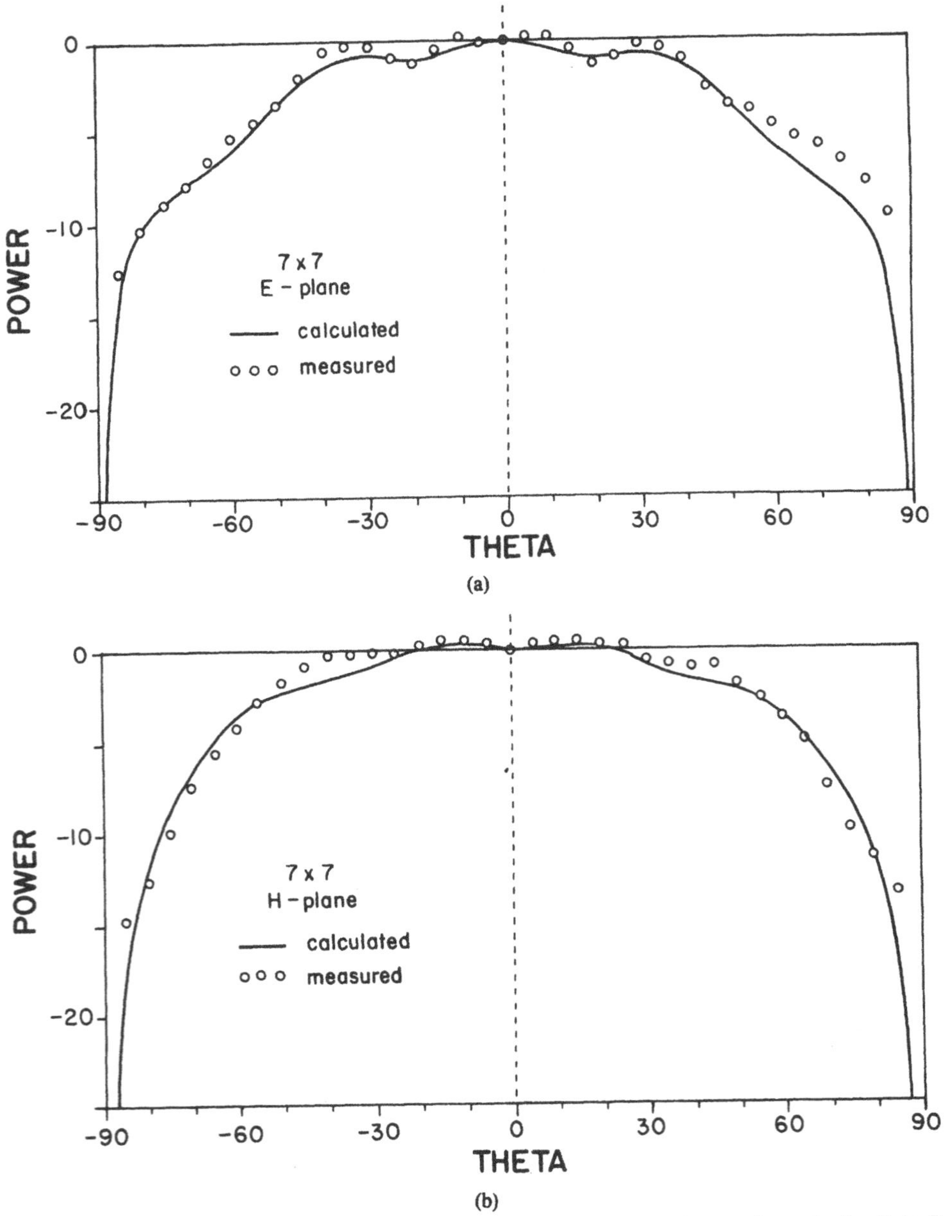

Fig. 11. (a) *E*-plane. (b) *H*-plane measured and calculated active element patterns (center element) of a 7 × 7 patch array. $\epsilon_r = 2.55$, $d = 0.16$ cm, $a = b = 3.45$ cm, $L = 2.0$ cm, $W = 3.0$ cm, $X_p = 0.55$ cm, $Y_p = 0$, $f = 4.35$ GHz.

References

[1] R. J. Mailloux, J. F. McIlvenna, and N. P. Kernweis, "Microstrip array technology," *IEEE Trans. Antennas Propagat.*, vol. AP-29, pp. 25–37, Jan. 1981.

[2] D. M. Pozar, "A microstrip antenna aperture coupled to a microstrip line." *Electron. Lett.*, vol. 21, pp. 49–50, Jan. 17, 1985.

[3] ——, "Input impedance and mutual coupling of rectangular microstrip antennas," *IEEE Trans. Antennas Propagat.*, vol. AP-30, pp. 1191–1196, Nov. 1982.

[4] ——, "Considerations for millimeter wave printed antennas," *IEEE Trans. Antennas Propagat.*, vol. AP-31, pp. 740–747, Sept. 1983.

[5] P. B. Katehi and N. G. Alexopoulos, "On the modeling of electromagnetically coupled microstrip antennas—The printed strip dipole," *IEEE Trans. Antennas Propagat.*, vol. AP-32, pp. 1179–1186, Nov. 1984.

[6] D. M. Pozar and D. H. Schaubert, "Scan blindness in infinite phased arrays of printed dipoles," *IEEE Trans. Antennas Propagat.*, vol. AP-32, pp. 602–610, June 1984.

[7] ——, "Analysis of an infinite array of rectangular microstrip patches with idealized probe feeds," *IEEE Trans. Antennas Propagat.*, vol. AP-32, pp. 1101–1107, Oct. 1984.

[8] C. C. Liu, J. Shmoys, A. Hessel, J. D. Hanfling and J. M. Usoff, "Plane wave reflection from microstrip-patch arrays—Theory and experiment," *IEEE Trans. Antennas Propagat.*, vol. AP-33, pp. 426–435, April 1985.

[9] J. Castaneda and N. G. Alexopoulos, "Infinite arrays of microstrip dipoles with a superstrate (cover) layer," in *1985 IEEE Antennas Propagat. Soc. Symp. Dig.*, pp. 713–717.

[10] D. M. Pozar, "Analysis of finite phased arrays of printed dipoles," *IEEE Trans. Antennas Propagat.*, vol. AP-33, pp. 1045–1053, Oct. 1985.

[11] ——, "General relations for a phased array of printed antennas derived from infinite current sheets," *IEEE Trans. Antennas Propagat.*, vol. AP-33, pp. 498–504, May 1985.

[12] ——, "Improved computational efficiency for the moment method solution of printed dipoles and patches," *Electromagn.*, vol. 3, pp. 299–309, Sept. 1984.

Analysis of a Series-Fed Aperture-Coupled Patch Array Antenna

Chen Wu, Jian Wang, Russell Fralich and John Litva

KEY TERMS
Patch antenna, aperture coupling, spectral domain analysis, antenna feed

ABSTRACT
A full-wave spectral domain model of a series-fed, traveling-wave aperture-coupled patch array antenna is presented. This configuration allows for performance optimization by adjustment of the aperture size, aperture location, and patch size. The theory successfully predicts the input impedance and far-field patterns of various C-band configurations.

I. INTRODUCTION

The feed in a large microstrip antenna array can be a source of problems. A large array of coax-fed antennas requires precise attachment of connectors to perhaps a thousand or more elements, whereas in a coplanar-fed network, the array performance is degraded due to spurious feedline radiation. The aperture-coupled patch array overcomes these difficulties. Such a design allows radiating elements and feedlines to occupy different substrates and isolates the radiation from the feed network by means of a common ground sheet. A series-fed array of printed dipoles has been previously developed [1]. However, in this array only the dipole offset location from the feedline can be used to optimize the array performance. This article presents an array in which the coupling apertures' position and size, as well as the patch-antenna size, can all be varied to adjust the coupling level to each patch, thereby allowing greater array design flexibility. A 1×4 series-fed traveling-wave aperture-coupled patch array has been developed. The theory, which uses the grounded dielectric slab Green's functions and moment method, successfully predicts the input impedance and far-field patterns of the array.

II. ANTENNA CONFIGURATION

The array configuration is shown in Figure 1. Each patch is identical with dimensions (L_p, W_p) and two equally spaced feedline guide wavelengths (λ_g) apart from center to center. This element separation corresponds to about $0.7\lambda_0$, which is sufficient to prevent grating lobes.

The transmission and reflection coefficients along the microstrip feedline at the kth aperture center, at $(x_{ap,k} y_{ap,k})$, are denoted as T_k and R_k, respectively, and the following recursion relations are used:

$$T_k = T'_{k-1} e^{-jk_f(y_{ap,k} - y_{ap,k-1})}, \quad (1)$$

$$R_k = R'_{k-1} e^{jk_f(y_{ap,k} - y_{ap,k-1})}, \quad (2)$$

where k_f is the propagation constant of the traveling-wave current mode along the feedline, $k = 1, 2, 3, \ldots, N$, and N is the number of elements in the array.

III. NUMERICAL DESCRIPTION

A full-wave spectral domain model [2] was previously developed to analyze an aperture-coupled single-element patch antenna. However, the feedline model [2] could not be used to analyze a series-fed aperture-coupled array. In this article, the feedline model was extended by calculating the reflection and transmission coefficients at each aperture on feedline. Therefore, the spectral domain model and moment method can be applied to analyze the series-fed aperture-coupled patch array.

As shown in Figure 1, the array can be divided into two regions: One lies above ground plane (denoted with a superscript p) and another is below the ground plane (superscript f). The following description includes the number of current modes and all matrix sizes necessary to successfully analyze a 1×4 series-fed traveling-wave array.

(*a*) *Surface Currents.* Since the array is operating near the resonant frequency, the surface electric current on the kth patch can be represented as a series of entire-domain basis functions,

$$\bar{\mathbf{J}}_k^p = \frac{1}{W_p} \sum_{i=1}^{N_p} I_i^p(k) \times \sin\left[\frac{i\pi}{L_p}\left(y - \left(y_p(k) - \frac{L_p}{2}\right)\right)\right] \hat{\mathbf{y}}, \quad (3)$$

where

$$x_p(k) - \frac{W_p}{2} \le x \le x_p(k) + \frac{W_p}{2},$$

$$y_p(k) - \frac{L_p}{2} \le y \le y_p(k) + \frac{L_p}{2},$$

$$N_p = 5.$$

By invoking the equivalence principle, the aperture at $(x_{ap,k}, y_{ap,k})$ with length and width $(L_{ap}(k), W_{ap}(k))$ can be replaced by magnetic surface currents located just above and below the ground plane. The magnetic surface current value

Reprinted with permission from *Microwave and Optical Tech. Lett.*, C. Wu, J. Wang, R. Fralich and J. Litva, "Analysis of a Series-Fed Aperture-Coupled Patch Array Antenna," vol. 4, no. 3, pp. 110–113, Feb. 1991. © John Wiley and Sons.

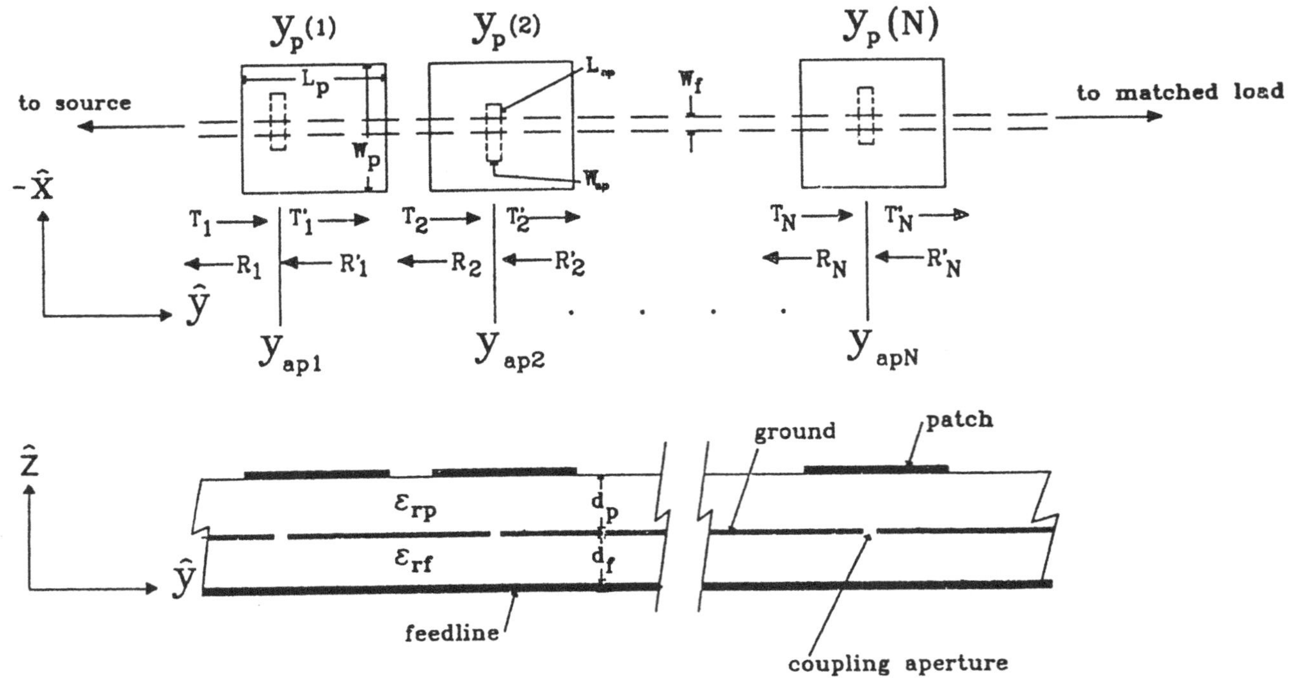

Figure 1 The configuration of a series-fed aperture-coupled patch array

at the kth aperture was modeled using subdomain piecewise sinusoidal (PWS) modes,

$$M(k) = \frac{1}{W_{ap}(k)} \sum_{i=1}^{N_{ap}} V_i(k) \times \frac{\sin[k_{ap}(h_{ap}(k) - |x - x_i(k)|)]}{\sin[k_{ap}h_{ap}(k)]}, \tag{4}$$

where

$$h_{ap}(k) = L_{ap}(k)/(N_{ap} + 1),$$

$$x_i(k) = x_{ap,k} - \frac{L_{ap}(k)}{2} + ih_{ap}(k),$$

$$x_i(k) - h_{ap}(k) \le x \le x_i(i) + h_{ap}(k),$$

$$y_{ap,k} - \frac{W_{ap}(k)}{2} \le y \le y_{ap,k} + \frac{W_{ap}(k)}{2},$$

$$N_{ap} = 4,$$

$$k_{ap} = (k_f + k_p)/2$$

(k_p is the wavenumber in region p).

To ensure a continuous tangential electric field across the aperture, the two magnetic surface currents have the same magnitude but are aligned in opposite directions:

$$\overline{\mathbf{M}}^p(k) = M(k)\hat{\mathbf{x}}, \quad \text{above ground plane}, \tag{5}$$

$$\overline{\mathbf{M}}^f(k) = -M(k)\hat{\mathbf{x}}, \quad \text{below ground plane}. \tag{6}$$

It also must be noted that all apertures were identical in size in this article.

The electric surface current on feedline, whose width is W_f, can be expanded as a superposition of a traveling-wave current and some piecewise sinusoidal (PWS) modes near the each aperture to account for current discontinuities. The surface current on the feedline proximate to the kth aperture can be written as

$$\overline{\mathbf{J}}_f^k = \overline{\mathbf{J}}_{tra}^k(k) + \overline{\mathbf{J}}_{pws}^k(k), \tag{7}$$

where

$$\overline{\mathbf{J}}_{tra}^k(k) = \frac{1}{W_f}[(T_k e^{-jk_f(y-y_{k-1})} + R_k e^{jk_f(y-y_{k-1})}) \times (1 - u(y - y_{ap,k})) + (T'_k e^{-jk_f(y-y_{k-1})} + R'_k e^{jk_f(y-y_{k-1})})u(y - y_{ap,k})]\hat{\mathbf{y}},$$

$$u(t) = \begin{bmatrix} 1, & t \ge 0 \\ 0, & t = 0 \end{bmatrix}, \quad y_{ap,k-1} \le y \le y_{ap,k+1},$$

and

$$\overline{\mathbf{J}}_{pws}^k(k) = \frac{1}{W_f} \sum_{i=1}^{N_f} I_i^f(k) \times \frac{\sin[k_f(h_f(k) - |y - y_i(k)|)]}{\sin[k_f h_f(k)]} \hat{\mathbf{y}}$$

$$h_f = L_f/(N_f + 1), \qquad L_f = \lambda_g/2,$$

$$y_i(k) = y_{ap,k} + i \cdot h_f - L_f/2,$$

$$N_f = 7,$$

assuming that the amplitude of the incident wave along feedline is $1/W_f$ and $T_1 = 1$. Since the array under analysis is operated in a traveling-wave configuration, $R'_N = 0$.

(*b*) *Moment Method Solution.* Using the above surface currents, the y-directed electric fields and the x-directed magnetic fields in the two regions (p and f) can be expressed as

$$E_y^p = \sum_{k=1}^{N} [E_y^p(\bar{\mathbf{J}}_k^p) + E_y^p(\overline{\mathbf{M}}_k^p)],$$

$$E_y^f = \sum_{k=1}^{N} [E_y^f(\bar{\mathbf{J}}_k^f) + E_y^f(\overline{\mathbf{M}}_k^f)],$$

$$H_x^p = \sum_{k=1}^{N} [H_x^p(\bar{\mathbf{J}}_k^p) + H_x^p(\overline{\mathbf{M}}_k^p)],$$

$$H_x^f = \sum_{k=1}^{N} [H_x^f(\bar{\mathbf{J}}_k^f) + H_x^f(\overline{\mathbf{M}}_k^f)],$$

By using the Galerkin method and applying the following boundary conditions: (1) the tangential magnetic field is continuous across the coupling apertures; (2) the total tangential electric field is zero on the feedline; (3) the total tangential electric field is zero on the patches; the problem can be described in matrix form as

$$\begin{bmatrix} [C^{ij}_{\mathrm{ap}_{ul}}] & [-C^{ij}_{af_{un}}] & [C^{ij}_{cs_u}] & [Y^{ij}_{u_{uv}} + Y^{ij}_{d_{uv}}] \\ [0] & [Z^{ij}_{mn}] & [Z^{ij}_{cs_m}] & [-T^{ij}_{fa_{mv}}] \\ [Z^{ij}_{kl}] & [0] & [0] & [T^{ij}_{pa_{kv}}] \\ [0] & [0] & [F] & [0] \end{bmatrix} \times \begin{bmatrix} [I_p] \\ [I_f] \\ [\mathrm{TR}] \\ [V_{\mathrm{ap}}] \end{bmatrix} = \begin{bmatrix} 0 \\ \cdot \\ 0 \\ \cdot \\ 2 \\ \cdot \\ 0 \end{bmatrix} \quad 18N + 1, \tag{8}$$

where $i, j, = 1, 2, 3, \ldots, N$; $m = 0, 1, 2, 3, 4, 5, 6, 7, 8$; $n = 1, 2, 3, 4, 5, 6, 7$; $k, l = 1, 2, 3, 4, 5$, and $u, v = 1, 2, 3, 4$. $[I_p]$ are the coefficients of the entire domain patch current basis modes, $[I_f]$ are the coefficients of the PWS feedline current modes and $[V_{\mathrm{ap}}]$ are the coefficients of PWS aperture modes.

Matrix [TR] describes the current transmission and reflection coefficients along the feedline: $[\mathrm{TR}] = [a_1\ b_1\ c_1\ d_1 \cdots a_N\ b_N\ c_N\ d_N]^t$, where t indicates *transposition*, and $a_i = R_i + T_i$, $b_i = R_i - T_i$, $c_i = R_i' + T_i'$, $d_i = R_i' - T_i'$. Matrix [F] describes the relationship between the current reflection and transmission coefficients along the feedline:

$$[\mathrm{F}] = \begin{bmatrix} 1 & -1 & 0 & 0 & & & \\ 0 & 0 & [f_{21}] & 0 & & \mathbf{0} & \\ 0 & 0 & 0 & [f_{32}] & & & \\ & & & & \ddots & & \\ & \mathbf{0} & & & & & \\ & & & & [f_{N,N-1}] & 0 & 0 \\ & & & & 0 & 1 & 1 \end{bmatrix}, \tag{9}$$

$$[f_{ij}] = \begin{bmatrix} 1 & 1 & -e^{-jt_{ij}} & -e^{-jt_{ij}} \\ 1 & -1 & -e^{jt_{ij}} & e^{-jt_{ij}} \end{bmatrix} \tag{10}$$

where $t_{ij} = k_f(y_{\mathrm{ap},i} - y_{\mathrm{ap},j})$, $i = 1, 2, 3, 4 \ldots N$, $j = i - 1$. The submatrices $[C_{\mathrm{ap}}]$ and $[T_{pa}]$ represent coupling between the patches and the apertures, while $[-C_{af}]$ and $[T_{fa}]$ represent the interaction between the feedline PWS modes and the apertures. $[C_{cs}]$ contains the interactions between the feedline traveling-wave currents and the apertures. $[Z_{kl}]$ and $[Z_{mn}]$ are moment-method impedance matrices of the entire-domain modes on the patches, and the PWS modes on the feedline. $[Z_{cs}]$ is the moment-method impedance matrix of the traveling wave and the PWS feedline modes. $[Y_u]$ and $[Y_d]$ are aperture admittance matrices.

Each of the matrix elements can be expressed by a spectral-domain Green's function which involves a six-fold complex integration [2]. It may be noted that in numerical calculation the underlying assumptions are that (*i*) the feedline modes couple only to the closest aperture and not to aperture of any other element, and (*ii*) the interelement coupling is only included among the patches (i.e., $\{Z_{kl}^{ij} \neq 0 \text{ when } i \neq j\}$).

IV. COMPARISON OF EXPERIMENTAL AND THEORETICAL RESULTS

Figure 2 shows measured and calculated input impedances for a 1 × 4 series-fed, aperture-coupled patch array for aperture lengths of 5.5 and 10 mm, respectively. In these examples, all four apertures are coincident with the center of the respective patches. The spectral domain model accurately predicted the input impedance characteristics for both configurations. The far-field E-plane pattern at the center frequency was also successfully predicted, as shown in Figure 3.

In the moment-method procedure, five entire-domain basis functions were used on each patch, four PWS modes described the aperture magnetic currents, and seven PWS modes were applied on the feedline underneath each aperture to account for current discontinuities. These current models resulted in not only an excellent agreement between theoretical and experiment results, but also in a convergent solution. If more basis functions were used, the results would generally vary by less than 3%.

V. CONCLUSIONS

The feedline model has been extended and a generalized analysis of a series-fed, traveling-wave, aperture-coupled patch antenna array has been described. Coupled integral equations incorporating the spectral-domain Green's function are used to predict the array input impedance and far-field pattern. This analysis is very useful for designing series-fed aperture-coupled patch arrays, especially when implementing array pattern synthesis.

ACKNOWLEDGMENT

The authors would like to thank Sharon Aspden of Rogers Corporation, Chandler, Arizona, USA for providing the substrates used in this article under the auspices of their University Assistance Program.

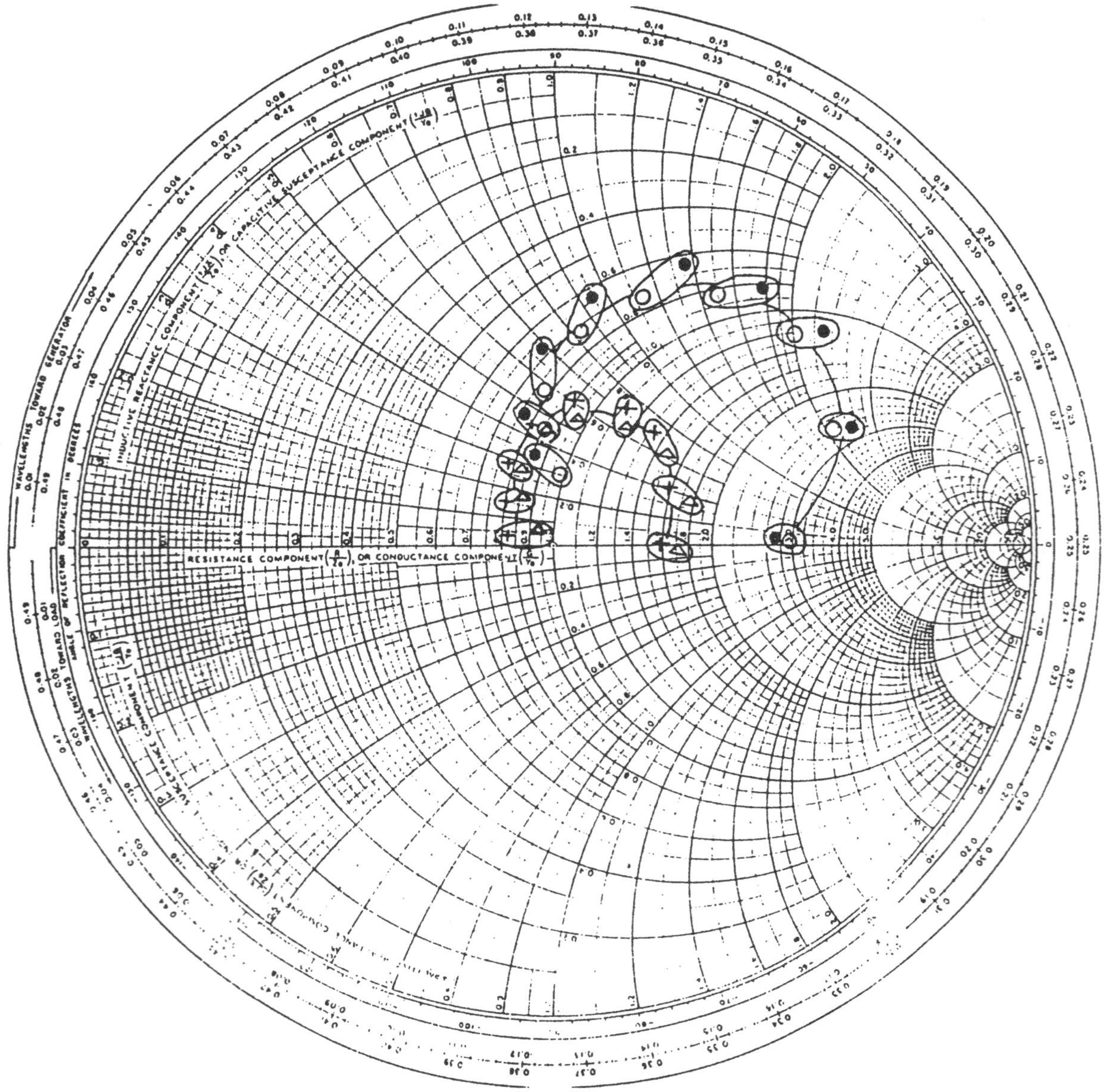

Figure 2 Input impedance of 1 × 4 series-fed aperture-coupled patch array L_p = 17.8 mm, W_p = 25.4 mm, W_{ap} = 0.8 mm, ε_{rp} = 2.33, d_p = 3.175 mm, ε_{rf} = 10.2, d_f = 2.54 mm, W_f = 2.35 mm. (a) L_{ap} = 5.5 mm, triangles—measured, crosses—calculated; (b) L_{ap} = 10.0 mm, closed circles—measured, open circles—calculated

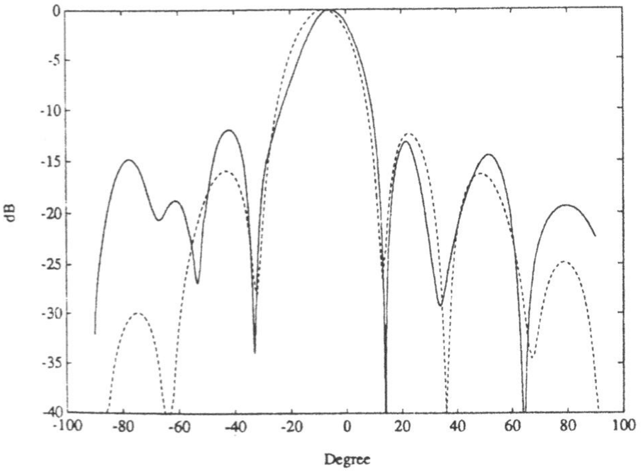

Figure 3 *E*-plane of four-element series-fed aperture-coupled patch array. Solid line—measured, dashed line—calculated, frequency = 4.8 (GHz)

REFERENCES

1. N. K. Das and D. M. Pozar, "Analysis and Design of Series-Fed Arrays of Printed Dipoles Proximity-Coupled to a Perpendicular Microstripline," *IEEE Trans. Antennas Propagat.*, Vol. AP-37, 1989, pp. 435–444.
2. P. L. Sullivan and D. H. Schaubert, "Analysis of an Aperture-Coupled Microstrip Antenna," *IEEE Trans. Antennas Propagat.*, Vol. AP-34, 1986, pp. 977–984.
3. D. M. Pozar, "A Reciprocity Method of Analysis for Printed Slot and Slot-Coupled Microstrip Antennas," *IEEE Trans. Antennas Propagat.*, Vol. AP-34, 1986, pp. 1439–1446.

Performance of Probe-Fed Microstrip-Patch Element Phased Arrays

CHUNG-CYTZ LIU, ALEXANDER HESSEL, LIFE FELLOW, IEEE, AND JERRY SHMOYS, SENIOR MEMBER, IEEE

***Abstract*—The performance trends of large phased arrays with probe-fed microstrip patch elements are discussed. In view of space constraints, this paper is limited primarily to single probe (unbalanced feed) for rectangular grid arrays with rectangular patch geometry. For balanced feed treatment, the reader is referred to Liu et al. in *Proceedings Phased Array 1985 Symposium*. The effects of substrate thickness and permittivity, patch dimensions and element spacing are considered. Pronounced surface and leaky wave effects which tend to limit the scan volume are found. Interaction between the patch resonance at broadside scan and the leaky wave resonances leads to an *E*-plane scan bandwidth trade-off. In the *H*-plane scan a transverse electric (TE)-type guided wave resonance is found which becomes pronounced when the *H*-plane patch dimension is a large fraction of the *H*-plane element spacing. This leads to a limitation of the *H*-plane patch dimension in the rectangular grid arrays. Based on the performance trends, a proposed element design procedure is presented and illustrated by design of an element suitable for wide angle scan over ±5 percent frequency band.**

I. Introduction

MICROSTRIP-PATCH array elements gained popularity because of their suitability for lightweight space and monolithic millimeter wave applications. The first trade-off studies of microstrip arrays appeared in [2]. While the analysis in [2] is limited to the *H*-plane scan of strip elements, a subsequent *E*-plane scan trade-off study of such phased arrays was given in [3]. An analysis of rectangular microstrip-patch phased arrays along with some numerical examples of scan performance was given in [4]. In the discussion we shall comment on the relation of our results to those of [4].

Following an outline of the analysis, we shall discuss the numerical results that are grouped in such a way as to exhibit the influence of array parameters on the scan and frequency performance trade-offs which, in turn, lead to the element design.

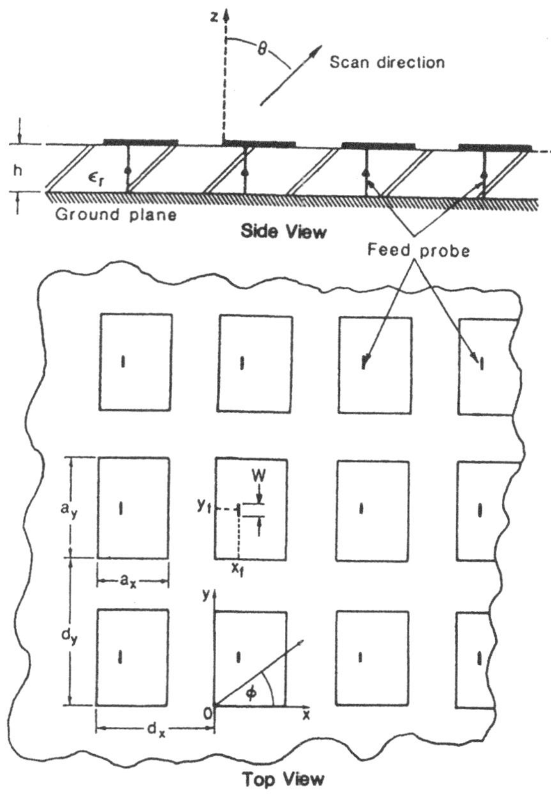

Fig. 1. Geometry of phased array of rectangular microstrip-patch elements.

II. Array Model and Outline of Analysis

The phased array model considered is shown in Fig. 1. The top of a lossless, grounded dielectric substrate of thickness h, small compared to free space wavelength, is periodically coated by perfectly conducting rectangular patches of negligible thickness in a rectangular grid arrangement. The patches are fed by probes simulated by current ribbons of a uniform z directed density. The equi-amplitude feed currents are pro-made of an equivalence principle that permits a breakup of analysis into two simpler problems. In the first, the apertures are absent. The fields generated in the resulting parallel-plate guide by the doubly periodic array of probes is determined in a gressively phased in both x and y directions. In order to carry out properly the computer simulation of a probe-fed microstrip phased array element performance, two convergence difficulties in the numerical analysis had to be overcome. The first is due to the rapid current variation on the underside of the patches in the vicinity of the probes. The second is caused by the edge condition, i.e., the current singularities at the patch edges giving rise to a slow convergence of the modal series involved, as well as to the necessity of inversion of large matrices. After removing these difficulties, the convergence of the numerical solution was examined and shown adequate. The approaches to remove the above-mentioned difficulties have been previously discussed in [3] for the technique of extraction of the rapid current variation near the probes, and in [5] for the overall convergence acceleration technique.

The present paper combines these two aspects. To extract the rapid current variation in the vicinity of the probes, use is

Reprinted from *IEEE Trans. Antennas Propaga.*, vol. 36, pp. 1501–1509, Nov. 1988.

unit cell by considering either x or y as the direction of propagation. The resulting modal series with a single index permits rapid evaluation of both the (partial) probe reactances and of the current distribution on the underside of the top plate. In the second problem, we remove the probes, restore the apertures, and place in them the negative of the current distribution $\mathbf{J}_{eq}$ obtained from the first problem. The analysis is now carried out in a unit cell, similarly as in [3], with respect to z as the direction of propagation, in terms of standard Floquet modes. The fields of the two problems are now linearly superposed to yield the desired overall solution. For details see [6]. For an unbalanced feed, the active probe impedance is evaluated via

$$Z_a = -\int_{\text{probe}} E_z \, dz/I, \tag{1}$$

where I denotes the total probe current and E_z is the z component of electric field arising from superposition of contributions from problems one and two.

The form of Galerkin's equations for the second problem is

$$\sum_j \hat{Z}_{ij} C_j = V_i \tag{2}$$

where C_j are the unknown amplitudes of the various basis functions $\mathbf{J}_j$ used to represent the patch current distribution and $\hat{Z}_{ij}$ are the impedance matrix elements of the form

$$\hat{Z}^{q,p}_{\ell m} = \sum_{\mathbf{n},\alpha} \frac{(\mathbf{J}^q_{\mathbf{l}}, \mathbf{e}^{\alpha}_{\mathbf{n}})(\mathbf{J}^p_{\mathbf{m}}, \mathbf{e}^{\alpha}_{\mathbf{n}})^*}{\overleftarrow{Y}^{\alpha}_{\mathbf{n}} + \overrightarrow{Y}^{\alpha}_{\mathbf{n}}} \tag{3}$$

where the single letter vector subscripts represent double indices referring to the spatial variation, and $q = (1, 2)$ and $p = (1, 2)$ denote the x or y direction of the vector basis functions $\mathbf{J}_m$; $\mathbf{e}^{\alpha}_{\mathbf{n}}$ denote the Floquet mode functions, while $\overleftarrow{Y}^{\alpha}_{n}$ and $\overrightarrow{Y}^{\alpha}_{\mathbf{n}}$ stand for the transverse electric (TE) ($\alpha = 1$) and transverse magnetic (TM) ($\alpha = 2$) Floquet mode admittances "looking" from the patch toward and away from the ground plane, respectively; the inner products $(\mathbf{a}, \mathbf{b})$ are integrals of $\mathbf{a} \cdot \mathbf{b}$ over the patch surface. The inhomogeneous terms V_i are derived from the equivalent source currents $\mathbf{J}_{eq}$ in the aperture location and are of the form:

$$V_i = \sum_{\mathbf{n},\alpha} \frac{(\mathbf{J}_{eq}, \mathbf{e}^{\alpha *}_{\mathbf{n}})(\mathbf{e}^{\alpha}_{\mathbf{n}}, \mathbf{J}_i)}{\overleftarrow{\mathbf{Y}}^{\alpha}_{\mathbf{n}} + \overrightarrow{\mathbf{Y}}^{\alpha}_{\mathbf{n}}}. \tag{4}$$

The basis functions $\mathbf{J}_i$, i.e.,

$$J^1_{m_x, m_y} = \frac{\sin (m_x \pi x/a_x)}{\sqrt{1-(1-2x/a_x)^2}} \frac{\cos (m_y \pi y/a_y)}{\sqrt{1-(1-2y/a_y)^2}} \mathbf{x}_0,$$

$$m_x = 0, 1 \cdots M_x \quad m_y = 0, 1 \cdots M_y \tag{5a}$$

$$J^2_{m_x, m_y} = \frac{\cos (m_x \pi x/a_x)}{\sqrt{1-(1-2x/a_x)^2}} \frac{\sin (m_y \pi y/a_y)}{\sqrt{1-(1-2y/a_y)^2}} \mathbf{y}_0 \tag{5b}$$

satisfy the edge conditions. The selection of the set of $\mathbf{J}_i - s$ used in computations (typically 52 $J_i - s$) was already discussed in detail in [5]. The set of Floquet wavenumbers was delimited by $-30 \leq n_x, n_y \leq +30$. The rate of convergence of the double modal series in (3) was exhibited in [5], where it was found that it requires acceleration. To this end, as in [5], Poisson summation is employed.

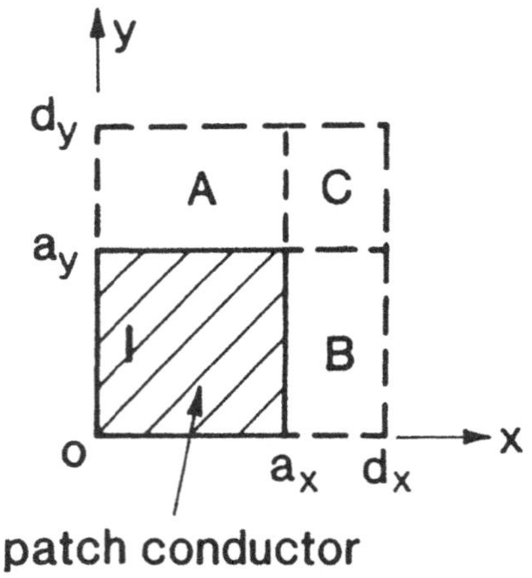

Fig. 2. Subdivision of the unit cell.

The evaluation of V_i involves an expansion, in terms of Floquet modes, of the equivalent sources generated in the apertures from problem one. Since the aperture is L-shaped, it had to be divided into regions (see Fig. 2). For improved convergence of the series involved in the evaluation of V_i, different representations of the equivalent sources had to be used in different regions. In regions A and C, in problem one, propagation in y was employed, for which the higher parallel plate guide modes decay exponentially along y, while in B and C propagation in x was used. In region C either representation yields good convergence and this feature was utilized for checking the required number of terms of the series in problem one, for sufficient numerical accuracy.

III. Performance Trends and Design Criteria

A. Patch Size

Let us now consider the performance trends. To begin with, we find that it is desirable to resonate the patch at center frequency at, say, broadside scan [3]. This raises the level of the active resistance R_a to a value acceptable for matching, since a nonresonant value of R_a may be as low as a few ohms. Moreover, we shall see that the broadside patch resonance exerts a major influence on the scan variation of active impedance.

The three curves shown in Fig. 3 exhibit the dependence of broadside R_a on patch size, for a square patch and a square lattice, with substrate thickness h/λ as a parameter. For a fixed value of h/λ the patch is resonated by selecting its linear dimension a for *maximum* R_a. If the patch dimensions were fixed, a resonance could be induced by varying the substrate thickness. One observes that the *thinner* the substrate, the *sharper* is the resonance and the *higher* the peak value of R_a. Conversely, the broadside Q *is lowered* with increasing of substrate thickness, indicating a larger broadside bandwidth. We also notice that the thinner the substrate, the larger is the resonance patch size.

Fig. 4 indicates that, in order to resonate a rectangular patch at broadside scan at a given frequency, an increase of the H-

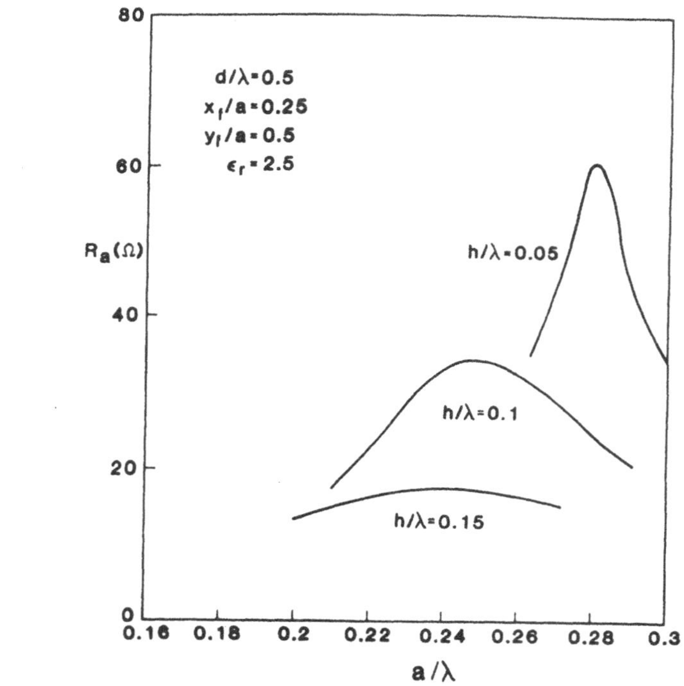

Fig. 3. Active resistance at broadside scan; square patch, square grid.

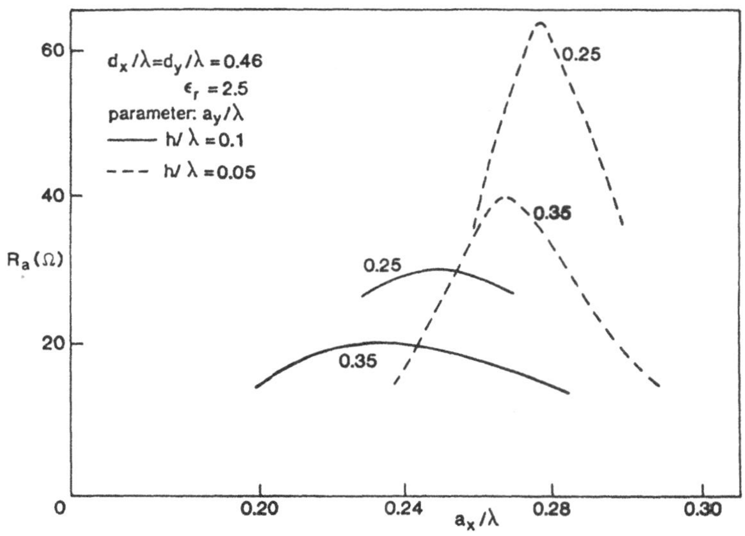

Fig. 4. Active resistance at broadside scan.

plane patch dimension results in reduction of the E-plane dimension and vice versa. In all cases considered $x_f = a_x/4$, $y_f = a_y/2$ (see Fig. 1).

B. Active Resistance for E-Plane Scan-Guided Wave Effects

The active resistance turns out to be a sensitive indicator not only of the bandwidth but also of the scan performance. The more gradual the average R_a variation with scan, the slower is the average active reactance scan variation and the more gradual is the mismatch scan loss. The level of R_a also indicates whether the element is suitable for a match into, say, a 50 Ω feed line. For these reasons most of the physical considerations of this paper are concentrated on R_a and its scan and frequency variation. In the final analysis, however, for design purposes, the active reflection coefficient must be, and is, considered. Fig. 5 shows a parametric dependence of the E-plane scan variation of R_a on the substrate thickness h/λ, for a square grid array with 0.5 λ spacing and a square patch 0.25 λ on a side. The dielectric constant is 2.5. For each value of h/λ one observes a resonant variation of R_a, flat near broadside rising to a peak followed by a sharper drop to a low level region. With increasing substrate thickness, the peaks and the associated low level regions drift toward broadside. The peaks of R_a have been traced to guided wave effects by examining the reciprocal $|\hat{Z}|^{-1}$ of the determinant of the impedance

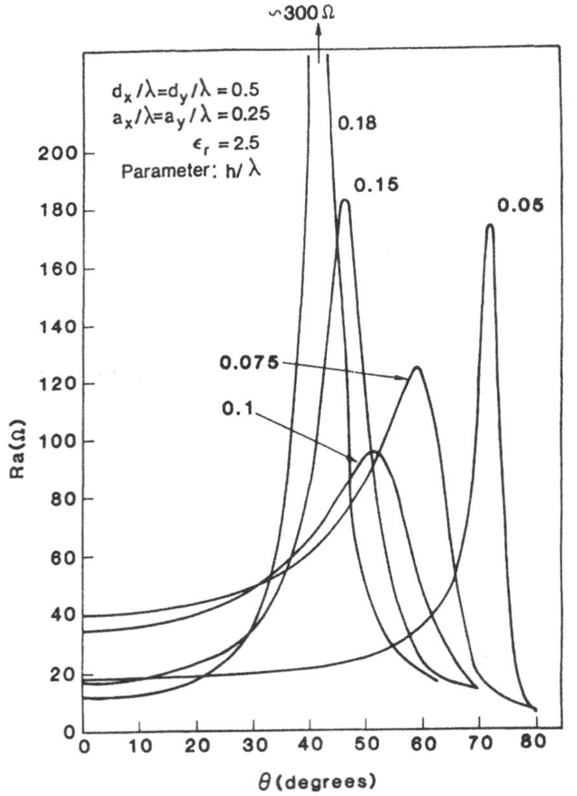

Fig. 5. *E*-plane active resistance versus scan angle (square patch, square grid).

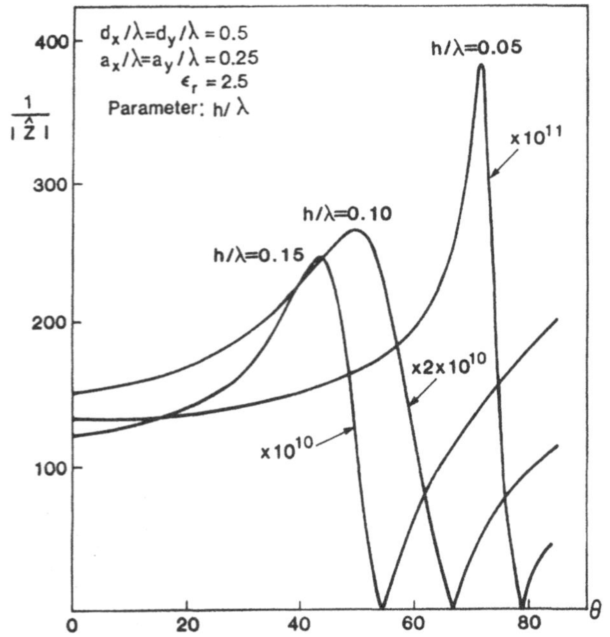

Fig. 6. Reciprocal of the determinant of the impedance matrix. Appropriate multiplication factor is indicated on each curve.

matrix ($\hat{Z}$) as a function of scan. This is shown in Fig. 6. It is seen that the peak locations and the Q_s values[1] in Figs. 5 and 6 are closely correlated. These guided waves, however, are no longer the "unperturbed" TM surface waves supported by the unloaded, grounded dielectric substrate slab, but (complex) leaky waves guided by the periodic microstrip-patch-loaded array structure as evidenced by the finite peak values of $|\hat{Z}|^{-1}$ and the finite Q_s in Fig. 6.

[1]Here Q_s denotes the Q of a resonance versus sin θ and not versus frequency.

The *E*-plane resonant peaks of R_a appear whenever a phase-match condition occurs between the guided leaky wave and a relevant grating lobe of the steering phase, i.e.,

$$\sin\theta_{LW} = \left|\frac{\beta_{LW}}{k} - \frac{\lambda}{d_x}\right|. \tag{6}$$

In (6), β_{LW} is the leaky wave phase constant. The leaky wave attenuation constant α determines the value of Q_s via

$$Q_s = \frac{\left(\beta_{LW} - \frac{2\pi}{d_x}\right)}{2\alpha} = \frac{\frac{\beta_{LW}}{k} - \frac{\lambda}{d_x}}{\alpha\lambda/\pi}. \tag{7}$$

A discussion of the nature of periodic structure leaky waves is found in [7]. The *E*-plane leaky wave peaks of R_a occur closer to broadside than the associated valley regions. These low level (but nonzero) R_a regions are located near the *E*-plane scan angles for which

$$\sin\theta_{SW} = \left|\frac{\beta_{SW}}{k} - \frac{\lambda}{d_x}\right|, \tag{8}$$

where β_{SW} is the TM surface wave phase constant in the absence of patch loading. Condition (8) deduced in [4] is tantamount to a phase-match between a grating lobe and the TM surface wave on the unloaded grounded dielectric substrate slab. (While the present analysis shows that R_a remains finite at θ_{SW}, albeit low, and X_a is bounded, by contrast, [4] invariably predicts a total reflection when (8) is satisfied.) Thus, during an *E*-plane scan, the loading of the dielectric substrate manifests itself in the appearance of guided wave resonances of the leaky type, as well as in the low level R_a values near θ_{SW}.

Fig. 5 also exhibits another interesting feature: the peak and the associated Q_s values are lowest for $h/\lambda = 0.1$. For this substrate thickness, the patch is close to its broadside resonance, as seen from Fig. 3. Both the peak and Q_s values rise as the patch is detuned by change of substrate thickness to a larger or smaller value. This is indicative of a strong influence of the broadside patch resonance on scan behavior. A resonant patch tends to damp the leaky wave peaks. On the other hand, when the patch is detuned, its loading is light and the scan variation of R_a becomes rapid. Furthermore, we shall see that the higher the broadside Q, i.e., the smaller the bandwidth, the slower will be the *E*-plane scan variation of R_a and the flatter the *E*-plane scan performance. We now exhibit this scan-bandwidth trade-off by comparing the frequency and scan performance of two arrays. The first (Fig. 7) has a smaller broadside bandwidth than the second (Fig. 8). Both arrays have a square grid with 0.5 λ spacing and ϵ_r of 2.5. In each case, the patch is resonated at broadside for maximum R_a at center frequency.

Fig. 7 shows a Smith chart for the unmatched active impedance normalized to 50 Ω for scan up to 60° off broadside

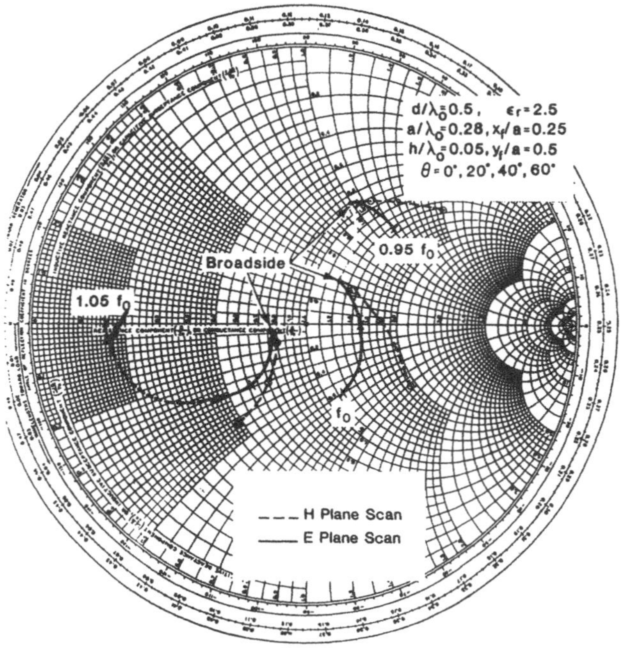

Fig. 7. Unmatched active impedance.

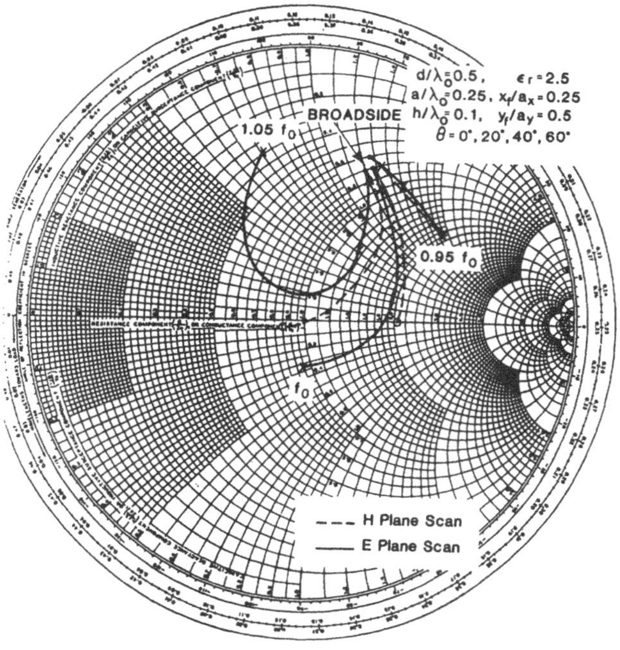

Fig. 8. Unmatched active impedance.

over ± 5 percent frequency band. We observe that the locus of broadside scan points is spread in an arc, but we have a reasonably tight unmatched scan locus. One also observes that the H-plane scan loci track those of the E-plane. The corresponding Smith chart for an identical array, but with 0.1 λ substrate thickness, is shown in Fig. 8. The broadside Q is now lower and bandwidth is increased, as evidenced by clustering of the broadside points. On the other hand, the scan variation is more rapid than for $h/\lambda = 0.05$. This is a clear manifestation of the scan-bandwidth trade-off.

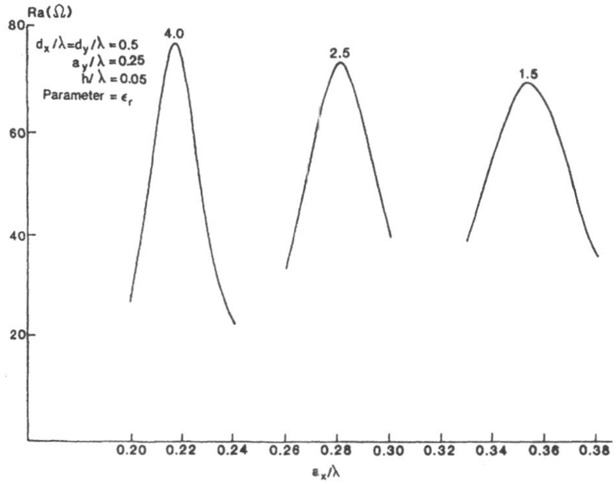

Fig. 9. Broadside Q for different choices of ϵ_r.

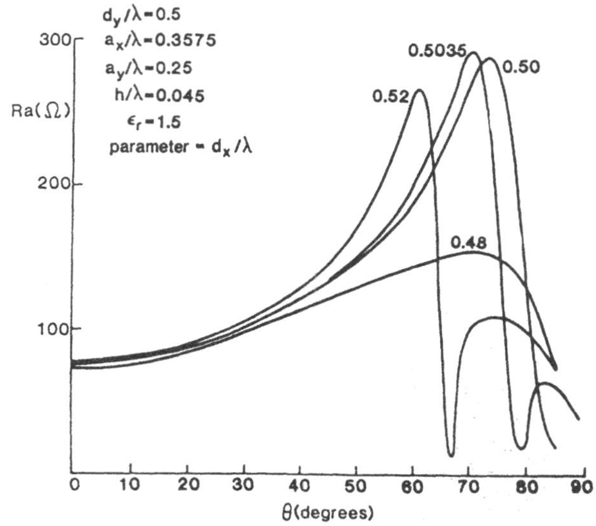

Fig. 10. Active resistance for E-plane scan, $\epsilon_r = 1.5$, $h/\lambda = 0.045$.

We now bring further evidence of this trade-off. Fig. 9 shows three broadside resonance tuning curves of R_a versus E-plane patch dimension a_x/λ. Each curve corresponds to a different value of ϵ_r. One observes that the highest Q is for $\epsilon_r = 4$ and the lowest for $\epsilon_r = 1.5$.

After what has been said, we expect for larger values of dielectric constant, where the broadside Q is highest, a more gradual scan variation. To verify this, we resonate the patch at broadside scan (for maximum R_a) for each ϵ_r and evaluate the scan variation. We also vary parametrically the E-plane spacing to simultaneously exhibit the scan dependence on d_x/λ.

Fig. 10 shows the case of $\epsilon_r = 1.5$. We observe a rapid scan variation. Also, as the E-plane spacing increases, the leaky wave peaks and surface wave dips migrate toward broadside reducing the usable scan sector. Fig. 11 is for $\epsilon_r = 2.5$. In this case, the broadside Q is higher than for $\epsilon_r = 1.5$ and the E-plane scan variation is more gradual. In Fig. 12, for $\epsilon_r = 4$, the broadside Q is the highest of the three and the E-plane scan variation is very smooth.

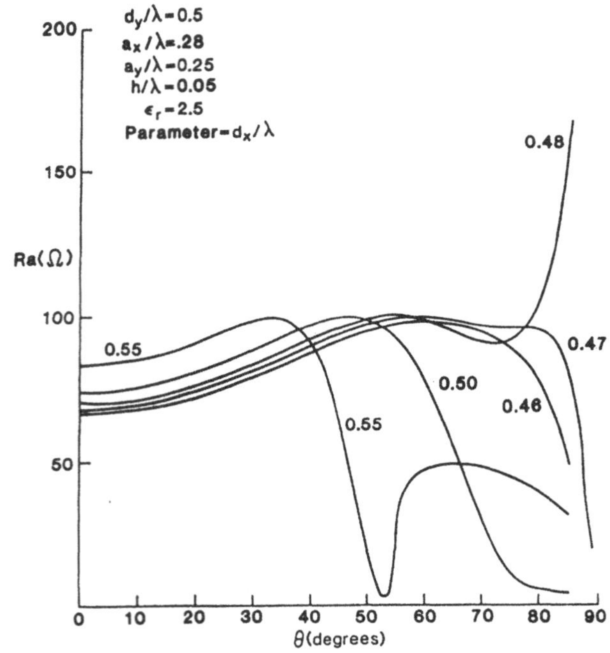

Fig. 11. Active resistance for E-plane scan, $\epsilon_r = 2.5$, $h/\lambda = 0.05$.

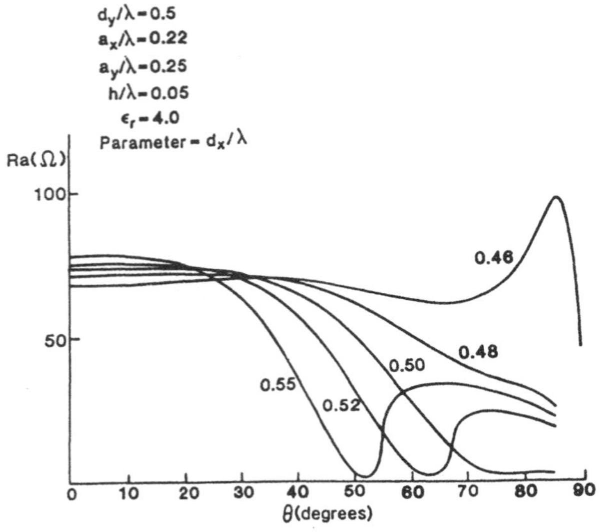

Fig. 12. Active resistance for E-plane scan, $\epsilon_r = 4.0$, $h/\lambda = 0.05$.

C. Active Resistance for the H-Plane Scan and Cross Polarization

An interesting feature appears in Figs. 7 and 8 in that the H-plane scan locus approximately tracks that for the E-plane. In order to shed light on this behavior, Fig. 13 shows the H-plane scan variation of R_a, with h/λ as a parameter, with the superimposed dashed E-plane variation for $h/\lambda = 0.05$. For other values of h/λ a comparison should be made between Figs. 5 and 13. It is observed that the H- and E-plane behavior of R_a is similar, and, in particular, the locations of the corresponding E- and H-plane peaks are quite close. Since the E-plane leaky wave resonances arise from TM guided waves, we conclude that for the array in question the H-plane

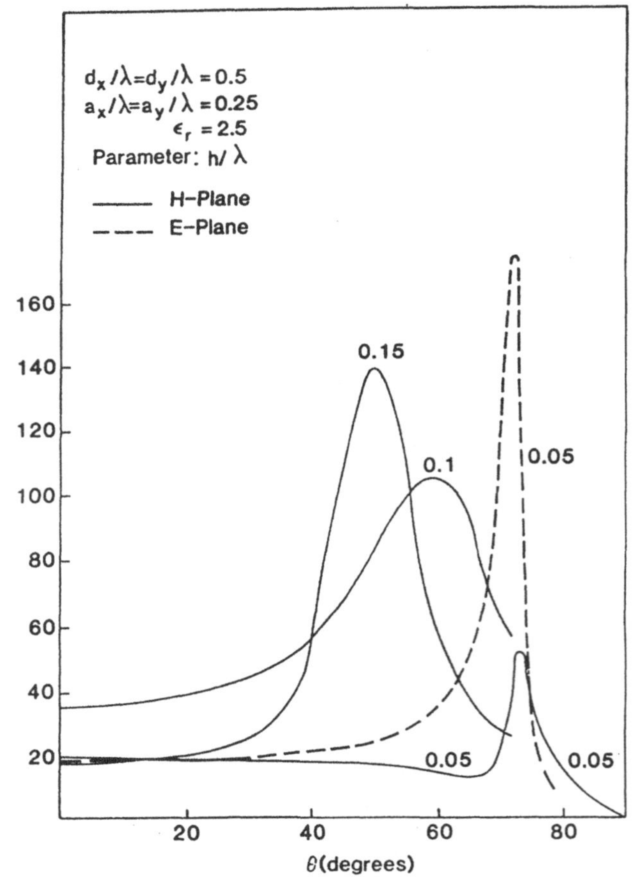

Fig. 13. Active resistance versus scan, H-plane. E- and H-plane peak locations track.

resonances are likewise caused by the TM guided waves excited by the asymmetrically located probe feed and, since in Fig. 13 $d_x = d_y$ and $a_x = a_y$, the surface wave dips and leaky wave resonant peaks of R_a in the H-plane scan are expected to follow those in the E-plane (for unbalanced feed).

Fig. 14 shows the H-plane cross polarization versus scan, with a_y/λ as a parameter, for $h/\lambda = 0.05$, $d/\lambda = 0.50$. It is observed that cross polarization increases with a_y.

We now come to the guidelines for choosing the H-plane patch dimension. To this end we have selected a balanced feed arrangement in order to prove a point. A similar H-plane scan variation is found for the unbalanced feed case. For the balanced feed, the location of the electric walls in the planes of array symmetry prevents for our dimensions, any TM guided waves from propagation along the H-plane direction. Therefore, no TM guided wave generated blindspots are possible. In spite of this, when the H-plane patch dimension a_y is large, leaky wave resonances appear. Thus, these resonances correspond to TE leaky waves for which the E-field is perpendicular to the electric walls. In Fig. 15, when the H-plane patch dimension is 0.25 λ, the scan variation of R_a is smooth, but for 0.35 λ a large TE resonant peak is observed. For $a_y = 0.4\ \lambda_0$ the resonant peak moves out of visible space. This behavior indicates that it is advisable to select the H-plane patch dimension at about 50 percent of the H-plane element spacing. There is another drawback to selecting large a_y for unbalanced feed, since the H-plane cross polarization increases.

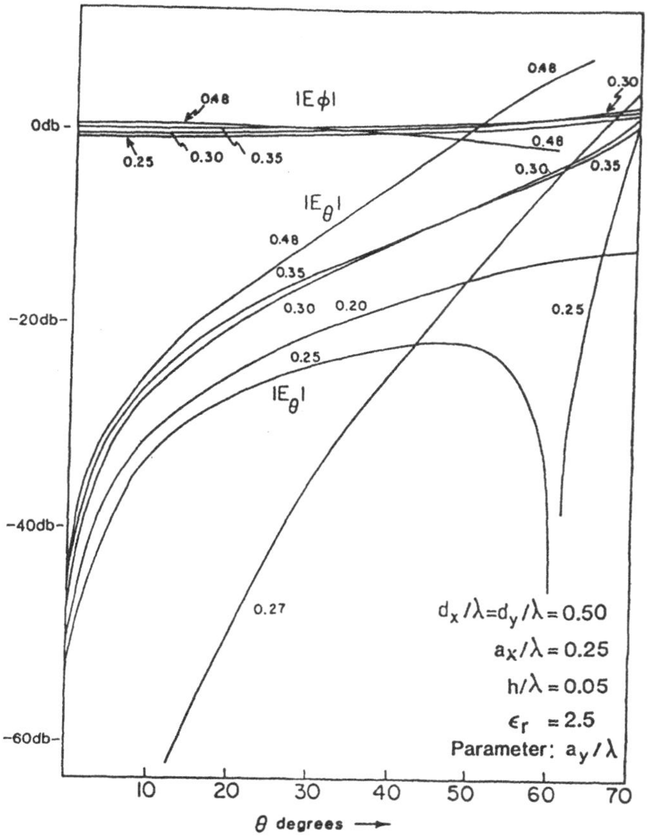

Fig. 14. *H*-plane scan cross-polarization. Influence of *H*-plane patch dimension.

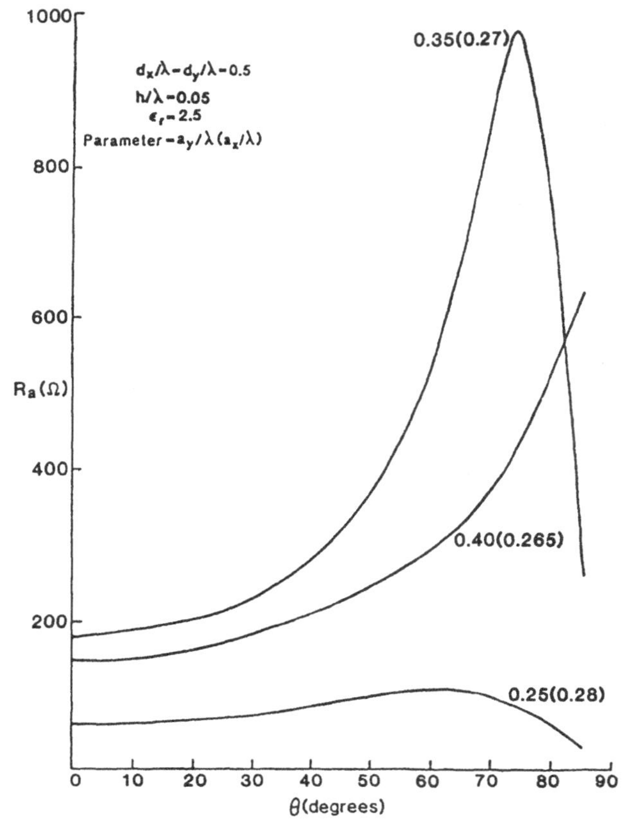

Fig. 15. Active resistance for *H*-plane scan. Balanced feed.

D. Computer-Aided Design Procedure

Based on the trade-offs discussed above, we can formulate a computer-aided design procedure for a microstrip patch phased array element with unbalanced feed. A similar procedure can be used for the design of an element with balanced feed.

1) Select ϵ_r in the range $2 \leq \epsilon_r \leq 5$:
   a) too high ϵ_r gives rise to narrow bandwidth (but smooth scan variation);
   b) too low ϵ_r gives rise to a large patch and thus to a small aperture with a resulting light loading and a rapid scan variation.
2) Select a low value of h/λ, say 0.05:
   a) this tends to reduce the scan variation at the expense of bandwidth;
   b) it also admits larger element spacings.
3) Determine the lattice to allow a desired blind-spot-free *E*-plane coverage, include sufficient margin for finite slope of the surface wave blind-spot and to avoid *H*-plane grating lobes.
4) Choose $a_y < a_x (a_y \approx d_y/2)$ (larger a_y may give rise to TE *H*-plane resonances).
5) Resonate element at broadside (or other scan angle).
6) Check the broadside bandwidth of the magnitude of active reflection coefficient $|\Gamma(f)|$.
7) If bandwidth is adequate, evaluate scan performance.
8) If bandwidth is too small, increase h/λ (or reduce ϵ) and reiterate.

The proposed design procedure is illustrated by way of an example where an unbalanced feed was assumed. The design was not optimized by iterative trials and crosspolarization level was not checked.

Consider a desired coverage of 50° off broadside in both the *E*- and *H*-planes over ±5 percent band. The value $\epsilon_r = 2.5$ was chosen and, in view of the wide angle coverage requirement in both planes, a square lattice with element spacing of 0.46 wavelength at center frequency was selected. The *H*-plane patch dimension was chosen 0.25 λ at center frequency.

To determine the substrate thickness, the patch was resonated by adjusting the *E*-plane patch dimension for a number of selected values of h/λ at center frequency. At this point we have the unmatched element geometry for each substrate thickness. Each array is now broadside matched and the magnitude of active reflection coefficient at broadside scan is computed as a function of frequency (see Fig. 16). We observe the frequency variation of $|\Gamma|$ with substrate thickness as a parameter. For $h/\lambda = 0.03$, the curve is steep indicating a narrow bandwidth. With increasing h/λ the broadside bandwidth increases. The results are summarized in the inset in Fig. 16. We see that the choice of $h/\lambda = 0.05$ at center frequency yields the desired broadside bandwidth of about 10-percent with voltage standing-wave ratio (VSWR) = 2:1. The *E*-plane patch dimension for this substrate thickness is 0.28 λ at center frequency. Having decided on the element geometry and the matching network, we can evaluate the scan performance. This is shown in Fig. 17. It is seen that while the *E*-plane

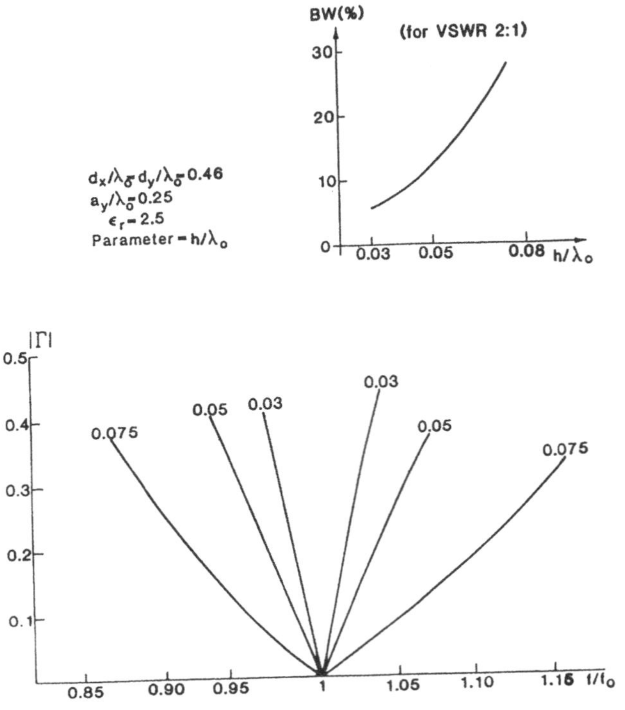

Fig. 16. Broadside bandwidth.

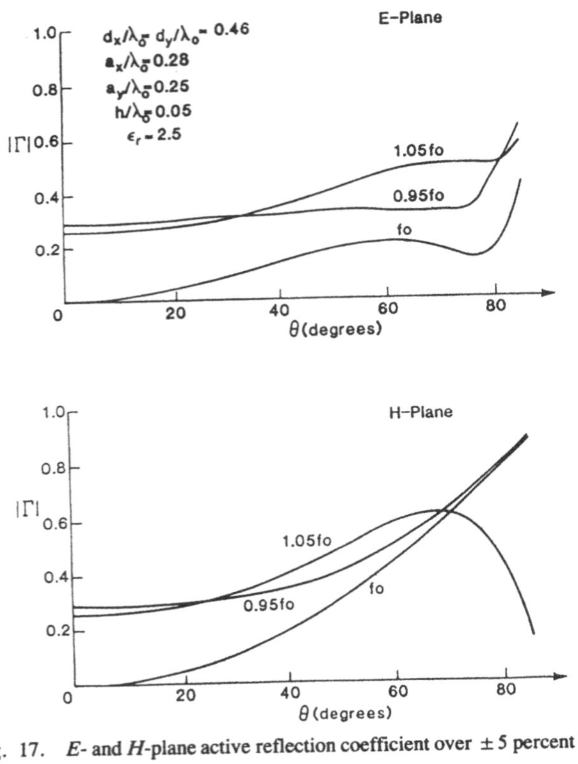

Fig. 17. *E*- and *H*-plane active reflection coefficient over ± 5 percent band.

scan VSWR at 50° is 2.2 at high frequency, the *H*-plane VSWR at high frequency is about 3 at $\theta = 50°$.

IV. Comments on Analysis in [4]

The discrepancies between our results and those of [4] stem from the following approximations in the method of analysis of [4].

1) Failure to extract the rapid current variation of the patch current in the vicinity of the probes, causing slow convergence of the patch current solution for not too thin substrates.

2)The neglect of the direct probe radiation influence on Z_a becomes important near the array blindness. The *E*-plane scan results of [4] are acceptable, but, as a consequence of the neglect of the direct probe contribution, (which also contains a surface wave contribution) [4] predicts a spurious *H*-plane blind spot at θ_{SW} where our results show a smooth, nonresonant behavior.

3) Use of sinusoidal basis functions that do not explicitly enforce the edge condition. This choice increases the required size of impedance matrix, and limits the practicality of the solution to thin, say, $h/\lambda < 0.03$ substrates. Nonetheless, [4] represents an important contribution to understanding the performance of microstrip patch phased arrays.

V. Summary and Conclusion

1) An accurate numerical solution for the scan performance of idealized probe-fed, periodic microstrip-patch phased arrays was presented.
2) Using the current basis (5), the solution is applicable to substrate thicknesses not less than, say, 0.03 λ. For $h < 0.03\ \lambda_0$ the use of sinusoidal basis is more appropriate [8].
3) Results of parametric studies disclose major performance trends.
4) They indicate that the scan and frequency performance require tighter *E*-plane spacings than in arrays of apertures in a ground plane.
5) The *E*-plane scan performance is limited by resonances of two types:
   a) a TM surface wave resonance in the neighborhood of

θ_{SW} of (8) as predicted by Pozar and Schaubert, but not causing a total reflection;

b) a leaky wave resonance closer to broadside than the surface wave blind spot.

While for a given ϵ_r and h/λ, the E-plane surface wave resonances can only be avoided by tightening the E-plane spacing (or by additional suppressors), leaky wave resonances can be damped by resonating the patch, say, at broadside with sufficiently high Q, as to yield an acceptable E-plane scan performance. The latter gives rise to characteristic scan-bandwidth trade-offs.

6) For H-plane scan with single-probe excitation, no TM surface wave blind spots are evident.
7) For patches closely spaced in the H-plane, TE type resonances are found in the H-plane scan in both balanced and unbalanced feed arrays.
8) For an unbalanced excitation a compromise between the mismatch and cross-polarization H-plane scan performance is required.
9) An element design procedure is presented along with an illustrative example.

References

[1] C. C. Liu, J. Shmoys, and A. Hessel, "Performance of probe-fed microstrip-patch phased array," in *Proc. Phased Array 1985 Symp.*.

[2] G. J. Laughlin, John Hopkins Univ., Appl. Phys. Lab., Tech. Memo TG-1278, AD-A015090, July 1975.

[3] C. C. Liu, J. Shmoys and A. Hessel, "*E*-plane performance trade-offs in two-dimensional microstrip-patch element phased arrays," *IEEE Trans. Antennas Propagat.*, vol. AP-30, pp. 1201–1206, Nov. 1982.

[4] D. M. Pozar and D. H. Schaubert, "Analysis of an infinite array of rectangular microstrip patches with idealized probe feeds," *IEEE Trans. Antennas Propagat.*, vol. AP-32, pp. 1101–1107, Oct. 1984.

[5] C. C. Liu *et al.*, "Plane wave reflection from microstrip-patch arrays: Theory and experiment," *IEEE Trans. Antennas Propagat.*, vol. AP-33, pp. 426–435, Apr. 1985.

[6] C. C. Liu, "Performance of probe-fed microstrip-patch element phased arrays," Ph.D. dissertation, Polytechnic Univ., Farmingdale, NY, 1986.

[7] A. Hessel, "General characterization of traveling-wave antennas," in *Antenna Theory*, pt. 2, R. Collin and F. J. Zucker, Eds. New York: McGraw-Hill, 1969, ch. 19.

[8] J. Schuss and J. D. Hanfling, "Synthesis of array feeds for low sidelobes and wide instantaneous bandwidths in space-fed phase steered antennas," in *Proc. Phased Array 1985 Symp.*.

Analysis of Infinite Arrays of One- and Two-Probe-Fed Circular Patches

JAMES T. ABERLE AND DAVID M. POZAR, FELLOW, IEEE

Abstract—An analysis of the radiation properties of infinite phased arrays of one- or two-probe-fed circular microstrip patches using a rigorous Green's function/moment method approach is presented. Two ways of treating the connection of a vertical feed probe to the patch are considered: an idealized probe feed model, which is useful for antennas printed on thin substrates or to predict general performance trends, and a more rigorous treatment that overcomes the deficiencies of the simpler model but requires larger amounts of computer time. Comparisons of both feed models with measured data from waveguide simulators are used to validate the theory for both one- and two-probe-fed elements.

I. Introduction

PHASED ARRAY ANTENNAS have yet to attain widespread use, primarily because the cost of implementing such antennas continues to be very high [1], but one promising approach to reducing the cost of phased arrays is the use of printed antenna technology. A variety of techniques have been used to analyze such antennas, the most rigorous of which utilize numerical techniques, such as the moment method. Recently, infinite phased arrays of printed dipoles and rectangular microstrip patches have been analyzed using this technique [2]–[10]. Experimental work with arrays of circular microstrip patches has also been reported [11].

The characteristics of a printed antenna array are determined by the choice of radiating element and feed, the substrate dielectric constant and thickness, and the element grid shape and spacing. But it will be seen that the scan performance of a circular patch array is very similar to that of a rectangular patch array, for the same substrate parameters and element grid. This is further confirmation of the premise that the dominant characteristics of printed antenna arrays are due to these factors, rather than the particular radiating element or feed used [3], [12].

In [3] the coaxial probe feed is modeled as a short electric line current, which excites current on the patch. The moment method is then used to determine the unknown patch current, and the patch input impedance is found as the reaction of the electric field scattered by the patch with the probe current. No attempt is made to model the rapidly varying patch current near the feed point, nor is the boundary condition that the tangential electric field must vanish on the probe feed enforced. Hence, the probe self-impedance as well as direct radiation from the probe is ignored. Nevertheless, this idealized probe feed model has been shown to give good agreement with measured data for both isolated patches [13] and patch arrays [3], as long as the substrate is no thicker than about 0.02 λ_0. In addition, this model is useful for predicting general trends in the behavior of patch radiators, even for arrays printed on thicker substrates.

In [4]–[7] and [9] the probe feeds are more rigorously modeled using a two-step technique, which avoids the problem of patch current singularities near the probe/patch interface. A treatment of two-probe-fed patches is given in [4], and in [6] and [7], the necessity of treating patches fed by more than one probe as multiport devices is shown. The approaches of [14] and [15] also overcome deficiencies of the idealized feed model for isolated patches, but none of the above analyses seem to be readily extendable to infinite arrays of circular patches. Hence, the extension of [3] to circular patches fed by one or two probes represents a first step in the solution of infinite arrays of circular patches.

A new, more rigorous treatment of probe-fed patch radiators, which overcomes the deficiencies of the idealized feed model, has recently been proposed [8], [10], and in this paper is applied to infinite arrays of probe-fed circular patches. The driving source is assumed to be a 1-V delta-gap generator at the base of the probe. Current on both the feed and patch is determined using the moment method. One of the basis functions used in the solution is a special attachment mode that exists on both the feed and patch. It insures current continuity from the feed to the patch and models the rapidly varying patch current near the feed point. The input impedance is then found as the reciprocal of the terminal current at the base of the probe feed. In both treatments presented here, the periodic grounded slab Green's function is used, limiting the analysis to one unit cell of the array.

Sections II and III present the theory of infinite arrays of one- and two-probe-fed circular patches using the idealized and improved probe feed models, respectively. In Section IV, both theories are compared to measured results from waveguides simulators. In Section V, relevant numerical results from the two theories are compared. Finally, conclusions are made in Section VI.

Reprinted from *IEEE Trans. Antennas Propaga.*, vol. 38, no. 4, pp. 421–432, April 1990.

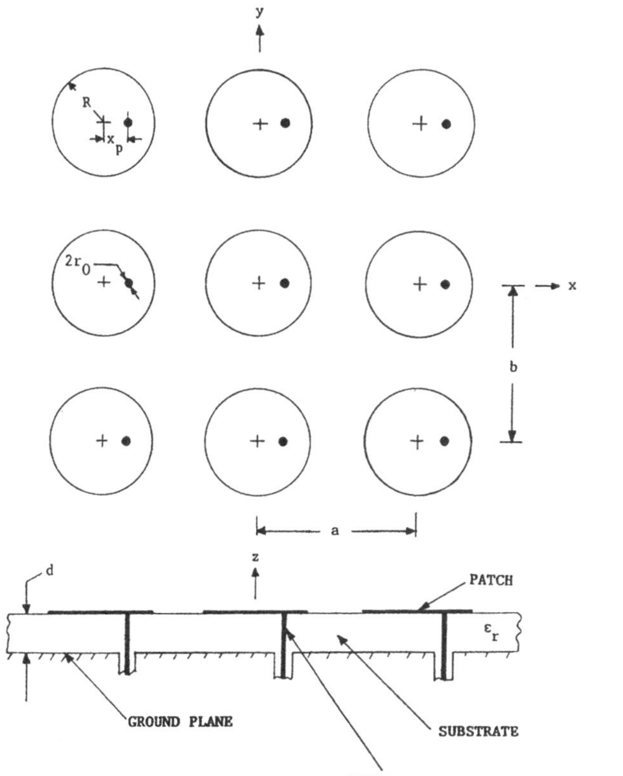

Fig. 1. Geometry of an infinite array of one-probe-fed circular microstrip patches. The ground plane is in the $z = 0$ plane.

II. The Idealized Probe Feed Model

A. Theory of One-Probe-Fed Circular Patch Infinite Arrays

The idealized probe feed model solution for infinite arrays of circular patches is an extension of the approach described in [2] and [3] for infinite arrays of other printed elements. Therefore, only a brief summary of the method will be presented here.

The geometry of an infinite, periodic planar array of identical one-probe-fed circular patches printed on a grounded dielectric substrate and probe-fed via the center conductor of a coaxial transmission line is shown in Fig. 1. If the $m = 0$, $n = 0$ patch is assumed to be centered at the origin, then the coordinates of the center of the m, nth patch are $x = ma$ and $y = nb$. The feed connection point for the m, nth element is located at $x = ma + x_p$ and $y = nb + y_p$. (In Fig. 1, $y_p = 0$). The electric field components from a horizontal current element on, and a vertical current element in, a grounded dielectric slab are needed for the calculation of the moment method impedance matrix and voltage vector elements. Eight of the nine components of the dyadic Green's function for the grounded dielectric slab were used in [3], but all nine components are needed in this paper to implement the improved probe feed model (see Section III).

The dielectric slab Green's function can be readily extended to the case of an infinite periodic array of current elements. This "periodic" Green's function can be written in the form of an infinite double sum of an infinite double integral over the spatial transform (spectral) coordinates, k_x and k_y. A theorem involving the Poisson sum formula can be used to reduce this periodic Green's function to an infinite double sum over discrete values of the spectral coordinates [2]. For notational convenience, it becomes useful to define a dyadic representation of the Green's function as

$$\bar{\bar{E}} = \hat{x}E_{xx}\hat{x} + \hat{x}E_{xy}\hat{y} + \hat{x}E_{xz}\hat{z} + \hat{y}E_{yx}\hat{x} + \hat{y}E_{yy}\hat{y} + \hat{y}E_{yz}\hat{z} + \hat{z}E_{zx}\hat{x} + \hat{z}E_{xy}\hat{y} + \hat{z}E_{zz}\hat{z}, \quad (1)$$

where

$$E_{pq}(x, y, z|x_0, y_0, z_0) = \frac{1}{ab}\sum_{m=-\infty}^{\infty}\sum_{n=-\infty}^{\infty} G_{pq}(-k_x, -k_y, z|z_0) \cdot e^{-jk_x(x-x_0)}e^{-jk_y(y-y_0)}. \quad (2)$$

$E_{pq}(x, y, z|x_0, y_0, z_0)$ is the $\hat{p}$-directed electric field at the point (x, y, z) due to an infinite periodic phased array of infinitesimal $\hat{q}$-directed current elements, where (x_0, y_0, z_0) is the location of the $m = 0$, $n = 0$ current element, and

$$k_x = \frac{2\pi m}{a} + k_0 u, \quad (3)$$

$$k_y = \frac{2\pi n}{b} + k_0 v, \quad (4)$$

$$u = \sin\theta\cos\phi, \quad (5)$$

$$v = \sin\theta\sin\phi. \quad (6)$$

The elements are phased for scanning to (θ, ϕ). The details of the Green's function components are given in [10, appendix].

The electric field integral equation (EFIE) representing the boundary condition that the total tangential electric field must vanish on the patch can be written as

$$\bar{E}^{\text{inc}}_{\tan}(x, y, d) = -\bar{E}^{\text{scat}}_{\tan}(x, y, d) = -\int\int_{S_0} \bar{\bar{E}}(x, y, x|x_0, y_0, d) \cdot \bar{J}_s(x_0, y_0)\, dx_0\, dy_0, \quad (7)$$

where $\bar{E}^{\text{inc}}$ is the incident electric field due to the probe feed, $\bar{E}^{\text{scat}}$ is the electric field scattered by the patch, S_0 is the patch surface, $\bar{\bar{E}}$ is the dyadic Green's function, and $\bar{J}_s$ is the unknown vector electric surface current density on the patch.

The surface current density $\bar{J}_s$ is expanded in a set of vector basis functions as

$$\bar{J}_s(x, y) = \sum_j I_j \bar{J}^P_j(x, y). \quad (8)$$

As will be made explicit later, the essential difference between the circular and rectangular patch solutions is exhibited in the choice of the expansion/testing modes. A Galerkin solution in the spectral domain is employed, as in [3], to reduce (7) to a matrix equation for the unknown coefficients I_j. We then have

$$[I] = [Z]^{-1}[V^t], \tag{9}$$

where the impedance matrix elements are given by [3]

$$\begin{aligned} z_{ij} = \frac{-1}{ab} \sum_{m=-\infty}^{\infty} \sum_{n=-\infty}^{\infty} &\{F_{xi}(k_x, k_y) \\ &\cdot G_{xx}(k_x, k_y, d|d) F_{xj}(-k_x, -k_y) \\ &+ F_{xi}(k_x, k_y) G_{xy}(k_x, k_y, d|d) F_{yj}(-k_x, -k_y) \\ &+ F_{yi}(k_x, k_y) G_{yx}(k_x, k_y, d|d) F_{xj}(-k_x, -k_y) \\ &+ F_{yi}(k_x, k_y) G_{yy}(k_x, k_y, d|d) F_{yj}(-k_x, -k_y)\}, \end{aligned} \tag{10}$$

and

$$\begin{aligned} V_i^t = \frac{-1}{ab} \sum_{m=-\infty}^{\infty} \sum_{n=-\infty}^{\infty} &\{F_{xi}(k_x, k_y) R_x(k_x, k_y) \\ &+ F_{yi}(k_x, k_y) R_y(k_x, k_y)\} e^{jk_x x_p} e^{jk_y y_p} \end{aligned} \tag{11}$$

is a voltage element representing the field from the probe integrated over a test mode on the patch, where

$$\bar{F}_i(k_x, k_y) = \iint_{\text{patch}} \bar{J}_i^P(x_0, y_0) e^{-jk_x x_0} e^{-jk_y y_0} \, dx_0 \, dy_0, \tag{12}$$

$$R_x(k_x, k_y) = \int_0^d G_{xz}(k_x, k_y, d|z_0) \, dz_0, \tag{13}$$

$$R_y(k_x, k_y) = \int_0^d G_{yz}(k_x, k_y, d|z_0) \, dz_0. \tag{14}$$

Note that the Fourier transforms defined here are the conjugates of those in [3].

The input impedance of the patch is calculated as the reaction of the patch scattered electric field with the current on the probe. In terms of the matrix elements,

$$Z_{\text{in}} = -\sum_j I_j V_j^e, \tag{15}$$

where

$$\begin{aligned} V_i^e = \frac{-1}{ab} \sum_m \sum_n &\{F_{xi}(-k_x, -k_y) R_x(k_x, k_y) \\ &+ F_{yi}(-k_x, -k_y) R_y(k_x, k_y)\} e^{-jk_x x_p} e^{-jk_y y_p} \end{aligned} \tag{16}$$

represents the field from an expansion mode on the patch integrated over the probe.

The scan performance of the one-probe-fed patch element is evaluated by calculating the reflection coefficient versus scan angle using

$$R(\theta, \phi) = \frac{Z_{\text{in}}(\theta, \phi) - Z_{\text{in}}(0, 0)}{Z_{\text{in}}(\theta, \phi) + Z_{\text{in}}^*(0, 0)}, \tag{17}$$

where $Z_{\text{in}}(0, 0)$ is the input impedance of the patch at broadside. Usually, the patch is used at or near its resonance, i.e., its input impedance at broadside is purely resistive. The normalized active element gain (for the case of no grating lobes in visible space) is given by

$$G_e(\theta, \phi) = \{1 - |R(\theta, \phi)|^2\} \cos\theta. \tag{18}$$

Since the patch is conjugate-matched to its broadside impedance, the reflection coefficient is zero at broadside and increases as the beam is scanned due to the patch impedance variation.

B. Theory of Two-Probe-Fed Circular Patch Infinite Arrays

The printed element in the infinite array environment fed with two probes is a two-port network which is described by a port impedance matrix $[Z^p]$. In [6] it was shown that patch radiators balanced-fed with reactive power dividers can exhibit some pernicious characteristics that would be missed if the patch were treated as a single input port device.

The elements of the patch port impedance matrix are calculated as

$$Z_{kl}^p = \frac{-1}{(I_k^{\text{inc}})(I_l^{\text{inc}})} \iiint_{V_0} \bar{E}_l^{\text{scat}} \cdot \bar{J}_k^{\text{inc}} \, dV_0; \qquad k, l = 1, 2, \tag{19}$$

where I_k^{inc} is the terminal current at the kth probe, $\bar{E}_k^{\text{scat}}$ is the electric field due to the patch surface current resulting from the kth probe, and $\bar{J}_k^{\text{inc}}$ is the current density on the kth probe. The superscript p is used here and in Section III-B to differentiate between port quantities and moment method matrix elements.

As for the one-probe-fed patch, we formulate a moment method solution to the integral equation representing the vanishing of the total electric field on the patch conductor using a Galerkin method in the spectral domain. We obtain a matrix equation which can be written as

$$[Z][I_k] = [V_k^t]; \qquad k = 1, 2, \tag{20}$$

where the elements of $[Z]$ are given by (10) and the elements of $[V_k^t]$ are given by

$$\begin{aligned} V_{ki}^t = \frac{-1}{ab} \sum_m \sum_n &\{F_{xi}(k_x, k_y) R_x(k_x, k_y) \\ &+ F_{yi}(k_x, k_y) R_y(k_x, k_y)\} e^{jk_x x_{pk}} e^{jk_y y_{pk}}; \\ &k = 1, 2. \end{aligned} \tag{21}$$

Notice that the matrix $[Z]$ is independent of the probe feeds and depends only on the patch and substrate parameters.

We can now rewrite (19) as

$$Z_{kl}^p = -\sum_j I_{lj} V_{kj}^e; \qquad k, l = 1, 2, \tag{22}$$

where

$$\begin{aligned} V_{kj}^e = \frac{-1}{ab} \sum_m \sum_n &\{F_{xi}(-k_x, -k_y) R_x(k_x, k_y) \\ &+ F_{yi}(-k_x, -k_y) R_y(k_x, k_y)\} e^{-jk_x x_{pk}} e^{-jk_y y_{pk}}. \end{aligned} \tag{23}$$

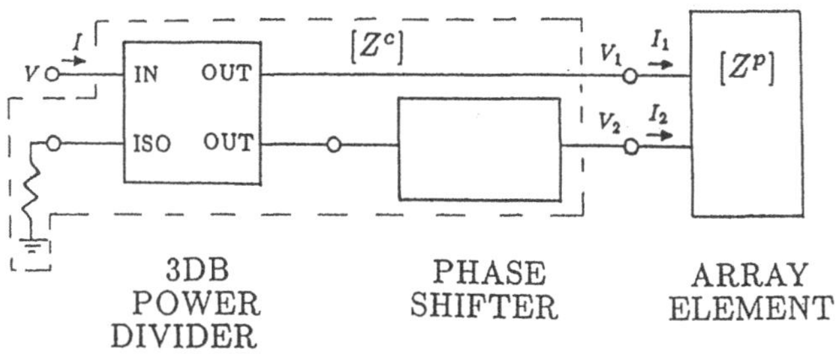

Fig. 2. Feed network configuration for two-probe-fed patch array element.

Clearly, (22) is the extension of (15) to the case of a two-probe-fed array element.

Once the two-port element impedance matrix has been calculated, further investigation of the properties of the two-probe-fed radiator depends on the network chosen to feed the element. Consider the feed network shown in Fig. 2, which consists of an ideal 3 dB power divider and an ideal phase-shifter. The power divider has an isolated port which is terminated in a matched load. Both power division and phase-shifting may actually take place within a single device, such as a branch-line or rat-race coupler. While more complicated than the feed network considered in [6], this feed network has isolation and avoids the parasitic mode effect described in [6]. For the two-probe-fed patch connected to the directional coupler/phase-shifter, the terminal voltages and currents are related by

$$\begin{bmatrix} V_1 \\ V_2 \end{bmatrix} = [Z^p] \begin{bmatrix} I_1 \\ I_2 \end{bmatrix}, \tag{24}$$

$$\begin{bmatrix} V \\ V_1 \\ V_2 \end{bmatrix} = [Z^c] \begin{bmatrix} I \\ -I_1 \\ -I_2 \end{bmatrix}, \tag{25}$$

where $[Z^p]$ is the two-port unit cell impedance matrix and $[Z^c]$ is the three-port impedance matrix of the terminated power-divider network. The impedance looking into each probe of the patch may then be found from a straightforward network analysis as

$$Z_1 = \frac{V_1}{I_1}, \tag{26}$$

$$Z_2 = \frac{V_2}{I_2}. \tag{27}$$

Note that these impedances depend not only on the patch but also on the feed network. The reflection coefficients at each probe are given by

$$R_1 = \frac{Z_1 - Z_0}{Z_1 + Z_0}, \tag{28}$$

$$R_2 = \frac{Z_2 - Z_0}{Z_2 + Z_0}, \tag{29}$$

where Z_0 is the characteristic impedance of the feed network. The reflection coefficient looking into the feed network is given by

$$R(\theta, \phi) = \frac{1}{2}\{R_1(\theta, \phi) + R_2(\theta, \phi)e^{-j2\Omega}\}, \tag{30}$$

where Ω is the phase difference between the probes.

For a two-probe-fed patch, mismatch at the probes results not only in mismatch at the input of the feed network, but possibly also in power dissipation in the feed network. For an ideal feed network, the relative power dissipated in the termination of the isolated port is given by

$$P_d(\theta, \phi) = \frac{1}{4}\left|R_1(\theta, \phi) - R_2(\theta, \phi)e^{-j2\Omega}\right|^2. \tag{31}$$

For two-probe-fed patches, the scanning performance is most conveniently evaluated in terms of the normalized active element gain pattern given by

$$G_e(\theta, \phi) = \{1 - |R(\theta, \phi)|^2\}\eta(\theta, \phi)\cos\theta, \tag{32}$$

where η is the feed network efficiency which for the ideal feed network is given by

$$\eta(\theta, \phi) = 1 - \frac{P_d(\theta, \phi)}{1 - |R(\theta, \phi)|^2}. \tag{33}$$

This accounts for power loss in the termination of the power divider.

The characteristic impedance of the feed network Z_0 can be adjusted so that, for a resonant patch, the reflection coefficients at the probes are zero at broadside. Hence, the normalized active element gain will be unity at broadside (for an ideal feed network) and will decrease away from broadside due to the patch impedance variation, the $\cos\theta$ factor, and the feed network efficiency.

C. The Expansion/Test Modes

The unknown electric surface current density on the circular patch is expanded in a set of entire domain basis (EDB) functions. In terms of the cylindrical coordinates ρ and ϕ, these functions are given by

$$\bar{J}_i^P(\rho, \phi) = \hat{\rho}\beta_{mn}J'_m(\beta_{mn}\rho)\cos m\phi - \hat{\phi}\frac{m}{\rho}J_m(\beta_{mn}\rho)\sin m\phi, \tag{34}$$

where the prime denotes differentiation with respect to the argument and $J'_m(\beta_{mn}R) = 0$, i.e., $\beta_{mn}R$ is the nth zero of $J'_m(x)$. The index i is used to reduce the set of basis functions to a one-dimensional sequence; the indices m and n are then determined by i. Note that these expansion modes correspond to one of the two sets of TM_{mn} modes of a circular cavity formed by electric walls on top and bottom and a magnetic wall around its circumference. This mode set corresponds to a wave function with even symmetry. For an array with elements fed at $\phi = 0$ (i.e., along the x-axis) and scanned only in the E-plane this set of EDB functions will suffice. However, in order to treat arbitrary feed positions and scan angles, it is necessary to use an additional set of EDB functions corresponding to a cavity wave function possessing odd symmetry. This set is given by

$$\bar{J}_i^P(\rho, \phi) = \hat{\rho}\beta_{mn}J'_m(\beta_{mn}\rho)\sin m\phi + \hat{\phi}\frac{m}{\rho}J_m(\beta_{mn}\rho)\cos m\phi. \quad (35)$$

The EDB functions given by (34) will be referred to as the "principal" set and those given by (35) as the "orthogonal" set.

The Fourier transforms used here are defined as in (12) and are given in closed form as

$$\begin{aligned}\bar{F}_i(k_x, k_y) = 2\pi R j^{-m+1}J_m(\beta_{mn}R) \\ \cdot \Bigg\{\hat{x}\left[\cos\alpha\cos m\alpha\frac{\beta_{mn}^2}{\beta_{mn}^2-\beta^2}J'_m(\beta R) + m\sin\alpha\sin m\alpha\frac{J_m(\beta R)}{\beta R}\right] \\ \cdot\hat{y}\left[\sin\alpha\cos m\alpha\frac{\beta_{mn}^2}{\beta_{mn}^2-\beta^2}J'_m(\beta R) - m\cos\alpha\sin m\alpha\frac{J_m(\beta R)}{\beta R}\right]\Bigg\},\end{aligned} \quad (36)$$

for the principal set, and

$$\begin{aligned}\bar{F}_i(k_x, k_y) = 2\pi R j^{-m+1}J_m(\beta_{mn}R) \\ \cdot \Bigg\{\hat{x}\left[\cos\alpha\sin m\alpha\frac{\beta_{mn}^2}{\beta_{mn}^2-\beta^2}J'_m(\beta R) - m\sin\alpha\cos m\alpha\frac{J_m(\beta R)}{\beta R}\right] \\ +\hat{y}\left[\sin\alpha\sin m\alpha\frac{\beta_{mn}^2}{\beta_{mn}^2-\beta^2}J'_m(\beta R) + m\cos\alpha\cos m\alpha\frac{J_m(\beta R)}{\beta R}\right]\Bigg\},\end{aligned} \quad (37)$$

for the orthogonal set, where

$$\beta = \sqrt{k_x^2 + k_y^2}, \quad (38)$$

$$\alpha = \arctan\left(\frac{k_y}{k_x}\right). \quad (39)$$

Because the basis functions given by (34) and (35) are real, we have

$$\bar{F}_i(-k_x, -k_y) = \bar{F}_i^*(k_x, k_y). \quad (40)$$

Numerical convergence checks and comparison of numerical results with measurements of both isolated circular patches and circular patches in waveguide simulators are used to determine the choice of expansion modes. Near its resonance, the patch acts very much like a resonant cavity, and hence, the resonant mode current dominates the total current [14]. This is why the idealized probe feed model works well for thin substrates, when only a few EDB modes are used. If too many modes are used, they will try to model the rapidly varying current near the probe/patch interface, and the calculated impedance will be too inductive. For linearly polarized elements scanned in the E-plane, the TM_{11} and the TM_{12} modes from the principal set yield good results. For off-E-plane scanning, the TM_{11} mode from the orthogonal set is included. This mode gives rise to cross-polarized field components and is excited by the asymmetry inherent in off-E-plane scanning. For circularly-polarized elements, the TM_{11} and TM_{12} modes from both principal and orthogonal sets are used. Thus, the idealized feed model calculations presented in the following sections use from two to four expansion modes.

III. The Improved Probe Feed Model

A. Theory of One-Probe-Fed Circular Patch Infinite Arrays

The analytic approach presented in this section is an extension of the method presented in [10] for rectangular patches. Therefore, only a brief summary will be presented here.

In the more rigorous treatment of probe-fed printed radiators, a delta-gap voltage generator is assumed to exist at the base of the probe. Hence, the EFIE representing the boundary condition that the total tangential electric field must vanish on both the feed and patch can be written as

$$\bar{E}_{\tan}^{\text{inc}} = -\bar{E}_{\tan}^{\text{scat}} \quad \text{on feed and patch,} \quad (41)$$

where

$$\bar{E}^{\text{inc}}(x, y, z) = \hat{z}\delta(z); x = x_p, y = y_p, \quad (42)$$

$$\bar{E}^{\text{scat}}(x, y, z) = \iiint \bar{\bar{E}}(x, y, z|x_0, y_0, z_0) \cdot \bar{J}(x_0, y_0, z_0)\,dx_0\,dy_0\,dz_0, \quad (43)$$

$\bar{\bar{E}}$ is the dyadic Green's function, and $\bar{J}$ is the unknown vector electric volume current density on the feed and patch. The current density on the feed and patch is expanded in a set of vector basis functions as

$$\bar{J}(x, y, z) = \sum_{j=1}^{N} I_j\bar{J}_j(x, y, z). \quad (44)$$

Each expansion mode may have two pieces, one of which exists on the feed, the other on the patch, i.e.,

$$\bar{J}_j(x, y, z) = \bar{J}_j^F(x, y, z) + \bar{J}_j^P(x, y, z); \quad j = 1, 2, \cdots, N, \quad (45)$$

where $\bar{J}_j^F$ is the part of the jth expansion mode existing on the feed, and $\bar{J}_j^P$ is the part of the jth expansion mode existing on the patch.

Since the feed radius is small, the probe is treated using a filamentary current approximation, as described in [10]. The field due to $\bar{J}_j^F$ is assumed to arise from an infinitesimally thin current filament, but $\bar{J}_j^F$ itself is assumed to exist on a cylinder of radius r_0.

Again, Galerkin's method is used to discretize the EFIE, yielding the matrix equation (9) for the unknown coefficients I_j. In this treatment, the voltage vector elements are given by the value of the corresponding expansion mode current at

$z = 0$, and the impedance matrix elements are given by

$$z_{ij} = z_{ij}^{FF} + z_{ij}^{FP} + z_{ij}^{PF} + z_{ij}^{PP}, \quad (46)$$

where

$$z_{ij}^{rs} \equiv -\iiint \bar{E}_j^s(x, y, z) \cdot \bar{J}_i^r(x, y, z)\, dx\, dy\, dz, \quad (47)$$

and $\bar{E}_j^s$ is the field due to the part of expansion mode j on structure s (i.e., either the feed or patch).

In order to continue with this development, it is now necessary to choose a set of modes to expand the unknown currents on the feed and patch. The first mode is an attachment mode which exists on both the feed and patch, and so insures continuity of current from the feed to the patch. The part of the attachment mode on the feed is taken to be a unit current, constant over $0 < z < d$. The patch part of the attachment mode is found by solving the boundary value problem of a cavity formed by electric walls on top and bottom and a magnetic wall around its circumference (the usual cavity model of a circular microstrip patch), and fed by a uniform current filament. We expect that this mode will accelerate the convergence of the moment method solution by modeling the rapid variation in patch current near the feed connection point. This part of the attachment mode is given by

$$\bar{J}_1^P(\rho, \phi) = \frac{-k}{4}\left[\frac{N_1(k\rho')}{\rho'}\{\hat{\rho}\rho - \rho_p \cos(\phi - \phi_p) + \hat{\phi}\rho_p \sin(\phi - \phi_p)\} + \sum_{m=0}^{\infty} \epsilon_m \cdot \left\{\frac{N_m'(kR) J_m(k\rho_p)}{J_m'(kR)}\right\} \left\{\hat{\rho} J_m'(k\rho)\cos m(\phi - \phi_p) - \hat{\phi}\frac{m}{k\rho} J_m(k\rho) \sin m(\phi - \phi_p)\right\}\right], \quad (48)$$

where

$$k = \sqrt{\epsilon_r} k_0, \quad (49)$$

$$\rho' = \sqrt{\rho^2 + \rho_p^2 - 2\rho\rho_p \cos(\phi - \phi_p)}, \quad (50)$$

$$\rho_p = \sqrt{x_p^2 + y_p^2}, \quad (51)$$

$$\phi_p = \arctan\left(\frac{y_p}{x_p}\right). \quad (52)$$

This current mode is the circular patch analog to the rectangular patch attachment mode current given by (32) of [10].

Near the TM_{11} dominant mode resonance, the attachment mode closely resembles the TM_{11} mode current, except that it exhibits a singularity at the feed connection point. It can be shown that the charge build-up at the probe tip associated with the discontinuous (at $z = d$) $\hat{z}$-directed feed current is exactly cancelled by a charge build-up associated with the patch current of (48) at $\rho' = 0$. The Fourier transform of (48) may be evaluated in closed form as

$$\bar{F}_1(k_x, k_y) = -\sum_{m=0}^{\infty} \epsilon_m j^{-m+1} \cdot \left[\left\{\frac{kJ_m(k\rho_p)J_m'(\beta R) - \beta J_m'(kR) J_m(\beta\rho_p)}{(k^2 - \beta^2) J_m'(kR)}\right\} \cdot \cos m(\alpha - \phi_p)\{\hat{x}\cos\alpha + \hat{y}\sin\alpha\} + \left\{\frac{mJ_m(k\rho_p)J_m(\beta R)}{k\beta R J_m'(kR)}\right\} \cdot \sin m(\alpha - \phi_p)\{\hat{x}\sin\alpha - \hat{y}\cos\alpha\}\right], \quad (53)$$

where β and α are given by (38) and (39), respectively. Since the attachment mode current is real, (40) holds for this mode as well. The remaining expansion modes are the two sets of EDB functions described in Section II-C.

From (47), it can be seen that if either mode i does not have a piece on structure r or mode j does not have a piece on structure s, then the matrix subelement z_{ij}^{rs} vanishes. Hence, the structure of the impedance matrix generated by the expansion modes described above is given by [10, eq. (45)]. The voltage vector is given by

$$[V^t] = \begin{bmatrix} 1 \\ 0 \\ \vdots \\ 0 \end{bmatrix}. \quad (54)$$

The matrix equation (9) is then solved, and the input impedance looking into the patch is calculated as

$$Z_{\text{in}} = \frac{1}{I_1}. \quad (55)$$

As in the idealized probe feed model, the scan performance of the patch element can be evaluated using (17).

B. Theory of Two-Probe-Fed Circular Patch Infinite Arrays

The two-port admittance matrix $[Y^P]$ is used to describe a two-probe-fed printed element in the improved probe feed model analysis. The elements of the matrix are defined as

$$y_{11}^p = \left.\frac{I_1^p}{V_1^p}\right|_{V_2^p=0}, \quad (56)$$

$$y_{12}^p = \left.\frac{I_1^p}{V_2^p}\right|_{V_1^p=0}, \quad (57)$$

$$y_{21}^p = \left.\frac{I_2^p}{V_1^p}\right|_{V_2^p=0}, \quad (58)$$

$$y_{22}^p = \left.\frac{I_2^p}{V_2^p}\right|_{V_1^p=0}, \quad (59)$$

where V_k^p and I_k^p are the terminal voltage and current, respectively at the kth feed.

The EFIE representing the boundary condition that the total tangential electric field must vanish on the feed and patch conductors when either feed is excited can be written as

$$\bar{E}_k^{\text{inc}} = -\bar{E}_k^{\text{scat}} \quad \text{on feeds and patch;} \qquad k = 1, 2, \tag{60}$$

where

$$E_k^{\text{inc}}(x, y, z) = \hat{z}\delta(z); x = x_{pk}, y = y_{pk}, \tag{61}$$

$$\bar{E}_k^{\text{scat}}(x, y, z) = \iiint \bar{E}(x, y, z|x_0, y_0, z_0) \cdot \bar{J}_k(x_0, y_0, z_0)\, dx_0\, dy_0\, dz_0, \tag{62}$$

and $\bar{J}_k$ is the vector electric volume current density on the feeds and patch when the kth feed is excited.

As for the one-probe-fed case, the unknown current density is expanded in a set of vector basis functions, and Galerkin's method is used to discretize the EFIE yielding the matrix equation (20), where the impedance matrix elements are given again by (46).

For the two-probe-fed case, two attachment modes are used, each of which exists on one feed and the patch. The patch part of each expansion mode can be obtained by generalizing (48). Additional expansion modes on the patch are once again drawn from the two sets described in Section II-C. The impedance matrix generated by this set of expansion modes has the structure given by [10, eq. (63)].

The voltage vectors are given by

$$[V_1^t] = \begin{bmatrix} 1 \\ 0 \\ \vdots \\ 0 \end{bmatrix}, \tag{63}$$

$$[V_2^t] = \begin{bmatrix} 0 \\ 1 \\ \vdots \\ 0 \end{bmatrix}, \tag{64}$$

The matrix equation (20) is then solved for the current coefficients. Notice that I_{11} and I_{12} are the port currents I_1^p and I_2^p when port 1 is excited by a 1-V generator and port 2 is short-circuited. Further, I_{21} and I_{22} are the port currents I_1^p and I_2^p when port 2 is excited by a 1-V generator and port 1 is short circuited. Hence, the two-port admittance matrix may be evaluated as

$$[Y^p] = \begin{bmatrix} I_{11} & I_{21} \\ I_{12} & I_{22} \end{bmatrix}. \tag{65}$$

The two-port impedance matrix $[Z^p]$ can then be obtained by inverting $[Y^p]$. The properties of the two-probe-fed patch radiator are then determined by this matrix and the feeding network described in Section II-B.

C. Convergence Considerations

Both solutions to the problem of infinite arrays of probe-fed circular patches described in this paper are formulated in terms of matrix elements which are expressed as infinite summations. Each term in the summation corresponds to a Floquet mode of the infinite periodic structure. In order to calculate these matrix elements with good accuracy, it is necessary to use about 101 terms in each series (i.e., $-50 < m, n < 50$) for the idealized probe feed model solution. However, in the more rigorous solution, it is necessary to use about 501 terms in the series involving the self-reaction of the attachment mode (i.e., z_{11} for the one-probe-fed case, and z_{11}, z_{12}, z_{21} and z_{22} for the two-probe-fed case). The reason for this is that the attachment mode used in the rigorous solution has a very wide spectral bandwidth since it models both the rapidly-varying patch current near the feed point and the discontinuous feed current.

An appropriate set of expansion modes from Section II-C must be chosen for use in the more rigorous solution. Through numerical convergence checks, it is found that the TM_{1n} modes from the principal set with $n = 1, 2, 3, \cdots, 8$ yield good results for linearly polarized elements scanned in the E-plane. For off E-plane scanning, the TM_{11} mode from the orthogonal set is also included. For circularly polarized elements, the TM_{1n}, $n = 2, 3, 4, \cdots, 8$ modes from the orthogonal set are also included.

A third convergence issue which must be addressed in the improved feed model is the number of terms required in the calculation of the series for the patch part of the attachment mode. As long as the patch is not fed near the edge, the series converges quite rapidly, and ten terms were used to obtain the results presented in the next sections.

IV. Waveguide Simulator Results

The waveguide simulator provides a convenient way to evaluate the performance of a radiating element in the infinite array environment [16]. In Figs. 3 and 4, one-probe-fed patch infinite array theory is compared with measurements from a waveguide simulator. In Fig. 3, the substrate is relatively thin, and both models yield good agreement with measured data. The relative dielectric constant used in the calculation is $\epsilon_r = 2.28$, which is outside the quoted tolerance of $\epsilon_r = 2.33 \pm 0.02$. However, the bandwidth of this radiator is extremely narrow, and other tolerances (e.g., patch radius) may be affecting the result. None of the other cases reported here required adjustment of the dielectric constant. The simulator is standard S-band waveguide operating in the TE_{10} mode. Although the TE_{20} mode may propagate above 4.16 GHz, it will not be excited because of the element configuration used. On the other hand, the TE_{01} mode could be excited by the unsymmetric patch radiators above 4.41 GHz. However, there was no indication that this occurred in the observed frequency range. Thus, scanning in the H-plane is simulated, and at 4.0 GHz, the scan angle is 31.34°. For one-probe-fed patches, the probe feed does not image correctly in the simulator. However, this does not seem to be a significant limitation, at least for thin substrates, because the position of the feed probe, relative to other patches in the vicinity, has

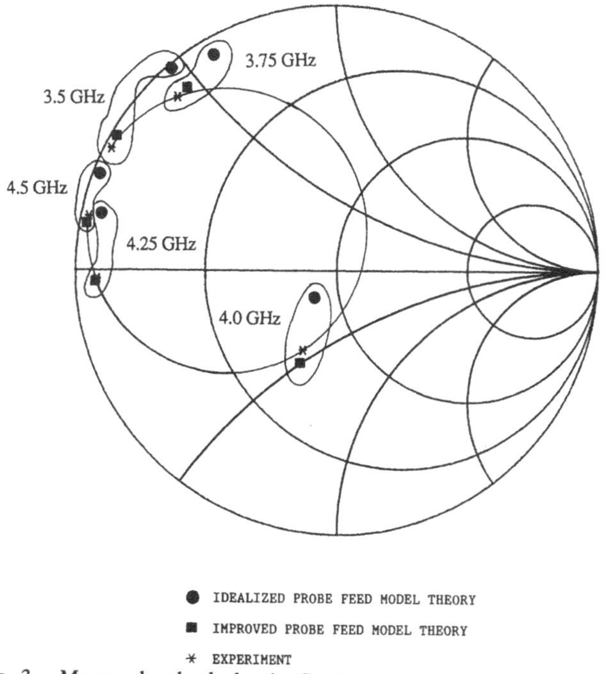

Fig. 3. Measured and calculated reflection coefficient magnitude of a one-probe-fed circular patch in an H-plane waveguide simulator ($\theta = 31.34°$ at 4 GHz, $\epsilon_r = 2.33$, $d = 0.079$ cm, $R = 1.429$ cm, $x_p = 0.5$ cm, $r_0 = 0.045$ cm, $a = 3.40$ cm, $b = 3.61$ cm).

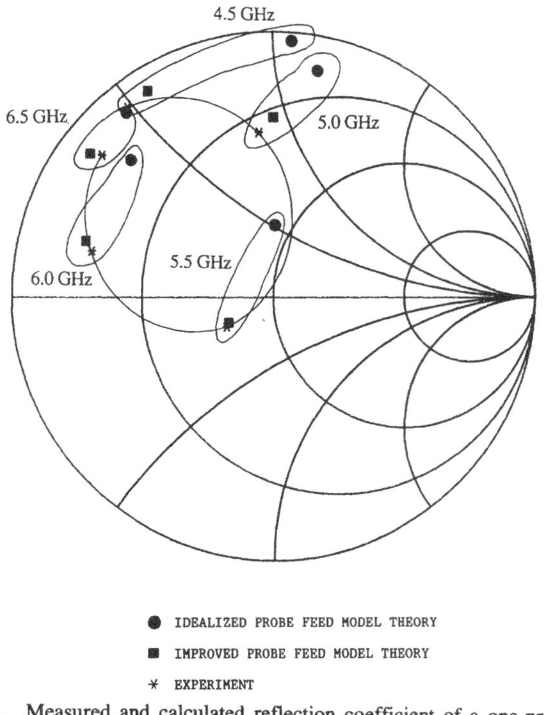

Fig. 4. Measured and calculated reflection coefficient of a one-probe-fed circular patch in an H-plane waveguide simulator ($\theta = 35.00°$ at 5.5 GHz, $\epsilon_r = 2.2$, $d = 0.159$ cm, $R = 1.0$ cm, $x_p = 0.4$ cm, $r_0 = 0.045$ cm, $d_x = 2.215$ cm, $d_y = 2.378$ cm).

little effect on the array performance [3], [13]. In Fig. 4, the substrate is somewhat thicker, and it is clear that the improved feed model generates significantly better agreement with the measured data. In this case, a C-band simulator operating in the TE_{10} mode was used, and at 5.5 GHz the simulated scan angle is 35.00° in the H-plane.

In Figs. 5 and 6, two-probe-fed patch infinite array theory

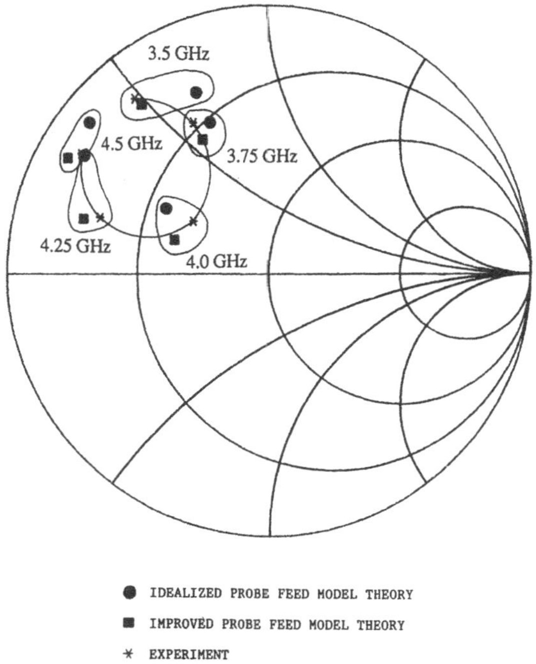

Fig. 5. Measured and calculated S_{11} of a two-probe-fed circular patch in an H-plane waveguide simulator ($\theta = 31.34°$ at 4 GHz, $\epsilon_r = 2.2$, $d = 0.159$ cm, $R = 1.429$ cm, $x_{p1} = -x_{p2} = 0.5$ cm, $r_0 = 0.045$ cm, $a = 3.40$ cm, $b = 3.61$ cm).

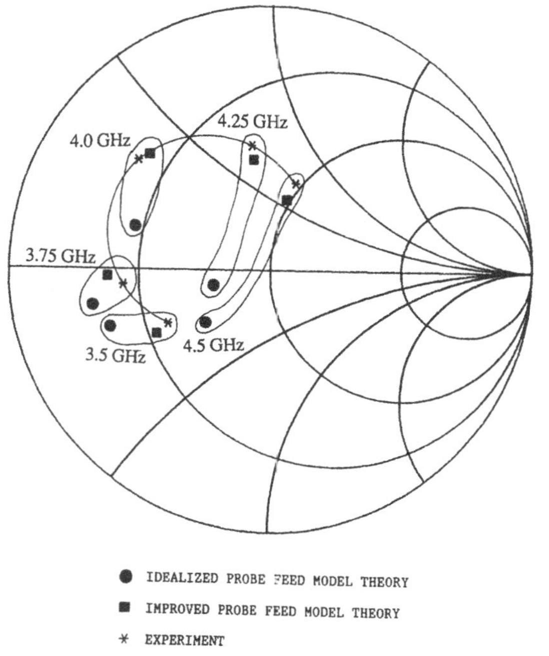

Fig. 6. Measured and calculated S_{21} of a two-probe-fed circular patch array in an H-plane waveguide simulator (same parameters as Fig. 5).

is compared with measurements from the S-band waveguide simulator; the patch is treated as a two port device, and its scattering parameters are measured. In general, the printed array element is a nonreciprocal device because of the asymmetry introduced by beam scanning. However, in the H-plane simulator, the balanced two-probe-fed patch retains its symmetry and is a reciprocal two-port. As can be seen, both models yield reasonable results for S_{11} (Fig. 5), but the more rigorous feed model yields much closer agreement with measured data for S_{21} (Fig. 6). This is primarily due to the fact that the idealized feed model does not implicitly include the probe-to-probe coupling that exists in a two-probe-fed patch. An *ad hoc* correction to the idealized feed model has been made to account for this coupling, by calculating the probe mutual impedance (in the infinite array environment) and adding this to the off-diagonal elements of the patch port impedance matrix. Another refinement that has been incorporated in the idealized feed model is to calculate the real part of the probe self-impedance and add this to the input impedance of the patch to account for the probe radiation. These refinements have been found to work fairly well for thin substrates, but even so it is clear that the improved feed model yields significantly better agreement with experiment.

It is apparent from these results that use of the improved feed model for probe-fed printed radiators is highly desirable. However, the improved feed model requires significantly more computer time than does the idealized feed model. This is primarily due to the number of terms required for accurate calculation of the Floquet mode summations for the matrix elements. For example, about 17 h VAX 11/750 CPU time were needed to compute the improved feed model data for Fig. 4, but less than 14 min were required to compute the idealized feed model data.

V. Numerical Results

In this section two numerical examples are studied. The first is a one-probe-fed linearly polarized circular patch infinite array element, and the second is a two-probe-fed circularly polarized circular patch infinite array element.

Fig. 7 shows the broadside-matched reflection coefficient magnitude in three planes of scan for the one-probe-fed circular patch array calculated in one-degree increments using both the idealized and the more rigorous probe feed models. (The "*D*-plane" is an intercardinal plane with $\phi = 45°$.) The array is printed on a substrate with $\epsilon_r = 2.55$ and $d = 0.06\ \lambda_0$. Since the element spacing along the x-axis is $0.51\ \lambda_0$, *E*-plane scanning to $\theta = 73.90°$ brings a grating lobe into real space. The condition for propagation of the TM_0 surface wave mode (discussed in [2] and [3]) is satisfied for $\theta = 68.90°$ in the *E*-plane and $\theta = 76.43°$ in the *H*-plane. The idealized probe feed model predicts total reflection at these scan angles, while the more rigorous feed model predicts a large (but not unity) reflection coefficient at the predicted angle in the *E*-plane, but not in the *H*-plane. Hence, the results of the improved feed mode are seen to be in agreement with those of [9].

Comparison of Fig. 7 with [10, fig. 7], which is for a rectangular patch array with the same substrate and array grid parameters, demonstrates the similarity between circular and rectangular patch arrays. This result agrees with the conclusions of [12], where a simple current sheet model was used to show that the dominant scanning characteristics of a printed phased array depend more on the substrate and element grid than on the particular element type. In addition, the fact that the two results in Fig. 7 are in close agreement supports the argument in [3] that the idealized feed model can be used to accurately predict the scan performance of such probe-fed arrays.

Fig. 8 shows the cross-polarization level computed using Ludwig's third definition [17] versus scan angle for the array of Fig. 7. The radiated field of the array is taken to be the $m = n = 0$ Floquet mode. Since this is no longer true near endfire, the polarization calculation is not meaningful for θ close to 90°. The cross-polarized field is zero in the *E*-plane, and the cross-polarization levels in the *D*- and *H*-planes predicted using the two feed models are fairly close (especially in the *D*-plane). These cross-polarization results are similar to the results obtained for the corresponding rectangular patch array [10, fig. 8].

For two-probe-fed patch arrays, the most convenient measure of scan performance is the active element gain pattern, (32), since it includes both power reflected out of and dissipated in the feed network. Two feed configurations are of particular interest:

1) balanced feed for linear polarization with reduced cross polarization,
2) orthogonal feed for circular polarization.

For the balanced feed, the probes are at $(S, 0)$ and $(-S, 0)$, and $\Omega = 180°$. For the circularly polarized feed, the probes are at $(S, 0)$ and $(0, S)$, and $\Omega = \pm 90°$ (depending on the desired rotation of the circular polarization). In general, for a two-probe-fed patch array element, $G_e(\theta, \phi)$ will not be equal to $G_e(\theta, \phi + \pi)$ (unlike the one-probe-fed patch array), because of the asymmetry of the radiating element. However, the difference is quite small. In the special case of a balanced feed, the element is symmetric, and $G_e(\theta, \phi)$ is then equal to $G_e(\theta, \phi + \pi)$.

Fig. 9 shows the active element gain in three planes of scan for the circularly polarized circular patch array calculated using the rigorous feed model. The array is printed on a substrate with $\epsilon_r = 2.55$, $d = 0.02\ \lambda_0$. The active element gain computed for the array using the idealized feed model is virtually identical to Fig. 9. Fig. 10 shows the axial ratio in decibels versus scan angle for the same array calculated using the more rigorous feed model. There is a dramatic difference in axial ratio levels predicted by the two feed models. Even at broadside, the rigorous feed model predicts a transmitted field which is not perfectly circularly polarized, in agreement with [7]. However, the idealized feed model predicts 0 dB axial ratio at broadside. The reason for this discrepancy is that the idealized feed model dramatically underestimates the amount of cross-polarized current excited by each of the probes in this configuration. Since the cross-polarized currents look different for different feeds, the two orthogonal polarizations are unbalanced. Much higher quality circular polarization can be obtained by feeding the patch with balanced pairs of probes [7].

Results for the equivalent square patch array are given by [10, figs. 12 and 13]. Again, it is apparent that the substrate and array grid parameters, not the element shape, are the primary factors controling the scan performance of the array.

VI. Conclusion

This paper has presented two solutions to the problem of an infinite array of circular microstrip patches fed by one or two probes. The idealized probe feed model, while pos-

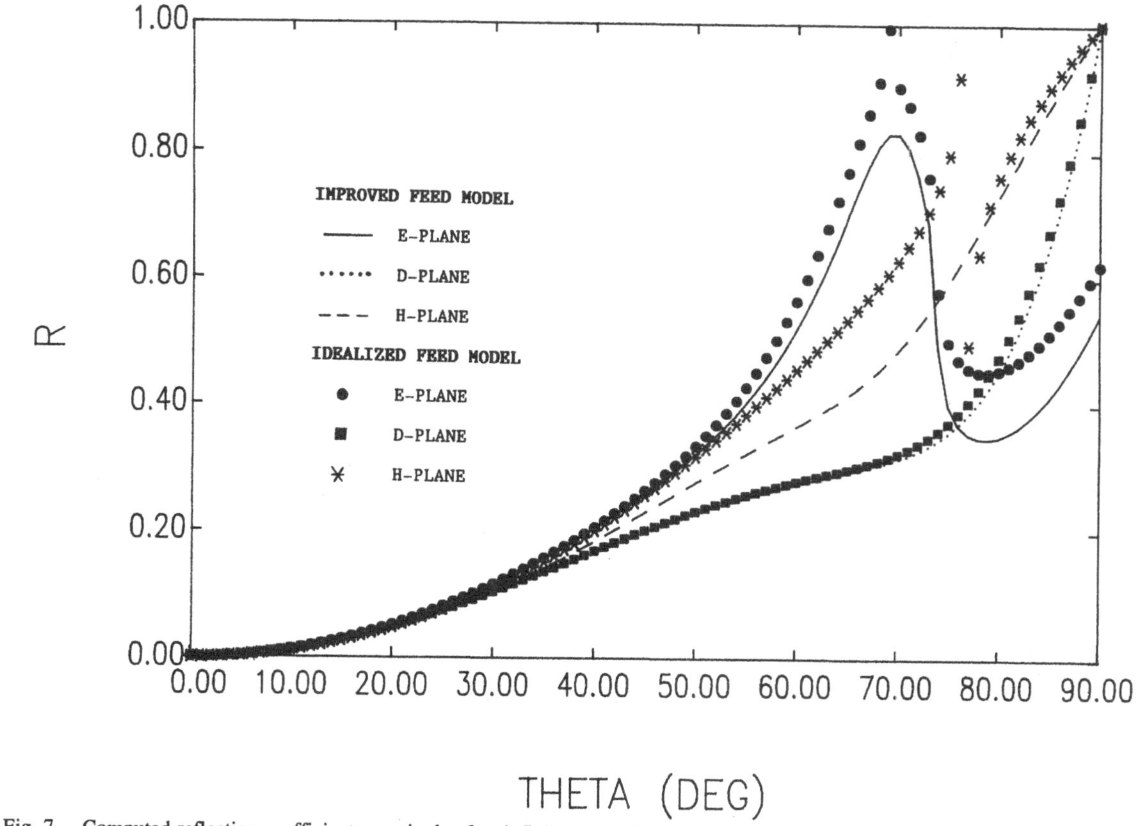

Fig. 7. Computed reflection coefficient magnitude of an infinite array of one-probe-fed circular patches ($\epsilon_r = 2.55$, $d = 0.06\ \lambda_0$, $R = 0.166\ \lambda_0$, $x_p = 0.083\ \lambda_0$, $r_0 = 0.0004\ \lambda_0$, $a = 0.51\ \lambda_0$, $b = 0.5\ \lambda_0$).

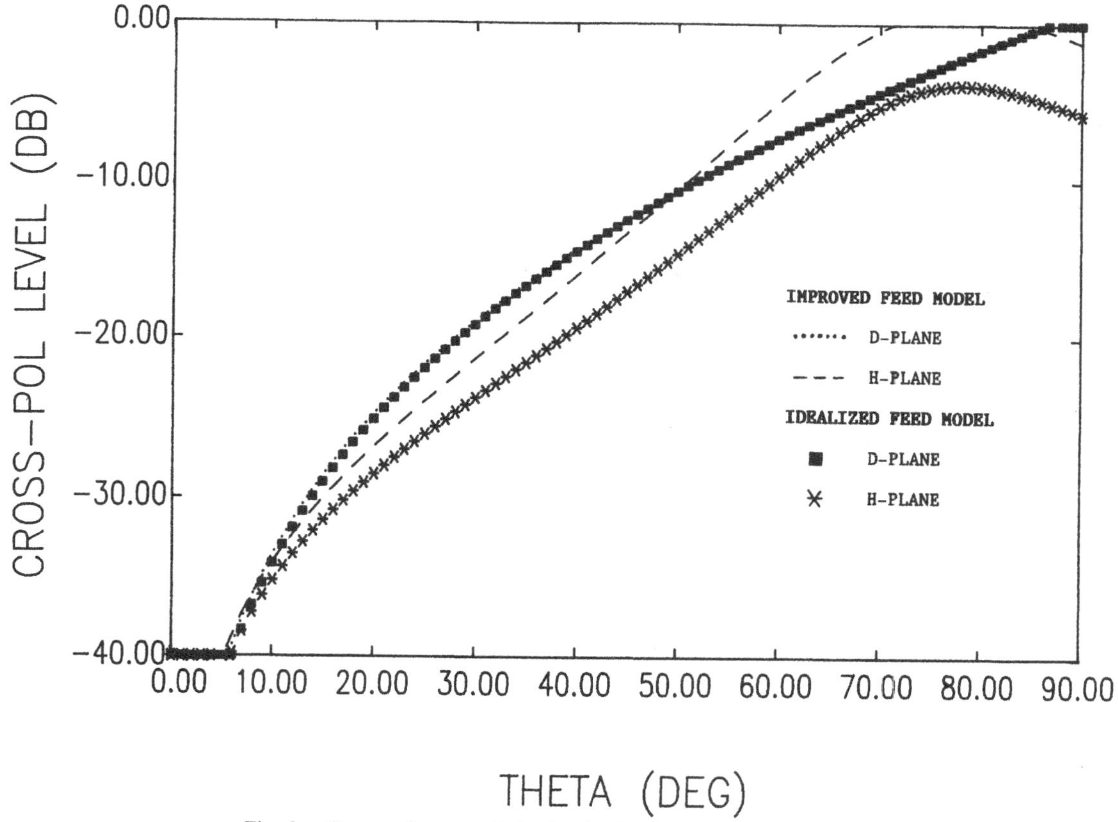

Fig. 8. Computed cross-polarization level for the array of Fig. 7.

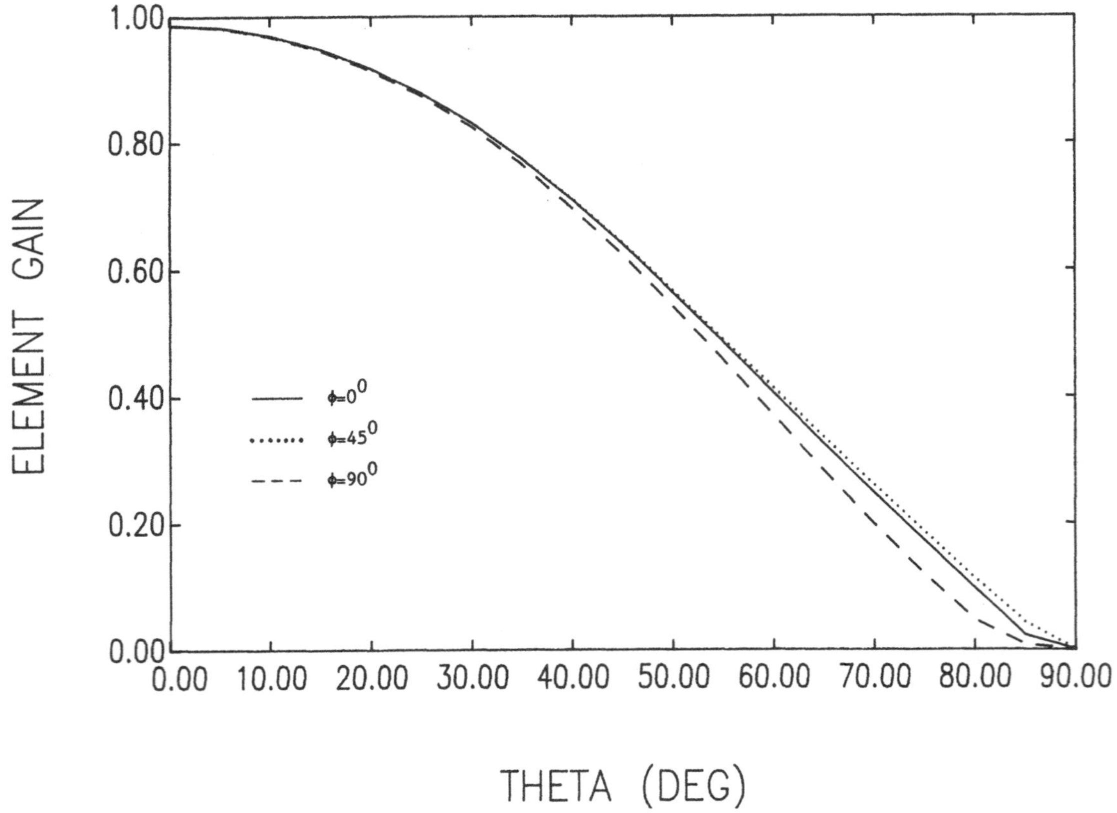

Fig. 9. Computed normalized active element gain of an infinite array of two-probe-fed circularly polarized patches ($\epsilon_r = 2.55$, $d = 0.02\ \lambda_0$, $R = 0.177\ \lambda_0$, $S = 0.0885\ \lambda_0$, $r_0 = 0.0004\ \lambda_0$, $\Omega = 90°$, $a = b = 0.5\ \lambda_0$).

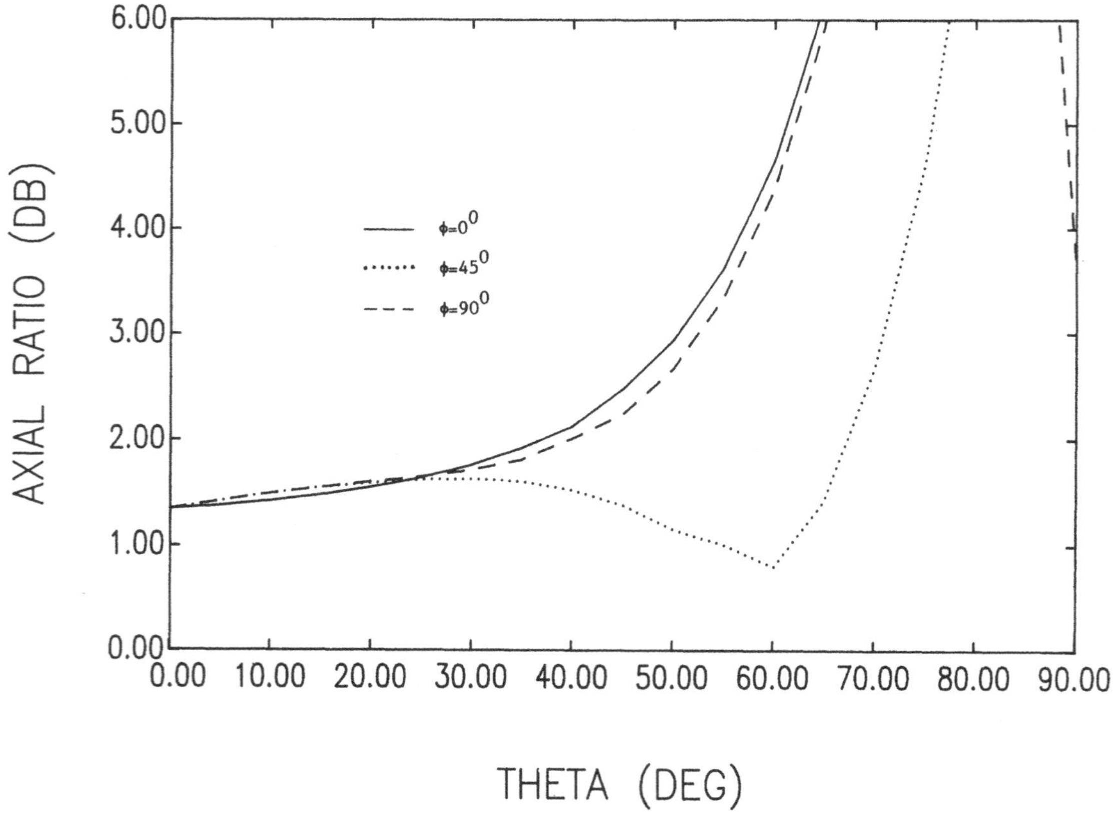

Fig. 10. Computed axial ratio for the array of Fig. 9.

sessing certain deficiencies, has been shown to be useful in predicting the scan performance of linearly-polarized patch arrays. The more rigorous probe feed model was developed in order to overcome the deficiencies in the idealized probe feed model, and generates excellent agreement with simulator measurements. In addition, this rigorous treatment seems to have some advantages over other published analyses, including applicability to a wide range of printed radiator problems. Unfortunately, this more rigorous approach requires a dramatic increase in the amount of computer time for the calculation. Some progress has been made in reducing this time, but work is continuing on the development of techniques to overcome this drawback.

References

[1] R. J. Mailloux, "Phased array theory and technology," *Proc. IEEE*, vol. 70, pp. 246-291, Mar. 1982.

[2] D. M. Pozar and D. H. Schaubert, "Scan blindness in infinite phased arrays of printed dipoles," *IEEE Trans. Antennas Propagat.*, vol. AP-32, pp. 602-610, June 1984.

[3] ——, "Analysis of an infinite array of rectangular microstrip patches with idealized probe feeds," *IEEE Trans. Antennas Propagat.*, vol. AP-32, pp. 1101-1107, Oct. 1984.

[4] C. C. Liu, A. Hessel, and J. Shmoys, "Performance of probe-fed microstrip-patch element phased arrays," presented at 1985 Phased Arrays Symp., RADC-TR-85-170, Sept. 1985.

[5] J. J. Schuss, J. D. Hanfling, R. J. Conti, and B. R. Herrick, "Numerical design of millimeter wave patch radiator arrays," presented at 1985 Phased Arrays Symp., RADC-TR-85-170, Sept. 1985.

[6] J. J. Schuss and J. D. Hanfling, "Nonreciprocity and scan blindness in phased arrays using balanced-fed radiators," *IEEE Trans. Antennas Propagat.*, vol. AP-35, pp. 134-138, Feb. 1987.

[7] J. J. Schuss and R. L. Bauer, "Axial ratio of balanced and unbalanced fed circularly polarized patch radiator arrays," in *IEEE Antennas Propagat. Soc. Symp. Dig.*, vol. 1, June 1987, pp. 286-289.

[8] J. T. Aberle and D. M. Pozar, "An improved probe feed model for printed patch antennas and arrays," in *IEEE Antennas Propagat. Soc. Symp. Dig.*, vol. II, June 1988, pp. 438-440.

[9] C. C. Liu, A. Hessel, and J. Shmoys, "Performance of probe-fed microstrip-patch element phased arrays," *IEEE Trans. Antennas Propagat.*, vol. 36, pp. 1501-1509, Nov. 1988.

[10] J. T. Aberle and D. M. Pozar, "Analysis of infinite arrays of probe-fed rectangular microstrip patches using a rigorous feed model," *Proc. Inst. Elec. Eng.*, vol. 136, pp. 110-119, Apr. 1989.

[11] K. Solbach, "Phased array simulation using circular patch radiators," *IEEE Trans. Antennas Propagat.*, vol. AP-34, pp. 1053-1058, Aug. 1986.

[12] D.M. Pozar, "General relations for a phased array of printed antennas derived from infinite current sheets," *IEEE Trans. Antennas Propagat.*, vol. AP-33, pp. 498-504, May 1985.

[13] ——, "Input impedance and mutual coupling of rectangular microstrip antennas," *IEEE Trans. Antennas Propagat.*, vol. AP-30, pp. 1191-1196, Nov. 1982.

[14] J. R. Mosig and F. E. Gardiol, "General integral equation formulation for microstrip antennas and scatterers," *Proc. Inst. Elec. Eng.*, vol. 132, pt. H, pp. 424-434, Dec. 1985.

[15] W. C. Chew and J. A. Kong, "Analysis of a circular microstrip disk antenna with a thick dielectric substrate," *IEEE Trans. Antennas Propagat.*, vol. AP-29, pp. 68-76, Jan. 1981.

[16] H. A. Wheeler, "A survey of the simulator technique for designing a radiating element in a phased-array antenna," in *Proc. 1970 Symp., Phased Array Antennas*, A. A. Oliner and G. H. Knittel, Eds. Dedham, MA: Artech House, 1972.

[17] A. W. Rudge, K. Milne, A. D. Olver, P. Knight, Eds., *The Handbook of Antenna Design*, vol. 1. UK: Peter Peregrinus, 1982.

Modelling of Wideband Proximity Coupled Microstrip Array Elements

J. S. Herd

Indexing terms: Antennas, Microstrip

A full wave analysis of infinite arrays of proximity coupled rectangular microstrip antennas is described. The broadband scanning properties of the arrays are numerically analysed in conjunction with experimental measurements to validate the analysis.

Introduction: A primary limitation of the conventional microstrip antenna is an inherently narrow bandwidth. This can be improved by increasing the substrate thickness. Feedline radiation and probe reactance make directly coupled microstrip lines or coaxial probes impractical for substrates that are thickner than approximately 0·05 dielectric wavelengths.

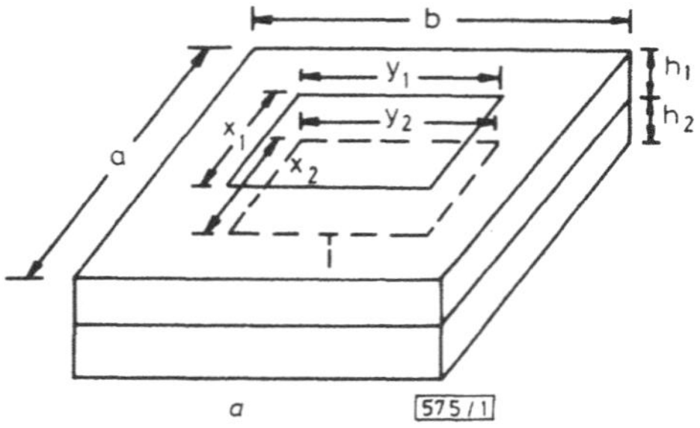

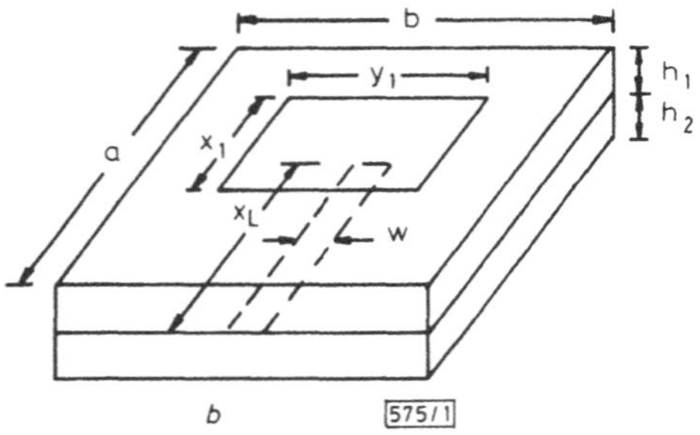

Fig. 1 *Proximity coupled element geometries*
a Probe feed
b Microstrip line feed

As an alternative, proximity coupling from an embedded microstrip line or patch can be used to excite a microstrip element as shown in Figs. 1*a* and *b*. These types of feeds relieve the constraint on total dielectric, and thus make wider bandwidths possible. Experimental results for isolated proximity coupled elements with bandwidths of up to 26% have been reported,[1,2] but so far no studies have been done to assess the effects of inter-element coupling in a large array.

An analysis that includes the array interactions, and incorporates them in the design of proximity coupled arrays for wide scanning bandwidth is described.

Theory: The infinite array spectral dyadic Green's function was used because it includes all effects caused by the dielectrics and the inter-element coupling. A method of moments was applied to the two element geometries in Fig. 1. The resulting matrix equation was then solved to obtain the unknown current distributions on the patches and feedlines. Fig. 1*a* shows an idealised vertical probe as in Reference 3. Entire domain modes were used to represent the current distribution on the patches. Fig. 1*b* shows an element for which a semi-infinite travelling wave mode was used to efficiently model the incident and reflected currents on the microstrip line, with piecewise sinusoidal modes on the microstrip line in the vicinity of the patch to model the nonpropagating currents near the open end of the line. The three types of modes are shown in Fig. 2.

A technique described in Reference 4 was used to overcome the problem of microstrip feedlines in an infinite planar array.

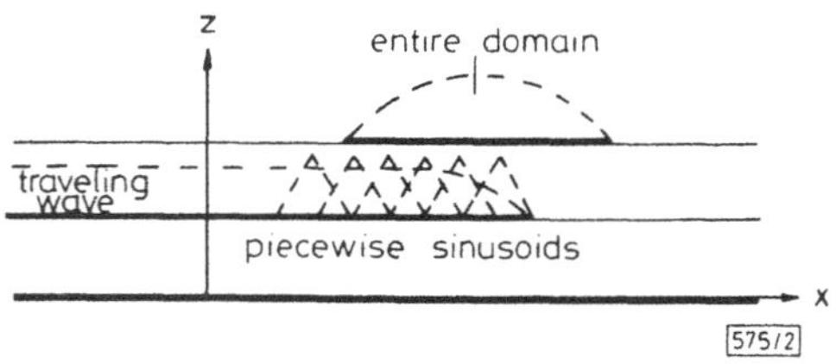

Fig. 2 *Moment method expansion modes*

In order to replicate travelling wave currents on the microstrip feedlines, it was necessary to theoretically extend the lines outside of the small unit cell of the infinite array which is approximately one-half wavelength per side. This was accomplished by observing that a feedline couples very little to patches and lines outside of the cell it feeds, as long as it does not pass within a few board thicknesses. As a consequence, the reactions between the line and the patch it feeds can be computed as if it were a single element fed by a semi-infinite line. All other reactions are computed in the presence of the infinite

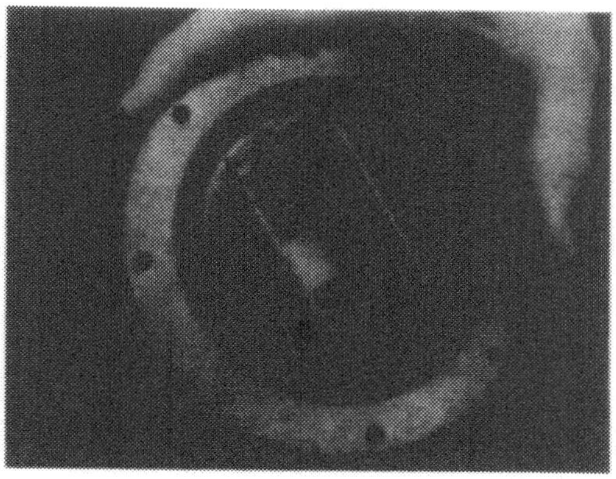

Fig. 3 *Infinite array waveguide simulator*

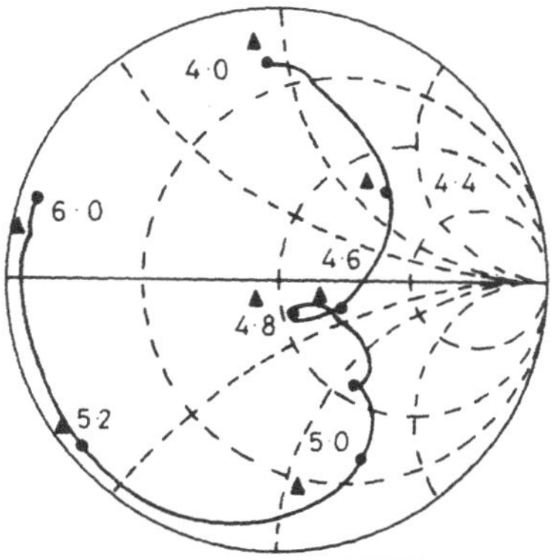

Fig. 4 *Probe feed proximity coupled element in waveguide simulator*

$h_1 = h_2 = 0{\cdot}159\,\text{cm};\ \varepsilon_1 = \varepsilon_2 = 2{\cdot}33;\ a = 3{\cdot}35\,\text{cm};\ b = 3{\cdot}60\,\text{cm}$
$x_1 = y_1 = 1{\cdot}80\,\text{cm};\ x_2 = y_2 = 1{\cdot}95\,\text{cm}$
● measured
▲ predicted

array because of the non-negligible coupling between unit cells.

Experiment: To validate the theory, experimental measurements were made using an infinite array waveguide simulator and an 11 × 11 array, Fig. 3 shows the waveguide simulator. Figs. 4 and 5 show the measured and computed input impedances for the two element types. Fig. 6 shows an 11 × 11 array of type 1a elements, and Figs. 7*a* and *b* show the measured and computed scanning impedance of the array.

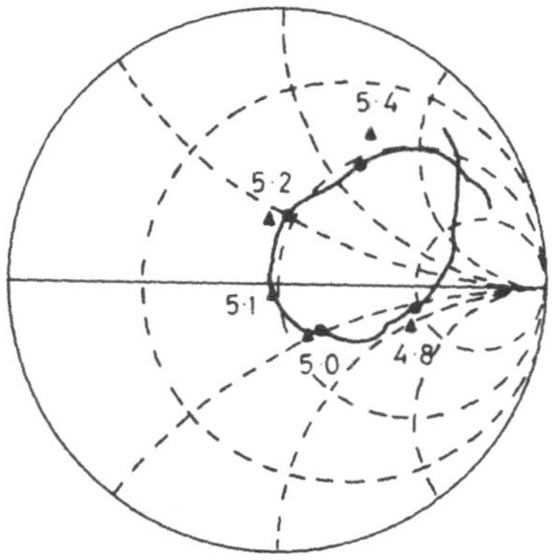

Fig. 5 *Microstrip line feed proximity coupled element in waveguide simulator*

$h_1 = h_2 = 0{\cdot}159\,\text{cm};\ \varepsilon_1 = \varepsilon_2 = 2{\cdot}33;\ a = 3{\cdot}35\,\text{cm};\ b = 3{\cdot}60\,\text{cm}$
$x_1 = y_1 = 1{\cdot}70\,\text{cm};\ x_L = 1{\cdot}80\,\text{cm};\ w = 0{\cdot}40\,\text{cm}$
● measured
▲ predicted

Conclusions: A method has been derived for accurately predicting the input impedance of proximity coupled modelling microstrip feedlines in an infinite array was developed and validated. It is concluded that all mutual coupling effects should be included in the design of proximity coupled arrays for wide scanning bandwidth.

Fig. 6 *11 × 11 proximity coupled array*

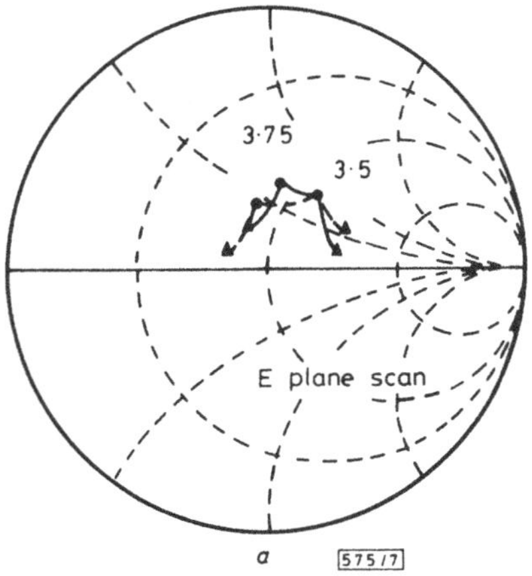

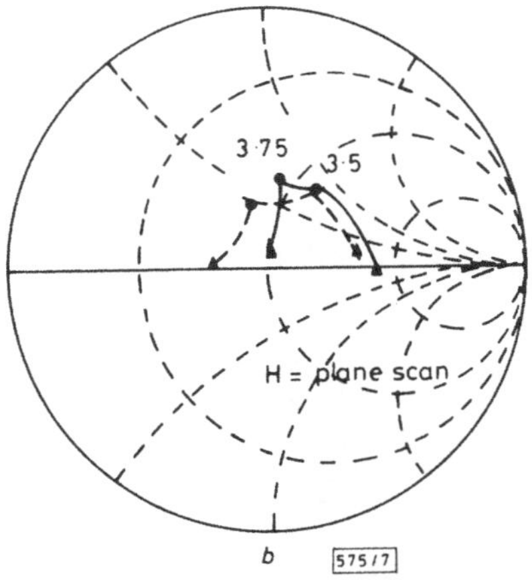

Fig. 7 *Scanning impedance of 11 × 11 array*
$h_1 = h_2 = 0{\cdot}159$ cm; $\varepsilon_1 = \varepsilon_2 = 2{\cdot}33$; $a = 3{\cdot}57$ cm; $b = 3{\cdot}97$ cm
$x_1 = 2{\cdot}14$ cm; $y_1 = 3{\cdot}19$ cm; $x_2 = 2{\cdot}30$ cm; $y_2 = 3{\cdot}43$ cm
- - - - predicted
——— measured
● 0° scan angle
▲ 50° scan angle
a *E*-plane scan
b *H*-plane scan

Acknowledgment: The author wishes to thank Robert Mailloux and David Pozar for their valuable suggestions and comments.

References

1 SABBAN, A.: 'A new broadband stacked two-layer microstrip antenna'. IEEE AP-S Int. Symp. Digest, 1983, pp. 63–66

2 COCK, R., and CHRISTANDOULOUS, C.: 'Design of a two layer, capacitively coupled microstrip element for broadband applications'. IEEE AP-S Int. Symp. Digest, 1987, pp. 936–939

3 POZAR, D. M., and SCHAUBERT, D. H.: 'Analysis of an infinite array of rectangular microstrip patches with idealized probe feeds', *IEEE Trans.*, 1984, **AP-32**, pp. 1101–1107

4 HERD, J. S.: 'Scanning impedance of proximity coupled rectangular microstrip antenna arrays'. PhD Dissertation, University of Massachusetts, 1989

Scanning Characteristics of Infinite Arrays of Printed Antenna Subarrays

David M. Pozar, *Fellow, IEEE*

Abstract—**In this paper the analysis and scanning characteristics of several different types of infinite arrays composed of subarrayed printed dipole and microstrip patch elements are presented. The analysis is based on full-wave moment method theory, and includes mutual coupling between elements in the subarray as well as between subarrays. The effect of subarraying on scan blindness is demonstrated for arrays using two-element subarrays of printed dipoles and microstrip patches, and it is found that subarraying can eliminate or reduce mismatch due to scan blindness. Results are also given for the amount of power radiated in grating lobes, which in many cases is found to be the primary factor that limits the practical scan range of subarrayed arrays. This paper also considers the effect of a subarray composed of one driven element and one parasitic element, and the use of a four-element synchronous subarray of microstrip patches to generate circular polarization. Data are given for impedance mismatch, power radiated into grating lobes, and the axial ratio; both square and rectangular patches are considered, and it is found that rectangular patches give better performance for all characteristics. Finally, results are shown for an infinite array of seven-element hexagonal subarrays of printed dipoles, and it is found that the large spacing between subarrays leads to a limited scan range.**

I. Introduction

ALMOST complete control of beam scanning, pattern shape, and polarization can be obtained from a phased array of closely spaced antenna elements, with individual amplitude, phase, and polarization control for each element. Such an antenna is very attractive for many radar and communications systems, but its high cost generally precludes its use in all but the most critical applications. Since array cost is proportional to the number of elements in the array, one way to reduce the expense of an array with a given gain requirement is to group the elements into contiguous subarrays, with excitations and polarization control applied at the subarray level. This technique can be used for both scanning and fixed-beam arrays [1], [2]. Besides the cost savings, subarraying can also be used to advantage in reducing feed losses, or for incorporating time-delay networks to improve array bandwidth. The disadvantages of subarraying are that it usually introduces grating lobes that cause a decrease in the scan range, and it may cause an increase in sidelobe levels. In a limited scan application where low sidelobes are not required, however, subarraying can be used very effectively. It is also possible to use subarrays with amplitude weighting to reduce the level of sidelobes or grating lobes [2].

Printed antenna arrays potentially offer further cost savings through the use of photolithographic fabrication techniques and compatability with microwave integrated circuit technology [3]. Characteristics of large arrays of printed dipole and microstrip patch elements have been presented in the literature [4]–[6], but the effect of subarraying on the performance of such antennas has not been investigated. We would expect that the grating lobe characteristics of a subarrayed printed array will be determined by the element spacing and the size of the subarray, just as for other types of arrays. But large printed arrays are plagued by intrinsic scan blindness problems, and the effect of subarraying on scan blindness was not clear before this work was undertaken. In addition, a rigorous analysis of subarrayed printed antennas can be used to study the characteristics of other interesting geometries, such as circularly polarized synchronous subarrays [7], [8], or subarrays with parasitic or dummy elements [9].

In this paper we present the analysis and scanning characteristics of several different types of infinite arrays composed of subarrayed printed dipole and microstrip patch elements. The analysis is an extension of the full-wave moment method theory used in [4]–[6], and includes the full effects of mutual coupling between all elements in the array. The four specific subarray geometries treated in this paper are described below.

We first study the effect of subarraying on scan blindness by considering the most basic case of a printed dipole or microstrip patch array composed of two-element subarrays, as shown in Fig. 1(a). As discussed in [4], scan blindness in infinite arrays of printed dipoles can be directly traced to the forced resonance of a surface wave on the dielectric substrate caused by phase matching with a nonpropagating Floquet mode. In an array with $\lambda/2$ element spacing, the blindness angle occurs before the onset of the grating lobe at 90°, and the blindness angle moves towards broadside as the element spacing is increased. In an array composed of two-element subarrays with an element spacing of $\lambda/2$, the spacing between subarrays will be λ, so it might appear that blindness will be more severe (closer to broadside) than in the nonsubarrayed case. But it will be seen from results presented in Section III that a properly spaced subarray can almost completely eliminated the unity (or near unity) reflection coefficient that is normally associated with the scan blindness in a standard printed array. Results will also be given for the amount of power radiated in grating lobes.

Another interesting geometry is an array with subarrays

Reprinted from *IEEE Trans. Antennas Propaga.*, vol. 40, no. 6, pp. 666–674, June 1992.

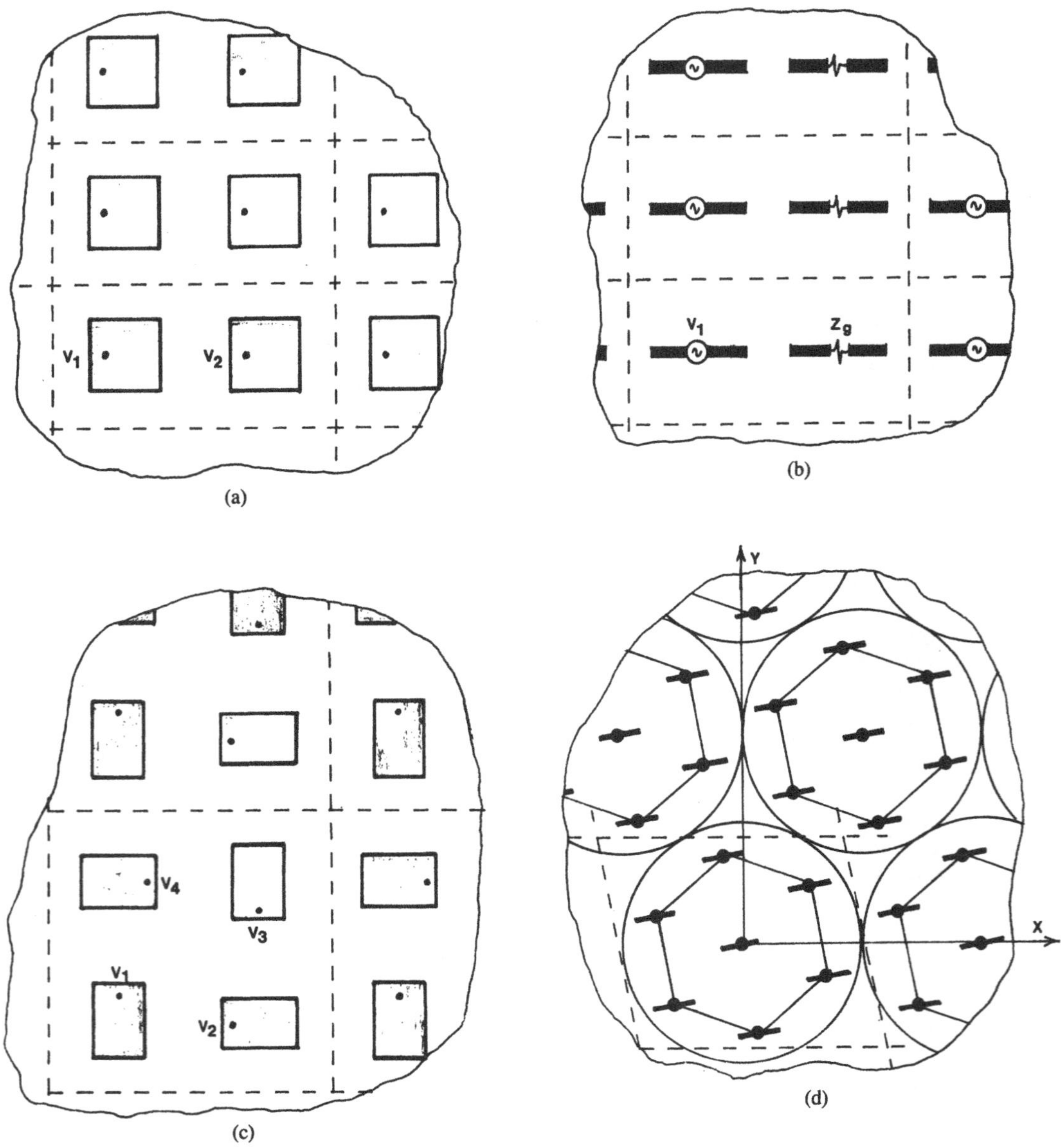

Fig. 1. Subarray geometries treated in this paper. (a) A two-element subarray of microstrip patch antennas. (b) A two-element printed dipole subarray with a parasitic element. (c) A circularly polarized synchronous subarray of microstrip patches. (d) A seven-element hexagonal subarray of printed dipoles.

composed of one driven element and one parasitic element, as shown in Fig. 1(b). Parasitic or dummy elements are sometimes used in arrays, with a close element spacing to provide proximity coupling from driven elements [9]. Since such elements do not require a feed, they do not significantly add to the cost of the array. In view of the advantageous effect that subarraying has on scan blindness, when all elements of a subarray are driven, it is legitimate to investigate whether the same advantage can be realized when one of the elements of the subarray is a parasitic element (either resistively loaded or shorted). For this case, we have found that the parasitic element does act to prevent a complete scan blindness, but the improvement is not as good as can be achieved when both subarray elements are fed.

The standard way to implement a circularly polarized array is to use circularly polarized elements, but this can significantly increase the complexity of the array feed network, as well as the array cost. An interesting alternative approach is to use a synchronous subarray composed of linearly polarized elements, fed in phase rotation [7], [8]. A four-element version of such a subarray is shown in Fig. 1(c). Some of the basic characteristics of the synchronous subarray, such as polarization and grating lobe properties in small arrays in the absence of mutual coupling, have been presented in the literature, but there is quite a variety of unanswered questions concerning the use of synchronous subarrays as elements in large arrays. In Section III we present data on scan blindness and input impedance, axial ratio, and power radiated by grating lobes for such arrays as a function of scan angle.

Finally, we consider hexagonal subarrays of printed dipole elements, as shown in Fig. 1(d). The hexagonal subarray can be thought of as a generalization of a triangular array grid, and allows an increase in the element spacing before grating lobes occur, when compared to an array with a rectangular grid [10]. Here we give results for an infinite array of seven-element hexagonal subarrays, including the effect of mutual coupling. We find that the large spacing between subarrays results in a rather limited scan range.

II. Analysis

A large variety of printed antenna element types are of practical interest [3], but here we consider only arrays of printed dipoles and probe-fed rectangular patches. These elements are familiar, and will serve to illustrate the essential properties of subarrayed arrays; it can be expected that subarrayed arrays using printed elements of a different type, but the same spacings and substrate parameters, will have very similar scan characteristics. Here we use the infinite array approximation, for the usual reasons of analytical convenience and accuracy in modeling large arrays. The solution is full-wave, employing the exact Green's function for the dielectric substrate and a moment method procedure, so all mutual coupling effects between and within subarrays are included. The dipole elements are assumed to be fed with ideal gap voltage generators [4], while the probe-fed patches use an idealized feed model consisting of a current probe [5]. This patch feed model does not rigorously model the flow of current from the probe to the patch, but it has been found to give good results for patches on thin substrates, and for predicting the trends in the scanning performance of patch arrays.

The theory of the subarray solution is similar in form to the infinite array theory described in [4], [5], [9], but because those solutions do not involve unit cells with multiple ports, it is worthwhile to briefly describe the present solution. In addition, we only present the formulation for probe-fed patches, as the printed dipole solution can easily be obtained from this theory. The geometry of an infinite array of subarrays is shown in Fig. 2. It is assumed that each subarray fills one unit cell, and that all subarrays are identical in structure and excitation, except for a uniform phase shift given by

$$e^{-jk_0(mau+nbv)} \tag{1}$$

where m, n are the indices of the unit cell, a, b are the spacings of the subarrays in the x, y directions, respectively, and u, v are the direction cosines for scanning at the angle θ,ϕ:

$$u = \sin\theta\cos\phi \tag{2a}$$

$$v = \sin\theta\sin\phi. \tag{2b}$$

Now assume that there are N elements in each subarray, with coordinates of their centers given by x_k, y_k, for $1 \leq k \leq N$. We will expand the surface current density on each element in the same set of M expansion functions, $\bar{J}_j(x, y)$, for $1 \leq j \leq NM$, so there are a total of NM expansion modes for each unit cell, or subarray. Then the spectral domain expression for the electric field generated by the infinite array of subarrays can be expressed as

$$\bar{E}(x, y, d) = \frac{1}{ab}\sum_{m=-\infty}^{\infty}\sum_{n=-\infty}^{\infty}\bar{\bar{G}}(k_{xm}, k_{yn})$$

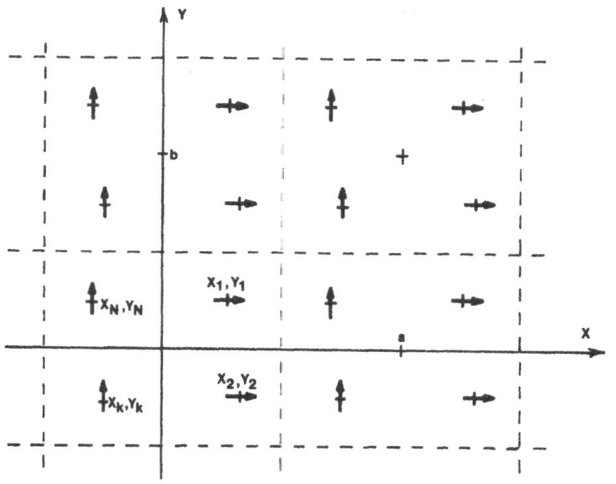

Fig. 2. Geometry of an infinite array of contiguous subarrays. Each subarray is composed of N printed antenna elements with center coordinates x_k, y_k, for $k = 1, 2, \cdots, N$. The spacing between subarrays is a, b. The substrate has a dielectric constant ϵ_r and a thickness d.

$$\cdot\sum_{j=1}^{NM} I_j\bar{F}_j^*(k_{xm}, k_{ym})e^{-jk_{xm}(x-x_j)}e^{-jk_{yn}(y-y_j)} \tag{3}$$

where $\bar{\bar{G}}$ is the dyadic Green's function in the spectral domain representing E_{xx}, E_{xy}, E_{yx} and E_{yy} for the dielectric slab (as defined in [9]), I_j is the (unknown) amplitude of the jth expansion function, and $\bar{F}_j$ is the Fourier transform of the jth expansion function defined as

$$\bar{F}_j(k_x, k_y) = \int_x\int_y \bar{J}_j(x, y)e^{-jk_x x}e^{-jk_y y}dx\,dy. \tag{4}$$

Equation (3) represents the periodic field of the array in terms of an expansion of Floquet modes, where k_{xm} and k_{yn} are the Floquet mode phase constants defined as

$$k_{xm} = \frac{2m\pi}{a} + k_0 u \tag{5a}$$

$$k_{yn} = \frac{2n\pi}{b} + k_0 v \tag{5b}$$

for a rectangular array grid. For a triangular grid, k_{yn} is replaced by

$$k_{yn} = \frac{2n\pi}{b} - \frac{2m\pi}{a\tan\alpha} + k_0 v \tag{5c}$$

where α is the skew angle of the array grid relative to the x-axis ($a = 90°$ for a rectangular grid).

Evaluating the electric field in (3) on the surface of the substrate and equating this to the negative of the tangential field radiated by the kth feed probe and testing with the $\bar{J}_i$ functions yields the following moment method equation for the unknown expansion mode coefficients:

$$[Z][I] = [V^t] \tag{6}$$

where the moment method impedance matrix elements are defined as

$$Z_{ij} = \frac{-1}{ab} \sum_{m=-\infty}^{\infty} \sum_{n=-\infty}^{\infty} \bar{F}_i(k_{xm}, k_{yn}) \cdot \bar{\bar{G}}(k_{xm}, k_{yn}) \cdot \bar{F}_j^*(k_{xm}, k_{yn}) e^{-jk_{xm}(x_i - x_j)} \cdot e^{-jk_{yn}(y_i - y_j)} \tag{7}$$

and the voltage vector elements for the ith test mode centered at x_i, y_i are defined as

$$V_{ik}^t = \frac{-1}{ab} \sum_{m=\infty}^{\infty} \sum_{n=-\infty}^{\infty} \bar{F}_i(k_{xm}, k_{yn}) \cdot \bar{G}_z(k_{xm}, k_{yn}) e^{-jk_{xm}(x_i - x_k^p)} e^{-jk_{yn}(y_i - y_k^p)} \tag{8}$$

where $\bar{G}_z$ is the Green's function vector in the spectral domain representing the electric fields, E_{xz} and E_{yz}, radiated by the kth z-directed unit-amplitude current probe located at x_k^p, y_k^p.

Now define port voltages, V_i^p, and port currents, I_i^p, at each of the N patches of a subarray. These port quantities are related by the short-circuit port impedance matrix $[Z^p]$:

$$[V^p] = [Z^p][I_p] \tag{9}$$

where the port impedance matrix elements are given by

$$Z_{ij}^p = -\sum_{k=1}^{NM} V_{ki}^e I_k^i. \tag{10}$$

In (10), I_k^i is the kth expansion mode coefficient when the ith feed probe is driven with a 1 A source, and all the other probes are open circuited. The voltage vector elements in (10) are defined as

$$V_{jk}^e = \frac{-1}{ab} \sum_{m=-\infty}^{\infty} \sum_{n=-\infty}^{\infty} \bar{G}_z(k_{xm}, k_{yn}) \cdot \bar{F}_j^*(k_{xm}, k_{yn}) e^{jk_{xm}(x_j - x_k^p)} e^{jk_{yn}(y_j - k_k^p)}. \tag{11}$$

Now if we assume Thevenin-type sources, with a voltage source V_{gk} at the kth element and a series generator impedance Z_g, then the port currents can be found as

$$[I^p] = \{[Z^p] + Z_g[U]\}^{-1}[V_g] \tag{12}$$

where $[U]$ is the identity matrix. The port voltages can then be found from

$$[V^p] = [V_g] - Z_g[I^p] \tag{13}$$

and so the input impedance seen at the kth port, when the subarray is excited by the voltage $[V_g]$, is

$$Z_{\text{in}}^k = \frac{V_k^p}{I_k^p}. \tag{14}$$

Because the generator voltages, port voltages, and port currents at corresponding elements in different subarrays are identical except for the phase shift defined in (1), the input impedances of corresponding elements in different subarrays will be identical. The input impedances of elements within the same subarray, however, will generally be different, and will have different variations with scan angle. In fact, it is possible for the input impedance of a specific element to have a negative real part, which simply means that power is being coupled out of that port, but into other ports of the subarray; the total power into the subarray will never be negative. A reflection coefficient can also be calculated at each element of the subarray, and is computed assuming a conjugate impedance match for broadside scan. Since the reflection coefficient at each element of a subarray will generally have a slightly different value, we often compute and plot the average reflection coefficient to avoid the need to present a large amount of data. The average reflection coefficient corresponds to the reflection coefficient that would be seen if an equal-split power divider were used to feed the elements of a subarray, thus there is a physical basis for defining such a quantity.

When all subarray elements are driven, the expansion mode coefficients can be found as a superposition of the responses due to individual probe excitations:

$$[I] = [Z]^{-1}[V^T] \tag{15}$$

where V_i^T is the total voltage vector excitation due to all the feed probes when the subarray is excited by $[V_g]$:

$$V_i^T = \sum_{k=1}^{N} I_k^p V_{ik}^t. \tag{16}$$

We are also interested in the power radiated by the array, which can be computed for each unit cell as

$$P = -\text{Re} \int_x \int_y \bar{J}^* \cdot \bar{E}\, dx\, dy = -\text{Re} \sum_{m=-\infty}^{\infty} \sum_{n=-\infty}^{\infty} \sum_{i=1}^{NM} I_i^* \bar{F}_i(k_{xm}, k_{yn}) \cdot \bar{\bar{G}}(k_{xm}, k_{yn}) \cdot \sum_{j=1}^{NM} I_j \bar{F}_j^*(k_{xm}, k_{yn}) e^{-jk_{xm}(x_i - x_j)} e^{-jk_{yn}(y_i - y_j)}. \tag{17}$$

The only terms in (17) that contribute to the real radiated power correspond to propagating Floquet modes, which include the main beam ($m = n = 0$ mode), and any grating lobes. If we defined P_{oo} as the real power radiated by the $m = n = 0$ Floquet mode, then we can compute a beam efficiency, η, as

$$\eta = \frac{P_{oo}}{P}. \tag{18}$$

The term "beam efficiency" is usually used to denote the ratio of power radiated through a specified beamwidth of the main beam of an antenna to the total radiated power. Since the beamwidth of an infinite array is zero, it appears that the use of this term for η is consistent with the usual definition of beam efficiency.

We also note that, in the absence of dissipative losses, the power radiated by a unit cell can also be computed from

$$P = \text{Re} \sum_{k=1}^{N} V_k^p I_k^{p*} \tag{19}$$

which can be used as a consistency check by comparing with the result obtained from (17).

Piecewise sinusoids were used on the printed dipole elements to expand x-directed current (typically three modes), while entire domain expansion modes were used to expand both the x- and y-directed currents on the microstrip patch elements (typically four modes). The Fourier transforms for these modes, and all the necessary Green's functions referred to above, can be found in the literature [9].

III. Results

A. The Effect of Subarraying on Scan Blindness

The effect of subarraying on the scan blindness condition in a large phased array of printed antenna elements can be demonstrated with a two-element subarray of the type shown in Fig. 1(a). We first consider printed dipoles, where one dipole is located at $x_1 = y_1 = 0$ in the reference unit cell, and the other dipole is located at $x_2, y_2 = 0.4\lambda$ (see Fig. 2). The subarray spacing is $a = b = 0.8\lambda$, and both dipoles are driven with voltage generators of $V_1 = V_2 = 1 + j0\,v$. The magnitude of the average reflection coefficient for the two dipoles is plotted versus E-plane scan angle in Fig. 3, for $x_2 = 0$, 0.1λ, 0.2λ, 0.3λ, and 0.4λ. The case for $x_2 = 0$ corresponds to a standard (nonsubarrayed) array with $a = 0.8\lambda$ and $b = 0.4\lambda$, so we expect to see a blindness in the E-plane, which in this case occurs at $\theta = 13°$ and results in a unity reflection coefficient magnitude. As x_2 is increased from 0.1 to 0.4λ, the reflection coefficient magnitude in the vicinity of 13° still shows a peak, but the peak value drops as the spacing increases. For a spacing of 0.4λ the blindness is almost completely removed. Of course, because of the subarray spacing of 0.8λ, a grating lobe occurs for scan angles beyond 14°, so there will be some reduction in the power radiated by the main beam of the array; this will be discussed further in the next example. This subarray has no effect on the H-plane scan characteristics.

The reason for the reduction in the scan blindness effect due to subarraying can be qualitatively explained in terms of phase interference between the excitation of the array and the surface wave of the substrate. In the nonsubarray case, blindness occurs when the transverse propagation constant of a certain Floquet mode equals the propagation constant, β_{sw}, of a surface wave of the dielectric substrate [4],

$$\frac{k_{xm}^2 + k_{yn}^2}{k_0^2} = \left(\frac{m\lambda}{a} + u\right)^2 + \left(\frac{n\lambda}{b} + v\right)^2 = \left(\frac{\beta_{sw}}{k_o}\right)^2 \tag{20}$$

which typically occurs for the $m = -1, n = 0$ Floquet mode. Physically what happens is that, at blindness, the transverse linear phase progression associated with this Floquet mode just matches the phase progression of a surface wave of the dielectric sheet; like the surface wave field, this

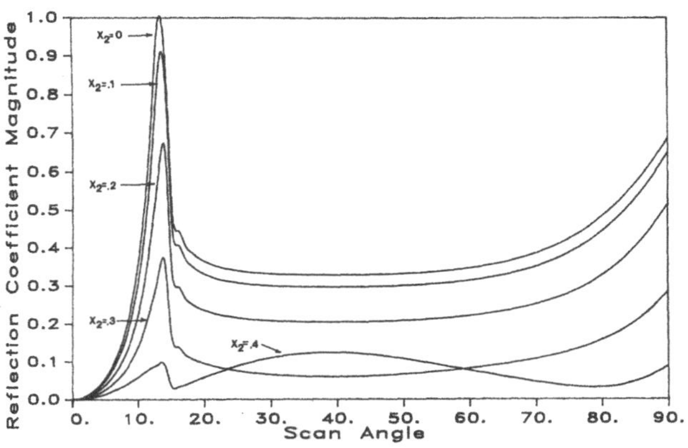

Fig. 3. Average reflection coefficient magnitude versus E-plane scan angle for an infinite array of two-element printed dipole subarrays. The first dipole is centered at $x_1 = y_1 = 0$; the second dipole is centered at $x_2, y_2 = 0$. Substrate dielectric constant $= 2.55$, substrate thickness $= 0.05\lambda$, dipole length $= 0.342\lambda$, dipole width $= 0.01\lambda$, $a = b = 0.8\lambda$, $Z_g = 8\Omega$. The dipole excitations are $V_1 = V_2 = 1 + j0$.

Floquet mode can propagate in the dielectric substrate, but not in the air region above the array (assuming no grating lobes are present). The addition of a second dipole in the unit cell, cophased with the first dipole but offset from it, upsets this linear phase progression and reduces the coupling to the surface wave field.

B. The Effect of Increasing the Number of Elements in a Subarray

We next consider the effect of the number of elements in a subarray, and examine the beam efficiency, for a microstrip patch array. The reflection coefficient magnitude and beam efficiency are plotted versus E-plane scan angle in Figs. 4 and 5 for an infinite array with either one, two, or three rectangular microstrip patch elements per subarray. The element spacing in each case is 0.5λ, so $a = 0.5\lambda$, 1.0λ, or 1.5λ, for $N = 1$, 2, or 3, respectively, and $b = 0.5\lambda$ for all cases. The subarray elements are fed in phase, with equal amplitudes. In Fig. 4, the reflection coefficient magnitude for the $N = 1$ case shows a scan blindness at $\theta = 76°$; for $N = 2$ the scan blindness that would normally appear at $\theta = 2°$ and $\theta = 76°$ are seen to be eliminated, and similarly for $N = 3$. The price paid for this improvement is shown in the beam efficiency plots of Fig. 5. For $N = 1$, there is no grating lobe in visible space, so the beam efficiency is unity for $\theta < 90°$. For $N = 2$, the subarray spacing is λ, so a grating lobe begins to propagate for $\theta > 0$, causing the beam efficiency to decrease as the scan angle increases; the power in the main beam is reduced to half for a scan angle of about 32°. For $N = 3$, the subarray spacing is 1.5λ, causing two grating lobes to appear for all scan angles greater than zero in the E-plane; the main beam power is reduced by half for a scan angle of about 20°. Thus we see that blindness effects can be eliminated or greatly reduced by subarraying, but only at the expense of main beam power being lost to grating lobes; in a limited-scan application, this may be an acceptable compromise.

The grating lobe and blindness effects associated with these

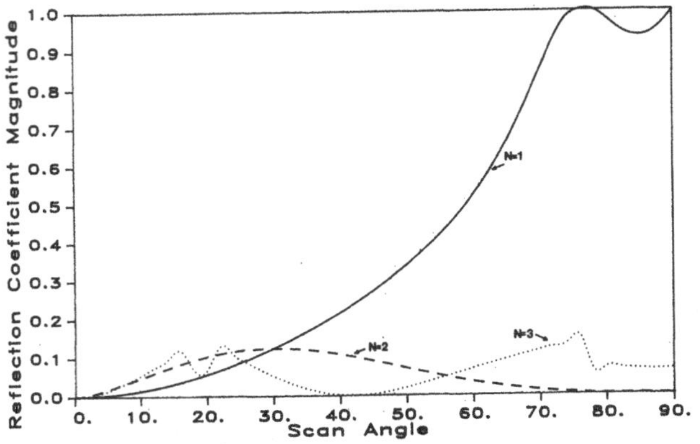

Fig. 4. Reflection coefficient magnitude versus E-plane scan angle for an infinite array of microstrip patch subarrays, for $N = 1, 2, 3$ elements per subarray. The element spacing is 0.5λ in both directions; the subarray is along the E-plane. For $N = 1$, $a = 0.5\lambda$; for $N = 2$, $a = 1.0\lambda$; for $N = 3$, $a = 1.5\lambda$; $b = 0.5\lambda$ for all cases. Substrate dielectric constant = 2.55, substrate thickness = 0.06λ, patch length = 0.28λ, patch width = 0.3λ. The patch excitations are $V_1 = V_2 = V_3 = 1 + j0$.

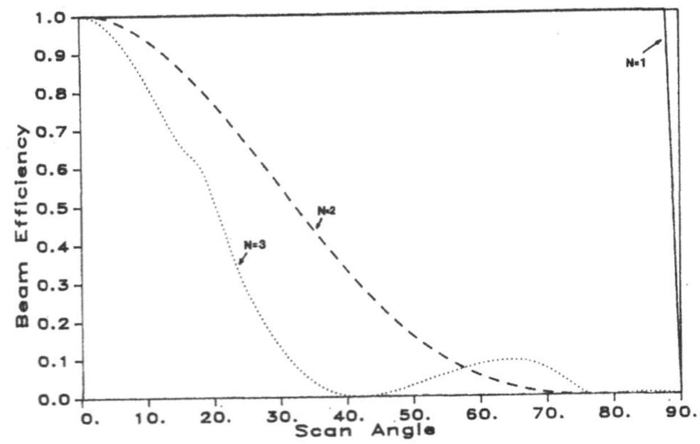

Fig. 5. Beam efficiency versus E-plane scan angle for the case of Fig. 4.

arrays can be clearly understood with the aid of a grating lobe and surface wave circle diagram [4], shown in Fig. 6 for $a = 1.5\lambda$ case. The large E-plane element spacing causes two grating lobes to be potentially present for all E-plane scan angles. For E-plane scan angles less than 19°, the $m = \pm 1$ circles intersect visible space; for scan angles greater than 19° the $m = \pm 1$ and $m = \pm 2$ circles intersect visible space. But the grating lobe power is reduced by the subarray element pattern; at broadside scan, the subarray pattern has nulls at precisely the locations where grating lobes would occur, and so there is no power lost to grating lobes. This occurs for both the $N = 2$ and $N = 3$ cases, or for any subarrayed array having uniform element spacing. Of course, perfect cancellation of a grating lobe by a subarray pattern null only occurs for infinite arrays, where the beamwidths are zero. In a finite array the subarray pattern null will split the grating lobe into two lobes, with some resulting power loss. The surface wave circles plotted with each grating lobe circle allow the prediction of scan blindness; the intersection of surface wave circles associated with the $m = \pm 1$ and $m = \pm 2$ circles lead to the potential

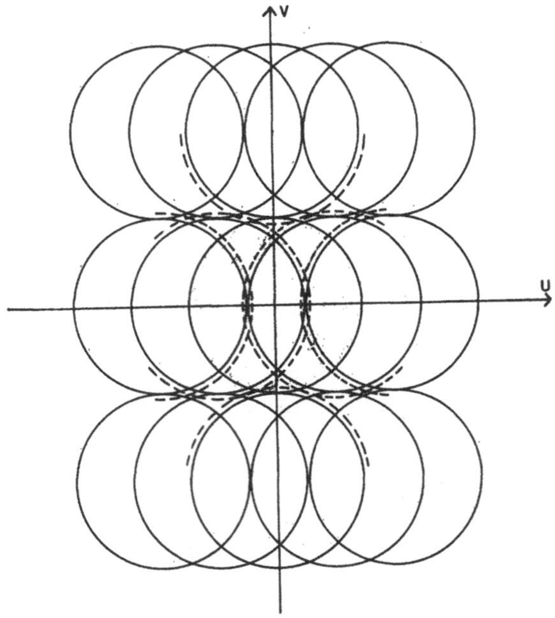

Fig. 6. Grating lobe and surface wave circle diagram for the case of Fig. 4 with $a = 1.5\lambda$. Visible space is indicated by the shaded region; the surface wave circles are dashed.

blindness at 21° and 18°. Blindness can also occur for other scan planes.

C. Subarrays with Parasitic Elements

We next consider the effect of a subarray with a parasitic element, and choose as an example a subarray with two printed dipole elements arranged along the E-plane of the array, as shown in Fig. 1(b). In this case, the E-plane subarray spacing is $a = 0.8\lambda$ and the H-plane spacing is $b = 0.5\lambda$; the spacing between the elements in the subarray is 0.4λ. Fig. 7 shows the reflection coefficient magnitude versus E-plane scan for three cases. As a reference, we show the reflection coefficient for a subarray with only one element; it has a complete blindness at $\theta = 13°$. The second result in Fig. 7 is for the two-element subarray when both dipoles are driven in phase. As in the case of Fig. 3, the blindness at $\theta = 13°$ is eliminated. The third curve in Fig. 7 shows the reflection coefficient for a two-element subarray when one element is fed, and the other is terminated in a 8Ω load. Notice that the peak at $\theta = 13°$ has a value of about 0.8, which indicates some reduction in blindness, but not a complete removal. Data was also run for a shorted parasitic dipole, and the response showed only a slight variation from the 8Ω case. Since some power is lost in the loaded dipole, it would be preferable to use shorted dipoles if a large number of parasitic elements were to be used in an array. The use of more parasitic elements, or closer spacings, may further reduce the blindness effect, but this has not been studied extensively.

D. Circularly Polarized Synchronous Subarrays

As discussed in [7] and [8], synchronous subarrays of linearly polarized elements can be used to generate circular polarization, avoiding the complexity and cost associated

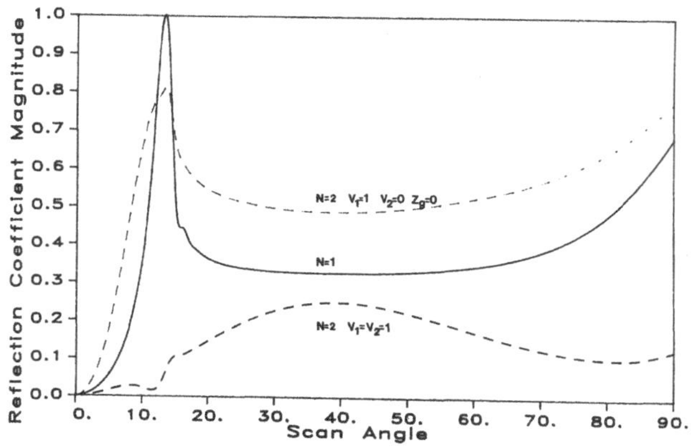

Fig. 7. Average reflection coefficient magnitude versus E-plane scan angle for an infinite array of printed dipole subarrays with a parasitic element. The first dipole is centered at $x_1 = y_1 = 0$; the second dipole is centered at $x_2 = 0.4\lambda$, $y_2 = 0$. Substrate dielectric constant = 2.55, substrate thickness = 0.05λ, dipole length = 0.342λ, dipole width = 0.01λ, $a = 0.8\lambda$, $b = 0.5\lambda$, $V_1 = 1 + j0$.

with circularly polarized elements. Typically a 2 × 2 subarray is used, as shown in Fig. 1(c), with phasings of $(1 + j0)$, $(0 + j)$, $(-1 + j0)$, $(0 - j)$ for the four elements. For a single synchronous subarray, fairly high grating lobes can appear in the diagonal planes [8], while the array factor for a large array of synchronous subarrays should reduce these lobes. Because of the subarray size, the spacing between subarrays is usually on the order of a wavelength, so grating lobe effects will also be important for beam positions scanned off broadside. Another consideration, for both the single subarray case and the array of subarrays case, is the difference in performance for arrays using rectangular patches versus arrays using square patches. In the case of square patches, we expect that mutual coupling between similarly polarized modes of adjacent patches will lead to significant cross polarization [11]; this effect should be reduced in the case of rectangular patches since similarly polarized modes on adjacent patches will not both be resonant, so coupling to the cross-polarized mode will be reduced.

Figs. 8–10 show the scan characteristics of a four-element synchronous subarray of probe-fed microstrip patches. Data are shown for scan planes at $\phi = 0°$ and $\phi = 45°$, and for square and rectangular patches. The element spacing in both planes is 0.5λ, so the spacing between subarrays is λ. Fig. 8 shows the average reflection coefficient magnitude versus scan angle, where it can be seen that a complete scan blindness does not occur, but there are some local peaks where blindness would normally occur at $\theta = 7°$ in the $\phi = 0°$ plane, and at $\theta = 24°$ in the $\phi = 45°$ plane. Also note that the mismatch is generally less for rectangular patches than for square patches. Fig. 9 shows the beam efficiency for the same array. A grating lobe starts to propagate for $\theta > 0$, so for both rectangular and square patches the main beam power decreases steadily with scan angle, although there is only about a 30% loss for scan angles up to 20°. Again, the beam efficiencies are generally better for rectangular patches. Note that the beam efficiency in the diagonal plane is not much different than the beam efficiency in the $\phi = 0$ plane, indicating that the array factor does

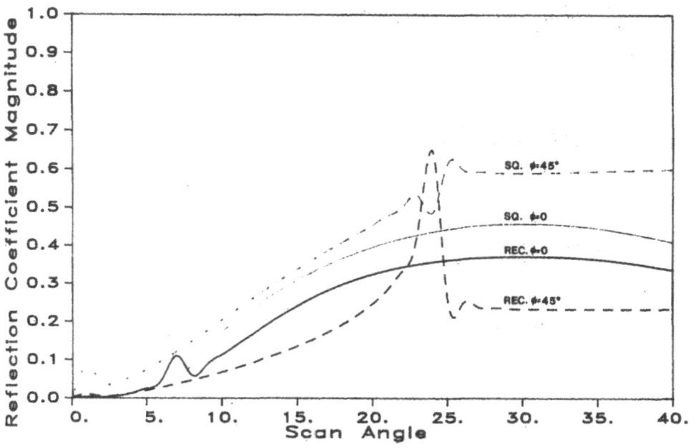

Fig. 8. Average reflection coefficient versus scan angle in the $\phi = 0°$ and $\phi = 45°$ planes for an infinite array of four-element circularly polarized synchronous subarrays of microstrip patches. The spacing between the subarray elements is 0.5λ, and the spacing between subarrays is $a = b = 1\lambda$. Substrate dielectric constant = 2.55, substrate thickness = 0.03λ, patch length = 0.3λ. Data are shown for rectangular patches (width = 0.25λ), and for square patches (width = 0.3λ). The patch excitations are $V_1 = 1 + j0$, $V_2 = 0 + j$, $V_3 = -1 + j0$, $V_4 = 0 - j$.

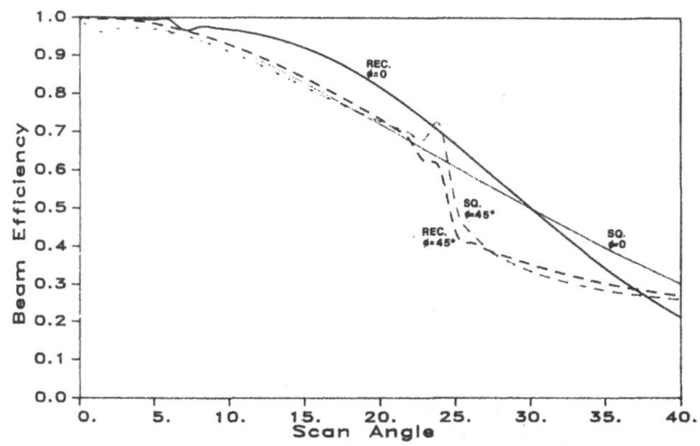

Fig. 9. Beam efficiency versus scan angle for the case of Fig. 8.

indeed reduce the grating lobes in that plane. Fig. 10 shows the axial ratio for the array. Note that perfect circular polarization is obtained for broadside scan, for both rectangular and square patches (this is essentially a result of symmetry). Scanning off broadside in either plane degrades the axial ratio, with more degradation occurring for square patches than for the rectangular patch case. Near the blindness angles of $\theta = 7°$ (for $\phi = 0$) and $\theta = 24°$ (for $\phi = 45°$) there are peaks in the axial ratio. This is because blindness is polarization sensitive, and only one of the linear components of the circularly polarized radiated field is affected by the blindness, causing an imbalance. But it appears that an axial ratio of less than 2 dB can be obtained over a scan range of at least 25°. Circularly polarized synchronous subarrays can also be constructed with circularly polarized elements, with a possible improvement in performance, but this seems to defeat the advantage of simplicity obtained when linearly polarized elements are used.

E. Characteristics of Hexagonal Subarrays

The hexagonal subarray arrangement shown in Fig. 1(d) contains seven elements in each subarray, so the spacing

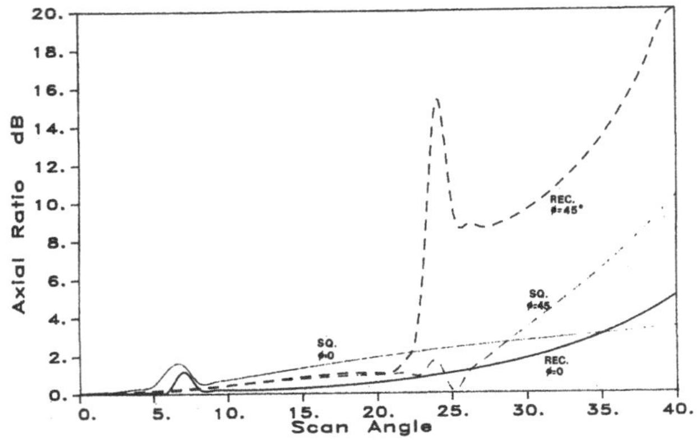

Fig. 10. Axial ratio versus scan angle for the case of Fig. 8.

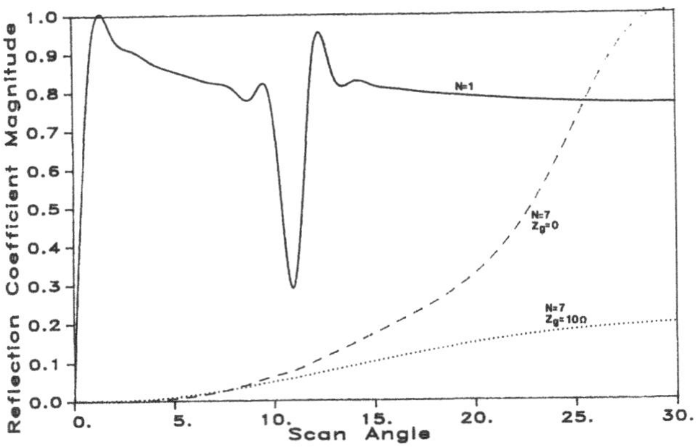

Fig. 11. Average reflection coefficient versus E-plane scan angle for an infinite array of seven-element hexagonal subarrays of printed dipoles. The spacings between the subarrays are $a = 1.1548\lambda$, $b = 1\lambda$; the array grid angle is 60°. Substrate dielectric constant = 2.55, substrate thickness = 0.05λ, dipole length = 0.342λ, dipole width = 0.01λ. The dipole excitations are $V_1 = V_2 = V_3 = V_4 = V_5 = V_6 = V_7 = 1 + j0$.

between subarrays is fairly large. As an example, we consider here a relatively closely spaced seven-element printed dipole subarray with $a = 1.1548\lambda$ and $b = 1\lambda$; the E-plane spacing between parallel rows of dipoles is then 0.33λ, and the H-plane spacing between adjacent dipoles is 0.44λ. The coordinates of the seven dipoles in the reference unit cell are given in Table I.

Fig. 11 shows the average reflection coefficient versus scan angle in the $\phi = 0°$ plane for this array. For comparison, the reflection coefficient magnitude for the $N = 1$ case is also shown. This corresponds to a standard array with the same values of a, b, and α, and the results show scan blindness at about $\theta = 1.3°$ and $\theta = 11°$. For the $N = 7$ subarray case, results are shown for $Z_g = 0$ and $Z_g = 10\Omega$; in both cases the scan blindnesses have been eliminated for scan angles below 30°. As expected, the $Z_g = 10\Omega$ case gives a better impedance match, but of course some reflected power is being dissipated in the generator resistance (this example demonstrates that array performance is affected by the generator impedance in a multiport subarray). In addition, as in the cases discussed above, the amount of power radiated in grating lobes increases with the scan angle. Thus, the hexagonal subarray of this example is limited to scan angles perhaps less than 25°; larger spacings will further reduce the scan range.

The grating lobe diagram for this array is shown in Fig. 12; for clarity, the surface wave circles have not been included. The diagram shows that at least two different pairs of grating lobe circles cover visible space for the $\phi = 0°$ and $\phi = 45°$ scan planes. As in the rectangular grid subarray case, the subarray element pattern produces nulls that eliminate (infinite array case), or reduce (finite array case), grating lobe effects for broadside scan.

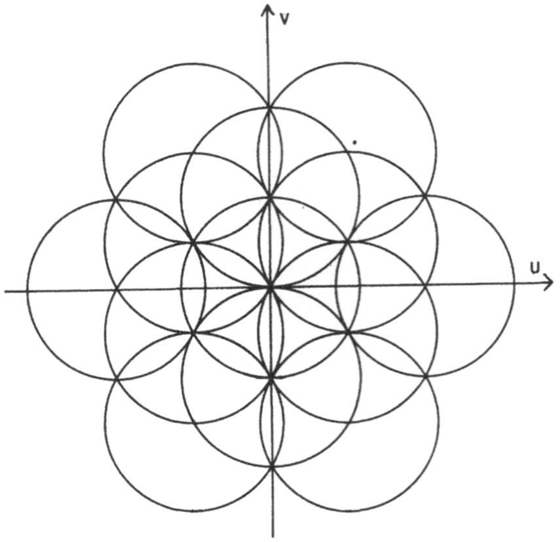

Fig. 12. Grating lobe diagram for the hexagonal subarray of Fig. 11. Visible space is indicated by the shaded region.

TABLE I
DIPOLE COORDINATES FOR HEXAGONAL SUBARRAY

i	x_i/λ	y_i/λ
1	0.000	0.000
2	0.330	0.286
3	−0.083	0.429
4	−0.412	0.143
5	−0.330	−0.286
6	0.083	−0.429
7	0.412	−0.143

IV. CONCLUSION

In this paper the use of subarrays in infinite arrays of printed antennas has been investigated in regard to the scan blindness effect, impedance mismatch, grating lobes, the use of parasitic elements, and the use of circularly polarized synchronous subarrays. Results were presented for several specific examples of subarrayed arrays of printed dipoles and probe-fed microstrip patches. We have found that subarrays can reduce or almost completely eliminate scan blindness, but the price paid for this improvement is the loss of power to grating lobes. We have quantified both the impedance mismatch and the beam efficiency for several cases, and conclude that subarraying can be successfully used to reduce the cost of a large phased array antenna if a wide scan range is not required. The use of subarrays with parasitic or dummy elements has also been found to reduce the mismatch caused by scan blindness, but to a lesser degree than the case where all subarray elements are driven in phase. The performance of an infinite array of synchronous subarrays was found to be quite good over a limited scan range, in terms of impedance matching, beam efficiency, and axial ratio. The scan performance of an array with hexagonal subarrays was shown to be acceptable over a somewhat reduced scanning range, primarily because of the relatively large subarray spacing associated with this configuration.

As demonstrated by the variety of examples considered in this paper, the present analysis is general enough to allow the study of a large class of problems involving subarrays. Additional problems that could be treated include the effect of amplitude tapering at the subarray level, and the use of overlapping subarrays.

Acknowledgment

The author would like to thank John Keevil of General Electric AstroSpace Division for discussions that led to the study of these problems, and Dan Schaubert of the University of Massachusetts for useful discussions on the scan blindness effect.

References

[1] R. J. Mailloux, "Grating lobe characteristics of arrays with uniformly illuminated contiguous subarrays," in *IEEE Int. Symp. Antennas Propagat. Dig.*, Boston, MA, 1984, pp. 511–514.

[2] R. L. Haupt, "Reducing grating lobes due to subarray amplitude tapering," *IEEE Trans. Antennas Propagat.*, vol. AP-33, pp. 846–850, Aug. 1985.

[3] D. M. Pozar, "Microstrip antennas," *Proc. IEEE*, vol. 80, pp. 79–91, Jan. 1992.

[4] D. M. Pozar and D. H. Schaubert, "Scan blindness in infinite phased arrays of printed dipoles," *IEEE Trans. Antennas, Propagat.*, vol. AP-32, pp. 602–610, June 1984.

[5] —, "Analysis of an infinite array of rectangular microstrip patches with idealized probe feeds," *IEEE Trans. Antennas Propagat.*, vol. AP-32, pp. 1101–1107, Oct. 1984.

[6] J. T. Aberle and D. M. Pozar, "Analysis of infinite arrays of one- and two-probe-fed circular patches," *IEEE Trans. Antennas Propagat.*, vol. 38, pp. 421–432, Apr. 1990.

[7] T. Teshirogi, M. Tanaka, and W. Cujo, "Wideband circularly polarized array antenna with sequential rotations and phase shifts of elements." in *Proc. Int. Symp. Antennas Propagat.*, Kyoto, Japan, 1985, pp. 117–120.

[8] P. S. Hall, J. Huang, E. Rammos, and A. Roederer, "Gain of circularly polarized arrays composed of linearly polarized elements," *Electron. Lett.*, vol. 25, pp. 124–125, Jan. 19, 1989.

[9] J. R. James and P. S. Hall, *Handbook of Microstrip Antennas.* London, U.K.: Peregrinus, 1989.

[10] A. J. Zaman, E. Rho, R. Q. Lee, and M. L. Zimmerman, "Grating lobe characteristics of a hexagonal array of subarrays," *IEEE Antennas Propagat. Symp. Dig.*, Ontario, 1991, pp. 1140–1143.

[11] D. M. Pozar and B. Kaufman, "Design considerations for low sidelobe microstrip arrays," *IEEE Trans. Antennas Propagat.*, vol. 38, pp. 1176–1185, Aug. 1990.

Chapter 8
Other Topics

In this chapter we give some coverage of a variety of topics related to microstrip antenna design, topics not included in the preceding chapters. They include the application of microstrip antenna technology to commercial products and systems, conformal antenna designs, radar cross section and the effect of substrate anisotropy, microstrip reflectarrays, and monolithic antenna designs.

We begin this chapter with the paper by Huang, written especially for this reprint volume to give an overview of the commercial applications of microstrip antennas. Many of these applications are benefiting from the defense-driven development of microstrip antenna technology that occurred during the 1980s. While cost was generally not a primary concern in the context of defense applications, the shift in the 1990s to commercial applications, such as direct broadcast television and personal communications systems, has made cost a major consideration in the market-driven application of microstrip antennas. Although microstrip antennas were advertised for many years as being a low-cost technology, the validity of this claim was often undermined by the high cost of substrate materials, which could be as high as $100 per square foot—a price that makes commercial use of such antennas impractical. For this reason many workers have explored low-cost ways of fabricating microstrip antennas and arrays. The paper by Daniel, Penard, Nedelec, and Mutzig, for example, concludes that economical microstrip arrays can be made with the usual PTFE substrate material only in the 20–30 GHz range (where the substrate size is physically fairly small and losses are not too severe), and that the array should be made using a single substrate layer with a coplanar feed network. The paper by Henderson and James gives a comparative study of the performance and cost tradeoffs between a microstrip array and a prime focus parabola for DBS reception at 12 GHz. The array materials cost was minimized by using a foam substrate with thin plastic films for the metalization, but the performance of the array was well below that of the reflector. A further cost-conscious application of microstrip antenna technology is described in the paper by Huang and Densmore, which describes a novel Yagi-type microstrip array design for a low-profile mobile satellite vehicle antenna. This antenna is intended to lie flat on the roof of a vehicle with a beam scanned low in elevation, and is mechanically steered in azimuth. Another area of development for commercial applications is the use of microstrip antennas of reduced size for operation around 900 MHz. The paper by Kuboyama, Imai, Hirasawa, and Fujimoto describes a post-loaded element for use in a pocket pager operating at about 930 MHz.

One of the unique features of microstrip antennas is their ability to be applied conformally to curved structures, while maintaining a low profile [1]–[4]. In many cases the curvature of the structure is large enough to be considered locally flat for purposes of antenna design, but if the antenna extends around the structure a more sophisticated treatment is necessary. An example of this problem is discussed in the paper by Jayakumar, Garg, Sarap, and Lal, for omnidirectional wraparound antennas mounted on circular cylinders. This paper contains useful design data on the number of circumferential elements required for a given azimuthal pattern ripple.

The radar cross-section of an antenna is important for radar and ECM applications, and deserves some mention here. Because of the high-Q nature of the microstrip antenna element, its RCS response exhibits large reflections at frequencies where either the dominant mode or higher order modes are resonant [5]. In addition, the effect of loading through the feed of the element can appreciably reduce or shift these RCS peaks. The paper by Pozar describes a full-wave moment method solution for the RCS of a probe-fed rectangular patch antenna, including the effect of both resistive and reactive loading. Similar solutions have been carried out for circular patches [6]. Also considered in the paper by Pozar is the effect of substrate anisotropy. Most composite dielectric substrates have some degree of anisotropy, usually of the uniaxial form. When the degree of anisotropy is small, the primary effect is to cause a shift in the resonant frequency of the microstrip element, a shift which can be important in view of the typically narrow bandwidth of microstrip antennas. More general anisotropic materials include ferrites which, when magnetically biased and used as antenna substrates, have been found to provide some unique properties. These properties include frequency tuning by changing the bias field strength [7], [8], obtaining circular polarization with a single feed point [9], [10], and dynamically impedance-matching a phased array versus scan angle [9], [11].

A significant hindrance to the implementation of high-gain microstrip arrays is the loss associated with a microstrip feed network. This loss is especially apparent in comparison with a reflector antenna, for gains in excess of about 28 dB. One way to overcome the inherent loss of microstrip transmission lines is to take a lesson from the reflector antenna and space-feed the array, which can be conveniently implemented in the form of the microstrip reflectarray [12], [13]. The reflectarray consists of a flat planar array of microstrip patches illuminated by a feed spaced above the array. The reflected phase from each array element is adjusted by design to simulate a constant phase front, thereby producing a main beam in a specified direction. The reflected field phase from an element can be controlled using a tuning stub of variable length, or by simply adjusting the resonant length of the element itself. This latter approach has been successfully implemented in a C-band reflectarray, as discussed in the paper by Targonski and Pozar.

Finally, we mention the subject of active or monolithic arrays, where microstrip antennas are integrated with amplifiers, phase shifters, or mixers. The objective of such fabrication is to exploit the potential cost savings associated with microwave integrated circuit technology, especially for applications such as large phased arrays and high-volume commercial products. Although attempts at such integration have been underway for many years, it has only been recently that successful integrated active antenna prototypes have been demonstrated [14], [15]. Most of these designs use hybrid integration, where individual MIC amplifiers, phase shifters, and switches are incorporated on a mother substrate containing the antenna elements. The paper by Sanzgiri, Pottenger, Bostrom, Denniston, and Lee represents an excellent example of this type of architecture, for a limited-scan 4x4 subarray using aperture coupled microstrip elements at 29 GHz. A more ambitious goal is to integrate all circuit and antenna functions monolithically on a single substrate. One example of progress in this direction is described in the paper by Ohmine, Kashiwa, Ishikawa, Iida, and Matsunaga, where a single aperture-coupled antenna element was integrated with a receiver front-end circuit at 40 GHz. Much work, however, remains to be done in the area of integrated phased array antennas, but recently proposed large-scale commercial satellite systems may provide a massive market-driven effort in this area.

References

[1] R. E. Munson, "Conformal microstrip antennas and microstrip phased arrays," *IEEE Trans. Antennas and Prop.*, vol. AP-22, pp. 74–78, Jan. 1974.

[2] S. Fonseca and A. Giarola, "Analysis of microstrip wraparound antennas using dyadic Green's functions," *IEEE Trans. Antennas and Prop.*, vol. AP-31, pp. 248–253, Mar. 1983

[3] J. Ashkenazy, S. Shtrikman, and D. Treves, "Electric surface current model for the analysis of microstrip antennas on cylindrical bodies," *IEEE Trans. Antennas and Prop.*, vol. AP-33, pp. 295–300, Mar. 1985.

[4] E. V. Sohtell and J. P. Starski, "Cylindrical microstrip patch phased array antenna—Chalscan C," *Military Microwaves Conf. Proceedings*, pp. 317–322, June 1986.

[5] E. H. Newman and D. Forrai, "Scattering from a microstrip patch," *IEEE Trans. Antennas and Prop.*, vol. AP-35, pp. 245–251, Mar. 1987.

[6] J. T. Aberle, D. M. Pozar, and C. R. Birtcher, "Radiation and scattering from probe-fed microstrip patch antennas," *IEEE Trans. Antenna and Prop.*, vol. 39, pp. 1691–1696, Dec. 1991.

[7] D. M. Pozar and V. Sanchez, "Magnetic tuning of a microstrip antenna on a ferrite substrate," *Electronics Letters*, vol. 24, pp. 729–731, June 1988.

[8] P. J. Rainville and F. J. Harackiewicz, "Magnetic tuning of a microstrip patch antenna fabricated on a ferrite film," *IEEE Microwave and Guided Wave Letters*, vol. 2, pp. 483–485, Dec. 1992.

[9] D. M. Pozar, "Radiation and scattering characteristics of microstrip antennas on normally biased ferrite substrates," *IEEE Trans. Antennas and Prop.*, vol. 40, pp. 1084–1092, Sept. 1992.

[10] J. S. Roy, P. Vaudon, A. Reineix, F. Jecko, and B. Jecko, "Circularly polarized far-fields of an axially magnetized circular ferrite microstrip antenna," *Microwave and Optical Technology Letters*, vol. 5, pp. 228–230, May 1992.

[11] H. Y. Yang and J. A Casteneda, "Infinite phased arrays of microstrip antenna on generalized anisotropic substrates," *Electromagnetics*, vol. 11, pp. 107–124, 1991.

[12] J. Huang, "Microstrip reflectarray," *IEEE Symp. on Antennas and Propagation Digest*, pp. 612–615, June 1991.

[13] D. C. Chang and M. C. Huang, "Microstrip reflectarray antenna with offset feed," *Electronics Letters*, vol. 28, pp. 1489–1491, July 1992.

[14] J. Huang, D. Rascoe, A. L. Riley, V. Lubecke, and L. Duffy, "A Ka-band MMIC phased array antenna," *IEEE Int'l Antennas and Propagation Symp. Digest*, pp. 1212–1215, June 1989.

[15] J. D. Hanfling, J. J. Schuss, and R. E. Morrow, "A dual polarized patch radiator for phased arrays," *IEEE Int'l Antennas and Propagation Symp. Digest*, pp. 1216–1219, June 1989.

Microstrip Antennas for Commercial Applications

JOHN HUANG
JET PROPULSION LABORATORY
CALIFORNIA INSTITUTE OF TECHNOLOGY
PASADENA, CA 91109

Abstract—**The microstrip antenna, because of its small size, light weight, low profile, and low manufacturing cost, is finding increasing applications in the commercial sector of the industry. This article discusses briefly some of the antenna's technical features and presents several important examples of the antenna's commercial applications, such as mobile satellite communications, direct broadcast satellite services, global positioning system, and medical hyperthermia usage.**

I. Introduction

Since the invention of the microstrip antenna [1], [2] four decades ago, the demand for its application has been increasing rapidly, especially within the last two decades. However, these applications have been in demand mostly by the Department of Defense. Because of their extremely thin profile (0.01 to 0.05 wavelength), printed microstrip antennas have found heavy applications in military aircraft, missiles, rockets, and satellites. In the commercial sector, the adoption of the microstrip antenna has not been as rapid, primarily due to the cost and the lack of demand. During the earlier years, the costs of the microstrip antenna's substrate material, design, and manufacturing processes were considered non-competitive when compared to the monopole, helix, horn, or parabolic reflector. In addition, the earlier popular terrestrial communication system's configuration and environment did not warrant the use of the microstrip antenna. During the last decade, however, the cost to develop and manufacture the microstrip antenna has dropped significantly, because of the maturity of microstrip antenna technology, the reduction in cost of substrate material and manufacturing process, and the simplified design process using newly developed computer-aided design (CAD) tools. Furthermore, the current satellite communication applications benefit greatly from the small size and low profile of the microstrip antenna. For example, in a UHF cellular terrestrial communication system, with plenty of RF power and antenna real estate for its base stations, the mobile unit can perform adequately with a simple low-gain monopole antenna without any concern about its system gain margin being too small. However, in developing L-band mobile satellite communications [3], because of the limited spacecraft solar-battery power and spacecraft antenna size, the mobile vehicle terminal requires a higher gain antenna (on the order of 10 dBi) to ensure an adequate system link margin. To cover a wide elevation angular region from 20 degrees to 60 degrees above the horizon with such a gain, an antenna such as a horn, helix, or monopole array will be too bulky to be mounted on top of a passenger automobile. A low-profile printed microstrip array not only offers an aesthetically pleasing appearance, but also yields a low manufacturing cost, especially when produced in large quantities.

This article discusses some of the microstrip antenna's technical features, its advantages, and disadvantages. Materials and fabrication techniques for commercial applications will also be briefly presented. Microstrip antenna applications in the areas of mobile satellite communications, the Direct Broadcast Satellite (DBS) system, and the Global Positioning System (GPS) will be highlighted. Nonsatellite-based applications, such as medical hyperthermia and remote sensing, will also be discussed.

II. Technical Background

This section presents the technical background of the microstrip antenna. The discussion is separated into three areas: features of the microstrip antenna, computer-aided design and fabrication, and materials.

II.1 Features of the microstrip antenna

A microstrip antenna [4], [5], as shown in Figure 1, consists of a radiating metallic patch or an array of patches on one side of a thin, nonconducting, supporting substrate panel (thickness is 0.01 to 0.05 free-space wavelength) with a ground plane on the other side of the panel. The metallic patch is normally made of copper foil or is copper-foil plated with a corrosion resistant metal, such as gold, tin, or nickel. Each patch can be made into a variety of shapes, with the most popular shapes being rectangular and circular. The supporting substrate material for an array application must be low in insertion loss with a loss tangent of less than 0.002. Generally, substrate materials [4] can be separated into three categories:

(1) With a relative dielectric constant in the range of 1.0 to 2.0, the material can be air, polystyrene foam, or honeycomb.
(2) With a relative dielectric constant in the range of 2.0 to 4.0, materials consist mostly of teflon.

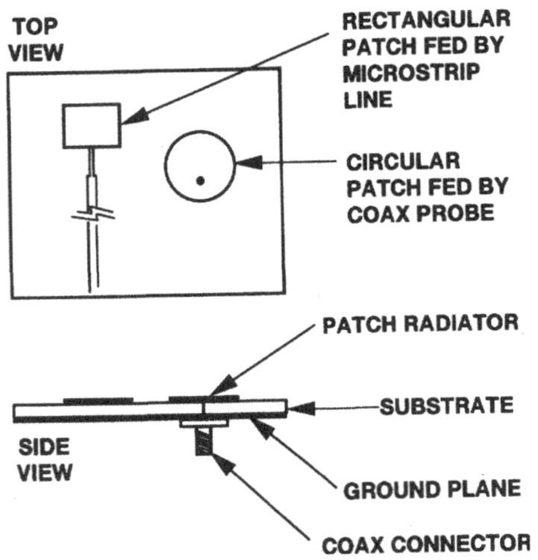

Fig. 1. Rectangular and circular microstrip patch antenna configurations.

(3) With a relative dielectric constant between 4 and 10, the material can be made of ceramic, quartz, or alumina.

Generally, the trend is that the lower the substrate dielectric constant is, the lower the antenna cost.

A single microstrip patch can be excited either by a coaxial probe or by a microstrip transmission line, as shown in Figure 1. For an array of microstrip patches, the patches can be combined either with microstrip lines located on the same side of the patches or with microstrip or strip lines designed on a separate layer placed behind the ground plane. For the separate-layer configuration, each patch and its feed line are electrically connected either by a thin metallic post or by an aperture-coupling slot [6]. In an array configuration, tens or hundreds of patch elements can be fabricated by a single low-cost etching process, and each single patch element of the array does not need to be fabricated individually (as many other types of radiating elements do), leading to a lower antenna manufacturing cost.

There are advantages as well as disadvantages associated with the microstrip antenna. By understanding these, one can readily design a microstrip antenna with the optimum efficiency, minimum risk, and lowest cost for a particular commercial application. The advantages of microstrip antennas when compared to conventional antennas (helix, horn, reflector, etc.) are:

- The extreme low profile of the microstrip antenna makes it lightweight and it occupies very little volume of the structure or vehicle on which it is mounted. It can be conformably mounted onto a curved surface so that it is aesthetically appealing and aerodynamically sound.
- The antenna (when produced in large quantities) or an array with many patch elements can be fabricated with a simple etching process, which leads to greatly reduced fabrication cost.
- Dual-frequency operation is possible by using either dual-stacked patches [7] or a patch with a loaded diode [8] or a stub [9].

The disadvantages of the microstrip antenna are:

- The antenna has a narrow bandwidth (generally less than 3%). However, with technology advancement, 15 to 20 % bandwidths have been achieved with techniques that use multiple stacked patches, thicker substrates, external matching circuits [10], a sequential rotation arrangement [11], [12], parasitic coupling [13], and similar configurations.
- Because of the small separation between the radiating patch and its ground plane (equivalent to small separation between two electrodes), the microstrip antenna can handle relatively low RF power. Generally, a few hundred watts of power or less is considered safe. However, depending on the substrate and metal thicknesses, and the frequency of operation, a few kilowatts of power for microstrip lines at X-band have been reported [14]. It should be noted that for space applications, the power-handling capability is generally less than that for ground application, due to multipacting breakdown [15].
- The microstrip array generally has a larger ohmic insertion loss than other types of antennas of equivalent aperture size. This ohmic loss mostly occurs in the dielectric substrate and the metallic conductor of the microstrip line power dividing circuit. It should be noted that a single microstrip patch element incurs very little loss because it is only one-half wavelength long. The loss in the power-dividing circuit of a microstrip array can be minimized by using series feed techniques [5], [16], waveguide and microstrip combined power dividers, honeycomb substrate, and similar techniques.

II.2 Computer-aided design (CAD) and fabrication

The choices of CAD software 15 to 20 years ago were limited and expensive. As a result, the "cut and try" method was relied on as a common solution to the design of microstrip circuits, which resulted in longer development time, higher antenna cost, and inaccuracy in performance prediction. This situation, however, has changed radically since the performance of personal computers has improved significantly and a variety of CAD tools are available. A microstrip array is generally composed of two components: (1) the power dividing microstrip lines and (2) the radiating microstrip patch elements. Most of the CAD software tools have placed an emphasis on the design of microstrip lines, because they were developed not just for microstrip antennas, but geared more toward the circuit design of transmitter and receiver components, such as filters, hybrids, and couplers. Examples of these tools are Touchstone, SuperCompact, Midas, LINMAC, and many others. Only a few CAD software tools were developed to handle both microstrip circuit lines and microstrip patch elements. The few that are known by the author are Micpatch, developed in Switzerland, using the transmission-line multiport technique; Ensemble, developed by Boulder Microwave Inc. in Colorado, using the integral equation technique [17]; and the Micropatch package, developed by the University of Colorado, employing the multiport segmentation technique [18]. These packages not only aid in the array's circuit layout design, but also calculate the antenna input impedance, radiation patterns, and antenna efficiencies. Mutual coupling effects between patch elements are included in the analysis.

It should be pointed out that, because microstrip lines have relatively wider bandwidth (more than 10% bandwidth), all the above CAD tools can generally design the power-dividing circuit of the microstrip array correctly the first time. However, because of the narrow band characteristics of the patch elements (less than 3% bandwidth), the CAD design will generally take one or two iterations for the actual patch element to resonate at the desired frequency. The inaccuracy in specifying the material parameters (i.e., substrate dielectric constant) by the manufacturers, and unknown fabrication tolerances, are two major contributors to the required iteration steps in designing the patch element. Nevertheless, the CAD tools can help the designer to complete a microstrip array design in several weeks, while the old "cut and try" method may take several months. Especially at millimeter wave frequencies, the "cut and try" method, because of the extreme small patch size, cannot be accomplished by using a knife. Every iteration needs to be done by a complete etching process. Thus, without CAD, antenna development can become very costly and time-consuming.

Once the CAD design is completed, the dimension data of the microstrip array is used to generate photographic film by using very accurate laser optics techniques. This film is then used with chemicals to etch off the copper from the copper-clad substrate panel to form the desired microstrip array antenna. One etching process can yield hundreds or thousands of patch elements on a single panel. For example, if one desires to mass produce thousands of microstrip array antennas with each antenna consisting of hundreds of patch elements, the automated photographic and etching processes [5] can accomplish the fabrication accurately in several days. This is not possible with other types of antenna elements such as the horn and helix. This automated microstrip antenna fabrication process can, therefore, lead to low manufacturing cost.

II.3 Material

The purpose of the substrate material of a microstrip antenna is primarily to provide mechanical support for the radiating patch elements. There are a variety of types of substrate materials. As discussed in Section II.1, the relative dielectric constant of these materials can be anywhere from 1 to 10. The most popular type of material is teflon-based, with a relative dielectric constant between 2 and 3. This teflon-based material, also called PTFE (polytetrafluoroethylene), has a structural form very similar to the fiberglass material used for digital circuit boards [but has a much lower insertion loss]. The selection of the correct material for a microstrip antenna should be based on cost, insertion loss, thermal stability, dielectric constant, and similar considerations. For commercial application, cost is one of the most important criteria in determining the substrate type. For example, a single patch antenna or an array of a few elements may be fabricated on low-cost fiberglass material at L-band frequency, while a 50-element array at 15 GHz may have to use higher-cost, but lower-loss, teflon-based material. For a large number of array elements at lower microwave frequencies, a nonmetallic honeycomb panel may be used as substrate to minimize insertion loss, antenna weight, and cost. A detailed discussion of substrate materials can be found in [5].

III. Applications for Mobile Satellite Communications

One of the most prominent communication systems that will utilize the microstrip antenna is the mobile satellite system. The examples given in Section I have clearly explained the reasons for applying microstrip antennas in a satellite-based communication system. The current terrestrial cellular system cannot provide complete coverage within a large global region, such as the United States. As a result, mobile-to-mobile communication would not be available in rural areas where no cellular station exists. A satellite-based system can fulfill this need by using either a few sets of fixed geostationary satellites or a large number of low Earth-orbiting satellites. Two examples of the geostationary satellite systems are the already implemented International Maritime Satellite System (INMARSAT) [19] and the developing Mobile Satellite (MSAT) [3] system. Both systems operate at L-band frequencies. The INMARSAT has several different versions (Standards A, B, C, and D) of Earth terminals that use reflector-type antennas and are mostly intended for sea-going vessel applications. However, its recently developed Standard-M terminal, which is intended for land application, uses a briefcase-size microstrip array antenna. This antenna, developed by Glocom Corporation of Rockville, Maryland, uses six circular patches and provides 14.5 dB of gain.

The MSAT system has been studied internationally by entities such as Canada, Japan, the European Space Agency (ESA), Australia, and the United States. Because of the satellite's limited power, limited antenna size, and high altitude, its ground terminal requires a medium-gain ($\sim$10 dBi gain), rather than a low-gain, antenna to satisfy the communication link requirements. Such a medium-gain antenna with a directive beam requires a satellite tracking capability to keep its beam pointed at the satellite while the vehicle is moving about. Two types of medium-gain antennas that use microstrip patches have been developed. One is the electronically steered phased array and the other is the mechanically steered array. Figure 2 shows two different phased arrays, developed by Ball Corporation [20] of United States, and the Toyota Central R&D Laboratories of Japan [21]. Both antennas use dual-stacked patches to cover both the transmitting and receiving frequency bands with 19 elements. Both antennas have a size of approximately 50 cm in diameter and 4 cm in height. Ball's study indicates that the phased array antenna, when produced in a volume of 50,000 units in a five-year period, would have a manufacturing cost of about U.S. $2000 per unit. Figure 3 presents a photograph of the Japanese antenna installed on a passenger car. A similar phased-array antenna system, developed by Ball Corporation,

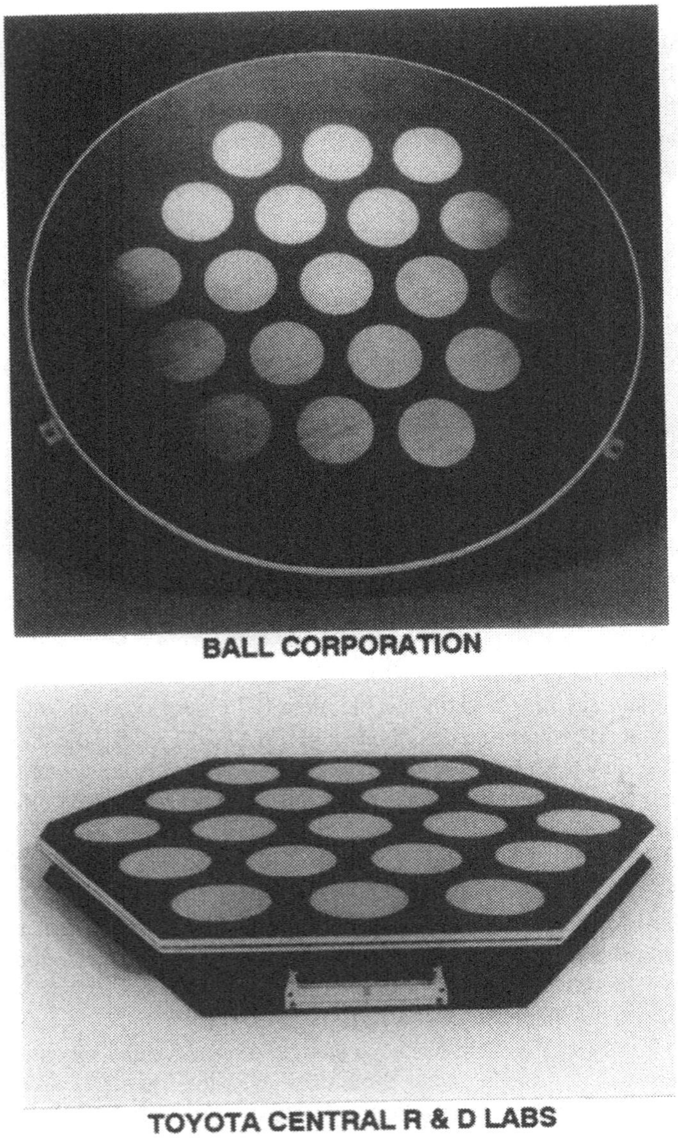

Fig. 2. Phased array antennas for mobile satellite communications.

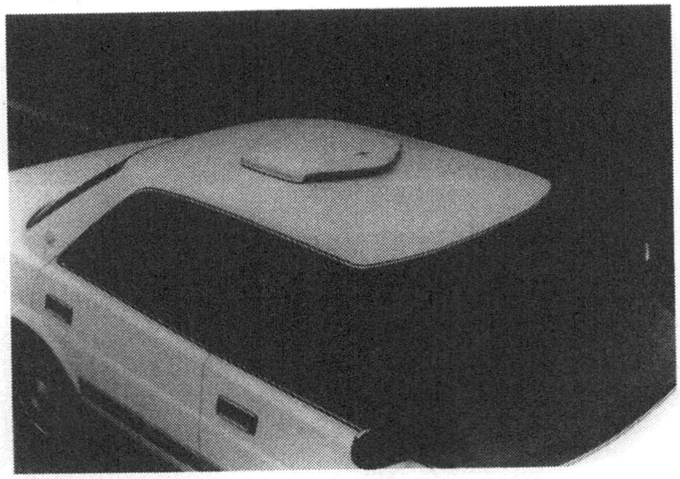

Fig. 3. Phased array installed on a passenger car. (Courtesy of the Toyota Central R & D Labs)

Fig. 4. Mechanically steered low-profile microstrip Yagi array for mobile satellite communications.

has been installed on Boeing aircraft for commercial aeronautical satellite communications. Regarding the mechanically steered antenna, the Jet Propulsion Laboratory (JPL) of United States has developed a unique low-profile and efficient antenna system. This is the mechanically steered microstrip Yagi array [22], shown in Figure 4. The array has a diameter of 48 cm and a height of 4 cm. For the same production quantity as the phased array, JPL has estimated that this mechanical antenna would cost approximately U.S. $400 per unit to the manufacturer. For both the phased array and the mechanically steered antennas, the microstrip radiator has demonstrated the capability of minimizing the antenna's size and cost. Most of the costs in either antenna system are attributed to the beam scanning and tracking components, not to the radiator portion.

Low Earth-orbiting satellite systems, such as the L-band IRIDIUM [23] system proposed by Motorola, Inc., would require multiple high-gain microstrip phased arrays on each satellite and low-gain omnidirectional antennas on the ground terminals. Each of the satellite's phased arrays has hundreds of microstrip patches. By printing all of these patches on a single flat panel, the fabrication process will lead to lower cost; the antenna system will require less volume and contribute less mass on the satellite. For the ground terminal, a low-gain omnidirectional antenna, such as a quadrifilar helix, will suffice. However, if antenna comformability to the vehicle's roof is required, the microstrip patch can certainly be used. A single patch excited at a fundamental mode or a higher-order mode can be considered. Circularly polarized, higher-order mode circular patch antennas [24] have demonstrated good omnidirectional coverage in the azimuth plane and optimum sectorial coverage in the elevation plane. A photograph of a fourth-order mode (TM_{41}) patch is shown in Figure 5, which gives calculated elevation patterns of three different patch modes.

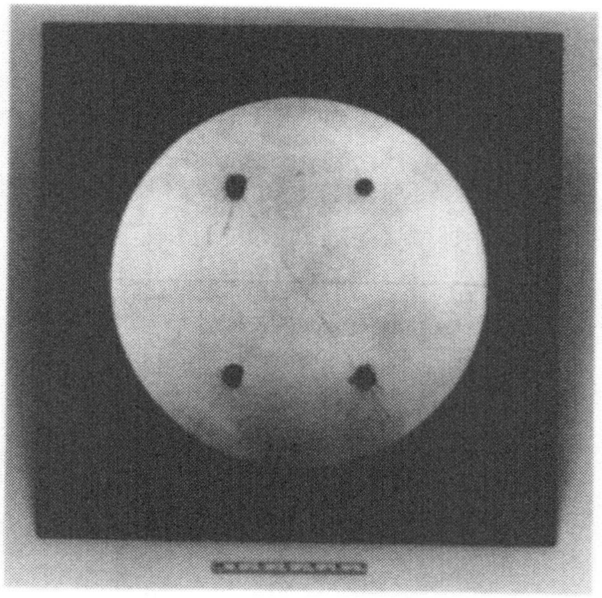

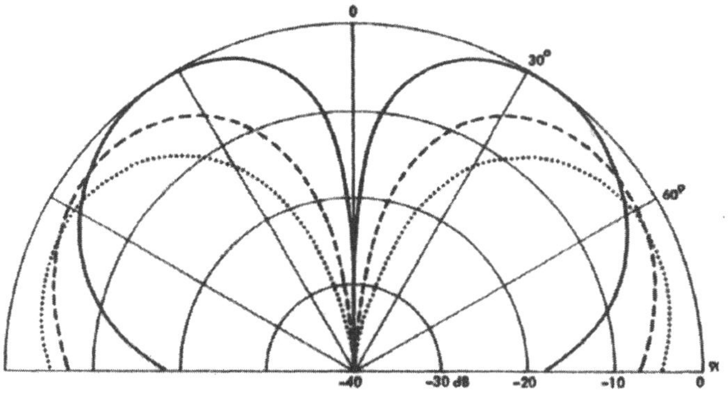

	MODE	RELATIVE DIELECTRIC CONSTANT	PEAK DIRECTIVITY	PEAK DIRECTION FROM ZENITH	RADIATOR DIAMETER
————	TM_{21}	1.25	6.9 dBi	35°	$0.9\lambda_0$
------	TM_{31}	2.2	4.6	54°	0.93
....	TM_{41}	4.2	4.0	69°	0.88

Fig. 5. Higher-order-mode circularly polarized microstrip antenna. Photo shows the S-band fourth-order-mode patch.

IV. Applications for Global Positioning System (GPS)

The satellite-based GPS, initiated thirty years ago for military missions, has grown to have significant commercial applications. It has twenty-four satellites circling the Earth every twelve hours at an altitude of 20,200 km. Each satellite transmits codes continuously at two frequencies in L-band. At any time, four of these satellites can together help a user on the ground to determine his precise position with an accuracy of 15 meters and the time to an accuracy of 100 nanoseconds. Thousands of these GPS ground terminal units were used by the U.S. Army in the Persian Gulf War for position finding, navigation, artillery fire control, and so forth. It is expected that millions of similar units will be used commercially by the general population for land vehicles, aircraft, and maritime vessels to determine their positions and directions.

The antenna needed for the ground terminal is a circularly polarized, omnidirectional, wide-beam, low-gain antenna. To minimize the antenna's size, mass, and costs at L-band, the microstrip patch radiator becomes the best candidate. Ball Corporation has developed a ceramic-loaded microstrip element (shown in Figure 6) that has a size of 5 cm × 5 cm, a thickness of 0.8 cm, and a weight of 85 grams. A higher dielectric constant ceramic-loaded substrate is used to reduce patch size and to increase beamwidth. The dual-stacked-patch technology is employed to achieve the required two L-band frequencies of the GPS system. Toyota Central R&D Laboratories of Japan has also developed a microstrip GPS antenna (illustrated in Figure 7) that uses a stub-loading technology [25] to achieve the dual-frequency capability with a single patch. This antenna was developed to be mounted on Toyota passenger automobiles for GPS navigational purposes. Figures 6 and 7 clearly show that it will be difficult to find another type of antenna to supersede the microstrip antenna's performance in terms of cost, mass, and size for GPS application.

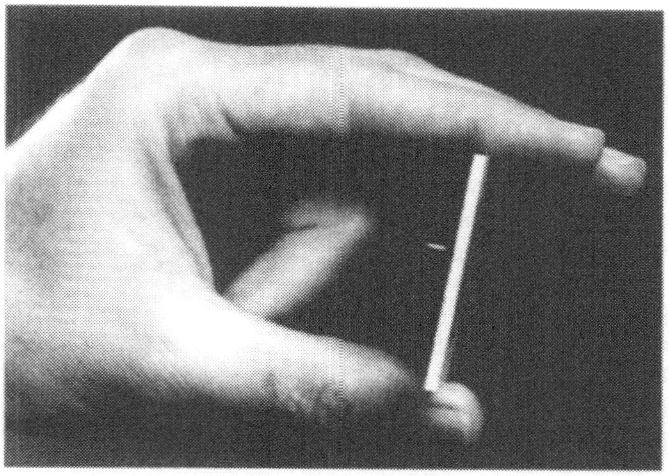

Fig. 6. Microstrip patch antenna for GPS application. (Courtesy of the Ball Corporation)

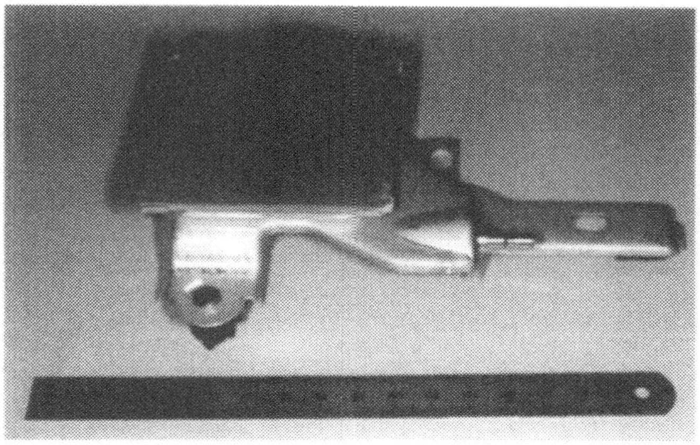

Fig. 7. Microstrip patch antenna for GPS application. (Courtesy of the Toyota Central R & D Labs)

V. Applications for Direct Broadcast Satellite (DBS) Systems

The DBS system has been providing television service to the general public in many countries. Its ground-user antenna is required to have a high gain of about 30 dBi and to operate at the frequency of about 12 GHz. Conventional parabolic reflector antennas have been adopted currently by many users. However, the parabolic reflector, because of its curved bulky structure, cannot be easily hidden or surface-mounted onto an existing building, and thus requires separate real estate for installation. As a result, a number of flat microstrip array antennas have been developed for DBS with the features of being lightweight, easily installed on a building wall, aesthetically appealing, and low in manufacturing cost. Another advantage of using a flat low-profile antenna is that its performance is less affected by wind or snow than a parabolic reflector.

Japanese companies have dominated the development of flat DBS antennas. The microstrip-type flat antennas developed by several Japanese companies are briefly discussed as follows. Yagi Antenna Corporation developed an array [26] with 1024 circular patch elements, which has an aperture size of 48 cm × 64 cm and a peak gain of about 33 dBi. NHK Science and Technical Research Laboratories has also developed a flat microstrip antenna [27] with increased bandwidth and efficiency. The antenna consists of 512 square patch elements with a size of 32 cm × 64 cm and a peak gain of 34 dBi. Other Japanese companies that have developed similar DBS microstrip arrays are DX Antenna [28], Matsushita Electric Works [29], and Sony Corporation.

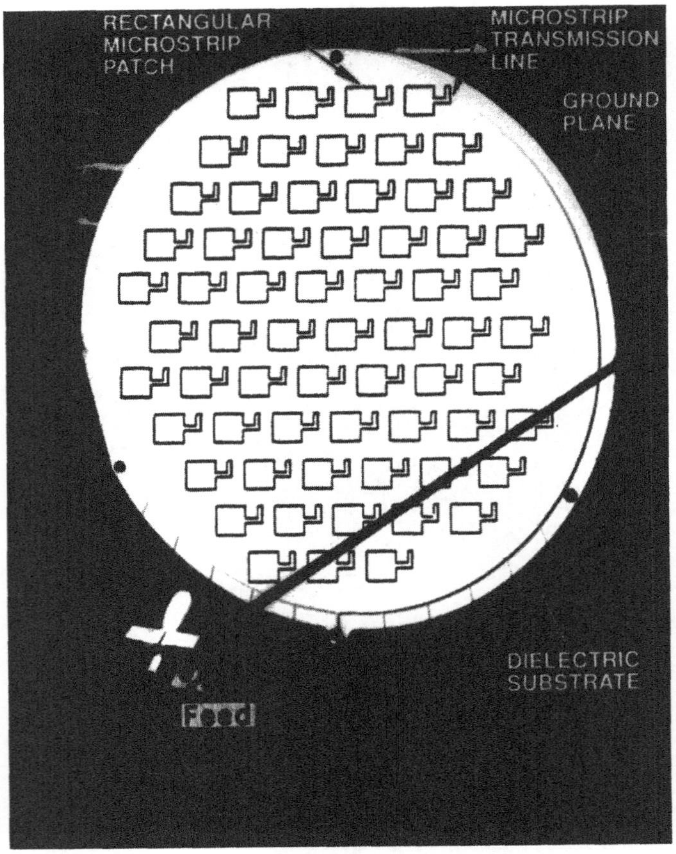

Fig. 8. Microstrip reflectarray surface-mounted on building's wall for DBS application.

One recently proposed antenna that may become a competitor in the DBS terminal application is the microstrip reflectarray [30], which combines the technologies of the parabolic dish and the array. The feed of the reflectarray (as shown in Figure 8) illuminates the flat aperture that has many printed microstrip patch elements. These identical patch elements, having different phasor delay lines, are not connected by any power division lines. The re-radiated waves from these elements will be coherently directed toward a desired direction in accordance with the predesigned phasor delay lines. Unlike the curved parabolic reflector, the flat reflecting surface of the reflectarray can be surface-mounted onto a building or a house wall, and thus occupy an insignificant amount of real estate. Moreover, unlike conventional array antennas, the reflectarray, without any power divider, has very little ohmic insertion loss, which helps the system's gain-over-temperature (G/T) performance when used as a receiving antenna.

The above fixed-beam microstrip array antennas are intended for a DBS system with a fixed ground terminal, such as a residential house. A DBS television receiving system is also applicable to mobile ground terminals, such as automobiles or trains. Toyota Central R&D Laboratories of Japan has recently developed a mechanically steered microstrip array for train applications. This antenna, shown in Figure 9, is a rectangular planar array that generates a fan beam with a wide beam in the elevation plane and a narrow beam in the azimuth plane. Therefore, mechanical steering in the azimuth plane is required to point its narrow beam at the satellite while the train is moving about. This planar array [31] has ninety-six circular patch elements excited by an array of helices and fed by rectangular waveguides. It provides a minimum of 27.5 dBi gain at Ku-band frequency. The complete antenna system has a diameter of 60 cm and a height of 13 cm.

NASA/JPL and the U.S. Information Agency/Voice of America (USIA/VOA) are jointly developing a Direct Broadcast Satellite Radio (DBSR) system, which will provide audio services (AM, FM, etc.) to reach a variety of radio receivers (fixed, portable, and mobile). The low-frequency audio signal is modulated onto a microwave frequency (i.e., S-band), and transmitted to the users via the satellite. When successfully developed, it is expected that millions of radio receivers will be in demand by the general population. Two types of microstrip antennas [32] have been studied for the DBSR system at the S-band frequency. One is the omni-directional higher-order-mode circular patch [24] that was discussed earlier. It is intended as an outdoor low-gain antenna. The second one is a four-element medium-gain patch array shown in Figure 10. It provides circular polarization with 12 dBi of gain. The sequential rotation technique [11], [12] is used to arrange the four elements to achieve good circular polarization over a relatively wide bandwidth. This four-element medium-gain patch array is intended as an indoor receiving antenna, and its extra gain, as compared with the outdoor low-

Fig. 9. Mechanically steered microstrip array antenna for mounting on trains for DBS application [31].

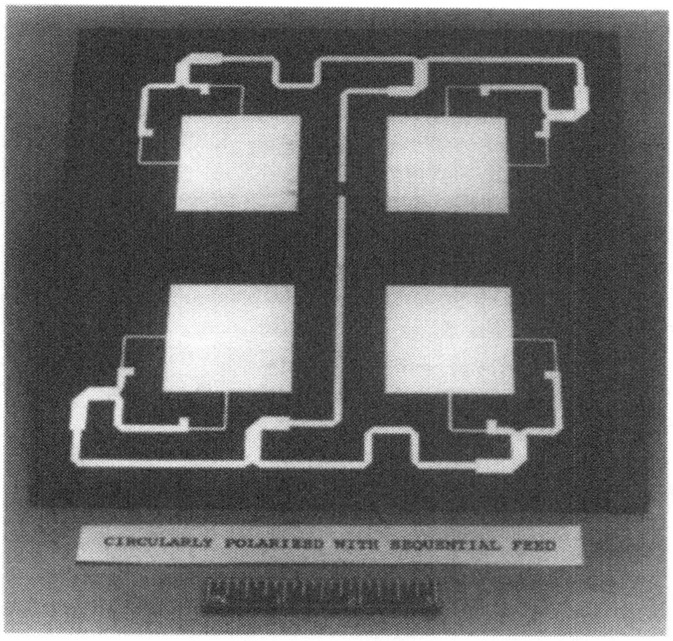

Fig. 10. S-band microstrip array with briefcase size for Direct Broadcast Satellite Radio (DBSR) service.

gain antenna, is used to compensate the RF loss incurred through the building structures.

VI. Nonsatellite-Based Applications

In addition to applications for satellite communications discussed in the previous sections, the microstrip antenna has also been found to be useful in many other areas. Because of its conformal mounting capability, microstrip antennas have been implemented on commercial aircraft for the purposes of altimetry, collision avoidance, remote sensing, and similar applications. Small-size microstrip arrays have also been developed for the automobile collision avoidance system [33] at millimeter frequencies, as well as for microwave sensing alarm system. The medical hyperthermia application is another exciting area in which the microstrip antenna has been found to be useful. Among these nonsatellite-based applications, two important areas—remote sensing and medical hyperthermia—are selected for further discussion in the following paragraphs.

In the area of remote sensing, the Synthetic Aperture Radar (SAR) technique has been used to determine ground soil grades, vegetation type, ocean wave speed and direction, and has significant commercial impact on the general population in terms of agriculture and weather prediction. For example, a C-band interferometric SAR [34], recently developed by NASA/JPL, has successfully measured the characteristics of the ocean waves, which contributed to the prediction of weather. A McDonnell Douglass DC-8 aircraft has been used as the radar platform. Two identical microstrip arrays, separated by a prescribed distance to properly perform the interferometric function, are flush-mounted on one side of the DC-8 fuselage. Each array [16] has a rectangular aperture and generates a fan-shaped beam in the broadside direction. A photograph of the microstrip array (Figure 11) shows the two rows of square microstrip patches with a total of 76 elements. These two rows are excited with opposite phases to achieve the required low cross-polarization. The relatively large number of elements are fed serially to minimize insertion loss and, thus, to achieve good efficiency. This array has

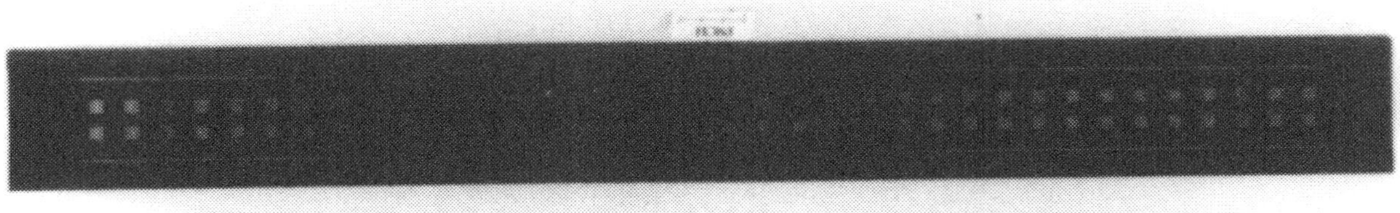

Fig. 11. Microstrip array for aircraft interferometric radar remote sensing application.

achieved less than −30 dB of cross-polarization in the main beam region and an overall efficiency of 72% with a peak gain of 23.8 dBi.

In the medical area, microwave energy has been found to be one of the most effective forms of inducing hyperthermia in treating malignant tumors. Difficulties have been experienced in heating deep-lying tissues and heating a relatively large volume of tissue. In general, the desired characteristics of a microwave applicator include an effective deposition of the energy in a defined tissue volume, good impedance matching, and conformality to the surface being treated, so that minimum leakage of microwave energy will occur outside of the treated area. In addition, the radiator should be lightweight and rugged with an easy-to-handle design. The microstrip patch radiator seems to meet all these requirements. The early designs of microstrip radiators for hyperthermia applications were printed dipoles [35] and an annular ring [36] at an S-band frequency. A more recent design involves the use of a circular microstrip disk [37] at an L-band frequency. Two coupled microstrip lines with adjustable separation [38] have been used at a UHF frequency to measure the temperature gradient inside the human body. An interesting and unique application of the microstrip radiator is the flexible patch applicator [39] at 430 MHz. Figure 12 shows how the flexible microstrip patch can be applied to a curved surface for maximum efficiency.

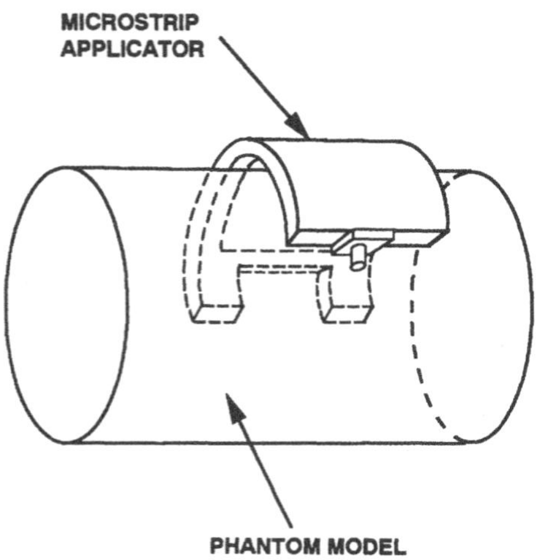

Fig. 12. Flexible microstrip patch mounted on a curved surface for hyperthermia medical application [39].

VII. Conclusion

In the past decade, because of its low profile, small size, and low manufacturing cost, the microstrip antenna has been found to be in significant demand for commercial applications. Especially in the area of satellite communications, the demand for microstrip antennas is most evident. In other areas, such as remote sensing, medical usage, automobile collision avoidance, and aircraft systems, the microstrip antenna has also found important applications. Research and development in the microstrip antenna area should be continued to improve performance parameters, such as the bandwidth and insertion loss. It is believed that this small-size antenna will continue to benefit the human race for many future years.

Acknowledgment

The author gratefully acknowledges Martin Huisjen and Brian Cox of Ball Corporation, Kazuo Sato of Toyota Central R&D Laboratories, and Harry Chen of TRW, for supplying the photographs and technical information on some of the antennas presented here. Thanks are also extended to Ronald Pogorzelski and Ann Tulinsteff for reviewing this article. A portion of the research described in this paper was carried out by the Jet Propulsion Laboratory, California Institute of Technology, under contract with the National Aeronautics and Space Administration.

References

[1] G. A. Deschamp, "Microstrip microwave antennas," Proc. 1953 Antenna Applications Symp., Sept. 1953.

[2] H. Gutton and G. Baissinot, "Flat aerial for ultra high frequencies," French patent No. 703113, 1955.

[3] W. Rafferty, K. Dessouky, and M. Sue, "NASA's Mobile Satellite Development Program," *Proceedings of the Mobile Satellite Conference,* pp.11–22, May 1988.

[4] I. J. Bahl and P. Bhartia, *Microstrip Antennas,* Dedham, Mass.: Artech House, 1980.

[5] J. R. James and P. S. Hall, *Handbook of Microstrip Antennas,* IEE, London: Peter Peregrinus Ltd., 1989.

[6] D. M. Pozar, "A microstrip antenna aperture coupled to a microstrip line," *Electronics Letters,* vol.21, pp.49–50, Jan. 1985.

[7] S. A. Long and W. D. Walton, "A dual frequency stacked circular disc antenna," *IEEE Trans. Antennas and Propag.,* AP-27, pp. 270–273, 1979.

[8] P. Bhartia and I. J. Bahl, "A frequency agile microstrip antenna," *IEEE AP-S Int. Symp. Digest,* pp. 304–307, 1982.

[9] J. Kerr, "Terminated microstrip antenna," Proc. 1978 Antenna Applications Symp., Sept. 1978.

[10] H. F. Pues and A. R. Van De Capelle, "An impedance-matching technique for increasing the bandwidth of microstrip antennas," *IEEE Trans. Antennas and Propag.*, vol. 37, pp. 1345–1354, Nov. 1989.

[11] T. Teshirogi, M. Tanaka, and W. Chujo, "Wideband circularly polarized array antenna with sequential rotations and phase shifts of elements," *Proc. Int'l Symp. Antennas and Propag.*, pp. 117–120, 1985.

[12] J. Huang, "A technique for an array to generate circular polarization with linearly polarized elements," *IEEE Trans. Antennas and Propag.*, AP-34, pp. 1113–1124, Sept. 1986.

[13] C. Wood, "Improved bandwidth of microstrip antennas using parasitic elements," *IEE Proc.*, 127H, pp. 231–234, 1980.

[14] H. Howe, Jr., "Stripline is alive and well," *Microwave J.*, vol. 14, p. 25, July 1971.

[15] R. Woo, "Final report on RF voltage breakdown in coaxial transmission lines," *JPL Technical Report 32-1500,* Oct. 1970.

[16] J. Huang, "A parallel-series-fed microstrip array with high efficiency and low cross-polarization," *Microwave and Optical Technology Letters,* vol. 5, pp. 230–233, May 1992.

[17] D. C. Chang, D. I. Wu, and J. C. Moore, "Modeling and computer-aided design for microstrip antenna array," *European Space Agency Workshop on Active Antennas,* June 1992.

[18] K. C. Gupta, R. Garg, and R. Chadha, *Computer-aided Design of Microwave Circuits,* Dedham, Mass.: Artech House, 1981.

[19] A. da Silva Curiel, "The first generation INMARSAT system," *IEEE 3rd Int'l Conf. on Satellite System for Mobile Communications and Navigation,* pp. 1–7, June 1983.

[20] F. Schmidt, "MSAT mobile electronically steered phased array antenna development," *Proceedings of the Mobile Satellite Conference,* pp. 217–222, May 1988.

[21] K. Sato, K. Nishikawa, and T. Hirako, "Development and field experiments of phased array antenna for land vehicle satellite communications," *IEEE AP-S Symp. Digest,* pp. 1073–1076, July 1992.

[22] J. Huang and A. Densmore, "Microstrip Yagi array antenna for mobile satellite vehicle application," *IEEE Trans. Antennas and Propag.,* Vol. 39, pp. 1024–1030, July 1991.

[23] J. J. Schuss et al., "Design of the IRIDIUM phased array antennas," *IEEE AP-S Symp. Digest,* pp. 218–221, June 1993.

[24] J. Huang, "Circularly polarized conical patterns from circular microstrip antennas," *IEEE Trans. Antennas and Propag.,* vol. AP-32, pp. 991–994, Sept. 1984.

[25] Y. Murakami et al., "Dual-band circularly polarized stub loaded microstrip antennas," *Proceedings of ISAP,* vol. 3, pp. 793–796, 1992.

[26] M. Haneishi, "A circularly polarized SHF planar array composed of microstrip pairs-element," *Int'l Symp. Antennas and Propag.,* pp. 125–128, 1985.

[27] K. Ohmaru and T. Murata, "A planar array antenna for satellite broadcasting reception," *Proc. Int'l Microwave Symp.,* pp. 115–120, July 1987.

[28] T. Watanabe, T. Fujita, and F. Deguchi, "Microwave planar array antenna design," *ITE Japan Technical Report, RE87-4,* 1987.

[29] K. Ito, K. Ohmaru, and Y. Konishi, "Planar antennas for satellite reception," *IEEE Trans. on Broadcasting,* vol. 34, pp. 457–464, Dec. 1988.

[30] J. Huang, "Microstrip reflectarray," *IEEE AP-S Symp. Digest,* pp. 612–615, June 1991.

[31] T. Watanabe et al., "Helical array antenna fed by waveguide," *Proceedings IEICE Spring Conf.,* Mar. 1994.

[32] T. K. Wu and J. Huang, "Low Cost Antennas for Direct Broadcast Satellite Radio," *IEEE AP-S Symp.,* pp. 1008–1011, June 1994.

[33] C. P. Wen et al., "Single Substrate Microwave Radar Transceiver including Flip-chip Integrated Circuits," *U.S. Patent No. 5115245,* May 1992.

[34] R. M. Goldstein and H. A. Zebker, "Interferometric radar measurement of ocean surface currents," *Nature,* vol. 328, pp. 707–709, Aug. 1987.

[35] F. Sterzer et al., "Microwave apparatus for the treatment of cancer," *Micro-wave J.,* vol. 23, pp. 39–44, Jan. 1980.

[36] I. J. Bahl, S. S. Stuchly, and M. A. Stuchly, "A new microstrip radiator for medical applications," *IEEE Trans. Microwave Theory and Techniques,* vol. MTT-28, pp. 1464–1468, Dec. 1980.

[37] R. De Leo, G. Cerri, and F. Moglie, "Microstrip patch applicators," *IEEE AP-S Symp. Digest,* pp. 524–527, June 1989.

[38] D. Kobayashi et al., "Coupled microstrip applicator for hyperthermia applications," *IEEE AP-S Symp. Digest,* pp. 532–535, June 1989.

[39] H. Kobayashi et al., "Flexible microstrip patch applicator for hyperthermia," *IEEE AP-S Symp. Digest,* pp. 536–539, June 1989.

DESIGN OF LOW COST PRINTED ANTENNA ARRAYS

J. P. DANIEL, Professor, Université de Rennes I, 35042 RENNES CEDEX, FRANCE

E. PENARD, Engineer | *CNET LANNION B*
M. NEDELEC, Tint | *Department MER/STA*
J. P. MUTZIG, Engineer | *22301 LANNION, FRANCE*

INTRODUCTION

Microstrip antenna arrays are often presented as low-cost elements for small communication or radar systems when narrow bandwith (less than 5 %) and medium gain (15 to 25 dB) are needed.

In fact to main conditions must be kept in mind when cheap antennas are required :

- first high frequency bands (20-30 GHz for instance) are well suited, because they require a rather small surface of usual printed circuit such as PTFE,
- secondly simple radiating structures and corporate feed, printed on the same side, are highly desirable.

Following the previous requirements some arrays have been investigated and tested in the 20/30 GHz band. Some of them have been already described [1]. Here two new aspects are considered : reduction of cross polarization using an alternate location of the elements on the feeding line and reduction of side lobe level using simple taper for linear serie's array. Moreover to avoid tapered distribution on two dimensional array, cross-fed structure including only four quarter wavelength matching sections lead to low side lobes in E and H planes.

1 - *RADIATION OF CORNER FED SQUARE PATCHES*

1.1 - Theory

The corner fed square patches have been choosen because they provide a high input impedance well suited for serie's array. When the patch is excited at one corner (fig. 1.1), the cavity model [2] shows that the main part of the internal field is the sum of two degenerated modes with equal amplitudes i.e. modes (0,1) and (1,0). If the higher modes are neglected the E_x and E_y fields along the edges exhibit the variations shown on (fig. 1.2). The far field is linearly polarized either in the E plane ($\phi = 0$) or in the H plane ($\phi = 90°$). For instance in the H plane, only the E_ϕ component arises.

As far as we want to know the cross polarization level, higher modes must be considered. The next mode (1,1) adds a contribution with a magnetic line distribution plotted on fig. 1.3. It must be noticed that M_1 is much more smaller than M_0 but gives the E_θ component. E_θ exhibits a null for $\theta = 0$ and increases with θ.

1.2 - Example of linear arrays (A_1 and A_2)

When the spacing along the feeding line equals one guided wavelength, the different elements are uniformly excited (fig. 2.1). Here the ten elements array is printed on a usual PTFE substrate ($\varepsilon_r = 2.17$ and thickness = 0.38 mm). The quarter wavelength transformer enables a good matching to the 50 Ω coaxial output.

As expected the H plane diagram exhibits the well known -13 dB first side lobe ; the cross polarization component is very low at $\theta = 0$ and then quickly increases with θ to -17 dB (fig. 2.2).

In order to reduce the cross-polarization level a new design has been investigated. Element spacing is now half wavelength but each two patches is fed on the out of phase part so equiphase excitation can be kept (fig. 3.1).

As a result copolar components (E_{ϕ_0} for the H-plane) add in phase when cross polar components (E_{θ_1}) add out of phase, and then the cross polarization level is lowered. The diagram plotted on fig. 3.2 shows that the cross polarization is reduced to -28 dB instead of the -17 dB. Due to the

Reprinted with permission from *Proc. ISAP,* J. P. Daniel, E. Penard, M. Nedelec and J. P. Mutzig, "Design of Low Cost Printed Antenna Arrays," pp. 121–124, Aug. 1985.

smaller length of the array the half-power beam width is increased and twenty elements with $\lambda/2$ spacing would give the same directivity.

2 - *TAPERED LINEAR SERIE'S ARRAY.*

2.1 - Theory

The previous arrays were uniformly excited. High side lobes are the consequences of this illumination. The idea was to design a non uniform amplitude distribution while keeping the simplicity of the previous serie's feeding.

Let us consider a linear array with one guided wavelength spacing ; as the radiating elements are identical, impedance transformers are necessary to get the given amplitude current. To do so a two step quarter wave transformer can be used in each cell (fig. 4.1). The transformed admittance Y_i in the π'_i plane is given by

$$Y_i = \left(\frac{Y_{L_2}}{Y_{L_1}}\right) Y_{i+1} = n_i Y_{i+1} \qquad (1)$$

where Y_{i+1} is the admittance at node i+1

Y_{L_2} and Y_{L_1} are the characteristic admittance of each quarter-wavelength transformer.

If the inputvoltage leads to a unit current in the first element, then the current distribution is readily obtained with the following relations :

$$I_o = 1 \; ; \; I_1 = n_1 \; ; \; I_i = n_i \, n_{i-1} \ldots n_1 \qquad (2)$$

one can deduce step by step the various ratio n_i from the known I_o, I_1, I_i weight.

2.2 - Results (array A_3)

A ten elements array have been built (fig. 4.2). The requirement was to get side lobe level lower than -20 dB. Only eight transformers were used because Y_{L_2} was choosen equal to the characteristic admittance of the half wavelength following line. Taking a characteristic impedance of the main line near 100 Ω, the various transformers exhibit impedance value between 75 to 95 Ω which are easily realized with microstrip lines. On fig. 4.3 the experimental H plane diagram is plotted ; it shows that the side lobe level is lower that -20 dB while the gain and the input impedance matching remain very similar to the uniform array.

3 - *CROSS-FED ARRAY (A_4).*

A square array of uniform excited elements would give -13 dB side lobes in the E and H planes. The cross-fed array is very attractive because, first its geometry gives a low side lobe level and secondly the corporate feed is very simple [1].

An array of 50 elements was designed at 20 GHz (figure 5.1). Quarter wave length transformer on each branch of the cross are calculated in such a way that the input power is equally devided between each patch and that the input impedance is matched to the 50 Ω coaxial connector. Because the width of the line for one section transformer was too wide we had to use a three section transformer for the vertical branch.

The radiation pattern is given fig. 5.2 for the H plane and one can note the very low side lobes (lower than -25 dB in both E and H plane). The measured gain is 23 dB, VSWR lower than 1.8 between 19.5 and 20.4 GHz.

CONCLUSION

Microstrip antenna arrays are very attractive for low cost communication system with medium gain (15 to 25 dB). High frequency band (20/30 GHz) seems to be well suited because rather small surfaces are needed and technical processes (photo etching) remain fairly simple. Reduction of cost requires also a simple design of one side printed antenna ; either for cross polarization and low side lobes performances it has been shown that the serie's structure brings a simple and feasible solution in the 20/30 GHz.

REFERENCES

[1] J. P. DANIEL, J. P. MUTZIG, M. NEDELEC, E. PENARD
"Réseau d'antennes imprimées dans la bande 20/30 GHz"
Onde Electrique, Janvier-février 1985 vol. 65 n° 1 pp 35-41.

[2] Y. T. LO, D. SOLOMON, W. F. RICHARD
"Theory and experiment on microstrip antennas"
IEEE Trans, march 1979, AP 27 pp 137.

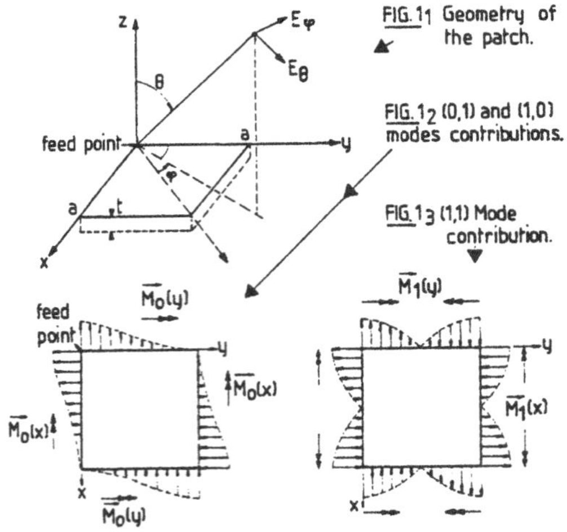

GEOMETRY AND MAGNETIC CURRENTS OF THE CORNER FED PATCH.

FIG. 2_1 Uniform linear array A_1 (F_o = 21.3 GHz)

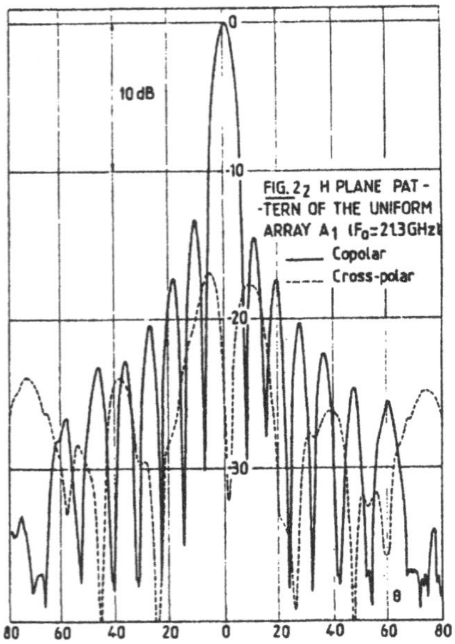

FIG. 2_2 H PLANE PATTERN OF THE UNIFORM ARRAY A_1 (F_0=21.3GHz)

FIG. 3_1 Uniform array A_2 with alternate elements

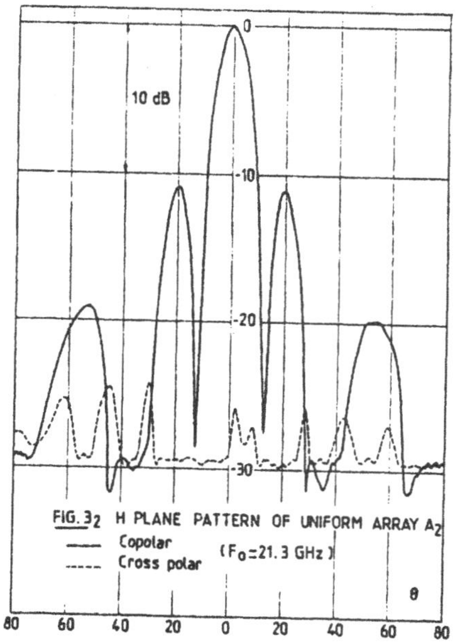

FIG. 3_2 H PLANE PATTERN OF UNIFORM ARRAY A_2
—— Copolar
---- Cross polar
(F_o=21.3 GHz)

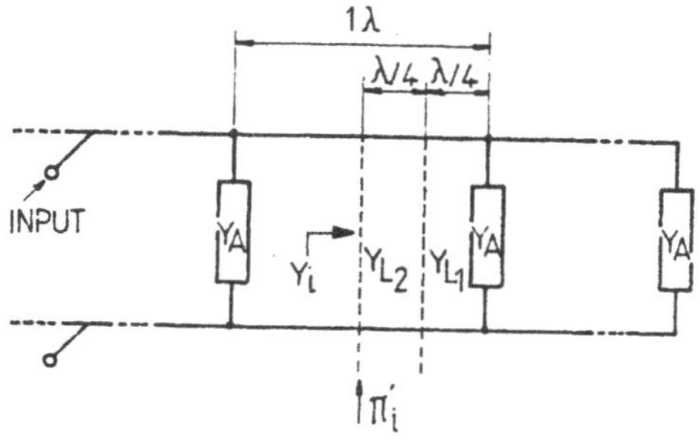

FIG. 4_1 STRUCTURE OF A TAPERED SERIE'S FED ARRAY.

FIG. 4_2 Tapered array A_3

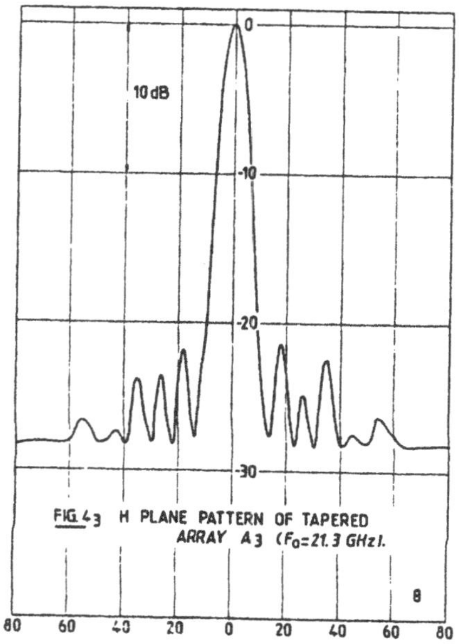

FIG. 4_3 H PLANE PATTERN OF TAPERED ARRAY A_3 (F_o=21.3 GHz).

FIG. 5_1 Cross fed array with coaxial output A_4

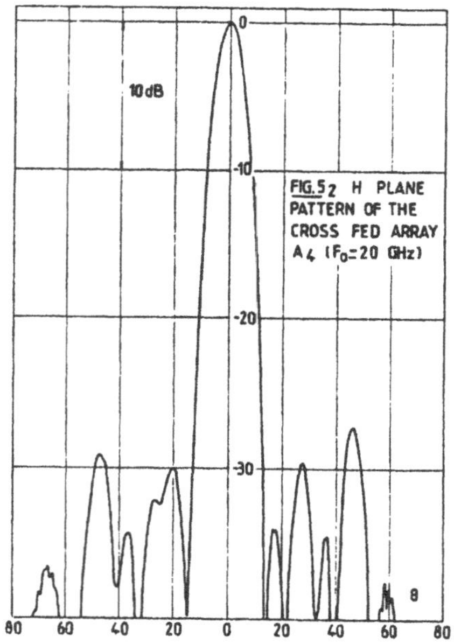

FIG. 5_2 H PLANE PATTERN OF THE CROSS FED ARRAY A_4 (F_o=20 GHz)

Low-cost flat-plate array with squinted beam for DBS reception

A. Henderson, BSc, PhD, CEng, MIEE
Prof. J.R. James, BSc, PhD, DSc, FEng, FIMA, FIERE, FIEE

Indexing terms: Antennas (microstrip), Satellite links and space communications

Abstract: There is a great deal of interest in the replacement of reflector antennas by low-cost wall-mounted flat-plate antennas for the reception of satellite signals. The paper describes the extent to which combline microstrip technology can be used to obtain a squinted, circularly polarised beam at very low cost. The antenna presented consists of a simple travelling wave printed structure comprising a foam/metal sandwich with metal strip polariser. Performance parameters such as bandwidth, dissipative loss and pattern quality are examined and the effect of manufacturing tolerances are discussed. Finally, comparisons are made with reflector antennas and other flat-plate designs, with particular reference to the DBS specification; the possibility of introducing beam control capabilities into the array is considered. It is concluded that the development of suitable low-cost substrates is a key issue and electronic beam control raises problems of a more fundamental nature.

1 Introduction

Attention is now being concentrated internationally on the development of flat-plate antenna arrays which are capable of receiving satellite transmissions and can be deployed on a variety of buildings. Typical applications include data transmission and direct broadcast satellite TV(DBS). These installations will necessarily be of low cost, and the market potential is considerable. In general, the technical specifications can be met by reflector antennas [1] provided that they can be positioned on a suitable wall, but other factors, such as environmental objections and performance degradation due to wind damage and weather effects [2], have led to intensified research into thin planar arrays.

In the long term, it is important that these planar arrays be designed with some degree of electronic beam control to accommodate the variety of geographical wall locations and also be capable of addressing a selection of satellites. The development of such a sophisticated, yet low-cost, array would appear to be some years away and, owing to its complexity, is only likely to be achieved in several stages. In the immediate future it is realistic to aim for a simpler array with a fixed beam squint in elevation, and some beam control in azimuth. Here we demonstrate the extent to which microstrip technology can meet some of these important requirements, such as low cost, a squinted beam and circular polarisation, prior to assessing its potential for further research into beam control aspects. The demand for a domestic array for DBS reception is an immediate technological challenge and the objective of the present research is to assess the feasibility of meeting this particular specification.

The paper commences in Section 2 with a summary of the DBS antenna requirement, as applied to the flat-plate array, and then Section 3 gives a detailed description of the construction and design of the microstrip array, along with appropriate measured results. Section 4 critically examines the overall performance of the microstrip antenna and compares this to other flat-plate design options. Finally, conclusions are stated in Section 5 about the suitability of the existing microstrip array for the DBS market, and areas for future research are identified.

2 DBS flat-plate requirements

The antenna requirements for DBS reception were laid down in 1977 by the WARC Conference [3] but, since that time, further improvements in receiver noise figures have allowed some relaxation in the antenna gain performance [4] for the UK allocated band of 11.7–12.1 GHz. A detailed breakdown of the down-link power budget will be given in Section 4.2 indicating the relationships between receiver noise figures and antenna gain requirement. The sidelobe envelope and crosspolarisation levels were also specified to minimise interference from other satellites, and are given in Fig. 12 for both the original WARC specification and a relaxed version which applies only to the case of UK reception from a UK satellite. Overall the antenna electrical specification is very stringent for a low budget microstrip array. The principal physical requirements for the array are that it is light to handle and of reasonable proportions for fixing to a wall. An in-built beam squint of about 24° allows vertical fixing with some slight adjustment for different geographical locations and azimuth optimisation.

By considering the design options available for planar structures, it is evident that many of the desired features of the DBS array are incompatible with known technologies. For instance, excellent planar arrays have been built with a boresight beam using high quality overlaid substrate [5] and suspended microstrip [6], all based on parallel fed elements. To convert these to a squinted beam design would require a much more complicated feeder topology. Furthermore there will be little prospect of devising a scanned system without incorporating

Reprinted with permission from *Proc. IEE,* A. Henderson and J. R. James, "Low-Cost Flat-Plate Array with Squinted Beam for DBS Reception," vol. 134, pt. H, pp. 509–514, Dec. 1987.

separate phasing for each element in the array. The design presented here, and elsewhere [7, 8], circumvents these difficulties by using a travelling-wave design on a cheap foam substrate but the inherent frequency scanning loss at the band edges is an obvious penalty.

3 Constructional details and design

3.1 Construction

Most of the previously published flat-plate arrays [5, 9–11] have been manufactured from expensive microwave plastic substrates, and to ensure that any array can compete in costs with mass-produced dishes it is essential that cheaper materials and simple manufacturing processes are chosen. The microstrip antenna presented here is manufactured from a sandwich structure of foam materials (see Fig. 1) and has a clear cost advantage over present plastic materials of about 10 : 1 in substrate costs.

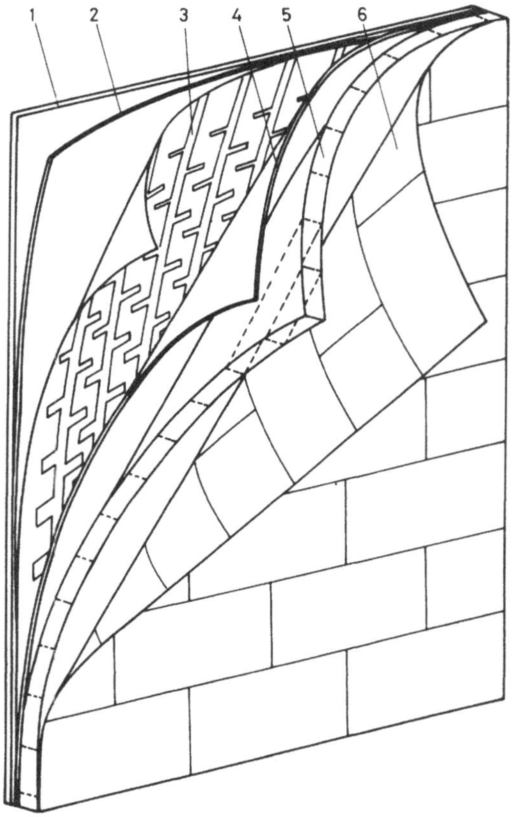

Fig. 1 *Laminar construction of foam microstrip array*

1 Metal groundplane 2 Foam substrate 3 Combline antenna elements 4 Foam spacer 5 Polarising grid 6 Weather-proof cover

The microstrip antenna is constructed from a thin layer of high quality foam substrate ($\varepsilon_r = 1.06$) and the radiating copper elements are etched from a layer of very thin copper-coated plastic film, which is then adhered to the foam sheet. The full-size DBS array is built up from four subarray units fed individually by coaxial cables, each subarray unit consisting of a fully integrated package using 16 comblines with a foam triplate feed. Because the combs are linearly polarised, a polarising sheet is required to match into the circularly polarised transmissions, which can easily be rotated to change the hand of polarisation. For the preliminary experiments, a simple metal plate-type polariser [12] was used, although it is feasible to design a flat printed sheet for this purpose, which would be more suitable for mass production. Finally the antenna can be protected from rainfall by a thin plastic box chosen to match the colour of the house walls.

3.2 Antenna design procedure

The subarray unit was optimised by a combination of CAD and practical testing procedures. To facilitate this process, the antenna was built up in several design stages from individually optimised subcomponents, as follows:

First, a single combline was designed [16] for maximum overall gain and pattern quality for the 11.7–12.1 GHz band. Secondly, four of these combs were combined into a subgroup and fed by an integrated power splitter. This subgroup corresponds to the dashed box in Fig. 2. Testing on this subgroup enabled the foam power splitter and the unwanted radiation at the microstrip/triplate discontinuity to be examined in detail, along with any mutual coupling effects between the combline elemental arrays. The subarray unit was then printed off from four subgroup sections with a 16 : 1 power splitter as shown in Fig. 3, and a metal plate polariser was designed for this unit, taking into account the squinted main beam angle and the proximity to the subarray. Finally measurements were made for four subarrays [17] connected by coaxial cables in a 4 × 1 configuration suitable for the DBS application.

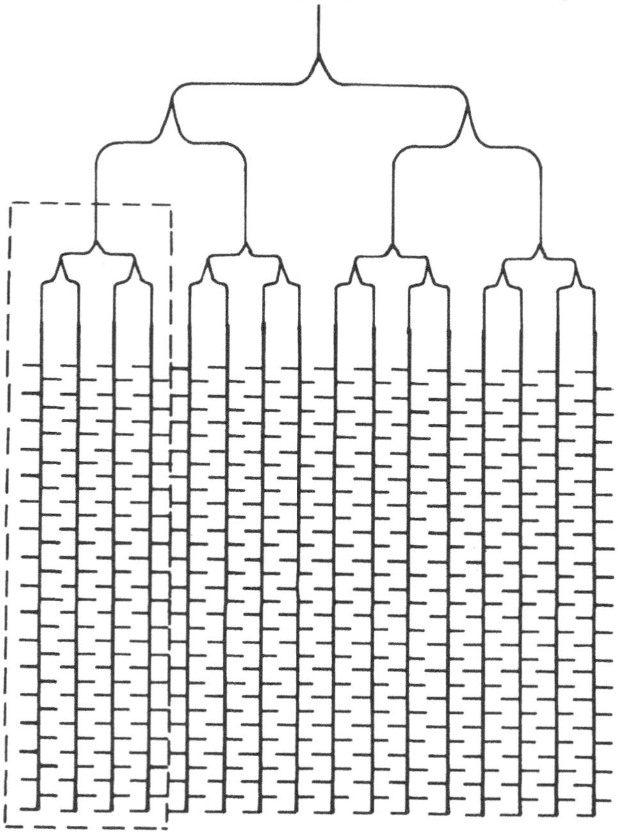

Fig. 2 *Photolithographic mask for subarray unit showing first 33 fingers*

The overall dimensions of the unit are 60 × 46 cm. The dashed box is a subgroup section

3.2.1 Combline array: The initial layout of the single combline array is generated such that the overall antenna gain is maximised over the 11.7–12.1 GHz band, taking into account the scanning losses and line dissipation as a function of antenna length and line impedance. Computations indicate that a minimum of 40 fingers using a

4 mm combline width realises the best overall gain figure for 1.59 mm foam substrate (see Fig. 4). The final design did, however, require a certain amount of cut-and-try experimentation to achieve an acceptable sidelobe level.

Fig. 3 *Subarray unit with 16 comblines and integrated feed*

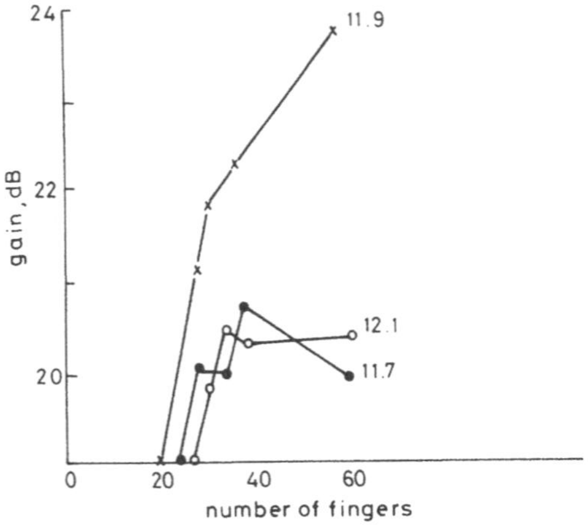

Fig. 4 *Computed overall gain of combline array with 4 mm feed width and a line attenuation of 0.1 dB/guide wavelength, for 11.7, 11.9 and 12.1 GHz*

Frequency scanning losses are included

Detailed measurements of the combline are given in Fig. 5 for 11.9 GHz, indicating (−13, −12) dB sidelobe levels with a fairly high level of cross polarisation of −17.5 dB, principally due to the wide terminal finger elements associated with the low permittivity substrate.

3.2.2 Four-comb group and triplate feed: The triplate feed was the most difficult subcomponent to optimise, requiring careful shaping of the power splitter tracks to minimise both unwanted mode generation within the feed and the dissipative loss. The final track layouts of the transmission through each track of the feed were made, and the power transmission characteristics are given in Fig. 6 for two of the outputs. Small differences in the power delivery to each track are due to manufacturing

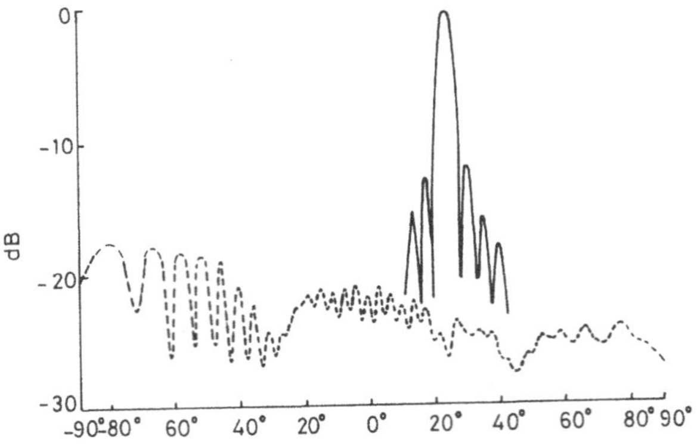

Fig. 5 *H-plane radiation pattern of 40-finger combline*

- - - - crosspolar

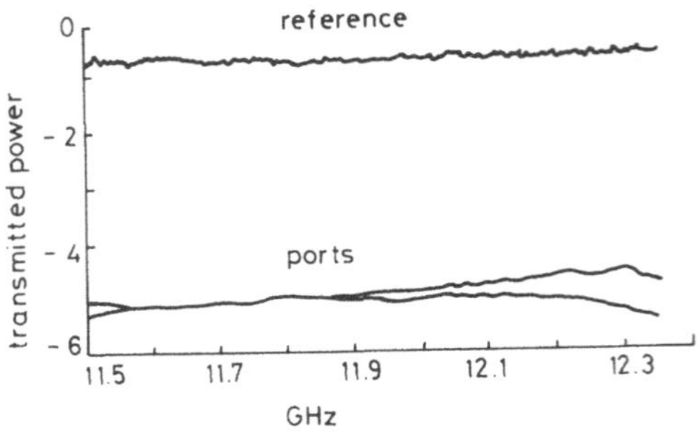

Fig. 6 *Transmission characteristics of 4 : 1 triplate splitter through two ports*

tolerances in the feed and mutual coupling between the four tracks. An average insertion loss of 0.4 dB is indicated from these figures although this was worsened by 1.0 dB when the combline elemental arrays were attached to form a four comb group. This was due to the power mismatch at the triplate/air boundary and was reduced by tapering the copper track in the vicinity of the junction and turning up the top plate of the feed to form a small horn-line structure. The input characteristics of the four-comb group are given in Fig. 7 showing a good

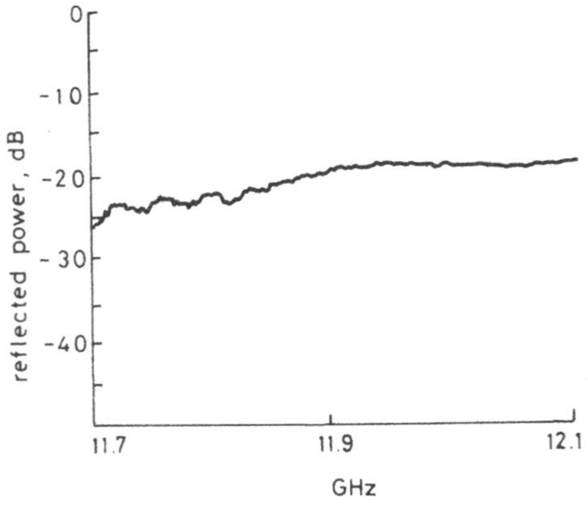

Fig. 7 *Input characteristics of 4-comb group*

power match over the band. The *E*-plane radiation patterns for the four-comb group are given in Fig. 8 and the symmetry in the main beam confirms an adequate power distribution within the feed. *H*-plane co-polar patterns are similar to those of a single combline. Gain measurements indicate an insertion loss of 1.0 dB for the triplate feed, and a full breakdown of the gain budget is given in Section 4.

3.2.3 The subarray unit: The measured *E*-plane radiation pattern [15] given in Fig. 9 showing (−11,

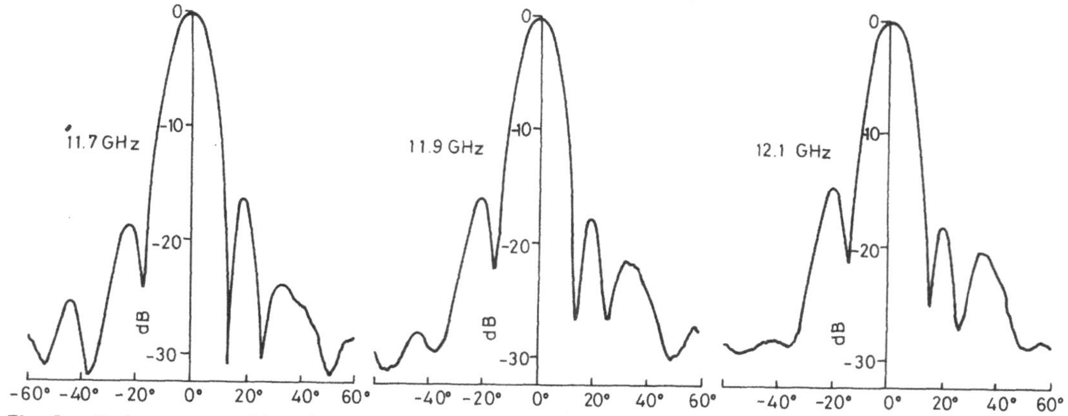

Fig. 8 *E-plane patterns of 4-comb group*

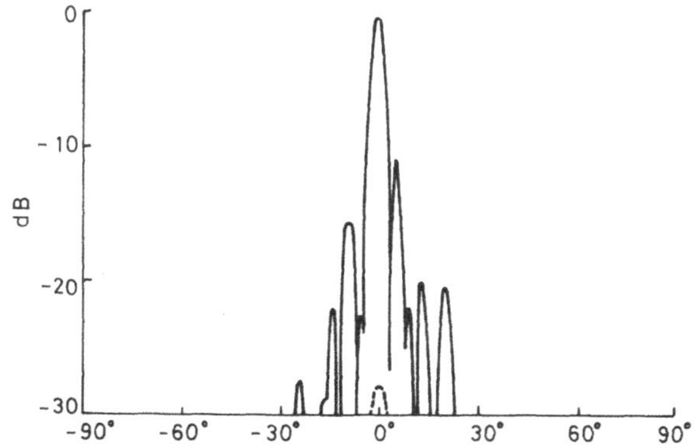

Fig. 9 *E-plane of subarray at 11.9 GHz*
—— copolar, - - - - crosspolar

−15) dB sidelobe levels and a −27 dB crosspolarisation level. The overall gain was 30 dB at 11.9 GHz, with 2 dB frequency scanning losses at the band edges, indicating a 3 dB insertion loss in the 16 : 1 triplate feed system.

3.2.4 The polarising grid: The metal plate polariser, Fig. 10, was chosen not only for its simplicity of con-

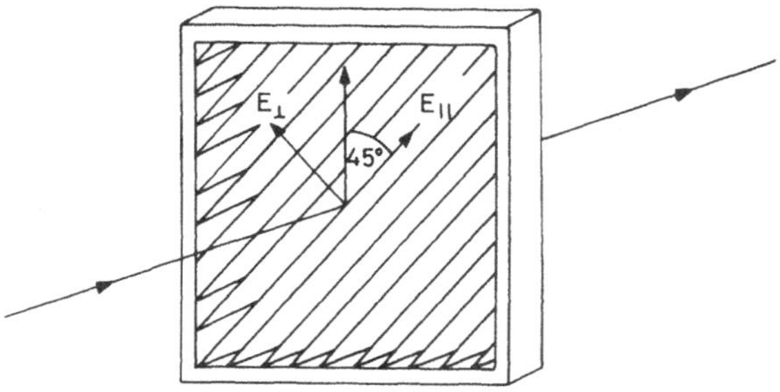

Fig. 10 *Metal-plate polariser*

struction and design but also for its acceptable bandwidth performance at main beam. Wire-grid sandwich structures as described by Warren [13] are preferable structurally for manufacturing reasons, but tests on a single layer type indicated a very narrow band response, unsuitable for the DBS application. A wider response could be obtained by using multilayer designs, typically involving meander lines [14], but there may be some penalty to pay in terms of increased insertion loss, and this would have to be determined by further research.

For a normally incident circularly polarised wave, the design of the plate polariser requires that the $E_{\parallel}$ component, see Fig. 10, is transmitted through the metal (waveguide) plates with a phase shift of 90° relative to the $E_{\perp}$ component. The dimensions of the plate are chosen such that $b = 0.607\ \lambda_0$ and $a = 0.655\ \lambda_0$ (*a* is the spacing between the plates and *b* the width of the plates) when the near-field effects at the plate edges are taken into account. To retain these phase conditions, the entire metal plate assembly should be tilted by an angle of 24° for the squinted beam antenna. Instead, the plate structure was laid in a plane conformal to the antenna elements, but each plate was tilted by 17.5° to take this into account, and the dimensions of the plates were modified slightly by cut-and-try methods to ensure the correct phasing conditions. The correction of 17.5° was derived from elementary geometrical relationships. Even so, the conformal construction does compromise the basic polariser design, by introducing progressive phase errors towards the edges of the aperture. Additionally, the polariser is located within the near-field region of the antenna itself, and measurements were necessary to determine the optimum separation distance for the best axial ratio and minimum insertion loss.

Fig. 11 gives the measured *E*- and *H*-plane patterns for the subarray/polariser unit at 11.9 GHz. An axial ratio of 0.5–0.8 dB was obtained over the band in the main beam although this worsened considerably in the sidelobe regions. An insertion loss of 0.5–0.7 dB was obtained for an optimum separation distance of 1 cm between the antenna and the polariser. It was concluded that a simple conformal polariser could be constructed for operation in close proximity to the antenna, provided that crosspolarisation levels are not too stringent at wider angles. This will be discussed further in Section 4 with reference to the DBS specification.

3.2.5 The large DBS array: Four subarray units were constructed, and their individual gain performances compared in Table 1, which gives a clear indication of the

Table 1: Gain of four identical subarray units showing tolerance effect

Subarray	1	2	3	4	
Gain (dB)	29.5	30.0	28.5	29.5	average = 29.4 dB

manufacturing tolerances. To minimise scanning loss over the frequency band, it was decided to arrange the four subarrays in a 4 × 1 matrix rather than a 2 × 2 matrix thus giving a narrower beam in azimuth than elevation and reducing any interference effects between satellite signals. Measurements of the gain and radiation patterns of the large array were made without the polarisers, and were subject to limitations on accuracy due to

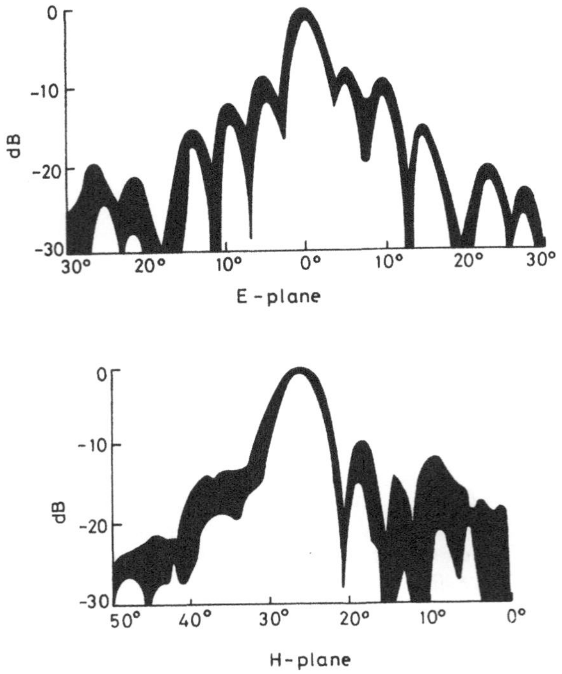

Fig. 11 *Circularly-polarised patterns at 11.9 GHz for subarray fitted with metal-plate polariser*

ground reflections and beam alignment errors; the results for the *E*-plane are given in Fig. 12. A gain figure of 33 dB was obtained at 11.9 GHz with 2 dB scanning losses at the band edges.

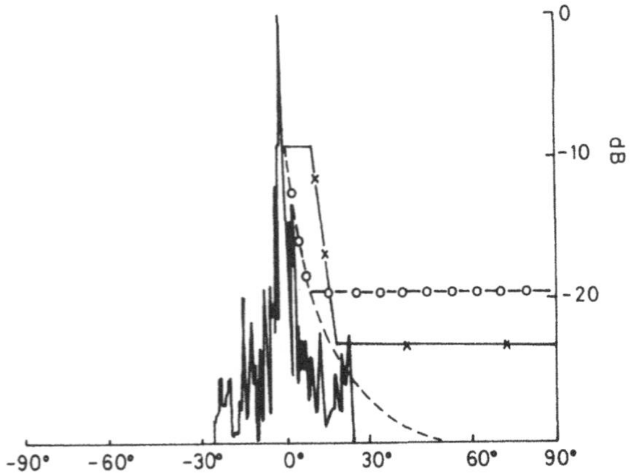

Fig. 12 *E-plane pattern for large DBS array, showing DBS antenna specifications for reception*

- - - - WARC specification for copolar, -x-x-x UK specification for cross-polar, -O-O-O UK specification for copolar.

4 Overall antenna performance

4.1 Gain budget and efficiency

Table 2 gives a budget of the gain performance of the large array and its subcomponents at 11.9 GHz, without polarisers. A 2 dB gain reduction is experienced at the band edges owing to the travelling wave design.

The 7.1 dB loss of gain is excessive and can be attributed to several sources (see Table 3). The overall antenna efficiency of about 20% is very poor when compared to 60–70% for a small parabolic dish. An examination of the gain penalties involved indicate that the use of a

Table 2: Gain budget for large array at 11.9 GHz

	Measured gain, dB	Predicted gain, dB	Shortfall in gain, dB	Cause
Single combline	21.0	*a* 22.1	1.1	Nonuniform current distribution
Four-comb subgroup with 4 : 1 feed	26.0	*b* 28.1 (= *a* + 6 dB)	2.1	1 dB due to feed losses*
Subarray	30.0	*c* 34.1 (= *b* + 6 dB)	4.1	3 dB due to corporate feed
Large array	33.0	*d* 40.1 (= *c* + 6 dB)	7.1	3 dB in cables and manufacturing tolerances

* The feed losses in the 4 : 1 splitter can be broken down into 0.5 dB at the triplate/microstrip junction, 0.4 dB dissipative loss in the tracks and 0.1 dB power mismatch.

Table 3: Sources of gain reduction for large array

(*a*) Cable loss and foam combiner	= 1.5 dB min	3 dB in total
(*b*) Variation in gain due to manufacturing	= 1.5 dB max	
(*c*) Triplate feed: (i) dissipation	= 1.5 dB	
(ii) mismatch	= 1.5 dB	
(*d*) Errors in combline design, either adhesive losses or inaccuracies in radiation mechanisms	= 1.1 dB	
Total	= 7.1 dB	

cheap foam sandwich structure has contributed about 2 dB by way of tolerance errors and power mismatches in feeds and at junctions. The polarising grids contribute an additional 0.5–0.7 dB loss.

4.2 Noise figures and pattern quality

As shown in Section 4.1, the microstrip DBS antenna has high dissipative losses and the possibility of additional system penalties due to increased thermal noise is now examined. Rainger *et al.* [15] have presented a comprehensive downlink power budget for DBS reception in the UK from which the systems performance of the microstrip array can be projected. The figures are summarised in Table 4 for the large array and two sizes of

Table 4: Comparison of antenna types and projected carrier-to-noise ratios (C/N) at 11.9 GHz: NF = noise figure

0.9 metre dish	0.6 metre dish	Large microstrip array	
0.5	0.5	5.2	Antenna dissipation, dB
−4.56	−8.07	−4.50	Aperture capture area, dB m^2
38.0	35.5	32.3	Antenna gain, dB
			Total *C/N*, dB:
14.0	10.5	9.0	8 dB NF receiver
17.5	14.0	12.2	5 dB NF receiver
20.7	17.2	14.9	2.5 dB NF receiver

These figures assume [15]
(*a*) 1.4 dB rain loss
(*b*) 0.3 dB galactic noise
(*c*) 160 W transmitted power
(*d*) antenna is modelled as a simple attenuator situated before receiver.
For C-MAC system, picture qualities are 'fair' for $C/N > 9$ and 'good' for $C/N > 12$

reflector dish. The high attenuation of the printed array is only a significant noise source when a low-noise receiver is used.

The microstrip array at midband operation is capable

of satisfying the carrier to noise (C/N) criterion provided a low-noise receiver is used. When frequency scanning losses are taken into account at the band edges, the grade of performance is not acceptable for a UK domestic system, and some amplification would be necessary to boost the C/N ratio by introducing FET devices in advance of the feed structure.

For reception in the UK, the antenna co-polar pattern, Fig. 12, meets the UK pattern specification in the E-plane but the crosspolarisation performance, derived from axial ratio measurements, is poor in the H-plane, Fig. 11, particularly at wide angles where ground reflections could introduce interference from other satellites. This isolation problem is chiefly a consequence of using cheap foam materials with a squinted beam design and it may be necessary to apply further effort to the redesign of the polarising sheet and combline elements for an improved wider angle response. Alternatively one should consider other forms of flat-plate array with cleaner patterns, but, as described below, their full capabilities when optimised are still unclear.

4.3 Alternative flat-plate designs

To date, information on the axial ratios of various flat-plate arrays is sparse, particularly for those using polarising grids. The suspended-substrate array of Rammos *et al.* [6] has an acceptable gain figure of 35.2 dBi over the UK DBS band and a −40 dB crosspolarisation level is quoted for the linearly polarised version. This array, with further redesign, should be capable of using circularly polarised (CP) elements directly, but figures are not generally available for this configuration as yet. As mentioned previously, this array, like all parallel fed arrays, has the disadvantage of being more difficult to adapt for beam steering operations than a microstrip travelling-wave type. Nevertheless, provided the necessary tolerances and CP characteristics can be controlled, the Rammos array is a good proposition for the immediate fixed beam DBS market, although generally it will not be conformal to the wall of the building.

Recently, a series of fixed beam arrays have been published using improved microstrip CP elements configured either in pairs [10, 11] or with parasitic overlays [9]. These have a wider CP bandwidth performance and lower axial ratios at wide angle, and as such deserve further examination for the UK DBS requirements. The arrays reported to date have lower gains than that demanded by the UK DBS specification and have been constructed on expensive substrates; some degradation seems likely should these techniques be applied to higher gain squinted arrays on cheap foam substrates.

5 Conclusions

This investigation has established that the domestic DBS requirement for a fixed squinted beam conformal array is indeed a challenging specification when severe substrate cost constraints are applied. The limitations in performance observed in the foam microstrip array such as lower efficiency, poorer pattern quality and scanning loss are a consequence of the array topology, the squinted beam and the low permittivity substrate. The poor pattern quality for this squinted beam design remains a critical issue requiring much more research into the wide angle characteristics of the combline and the polariser for this particular antenna. The tolerance effects associated with foam substrates subjected to extremes of temperature are also important, and the final choice of antenna configuration for a commercial product will depend largely on this factor.

For these reasons, it seems likely that the first generation of flat-plate arrays will have a conventional boresight design with antenna elements printed onto plastic substrate(s) rather than onto foam and possibly employing low loss feeding structures, such as waveguide, to meet gain and pattern quality specifications. The development of a cheap plastic material with temperature stability is imperative enabling a greater design flexibility and control in manufacture. One manufactured flat plate array has already been publicised [18] and it seems only a matter of time before these devices appear on the market.

Some provision for electronic beam scanning by incorporating active devices into the feed structures would be a major breakthrough, enabling the array to be installed fully conformal to the wall. The difficulties encountered in this investigation with controlling pattern quality for squinted beams raise some doubts as to whether this objective can be achieved without considerable effort and it will remain a challenge for the next few years.

6 Acknowledgments

We would like to thank P. Haskins of RMCS for the detailed measurements undertaken during this project, and acknowledge the financial support of the British Broadcasting Corporation during the initial phase of our work.

7 References

1 CLARRICOATS, P.J.B.: 'Reflector antennas for satellite systems — a review'. 14th European Microwave Conference, Liége, Belgium, 10–13 September 1984, pp. 9–20

2 ITOH, K., OGAWA, Y., OHMIYA, M., and SASAKI, M.: 'Adaptability experiments of satellite broadcasting antenna systems in snowy district'. Proc. ISAP, Kyoto, Japan, 1985, pp. 201–204

3 International Telecommunications Union Radio Regulations, Appendix 30, Geneva, Switzerland, 1979

4 MILLARD, G. H., and GANDY, C.: 'Antennas for direct reception of broadcasts from satellites', Proc. ICAP, Norwich, England, 1983, pp. 365–368

5 DUBOST, G., and VINATIER, C.: 'Large bandwidth and high gain array of flat folded dipoles acting at 12 GHz', Proc. ICAP, Norwich, 1983, pp. 145–149

6 RAMMOS, E.: 'Suspended substrate line antenna fits 12 GHz satellite application', *Microwave Syst. News*, 1984, **14**, pp. 111–125

7 HENDERSON, A., JAMES, J.R., HALL, P.S., STOTT, J.H., and BOARDMAN, D.H.: 'Investigation of a cost-constrained 12 GHz "flat-plate" antenna for DBS', Proc. ICAP, Warwick, England, 1985, pp. 108–112

8 HENDERSON, A., and JAMES, J.R.: 'Improved microstrip flat-plate array for domestic reception', IEEE AP-S International Symposium Digest, Philadelphia, June 8–13, 1986, pp. 565–568

9 HORI, T., and NAKAJIMA, N.: 'Broadcast circularly polarised microstrip array with co-planar feed', *Trans. Inst. Electron. & Commun. Eng. Jpn. Part B*, 1985, **68**, pp. 515–522 (in Japanese)

10 HANEISHI, M., and TAKAZAWA, H. 'Broadband CP planar array composed of a pair of dielectric resonator antennas', *Electron. Lett.*, 1985, **21**, (10), pp. 437–439

11 HANEISHI, M., YOSHIDA, S., and GOTO, N.: 'A broadband microstrip array composed of circularly polarised microstrip antenna', IEEE AP-S, International Symposium Digest, May, 160–163

12 LAIT, A.J.: 'Broadband circular polarisers', *Marconi Rev.*, 1969, **32**, pp. 159–184

13 WARREN, K.A.J.: 'A planar antenna circular polarisation converted utilising printed circuit technology', *Marconi Rev.*, 1980, **43**, pp. 176–184

14 YOUNG, L., ROBINSON, L.A., and HACKING, C.A.: 'Meander-line polariser', *IEEE Trans*, 1973, **AP-21**, pp. 376–378

15 RAINGER, P. *et al.*: 'Satellite broadcasting' (John Wiley and Sons, Chichester, 1985)

16 JAMES, J. R., HALL, P.S., and WOOD, C.: 'Microstrip antenna theory and design' (Peter Peregrinus Ltd, Stevenage, UK, 1981) chap. 5

17 FORREST, J.R.: 'Assessing antennas for small SATCOM terminals', *Microwave Syst. News*, 1981, **11**, pp. 77–100.

18 'Flat is now beautiful', *Electron. and Power*, 1987, **134**, (2), p. 227

Microstrip Yagi Array Antenna for Mobile Satellite Vehicle Application

John Huang, *Member, IEEE,* and Arthur C. Densmore

***Abstract*—A new antenna structure is formed by combining the Yagi-Uda array concept and the microstrip radiator technique. This antenna, called the microstrip Yagi array, has been developed for the mobile satellite (MSAT) system as a low-profile, low-cost, and mechanically steered medium-gain land-vehicle antenna. With antenna's active patches (driven elements) and parasitic patches (reflector and director elements) located on the same horizontal plane, the main beam of the array can be tilted, by the effect of mutual coupling, in the elevation direction to provide optimum coverage for Continental U.S. users. Because the parasitic patches are not connected to any of the lossy RF power distributing circuit, while still contributing to achieve the MSAT required antenna gain, the antenna becomes a very efficient radiating system. With the complete monopulse beamforming and power distributing circuits etched on a single thin stripline board underneath the microstrip Yagi array, the overall L-band antenna system has achieved a very low profile for vehicle's rooftop mounting, as well as a low manufacturing cost. Experimental results are presented to demonstrate the excellent performance of this microstrip Yagi antenna.**

Introduction

A MAJOR element of the mobile satellite (MSAT) program has been the development of several types of medium-gain circularly polarized L-band vehicle antennas. Prior to the microstrip Yagi development, two types of medium-gain antennas had been successfully developed in the United States. One type is the electronically steered planar phased arrays [1] developed separately by Ball Aerospace Corporation [2] and Teledyne Ryan Electronics [3] under contract with the Jet Propulsion Laboratory (JPL). The other is the mechanically steered 1 × 4 tilted microstrip patch array developed by JPL [4]. The phased array offers the advantages of low profile (1-in thick) and beam agility at the expense of higher manufacturing cost (several thousand dollars per unit). On the other hand, the mechanically steered 1 × 4 array antenna offers a low manufacturing cost (approximately $600 per unit), while it suffers from a high profile (6-in tall).

To combat the disadvantages of high cost or high profile of the above antennas, a new antenna system has been introduced and developed, which not only offers both the advantages of low profile and low cost, but shows excellent antenna efficiency as well. This is the mechanically steered planar microstrip Yagi array. A single "row" of microstrip Yagi array [5], as shown in Fig. 1, consists of a single driven element patch, a single parasitic reflector patch, and two or three director patches. The beam peak of this array, due to the effect of mutual coupling and Yagi-Uda principle [6], [7], can be tilted away from the broadside direction and toward the direction of the director patches. To achieve the MSAT required gain of 10 dBic within the angular region between 20° and 60° elevations, four rows of microstrip Yagi arrays are needed as configured in Fig. 2. Because only four driven element patches are directly connected to and driven by the RF power distributing circuitry, the complexity and RF insertion loss of the power distributing circuit are dramatically reduced as compared to a conventional planar array. This results in improved antenna gain and system noise temperature.

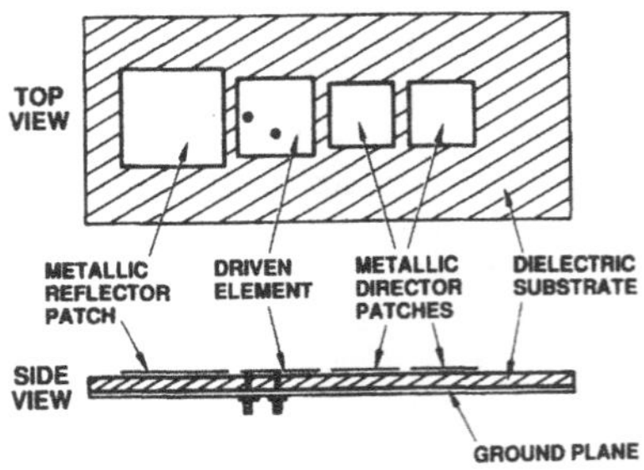

Fig. 1. Microstrip Yagi array configuration.

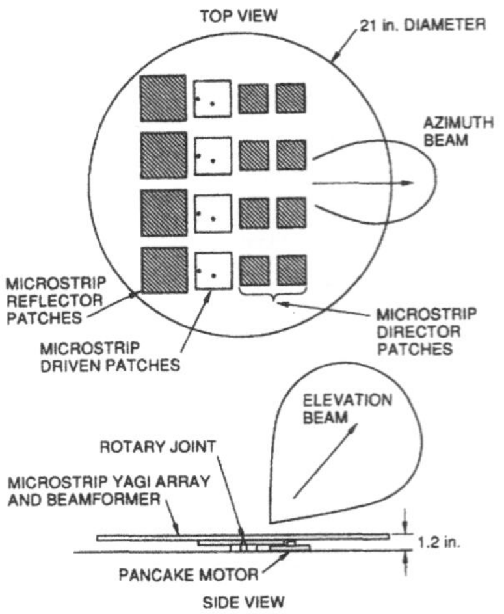

Fig. 2. Planar microstrip Yagi array concept for MSAT application.

As illustrated in Fig. 2, in addition to the microstrip Yagi, two other major components of the antenna system are the beamformer and the mechanical rotating platform. The beamformer provides the functions of monopulse tracking and RF

Reprinted from *IEEE Trans. Antennas Propaga.,* vol. 39, no. 7, pp. 1024–1030, July 1991.

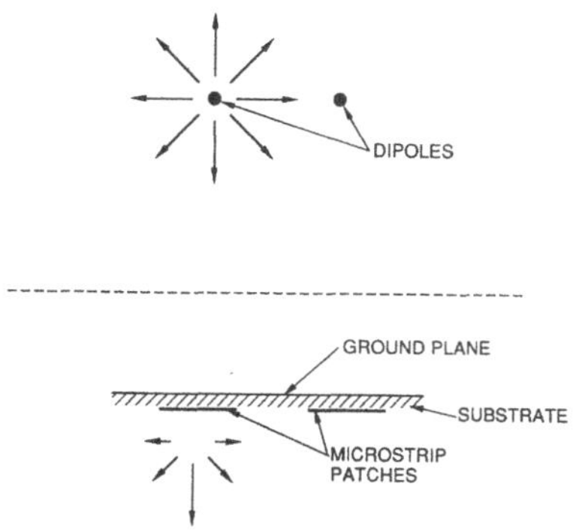

Fig. 3. Radiation mechanisms of dipoles and microstrip patches.

power distribution, and is constructed on a single layer of thin stripline board. This beamformer circuitry is uniquely designed for the communication antenna to provide simultaneous transmit and receive signals. The mechanical rotating platform accomplishes the functions of azimuth rotation by a thin pancake motor (thickness = 0.5 in) whose motion is controlled by the monopulse signal. The purpose of this article is to provide a detailed description of the microstrip Yagi array and its measured performance results.

Principle of Microstrip Yagi Array

The microstrip Yagi antenna, as depicted in Fig. 1, consists of a driven patch element and a few parasitically coupled reflector and director patch elements. It utilizes a similar principle as a conventional Yagi–Uda dipole array where the electromagnetic energy is coupled from the driven element dipole through space into the parasitic dipoles and then reradiated to form a directional beam. In a microstrip Yagi array, however, the electromagnetic energy is coupled from the driven patch to the parasitic patches not only through space but also by surface waves in the substrate. Because a dipole radiates RF power equally around its axis, it can couple relatively strong power to its neighbors as shown in Fig. 3, even when the spacings between dipoles are large. This spacing can be a fraction or multiple of a wavelength. The microstrip patch, on the other hand, radiates primarily in its broadside direction, as shown in Fig. 3, and very little along the direction of the ground plane. This is why a microstrip patch generally couples very little energy [8] to its neighboring patches, especially when the patches' separation becomes large. As a consequence, in order for the microstrip patches to function similarly to the Yagi dipoles, the adjacent patches need to be placed very closely to each other so that a significant amount of coupling can be obtained through surface waves in the substrate. It was found experimentally [5] that the gap distance between two patch edges should be equal to or less than the dielectric substrate thickness. With such a small gap between the two patches (almost touching each other), the coupling mechanism must be primarily a surface wave effect rather than through radiation. This is because that a significant amount of edge field that starts propagating from the dielectric substrate of its originating patch will continue propagating through the substrate into the next "almost touching" patch. The amount of edge field that radiates out into space from the small gap does not have a mechanism to radiate back into this same gap. Since the amount of surface wave excitation is a strong function of the dielectric constant and substrate thickness, the pattern shape of the microstrip Yagi is also a function of these two parameters. One other important factor that also determines the pattern shape and beam direction is the sizes of the parasitic reflector and director patches. When the size of the parasitic patch is different from the resonant dimension of the driven element, a reactance component is added to its radiation impedance at the resonant frequency of the driven element. This reactance introduces an added phase component to its radiation field. It is this phase, together with the separation delay phase (phase velocity of the surface wave) between patches, that contributes to the microstrip Yagi array's beam tilt from its broadside direction. Since the microstrip antenna always requires a ground plane and its radiation is primarily in the patch's broadside direction, the beam tilt direction of the microstrip Yagi can never become completely endfire. To summarize, this beam tilt angle is a function of dielectric constant [5], substrate thickness, patch separation, gap distance, and sizes of the parasitic patches. All these parameters are related to each other. However, in general, the patch separation and parasitic patch sizes govern the relative phase, while the gap distance and dielectric constant are strong factors that determine the amplitude of the coupling to each parasitic element.

It was found that the performance of the microstrip Yagi array is not strongly impacted when a second reflector patch and a third director patch are added to the array. This is not the case in a dipole Yagi array where two reflectors and more than ten directors have been used to enhance the endfire directivity. This is because, as explained previously, the coupling mechanism in the microstrip array is primarily a surface wave effect. As an example, the amount of energy that couples into the second director of the microstrip Yagi is significantly reduced after most of the energy coupled into the first director patch has been radiated. In the case of the dipole Yagi, not only the first director can couple energy into the second director, but the driven element dipole also couples into the second director through space. Therefore, the amplitude decay in the director elements of the dipole Yagi happens much more slowly than that of the microstrip Yagi.

Although the configuration and coupling mechanism of the microstrip Yagi and the dipole Yagi are different, the design rules [9] for these two types of Yagi are very similar. It was empirically determined that the dimension ratio of the reflector patch to the driven element patch is somewhere between 1.1 to 1.3, depending on the substrate thickness and dielectric constant. The dimension ratio between the director patch and the driven element patch should be between 0.8 and 0.95. The distance between the centers of the reflector and the driven elements is about 0.35 free-space wavelength, while the separation between the centers of the director and driven elements should be approximately 0.3 free-space wavelength. Because of the size of the microstrip patches, the microstrip Yagi array has a few limitations in design not present for the dipole Yagi arrays. For example, the dielectric constant of the substrate material can neither be too low nor too high. If the relative dielectric constant is very low (less than 1.5), the patch size becomes larger than 0.35 free-space wavelength and consequently the separation distance that is required above can no longer be held. On the other hand, if the relative dielectric constant is too high (more than five), the patches will be very small and the gaps between patches become large in order to hold the above required element separation. As a result, the coupling energy will be insufficient for the Yagi effect to function.

Several important and interesting characteristics of the microstrip Yagi antenna have been found through experiments. First, both linearly and circularly polarized waves can be produced from the microstrip Yagi array [5]. For linear polarization, either E-plane coupled or H-plane coupled microstrip Yagis are feasible. The E-plane coupled array, however, shows a stronger coupled field than the H-plane coupled array since E-plane field (Dirichlet boundary condition) can travel much more strongly along the ground plane than the H-plane field (Neumann boundary condition). As a result, the E-plane coupled microstrip Yagi array is able to tilt its beam much closer to the ground plane than the H-plane coupled array. Second, if good circular polarization (CP) is desired in the microstrip Yagi's main beam region, the phase difference between the two orthogonal feeds needs to deviate from the conventionally required 90°. This is due to the fact that the main beam is no longer in the broadside direction, and that the E-plane and H-plane coupled fields have different coupled phase terms. Fig. 4(a) shows the pattern of a circularly polarized microstrip Yagi array where the two orthogonal feed probes are 90° apart in phase. Fig. 4(b) presents the pattern of the same array with improved CP when the two probes are fed 115° apart in phases. This 115° phase differential was optimized experimentally for the particular MSAT array. A differently designed microstrip Yagi array will require a different phase differential value for optimized CP quality. The third interesting finding is that, because the E-plane array has a much stronger surface wave coupling than the H-plane array, the input impedance of the E-plane feed is significantly affected by the parasitic patches. As a result, to achieve the nominal 50-Ω input impedance match, the position of the E-plane probe in the driven patch of the Yagi array needs to be readjusted from that of the classical single microstrip patch. This readjusted new feed position has been found to be closer to the edge of the patch than its old position. The amount of adjustment in the feed position will be presented in the next section. Finally, it was not surprising to find that the microstrip Yagi array has a wider input impedance bandwidth than the driven element by itself. This is because the parasitic directors and reflector, having different sizes from the driven element, cause a "log-periodic frequency independent" phenomenon to occur. The curve in the Smith chart of Fig. 5(a) gives the input impedance as a function of frequency for a single patch (2.2-in square) with relative dielectric constant of 2.5 and substrate thickness of 0.25 in. It shows a single-loop response with a relatively narrow bandwidth. Fig. 5(b) gives the input impedance curve of the same element with parasitically coupled director and reflector patches (dimensions are given in

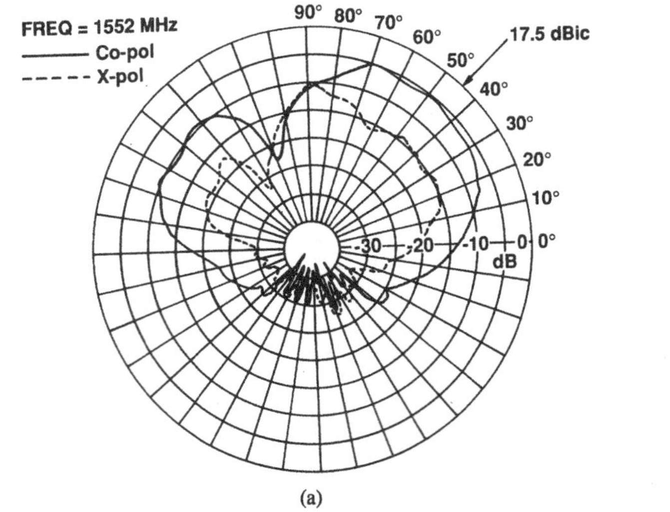

(a)

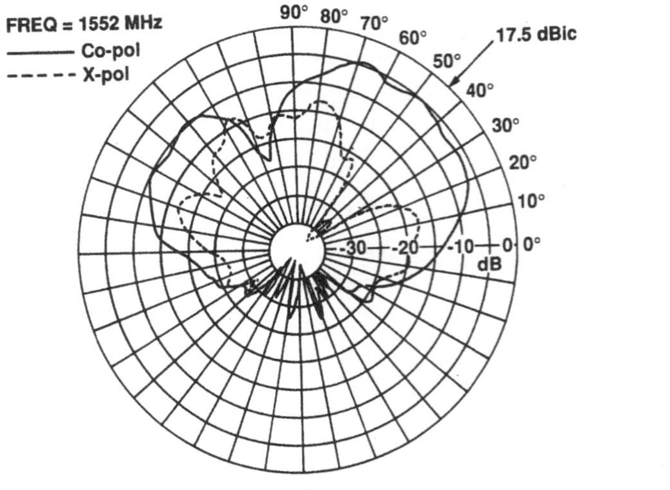

(b)

Fig. 4. Measured elevation patterns of circularly polarized microstrip Yagi array, two feeds are separated. (a) 90° in phase differential. (b) 115° in phase differential.

next section). There are two input probes connected to the driven-element patch for circular polarization. One is called the H-plane coupled probe, while the other is the E-plane coupled probe. This input impedance is measured at the single input probe for the H-plane coupled microstrip Yagi array while the E-plane coupled input probe is terminated with a 50-Ω load. The curve shows a double-loop response which has a double resonance and covers both the MSAT's transmit and receive frequencies. Similar double-loop response is achieved for the E-plane coupled array input impedance as shown in Fig. 5(c).

For possible other future applications, it is important to briefly point out here the essential empirical design steps for this microstrip Yagi array. The first step involves several single-driven-element patches having different dielectric constants and each with two orthogonal feeds are fabricated and tested. Both feeds of each patch are impedance matched at the center frequency of the band in interest. The second step encompasses several microstrip Yagi arrays that are fabricated using the above driven-element patches. Each array's parasitic reflector and director patches are designed by using the design rules given previously. Range tests are then per-

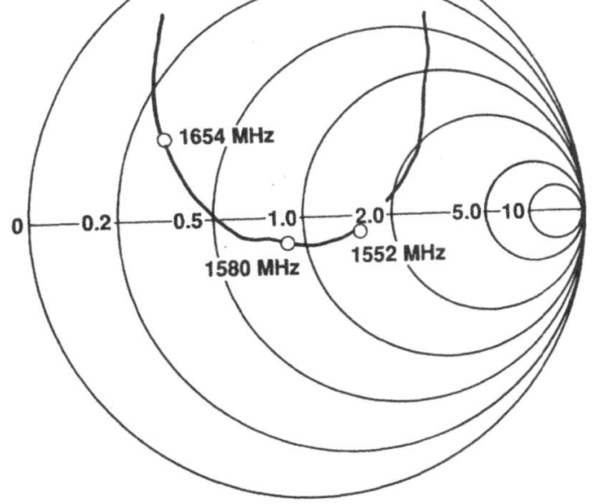

(a)

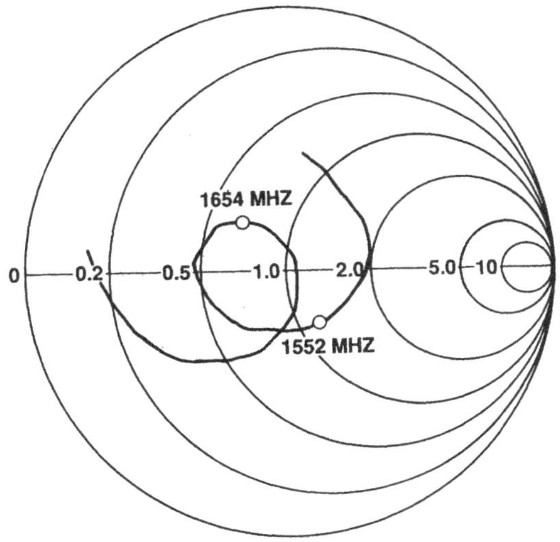

(b)

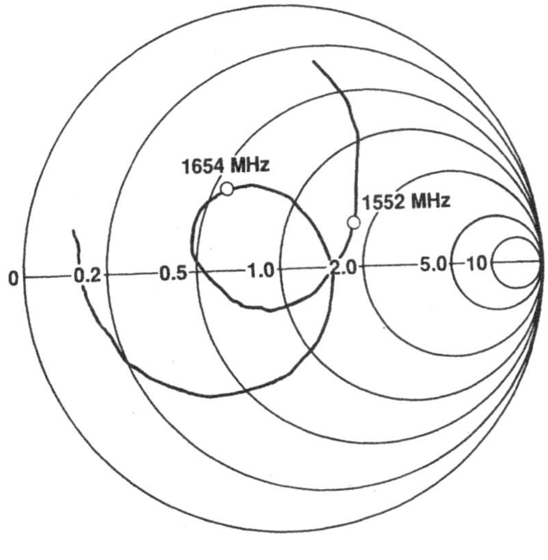

(c)

Fig. 5. Measured input VSWR of (a) a single microstrip patch; (b) *H*-plane coupled microstrip Yagi; (c) *E*-plane coupled microstrip Yagi.

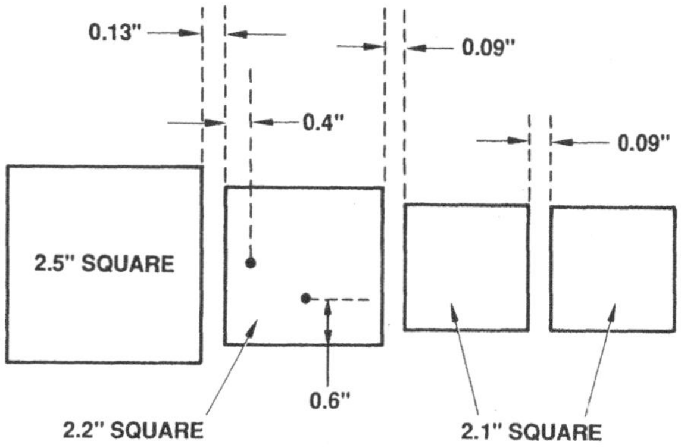

Fig. 6. Dimensions of a single row of the MSAT microstrip Yagi array.

formed to find the array that gives the best co-pol pattern in terms of the beam tilt angle and beamwidth. For these measurements, the two feeds on each driven-element patch are connected by a 90°-hybrid. Third step, the two feed probes of the above selected array are retuned by moving them along the respective center lines on the driven-element patch to find the locations for optimum impedance match. Remember that the input impedance match of the single driven-element patch has been altered by the parasitic patches through strong couplings. Final step, range tests are performed again to generate both the co-pol and cross-pol patterns. The cross-pol level is minimized at the required main beam tilt angle by adjusting the phase differential between the two orthogonal feeds. In other words, the phase differential between the two feeds is no longer 90°. Certainly, accurate mathematical tools, such as the method of moments, should be developed in the future to ease the above design procedures.

Description and Performance of MSAT Microstrip Yagi Array

The MSAT system requires its medium-gain vehicle antennas to provide a minimum gain of 10 dBic (dB with respect to isotropic circularly polarized radiation) between 20° and 60° elevation angles with a circularly polarized radiation. The antenna must transmit at 1.654 GHz while simultaneously receiving at 1.552 GHz. An automatic satellite tracking mechanism must keep pointing its main beam toward the satellite as the vehicle moves about.

In order to meet the above MSAT specified minimum gain requirement, four identical rows of microstrip Yagi arrays with a row separation of 4.5 in (0.61 wavelength at 1.6 GHz) are designed. All microstrip patches are etched on a single board of dielectric material with a thickness of 0.25 in, relative dielectric constant of 2.5, and board size of 10 × 17 in. The detailed microstrip patch dimensions of each row are given in Fig. 6 where it should be noted that the two orthogonal probes are located at different distances from the edges of the driven patch for an optimum impedance match. A 1/8-in thick stripline beamformer board, located flush below the array, is designed to provide power division with uniform excitations through feed probes to the four driven elements of the 4-row Yagi array. This beamformer also provides the required monopulse tracking circuitry. Since

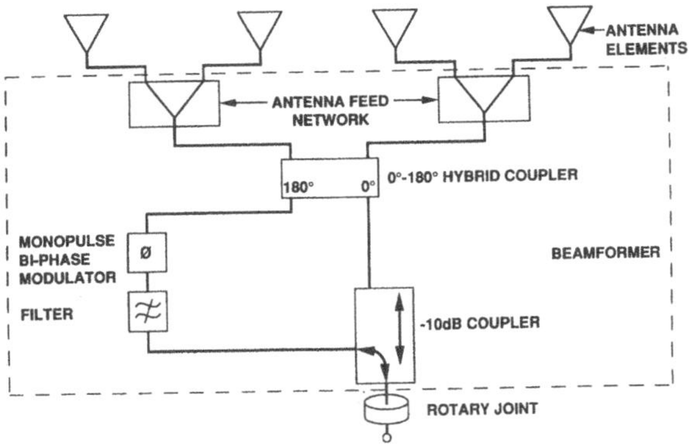

Fig. 7. Block diagram of the MSAT microstrip Yagi array antenna system.

Fig. 8. Integrated MSAT microstrip Yagi array antenna system; radome height = 1.5 in; diameter = 21 in.

both the microstrip Yagi array and the beamformer boards can be fabricated with a simple etching process on two thin layers of dielectric boards, this lends to an antenna system with relatively low profile and low manufacturing cost. The RF block diagram of the overall antenna system is shown in Fig. 7.

The planar microstrip Yagi array and the beamformer boards are bonded together and mounted onto a half-in thick pancake motor that is controlled by the monopulse feedback pointing error signal to track the satellite. The integrated antenna system, as shown in Fig. 8, is covered by a 0.04-in thick radome having a height of 1.5 in and diameter of 21 in. It has been found that the elevation radiation pattern is quite sensitive to the microstrip array ground plane dimension immediately in front of the director patches. This is due to the fact that the beam is tilted in elevation by the Yagi effect to the low elevation angle close to the ground plane where strong edge effect is critical to the elevation pattern shape. A 1.5-in ground plane extension in front of the last director patch was found to yield the optimum elevation pattern performance for MSAT. Presented in Figs. 9(a) and (b) are the measured antenna elevation patterns at the receive and transmit frequencies, respectively. Fig. 10 gives the sum and difference azimuth patterns measured at 40° elevation angle and at the receive frequency. The azimuth pattern for the transmit frequency at 40° elevation angle is given in Fig. 11. The antenna gains (referenced at the rotary joint) measured at various elevation angles are listed in Table I. Except for the

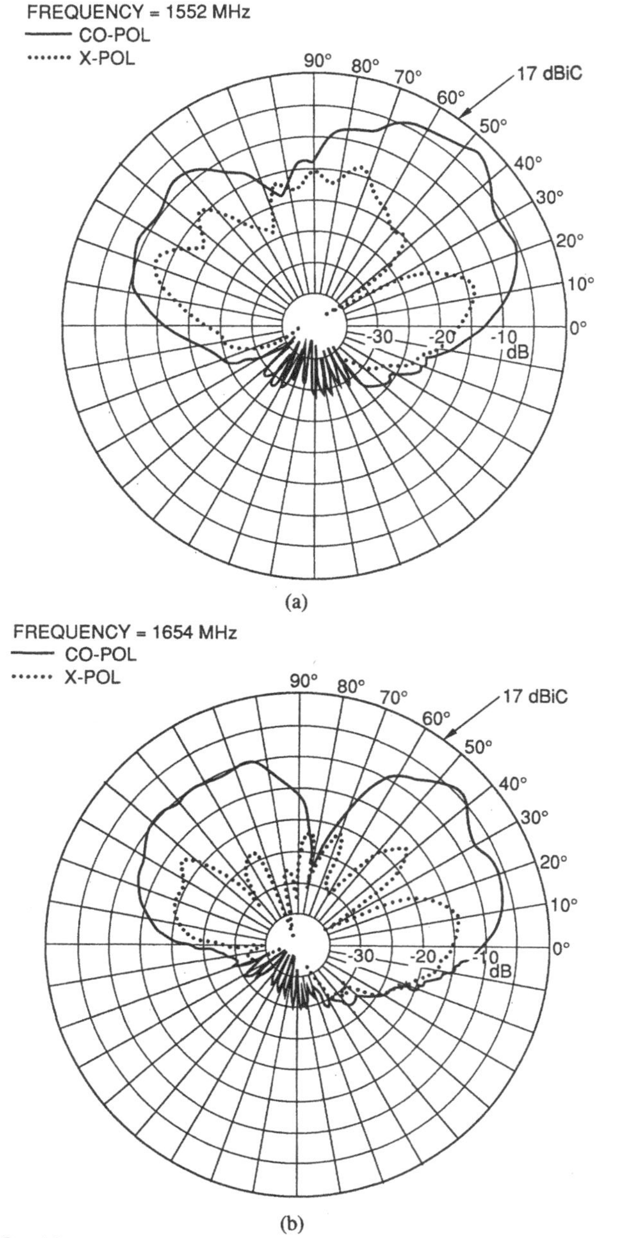

Fig. 9. Measured elevation patterns of the MSAT microstrip Yagi array. (a) Frequency = 1552 MHz. (b) Frequency = 1654 MHz.

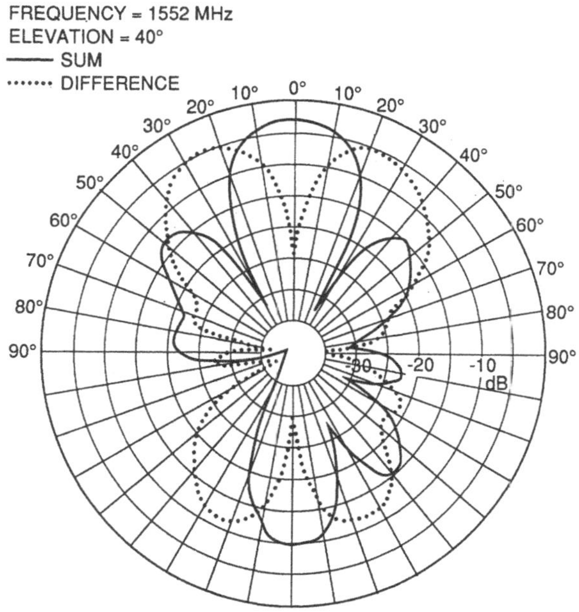

Fig. 10. Measured azimuth sum and difference patterns at frequency of 1552 MHz.

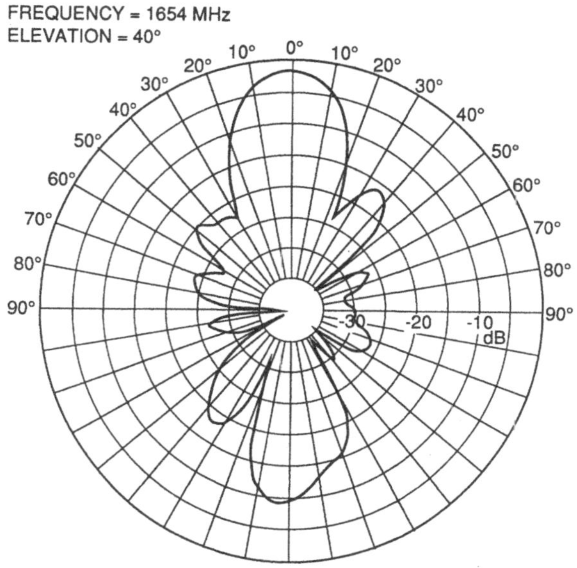

Fig. 11. Measured azimuth pattern at frequency of 1654 MHz.

9.5 dBic of gain measured within a small elevation angular region at 60° of the transmit frequency, the antenna has achieved the required 10 dBic of gain throughout the specified coverage region. Due to the low insertion loss of the complete antenna system (0.8 dB between the rotary joint and radiator), excellent antenna G/T (gain over system noise temperature) ratio has been achieved. At 40° elevation angle, the measured antenna system noise temperature is 295°K and the antenna gain is 13.0 dBi, which give a G/T ratio of −11.7 dB/°K. The measured input mismatches at each one of the two orthogonal feeds of the driven patch are given in Figs. 5(b) and 5(c).

TABLE I
MEASURED ANTENNA GAIN AT VARIOUS ELEVATION ANGLES

Elevation Angle	Gain at 1552 MHz (dBic)	Gain at 1654 MHz (dBic)
20°	11.0	11.3
30°	10.7	10.0
40°	13.0	11.7
50°	14.3	12.0
60°	13.1	9.5

From the elevation patterns, one noticeable shortcoming of the microstrip Yagi array is its high back lobe. This is also the case for the conventional dipole Yagi array. Fortunately, for MSAT application, investigation has lead to the conclusion that this high back lobe will not present any adverse effect to the overall system in terms of tracking, multipath and satellite interference. The reason that the antenna will not track the satellite with its back lobe is that the difference pattern of the back lobe has a reverse pointing error slope from that of the front lobe. The slopes of the front and back lobes are opposite since the wavefront phase lead/lag as seen from a particular half of the antenna with a particular sense of rotation of the antenna in azimuth is opposite when the wavefront approaches the array from one side as opposed to the other. The high back lobe certainly will present worse multipath effect than a low back lobe antenna. However, due to the fact that the medium-gain antenna has a much narrower azimuth beamwidth than an omnidirectional low-gain antenna and therefore, being a spatial filter, the multipath effect due to the high backlobe of the Yagi array will not have any detrimental impact on the overall system. For the issue of multisatellite interference, as long as all the vehicles are located well away from the Earth's equator, such as in the U.S., no antenna backlobe will be pointed toward any geostationary satellite during communication. Although during acquisition period, the backlobe can be directed at other geostationary satellites, the transmit power will be automatically turned off during this period to avoid interference. Nevertheless, even though no severe system impact is created, power will be wasted in the high back lobe. This waste in power, however, is overcome by the excellent efficiency of the Yagi array. With only 0.8 dB of RF loss in the entire antenna system, this low-profile antenna has provided excellent antenna gain and G/T for the MSAT system.

A stand-alone closed-loop monopulse system to track the satellite's pilot signal will not function well during signal outage caused by building blockage or foliage shadowing. To combat this problem, an open-loop sensor has been augmented onto the closed-loop monopulse to form a robust hybrid tracking system. The open-loop sensor currently being used is a single-axis angular rate sensor [10]. With this hybrid tracking system, the Yagi antenna has functioned flawlessly in recent field tests.

Since two major components of the antenna system, the microstrip Yagi array and the beamformer, can be manufactured by simple etching processes, a relatively low cost antenna system has been realized. The production cost of the antenna is lower than that estimated for the mechanically steered 1×4 tilted microstrip array [11]. For a 10 000-units-per-year production, The manufacturing cost of the microstrip Yagi array is expected to be around $450 per unit. The major cost components of this antenna are the dielectric circuit board material, open-loop angular rate sensor, pancake motor, rotary joint, and assembly labor.

Conclusion

A mechanically steered antenna with low-cost, low profile, and low insertion loss has been successfully developed for the MSAT system. This antenna combines the excellent efficiency of the microstrip Yagi concept and an elegant design of the monopulse beamformer to achieve the required antenna gain and high G/T ratio. Experiment has indicated that, with a slightly different array design, the microstrip Yagi antenna may also provide the required performance for the Canadian region where a lower elevation coverage is encountered.

Acknowledgment

The authors wish to express appreciation to Dr. William Rafferty, Dr. Ken Woo, and Mr. Richard Emerson for their support and encouragement, and to Mr. Cosme Chavez for his dedicated assistance in all the measurements.

References

[1] J. Huang, "L-band phased array antennas for mobile satellite communications," in *IEEE Veh. Technol. Conf. Rec.*, June 1987, pp. 113–117.

[2] Ball Aerospace Systems Group, "Electronically steered phased array vehicle antenna for MSAT-X," JPL MSAT-X Rep. 151–153, Dec. 1987.

[3] Teledyne Ryan Electrons, "Electronically steered phased array vehicle antenna for MSAT-X," JPL MSAT-X Rep. 154, Mar. 1988.

[4] V. Jamnejad, "A mechanically steered monopulse tracking antenna for PiFEx: RF system design," *JPL MSAT-X Quart.*, no. 13, 1988.

[5] J. Huang, "Planar microstrip Yagi array antenna," in *IEEE Antennas Propagat. Soc./Ursi Symposium Dig.*, June 1989, pp. 894–897.

[6] S. Uda, "Wireless beam of short electric waves," *J. Inst. Elec. Eng.*, no. 452, pp. 273–282, Mar. 1926; no. 472, pp. 1209–1219, Nov. 1927.

[7] H. Yagi, "Beam transmission of ultra short waves," *Proc. IRE*, vol. 16, pp. 715–741, June 1928.

[8] R. J. Mailloux *et al.*, "Microstrip array technology," *IEEE Trans. Antennas Propagat.*, vol. AP-29, p. 29, Jan. 1981.

[9] C. A. Chen and D. K. Cheng, "Optimum element lengths for Yagi-Uda arrays," *IEEE Trans. Antennas Propagat.*, vol. AP-23, Jan. 1975.

[10] D. Bell, "Vehicle antenna open-loop pointing," *JPL MSAT-X Quart.*, no. 6, May 1986.

[11] B. Jackson, "Cost study for MSAT-X vehicle communication system," JPL Rep. D-7210, Feb. 1990.

POST LOADED MICROSTRIP ANTENNA FOR POCKET SIZE EQUIPMENT AT UHF

Haruhiro KUBOYAMA, Shigeru IMAI,
Kazuhiro HIRASAWA and Kyohei FUJIMOTO

Institute of Applied Physics, The University of Tsukuba
Ibaraki-ken 305, Japan

Introduction

As the demand for personal communications increases, the demand for small antennas increases as well. A post-loaded rectangular microstrip antenna (P-MSA) has been developed for small size UHF equipment. The P-MSA is an antenna modified from an ordinary quarter-wave rectangular microstrip antenna (O-MSA) by replacing the shorted end with some reactance posts, and is designed to have higher gain than the O-MSA. The purpose of this paper is to show the analysis and design of the P-MSA with three posts loaded, placing special emphasis on the application to pocket size UHF equipment such as paging receivers. Since the antenna is intended to be placed in a very limited area inside small equipment, the ground plane is made almost the same size as the radiating patch. The antenna characteristics such as resonance frequency and gain are investigated with respect to the antenna dimensions, and some interesting design parameters are found, such as the relation of the resonance frequency with the patch width and the effect of the length of the ground plane upon gain. The process to obtain the maximum gain is discussed for the antenna, whose size is limited to fit small equipment.

Antenna structure

The antenna structure of the P-MSA is shown in Fig. 1. The P-MSA shown here has three posts replacing the shorted end of the O-MSA and an extended patch of length $\underline{a}$, which forms a radiating aperture at its end. The location of the posts placed on the side of the patch, length $\underline{b}$, essentially determines the resonance frequency. The posts act as reactance loading in the two-plate guided line, so that the length of the antenna becomes shorter than that of the O-MSA [1].

Resonance frequency and band width

Input impedance characteristics with respect to length $\underline{b}$ and width $\underline{w}$ are shown in Fig. 2. The matching condition can be easily obtained by adjusting the position of the feed point. Length $\underline{b}$ is considered to be the most important factor to determine the resonance frequency [2], but it was found experimentally that the resonance frequency $\underline{Fr}$ is related to both $\underline{b}$ and $\underline{w}$ and is given by the following equation, provided that $\underline{w}$ is equal or smaller than a quarter wavelength:

$$Fr = \frac{C}{4 \times (b + \alpha w)} \times \frac{1}{\sqrt{\varepsilon_e}} \qquad (1)$$

where $\underline{C}$ is the velocity of light and ε_e is an effective

Reprinted with permission from *Proc. ISAP*, H. Kuboyama, S. Imai, K. Hirasawa and K. Fujimoto, "Post Loaded Microstrip Antenna for Pocket Size Equipment at UHF," pp. 433–436, Aug. 1985.

permittivity[3] in which the effect of the fringing field along the edge is taken into account. The resonance frequency Fr calculated by taking $\alpha = 0.22$ in Eq. (1) agrees very well with measured results as shown in Fig. 3. Length b can be reduced by widening w for the same resonance frequency, which implies effectively shortening the antenna length. Length b of the P-MSA becomes about 8% shorter than that of the O-MSA. The relative bandwidth increases in proportion to w, although the variation is small. The extension of the gorund plane length ΔZ makes the resonance frequency lower as shown in Fig. 4. This can be attributed to the effect of fringing field which enlarges the radiation aperture as the length ΔZ becomes longer. However, the effect of the fringing field terminates near ΔZ = 10 mm and there is almost no frequency variation over ΔZ = 10 mm. As discussed later, ΔZ is better to have six times the thickness of the substrate.

Radiation patterns and gain

Radiation patterns in the horizontal (X-Y) plane are shown in Fig. 5-a, where the vertical electric field pattern is seen nearly omnidirectional and the horizontal electric field is a figure-eight shape, as expected. It is interesting to know the performance of the antenna for pocket size equipment being placed near a human body, since pocket size equipment is often operated in the bearer's pocket. Variation of the radiation pattern caused by the human body effect is shown in Fig. 5-b. The maximum level can be seen in the figure to be slightly larger than that in the free space. This is resulted from the magnetic current radiation superposed with its image radiation [4].

Gain with respect to a is shown in Fig. 6, where gain increases in proportion to a under the condition that a is longer than b/4. Gain against ΔZ is shown in Fig. 7. The gain increases with ΔZ, but up to $\Delta Z \simeq 5$ mm for $t = 0.8$ and up to $\Delta Z \simeq 10$ mm for $t = 1.6$. It can be said that $\Delta Z = 6t$ is long enough to obtain the maximum gain. Fig. 8 illustrates gain versus t, where the gain shows an increase up to $t = \lambda_g/128$ (λ_g: the wavelength in the microstrip structure) and becomes constant for $t > \lambda_g/128$. Thus t is chosen to be greater than or equal to $\lambda_g/128$. It is often encountered in practice that the antenna mount is limited within a small volume [LxWxT], and yet the antenna gain is desired to be as higher as possible. The procedure to choose the best design parameters for such a case is as follows;

(1) Set $t = \lambda_g/128$ when $T \geq \lambda_g/128$. Set $t = T$ otherwise.
(2) Set $w = \lambda_g/128$ when $W \geq \lambda_g/4$. Set $w = W$ otherwise.
(3) Find b from Eq. (1) after Fr and w are determined.
(4) Choose $\Delta Z = 6t$ whenever it is possible.
(5) Finally $a = L - (b+2\Delta Z)$.

The maximum gain can be obtained from the parameters chosen in the above five steps.

Conclusions

A post loaded microstrip antenna for UHF pocket size equipment has been introduced and the performance of the resonance frequency, bandwidth, radiation pattern and gain are investigated. Design parameters for resonance frequency and gain are obtained. A design process to get the maximum gain for the P-MSA with three posts loaded, under the limitation of a small volume [LxWxT], has been derived. The maximum radiation level of the P-MSA placed near or on a human body has shown an increase as compared with that in the free space. The P-MSA here is discussed mostly for application to pocket size equipment, but this type of antenna may also be useful for other equipment for which small antennas are required.

References

[1] James,J.R., Hall,P.S. & Wood,C.: "Microstrip Antenna Theory and Design", Sec.4.6, pp.102-105, Peter Peregrinus Ltd. (1981).

[2] Bahl,I.J. & Bhartia,P.: "Microstrip Antenna", chap.2, pp.32-84, Artech House (1980).

[3] Schneider,M.V.: "Microstrip dispersion", Proc. IEEE, vol.60, pp.144-146 (Jan. 1972).

[4] "Antenna engineering handbook", Sec.7.4, pp.319-321, OHM-Sha Co., Tokyo, Japan (in Japanese) (1980).

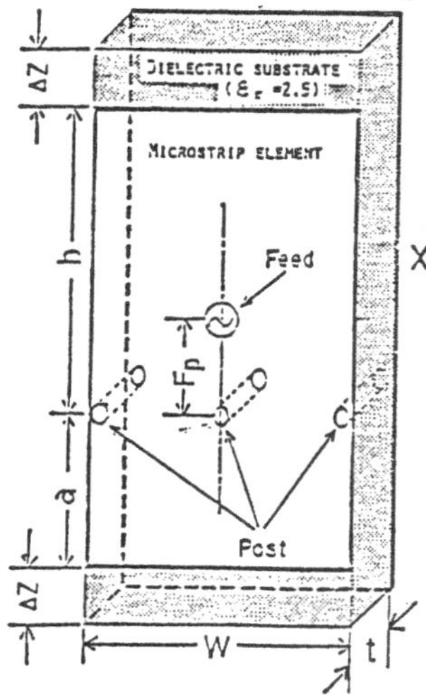

Fig. 1 Antenna structure

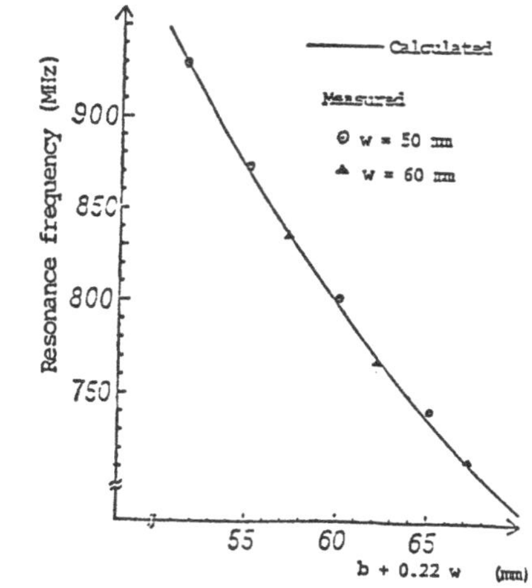

Fig. 3 Resonance frequency versus antenna dimension
a=25 mm,Fp=4 mm
t=0.8 mm,ΔZ=5 mm

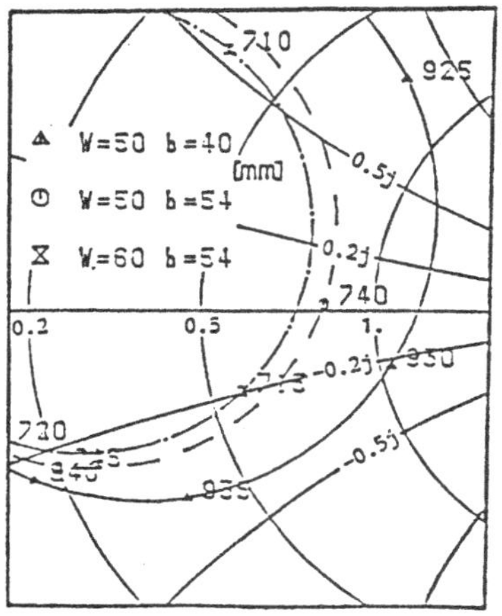

Fig. 2 Input impedance
Z_0=50 ohm,a=25 mm
Fp=4 mm,t=0.8 mm,ΔZ=5 mm

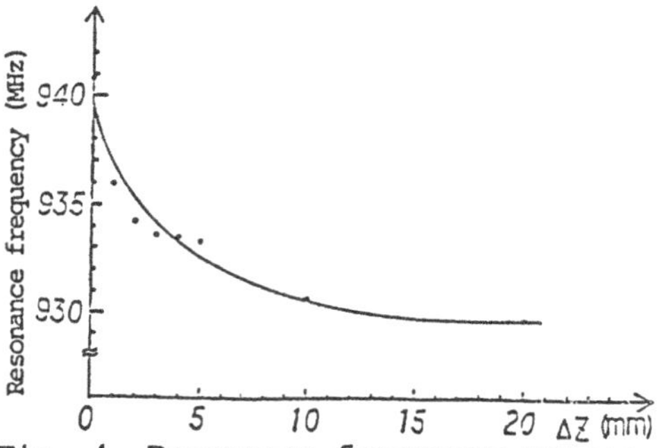

Fig. 4 Resonance frequency versus the extension of the ground plane length ΔZ
a=25 mm,b=46 mm,Fp=4 mm
t=0.8 mm,w=30 mm

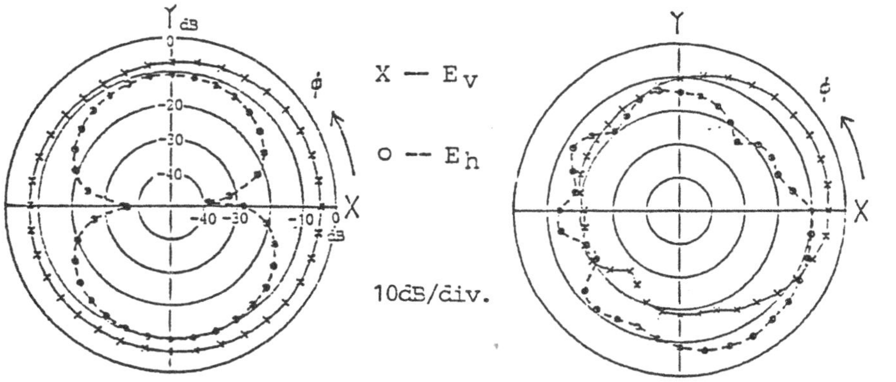

5-a In the free space 5-b On a human body(chest)

Fig. 5 Radiation patterns normalized with half-wavelength dipole level
a=17 mm,b=50 mm,Fp=6 mm,t=1.2 mm,w=30 mm,ΔZ=5 mm

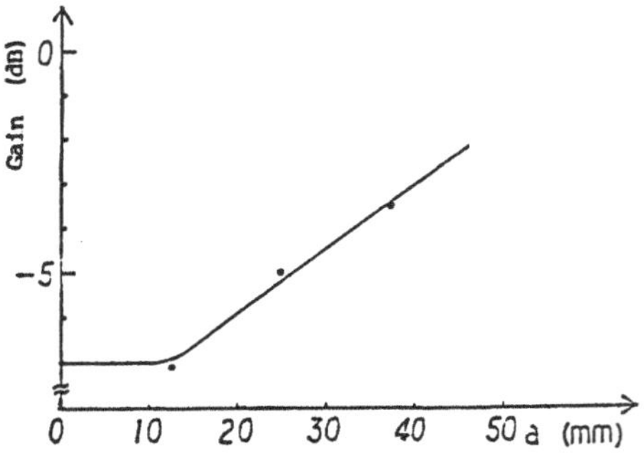

Fig. 6 Gain versus length a
b=46 mm,Fp=4 mm,t=0.8 mm
w=30 mm,ΔZ=5 mm

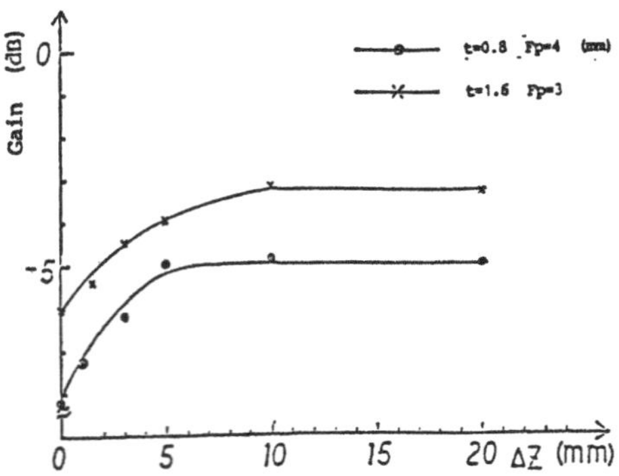

Fig. 7 Gain versus the extension of the ground plane length ΔZ
a=25 mm,b=46 mm,w=30 mm

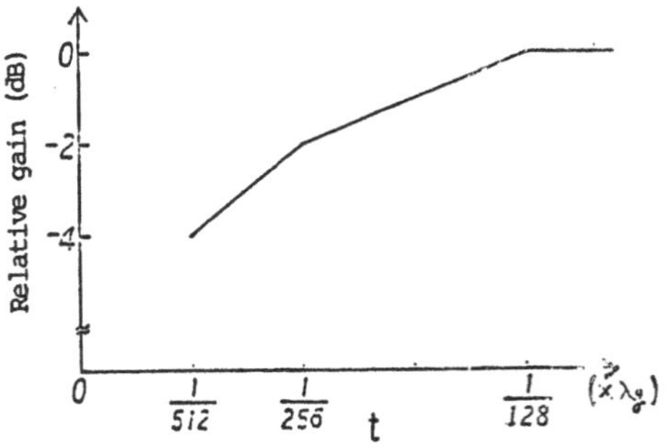

Fig. 8 Relative gain versus the thickness t of the substrate
a=25 mm,w=30 mm

A Conformal Cylindrical Microstrip Array for Producing Omnidirectional Radiation Pattern

IMMANUAL JAYAKUMAR, RAMESH GARG, MEMBER, IEEE,
B. K. SARAP AND BHAGWAN LAL

Abstract—**Radiation patterns of rectangular microstrip patches arranged circumferentially on a circular cylinder have been computed. Both the axial patches and circumferential patches have been used. It has been found that the number of circumferential patches required for a given ripple in gain pattern is considerably less, thus simplifying the feed network. Comparison with experimental measurements shows good agreement.**

I. Introduction

A number of investigations [1]–[7] have analyzed rectangular microstrip patches mounted on cylindrical surfaces, because of their importance in aerospace applications. In most of these applications like radio guidance and control of missiles, telemetry and telecommand of satellites, and control of air traffic at airports, the requirement is an omnidirectional pattern in the circumferential plane of the cylinder. This type of radiation pattern can be obtained by wrapping a microstripline around the circumference of the cylinder and feeding it at a number of points distributed uniformly along the circumference of the cylinder [1]. The other approach is to use a number of discrete radiators arrayed along the circumference and excited in same amplitude and phase [7]. The latter scheme has been implemented in this communication. It has been found that the feed network is simpler for this case as fewer feed points are required. This will give rise to a considerable savings in space and weight of the antenna.

II. Analysis

Radiation from a microstrip antenna mounted on a conducting circular cylinder can be analyzed in terms of line sources of magnetic currents in front of the cylinder, the analysis for which is available in [8]. The presence of a dielectric substrate between the cylinder and the magnetic current leaves the radiation pattern largely unchanged as shown in [6].

The rectangular patches in the cylindrical array can be excited in two different modes. In one mode the radiating slots are oriented axially along the cylinder and the patch is called an axial patch. In the other mode of excitation the radiating slots are oriented along the circumference and the corresponding patch is called a circumferential patch. These are shown in Fig. 1 along with the excitation.

The radiation pattern for a circumferential slot on a cylinder of radius a is given as [8]

$$E_\theta(\theta=\pi/2)=A\sum_{n=-\infty}^{\infty}\frac{j^n e^{jn(\Phi-\Phi_0)}}{H_n^{(2)}(k_0a)}\,\frac{\sin\,(n\alpha/2)}{n\alpha/2} \qquad (1)$$

where α is the angle subtended by the slot at the center of the cylinder, Φ_0 is the angular displacement of the midpoint of the radiating slot, and $k_0=2\pi/\lambda_0$. Similarly, the radiation pattern for an axial slot on a cylinder will be [8]

$$E_\Phi(\theta=\pi/2)=B\sum_{n=-\infty}^{\infty}\frac{j^n e^{jn(\Phi-\Phi_0)}}{H_n^{(2)\prime}(k_0a)} \qquad (1)$$

where Φ_0 gives the location of the slot, and the prime indicates derivative with respect to the argument. A and B are constants independent of Φ. In order to determine the electric field in the radiation zone due to a single patch or an array of patches on a conducting cylinder the principle of superposition can be used.

III. Computational Results

A computer program was written to evaluate (1) and (2), which involve an infinite series summation of the Hankel function of the second kind and its derivative. With the cylinder diameter and frequency increasing the argument of the Hankel function increases, and therefore the number of terms that must be summed before convergence is assured increases. Two different cylinder diameters and two different frequencies were used. These were 0.5 m and 1.0 m, and 550 MHz and 2.25 GHz, respectively. The argument k_0a of the Hankel function varied from 2.8798 to 23.562, corresponding to 550 MHz, 0.5 m diameter; and 2.25 GHz, 1.0 m diameter, respectively.

Subroutines were written to calculate Bessel functions of the first and second kind. In the computation of the Bessel functions of first kind, asymptotic formulas for the large and small argument cases were employed, and also the normal series summation formula when arguments and orders were comparable was used. In the computation of the Bessel functions of the second kind, the polynomial approximation formulas were employed.

An error criterion was used to terminate the infinite summation. Since the maximum absolute value was of the order of 0.8, any term whose contribution was less than 0.001 was neglected. For the argument of Hankel function varying from 2.8798 to 23.562 the maximum value of n increased from 9 to 47 when the above error criterion was applied. It is comparable to $2k_0a$. The computer program was executed on a Burroughs System.

The radiation pattern for a single axial slot displaced by $\Phi_0 = 20°$ on a cylinder has been computed for the frequency of 550 MHz and cylinder diameter of 0.5 m. It agrees closely with that given in [9]. The radiation patterns for 550 MHz with $2a = 0.5$ m and 1.0 m and also for 2.25 GHz with $2a = 0.5$ m and 1.0 m have been computed both for circumferential and axial patches. It was seen that the number of radiating elements on the cylinder determined the level of ripple in the gain. The results of the computation are summarized in Table I. As may be seen from this table, there is a drastic reduction in

Reprinted from *IEEE Trans. Antennas Propaga.*, vol. AP-34, pp. 1258–1261, Oct. 1986.

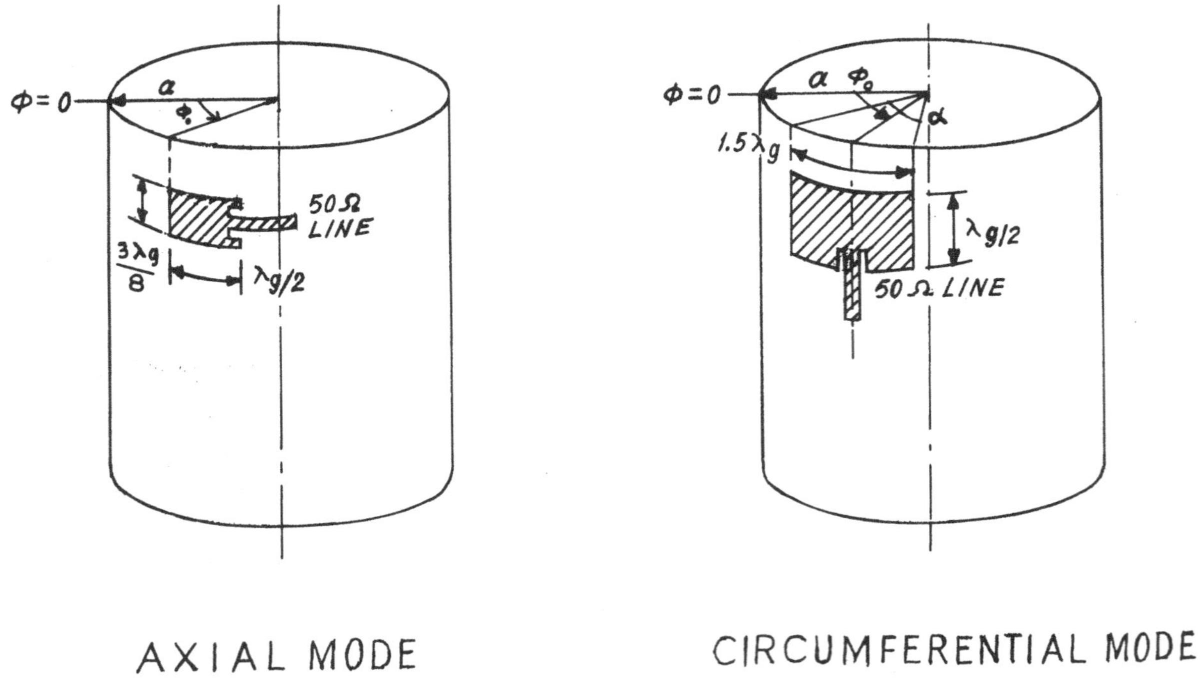

Fig. 1. Axial mode and the circumferential mode of excitations of a rectangular patch microstrip antenna.

TABLE I
COMPARATIVE RIPPLE (dB) IN THE GAIN PATTERNS OF A CYLINDRICAL ARRAY CONSISTING OF CIRCUMFERENTIAL PATCHES AND AXIAL PATCHES

Number of Elements	Gain Ripple (dB)							
	Circumferential patches				Axial patches			
	Cylinder dia. = 1m		Cylinder dia. = 0.5m		Cylinder dia. = 1m		Cylinder dia. = 0.5m	
	Operating Freq. MHz		Operating Freq. MHz		Operating Freq. MHz		Operating Freq. MHz	
	550	2250	550	2250	550	2250	550	2250
2	±10.9	±21.6	±4.5	±16.7	±7.16	±10.53	±4.72	±10.17
3	± 4.3	±10.8	±0.01	± 7.94	±4.86	±11.27	±3.33	± 8.10
4	± 0.78	± 7.4	-	± 4.46	±3.23	±17.74	±2.0	±11.97
6	± 0.13	± 5.5	-	± 2.28	±4.34	± 4.54	±0.14	± 8.09
8	-	± 8.04	-	± 7.0	±0.19	±13.26	±0.01	±10.36
12	-	± 4.6	-	± 0.31	±0.01	±10.59	-	± 3.75
16	-	± 0.11	-	-	-	± 6.81	-	± 0.01
32	-	-	-	-	-	± 0.02	-	-

the number of radiating elements when circumferential patches are employed to obtain a given ripple from a cylinder of given diameter. This is due to the fact that the radiating slots for the circumferential patches have been taken to be 1.5 λ_g long, and thereby cover a larger extent on the cylinder. Mutual coupling between the patches and the effect of periodicity on element pattern have been neglected in calculating radiation patterns. The effect of periodicity is expected to be small for smaller values of $k_0 a$ and larger size of radiating slot [10].

IV. Experimental Verification

Experiments have been conducted to confirm theoretical predictions. These were carried out for patches resonant at 2.25 GHz and mounted on a 0.5 m diameter cylinder. The substrate used had dielectric thickness 0.159 cm and dielectric constant $\epsilon_r = 2.50$. The patches were excited by 50 ohm microstrip. For feeding the array, a corporate feed network was employed, which was fabricated on the same substrate and mounted on the cylinder along with the patches. The patches were designed for resonant frequency and input

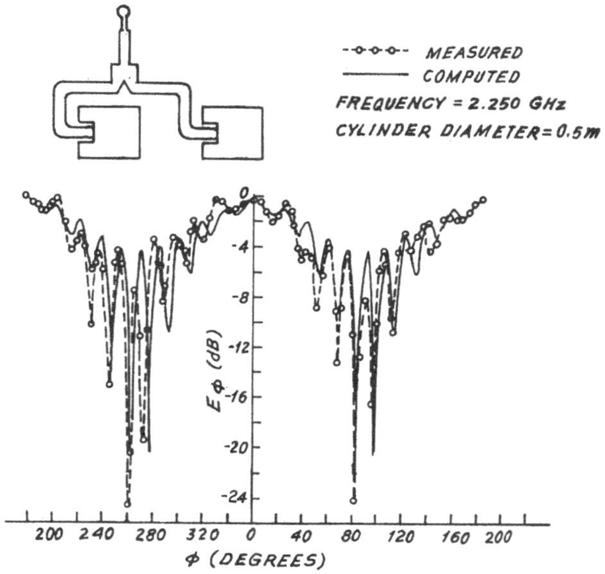

Fig. 2. Comparison of computed and measured radiation patterns in $\theta = 90°$ plane for two axial patches mounted diametrically opposite on a circular cylinder.

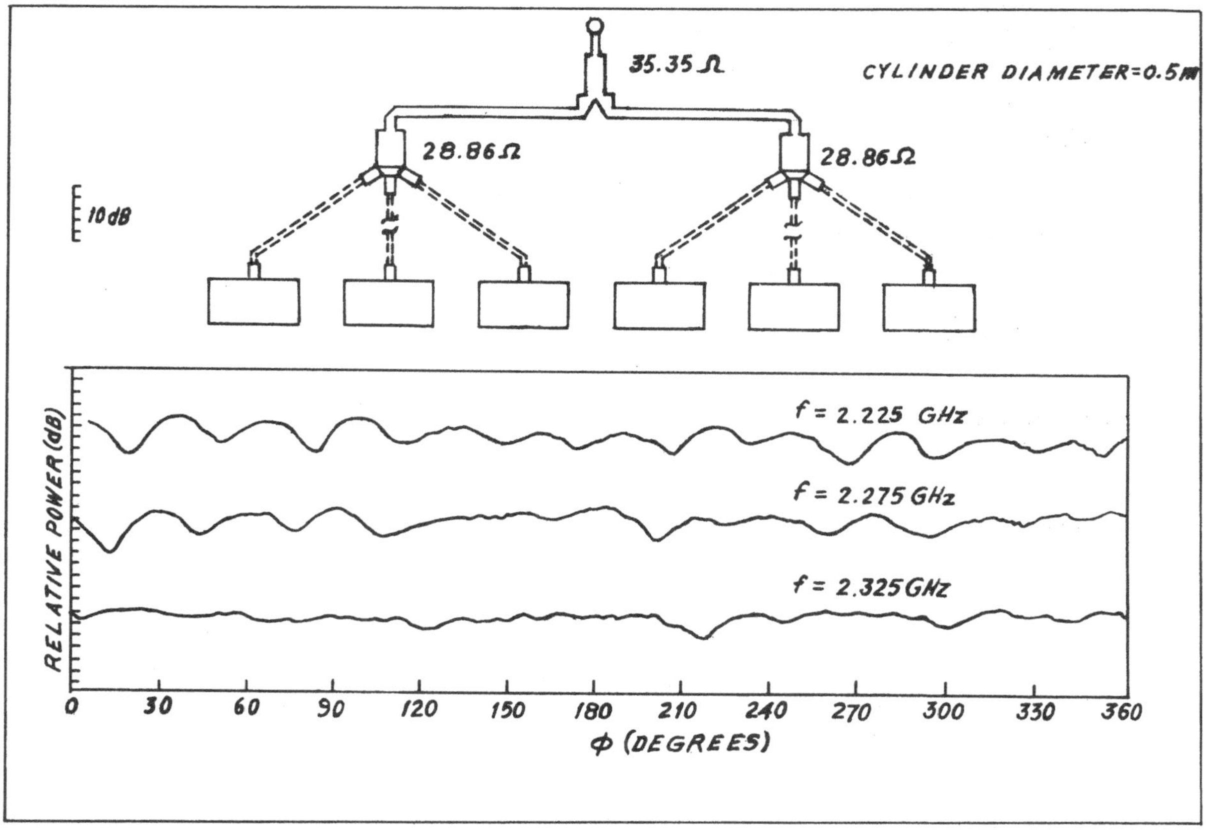

Fig. 3. Measured radiation patterns of a six-element circumferential patch array mounted on a circular cylinder.

impedance using the cavity model for planar rectangular patches [11]. The change in these characteristics when the patch was mounted on the cylinder was found to be negligible, as predicted in [2].

A radiation pattern in the circumferential plane of the cylinder for a two-element array employing axial patches is shown in Fig. 2. Feed arrangement and the opened-out view of the array is shown as inset there. Comparison with the computed result shows good agreement, thus verifying the approach used.

For the final omnidirectional array circumferential patches were used because of the simpler feed arrangement and because fewer patches are required for a given ripple in gain. The length of the radiating slot was 1.5 λ_g at 2.225 GHz. Radiation patterns for six element array, fed in phase, are shown in Fig. 3. The corporate feed structure employing 1:3 and 1:2 power dividers is also shown there in the opened-out view of the array. Comparison with the computed results for radiation pattern show that the agreement is better at a frequency of 2.325 GHz, and not at the design frequency. The shift in frequency could be attributed to the change in element radiation pattern due to the effect of substrate and periodicity. However, it emphasizes the fact that an omnidirectional radiation pattern can be obtained using fewer circumferential patches each of length 1.5 λ_g. For comparison, a wraparound antenna suggested in [1] would have required 38 feed points thus increasing the size of feed network considerably.

References

[1] R. E. Munson, "Conformal microstrip antennas and microstrip phased arrays," *IEEE Trans. Antennas Propagat.*, vol. AP-22, pp. 74–78, 1974.

[2] C. M. Krowne, "Cylindrical rectangular microstrip antenna," *IEEE Trans. Antennas Propagat.*, vol. AP-31, pp. 194–199, 1983.

[3] R. E. Post and D. T. Stephenson, "The design of a microstrip antenna array for a UHF space telemetry link," *IEEE Trans. Antennas Propagat.*, vol. AP-29, pp. 129–134, 1981.

[4] S. B. Fonseca and J. Giarola, "Analysis of microstrip wraparound antennas using dyadic Green's functions," *IEEE Trans. Antennas Propagat.*, vol. AP-31, pp. 248–253, 1983.

[5] K. R. Jacobsen, "The radiation from rectangular microstrip antennas mounted on two-dimensional objects," *IEEE Trans. Antenna Propagat.*, vol. AP-32, pp. 1255–1259, 1984.

[6] J. Ashkenazy, S. Shtrikman, and D. Treves, "Electric surface current model for the analysis of microstrip antennas on cylindrical bodies," *IEEE Trans. Antenna Propagat.*, vol. AP-33, pp. 295–300, 1985.

[7] T. S. Campbell *et al.*, "The design and fabrication of microstrip omnidirectional array antennas for aerospace applications," NTIS Tech. Rep. N77-13332, Nov. 1976.

[8] R. F. Harrington, *Time-Harmonic Electromagnetic Fields.* New York: McGraw-Hill, 1961.

[9] G. Markov, *Antenna.* Moscow: Progress Publishers, 1965.

[10] J. C. Herper, A. Hessel, and B. Tomasic, "Element pattern of an axial dipole in a cylindrical phased array, Part I: Theory," *IEEE Trans. Antenna Propagat.*, vol. AP-33, pp. 259–272, 1985.

[11] I. J. Bahl and P. Bhartia, *Microstrip Antennas.* Dedham, MA: Artech House, 1980.

Radiation and Scattering from a Microstrip Patch on a Uniaxial Substrate

DAVID M. POZAR, MEMBER, IEEE

Abstract—The problem of a rectangular microstrip antenna printed on a uniaxially anisotropic substrate is treated. The effect of anisotropy on the resonant frequency and surface wave excitation of the antenna is considered, and the radar cross section (RCS) of the antenna is calculated. The RCS calculation includes the effect of the load impedance (antenna mode scattering). Results for the resonant frequency of a patch on a uniaxial substrate are compared with measurements, and the RCS of a patch on an isotropic substrate is compared with measurements. The derivation of the uniaxial Green's function in spectral form, the associated moment method analysis for the input impedance and scattering of the microstrip patch, and the expressions for the far-zone fields of a source on a uniaxial substrate are presented.

I. Introduction

THE STUDY OF anisotropic microwave substrates is of interest for at least two reasons. First, as the review article [1] points out, many practical substrates have a significant amount of anisotropy that can affect the performance of printed circuits and antennas, and thus accurate characterization and design must account for this effect. Second, it is possible that the use of such materials may have a beneficial effect on circuit or antenna (either radiation or scattering) performance. An example is the use of an anisotropic substrate to achieve equal even and odd mode phase velocities for microstrip coupled line directional couplers [2].

As the references of [1] indicate, a considerable amount of work on the characterization and application of anisotropic substrates for microwave transmission lines and circuits has been carried out, but there has been little effort in this direction for antenna applications. The present paper thus treats the problem of a rectangular microstrip antenna on a uniaxial dielectric substrate; both radiation and scattering radar cross section (RCS) are discussed. Since the effect of a load impedance on the RCS of a microstrip antenna has not yet been treated even for an isotropic substrate, a number of results pertaining to this case are presented. Measured resonant frequency of a microstrip patch on a uniaxial substrate is compared with calculations, and the measured RCS of a microstrip antenna on an isotropic substrate is compared with calculations.

The spectral domain Green's functions for a variety of planar anisotropic geometries have been derived and presented in the literature [1], [3], and [4]. Although these results may be general enough to yield the required Green's function for the uniaxial case of the present work, the specialization to the uniaxial case is derived here and shown to be quite simple in form—actually no more complicated than the corresponding result for the isotropic case [5]. The next section presents the Green's function derivation, and discusses the moment method formulation for the input impedance and radar cross section of the microstrip antenna. Expressions for the far-zone fields of the printed antenna are also presented. Various results for radiation and scattering are presented in Section III, with some comparisons with measurements.

II. Theory

A. The Green's Function of a Uniaxial Substrate in Spectral Form

In this section the Green's function for the uniaxial dielectric substrate is derived. The relevant geometry is shown in Fig. 1, where a unit infinitesimal current source in the $\hat{x}$ direction is located at (x_0, y_0, d). The dielectric is characterized by a permittivity tensor of the form

$$\bar{\bar{\epsilon}}=\epsilon_0\begin{bmatrix}\epsilon_x & 0 & 0\\ 0 & \epsilon_x & 0\\ 0 & 0 & \epsilon_z\end{bmatrix}, \tag{1}$$

and it is desired to find the resulting E_x, E_y, and E_z fields.

While it is possible to introduce vector potentials, it is actually simpler to work directly with wave equations for E_z and H_z, and find the transverse fields from these field components. Thus, Maxwell's equations,

$$\nabla\times\bar{E}=-j\omega\mu_0\bar{H}, \tag{2a}$$

$$\nabla\times\bar{H}=j\omega\bar{\bar{\epsilon}}\cdot\bar{E}, \tag{2b}$$

for the fields inside the dielectric region can be solved simultaneously for the following wave equations:

$$-\nabla\times\nabla\times\bar{E}+\omega^2\mu_0\bar{\bar{\epsilon}}\cdot\bar{E}=0 \tag{3a}$$

$$-\nabla\times(\bar{\bar{\epsilon}})^{-1}\cdot(\nabla\times\bar{H})+\omega^2\mu_0\bar{H}=0. \tag{3b}$$

Applying the divergence condition that $\nabla\cdot(\bar{\bar{\epsilon}}\cdot\bar{E})=0$ to the z-component of (3a) yields a wave equation for E_z:

$$\frac{\partial^2 E_z}{\partial x^2}+\frac{\partial^2 E_z}{\partial y^2}+\frac{\epsilon_z}{\epsilon_x}\frac{\partial^2 E_z}{\partial z^2}+\epsilon_z k_0^2 E_z=0, \tag{4a}$$

where $k_0=\omega\sqrt{\mu_0\epsilon_0}=2\pi/\lambda_0$. Similarly, applying $\nabla\cdot\bar{H}=0$

Reprinted from *IEEE Trans. Antennas Propaga.*, vol. AP-35, pp. 613–621, June 1987.

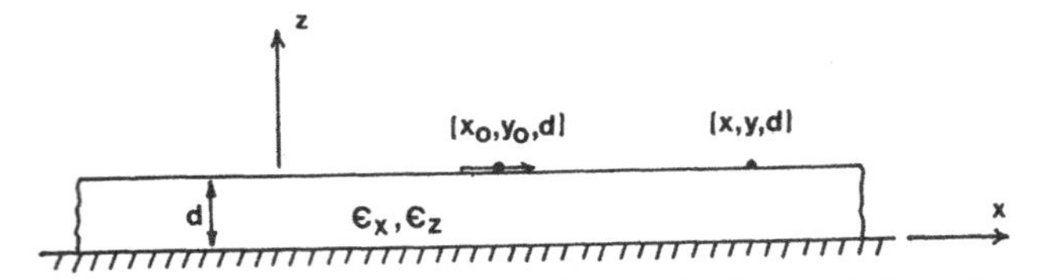

Fig. 1. Geometry of an infinitesimal $\hat{x}$-directed unit current element on a uniaxial dielectric substrate.

to the z-component of (3b) yields a wave equation for H_z:

$$\frac{\partial^2 H_z}{\partial x^2}+\frac{\partial^2 H_z}{\partial y^2}+\frac{\partial^2 H_z}{\partial z^2}+\epsilon_x k_0^2 H_z=0. \tag{4b}$$

By assuming plane wave propagation of the form $e^{\pm jk_x x}e^{\pm jk_y y}e^{\pm jk_z z}$ the dispersion relations may be obtained as

$$k_z^2=\epsilon_x k_0^2-\beta^2=k_a^2 \tag{5a}$$

for H_z (transverse electric (TE) waves), and

$$k_z^2=\epsilon_x k_0^2-\epsilon_x\beta^2/\epsilon_z=k_b^2 \tag{5b}$$

for E_z (transverse magnetic (TM) waves), where

$$\beta^2=k_x^2+k_y^2. \tag{6}$$

Observe that there exist distinct propagation constants for the TE and TM waves in the uniaxial dielectric. The above results can be specialized to the air region ($z > d$) by allowing $\epsilon_x = \epsilon_z = 1$, in which case both the TE and TM propagation constants reduce to

$$k_z^2=k_0^2-\beta^2=k_2^2. \tag{7}$$

We now define a Fourier transform pair as

$$E(x, y, z)=\frac{1}{4\pi^2}\iint_{-\infty}^{\infty}\tilde{E}(k_x, k_y, z)e^{jk_x x}e^{jk_y y}\,dk_x\,dk_y \tag{8a}$$

$$\tilde{E}(k_x, k_y, z)=\iint_{-\infty}^{\infty}E(x, y, z)e^{-jk_x x}e^{-jk_y y}\,dx\,dy. \tag{8b}$$

In the transform domain, the transverse fields can be written in terms of $\tilde{E}_z$ and $\tilde{H}_z$ as

$$\left(\epsilon_x k_0^2+\frac{\partial^2}{\partial z^2}\right)\tilde{E}_x=jk_x\frac{\partial}{\partial z}\tilde{E}_z+\omega\mu_0 k_y\tilde{H}_z \tag{9a}$$

$$\left(\epsilon_x k_0^2+\frac{\partial^2}{\partial z^2}\right)\tilde{E}_y=jk_y\frac{\partial}{\partial z}\tilde{E}_z-\omega\mu_0 k_x\tilde{H}_z \tag{9b}$$

$$\left(\epsilon_x k_0^2+\frac{\partial^2}{\partial z^2}\right)\tilde{H}_x=jk_x\frac{\partial}{\partial z}\tilde{H}_z-\omega\epsilon_0\epsilon_x k_y\tilde{E}_z \tag{9c}$$

$$\left(\epsilon_x k_0^2+\frac{\partial^2}{\partial z^2}\right)\tilde{H}_y=jk_y\frac{\partial}{\partial z}\tilde{H}_z+\omega\epsilon_0\epsilon_x k_x\tilde{E}_z. \tag{9d}$$

In the above, $\partial^2/\partial z^2 = -k_z^2$, but k_z may be either k_a or k_b in the dielectric region, and k^2 in the air region.

The general form of $\tilde{E}_z$ and $\tilde{H}_z$ is

$$\tilde{E}_z=Ae^{-jk_2 z},\qquad \text{for } z>d \tag{10a}$$

$$\tilde{H}_z=Be^{-jk_2 z},\qquad \text{for } z>d \tag{10b}$$

$$\tilde{E}_z=C\cos k_b z+D\sin k_b z,\qquad \text{for } 0<z<d \tag{11a}$$

$$\tilde{H}_z=E\sin k_a z+F\cos k_a z,\qquad \text{for } 0<z<d. \tag{11b}$$

The transverse field components of (9a)–(9d) can then be rewritten as

$$\tilde{E}_x=\frac{jk_x\epsilon_z}{\epsilon_x\beta^2}\frac{\partial}{\partial z}\tilde{E}_z+\frac{\omega\mu_0 k_y}{\beta^2}\tilde{H}_z \tag{12a}$$

$$\tilde{E}_y=\frac{jk_y\epsilon_z}{\epsilon_x\beta^2}\frac{\partial}{\partial z}\tilde{E}_z-\frac{\omega\mu_0 k_x}{\beta^2}\tilde{H}_z \tag{12b}$$

$$\tilde{H}_x=\frac{jk_x}{\beta^2}\frac{\partial}{\partial z}\tilde{H}_z-\frac{\omega\epsilon_0\epsilon_z k_y}{\beta^2}\tilde{E}_z \tag{12c}$$

$$\tilde{H}_y=\frac{jk_y}{\beta^2}\frac{\partial}{\partial z}\tilde{H}_z+\frac{\omega\epsilon_0\epsilon_z k_x}{\beta^2}\tilde{E}_z, \tag{12d}$$

which can be verified by substitution of (12) into (9), and using (10) and (11). Applying (12a) and (12b) to (11a) and (11b) to enforce the boundary condition that $E_x = E_y = 0$ at $z = 0$ yields $D = F = 0$. There then remain four constants (A, B, C, E) to be evaluated by the continuity of E_x, E_y, and H_x at $z = d$, and a jump condition in H_y at $z = d$. After some straightforward algebraic manipulation, the results for $\tilde{E}_z$ and $\tilde{H}_z$ are, for $z > d_1$,

$$\tilde{E}_z=\frac{k_x k_b\sin k_b d}{j\omega\epsilon_0 T_m}e^{-jk_2(z-d)}e^{-jk_x x_0}e^{-jk_y y_0} \tag{13a}$$

$$\tilde{H}_z=\frac{-jk_y\sin k_a d}{T_e}e^{-jk_2(z-d)}e^{-jk_x x_0}e^{-jk_y y_0}, \tag{13b}$$

where

$$T_m=\epsilon_x k_2\cos k_b d+jk_b\sin k_b d \tag{14a}$$

$$T_e=k_a\cos k_a d+jk_2\sin k_a d. \tag{14b}$$

The zeros of the T_m and T_e functions correspond to TM and TE surface wave poles, respectively.

Using (12a), (12b), (13a), (13b), and (8a), the transverse fields at (x, y, d) can be written as

$$E_x(x, y, d)=\frac{-jZ_0}{4\pi^2 k_0}\iint_{-\infty}^{\infty}Q_{xx}(k_x, k_y)\cdot e^{jk_x(x-x_0)}e^{jk_y(y-y_0)}\,dk_x\,dk_y \tag{15a}$$

$$E_y(x, y, d)=\frac{-jZ_0}{4\pi^2 k_0}\iint_{-\infty}^{\infty}Q_{yx}(k_x, k_y)\cdot e^{jk_x(x-x_0)}e^{jk_y(y-y_0)}\,dk_x\,dk_y, \tag{15b}$$

where

$$Q_{xx}=\frac{k_x^2 k_2 k_b \sin k_b d}{\beta^2 T_m}+\frac{k_0^2 k_y^2 \sin k_a d}{\beta^2 T_e}, \quad (16a)$$

$$Q_{yx}=\frac{k_x k_y k_2 k_b \sin k_b d}{\beta^2 T_m}+\frac{k_0^2 k_x k_y \sin k_a d}{\beta^2 T_e}, \quad (16b)$$

and $Z_0 = \sqrt{\mu_0/\epsilon_0}$. We will also require the fields due to a $\hat{y}$-directed current element, to allow both $\hat{x}$ and $\hat{y}$ current expansion on the patch. These results can be obtained from the above case by a change of coordinates to give

$$Q_{yy}=\frac{k_y^2 k_2 k_b \sin k_b d}{\beta^2 T_m}+\frac{k_0^2 k_x \sin k_a d}{\beta^2 T_e} \quad (17a)$$

$$Q_{xy}=Q_{yx}. \quad (17b)$$

The above results reduce to the appropriate isotropic results of [5] or [6] for $\epsilon_x = \epsilon_z = \epsilon_r$. It is interesting that the expressions for the uniaxially anisotropic case are no more complicated than the corresponding isotropic results; this is apparently due to the fact that the two eigenmodes of the uniaxial case couple individually to the TM and TE waves.

B. Far-Zone Fields for a Source on a Uniaxial Substrate

For the calculation of antenna patterns or radar cross section, a stationary phase evaluation of the infinite Fourier transform integrals yields convenient and useful results. We will first consider the fields due to an $\hat{x}$-directed source; the fields from $\hat{y}$-directed currents can then be obtained by a coordinate transformation.

From the analysis of Section II-A, the exact electric fields at (x, y, z) in the air region $(z > d)$ due to a unit $\hat{x}$-directed current source at (x_0, y_0, d) can be written as

$$E_i(x, y, z)=\frac{-jZ_0}{4\pi^2 k_0}\iint_{-\infty}^{\infty} P_{ix}(k_x, k_y)e^{jk_x(x-x_0)} \cdot e^{jk_y(y-y_0)}e^{-jk_2(z-d)}\, dk_x\, dk_y, \quad (18)$$

where the index $i = x, y$, or z, and the P_{ix} are defined as

$$P_{xx}=\frac{k_b k_2 k_x^2 \sin k_b d}{\beta^2 T_m}+\frac{k_0^2 k_y^2 \sin k_a d}{\beta^2 T_e} \quad (19a)$$

$$P_{yx}=\frac{k_b k_2 k_x k_y \sin k_b d}{\beta^2 T_m}-\frac{k_0^2 k_x k_y \sin k_a d}{\beta^2 T_e} \quad (19b)$$

$$P_{zx}=\frac{k_b k_x \sin k_b d}{T_m}. \quad (19c)$$

Stationary phase evaluation of the above integrals and conversion to spherical coordinates then yields

$$E_\theta(r, \theta, \phi)=\frac{-Z_0}{2\pi r} e^{-jkr}e^{jk_x x_0}e^{jk_y y_0}e^{jk_2 d}P_\theta \cos\theta \cos\phi \quad (20a)$$

$$E_\phi(r, \theta, \phi)=\frac{-Z_0}{2\pi r} e^{-jkr}e^{jk_x x_0}e^{jk_y y_0}e^{jk_2 d}P_\phi \cos\theta \sin\phi, \quad (20b)$$

where

$$P_\theta=\frac{-k_0 k_b \sin k_b d}{T_m}, \quad (21a)$$

$$P_\phi=\frac{k_0^2 \sin k_a d}{T_e}. \quad (21b)$$

In (20) and (21) (and in k_a and k_b in these expressions), k_x and k_y are evaluated at the stationary phase point as

$$k_x=k_0 \sin\theta \cos\phi, \quad (22a)$$

$$k_y=k_0 \sin\theta \sin\phi. \quad (22b)$$

Integration over x_0 and y_0 in (20) will give the fields of an expansion mode; this is actually multiplication by the Fourier transform of the mode.

C. The Moment Method Solution for the Antenna Characteristics

Now that we have the necessary Green's function components, it is relatively straightforward to formulate the moment method solution for the usual antenna characteristics. We assume the patch is probe-fed, as shown in Fig. 2, although other feeding techniques could be implemented. As discussed in [5] and [6], the probe feed model is somewhat idealized, but is reasonably valid for thin substrates.

The surface current density on the patch is expanded in a set of N expansion modes $\bar{J}_n(x, y)$ with unknown coefficients I_n:

$$\bar{J}(x, y)=\sum_{n=1}^{N} I_n \bar{J}_n(x, y). \quad (23)$$

The nth expansion mode may represent current flow in either the $\hat{x}$ or $\hat{y}$ direction. For the present work, entire domain expansion modes have been selected, with the form

$$\bar{J}_n(x, y)=\frac{\hat{x}}{W} \sin\frac{k\pi}{L}(x+L/2) \cos\frac{l\pi}{W}(y+W/2) \quad (24a)$$

for $\hat{x}$-directed modes, while the $\hat{y}$-directed modes have the form

$$\bar{J}_n(x, y)=\frac{\hat{y}}{L} \cos\frac{k\pi}{L}(x+L/2) \sin\frac{l\pi}{W}(y+W/2). \quad (24b)$$

The Fourier transform of these modes can be defined, consistent with (8a), as

$$\bar{F}_n(k_x, k_y)=\int_x\int_y \bar{J}_n(x, y)e^{-jk_x x}e^{-jk_y y}\, dx\, dy, \quad (25)$$

and can be easily evaluated in closed form.

The electric field integral equation, which enforces the boundary condition that the incident field from the probe plus the scattered field from the patch currents must vanish on the

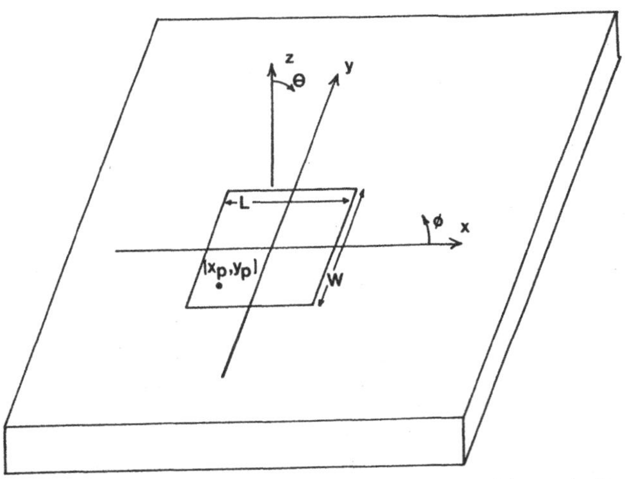

Fig. 2. Geometry of a rectangular microstrip antenna with a probe feed.

patch surface, can then be discretized into the usual matrix form as

$$[Z][I]=[V^p]. \tag{26}$$

The impedance matrix elements are

$$Z_{mn}=-\int_s \bar{J}_m \cdot \bar{E}_n \, ds=\frac{jZ_0}{4\pi^2 k_0} \cdot \iint_{-\infty}^{\infty} \bar{F}_m^* \cdot \bar{\bar{Q}} \cdot \bar{F}_n \, dk_x \, dk_y, \tag{27}$$

where $\bar{E}_m$ is the electric field generated by the nth expansion mode, and $\bar{\bar{Q}}$ is a dyad defined for notational convenience as

$$\bar{\bar{Q}}=\hat{x}Q_{xx}\hat{x}+\hat{x}Q_{xy}\hat{y}+\hat{y}Q_{yx}\hat{x}+\hat{y}Q_{yy}\hat{y}. \tag{28}$$

The voltage vector elements are

$$V_m^p=\int_s \bar{E}^p \cdot \bar{J}_m \, ds=\int_{z=0}^{d} \bar{J}^p \cdot \bar{E}_m \, dz = \frac{-jZ_0}{4\pi^2 k_0} \iint_{-\infty}^{\infty} [Q_{zx}\hat{x}+Q_{zy}\hat{y}] \cdot \bar{F}_m e^{jk_x x_p} e^{jk_y y_p} \, dk_x \, dk_y, \tag{29}$$

where $\bar{E}^p$ represents the incident electric field caused by $\bar{J}^p$, a unit current source on the probe. In (29), Q_{zx} and Q_{zy} represent the E_z field due to either an $\hat{x}$ or $\hat{y}$ current, respectively, integrated from $z = 0$ to d. From the analysis of Section II-A, these terms can be derived as

$$Q_{zx}=\frac{j\epsilon_x k_2 k_x \sin k_b d}{\epsilon_z k_b T_m}, \tag{30a}$$

$$Q_{zy}=\frac{j\epsilon_x k_2 k_y \sin k_b d}{\epsilon_z k_b T_m}. \tag{30b}$$

The reciprocity theorem [7] has been used in (29).

As discussed in [5], the infinite integrals in (27) and (29) can be simplified by converting to polar coordinates. Nevertheless, numerical integration is still required, which results in long computing times. Some savings in computer time can be obtained by integrating the entire $[Z]$ matrix or $[V^p]$ vector at the same time, to avoid recalculations of the Q functions [8].

Once the $[Z]$ and $[V^p]$ matrices are calculated, the expansion mode coefficients can be found via (26). The input impedance is then

$$Z_{in}=-\int_{z=0}^{d} E_z \, dz=-[I]^t[V^p], \tag{31}$$

where E_z is the total $\hat{z}$ field generated by the patch currents. The probe self-impedance is neglected here.

Another quantity of interest is the amount of surface wave power excited on the dielectric slab. As in [8], a radiation efficiency can be defined as

$$e=\frac{P_{\rm rad}}{P_{\rm rad}+P_{\rm sw}}, \tag{32}$$

where $P_{\rm rad}$ is the power radiated by the antenna and $P_{\rm sw}$ is the power delivered to surface waves. These quantities can be calculated as

$$P_{\rm rad}=[I]^{t*}[R][I], \tag{33a}$$

$$P_{\rm sw}=[I]^{t*}[R^{\rm sw}][I], \tag{33b}$$

where R_{mn} = Re (Z_{mn}) is the real part of the impedance matrix elements (space wave and surface wave), and $R_{mn}^{\rm sw}$ is the contribution from the surface wave fields, which can be evaluated from the residue of the surface wave pole, or poles.

D. The Moment Method Solution for the Radar Cross Section

The radar cross section of a microstrip patch has recently been treated [9], although the effect of the terminating impedance (antenna mode scattering) was not considered. Besides [9] and [10], there has been very little work on the RCS of printed antennas in the literature. In this section, we will develop the theory for the radar cross section of a rectangular patch antenna on a uniaxial substrate, including the effect of the terminating impedance. Only monostatic scattering is considered here, but the bistatic case is a trivial extension.

Let the microstrip patch antenna of Fig. 2 be illuminated by a unit amplitude plane wave of the form,

$$\bar{E}^i=\bar{E}_0 e^{-j\bar{k}\cdot\bar{r}}, \qquad |\bar{E}_0|=1, \tag{34}$$

from the direction θ, ϕ. In the absence of the patch antenna, a plane wave $\bar{E}^r$ will be specularly reflected from the grounded dielectric slab, so that $\bar{E}^i + \bar{E}^r$ will satisfy the appropriate boundary conditions at the air-dielectric interface and at the ground plane.

In the presence of the terminated patch antenna, the above fields will be augmented by $\bar{E}^s$, the field due to the patch currents, and $\bar{E}^p$, the field due to the current flowing on the probe. If $\bar{E}^s$ and $\bar{E}^p$ are computed in the presence of the grounded dielectric slab, through the use of the Green's functions of Section II-A, then these fields will also satisfy the appropriate boundary conditions on the grounded dielectric slab. The remaining boundary condition is that the total field must vanish on the patch surface S:

$$\bar{E}^i+\bar{E}^r+\bar{E}^s+\bar{E}^p=0, \qquad \text{on S.} \tag{35}$$

The scattered field from the patch is

$$\bar{E}^s=\sum_{n=1}^{N} I_n\bar{E}_n, \tag{36}$$

where $\bar{E}_n$ is the field from the nth expansion mode. Using this in (35), and weighting the resulting equation with $\bar{J}_m$ yields the following result:

$$\int_s \bar{J}_m \cdot (\bar{E}^i+E^r)\, ds = -\sum_{n=1}^{N} I_n \cdot \int_s \bar{J}_m \cdot \bar{E}_n\, ds - \int_s \bar{J}_m \cdot \bar{E}^p\, ds. \tag{37}$$

This can be rewritten, using the definition of $[Z]$ from (27) and $[V^p]$ from (29) (in (29), V_m^p is defined assuming 1 A on the probe), as

$$[V^i]=[Z][I]-I_L[V^p], \tag{38}$$

where I_L is the current through the load, and hence through the probe feed, and $[V^i]$ is an excitation vector with elements,

$$V_m^i=\int_s \bar{J}_m \cdot (\bar{E}^i+\bar{E}^r)\, ds = \int_v \bar{E}_m \cdot \bar{J}^i\, dv = \frac{4\pi\bar{E}_m \cdot \bar{E}_0}{j\omega\mu_0}, \tag{39}$$

where $4\pi r/j\omega\mu_0$ is the required strength of an infinitesimal dipole source to produce a unit amplitude plane wave. In (39), the reciprocity theorem was used to evaluate V_m^i in terms of $\bar{E}_m$, the far-zone field of mode m (in the presence of the grounded dielectric slab), and $\bar{E}_0$, the incident plane wave amplitude vector.

The load or probe current can be determined as

$$I_L=\frac{V_L}{Z_L}=\frac{1}{Z_L}[I]^t[V^p], \tag{40}$$

where Z_L is the load impedance connected to the probe-fed patch antenna. Substituting (40) for I_L in (38) allows the unknown patch currents to be solved as

$$[I]=\{[Z]-\frac{1}{Z_L}[V^p][V^p]^t\}^{-1}[V^i]. \tag{41}$$

The scattered far-zone field from the patch can then be found from (36) and (20), and the radar cross section computed from

$$\sigma_{\theta\theta}=4\pi r^2|E_\theta^s|^2, \tag{42}$$

assuming $\bar{E}_0$ is θ polarized and of unit magnitude. Other polarizations can be similarly found.

The qualitative behavior of the effect of the load impedance on the RCS can be inferred by considering the one-mode form

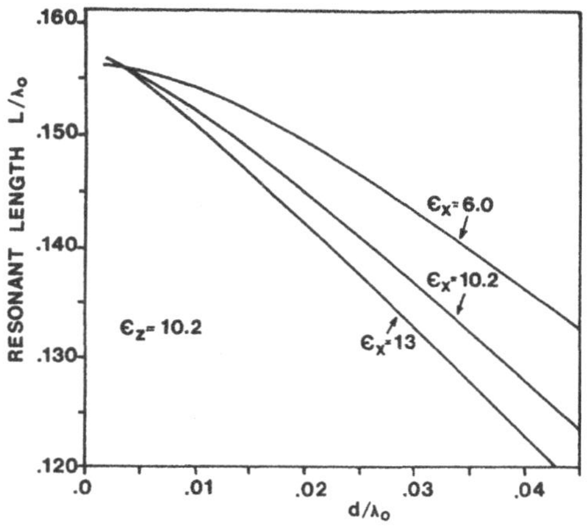

Fig. 3. The effect of substrate anisotropy on the resonant length of a rectangular patch antenna.

of (41):

$$I_1=\frac{V_1^i}{Z_{11}-(V_1^p)^2/Z_L}. \tag{43}$$

Since the scattered field in (42) is proportional to the patch current, (43) shows that the RCS will be greatest for an open circuit load ($Z_L = \infty$), assuming Z_{11} is close to resonance so that $|Z_{11}|$ is near a minimum. Similarly, the RCS will be the smallest for a short-circuited patch ($Z_L = 0$).

III. Results

A. Antenna Results

In this section we present some results on the effect of substrate anisotropy on printed antenna performance. While the previously described analysis is general enough to treat both positive uniaxial ($\epsilon_z > \epsilon_x$) and negative uniaxial ($\epsilon_z < \epsilon_x$) cases, as well as a wide range of anisotropy ratios (AR $= \epsilon_z/\epsilon_x$), most substrates in common use are negative uniaxial with anisotropy ratios less than 1.4 or so. As a practical example, then, we will consider Epsilam-10 material, which has been quoted in [1] as having $\epsilon_x = 13.0$, $\epsilon_z = 10.2$.

Fig. 3 shows the resonant length of a rectangular microstrip patch of width 0.23 λ_0 on a substrate with $\epsilon_z = 10.2$, versus substrate thickness, for various values of ϵ_x. The curve for $\epsilon_x = 10.2$ is then the result assuming the substrate was isotropic with $\epsilon_r = 10.2$. The curve for $\epsilon_x = 13.0$ is the result when the uniaxial anisotropy is included. Observe that the effect of the $\epsilon_x = 13.0$ anisotropy is to decrease the resonant length, or to increase the effective dielectric constant of the substrate, and that the amount of this change increases with increasing substrate thickness. That the effect of anisotropy increases with substrate thickness is not unexpected, since the electric field under the patch, while predominantly $\hat{z}$-directed for thin substrates, becomes more transversely directed as d increases.

The data also show that the shift in resonant frequency is about equal to the bandwidth of the antenna [11]—for example, for $d = 0.025\ \lambda_0$ the bandwidth is about 2 percent, while the lowering in resonant frequency due to $\epsilon_x = 13.0$, instead of $\epsilon_x = 10.2$, is about 2 percent. The practical implication is that an antenna designed for a specific operating frequency may actually resonate outside of the antenna bandwidth if the substrate anisotropy is not considered. Fig. 3

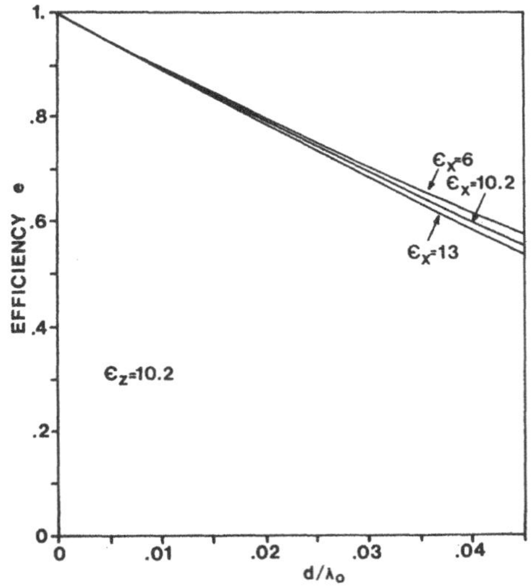

Fig. 4. Radiation efficiency (based on power lost to surface waves) for a rectangular patch antenna on a uniaxial substrate.

TABLE I

d cm	d/λ_o	L cm	W cm	x_p cm	y_p cm	measured resonant frequency GHz	calculated resonant frequency ϵ_x=10.2 ϵ_z=10.2	% error	ϵ_x=13.0 ϵ_z=10.2	% error	calc. band-width
0.127	0.01	2.00	3.00	0.350	0.	2.264	2.285	1.0	2.268	0.2	0.6%
0.127	0.02	0.95	1.50	0.155	0.	4.495	4.58	1.9	4.52	0.6	1.4%
0.254	0.02	1.90	3.00	0.300	0.	2.242	2.29	2.2	2.26	0.8	1.2%

also shows the result for $\epsilon_x = 6.0$, which is seen to increase the resonant length, or lower the effective dielectric constant.

Fig. 4 shows the radiation efficiency, based on power lost to surface waves as defined in (32), for the same geometry as Fig. 3 ($\epsilon_z = 10.2$, $W = 0.23\ \lambda_0$). It is seen that changing ϵ_x has little effect, and that the downward trend of e with increasing substrate thickness is the same as that observed for isotropic substrates [11]. The fact that there is such little difference in e for various values of ϵ_x is probably because the only surface wave propagating, for this range of substrate thickness, is the TM_0 mode, which has a dominant E_z field.

Measurements were made on a set of rectangular microstrip antennas printed on Epsilam-10 substrates, with probe feeds. Calculations of the resonant frequency were then made with and without the anisotropy effect. The results are shown in Table I.

The inclusion of the substrate anisotropy is seen to give a resonant frequency much closer to the measured value, while its neglect gives a resonant frequency well outside the bandwidth of the antenna.

An interesting possible application of a uniaxial material would be to employ it as either a substrate or superstrate to achieve circular polarization from a printed antenna that would normally radiate linear polarization. This idea is related to the well-known quarter-wave plate [7] of optics, where a uniaxial material is used to convert linear to circular polarization. Thus, a uniaxial superstrate, with its optic axis parallel to the antenna substrate and oriented 45° from the electric field polarization of the patch (resonant direction) should generate circular polarization in a way similar to a meander line polarizer. A possible drawback might be the excitation of large amounts of surface wave power, since the required superstrate thickness would be rather large, electrically. Lifting the superstrate off the antenna substrate would help to alleviate this problem.

The more interesting possibility, however, would be to generate circular polarization solely through the use of a uniaxial substrate. Unfortunately, this does not appear to be practical, as a simple cavity-model type of analysis shows that the predominant electric field under the patch is $\hat{z}$-directed, even for a misaligned optic axis. Thus, the substrate would have to be quite thick ($d > 0.25\ \lambda_d$) electrically for a significant amount of transverse electric field to be present, and for the necessary phase shift to be generated between the two eigenmodes of the material.

B. Radar Cross-Section Results

As discussed in [9], the calculation of the RCS properties of a microstrip antenna is very time consuming because broadband data is generally required, and many expansion modes are then needed. The present solution is even more time consuming than that of [9], because the voltage vector elements of (29) must be calculated in addition to the impedance matrix elements of (27), in order to treat the feed termination effect. (In [9], the patch was assumed to be open-circuited, so the voltage vector elements were not needed.)

Most of the results to follow deal with isotropic substrates, for two reasons. First, there has been very little work reported

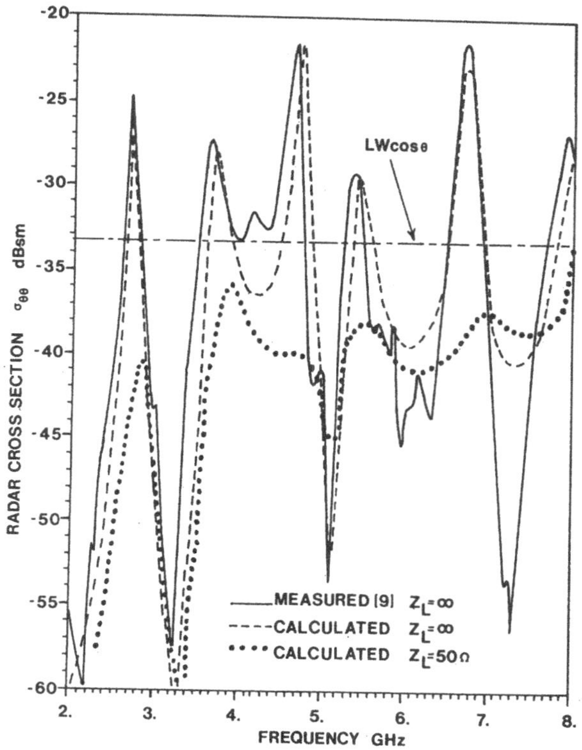

Fig. 5. Radar cross section versus frequency of a rectangular microstrip patch on an isotropic substrate. Measured data for an open-circuited patch ($Z_L = \infty$); calculated data for $Z_L = \infty$, and $Z_L = 50\ \Omega$. $L = 3.66$ cm, $W = 2.60$ cm, $\epsilon_x = \epsilon_z = 2.17$, $d = 0.158$ cm, $x_p = 0.0163$ cm, $y_p = 0.011$ cm, $\theta = 60°$, $\phi = 45°$.

in the literature on the RCS of printed antennas ([9], [10]), and no theory on the effect of the load termination. Second, as the data will show, the effect of the uniaxial substrate is to merely cause a slight shift in the location of the resonant peaks of the RCS frequency response.

Fig. 5 shows the RCS of a rectangular microstrip antenna on an isotropic ($\epsilon_r = 2.17$) substrate, over a wide frequency band. The measured data is from [9], for an open-circuited patch (no feed). Calculations are plotted for the open-circuit case ($Z_L = \infty$), and for the terminated case ($Z_L = 50\ \Omega$), and were made using 15 expansion modes. First note that the peaks in the RCS response can be identified with the resonances of the patch as given by the simple cavity model [9], and thus the response in the vicinity of a given peak can be traced to the response of a specific expansion mode of the form of (24a) or (24b). This is another reason for using the entire domain type of expansion modes. The incidence/scattering angles for the data of Fig. 5 are $\theta = 60°$, $\phi = 45°$, so both $\hat{x}$ and $\hat{y}$ modes are excited on the patch. If ϕ were chosen as either 0° or 90°, then either the $\hat{y}$ or $\hat{x}$ mode responses would be absent. It is seen that the resonant peaks are well above the cross-sectional area, $LW \cos\theta$, of the patch.

The effect of a 50 Ω load impedance is seen to reduce the peak RCS values by 10–20 dB. The location of the feed point is important; here it was chosen near a corner of the patch, so that the feed would influence virtually all of the modes. If the feed were located at $y_p = 0$, however (as it often is for an antenna operating in the dominant $\hat{x}$ mode), it would not couple to many of the $\hat{y}$ modes, and thus not affect the peak RCS of those modes.

To further study the effect of the load impedance on the RCS of the microstrip antenna, Fig. 6 shows the RCS at $f = 2700$ MHz for the patch of Fig. 5, for various load impedance magnitudes and phases. This frequency corresponds to the first resonant peak in Fig. 5, and the incidence angle is selected as $\theta = \phi = 0°$ for maximum response. The data of Fig. 6 show that the RCS is maximum for an open-circuit load, and generally a minimum for a short-circuit load, in agreement with the argument near the end of Section II-D. The phase of the load impedance does not appear to have a large effect on the RCS response, although the dip for $Z_L \approx -j11\ \Omega$ does seem to indicate that the RCS can be substantially reduced at a specific frequency by creating a resonance with an appropriate load impedance. Such a load would probably have a deleterious effect on the antenna performance, however.

The data of Figs. 5 and 6 suggest some obvious methods of RCS reduction for microstrip antennas. Assume, for example, that the patch antenna is expected to operate at a single frequency, with a linear polarization. There is nothing obvious that can be done to reduce the RCS of the antenna at its operating frequency without sacrificing antenna performance (at least the bandwidth of the RCS peak will be quite narrow). Out of band, however, a number of things can be done. First, shorting pins can be placed along the $x = 0$ line of Fig. 2, to show a low-impedance to higher order (odd) $\hat{x}$ modes, and to a number of the $\hat{y}$ modes, without substantially affecting the dominant antenna mode. Next, the antenna feed should be placed off of a symmetry plane. Thus, x_p can be adjusted for an impedance match to the antenna, but y_p should be placed close to $y_p = -W/2$. In this way the loading of whatever is

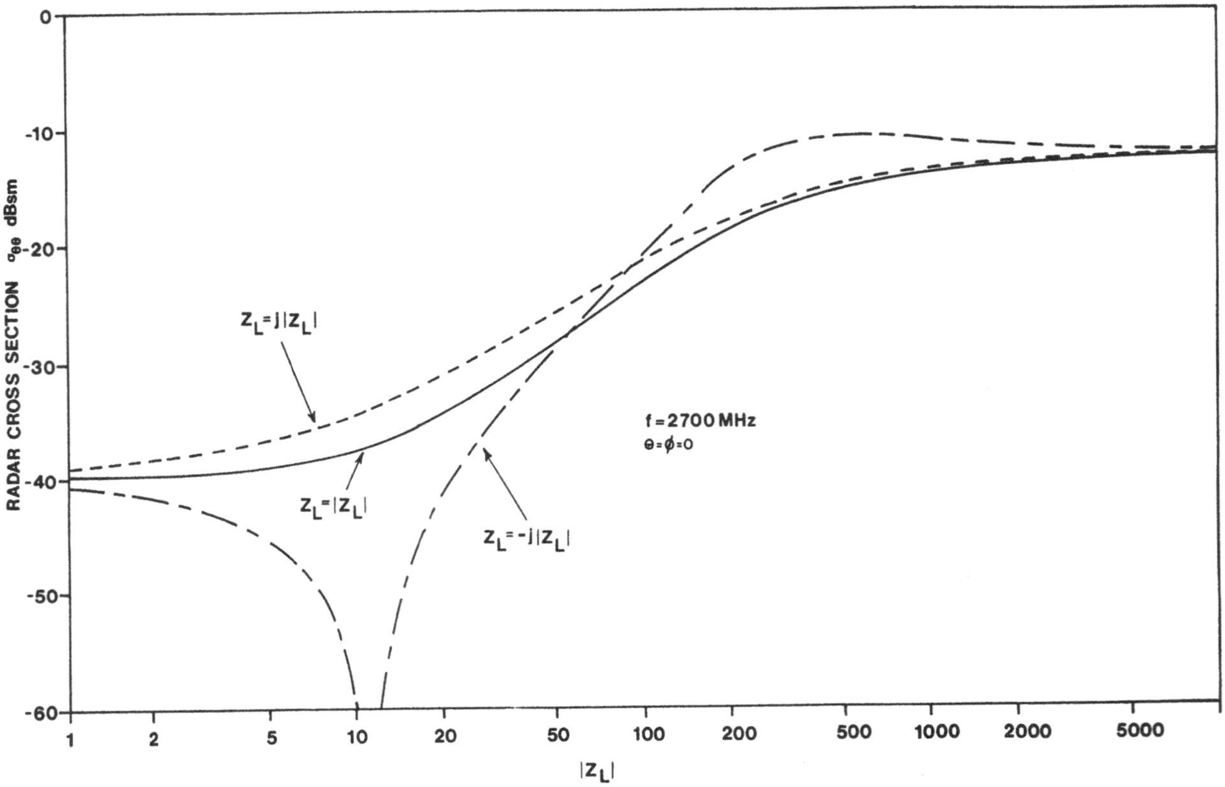

Fig. 6. Radar cross section of the microstrip antenna of Fig. 5, at f = 2700 MHz, for various magnitudes and phases of the load impedance Z_L. $\theta = \phi = 0°$.

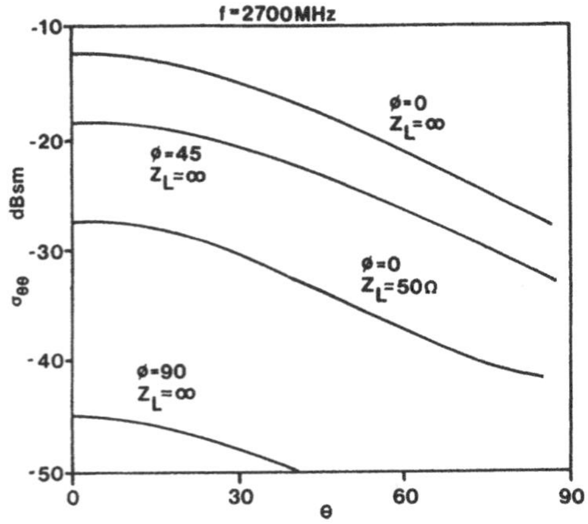

(a)

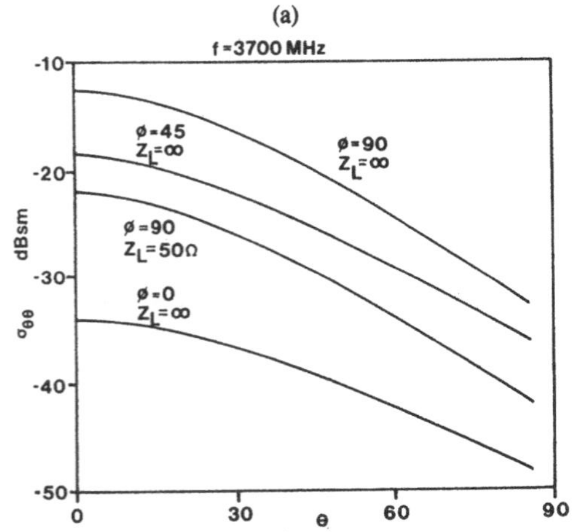

(b)

Fig. 7. Radar cross section of the microstrip antenna of Fig. 5, versus θ, for various values of azimuth angle, ϕ, and load impedance Z_L. (a) f = 2700 MHz. (b) f = 3700 MHz.

connected to the antenna will have the most beneficial effect. If the extra complexity can be tolerated, a narrow bandpass filter can be inserted between the microstrip antenna and the rest of the system. This filter should be wide enough to pass the desired signal to or from the antenna, and outside of this band show a low impedance to the antenna to reduce the out-of-band RCS.

Besides the frequency response, one might also be interested in the variation of RCS with angle. Figs. 7(a) and 7(b) show the monostatic RCS patterns, versus θ, for two frequencies and various load impedances Z_L and azimuth angles ϕ. The frequency 2700 MHz corresponds to the lowest order $\hat{x}$ mode, so the response for $\phi = 0$ is much greater than for $\phi = 90°$. Similarly, the frequency of 3700 MHz of Fig. 7(b) corresponds to the lowest order $\hat{y}$ mode, so the response is strongest for $\phi = 90°$. Changing ϕ to 45° reduces the RCS by about 6 dB, as expected. The variation with θ is smooth, and is essentially the same as the radiation pattern of the patch antenna. Higher order modes, however, may have multilobed patterns with nulls at certain angles.

The effect of a uniaxial substrate is considered in Figs. 8(a) and 8(b). The data of Fig. 8(a) are for an isotropic substrate with $\epsilon_x = \epsilon_z = 10.2$, and shows the first two peaks in the RCS of a patch antenna for $Z_L = \infty$ and $Z_L = 50\ \Omega$ terminations. As before, the 50 Ω load is seen to substantially reduce the RCS peaks. Fig. 8(b) is for the same geometry, but with $\epsilon_x = 13.0$, $\epsilon_z = 10.2$. The only noticeable change is to cause a slight lowering in the resonant frequency of the peaks. As suggested previously, the lack of any more substantial effect is probably due to the fact that the substrate is quite thin, electrically. More complicated media, such as biaxial, gyro-

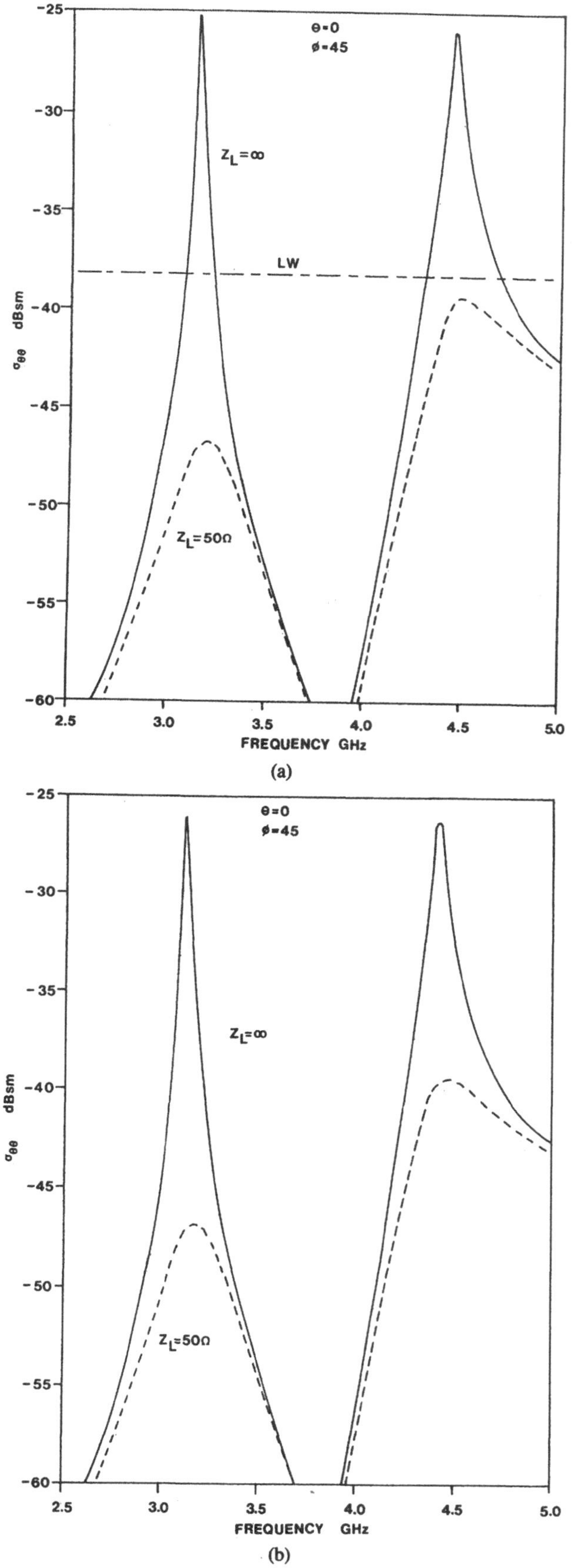

Fig. 8. Radar cross section of a microstrip antenna. $L = 1.0$ cm, $W = 1.5$ cm, $d = 0.127$ cm, $x_p = -L/4, y_p = -W/4, \theta = 0, \phi = 45°$. (a) $\epsilon_x = \epsilon_z = 10.2$. (b) $\epsilon_x = 13.0$, $\epsilon_z = 10.2$.

tropic, or bianisotropic, or the use of an anisotropic superstrate, may have more interesting effects.

IV. Conclusion

The effect of a uniaxial substrate on the radiation and scattering of a microstrip patch has been studied. The Green's function for a uniaxial substrate has been derived in the spectral domain, and the far-zone fields of a source on such a substrate have been presented. The primary effect of a uniaxial anisotropy is seen to be a change in the effective dielectric constant of the material, leading to a shift in the resonant frequency of the antenna. Calculated resonant frequencies have been compared with measurements.

The radar cross section of a microstrip antenna, including the effect of the feed termination, has been presented in terms of its dependence on frequency, load impedance, scattering angle, and anisotropy. The RCS of an open-circuited patch antenna on an isotropic substrate was compared with measurements.

Acknowledgment

The author would like to thank Dr. E. H. Newman of The Ohio State University ElectroScience Laboratory for providing the measured data of Fig. 5.

References

[1] N. G. Alexopoulos, "Integrated-circuit structures on anisotropic substrates," *IEEE Trans. Microwave Theory Tech.*, vol. MTT-33, pp. 847–881, Oct. 1985.

[2] N. G. Alexopoulos and S. A. Maas, "Characteristics of microstrip directional couplers on anisotropic substrates," *IEEE Trans. Microwave Theory Tech.*, vol. MTT-30, pp. 1267–1270, Aug. 1982.

[3] C. M. Krowne, "Green's function in the spectral domain for biaxial and uniaxial anisotropic planar dielectric structures," *IEEE Trans. Antennas Propagat.*, vol. AP-32, pp. 1273–1281, Dec. 1984.

[4] ——, "Determination of the Green's function in the spectral domain using a matrix method: Application to radiators or resonators immersed in a complex anisotropic layered medium," *IEEE Trans. Antennas Propagat.*, vol. AP-34, pp. 247–253, Feb. 1986.

[5] D. M. Pozar, "Input impedance and mutual coupling of rectangular microstrip antennas," *IEEE Trans. Antennas Propagat.*, vol. AP-30, pp. 1191–1196, Nov. 1982.

[6] D. M. Pozar and D. H. Schaubert, "Analysis of an infinite array of rectangular microstrip patches with idealized probe feeds," *IEEE Trans. Antennas Propagat.*, vol. AP-32, pp. 1101–1107, Oct. 1984.

[7] J. A. Kong, *Electromagnetic Wave Theory*. New York: Wiley Interscience, 1986.

[8] D. M. Pozar, "Finite phased arrays of rectangular microstrip antennas," *IEEE Trans. Antennas Propagat.*, vol. AP-34, pp. 658–665, May 1986.

[9] E. H. Newman and Y. Yeh, "Scattering from a microstrip patch," presented at Nat. Radio Sci. Meet., June 1986.

[10] R. V. McGahan and B. Rama Rao, "Electromagnetic backscatter from microstrip arrays: Theory and measurement," in *1985 Phased Arrays Symp. Proc.*, RADC-TR-85-171.

[11] D. M. Pozar, "Considerations for millimeter wave printed antennas," *IEEE Trans. Antennas Propagat.*, vol. AP-31, pp. 740–747, Sept. 1983.

Analysis and Design of a Microstrip Reflectarray Using Patches of Variable Size

S.D. Targonski* and D.M. Pozar
Department of Electrical and Computer Engineering
University of Massachusetts
Amherst, MA 01003

Introduction

In a great number of microwave applications a highly directive antenna with a main beam scanned to a certain angle is required. To achieve this a certain aperture illumination with progressive phasing is used. The two primary ways to do this are reflectors and arrays. The reflector antenna uses its geometry to create the desired phase across the aperture, while the array employs distinct elements fed with progressive phasing. Reflector antennas are advantageous in the fact that they typically exhibit large bandwidth and low loss. The main disadvantage of the reflector is the geometrical constraint it imposes on the design. The most popular reflector, the parabolic reflector, also exhibits inherently high cross polarization levels. Microstrip patch arrays are lightweight, low-profile antennas that are capable of low cross polarization levels but typically have small bandwidth and fairly large loss at microwave frequencies.

Obviously, it would be beneficial to combine some of the more attractive features of reflectors and arrays. This is accomplished by the reflectarray [1]. In the reflectarray, a primary source illuminates a reflecting surface, producing a specularly reflected field, as in the case of a reflector antenna. However, the reflecting surface is covered with an array of radiating elements that produce a scattered field. These elements can be tuned to produce the required phasing over the aperture, thereby eliminating the geometrical constraint of ordinary reflectors.

In [2], the author designed a microstrip reflectarray using patches with an attached stub for phasing as radiating elements. In [3], the authors showed that the same type of phase control could be achieved by using patches of variable resonant length. This technique eliminates the need for a triangular grid spacing, and produces a larger bandwidth since the bandwidth of the stubs is no longer a factor.

In this paper the steps taken in the design of a microstrip reflectarray using patches of variable size are outlined. Measured and theoretical results are shown for the finished design, and several important performance criteria are compared with the microstrip reflectarray of [2].

Design of the Microstrip Reflectarray

The basis for the design of the microstrip reflectarray is the solution for the phase of the reflected field from an infinite array of microstrip patches illuminated by a plane wave [3]. Several restrictions are then employed in using this data for the design, as in [2]. The reflectarray must be large enough so that the infinite array solution is valid, and the F/D ratio must be large enough so that the incident field is locally a plane wave. The assumption that each element in the reflectarray scatters with the same phase as the infinite array is then invoked. Results shown in [2] support the use of this approximation.

Reprinted from *IEEE Symposium on Antennas and Propagation Digest,* vol. 3, pp. 1820–1823, June 1994.

Figure 1 shows several curves of phase versus patch length for different angles of incidence. Notice that the curves do not differ greatly until the incidence angle approaches 40 degrees. This brings up an important point. The phase computed in Figure 1 is the phase in the direction of the incidence angle. The fact that the phase curves start to differ for incidence angles greater than 40 degrees suggests that this design approach may not work if trying to scan the main beam to angles far off broadside.

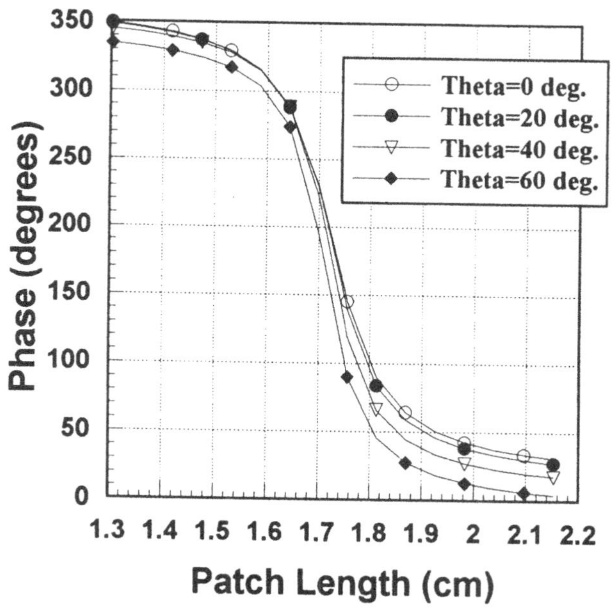

Figure 1. Reflected field phase versus patch length for an infinite array of microstrip patches with plane wave incidence at various angles.

The design equation for the reflection coefficient phase for each element is

$$\phi_i = k_o(R_i - \mathbf{r}_i \cdot \mathbf{r}_o) \quad (1)$$

where ϕ_i is the reflection coefficient phase, R_i is the distance from the phase center of the feed to each element, $\mathbf{r}_i$ is the vector from the center of the reflectarray to each element, and $\mathbf{r}_o$ is the unit vector in the main beam direction. The length of each element is then adjusted to produce the desired phase.

An interesting effect occurs if the specularly reflected field is included in the analysis of [3]. At the design frequency of the reflectarray, the scattered field from the patches is of nearly equal amplitude and has an approximate phase difference of 180 degrees as compared to the specularly reflected field. This effectively eliminates the specular reflection, and there is no loss of efficiency due to the specular reflection.

Results for the Microstrip Reflectarray

Using the design procedures described above, a microstrip reflectarray was designed and fabricated with the same parameters of [2] but using patches of variable size. Figure 2 shows the measured and theoretical E-plane patterns for the reflectarray taken at 5.305 GHz. The main beam is scanned to 25 degrees with a beamwidth of 7.5 degrees. The peak sidelobe level was measured to be 20.5dB down from the main beam amplitude. The results are in good agreement with theoretical calculations. The measured and computed H-plane patterns are shown in Figure 3. The agreement between computed and measured patterns is excellent, and is better than in the H-plane case since aperture blockage effects are reduced.

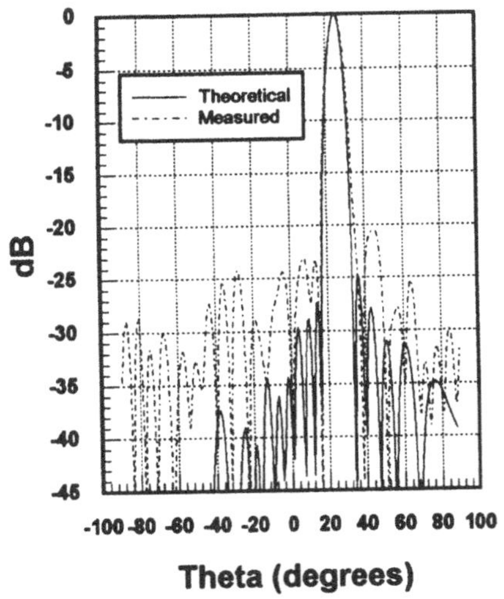

Figure 2. Measured and theoretical E-plane patterns at 5.305 GHz.

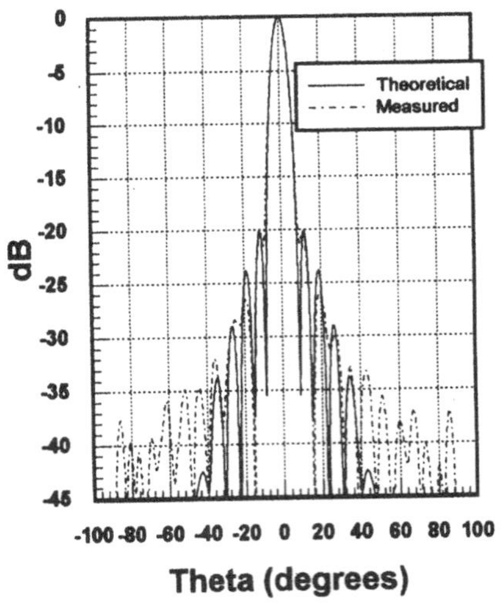

Figure 3. Measured and theoretical H-plane patterns at 5.305 GHz.

Table 1 shows a comparison between the reflectarray described in this paper and the reflectarray of [2]. Several important performance criteria of the two different designs are compared. The two designs exhibit similar performance in most areas, however, the reflectarray using variable size patches has a slightly larger bandwidth, which is expected because of the absence of tuning stubs. The increase in bandwidth may also result from the fact that the reflectarray consists of many different sized patches, each operating at a slightly different resonant frequency. The bandwidth is defined in this case as the gain bandwidth 1dB down from the peak gain, and the reflectarray of [2] exhibits a bandwidth of 3.7% while the reflectarray using variable size patches has a 4.6% bandwidth.

Table 1. Microstrip reflectarray performance comparison.

	Reflectarray Using Tuning Stubs [2]	Reflectarray Using Variable Size Patches
Peak Gain	27.1 dBi	27.0 dBi
E-plane Beamwidth	7.7 degrees	7.5 degrees
H-plane Beamwidth	7.7 degrees	7.5 degrees
E-plane Sidelobe level	-22 dB	-20.5 dB
H-plane Sidelobe level	-18 dB	-20 dB
Gain bandwidth	3.7 %	4.6 %
E-plane cross polarization level	-29 dB	-40 dB

In conclusion, a microstrip reflectarray using variable size patches was designed and tested. It exhibited similar performance to the microstrip reflectarray using patches with tuning stubs attached, with the additional advantage of having less restriction on the array grid spacing. It also exhibited a slightly larger bandwidth.

References:

[1] R.G. Malech, "The Reflectarray Antenna System", 12th Annual Antenna Symposium, USAF Antenna Research and Development Program 1, University of Illinois, 1962.

[2] T.A. Metzler, Design and Analysis of a Microstrip Reflectarray, Ph.D. dissertation, University of Massachusetts, September 1992.

[3] D.M. Pozar and T.A. Metzler, "Analysis of a Reflectarray Antenna Using Microstrip Patches of Variable Size", *Electronics Letters*, vol. 29, pp. 657-658, April 1993.

ACTIVE SUBARRAY MODULE DEVELOPMENT FOR Ka BAND SATELLITE COMMUNICATION SYSTEMS

Shashi Sanzgiri*, Warren Pottenger
Doug Bostrom and David Denniston
Texas Instruments, Mckinney, Texas
Richard Lee
NASA Lewis, Cleveland, Ohio

1.0 Introduction

Due to the demand for increased satellite communication capacity, new satellite communication technologies operating at Ka band are being investigated in the United States and abroad. The recent launching of ACTS communication satellite by NASA and on-going communication experiments at Ka band, will provide momentum for future launching of Ka band satellites. These satellites will provide wide band communication channels to fixed and mobile terminals. For aircraft to satellite communications, conformal electronically scanned phased arrays are preferred over reflector antennas for improved satellite tracking and to reduce life cycle cost, size, weight and aerodynamic drag.

This paper describes development of a low profile 4 x 4 element subarray module which can be used as a building block for large arrays. The program was funded by NASA Lewis as a phased array packaging technology program. We have delivered three such modules to NASA. These modules will be used in an array configuration to demonstrate mobile satellite communication link between ACTS satellite and the aircraft.

2.0 Subarray Module Description

The subarray module consists of a 4 x 4 element array with elements spaced .33" apart. The module provides transmit only function at 29.6 GHz with a bandwidth of 5%. The subarray is designed to scan over +/- 30 ^{0} with less than 3 dB drop in ERP. The goal was to provide 75 watts ERP at broadside from each module.

The module utilizes a 'tile' array architecture in which the radiating element layer, MMIC device layer, and signal distribution layer are parallel to each other . The layers are interconnected using wire bonds and electromagnetic coupling. The description of each of these layers is given later on.

The module block diagram is shown in Figure 1 and its form factor is shown in Figure 2. The module size is 1.3" x 1.3" x 0.3". The RF portion of the module consists of a 16 way power divider, 16 MMIC phase shifters, 16 MMIC consists of a serial data bus and ASIC chips that perform serial to parallel data conversion for controlling phase shifters and power amplifier chips. The module consists of two ASICs each controlling half of the module. The control subsystem also consists of drain bias switches, which are used to turn 'on' or 'off' any selected amplifiers. The DC bias portion of the module provide regulated and selected bias levels to the gate and drain of the power amplifiers. This part of the system also includes appropriate circuits for stabilizing the power amplifier performance.

Reprinted from *IEEE Symposium on Antennas and Propagation Digest,* vol. 2, pp. 860–863, June 1994.

3.0 Layer Description

Figure 3 shows the module assembly with the lid removed. The module consists of three main layers: radiating element layer, device layer and signal distribution layer. The radiating element layer consists of 16 microstrip patch elements. To provide low resistance thermal paths for heat dissipation from the power amplifiers, the patch elements are embedded in the cavities formed in the module housing. This cavity backed construction also provides good bandwidth and scan angle performance. The isolated element provided a bandwidth of 8 % for 2:1 VSWR and 6 dB gain. The cavity backed element is excited through a slot by a microstrip line carrying the output signal from the power amplifier.

The device layer consists of MMIC power amplifiers and phase shifters. The power amplifier is a three stage 50µm-100µm-200µm PHEMPT device which delivers 100 mW at 23 dB gain and 30 % efficiency . The phase shifter provides 4 bits of switched line lengths. The switching device is GaAs PIN diode. The insertion loss of the phase shifter is about 4.5 dB at room temperature. Two phase shifters and two amplifier chips along with the RF,control and bias lines are assembled on a carrier plate. The subarray module consists of eight such carrier plates. The carrier plates are epoxied to the housing floor. The slots on the ground plane side of the carrier plate serve as the coupling medium between the amplifier output lines and the patch elements.

The signal distribution layer consists of a multilayer RF,DC and logic distribution network. The RF distribution consists of shielded microstrip transmission line. On top of this transmission line is attached a seven layer DC/Logic distribution board constructed of 1 mil thick polyamide layers. The signal distribution layer also incorporates ASIC chips, voltage regulators, bias adjust resistors , drain switches and amplifier stabilizing circuits.

4.0 Module Performance

Three subarray modules were fabricated on the NASA program. The modules have been tested individually and as a two module (4 x 8 element) array. Figure 4a shows the broadside pattern and Figure 4b shows pattern scanned to 30^0 in the H-plane for single subarray module. We have realized 77 watts ERP at 29.6 GHz for such a module.

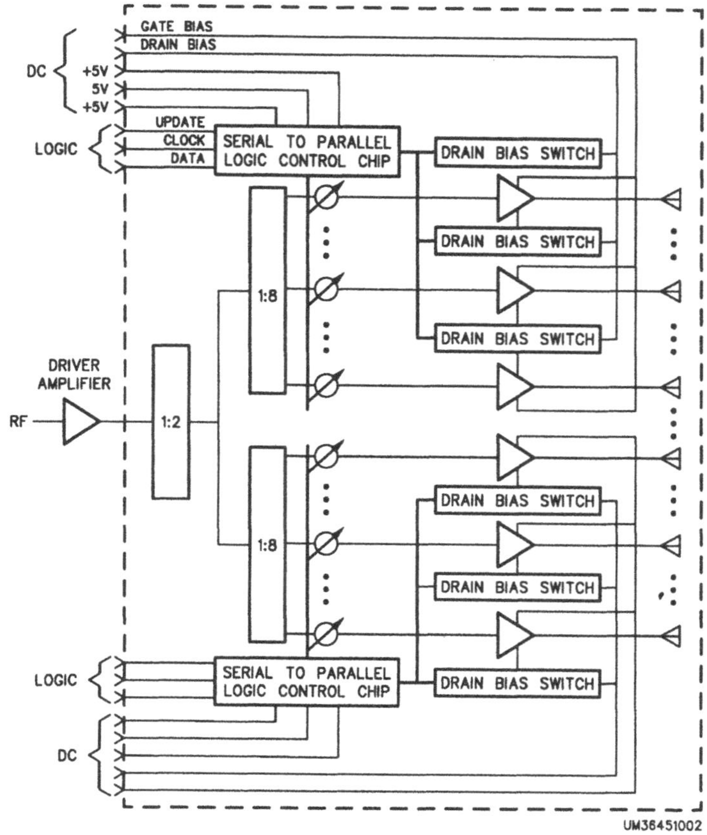

Figure 1. Active subarray module block diagram

Figure 2. Active 16-element subarray module

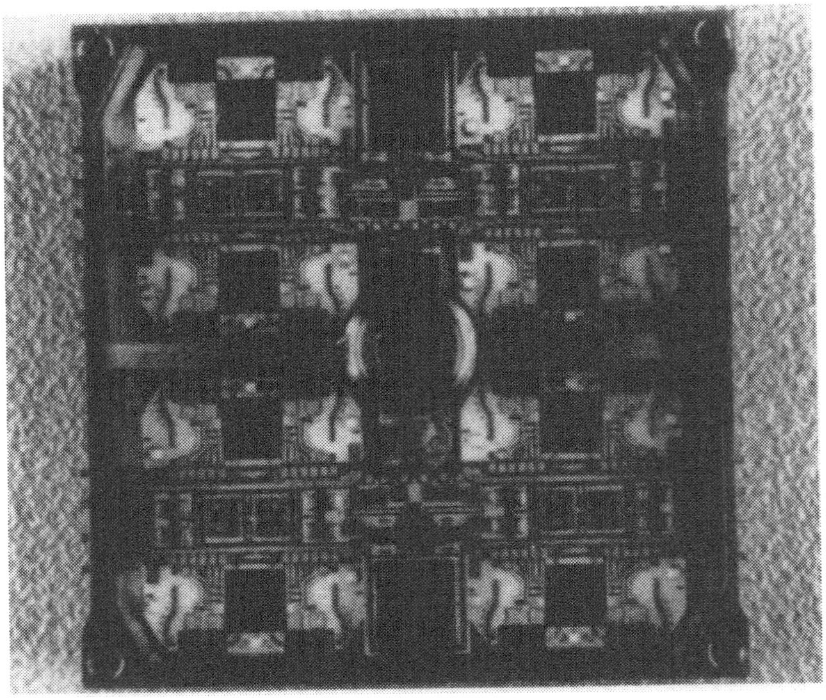

Figure 3. Active subarray module layout

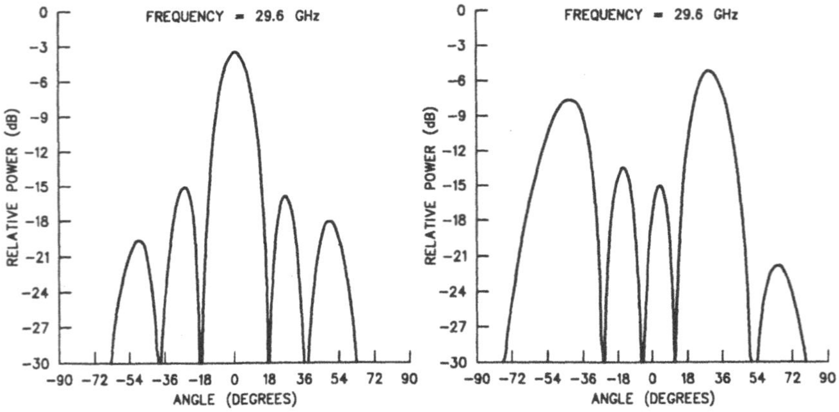

(a) 0° scan

(b) 30° scan

Figure 4. Active subarray patterns

AN MMIC APERTURE–COUPLED MICROSTRIP ANTENNA IN THE 40GHz BAND

Hiroyuki OHMINE, Takuo KASHIWA, Takahide ISHIKAWA,
Akio IIDA and Makoto MATSUNAGA
Mitsubishi Electric Corporation
Ofuna,Kamakura–city, Japan 247

1. Introduction

Recently, in millimeterwave and microwave applications, several concepts of monolithic antennas with which active solid–state devices and feed circuits are integrated have been reported [1]–[4]. This paper presents the fabrication and test results of an MMIC aperture–coupled microstrip antenna (MSA) integrated with a mixer and an amplifier. The characteristics of the antenna radiator were evaluated separately and a good radiation efficiency of 89% was obtained in the 40GHz band.

2. Configuration

Fig.1 shows the photographs of a fabricated aperture–coupled MSA with GaAs monolithic circuits, a mixer with a conversion loss of 9.5dB [5] and an amplifier with a gain of 9dB [6]. Fig.2 shows the configuration of the aperture–coupled MSA [7]–[9]. The coupling between a feed line and a radiating element is accomplished by an aperture in the ground plane. In this case, quartz (dielectric permittivity ε_r=3.8) and GaAs (ε_r=12.9) are used for the substrates of the radiating element and the feed circuit, respectively.

The major advantage of this configuration is that the reverse side of the antenna radiator is used for the active and feed components. Another advantage is that the design matches the antenna substrate (low ε_r) to the electrical characteristics such as high radiation efficiency and wide bandwidth. In addition, the ground plane isolates feed circuits from radiating elements and results in reducing undesirable coupling [7], [9].

3. Input Impedance

The followings show the characteristics of the antenna radiator exclusive of GaAs MMIC circuits. Fig.3 shows the measured input impedance of the antenna radiator for various open–stub lengths as a function of frequency in a Smith chart. The measurements were made with an HP–8510B network analyzer. In this figure, when the stub length is approximately 1/4 λg, where λg is the wave–length in the dielectric substrate, the locus of the impedance rotates around the real axis and an impedance match is achieved.

The measured resonant frequency versus the slot length is plotted in Fig.4. It is seen from this figure that resonant frequency decreases with increasing the slot length [8]. Also plotted in Fig.4 is the input resistance at resonance versus the slot length. As the slot length is reduced the input resistance decreases. From this figure, the slot length can be determined to achieve a matching condition. For comparison the calculated resonant frequency of circular MSA without slot based on the cavity model is plotted in Fig.4. From this figure, the resonant frequency of the antenna is determined primarily by

Reprinted with permission from *Proc. ISAP,* H. Ohmine, T. Kashiwa, T. Ishikawa, A. Iida and M. Matsunaga, "An MMIC Aperture-Coupled Microstrip Antenna in the 40 GHz Band," pp. 1105–1108, Sept. 1992.

the patch length, but it is affected slightly by the slot length.

The calculated and measured bandwidth of this antenna are shown in Fig.5. Calculated values are obtained from the mode expansion analysis in circular MSA [10]. The measured bandwidths (BW) that satisfy VSWR≦1.5, 2, and 3 are obtained 1.5%, 3%, and 5.5%, respectively. Measured values are in good agreement with calculated values. It is clear from this figure that equal bandwidth is obtained compared to circular MSA without slot.

4. Efficiency

Fig.6 shows the measured radiation patterns in the (a) E–plane and (b) H–plane. Cross–polarization is below 15dB in the E–plane and H–plane. The front–to–back ratio is consistently below 16dB. In the calculation, the finite ground plane is taken into consideration. Calculated values are in relatively good agreement with measured values, with some distortions being due to diffractions from antenna fixture. The measured gain of this antenna is 6dB at boresite including loss of 1.5dB of the feed line. The directive gain of this calculated pattern is 8dBi. Therefore, the antenna radiation efficiency of approximately 89% (directive gain minus measured gain) is obtained by excluding the above loss.

5. Conclusion

An MMIC aperture–coupled microstrip antenna integrated with a mixer and an amplifier has been developed and its high efficient characteristics have been verified in the 40GHz band. The antenna has reasonable gain, bandwidth, low cross–polarization and back–radiation.

References

[1]J.F.Millvenna," Monolithic Phased Arrays for EHF Communications Terminals," Microwave Journal, pp.113–125, Mar. 1988.

[2]D.M.Pozar and D.H.Schaubert," Comparison of Architecture for Monolithic Phased Array Antennas," Microwave Journal, pp.93–104, Mar. 1986.

[3]R.J.Mailoux," Phased Array Architecture for mm–Wave Active Arrays," Micro–wave Journal, pp.117–120, July 1986.

[4]G.Falciasecca, et al.," Feasibility of an EHF(40/50GHz) Mobile Satellite System Using Highly Inclined Orbits," International Mobile Satellite Confe–rence, Ottawa, pp.131–135, 1990.

[5]K.Itoh, et al.," A 40GHz Band Monolithic Even Harmonic Mixer with an Anti–parallel Diode Pair," 1990 MTT–S Int. Microwave Symp. Dig., pp.879–882.

[6]K.Seino, et al.," A Broadband MMIC Amplifier Module Operating at Ka Band," Natl. Conv. Rec., IEICE Japan, C–66, 1991.

[7]D.M.pozar," Microstrip Antenna Aperture Coupled to a Microstrip Line," Electronics Letters, vol.21,no.2, pp.49–50, 1985.

[8]P.L.Sullivan and D.H.Schaubert," Analysis of Aperture Coupled Microstrip An–tenna," IEEE Trans. Antennas Propagat., vol.AP–34, pp.997–984, Aug. 1986.

[9]D.M.pozar," A reciprocity method of analysis for printed slot and slot–coupled microstrip antennas," IEEE Trans. Antennas Propagat., vol.AP–34, pp.1439–1446, Dec. 1986.

[10]I.J.Bahl and P.Bhartia, *Microstrip Antennas*, Dedham, MA:Artech House, 1981.

(a) Circular microstrip antenna.

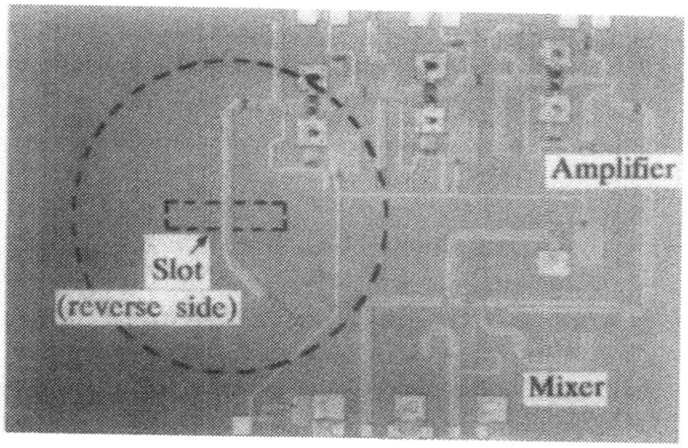

(b) Mixer and amplifier.
(chip size is 3.3mm×5mm)

Fig.1. Photograph of a fabricated aperture–coupled microstrip antenna and GaAs monolithic circuits.

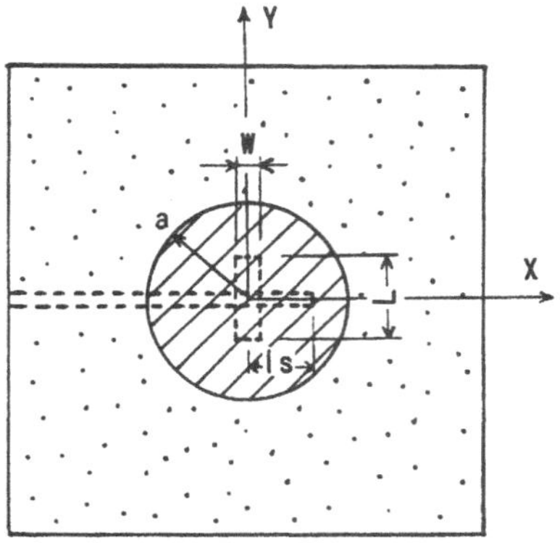

(a) Top view.

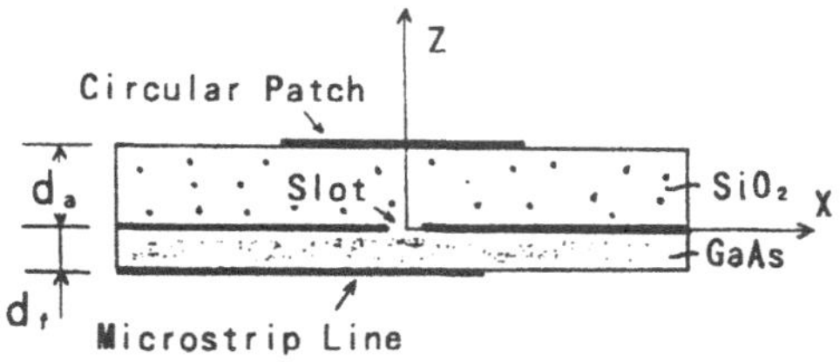

(b) Cross–sectional view.

Fig.2. Configuration of an aperture–coupled microstrip antenna.

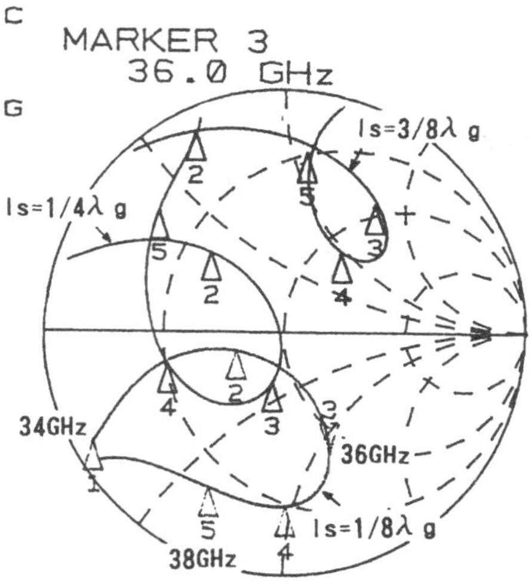

Fig.3. Measured input impedance as a function of the open–stub length. $a=0.132\lambda o, L=0.108\lambda o, W=0.024\lambda o$.

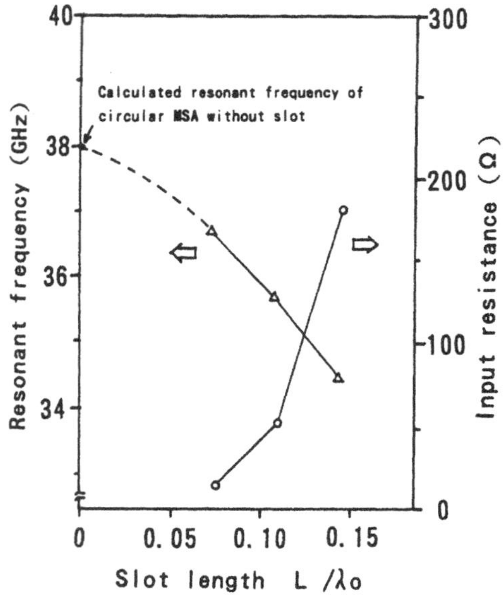

Fig.4. Measured resonant frequency and input resistance at resonance versus slot length. a=0.132λo,W=0.024λo.

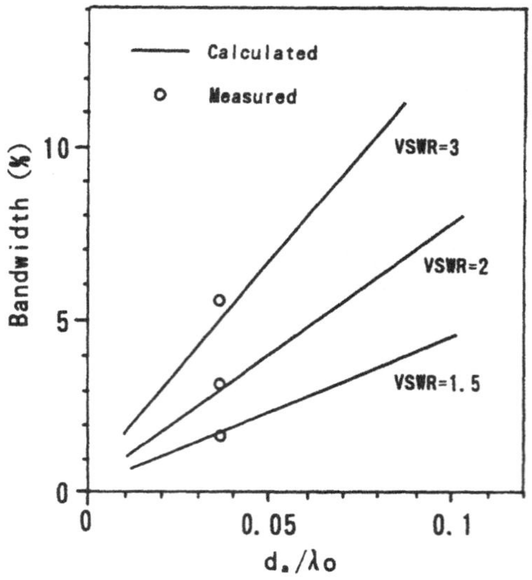

Fig.5. Calculated and measured bandwidth versus thickness of quartz substrate.

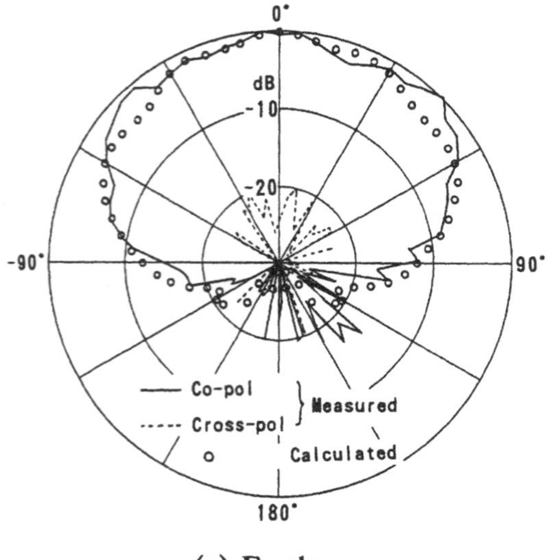

(a) E−plane

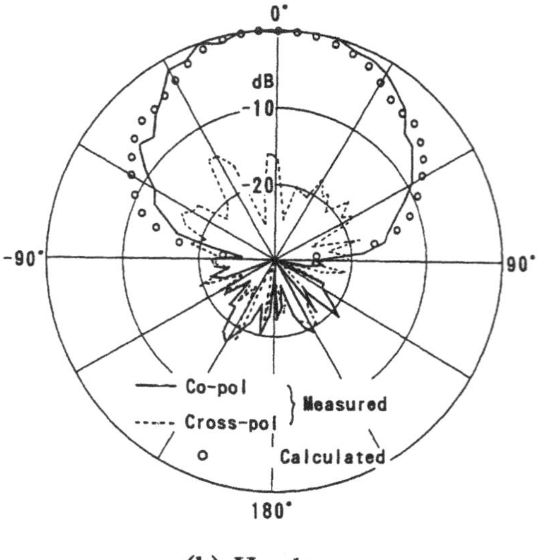

(b) H−plane

Fig.6. Calculated and measured radiation patterns. a=0.126λo,L=0.108λo,W=0.024λo,f=38GHz.

Author Index

A
Aberle, James T., 344
Adrian, A., 128
Alexópoulos, Nicólaos G., 223
Arai, Hiroyuki, 145

B
Bahl, Inder J., 95
Bhartia, Prakash, 95
Bi, Zhi-Qiang, 259
Bobinchak, J., 180
Bostrom, Doug, 417

C
Capelle, Antoine R. van de, 167, 205
Carver, Keith R., 3
Chang, Esin, 73
Chow, Francis Y. M., 274
Chujo, Wataru, 136, 145
Citerne, J., 26
Croq, Frederic, 185

D
Daniel, J. P., 26, 212, 215, 380
Das, Nirod K., 249
Davidovitz, Marat, 90
Denniston, David, 417
Densmore, Arthur C., 390
Drissi, M., 26
Dubost, G., 26

E
Edimo, E., 142

F
Farrar, Frederick G., 148
Fralich, Russell, 331
Fujimoto, Kyohei, 397
Fujise, Masayuki, 145

G
Gardiol, F. E., 232
Garg, Ramesh, 217, 401
Goto, Naohisa, 145
Gupta, Kuldip C., 117

H
Hall, C. M., 280
Hall, P. S., 107, 140, 176, 192, 280
Hayes, Scott T., 148
Henderson, A., 384
Herd, J. S., 356
Hessel, Alexander, 335
Himdi, M., 212, 215
Hirasawa, Kazuhiro, 397
Huang, John, 100, 140, 305, 371, 390

I
Iida, Akio, 421
Imai, Shigeru, 397
Ishikawa, Takahide, 421
Iwasaki, Hisao, 124

J
Jackson, David R., 223
James, J. R., 51, 384
Jayakumar, Immanual, 401
Jones, Bevan B., 274

K
Kashiwa, Takuo, 421
Kaufman, Barry, 178, 295
Kawabata, Kazuaki, 124
Kishk, A. A., 79
Kuboyama, Haruhiro, 397

L
Lal, Bhagwan, 401
Lee, K. F., 87, 180
Lee, R. Q., 180
Lee, Richard, 417
Levine, Ely, 287
Litva, John, 259, 331
Liu, Chung-Cytz, 335
Long, Stuart A., 73
Luk, K. M., 87

M
Malamud, Gabi, 287
Matsunaga, Makoto, 421
Mink, James W., 3
Mosig, J. R., 232
Munson, Robert E., 68
Mutzig, J. P., 380

N
Nakano, Masayuki, 145
Nedelec, M., 380

O
Ohmine, Hiroyuki, 421

P
Palanisamy, V., 217
Penard, E., 380
Pottenger, Warren, 417
Pozar, David M., 51, 130, 157, 178, 185, 241, 249, 295, 311, 323, 344, 359, 405, 414
Pues, Hugo F., 167, 205

R
Rammos, E., 140
Richards, William F., 73
Roederer, A., 140

S
Sanzgiri, Shashi, 417
Sarap, B. K., 401
Sawada, Hisashi, 124
Schaubert, Daniel H., 59, 128, 148, 269
Seeto, Anthony W., 274
Shafai, Lotfollah, 79
Sharaiha, A., 142
Sharma, P. C., 117
Shmoys, Jerry, 335
Shtrikman, Shmuel, 287
Sindoris, Arthur, 148
Solbach, Klaus, 317
Splitt, Georg, 90
Stuchly, Stanislaw S., 95

T
Tam, P. Y., 87
Tanaka, Masato, 136
Targonski, Stephen D., 130, 414
Terret, C., 26, 142, 215
Teshirogi, Tasuku, 136
Thouroude, D., 212
Treves, David, 287

V
Van de Capelle, Antoine R., 167, 205

W
Wang, Jian, 331
Wu, Chen, 259, 331
Wu, Ke-Li, 259

Z
Zürcher, J.-F., 182

Subject Index

A

Active element match, 318–319
Active input impedance, 325, 326
Active resistance for E-Plane scan, 337–338
Active resistance for H-plane scan, 340–341
Active subarray module development, 417–420
Analysis techniques, for microstrip elements, 6–12
Anisotropy, 4–5
  uniaxial, 409
Aperture coupling, 331–334, 421–424
Arrays
  analysis, 309–355
  architectures, 269
  and CAD, 51–56
  circuit representations of, 274–275
  circularly polarized, 112–116
  comb-line, 194, 385–386
  conformal cylindrical, 401–404
  corporate-fed, 162–163
  description, 305–307
  design, 267, 280–286
  design and performance, 197–200
  design considerations for, 295–304
  directivity of, 292–294
  efficiency, 283
  elements, modeling, 356–358
  with feed network, 290–292
  fixed beam, 54–55
  four-element, 290–292
  gain, 283
  gain of, 292–294
  gain of circulary polarized, 140–141
  infinite, 346–350
  large, 292–294
  linearly polarized, 140–141
  log-periodic, 162, 192–201
  low sidelobe, 295–304
  microstrip dipoles, 47
  microstrip line, 112
  model, 335–336
  off-axis radiation, 137–138
  optimization, 283–284
  optimum form, 194–195
  parallel-series-fed, 305–308
  phased, 55, 68–72, 335–343
  planar, 42–43
  planar sequential, 137–138
  sequentially rotated, 112–114
  series, 193
  series-fed, 162–163, 274–279
  with squinted beam, 384–389
  techniques, 269–273
  wide-bandwidth patch, 195–196
  Yagi, 162
Axial ratio, 163–164

B

Ball Corporation, phased array antenna, 373
Bandwidth
  axial ratio, 163–165
  definitions, 157–158
  enlargement factor, 168–170
  fundamental limits, 158–159
  impedance, 158–160
  increasing, 157–166, 167–175, 178–179
  multioctave, 192–201
  and multiple resonances, 160–161
  narrow, 297–298
  pattern, 161–162
  performance improvement of, 200
  and phase accuracy, 297–298
  and resonance frequency, 397–398
Beam efficiency, 362–365
Beam, squinted, 384–389
Broad-band antenna, 29, 36

C

CAD (computer-aided design), 6
  commercial software, 53
  and fabrication, 372–373
  procedure for unbalanced feed array, 341–342
  simplified process in development, 371
  software for microstrip arrays, 309
Cavity models, 35, 212–214, 215–216
  CAD-oriented, 212–214, 215–216
  and effective loss tangent, 28
  modal-expansion, 7
Circular patch antennas, 79–86, 87–89
  antenna configuration, 372
  crosspolarisation, 87–89
  feed location, 82
  radiation efficiency of, 79–86
  radiators, 317–322
  two-probe-fed, 344–355
Circularly polarized elements, 105–106, 107–116, 117–123, 130–135, 136–139
Coaxial feed modeling, 261–262
Coax-to-microstrip transistion, 263–264

Comb-line array, 385–386
Coplanar corporate feed effects, 280–285
Corner fed patch, magnetic currents of, 382
Corner fed square patches, radiation, 380–381
Corporate feed effects, 280–285
Corporate feed loss mechanisms, 280–281
Coupling, measurements, 318
Cross-fed array, 381
  with coaxial output, 383
Cross polarization, 299–300, 302, 305, 308
  H-plane scan and, 340–341
Crossed slots, aperture coupling, 130–131

D

DBS (Direct Broadcast Satellite), 371
  flat-plate requirements, 384–385
  large array, 387–388
  system, 376–377
DBSR (Direct Broadcast Satellite Radio), 376–377
Diffraction effects, 300–302
Dipoles, 255–256
  coupled, 38–39
  dual-polarized, 38
  electromagnetic, 38–39
  flat-folded, 37, 38
  mono-polarized, 37
  radiation mechanisms, 391
Dispersive boundary condition, 262
Dissipation loss of active element, 320–321
Dual-beam patterns, 43
Dual elements, 105–106, 107–116

E

Electrically thick antennas, 73–78
  bandwidth, 75–76
  impedance, 74
  radiation pattern, 76–77
  resonant frequency, 74
Element match, 302–304
Element patterns, 325–326
Element positioning errors, 300–302
Elements, single-fed CP, 164
E-plane patterns, of four-comb group, 387
Excitation
  and input impedance, 236–237
  optimization of, 276–279
  treatment, 260–261

F

Far-zone fields, 407
Feeding
  of double-layered antenna, 145–147
  dual polarized antenna, 142–144
  error analysis of, 146–147
  geometry, 284–285
  location, 119–120
  multi-point, 109–110
  of self-diplexing antenna, 145–147
  single point, 110
  singly-fed proximity coupled, 124–127
  techniques, 57–58, 61–66
Feed network, 287–294
  isolation, 302–303
  radiation, 298, 302
Feed radiation, 270–271
Feedlines, dual, 130–131
Finite difference method, 259–266
Finite element technique, 12
Finite ground plane effect, 100–104
Finite phased arrays, 323–330
Flat-plate array, 384–389
  alternative designs, 389
  construction, 385–386
Flexible patch, on curved surface, 378
Four-comb group, 386–387
Frequency tuning, 66
Full-wave analyses, 52

G

GPS (global positioning system), 371, 375
Green's functions, 223–225, 311–312, 362
  for Hertzian dipole, 223–224
  interpolation among, 235
  of uniaxial substrate, 405–413
Ground plane size, 85

H

H-plane radiation pattern, of comb-line, 386
Hexagonal subarrays, 365–366
Higher order mode operation, 66–67
Hyperthermia medical application, 378

I

Impedance, input, 236–237, 311–316
Impedance-matching techniques, 167–175
Infinite arrays, 344–355
  one-probe-fed circular patch, 345–350
  scanning characteristics, 359–367
Interferometric radar application, 377

K

Ka Band satellite communication systems, 417–420

L

Linear system, resolution of, 235
Line loss, 280, 287–290
Loop-complementary structure, 33
Losses, relevance of, 235–236
Loss tangent determination, 28
Low cost arrays, 380–383, 384–389

M

Matrix condition, 236
Matsushita Electric Works, 376
Metal-plate polarizer, 387, 388
Microstrip
  basic elements, 57–58
  characteristics, 59–60
  covered, 255
  defined, 3
  patch, arrays, 335–343
  radiator, 69–70
  spirals, 111
Microstrip antenna
  circular, 16
  dipole, 21–22
  features of, 371–372
  rectangular, 13–14
Microstrip dipole, 255–256
Microstrip elements
  bandwidth, 19–21
  circularly polarized, 17–19
  efficiency, 19–21
  quality factor, 19–21
Microstrip line, 255
  inclined covered, 255–256
MMIC aperture-coupled antenna, 421–424
MMIC power amplifiers, 418
Mobile satellite communications, 145–147
  applications, 373–374, 390–396
Modal expansion technique, 101–102
Model, rectangular antenna, 205–211
Moment method, 11–12, 233–235
  expansion modes, 356
  formulation, 312–313
Moment method solution, 243–245, 407–409
  for radar cross section, 408–409
Mutual coupling, 298–299, 302, 309–355

N

NASA/JPL, 376
NHK Science and Technical Laboratories, 376
Noise figures, for DBS antenna, 388
Nonsatellite-based applications, 377–378
Numerical analysis techniques, 11

O

One-probe-fed arrays, theory, 345–346, 348–349
One-probe-fed circular patch, 345–347
Orthogonal polarizations, 108

P

Parallel feed configuration, 133–134
Parasitic element in subarray, 360
Patch
  antenna, 245–246, 331–334
  arbitrarily shaped, 217–222
  array, 317–318
  array geometry, 276
  circuit representations of, 274–275
  circular, 11, 17
  circularly polarised, 108–109
  with coaxial feed, 27
  dual polarised, 37, 110–111
  experimental characterization, 275–276
  with microstrip feed, 27
  parasitic, 251–253
  probe compensation in thick, 176–177
  radiation, 237–240
  rectangular, *see* rectangular patch
  size, 336–337
  slot-fed, 34, 37
  slot-loaded, 37
  slotted, 237
  SSFIP, 182–184
  variable size, 414–416
Phase accuracies, 295–296, 301–302
Phased array, 323
  geometry, 335
  for passenger car, 375
  simulation, 317–322
Phase shift, of elements, 136–139
Photolithographic mask, for subarray, 385
Planar radiating sources, basic, 26–27
Polarization
  of antenna systems 107–116
  circular, 128–129, 249–250
  cross, 299–300, 302
  dual, 128–129
  linear, 131
  low cross, 305–308
Post-loaded antenna, 397–400
Printed dipole subarray, 360
Probe compensation, in thick patches, 176–177
Probe-fed arrays, 335–343
Probe-fed EMCP antennas, 29
Probe feed model, 345–346
  improved, 348–349
Proximity coupled arrays, 178–179, 356–358, 360
Proximity coupled element, 356

R

Radar cross-section, 410–413
Radiated field, asymptotic expressions for, 237
Radiating edge feeds, models, 28
Radiation
  characteristics, 59, 118, 120
  from feed network, 298
  field, 225
  into the horizon, 229–230
  loss, 280–281
  from microstrip lines, 287–290
Radiation pattern, 76–77, 101–104, 119, 120, 237–238

Radiation pattern, (cont.)
effects, 281–282
H-plane, 386
performance improvement, 200
Radiator microstrip, 69–70
Radiators, circular patch, 317–322
Radome, height, 394
Reciprocity method of analysis, 241–249
Rectangular patch antenna, 7, 212–214, 237, 323–330
on circular cylinder, 401–404
configuration, 372
mutual coupling of, 311–316
radiation patterns, 10, 401–404
variations on, 14
wall admittance of 9
Reduced analyses, defined, 52
Reflectarray, design, 414–416
Resonance conditions, 227–228, 236
Rotation, sequential, 164–165

S
Scan blindness, 337–339, 363, 367
Scanning characteristics of infinite arrays, 359–367
Scatterers, 232–240
Scattering analysis, multiport, 249–258
Scattering, from uniaxial substrate, 405–413
Segmentation technique, 217–222
Self-diplexing antenna, isolation characteristics of, 145–146
Sequential rotations, 136–139
Series-fed patch array antenna, 331–334
Series feed configuration, 131–132
Shaped patterns, synthesis of, 274–279
Sidelobe, performance, 295–296
Simulation, using circular patch radiators, 317–322
Slot-fed antenna, 36
Slot, microstripline-fed, 243–245
Slots, bandwidth, 34
Small array, 317–318
Sony Corporation, 376
Spectral domain analysis, 331–334
Spirals, microstrip, 111
Splitter, triplate
Squinted beam for DBS reception, 384–389
SSFIP (strip-slot-foam-inverted patch), 182–184
Stacked-microstrip antennas, 29
Stripline, defined, 3
Subarray module
description, 417–418
performance, 418–419
Subarrays
circularly polarized, 364–365
with comb-lines, 386
geometry of contiguous, 361
hexagonal, 365–366
increasing elements in, 363–364
integrated feed, 386
with metal-plate polarizer, 388
optimization by CAD, 385
with parasitic elements, 364
photolithographic mask for, 385
synchronous, 364–365
scan blindness, effect on, 363
Substrate
available microwave, 4, 6
material as mechanical support, 373
optimum thickness of, 231
PCA properties, effect on, 226–227
permittivity, 84
special, 111–112
surface wave scattering, 282–283
thickness, 84
two-layer, 90–94
Surface wave
effects, 270–271
elimination of, 228–229
excitation, 60
relevance of, 235–236
scattering, 282–283
structure, 225–226
Synthesis of shaped patterns, 274–279

T
Tapered array, 383
Tapered linear series array, 381
Toyota Central phased array antennas, 373, 376
Transmission-line model, 6, 205–211
analysis, 29
radiating slot, 33
verification, 210
Triplate, defined, 3
Triplate feed, 387–388
Two-layer substrates, 90–94
Two-probe-fed arrays, theory, 346–347, 349–350
Two-probe-fed circular patches, 344–355

U
Uniaxial substrate, radiation and scattering, 405–413

V
Vehicle application of Yagi array antenna, 390–396

W
Waveguide simulator, 350–353
impedance measurements, 319–321
infinite array, 357
Wideband arrays, 356–358

Y
Yagi Antenna Corporation, 375
Yagi array, 390–396
description, 393–394
for mobile satellite communications, 374

Editors' Biographies

DAVID M. POZAR received the BSEE degree in 1975 and the MSEE degree in 1976 from the University of Akron, and the PhD degree in 1980 from Ohio State University. During the course of his undergraduate studies he spent one year as an engineering co-op assistant at the National Security Agency, Fort Meade, Maryland. He was a graduate research assistant at the ElectroScience Lab of Ohio State University while pursuing the PhD degree, and became a research assistant upon completion of the PhD degree. He joined the faculty at the University of Massachusetts in 1980, and in 1989 became professor of electrical and computer engineering. In 1988 he spent a six-month sabbatical leave at Ecole Polytechnique Federale de Lausanne, in Lausanne, Switzerland, participating in several short courses and workshops on printed antennas.

Professor Pozar is a fellow of the IEEE, belonging to the Antennas and Propagation Society and the Microwave Theory and Techniques Society. He has served as an associate editor of the IEEE Transactions on Antennas and Propagation (1983–1986 and 1989–1992), as a member of the IEEE AP-S AdCom (1989–1991), and as an associate editor of the IEEE AP-S newsletter (1982–1984). In 1981 he received the "Outstanding Professor for 1981" award from Eta Kappa Nu, the student honor society. In 1984 he received an NSF Presidential Young Investigator Award, and the "Keys to the Future Award" from the IEEE Antennas and Propagation Society. In 1985 he received the University of Massachusetts Engineering Alumni Association "Outstanding Junior Faculty Award." In 1986 he received the R.W.P. King Best Paper Award from the IEEE Antennas and Propagation Society. In 1987 he received the URSI Issac Koga Gold Medal for his work on printed antennas and phased arrays. He again received the R.W.P. King Best Paper Award in 1988. In 1989 he received the United Technologies Corporation "Outstanding Teaching Award." He presently serves as a distinguished lecturer for the IEEE Antennas and Propagation Society.

Professor Pozar has published over 100 papers on microstrip antennas and phased arrays, and is the author of *Antenna Design Using Personal Computers* (Artech House, 1985), and *Microwave Engineering* (Addison-Wesley, 1990). He is also the author of *PCAAD,* a software package for personal computer-aided antenna design.

DANIEL H. SCHAUBERT received the BS, MS, and PhD degrees in electrical engineering from the University of Illinois. From 1974 to 1980 he was a senior research and development engineer at Harry Diamond Laboratories, Adelphi, Maryland. From 1980 to 1982 he was project engineer and program manager at the Bureau of Radiological Health, Rockville, Maryland. Since 1982 he has been at the University of Massachusetts, Amherst, where he is currently professor of electrical and computer engineering. He became head of the department in 1994. His current research interests include antenna design and analysis, and numerical solution of electromagnetic scattering problems. He is particularly interested in phased arrays at microwave and millimeter-wave frequencies, in wide-bandwidth radiation, and in problems involving dielectric media.

Dr. Schaubert is active in the IEEE Antennas and Propagation Society, having served twice as a member of the administrative committee and as chair of the membership committee and the Washington, D.C. chapter. He was editor of the AP-S newsletter and secretary-treasurer of the Society, and is administrator of the Distinguished Lecturers Program. He received the 1986 Eta Kappa Nu teaching award. He has served on the technical review committee for the Joint Services Electronics Program and several ad hoc panels and review boards; he is a member of Eta Kappa Nu and Tau Beta Pi.

Printed and bound by CPI Group (UK) Ltd, Croydon, CR0 4YY
04/07/2022
03133467-0004